Freshwater Mussels of Alabama and the Mobile Basin in Georgia, Mississippi and Tennessee

Freshwater Mussels of Alabama and the Mobile Basin in Georgia, Mississippi and Tennessee

James D. Williams,
Arthur E. Bogan and
Jeffrey T. Garner

Foreword by
E.O. Wilson

With special thanks to
Sherry L. Bostick
Richard T. Bryant

Sponsored by
Alabama Department of Conservation and Natural Resources, Division of Wildlife
and Freshwater Fisheries

The University of Alabama Press / Tuscaloosa

∞
The paper on which this book is printed meets the minimum requirements of American National Standard for Information Science—Permanence of Paper for Printed Library Materials, ANSI Z39.48-1984.

Library of Congress Cataloging-in-Publication Data

Williams, James D. (James David), 1941–
  Freshwater mussels of Alabama and the Mobile Basin in Georgia, Mississippi, and Tennessee / James D. Williams, Arthur E. Bogan, and Jeffrey T. Garner; foreword by E.O. Wilson.
      p. cm.
  Includes bibliographical references and index.
  ISBN-13: 978-0-8173-1613-6 (cloth : alk. paper)
  ISBN-10: 0-8173-1613-2 (alk. paper)
  1. Freshwater mussels—Alabama. 2. Freshwater mussels—Alabama—Mobile Bay Watershed. I. Bogan, Arthur E., 1950– II. Garner, Jeffrey T., 1965– III. Title.
  QL430.6.W55 2007
  594'.4--dc22
                    2007017010

Publication of this book was made possible by support from Alabama Department of Conservation and Natural Resources, Division of Wildlife and Freshwater Fisheries and Auburn University.

We take great pleasure in dedicating this book to

Herbert Huntington Smith (1851–1919)

and

Herbert D. Athearn (1923–)

Without their efforts to document the molluscan fauna of Alabama and adjacent states during the twentieth century, our knowledge and understanding of freshwater mussels of the region would be vastly diminished.

CONTENTS

Foreword

At first glance its title might easily give the impression that this book is a specialized treatise on an obscure group of organisms in a relatively restricted part of their range. Nothing could be farther from the truth. James D. Williams and his coauthors have thrown a searchlight on one of the richest and most imperiled assemblages in North America. The mussel fauna of Alabama and in the Mobile Basin of adjacent states comprises dozens of invertebrate equivalents of the Ivorybilled Woodpecker which, when their images are magnified, project an equal mystery and beauty.

Ever since I was a university student, already steeped in the natural history of my native state, I knew there was something extraordinary about the freshwater mussels of Alabama. They are extremely diverse, I knew. Now I realize, from the lucid and thorough monograph, that Alabama's 178 species are the most diverse in North America. I also knew a long time ago that the extensive damming of Alabama's rivers had extinguished a great many of these species, some unique in their anatomy and habitat adaptation. Now from these pages I learn that many of the 6 percent of U.S. species known to be extinct were in Alabama, and that today 48 species of those known from the state are federally listed as imperiled.

Mussels are not dismissible, even by those who have little interest in the natural world. These beautiful little animals are part of the economic and cultural history of Alabama and its neighboring states. Their presence is a signature of healthy aquatic ecosystems, to which they contribute as living water filters. Their great diversity, albeit ravaged by habitat destruction, makes them a regional treasure. Alabama, if measured by the standard of biological wealth, and it should be, deserves to be called the Aquatic State.

It has been my experience during many years spent in systematic biology and conservation that it is difficult for the public, indeed difficult even for biologists, to respond to statistics and other generalizations about valuable faunas alone. People do care about species of wildlife, however, if they can see a picture of it, know its name, and read what is known of its distribution and natural history. In addition to their contribution in mussel biology, this is what the authors have given us.

– Edward O. Wilson

Acknowledgments

A faunal treatise typically requires the assistance of scores of individuals, institutions and agencies during the course of its preparation. This book is no exception. The enthusiastic encouragement of colleagues who gave generously of their time and resources was a real inspiration in the completion of this project. In the following paragraphs we would like to express our appreciation to those who have contributed to the production of this book. We sincerely hope that no one was inadvertently omitted from these acknowledgments. However, in a work of this size and duration, with the large number of people from whom we received assistance, omission of one or more individuals is a real possibility. To those we extend a sincere apology. Any errors found within this volume are strictly the responsibility of the authors.

Without the assistance, knowledge, support and efficiency of Sherry Bostick this book would never have come to fruition. Her tolerance of and patience with we three authors is a demonstration of her virtues, especially in light of our numerous formatting, editing, grammatical and punctuation errors, some of which she had to correct more than once because we failed to learn from our mistakes the first time. Revelation of our errors and subsequent instruction was always imparted with patience and humor. Sherry was responsible for taking a large series of seemingly random texts, photographs, figures and maps cast her way and assembling them into this sizeable volume. In addition to correcting our mistakes and assembling the book, Sherry worked long hours during museum trips, entering literally tens of thousands of museum records into a database with an accuracy and speediness that was truly amazing. Days spent in often dark and dusty museums were almost always endured in good spirits even though there was little to look forward to at the end of the day except some exotic cuisine, modest libation and further quality time with us. She also scanned and cleaned all figures herein reproduced. The contributions of Sherry Bostick to this volume cannot be overemphasized. We are all considerably better authors for being associated with her for this extended endeavor.

We were very fortunate to have the assistance of Richard (Dick) T. Bryant who photographed all species of Alabama mussels, including males and females of the sexually dimorphic species. His photographs are outstanding and furnish an excellent record of the mussels of Alabama.

Cindy Bogan has been involved with this project since its inception, providing moral and logistical support. She reviewed many parts of this volume, always delivering thoughtful insights and pointed questions. In addition to contributing occasional assistance with field work, Lisa and Jacob Garner are gratefully acknowledged for their patience with and tolerance of Jeff's frequent physical and extended mental absence. Jeff would also like to express sincere appreciation to Dr. Paul Yokley, Jr., for a kindly and knowledgeable welcome to the study of malacology.

Logistics

This book is a product of nearly two decades of data accumulation and assembly. Much of the funding during the last five years, as well as funds to offset production costs, were provided by the federal Wildlife Conservation and Restoration Program through the Alabama Division of Wildlife and Freshwater Fisheries (ADWFF). Completion of any major undertaking requires considerable administrative support. We were fortunate to have excellent assistance during the last five years of the project. Nick Nichols and Stan Cook, ADWFF, Montgomery, Alabama, were instrumental in getting Wildlife Conservation and Restoration Program funding for the project and following it through to its completion. Dr. Elise Irwin, Auburn University, Auburn, Alabama, served as project coordinator and provided assistance throughout the project. Mary Lou Smith, Auburn University, supplied purchasing and contract support and expert guidance on numerous procedural and fiscal questions. Paul Kittle and Tom Haggerty at the University of North Alabama graciously provided laboratory space for much of the work on mussel soft anatomy.

Text Review

We were most fortunate to have an exceptional group of reviewers. The text was provided to reviewers based on their expertise with mussels of a particular taxonomic group or a particular geographic region. All text was reviewed by two or more colleagues who pointed out our errors as well as suggested additions and deletions. We especially appreciate their patience and attention to large volumes of material that was often accompanied by a request for a quick turnaround time. We offer our special thanks to Steve Ahlstedt, Bob Butler, Kevin Cummings, Sam Fuller (deceased), Mike Gangloff, Jim Godwin, Wendell Haag, John Harris, Paul Hartfield, Bob Howells, Don Hubbs, Paul Johnson, Jess Jones, Stuart McGregor, Russ Minton, Paul Parmalee (deceased), Malcolm Pierson (deceased), Andy Rindsberg, Kevin Roe, Doug Smith, Jamie Smith, Caryn Vaughn and Tom Watters. Text of this volume was greatly improved by the expert assistance of our copy editors, Lisa Kelly and Susan Harris.

Museum Support

We are especially grateful to the museums, their curators, collection managers and other personnel, as well as librarians and archivists, who provided us with convenient access to mussel collections, catalog data, lab equipment and library resources. They were always congenial and accommodating to us during our visits. They also responded to requests for loan material and provided clarification of locality data and catalog information. Their support represents a major contribution to this effort, and we express our sincere thanks to the following institutions and individuals: Academy of Natural Sciences of Philadelphia—George Davis, Dan Graf, Gary Rosenberg, Earle Spamer, Eileen Mathias; American Museum of Natural History, New York—Mary DeJong, Paula Mikkelsen; Auburn University Museum—Mike Gangloff, Michael Buntin; Canadian Museum of Nature, Ottawa, Canada—Jean-Marc Gagnon; Carnegie Museum of Natural History—John Rawlins, Tim Pearce; Cincinnati Museum of Natural History—Jeff Davis; Columbus Museum, Columbus, Georgia—Frank Schnell; Delaware Museum of Natural History—Jeff Halfpenny, Al Chadwick; Field Museum of Natural History, Chicago—Rütiger Bieler; Florida Museum of Natural History, University of Florida—Kurt Auffenberg, Gustav Paulay, John Slapcinsky, Fred Thompson; Illinois Natural History Survey—Kevin Cummings, Christine Mayer; Los Angeles County Museum—Lindsey Groves; McClung Museum, University of Tennessee—Paul Parmalee (deceased); Milwaukee Public Museum—Joan Jass; Mississippi Museum of Natural Science—Libby Hartfield, Bob Jones, Todd Slack; Museum of Comparative Zoology, Harvard University—Ken Boss, Richard Johnson, Ruth Turner (deceased); Museum of Fluviatile Mollusks—Herbert Athearn; Muséum national d'Histoire naturelle, Paris, France—Philippe Bouchet, Virginie Héros; Nationaal Natuurhistorisch Museum, Leiden, The Netherlands—Edmund Gittenberg, Jeroen Goud; Natural History Museum, Bern, Switzerland—Margret Gosteli; North Carolina State Museum of Natural Sciences—Betsy Bennett, Steve Busack, Suzanne Cooper, Jamie Smith; North Carolina State University, College of Veterinary Medicine—Jay Levine; North Carolina State University, D.H. Hill Library—Jonathan Underwood; Ohio State University Museum—Kathy Borror, David Stansbery, Tom Watters; Senckenberg Forschungsinstitut und Naturmuseum, Frankfurt am Main, Germany—Ronald Janssen, Eike Neubert; U.S. National Museum of Natural History, Smithsonian Institution—Cheryl Bright, Paul Greenhall, Robert Hershler, Tyjuana Nickens, Leslie Overstreet, David Steere, Daria Wingreen; University of Alabama, Department of Biological Sciences—David Campbell, Chuck Lydeard; University of Alabama Libraries—Marina Klarić; University of Michigan, Museum of Zoology—Liath Appleton, Jack Burch, Diarmaid O'Foighil, Randy Hoeh, Renee Sherman; University of North Alabama—Paul Kittle, Tom Haggerty; University of North Dakota, Grand Forks—Joseph Hartman; and Wagner Free Institute of Science—Susan Glassman.

Field Assistance

Mussels from Alabama streams are fairly well represented in museum collections. However, much of the museum material is decades old and new information on current distributions was needed. Numerous individuals assisted the authors with mussel sampling in Alabama and the Mobile Basin, sometimes in streams where conditions were less than pleasant, not to mention safe. Without the diligent sampling efforts of these individuals, recent distribution and conservation information would be incomplete. We appreciate the efforts of Steve Ahlstedt, Moez Ali, Dave Armstrong, Darryl Askew, Lee Bain, Dick Biggins, Holly Blalock-Herod, Jayne Brim Box, Michael Buntin, Bob Butler, Holly Cabrera, Andre Daniels, Bryce Daniels, Doug Darr, Phil Ekema, Ryan Evans, Keith Floyd, Andy Ford, Pam Fuller, Sam Fuller (deceased), Mike Gangloff, Jacob Garner, Lisa Garner, Traci George, Deanna Gilchrist, Alan Gooch, Chris Greene, Wendell Haag, Jay Haffner, Tom Haggerty, Dennis Haney, April Hargis, Paul Hartfield, Phillip Henderson, Jeff Herod, Amy Hester, Eric Hill, Gary Hill, Tim Hogan, Chuck Howard, Don Hubbs, Mark Hughes, Rob Hurt, Elise Irwin, P.J. Jessie, Judy Johnson, Paul Johnson, Carol Johnston, Alan Jones, Philip Kilpatrick, Ricardo Lattimore, Bob Lewis, Jay Lowery, Chuck Lydeard, Monte McGregor, Stuart McGregor, Jerry Moss, Christine O'Brien, Noel Ocampo (deceased), Pat O'Neil, Malcolm Pierson (deceased), Jim Piper, Sandy Pursifull, Shane Ruessler, Jeanne Serb, Rob Shearer, Doug Shelton, Tom Shepard, Jeffrey Sides, Brett Smith, Brooke Smith, Andrew Suddith, Austin Suddith, Ken Weathers, Doug Weaver, Paul Yokley, Jr. and Joe Zolczynski. Logistical assistance for field-related activities was provided by Freda McCrary and Brenda Morrison, ADWFF; Jack Pounders, Southeastern Divers Incorporated; and Larry Neill, Tennessee Valley Authority.

Information

Throughout the course of this project many individuals provided specimens, unpublished data, reports, publications and advice on a variety of topics involving Alabama and Mobile Basin mussels. This information was provided via personal communication at meetings and workshops, e-mail correspondence and sometimes lengthy phone conversations. We are indebted to the following individuals who gave freely of their time and expertise to assist in the completion of this volume: Steve Ahlstedt, Bob Baker, Dick Biggins, Philippe Bouchet, Michael Buntin, Bob Butler, Stephanie Chance, Ashley Dumas, Bill Eschmeyer, Ryan Evans, Mike Gangloff, Wendell Haag, Fred Harders, Paul Hartfield, Don Hubbs, John Jenkinson, Paul Johnson, Jess Jones, Jim Layzer, Harry Lee, Chuck Lydeard, Henry McCullagh, Stuart McGregor, Dick Neves, Sabrina Novak, Evan Peacock, Richard Petit, Malcolm Pierson (deceased), Sandy Pursifull, Shane Ruessler, Doug Shelton, Peggy Shute, Damien Simbek, Carson Stringfellow, Mel Warren and Paul Yokley, Jr.

Genetics Data

In an effort to resolve taxonomic problems, both at the species and genus levels, we obtained results of genetic analyses from several molecular biologists. These individuals not only furnished published and unpublished data but also provided assistance in interpreting results, allowing us to make more informed taxonomic decisions. We wish to thank the following individuals: Dave Berg, David Campbell, Curt Elderkin, Randy Hoeh, Karen Kandl, Steve Karl, Todd Levine, Chuck Lydeard, Morgan Raley, Kevin Roe and Jeanne Serb.

Photographs/Artwork

We extend our gratitude to individuals and institutions that contributed photographs and illustrations for use in this volume. Photographs and artwork are critical to the production of any book on a group of animals, especially mussels with their interesting shell shapes, sculpture and color. Photographs of conglutinates, mantle displays and other soft parts were generously provided by Chris Barnhart, Jayne Brim Box, Noel Burkhead, William Henley, Alex Huryn, Paul Johnson, Eike Neubert, Dick Neves, Christine O'Brien, Bill Roston and Shane Ruessler. Zebra Mussel photographs were supplied by Jerrie Nichols and Don Schloesser. In some instances photography proved difficult or impossible to illustrate selected features of mussel anatomy and morphology; in these situations we secured the services of Susan Trammell who created excellent illustrations. Malcolm Pierson (deceased) provided beautiful color slides from his collection of Alabama stream photographs. We also thank Merrily Harris, Hoole Special Collections Library Archives at The University of Alabama, Tuscaloosa, and Bill Tharpe, Alabama Power Company, Corporate Archives, Birmingham, Alabama, and their institutions, for permission to use historical photographs of Alabama streams. Jonathan Raine, North Carolina State Museum of Natural Sciences (NCSM), assisted with the scanning of some figures.

Mapping Assistance

The art and science of producing good maps is essential to providing a visual expression of geographic data. This task was most ably accomplished by the following individuals: Gareth Turner revised distribution and introductory maps; David Coley created introductory maps; and Britton Wilson created drafts of distribution maps for all species. The distribution base map was generously provided by the Geological Survey of Alabama (GSA).

Institutional Abbreviations

The following abbreviations are used for state and federal agencies, nongovernmental organizations and institutions housing permanent mussel collections referenced in the text. Although some museum collections may have more than one abbreviation currently in use, the acronym given below is the one utilized in this volume.

ADCNR   Alabama Department of Conservation and Natural Resources
ADWFF   Alabama Division of Wildlife and Freshwater Fisheries
AFS   American Fisheries Society, Bethesda, Maryland
ALNHP   Alabama Natural Heritage Program, Auburn University, Auburn
AMNH   American Museum of Natural History, New York, New York
ANSP   Academy of Natural Sciences, Philadelphia, Pennsylvania
AUM   Auburn University Museum, Auburn, Alabama
BMNH   British Museum of Natural History, London, England
CM   Cincinnati Museum of Natural History, Cincinnati, Ohio
CMNH   Carnegie Museum of Natural History, Pittsburgh, Pennsylvania
CMNML   Canadian Museum of Nature, Mollusks, Ottawa, Ontario (formerly National Museums of Canada [NMC])
DMNH   Delaware Museum of Natural History, Wilmington
FMNH   Field Museum of Natural History, Chicago, Illinois
GSA   Geological Survey of Alabama, Tuscaloosa
ICZN   International Commission of Zoological Nomenclature, The Natural History Museum, London, England
INHS   Illinois Natural History Survey, Champaign
IUCN   International Union for Conservation of Nature and Natural Resources
MCZ   Museum of Comparative Zoology, Harvard University, Cambridge, Massachusetts
MFM   Museum of Fluviatile Mollusks, Cleveland, Tennessee (currently housed at the North Carolina State Museum of Natural Sciences, Raleigh)
MMNS   Mississippi Museum of Natural Science, Jackson
MNHN   Muséum national d'Histoire naturelle, Paris, France
MPM   Milwaukee Public Museum, Milwaukee, Wisconsin
NCSM   North Carolina State Museum of Natural Sciences, Raleigh
OSUM   Ohio State University Museum, Columbus
RMNH   Rijksmuseum van Natuurlijke Historie, Leiden, The Netherlands (now known as Nationaal Natuurhistorisch Museum [NNM])
SMF   Senckenberg Forschungsinstitut und Naturmuseum, Frankfurt am Main, Germany
TNARI   Tennessee Aquarium Research Institute (formerly Southeastern Aquatic Research Institute)
TNC   The Nature Conservancy
TVA   Tennessee Valley Authority
UAUC   University of Alabama Unionid Collection, Tuscaloosa
UF   University of Florida, Florida Museum of Natural History, Gainesville
UMMZ   University of Michigan Museum of Zoology, Ann Arbor
UNA   University of North Alabama, Florence
USACE   U.S. Army Corps of Engineers
USFS   U.S. Forest Service, Department of Agriculture
USFWS   U.S. Fish and Wildlife Service, Department of the Interior
USGS   U.S. Geological Survey, Department of the Interior
USNM   U.S. National Museum of Natural History, Smithsonian Institution, Washington, DC
UT   University of Tennessee, McClung Museum, Paul W. Parmalee Malacology Collection, Knoxville
WFIS   Wagner Free Institute of Science, Philadelphia, Pennsylvania

Chapter 1
Introduction

To those individuals who have never worked with unionoid mussels, these animals may appear to be about as boring as any organism could possibly be. After all, they have no head and only one foot and spend most of their lives in the bottom of a stream or lake with only their posterior ends exposed. Mussels are poorly understood, but they are in fact very interesting animals. In addition to their diversity and complex life histories, they are believed to have the longest lifespan of any living freshwater invertebrate in the world, some living more than 100 years.

One might well ask why a book should be written about Alabama freshwater mussels. This volume was driven by a need for information on the taxonomy, biology, ecology, zoogeography and conservation of unionoid mussels of the southeastern U.S. Mussel diversity in Alabama and the Mobile Basin is higher than that of any other area in the U.S. Approximately 60 percent of all U.S. mussel species are known from Alabama, which is reason enough to produce a treatise on this fauna. The demise of the North American mussels during the twentieth century due to habitat alterations represents an equally compelling reason for this book.

The southeastern U.S. is the center of aquatic biodiversity in North America, and Alabama is near the epicenter. The numbers of mussels, snails, crayfish and aquatic turtles in Alabama exceed those of all other states, and Alabama is second only to Tennessee in the number of freshwater fish species (Etnier and Starnes, 1993; Lydeard and Mayden, 1995; Boschung and Mayden, 2004). There are currently 2 families and about 300 species of mussels recognized in the U.S. and Canada (Turgeon et al., 1998). Alabama has members of both families, Margaritiferidae and Unionidae, including 178 species representing 43 genera. Only 7 genera found in the U.S. and Canada are not known from the state. Other southeastern states with high mussel diversity include Tennessee with 129 species (Parmalee and Bogan, 1998), Georgia with 123 species and Kentucky with 104 species (Cicerello and Schuster, 2003).

Many mussels are aesthetically pleasing to the eye. The exterior of the shell, or periostracum, of many species is dark olive to brown or black, but some are very colorful. For example, various members of *Epioblasma*, *Lampsilis* and *Villosa* are shades of yellow or green with dark green bands radiating across the shell disk. The shell interior can be even more colorful, with pearly nacre ranging from white or bluish white to purple, pink, orange, salmon and peach. Many mussels

have oddly shaped shells, some adorned with ridges, knobs, pustules or spines. It was likely the odd shapes and colorful shells that made mussels attractive to the naturalists and conchologists in the 1800s and early 1900s. While there are still a few conchologists that buy, sell and trade mussel shells, most deal primarily in marine shells.

Mussels have been utilized for a variety of purposes. Native Americans harvested them for food in large quantities, as suggested by the shell middens often associated with settlements along large rivers. The shells were used for a variety of purposes, including use as ornaments, jewelry and tools. Research on mussel shells from Native American middens has provided biologists with interesting insights into the prehistoric freshwater mussel communities. Some mussel species are currently rare or absent from a particular river but have been found to be common in middens. One recently described mussel species was discovered in Native American middens but went extinct before its collection by naturalists.

From the mid- to late 1800s early American settlers harvested mussels intermittently for pearls. During the late 1800s and first half of the 1900s mussel shell was harvested for production of pearl buttons. The button industry became obsolete with the development of plastics. In recent decades the interest in mussels has centered on the harvest of shell for the cultured pearl industry. This industry is highly unpredictable and typically goes through cycles depending on market demand. The most recent boom period was in the late 1980s and early 1990s when high-quality shell sold for up to $9.00 per pound. Most of the shell was harvested from reservoirs and tailwaters of large rivers and was composed primarily of common, widespread, thick-shelled species.

Mussels received considerable attention from the 1820s to the early 1900s when naturalists raced to describe this highly diverse and conchologically variable fauna. Most of the descriptions were based on shell characters with little or no study of soft anatomy. The individuals publishing the descriptions adhered to the prevailing typological species concept, which resulted in the fauna being overly described, creating hundreds of species-level synonyms.

During the past three decades great strides have been made in unraveling the biology, phylogeny and distribution of mussels in the southeastern U.S. However, much remains to be learned, as is evidenced by the recent discovery and description of a new genus, *Hamiota* Roe and Hartfield, 2005, and a new species,

Pleurobema athearni Gangloff et al., 2006, both of which occur in Alabama. Application of molecular genetic techniques to phylogenetic problems has been and will continue to be a useful tool in resolving taxonomic relationships at the genus and species levels. This will be essential in assisting conservation biologists in making management decisions. It should be noted that our list of mussel species is in a state of flux, due primarily to application of molecular techniques to the fauna. Taxonomy employed in this book is a reflection of current understanding and will undoubtedly be subject to future changes.

The incredible aquatic diversity of the southeastern U.S. was drastically reduced during the past century. Mussel harvest resulted in localized depletions of populations of some species, but it was other anthropogenic factors—including dams, channelization, industrial and municipal pollution, erosion and silt—that destroyed riverine habitat and the associated aquatic fauna. The introduction of nonindigenous species has also been a factor in the demise of native species. Most mussels in this region have suffered from substantial habitat loss, and many species are now confined to small portions of their former ranges.

Today much of the remaining mussel diversity in Alabama and the Mobile Basin is in imminent danger of extinction due primarily to habitat loss. However, recent advances in propagation and husbandry techniques may afford protection from immediate extinction for some species and public concern for protection and management of aquatic habitats appears to be on the increase. More intensive restoration efforts are needed or species will continue to be lost. While it is too late for the species that became extinct during the past century, there is still time to rescue a significant percentage of our mussel fauna. A Herculean effort, with new and innovative approaches to aquatic ecosystem conservation, will be required to stem the tide of mussel extinctions in Alabama and the remainder of the U.S. One of the purposes of this volume is to provide a foundation on which future conservation and research efforts can be based.

Chapter 2
Mussel Studies in Alabama

Natural history explorations of Alabama and the Mobile Basin—geographically remote from museums, libraries and the early budding cultural centers of the northeastern U.S.—did not begin in earnest until the early 1800s. Discovery of the diverse southeastern flora and fauna by natural historians prompted intense interest in the U.S. as well as abroad. Mollusks were among the first organisms to attract the attention of naturalists. Empty dry shells of mussels and snails required little or no effort to prepare as colorful and interesting natural history objects. The fact that they were abundant, with tremendous diversity of shapes and sizes, and easily transported resulted in large numbers being shipped to institutions and individual collectors in the northeastern U.S. The first shipments of mussel shells from the southeast apparently arrived in the northeast during the late 1820s, and species descriptions soon followed.

The first publications on mussels from Alabama and the Mobile Basin were descriptions of new species that appeared in 1830. The most active workers in Alabama during the 1830s were Isaac Lea and Timothy Abbott Conrad, both of whom lived in Philadelphia, Pennsylvania, and were very active in the Academy of Natural Sciences of Philadelphia (ANSP), where they worked on fossil and recent mollusks. Initially, there was a great deal of admiration and respect between these two natural historians. However, it did not take more than a couple of years before a rivalry developed, resulting in an intense competition for each man to lay claim to the new species of mussels and describe them before his competitor. This competition persisted for decades and along with it came copious criticism of each other's work.

Isaac Lea performed little field work but was the most prolific describer of freshwater mollusks of the U.S., using material shipped to him by others (including Reverend G. White, Bishop S. Elliott, W.H. De Camp, W. Spillman, E.R. Showalter, G. Hallenbeck, J. Postell, Judge C. Tait, Professor Tuomey, L.B. Thornton, Dr. Budd, W. Gesner, B. Pybas and C.M. Wheatley). Lea published a total of 279 titles, 239 of which dealt with mollusks (Scudder, 1885). His first malacological paper was published in 1828, and he continued publishing until late in life, with his last paper appearing in 1876. Lea described a total of 1,842 new mollusk species; of these, 851 were of recent and fossil unioniform bivalves, including 180 recent taxa from Alabama and the Mobile Basin.

Lea's malacological collections were left primarily to the Smithsonian Institution, U.S. National Museum of Natural History (USNM), including most of his figured and type specimens (Johnson, 1956, 1971, 1974, 1980, 1998; Johnson and Baker, 1973; Boyko and Sage, 1996). Several other museums contain significant collections of Lea specimens, including the ANSP, Harvard University's Museum of Comparative Zoology (MCZ) and some European museums.

Isaac Lea (1792–1886). Photograph from *The published writings of Isaac Lea, LL.D.* (Scudder, 1885).

Most of Lea's mussel descriptions appeared in the *Transactions* and the *Proceedings of the American Philosophical Society* and in the *Journal* and the *Proceedings of the Academy of Natural Sciences of Philadelphia*. He privately reprinted papers from the *Transactions* and the *Journal* in his *Observations on the Genus Unio* (Bogan and Bogan, 2002), which amounted to 13 volumes (1834–1874) and 3 indices (Lea, 1867a, 1869b, 1874f). Only 250 copies of the *Observations* were published. Lea listed the described species of freshwater bivalves and provided his own classification in a synopsis, of which four editions were published (Lea, 1836, 1838b, 1852a, 1870). He chose to place most species in the genera *Unio* Retzius, 1788, *Anodonta* Lamarck, 1799, and *Margaritana* Schumacher, 1816, until the anatomy of described species was better understood.

A major upheaval occurred in Lea's life with Conrad's publication of *A Synopsis of the Family of Naiades of North America* (1853). In his synopsis Conrad corrected some of Lea's dates of publication of named taxa and gave priority of many to the pioneering naturalist Constantine Samuel Rafinesque. Lea claimed the date for a name as the date on which that name was read before a scientific society (e.g., ANSP and American Philosophical Society), while Conrad claimed the date for a name to be the date when the description appeared in print. This irritated Lea to the extent that he published two editions of a "Rectification" of Conrad's synopsis (Lea, 1854, 1872a), in which he claimed to correct erroneous dates provided by Conrad. Lea then began to publish short, mostly four to six line, descriptions of new species in Latin, accompanied by type locality and the collector of the specimens. This established the earliest date for Lea's species names. He would subsequently publish a full description with a figure for most species. Later in his career Lea added brief notes on unioniform anatomy and glochidia.

Conrad's views differed from those of Lea with regard to the establishment of species name priority long before his 1853 synopsis. Conrad (1834b) remarked, "If in any instance I have thus introduced a shell of which a *published* description already exists, I will cheerfully resign the name I may have given it and 'render unto Caesar the things which are Caesar's'; but where a claim is made upon such shells, merely because the descriptions may have been read to the members of some institution, I shall certainly not feel myself compelled to make any such restitution. The day of publication can alone decide the question of this kind, else a naturalist might describe fifty species at random, read them at a meeting of some scientific institution, and then cull out at his leisure a half dozen new ones, and claim them as his own six years afterwards, although perhaps they had been published under other names, five years before his descriptions appeared. Other reasons could be given, but it is unnecessary to multiply them in so plain a case."

Timothy Conrad was occasionally the benefactor of material shipped to the ANSP, but he also took extended trips to collect fossil and recent mollusks for himself and other naturalists, many of whom contributed funds to support his travels. One such trip

was an expedition to the southeastern U.S. with the ultimate goal of reaching Claiborne, Alabama, where he was the invited guest of Judge Charles Tait. Conrad reached Wilmington, North Carolina, by schooner in December 1832 and made his way over land and by steamer to Claiborne, by way of Columbus, Georgia, in February 1833 (Wheeler, 1935).

Conrad conducted initial explorations in Alabama in March and April 1833 along the Alabama, Tombigbee and Black Warrior rivers. These trips were on steamers that hauled cargo and passengers between various river ports. In May 1833 he embarked on a steamer bound for Tuscaloosa, where he began an extended six-week trip to explore the headwaters of the Black Warrior River and a reach of the Tennessee River and its tributaries in northern Alabama. This trip was very successful, and Conrad collected large numbers of shells from which he later described new species. He returned to Claiborne and continued to collect specimens and write before departing for Mobile, in February 1834, where he boarded a schooner and returned to New York. Data and specimens collected on this trip formed the basis of several classic papers by Conrad on fossil and living mollusks (Wheeler, 1935).

Timothy Abbott Conrad (1803–1877). Photograph from Wheeler (1935), with permission of the Paleontological Research Institution, Ithaca, New York.

In 1834 Conrad published two papers describing a total of 32 new unionid species, all occurring in Alabama and most recognized as valid today (Conrad, 1834a, 1834b). Conrad (1834b) provided a brief table of synonymy for many of the mussels previously described, as well as some discussion of freshwater mussel distribution by river drainage, noting differences between the Tennessee River drainage and Mobile Basin faunas. Conrad (1834b) also included colored plates of mussel shells.

The first attempt to provide a comprehensive overview of freshwater and terrestrial mollusks of Alabama was published in the *Alabama Geological Survey Report of Progress for 1876* by James Lewis. This publication provided a complete list of species with distributional annotations and an appendix with a brief discussion of possible synonyms for selected species. The list included a total of 256 species that were arranged in 3 genera, *Unio* (238 species), *Margaritana* (13 species) and *Anodonta* (5 species). Lewis, a conchologist who lived in Mohawk, New York, based this report on shells collected from Alabama and adjacent states by several conchologists and naturalists. One individual singled out by Lewis for his contribution prior to 1861 was Dr. E.R. Showalter of Mobile, formerly of Uniontown, who provided specimens and notes on geographical distribution. Other individuals contributing shells from Alabama and acknowledged by Lewis included Truman H. Aldrich of Montevallo, formerly of Selma; L.B. Thornton, Esq., and B. Pybas, both residents of Tuscumbia; and Dr. Eugene Allen Smith, Alabama State Geologist in Tuscaloosa. Lewis also recognized several individuals for their efforts collecting shells in the Chattahoochee River drainage prior to 1861, including Hugh M. Neisler, Garrett Hallenbeck and William Gesner, all from Columbus, Georgia, as well as the Right Reverend Stephen Elliott of Savannah, Georgia (Lewis, 1876).

Following his overview of Alabama mollusks, Lewis (1877) revisited the Alabama mussel fauna in a publication on the Unionidae of Ohio and Alabama. This publication represents the first attempt to understand the geographic distribution and evolutionary relationships of the Alabama mussel fauna. It also includes a more thorough discussion of possible synonyms than the preceding work. Using conchological characters, Lewis identified several species groups as closely related. Some of the taxonomic problems recognized by Lewis (e.g., species currently recognized in the genus *Toxolasma*) have yet to be resolved.

Between 1883 and 1934 Dr. Samuel Hart Wright (1825–1905) and his son Berlin Hart Wright (1851–1940) described 52 species of mollusks from Florida and adjacent states, including Alabama (Johnson, 1967b). Samuel H. Wright also published a short note

on the similarity of the mussels in peninsular Florida to those of adjacent states (S.H. Wright, 1891), which represents one of the earliest attempts to report on the geographic distribution of the Gulf Coast mussel fauna. Berlin H. Wright described several species of mussels from Alabama in the late 1800s and early 1900s. He developed a checklist of North American Unionidae based on Lea's synopsis with the addition of species described since 1870, listing species under *Anodonta*, *Margaritana* and *Unio* (B.H. Wright, 1888). In 1902 B.H. Wright, along with Bryant Walker, produced a privately published checklist of North American unionoids (Wright and Walker, 1902). This was based primarily on Charles Torrey Simpson's 1900 synopsis but also included species described since the publication of that volume.

Charles Torrey Simpson (1846–1932). Photograph from *Florida's Pioneer Naturalist* (Rothra, 1995).

In 1900 Charles Torrey Simpson published two papers that included the mussel fauna of Alabama and the Mobile Basin. The comprehensive *Synopsis of the Naiades, or Pearly Fresh-Water Mussels* (Simpson, 1900b) provided information on most species throughout the world. Publication of this volume was suggested and supported by the Reverend and Mrs. L.T. Chamberlain, son-in-law and daughter of Isaac Lea. It was to be essentially the fifth edition of Lea's synopsis. It also represented an attempt to split the genus *Unio* into recognizable lineages based in part on soft anatomy, especially that of gravid females. This represented a major step forward in understanding the evolutionary relationships of the Unionidae. Simpson (1900a) described several new mussel species and provided figures and additional conchological characters for some that were previously described but not illustrated. This work included descriptions of new species from Alabama and the first illustrations of several species previously described from the state. In 1914 Simpson published a three-part monograph entitled *A Descriptive Catalogue of the Naiades, or Pearly Fresh-water Mussels*. It was similar to the earlier publication (Simpson, 1900b) but included species described since 1900 as well as information on shell morphology and distribution, including the type locality for each species.

Herbert Huntington Smith (1851–1919). Photograph from Hoole Special Collections Library, The University of Alabama, Tuscaloosa.

Herbert Huntington Smith arrived in Alabama in 1901 and remained in the state until his untimely death at the age of 68 in 1919, when he was killed in a train accident on the University of Alabama campus. By all accounts Smith was an outstanding zoological collector and field biologist. From 1901 to 1909 he was employed by various individuals and institutions to conduct field work in Alabama and adjacent states. During much of this time he was working for a group of conchologists, whom he fondly referred to as "The Syndicate," that paid a portion of his salary and provided some funding for travel expenses in return for mollusks collected from Alabama and surrounding states. The number of individuals in The Syndicate varied from year to year but included such notables as Truman H. Aldrich, George H. Clapp, John B. Henderson, Jr., Henry A. Pilsbry and Bryant Walker. In 1909 Smith began work as curator of the Alabama Museum of Natural History, but he continued to collect mollusks until his death. His unionid collections are housed today in museums around the U.S. and serve as the foundational record of historical diversity and abundance in Alabama.

Bryant Walker (1856–1936), a wealthy Detroit lawyer, was a private shell collector and prominent member of The Syndicate. His support of freshwater mollusk work in the early 1900s represents a major contribution to southeastern malacology. During this period he financed the publication of Simpson's 1914 catalog of freshwater mussels, a landmark volume. Walker's personal mollusk collection was donated to the University of Michigan Museum of Zoology (UMMZ).

Anson A. Hinkley (1857–1920) of Aledo, Illinois, made a mollusk collecting trip to Alabama in fall 1903 and returned with some of his "conchological friends" during fall 1904. The second trip also included areas of Mississippi. They traveled by railroad, and most of their collecting was in streams crossed by rail lines or near train stops and transfers. Results of their collecting efforts were published in two papers (Hinkley, 1904, 1906). It is interesting to note that Hinkley (1904) placed all mussel species in *Unio* or *Anodonta*. However, in the second paper Hinkley (1906) adopted the generic classification proposed by Simpson (1900b). Both papers provided important notes on distribution and abundance.

Arnold Edward Ortmann (1863–1927) worked as curator of invertebrate zoology at the Carnegie Museum of Natural History (CMNH) in Pittsburgh, Pennsylvania, from 1903 until his death in 1927. From 1909 to 1927 he also taught classes at the University of Pittsburgh. During this period he made significant contributions to the biology, ecology, evolution, zoogeography and conservation of aquatic invertebrates, especially unionid mollusks. In particular, his work on soft anatomy of unionids was a major advancement beyond the efforts of previous workers and has proved to be a lasting contribution to our knowledge of this group. Much of Ortmann's unionid mollusk research centered on the fauna of the Tennessee, Cumberland and Ohio rivers. His paper on the mussel fauna of the Tennessee River drainage downstream of Walden Gorge serves as a baseline for modern comparisons with the preimpoundment fauna (Ortmann, 1925), and his discussion of the loss of the Muscle Shoals fauna when the Tennessee River was impounded was one of the first proposals for protection of mollusk habitat. Ortmann also published on Mobile Basin and Gulf Coast unionids in a series of papers on taxonomy and soft anatomy (e.g., Ortmann, 1913b, 1914, 1915, 1921, 1923a, 1923b, 1924a).

Lorraine Scriven Frierson (1861–1933) was an amateur conchologist who lived in Louisiana and worked primarily on southeastern mussels. Between 1900 and 1927 he addressed taxonomic problems and described new species of mussels, several from Alabama. Frierson (1927) published a classification and checklist of North American mussels that was his most important work and included a synonymy and numerous generic realignments as well as the description of new taxa.

> "The fact that Conrad made mistakes at times, can have no weight, for no author is free from these" Frierson (1914).

Henry van der Schalie (1907–1986) was curator of mollusks at the UMMZ and actively worked on Alabama unionids during the 1930s. Much of his work was based on material collected from various parts of Alabama and deposited in the UMMZ. He traveled to Alabama for work in the Cahaba River, which resulted in a published survey (van der Schalie, 1938a). Other important work included papers on the faunas of the Tombigbee and lower Tennessee rivers prior to modern habitat alterations (van der Schalie, 1938c, 1938d).

In 1940 van der Schalie published a paper on the ecology and diversity of mussels in the Chipola River in northwestern Florida and headwater tributaries in Houston County, Alabama. This publication appeared in 1940, but most of the material on which it was based was collected between 1915 and 1918 by C.A. Burke and J. Burke, brothers who made many collections in southeastern Alabama and northwestern Florida for H.H. Smith. Approximately 40 years later, van der Schalie (1981a, 1981b) again published on Alabama mollusks. These papers were stimulated by environmental battles resulting from legal challenges to the proposed construction of the Tennessee Tombigbee Waterway.

William J. Clench (1897–1984) and P. Sheldon Remington, Jr., (1899–1975) collected freshwater and terrestrial mollusks on a 1924 expedition across the southeast funded by the UMMZ and Bryant Walker (Remington and Clench, 1925). They collected unionids from Huntsville, Alabama, to Chattanooga, Tennessee. The specimens are now housed in the UMMZ.

Clench and Ruth D. Turner (1914–2000), of the MCZ, reported on the freshwater mollusks of Alabama, Georgia and Florida, from the Escambia River drainage east to the Suwannee River drainage (Clench, 1955; Clench and Turner, 1956). This survey came about as a result of the planned construction of Jim Woodruff Dam on the Apalachicola River by the U.S. Army Corps of Engineers (USACE). With the realization that this dam would obliterate most of the upstream riverine fauna, including lower reaches of the Chattahoochee River in Alabama, a group of biologists from the Florida Museum of Natural History coordinated a biological inventory of the area. An additional grant from Harvard University extended the scope of the survey to include drainages from the Escambia River to the Suwannee River (Clench and Turner, 1956). This represented the most intensive sampling of eastern Gulf Coast drainages since the collections made by the Burke brothers in the early 1900s.

Most of the mussel work performed by the Tennessee Valley Authority (TVA) since its inception in 1933 has been related to the conservation of mussel resources. Much of the early work was unpublished but provides insight into changes in mussel populations caused by impoundment of the river (e.g., Cahn, 1936; TVA, 1964, 1966). Billy G. Isom and colleagues published results of several mussel surveys of Tennessee River tributaries (Isom 1968; Isom and Yokley, 1968, 1973; Isom et al., 1973) and an assessment of Tennessee River commercial mussel resources (Isom, 1969). The last record of many riverine species in the Tennessee River proper is found in Gooch et al. (1979), a TVA report that was released only in draft form. During the 1980s TVA initiated the Cumberlandian Mollusk Conservation Program, which included surveys as well as studies on life history and ecology. Paint Rock and Elk rivers were included in the program. People involved in these studies who worked in Alabama were Steven A. Ahlstedt, Billy G. Isom, John J. Jenkinson, Leroy M. Koch and Charles F. Saylor. Important work on in vitro culture of mussels was carried out in Muscle Shoals by Isom, in collaboration with Robert G. Hudson of Presbyterian College, Clinton, South Carolina (Isom and Hudson, 1982; Hudson and Isom 1984). This work was later put to practical use by Donald C. Wade, Damien J. Simbek and others utilizing newly cultured juvenile mussels in toxicity tests.

In 1955 Herbert D. Athearn moved from Boston, Massachusetts, to Cleveland, Tennessee, located near the headwaters of the Mobile Basin. Upon his arrival he began to collect aquatic mollusks throughout the southeastern states, including all parts of Alabama, continuing at a steady pace until 2002. His personal collection, the Museum of Fluviatile Mollusks (MFM), was maintained in his home until 2007, when it was donated to the North Carolina State Museum of Natural Sciences (NCSM). Athearn collected mussels as they were slipping into extinction with impoundment of various rivers, at the same time discovering new species in areas that had been previously sampled. Among Athearn's discoveries were three new species, *Alasmidonta mccordi*, *Lampsilis haddletoni* and *Villosa choctawensis* (Athearn, 1964), all described from Alabama. Athearn's vast field experience provided a tremendous foundation for the study of southeastern aquatic mollusk distribution and ecology.

Herbert D. Athearn (1923–). Photograph from *The Dixie Ranger*, a newsletter of The Dixie Company, a stove manufacturer in Cleveland, Tennessee, Volume 2(11), February 1959.

In the late 1950s an important study of commercial mussel stocks in the Tennessee River was carried out by George D. Scruggs, Jr., of the U.S. Fish and Wildlife Service (USFWS) (Scruggs, 1960). Wheeler Reservoir was the major focus of the study. This paper allowed documentation of faunal changes in the Tennessee River reservoirs as the mussel community stabilized following impoundment. During the 1970s the USFWS took on an instrumental role in detailing distributional patterns of mussels in Alabama, with special emphasis on imperiled species for possible protection under the federal Endangered Species Act. In addition to surveys performed by USFWS personnel, numerous studies were funded by the agency. From the 1970s through the turn of the century, field offices that had influence on mussel studies in Alabama and the Mobile Basin include those in Daphne, Alabama, Jackson, Mississippi, and Asheville, North Carolina. Personnel involved in the studies include Richard G. Biggins, Robert Bowker, Robert S. Butler, Paul D. Hartfield, Jeff Powell and James Stewart.

In the 1960s Paul Yokley, Jr., professor of biology at the University of North Alabama (UNA), completed a doctoral degree at Ohio State University under Dr. David H. Stansbery. His dissertation research, a gross and histological examination of *Pleurobema cordatum* anatomy (Yokley, 1968), is one of the most detailed assessments of a freshwater mussel's anatomy to date. Students of Yokley who later worked in the field of malacology include Zachary H. Bowen, Jeffrey T. Garner, Charles H. Gooch, Michael A. Hoggarth, Stuart W. McGregor, Terry D. Richardson, Jeff Selby, Damien J. Simbek and Donald C. Wade. Yokley (1972) provided a detailed account of the life history of *P. cordatum* that was one of the first to incorporate histological examination of gonads with brooding period and host fish determination. Yokley also surveyed many Alabama rivers and creeks for mussels. Most material from these surveys was deposited in the Ohio State University Museum (OSUM). During the late 1980s Yokley, along with Thomas M. Haggerty and several students, including Garner, Bowen and Alan Gooch, worked toward furthering in vitro mussel culture techniques previously developed by TVA.

Beginning in the 1960s Richard I. Johnson, MCZ, published a number of papers on unionids that provided some clarification on distributions of selected species in Gulf Coast drainages, including those of Alabama (Johnson, 1967a, 1968, 1969a). Johnson also included references to the Gulf Coast fauna in papers dealing with the Atlantic Coast region (Johnson, 1970) and peninsular Florida (Johnson, 1972b). He published monographs on *Medionidus* (Johnson, 1977) and *Epioblasma* (Johnson, 1978), as well as described *Utterbackia peggyae* (Johnson, 1965) and *Margaritifera marrianae* (Johnson, 1983). Additionally, Johnson published a series of useful type catalogs of various authors and institutions (see Appendix).

Paul W. Parmalee (1927–2006) came to the McClung Museum, University of Tennessee (UT), Knoxville, in 1973 to establish a zooarchaeology program and began working on the mussel fauna of Tennessee. His efforts include *The Freshwater Mussels of Tennessee* (Parmalee and Bogan, 1998). He was involved with the identification of animal remains, including mussels, from archaeological middens and rock shelters along the Tennessee River (Parmalee, 1994). Parmalee's students, including Arthur E. Bogan and Neil D. Robison, have provided identification of archaeological shell materials from Alabama (Robison, 1983; Robison and Bogan, unpublished data). Arthur Bogan incidentally collected unionids during his surveys of the gastropods of the Cahaba and Coosa River drainages (Bogan et al., 1995). The most recent summary of the freshwater bivalve fauna of the Tennessee River in northern Alabama, comparing the archaeological record with pre- and postimpoundment records, was produced by Parmalee and another of his students, Mark H. Hughes (Hughes and Parmalee, 1999).

In August 1970 James D. Williams arrived at the Mississippi State College for Women in Columbus. Together with Glenn H. Clemmer, an ichthyologist at Mississippi State University in Starkville, and their students, Williams began to sample fish and mussels in the Tombigbee River drainage in western Alabama and eastern Mississippi. The inspiration for the intensive sampling was the impending destruction of the Tombigbee River by the USACE with the construction of the Tennessee Tombigbee Waterway. This waterway project not only destroyed the largest remaining unimpounded river in the Mobile Basin, it connected the river to the Tennessee River drainage via an overland canal. In the spring of 1971 Dr. David H. Stansbery visited the biologists in Mississippi and made arrangements to have the Tombigbee mussel collections shipped to the OSUM in Columbus, Ohio. Dr. Stansbery was very excited to see species of *Pleurobema*, *Epioblasma* and *Quadrula*, including several that had not been collected in almost 50 years. The Tombigbee River sampling continued until 1972.

In August 1972 Williams moved to the Tuskegee Institute, Tuskegee, Alabama. Soon after his arrival Williams and a graduate student, Charles S. Baldwin, began to sample mussels in the Cahaba River drainage, including most of the stations visited by van der Schalie in 1938. With encouragement from Samuel L.H. Fuller, this project evolved into an M.S. thesis (Baldwin, 1973). During his tenure at the Tuskegee Institute, Williams and a colleague, Randall Grace, continued to work on Tombigbee River mussels. The bulk of the mussel specimens collected by Williams from the Cahaba and Tombigbee rivers was deposited in the OSUM.

After leaving the Tuskegee Institute in 1974 Williams worked in the USFWS Endangered Species Office in Washington, DC, where he remained until 1987. Working in the Washington area provided Williams with the opportunity to spend time in the mollusk collection in the Smithsonian Institution. Access to the type collection in the USNM was essential to resolving the synonymies of most of the Alabama and Mobile Basin mussel fauna, which led to the initiation and eventual publication of this volume.

In July 1987 Williams moved to the USFWS National Fisheries Research Center, currently a U.S. Geological Survey (USGS) research lab, in Gainesville, Florida. During the 1990s Williams and research associates surveyed mussels in most of the eastern Gulf Coast drainages of Alabama, Florida and Georgia, from the Apalachicola Basin west to the Escambia River drainage. The sampling effort in the Apalachicola Basin, by far the largest of the eastern Gulf Coast drainages, was under the leadership of Jayne Brim Box and involved numerous graduate students (Brim Box and Williams, 2000). During the course of the Apalachicola Basin survey several ancillary projects developed. One project, an M.S. thesis by Christine A. O'Brien, examined the reproductive biology of four rare mussels (O'Brien and Williams, 2002). The Choctawhatchee River, another eastern Gulf Coast drainage, was surveyed by Holly N. Blalock-Herod, Jeff J. Herod, Stuart W. McGregor and Williams (Blalock-Herod et al., 2005). Almost all Gulf Coast drainage mussels collected by Williams and the students were deposited in the Florida Museum of Natural History at the University of Florida (UF) in Gainesville.

Graduate students in several departments at Auburn University have produced theses and dissertations regarding mussels. Mussel surveys by John C. Hurd (1974) in the Coosa River drainage and by John J. Jenkinson (1973) in tributaries of the Tallapoosa and Chattahoochee rivers provided important distributional information prior to modern extirpation of many species from those streams. More recently Judy A. Johnson surveyed the Tallapoosa River drainage (Johnson, 1997; Johnson and DeVries, 2002), and Mike M. Gangloff (2003) resurveyed Coosa River tributaries examined earlier by Hurd. During the survey Gangloff discovered a previously undetected species of *Pleurobema* (Gangloff et al., 2006). Other recent work by Auburn students includes an analysis of the commercial mussel fishery in Wheeler Reservoir by Zachary H. Bowen (1993) and a study of shell growth, behavior and the effects of suspended sediment on *Amblema plicata* by Chuck Howard (1999). Elise R. Irwin, professor at Auburn University, investigated imperiled Tallapoosa River species, including mussels (Irwin et al., 1998).

During the 1980s and 1990s the Mississippi Museum of Natural Science (MMNS) produced a series of survey and status reports on imperiled Tombigbee River species. Most of the work was performed by museum personnel, including Paul D. Hartfield, Robert L. Jones, Charles L. Knight and Terry Majure (Hartfield and Jones, 1989, 1990; Jones, 1991; Jones et al., 1996; Jones and Majure, 1999), but other work was done by contractors (e.g., Pierson, 1991a, 1991b).

Andrew C. Miller and Barry S. Payne from the USACE Waterways Experiment Station, Vicksburg, Mississippi, commenced working in Alabama and the Mobile Basin in the 1980s. Most of the work consisted of mussel surveys and assessments in various reaches of navigable waterways (Miller et al., 1987; Miller, 1994, 2000, 2001). The bulk of the work carried out by the USACE was conducted as part of activities related to the maintenance of navigational channels.

At the University of Alabama in Huntsville, during the late 1980s and early 1990s, Jeffrey T. Garner worked on a master's thesis that described the annual reproductive cycle of *Quadrula metanevra* (Garner, 1993; Garner et al., 1999). This work was carried out under the guidance of Richard F. Modlin, with instruction and assistance from Paul Yokley, Jr., and Tom Haggerty at the UNA.

In the 1990s Charles Lydeard arrived at the University of Alabama, Tuscaloosa, and began a molecular phylogenetics program with primary emphasis on freshwater mussels and snails. Numerous dissertations and publications have been produced by Lydeard, his students and postdoctoral associates (e.g., Lydeard et al., 1996; Roe and Lydeard, 1998; Lydeard et al., 2000; Roe et al., 2001; Buhay and Haag, 2003; Buhay et al., 2003; Roe and Hoeh, 2003; Serb et al., 2003; Campbell et al., 2005; Roe and Hartfield, 2005). Their research has shed new light on old taxonomic questions. Students and postdoctoral fellows who studied mussels under Lydeard include Jennifer E. Buhay, David C. Campbell, Russ L. Minton, Kevin J. Roe and Jeanne M. Serb.

In 1996 Paul D. Johnson began life history studies on Mobile Basin mussels at the Southeastern Aquatic Research Institute in Cohutta, Georgia. The institute is part of the Tennessee Aquarium in Chattanooga and is now called the Tennessee Aquarium Research Institute (TNARI). Most of the life history work was part of an effort to perfect propagation techniques for imperiled mollusks. In addition to life history research, mussel surveys were carried out in selected streams, including a comprehensive comparison of recent and historical mussel assemblages in the upper Coosa River drainage with assistants Ryan Evans and Sabrina Novak (Johnson and Evans, 2000; Novak, 2004).

During the past decade biologists at other universities in Alabama and Georgia have been involved in mussel research. Carson Stringfellow, a graduate student at Columbus State University, Columbus, Georgia, under the direction of George E. Stanton, surveyed the Chattahoochee River drainage in the vicinity of Columbus, Georgia (Stringfellow, 1997). More recently a graduate student, Megan M. Pilarczyk, at Troy University, under the supervision of Paul M. Stewart, has examined the biology and distribution of mussels in Gulf Coast drainages (Pilarczyk et al., 2005; Pilarczyk et al., 2006).

The four national forests in Alabama contain streams with healthy mussel populations. During the past two decades U.S. Forest Service (USFS) personnel have been important contributors to Alabama malacology. Wendell R. Haag and Melvin L. Warren, Jr., of the USFS Forest Hydrology Laboratory, Oxford, Mississippi, have conducted significant life history and ecological studies of mussels in the national forests of Alabama (Haag and Warren, 1997, 1998; Haag et al., 1999; Haag and Warren, 2000). Leigh A. McDougal and April Hargis have also contributed to national forest mussel assessments.

In the 1990s Stuart W. McGregor, with the Geological Survey of Alabama (GSA), began a series of comprehensive mussel surveys of selected streams in Alabama and the Mobile Basin. Other GSA personnel assisting McGregor with mussel surveys include Maurice F. Mettee, Patrick E. O'Neil and Tom E. Shepard. Numerous survey reports and publications have been produced by GSA personnel (e.g., McGregor, 1992; McGregor and Pierson, 1999; McGregor et al., 1999; McGregor et al., 2002; McGregor and Garner, 2004; McGregor and Haag, 2004).

Mussel surveys in streams of Alabama and the Mobile Basin have been performed by several nongovernmental agencies and individuals. A number of streams have been surveyed by J. Malcolm Pierson (deceased), with the Alabama Power Company; some of these surveys have resulted in the discovery of previously unknown populations of imperiled species, as well as new distribution records. Pierson also conducted status surveys for *Pleurobema decisum*, *Pleurobema taitianum* and *Quadrula stapes* (Pierson, 1991a, 1991b). Jim C. Godwin, with The Nature Conservancy (TNC) Natural Heritage Program, has surveyed a number of Alabama streams, including Paint Rock River. Various streams have been surveyed by W. Henry McCullagh, Harry Lee and Doug N. Shelton. In 2002 McCullagh and his associates published a survey of the Sipsey River and compared historical and recent distributions. Deborah Wills (1995) privately published an account of mussels in the Tennessee River in the vicinity of Decatur, Alabama. The publication includes numerous color photographs of mussel shells.

In response to intense commercial harvest pressure on mussels, the Alabama Department of Conservation and Natural Resources (ADCNR) initiated a new management strategy that included a Mussel Management Supervisor position within the Game and Fish Division (now Division of Wildlife and Freshwater Fisheries). The function of the program was to provide a mechanism for monitoring and managing mussels in the state, as well as perform basic research on their biology. The position was funded with revenue generated from a tax on commercial mussels, and a biologist, Jeffrey T. Garner, was hired to fill the position in 1995. In addition to monitoring mussel harvest and population densities in areas of commercial interest, Garner has conducted basic faunal surveys in many areas of the state. He has also studied the annual gametogenic cycles of selected mussel species in collaboration with Tom Haggerty (Haggerty et al., 1995; Garner et al., 1999; Haggerty and Garner, 2000; Haggerty et al., 2005).

Work of the investigators reviewed above spans 175 years and provides a substantial background on which future conservation efforts and research questions can be based. The increased interest in mussel biology, conservation, ecology and phylogenetics during the past three decades is encouraging. Many individuals from state and federal agencies, universities, nongovernment institutions and the public have made significant contributions to the research on mussels in Alabama and adjacent states. The ADCNR has demonstrated a commitment to increase conservation efforts by establishing the Alabama Aquatic Biodiversity Center in Marion, Alabama. The function of the facility is to propagate mussels and other nongame aquatic species for release into the wild. Paul D. Johnson was selected as the first supervisor of the facility in 2005. It will take sustained effort into the foreseeable future to ensure the survival of the mussel fauna of Alabama and the Mobile Basin.

Chapter 3
Inland Waters of Alabama and the Mobile Basin

Physiography

The landmass within the boundary of Alabama and the Mobile Basin extends from the southern terminus of the Appalachian Mountains and Interior Low Plateau south to the Gulf of Mexico. This area encompasses a variety of geologic formations, physiographic provinces, soil types and hydrologic conditions. Streams draining this diverse landscape vary greatly in substrate, gradient, width, depth, chemistry and volume. Alabama and the Mobile Basin include sections of six physiographic provinces: the Appalachian Plateau, Blue Ridge, Coastal Plain, Interior Low Plateau, Piedmont and Valley and Ridge (Figure 3.1). The geology and hydrology combine to produce a variety of aquatic habitats that support a remarkable diversity of aquatic organisms. The aquatic biodiversity of Alabama and the Mobile Basin exceeds that of any other area of comparable size in North America.

The Interior Low Plateau province in northern Alabama encompasses most of Lauderdale, Limestone and Madison counties north of the Tennessee River, and the northern portions of Colbert, Lawrence and Morgan counties south of the river. In Alabama its western margin is bound by the Coastal Plain and its southern and eastern margins by the Appalachian Plateau. It is underlain primarily by limestone and dolostone, with sandstone tablelands. The Alabama portion of the Interior Low Plateau is entirely within the Tennessee River drainage.

The Appalachian Plateau province occupies a large section of north-central and northeastern Alabama. It is the second largest physiographic region in the state. The Appalachian Plateau is bound on the west and south by the Coastal Plain, on the east by the Valley and Ridge and on the north by the Interior Low Plateau. Geology of the Appalachian Plateau ranges from shale and sandstone in the southern portion to limestone lowlands with shale and sandstone tablelands in the northern portion. The Alabama portion of the Appalachian Plateau is drained by the Tennessee River in the northeast, with most of the remainder drained by the Black Warrior River. The Black Warrior River drainage is shaped like a shallow saucer, confined along its southwestern margin by the Sand Hills, with one outflow at Tuscaloosa. Along its southeastern margin a small section of the Appalachian Plateau lies within the drainage of Big Wills Creek and the Little River, both tributaries of the Coosa River. The Sipsey and Buttahatchee rivers also have their headwaters along the western edge of the Appalachian Plateau.

In Alabama and the Mobile Basin, the Valley and Ridge province extends in a diagonal band from central Alabama northeastward into northwestern Georgia and southeastern Tennessee. It lies adjacent to the Piedmont on the east, with the exception of a small area in northern Georgia where it abuts the Blue Ridge. It is bounded by the Interior Low Plateau to the west and by the Coastal Plain to the south. The Valley and Ridge province is characterized by a series of approximately parallel ridges and valleys of various widths oriented northeast to southwest. Geology of the province is mainly limestone and dolostone in the valley floors with sandstone and shale ridges. The Coosa River is the primary drainage of the Alabama portion of the Valley and Ridge. It occupies a broad alluvial valley with sand, gravel and cobble substrates periodically interrupted by outcrops of limestone and dolostone. The remainder of the Valley and Ridge is drained by the Cahaba River to the southwest, and a few headwater tributaries of the Black Warrior River lie along the western margin.

The Piedmont province is located in east-central Alabama and extends eastward into Georgia. Its northern margin is adjacent to the Valley and Ridge province, and its southern boundary is adjacent to the Coastal Plain. In northwestern Georgia the Piedmont adjoins the southernmost extent of the Blue Ridge province. Geology of the Piedmont is variable, consisting of schists, gneisses, granites and slates, with narrow belts of quartzite and marble. The highest point in Alabama, Cheaha Mountain with an elevation of 734 m, lies within the Piedmont near its border with the Valley and Ridge province. The principal drainage of the Piedmont is the Tallapoosa River, with smaller portions drained by the Coosa and Chattahoochee rivers.

Of the six physiographic provinces that comprise Alabama and the Mobile Basin, the Blue Ridge is the smallest. It is located in northwestern Georgia near the Tennessee and North Carolina border and has the highest elevations in the Mobile Basin, up to 1,200 m. The Blue Ridge represents the remnants of a former highland area that is older than the adjacent Piedmont to the south and Valley and Ridge to the west. The Blue Ridge is underlain by igneous and metamorphic rocks. The Mobile Basin portion of the Blue Ridge is drained by the Conasauga, Coosawattee and Etowah rivers. These headwater tributaries of the Coosa River are characterized by moderate to steep gradient with predominantly cobble and gravel substrates.

The Coastal Plain is the largest of the six physiographic provinces in Alabama and the Mobile

Basin, covering the southern and western portions of the area. The inner margin of the Coastal Plain is bordered by the Interior Low Plateau, Appalachian Plateau, Valley and Ridge and Piedmont. River drainages of the Coastal Plain include portions of the Tallapoosa; the lower reaches of the Coosa, Cahaba and Black Warrior rivers; almost all of the Tombigbee River; and all of the Alabama River. This province is underlain by Mesozoic and Cenozoic sediments that

consist of chalk, marl, claystone, soft limestone, gravel and unconsolidated sands and clays. The contact between the Coastal Plain and the adjacent upland provinces is referred to as the Fall Line (Figure 3.1). This feature derives its name from the steep gradient typically found in stream channels where they approach or cross this boundary. Stream reaches near the Fall Line are characterized by shoals and waterfalls of varying heights.

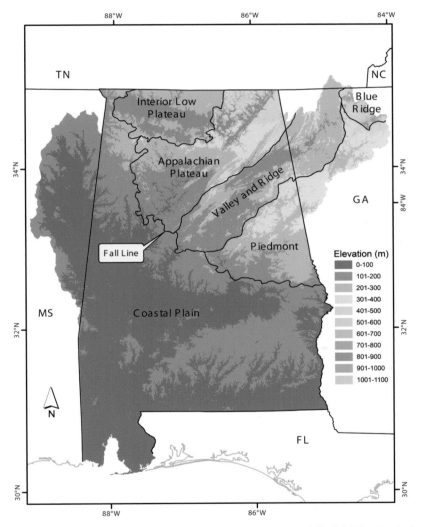

Figure 3.1. Physiographic provinces and range of elevations of Alabama and the Mobile Basin in Alabama, Georgia, Mississippi and Tennessee.

Rivers of Alabama and the Mobile Basin

The region addressed in this book encompasses Alabama and parts of three adjacent states, Georgia, Mississippi and Tennessee (Figure 3.2). Parts of three major river basins lie within the boundaries of Alabama: the Tennessee River drainage of the Mississippi Basin, most of the Mobile Basin, and the western reaches of the Apalachicola Basin. Six smaller

drainages cover the southern portion of the state. None of the drainages lie completely within Alabama. The Tennessee River enters the state from Tennessee in the northeastern corner and reenters Tennessee from northwestern Alabama, flowing northward to the Ohio River. Eastern reaches of the Mobile Basin originate in northwestern Georgia and southeastern Tennessee, and western reaches of the basin arise in north-central

Alabama and northeastern Mississippi. Waters of the Mobile Basin empty into Mobile Bay on the Gulf of Mexico in southwestern Alabama. The westernmost river drainage of the Apalachicola Basin, the Chattahoochee River, begins in northeastern Georgia and flows southwestward, then southward, forming the Alabama and Georgia state line. The Choctawhatchee, Yellow, Blackwater, Escambia and Perdido River drainages arise in Alabama and flow across the Florida panhandle before emptying into the Gulf of Mexico.

The Escatawpa River drainage originates in extreme southwestern Alabama, then flows into Mississippi where it joins the Pascagoula River before emptying into the Gulf of Mexico. Relationships of Gulf Coast drainages are presented in a schematic diagram in Figure 3.3. All waters that flow through Alabama eventually empty into the Gulf of Mexico. However, their drainage histories and evolution have resulted in very different aquatic faunas, characterized by high levels of endemism in most drainages.

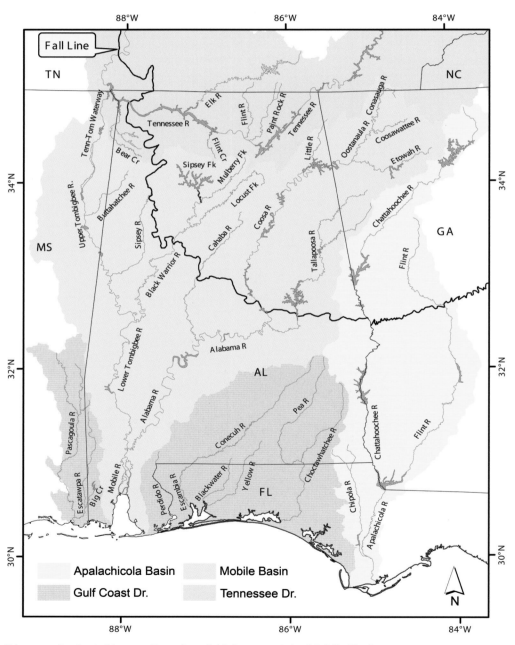

Figure 3.2. Rivers and selected large tributaries of Alabama and the Mobile Basin.

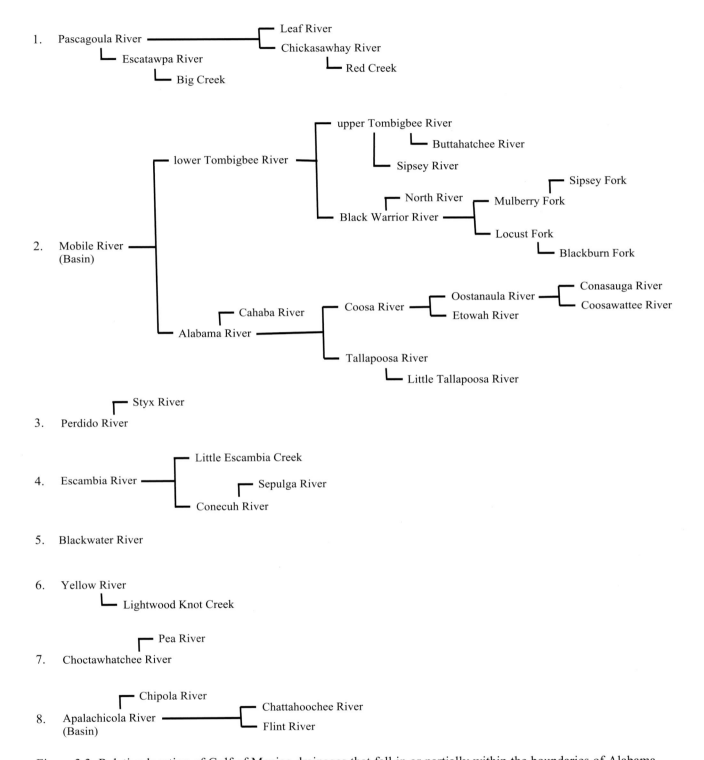

Figure 3.3. Relative location of Gulf of Mexico drainages that fall in or partially within the boundaries of Alabama, listed west to east. The Chickasawhay, Leaf and Pascagoula rivers are in Mississippi, but some tributaries of the Chickasawhay River originate in Alabama. The Apalachicola River in Florida and the Flint River in Georgia are also outside of Alabama. These rivers are included for completeness.

Tennessee River

The Tennessee River, with a drainage area of approximately 110,000 km^2, is the largest tributary of the Ohio River and one of the largest river systems in the southeastern U.S. It drains portions of seven states, Alabama, Georgia, Kentucky, Mississippi, Tennessee, North Carolina and Virginia, and five physiographic provinces, Appalachian Plateau, Blue Ridge, Coastal Plain, Interior Low Plateau and Valley and Ridge (Zurawski, 1978). From its headwaters in southwestern Virginia, western North Carolina and eastern Tennessee, the Tennessee River flows southwestward into Alabama.

The Alabama reach of the Tennessee River, about 322 km long, drains the northern portion of the state (Figure 3.4). It enters Alabama in the extreme northeastern corner and follows a southeastern course in the Sequatchie Valley before turning northwestward near Guntersville in southeastern Marshall County. Prior to its impoundment this reach was characterized by a wide channel with islands and shoals (Hall and Hall, 1916). The valley along this reach is relatively narrow and most tributaries relatively short. After an abrupt turn to the northwest the river valley is wider, with considerably longer tributaries (e.g., Cypress Creek, Big Nance Creek, Flint Creek, Elk River, Flint River and Paint Rock River).

Perhaps the most unique feature of the preimpoundment Tennessee River was the extensive Greater Muscle Shoals, which extended from above the mouth of the Elk River to below the mouth of Bear Creek, a distance of about 85 km with a drop in bed elevation of approximately 43 m (Figure 3.5) (Hall and Hall, 1916; Isom, 1969; Garner and McGregor, 2001). In this reach there were many shallow areas where it was possible to ford the river during normal flow conditions (Figure 3.6).

Much of the Tennessee River flowed over bedrock before its impoundment, though the riverbed has unconsolidated sediments deposited in many areas. During construction of dams in this reach the engineers had difficulty finding sufficient quantities of gravel to mix concrete (Moneymaker, 1941; TVA, 1949). The carbonate terrain of the Tennessee River Valley is conducive to an abundance of springs, many of them large, as well as sinkhole ponds and swamps. These contribute to the maintenance of water quality and quantity during periods of drought.

The Tennessee River reenters Tennessee at the extreme northwestern corner of Alabama, where it forms a common border with the state of Mississippi. From there it flows northward to empty into the Ohio River in southwestern Kentucky.

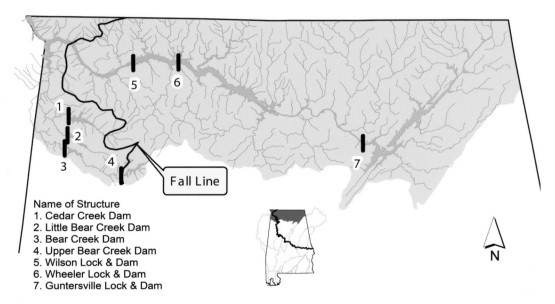

Name of Structure
1. Cedar Creek Dam
2. Little Bear Creek Dam
3. Bear Creek Dam
4. Upper Bear Creek Dam
5. Wilson Lock & Dam
6. Wheeler Lock & Dam
7. Guntersville Lock & Dam

Figure 3.4. Tennessee River drainage in northern Alabama. The numbered black bars represent major dams in the drainage. Dams 1 to 4 are in the Bear Creek system, and dams 5 to 7 are on the Tennessee River proper.

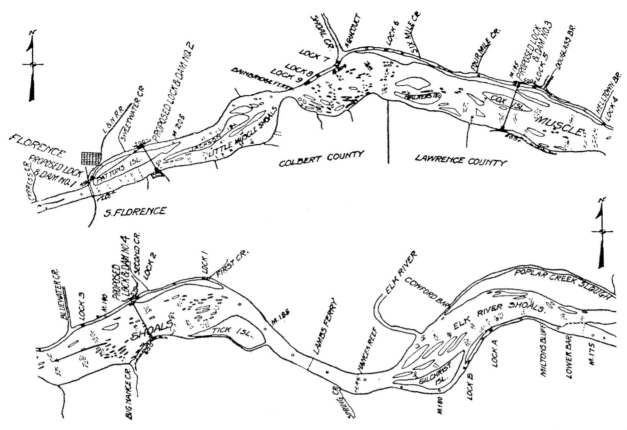

Figure 3.5. A portion of a 1913 USACE map of the Tennessee River, Muscle Shoals section, showing "proposed improvements on the Tennessee River at Muscle Shoals" (Hall and Hall, 1916) (Colbert Shoals not included).

Figure 3.6. Horse-drawn wagon fording Muscle Shoals, on the Tennessee River prior to its impoundment, pre-1929. Photograph by Roland Harper, from the W.S. Hoole Special Collections Library, The University of Alabama.

Pascagoula River

The Pascagoula River originates with the confluence of the Chickasawhay and Leaf rivers in southeastern Mississippi and enters the Gulf of Mexico just west of the Alabama and Mississippi state line. It has a drainage area of 24,600 km². No part of the main channels of the Chickasawhay, Leaf or Pascagoula rivers are in Alabama, but tributaries of the Chickasawhay and Pascagoula originate in the southwestern corner of the state. The Chickasawhay tributary, Red Creek, is located in northwestern Washington County and southwestern Choctaw County. Access to Red Creek is limited, but at an accessible site in the middle reach of the stream the channel was observed to be somewhat incised, 2 m to 3 m wide, with a sand and sandy clay bottom.

The largest eastern tributary of the Pascagoula River is the Escatawpa River, which has its headwaters in extreme western Mobile and Washington counties in southwestern Alabama. The drainage area of the Escatawpa River in Alabama is approximately 1,300 km², not including the Big Creek drainage. Big Creek in western Mobile County, with a drainage area of about 565 km², is the largest Alabama tributary of the Escatawpa River (USACE, 1985a). The Escatawpa River and Big Creek are characterized by wide, sand-bottomed channels with tannic water (Figure 3.7). There are no impoundments on the Escatawpa River proper, but there is a large water supply impoundment on Big Creek. There are no known historic occurrences of mussels in the Escatawpa River drainage in Alabama. However, *Utterbackia imbecillis* and *Pyganodon grandis* are known from one site on the Big Creek impoundment. It appears that these records represent recent introductions, most likely as glochidia on introduced fish.

Figure 3.7. Escatawpa River, U.S. Highway 98 near the Alabama and Mississippi state line, Mobile County, Alabama, 27 October 2000. Photograph by J.D. Williams.

Mobile Basin

The Mobile Basin is the largest basin emptying into the Gulf of Mexico east of the Mississippi River. Its drainage area is approximately 113,000 km² and includes most of Alabama and portions of Georgia, Mississippi and Tennessee (USACE, 1985a). Upstream of the Mobile River the basin can be divided into eastern and western sub-basins that are almost equal in area. The major rivers in the eastern sub-basin include the Coosa, Tallapoosa, Cahaba and Alabama. There are only two large rivers in the western sub-basin, the Black Warrior and Tombigbee.

The Mobile River is formed at the junction of the Alabama and Tombigbee rivers and flows approximately 50 km to Mobile Bay. Much of its course is characterized by the wide, interconnected floodplain of the Mobile Delta, which ranges in width from 15 km to 20 km. Major distributaries of the Mobile Delta include the Tensaw, Blakeley, Apalachee, Spanish and Middle rivers. The Mobile River is tidally influenced, with detectable tide cycles along its entire length during periods of average to low flow. Streambed gradient is low in the river channel, and substrates primarily consist of mixtures of sand and mud with organic debris. Substrates in backwater areas are usually mud with large quantities of organic debris. Gravel is present in upstream reaches of some tributaries but is uncommon in the main channel of the river. The Mobile River is typically not subjected to navigational dredging, and there are no impoundments. The Mobile Delta remains in a relatively natural state, with a large portion set aside as various nature preserves.

Tombigbee River

The Tombigbee River is one of the longest rivers on the eastern Gulf Coast (Figure 3.8). For purposes of discussion and administration, many aquatic biologists, regulatory agencies and natural resources managers divide the river into two smaller units. From the headwaters in northeastern Mississippi downstream to its junction with the Black Warrior River at Demopolis, Alabama, it is often referred to as the upper Tombigbee River. The remainder of the Tombigbee, from the mouth of the Black Warrior River downstream to its junction with the Alabama River, is generally referred to as the lower Tombigbee River. While these divisions have been recognized for many years and are extremely useful, they are not official or locally used names and do not appear on topographic maps. Division of the Tombigbee River into upper and lower reaches is especially useful in the discussion of aquatic organism distributions and is utilized in this volume.

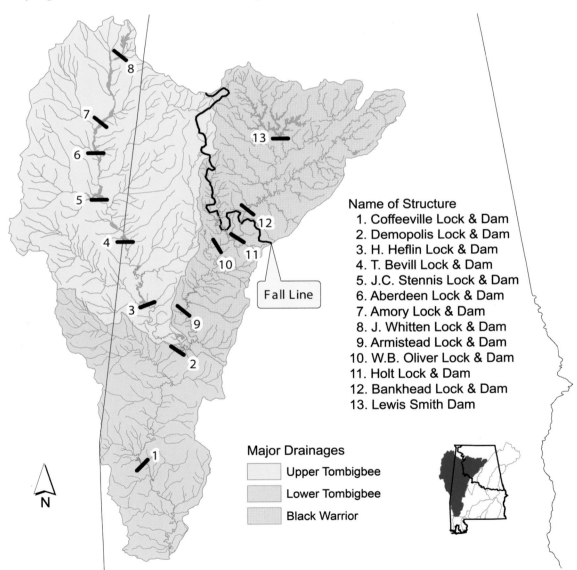

Name of Structure
1. Coffeeville Lock & Dam
2. Demopolis Lock & Dam
3. H. Heflin Lock & Dam
4. T. Bevill Lock & Dam
5. J.C. Stennis Lock & Dam
6. Aberdeen Lock & Dam
7. Amory Lock & Dam
8. J. Whitten Lock & Dam
9. Armistead Lock & Dam
10. W.B. Oliver Lock & Dam
11. Holt Lock & Dam
12. Bankhead Lock & Dam
13. Lewis Smith Dam

Major Drainages
Upper Tombigbee
Lower Tombigbee
Black Warrior

Figure 3.8. Western Mobile Basin in Alabama and Mississippi. The numbered black bars represent major dams on the Tombigbee and Black Warrior rivers. Dams 1 and 2 are on the lower Tombigbee River, dams 3 to 8 are on the upper Tombigbee River, and dams 9 to 13 are in the Black Warrior River drainage.

The headwaters of the Tombigbee River arise in the Coastal Plain of Prentiss and Tishomingo counties in the northeastern corner of Mississippi. Its course in Mississippi is generally southern, paralleling the Mississippi and Alabama state line, before turning southeastward near Columbus, Mississippi, and crossing the state line near Pickensville, Alabama. The Tombigbee River continues a southeastern course to the junction with the Black Warrior River just north of Demopolis, Alabama. The upper Tombigbee River is about 428 km long and has a drainage area of approximately 23,490 km^2 (USACE, 1985a). Large tributaries in the upper Tombigbee River drainage include the Noxubee River (3,673 km^2), Tibbee Creek (2,893 km^2), Buttahatchee River (2,253 km^2), Luxapallila Creek (2,059 km^2) and Sipsey River (2,044 km^2) (USACE, 1985a).

Streams in the upper Tombigbee River drainage predominantly have low to moderate gradient, with pool and riffle habitats (Figure 3.9). Substrates are typically mixtures of sandy clay, sand and gravel overlying chalk and marl, which are often locally exposed. Exposed chalk and marl are more prevalent in the Black Belt Prairie than other parts of the drainage. Mackeys Creek had a substrate of Paleozoic bedrock with patches of gravel prior to its destruction by the USACE as part of the divide cut section of the Tennessee Tombigbee Waterway. Some of the headwaters of northeastern tributaries (e.g., Buttahatchee River) are deeply incised, cutting through overlying sand and gravel formations to Paleozoic bedrock.

Prior to the 1970s there were no dams on the main channel of the upper Tombigbee River. Beginning in the mid-1970s the USACE initiated construction of the Tennessee Tombigbee Waterway, a project designed to connect the Tombigbee and Tennessee rivers through a series of impoundments and a barge canal. This project impounded and canalized the entire main channel of the Tombigbee River, connecting the two drainages via the Yellow Creek embayment behind Pickwick Dam on the Tennessee River. This project resulted in the destruction of the last large unimpounded river in the Mobile Basin, along with its riverine fauna (J.D. Williams, personal observation), but fell far short of expectations with regard to utilization as a waterway.

Figure 3.9. Upper Tombigbee River about 6.5 km southwest of Pickensville, Pickens County, Alabama, August 1974. Photograph by J.D. Williams.

The lower Tombigbee River flows southward from its junction with the Black Warrior River to its junction with the Alabama River. Its entire watershed is within the Coastal Plain, flowing across the Black Belt Prairie, Chunnenuggee Hills, Southern Red Hills, Lime Hills and Southern Pine Hills. Diverse geologic strata produce a variety of stream habitats. Streams typically have low gradient and substrates of sandy mud, sand or gravel (Figure 3.10). Some streams may have reaches of steeper gradient and rocky substrates in areas where they are underlain by claystone, marl, sandy limestone or calcareous sandstone. Groundwater is sufficient to maintain flow during droughts, but some streams may be reduced to isolated pools. The lower Tombigbee River is about 275 km long and has a drainage area of approximately 12,460 km^2. There are several large tributaries to the lower Tombigbee River, including the Sucarnoochee River (2,523 km^2), Chickasaw Bogue (891 km^2), Okatuppa Creek (808 km^2), Tuckabum Creek (666 km^2), Satilpa Creek (560 km^2) and Santa Bogue (471 km^2) (USACE, 1985a).

Figure 3.10. Lone Brothers Bar, lower Tombigbee River prior to its impoundment, looking upstream, Marengo and Sumter counties, Alabama, 13 October 1908. Photograph by Roland Harper, from the W.S. Hoole Special Collections Library, The University of Alabama.

Modification of the lower Tombigbee River began prior to the Civil War. Early alterations included dynamiting shoals and removing snags, which enabled navigation between St. Stephens, Washington County, and Tuscaloosa, Tuscaloosa County, during low flow periods. Impoundment of the lower Tombigbee River proper began in the early 1900s with the construction of four low locks and dams between Jackson and Demopolis to open the river to barge traffic (Hall and Hall, 1916). These were replaced with higher locks and dams during the mid-1900s. Conversion of the lower Tombigbee River into a barge canal was completed by cutting off two meander loops between Coffeeville Lock and Dam and the confluence with the Alabama River.

Black Warrior River

The Black Warrior River drainage in north-central Alabama lies entirely within the borders of the state (Figure 3.8). Its headwater tributaries originate on the Appalachian Plateau (Figures 3.11, 3.12), with the exception of a small area along the eastern edge of the drainage within the Valley and Ridge province. After the Black Warrior River crosses the Fall Line near Tuscaloosa it continues its southward course to join the Tombigbee River at Demopolis. The drainage downstream of Tuscaloosa lies almost entirely on the Coastal Plain. The Black Warrior River is about 280 km long and has a drainage area of approximately 16,250 km^2. Tributaries with the largest drainage areas are the headwater rivers, Mulberry Fork (6,133 km^2) and Locust Fork (3,131 km^2). Other large watersheds within the Mulberry and Locust Fork drainages include the Sipsey Fork (2,585 km^2), Lost Creek (900 km^2) and Blackburn Fork (495 km^2), which are also above the Fall Line. The largest tributary below the Fall Line is Big Sandy Creek with a drainage area of 461 km^2. North River is an interesting watershed in that the main channel of the river lies along the Fall Line and most tributaries on its east side are above and most tributaries on the west side are below the Fall Line.

Figure 3.11. Clear Creek Falls, upper Black Warrior River drainage, prior to its impoundment behind Lewis Smith Dam, Winston County, Alabama, 18 November 1959. Photograph courtesy Alabama Power Company Archives.

Figure 3.12. Sipsey Fork prior to its impoundment behind Lewis Smith Dam, looking upstream, Winston or Walker County, Alabama, late 1950s. Photograph courtesy Alabama Power Company Archives.

In the Black Warrior River drainage, streams above the Fall Line typically have a moderate gradient and predominantly bedrock substrates, composed of sandstone, shale and slate in areas with flow, and silt, sand and gravel in pools (Figures 3.13, 3.14). In some watersheds the stream channels are incised more than 50 m below the surface of the sandstone and shale plateau landscape. Stream flow can be highly variable in these streams due to the lack of groundwater to sustain flow during periods of low precipitation. A portion of the watersheds of streams along the extreme northeastern side of the drainage, in the Valley and Ridge province, originate in areas with limestone, dolomite, calcareous shale and chert.

Figure 3.13. Cahaba Lily (*Hymenocallis coronaria*), Black Warrior River prior to its impoundment, Squaw Shoals, Tuscaloosa County, Alabama, 4 June 1913. Photograph by Roland Harper, from the W.S. Hoole Special Collections Library, The University of Alabama.

Figure 3.14. Black Warrior River, preconstruction site of the western abutment for Lock and Dam 17 near the Tuscaloosa and Jefferson county line, Alabama, early 1900s. Photograph by Roland Harper, from the W.S. Hoole Special Collections Library, The University of Alabama.

The main channel of the Black Warrior River was altered between 1895 and 1915 with construction of low locks and dams (Hall and Hall, 1916). These were gradually replaced with four major navigation dams present today. There is also a major dam on the Sipsey Fork, constructed by the Alabama Power Company in 1961. Recreation and water supply impoundments are present on the North River (constructed in the mid-1960s) in Tuscaloosa County, and the Blackburn Fork of the Little Warrior River (constructed about 1938) in Blount County. In addition to the impoundments, water quality declines have been brought about by surface mining with resultant acid mine drainage and increased sedimentation in many watersheds of the Black Warrior River drainage.

Alabama River

The Alabama River, approximately 490 km in length, originates at the junction of the Coosa and Tallapoosa rivers, about 10 km north of Montgomery in south-central Alabama (Figure 3.15). It has a strongly meandering course westward to the vicinity of Selma, where it turns southwestward and flows to the Tombigbee River to form the Mobile River about 50 km north of Mobile Bay (Figure 3.16). The main channel of the Alabama River lies in an alluvial floodplain of varying widths for its entire length, which is entirely within the Coastal Plain province. Most northern tributaries upstream of Selma arise in the Fall Line Hills. They typically have sand and gravel substrates with low to moderate gradient, characterized by riffle, run and pool habitats. The southern tributaries of that reach arise in the Black Belt Prairie, an area underlain by chalk and marl deposits with a veneer of black soil rich in organic matter. The character of these tributaries varies depending on the local situation, but they typically have chalk or marl bottoms with variable cover of sand and gravel. Streams of the Black Belt Prairie are typically low gradient and have little or no flow during drier periods of the year due to the lack of adequate groundwater discharge.

Figure 3.15. Parker Island, Elmore County, Alabama, looking northwest, 11 August 1967. The Coosa River enters the photograph from the lower right and the Tallapoosa River from the bottom right, joining to form the Alabama River. A branch of the Tallapoosa River in the lower portion of the photograph forms the southeastern border of the island. The Alabama River exits the photograph near the center left and reappears in a meander in the upper left corner. The outlet channel from the Walter Bouldin hydropower facility is on the upper right. Photograph courtesy Alabama Power Company Archives.

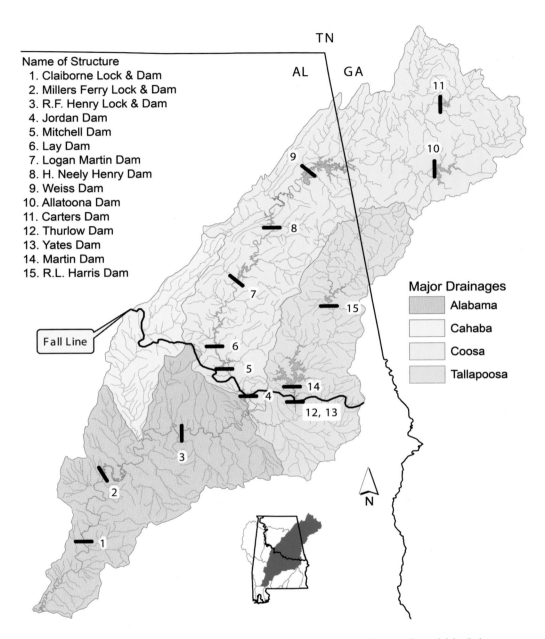

Name of Structure
 1. Claiborne Lock & Dam
 2. Millers Ferry Lock & Dam
 3. R.F. Henry Lock & Dam
 4. Jordan Dam
 5. Mitchell Dam
 6. Lay Dam
 7. Logan Martin Dam
 8. H. Neely Henry Dam
 9. Weiss Dam
10. Allatoona Dam
11. Carters Dam
12. Thurlow Dam
13. Yates Dam
14. Martin Dam
15. R.L. Harris Dam

Major Drainages
Alabama
Cahaba
Coosa
Tallapoosa

Figure 3.16. Eastern Mobile Basin in Alabama, Georgia and Tennessee. The numbered black bars represent major dams on the Alabama, Coosa and Tallapoosa rivers. Dams 1 to 3 are on the Alabama River, dams 4 to 11 are on the Coosa River, and dams 12 to 15 are on the Tallapoosa River.

Near Selma the Alabama River is joined by the Cahaba River. Downstream of Selma the southwestward course of the river cuts across most of Alabama's Coastal Plain physiographic districts, including the Fall Line Hills, Black Belt Prairie, Chunnenuggee Hills, Southern Red Hills, Lime Hills and Southern Pine Hills physiographic districts. Geological formations in these districts result in a variety of habitats that differ in substrate, gradient and water volume. The Cahaba River (drainage area 4,727 km^2) is the largest tributary of the Alabama River.

Other tributaries with large drainage areas include Bogue Chitto (943 km^2), Catoma Creek (880 km^2), Big Flat Creek (800 km^2), Big Swamp Creek (725 km^2), Big Mulberry Creek (715 km^2), Pintlalla Creek (681 km^2), Beaver Creek (665 km^2) and Limestone Creek (458 km^2).

The Alabama River was not impounded (Figure 3.17) until the late 1960s when the USACE initiated construction of Claiborne, Millers Ferry and R.F. Henry locks and dams. They were constructed primarily for navigation and destroyed all shoal habitat in the

Alabama River proper. Claiborne Lock and Dam, the lowermost structure, has a fixed crest spillway and no power generation facilities. The remaining two structures have gated spillways to regulate the water level and the capacity to generate small amounts of hydropower. The uppermost structure, R.F. Henry Lock and Dam, impounds the river upstream to above the confluence of the Coosa and Tallapoosa rivers.

Figure 3.17. Alabama River, looking downstream from Bridgeport Landing, prior to its impoundment behind Millers Ferry Dam, Wilcox County, Alabama, 30 August 1922. Photograph by Roland Harper, from the W.S. Hoole Special Collections Library, The University of Alabama.

Cahaba River

The Cahaba River drainage is located in central Alabama and lies entirely within the state. It originates in the Valley and Ridge province northeast of Birmingham and flows in a southwestern direction to the Fall Line. It crosses the Fall Line at Centreville (Figure 3.18) and continues south on the Coastal Plain to its confluence with the Alabama River, 12 km southwest of Selma. The Cahaba River is approximately 314 km long, of which 184 km lies above the Fall Line. It has a drainage area of 4,725 km^2 with approximately 55 percent above the Fall Line (USACE, 1985a).

Above the Fall Line the Cahaba River and its tributaries drain the limestone and dolostone formations of the Valley and Ridge province. These streams are characterized by moderate to steep gradient with pools of varying length interrupted by bedrock shoals. Gradient of the 47 km reach upstream of Centreville (i.e., above the Fall Line) is approximately 0.9 m per km (Hall and Hall, 1916). The largest tributary above the Fall Line is the Little Cahaba River, which enters the Cahaba River about 16 km upstream of Centreville.

The Little Cahaba River has a drainage area of 685 km^2. Other large tributaries situated above the Fall Line include Buck and Schultz creeks, each with a drainage area of approximately 190 km^2.

Downstream of Centreville the character of the river changes abruptly. The Coastal Plain reach of the Cahaba River is characterized by reduced gradient and substrates composed predominantly of sand and gravel. The drop in elevation in this 140 km reach is about 0.3 m per km (Hall and Hall, 1916). The river channel is characterized by long pools separated by shallow, swift riffles and sand and gravel bars. The lowermost reach of the Cahaba River flows through the Demopolis Chalk of the Black Belt Prairie. Substrates in this reach are often chalky bedrock overlain with varying amounts of sand and gravel, and the banks in some areas are steep and chalky. The largest Coastal Plain tributary is Oakmulgee Creek, which enters the Cahaba River about 40 km above its mouth and has a drainage area of approximately 610 km^2. Other Coastal Plain tributaries range from 60 km^2 to 90 km^2 in drainage area.

Figure 3.18. Cahaba River, looking upstream from the U.S. Highway 82 bypass bridge north of Centreville, Bibb County, Alabama, July 2001. Photograph by J.M. Pierson.

The Cahaba and Little Cahaba rivers proper have escaped channelization and large dams with the exception of Purdy Reservoir, a water supply impoundment on the Little Cahaba River east of Birmingham on the Jefferson and Shelby county line. However, there are other smaller impoundments in the Cahaba River drainage both above and below the Fall Line.

Coosa River

Of the seven river drainages that comprise the Mobile Basin, the Coosa River is the most complex in terms of physiography and geology, which combine to produce a variety of aquatic habitats. The Coosa River is about 460 km long and has a drainage area of approximately 26,335 km^2. Its headwaters originate in the foothills of the Appalachians in southeastern Tennessee and northwestern Georgia and drain portions of the Blue Ridge, Piedmont and Valley and Ridge physiographic provinces. Elevations of more than 1,200 m in Blue Ridge headwater tributaries are the highest in the entire Mobile Basin. The Coosa River proper originates with the junction of the Etowah and Oostanaula rivers at Rome, Georgia. Five small rivers—Conasauga, Jacks, Coosawattee, Cartecay and Ellijay—plus numerous creeks come together to form the Oostanaula River, which has a drainage area of 5,569 km^2 (USACE, 1985b). The Etowah River, with larger tributaries including Little River, Amicalola, Euharlee and Pumpkinvine creeks, has a drainage area of 4,817 km^2 (USACE, 1985b).

From Rome, Georgia, the Coosa River meanders southwestward in the Valley and Ridge province for a distance of about 50 km to the Alabama and Georgia state line. It continues its course southwestward through an alluvial valley to near the mouth of Waxahatchee Creek in the northeastern corner of Chilton County, where it passes onto the Piedmont. Prior to its impoundment, the reach from Rome to northeastern Chilton County was characterized by a moderate gradient with long pools interrupted by gravel, cobble and bedrock shoals (Figures 3.19, 3.20). Preimpoundment photographs reveal extensive gravel bars in this reach, an unusual habitat above the Fall Line in the Mobile Basin.

The reach of the Coosa River that cuts southeastward across the Piedmont is about 75 km in length. The preimpoundment Piedmont reach was distinctly different from the Valley and Ridge reach in having a steeper gradient with predominantly bedrock substrates (Figures 3.21, 3.22, 3.23). Some of the larger Coosa River tributaries in Alabama include Terrapin Creek (736 km^2), Big Canoe Creek (717 km^2), Ohatchee Creek (578 km^2), Kelly Creek (539 km^2), Tallaseehatchee Creek (515 km^2), Talladega Creek (453 km^2) and Choccolocco Creek (479 km^2) (Figure 3.24). The Coosa River passes onto the Coastal Plain near Wetumpka and flows approximately 18 km to join the Tallapoosa River and forms the Alabama River.

Figure 3.19. Coosa River, Broken Arrow Bar and old Lock and Dam 4, prior to its impoundment behind Logan Martin Dam, approximately 5 km northwest of Lincoln, Talladega County, Alabama, about 1962. Photograph courtesy Alabama Power Company Archives.

Figure 3.20. Coosa River, mouth of Choccolocco Creek, after reservoir clearing prior to its impoundment behind Logan Martin Dam, approximately 10 km southwest of Lincoln, Talladega County, Alabama, about 1962. Photograph courtesy Alabama Power Company Archives.

Figure 3.21. Coosa River at Butting Ram Shoals, looking upstream from the west bank, prior to its impoundment. Large island in left center of the photograph is located at approximately river mile 36.7. Photograph from early 1900s, courtesy Alabama Power Company Archives.

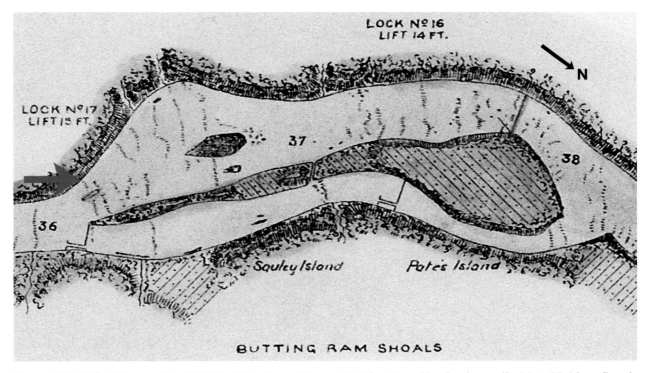

Figure 3.22. USACE map, dated 1889, of the Coosa River at Butting Ram Shoals, river mile 36 to 38 (river flow is from right to left), showing proposed locks and dams. Arrow at left pointing upstream indicates the approximate site and direction from which the photograph in Figure 3.21 was taken.

Figure 3.23. Coosa River, Mitchell Dam site prior to construction, Chilton and Coosa counties, Alabama, about 1920. Photograph courtesy Alabama Power Company Archives.

Figure 3.24. Jackson Shoals, Choccolocco Creek, tributary to the Coosa River, downstream of a small hydropower dam, Talladega County, Alabama, about 1915. Photograph courtesy Alabama Power Company Archives.

Alteration of the Coosa River began in the 1890s with construction of three low dams to open the river to barge traffic between Rome, Georgia, and Talladega County, Alabama. In the early 1900s three dams were constructed on the river: Lay Dam (1914), Mitchell Dam (1923) and Jordan Dam (1929). These were the first large hydropower dams on the river. They are located in the Piedmont reach between Coosa River miles 18 and 51, an area characterized by a steep gradient and predominantly bedrock substrate. This reach was often referred to as "The Devil's Staircase." Additional Coosa River hydropower dams were constructed in the Valley and Ridge province: Weiss Dam (1961), Logan Martin Dam (1964) and H. Neely Henry Dam (1966). None of the hydropower dams have lock facilities (Hall and Hall, 1916; Ackerman, 1950; USACE, 1985a; Jackson, 1997).

Tallapoosa River

The Tallapoosa River is located in east Alabama and west Georgia and is the easternmost river system in the Mobile Basin. The upper two-thirds of the basin lies on the Piedmont and the lower one-third is on the Coastal Plain. Geology within the Piedmont generally consists of metamorphic rocks including schists, gneisses, granites and slates, with narrow belts of sandstone and dolomite. Portions of the Piedmont have elevations up to 425 m along the divide between the Tallapoosa and Coosa River drainages. The Tallapoosa River proper and lower reaches of large tributaries are incised 30 m to 100 m below the surrounding plateau. The streambed is typically bedrock with rubble or small amounts of sand, silt and gravel. Most streams are characterized by series of pools and riffles.

Figure 3.25. The Fall Line on the Tallapoosa River at Tallassee, Elmore and Tallapoosa counties, Alabama. The low dam visible in the upper right corner was later replaced by Thurlow Dam, which was constructed in the 1920s. Photograph courtesy Alabama Power Company Archives.

The Fall Line is the most important physiographic feature of the Tallapoosa River drainage. Most of the rapids and waterfalls that characterize the Fall Line were inundated behind Thurlow Dam (Figure 3.25), which was completed by 1930 (Jackson, 1995). The streambed elevation drops approximately 15 m over a distance of 120 m in the Thurlow Dam area. Yates and Martin dams, located approximately 5 and 18 miles upstream of Thurlow Dam, respectively, inundate other rapids and falls on the Tallapoosa River.

The Coastal Plain portion of the Tallapoosa River drainage is underlain by Cretaceous sediments and is traversed by two physiographic subdivisions, the Fall Line Hills and Black Belt Prairie. The Fall Line Hills subdivision to the north, which varies in width from 25 km to 45 km, is developed on sand, clay and gravel. The streams in this area have sand to gravel substrates, occasionally mud, and slow to moderate currents. Exceptions are the rocky streams just below the Fall Line that have eroded through the sand and gravel formations and are flowing over the underlying Paleozoic bedrock. The Black Belt Prairie subdivision to the south is developed on the chalky marl of the Selma formation. During normal flow conditions streams in this area typically have little or no current due to low relief and reduced volumes of groundwater. The near absence of groundwater is also reflected in the intermittent nature of many streams in this area.

Perdido River

The Perdido River lies west of the Escambia River drainage and east of Mobile Bay, completely within the Coastal Plain. It originates along the border of Baldwin and Escambia counties in southwestern Alabama. The Perdido River is the boundary between Baldwin County, Alabama, and Escambia County, Florida, for most of its length. It flows southward about 103 km before emptying into Perdido Bay on the Gulf of Mexico. Its drainage area is approximately 2,400 km^2. Streams in upper reaches have narrow, incised channels. Downstream the relief gradually decreases before the river terminates in coastal marshes. Streams in the drainage typically have sand or sandy clay substrates and tannic, darkly stained water. The largest Perdido River tributary is the Styx River, which has a drainage area of about 676 km^2 and enters the river from the west (USACE, 1985b). There are no dams on the Perdido River or its larger tributaries.

Perdido River habitat superficially appears suitable for mussels, but it is one of only three drainages in Alabama in which they appear to be absent. A possible explanation for this absence is a combination of oligotrophic conditions and acidic water. Another explanation could be the relatively recent inundation of the river during a sea level highstand about three million years ago, as well as partial inundation during the Pleistocene. However, this does not explain the absence of *Corbicula fluminea*, which colonized almost all streams of Alabama during the 1900s.

Conecuh River

The Escambia and Conecuh rivers are the same stream with the upstream portion in Alabama known as the Conecuh River and the downstream portion in Florida known as the Escambia River. The Conecuh River becomes the Escambia River near its junction with Little Escambia Creek, immediately south of the Alabama and Florida state line. The Conecuh River drainage is located in south-central Alabama and is confined to the Coastal Plain (Figure 3.26). In the headwaters of the Conecuh River, near Union Springs, it flows through relatively high, hilly terrain reaching elevations of about 150 m. Its course is generally southwestern for about 311 km to the Alabama and Florida state line. The Escambia River continues southward in Florida for about 93 km to Escambia Bay, an arm of Pensacola Bay.

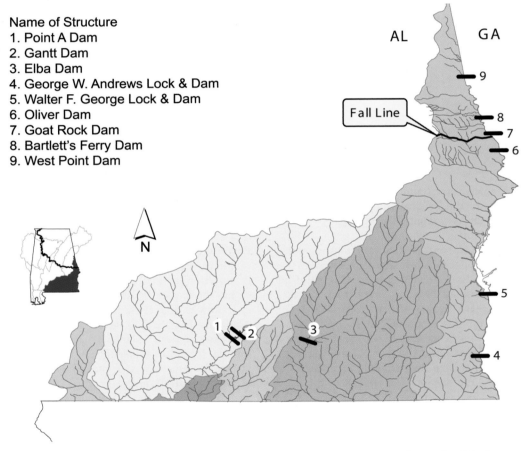

Name of Structure
1. Point A Dam
2. Gantt Dam
3. Elba Dam
4. George W. Andrews Lock & Dam
5. Walter F. George Lock & Dam
6. Oliver Dam
7. Goat Rock Dam
8. Bartlett's Ferry Dam
9. West Point Dam

Figure 3.26. Eastern Gulf Coast drainages in Alabama, from west to east: Conecuh, Blackwater, Yellow, Choctawhatchee, Chipola and Chattahoochee rivers. The numbered black bars represent approximate locations of major dams on the Conecuh River (dams 1 and 2), Pea River (dam 3; Pea River is a tributary of the Choctawhatchee River) and Chattahoochee River (dams 4 to 9). Several small mill dams and diversions in the vicinity of Columbus, Georgia, are not shown.

Drainage area of the Conecuh River is about 8,671 km^2 (USACE, 1985a). The Conecuh River is generally characterized by a low to moderate gradient with low banks and predominantly sand and gravel substrates (Figure 3.27). In the vicinity of the Alabama and Florida state line the Escambia River width ranges from 24 m to 37 m and depth ranges from 1.2 m to 1.8 m (Wurtz and Roback, 1955). There are numerous tributaries in the drainage; the largest are the Sepulga River with a drainage area of 2,714 km^2 and Patsaliga Creek with a drainage area of 1,554 km^2 (USACE, 1985a). Both streams enter the Conecuh River from the north. There are two impoundments on the Conecuh River, located north of Andalusia in Covington County. The two dams, Point A Dam and Gantt Dam, were constructed in 1923 and are for hydropower generation.

Figure 3.27. Conecuh River just upstream of County Highway 4, east of Brewton, Escambia County, Alabama, 30 June 1995. Photograph by J.D. Williams.

Blackwater River

The Blackwater River, located between the Escambia and Yellow rivers, is the smallest of the Gulf Coast rivers in Alabama (Figure 3.26). It originates in eastern Escambia and southwestern Covington counties, Alabama, and flows southwest into Blackwater Bay, an arm of Pensacola Bay, Florida. It has a total length of 94 km and a drainage area of approximately 2,227 km^2 (USACE, 1985b) and is confined to the Coastal Plain. The river is a blackwater stream characterized by tannic, darkly stained water with low concentrations of minerals and nutrients. Substrates of the Blackwater River are coarse, white sand or mixtures of sand and quartz gravel (Beck, 1965). There are no impoundments on the Blackwater River proper. It is one of only three drainages in Alabama in which there are no known occurrences of mussels.

Yellow River

The Yellow River arises in northern Covington and southeastern Crenshaw counties in south-central Alabama (Figure 3.26). It flows southwest for 185 km entirely within the Coastal Plain, then empties into Blackwater Bay, Florida. The Alabama reach of the river is about 80 km in length with a drainage area of 1,248 km^2, almost entirely in Covington County (USACE, 1985b). The river is swift, shallow and narrow in its upper reaches, becoming deeper and more sluggish downstream. Substrates are predominantly fine to coarse sand or sandy clay, with occasional patches of gravel. The two largest Alabama tributaries of the Yellow River are Five Runs Creek (319 km^2) and Lightwood Knot Creek (306 km^2) (USACE, 1985b). Lightwood Knot Creek has an impoundment, Frank Jackson Reservoir.

Lake Jackson in Florala, Alabama, is a large sinkhole lake that straddles the Alabama and Florida state line in the Yellow River drainage. There are no native mussels known from the lake. However, it appears that *Utterbackia imbecillis* was introduced sometime during the past decade. The introduction most likely occurred via fish introductions. It is not known if it has become established as a reproducing population.

Choctawhatchee River

The Choctawhatchee River drainage is located on the Coastal Plain in southeastern Alabama (Figure 3.26). The Choctawhatchee River proper originates at the junction of the East Fork and West Fork Choctawhatchee rivers southwest of Eufaula, Alabama. It flows generally southward and crosses the Florida panhandle before emptying into Choctawhatchee Bay, Florida. The Choctawhatchee River is the third largest eastern Gulf Coast drainage in Alabama. It has a total length of 288 km, of which 148 km lie in the state. The total drainage area is 12,030 km^2, of which 8,179 km^2 are in Alabama (USACE, 1985a, 1985b). Streams of the Choctawhatchee River drainage are characterized by sand and gravel substrates with occasional rocky outcrops and slow to moderate flow (Figure 3.28).

The Pea River is the largest Choctawhatchee River tributary, approximately 245 km long with a drainage area of 4,027 km^2, which is about one-third of the total Choctawhatchee River drainage. From its headwaters in Barbour and Bullock counties, Alabama, the Pea River flows generally southward before turning east to join the Choctawhatchee River near Geneva, Alabama. Other major Choctawhatchee tributaries in Alabama include Claybank Creek (609 km^2), Double Bridges Creek (505 km^2) and Little Choctawhatchee River (414 km^2). The only large impoundment in the Choctawhatchee River drainage is on the Pea River at Elba, Alabama. This dam was built in 1903 and is about 10 m high and is a barrier to upstream fish migration.

Figure 3.28. West Fork Choctawhatchee River on Alabama Highway 27, about 9.7 km southeast of Ozark, Dale County, Alabama, 5 May 1986. Photograph by J.D. Williams.

Apalachicola Basin

The Apalachicola Basin, located in eastern Alabama, western Georgia and northern Florida, has three principal tributaries: the Chattahoochee, Flint and Chipola rivers (Figure 3.2). It is the second largest Gulf Coast drainage east of the Mississippi River. The northern one-third of the basin drains the Blue Ridge and Piedmont provinces, and the southern two-thirds is entirely on the Coastal Plain. Drainage area of the entire basin is approximately 50,760 km^2. The combined length of the Chattahoochee and Apalachicola rivers, from the Chattahoochee's headwaters in the mountains of northeastern Georgia to the Apalachicola's mouth in Apalachicola Bay, Florida, is approximately 872 km (USACE, 1985a, 1985b).

Chattahoochee River

The Chattahoochee River, the largest tributary of the Apalachicola River, encompasses portions of Alabama, Florida and Georgia, including parts of three physiographic provinces: Blue Ridge, Piedmont and Coastal Plain. It flows southwestward from its origin in northeastern Georgia to the Alabama state line, then southward to form the boundary between Alabama and Georgia for a distance of 277 km. The Chattahoochee River drainage area is approximately 22,710 km^2, almost half the total drainage area of the Apalachicola Basin. The streambed elevation of the Chattahoochee River drops 340 m over its length. The river is regulated for most of its course by five large USACE dams and nine nonfederal dams of various sizes (Figure 3.26, smaller dams not shown).

The Chattahoochee River drainage from just south of West Point, Georgia, downstream to Columbus, Georgia, and Phenix City, Alabama, is in the Piedmont. Streams in this drainage are characterized by moderate gradient with pool and riffle habitat and bedrock bottoms (Figure 3.29).

From the Fall Line downstream to the Apalachicola River, the Chattahoochee River is impounded by three dams, which have converted it into a series of navigation pools (Figures 3.30, 3.31). Tributaries on the Coastal Plain typically have lower gradient, with sandy clay or sand and gravel bottoms and occasional outcrops of bedrock. There are five large tributaries to the Chattahoochee River on the Coastal Plain in Alabama: Cowikee Creek (1,204 km^2), Uchee Creek (865 km^2), Abbie Creek (513 km^2), Omusee Creek (461 km^2) and Hatchechubbee Creek (394 km^2). Chattahoochee tributaries above the Fall Line are considerably smaller in drainage area.

Figure 3.29. Chattahoochee River at the Fall Line, Columbus, Muscogee County, Georgia, and Phenix City, Russell County, Alabama, 4 May 1986. Photograph by J.D. Williams.

Figure 3.30. George W. Andrews Lock and Dam on the Chattahoochee River, Houston County, Alabama, and Early County, Georgia, February 1999. Photograph by J.D. Williams.

Figure 3.31. Impounded Chattahoochee River at upper end of reservoir behind George W. Andrews Lock and Dam, looking upstream from the Alabama Highway 10 bridge, Henry County, Alabama, during a reservoir drawdown in October 2000. Photograph by J.D. Williams.

The Chattahoochee River was one of the first southeastern rivers to undergo drastic anthropogenic alterations. These began in the early 1800s with removal of snags, construction of rock wing dams and blasting of shoals to open the river for navigation. Poor agricultural and timber harvest practices resulted in excessive loads of sediment being deposited in the river, leaving it in a drastically altered state (Glenn, 1911; Thurston, 1973; Brim Box and Williams, 2000). The first dam on the Chattahoochee River was constructed in 1834 at Columbus, Georgia, and was built to provide water power to industrial mills along the river. It was followed by construction of larger dams in the 1900s for hydroelectric power generation and flood control. The largest dams directly affecting the Alabama reach of the Chattahoochee River were built by the USACE. Three were built primarily for navigation—Jim Woodruff Lock and Dam (1954), located in Florida just south of the Alabama state line (not included on Figure 3.26), George W. Andrews Lock and Dam (1963) and Walter F. George Lock and Dam (1963)—and one, West Point Dam (1975), for hydropower generation and flood control. There are three smaller hydropower dams downstream of West Point Dam: Oliver, Goat Rock and Bartlett's Ferry dams (Figure 3.26).

Chipola River

Big and Cowarts creeks, the headwaters of the Chipola River, originate in Houston County in extreme southeastern Alabama and follow a southward course to the Alabama and Florida state line (Figure 3.26). The Chipola River continues southward to join the Apalachicola River in the Florida panhandle. The Chipola River lies entirely on the Coastal Plain; it is approximately 150 km long with a drainage area of 3,212 km[2]. Big and Cowarts creeks are low gradient streams with sand to sandy mud substrates, with occasional localized areas of gravel and compact clay. Several of the headwater streams are spring-fed and maintain consistent flow during periods of low rainfall and typically have low turbidity.

Interbasin Connections

Within the boundaries of Alabama and the Mobile Basin there are two man-made canals joining waters of different drainages across their common divide. These connections remove barriers to dispersal of aquatic fauna into adjacent drainages. The connections are between the Conecuh and Tallapoosa River drainages and the Tennessee and Tombigbee River drainages. While these connections are known to have allowed movement of fishes into adjacent drainages (Williams, 1965; Etnier and Starnes, 1993), as yet no mussels are known to have crossed the divides.

The interbasin connection between the Conecuh and Tallapoosa River drainages is located on the eastern side of Union Springs, Bullock County, Alabama, and dates back to the 1870s. The purpose of the water diversion was to power a grist mill. The drainage divide is a steep north-facing ridge, with an elevation of approximately 150 m, and consists of sand and clay underlain by chalky marl (Monroe, 1941). This connection involves a one-way diversion of water from the extreme headwaters of the Conecuh River into Old Town Creek of the Tallapoosa River drainage. This is accomplished by a small man-made ditch (1 m to 3 m wide). At the point where the river crosses the divide, a waterfall approximately 10 m high is formed and is locally known as Conecuh Falls (Monroe, 1941; Williams, 1965).

In eastern Mississippi the headwaters of the Tombigbee River are connected to Yellow Creek, a tributary of the Tennessee River, by the Tennessee Tombigbee Waterway, a barge canal that was completed in 1985. Operation of the barge canal permits free movement of fishes in both directions through a series of locks. Movement of fishes from the Tombigbee River drainage into the Tennessee River drainage has been documented (Etnier and Starnes, 1993), but no mussels are known to have crossed the divide between these two systems. The movement of fishes infested with glochidia through the barge canal remains a possibility.

Chapter 4
Mussel Distribution in Alabama and the Mobile Basin

Approximately 300 mussel species in 2 families, Margaritiferidae and Unionidae, occur in the U.S. and Canada (Turgeon et al., 1998). Most of this diversity is in the eastern U.S. but is particularly concentrated in the southeast. Other aquatic groups (e.g., snails, crayfish, fishes and turtles) also have high levels of diversity in the southeastern U.S. (Lydeard and Mayden, 1995). There are 178 mussel species in 43 genera reported from Alabama and the Mobile Basin. All of these occurred within the boundary of Alabama, which has the highest number of mussel species reported for a single state in the U.S., followed by Tennessee with 129 species (Parmalee and Bogan, 1998), Georgia with 123 species and Kentucky with 104 species (Cicerello and Schuster, 2003). This exceptional diversity is the product of an old and geologically stable land mass, which has not been glaciated and is especially rich in river systems and physiographic provinces.

Mussels, like other freshwater organisms, are isolated by freshwater drainages. Natural dispersal between drainages requires aquatic connections, usually brought about by geological events such as stream capture, subsidence and uplift that alter surface elevations, which may in turn reshape drainage patterns. Also, climatic changes that produce glacial cycles raise and lower sea levels, resulting in the connection or isolation of adjacent drainages. Lower sea levels result in downstream extension of rivers that may merge with streams of adjacent drainages, uniting formerly isolated aquatic faunas. Higher sea levels during interglacial periods flood lower reaches of rivers, which can isolate portions of a drainage or entirely inundate small coastal drainages. These scenarios have likely played a role in the evolution of the aquatic fauna in Alabama and the Mobile Basin.

Mussels in Alabama and the Mobile Basin vary in their distribution patterns. Some are widespread, while others are endemic to a single river system. Based on these distribution patterns there appears to be four distinct geographic mussel assemblages, those of the Tennessee River drainage, Mobile Basin, Gulf Coast drainages and Apalachicola Basin. Each assemblage has shared species but is characterized by a group of endemics that is useful in delineating assemblage boundaries. Factors that may have played a role in the number of endemics in a given assemblage include drainage size, relationship to other drainages and period of isolation. Fish assemblages have also been recognized from these same four regions (Boschung and Mayden, 2004).

There is a variety of distribution patterns among the four geographic assemblages (Table 4.1). The Margaritiferidae is represented by only two genera, *Cumberlandia* in the Tennessee River drainage and *Margaritifera* in Gulf Coast drainages and the Mobile Basin. Of the Unionidae's 41 genera, 14 occur in all of the mussel assemblages. Seven genera occur in three of the assemblages, exhibiting three distributional patterns: *Obovaria* and *Ptychobranchus* have representatives in all but the Apalachicola Basin; *Alasmidonta* and *Lasmigona* have representatives in all but the Gulf Coast drainages; and *Anodontoides*, *Glebula* and *Hamiota* have representatives in all but the Tennessee River drainage. Of the 10 genera that occur in only 2 of the assemblages, all are shared between the Tennessee River drainage and the Mobile Basin, the most diverse river drainages. Nine genera occur only in the Tennessee River drainage, but *Lemiox* is the only endemic genus (the others also occur in additional Interior Basin drainages). *Elliptoideus* is the only genus endemic to the Apalachicola Basin. There are no endemic unionid genera in the Mobile Basin and Gulf Coast drainages between the Mobile and Apalachicola rivers.

Comparison of historical assemblages is made difficult by the presence of adventive populations following the impoundment of major rivers, as well as by accidental and intentional introductions. Four species appear to be adventive in Alabama reaches of the Tennessee River since its impoundment: *Anodonta suborbiculata*, *Arcidens confragosus*, *Lasmigona complanata* and *Utterbackia imbecillis* (Garner and McGregor, 2001). An additional species, *Quadrula apiculata*, was apparently intentionally introduced into the Tennessee River (Parmalee and Bogan, 1998). The distributional history of *Toxolasma parvum* in the Mobile Basin and Gulf Coast drainages is not well understood, but the species may be adventive in those systems. *Anodonta suborbiculata* was not found in the Mobile Basin until the 1970s. It is also known from two impoundments of the Conecuh River, where it appears to be introduced.

Several mussel species (e.g., *Ligumia subrostrata*, *Toxolasma parvum* and *Utterbackia imbecillis*) have been observed to inhabit commercial fish culture ponds. At least one verifiable case of the introduction of these three species via cultured fish shipped from a fish farm in the Black Warrior River drainage, Alabama, to the Flint River drainage, Georgia, has been observed (R.C. Stringfellow, personal communication).

Table 4.1. Distribution of 43 genera across 4 mussel assemblages in Alabama and the Mobile Basin. TN = Tennessee River drainage; MB = Mobile Basin; GC = Gulf Coast drainages; ACF = Apalachicola, Chattahoochee and Flint River drainages (Apalachicola Basin). *Plectomerus* and *Uniomerus* are not known from the Alabama reaches of the Tennessee River.

	TN	MB	GC	ACF
Margaritiferidae				
Cumberlandia	X			
Margaritifera		X	X	
Unionidae				
Actinonaias	X			
Alasmidonta	X	X		X
Amblema	X	X	X	X
Anodonta	X	X	X	X
Anodontoides		X	X	X
Arcidens	X	X		
Cyclonaias	X			
Cyprogenia	X			
Dromus	X			
Ellipsaria	X	X		
Elliptio	X	X	X	X
Elliptoideus				X
Epioblasma	X	X		
Fusconaia	X	X	X	X
Glebula		X	X	X
Hamiota		X	X	X
Hemistena	X			
Lampsilis	X	X	X	X
Lasmigona	X	X		X
Lemiox	X			
Leptodea	X	X		
Ligumia	X	X		
Medionidus	X	X	X	X
Megalonaias	X	X	X	X
Obliquaria	X	X		
Obovaria	X	X	X	
Pegias	X			
Plectomerus	X	X		
Plethobasus	X			
Pleurobema	X	X	X	X
Pleuronaia	X			
Potamilus	X	X		
Ptychobranchus	X	X	X	
Pyganodon	X	X	X	X
Quadrula	X	X	X	X
Strophitus	X	X		
Toxolasma	X	X	X	X
Truncilla	X	X		
Uniomerus	X	X	X	X
Utterbackia	X	X	X	X
Villosa	X	X	X	X
Total Taxa	38	32	20	20

Of the 178 mussel species known from Alabama and the Mobile Basin, only *Elliptio crassidens*, *Megalonaias nervosa*, *Pyganodon grandis* and *Utterbackia imbecillis* belong to all four of the mussel assemblages. These species are also widespread elsewhere in the Mississippi Basin and Gulf Coast drainages (Table 4.2).

There are 93 mussel species known from Alabama reaches of the Tennessee River drainage, which makes it the most diverse assemblage in Alabama and the Mobile Basin. This exceptionally high diversity is the result of the occurrence of two overlapping aquatic faunal provinces, each represented by a large number of species. The highly endemic upper Tennessee River fauna, often referred to as the Cumberlandian fauna, meets the more widespread mussel fauna of the Mississippi Basin at greater Muscle Shoals, which extends from above the mouth of Elk River (Elk River Shoals) downstream to below the mouth of Bear Creek (Colbert Shoals), a distance of about 85 km (Isom, 1969; Garner and McGregor, 2001). Based on archaeological evidence, members of the Cumberlandian fauna once extended downstream at least to the mouth of the Duck River, with some species of *Epioblasma* extending to the mouth of the Tennessee River. Native fish diversity in Alabama reaches of the Tennessee River drainage is likewise high, 162 native species (Boschung and Mayden, 2004).

Eighteen mussel species are shared between the Tennessee River drainage in Alabama and the Mobile Basin, 13 of which (*Amblema plicata*, *Ellipsaria lineolata*, *Elliptio crassidens*, *Fusconaia ebena*, *Lampsilis teres*, *Leptodea fragilis*, *Ligumia recta*, *Megalonaias nervosa*, *Obliquaria reflexa*, *Pyganodon grandis*, *Quadrula metanevra*, *Quadrula verrucosa* and *Truncilla donaciformis*) appear to have been present historically, with the remainder arriving since modern perturbations to river habitats. These 13 species are widespread in the Mississippi Basin and several occur in Gulf Coast drainages.

The genus *Epioblasma* is particularly diverse in the Tennessee River drainage, with 18 species known from the Alabama reach of the river (Table 4.2). This represents 19 percent of all species known from the drainage. Unfortunately, *Epioblasma* species have not fared well with destruction of their shoal habitat. Eleven species are extinct, and five are extirpated from the state. The only extant species of the genus in Alabama are *Epioblasma brevidens* and *Epioblasma triquetra*.

The Mobile Basin has a total of 73 mussel species, which is the highest among Gulf Coast drainages, with the exception of the Mississippi Basin. This drainage also has an exceptionally high level of endemism, with 34 species (47 percent of the total). The Mobile Basin fish fauna, 178 native species, is likewise highly diverse; however, the level of fish endemism (42 species) is only 24 percent (Boschung and Mayden, 2004).

Mussel diversity within the Mobile Basin is distributed unevenly between eastern and western reaches of the drainage. Western reaches have 57 species, 78 percent of the total, whereas eastern reaches have 67 species, or 92 percent of the total. Most of the difference in diversity between the sub-basins can be attributed to the different levels of endemism. Of the 34 species endemic to the Mobile Basin, 15 are found in both eastern and western reaches. Of the remainder, 15 are known only from the eastern and 4 only from the western reaches of the basin. The genus *Pleurobema* accounts for much of the Mobile Basin endemism. Seven of the 15 species (47 percent) endemic to the eastern reaches of the basin are species of *Pleurobema* known only from the Coosa River drainage.

Alabama Gulf Coast drainages consist of portions of six rivers: the Pascagoula, Perdido, Conecuh, Blackwater, Yellow and Choctawhatchee. The entire drainage area of each of these rivers lies in the Coastal Plain physiographic province. Adjacent basins, the Mobile and Apalachicola, both have headwater rivers that arise in the geologically older upland physiographic provinces above the Fall Line. Two of the drainages, Perdido and Blackwater, apparently do not support any freshwater aquatic mollusks. A third drainage, the Pascagoula River, is unusual in that in Mississippi it supports 33 species of mussels (Jones et al., 2005), but its tributaries in Alabama have a depauperate fauna. Alabama tributaries of the Pascagoula River drainage are the Escatawpa River and its major tributary, Big Creek, from which the only mussel records are of *Pyganodon grandis* and *Utterbackia imbecillis* that were possibly introduced into Big Creek Reservoir. Red Creek, a tributary of the Pascagoula River in Washington County, Alabama, supports populations of three mussel species, *Lampsilis*

straminea, *Villosa lienosa* and *Villosa vibex*. Red Creek is frequently overlooked because it is not associated with the Escatawpa River, instead flowing into the Chickasawhay River in Mississippi.

The remaining Alabama Gulf Coast drainages—the Conecuh, Yellow and Choctawhatchee rivers—have a cumulative total of 32 species (Table 4.2). Of those, 11 (34 percent) are endemic. Distribution of endemics varies across the drainages, but over half occur in all three (*Hamiota australis*, *Pleurobema strodeanum*, *Ptychobranchus jonesi*, *Quadrula succissa*, *Toxolasma* sp. and *Villosa choctawensis*). Three species are endemic to a single drainage: *Fusconaia burkei* and *Obovaria haddletoni* in the Choctawhatchee River drainage, and *Fusconaia rotulata* in the Conecuh River drainage. The remaining two species have unusual distribution patterns, with *Elliptio mcmichaeli* occurring in the Choctawhatchee and Escambia River drainages and *Fusconaia escambia* in the Escambia and Yellow River drainages.

There are 30 species known from Alabama reaches of the Chattahoochee and Chipola rivers, which are part of the Apalachicola Basin (Table 4.2). Brim Box and Williams (2000) reported 33 species from the Apalachicola Basin. Two of the species (*Anodonta heardi* and *Amblema neislerii*) are not known to occur in the Alabama portion of the basin but are herein included as species of hypothetical occurrence. Based on genetic analyses, *Strophitus subvexus* reported by Brim Box and Williams (2000) from the Apalachicola Basin was reidentified as *Anodontoides radiatus*. Of the 30 species known from Alabama reaches of the basin, all but *Utterbackia peggyae* occur in the Chattahoochee River drainage. Eleven species (37 percent) are endemic to the Apalachicola Basin, all of which historically occurred in Alabama.

There are two interbasin connections in Alabama and the Mobile Basin. Mussel dispersal is likely to occur in one, the Tennessee Tombigbee Waterway. This connection between the upper Tombigbee River and the Tennessee River in northeastern Mississippi has resulted in interbasin dispersal of fishes but it is not known if mussels have successfully crossed the divide. The other interbasin connection is between the Conecuh and Tallapoosa River drainages, located on the eastern side of Union Springs, Bullock County, Alabama. This connection is a one-way diversion of water from the headwaters of the Conecuh River into Old Town Creek of the Tallapoosa River drainage. This diversion involves only a small part of the Conecuh River and has been in place for over a century. It does not appear to have resulted in an interbasin transfer of mussels.

Table 4.2. Distribution of mussels in the major river drainages of Alabama and the Mobile Basin. Bold vertical lines distinguish the four geographic mussel assemblages: 1. Tennessee River drainage (TEN = Tennessee River); 2. Mobile Basin (MOB = Mobile River, LTM = lower Tombigbee River, UTM = upper Tombigbee River, BLW = Black Warrior River, ALA = Alabama River, CAH = Cahaba River, COS = Coosa River, TAL = Tallapoosa River); 3. Gulf Coast drainages (PAS = Pascagoula River, CON = Conecuh River, YEL = Yellow River, CHO = Choctawhatchee River); 4. Apalachicola Basin (CHP = Chipola River, CHT = Chattahoochee River).

	TEN	MOB	LTM	UTM	BLW	ALA	CAH	COS	TAL	PAS	CON	YEL	CHO	CHP	CHT
Margaritiferidae															
Cumberlandia monodonta	X														
Margaritifera marrianae			X	X		X					X				
Unionidae															
Actinonaias ligamentina	X														
Actinonaias pectorosa	X														
Alasmidonta marginata	X														
Alasmidonta mccordi								X							
Alasmidonta triangulata														X	
Alasmidonta viridis	X														
Amblema elliottii							X	X							
Amblema plicata	X		X	X	X	X	X	X	X		X				
Anodonta suborbiculata	X		X	X	X	X	X	X	X		X				
Anodonta sp.				X							X				
Anodontoides radiatus			X	X	X	X	X	X	X		X		X	X	X
Arcidens confragosus	X	X	X	X	X	X		X							
Cyclonaias tuberculata	X														
Cyprogenia stegaria	X														
Dromus dromas	X														
Ellipsaria lineolata	X		X	X	X	X	X	X	X						
Elliptio arca			X	X	X	X	X	X	X						
Elliptio arctata			X	X	X	X	X	X	X		X		X	X	X
Elliptio chipolaensis														X	X
Elliptio crassidens	X	X	X	X	X	X	X	X	X		X	X	X	X	X
Elliptio dilatata	X														
Elliptio fraterna															X

	TEN	MOB	LTM	UTM	BLW	ALA	CAH	COS	TAL	PAS	CON	YEL	CHO	CHP	CHT
Elliptio fumata														X	X
Elliptio mcmichaeli											X		X		
Elliptio nigella														X	X
Elliptio pullata											X	X	X	X	X
Elliptio purpurella														X	X
Elliptoideus sloatianus															X
Epioblasma arcaeformis	X														
Epioblasma biemarginata	X														
Epioblasma brevidens	X														
Epioblasma capsaeformis	X														
Epioblasma cincinnatiensis	X														
Epioblasma flexuosa	X														
Epioblasma florentina	X														
Epioblasma haysiana	X														
Epioblasma lenior	X														
Epioblasma lewisii	X														
Epioblasma metastriata					X		X	X							
Epioblasma obliquata	X														
Epioblasma othcaloogensis								X							
Epioblasma penita			X	X	X	X	X	X							
Epioblasma personata	X														
Epioblasma propinqua	X														
Epioblasma stewardsonii	X														
Epioblasma torulosa	X														
Epioblasma triquetra	X														
Epioblasma turgidula	X														
Epioblasma sp. cf. capsaeformis	X														
Fusconaia apalachicola															X
Fusconaia burkei													X		
Fusconaia cerina		X	X	X	X	X	X	X	X						
Fusconaia cor	X														
Fusconaia cuneolus	X														

	TEN	MOB	LTM	UTM	BLW	ALA	CAH	COS	TAL	PAS	CON	YEL	CHO	CHP	CHT
Fusconaia ebena	X	X	X	X	X	X	X	X							
Fusconaia escambia											X	X			
Fusconaia rotulata											X				
Fusconaia subrotunda	X														
Glebula rotundata		X	X			X					X				
Hamiota altilis						X	X	X	X		X				
Hamiota australis												X	X		
Hamiota perovalis			X	X	X	X									
Hamiota subangulata														X	X
Hemistena lata	X														
Lampsilis abrupta	X														
Lampsilis binominata															X
Lampsilis fasciola	X														
Lampsilis floridensis											X		X		X
Lampsilis ornata			X	X	X	X	X	X	X		X				
Lampsilis ovata	X														
Lampsilis straminea	X		X	X	X	X	X	X	X	X	X	X	X	X	X
Lampsilis teres		X	X	X	X	X	X	X	X						
Lampsilis virescens	X				X	X	X	X	X						
Lasmigona alabamensis				X											
Lasmigona complanata	X														
Lasmigona costata	X														
Lasmigona etowaensis					X		X	X							
Lasmigona holstonia	X														
Lasmigona subviridis	X														X
Lemiox rimosus	X														
Leptodea fragilis			X	X	X	X	X	X	X						
Leptodea leptodon	X			X	X	X									
Ligumia recta	X			X	X	X	X	X	X						
Ligumia subrostrata		X													
Medionidus acutissimus			X	X	X	X	X	X	X		X	X	X		
Medionidus conradicus	X														

	TEN	MOB	LTM	UTM	BLW	ALA	CAH	COS	TAL	PAS	CON	YEL	CHO	CHP	CHT
Medionidus parvulus							X	X							X
Medionidus penicillatus														X	X
Megalonaias nervosa	X	X	X	X	X	X	X	X	X		X				
Obliquaria reflexa	X	X	X	X	X	X	X	X	X						
Obovaria haddletoni													X		
Obovaria jacksoniana			X	X	X	X	X								
Obovaria olivaria	X														
Obovaria retusa	X														
Obovaria subrotunda	X														
Obovaria unicolor			X	X	X	X	X	X							
Pegias fabula	X														
Plectomerus dombeyanus		X	X	X	X	X	X	X							
Plethobasus cicatricosus	X														
Plethobasus cooperianus	X														
Plethobasus cyphyus	X														
Pleurobema athearni								X							
Pleurobema clava	X														
Pleurobema cordatum	X														
Pleurobema curtum				X											
Pleurobema decisum			X	X	X	X	X	X	X						
Pleurobema fibuloides								X							
Pleurobema georgianum								X							
Pleurobema hanleyianum								X							
Pleurobema hartmanianum								X							
Pleurobema marshalli			X	X											
Pleurobema oviforme	X														
Pleurobema perovatum			X	X	X	X	X		X						
Pleurobema plenum	X														
Pleurobema pyriforme														X	X
Pleurobema rubellum					X		X								
Pleurobema rubrum	X														
Pleurobema sintoxia	X														

	TEN	MOB	LTM	UTM	BLW	ALA	CAH	COS	TAL	PAS	CON	YEL	CHO	CHP	CHT
Pleurobema stabilis								X							
Pleurobema strodeanum											X	X	X		
Pleurobema taitianum			X	X	X	X	X								
Pleurobema verum						X	X								
Pleuronaia barnesiana	X														
Pleuronaia dolabelloides	X														
Potamilus alatus	X														
Potamilus inflatus			X	X	X	X	X								
Potamilus ohiensis	X														
Potamilus purpuratus			X	X	X	X	X	X							
Ptychobranchus fasciolaris	X														
Ptychobranchus foremanianus						X	X	X							
Ptychobranchus greenii				X	X										
Ptychobranchus jonesi											X	X	X		
Ptychobranchus subtentum	X														
Pyganodon cataracta									X						X
Pyganodon grandis	X	X	X	X	X	X	X	X	X	X	X		X	X	X
Quadrula apiculata	X	X	X	X	X	X	X	X	X						
Quadrula asperata			X	X	X	X	X	X	X						
Quadrula cylindrica	X														
Quadrula infucata														X	X
Quadrula intermedia	X														
Quadrula kieneriana								X							
Quadrula metanevra	X		X	X	X	X	X	X							
Quadrula nobilis		X	X	X	X	X	X	X	X						
Quadrula pustulosa	X														
Quadrula quadrula	X														
Quadrula rumphiana			X	X	X	X	X	X	X						
Quadrula sparsa	X					X									
Quadrula stapes				X	X										
Quadrula succissa											X	X	X		
Quadrula verrucosa	X		X	X	X	X	X	X	X						

	TEN	MOB	LTM	UTM	BLW	ALA	CAH	COS	TAL	PAS	CON	YEL	CHO	CHP	CHT
Strophitus connasaugaensis						X	X	X	X						
Strophitus subvexus			X	X	X										
Strophitus undulatus	X														
Toxolasma corvunculus				X	X	X	X	X	X						
Toxolasma cylindrellus	X														
Toxolasma lividum	X														
Toxolasma parvum	X	X	X	X	X	X	X	X	X		X				
Toxolasma paulum														X	X
Toxolasma sp.												X	X		
Truncilla donaciformis	X			X	X	X	X	X	X		X	X			
Truncilla truncata	X														
Uniomerus columbensis														X	X
Uniomerus tetralasmus			X	X		X	X	X	X		X	X	X	X	X
Utterbackia imbecillis	X	X	X	X	X	X	X	X	X	X	X	X	X	X	X
Utterbackia peggyae													X	X	
Villosa choctawensis											X	X	X		
Villosa fabalis	X														
Villosa iris	X														
Villosa lienosa			X	X	X	X	X	X	X	X	X	X	X	X	X
Villosa nebulosa					X		X	X							
Villosa taeniata	X														
Villosa trabalis	X														
Villosa umbrans								X							
Villosa vanuxemensis	X														
Villosa vibex			X	X	X	X	X	X	X	X	X	X	X	X	X
Villosa villosa											X			X	X
Total Taxa	93	14	40	51	51	51	50	54	36	5	30	15	21	20	29

Chapter 5
Mussels in the Archaeological Record

Mollusk collections in natural history museums provide a record of the mussel fauna but date only to the early 1800s. The archaeological record provides documentation of species that once occurred in localized areas and their use by Native Americans prior to the arrival of European settlers. Faunal remains recovered from archaeological excavations often contain vast quantities of mussel shell (e.g., Morrison, 1942; Warren, 1975). They provide a great source of data on prehistoric faunal distribution and relative abundance. These records are very important in understanding species composition in areas where the fauna has been decimated during past centuries due to the destruction of native habitats, especially since many rivers were not surveyed prior to these changes. This type of data has been used to better evaluate faunal changes in various rivers (e.g., Parmalee and Bogan, 1998). The archaeological mussel record of Alabama and the Mobile Basin is herein used to examine previous mussel diversity.

The rich archaeological mussel record of the Tennessee River has been studied more intensely than that of other rivers in Alabama and the Mobile Basin. Hughes and Parmalee (1999) summarized Tennessee River archaeological data. Studies from the Alabama reach of the Tennessee River include Morrison (1942), Warren (1975), Curren et al. (1977), Hanley (1983) and Parmalee (1994), as well as unpublished data. Clench (1974) provided additional mussel information from Russell Cave, somewhat removed from the main channel of the river but containing species that were probably collected from the Tennessee River. Bogan (1990) compared mussel diversity and evenness among archaeological sites in the Tennessee River drainage and other Interior drainages.

In the Alabama reach of the Tennessee River, 69 mussel species have been identified from archaeological sites (Table 5.1). A detailed comparison of the Tennessee River drainage archaeological, preimpoundment and postimpoundment mussel fauna was presented by Hughes and Parmalee (1999). Three species—*Epioblasma lewisii*, *Quadrula sparsa* and *Villosa fabalis*—are known from Alabama only in the archaeological record. Eleven species were reported in historical preimpoundment collections from the Tennessee River in Alabama but are not in the archaeological record. These are *Epioblasma turgidula*, *Fusconaia ebena*, *Hemistena lata*, *Lampsilis teres*, *Leptodea leptodon*, *Megalonaias nervosa*, *Obovaria olivaria*, *Quadrula quadrula*, *Quadrula verrucosa*, *Toxolasma parvum* and *Truncilla donaciformis*.

Whether they were overlooked or ignored by Native Americans or colonized the Tennessee River subsequent to deposition of prehistoric shell middens is unclear. If colonization is the explanation for the appearance of these species since deposition of the shell middens, all would have colonized from downstream reaches of the Tennessee River, with the exception of *E. turgidula*. It could have colonized the Alabama reach from upper portions of the Tennessee River drainage as well as from the Elk River. Four species—*Anodonta suborbiculata*, *Arcidens confragosus*, *Lasmigona complanata* and *Utterbackia imbecillis*—have apparently colonized Alabama from lower reaches of the Tennessee River since it was impounded (Garner and McGregor, 2001).

The largest collection of archaeological mussel shell excavated from Tennessee River sites in Alabama is from Widows Creek (Warren, 1975). The shell midden is located in Jackson County on a natural levee along the right-descending bank of the Tennessee River, at the mouth of Widows Creek. Warren (1975) identified 59,809 valves representing 52 mussel species (adjusted to taxonomy herein recognized). The estimated age of the shell and associated artifacts range from AD 350 to about AD 800. Based on modern ecology of the species recovered, the site is believed to have had shoal habitat (Warren, 1975).

The second largest collection of archaeological mussel shell from the Tennessee River in Alabama is from a study of seven shell midden sites between Tennessee River miles 226 and 242, downstream of Muscle Shoals. The excavations, the first major archaeological study in the area, were made just prior to their impoundment behind Pickwick Dam (Morrison, 1942). A combined total of 31,950 valves representing 53 species (adjusted to taxonomy herein recognized) was reported. Deposition was estimated to have occurred from 3000 BC to AD 1000.

Mussel shells from prehistoric middens are usually unmodified (Morrison, 1942). However, Warren (1975) figured several valves with man-made holes through them, the utility of which remains unknown. In other contexts, such as burials, human-modified mussel shells have been found. Late prehistoric burials located in eastern Tennessee have contained a variety of modified shells, including shell spoons carefully shaped from valves of *Lampsilis ovata* and elongate shells (e.g., *Elliptio dilatata* and *Ligumia recta*) with the ventral margin cut or ground to a straight edge (Bogan, 1980; Bogan and Polhemus, 1987).

Table 5.1. Freshwater mussels reported from archaeological (before AD 1400) sites in river drainages of Alabama and the Mobile Basin in Georgia and Mississippi (adjusted to taxonomy herein recognized).

	Tennessee River	Tombigbee River	Alabama River	Etowah River	Chattahoochee River
Margaritiferidae					
Cumberlandia monodonta	X				
Unionidae					
Actinonaias ligamentina	X				
Actinonaias pectorosa	X				
Alasmidonta marginata	X				
Alasmidonta viridis	X				
Amblema elliottii				X	
Amblema plicata	X	X	X		
Cyclonaias tuberculata	X				
Cyprogenia stegaria	X				
Dromus dromas	X				
Ellipsaria lineolata	X	X	X		
Elliptio arca		X	X	X	
Elliptio arctata		X		X	
Elliptio crassidens	X	X	X	X	X
Elliptio dilatata	X				
Elliptio fumata					X
Elliptio pullata					X
Elliptoideus sloatianus					X
Epioblasma arcaeformis	X				
Epioblasma biemarginata	X				
Epioblasma brevidens	X				
Epioblasma capsaeformis	X				
Epioblasma flexuosa	X				
Epioblasma florentina	X				
Epioblasma haysiana	X				
Epioblasma lewisii	X				
Epioblasma metastriata			X	X	
Epioblasma obliquata	X				
Epioblasma penita		X			
Epioblasma personata	X				
Epioblasma propinqua	X				
Epioblasma stewardsonii	X				
Epioblasma torulosa	X				
Epioblasma triquetra	X				
Fusconaia apalachicola					X
Fusconaia cerina		X	X		
Fusconaia cor	X				
Fusconaia cuneolus	X				
Fusconaia ebena		X	X		
Fusconaia subrotunda	X				
Hamiota altilis				X	
Hamiota perovalis		X			
Lampsilis abrupta	X				
Lampsilis fasciola	X				
Lampsilis ornata		X	X		
Lampsilis ovata	X				

	Tennessee River	Tombigbee River	Alabama River	Etowah River	Chattahoochee River
Lampsilis straminea		X	X		
Lampsilis teres		X			
Lampsilis virescens	X				
Lasmigona alabamensis		X			
Lasmigona costata	X				
Lemiox rimosus	X				
Leptodea fragilis	X	X	X	X	
Ligumia recta	X	X	X	X	
Medionidus acutissimus				X	
Medionidus conradicus	X				
Megalonaias nervosa		X		X	
Obliquaria reflexa	X	X	X		
Obovaria jacksoniana		X			
Obovaria retusa	X				
Obovaria subrotunda	X				
Obovaria unicolor		X	X	X	
Plectomerus dombeyanus		X	X		
Plethobasus cicatricosus	X				
Plethobasus cooperianus	X				
Plethobasus cyphyus	X				
Pleurobema clava	X				
Pleurobema cordatum	X				
Pleurobema decisum		X	X	X	
Pleurobema georgianum				X	
Pleurobema hartmanianum				X	
Pleurobema marshalli		X			
Pleurobema oviforme	X				
Pleurobema perovatum		X			
Pleurobema plenum	X				
Pleurobema pyriforme					X
Pleurobema rubrum	X				
Pleurobema sintoxia	X				
Pleurobema stabilis				X	
Pleurobema taitianum		X	X		
Pleuronaia barnesiana	X				
Pleuronaia dolabelloides	X				
Potamilus alatus	X				
Potamilus purpuratus		X		X	
Ptychobranchus fasciolaris	X				
Ptychobranchus foremanianus				X	
Ptychobranchus subtentum	X				
Pyganodon grandis	X				
Quadrula apiculata		X			
Quadrula asperata		X	X	X	
Quadrula cylindrica	X				
Quadrula infucata					X
Quadrula intermedia	X				
Quadrula metanevra	X	X	X		
Quadrula nobilis		X	X		
Quadrula pustulosa	X				
Quadrula rumphiana		X			
Quadrula sparsa	X				

	Tennessee River	Tombigbee River	Alabama River	Etowah River	Chattahoochee River
Quadrula stapes		X	X		
Quadrula verrucosa		X	X	X	
Strophitus connasaugaensis				X	
Strophitus subvexus		X			
Strophitus undulatus	X				
Toxolasma corvunculus				X	
Toxolasma lividum	X				
Toxolasma parvum		X			
Truncilla donaciformis		X	X		
Truncilla truncata	X				
Uniomerus tetralasmus		X			
Villosa fabalis	X				
Villosa iris	X				
Villosa lienosa		X			
Villosa taeniata	X				
Villosa trabalis	X				
Villosa umbrans				X	
Villosa vanuxemensis	X				
Total Taxa	**69**	**37**	**22**	**22**	**7**

The importance of mussels in Native American diets has been a point of disagreement. Parmalee and Klippel (1974) examined caloric values of mussel tissues (*Actinonaias ligamentina* and *Potamilus alatus*) and compared them with values of other animals commonly associated with Native American middens, such as White-tailed Deer, Wild Turkey and rabbit. These analyses suggested that mussels are not particularly nutritious. Mussels contain far fewer calories than other meats and may have provided only a supplement to the Native American diet.

The archaeological mussel record of the Mobile Basin is less studied than that of the Tennessee River drainage. Numerous exposed sites along rivers and creeks have not been studied or have received only superficial attention. There are archaeological mussel records from 16 sites in the Tombigbee, Etowah and Alabama River drainages.

The Tombigbee River is best represented in the archaeological record with nine sites reported in the literature (Rummel, 1980; Woodrick, 1981, 1983; Hanley, 1982, 1983, 1984a, 1984b; Reitz, 1987; Hartfield, 1990; Peacock, 1998, 2000). A total of 37 species was identified in these studies (Table 5.1). An additional site, in Clarke County, Alabama, is currently being investigated (A.A. Dumas, personal communication). Mussel diversity in the Tombigbee River proper has been greatly reduced by the construction of the Tennessee Tombigbee Waterway.

The only archaeological mussel assemblage reported from the Coosa River drainage is from Etowah Mounds along the Etowah River, Bartow County,

Georgia (van der Schalie and Parmalee, 1960). A total of 22 species (adjusted to taxonomy herein recognized) was identified from this site, plus unidentified species of *Pleurobema* and *Quadrula* (Table 5.1). The Etowah Mounds study is important because the Etowah River was poorly sampled historically, prior to extirpation of most mussel species.

Archaeological mussel remains have been reported from six sites along the Alabama River. One site is located at Eureka Landing in Monroe County, from which seven species were identified (Wood et al., 1987). From 4 sites in Lowndes County and a single site in Autauga County a total of 19 species was recovered (Dickens, 1971; Womochel, 1982). Thirteen species were identified during casual observations at a prehistoric shell midden along the Alabama River in Montgomery County, just upstream of the Alabama River Parkway (J.T. Garner, personal observation). The combined list from these studies, 22 species, appears to represent a small portion of the Alabama River fauna (Table 5.1).

None of the studies of archaeological sites along Alabama reaches of the Chattahoochee River have included mussel remains. However, six species were reported in the description of *Fusconaia apalachicola*, which was described from archaeological remains (Williams and Fradkin, 1999) (Table 5.1). The material was from a casual collection at the mouth of Omussee Creek, Houston County, Alabama (F.T. Schnell, personal communication). An additional species, *Elliptio fumata*, was located in archaeological samples housed in the Columbus Museum, Columbus, Georgia.

Chapter 6
Commercial Utilization of Mussels

Native Americans apparently treasured pearls of freshwater mussels, and "immense quantities" have been unearthed from mounds along the Mississippi River (Kunz, 1898a). However, following the colonization of North America by Europeans, these objects of value were ignored—until 1857, when discovery of a valuable pearl in New Jersey set off the first wave of intense exploitation of the resource. Within one year pearls were sent to Tiffany & Co. in New York from "nearly every state." Harvest pressures were periodic and usually regional in scope. Kunz stated that "every ten years or so" renewed interest would arise in various regions, associated with discoveries of valuable pearls that would bring on "a fresh campaign of ignorant extermination" for several summers until the resource was once again exhausted. The practice of harvesting mussels strictly for pearls appears to have diminished greatly with the advent of the pearl button industry. Harvest of mussels for their shell proved to be a much more dependable source of income. However, pearls continued to be collected as a byproduct of shell harvest (Figure 6.1).

Figure 6.1. Natural pearls taken from a variety of mussel species. The largest pearl near the center is 10 mm in length. © Richard T. Bryant.

John F. Boepple immigrated to the U.S. from Germany in 1887, settled in Muscatine, Iowa, and started America's first freshwater pearl button company in 1891. By 1898 there were 11 plants in Iowa and Illinois producing finished buttons and an additional 38 "saw works" that cut shell to produce button blanks (Figures 6.2, 6.3) to be completed at other factories (Smith, 1899). Button factories eventually sprang up in other parts of the country, but Muscatine remained the center of the industry, and shell was shipped there from other states via the railroad. The earliest mention of Alabama shell harvest in the literature was around 1904 (Claassen, 1994). In 1931 Alabama ranked sixth in the total mussel harvest from 13 states with more than 37 million pounds sold (Johnson, 1934). The Tennessee River was a major source of shell, while the Tombigbee River was a minor contributor (Coker, 1919). The Alabama River was exploited later (Claassen, 1994). By 1942 the Tennessee River was the primary producer of mussel shell in the U.S. (Scruggs, 1960).

Commercial harvest of freshwater mussel shell ceased in Alabama reaches of the Tennessee River following its impoundment, which was completed in 1939. The assumption was made that populations were destroyed with impoundment. However, in 1945 a few unemployed former mussel harvesters investigated the old beds in the upper reaches of Wheeler Reservoir and found mussels to be abundant (Table 6.1). The industry was revived "almost overnight" (Bryan and Miller, 1953).

Table 6.1. Summary of shell harvested from TVA reservoirs in northern Alabama between 1945 and 1962. Data from an unpublished TVA report entitled "Summary and Report of Tennessee River Mussel Survey in Alabama, July 1–November 30, 1963," dated 7 February 1964.

Calendar Year	Tons	Dollars per Ton	Total Dollar Value
1945	220	39.36	8,660
1946	1,275	38.26	48,781
1947	1,610	53.13	85,540
1948	2,663	43.27	115,229
1949	1,570	35.00	54,950
1950	3,135	30.00	94,050
1951	2,491	40.00	99,640
1952	4,124	45.00	185,580
1953	6,390	55.15	352,388
1954	6,815	42.00	286,270
1955	6,283	44.00	276,452
1956	3,805	59.00	224,495
1957	4,125	76.00	313,500
1958	2,330	60.00	139,800
1959	4,159	69.46	288,911
1960	6,728	123.18	828,768
1961	5,549	127.94	709,943
1962	2,864	154.63	442,859

Use of mussel shell for button production continued into the 1960s, but at a declining rate as plastic buttons replaced pearl buttons beginning in the 1940s. In the mid-1950s an export market developed for North American mussel shell to be used in the production of nuclei for cultured pearls (Isom, 1969). This new market maintained the shell harvest business as the button industry waned.

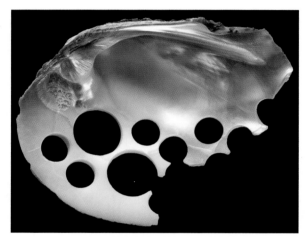

Figure 6.2. *Megalonaias nervosa* shell that was used to produce button blanks. © Richard T. Bryant.

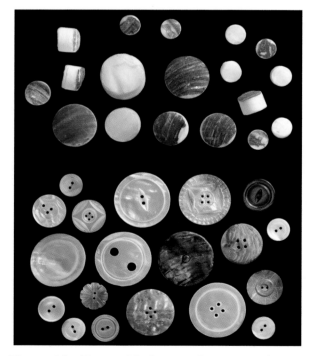

Figure 6.3. Button blanks cut from mussel shells (above) and finished buttons (below). © Richard T. Bryant.

In the 1950s and 1960s mussel shell continued to be harvested from the Tennessee River. Between 1952 and 1962 the Alabama reaches of the Tennessee River contributed 49 percent to 79 percent to the river's total harvest (Isom, 1969). A detailed evaluation of the mussel fishery on the Tennessee River's Wheeler Reservoir in Alabama estimated harvest from the reservoir to be 570 tons in 1991 and 1992 (Bowen et al., 1994). Most of the harvest was composed of *Megalonaias nervosa* (45 percent), *Pleurobema cordatum* (25 percent) and *Ellipsaria lineolata* (10 percent). The study was carried out during a period of high shell demand, during which a peak price of $9.00 per pound ($19.80 per kg) was reported for top quality *M. nervosa* shell. Estimated wholesale value of the 12-month catch was $2,119,921.

Reporting of shell purchased from commercial harvesters has been required by the state of Alabama since the 1960s. However, until the 1990s there was no systematic structure for submission or maintenance of the records, and harvest data from 1963 to 1993 are difficult to assess. A resurgence of harvest pressure during the late 1980s and early 1990s (though on a smaller scale than what occurred from the 1940s to 1960s) prompted the establishment of a standardized receipt system for commercial mussel shell sold in Alabama. The first complete year for which receipt records are available is 1994, during which a total harvest of 530 tons was reported. A steep decline in market demand for shell began in late 1996, and harvest reached its lowest level in 1999 when a total of only 22 tons was reported. From 2000 through 2004 the market was variable, with annual reported harvest ranging from 75 tons to 263 tons. Note that numbers from 1994 to 2004 represent those reported on receipts. There is ample anecdotal evidence that a considerable amount of illegally harvested undersized shell was exported from Alabama to states with smaller size limits without being reported.

The future of commercial harvest in Alabama remains dependent on market demand for shell. As of 2007 commercial mussel stocks were high in most areas of Alabama rivers that are open to harvest. Current size limit regulations are conservative and should ensure a renewable resource.

Harvest Methods

Methods used by early pearl hunters to gather mussels were "the simplest and most primitive"—they simply waded into shallow water and picked them up by hand (Kunz, 1898a). The development of the pearl button industry created a need for greater quantities of shell to be harvested, and more efficient methods were devised. Early methods included tongs, hand rakes and rakes operated by means of a windlass (Smith, 1899). The tongs were similar to those used to harvest oysters, constructed of two opposable rakes, and could be used in water up to 4.5 m deep. Hand rakes varied in pattern but basically consisted of a wire basket attached to the back of an iron rake with handles 4.3 m to 6.1 m long. Rakes pulled along the bottom with a rope attached to a windlass were usually larger than hand rakes. Also in operation at Muscatine, Iowa, during the 1890s was at least one steam-powered scow, which used a steam-powered dredge to harvest mussels (Smith, 1899).

In the spring of 1897 the crowfoot brail was developed and soon became the primary method of harvest due to its "cheapness and efficiency" (Smith, 1899). A brail is comprised of a series of four-pronged hooks (Figure 6.4) made of stout wire attached to a horizontal bar with rope or chains. The contraption is lowered from a boat (Figure 6.5) and pulled along the river bottom; mussels close their valves on the prongs (mussels are generally positioned in the sediment with only their posterior ends exposed and valves slightly agape) and are pulled from the sediment. The mussels usually remain attached to the brail hooks until forcibly removed (Figure 6.6). Brail boats typically use the river current for power, propelled by an underwater sail called a "mule" (Smith, 1899; Parmalee and Bogan, 1998).

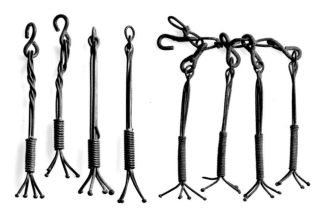

Figure 6.4. Brail hooks used in the harvest of mussels. © Richard T. Bryant.

Figure 6.5. Mussel brail boat on the Tombigbee River, Alabama, August 1974. Photograph by J.D. Williams.

During the late 1960s harvesters began using dive gear to aid in gathering mussels (Parmalee and Bogan, 1998). Using this method, mussels are simply taken by hand by a diver supplied with air from tanks or, more typically, a boat-mounted compressor. By the 1990s this became the primary method of harvest. It is more efficient than use of a brail because noncommercial species and undersized mussels can be passed over without handling.

Figure 6.6. Mussels attached to crowfoot brail hooks. Photograph by L.M. Koch.

Commercial Species

Fusconaia ebena was the most important species to the pearl button industry. *Fusconaia subrotunda, Lampsilis abrupta, Lampsilis higginsi, Lampsilis teres, Ligumia recta, Megalonaias nervosa* and *Obovaria olivaria* were other significant species (Smith, 1899; Coker, 1919). Pustulose species were initially rejected by the industry but came into use as more desirable species became scarce. Commonly harvested pustulose species included *Plethobasus cooperianus, Quadrula pustulosa* and *Quadrula quadrula* (Coker, 1919). Some species, such as *Ellipsaria lineolata, Quadrula nodulata* and *Quadrula verrucosa*, were suitable for button production but were too scarce to make up significant portions of the harvest (Smith, 1899; Coker, 1919). Other species were used by the industry but had thin shells, produced excessive waste or had nacre that was too hard or brittle. This group included *Amblema plicata, Fusconaia flava, Lampsilis cardium* Rafinesque, 1820, *Plethobasus cyphyus* and *Potamilus capax* (Smith, 1899; Coker, 1919). The shell of *Plethobasus cyphyus* is so hard that it actually damaged the saws used to cut the blanks (Coker, 1919).

Some species were of variable quality, depending on their river of origin, such as *Actinonaias ligamentina* and *Lampsilis siliquoidea*. The latter was not used by the industry until populations in Lake Pepin and the St. Croix River, Minnesota and Wisconsin, were found to be of a form in which buttons could be cut from "practically the entire surface" (Coker, 1919). *Pleurobema cordatum*, along with *Pleurobema plenum* and *Pleurobema rubrum*, were initially considered

species that were useful but less desirable. However, they eventually became the most significant species harvested from some reaches of the Tennessee River (Scruggs, 1960).

As of 2007, most of the commercial harvest exported for the cultured pearl industry was comprised of *Fusconaia ebena* and *Megalonaias nervosa*. Other species that make up a substantial portion of the harvest include *Amblema plicata* and *Quadrula quadrula*. *Pleurobema cordatum* was initially an important species for the cultured pearl industry, but declined as its abundance diminished due to harvest outpacing reproduction. Failure of recruitment to maintain population levels was due to changes in habitat and possibly the availability of host fishes following impoundment of the river (Scruggs, 1960).

Regardless of the mussel species, they are treated the same in production of cultured pearl nuclei. The harvested shell is cut into cubes that are ground into rounded, polished nuclei, sometimes called "seed pearls" (Figure 6.7). The nuclei are inserted into a variety of bivalve species (e.g., pearl oysters) to produce cultured pearls (Landman et al., 2001). Development of the black pearl market created demand for old, large shells, which often have discolored nacre (e.g., *Megalonaias nervosa* residing in the river channel commonly called "river boards" by harvesters). Since the pearls are dark in color, nuclei made from discolored nacre can be used for their culture. *Potamilus alatus* and other purple-nacred species are periodically harvested to meet demand for purple shell to make jewelry and novelties.

Figure 6.7. Cubes of mussel shell (upper left) used in production of rounded nuclei, often referred to as "seed pearls," ready for insertion into a pearl oyster to produce a cultured pearl. © Richard T. Bryant.

Chapter 7
Mussel Conservation in Alabama

As currently understood, the continental U.S. harbors the world's most diverse freshwater mussel fauna. During the past century this fauna, about 300 taxa, has experienced a greater decline than any other wide-ranging group of organisms (Strayer et al., 2004). Approximately 25 percent of mussels in the U.S. are currently listed as federally endangered or threatened, and approximately 50 percent of remaining taxa are species of special conservation concern (Williams et al., 1993). The U.S. Fish and Wildlife Service (USFWS) currently considers 6 percent of mussels to be extinct. The southeastern region of the U.S. contains about 94 percent of the nation's mussel taxa, including more than 95 percent of the 70 federally protected mussels. Forty-eight federally protected species are known from Alabama, more than any other state (Table 7.1). The Nature Conservancy (TNC) recognizes approximately 52 percent of U.S. mussels as extinct or imperiled, compared to only 8 percent of birds and mammals (TNC, 1997). Lydeard and Mayden (1995) called attention to the extraordinary diversity and imperilment of the southeastern U.S. aquatic fauna and lamented its lack of consideration as a worldwide diversity hot spot. The loss of mussel diversity in the southeast is part of a global loss of nonmarine mollusks (Neves et al., 1997; Lydeard et al., 2004).

Aquatic and terrestrial habitat modifications during the past century have left no watershed untouched. Anthropogenic activities, including impoundment, channelization, construction of impervious surfaces in watersheds (altering surface water runoff), channel destabilization, contamination, eutrophication and sedimentation, continue to destroy mussel habitat, resulting in diminishing populations for most species. Some species have been reduced to small isolated populations in short stream reaches and may consist entirely of old animals with limited reproduction and recruitment. The same anthropogenic alterations have also reduced native fish populations, sometimes making glochidial hosts unavailable and further hampering successful mussel recruitment. Without a coordinated effort among federal, state and local governments and the conservation community, it is likely that a significant loss of the remaining fauna will occur in the near future.

Decline of the mussel fauna began more than a century ago with pollution and the impoundment of many rivers. Excessive collection of mussels for pearl harvest contributed to the reduction of mussel populations in localized reaches of small rivers and creeks. Commercial mussel harvest for the pearl button industry also depleted local stocks in some rivers of the Mississippi Basin (Smith, 1899; Coker, 1919; Anthony and Downing, 2001).

Dwindling mussel populations during the late 1800s did not go unnoticed. Kunz (1898a, 1898b) conducted a nationwide survey via correspondence on the status of freshwater pearl fisheries for the U.S. Fish Commission. He described the slaughter of mussels by pearl hunters during "periods of excitement." These periods usually followed news or rumors of valuable pearls being discovered in a particular region or stream. This activity began in the 1850s and continued through the early 1900s. With the survey, Kunz found that approximately three-fourths of the respondents described partial or complete exhaustion of mussel beds in widely scattered localities throughout the eastern U.S. None from Alabama were mentioned, but there was a report of pearl hunters depleting mussel beds in the Oostanaula and Etowah rivers in the Mobile Basin near Rome, Georgia.

Simpson (1899) called attention to the need for protection of mussels. He identified the problems caused by pearl hunters and the developing pearl button industry as well as municipal and industrial wastes. Simpson also pointed to unusual problems, such as hogs that forage in streams and feed on mussels. Smith (1899) provided a detailed account of the pearl button industry in the U.S. and also recommended protection of mussel resources. He proposed that harvest of small mussels be prohibited and a minimum legal size for each commercially important species be prescribed by law. Smith recommended a closed season immediately prior to and during the spawning period of harvested species. He also identified damage to mussel beds by sewage and factory refuse and called for regulations to prohibit the dumping of waste into streams.

During the 1890s the developing pearl button industry quickly surpassed pearl hunters as the primary exploitation of mussel resources. Coker (1914a) reviewed the condition of the industry and provided recommendations for the protection of mussels. He considered the depletion so severe that national legislation was recommended to remedy the problem. Coker reported that 20 states were participating in shell harvest, but only 3 had taken actions to protect the resource. He recommended restrictions on the method of harvest and size of shells harvested and suggested rotating closure of certain regions to allow for recovery. This legislation was never implemented at the national level. Coker (1914b) pointed out that Mississippi River dams, in addition to over-harvest, negatively impacted the mussel resource. Coker (1916) again reminded fishery managers of the rapid decline of mussel populations and called for immediate action to reverse the trend. He suggested that in addition to the protection of existing mussel populations, propagation might be needed to restore populations in some streams. Artificial propagation of mussels began soon after the completion of the U.S. Bureau of Fisheries Biological Station at Fairport, Iowa, in 1910 (Coker, 1921). Mussel research continued at the station into the 1930s.

Ellis (1931a) recognized the continued reduction of mussel populations and identified environmental perturbations, such as siltation and municipal and industrial wastes, that were exacerbating the problem. He also pointed out that these disturbances impacted mussel host fish populations. Diminishing mussel populations resulted in increased value of the resource. Ellis (1931b) reported on the effect of river impoundment on mussel populations, specifically their suffocation caused by settlement of silt from the water column. Ellis (1941) expounded on the problems created by impoundments on riverine habitats and their associated fauna.

Ortmann (1924b) provided one of the first mollusk conservation publications during the early 1900s with a discussion of the mussel fauna decline in the Tennessee River at Muscle Shoals, Alabama. This was caused by the destruction of shoal habitat following construction of Wilson Dam. Ortmann (1924b, 1925) explained factors resulting in this unusually diverse fauna, which is a product of two different faunas, known as "Cumberlandian" and "Ohioan," that overlap in this extensive reach of large river shoal habitat. Ortmann (1924b) lamented the destruction of this natural resource and suggested that the unique features of the "mussel shoals" deserved to be kept intact and preserved as a "natural monument," which would be "second only to very few other monuments of the United States." Ortmann (1925) provided a comprehensive list of mussels in the Alabama reach of the Tennessee River prior to its impoundment. This allowed documentation of the decimation of what was arguably the most locally diverse freshwater mussel fauna in the world by the subsequent surveys of Stansbery (1964) and Garner and McGregor (2001). Garner and McGregor reported a cumulative total of 79 species, including 10 adventive species, from the Muscle Shoals area, with 39 species extant in the late 1990s. Two additional species, *Cyprogenia stegaria* and *Pleurobema sintoxia*, have been encountered in that river reach since that assessment was published, bringing the current total to 41 extant species (J.T. Garner, personal observation).

Jones (1938) provided an enlightened view of the problem of aquatic and terrestrial habitat destruction in the state of Alabama. In addition to stream pollution

and siltation, he singled out deforestation and the resulting diminished recharge of the water table as a major problem. Jones identified lower groundwater levels as a serious threat to streams, pointing out that many formerly perennial springs and small streams had become intermittent. Factors contributing to the depletion of mussels in the eastern U.S. were summarized by van der Schalie (1938b), who cited dams, sedimentation, industrial development, intensive farming and clearing of ground cover as problems.

A study of the dwindling Tennessee River commercial mussel stocks by George D. Scruggs, USFWS, was carried out in 1956 and 1957 (Scruggs, 1960). This was perhaps the first use of scuba to assess mussel populations. Scruggs determined that the reduction was brought about not by excessive harvest but by lack of recruitment resulting from unsuitable habitat conditions caused by impoundment of the river.

In the late 1960s national concern over the decline of aquatic and terrestrial fauna resulted in the passage of legislation to protect threatened and endangered species. In 1968 the American Malacological Union (now the American Malacological Society) convened a symposium on rare and endangered mollusks of North America at their annual meeting, in Corpus Christi, Texas. Results of the symposium were published in *Malacologia* and served as a modern foundation for the evaluation of mussel conservation status (Athearn, 1970; Heard, 1970a; Stansbery, 1970a). This compilation was the first attempt to develop a list of endangered and extinct mussels. The USFWS held another symposium on rare and endangered mollusks in 1971. A paper by Stansbery (1971) identified additional taxa from the southeastern U.S. as candidates for a national list of endangered species.

The federal Endangered Species Act (ESA) of 1973 was the most proactive legislation ever enacted to conserve endangered animals and plants (Williams, 1976). Regulations implementing this legislation gave the responsibility of listing and recovery of freshwater and terrestrial species to the USFWS. The first 23 mussel taxa were added to the federal endangered species list in 1976 in response to a request by the Fund for Animals that resulted in the addition of 159 animal species (USFWS, 1976). Following this initial listing, species were added to the list at a much slower pace (Figure 7.1).

In response to the ESA, the southeastern USFWS field stations initiated programs for listing and recovery of protected species. Species that occur in Alabama have been assigned to various USFWS field offices. The offices in Daphne, Alabama, Jackson, Mississippi, Jacksonville, Florida, and Panama City, Florida, have been responsible for mussel species in Gulf Coast drainages, including the Mobile Basin. The Asheville, North Carolina, and Cookeville, Tennessee, field offices have dealt with most Tennessee River drainage species that occur in Alabama. Responsibilities of these offices include listing activities, drafting recovery plans, designating critical habitat and issuing biological opinions in support of species protection. During the 1980s and 1990s the field offices in Asheville, Jackson and Jacksonville listed more than 50 mussel species, most of which occur in Alabama.

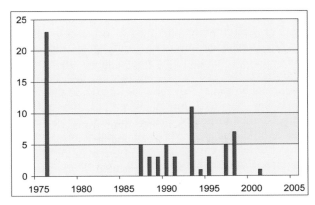

Figure 7.1. Number of mussels listed per year under the federal Endangered Species Act.

A number of recovery plans were drafted beginning in the 1970s and are available on the USFWS website. One recovery plan developed by the Jackson Field Office took an ecosystem approach for the Mobile Basin and included 17 mussel species (USFWS, 2000). Two additional recovery plans included multiple mussel species, the first for seven Apalachicola Basin species (USFWS, 2003) and the second for five Cumberlandian species (USFWS, 2004).

In addition to directly applying conservation measures, the USFWS has been a major funding source for studies by other agencies and institutions. These have been instrumental in developing biological, ecological and taxonomic information necessary for mussel conservation and the technology and knowledge for mussel propagation.

The Convention on International Trade in Endangered Species of Wild Flora and Fauna (CITES) was enacted in 1975. It is an agreement among governments that provides a means of controlling export and import of rare species. There are two appendices to the convention. Species on Appendix I require permits from both the exporting and importing countries, and those on Appendix II require a permit from only the exporting country. Seventeen species known from Alabama are currently listed on Appendix I and three species are on Appendix II. The species are listed on the CITES website.

During the 1970s state natural resource agencies also developed lists of endangered and threatened species. In Alabama the first evaluation of rare and endangered animals occurred at a symposium convened at Birmingham-Southern College in 1972, and the results were published by the Alabama Department of Conservation and Natural Resources (ADCNR) (Keeler, 1972). However, no invertebrates were included in the 1972 symposium. A second symposium, sponsored by ADCNR and the Alabama Museum of Natural History, convened in Tuscaloosa at the University of Alabama in 1975 and included an evaluation of the conservation status of Alabama mussels (Stansbery, 1976). A subsequent conference at Auburn University in 1983 included only vertebrates.

Beginning in the 1970s nongovernmental organizations and individuals also began to independently publish lists of rare and/or extinct species. Opler (1977) and Palmer (1985) provided lists of extinct species, which included 8 and 15 mussel taxa known from Alabama, respectively. Harris (1990) published a review of rare invertebrates in Alabama, which included a list of endangered and threatened mussels based primarily on the federal list and information provided by Stansbery (1976). Turgeon et al. (1988) provided a summary of endangered, threatened and extinct mussels. This review provided an up-to-date summary but was based primarily on the federal endangered species list. The summary was updated in Turgeon et al. (1998).

During the 1990s several publications evaluated the conservation status of mussels. Williams et al. (1993) published a comprehensive review that included all mussel species in the U.S. and Canada. A broader perspective of mussel extinction was authored by Bogan (1993, 1998, 2006). Lydeard et al. (1999) provided a review of the conservation status of Alabama mussels. A national strategy for the conservation of mussels was developed by the National Native Mussel Conservation Committee (National Native Mussel Conservation Committee, 1998). This strategy outlined specific actions to prevent further extinctions and recover declining populations.

The Tennessee Valley Authority (TVA) has been involved in Tennessee River mussel conservation for several decades. Early activities revolved around commercial mussel resources. Shelling activity was monitored as early as 1936 in Wheeler Reservoir (Cahn, 1936). Monitoring of harvest continued with the "rebirth" of the vocation in 1945 (Bryan and Miller, 1953). Isom (1969) published a summary of Tennessee River mussel resources. Recommendations by TVA in the 1960s resulted in the establishment of mussel sanctuaries in tailwaters of Tennessee River dams by the Alabama Game and Fish Division (now the Alabama Division of Wildlife and Freshwater Fisheries, ADWFF) in 1967 (unpublished TVA memo dated 9 October 1964; F. Harders, personal communication). A study of the reproductive biology of *Pleurobema cordatum*, the most important commercial species at the time, was supported by TVA in the late 1960s (Yokley, 1968, 1972).

In the 1970s TVA redirected its mussel concerns to endangered species. A natural heritage program was initiated in 1976, with a database of federal and state protected species, to support the management of biodiversity on TVA lands (P.W. Shute, personal communication). In response to a USFWS jeopardy biological opinion on construction of Columbia Dam (Duck River, Tennessee), TVA initiated the Cumberlandian Mollusk Conservation Program (CMCP) (Jenkinson, 1980, 1982). Aspects of the CMCP included surveys of Tennessee River drainage streams, including Elk and Paint Rock rivers, which lie partially in Alabama, and mussel life history work. Another TVA research program was aimed at producing juvenile mussels using in vitro culture (Isom and Hudson, 1982; Hudson and Isom, 1984). These efforts took place at a laboratory in Muscle Shoals. Culture of mussels without utilizing a host fish was accomplished, but rearing of juveniles in laboratory settings proved elusive. Later, propagated juveniles of common species were used in toxicity tests at a laboratory at Browns Ferry Nuclear Plant in Limestone County, Alabama (Wade et al., 1993).

In 1989 TNC established the Alabama Natural Heritage Program (ALNHP) to identify significant communities of rare and endangered species. The ALNHP database includes mussel records from surveys performed by program personnel as well as records compiled from other workers. The ALNHP database is also used by other conservation organizations and state and federal agencies for establishing conservation priorities.

The Alabama chapter of TNC has been protecting parcels of land important to imperiled species since 1999, but the national organization bought its first property in the state in 1967 (J. Danter, personal communication). Many TNC projects directly benefit the mussel fauna. A TNC Freshwater Initiative, chartered in 1999, resulted in emphasis being placed on the Cahaba and Paint Rock rivers in particular. In a national review of biologically significant rivers, TNC recognized 26 in Alabama as "critical watersheds" (TNC, 1997). In November 2004 the "Marvel Slab" on the Cahaba River, Bibb County, Alabama, was removed in a multi-agency effort led by the Alabama chapter of TNC. This structure was a low-water concrete bridge that resulted in a barrier to fish passage for much of the year.

A second state heritage program was begun by the State Lands Division of ADCNR in late 1996. They also maintain a species database, which provides support for the state's Forever Wild land acquisition program.

Escalation of demand for mussels during the early 1990s prompted ADCNR to respond with conservation and management actions. Regulations were adopted to establish a list of commercially harvestable species and size limits, as well as to limit days of harvest to weekdays. A detailed, standardized receipt system was put in place to better monitor the harvest. The receipts are completed by the buyer at the time of the transaction with the harvester. Information required on the receipts includes names of the harvester and buyer, date of the transaction, pounds of each species harvested, river or reservoir from which they were taken and total amount paid for the shell. License fees were increased and a tax on the harvest instituted, producing revenues to fund a mussel management supervisor position. Initial efforts of the program included monitoring of mussel harvest using receipt data and routine assessment of selected mussel beds subject to heavy harvest pressures. With a reduction in commercial mussel harvest in late 1996, more emphasis was placed on surveys and assessments of imperiled species.

In 1996 the Southeastern Aquatic Research Institute was founded by the Tennessee Aquarium in Chattanooga, Tennessee. One of the first projects of the institute was to convene a conference of specialists to review the conservation status of the southeastern aquatic fauna (Benz and Collins, 1997). Proceedings of the conference included a chapter on the conservation status of aquatic mollusks (Neves et al., 1997), which called attention to causal factors of aquatic mollusk losses in Alabama and other southeastern states.

In 2000 the Southeastern Aquatic Research Institute was moved to a former USFWS fish hatchery in Cohutta, Georgia. The name was later changed to the Tennessee Aquarium Research Institute (TNARI). An emphasis of the institute was life history investigations and propagation of rare mussels and snails, along with surveys to document their distributions and locate populations for brood stock. The effort focused on the Mobile Basin and Tennessee River drainage faunas, and progeny resulting from the propagation program have been used to augment populations in various Alabama rivers. Juvenile mussels have also been produced at Tennessee Technological University, Cookeville, and released into the Alabama reach of the Tennessee River.

In 2001, at the request of ADCNR, USFWS designated a Nonessential Experimental Population for 16 mussel species in the Tennessee River downstream of Wilson Dam. This effort was led by the USFWS Asheville, North Carolina, field office. In 2003 pilot populations of *Dromus dromas* from Clinch River, Tennessee, and *Epioblasma* sp. cf. *capsaeformis* and *Lemiox rimosus* from Duck River, Tennessee, were reintroduced to the area by ADCNR, with the assistance of the Tennessee Wildlife Resources Agency. Observation of substantial survival of these reintroduced mussels in 2005, 2006 and 2007 suggested that habitat is suitable for reintroductions of those, and potentially other, species.

In 2002 an imperiled species conference convened at Auburn University to update results of the 1983 conference, and freshwater mollusks were added. Results of this meeting were published in a four-volume set entitled *Alabama Wildlife*. The set includes a checklist of species (Mirarchi, 2004), species accounts for aquatic species of conservation concern (Mirarchi et al., 2004), species accounts for terrestrial species of conservation concern (Mirarchi et al., 2004) and conservation recommendations (Mirarchi et al., 2004). These volumes represent the first step in the development of a Comprehensive Wildlife Conservation Strategy by the ADCNR, which will be used to guide future nongame conservation programs in the state.

The future of mussel conservation in Alabama is promising. Growing interest in nongame animals by the public, state and federal governments and nongovernmental organizations (e.g., Cahaba River Society, TNC and World Wildlife Fund) is reflected in increased funding for conservation and habitat protection. In 2005 ADCNR opened the Alabama Aquatic Biodiversity Center in Marion, Alabama. The purpose of this facility is to propagate mussels and other nongame aquatic species for release into the wild in order to augment existing populations and reintroduce species into stream segments where conservation and restoration efforts have improved habitat. The most favorable site for reintroductions of mussels, as well as snails, in the Mobile Basin is in tailwaters of Jordan Dam on the Coosa River. Minimum flows from the dam by the Alabama Power Company have returned the habitat in this 11-km free-flowing reach to a semblance of historical conditions. The first reintroductions in this reach took place in 2005, involving *Epioblasma penita* and *Medionidus acutissimus* juveniles cultured at TNARI. Parent stock for the cultured mussels came from the Buttahatchee River and Holly Creek of the upper Coosa River drainage, respectively (P.D. Johnson, personal communication). The status of Alabama's mussel fauna will remain critical for the foreseeable future, but these and similar conservation measures may prevent extinction of some species until large-scale habitat restoration provides additional secure habitats for the recovery of aquatic fauna and flora.

Table 7.1. Alabama mussel taxa (48) listed as endangered (42) or threatened (6) under provisions of the federal Endangered Species Act as of January 2006.

	USFWS Status	USFWS Year Listed
Cyprogenia stegaria	E	1990
Dromus dromas	E	1976
Elliptio chipolaensis	T	1998
Elliptoideus sloatianus	T	1998
Epioblasma brevidens	E	1997
Epioblasma capsaeformis	E	1997
Epioblasma florentina florentina	E	1976
Epioblasma florentina walkeri	E	1977
Epioblasma metastriata	E	1993
Epioblasma obliquata obliquata	E	1990
Epioblasma othcaloogensis	E	1993
Epioblasma penita	E	1987
Epioblasma torulosa torulosa	E	1976
Epioblasma turgidula	E	1976
Fusconaia cor	E	1976
Fusconaia cuneolus	E	1976
Hamiota altilis	T	1993
Hamiota perovalis	T	1993
Hamiota subangulata	E	1998
Hemistena lata	E	1989
Lampsilis abrupta	E	1976
Lampsilis virescens	E	1976
Lemiox rimosus	E	1976
Leptodea leptodon	E	2001
Medionidus acutissimus	T	1993
Medionidus parvulus	E	1993
Medionidus penicillatus	E	1998
Obovaria retusa	E	1989
Pegias fabula	E	1988
Plethobasus cicatricosus	E	1976
Plethobasus cooperianus	E	1976
Pleurobema clava	E	1993
Pleurobema curtum	E	1987
Pleurobema decisum	E	1993
Pleurobema furvum (= rubellum)	E	1993
Pleurobema georgianum	E	1993
Pleurobema marshalli	E	1987
Pleurobema perovatum	E	1993
Pleurobema plenum	E	1976
Pleurobema pyriforme	E	1998
Pleurobema taitianum	E	1987
Potamilus inflatus	T	1990
Ptychobranchus greenii	E	1993
Quadrula intermedia	E	1976
Quadrula sparsa	E	1976
Quadrula stapes	E	1987
Toxolasma cylindrellus	E	1976
Villosa trabalis	E	2001

Chapter 8
Ecology and Biology of Mussels

Worldwide there are about 19 bivalve families that occur partially or entirely in fresh water (Bogan, 1993, in press). Unioniform bivalves are an entirely freshwater group and are unique among the Bivalvia in possessing an obligate parasitic larval stage. In the Unionoidea the larval stage is the glochidium, most of which use a fish for a host (Wächtler et al., 2001). Unionoid parasitism has evolved in close association with the fish fauna (Watters, 1992). Utilization of a host provides a secure environment for early development and serves as an active means of dispersal.

In North America, unionoid bivalves, commonly called mussels, occur in a variety of aquatic habitats from near the Arctic Circle south to the tropics. Under such diverse conditions they have evolved varied morphology, reproductive traits and life history strategies.

Ecology and Habitat

Few studies have been concerned with the functional role of mussels in aquatic ecosystems. Such a study was carried out on *Actinonaias ligamentina* and *Amblema plicata* in controlled mesocosms (Vaughn et al., 2004). The study found that mussels can turn over a substantial proportion of the water column even at low densities and that populations have the potential to influence ecosystem processes, including community respiration, water column ammonia, nitrate and phosphorus concentrations and algal clearance rates.

Mussel ecology and factors limiting their distribution have been fertile ground for observation and speculation since some of the first works by early naturalists. More recently such commentary has been supported with detailed analyses. Studies of mussel microhabitats have included such factors as substrate composition, water depth, water temperature, bottom roughness, current velocity, variation in current, distance from shore and turbidity (Sickle, 1980; Strayer, 1981; Salmon and Green, 1983; Stern, 1983; Holland-Bartels, 1990; Strayer, 1993; Layzer and Madison, 1995; Hornbach et al., 1996; McRae et al., 2004). The primary conclusions of these studies were that there are no strong correlations between mussel distribution and various microhabitat variables. Strayer and Ralley (1993) stated that "physical aspects of microhabitats are of limited use in predicting the occurrence and species composition of unionaceans in running waters and raise serious doubts about the utility of using such an approach to predict the distribution and abundance of unionacean species." Vannote and Minshall (1982) studied the ecology of *Margaritifera*

falcata (Gould, 1850) in the Salmon River Canyon, Idaho, and hypothesized that this species achieves maximum density and age in areas where large boulders stabilized cobble and gravel substrates. Shear stress has been found to be a factor in location of mussels within a stream. Areas with low shear stress were found to be best for mussels (Layzer and Madison, 1995), and such stresses in modified flow downstream of some flood-control dams are believed to sweep newly metamorphosed juveniles downstream before settling (Hardison and Layzer, 2001).

Juvenile mussels have been reported to have clumped distributions, with greater densities downstream of rocks in riffles and runs, and to be positively correlated with the occurrence of sphaeriid clams (Neves and Widlak, 1987). However, as juveniles increase in age they are more likely to be encountered with adults of the same species. Juvenile mussels of some species attach to pebbles and coarse gravel with byssal threads (Isely, 1911; Howard and Anson, 1922). This has been reported in swift current and is presumably to aid in preventing juveniles from being swept downstream. Neves and Widlak (1987) did not report any juveniles with byssal threads, but this may have been due to the collecting methods used.

Differences in large-scale distribution patterns of North American mussel species were commented upon early by Rafinesque (1820), Conrad (1834b), Simpson (1893) and Ortmann (1913a). They noted how depauperate the European fauna was compared with that of the North American Interior Basin. Conrad (1834b) observed general differences in shell morphology, such as the absence of heavily plicate and tuberculate species from Atlantic Coast drainages.

On a regional scale, van der Schalie (1938c) documented stream size as the primary factor determining mussel faunas. Surface geology and stream size were identified as major environmental features controlling mussel distributions in Michigan (Strayer, 1983). Surface geology also regulates hydrology, slope and turbidity of streams, which impact mussel ecology. Strayer (1983) found diversity to be additive from headwaters to big river habitat, but stream size alone was not sufficient to predict mussel diversity. However, the combination of drainage area with surface geology was found to be a good predictor of diversity in the River Raisin, Michigan (McRae et al., 2004). The relationships of fish and mussel diversity and drainage area were analyzed by Watters (1992, 1993b). This study reported that in large rivers mussel diversity appears directly related to fish diversity, but in small streams mussel diversity is related to drainage area. In Bavarian streams mussel metabolic rate was found to be an important factor in mussel distribution, but potential glochidial host distribution did not explain mussel distribution (Bauer et al., 1991).

Strayer (1993) examined environmental factors on a scale of 1 km to 10 km and found that macrohabitat variables were less effective in predicting mussel community structure than surface geology and stream size, but stream size and presence or absence of tidal influence were the most effective predictors of mussel distribution in lower reaches of the Hudson River, New York. Strayer (1993) suggested that unmeasured or smaller-scale processes may determine mussel distribution in Atlantic Coast drainages. In additional work on the Hudson River, Strayer et al. (1994) concluded that patchy mussel distribution was not explained by the environmental data collected, expressing doubt as to "whether it is worthwhile to focus on such variables in future studies of unionid ecology."

Water velocity was determined to be an important factor in mussel community composition, making mussel distributions irregular (Vaughn, 1997). Dispersal of glochidia on fishes provides the dispersal mechanism for distributing mussels within and between the irregular habitats. Strayer (1999b) concluded that the patchiness of mussel distribution is best explained by flow refuges found under flood and high flow conditions.

Haag and Warren (1998) included mussel reproductive strategy, along with fish fauna and ecological factors, in an attempt to predict mussel community structure. Fish community variability and mussel reproductive strategy were found to better explain mussel diversity than ecological parameters. A later study confirmed that local mussel species richness is constrained by the number of available glochidial hosts, but undetermined regional processes may also be involved (Vaughn and Taylor, 2000).

Disturbance and Effects on Distribution

Detrimental effects of anthropogenic activities—including timber harvest, agriculture, channelization, dredging, impoundment, mining, and industrial and domestic pollution—on aquatic habitats and mussels have been noted since the 1800s (Lyell, 1849; Higgins, 1858; Rhoads, 1899; Ortmann, 1909c, 1918) (Figures 8.1, 8.2, 8.3). Silt has been identified as a major culprit in mussel declines (e.g., Kunz, 1898a; Ellis, 1936, 1937; Clench, 1955; Matteson, 1955; Imlay, 1972; Way et al., 1989). The U.S. Environmental Protection Agency (1990) reported sedimentation to be the number one pollutant in U.S. rivers.

Figure 8.1. USACE channelization of the lower reach of Luxapalila Creek, near its confluence with the Tombigbee River, Lowndes County, Mississippi, September 1971. Photograph by J.D. Williams.

Figure 8.2. USACE dredge pipe containing mussels, following channel maintenance dredging of the Alabama River, Claiborne Landing, Monroe County, Alabama, October 1973. Photograph by J.D. Williams.

Figure 8.3. USACE Jim Woodruff Dam, Apalachicola River, Jackson County, Florida, impounds the Chattahoochee River in southeastern Alabama, July 1993. Photograph by J.D. Williams.

Reservoirs are inhospitable habitats for most mussel species, and faunas of formerly free-flowing rivers have not remained intact (van der Schalie, 1938b, 1938d; Bates, 1962; Williams et al., 1992; Hughes and Parmalee, 1999). Problems facing the unionid fauna in a reservoir are increased sedimentation, low dissolved oxygen levels and loss of fish hosts, which inhibits recruitment. Reservoirs also are subject to invasion by species tolerant of disturbed habitats (Bates, 1962). Dams have been recognized as a barrier to fish migration. Even small dams will impede fish movements and thus directly affect upstream dispersal of unionids (Watters, 1996b).

Areas downstream of dams are also negatively affected. Cahn (1936) surveyed the mussel fauna downstream of Norris Dam on the Clinch River, Tennessee, and reported total mussel extirpation from the area due to decreased water temperature and dissolved oxygen, as well as scouring effects of discharge from the dam. Long-term effects of reservoirs on the distribution of mussels downstream of dams were carefully documented in Oklahoma by Vaughn and Taylor (1999). The greatest impact was just below the dam, with the fauna beginning to recover with greater distance from the dam and the rarest species occurring farthest downstream of the dam. A similar pattern of recovery downstream of impoundments was reported for Bear Creek in northwestern Alabama and northeastern Mississippi (McGregor and Garner, 2004). Factors attributed to such a recovery gradient include physical stresses, habitat change, availability of food and host fishes, stability of flow and water temperature (Vaughn and Taylor, 1999).

Sedimentation has been determined to be a major factor in habitat destruction, resulting in corresponding shifts in mussel fauna (Brim Box and Mossa, 1999) (Figure 8.4). However, it may take years for the effects to become obvious. Sedimentation has wide-ranging effects on mussels, including inhibition of feeding and reproduction as well as suffocation when they are covered. Another substantial factor in disruption of mussel habitat is in-stream and near-stream sand and gravel mining in the southeastern U.S. (Hartfield, 1993). Such activities result in headcutting, which affects stream slope and erosional regime far upstream of the mining. River channel morphology is related to sediment load and directly affects fish abundance, diversity and reproduction, as well as ecology and dispersal. However, there is much uncertainty regarding the association of mussels and sediment. Poole and Downing (2004) linked mussel declines with watershed characteristics, lack of riparian forest, excessive siltation and the most intensive agricultural practices. They found the mussel fauna to do well in river reaches with greater than 50 percent riparian woodland buffer.

They also observed that substrate heterogeneity was important to mussel fauna health.

Figure 8.4. Erosion in Providence Canyon State Park, Stewart County, Georgia, which began in the 1840s and continues to deposit sediment in the Chattahoochee River via Turner Creek, November 1996. Photograph by J.D. Williams.

The effects of acid mine drainage on freshwater mussels have been well documented (Ortmann, 1909c, 1918; Soucek et al., 2003) (Figure 8.5). In the Powell River, Virginia, acute toxicity due to the combination of acid and dissolved metals combine with urban runoff and nutrients to negatively affect mussel populations (Soucek et al., 2003). Mussel fauna declines in the Little South Fork Cumberland River, Tennessee and Kentucky, have been attributed to pollutants from oil extraction activities in upper reaches and surface coal mining in lower reaches (Warren and Haag, 2005). Effluents from sewage treatment plants negatively impact mussel populations for several kilometers downstream of discharges (Fuller, 1974; Stansbery and Stein, 1976; Horne and McIntosh, 1979; Goudreau et al., 1993). Specific factors determined to be problematic include low dissolved oxygen as well as increased levels of chlorine, ammonia and monochloramines.

Figure 8.5. Coal strip mine in Fayette County, Alabama, located in the Black Warrior River drainage, October 1973. Photograph by J.D. Williams.

Severe droughts, often overlooked as having detrimental effects on mussel populations, may have significant impacts on survival of populations. Bogan (1993) reported that three species of mussels became extinct when a river dried in Israel. Severe droughts limit discharge, affecting dissolved oxygen levels and water temperatures, which leads to impaired metabolism and reproduction and results in stranding, subjecting mussels to predation and desiccation. Levels of mortality associated with stranding vary among species, and appears to be related to stream size. Depressions under woody debris in a stream channel may provide refuge for mussels when a stream dries (Gagnon et al., 2004; Golladay et al., 2004).

Mussel Behavior

Some mussel species have been observed to be more motile than others. Some appear to remain sedentary for their whole lives. Patterns of horizontal migration have been observed among various species and the causes have been a source of speculation. Mussel movement (Figure 8.6) has been observed to be stimulated by unfavorable conditions, with water depth being an important variable (Isely, 1914; Allen, 1922; van der Schalie, 1938b; Gagnon et al., 2004; Golladay et al., 2004). *Amblema plicata* have been reported to move little in water approximately 1 m deep but to migrate to deeper water when placed in shallow areas (Isely, 1914). Onset of winter has also been reported to stimulate mussels to migrate to deeper water (Evermann and Clark, 1918). However, such seasonal migration was not reported by van der Schalie (1938c) or Matteson (1955). Other factors reported to stimulate horizontal movement of mussels include the search for food (Bovbjerg, 1957), dispersal from crowded conditions (Kat, 1982) and congregation for spawning (Downing et al., 1993; Amyot and Downing, 1998). Amyot and Downing (1997) reported no difference in

movements of *Elliptio complanata* between night and day, which contradicts the finding of Imlay (1968).

Deliberate orientation of mussels in the substrate, relative to current direction, has long been observed. Numerous authors have noted that mussels orient their posterior ends into the current (e.g., Allen, 1922; Baker, 1928a; Matteson, 1948; Clarke, 1973; Maio and Corkum, 1997). Sex and shell inflation were found to be significantly correlated to orientation of *Lampsilis siliquoidea* (Barnes, 1823) in Ohio streams, with females oriented downstream and less inflated males upstream (Horn, 1983). Mussels living in areas prone to flood events have been reported to have tendencies to orient parallel to the current, whereas orientation of those in more stable rivers is less predictable (Maio and Corkum, 1997). Mussels in more stable habitats have been reported to grow larger than those in unstable habitats, but shell size is not related to orientation and burrowing depth is not related to hydrological events or stability of flow (Maio and Corkum, 1997).

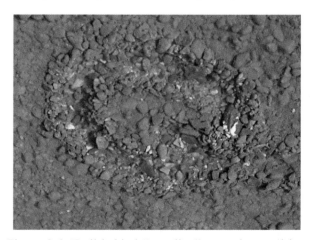

Figure 8.6. Trail behind "crawling" mussel, gravel bar on the Tombigbee River, Lowndes County, Mississippi, November 1971. Photograph by J.D. Williams.

Results of studies on vertical movement of mussels in the substrate are variable. Several studies have reported burrowing activity to be correlated with water temperature and season (van der Schalie, 1938c; Amyot and Downing, 1997). However, Matteson (1955) found mussels not to "embed significantly" during winter, at least in lentic habitats. Watters et al. (2001) experimented with eight mussel species to determine vertical migration patterns and found that they surfaced in the spring and burrowed in the fall in relation to water temperature and day length. There were two modifications of this pattern: some came up and stayed above the surface for the summer, while others came up to spawn, reburied and did not resurface until the fall. These behavior patterns were not correlated with taxonomy at the family or subfamily level. Perles et al. (2003) reported vertical movement of *Lampsilis siliquoidea* in Arkansas to be strongly correlated with day length and less strongly with water temperature. In Finland *Anodonta piscinalis* Nilsson, 1822, was noted to be more visible during its June spawning period than later in the summer, but was more active in August (Saarinen and Taskinen, 2003). Some species were found to burrow in response to emersion during droughts, but others migrated horizontally with the receding waterline to remain submerged (Gagnon et al., 2004).

Burrowing mussels affect local stream nutrient dynamics by egestion of feces and pseudofeces (Vaughn et al., 2004). Bioturbation of the substrate by vertical and horizontal movement of mussels increases substrate water and dissolved oxygen and releases nutrients into the water column from the substrate. The presence of mussels also serves to stabilize the substrate, increasing habitat stability (McCall et al., 1979; Vaughn and Hakenkamp, 2001). Trueman (1968) detailed the biomechanics of burrowing behavior of mussels.

Mussel Interactions with Other Species

Mussels serve as food for a variety of animals, including crayfish, some fishes (e.g., *Aplodinotus grunniens*, *Amieurus nebulosus*, *Ictalurus punctatus*, and *Pylodictis olivaris*), amphibians (salamanders) and reptiles (map turtles) (Fuller, 1974; Vogt, 1981; Shively and Vidrine, 1984). Other animals that include mussels among their prey are some birds (e.g., waterfowl, crows, Limpkin, Snail Kite and Boat-tailed Grackle) and mammals (e.g., River Otter and Mink) (Fuller, 1974; Apgar, 1887; Convey et al., 1989; Watters, 1995). Muskrat predation may have an effect on species and size composition of a mussel community (Neves and Odum, 1989; Jokela and Mutilainen, 1995; Tyrell and Hornbach, 1998; Diggins and Stewart, 2000). Zimmerman et al. (2003) reported flatworms, *Macrostomum* sp. (Platyhelminthes), to be voracious predators on newly metamorphosed juvenile mussels held in captivity.

Effects of nonindigenous species on some of the vertebrate groups are fairly well understood. Clavero and Garcia-Berthou (2005) analyzed the IUCN Red List of Threatened Species with regard to the role of nonindigenous species in animal extinctions. They concluded that nonindigenous species are the leading cause of worldwide bird extinctions and the second major cause of worldwide fish and mammal extinctions. However, Gurevitch and Padilla (2004) suggested that additional data are required for critical evaluation of the effect of nonindigenous species on native species extinctions. Different nonindigenous species appear to impact native mussels to varying degrees.

Corbicula fluminea was introduced to the U.S. from Asia and has spread through many of the river systems in North America (McMahon, 1982; Counts, 1986). Some literature has suggested competition between *C. fluminea* and declining native mussels (e.g., Fuller and Imlay, 1976; Kraemer, 1979; Clarke 1986, 1988). These studies simply correlate the decline of native species with the explosive expansion of *C. fluminea*. However, direct cause-and-effect evidence of this relationship is weak (Strayer, 1999a). Others have compared the biology and physiology of *C. fluminea* to that of native mussels to determine if there is evidence of competition (Leff et al., 1990; McMahon, 1999). Since *C. fluminea* does not use a host to complete its life cycle there is no competition with mussels for hosts. *Corbicula fluminea* filters bacteria and algae from the water column at a higher rate than do mussels. However, native mussel densities often remain high in areas with *C. fluminea* (Leff et al., 1990; McMahon, 1999). *Corbicula fluminea* has been reported to have a significant impact on stream carbon cycles and to influence organic material concentrations in the substrate and may have an impact on the standing organic matter (Hakenkamp and Palmer, 1999). It is capable of pedal feeding when water column food levels are low (McMahon, 1999). Mussels have been reported to utilize pedal feeding as juveniles, but there is no indication that they do it as adults (Yeager et al., 1994; Gatenby et al., 1997).

Dreissena polymorpha and *Dreissena bugensis* Andrusov, 1897, have been introduced into North America from the Black and Caspian seas. *Dreissena polymorpha* has rapidly spread across eastern North America, but *D. bugensis* appears to be confined to the Great Lakes and St. Lawrence River. The spread of *D. polymorpha* has had dramatic effects on native mussels in some areas. Populations of this species can reach densities of up to 750,000 per m^2 and cause severe reductions in phytoplankton and zooplankton populations, as they have in the Great Lakes (Kovalak et al., 1993). These small, byssally attached bivalves have proven to have a major impact on the native mussel populations simply by competing for living space and attaching directly to shells of living mussels (Figures 8.7, 8.8). This has resulted in local extirpation of mussel populations or severe stress on surviving individuals (e.g., Hebert et al., 1991; Mackie, 1991; Hunter and Bailey, 1992; Haag et al., 1993; Gillis and Mackie, 1994; Nalepa, 1994; Lavrentyev et al., 1995; Ricciardi et al., 1996, 1998; Strayer and Smith, 1996; Baker and Hornbach, 1997; Martel et al., 2001). The native mussel shell seems to be a preferred settling site for *D. polymorpha* veligers (Ricciardi et al., 1996). Martel et al. (2001) found that decline in mussel populations began after attachment of *D. polymorpha*, and that once fouled, mussel species were extirpated in four to seven years. Strayer (1999a) observed that mussel extirpation by *D. polymorpha* is due to fouling of the shell and competition for food.

Figure 8.7. Freshwater mussels encrusted with *Dreissena polymorpha*, from Presque Isle, Lake Erie, Erie County, Pennsylvania, July 1991. Photograph by D.W. Schlosser.

Figure 8.8. *Amblema plicata* encrusted with *Dreissena polymorpha*, from the western basin of Lake Erie, near Monroe, Monroe County, Michigan, November 1989. Photograph by S.J. Nichols.

A nonindigenous gastropod, *Potamopyrgus antipodarum* (Gray, 1843) (Hydrobiidae), has been introduced into the Snake River Basin in the northwestern U.S. Localized populations of 20,000 to 500,000 per m^2 have been reported (Hall et al., 2003). High densities of this species are having a major impact on local ecology. Though *P. antipodarum* has not been reported from the southeast, conditions in some streams in the region appear suitable for the species. Another potential problem is the introduction of native mussel species outside of their historic ranges, which may occur via introductions of fishes infested with glochidia or physical transplantation of adults (Bogan et al., 2002).

Introduction of molluscivorous fishes represents a potential threat to native mollusk populations. Native fishes introduced and established in drainages outside of their native ranges in the U.S. and known to prey on mussels include *Ictalurus furcatus* (Lesueur, 1819) (Blue Catfish) and *Pylodictis olivaris* (Rafinesque, 1818) (Flathead Catfish) (Ictaluridae). An Asian molluscivore introduced to the U.S., *Mylopharyngodon piceus* (Richardson, 1846) (Black Carp) (Cyprinidae), is used by the aquaculture industry to control snails in commercial fish ponds. This species has escaped into the open waters of the Mississippi Basin and has the potential to do great harm to mussel and snail populations (Nico et al., 2005).

Parasites and Diseases

A variety of organisms are commensal or parasitic on mussels, but information on such relationships is limited (Fuller, 1974). Groups known to share a relationship with mussels include bacteria, Protozoa, Platyhelminthes, Nematoda, Annelida, Tardigrada and Arthropoda (e.g., unionicolid mites and insects such as chironomid larvae) (Roback et al., 1980; Benz and Curren, 1997; Chittick et al., 2001). Unioinicolid mites are common parasites of unionid gills and mantle but have not been reported in margaritiferids (Mitchell, 1955; Vidrine, 1996a–e; Gledhill and Vidrine, 2002). These mites have a parasitic stage that infests larvae of midges (Chironimidae) (Edwards and Dimock, 1996; Gledhill and Vidrine, 2002). Many mite species appear to be host-specific or restricted to a small suite of species, whereas others are host generalists (Vidrine, 1996a; Gledhill and Vidrine, 2002).

Aspidogastrid trematodes (Platyhelminthes) are common parasites of unionids but have not been reported in North American margaritiferids (Smith, 2001). Species identified from mussels include *Aspidogaster concicola* von Baer, 1826, *Cotylapsis insignis* Leidy, 1857, *Cotylapsis cokeri* Barker and Parsons, 1914, *Cotylapsis stunkarki* Rumbod, 1928, *Cotylogaster occidentalis* Nickerson, 1902, and *Lopotaspis interiora* Ward and Hopkins, 1931 (Hendrix et al., 1985). The site of infection within a mussel varies with the species infected (Benz and Curren, 1997). Heavy loads of trematodes have been reported to replace practically all gonadal tissue, effectively castrating affected individuals (Sterki, 1898a).

Little is known about natural bacteria associated with mussels. The fact that bacteria make up a significant portion of the mussel diet makes studies of bacterial flora difficult (Šyvokienė et al., 1987, 1988; Šyvokienė, 1988; Nichols and Garling, 2000; Christian et al., 2004). Chittick et al. (2001) reported a total of 18 species of aerobic bacteria to comprise the flora of *Elliptio complanata* digestive glands. However, bacteria have been reported to have detrimental effects on glochidia while still in the marsupium (Ellis, 1929) and to retard growth when present in large quantities (Imlay and Paige, 1972).

Clinal Variation in Shell Shape

Shell shape has long been recognized to vary among populations of the same mussel species in differing habitats (Figure 8.9), and these populations were once perceived as closely related species. Variation in shell morphology between rivers and lakes was studied in Europe by Hueber (1871). Studies on the relation of shell shape and stream influence in Europe were continued by Hazay (1881), Wallengren (1905), Sell (1908), Buchner (1910) and Israel (1910, 1911, 1913). These were followed by a law proposed by Haas and Schwarz (1913) stating that "the *same* types under the same biological (ecological) conditions produce the same variants; *different* types under like conditions produce convergent (parallel) local variants. In the case of a sufficiently lengthy isolation the local variants subject to biologically similar environments, may become constant or fixed local forms" (translation by Grier, 1920).

Figure 8.9. Variation in shell shape of *Fusconaia cerina* from large river (left) and creek (right) habitats. © Richard T. Bryant.

Similar observations were being made in the U.S. (e.g., Wilson and Clark, 1912, 1914; Coker, 1914a; Danglade, 1914; Utterback, 1916b, 1917). Ortmann (1909a) began making observations on shell inflation and stream position and followed with more detailed observations on the upper Tennessee River drainage fauna (Ortmann, 1918). Ortmann (1920) published his classic paper on the correlation of shape and station in freshwater mussels, which included the general law that "in the larger rivers, these shells are more convex and swollen; in the headwaters, they are flat and compressed; and in intermediate parts, the intergrades between the extremes are found." Ortmann (1920) added a corollary: "The obesity [inflation] of the shell changes a little with age, so that young shells, in the average, are more swollen than old ones." With the decrease in inflation there is often a correlated increase in shell length. Some of the tuberculate species found in large rivers often have reduced or absent tubercles in headwater streams. Ortmann noted that not all species

follow these clinal variations (e.g., it occurs in none of the unionids in Atlantic Coast drainages). Ortmann (1920) reported that most of the species exhibiting variation from compressed headwater form to inflated big-river form were "primitive" short-term brooders such as *Amblema*, *Fusconaia*, *Pleurobema* and *Quadrula*. However, Ortmann (1920) found some of the "comparatively primitive" long-term brooders such as *Dromus* and *Obovaria* to also demonstrate such clinal variation.

Grier (1920) worked under Ortmann comparing morphology in a suite of mussel species from Lake Erie and the upper Ohio River. Lake Erie is of Pleistocene origin so the fauna is relatively young. Lake Erie specimens showed greater inflation than those of the Ohio River. Grier recognized a lake effect with greater inflation and less shell height; the same species from the Ohio River were less inflated with greater shell height. He attributed these variations to differences between lake and river environments including water temperature, current, sediment and food resources. The findings of Ortmann and Grier were supported by Ball (1922), Baker (1926), Grier and Mueller (1926), Brown et al. (1938) and van der Schalie (1938c).

Lampsilis radiata (Gmelin, 1791) in Canadian lakes were reported to vary in shell size, thickness and weight in relation to turbulence, water depth, alkalinity, pH and sodium chloride levels (Green, 1972). Later studies continued to examine the relation of shell shape to environmental variables (Eager, 1978; Horn and Porter, 1981; Mackie and Topping, 1988; but see Kesler and Bailey, 1993). *Lampsilis radiata* was reported to grow faster in sand than mud (Kat, 1982; Hinch et al., 1986). Morris and Corkum (1999) found faunal differences between rivers with forested versus grassy riparian areas. Morris and Corkum (1999) also reported *Lasmigona complanata* living in a forested area to grow slower throughout life and reach maximum size at an older age than individuals in a grassy area.

Overall shell morphology has been recognized as performing a functional role in stabilizing a mussel in the substrate (Savazzi and Yao, 1992). Watters (1994b) examined the specific function of shell sculpture with regard to its stabilization role.

Food and Feeding

Compared to many animal groups, little is known about the food and feeding of mussels. Fuller (1974) summarized the earlier information on this subject. Lefevre and Curtis (1910c, 1912), Wilson and Clark (1912), Allen (1914, 1921), Evermann and Clark (1918, 1920), Coker et al. (1921), Howard (1922) and Churchill and Lewis (1924) presented limited information based on preserved mussel stomach contents, which included mud, desmids, diatoms, rotifers, flagellates and other unicellular organisms.

Following his review of available literature, Fuller (1974) observed, "It might usefully be emphasized that the prevalent notion that mussels feed mainly on diatoms is a myth." Based on an analysis of carbon isotope ratios in algae—including diatoms—and other foodweb components, Nichols and Garling (2000) stated that mussels from both river and lake environments use primarily bacterial carbon, but algae provide key nutrients such as lipids and vitamins A and D. Bacterial clearance rates of three mussel species from riverine habitats have been reported to be higher than those of three species from lake habitats (Silverman et al., 1997). Bacterial clearance rates and incorporation of bacteria nutrients have been shown to be slower in *Toxolasma texasense* (Lea, 1857) than nonindigenous *Corbicula fluminea* and *Dreissena polymorpha* (Silverman et al., 1995). Overall clearance rates of *C. fluminea* have also been reported to be greater than those of *Elliptio complanata* (Lightfoot, 1786) (Leff et al., 1990). Mussels have been reported to be detritivores and obtain a large portion of their diet from the microbial biomass of fine particulate organic matter (FPOM) (Christian et al., 2004).

Particle selection takes place at the incurrent aperture, on the gills and on the labial palps, and cilia play an active role in movement of food particles within the mussel (Allen, 1914, 1921). Mussel species that occur in river habitats have been found to possess significantly more cilia than those from lake habitats (Silverman et al., 1997). Understanding of suspension-feeding mechanics in marine bivalves was enhanced by use of endoscopes (Beninger et al., 1992; Ward et al., 1993). Tankersley and Dimock (1993b) applied this technique to freshwater mussels in a study of functional gill morphology in *Pyganodon cataracta*.

Mussels have been described as filter or suspension feeders (Strayer et al., 1994; Brim Box and Mossa, 1999; Strayer, 1999b; McMahan and Bogan, 2001). However, Raikow and Hamilton (2001) pointed to a lack of data supporting these animals being purely suspension feeders. They found that 80 percent of nitrogen used by unionids is extracted from deposited FPOM and only 20 percent from suspended FPOM. Newly transformed juvenile mussels utilize pedal feeding, with a sweeping motion of the ciliated foot (Yeager et al., 1994) (Figure 8.10). This form of feeding occurs prior to development of the gills, palps and other structures used for particle capture and sorting in adults. Pedal feeding is common in marine bivalves (Gatenby et al., 1997).

Particle selection in mussels has been shown to be based on size and density and not type of material. Material not ingested is rejected as pseudofeces. Particle size ingested varies among species (Patterson, 1984, 1986; Leff et al., 1990; Baker and Levinton, 2003; Beck and Neves, 2003). Enzymes reported in

mussel stomachs include proteases, peptidases and lipases and pH has been reported to range from 6.7 in the esophagus to 6.0 in the style sac (Lomte, 1973). Cellulase has been reported to be an important enzyme of intermediary metabolism in detritivores such as mussels, and cellulase activity has been used as an indicator of bivalve condition (Johnson et al., 1998). Zinc exposure has been reported to depress cellulase activity, suggesting a mechanism by which elevated metal contaminants negatively affect mussels (Farris et al., 1989, 1994). It has been suggested that bacteria in freshwater mussel digestive tracts aid in digestion. Abundance of bacteria has been reported to increase during feeding, while bacterial diversity remains constant (Šyvokienė et al., 1987, 1988; Šyvokienė, 1988).

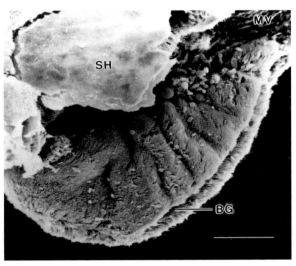

Figure 8.10. Ciliated foot of juvenile *Lampsilis cardium* used for pedal feeding. Byssal groove (BG), microvillar foot surface (MV), shell (SH). Note shell margin broken away to expose foot. Scale bar = 50 μm (Lasee, 1991).

Perfection of freshwater mussel propagation techniques and their growing use in conservation programs has led to the necessity of an artificial diet, primarily for juveniles. Addition of sediment has been found to be important for growth and survival of juveniles (Hudson and Isom, 1984). Gatenby et al. (1996) determined that a commercially available bacterial suspension was not an effective diet and noted the role of certain environmental bacteria in digestion. Gatenby et al. (1996, 1997) reported some success in rearing juvenile mussels on diatoms high in lipids.

Life Cycle Overview

The life cycle of freshwater mussels is very complex (Figure 8.11). They are ovoviviparous, and the larval stage, called a glochidium, is parasitic. Below is a brief summary of the life cycle. More detailed information is given under subsequent headings.

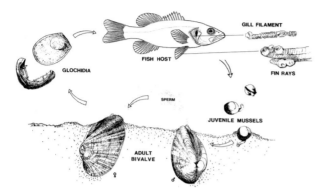

Figure 8.11. Generalized mussel life cycle.

Male freshwater mussels release sperm into the water column, it is taken in by the female via the incurrent aperture, and fertilization occurs internally, apparently in the suprabranchial chamber as ova are passed from the gonad to the marsupia (Lillie, 1901; Yokley, 1972). The fertilized eggs are deposited in interlamellar spaces of the gills, which are modified as marsupia, where they complete development into glochidia (Lefevre and Curtis, 1910b; Heard, 1975a). After discharge most glochidia are obligate parasites, utilizing various species of fish as hosts—with the exception of *Simpsonaias ambigua* (Say, 1825), which uses the *Necturus maculosus* (Mudpuppy) (Proteidae), an aquatic salamander (Howard, 1951). A few other species have been shown to use amphibian larvae as glochidial hosts in laboratory experiments (e.g., *Utterbackia imbecillis*) (Watters and O'Dee, 1997a). While encysted on the host, glochidia transform into juvenile mussels in which organs for a free-living existence are partially formed.

Life history strategies are highly variable among freshwater mussel species, but most can be placed into one of two broad groups: short-term brooders and long-term brooders (Schierholz, 1889; Sterki, 1895, 1898a; Lefevre and Curtis, 1910b; Ortmann, 1911a). The two groups differ in duration and seasonal timing of gamete production, spawning and glochidial brooding. Short-term brooders produce gametes over an extended period, typically from autumn to the following summer, and spawning occurs over an extended period, usually from late winter or early spring to early summer (Haggerty et al., 1995; Garner et al., 1999). Long-term brooders produce gametes over a much shorter period in late summer and/or autumn, followed by a short spawning period of as few as two to three weeks (Haggerty and Garner, 2000). Short-term brooders are typically gravid during spring and early summer,

releasing glochidia shortly after they are fully developed. Long-term brooders typically hold fully developed glochidia in the marsupia from late autumn to the following spring or summer. However, a few species fall into neither the typical short-term nor long-term brooding categories (e.g., *Megalonaias nervosa* is a short-term brooder but has a gametogenic cycle similar to those of long-term brooders, and its brooding period is late autumn and early winter instead of spring and summer) (Haggerty et al., 2005).

The long life span of mussels is an often overlooked aspect of their biology. Many species live 30 to 70 years, and *Margaritifera margaritifera* (Linnaeus, 1758) has been reported to live to about 200 years (Bauer, 1992; Zuiganov et al., 2000). This long life span is recorded as annuli in the shell. However, use of annuli to age shells has recently been found to be inaccurate (Downing et al., 1992). Analysis of annuli may in fact be underestimating actual shell age in some cases, and overestimating in others (Kesler and Downing, 1997). Growth equations have also been used to estimate mussel ages (Anthony and Downing, 2001). The increments in shell deposition have been used to estimate local water chemistry and water temperature changes over time by analyzing oxygen isotope ratios and metal content (Mutvei and Westermark, 2001).

Sex, Gametes and Genetics

Most members of the Unionidae and Margaritiferidae appear to be dioecious, but a few are known to be simultaneous hermaphrodites, at least in some populations (Sterki, 1898b; Tepe, 1943; Heard, 1970b, 1975a, 1979a; van der Schalie, 1970). However, occasional hermaphrodites, with varying combinations of oogenic and spermatogenic tissues, have been reported in approximately 30 predominately dioecious species (van der Schalie, 1966, 1969, 1970; Heard, 1970b; Haggerty et al., 1995; Garner et al., 1999). In Europe, *Margaritifera margaritifera* has been found to occur in populations with half female and half hermaphroditic individuals (Grande et al., 2001). A population of *Elliptio complanata* in Quebec, Canada, was reported to be composed of 20 percent males and females with the remainder hermaphroditic, and it was suggested that there may be protandrous individuals as well as sequential and simultaneous hermaphrodites in the population (Downing et al., 1989). This contrasted with the findings of Heard (1979a) and Kat (1983c) who reported no hermaphroditic individuals for *E. complanata*. Heard (1975a) classified dioecious unionids as female hermaphrodites and male hermaphrodites. Hoeh et al. (1995) looked at simultaneous hermaphroditism from a phylogenetic perspective and suggested that it has a fitness advantage over gonochorism.

Bivalve gametes are produced in small compartments of the gonad, which are the basic functional units of the organ. The compartments are variously referred to as acini (Yokley, 1972; Heard, 1979a), alveoli (Coe and Turner, 1938; Ropes and Stickney, 1965) and follicles (Loosanoff, 1937; Dudgeon and Morton, 1983). The terms may be used interchangeably between the sexes. Spermatogonia or oogonia, depending on the sex of an individual, are embedded in the epithelial walls of the acini. In hermaphrodites, male and female germ cells typically do not occur within the same acinus. In males spermatogenesis proceeds as successive divisions produce primary and secondary spermatocytes, spermatids and spermatozoa. As these divisions occur around the acinus periphery, spermatozoa are crowded toward the lumen (Haggerty et al., 1995; Garner et al., 1999). Another cell type, sperm morulae or multinucleated inclusions, is abundant in male gonads during certain periods of the year, but their fate and function have been a matter of conjecture (Heard, 1975a). In females, oogonia produce oocytes that are attached to the acinus wall by cytoplasmic stalks (Beams and Sekhon, 1966). They eventually break free as oocytes mature into ova (Haggerty et al., 1995; Garner et al., 1999).

Unionoid spermatozoa consist of a head, midpiece and tail (Figure 8.12). Sperm morphology has been described for some North American unionid species, including *Anodontoides ferussacianus* (Lea, 1834) (Edgar, 1965), *Ligumia subrostrata* (Trimble and Gaudin, 1975) and *Truncilla truncata* (Waller and Lasee, 1997). An ultrastructural study of spermatogenesis in the European *Anodonta cygnea* (Linnaeus, 1758) was provided by Rocha and Azevedo (1990). Unionoid sperm are considered primitive, and their morphology and ultrastructure have been used to trace the relationship between the Trigoniidae and Unioniformes (Healy, 1989, 1996).

Figure 8.12. *Elliptio fumata* spermatozoan, from the Apalachicola Basin, Georgia, 16 March 1995. Photograph by C.A. O'Brien.

Spermatozoa have been reported to be discharged in the form of spermatozeugmata in at least seven species (Utterback, 1915, 1931; Edgar, 1965; Lynn, 1994; Barnhart and Roberts, 1997; Waller and Lasee, 1997). Spermatozeugmata have been described as "hollow globular masses" (Utterback, 1931) or "sperm balls" (Edgar, 1965) (Figure 8.13). They are comprised of numerous spermatozoa with their heads closely packed together and buried in a body of undetermined composition (Edgar, 1965). The sperm flagella extend outward and beat synchronously, turning and moving the body through the water. Spermatozeugmata of *Anodonta suborbiculata* were reported to contain 3,600 spermatozoa (Barnhart and Roberts, 1997); those of *Truncilla truncata* were reported to be made up of 8,000 to 9,000 spermatozoa (Waller and Lasee, 1997).

Figure 8.13. Cross section of a *Medionidus penicillatus* spermatozeugmata, from Sawhatchee Creek, Terrell County, Georgia, 15 August 1995. Photograph by C.A. O'Brien.

Unionoid ova are round and surrounded by a vitelline membrane that remains intact through most of the embryonic development. Ova of unionoids have greater yolk volumes than do ova of marine species with planktonic larvae (McMahon and Bogan, 2001). Beams and Sekhon (1966) used electron microscopy to describe the development of the oocyte stalk and the mechanism of yolk deposition. Lillie (1895, 1901) provided a detailed account of the unionid ovum and its embryology following fertilization, including lineages of various cell lines.

Water temperature is believed to play a role in seasonal timing of gametogenic activities, but the influence of other environmental factors is poorly understood. Mussels have been found to cease gametogenesis downstream of dams that have cold-water discharge from the hypolimnion. However,

Megalonaias nervosa moved from a cold-water habitat to a warm-water reservoir were reported to reinitiate gametogenesis, with almost half of the transplanted individuals becoming hermaphroditic (Heinricher and Layzer, 1999).

Unionidae and Margaritiferidae have a diploid chromosome number of 38 (Jenkinson, 1976). Unlike most organisms, unionoid mussel offspring inherit male as well as female mitochondria. This phenomenon has been termed double uniparental inheritance (DUI) (Skibinski et al., 1994a, 1994b; Zouros et al., 1994a, 1994b; Hoeh et al., 1996). Zouros et al. (1994a) first described DUI in mytilid bivalves. It was subsequently documented in unionids (Hoeh et al., 1996; Liu et al., 1996; Hoeh et al., 2002) and finally recognized in the marine bivalve family Veneridae (Passamonti and Scali, 2001). In Unionoidea the male mitotype appears to be restricted to male gonadal tissue and sperm, with all somatic tissues and female gonads containing the female mitotype.

Phylogenetic analyses have demonstrated that DUI has produced two independent mitochondrial DNA lineages within a species, with male and female lineages diverging (Hoeh et al., 1996, 2002), but with the male lineage evolving at a faster rate (Krebs, 2004). Curole and Kocher (2002) examined the junction of the cytochrome c oxidase II (COII) and cytochrome c oxidase I (COI) genes and reported the male sequence has a 600 base pair extension that is absent in the female sequence. Assessment of DUI may represent an important avenue of research for examining unioniform phylogeny.

The age at sexual maturity has been studied for a small percentage of freshwater mussel species. Most studies have found that reproduction begins between three and five years of age (Stein, 1969; Yokley, 1972; Zale and Neves, 1982b; Weaver et al., 1991; Jirka and Neves, 1992; Haag and Staton, 2003). The prolific and often invasive *Sinanodonta woodiana* (Lea, 1834) was found to reach sexual maturity in less than one year (Dudgeon and Morton, 1983). A few species have been reported to reach sexual maturity at two years of age, including *Elliptio arca*, *Lampsilis ornata* (Haag and Staton, 2003), *Lampsilis teres* (Chamberlain, 1931) and *Utterbackia imbecillis* (Allen, 1924). In a population of *E. arca* and *L. ornata*, 100 percent of individuals examined had reached sexual maturity by age two. *Megalonaias nervosa* was not reported to begin reproduction until age eight in upper reaches of the Mississippi River (Woody and Holland-Bartels, 1993). Individual *Quadrula asperata* and *Quadrula pustulosa* may reach sexual maturity as early as three years, but Haag and Staton (2003) did not find 100 percent of females to be mature until age nine in the former and age seven in the latter. Age at sexual maturity has been reported to differ between the sexes in *Cumberlandia*

monodonta, with males maturing at four to five years and females maturing at five to seven years (Baird, 2000).

Fertilization success has been found to be variable among mussel species. Some species appear to routinely discharge conglutinates that contain varying percentages of undeveloped eggs (Haag and Staton, 2003). A fertilization rate of 95 percent has been reported for *Quadrula cylindrica strigillata* (Wright, 1898) (Yeager and Neves, 1986). Downing et al. (1993) reported reproductive failure of *Elliptio complanata* at densities less than 10 animals per m^2 and 100 percent fertilization rate only with densities of at least 40 animals per m^2.

Glochidium and Fecundity

The glochidium is the parasitic larval stage of unionoid bivalves and is unique to the group. Heard (1998a, 2000) examined the history of unionoid reproductive biology from Aristotle to Houghton (1862). Glochidium development and morphology was described and illustrated by Lillie (1895, 1901), Harms (1907), Herbers (1914) and Wood (1974). Haas (1933) summarized previous studies on glochidium development and morphology. Lea (1858d) figured and described glochidia from 38 species and later included descriptions of glochidia with some species descriptions. Hoggarth (1999) provided scanning electron micrographs and descriptions for many of the North American species. Hoggarth and Gaunt (1988) and Hoggarth (1993) investigated glochidial shape and mechanics of attachment, suggesting some shapes are more effective in anchoring glochidia to the host.

Glochidia vary among species with regard to shape and morphology, but there are three basic types: hooked, hookless and ligulate (ax-head shaped) (Lefevre and Curtis, 1910b; Arey, 1924). The most common glochidium is the hookless type, which has a subovate to subelliptical outline (Figure 8.14). The thickened ventral margin of hookless glochidia is composed primarily of cuticular material, but is roughened with microstylets, which are positioned in several rows parallel to the shell margin (Arey, 1924; O'Brien et al., 2003). The ventral margin of some *Elliptio* glochidia has a low, centrally located fold that resembles a poorly developed hook, but these species are considered to be of the hookless variety (O'Brien et al., 2003). Some *Epioblasma* have supernumerary hooks, which are microscopic, straight, sharply pointed hooks located along the ventral margin. Hooked glochidia tend to have heavier valves that are generally subtriangular in outline and are often larger than hookless glochidia (Arey, 1924; Bauer, 1994) (Figure 8.15). These species are equipped with a prominent styliform hook, which is centrally located on the ventral margin of each valve (Figure 8.16). The hook is adorned with microstylets that may continue onto the ventral shell margin and around the periphery to varying degrees, depending on species (Arey, 1924).

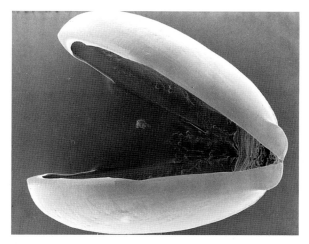

Figure 8.14. Hookless glochidium, *Villosa vibex*, Flint River, Georgia. Photograph by C.A. O'Brien.

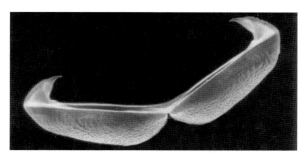

Figure 8.15. Hooked glochidium, *Lasmigona costata*, Milwaukee River, Washington County, Wisconsin (Clarke, 1985).

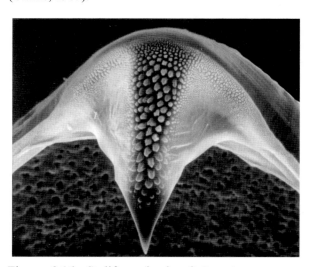

Figure 8.16. Styliform hook of *Lasmigona costata* glochidium, Milwaukee River, Washington County, Wisconsin (Clarke, 1985).

Ligulate glochidia are restricted to the unionid genus *Potamilus* (Figure 8.17). They have short hinge lines and more or less straight anterior and posterior sides which flare ventrally to meet the laterally expanded ventral margin. Glochidia of most *Potamilus* species have prominent lanceolate hooks on the lateral margins of the ventral flange. The ventral flange is adorned with microstylets. However, the microstylets do not extend up onto the lateral hooks. These species have a lateral gape to the extent that right and left valves make contact only along their dorsal and ventral margins.

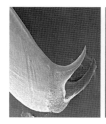

Figure 8.17. Ligulate glochidium, *Potamilus alatus*, Tennessee River near Decatur, Morgan County, Alabama. Photograph by C.A. O'Brien.

A glochidium is composed of two chitinous valves attached by an elastic dorsal hinge ligament and a single large adductor muscle that serves to close the valves (Figure 8.18). Once attached to a host the adductor remains contracted. It degenerates during metamorphosis and is replaced by anterior and posterior adductor muscles in juveniles (Wächtler et al., 2001). A series of sensory hairs, variable among species with regard to size, shape and position, are believed to be sensitive to chemical and possibly mechanical stimuli (Wächtler et al., 2001). Stimulation of the hairs results in closing of the valves. The glochidium mantle is epithelial and is believed to be involved in nutrient uptake during the parasitic stage (Wächtler et al., 2001; Fisher and Dimock, 2002a, 2002b). Fields of cilia are located on certain areas of the mantle surface and create water currents to pass oxygen and chemical stimuli (Wächtler et al., 2001). Glochidia of some species have a long larval thread, which is not homologous to the byssal thread of juveniles (Lefevre and Curtis, 1910b; Wächtler et al., 2001). The function of the larval thread is believed to aid in bringing glochidia into contact with a host. Threads of multiple glochidia often become entangled during discharge from the adult, which may make small clusters more visible to fish or form large web-like masses in which fish may become entangled (Haag and Warren, 1997; C.A. O'Brien, personal communication).

Glochidia attach to the host by clamping the two valves together around the host's tissues. Hooked glochidia primarily infest fins of hosts, and hookless and ligulate glochidia encyst primarily on gill filaments (Lefevre and Curtis, 1911; Arey, 1924; Jansen, 1991; Jansen and Hansen, 1991). Arey (1924) described the mechanics of attachment in detail. Surber (1912, 1915) discussed problems associated with attempts to identify glochidia taken from fish and provided details on their identification based on size and shape. Glochidia range in size from 50 µm to 450 µm in shell length (Bauer, 1994). Glochidium size appears to be independent of shell allometry as well as phylogeny, brooding strategy, marsupium position or habitat type. However, species that are host generalists tend to have larger glochidia (Bauer, 1994).

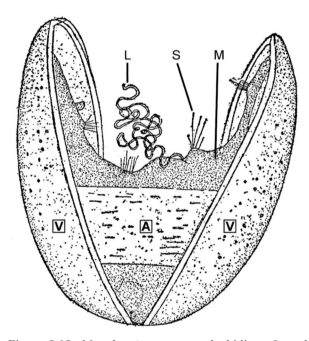

Figure 8.18. *Megalonaias nervosa* glochidium. Larval adductor muscle (A), larval thread (L), mantle (M), sensory hairs (S), valve (V) (Coker et al., 1921).

Fecundity has been estimated for only a few freshwater mussel species. Factors affecting fecundity are not well understood, but a few generalizations are apparent. Whereas glochidium size remains constant within a species, fecundity is positively related to size of the female, increasing exponentially in some species (Bauer, 1994; Haag and Staton, 2003). There is a trade-off between glochidium size and number of glochidia, with fecundity being highest in species with smallest glochidia. Margaritiferids have small glochidia relative to most unionids and have the highest reported numbers of glochidia. *Cumberlandia monodonta* has been reported to produce between 2,000,000 and 9,000,000 glochidia (Baird, 2000), and a European population of *Margaritifera margaritifera* has been reported to produce up to 17,000,000 glochidia per year (Young

and Williams, 1984). Fecundity of unionids has been reported to average from 9,647 glochidia per year in *Quadrula asperata* (Haag and Staton, 2003) to 750,000 per year in *Megalonaias nervosa* (Haggerty et al., 2005). However, with maximum adult size ranging from 35 mm to 280 mm, and glochidium sizes ranging from 50 µm to 450 µm (Bauer, 1994), the relationship between size and fecundity within the Unionidae is not clear. Some species with relatively large glochidia produce large numbers because adult females attain large sizes (e.g., *Megalonaias nervosa* glochidia may be up to 280 µm long, and females have been reported to average 750,000 per year) (Haggerty et al., 2005). There is some evidence that species that discharge glochidia bound in conglutinates produce fewer glochidia (Kat, 1984). More inclusive studies of fecundity and detailed comparisons involving life history traits, glochidium size and adult size are needed.

Brooding

Female unionoids brood developing embryos and glochidia in the marsupial portion of their gills. Most species produce a single brood per year and can be placed into one of two groups: short-term brooders or long-term brooders (Sterki, 1895; Ortmann, 1909b; Lefevre and Curtis, 1910b). Short-term brooders typically hold glochidia from late winter or spring to summer, releasing glochidia soon after they are fully developed. Long-term brooders usually hold fully developed glochidia in the marsupia from late autumn to the following spring or summer. Ortmann (1911a) coined the terms tachytictic for short-term brooders and bradytictic for long-term brooders. However, some species do not fall into either group (Heard, 1998b). It has been suggested that *Cumberlandia monodonta* produces two broods per year (Gordon and Smith, 1990) and that *Glebula rotundata* produces up to three broods per year (Parker et al., 1984).

There is some indication that the parent provides nourishment to glochidia during the brooding period. Schwartz and Dimock (2001) provided ultrastructural evidence that such nutrient transfer takes place in *Utterbackia imbecillis* and *Pyganodon cataracta*. Lamellar tissues of the marsupia initially have concentrations of glycogen that diminish as the brooding period progresses. Such glycogen concentrations have not been observed in nonmarsupial gills. Also, septa of the brood chambers are equipped with numerous mitochondria and microvilli, which suggest the capability of active transport into or out of the marsupia. Absorption of amoebocytes by embryos from the parent mussel across the gill membrane has also been reported (Pelseneer, 1935).

During the brooding period gills continue to function as feeding and respiratory organs even though the lumina of the water tubes are restricted by the glochidia with which they are filled. Species of *Anodonta*, as well as some other genera, partially compensate for the loss of water transport capacity by development of secondary water tubes, which disappear after discharge of the glochidia (Ortmann, 1911b; Tankersley and Dimock, 1992). Marsupial gills continue to contribute to filter-feeding efforts during the brooding period, even though flow of water around the outer demibranchs is impeded by their swollen condition (Tankersley and Dimock, 1993b).

At the end of the brooding period glochidium discharge occurs via the excurrent aperture, aided by rapid adductions of the valves and contractions of the brooding demibranchs (Tankersley and Dimock, 1993b). There have been reports of glochidia of some species being discharged through ruptures in the ventral gill margin (Ortmann, 1910a; Sterki, 1911; Richard et al., 1991), but there has been no absolute confirmation of this and there is no known mechanism for getting glochidia from the mantle cavity to the excurrent aperture if they emerge from the ventral margin of the gill. Tankersley and Dimock (1993b) observed glochidia discharge via the excurrent aperture in a species of *Pyganodon* that had previously been reported to discharge via gill rupture (Richard et al., 1991).

Host Attraction Strategies

Some unionoid species have derived elaborate methods of increasing opportunities for their glochidia to come into contact with hosts, whereas others appear to leave this important aspect of their life history to chance. One method of attracting potential hosts to glochidia is by packaging them into bundles that resemble food items, called conglutinates (Lea, 1858d; Ortmann, 1912a; Utterback, 1931; Morrison, 1973). Conglutinates are held together by their membranes, which have adhesive qualities, embedded in a mucilaginous matrix or held together by undeveloped eggs (Lefevre and Curtis, 1912; Haag and Staton, 2003; Haag and Warren, 2003). Conglutinates of *Pleurobema collina* (Conrad, 1837) have been estimated to contain 93 percent mature glochidia and 7 percent unfertilized eggs (Hove and Neves, 1994) and those of *Fusconaia cuneolus* have been reported to include 1 percent to 5 percent unfertilized eggs (Bruenderman and Neves, 1993). Conglutinates of *Strophitus undulatus* are composed of a spongy matrix in which glochidia are completely embedded but emerge from pores following discharge from the marsupium (Watters, 2002).

Each conglutinate consists of the entire contents of a single marsupial water tube. They may resemble insect larvae or pupae, fish eggs or larval fishes, worms, leeches or flatworms (Hartfield and Hartfield, 1996; Haag and Warren, 1997; Watters, 1999; Jones and Neves, 2002b; Jones et al., 2004) (Figure 8.19). Conglutinates are believed to aid in delivering glochidia

to potential host fishes. When bitten the conglutinate ruptures and glochidia are released into the fish's mouth and gill chamber, bringing them into contact with gill filaments, where they may attach and become encysted.

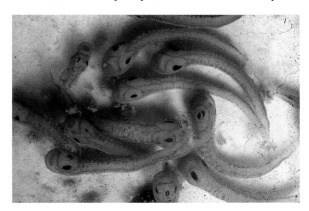

Figure 8.19. *Ptychobranchus fasciolaris* conglutinates that mimic larval fish. Photograph by W.N. Roston.

Marsupial contents are often prematurely aborted in the form of conglutinates when a gravid female is disturbed or stressed, primarily in short-term brooders (Lefevre and Curtis, 1912). These aborted conglutinates are often composed of early embryos or immature glochidia. Some species that discharge aborted conglutinates discharge glochidia singly or in irregular masses when they are mature. These conglutinates lose their cohesiveness as glochidia mature (Lefevre and Curtis, 1912). Descriptions of conglutinates were often reported without qualifying statements as to the development of glochidia contained within them. Therefore, the extent of conglutinate use for attracting glochidial hosts is not well understood for some species.

Mussels of the genus *Hamiota* have taken conglutinate production one step further. They package the entire contents of both gills as a superconglutinate (Haag et al., 1995, 1999; Hartfield and Butler, 1997; O'Brien and Brim Box, 1999; Roe et al., 2001). Superconglutinates are discharged into paired, transparent mucus tubes that are attached to each other and closed on the distal end. The proximal end of the mucus tube remains attached to the excurrent aperture of the adult female, at least initially. The mucus tubes may reach 2.5 m in length (Haag et al., 1995). Action of the water current causes the superconglutinate at the distal end to display a fish-like motion and elicit attack by predatory host fishes (Haag and Warren, 1999).

Some species of mussels, such as *Strophitus subvexus* and *Megalonaias nervosa*, discharge glochidia embedded in a mucus web, which may serve to entangle fish and increase the likelihood of attachment (Haag and Warren, 1997; C.A. O'Brien, personal communication). Many mussel species are believed to simply discharge their glochidia into the water column and leave the possibility of encounter with a suitable host to chance. This strategy is difficult to study, but adaptations to this mode of dispersal may include production of glochidia in large numbers, small glochidia that remain suspended in the water column for longer periods, glochidia capable of remaining viable for longer periods and lack of host specificity. However, none of these characteristics have been correlated to broadcast dispersal of glochidia.

One of the more unusual reproductive behaviors has been described for a European species, *Unio crassus* Retzius, 1788. Females have been observed to migrate to the water's edge and spurt streams of water, containing glochidia, through the air up to 1 m toward the center of the stream. As the water and glochidia splash on the surface, the resulting disturbance has been observed to attract potential glochidial host fishes (Vicentini, 2005).

Some mussel species, primarily long-term brooders, have taken a different approach to increasing probability of host infestation. They have evolved modifications to the posterioventral mantle margin that are used to lure potential hosts into their vicinity, after which glochidia are discharged singly or in loose masses. There are different types of mantle lures, including fleshy pads with microlures (*Epioblasma*) (Figure 13.8), mantle flaps (e.g., *Lampsilis*) (Figure 13.19), caruncles (*Toxolasma*) (Figure 13.42) and papillate mantle folds (e.g., *Villosa*) (Figure 13.43). These structures have evolved to resemble small fishes, crayfish, insect larvae and worms. Mantle displays have been reported to react to light (*Lampsilis*) and to pulse with contractions that move along their length (*Lampsilis*, *Ligumia* and *Villosa*) (Kraemer, 1970; Haag et al., 1999). *Villosa nebulosa* was reported to display its papillate mantle fold primarily at night, and *Villosa vibex* displays primarily during the day (Haag and Warren, 2000). It has also been suggested that mantle flaps serve to help aerate the gravid marsupia and create currents to increase sperm uptake (Utterback, 1931). Mantle lures are generally reduced or absent in males.

Other freshwater mussel species have posterioventral mantle modifications that have not been properly described and are poorly understood. *Medionidus acutissimus* displays a black mantle margin with a small white patch that it flickers rapidly (Haag and Warren, 2003). Additional species with apparent modifications to their posterioventral mantle margins include *Ellipsaria lineolata*, *Lemiox rimosus* and *Medionidus conradicus*. Species with rudimentary mantle modifications that are poorly understood include *Actinonaias ligamentina*, *Leptodea fragilis*, *Obovaria subrotunda* and *Potamilus alatus*. *Quadrula verrucosa*, a short-term brooder, displays a thickened, fleshy pad of mantle tissue on the posterioventral margin during

spring. The pad has numerous irregular wrinkles parallel to the margin, which may be adorned with irregular, flattened papillae.

Fish Hosts and Metamorphosis

Most glochidia require an intermediate host (e.g., Lefevre and Curtis, 1910b; Surber, 1912, 1913, 1915; Fuller, 1974; Watters, 1994a). The majority of mussel species utilize various species of fish, but *Simpsonaias ambigua* uses the external gills of *Necturus maculosus* (Mudpuppy) (Proteidae) (Howard, 1951). *Utterbackia imbecillis*, a known generalist with regard to glochidial hosts, was reported to transform on larvae of *Rana catesbeiana* (Bullfrog) (Ranidae), *Rana pipiens* (Northern Leopard Frog) (Ranidae) and *Ambystoma tigrinum* (Tiger Salamander) (Ambystomatidae) in laboratory experiments, as well as a number of nonnative fishes and amphibians (Watters and O'Dee, 1997a). However, there have been no reports of *U. imbecillis* using amphibian hosts in nature.

Glochidia attach to a host by clamping the two valves together on a gill filament or fin. Species with hookless and ligulate glochidia infest primarily the host's gills, but hooked glochidia may infest either gills or fins (Arey, 1924). Most species capable of attaching to fins are equipped with marginal hooks that aid in gripping the host. Infestation of fins occurs primarily at the distal margin, and gill infestation is heavier on the distal third of the filaments (Lefevre and Curtis, 1912). Once a glochidium is attached to a gill or fin the host's tissues grow around it until it is completely embedded within a cyst as the host's response to mechanical damage of the tissues (Arey, 1921) (Figure 8.20). Formation of the cyst occurs much more rapidly during gill infestations than during fin infestations. A glochidium may be fully embedded within a gill cyst in as little as 2 to 3 hours, whereas it takes from 6 to more than 24 hours for a fin-attached glochidium to become fully encysted (Young, 1911; Lefevre and Curtis, 1912; Arey, 1932b). Cysts of fin-attached glochidia are comprised of epithelial tissues, whereas glochidia attached to fins are embedded in cellular connective tissues, covered by epithelial tissue (Arey, 1932b). Glochidia which fail to infest a host perish unattached. *Villosa iris* and *Actinonaias pectorosa* glochidia have been reported to survive longer at lower temperatures (Zimmerman and Neves, 2002). Margaritiferid glochidia have been reported to survive in the water column for as long as 11 days (Murphy, 1942).

While inside the cyst the glochidium undergoes metamorphosis to the juvenile stage, with partially developed organs needed for a free-living existence (Coker et al., 1921). Changes in the glochidium mantle during the encystment period were detailed by Blystad (1923), who described three stages of metamorphosis: 1) encapsulation and encystment on the host;

2) mushroom body stage during which the host's tissues between the valves are digested; 3) formation of juvenile anatomical features. Fisher and Dimock (2002a) recognized two stages of metamorphosis: 1) the first four days, with disintegration of the larval adductor muscle and formation of the mushroom body; 2) the final days of metamorphosis, with development of juvenile structures including the stomach, intestine, digestive glands, foot, gill buds and nerve cords. Fisher and Dimock (2002a) reported high levels of DNA, RNA and protein synthesis during both stages of metamorphosis.

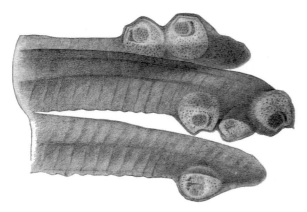

Figure 8.20. *Actinonaias ligamentina* glochidia encysted on gill filaments of *Ambloplites rupestris* (Rock Bass) (Centrarchidae) (Lefevre and Curtis, 1912).

There has been varied discussion about provision of nutrients to the glochidium by the host during the parasitic stage. The intercepted tissue clamped between the two valves as a glochidium attaches to the host quickly disappears, with the nutrients presumably absorbed by the glochidium (Lefevre and Curtis, 1912; Blystad, 1923; Arey, 1932c). Additionally, degeneration of the larval adductor muscle and mantle tissues, including the mushroom body, is believed to provide nutrients during transformation (Arey, 1932c; Fisher and Dimock, 2002b). The requirement of organic compounds, amino acids, vitamins and plasma for in vitro glochidium transformation suggests that some nutritional benefit is received from the host (Isom and Hudson, 1982). Most mussel species undergo little or no increase in size during encystment, but there are exceptions, such as *Potamilus ohiensis*, *Truncilla donaciformis* and *Margaritifera margaritifera* (Lefevre and Curtis, 1912; Coker et al., 1921; Arey, 1932c; Bauer, 1994). *Potamilus ohiensis* has been reported to increase to 40 times its original size during the encystment period with apparent secondary capillary growth in gill tissues surrounding the cyst, suggestive of provision of nutrients to the growing glochidium

(Arey, 1932b). Such secondary gill growths do not occur during encystment in all mussel species (Arey, 1932b). Blystad (1923) suggested that glochidia attached to gills are more dependent on nutrients from the host than those attached to fins, since gill-attached glochidia are more in contact with the lymph stream and they retain their larval mantle, which may play a digestive role during development.

Different species of freshwater mussel display varying degrees of glochidial host specificity. As Coker et al. (1921) eloquently stated, "Some are more catholic in their tastes than others, yet for any mussel there is a limited number of species of fish upon which it will attach and complete metamorphosis." Some appear to utilize a single species, genus or family of fishes, whereas others may use a wide variety of species across a number of families. For example, *Pyganodon grandis* has been reported to utilize at least 21 species representing 7 fish families. Adult mussels are not selective in their discharge of glochidia, and glochidia are not selective in which species they will infest. The specificity appears to be a factor of the host's immunity to some species but not others, with glochidia rejected from unsuitable hosts after they become attached. Arey (1932a) microscopically studied the host immune reaction to glochidial infestation. Glochidial cysts formed on immune fishes become larger and more irregular than those in which metamorphosis was to occur, with the size increase due to additional cellular connective tissues in the cyst wall. After the first day the cyst walls become thinner, followed by sloughing a day or more later. Many glochidia are destroyed by cytolysis prior to being sloughed from the host. Waller and Mitchell (1989) compared tissue reactions of *Lampsilis siliquoidea* glochidia infestations between a known host, *Sander vitreus* (Walleye) (Percidae), and a nonhost, *Cyprinus carpio* (Common Carp) (Cyprinidae). Encystment of glochidia on the host was complete in 6 hours while few glochidia on the nonhost became encysted and were sloughed within 60 hours.

Detailed study of glochidial hosts using populations of mussels and fishes from distant parts of their ranges has suggested that host fish relationships may be more complicated that previously believed. For example, *Villosa umbrans* has been found to vary in host use among populations, with some using sunfishes (*Lepomis*), some using sculpins (*Cottus*) and some using both (W.R. Haag and P.D. Johnson, personal communication). Also, there may be a temporal aspect to host fish suitability, as it has also been reported that fish eventually acquire immunity to glochidial infestations (Reuling, 1919; Arey, 1932a; Neves et al., 1985; Watters and O'Dee, 1996; O'Connell and Neves, 1999; Rogers and Dimock, 2003). Neves et al. (1985) summarized information on acquired host immunity to repeated glochidial exposure and infestation. Kirk and Layzer (1997) depressed the immune system of a non-host fish and were able to obtain metamorphosed juveniles. O'Connell and Neves (1999) documented an immune response in host fishes that had previously been infected with glochidia.

Duration of the parasitic period is highly variable within and among species. Species with small glochidia tend to have longer parasitic stages than those with larger glochidia (Bauer, 1994). Water temperature and other environmental factors play a role in the duration of the parasitic period. In general, length of the parasitic period increases with decreases in temperature, but this can vary among trials of the same species with temperature held constant. In experimental trials with *Villosa iris* glochidia on *Micropterus dolomieu* (Smallmouth Bass) (Centrarchidae) the transformation period varied from 38 to more than 105 days at constant temperature in the same trial (Zale and Neves, 1982a). Shortest reported parasitic periods were for *Ligumia recta* and *Ligumia subrostrata*, which transformed on *Micropterus* sp. in 7 days at 20.5° C. However, other infestations at the same temperature required longer periods to transform (Lefevre and Curtis, 1912). The longest reported transformation period involved *Actinonaias ligamentina* and *Ligumia recta* on "black bass and crappie" and lasted from 91 to 112 days (Lefevre and Curtis, 1912).

At the end of the parasitic period a newly transformed juvenile excysts, drops to the substrate and begins to grow (Figure 8.21). Excystment is apparently accomplished in part by activity of the juvenile within the cyst, as it has been observed to periodically move and push upon the cyst wall with the foot and by slightly opening and closing its valves (Young, 1911; Arey, 1932b). Young (1911) reported development of lymph-filled vacuoles in the cyst wall over time, which may weaken the cyst and aid in eventual excystment. However, Arey (1932b) noted no evident thinning or weakening of the wall, but recorded gross sloughing of gill cells as a possible aid in the excystment process. Waller and Mitchell (1989) reported excystment to begin with thinning of the capsule walls.

The juvenile mussel excysts from the host with a mouth, intestine, foot, two adductor muscles and rudimentary gills (Coker et al., 1921). It immediately begins to crawl once it is free from the cyst, using its long, ciliated foot. Some juvenile mussels eventually produce a byssal thread, with which they attach to the substrate (Lefevre and Curtis, 1912). Habits and habitats of juvenile mussels are subjects in need of further study. Isely (1911) concluded that juveniles occur in coarse gravel and pebble substrates in "fairly swift" water and migrate to areas with muddy substrates, yet there is no indication that a search of other habitats was performed. In a later study, which

encompassed riffle, run and pool habitats, juveniles were found to occur in greater numbers in riffles and runs, with greatest densities just downstream of boulders (Neves and Widlak, 1987). There appear to be no studies dealing with juvenile ecology in areas disjunct from riverine habitats.

Figure 8.21. Juvenile *Lampsilis cardium*, 21 days postexcystment. Note shell growth beyond the original glochidial shell. Scale bar = 100 µm (Lasee, 1991).

There are at least three species of freshwater mussels for which the glochidium has been reported to possibly be a facultative parasite. *Strophitus undulatus* was observed to undergo metamorphosis inside its elaborate conglutinate, with emerging juveniles having anterior and posterior adductor muscles, a ciliated foot, gill buds and other internal organs (Lefevre and Curtis, 1911, 1912). However, *S. undulatus* glochidia have also been found to undergo transformation on a number of fish hosts (Watters, 1994a). *Utterbackia imbecillis* glochidia have also been reported to be facultative parasites (Howard, 1914a, 1915). Though the observations of Howard (1914a) were convincing, experiments by Tucker (1928) did not duplicate his results and there have been no additional reports of this phenomenon in the species. Similar observations of direct development have been made for *Lasmigona subviridis* from a population in eastern North Carolina (Barfield and Watters, 1998). Watters et al. (1998) noted two different glochidium types in the marsupia of *Lasmigona costata* and noted that the round, weakly hooked or unhooked glochidia may be evidence of direct development.

Ellis and Ellis (1926) and Ellis (1929) reported achieving transformation of mussel glochidia in vitro but never published detailed methods. Isom and Hudson (1982) presented a method for in vitro culture for artificial transformation of glochidia to juvenile mussels using fish plasma along with other organic and inorganic compounds. Hudson and Isom (1984) documented the first success in rearing juveniles in a laboratory. Keller and Zam (1990) refined and simplified the in vitro culture technique. Uthaiwan et al. (2001, 2002) followed up on this method and found that best results came with using plasma from host fishes.

Physiology

McMahon and Bogan (2001) compiled a concise overview of the limited information on mussel physiology and compared it with that of *Corbicula fluminea*. Silverman et al. (1985, 1989) examined the role and formation of calcium concretions often found in gill connective tissues. Dietz et al. (2000) provided a detailed examination of mussel kidney function. Byrne (2000) reported on the interstitial tissue calcium concretions of a hyriid from Australia and suggested different concentrations may be phylogenetically controlled.

Anatomy Overview

Simple observations of mussel soft anatomy were provided by some of the earliest malacologists, including Rafinesque (1820), Barnes (1823), Agassiz (1852) and Lea (e.g., 1828, 1848a and later publications). However, a detailed anatomical description was not available until G.B. Simpson (1884) published a well-illustrated guide to the anatomy of *Pyganodon cataracta* that included external and internal shell landmarks, mantle and cilia as well as digestive, nervous, circulatory and reproductive systems. The intestine illustration included three segments between the stomach and anus, where most subsequent authors have shown only a simple tube. In the illustration of the nervous system were two structures referred to as the "auditory organ", which are apparently the statocysts. Girod (1889) provided a detailed illustrated guide for the dissection of European *Anodonta*.

Troschel (1847) was among the first to use soft anatomy characters for unioniform classification. The first Americans to use soft anatomy characters in classification were Stimpson (1851), Agassiz (1852), Sterki (1895, 1898a) and C.T. Simpson (1896, 1900b, 1914). Simpson (1896, 1900b) relied on marsupial characters to develop his classification of the Unioniformes. Comparative studies with illustrated gross soft anatomy were published in a series of classic papers by Ortmann (e.g., 1910a, 1910b, 1911b, 1912a, 1912b, 1913b, 1914, 1915, 1919), which used the information to separate genera and species into subfamilies. Detailed shell landmarks and soft anatomy of five Atlantic Coast mussel species, with carefully labeled digestive, circulatory and nervous systems, were produced by Reardon (1929). However, Reardon's figures were greatly reduced and are rather difficult to use. Fuller and Bereza (1974) and Heard (1974)

stressed the value of anatomical characters in taxonomy.

In a dissertation Yokley (1968) histologically examined the anatomy of *Pleurobema cordatum*, including the nervous, digestive, circulatory, excretory and reproductive systems as well as the gross anatomy of the animal. Kraemer (1970) provided information on the morphology and behavior of *Lampsilis* mantle flaps. Kokai (1974) examined the apertures of two species of *Quadrula* and found overlap in aperture size. Kokai (1976) later studied the aperture characters of 18 mussel species and observed the allometric increase in the number of papillae on the incurrent aperture of *Amblema plicata* with increasing shell length. The musculature of European *Anodonta* was described by Brück (1914), who identified four layers of muscles, in addition to the transverse muscles, in the foot. Brück (1914) also examined the histological structure of muscles and included good illustrations of the extent of each muscle and its placement in the foot and visceral mass. Smith (1983) provided anatomical information on margaritiferid mantle attachments, which were previously erroneously labeled as muscles.

Digestive System

Guthiel (1912) and Graham (1949) detailed the digestive system of European *Anodonta*. Graham noted the dorsal fold of the stomach and gave a three dimensional view of the organ. Kat (1983a, 1983b) described and illustrated the stomach floor in *Lampsilis* and *Anodonta*, noting distinction between the two, but did not discuss specific variation. Smith (1986) provided detailed illustrations of the margaritiferid stomach including the position of the digestive glands, typhlosoles and ridges and sorting areas on the stomach floor. Bogan and Hoeh (1995) illustrated the difference in the stomach floor between two *Utterbackia* species. The crystalline style is a flexible, gelatinous rod of glycoprotein that extends from the style sac into the stomach and impinges on the gastric shield, where it is believed to aid in digestion (Nelson, 1918; Lomte and Jadhav, 1980; Alyakrinskaya, 2001).

Guthiel (1912) illustrated sections of the intestine and revealed its histological structure and musculature. Other digestive structures identified by Guthiel (1912) include the esophagus, crystalline style and "liver gland" (i.e., digestive gland). Jegla and Greenberg (1968) suggested that the location of the intestine through the heart facilitated osmoregulation by filtration of fluid and excretion of metabolic wastes from the heart to the intestine. Within the ventricle, the intestine walls are much thinner and the intestinal typhlosole in this area has more abundant hemolymph sinuses (Narain and Singh, 1990). Myers and Franzen (1970) histologically examined the pericardial cavity and nephridia. Waste particles are broken down enzymatically, the metabolites of which are excreted by way of secondary vesicles into the nephridial lumen or pericardial cavity.

After the rectum transverses the heart, it extends posteriorly over the dorsal surface of the posterior adductor muscle. In the Margaritiferidae the anus is positioned on the dorsum of the posterior adductor muscle. In the Unionidae the rectum extends past the adductor muscle dorsum, and the anus is typically positioned on the posterior side of the posterior adductor muscle (Bogan, 1992).

Nervous System and Statocysts

The nervous system was included in a series of mussel illustrations by G.B. Simpson (1884). Splittstößer (1913) illustrated the nervous system of *Anodonta* including the statocysts. Kraemer (1967, 1984) investigated the pallial nerves of *Lampsilis cardium* and reported mantle tissues adjacent to the mantle flap to be proportionately richer in nerves than the remainder of the mantle. A ganglion was found just anterior of the mantle flap. *Toxolasma parvum* caruncle enervation was examined and figured by Kraemer (1984).

Statocysts are paired organs located within the mussel foot that are believed to be associated with equilibrium. Kraemer (1978) documented two statocysts, widely separated from each other and attached to the nerve connecting the cerebral and pedal ganglia (figured by G.B. Simpson, 1884). More information on the enervation of the statocysts of *Lampsilis cardium*, including photomicrographs, was provided by Kraemer (1984).

Gills and Marsupia

The four gills of mussels serve a variety of functions, including respiration, sorting and transport of food particles and as a site for storing calcium. The ventral margin of the inner gills has a food groove lined with cilia that moves food particles anterior to the labial palps. Gill structure was illustrated by Ortmann (1911a, 1919), and more recently internal gill structure was illustrated by Kilias (1956). Water enters the branchial chamber through the incurrent aperture and moves into the gill filaments via ostia. Inside the gill it is moved from transverse canals into larger water channels (i.e., water tubes), where it is sent dorsally into the suprabranchial chamber and eventually exits through the excurrent aperture. Ortmann (1911a) and Heard (1975a) noted that nonmarsupial gills have less crowded and thinner septa than marsupial gills. Female anodontine species develop secondary water tubes when they become gravid (Ortmann, 1912a). These secondary water tubes develop prior to deposition of eggs into the marsupium and disappear one to two weeks after glochidial discharge (Richard et al., 1991;

Tankersley and Dimock, 1992). Kays et al. (1990) published a detailed study of the ultrastructure of gill water canals and water channels of *Ligumia subrostrata*. Way et al. (1989) provided photomicrographs of the lateral frontal cilia of various freshwater bivalves, including mussels. Kays et al. (1990) examined the histochemistry of gills and movement of water through the water canals. Dietz (1985) presented an overview of ion regulation in the gills. Gardiner et al. (1991) reported that gill ostia are controlled by musculature and that gills are major sites of ion regulation. Richard et al. (1991), Tankersley and Dimock (1992, 1993a) and Tankersley (1996) documented the structure and function of the marsupial gill with respect to water flow around developing glochidia. Gills store calcium in the form of calcified, chitinous rods or concretions that are mobilized during reproduction, and serve as a calcium reservoir during hypoxia (Silverman et al., 1983, 1985, 1987, 1989).

The brooding of developing embryos and glochidia has long been recognized as occurring in the gills. In the Unionidae marsupia occupy various parts of the gills, but the portion occupied is consistent within each genus. Shape and location of the marsupium was used extensively by Simpson (1900b, 1914) as the basis of his classification of unioniform bivalves, and generic groups were named based on gill morphology (e.g., Tetragenae for genera with all four gills marsupial, Homogenae for genera with marsupia occupying the entire outer gill). General terms used to describe marsupial placement are tetragenous (all four gills), ectobranchous (outer gills only) and endobranchous (inner gills only, which does not occur in the Unionidae or Margaritiferidae). Gill septa in marsupial areas are more closely spaced than those in nonmarsupial areas (Ortmann, 1910c, 1911a, 1912a; Utterback, 1915). Peck (1877) first recognized the dimorphism and illustrated the crowded septa of marsupial gills but made no mention of them in the text (Heard, 1975a).

Heart, Pericardium and Hemolymph

The heart is located dorsal to the visceral mass, positioned against the shell hinge plate, between the anterior and posterior adductor muscles. A detailed histological study of heart structure was produced by Krug (1922). The heart consists of a ventricle and two symmetrical lateral auricles, all enclosed in a pericardial sac (Narain and Singh, 1990; McMahon and Bogan, 2001). The lower intestine passes through the center of the heart and the pericardium. Location of the lower intestine within the ventricle enhances excretory efficiency by allowing elimination of unwanted substances from ventricular blood directly into the intestinal lumen (Narain and Sing, 1990).

Mussel hemolymph is moved by heart action. It exits the heart from two aortas, across the gill capillary beds and through the kidneys before returning to the heart (Willem and Minne, 1899; Schwanecke, 1913; Ellis et al., 1931; Dundee, 1953; Brand, 1972; Narain, 1972; Hazleton and Isenberg, 1977). The hemolymph does not contain corpuscles or specialized respiratory pigment but does contain amoebocytes (Bonaventura and Bonaventura, 1983). Ellis et al. (1931) studied mussel hemolymph and reported three different types of amoebocytes.

Byssal Threads

A byssal thread, also known as a byssus, is a thin, elastic strand that extends from the foot and apparently serves to anchor the juvenile mussel to the substrate (Sterki, 1891a, 1891b; Frierson, 1903b; Isely, 1911). Byssal threads may reach lengths of 75 mm, with branches on the distal end where attachment to the substrate occurs (Sterki, 1891a, 1891b). They apparently occur in all major groups of the Unionidae and Margaritiferidae, but the number of species that produce them is poorly understood (Ziuganov et al., 1994; Smith, 2000). They were first reported by Kirtland (1840) and Anthony (1841). Only *Medionidus* has been documented as producing or retaining byssal threads as adults (Lea, 1856c).

Smith (2000) histologically examined the unionid byssal gland, which is a spherical complex and is persistent throughout life, located inside the visceral mass posterior to the cerebral and pleural ganglia. It is suspended by muscles and connected with the outside via a stem canal. There are no retractor muscles associated with the gland complex. Byssal glands appear to be absent in *Pyganodon* (Smith, 2000).

Chapter 9
Shell Morphology and Soft Anatomy

Shell Morphology

Freshwater mussels have two valves held together along their dorsal margins by a flexible, proteinaceous hinge ligament. When the adductor muscles are relaxed, the hinge ligament pulls the valves apart ventrally. The hinge ligament is typically dark brown to black in color and varies among species in length and thickness.

Valves of freshwater mussels consist of four layers. The thin, uncalcified, proteinaceous outer layer, the periostracum, protects the underlying layers from erosion and dissolution. Beneath the periostracum is the calcified prismatic layer, which is composed of minute calcium carbonate prisms oriented perpendicular to the shell surface. The inner nacreous layer is thickest and is composed of calcium carbonate (aragonite) in an organic matrix. The fourth layer, the hypostracum, is found only beneath muscle attachments. Since muscles pass through the mantle and attach directly to the shell, they produce their own nacre, which makes up the hypostracum. As the shell grows the posterior adductor muscle migrates in a posterioventral direction and leaves a path of hypostracum imbedded in the nacreous layer. This is evident in some species, especially those with thin shells. Detailed accounts of shell composition and structure were provided by Coker et al. (1921), Taylor et al. (1969), Petit et al. (1980) and Checa (2000).

The periostracum and prismatic layers are produced only along the margin of the mantle. Therefore, external damage to the shell disk and umbo is not repaired. The nacreous layer is laid down by the entire mantle surface and makes up the bulk of the shell. Injuries penetrating the nacreous layer may be repaired by the mantle, as new nacre is secreted. All four layers are produced throughout the life of a mussel, but shell deposition, and thus growth, slows dramatically once an individual reaches sexual maturity. A series of darkened, concentric rings called annuli are present on the external surface of many mussels (Figure 9.1). These features are thought to represent growth rests and are often used to estimate the age of a given specimen. Accuracy of age estimations based on external annuli may be unreliable in older individuals, but more accurate estimations may be obtained by examination of shell thin-sections (Neves and Moyer, 1988).

Freshwater mussel species vary greatly in size. In North America the smallest species, *Toxolasma parvum*, barely exceeds 30 mm in length, while the largest species, *Megalonaias nervosa*, often exceeds 200 mm in length and may approach 280 mm. Herein shell length is defined as the distance between the anterior and posterior margins, measured parallel to the hinge ligament (measured approximately parallel to the hinge plate in alate species). Shell height is defined as the greatest distance between dorsal and ventral margins, measured near the midpoint of the hinge ligament, perpendicular to the length measurement. Shell width is the greatest distance between the outer surfaces of the two valves. Valve thickness is highly variable among species, ranging from less than 1 mm (e.g., *Utterbackia imbecillis*) to more than 20 mm (e.g., large *Megalonaias nervosa*).

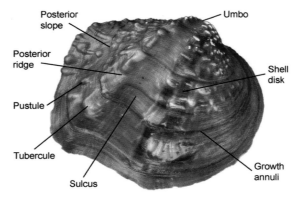

Figure 9.1. External morphology of a freshwater mussel shell.

Shell inflation refers to shell width in relation to height. It varies among species and among individuals within a species. This character is often important in identification, although the terms "inflated" and "compressed" are somewhat subjective.

Freshwater mussels are highly variable in shell morphology among species and among individuals within a species. The greatest variation within species is typically in individuals from different habitats and is expressed in shell shape, degree of inflation and amount of sculpture. Some species, primarily long-term brooders, are sexually dimorphic. In most long-term brooders dimorphism is expressed in degree of shell inflation posteriorly, with the female shell cavity expanded to accommodate gills filled with glochidia, which remain distended for much of the year. The most extreme cases of sexual dimorphism of the shell occurs in the genus *Epioblasma*, in which females have elaborate lobes or extrapallial swellings. In *Epioblasma* these are often erroneously referred to as marsupial swellings, but in fact accommodate modified mantle margins that are apparently used for glochidial host attraction displays and not marsupia. A wide variety of terms have been used to describe shapes of freshwater mussels. Descriptive shape terminology herein used is limited to the following: elliptical, oval (ovate), round (rotund), quadrate, rhomboidal, trapezoidal and triangular. Use of the prefix "sub" with any shape indicates that the shell outline approaches the shape being described.

Shell margins are important aspects of shape and are often significant in species identifications, but their boundaries are seldom defined. Herein margins were delineated based on lines established in relation to adductor muscle scars (Figure 9.2). A horizontal line positioned along the dorsal margins of the anterior and posterior adductor muscle scars served as a baseline. A vertical line perpendicular to the baseline was positioned along the interior margin of each adductor muscle scar. Shell margin outside of the anterior vertical line was defined as the anterior margin. Shell margin outside of the posterior vertical line was defined as the posterior margin. Dorsal and ventral margins lie between the two vertical lines. The umbo, which often projects above the dorsal margin, was not included as part of the dorsal margin description. Some species have one dorsal wing located posterior of the umbo (e.g., *Lasmigona complanata*), while a few have an additional dorsal wing anterior of the umbo (e.g., *Potamilus inflatus*). A poorly developed dorsal wing is present in some species (e.g., *Cyclonaias tuberculata*).

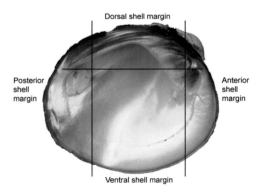

Figure 9.2. Lines that delineate the shell margins as herein applied.

Most freshwater mussels have a posterior ridge that extends diagonally from near the umbo to the posterioventral margin of the shell. The posterior ridge ranges from rounded to sharply angled. The portion of shell from the crest of the ridge to the posteriodorsal margin is referred to as the posterior slope (Figure 9.1). The posterior ridge of some species (e.g., *Epioblasma biemarginata*) may be doubled, resulting in a biangular margin distally (Figure 9.3). Some species have a shallow sulcus anterior of the posterior ridge, separating it from the remainder of the shell. Species with a wide posterior slope or dorsal wing may have a shallow sulcus posterior of the posterior ridge.

Figure 9.3. Doubled posterior ridge, resulting in a biangulate posterior margin on a male *Epioblasma biemarginata*.

The shell surface of freshwater mussels varies from smooth to highly sculptured. Sculpture includes small pustules or large tubercles (Figure 9.1), large knobs (Figure 9.4), corrugations (Figure 9.4) and plications (Figure 9.5). Shell sculpture is often confined to the posterior portion of the shell and may be most pronounced on the posterior ridge and slope. Shell sculpturing within a species often shows clinal variation from headwaters to large rivers.

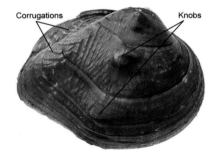

Figure 9.4. Corrugations and knobs on *Obliquaria reflexa*.

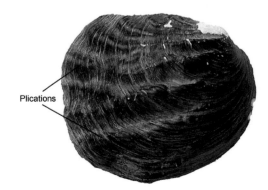

Figure 9.5. Plications on *Amblema elliottii*.

The shell umbo, sometimes referred to as the beak, is the oldest portion of the shell and is located at the anterior end of the hinge ligament (Figure 9.1). Its degree of inflation and elevation above the hinge line are highly variable among species. In some species the umbo is greatly inflated (e.g., *Lampsilis ornata*), whereas others have a compressed umbo (e.g., *Lasmigona alabamensis*). In some species the umbo is elevated well above the hinge line (e.g., *Pyganodon grandis*), but in others the umbo is barely elevated above the hinge line or not at all (e.g., *Utterbackia imbecillis*) (Figure 9.6). For species with umbos elevated above the hinge line, their orientation may be important in identification. Umbos exhibit minute sculpture comprised of concentric loops, ridges or corrugations, which may be adorned with small nodules, but this feature is often lost due to erosion (Figure 9.7).

Figure 9.7. Umbo sculpture, *Lasmigona complanata* (Clarke, 1985).

Figure 9.6. Umbo elevated above hinge line in *Pyganodon grandis* (upper shell) and flush with hinge line in *Utterbackia imbecillis* (lower shell).

Periostracum texture ranges from smooth to satiny, cloth-like or rough and may be important for identification. Texture often affects the luster of the shell surface, which may be shiny or dull. Periostracum color varies, ranging from yellow to green, brown or black, and is often adorned with darker rays, chevrons or other markings. Shell markings are typically some shade of green. Rays may be narrow or wide, continuous or interrupted, straight or wavy. The periostracum typically darkens with age, often obscuring shell markings.

The most prominent interior shell feature of most freshwater mussels is the hinge plate. It lies along the dorsal margin and includes pseudocardinal and lateral teeth (Figure 9.8). The teeth of one valve interlock with those of the other. They function to align the two valves when closing and to help counteract shear stress when closed. Hinge teeth vary greatly among taxa and are rudimentary (e.g., *Strophitus*) or absent (e.g., *Anodonta*) in a few genera. Pseudocardinal teeth are located anteriorly, typically under or just anterior of the umbo. They range from thick and triangular to thin and blade-like. Small accessory denticles may be present anterior and/or posterior to the pseudocardinal teeth, but their variability within species makes them of little use in identification. Lateral teeth lie posterior to the umbo and pseudocardinal teeth. They are elongate ridges, usually with two in the left valve and one in the right. Their thickness, length and degree of curvature may be helpful in identification. The area between the pseudocardinal and lateral teeth is the interdentum (Figure 9.8). It may be wide and flat (e.g., *Quadrula quadrula*) or very narrow, no wider than the thickness of the shell (e.g., *Potamilus ohiensis*). Length and width of the interdentum are often important in identification, but considerable variation is present in some species. Interdentum width is measured from dorsal to ventral and may be affected by extreme erosion of the umbo and hinge plate. Interdentum length is measured from the posterior end of the pseudocardinal teeth to the anterior end of the lateral teeth. Some genera (e.g., *Alasmidonta* and *Lasmigona*) have an interdental projection in the left valve.

The umbo cavity is a depression located inside the shell beneath the umbo (Figure 9.8). This feature may be deep and cave-like, especially in those species with a wide interdentum (e.g., *Fusconaia ebena*). Deep umbo cavities of some species are compressed with width greater than height. Umbo cavities are very shallow to almost nonexistent in some species (e.g., *Elliptio dilatata*).

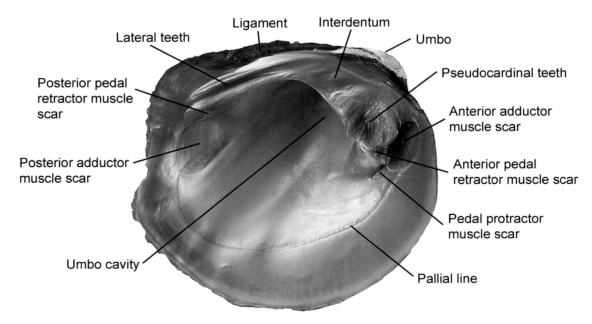

Figure 9.8. Internal morphology of a freshwater mussel shell.

Freshwater mussel valves have scars where muscles were attached (Figure 9.8). There are typically five prominent muscle scars in each valve, corresponding to the anterior and posterior adductor, anterior and posterior pedal retractor and pedal protractor muscles. The largest are the adductor muscle scars. Adductor muscle scars are located dorsally, with the anterior scar just ventral or anterior to the pseudocardinal teeth and the posterior scar just ventral to the posterior end of the lateral teeth. The anterior adductor scar is usually more deeply impressed than the posterior adductor scar and often has a rougher surface. The pedal protractor scar is small, located just posterioventral to the anterior adductor scar. The anterior pedal retractor scar is located adjacent to, and posterior of, the anterior adductor scar. The posterior pedal retractor scar is located adjacent to, and anterior or dorsal to, the posterior adductor scar. Pedal retractor and protractor scars may be entirely fused with their respective adductor muscle scars.

Groups of very small muscle scars formed by pallial and dorsal muscles are also present on the interior shell surface. These scars are in the form of small, shallow pits in the shell nacre. The pallial line is formed by pallial muscle scars (Figure 9.8). It originates just ventral to the anterior adductor scar and extends ventrally and posteriorly, lying concentric with, but removed from the ventral margin, and connects with the posterior adductor scar. Typically, individual scars of the pallial line are more deeply impressed anteriorly than posteriorly. Dorsal muscle scars are located in or near the umbo cavity and may be clearly visible or almost nonexistent. Dorsal muscle scars can be especially difficult to see in small specimens or in those with a deep umbo cavity. The Margaritiferidae have small but obvious mantle attachment scars on the inside of the shell disk between the anterior and posterior adductor scars, above the pallial line.

The pearly nacreous layer ranges from white or bluish white to shades of pink, salmon or purple. Nacre color may be consistent, making this feature helpful in species identification. However, some species are consistently white in portions of their ranges but variable in others (e.g., *Quadrula verrucosa* and *Obliquaria reflexa* are white in the Tennessee River drainage but white to reddish purple or purple in the Mobile Basin). Inner surfaces of valves are iridescent posteriorly, which tends to be much more apparent in species with white nacre.

Soft Anatomy

The body of a freshwater mussel is surrounded by a thin layer of tissue, the mantle, which lies flat against the inside of the shell (Figure 9.9). The mantle is attached to the shell by small dorsal and pallial muscles (Figures 9.10, 9.11). Dorsal muscles are located in the umbo cavity, and pallial muscles are located distally in a narrow line (pallial line) more or less concentric with, but removed from the ventral margin. Visceral mass is the term used for the collective assemblage of internal organs, including the digestive system, kidney and reproductive tissues, which are held within a thin sheath of muscle (Figure 9.9). The foot is a large, muscular, typically somewhat triangular organ situated anterioventrally, used for locomotion and/or anchoring in the substrate (Figures 9.9, 9.10, 9.11).

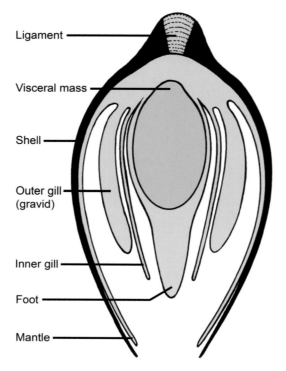

Figure 9.9. Schematic cross section of a freshwater mussel. Illustration by S. Trammell.

Anterior and posterior adductor muscles are located near the dorsal margin and serve to close the shell upon contraction. The much smaller anterior and posterior pedal retractor muscles are located near their respective adductor muscles. The pedal protractor is located near the anterior adductor muscle. The pedal protractor serves to extend the foot, and the pedal retractors withdraw the foot (Figure 9.10).

Freshwater mussels have one pair of demibranchs, often referred to as gills, on each side of the visceral mass (Figures 9.9, 9.11). Each demibranch is comprised of numerous long, thin filaments fused together to form sheets. Demibranchs of each side are joined along their dorsal margins. The proportion of the dorsal margin of each set of demibranchs that is connected to the visceral mass differs among species and within some species. Most are connected only anteriorly, being free for the remainder of the length of the visceral mass, then conjoined with the demibranchs of the opposite side posteriorly. Others are completely connected to the visceral mass. Dorsal gill margins are usually straight to slightly sinuous but may be slightly concave or slightly convex in some individuals. Ventral margins are

usually convex or elongate convex but may be slightly bilobed, especially the outer gills of long-term brooder females, in which the posterior lobe is typically marsupial.

The portions of gills used as marsupia vary among mussel taxa. Most species use outer gills only, but some use all four. Many of those that use only outer gills use only a portion of the gill. Gills of some species are greatly distended when gravid, whereas some are only slightly thickened (padded).

Labial palps are small, paired, usually somewhat triangular structures located on the left and right sides of each individual at the anterior end of the gills (Figure 9.11). They assist in food transfer from the gills to the oral groove, which leads to the mouth. Each pair of palps is fused basally and often partially fused dorsally.

Most unionoids have (from dorsal to ventral) supra-anal, excurrent and incurrent apertures (Figure 9.10). These structures are frequently incorrectly referred to as siphons. Mussel apertures function as siphons but differ morphologically. True siphons have mantle tissues fused on both ends of the opening, forming a tube, such as found in *Corbicula fluminea* (Figure 15.1). In margaritiferids and unionids mantle margins converge but are not fused, with the exception of the mantle bridge separating the excurrent and supra-anal apertures in most species. Incurrent apertures have papillate margins. Papillae vary among species and occasionally within species. Papilla types include simple, bifid, trifid and arborescent. Some species frequently have more than one type. Excurrent aperture margins may be papillate, crenulate or smooth. When papillae are present, they are much smaller than those of the incurrent aperture and are almost always simple. The supra-anal aperture margin is usually smooth but may be minutely crenulate in some species or individuals. The mantle wall within the apertures may differ in color from the remainder of the mantle. There is usually a marginal color band in the excurrent aperture. This band may be solid, mottled or with dark lines that are usually straight and perpendicular to the mantle margin but are sometimes interconnected to form a reticulated pattern. The excurrent and supra-anal apertures are usually separated by a mantle bridge connecting the mantle of the two sides (Figures 9.10, 9.11). The mantle bridge is of variable length, and the bridge in some individuals has irregular, small perforations.

Descriptions of gross soft anatomy of freshwater mussels were provided in a series of papers by Ortmann (1912a, 1913b, 1914, 1915, 1916a, 1921, 1923a, 1923b, 1924a) and Utterback (1915, 1916a, 1916b). There is a paucity of descriptive work on soft anatomy in recent literature.

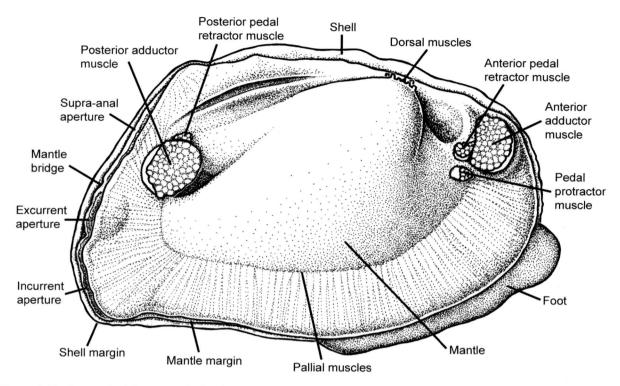

Figure 9.10. Anatomical features of a freshwater mussel with the right valve removed. Illustration by S. Trammell.

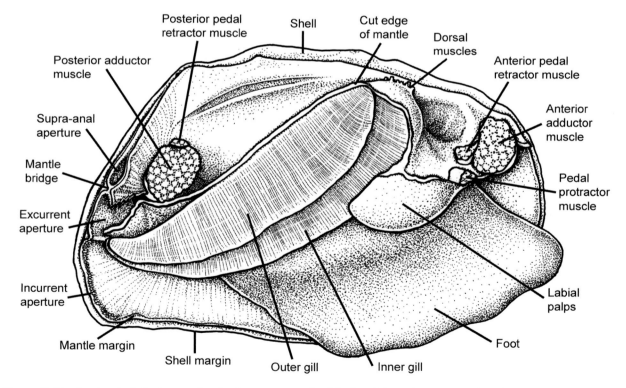

Figure 9.11. Anatomical features of a freshwater mussel with the right valve removed and the mantle cut away. Illustration by S. Trammell.

Chapter 10
Higher Classification of the Order Unioniformes

Background

Lea (1836, 1838b, 1852a, 1870) was first to devise a classification system for North American freshwater mussels. Simpson (1896, 1900b, 1914) revised and expanded the fourth edition of Lea's 1870 freshwater mussel synopsis. However, Simpson's classification departed from Lea's artificial scheme, which grouped species into very few genera based entirely on shell characters. Simpson recognized many of the genera proposed in the second half of the nineteenth century and introduced a number of new genera. Simpson (1896) based his initial classification on shell characters, but his larger descriptive catalog recognized two families, Unionidae Rafinesque, 1820, and Mutelidae Gray, 1847 (= Iridinidae Swainson, 1840), based on larval morphology as observed by von Ihering (1891). Unionidae have a glochidium larval stage, and Mutelidae have a lasidium larval stage. Simpson (1914) split the Unionidae into two subfamilies, Unioninae Rafinesque, 1820, which use the outer or all four gills as marsupia, and Hyriinae Swainson, 1840, which use only the inner gills as marsupia. The remainder of Simpson's classification, to the generic level, was based on position and shape of the marsupium. Ortmann (1912a, 1916a, 1916b, 1919) and Ortmann and Walker (1922a) relied more heavily on gross anatomy and proposed a slightly different classification of the order Unioniformes Rafinesque, 1820. The order was formerly named Unionoida Stoliczka, 1871.

Modell (1942), in his classification of freshwater bivalves, used umbo sculpture as the principal character on which he based his relationships and relegated shell characters, larval type and anatomical features to a secondary role. Subsequent publications of Modell (1949, 1964) represent continued efforts to refine his proposed classification.

Parodiz and Bonetto (1963) provided a classification of South American freshwater bivalves based on larval stage, with Unionoidea Rafinesque, 1820, having a glochidium and Mutelacea Gray, 1847 (= Etherioidea Deshayes, 1830), having a lasidium/haustorium. South American unioniforms were separated at the family level by Parodiz and Bonetto (1963) based on larval stage. They recognized that South American unionoids belong to the family Hyriidae Swainson, 1840. Worldwide, Parodiz and Bonetto (1963) recognized five families divided into two superfamilies, with the position of the Etheriidae Swainson, 1840, left in question because of a lack of information on the larval stage.

Haas (1969a) published a catalog of freshwater bivalve taxa of the world. However, his treatment of North American bivalves was 40 years out of date, relying on Frierson (1927) as the last comprehensive treatment. Haas (1969a) combined North American freshwater bivalves into a single superfamily and four families. Haas (1969a) recognized Margaritiferidae Henderson, 1929, but placed the hyriids as a subfamily of the Unionidae.

Haas (1969b) presented a list of fossil and recent genera of freshwater bivalves. There are differences between Haas (1969b) and Haas (1969a), in which he recognized 150 genera and 112 subgenera in 4 families, with the Unionidae divided into 6 subfamilies.

Starobogatov (1970) developed a freshwater mollusk classification that included the Unionoida (= Unioniformes). He introduced a large number of new higher taxa, recognizing the freshwater bivalves as part of the Order Actinodontida and splitting the Unionoida of others into 3 superfamilies, 12 families, 30 subfamilies and 37 tribes.

Heard and Guckert (1970) provided a higher classification of North American freshwater mussels based on marsupial characters, duration of brooding period and other reproductive characters. Although they included all of the currently recognized North American genera in their classification, failure to recognize multiple origins of the tetragenous condition led to problems with their classification. Heard and Guckert (1970) erected three new subfamilies and divided unionoids into the families Margaritiferidae, Amblemidae Rafinesque, 1820, Hyriidae and Unionidae, with the Margaritiferidae basal to the entire freshwater bivalve radiation.

Scarlato and Starobogatov (1979) united the Trigonoida Dall, 1889, and Unionoida in a single order Unioniformes. They recognized three superfamilies of unionoids, Etherioidea, Mullerioidea Deshayes, 1830, and Unionoidea. The superfamily Unionoidea included the recent families Margaritiferidae, Amblemidae, Unionidae, Lampsilidae von Ihering, 1901, and Hyriidae.

Skelton and Benton (1993) included the Unionoida under the order Trigonioida Dall, 1889, apparently following Scarlato and Starogatov (1979) but using a junior synonym, Trigonioida, in place of Unionoida Rafinesque, 1820 (Stoliczka, 1871). They did not recognize any superfamilies and only four families— Etheriidae, Margaritiferidae, Mutelidae and Unionidae. Skelton and Benton (1993) combined Mycetopodidae Gray, 1840, under Mutelidae and combined Hyriidae under Unionidae without explanation.

The classification herein adopted recognizes six families in the Unioniformes. Allocation of families into superfamilies is in a state of flux (Roe and Hoeh, 2003).

Order Unioniformes

The order Unioniformes Rafinesque, 1820 (= Unionoida Stoliczka, 1871), as currently recognized, contains approximately 800 species in 180 genera (Bogan, in press). Unioniformes has been split into two superfamilies, Unionoidea and Etherioidea, based primarily on larval morphology (Parodiz and Bonetto, 1963; Haas, 1969a, 1969b; Wächtler et al., 2001; Bogan, in press). There are six families in the Unioniformes, all restricted to freshwater. Five of the families—Hyriidae, Margaritiferidae, Unionidae, Iridinidae and Mycetopodidae—are supported as monophyletic clades at present, based on preliminary phylogenetic analyses including DNA sequences, shell morphology and anatomical data. The sixth family, Etheriidae, is paraphyletic (Bogan and Hoeh, 2000). Hyriids, margaritiferids, unionids, iridinids and mycetopodids typically have larvae that are obligate parasites on gills, fins and other external structures of hosts, usually fishes. Etheriids are presumed to have such a life cycle, but its larvae have never been reported. The name Unioniformes Rafinesque, 1820, was herein adopted over Unionoida Stoliczka, 1871, following the suggestion of Starobogatov (1991) and order level suffix usage (-iformes) in some other animal groups (e.g., fishes and birds). Unioniformes Rafinesque, 1820, is not newly introduced here but has been used for more than 20 years in the literature (e.g., Zatravkin and Starobogatov, 1984; Zatravkin and Bogatov, 1987; Zatravkin and Lobanov, 1987, 1989; Bogatov and Zatravkin, 1988; Starobogatov, 1992; Bogatov, 2001; Bogatov et al., 2002; Starobogatov et al., 2004).

Diagnosis

Restricted to freshwater; adults with paired valves; external ligament; valves nacreous with prismatic layer and periostracum, crystalline structure of valves aragonite; hinge teeth present or absent, when present represented by a combination of pseudocardinal and posterior lateral teeth or pseudotaxodont teeth; usually with two adductor muscles; larval stage typically parasitic, using fins, gills and other external structures of hosts, usually fishes.

Superfamily Unionoidea Rafinesque, 1820

Historically the superfamily Unionoidea has been composed of three families—Hyriidae, Margaritiferidae and Unionidae. However, recent phylogenetic analyses suggest inclusion of only Margaritiferidae and Unionidae in the superfamily (Roe and Hoeh, 2003). Unionoidea has a glochidium larval stage as opposed to the lasidium of Mycetopodidae and haustorium of Iridinidae. The Hyriidae also has a glochidial larval stage, but it is unclear if it is homologous to the glochidial stage of unionids and margaritiferids (Simpson, 1900b, 1914; Ortmann, 1912a; Parodiz and Bonetto, 1963; Haas, 1969a, 1969b; Wächtler et al., 2001).

Diagnosis

Ligamental notch located at posterior end of ligament, dorsal to lateral teeth, typically shallow and rounded ventrally as opposed to deep v-shaped notch in Etherioidea. Gills thin, often elongate, usually rounded ventrally, opposing gills fused posterior of foot, outer gill may be fused with mantle, inner gill may be fused with visceral mass; females with marsupia for developing glochidia occupying outer, inner or all four gills, or restricted to a portion of outer gill; mantle not fused to form true siphons.

If the family Hyriidae is found to belong to a superfamily other than Unionoidea, then the character of marsupial inner gills will be removed from the Unionoidea diagnosis.

Chapter 11
Explanation of Accounts

Family accounts for the Margaritiferidae and Unionidae include a brief introduction, diagnosis and general distribution, with a map. Accounts for each of the 43 genera that occur in Alabama and the Mobile Basin are also provided and contain an introduction, type species, diagnosis and synonymy.

Species accounts are included for all freshwater mussels known to occur, or formerly occur, in Alabama and the Mobile Basin of Georgia, Mississippi and Tennessee. Each species account was written to stand alone, which resulted in some repetition among accounts. Accounts are arranged in alphabetical order by genus, and species within a genus. The species accounts are assembled in the same sequence, and the format of the accounts is presented below.

Scientific and Common Names

Scientific names follow those recognized in Turgeon et al. (1998), except where subsequent taxonomic changes were published in peer-reviewed literature or supported by extensive comparison of museum material during the production of this volume (Table 11.1). Taxonomic changes based on recent genetic analyses were followed in cases that were supported by shell morphology, soft anatomy and/or zoogeographical evidence. Subspecies are discussed in the accounts of their respective species. Author and date of publication follow each scientific name in the heading of each account and are not repeated in the text of the account. The standard taxonomic practice of using parentheses around author and date for names originally described in a different genus was followed. Citations for all original descriptions of mollusks are in the Literature Cited. For nonmolluscan taxa, with the exception of glochidial hosts, author and date of publication are in the text, but citations for their original descriptions are not in the Literature Cited.

Table 11.1. Names of freshwater mussels herein used that are not in or deviate from those used by Turgeon et al. (1998). Taxonomic changes are explained in the genus and species accounts. Names preceding the equal symbol are those used in this volume. Names following the equal sign are those used in Turgeon et al. (1998).

Elliptio fumata = not in Turgeon et al. (1998)
Elliptio pullata = not in Turgeon et al. (1998)
Elliptio purpurella = not in Turgeon et al. (1998)
Fusconaia apalachicola = recently described species
Fusconaia burkei = *Quincuncina burkei*
Fusconaia rotulata = *Obovaria rotulata*

Hamiota altilis = *Lampsilis altilis*
Hamiota australis = *Lampsilis australis*
Hamiota perovalis = *Lampsilis perovalis*
Hamiota subangulata = *Lampsilis subangulata*
Lampsilis floridensis = not in Turgeon et al. (1998)
Lasmigona alabamensis = *Lasmigona complanata alabamensis*
Lasmigona etowaensis = not in Turgeon et al. (1998)
Obovaria haddletoni = *Lampsilis haddletoni*
Pleurobema athearni = recently described species
Pleurobema fibuloides = not in Turgeon et al. (1998)
Pleurobema hartmanianum = not in Turgeon et al. (1998)
Pleurobema stabilis = not in Turgeon et al. (1998)
Pleuronaia barnesiana = *Fusconaia barnesiana*
Pleuronaia dolabelloides = *Lexingtonia dolabelloides*
Ptychobranchus foremanianus = not in Turgeon et al. (1998)
Quadrula infucata = *Quincuncina infucata*
Quadrula kieneriana = not in Turgeon et al. (1998)
Quadrula nobilis = not in Turgeon et al. (1998)
Quadrula succissa = *Fusconaia succissa*
Quadrula verrucosa = *Tritogonia verrucosa*
Uniomerus columbensis = not in Turgeon et al. (1998)
Villosa umbrans = *Villosa vanuxemenis umbrans*

Common names also follow Turgeon et al. (1998) but are capitalized, as is the practice for some other groups of animals (e.g., birds, reptiles and amphibians). Capitalization helps avoid confusion by identifying standardized common names. For example, reference to a fragile papershell could apply to a number of thin-shelled species. However, by capitalizing Fragile Papershell, it can easily be recognized as the common name for *Leptodea fragilis*. Common names of nonmolluscan species are also capitalized.

Illustrations

A color photograph of the exterior of the right valve and the interior of the left valve are provided for each species. Shells are consistently positioned with dorsal margins of anterior and posterior adductor muscle scars on a horizontal plane. This resulted in some species appearing misaligned (e.g., *Lasmigona alabamensis*). Photographs of both sexes are presented for those species with pronounced sexual dimorphism. Species with subtle sexual dimorphism are illustrated with a single photograph. Additional photographs of some species are used to illustrate variation in shell morphology, nacre color or juvenile form. If available, locality data, collection date and museum catalog number are included for each figured specimen (note: for many specimens, collection locality and date are incomplete or absent). The distance measurements in the locality data of specimens illustrated are given as recorded on museum labels without conversion to the metric system. Museum acronyms are listed in the Institutional Abbreviations at the beginning of this book. Interpolated locality data for figured specimens are enclosed in brackets.

Shell Description

Shell descriptions are based on examination of specimens from Alabama and the Mobile Basin when possible. Shell descriptions are presented in telegraphic style and follow the same general sequence in each account.

Soft Anatomy Description

Soft anatomy descriptions are based on observations of fresh and preserved material. When fresh material was available, descriptions were made from sacrificed, unrelaxed individuals. Individuals were placed on their left sides and opened by severing the anterior and posterior adductor muscle groups. Observations were made on the right side of the animals. For protected, extremely rare or extinct species, previously preserved material was used. Preserved soft tissues are generally varying shades of tan, so colors of those specimens may differ from those observed in live individuals. A notation is provided at the end of the description for those based on preserved material. If no fresh or preserved material was available, published information was utilized when possible.

Fresh material was fixed in formalin and deposited in the North Carolina State Museum of Natural Sciences. Previously preserved material was examined from collections at the University of Alabama Unionid Collection, Tuscaloosa; Auburn University Museum; Florida Museum of Natural History, Gainesville; National Museum of Natural History, Leiden, Netherlands; and Ohio State University Museum, Columbus.

Soft anatomy descriptions are presented in telegraphic style and follow the same general sequence in each account. Because measurements mean little without context of dimensions of the individual, measurements of various structures are expressed as percentages of shell length or other appropriate structure (e.g., gill height given as a percentage of gill length).

Sample sizes and stream of origin are listed at the end of each soft anatomy description. Unless otherwise stated, specimens were collected in Alabama. County names are provided for specimens collected from creeks, since the same name has often been applied to

different creeks in disjunct parts of the state. For those species with only one specimen available, various measurements have been qualified with "approximate" since a range cannot be given. Because most material was collected from Alabama and adjacent states, wide-ranging species are represented by specimens from only a small portion of their ranges. Therefore, differences in observed characters from remote populations would not be unexpected. When characters differed from those reported in the literature, comments were added to the end of appropriate paragraphs.

Glochidium Description

Descriptions of glochidia are provided for species with published information or for which specimens were available. Glochidium description terminology, including measurements and shape, is based primarily on Hoggarth (1999) (Figure 11.1).

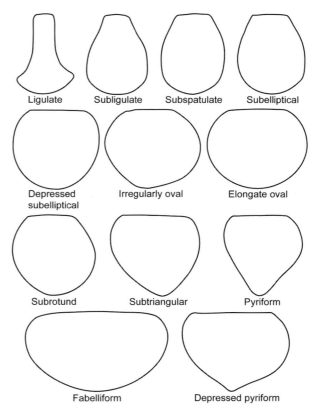

Figure 11.1. Glochidial shapes and terminology based on Hoggarth (1999). Illustration by S. Trammell.

Similar Species

A comparison with similar sympatric species is included in each account. Similar species may or may not be closely related. Characters given are primarily based on shell morphology, periostracum and nacre color, but obvious differences in soft anatomy are offered when relevant.

General Distribution

General distributions are based on written accounts combined with personal experience of the authors. Distributions are described generally from north to south. Within basins and along the Gulf Coast, distributions are given from east to west. Citations for species distributions are limited to those necessary to provide documentation for its range, and did not necessarily include all relevant citations.

Alabama and Mobile Basin Distribution

Distributions within Alabama and the Mobile Basin are based primarily on museum records and recent field work by the authors. Records were also extracted from published and other unpublished sources. A synopsis of recent (1990–2006) distribution is included in this section.

Ecology and Biology

General observations on habitat are given for most species based primarily on field experience of the authors. However, habitat preference of some species is poorly understood, especially those that became extinct during the early part of the twentieth century. For those species, inferences were made based on the streams where they occurred and what is known about closely related species.

Biological information was taken primarily from the literature. For most species, little is known beyond its period of gravidity and glochidial hosts. Glochidial hosts have been reported for various species based on observations of natural infestations as well as through laboratory trials. Reported natural infestations were typically not followed to excystment, so were never confirmed. These potential hosts are differentiated in the accounts. Glochidial hosts listed in the accounts represent the earliest reports for a given species, unless a subsequent report utilizing laboratory trials has confirmed its status as a host. For some species with no published biological information, inferences were made based on knowledge of closely related species.

Current Conservation Status and Protection

This section presents a review of conservation status based on published assessments. In order to give a brief history of the conservation status of each species, all assessments pertinent to Alabama are cited. State status is not given for Georgia, Mississippi or Tennessee, since the Mobile Basin includes only small portions of those states.

Remarks

This section covers a variety of subjects, including, but not restricted to, systematics and taxonomy, morphological differences among populations, distributional changes, archaeological records and erroneous records. Explanation of taxa names derived from Native American words is given. A brief biographical statement is presented for most patronyms.

Synonymy

The taxonomic history of freshwater mussels is fraught with difficulties, stemming from a high degree of shell variation within taxa coupled with overzealous and competitive efforts by early conchologists to describe the fauna. A primary synonymy—limited to all scientific names applied to a given species subsequent to its description, but not all of its generic combinations—is provided in each account. For species that were described from Alabama and the Mobile Basin, the first published illustration of the type specimen is included, except for those few in which the photograph quality was too poor for reproduction.

Distribution Map

In each account one of three base maps is utilized to illustrate distribution: 1) Alabama and the Mobile Basin; 2) Tennessee River drainage of Alabama; or 3) eastern Gulf Coast drainages of southern Alabama. Dots on the distribution maps represent compilations of museum records, recent field observations by the authors and written records. A single dot may represent multiple collections made in close proximity or collections from a given site on multiple dates. In some cases, when locality data for a specimen were imprecise a dot was placed on the map in the area from which it was most likely collected. For example, a museum record of *Pleurobema taitianum* collected from the Alabama River by Judge Tait was assumed to have been collected near Claiborne, Alabama, where he lived and collected extensively.

The distribution of a particular species portrayed on the map in its account may not match the major drainage information provided in Table 4.2. The only records of some species from a given drainage had incomplete locality data, preventing their inclusion on the map in their respective accounts. These were included in Table 4.2. Specimens reportedly collected from outside the accepted range of a given species are not plotted on the maps but are discussed in the Remarks of the respective species account.

Chapter 12
Family Margaritiferidae

The family Margaritiferidae Henderson, 1929, is Holarctic in distribution and currently represented by two genera, *Margaritifera* Schumacher, 1816, and *Cumberlandia* Ortmann, 1912b. These two genera occur in Alabama, each represented by a single species. Some authors recognize additional genera (e.g., Starobogatov, 1995; Smith, 2001). Historical problems due to various spellings of *Margaritifera* and Margaritiferidae were stabilized by the ICZN (1957: Opinion 495). However, the ICZN (1957: Opinion 495) gave the authorship of Margaritiferidae as Haas (1940) but overlooked the earliest use of the correctly spelled name by Henderson (1929). The type genus of the family is *Margaritifera* Schumacher, 1816.

Margaritiferidae has been considered the most primitive unioniform family (e.g., Ortmann, 1912a; Heard and Guckert, 1970; Roe and Hoeh, 2003). Fossil margaritiferids date from the Upper Cretaceous (Haas, 1969b; Watters, 2001).

Davis and Fuller (1981) suggested that Margaritiferidae be recognized as a subfamily of Unionidae. Smith and Wall (1984) and Smith (2001) argued for its family level status based on unique anatomical characters. Starobogatov (1995) proposed a revised classification of the family, recognizing nine genera. Smith (2001) proposed a revised classification of the family, recognizing three genera worldwide: *Pseudunio* Haas, 1910 (five species), *Margaritifera* (one species) and *Margaritinopsis* Haas, 1912 (six species). Neither Starobogatov (1995) nor Smith (2001) presented a phylogenetic analysis to support their proposed classification. Bogan et al. (2000) and Hoeh et al. (2001) used DNA analyses to provide evidence that Margaritiferidae is a monophyletic unit. Huff et al. (2004) reported the genus *Margaritifera* to be polyphyletic with *Cumberlandia* Ortmann, 1912b, sister to *Pseudunio auricularia* (Spengler, 1793) of southwestern Europe based on a phylogenetic analysis of DNA sequence data.

Diagnosis

Lateral mantle attachment scars present (Figure 12.1); hinge plate variably developed; umbo cavity shallow.

Diaphragm incomplete, formed by combination of gills and diaphragmatic septa on mantle lobes; gills not connected to mantle posteriorly; posterior end of labial palps separated from anterior end of inner gills by a large gap; incurrent and excurrent apertures poorly defined, excurrent and supra-anal apertures continuous; anus located on dorsal margin of posterior adductor muscle; gills lack water tubes, interlamellar connections irregularly scattered or form irregular rows oriented oblique to gill filaments, septa incomplete; all 4 gills marsupial; glochidium very small, without ventral hooks but with small, irregular teeth on ventral margin (Ortmann, 1912a, 1912b; Simpson, 1914; Haas, 1969b; Smith, 1983, 2001; Nezlin et al., 1994; Pekkarinen and Valovirta, 1996; Bogan, 2004).

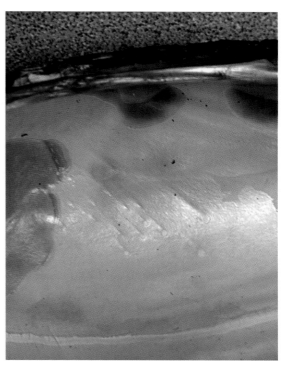

Figure 12.1. Mantle attachment scars on central portion of a margaritiferid valve. Photograph by A.E. Bogan.

Distribution

Two genera and five species of Margaritiferidae occur in North America (Turgeon et al., 1998). Both genera and two species are found in Alabama.

Margaritiferidae occurs in Morocco, northwestern Africa (van Damme, 1984), and from western Europe to northern Russia. In Asia margaritiferids are found in northeastern Myanmar, northwestern Vietnam, Japan and Siberia. In North America the family occurs along a portion of the western coast and sporadically in central, southeastern and eastern regions (Smith, 2001) (Figure 12.2).

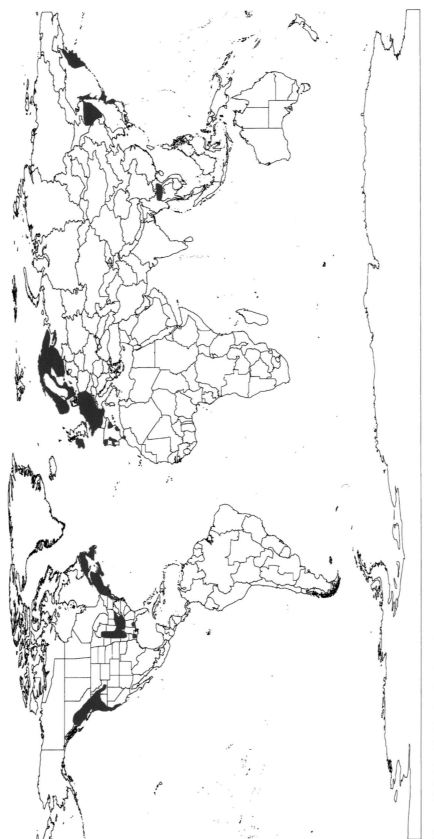

Figure 12.2. Worldwide distribution of the family Margaritiferidae.

Genus *Cumberlandia*

Cumberlandia Ortmann, 1912, is restricted to the Mississippi Basin, with the southeastern extent of its range in the Tennessee River drainages of Alabama. *Cumberlandia* is a monotypic genus (Turgeon et al., 1998).

Ortmann (1912b) erected the genus based on its gill septa, which are continuous and oriented oblique to gill filaments, as opposed to scattered and irregularly distributed as in *Margaritifera* Schumacher, 1816. *Cumberlandia* conglutinates are branched and feathery (Baird, 2000). Smith (2001) placed *Cumberlandia* in the synonymy of *Margaritinopsis* Haas, 1912, a genus from eastern Asia. However, recent genetic analyses do not support this proposed classification (Huff et al., 2004; A.E. Bogan and M.E. Raley, unpublished data). With taxonomic issues unresolved, the name *Cumberlandia* is herein retained.

Type Species
Unio monodonta Say, 1829

Diagnosis
Shell elongate; posterior ridge inflated; without sculpture; lateral teeth weak; lateral mantle attachment scars located centrally.

Interlamellar gill connections form irregular rows oriented oblique to gill filaments; mantle free around shell margin, not forming connections between the two mantle margins; incurrent and excurrent apertures elongate, supra-anal aperture absent; all four gills marsupial; glochidium very small, length 50–65 μm (Ortmann, 1912b).

Synonymy
None recognized.

Cumberlandia monodonta (Say, 1829)
Spectaclecase

 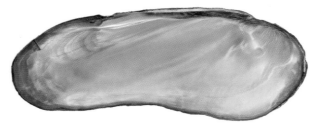

Cumberlandia monodonta – Length 138 mm, UF 370413. Tennessee River, river mile 255.8, Lauderdale County, Alabama, 7 June 1969. © Richard T. Bryant.

Shell Description

Length to 180 mm; thin when young, becoming thicker with age; somewhat compressed anteriorly, inflated along posterior ridge; outline elongate elliptical, arcuate; posterior margin rounded; anterior margin rounded; dorsal margin convex; ventral margin straight to concave; posterior ridge inflated, rounded; posterior slope flat; shallow radial depression anterior of posterior ridge; umbo low, may be elevated above hinge line, umbo sculpture well-developed concentric ridges; periostracum olive brown, darkening to black with age.

Pseudocardinal teeth small, tubercular, 1 in each valve in young individuals, older individuals with a single tooth in the right valve that fits into a depression in left valve; lateral teeth in young individuals moderately long, straight, 2 in left valve, 1 in right valve, lateral teeth of older individuals fused into thickened hinge line; umbo cavity shallow; nacre white, sometimes with a granular texture; lateral mantle attachment scars present centrally.

Soft Anatomy Description

Mantle creamy white to tan, outside of apertures grayish brown to rusty brown; visceral mass pearly white to creamy white; foot creamy white to tan, often darker distally. Lea (1863d) described the foot as being "at the anterior end" of the visceral mass, "not along the whole inferior part of the abdominal sack as usual".

Gills tan; dorsal margin straight to slightly sinuous, ventral margin straight to slightly convex; gill length 60–70% of shell length; gill height 20–25% of gill length; outer gill height 70–80% of inner gill height; gill dorsal margin only connected to visceral mass anteriorly; interlamellar septa continuous, obliquely oriented.

No gravid females were available for marsupium description. Ortmann (1912b) and Utterback (1915) reported that all four gills are marsupial.

Labial palps tan, usually with an irregular white area dorsally; elongate, straight to slightly convex dorsally, convex ventrally, usually bluntly pointed distally, occasionally rounded; palp length 20–30% of gill length; palp height 45–55% of palp length; distal 25–35% of palps bifurcate.

Continuous excurrent and supra-anal apertures considerably longer than incurrent aperture.

Incurrent aperture length 15–20% of shell length; creamy white within, rusty brown, gray or black basal to papillae, generally not presented in a regular pattern; papillae in 1 row, truncate, width variable, mostly wide, distal margins of larger papillae crenulate, occasional individuals without papillae on incurrent aperture, margin smooth but undulating; papillae usually some combination of creamy white, brown and gray.

Excurrent and supra-anal apertures continuous, not interrupted by mantle bridge; length 20–25% of shell length, occasionally greater; creamy white within, thin marginal color band dark brown to grayish brown or rusty brown; margin smooth to minutely crenulate. Ortmann (1912b) and Utterback (1915) reported the condition of continuous excurrent and supra-anal apertures as an absence of a supra-anal aperture.

Specimens examined: Tennessee River (n = 5).

Glochidium Description

Length 50–65 µm; height 52–65 µm; without styliform hooks; outline subrotund; dorsal margin straight, ventral, anterior and posterior margins convex (Howard, 1915; Surber, 1915; Baird, 2000).

Similar Species

Cumberlandia monodonta may superficially resemble *Ligumia recta*, but *C. monodonta* is usually more arcuate, has rudimentary pseudocardinal teeth and lateral teeth fused into a thickened hinge line in adults. *Cumberlandia monodonta* also has lateral mantle attachment scars, which are characteristic of the Margaritiferidae.

General Distribution

Cumberlandia monodonta occurs in the Mississippi Basin from southern Minnesota and Wisconsin (Cummings and Mayer, 1992) south to the Ouachita River drainage in south-central Arkansas, and in the Ohio River drainage from Ohio and West Virginia downstream to the mouth of the Ohio River, including some tributaries (Cicerello et al., 1991; Cummings and Mayer, 1992). *Cumberlandia monodonta* was historically widespread in the Cumberland River downstream of Cumberland Falls (Cicerello et al., 1991; Parmalee and Bogan, 1998) and is known from throughout the Tennessee River drainage (Neves, 1991; Parmalee and Bogan, 1998).

Alabama and Mobile Basin Distribution

Cumberlandia monodonta is confined to the Tennessee River drainage. It is known only from the Tennessee and Elk rivers.

Cumberlandia monodonta is now apparently restricted to Guntersville and Wilson Dam tailwaters.

Ecology and Biology

Cumberlandia monodonta inhabits medium to large rivers. It usually occurs in moderate to swift current. *Cumberlandia monodonta* is generally found under large flat rocks or in crevices among rocks but is occasionally encountered well buried in gravel substrates.

The life history of *Cumberlandia monodonta* in the Gasconade and Meramec rivers, Missouri, was reported by Baird (2000). That population had a 1:1 sex ratio, no sexual dimorphism in shell morphology and no hermaphroditic individuals. However, occasional hermaphrodites have been reported (van der Schalie, 1966). The age at sexual maturity in the Gasconade and Meramec rivers is four to five years for males and five to seven years for females. *Cumberlandia monodonta* is a long-term brooder. In the Gasconade and Meramec rivers females brooded eggs between September and December, and mature glochidia in April and May. Conglutinates are white, branched and feathery. Fecundity was reported to

range from approximately 2,000,000 to more than 9,000,000 (Baird, 2000). It has been suggested that *C. monodonta* may produce two broods per year (Howard, 1915; Gordon and Smith, 1990), but Baird (2000) reported no evidence of biannual reproduction.

Baird (2000) reported no successful transformations of *Cumberlandia monodonta* glochidia in laboratory host fish trials. However, natural infestations of *C. monodonta* were found on *Moxostoma macrolepidotum* (Shorthead Redhorse) (Catostomidae) and *Hybopsis amblops* (Bigeye Chub) (Cyprinidae) from the Meramec River, Missouri (Baird, 2000).

Current Conservation Status and Protection

Cumberlandia monodonta was listed as endangered throughout its range by Stansbery (1970a, 1971) and threatened throughout its range by Williams et al. (1993). In Alabama it was designated a species of special concern by Stansbery (1976), threatened by Lydeard et al. (1999) and a species of highest conservation concern by Garner et al. (2004). *Cumberlandia monodonta* was elevated to a candidate for protection under the federal Endangered Species Act in 2003.

Remarks

Cumberlandia monodonta was historically abundant in some reaches of the Tennessee River in Alabama. Hinkley (1906) reported as many as "two hundred being found under one slab". Wilson and Clark (1914) described a mussel bed in the Cumberland River in which *C. monodonta* was common until the species was nearly exterminated by being taken for fish bait.

Cumberlandia monodonta is rare in prehistoric shell middens in Alabama, with a single specimen reported from Muscle Shoals (Hughes and Parmalee, 1999). The scarcity of this species in middens may be due to its proclivity for living under large rocks, making it difficult to harvest, not an indication of prehistoric rarity.

Synonymy

Unio monodonta Say, 1829. Say, 1829:293; Say, 1830a:[no pagination], pl. 6
Type locality: Falls of the Ohio and Wabash rivers. Type specimen not found. Length not given in original description.

Unio soleniformis Lea, 1831. Lea, 1831:87, pl. 10, fig. 17; Lea, 1834b:97–99, pl. 10, fig. 17

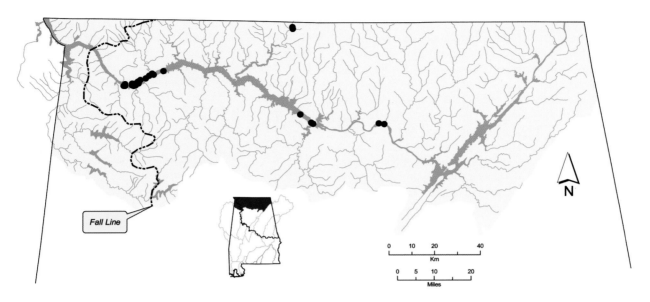

Distribution of *Cumberlandia monodonta* in the Tennessee River drainage of Alabama.

Genus *Margaritifera*

Margaritifera Schumacher, 1816, is Holarctic, with four species in North America (Turgeon et al., 1998). One species occurs in Atlantic Coast drainages, two in Gulf Coast drainages, and one in northern Pacific Coast drainages and a small portion of the northern Rocky Mountains. One widespread species, *Margaritifera margaritifera* (Linnaeus, 1758), is found both in Europe and northeastern North America. A single species is endemic to southern Alabama.

The spelling and authorship of *Margaritifera* were stabilized by the ICZN (1957: Opinion 495).

Type Species
Mya margaritifera Linnaeus, 1758

Diagnosis
Shell thick, compressed; lateral teeth variously developed; lateral mantle attachment scars located centrally on inside of shell.

Interlamellar gill connections scattered and irregularly distributed in indistinct rows oblique to filaments; mantle free around shell margin, not forming connections between the two edges; incurrent and excurrent apertures elongate, darkly pigmented, excurrent and supra-anal apertures continuous; gills without water tubes and septa; all four gills marsupial; glochidium very small, length 48–90 μm (Ortmann, 1912a; Simpson, 1914; Haas, 1969b; Nezlin et al., 1994; Pekkarinen and Valovirta, 1996).

Synonymy
Damalis Leach, 1847
Schalienaia Starobogatov, 1970

Margaritifera marrianae Johnson, 1983
Alabama Pearlshell

Margaritifera marrianae – Upper figure: length 86 mm, UF 269728. Hunters [Hunter] Creek, 8 miles southwest of Evergreen, Conecuh County, Alabama, February 1912. Lower figure: length 65 mm, UF 66135. Horse Creek near Luverne, Crenshaw County, Alabama, August 1915. © Richard T. Bryant.

Shell Description

Length to 96 mm; usually moderately thick, occasionally thin; moderately inflated; outline elliptical to trapezoidal, may be arcuate; posterior margin obliquely truncate to bluntly pointed; anterior margin rounded; dorsal margin slightly convex; ventral margin straight, may be slightly concave in older individuals; posterior ridge low, rounded; posterior slope flat to slightly concave; posterior 65% of shell sculptured with plications, increasing in size posterioventrally, some with reduced sculpture anterior of posterior ridge, plications radiate from posterior ridge onto posterior slope and disk, juveniles often with reduced sculpture; umbo broad, not inflated, elevated slightly above hinge line, umbo sculpture unknown; periostracum smooth, shiny on disk, somewhat roughened on posterior slope, dark olive to brown, becoming dark brown to black with age.

Pseudocardinal teeth blunt, triangular, moderately thick, 2 divergent teeth in left valve, 1 tooth in right valve; lateral teeth straight, moderately thick, 2 in left valve, 1 in right valve; interdentum moderately long, narrow; umbo cavity broad, shallow; nacre in juveniles somewhat bluish, white in adults.

Soft Anatomy Description

Mantle tan, outside of apertures dark brown with light brown bands perpendicular to margin; visceral mass creamy white or pearly white to tan; foot tan to light brown.

Gills tan to light brown; dorsal margin straight to slightly convex, ventral margin elongate convex; gill length 55–60% of shell length; gill height 45–55% of gill length; outer gill height 80–85% of inner gill height; gill dorsal margin only connected to visceral mass anteriorly; interlamellar gill connections present, more or less arranged in oblique rows.

No gravid females were available for marsupium description. Ortmann (1912a) reported members of the Margaritiferidae to characteristically use all four gills.

Labial palps tan; straight to slightly concave dorsally, convex ventrally, bluntly pointed distally; palp length 25–30% of gill length; palp height 35–50% of palp length; distal 20–30% of palps bifurcate.

Continuous excurrent and supra-anal apertures considerably longer than incurrent aperture.

Incurrent aperture length 15–25% of shell length; creamy white to tan within, dark brown, rusty brown or gray basal to papillae, may be mottled; papillae in 1 row, short, thick, simple and bifid, occasionally with few trifid, truncate distally, distal ends of some thicker papillae minutely crenulate; papillae usually some shade of brown, occasional individuals may have a few orange papillae.

Excurrent and supra-anal apertures continuous, not separated by mantle bridge or distinguished by marginal coloration, papillae or crenulations; length 30–35% of shell length; usually gray within, occasionally tan, marginal color band dark brown, may have lighter mottling; margin crenulate, crenulations often minute.

Specimens examined: Little Cedar Creek, Conecuh County (n = 4).

Glochidium Description

Glochidium unknown.

Similar Species

Superficially *Margaritifera marrianae* resembles *Quadrula verrucosa*. Both have plications on the posterior slope, but *Q. verrucosa* is usually heavily tuberculate over most of the shell while *M. marrianae* lacks tubercles and is typically without sculpture anteriorly. The posterior ridge of *M. marrianae* is low and rounded, while that of *Q. verrucosa* is prominent. These two species are only sympatric in a few tributaries of the Alabama River in Monroe County, Alabama.

General Distribution

Margaritifera marrianae is known only from a portion of the Conecuh River drainage in Conecuh and Crenshaw counties, Alabama, and localized populations in the Mobile Basin.

Alabama and Mobile Basin Distribution

In the Conecuh River drainage *Margaritifera marrianae* is known from several tributaries in Conecuh and Crenshaw counties. Mobile Basin populations are localized, occurring primarily in the Alabama River drainage, where they are restricted to the Big Flat and Limestone Creek systems, Monroe County. An additional population is known from the Buttahatchee River, Marion County, in the Tombigbee River drainage.

Margaritifera marrianae appears to be extant only in two small Conecuh River tributaries in Conecuh County (Pilarczyk et al., 2006).

Ecology and Biology

Margaritifera marrianae inhabits headwater streams with slow to moderate current in mixed sand and gravel substrates. It can occasionally be found in sandy mud. Frierson (1927) incorrectly reported "*Margaritifera hembeli*" (= *M. marrianae* in part) as occurring "only in water free from lime." To the contrary, *M. marrianae* often occurs in streams with abundant limestone and groundwater inflow. *Margaritifera marrianae* has been reported to occur in male-female pairs, with the male upstream of the female (Shelton, 1997). This was determined based on sexual dimorphism in shell morphology. However, other workers discount the visual recognition of sexual dimorphism in *Margaritifera* (R. Araujo, personal communication).

Margaritifera marrianae appears to be a long-term brooder. Gravid females have been reported in December (Shelton, 1997). Its glochidial hosts are unknown.

Current Conservation Status and Protection

Margaritifera hembeli (= *Margartifera marrianae* in part) was considered endangered throughout its range by Athearn (1970) and Stansbery (1971) and in Alabama by Stansbery (1976). Subsequent to recognition of *M. marrianae* as a valid species, it was recognized as endangered throughout its range by Williams et al. (1993) and in Alabama by Lydeard et al. (1999). Garner et al. (2004) designated *M. marrianae* a species of highest conservation concern in the state. It was elevated to a candidate for protection under the federal Endangered Species Act in 2003.

Remarks

After a review of the systematics and distribution of recent Margaritiferidae, Smith (2001) proposed a realignment of the species into three genera. This realignment resulted in *marrianae* being moved from the genus *Margaritifera* to *Pseudunio*. However, recent genetic analyses do not support this proposed classification (A.E. Bogan and M.E. Raley, unpublished data). With the matter not clearly resolved, the combination *Margaritifera marrianae* is herein retained.

The *Margaritifera marrianae* population in the Buttahatchee River is known from a single specimen, collected in 1909 by H.H. Smith. The specimen is in the Florida Museum of Natural History.

Margaritifera marrianae is a patronym for Marrian Geer Gleason Johnson, wife of Richard I. Johnson.

Synonymy

Margaritifera marrianae Johnson, 1983. Johnson, 1983:299–304, pl. 41 [original figure not herein reproduced]
Type locality: Hunters [Hunter] Creek, 8 miles southwest of Evergreen, Conecuh County, Alabama. Holotype, MCZ
    28491, length 83 mm.

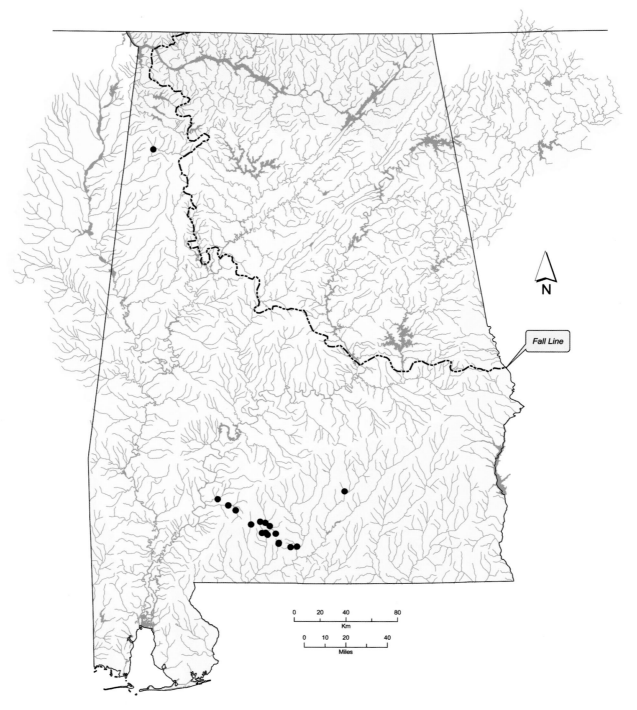

Distribution of *Margaritifera marrianae* in Alabama and the Mobile Basin.

Chapter 13
Family Unionidae

Globally there are approximately 142 genera recognized in the Unionidae Rafinesque, 1820 (Bogan, in press). In Turgeon et al. (1998), the Unionidae in North America, north of Mexico, was reported to be composed of 49 genera and 299 species and subspecies. Since publication of Turgeon et al. (1998) three genera—*Lexingtonia* Ortmann, 1914, *Quincuncina* Ortmann *in* Ortmann and Walker, 1922, and *Tritogonia* Agassiz, 1852—have been synonymized. One genus, *Pleuronaia* Frierson, 1927, has been elevated from synonymy, and one new genus, *Hamiota* Roe and Hartfield, 2005, has been described. Alabama has 176 species in 41 genera. The unionid fossil record dates from the Triassic (Haas, 1969b; Watters, 2001). The type genus of the family is *Unio* Retzius, 1788.

Phylogenetic relationships of many unionid genera are poorly understood at this time (McMahon and Bogan, 2001). There have been various organizations of genera within the Unionidae. Most classifications do not agree entirely but have some overlap, ranging from the conservative interpretation of Haas (1969a) to the overly inflated classification of Starobogatov (1970). Some classifications focus solely on North American genera which limits their utility. Recognition of subfamilies, if they exist, must consider the entire family and its worldwide distribution. Preliminary molecular analyses do not support any of the subfamily divisions proposed to date. For this reason subfamilies are not herein included. Concepts of groups such as the Lampsilinae, Anodontinae and Ambleminae are useful hypotheses but have no validity unless supported by a worldwide phylogenetic analysis, which should include conchological, anatomical, life history and genetic data.

Diagnosis

Shell valves approximately equal; external surface often sculptured; umbo typically sculptured; umbo cavity deep to very shallow; usually with 2 pseudocardinal teeth and 2 lateral teeth in left valve and 1 pseudocardinal tooth and 1 lateral tooth in right valve; hinge plate and teeth may be reduced or absent.

Diaphragm complete, formed only by gills; mantle margins drawn together by diaphragm but not united; incurrent and excurrent apertures separate, excurrent aperture usually closed dorsally by mantle bridge, typically with distinguishable supra-anal opening dorsal to excurrent aperture; anus typically located on posterior margin of posterior adductor muscle; gills always with water tubes, formed by interlamellar connections developed as continuous septa running parallel to gill filaments; females with marsupia for developing glochidia occupying outer or all 4 gills, or restricted to portion of outer gills; glochidia of various shapes and sizes, with or without ventral hooks (Ortmann, 1912a; Simpson, 1914; Haas, 1969b).

Distribution

The Unionidae has primarily a Holarctic distribution, including most of Europe and North America (Figure 13.1). It is also known from portions of northern and sub-Saharan Africa and northwestern Madagascar (Pilsbry and Bequaert, 1927; van Damme, 1984; Mandhal-Barth, 1988; Appleton, 1996; Daget, 1998). The family is represented throughout most of Asia, including the islands of Japan, the Philippines and Borneo, east to Java, but is absent from Australia, New Zealand and New Guinea (Haas, 1969a; Subba Rao, 1989; Bishai et al., 1999; Bába, 2000). In North America unionids occur from Alaska to Nova Scotia and south to Panama (Haas, 1969a).

Genera *Incertae Sedis*

Bariosta Rafinesque, 1831

Rafinesque described one species, *Bariosta ponderosus* Rafinesque, 1831, in the genus and referred *Obliquaria sinuata* Rafinesque, 1820, to it but did not designate a type species. The type locality for *B. ponderosus* was listed as "the lower Ohio and Mississippi." Herrmannsen (1846–1852) included *Bariosta* in his supplement but listed no species. Agassiz (1852) did not mention the genus or the species. Fischer (1886) listed *Unio ponderosus* (Rafinesque, 1831) as the type species of *Bariosta*. Simpson (1900b, 1914) listed *B. ponderosus* as an indeterminate unionid and erroneously included *Lampsilis diploderma* Rafinesque, 1831, and *Lampsilis? vittatis* Rafinesque, 1831, under *Bariosta*. Frierson (1914) listed *B. ponderosus* as the type species of *Bariosta* and claimed it was synonymous with *Unio crassidens* var. a Lamarck, 1819 (= *Unio trapezoides* Lea, 1831). Frierson (1914) stated that if *Bariosta* is different from *Amblema* Rafinesque, 1820, it would then belong in the synonymy of *Plectomerus* Conrad, 1853. Ortmann and Walker (1922a) discussed *Bariosta* and *B. ponderosus* under *Plectomerus trapezoides* Lea, 1831 (= *Plectomerus dombeyanus* (Valenciennes, 1827)). They claimed the species was not recognizable from the description and thus the genus was unrecognizable.

Frierson (1927) included *U. ponderosus* in the synonymy of *P. dombeyanus*. Haas (1969b) synonymized *Bariosta* with *Amblema* following Frierson (1927). No type specimen of *B. ponderosus* has been located (Johnson, 1973, 1980; Johnson and Baker, 1973). Johnson and Baker (1973) identified the lectotype of *O. sinuata* as *Elliptio dilatata* (Rafinesque, 1820). Due to the lack of type specimens and the inability to identify the type species, *Bariosta ponderosus* Rafinesque, 1831, and *Bariosta* Rafinesque, 1831, are herein considered *nomina dubia* and *Bariosta* a *nomen oblitum*.

Flexiplis Rafinesque, 1831

Flexiplis Rafinesque, 1831, was described as a subgenus of *Anodonta* Lamarck, 1799, in the remarks for *Anodonta digonota* Rafinesque, 1831, which is the type species by monotypy. Rafinesque listed the locality for *A. digonota* as Lake Erie. *Flexiplis* was not listed by Herrmannsen (1849–1852). Haas (1969a) placed *A. digonota* as a junior synonym of *Pyganodon cataracta* (Say, 1817). *Pyganodon cataracta* does not occur in Lake Erie, but *Pyganodon grandis* (Say, 1929) does and is variable in shell form. However, Haas (1969b) included *Flexiplis* as a junior synonym of *Lastena* Rafinesque, 1820. If the identification of the species *Anodonta* (*Flexiplis*) *digonota* Rafinesque, 1831, as a synonym of *P. grandis* is accepted, then *Flexiplis* is an unused senior synonym of *Pyganodon* Crosse and Fischer *in* Fischer and Crosse, 1894. However, *Anodonta digonota* Rafinesque, 1831, is herein considered a *nomen dubium* since it is not readily identifiable from the description and no type material is known. In this light, *Flexiplis* Rafinesque, 1831, is herein considered a *nomen dubium* and a *nomen oblitum*.

Gonamblus Rafinesque, 1831

Rafinesque (1831) erected this genus in the remarks for *Alasmodon ponderosum* Rafinesque, 1831, and stated that characters of the teeth should separate it from *Lasmigona* Rafinesque, 1831. *Alasmodon ponderosum* was the only species included in *Gonamblus* and is the type species by monotypy. *Gonamblus* was not listed by Herrmannsen (1846–1852). It was tentatively placed in the synonymy of *Plectomerus* Conrad, 1853, by Haas (1969b) based on a misunderstanding of Frierson (1927). No type specimens have been identified, and it is not recognizable from the description. Therefore, *Alasmodon ponderosum* Rafinesque, 1831, is herein considered a *nomen dubium* and *Gonamblus* Rafinesque, 1831, a *nomen dubium* and a *nomen oblitum*.

Leucosilla Rafinesque, 1831, *fide* Scudder, 1882

The genus name *Leucosilla* Rafinesque, 1831, was listed by Scudder (1882) and attributed to the compendium of Agassiz (1842–1846) but was never published by Rafinesque in any of his papers. It is a *nomen nudum* and a *nomen oblitum*.

Margarita Lea, 1836, *non* Leach, 1814 (Mollusca: Pteriidae)

Lea (1836) erected the genus *Margarita* in the first edition of his synopsis and included five subgenera but not a nominotypical subgenus. *Margarita* was not directly associated with any described species names, and Lea did not designate a type species for *Margarita*. This genus name is a junior primary homonym and is a *nomen nudum*.

Margaron Lea, 1852, *nomen novum* for *Margarita* Lea, 1836, *non* Leach, 1814

Lea (1852a) replaced the junior primary homonym *Margarita* Lea, 1836, with the new name *Margaron* Lea, 1852, in the third edition of his synopsis. He included seven subgenera with this name but did not include a nominotypical subgenus. Neither *Margarita* nor *Margaron* were used by Lea in direct association with a described species. *Margaron* was not listed by Herrmannsen (1849–1852). This replacement name as used by Lea is a *nomen nudum*.

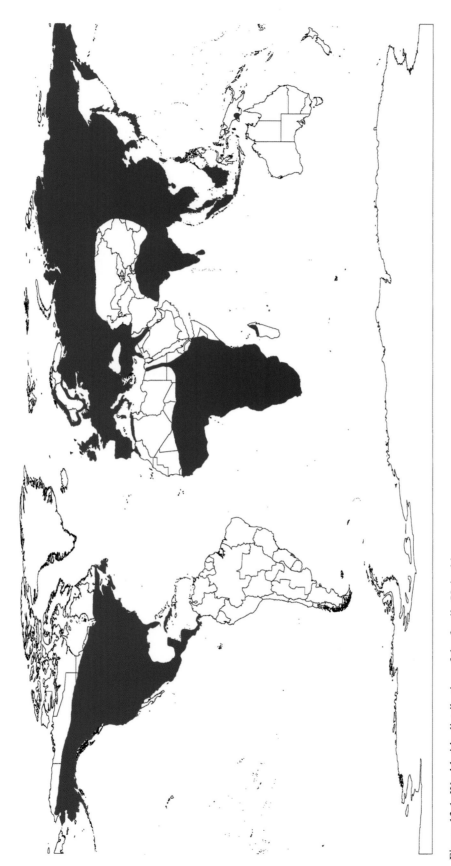

Figure 13.1. Worldwide distribution of the family Unionidae.

Genus *Actinonaias*

Actinonaias Crosse and Fischer *in* Fischer and Crosse, 1894, as herein used, includes two species found north of the Rio Grande (Turgeon et al., 1998). One species occurs in the Mississippi and Great Lakes basins, and one is endemic to the Tennessee and Cumberland River drainages. Both species are known from Alabama where they are restricted to the Tennessee River drainage. Several species, including the type species, are restricted to Mexico. Zoogeographic patterns and the disjunct distribution of *Actinonaias* species suggest that the genus is polyphyletic.

Haas (1969a) included seven Mexican species in *Actinonaias,* with the type species as *Unio sapotalensis* Lea, 1841, but placed *A. ligamentina* (Lamarck, 1819) and *A. pectorosa* (Conrad, 1834) in two different subgenera of *Lampsilis* Rafinesque, 1820. Ortmann (1911b) determined that *A. ligamentina* was not a *Lampsilis* based on soft anatomy.

Type Species
Unio sapotalensis Lea, 1841

Diagnosis
Shell oval to subelliptical; shell surface without sculpture; umbo positioned toward anterior end; sexual dimorphism subtle.

Inner lamellae of inner gills completely connected to visceral mass; outer gills marsupial; marsupium reniform in outline, restricted to posterior portion of gills, extended ventrally beyond original edge of gill when gravid; glochidium without styliform hooks (Ortmann, 1912a, 1919; Haas, 1969b).

Synonymy
Ortmanniana Frierson, 1927

Actinonaias ligamentina (Lamarck, 1819)
Mucket

Actinonaias ligamentina – Length 68 mm, UF 63705. Tennessee River, Lauderdale County, Alabama. © Richard T. Bryant.

Shell Description

Length to 155 mm; thick; compressed when young, becoming somewhat inflated with age; outline variable, from oval to elliptical; posterior margin rounded to bluntly pointed, females slightly more rounded than males; anterior margin broadly rounded; dorsal margin straight to slightly convex; ventral margin broadly rounded, almost straight in some individuals; posterior ridge low, rounded, angular near umbo; posterior slope steep dorsally, flattening posterioventrally; without sculpture; umbo broad, slightly inflated, elevated slightly above hinge line, umbo sculpture faint, irregular, double-looped ridges; periostracum smooth to irregularly roughened, tawny to greenish brown, darkening with age, often with broad, dark green rays that may become obscure in older individuals.

Pseudocardinal teeth thick, erect, striate, triangular, 2 divergent teeth in left valve, often separated dorsally, 1 tooth in right valve, often with accessory denticle anteriorly; lateral teeth thick, moderately long, straight to slightly curved, 2 in left valve, 1 in right valve; interdentum long, narrow; umbo cavity wide, moderately shallow; nacre white.

Soft Anatomy Description

Mantle tan, outside of apertures pale tan to rusty orange or grayish brown, mottled with gray or grayish brown, mottling usually in form of crude bands, perpendicular to mantle margin; visceral mass creamy white to tan; foot tan, often darker distally. Ortmann (1912a) reported the mantle margin just ventral to the incurrent aperture of females to be "slightly lamellate for a certain distance, with fine crenulations".

Gills tan; dorsal margin straight to slightly sinuous, ventral margin convex or elongate convex, outer gills may be slightly bilobed in females; gill length 50–60% of shell length; gill height 40–55% of gill length; outer gill height 70–95% of inner gill height; inner lamellae of inner gills completely connected to visceral mass.

Outer gills marsupial; glochidia held in approximately 50% of gill length, centrally located; marsupium padded when gravid; creamy white, distal margin gray. Lea (1863d) reported glochidia to be held in 66% of the outer gill, and Utterback (1916b) reported glochidia to be held across the entire gill. Lea (1863d) reported gravid marsupia to be "enormously distended" and extended well below the ventral gill margin. Ortmann (1912a) reported the distal edge of marsupial gills to be brownish or blackish but stated that distal coloration may be indistinct or lacking.

Labial palps creamy white to tan; straight to slightly convex dorsally, convex ventrally, bluntly pointed distally; palp length 20–30% of gill length; palp height usually 55–70% of palp length, occasionally greater; distal 25–50% of palps bifurcate.

Incurrent, excurrent and supra-anal apertures of variable lengths relative to one another.

Incurrent aperture length 10–15% of shell length; creamy white within, occasionally with grayish cast, usually with grayish brown to rusty orange basal to papillae, without pattern or with dark brown bands somewhat perpendicular to margin, converging proximally; papillae in 1–2 rows, mostly simple, often with a few bifid, occasionally with very few arborescent papillae; papillae usually some combination of tan, dark brown and rusty orange.

Excurrent aperture length 10–15% of shell length, occasionally less; usually creamy white within, occasionally grayish brown or with grayish cast, marginal color band usually pale tan to dark brown or rusty orange, usually mottled with dark brown that forms bands distally, bands perpendicular to margin; margin crenulate.

Supra-anal aperture length 10–20% of shell length; creamy white within, may have grayish or grayish brown cast, often with thin dark brown or grayish brown band marginally; margin smooth, but may be undulating; mantle bridge separating supra-anal and excurrent apertures imperforate, length 10–25% of supra-anal length.

Specimens examined: Clinch River, Tennessee (n = 3); Duck River, Tennessee (n = 3).

Glochidium Description

Length 220–232 µm; height 243–265 µm; without styliform hooks; outline subelliptical; dorsal margin slightly convex or undulate, ventral margin broadly rounded, anterior and posterior margins convex (Surber, 1912; Utterback, 1916b; Coker, et al., 1921; Merrick, 1930; Hoggarth, 1999). Jirka and Neves (1992) provided glochidial measurements that fall within ranges given here but the longer measurement was given as length and the shorter measurement as "width".

Similar Species

Actinonaias ligamentina resembles *Actinonaias pectorosa* but has a thicker shell, is more oval in outline and may be slightly less inflated. The periostracum of *A. ligamentina* often has uninterrupted rays, while those of *A. pectorosa* are usually interrupted.

Actinonaias ligamentina resembles male and young, uninflated female *Lampsilis abrupta*. However, *L. abrupta* has a steeper posterior slope. Also, the periostracum of *A. ligamentina* is typically greener and more heavily rayed than that of *L. abrupta*.

General Distribution

Actinonaias ligamentina occurs in some parts of the Great Lakes Basin, including portions of lakes Erie, Michigan and Ontario and some tributaries (La Rocque and Oughton, 1937; Dawley, 1947; Clarke, 1973, 1981a; Mathiak, 1979). It occurs in most of the Mississippi Basin from Minnesota (Dawley, 1947) south to Louisiana (Vidrine, 1993) and from headwaters of the Ohio River drainage in New York (Strayer et al., 1992) west to eastern Kansas (Murray and Leonard, 1962) and Nebraska (Hoke, 2000). *Actinonaias*

ligamentina is known from the Cumberland River drainage downstream of Cumberland Falls (Cicerello et al., 1991) and throughout the Tennessee River drainage (Ahlstedt, 1992a, 1992b; Parmalee and Bogan, 1998). There are several reports of *A. ligamentina* from the southern Hudson Bay Basin (Clarke, 1973). However, these records appear to have been based on misidentifications and were discussed by Clarke (1973) and Graf (1997).

Alabama and Mobile Basin Distribution

Actinonaias ligamentina is confined to the Tennessee River drainage. Most records are from the Tennessee River, but a few exist from tributaries. Ortmann (1925) reported it from lower Elk River, Limestone County.

A collection of weathered dead shells from Second Creek, Lauderdale County, is the only recent record of *Actinonaias ligamentina* from Alabama (S.W. McGregor, personal communication). In 1987 it was collected live from Shoal Creek just upstream of the Alabama and Tennessee state line in Lawrence County, Tennessee (J.T. Garner, personal observation).

Ecology and Biology

Actinonaias ligamentina typically inhabits large creeks and rivers, where it occurs in gravel and cobble substrates of shoals and runs. Ortmann (1919) described it as "best fitted for the rough parts, riffles with strong current and heavy gravel and rocks". It may also be found in sandy mud or gravel along stream margins. However, *A. ligamentina* can occur in some areas of large lakes (e.g., an embayment of Lake Michigan) (Coker et al., 1921). Baker (1898b) and Call (1900) suggested that it occurs in muddy habitats of rivers.

Actinonaias ligamentina is dioecious and has been reported to reach sexual maturity as young as four years of age (Jirka and Neves, 1992). In the New River, Virginia and West Virginia, gametogenesis begins during early autumn in both sexes, but spawning does not occur until the following July and August (Jirka and Neves, 1992). *Actinonaias ligamentina* is a long-term brooder, becoming gravid in August, with glochidia maturing by October and brooded until the following July (Conner, 1909; Ortmann, 1909b, 1919; Surber, 1912; Jirka and Neves, 1992). Overlapping periods of gravidity have been reported for *A. ligamentina* in the upper Mississippi River, with gravid females collected during every month except January (Coker et al., 1921). Glochidial discharge was reported to occur from the end of March until mid-May in the New River, Virginia and West Virginia (Jirka and Neves, 1992).

Actinonaias ligamentina has been reported to abort marsupial contents during unfavorable conditions, but not as readily as most short-term brooders (Lefevre and Curtis, 1912). Glochidial conglutinates of *A.*

ligamentina are white, elongate, broad, flattened and tapered to blunt points on both ends. They are held together by a clear mucilaginous matrix but lose their cohesion as glochidia mature (Lefevre and Curtis, 1912; Utterback, 1916b).

Actinonaias ligamentina is a glochidial host generalist. Fishes that have been demonstrated to serve as glochidial hosts in laboratory trials include *Ambloplites rupestris* (Rock Bass), *Lepomis cyanellus* (Green Sunfish), *Micropterus salmoides* (Largemouth Bass), *Pomoxis annularis* (White Crappie) and *Pomoxis nigromaculatus* (Black Crappie) (Centrarchidae); *Campostoma anomalum* (Central Stoneroller) and *Notropis buccatus* (Silverjaw Minnow) (Cyprinidae); *Morone chrysops* (White Bass) (Moronidae); and *Etheostoma tippecanoe* (Tippecanoe Darter), *Perca flavescens* (Yellow Perch) and *Sander canadensis* (Sauger) (Percidae) (Lefevre and Curtis, 1912; Coker et al., 1921; Watters et al., 1998). Additional fishes for which a natural infestation with *A. ligamentina* glochidia has been reported include *Lepomis macrochirus* (Bluegill) and *Micropterus dolomieu* (Smallmouth Bass) (Centrarchidae) (Coker et al., 1921). Coker et al. (1921) also reported natural infestations of *A. ligamentina* glochidia on *Anguilla rostrata* (American Eel) (Anguillidae) and *Noturus gyrinus* (Tadpole Madtom) (Ictaluridae) but deemed them to be insignificant. In a study of glochidial encystment, Young (1911) included *Lepomis humilis* (Orangespotted Sunfish) (Centrarchidae) and *Fundulus diaphanus* (Banded Killifish) (Fundulidae) in a suite of fishes used for potential hosts. Though *A. ligamentina* glochidia successfully encysted on *L. humilis* and *F. diaphanus*, it was unclear if transformation occurred.

Though the period of glochidial encystment has been reported to be as short as 13 days during spring and summer, it was reported to last as long as 16 weeks during an artificial infestation trial in winter (Lefevre and Curtis, 1912; Coker et al., 1921).

Lefevre and Curtis (1910b) were unable to find larval threads on *Actinonaias ligamentina* glochidia. However, juveniles with byssal threads as late as their second year have been reported (Utterback, 1916b; Coker et al., 1921). Larval threads of glochidia and byssal threads of juveniles are not believed to be homologous.

Current Conservation Status and Protection

Actinonaias ligamentina was reported as currently stable throughout its range by Williams et al. (1993). However, Stansbery (1976) and Lydeard et al. (1999) considered it endangered in Alabama, and Garner et al. (2004) designated it a species of highest conservation concern in the state.

Remarks

Actinonaias ligamentina occurred in the Tennessee River across north Alabama prehistorically, but archaeological evidence suggests that its abundance was variable. It was apparently widespread but not abundant in most of the Alabama reach (Morrison, 1942; Hughes and Parmalee, 1999), but comprised almost 8 percent of material recovered near Bridgeport in northeastern Alabama (Warren, 1975). Ortmann (1925) commented on the scarcity of *A. ligamentina* at Muscle Shoals compared to other areas within the Ohio River drainage. In more recent studies it was found to be very rare in tailwaters of Guntersville Dam (Isom, 1969). Bowen et al. (1994) reported it from Wheeler Reservoir but gave no indication of abundance. If extant in that reach, it is very rare (Garner and McGregor, 2001).

Actinonaias ligamentina in Alabama reaches of the Tennessee River is represented by a peculiar dwarfed form described as *Actinonaias ligamentina gibba* by Simpson (1900b). Another subspecies, *Actinonaias carinata* (= *ligamentina*) *orbis* was described from archaeological remains in northwestern Alabama by Morrison (1942). However, neither of these subspecies is currently recognized.

Actinonaias ligamentina was formerly included in the genus *Lampsilis* based on its close resemblance to *L. abrupta*. Ortmann (1911a) determined that "*Lampsilis ligamentina* is no *Lampsilis* at all" based on its lack of a mantle flap.

Actinonaias ligamentina was an important species in the pearl button industry during the late 1800s and first half of the 1900s. Wilson and Clark (1914) stated that *A. ligamentina* is "the most desirable form with which to stock the [Cumberland] river and extensive plantings from the falls to the mouth would greatly increase its value as a mussel stream."

Synonymy

Unio crassus Say, 1817, *non* Retzius, 1778. Say, 1817:[no pagination], pl. 1, fig. 8

Unio ligamentina Lamarck, 1819. Lamarck, 1819:72

Type locality: Ohio River. Holotype, MNHN, length 73 mm. Johnson (1969) restricted the type locality to the Ohio River, Cincinnati, Ohio.

Unio (Elliptio) fasciatus Rafinesque, 1820. Rafinesque, 1820:294

Unio (Elliptio) fasciatus var. *alternata* Rafinesque, 1820. Rafinesque, 1820:294

Unio (Elliptio) fasciatus var. *cuprea* Rafinesque, 1820. Rafinesque, 1820:294

Unio (Elliptio) fasciatus var. *nigrofasciata* Rafinesque, 1820. Rafinesque, 1820:294

Unio carinatus Barnes, 1823. Barnes, 1823:259, pl. 11, fig. 10
Unio ellipticus Barnes, 1823, *non* Rafinesque, 1820. Barnes, 1823:259, pl. 19, figs. 19a–d (outline figure)
Unio (Obliquaria) calendis Rafinesque, 1831. Rafinesque, 1831:3
Mya gravis Wood, 1856. Wood, 1856:199, pl. 1, fig. 6
Unio pinguis Lea, 1857. Lea, 1857c:84; Lea, 1858e:78, pl. 15, fig. 58; Lea, 1858i:78, pl. 15, fig. 58
Unio upsoni Marsh, 1880. Marsh, 1880:1
Lampsilis ligamentinus var. *gibbus* Simpson, 1900. Simpson, 1900b:540
Lampsilis ligamentina var. *nigrescens* Simpson, 1914. Simpson, 1914:82
Actinonaias carinata orbis Morrison, 1942. Morrison, 1942:361; Johnson, 1975:32, pl. 1, fig. 3
Type locality: [Tennessee River,] Florence, [Lauderdale County,] Alabama. Holotype, USNM 84998, length 66 mm.

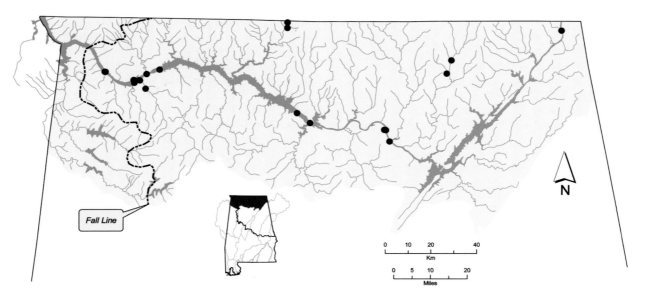

Distribution of *Actinonaias ligamentina* in the Tennessee River drainage of Alabama.

Actinonaias pectorosa (Conrad, 1834)
Pheasantshell

Actinonaias pectorosa – Length 76 mm, UF 63712. Tennessee River, Alabama. © Richard T. Bryant.

Shell Description

Length to 150 mm; thin when young, becoming moderately thick with age; moderately inflated; outline elliptical to oval; posterior margin narrowly rounded, may be obliquely truncate in young individuals; anterior margin narrowly rounded; dorsal margin straight to slightly convex; ventral margin straight to slightly convex; posterior ridge well-developed, rounded, less pronounced posterioventrally; posterior slope flat to slightly convex or concave; umbo broad, moderately inflated, elevated slightly above hinge line, umbo sculpture unknown; periostracum light olive to tawny, usually with wide, interrupted green rays, may become dark brown with age.

Pseudocardinal teeth moderately thick, erect, 2 somewhat triangular, slightly divergent teeth in left valve, 1 large, truncate tooth in right valve, older individuals often with well-developed accessory denticle anteriorly and posteriorly; lateral teeth short, thick, usually straight, occasionally slightly curved, 2 in left valve, 1 in right valve; interdentum moderately long, narrow to moderately wide; umbo cavity moderately deep, open; nacre white, may have salmon tint in umbo cavity.

Soft Anatomy Description

Mantle tan, outside of apertures pale rusty tan to rusty orange with grayish brown mottling in form of crude bands, perpendicular to margin; visceral mass pearly white to creamy white or light tan; foot tan, may be darker distally. Ortmann (1912a) reported the mantle margin just ventral to the incurrent aperture to be "slightly lamellar and indistinctly crenulated, and emphasized by a streak of black pigment".

Gills tan; dorsal margin slightly sinuous, ventral margin convex or elongate convex, outer gills may be slightly bilobed in females; gill length 50–60% of shell length; gill height 50–60% of gill length; outer gill height usually 70–90% of inner gill height; inner lamellae of inner gills completely connected to visceral mass.

Outer gills marsupial; glochidia held in posterior lobe of gill but not extending completely to posterior end, occupying approximately 35% of gill; marsupium somewhat triangular in outline, padded when gravid; creamy white, distal margin grayish brown.

Labial palps creamy white; straight to slightly concave dorsally, convex to deeply convex ventrally, bluntly pointed distally; palp length 20–30% of gill length; palp height 55–70% of palp length; distal 40–50% of palps bifurcate.

Incurrent aperture longer than or equal to excurrent aperture; supra-anal aperture longer or shorter than incurrent aperture.

Incurrent aperture length 15–20% of shell length; creamy white within, may have grayish cast, with band of rusty tan or rusty orange with brown mottling or specks basal to papillae; papillae in 2 rows, short, thick, blunt, mostly simple, with occasional bifid, some large papillae of inner row may be crenulate distally, approaching form of arborescent; papillae some combination of rusty orange, grayish brown and dark brown.

Excurrent aperture length 10–15% of shell length; creamy white within, sometimes with brownish or grayish cast, marginal color band rusty tan to rusty orange with brown specks or brown or grayish brown mottling in form of crude bands perpendicular to margin; margin crenulate.

Supra-anal aperture length 10–20% of shell length; creamy white to tan within, may have grayish cast, sometimes with rusty tan marginal band with grayish brown mottling that may be in form of crude bands perpendicular to margin; margin smooth; mantle bridge separating supra-anal and excurrent apertures imperforate, length 5–25% of supra-anal length.

Specimens examined: Clinch River, Tennessee (n = 3).

Glochidium Description

Length 244–253 µm; height 260–290 µm; without styliform hooks; outline subelliptical; dorsal margin straight, ventral margin broadly rounded, anterior and posterior margins convex (Ortmann, 1912a; Hoggarth, 1999).

Similar Species

The shell of *Actinonaias pectorosa* resembles that of *Actinonaias ligamentina*, but is thinner, usually more elongate and more inflated. The rays of *A. pectorosa* are usually interrupted, while those of *A. ligamentina* are often uninterrupted.

Small *Actinonaias pectorosa* may also resemble *Villosa taeniata* but have deeper umbo cavities. The posterior ridge of *A. pectorosa* is more pronounced than that of *V. taeniata*.

General Distribution

Actinonaias pectorosa was historically widespread in the Tennessee and Cumberland River drainages. It is known from most parts of the Cumberland River drainage but is confined to tributaries in middle and lower reaches of that system (Cicerello et al., 1991; Parmalee and Bogan, 1998). In the Tennessee River drainage *A. pectorosa* is known from headwaters in southwestern Virginia (Ahlstedt, 1992a, 1992b) downstream to Muscle Shoals (Ortmann, 1925), with a disjunct population in the Duck River, Tennessee (Parmalee and Bogan, 1998).

Alabama and Mobile Basin Distribution

Actinonaias pectorosa historically occurred across northern Alabama in the Tennessee River and some tributaries. However, all records from the Tennessee River proper are from Muscle Shoals. Isom and Yokley (1973) reported it from the Paint Rock River, but no museum records from that system were found.

There are no recent records or reports of *Actinonaias pectorosa* from Alabama. The most recent reported observation was from the Paint Rock River in 1965 (Isom and Yokley, 1973).

Ecology and Biology

Actinonaias pectorosa is primarily a species of large creeks to medium rivers. However, records from Muscle Shoals suggest its ability to occupy large river habitats under some circumstances. It usually occurs in shoals in sand and gravel swept clear of silt by current.

Actinonaias pectorosa is a long-term brooder, gravid from September through May (Ortmann, 1921). Fishes found to serve as its glochidial hosts in laboratory trials include *Ambloplites rupestris* (Rock Bass), *Micropterus dolomieu* (Smallmouth Bass), *Micropterus punctulatus* (Spotted Bass) and *Micropterus salmoides* (Largemouth Bass) (Centrarchidae); *Cottus carolinae* (Banded Sculpin) (Cottidae); and *Sander canadensis* (Sauger) (Percidae) (J.B. Layzer, personal communication). However, *A. rupestris* appears to be only a marginal host.

Current Conservation Status and Protection

Actinonaias pectorosa was listed as a species of special concern throughout its range by Williams et al. (1993). It was reported extirpated from Alabama by Lydeard et al. (1999) and Garner et al. (2004).

Remarks

Actinonaias pectorosa was collected from Muscle Shoals historically, but it was apparently rare in the Tennessee River proper. The only report of this species from prehistoric shell middens is from near Bridgeport in northeastern Alabama (Warren, 1975).

Synonymy

Unio pectorosus Conrad, 1834 (May). Conrad, 1834b:37, pl. 6, fig. 1

Type locality: Bank of Elk River, near its junction with the Tennessee [River], [Lauderdale County,] at Mussel [Muscle] Shoals, [Alabama]. Type specimen not found. Conrad (1834b) gave no measurement of the figured shell in the original description.

Unio perdix Lea, 1834 (August/September). Lea, 1834a:72, pl. 11, fig. 31; Lea, 1834b:184–185, pl. 2, fig. 31
Unio biangularis Lea, 1840. Lea, 1840:288
Unio biangulatus Lea, 1842. Lea, 1842b:197, pl. 9, fig. 6 [unjustified emendation of *Unio biangularis* Lea, 1840]; Lea, 1842c:35, pl. 9, fig. 35

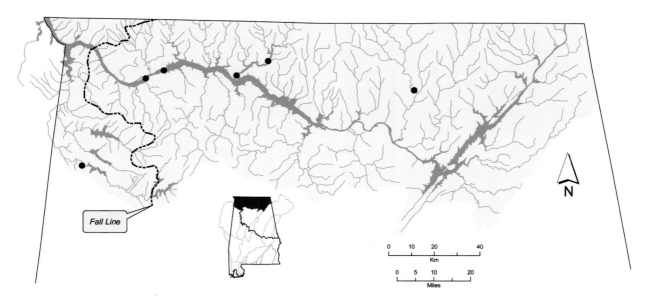

Distribution of *Actinonaias pectorosa* in the Tennessee River drainage of Alabama.

Genus *Alasmidonta*

Alasmidonta Say, 1818, occurs in the Great Lakes and Mississippi basins, Atlantic Coast drainages from Canada to Georgia, and Gulf Coast drainages in Florida, Georgia and Alabama. Twelve *Alasmidonta* species are recognized, all confined to North America (Turgeon et al., 1998). Four species are known from Alabama, one species each in the Apalachicola and Mobile basins, and two species in the Tennessee River drainage.

There are two additional specimens of *Alasmidonta* from the Mobile Basin in museum mollusk collections, one each in the ANSP and OSUM, both from the Etowah River, Georgia. One specimen, ANSP 126760 (Figure 13.2), is clearly not *Alasmidonta mccordi*. It was identified as *Alasmidonta varicosa* (Lamarck, 1819) by Clarke (1981b) with the notation

"locality probably incorrect" and no further explanation. This specimen, based on conchological characters, is more similar to *Alasmidonta triangulata* (Lea, 1858) of the Apalachicola Basin than to *A. varicosa*, which occurs in Atlantic Coast drainages. Based on the highly endemic mussel fauna of the Apalachicola Basin and the Coosa River drainage, this shell most likely represents an undescribed species endemic to the Mobile Basin.

The second specimen, OSUM 15113 (Figure 13.3), is most similar in appearance to *Alasmidonta mccordi*—the only recognized species of *Alasmidonta* from the Mobile Basin—and most likely represents that species. *Alasmidonta mccordi* was described based on a single, slightly misshapen specimen, which makes positive identification of this shell difficult.

Figure 13.2. *Alasmidonta* sp. Length 64 mm, ANSP 126760. Upper figure: outside right valve and inside left valve. Lower figure: inside right valve and outside left valve. Etowah River, Georgia, [1800s]. © Richard T. Bryant.

Figure 13.3. *Alasmidonta mccordi?* Length 44 mm, OSUM 15113. Etowah River, Georgia. Photograph by J.D. Williams.

Say (1818) erected *Alasmidonta* and considered it intermediate between *Unio* and *Anodonta* based on reduced dentition of its hinge plate. *Alasmidonta* is a replacement name for *Monodonta* Say, 1817, *non* Lamarck, 1799. This genus was included in a monograph of the tribe Alasmidontini by Clarke (1981b).

Type Species
Monodonta undulata Say, 1817

Diagnosis
Shell elliptical to rhomboidal; inflated; typically somewhat thin; posterior ridge well-developed; shell disk generally smooth, but posterior slope often sculptured with fine ridges; umbo sculpture coarse, often extending onto shell disk; pseudocardinal teeth poorly to moderately developed, lateral teeth rudimentary to absent, interdental projection on left valve with corresponding depression in hinge of right valve.

Soft tissues usually some shade of orange; inner lamellae of inner gills completely or partially attached to visceral mass; incurrent and excurrent apertures separate; outer gills with secondary water tubes; outer gills marsupial; marsupium distended when fully gravid; glochidium with styliform hooks (Ortmann, 1912a; Simpson, 1914; Haas, 1969a; Clarke, 1981b).

Synonymy
Monodonta Say, 1817, *non* Lamarck, 1799
Decurambis Rafinesque, 1831
Calceola Swainson, 1840, *non* Lamarck, 1799
Hemiodon Swainson, 1840
Uniopsis Swainson, 1840
Bullella Simpson, 1900
Pressodonta Simpson, 1900, *nomen novum* for
    Calceola Swainson, 1840
Rugifera Simpson, 1900
Prolasmidonta Ortmann, 1914
Jugosus Simpson, 1914
Alasmidens Clarke, 1981

Alasmidonta marginata Say, 1818
Elktoe

Alasmidonta marginata – Length 65 mm, UF 225812. Tennessee River, Tuscumbia, Colbert County, Alabama. © Richard T. Bryant.

Shell Description

Length to 79 mm; thin when young, becoming thick with age; inflated posteriorly; outline elongate rhomboidal; posterior margin obliquely truncate, bluntly pointed ventrally; anterior margin narrowly rounded; dorsal margin straight to slightly convex; ventral margin straight to slightly concave or slightly convex; posterior ridge high, sharply angled; posterior slope steep, sometimes adorned with weak ridges extending from posterior slope to dorsal margin at an oblique angle; umbo large, inflated, umbo sculpture heavy ridges, usually double-looped, angular on posterior ridge; periostracum yellowish to greenish brown, usually with numerous thin, dark green rays and darker green spots.

Pseudocardinal teeth rudimentary, thin, low, elongate, triangular when viewed in profile, 1 in left valve, 1–2 in right valve; left valve with prominent, irregular, compressed interdental projection; lateral teeth absent, but hinge line thickened; umbo cavity wide, moderately deep; nacre bluish white, may be tinged with pink or tan, especially in umbo cavity.

Soft Anatomy Description

Mantle creamy white, posterior 50% of mantle margin exterior surface with narrow band of irregular, somewhat rectangular, dark brown blotches; visceral mass creamy white; foot creamy white. Lea (1863d) reported the rectangular blotches on the exterior surface of the mantle to sometimes be fused into a continuous line ventrally. Utterback (1915) reported the foot to be orange and "very long and powerful".

Gills tan; dorsal margin straight, inner gill ventral margin elongate convex, outer gill ventral margin more deeply convex, somewhat acute distally; gill length approximately 70% of shell length; gill height approximately 45% of gill length; outer gill height equal to inner gill height; inner lamellae of inner gills completely connected to visceral mass.

No gravid females were available for marsupium description. Lea (1863d) reported outer gills to be marsupial, with glochidia held across the entire gill. Utterback (1915) and Clarke (1981b) reported gills to be greatly distended when gravid. Ortmann (1912a) reported gravid gills to range from yellowish white to brown, and Utterback (1915) reported them to be "bluish" when holding late embryos.

Labial palps creamy white; straight dorsally, convex ventrally, bluntly pointed distally; palp length approximately 15% of gill length; palp height approximately 65% of palp length; distal 50% of palps bifurcate. Clarke (1981b) reported labial palps to be pale brown.

Incurrent and supra-anal apertures of similar length. Excurrent aperture longer than incurrent and supra-anal apertures. Clarke (1981b) reported the supra-anal aperture to be much longer than incurrent and excurrent apertures and the incurrent aperture to be considerably longer than the excurrent aperture based on a single female specimen from the Clinch River, Tennessee.

Incurrent aperture length approximately 15% of shell length; creamy white within, brown basal to papillae; papillae in 2 rows, simple, short, thick, blunt; papillae grayish brown. Clarke (1981b) reported incurrent aperture papillae to be flattened and "partially fused along their length to the mantle edge".

Excurrent aperture length approximately 20% of shell length; creamy white within, marginal color band light brown, dark distally with a few slightly darker lines perpendicular to margin; margin minutely crenulate. Clarke (1981b) reported the excurrent aperture to be only 7% of shell length based on a single specimen. Utterback (1915) reported the excurrent aperture margin to have fine papillae.

Supra-anal aperture length approximately 15% of shell length; creamy white within, without marginal color band but coloration from outside of margin may show through translucent mantle; margin smooth; mantle bridge separating supra-anal and excurrent apertures imperforate, length approximately 35% of supra-anal length.

Specimen examined: Hiwassee River, Tennessee (n = 1) (specimen previously preserved).

Glochidium Description

Length 300–350 μm; height 346–380 μm; with short styliform hooks; outline pyriform; dorsal margin straight to slightly undulate, ventral margin narrowly pointed, anterior and posterior margins convex but becoming straight to slightly concave ventrally, posterior margin slightly more angular than anterior margin; larval thread present (Surber, 1912; Utterback, 1915; Ortmann, 1919; Coker et al., 1921; Clarke, 1981b; Hoggarth, 1999).

Similar Species

Alasmidonta marginata differs from *Epioblasma arcaeformis* and *Epioblasma triquetra* in attaining a larger size but having a thinner shell. *Alasmidonta marginata* has thin, lamellar pseudocardinal teeth, an interdental projection and rudimentary lateral teeth. Pseudocardinal teeth of *E. arcaeformis* and *E. triquetra* are more triangular, the lateral teeth are well-developed and they lack an interdental projection.

General Distribution

Alasmidonta marginata is known from eastern reaches of the Great Lakes drainage from the Ottawa River to Lake Michigan (La Rocque and Oughton, 1937; Burch, 1975a; Mathiak, 1979). It has also been reported from the Susquehanna and Hudson River drainages of the Atlantic Coast in New York and Pennsylvania (Ortmann, 1919; Clarke, 1981b). *Alasmidonta marginata* occurs in much of the Mississippi Basin from Wisconsin and Minnesota (Dawley, 1947; Mathiak, 1979) south to Arkansas (Vidrine, 1993), and from headwaters of the Ohio River drainage in New York (Strayer et al., 1992) west to eastern Kansas (Clarke, 1981b). It occurs in the Cumberland River drainage downstream of Cumberland Falls, Kentucky and Tennessee (Cicerello et al., 1991; Parmalee and Bogan, 1998; Cicerello and Laudermilk, 2001). *Alasmidonta marginata* is known from throughout the Tennessee River drainage in Alabama, Tennessee and Virginia (Ahlstedt, 1992a, 1992b; Parmalee and Bogan, 1998).

Alabama and Mobile Basin Distribution

Alasmidonta marginata occurred historically in the Tennessee River, but a paucity of records suggests that it was rare there. It was also found in several tributaries.

In 2004 a shell of *Alasmidonta marginata* was found in a muskrat midden on the Paint Rock River, Whitaker Nature Preserve, Jackson County, Alabama (D.N. Shelton, personal communication). Prior to that collection the most recent dated museum material was collected from Shoal Creek, Lauderdale County, in 1909.

Ecology and Biology

Alasmidonta marginata is primarily a species of large creeks to medium rivers, though records from Muscle Shoals indicate its ability to inhabit large rivers under some conditions (Ortmann, 1925). It may be found in a variety of substrates including combinations of silt, sand, gravel and cobble, but more commonly in gravel and cobble of shoal habitats (Ortmann, 1919; Buchanan, 1980).

Alasmidonta marginata is dioecious, though occasional hermaphrodites occur (van der Schalie, 1970). It is a long-term brooder, gravid from late July until the following June (Ortmann, 1909b, 1914, 1919; Utterback, 1915; Coker et al., 1921; Baker, 1928a). Fishes reported to serve as glochidial hosts for *A. marginata*, based on observations of natural infestations, include *Catostomus commersonii* (White Sucker), *Hypentelium nigricans* (Northern Hog Sucker) and *Moxostoma macrolepidotum* (Shorthead Redhorse) (Catostomidae); and *Ambloplites rupestris* (Rock Bass) and *Lepomis gulosus* (Warmouth) (Centrarchidae) (Howard and Anson, 1922).

Current Conservation Status and Protection

Alasmidonta marginata was listed as a species of special concern throughout its range by Williams et al. (1993) and in Alabama by Stansbery (1976) and Lydeard et al. (1999). Garner et al. (2004) classified it as extirpated from the state, though it was subsequently rediscovered in the Paint Rock River.

Remarks

Ortmann (1925) did not report *Alasmidonta marginata* from the Tennessee River proper downstream of Walden Gorge and suggested that its absence was due to its preference for smaller rivers. However, there was a single archaeological specimen reported from Muscle Shoals (Morrison, 1942).

Synonymy
Alasmidonta marginata Say, 1818. Say, 1818:459–460
Type locality: Scioto River, Chillicothe, Ohio.
Mya rugulosa Wood, 1828. Wood, 1828:182–183
Alasmodon (*Decurambis*) *scriptum* Rafinesque, 1831. Rafinesque, 1831:5
Margarita (*Margaritana*) *truncata* "Say" Lea, 1838. Lea, 1838b:135
Alasmidonta (*Decurambis*) *marginata susquehannae* Ortmann, 1919. Ortmann, 1919:187, pl. 12, fig. 4
Alasmidonta marginata var. *variabilis* F.C. Baker, 1928. F.C. Baker, 1928a:194, pl. 69, figs. 4–9

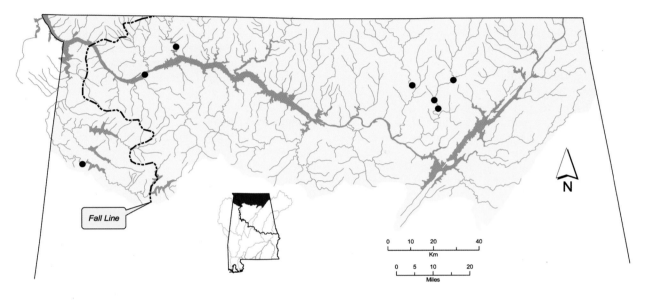

Distribution of *Alasmidonta marginata* in the Tennessee River drainage of Alabama.

Alasmidonta mccordi Athearn, 1964
Coosa Elktoe

Alasmidonta mccordi – Upper figure: outside right valve and inside left valve. Lower figure: inside right valve and outside left valve. Length 58 mm, CMNML 20094 (holotype). Coosa River, Ten Island Shoals, 3.2 miles south of Greenport, St. Clair County, Alabama, 2 August 1956. Note: The holotype is slightly misshapen. The shell anterior of the umbo is slightly depressed while the posterior one-third is expanded dorsally. © Richard T. Bryant.

Shell Description

Length to 58 mm; moderately thick; moderately inflated; outline oval (holotype slightly misshapen); posterior margin narrowly rounded; anterior margin rounded; dorsal margin slightly convex; ventral margin slightly convex; posterior ridge broadly rounded; posterior slope steep; umbo inflated, elevated above hinge line, turned slightly forward, umbo sculpture single-looped ridges; periostracum smooth, shiny, yellowish to greenish brown, with dark green rays of varying widths over entire disk.

Pseudocardinal teeth small, triangular, 1 in each valve (misshapen holotype with compressed pseudocardinal teeth); lateral teeth absent, but dorsal margin slightly thickened; interdental projection in left valve small, obscure; umbo cavity wide, moderately deep; nacre white.

Soft Anatomy Description

Soft anatomy unknown.

Glochidium Description

Glochidium unknown.

Similar Species

Alasmidonta mccordi bears a vague resemblance to *Lampsilis ornata*. However, *A. mccordi* is less inflated and has considerably less-developed pseudocardinal and lateral teeth.

General Distribution

Alasmidonta mccordi is known only from the Coosa River in Alabama.

Alabama and Mobile Basin Distribution

Alasmidonta mccordi is known only from its type, collected from the Coosa River, Ten Island Shoals, just below old Lock and Dam 2, 3.2 miles south of Greensport, St. Clair County, Alabama.

Ecology and Biology

The type specimen of *Alasmidonta mccordi* was collected alive from shallow, swift water in the Coosa River. The substrate was sand and gravel in an area thickly strewn with rock debris from Lock and Dam 2 (Athearn, 1964).

Alasmidonta mccordi was presumably a long-term brooder, gravid from late summer or autumn to the following summer. Its glochidial hosts are unknown.

Current Conservation Status and Protection

Alasmidonta mccordi was considered endangered by Athearn (1970) and Stansbery (1971, 1976). It was listed as extinct by Williams et al. (1993), Turgeon et al. (1998), Lydeard et al. (1999) and Garner et al. (2004).

Remarks

In addition to the type of *Alasmidonta mccordi*, there are two specimens of *Alasmidonta* from the Coosa River drainage, both from Etowah River, Georgia. However, shell characters of those individuals (shape, shell thickness and hinge plate) do not correspond exactly with those of *A. mccordi*, nor do they match each other (Figures 13.2, 13.3). The paucity of material, compounded by the poor condition of one individual and the misshapen shell of the *A. mccordi* holotype, precludes the specific identity of these specimens. Considering the large amount of shell material taken from the Coosa River drainage prior to destruction of its riverine habitat, it is surprising that more specimens of *Alasmidonta* were not collected.

In his systematic review of the tribe Alasmidontini, Clarke (1981b) was unable to align *Alasmidonta mccordi* with existing groups. He erected a new subgenus, *Alasmidens*, for this species based on conchological characters of the holotype, which is slightly misshapen.

Synonymy

Alasmidonta mccordi Athearn, 1964. Athearn, 1964:134, pl. 9, figs. a, b
Type locality: Coosa River, Ten Island Shoals, just below old Lock and Dam 2, 3.2 miles south of Greenport, St. Clair County, Alabama, 2 August 1956. Holotype, CMNML 20094, length 58 mm.

 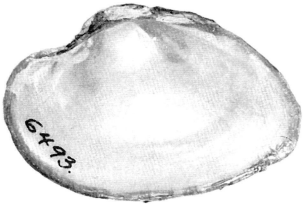

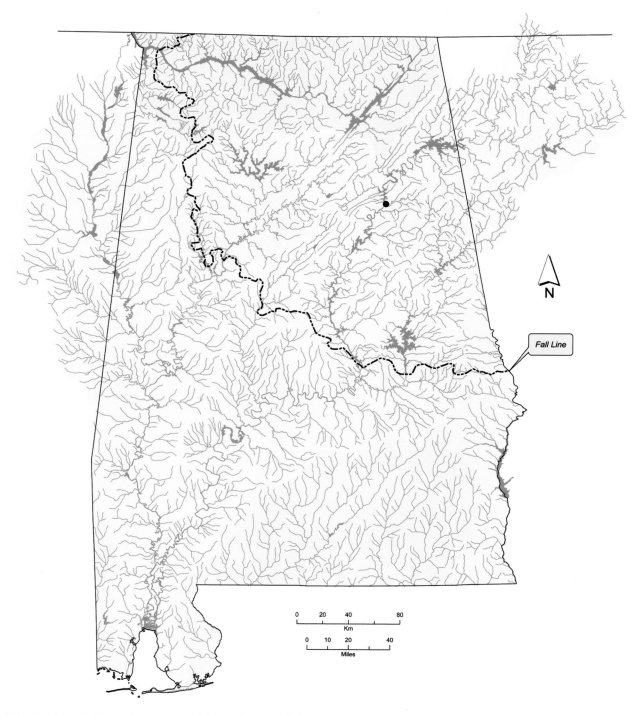

Distribution of *Alasmidonta mccordi* in Alabama and the Mobile Basin.

Alasmidonta triangulata (Lea, 1858)
Southern Elktoe

Alasmidonta triangulata – Upper figure: length 55 mm, MCZ 190391. Chipola River, Dead Lake, Chipola Park, 20 miles south of Blountstown, Calhoun County, Florida, 3 September 1954. Lower figure: length 39 mm, MCZ 254754. [Chattahoochee River,] Columbus, Muscogee County, Georgia. © Richard T. Bryant.

Shell Description

Length to 61 mm; moderately thin, thickened anteriorly; greatly inflated; outline subtriangular; posterior margin obliquely truncate, with blunt point posterioventrally; anterior margin rounded; dorsal margin straight; ventral margin convex; posterior ridge sharp; posterior slope steep, flat, often with fine wrinkles; umbo broad, triangular, inflated, elevated well above hinge line, umbo sculpture strong concentric ridges; periostracum yellowish brown to dark olive in young individuals, adults olive brown to black, some individuals with dark green rays that may become obscure with age.

Pseudocardinal teeth weak to moderately developed, 2 teeth in left valve, posterior tooth high, somewhat triangular, anterior tooth often low, rudimentary to well-developed, 1 subtriangular tooth in right valve, usually with flat crest, pseudocardinal tooth in right valve considerably larger than those in left valve; lateral teeth absent, but dorsal margin slightly thickened; interdental projection on left valve low, broad; umbo cavity wide, deep; nacre white.

Soft Anatomy Description

Mantle tan, with some rusty brown and grayish brown outside of apertures, a few irregular dark brown blotches on mantle exterior along posterior margin; visceral mass creamy white; foot pale tan.

Gills tan; dorsal margin straight, ventral margin convex; gill length approximately 65% of shell length; gill height approximately 50% of gill length; outer gill height equal to inner gill height; inner lamellae of inner gills completely connected to visceral mass.

No gravid females were available for marsupium description. Lea (1859f) reported outer gills to be marsupial, with glochidia held across the entire gill.

Labial palps pale tan; straight dorsally, convex ventrally, narrowly rounded distally; palp length approximately 40% of gill length; palp height approximately 70% of palp length; distal 50% of palps bifurcate.

Incurrent aperture longer than excurrent and supra-anal apertures; excurrent and supra-anal apertures of similar length.

Incurrent aperture length approximately 15% of shell length; pale tan and grayish brown within, pale rusty tan and dark brown basal to papillae; papillae simple, short, blunt; papillae pale rusty tan, some with dark brown edges basally.

Excurrent aperture length approximately 10% of shell length; pale tan within, marginal color band irregular, pale rusty tan and grayish brown; margin smooth but undulating.

Supra-anal aperture length approximately 10% of shell length; pale brown within, creamy white marginally; margin smooth; mantle bridge separating supra-anal and excurrent apertures imperforate, length approximately 25% of supra-anal length.

Specimen examined: Uchee Creek, Russell County (n = 1).

Glochidium Description

Glochidium unknown. Lea (1858d) provided a description of an "*Alasmidonta triangulata*" glochidium and commented that he "could not observe any hooks". Since hooks are characteristic of *Alasmidonta* glochidia, the description of Lea requires confirmation.

Similar Species

Alasmidonta triangulata has a distinctive shell but may superficially resemble some *Lampsilis binominata* in shape. However, the lateral teeth of *A. triangulata* are rudimentary compared to the well-developed lateral teeth of *L. binominata*. Also, the posterior ridge of *A. triangulata* is sharper and the periostracum is darker than those of *L. binominata*. The periostracum of *A. triangulata* is yellowish brown to olive brown or black, but that of *L. binominata* is always yellowish.

General Distribution

Alasmidonta triangulata is an Apalachicola Basin endemic, known from the Apalachicola, Chattahoochee, Chipola and Flint River drainages in Alabama, Georgia and Florida. Most records are from the Coastal Plain, but there are a few from the Piedmont in Georgia.

Alabama and Mobile Basin Distribution

Alasmidonta triangulata is restricted to the Chattahoochee River and some of its tributaries, where records are rare.

Alasmidonta triangulata is extant only in the Uchee Creek system, where it is rare.

Ecology and Biology

Alasmidonta triangulata inhabits large creeks and rivers in a variety of substrates, including various mixtures of mud, sand and gravel. It may be found in currents or pools.

Alasmidonta triangulata is presumably a long-term brooder, gravid from late summer or autumn to the following summer. Its glochidial hosts are unknown.

Current Conservation Status and Protection

Alasmidonta triangulata was reported as endangered throughout its range by Athearn (1970), Stansbery (1971) and Williams et al. (1993). In Alabama, it was listed as a species of special concern by Stansbery (1976), endangered by Lydeard et al. (1999) and a species of highest conservation concern by Garner et al. (2004).

Remarks

Alasmidonta triangulata was considered a synonym of *Alasmidonta undulata* by Clarke (1981b). However, Brim Box and Williams (2000) recognized it as valid and suggested that it is endemic to the Apalachicola Basin. Some question remains as to the taxonomic status of *Alasmidonta* in southern Atlantic Coast drainages since some individuals from the Savannah and Ogeechee River drainages are conchologically similar to *A. triangulata*.

Synonymy

Margaritana triangulata Lea, 1858. Lea, 1858b:138; Lea, 1859f:228, pl. 32, fig. 111; Lea, 1859g:46, pl. 32, fig. 111
Type locality: Upper Chattahoochee [River,] Georgia, Bishop Elliott; [Chattahoochee River,] Columbus, [Muscogee County,] Georgia, Dr. Boykin and J. Postell; Potato Creek, Georgia, Reverend G. White; Sawney's Creek, South Carolina, Dr. Blanding. Lectotype, USNM 86249, length 56 mm, designated by Johnson (1970), is from Columbus, Muscogee County, Georgia.

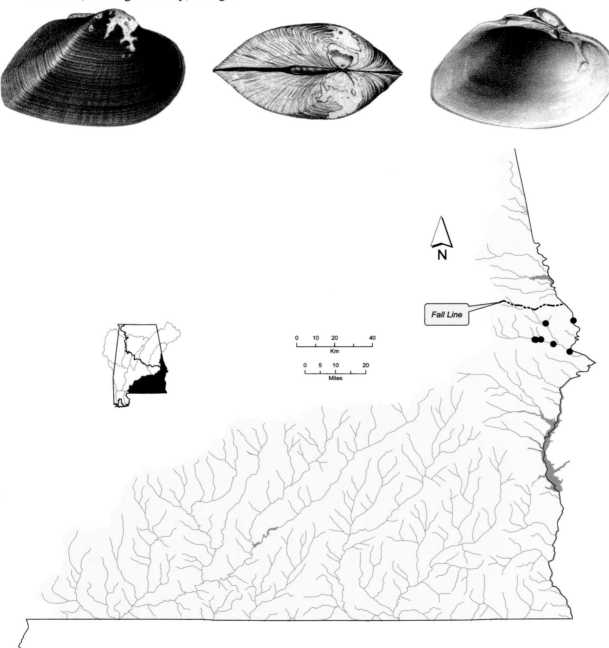

Distribution of *Alasmidonta triangulata* in the eastern Gulf Coast drainages of Alabama.

Alasmidonta viridis (Rafinesque, 1820)
Slippershell Mussel

Alasmidonta viridis – Length 39 mm, UMMZ 101327. Tuscumbia, Colbert County, Alabama. © Richard T. Bryant.

Shell Description

Length to 55 mm; moderately thin, thicker anteriorly; inflated; outline rhomboidal to trapezoidal; posterior margin obliquely truncate, with blunt point posterioventrally; anterior margin rounded; dorsal margin convex; ventral margin straight to slightly convex; posterior ridge high, rounded; posterior slope steep, flat to slightly concave; umbo moderately inflated, slightly elevated above hinge line, umbo sculpture heavy, irregular loops; periostracum greenish brown to tawny, with numerous wavy green rays, young individuals may be very pale, becoming darker with age, rays often become obscure.

Pseudocardinal teeth variable, 1–2 small, triangular teeth in left valve, anterior tooth may be rudimentary, 1 larger, raised, irregular tooth in right valve; lateral teeth absent, but dorsal margin thickened; interdental projection of left valve triangular, may be well-developed, may be fused with pseudocardinal tooth; umbo cavity moderately shallow; nacre dull white.

Soft Anatomy Description

Mantle tan, dark grayish brown outside of apertures, outside surface of mantle with dark brown margin in vicinity of apertures; visceral mass pale tan; foot tan. Clarke (1981b) reported the mantle to be grayish white, covered with a fine, reticulated pattern of small, pale orange-white patches. Lea (1863d) reported the mantle to be thickened marginally, with "numerous quadrate maculations [blotches] on the outer edges".

Gills tan; inner gill ventral margin almost straight, outer gill ventral margin convex; gill length approximately 60% of shell length; gill height approximately 50% of gill length; outer and inner gill height approximately equal; inner lamellae of inner gills attached to visceral mass only anteriorly. Clarke (1981b) reported gills to be pale yellowish brown.

No gravid females were available for marsupium description. Clarke (1981b) reported outer gills to be marsupial, thick, pad-like and pale brown when gravid. Lea (1863d) reported glochidia to be held across the entire outer gill.

Labial palps tan; slightly convex dorsally, convex ventrally, bluntly pointed distally; palp length approximately 15% of gill length; palp height approximately 70% of palp length; distal 70% of palps bifurcate.

Incurrent aperture longer than excurrent and supra-anal apertures; excurrent aperture longer than supra-anal aperture.

Incurrent aperture length approximately 20% of shell length; tan within, with irregular rusty tan basal to papillae; papillae in 2–3 rows, outer rows smaller, long, slender, simple; papillae pale rusty tan, some with gray edges.

Excurrent aperture length approximately 15% of shell length; tan within, margin with irregular dark brown lines, some parallel to margin; margin papillate, papillae small, simple; papillae grayish brown.

Supra-anal aperture length approximately 10% of shell length; tan within, with sparse grayish brown marginally; margin smooth; mantle bridge separating supra-anal and excurrent apertures imperforate, length approximately 150% of supra-anal length. Clarke (1981b) reported the mantle bridge to be short, but Ortmann (1912a) reported it to be almost as long as the excurrent aperture.

Specimen examined: Fowler Creek, Madison County (n = 1) (specimen previously preserved).

Glochidium Description

Length 286–320 µm; height 232–270 µm; with short styliform hooks; outline depressed pyriform; dorsal margin straight, ventral margin forming blunt point, anterior and posterior margins convex, anterior margin slightly more convex than posterior margin (Surber, 1912; Ortmann, 1914; Utterback, 1915; Clarke, 1981b; Hoggarth, 1999).

Similar Species

The shells of some *Alasmidonta viridis* resemble small *Strophitus undulatus*, but the former has better developed pseudocardinal teeth and a more pronounced posterior ridge. Some *A. viridis* are more swollen posteriorly than *S. undulatus*.

Alasmidonta viridis may resemble *Pegias fabula*. However, *P. fabula* has a biangulate posterior ridge that is usually adorned with irregular knobs, whereas the posterior ridge of *A. viridis* is not biangulate or knobbed. *Pegias fabula* is usually much more decorticated than *A. viridis*, sometimes completely devoid of periostracum.

General Distribution

Alasmidonta viridis is known to occur in the Great Lakes Basin from the Ottawa River west to Lake Michigan (Burch, 1975a; Mathiak, 1979; Clarke, 1981b). In the Mississippi Basin it occurs from southern Wisconsin (Cummings and Mayer, 1993) south to Arkansas (Harris and Gordon, 1990), and from the Ohio River drainage in Ohio (Cummings and Mayer, 1992) west to Missouri (Oesch, 1995) and Arkansas (Harris and Gordon, 1990). *Alasmidonta viridis* is found throughout the Cumberland River drainage, Kentucky and Tennessee (Parmalee and Bogan, 1998; Cicerello and Laudermilk, 2001). It is widespread in the Tennessee River drainage in Alabama, North Carolina, Tennessee and Virginia (Neves, 1991; Parmalee and Bogan, 1998).

Alabama and Mobile Basin Distribution

Alasmidonta viridis is limited to the Tennessee River drainage. There are a few museum specimens from the Tennessee River proper, collected before the river was impounded, but most do not have precise locality data. It is also known from several tributaries. Ortmann (1925) reported it from upper Elk River in Tennessee, suggesting that it probably occurred in the lower reaches of that river in Alabama.

The only recent reports of *Alasmidonta viridis* from Alabama are from Fowler Creek, Madison County, and Paint Rock River tributaries (McGregor and Shelton, 1995; Ahlstedt, 1998).

Ecology and Biology

Alasmidonta viridis is primarily a species of headwaters and tributary streams, though it has been reported from some medium to large rivers (Buchanan, 1980; Parmalee and Bogan, 1998) and lakes (Baker, 1928a). *Alasmidonta viridis* most often occurs in flowing water, in mixtures of sand and gravel swept free of silt by current. It may also be found associated with aquatic vegetation in mud and sand as long as current is continuous (Buchanan, 1980; Parmalee and Bogan, 1998). However, van der Schalie (1938c) reported *A. viridis* to occur in pools as well as riffles. It is often well-buried in the substrate (Baker, 1928a).

Alasmidonta viridis is a long-term brooder, gravid from September until the following spring or summer (Surber, 1912; Ortmann, 1914, 1921; Clarke, 1981b). They have been observed exposed on the substrate surface displaying a white mantle margin during January and February in the upper Little Tennessee River, North Carolina (S.A. Ahlstedt, personal communication). This behavior is presumably to attract glochidial hosts.

Observations of natural infestation of *Alasmidonta viridis* glochidia on *Cottus carolinae* (Banded Sculpin) (Cottidae) have been reported (Zale and Neves, 1982c). The glochidia were attached to the gills, which is atypical for anodontine species with hooked glochidia, which usually attach to the fins and tail of hosts. Clarke and Berg (1959) reported *A. viridis* glochidia to use *Cottus bairdii* (Mottled Sculpin) (Cottidae) and *Etheostoma nigrum* (Johnny Darter) (Percidae) as glochidial hosts, citing J.P.E. Morrison, personal communication, but giving no details.

Current Conservation Status and Protection

Alasmidonta viridis was listed as a species of special concern throughout its range by Williams et al. (1993) and in Alabama by Lydeard et al. (1999). It was listed as a species of highest conservation concern in Alabama by Garner et al. (2004).

Remarks

Ortmann (1925) regarded *Alasmidonta minor* (Lea, 1845) as a member of the Cumberlandian fauna, describing it as being found in the Tennessee and Cumberland rivers and the Kentucky River where it flows near the Cumberland River. *Alasmidonta minor* was later placed in the synonymy of *Alasmidonta viridis*, which is widespread outside the Cumberlandian region.

The preference of *Alasmidonta viridis* for small stream habitats is suggested by the paucity of material recovered from prehistoric middens along the Tennessee River. Only two individuals have been reported, both from near Muscle Shoals (Morrison, 1942; Hughes and Parmalee, 1999).

Synonymy

Unio viridis Rafinesque, 1820. Rafinesque, 1820:293

Type locality: Cincinnati, Ohio.

Comment: Lectotype selection (ANSP 20219) by Vanatta (1915) was subsequently found to be invalid. Clarke (1981) designated USNM 8626a as the neotype.

Unio calceolus Lea, 1828. Lea, 1828:265, pl. 4, fig. 1; Lea, 1834b:7, pl. 3, fig. 1

Alasmodonta truncata Conrad, 1834, *nomen nudum*. Conrad, 1834b:73

Margarita (*Margaritana*) *deltoidea* Lea, 1836, *nomen nudum*. Lea, 1836:44

Margaritana deltoidea Lea, 1838. Lea, 1838a:43, pl. 13, fig. 38; Lea, 1838c:43, pl. 13, fig. 38

Calceola angulata Swainson, 1840. Swainson, 1840:382

Margaritana minor Lea, 1845. Lea, 1845:166; Lea, 1848a:82, pl. 8, fig. 26; Lea, 1848b:82, pl. 8, fig. 26

Unio diversus Conrad, 1856. Conrad, 1856:172 (outline figure)

Type locality: Shoal Creek, [Lauderdale County,] northern Alabama, Professor T.P. Hatch. Type specimen not found.

Comment: Ortmann (1925) considered *Unio diversus* a spurious species and questioned its placement in *Alasmidonta*. This taxon likely belongs here (Parmalee and Bogan, 1998).

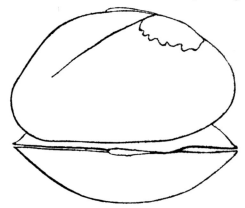

Alasmidonta calceolus danielsi Baker, 1928. Baker, 1928a:187, pl. 69, fig. 2; pl. 72, figs.7–11
Alasmidonta calceolus magnalacustris Baker, 1928. Baker, 1928a:188, pl. 69, fig. 3; pl. 72, figs.12–16

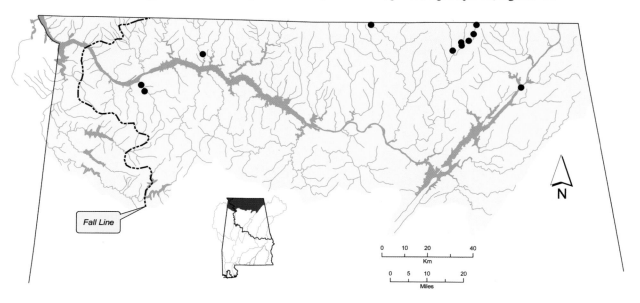

Distribution of *Alasmidonta viridis* in the Tennessee River drainage of Alabama.

Genus *Amblema*

Amblema Rafinesque, 1820, is confined to eastern North America, occurring throughout the Mississippi and Great Lakes basins and in eastern Gulf Coast drainages. It contains three species (Turgeon et al., 1998). Two species occur in Alabama, and the third hypothetically occurred in the state.

Frierson (1914) identified the type species, by elimination, as *Amblema costata* Rafinesque, 1820. Due to confusion between *Amblema* Rafinesque, 1819, and *Amblema* Rafinesque, 1820, this genus name has been used inconsistently. *Crenodonta* Schlüter, 1838, has also been applied to species currently recognized as *Amblema* and *Megalonaias* Utterback, 1915 (e.g., Clench and Turner, 1956). Clarke and Clench (1965) noted an earlier type species designation of *Unio securis* (Deshayes, 1830) (= *Unio securis* Lea, 1829) by Herrmannsen (1852) making *Crenodonta* Schlüter, 1838, a junior synonym of *Ellipsaria* Rafinesque, 1820. The ICZN (1968: Opinion 840) suppressed *Amblema* Rafinesque, 1819. *Amblema* Rafinesque, 1820, is a *nomen conservatum* (ICZN, 1968: Opinion 840). The genus *Cokeria* Marshall, 1916, was described from North Dakota, based on *C. southalli* Marshall, 1916, an aberrant *Amblema plicata* (Say, 1817). Thus, *Cokeria* is a junior synonym of *Amblema*.

Type Species

Amblema costata Rafinesque, 1820 (= *Unio plicatus* Say, 1817)

Diagnosis

Shell large, round to trapezoidal; umbo positioned anteriorly; posterior half of shell with plications or folds variable in number and development; anterior adductor muscle scar large, flat, heavily sculptured; pseudocardinal and lateral teeth well-developed.

Supra-anal aperture separated from excurrent aperture by very short mantle bridge, absent in some individuals; inner lamellae of inner gills connected to visceral mass only anteriorly; all 4 gills marsupial; marsupium not extended ventrally beyond original gill edge when gravid; glochidium without styliform hooks (Ortmann, 1912a; Utterback, 1915; Baker, 1928a; Haas, 1969b).

Synonymy

Cokeria Marshall, 1916

Amblema elliottii (Lea, 1856)
Coosa Fiveridge

 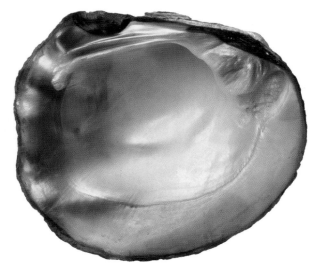

Amblema elliottii – Length 95 mm, UF 21252. Coahulla Creek, Cedar Ridge, Whitfield County, Georgia. © Richard T. Bryant.

Shell Description

Length to 145 mm; very thick; moderately inflated; outline subquadrate to subrotund; posterior margin obliquely truncate to very broadly rounded; anterior margin broadly rounded; dorsal margin straight but often raised into very slight wing; ventral margin broadly rounded; posterior ridge low, broadly rounded, somewhat obscured by sculpture; posterior slope wide, sculptured with regular, small, wide, often arcuate plications across its width, terminating on dorsal margin; center of disk sculptured with 3–5 well-developed, parallel plications, becoming larger and deeper with age; umbo slightly inflated, elevated above hinge line, located toward anterior margin, umbo sculpture a few corrugated ridges; periostracum greenish brown in young specimens, dark olive to black in adults.

Pseudocardinal teeth thick, rough, triangular, 2 divergent teeth in left valve, 1 tooth in right valve, often with accessory denticle; lateral teeth thick, short, slightly curved, 2 in left valve, 1 in right valve; interdentum short, wide; umbo cavity broad, moderately deep; nacre white.

Soft Anatomy Description

Mantle creamy white to tan, usually grayish brown outside of apertures; visceral mass tan, may be creamy white adjacent to foot; foot creamy white to tan.

Gills tan; dorsal margin straight to slightly sinuous, ventral margin convex to elongate convex; gill length usually 55–65% of shell length; gill height 40–50% of gill length; outer gill height 65–80% of inner gill height; inner lamellae of inner gills connected to visceral mass only anteriorly.

No gravid females were available for marsupium description; probably similar to *Amblema plicata*, with all four gills marsupial, marsupium slightly padded when gravid.

Labial palps tan; straight to slightly curved dorsally, convex ventrally, usually bluntly pointed distally, occasionally rounded; palp length 20–30% of gill length; palp height 50–70% of palp length; distal 30–45% of palps bifurcate.

Incurrent aperture usually longer than excurrent aperture, occasionally equal; supra-anal aperture usually longer than incurrent aperture, occasionally equal.

Incurrent aperture length 20–30% of shell length; creamy white to light tan within, usually gray or grayish brown basal to papillae; papillae in 2–3 rows, mostly simple, long, thick, occasionally with few bifid, inner row larger, more sparse; papillae usually some combination of creamy white, tan, rusty tan and rusty orange, often with grayish brown basally or between papillae.

Excurrent aperture length 15–25% of shell length; creamy white to light tan within, usually with marginal color band of rusty tan, rusty orange, grayish brown or light gray; margin usually with small, simple papillae, occasional specimens with crenulate margin; papillae creamy white, rusty tan, rusty orange, grayish tan or grayish brown.

Supra-anal aperture length 25–40% of shell length; creamy white within, without marginal color band; margin smooth; mantle bridge separating supra-anal and excurrent apertures imperforate, length approximately 5% of supra-anal length, may be slightly greater, occasionally absent.

Specimens examined: Big Canoe Creek, St. Clair County (n = 4); Coosa River (n = 3).

Glochidium Description

Glochidium unknown.

Similar Species

The shell of *Amblema elliottii* differs from that of *Amblema plicata* in having fewer and larger plications and being more rounded and less elongate.

Amblema elliottii can be distinguished from *Megalonaias nervosa* by its deeper, heavier plications, less wrinkled shell surface and a shallower umbo cavity. Subadult *M. nervosa* have small, chevron-shaped pustules that are absent in *A. elliottii*.

General Distribution

Amblema elliottii is a Mobile Basin endemic found in the Coosa River drainage of Alabama and Georgia and the upper Cahaba River.

Alabama and Mobile Basin Distribution

Amblema elliottii occurs throughout the Coosa River drainage of Alabama and Georgia as well as upper reaches of the Cahaba River.

Amblema elliottii is extant in most of its historical range. It appears to be declining in some localized areas and has been extirpated from some areas.

Ecology and Biology

Amblema elliottii occurs in large creeks to large rivers. It is found in slow to strong current in substrates of sand, gravel and cobble. *Amblema elliottii* occurs at depths of less than 1 m to more than 6 m.

Amblema elliottii is presumably a short-term brooder, gravid in spring and summer. Its glochidial hosts are unknown.

Current Conservation Status and Protection

Amblema elliottii was not recognized until the late 1990s (Mulvey et al., 1997). In Alabama Lydeard et al. (1999) listed it as a species with undetermined status, and Garner et al. (2004) designated it a species of moderate conservation concern.

Remarks

Amblema elliottii was in the synonymy of *Amblema plicata* until a genetic analysis by Mulvey et al. (1997) demonstrated its distinctiveness. This analysis also compared other populations of *Amblema* and found that *A. elliottii* is most closely related to *A. neislerii*, an Apalachicola Basin endemic.

Amblema elliottii and *Amblema plicata* converge in shell characters and may be difficult to distinguish, although they have been found to be genetically distinct (Mulvey et al., 1997). The two species appear to occur sympatrically at some localities above the Fall Line in the Cahaba River drainage. However, the taxonomic status of *Amblema* in the Cahaba River drainage has not been addressed. Some individuals downstream of Logan Martin Dam approach the shell form of *A. plicata*, which also poses questions regarding its occurrence in the Coosa River.

Amblema elliottii was named in honor of the "Right Reverend Stephen Elliott, of Georgia" who collected and shipped mussels from Georgia and adjacent states to his friend Isaac Lea (Lea, 1856d).

Synonymy

Unio elliottii Lea, 1856. Lea, 1856d:262; Lea, 1858e:54, pl. 7, fig. 37; Lea, 1858h:54, pl. 7, fig. 37
Type locality: Othcalooga [Oothkalooga] Creek, Gordon County, Georgia, Bishop Elliott. Lectotype, USNM 84019, length 128 mm, designated by Johnson (1974).

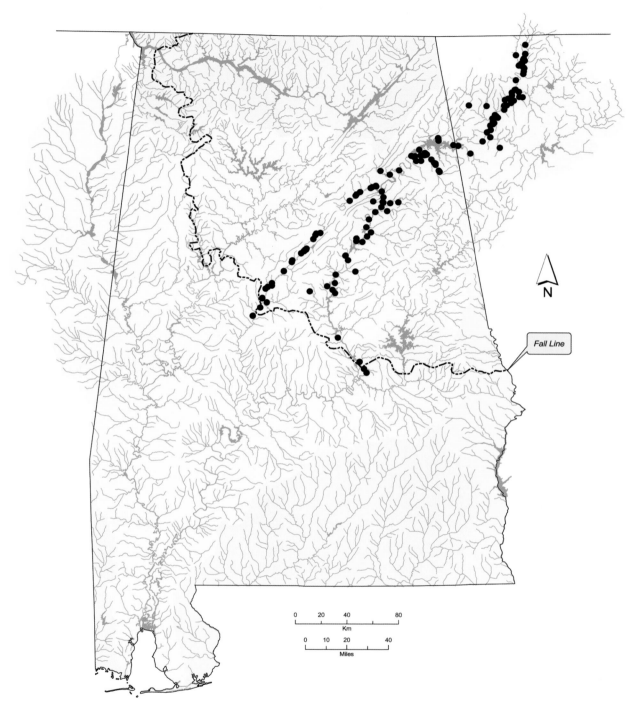

Distribution of *Amblema elliottii* in Alabama and the Mobile Basin.

Amblema plicata (Say, 1817)
Threeridge

Amblema plicata – Upper figure: length 107 mm, AUM 2107. Tennessee River, Buck Island, river mile 249.8, Lauderdale County, Alabama. Lower figure: length 76 mm, UMMZ 100815. Black Warrior River, Squaw Shoals, Jefferson County, Alabama, July 1910 or August 1911. © Richard T. Bryant.

Shell Description

Length to 178 mm; thick; moderately compressed to inflated, often less inflated in small streams; outline round to quadrate, may be somewhat rhomboidal or trapezoidal; posterior margin rounded to obliquely truncate; anterior margin broadly rounded; dorsal margin straight to slightly convex, often developed into slight dorsal wing posteriorly; ventral margin straight to broadly rounded; posterior ridge low, rounded, almost obscured by sculpture; posterior slope broad, flat, may have slight radial depression; posterior 65% of shell adorned with large, oblique plications, some individuals with additional wrinkles and small ridges on disk, especially those from tributary streams, posterior slope

usually with small, concentric plications extending from posterior ridge to dorsal margin at an oblique angle, less prominent in individuals from large rivers; umbo compressed to inflated, elevated above hinge line, may be turned slightly anteriad, sculpture a few irregular concentric ridges, nodulous on posterior ridge; periostracum dark green to tawny in young individuals, becoming dark brown or black with age, individuals from tributary streams often tawny throughout life.

Pseudocardinal teeth large, thick, rough, elevated, triangular, 2 divergent teeth in left valve, 1 tooth in right valve, usually with well-developed accessory denticle anteriorly; lateral teeth moderately long, straight, slightly serrated, 2 in left valve, 1 in right valve; interdentum short, moderately wide; umbo cavity deep; nacre white, occasionally with pink or purplish tint posteriorly.

Soft Anatomy Description

Mantle creamy white to light brown, may have slight golden cast, usually dark brown, gray or black outside of apertures; visceral mass creamy white to pearly white; foot creamy white to tan, may be slightly darker distally.

Gills creamy white to tan or light brown, may have slight golden cast; dorsal margin straight to slightly sinuous, ventral margin convex; gill length usually 50–60% of shell length; gill height usually 40–50% of gill length, occasionally greater; outer gill height usually 60–80% of inner gill height; inner lamellae of inner gills usually connected to visceral mass only anteriorly.

All 4 gills marsupial; glochidia held across entire gill except extreme anterior and posterior ends; marsupium slightly to moderately padded when gravid; creamy white. Lea (1863d) reported only outer gills to be marsupial. Utterback (1915) reported gravid gills to be "rich brown" when gravid with mature glochidia.

Labial palps creamy white to tan; broadly connected to mantle basally, straight to slightly concave or slightly convex dorsally, convex ventrally, rounded to bluntly pointed distally; palp length 15–45% of gill length; palp height usually 50–75% of palp length, occasionally greater; distal 25–45% of palps bifurcate. Utterback (1915) reported labial palps to be brown.

Incurrent aperture usually longer than excurrent aperture, occasionally equal; supra-anal aperture usually longer than incurrent aperture.

Incurrent aperture length 20–30% of shell length; creamy white to tan within, rarely with grayish cast, usually with band of variable shades of black, gray or brown basal to papillae; papillae in 1–2 rows, typically simple, occasionally with a few bifid, of variable length, short and thick to long and slender; papillae usually some combination of creamy white, tan, brown, dull orange and gray, may be pale or translucent distally. Utterback (1915) described incurrent aperture papillae as being small and "arboreal" (i.e., arborescent). Utterback (1915) reported incurrent aperture papillae of some individuals to be yellowish.

Excurrent aperture length 10–25% of shell length; creamy white within, usually with marginal color band that is translucent creamy white or of variable shades of black, gray or brown; margin crenulate or with small, simple papillae, crenulations may be minute; papillae tan, rusty brown or dull orange.

Supra-anal aperture length 20–35% of shell length; creamy white within, with or without thin, black or brown marginal band; margin smooth; mantle bridge separating supra-anal and excurrent apertures imperforate, length usually 5–10% of supra-anal length, mantle bridge sometimes absent.

Specimens examined: Bear Creek, Colbert County (n = 3); Bogue Chitto, Dallas County (n = 2); Conecuh River (n = 3); Estill Fork, Jackson County (n = 2); Sipsey River (n = 3); Tennessee River (n = 9); Tombigbee River (n = 3).

Glochidium Description

Length 185–210 μm; height 195–220 μm; without styliform hooks; outline subrotund; dorsal margin straight, ventral margin broadly rounded, anterior and posterior margins convex and equal; larval thread present (Lefevre and Curtis, 1910b; Ortmann, 1912a, 1914; Surber, 1912; Howard, 1914b; Utterback, 1915, 1916a; Coker et al., 1921).

Similar Species

The shell of *Amblema plicata* differs from that of *Amblema elliottii* in being more quadrate and elongate and having fewer plications.

Amblema plicata from large rivers can be distinguished from *Megalonaias nervosa* by having deeper plications on the shell disk, fewer wrinkles and a smooth umbo. Specimens of *A. plicata* from creeks are usually more wrinkled than those from large rivers but can still be separated from *M. nervosa* by a relatively smooth umbo.

Amblema plicata may resemble some *Arcidens confragosus* but has a thicker, less inflated shell, usually with deeper plications. Lateral teeth of *A. plicata* are well-developed, but those of *A. confragosus* are poorly developed.

General Distribution

Amblema plicata occurs in the Hudson Bay Basin (Dawley, 1947; Burch, 1975a). It also occurs in the Great Lakes Basin of southern Canada and north-central United States (La Rocque and Oughton, 1937; Burch, 1975a; Strayer et al., 1992). In the Mississippi Basin *A. plicata* occurs from Minnesota (Dawley, 1947) south to Louisiana (Vidrine, 1993), and from headwaters of the Ohio River in western New York (Strayer et al., 1992) west to eastern Kansas (Murray and Leonard, 1962) and Nebraska (Hoke, 2000). *Amblema plicata* is found in most of the Cumberland River drainage downstream of Cumberland Falls (Cicerello et al., 1991; Parmalee and Bogan, 1998) and throughout the Tennessee River drainage (Ahlstedt, 1992a, 1992b; Parmalee and Bogan, 1998). It occurs in Gulf Coast drainages from the Choctawhatchee River, Alabama and Florida, west to the Neuces River, Texas, including the Mobile Basin (Burch, 1975a; Butler, 1990; R.G. Howells, personal communication).

Alabama and Mobile Basin Distribution

Amblema plicata occurs in the Tennessee River drainage and Mobile Basin with the exceptions of upper Tallapoosa River drainage and possibly the Coosa River drainage. However, individuals that resemble *A. plicata* have recently been collected from the Coosa River downstream of Logan Martin Dam. Whether these represent an adventive population of *A. plicata* or aberrant individuals of *Amblema elliottii* remains unclear. Along the Gulf Coast, *A. plicata* occurs in the Conecuh River system. Though never collected from the Choctawhatchee River drainage in Alabama, it was collected from Florida reaches of that drainage during the early 1900s.

Amblema plicata remains widespread and common in the Tennessee River but has disappeared from many tributaries. It has also disappeared from much of the Mobile Basin and has a patchy distribution in the Conecuh River.

Ecology and Biology

Amblema plicata occurs in shoals and pools of large creeks and rivers. It is also found in impounded reaches of large rivers, including overbank habitats, as well as natural lakes. In riverine reaches it inhabits substrates of gravel and sand, but in overbank habitats of reservoirs it can be found in substrates composed of any combination of mud, sand, gravel and *Corbicula fluminea* shells. Though it is silt tolerant, it is less common in soft mud than in more solid substrates (e.g., sandy mud, gravel and *Corbicula* shells). *Amblema plicata* occurs at depths ranging from less than 1 m to more than 10 m.

Gut contents of *Amblema plicata* from the Wisconsin River, Wisconsin, have been examined (Bisbee, 1984). Green algae were most prevalent, but blue-green algae and diatoms were also abundant.

Amblema plicata reaches sexual maturity at four years of age (Stein, 1969). Trematode infestation has been reported to cause sterility in small percentages of some populations (Haag and Staton, 2003). Trematode infestation was not limited to large individuals.

Male *Amblema plicata* have been reported to discharge spermatozeugmata with an average diameter of 44.9 μm (Waller and Lasee, 1997). *Amblema plicata* is a short-term brooder, gravid from May or June to August in various parts of its range (Frierson, 1904; Surber, 1912; Ortmann, 1909b, 1912a, 1914; Utterback, 1915, 1916a; Coker et al., 1921; Holland-Bartels and Kammer, 1989; Howells, 2000). In the Sipsey River, Alabama, and Little Tallahatchie River, Mississippi, 71 percent and 77 percent of mature females have been reported gravid, respectively, during peak representative period (Haag and Staton, 2003). *Amblema plicata* marsupial contents were quantified by Haag and Staton (2003), who reported an average of 1.3

percent unfertilized eggs. Conglutinates of *A. plicata* have been reported as white, compressed and lanceolate, discharged in broken or loose, irregular masses (Ortmann, 1912a; Utterback, 1915). However, Haag and Staton (2003) found eggs and embryos to be unbound within the marsupia. Fecundity of *A. plicata* was reported to average 229,738 glochidia per year in the Sipsey River, Alabama, and 325,709 glochidia per year in the Tallahatchie River, Mississippi (Haag and Staton, 2003).

Amblema plicata is a generalist with regard to glochidial hosts. Fishes found to serve as hosts during laboratory trials include *Ambloplites rupestris* (Rock Bass), *Lepomis cyanellus* (Green Sunfish), *Lepomis gibbosus* (Pumpkinseed), *Lepomis macrochirus* (Bluegill), *Lepomis megalotis* (Longear Sunfish), *Micropterus salmoides* (Largemouth Bass), *Pomoxis annularis* (White Crappie) and *Pomoxis nigromaculatus* (Black Crappie) (Centrarchidae); *Lepisosteus platostomus* (Shortnose Gar) (Lepisosteidae); and *Perca flavescens* (Yellow Perch) (Percidae) (Howard, 1914b; Coker et al., 1921; Stein, 1968). Coker et al. (1921) also successfully transformed *A. plicata* glochidia on *Lepomis euryorus* McKay, 1881, which was later determined to be a hybrid of *L. cyanellus* and *L. gibbosus* (Trautman, 1981). In addition to some of the above, fishes reported to have been collected with natural infestations of *A. plicata* glochidia include *Hypentelium nigricans* (Northern Hog Sucker), *Moxostoma duquesnei* (Black Redhorse), *Moxostoma erythrurum* (Golden Redhorse) (Catostomidae); *Cyprinella spiloptera* (Spotfin Shiner), *Cyprinella whipplei* (Steelcolor Shiner), *Erimystax dissimilis* (Streamline Chub) and *Notropis atherinoides* (Emerald Shiner) (Cyprinidae); *Esox lucius* (Northern Pike) (Esocidae); *Hiodon tergisus* (Mooneye) (Hiodontidae); *Ictalurus punctatus* (Channel Catfish) (Ictaluridae); *Percina caprodes* (Logperch) and *Sander canadensis* (Sauger) (Percidae); and *Aplodinotus grunniens* (Freshwater Drum) (Sciaenidae) (Surber, 1913; Coker et al., 1921; Weiss and Layzer, 1995). Coker et al. (1921) also reported natural infestation of *Morone chrysops* (White Bass) (Moronidae) by *A. plicata* glochidia but deemed the record to be insignificant.

Amblema plicata has been reported to grow at different rates among differing habitat types (Stansbery, 1970b). Those in deep water habitats, with heavier silt accumulations, grow slower than those in shallower water with less silt accumulation.

Current Conservation Status and Protection

Amblema plicata was listed as currently stable throughout its range by Williams et al. (1993) and in Alabama by Lydeard et al. (1999). Garner et al. (2004)

designated *A. plicata* a species of low conservation concern in the state.

Amblema plicata is an important commercial species in Alabama and is protected by state regulations, the most important being a minimum size limit of $2^5/_8$ inches (67 mm) in shell height.

Remarks

Variation in *Amblema plicata* shell morphology can be extreme among individuals from different habitats (i.e., small streams versus large rivers), as well as among allopatric populations in similar sized rivers (Ortmann, 1920; Ball, 1922). Utterback (1915) recognized the early taxonomic problems that stemmed from variation in shell morphology, calling *Amblema* "a group of all sorts of inter-grading and puzzling forms".

Amblema plicata and *Amblema elliottii* converge in shell characters and may be difficult to distinguish, although they have been found to be genetically distinct (Mulvey et al., 1997). The two species appear to occur sympatrically at some localities in the Cahaba River drainage. However, the taxonomic status of *Amblema* in the Cahaba River drainage has not been addressed. Some *Amblema*, presumably *A. elliottii*, found in the Coosa River downstream of Logan Martin Dam approach the shell form of *A. plicata*. Additional genetic analysis is needed to precisely delineate the distribution of these two species.

Amblema plicata from the Tennessee River proper are generally inflated, with little sculpture other than primary plications on the shell disk and often rudimentary plications on the posterior slope. Specimens from tributaries are more compressed and have more secondary wrinkles and plications, including well-developed plications on the posterior slope. Periostracum color also differs, with those from the Tennessee River proper typically being dark green when young, becoming dark greenish brown or black with age, and those from tributaries being tawny when young, often darkening little with age.

Variation between small river and large river populations can also be seen in the Mobile Basin. *Amblema plicata* shells from the Tombigbee River proper resemble those from the Tennessee River proper. Specimens from tributaries usually have more numerous plications. However, specimens of *A. plicata* from upper Black Warrior River drainage differ in having plications that are very shallow or lacking and a more wrinkled shell surface, as well as being more pointed posteriorly. The upper Black Warrior specimens could represent a distinct species, since several other endemic unionids occur there, but there are no extant *Amblema* in the upper Black Warrior River drainage, precluding anatomical and genetic comparisons.

Amblema plicata from Gulf Coast drainages have more numerous primary plications and less inflated umbos than those of large rivers in the Tennessee River drainage and Mobile Basin. This form has been recognized as *Amblema perplicata*. Genetic analyses presented by Mulvey et al. (1997) suggest that *A. perplicata* is conspecific with *A. plicata*.

There appears to be differences in tolerance of impounded conditions between Tennessee River drainage and Mobile Basin populations of *Amblema plicata*. It may be locally common in overbank habitats of Tennessee River reservoirs, but is very rare or absent in similar habitats in the Mobile Basin.

Amblema plicata is uncommon in prehistoric shell middens across northern Alabama (Morrison, 1942; Warren, 1975; Hughes and Parmalee, 1999). However, Ortmann (1925) reported it to be abundant from Muscle Shoals upstream prior to impoundment. This suggests that *A. plicata* was rare in northern Alabama reaches of the Tennessee River prehistorically but increased in relative abundance by the early 1900s. This species remains abundant in some reaches of the Tennessee River drainage including some overbank areas of reservoirs.

Synonymy

Unio plicata Say, 1817. Say, 1817:[unpaginated, pages 11–12]
Type locality: Lake Erie. Haas (1930) designated a neotype, SMF 4305, length 133 mm, from the Ohio River.
Unio peruvianus Lamarck, 1819. Lamarck, 1819:71
Unio rariplicata Lamarck, 1819. Lamarck, 1819:71
Amblema costata Rafinesque, 1820. Rafinesque, 1820:315, pl. 82, figs. 13, 14
Unio undulatus Barnes, 1823. Barnes, 1823:120, pl. 2
Unio perplicatus Conrad, 1841. Conrad, 1841:19; Conrad, 1850:276, pl. 38, fig. 2

Unio atrocostatus Lea, 1845. Lea, 1845:163; Lea, 1848a:70, pl. 2, fig. 5; Lea, 1848b:44, pl. 2, fig. 5

Type locality: [Alabama River,] Claiborne, [Monroe County,] Alabama, Judge Tait; [Black Warrior River,] Tuscaloosa, [Tuscaloosa County,] Alabama, B.W. Budd, M.D.; [Red River,] Alexandria, [Rapides Parish,] Louisiana, J. Hale, M.D. Lectotype, USNM 84015, length 74 mm, designated by Johnson (1974), is from Claiborne, Monroe County, Alabama, not from Alexandria, Louisiana, as originally reported by Johnson (1974).

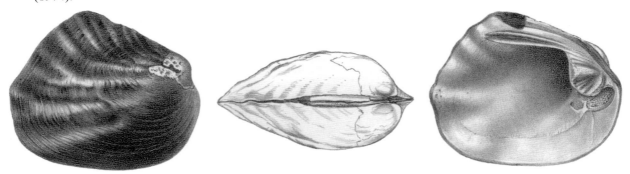

Unio hippopoeus Lea, 1845. Lea, 1845:163; Lea, 1848a:67, pl. 1, fig. 1; Lea, 1848b:67, pl. 1, fig. 1
Unio latecostatus Lea, 1845. Lea, 1845:163; Lea, 1848a:68, pl. 1, fig. 2; Lea, 1848b:42, pl. 1, fig. 2
Type locality: [Black Warrior River,] Tuscaloosa, [Tuscaloosa County,] Alabama, B.W. Budd, M.D. Type specimen not found, also not found by Johnson (1974).

Unio pearlensis Conrad, 1855. Conrad, 1855:256; Reeve, 1864:[no pagination], pl. 40, fig. 42
Unio brazosensis Lea, 1868. Lea, 1868a:144; Lea, 1868c:309, pl. 48, fig. 122; Lea, 1869a:69, pl. 48, fig. 122
Unio lincecumii Lea, 1868. Lea, 1868a:144; Lea, 1868c:312, pl. 49, fig. 125; Lea, 1869a:72, pl. 49, fig. 125
Unio pauciplicatus Lea, 1872. Lea, 1872b:156; Lea, 1874c:29, pl. 9, fig. 26; Lea, 1874e:33, pl. 9, fig. 26
Unio quintardii Cragin, 1887. Cragin, 1887:6; Pilsbry, 1892:131, pl. 7, figs. 1–3
Unio pilsbryi Marsh, 1891. Marsh, 1891:1; Pilsbry, 1892:131, pl. 8, figs. 7, 8
Cokeria southalli Marshall, 1916. Marshall, 1916:133

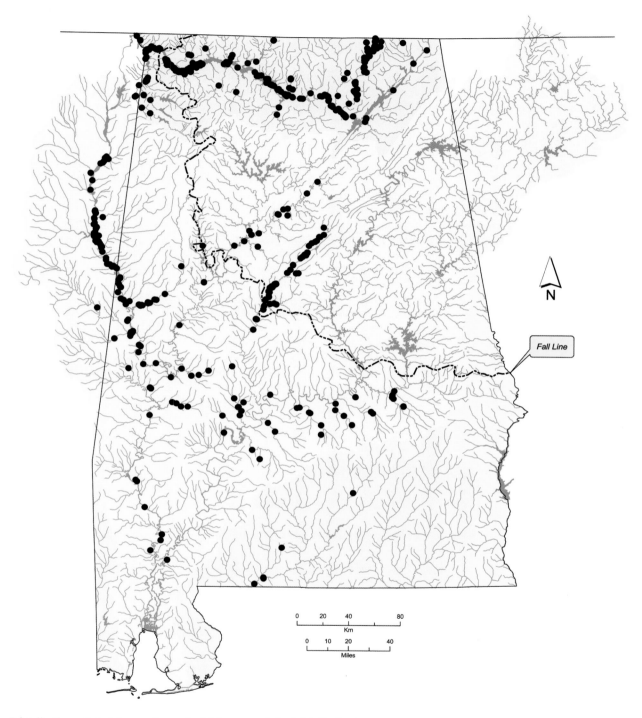

Distribution of *Amblema plicata* in Alabama and the Mobile Basin.

Genus *Anodonta*

The distribution of *Anodonta* Lamarck, 1799, is Holarctic, and the type species is from western Europe. In North America it occurs from Alaska and Canada to Mexico. Turgeon et al. (1998) listed 10 *Anodonta* species, 2 of which occur in Alabama. There is at least one additional undescribed species known from eastern Gulf Coast drainages, which is herein recognized based on shell morphology.

The number of *Anodonta* species worldwide is not clear due to poor delineation of species and varying interpretations of the species concept. Ortmann (1912a) observed that "in Europe the species-making in this group has gone beyond all the bounds of reason." *Anodonta* Lamarck, 1799, is a *nomen conservatum* (ICZN, 1926: Opinion 94; 1959: Opinion 561). Hoeh (1990) used morphological and allozymic data to produce a phylogeny that divided North American *Anodonta*, along with the European type species *A. cygnea* into three clades: *Anodonta*; *Pyganodon* Crosse and Fischer *in* Fischer and Crosse, 1894; and *Utterbackia* Baker, 1927. The latter two were elevated from subgeneric to generic status (Hoeh, 1990).

Type Species
Mytilus cygnea Linnaeus, 1758

Diagnosis
Shell thin; elliptical to oval; compressed to inflated; umbo sculpture double-looped or parallel bars; without hinge teeth.

Inner lamellae of inner gills connected to visceral mass only anteriorly; supra-anal aperture usually small, separated from excurrent aperture by wide mantle bridge (may be as long as or longer than either of the 2 apertures); outer gills marsupial; marsupium occupying entire gill, distended when gravid, secondary water tubes appear when gravid; glochidium with styliform hooks (Ortmann, 1912a; Simpson, 1914; Haas, 1969b).

Synonymy
Anodonta Lamarck, 1799, *nomen conservatum*
Cista Huebner, 1810
Edentula Nitzsch, 1820
Lipodonta Nitzsch, 1820
Anodontina Schlüter, 1838
Colletopterum Bourguignat, 1880
Pteranodon Fischer, 1886
Euanodonata Westerlund, 1890
Gabillotia Servain, 1890
Brachyanodon Crosse and Fischer *in* Fischer and Crosse, 1894
Arnoldina Hannibal, 1912
Nayadina De Gregorio, 1914
Utterbackiana Frierson, 1927
Euphrata Pallary, 1933
Anemina Haas, 1969

This synonymy does not include the plethora of Russian and Asian generic names described or elevated by Starobogatov (1970) or those described by subsequent Russian authors.

Anodonta suborbiculata Say, 1831
Flat Floater

Anodonta suborbiculata – Upper figure: length 124 mm, UF 376151. Coosa River, Weiss Reservoir, mouth of Big Cedar Creek, about 2 air miles east of Alabama and Georgia state line, river mile 258, Floyd County, Georgia, 20 August 1997. Lower figure: juvenile, length 44 mm, UF 375318. Patsaliga Creek, slough on impounded lower end, 7 air miles north of County Road 59 bridge, Covington County, Alabama, 24 July 1995. © Richard T. Bryant.

Shell Description

Length to 200 mm; thin; compressed; outline round to oval; posterior margin bluntly pointed to broadly rounded; anterior margin broadly rounded; dorsal margin straight to slightly convex; ventral margin broadly rounded; posterior ridge distinct but not prominent; posterior slope low to moderately steep, slightly concave, often with slight dorsal wing posteriorly, wing generally more pronounced in young individuals; umbo broad, moderately inflated, typically not elevated above hinge line, umbo sculpture irregular, nodulous ridges; periostracum smooth, often shiny, light yellow to dark brown, occasionally with weak green rays.

Pseudocardinal and lateral teeth absent; umbo cavity wide, shallow; nacre white, often with salmon tint.

Soft Anatomy Description

Mantle pale tan, thicker outside of pallial line, gray to grayish brown outside of apertures; visceral mass creamy white; foot pale tan.

Gills pale grayish brown; dorsal margin concave, ventral margin convex; gill length approximately 50% of shell length; gill height approximately 45% of gill length; outer gill height approximately 80% of inner gill height; inner lamellae of inner gills connected to visceral mass only anteriorly.

No gravid females were available for marsupium description.

Labial palps pale gray; straight dorsally, convex ventrally, bluntly pointed distally; palp length approximately 20% of gill length; palp height approximately 60% of palp length; distal 35% of palps bifurcate.

Incurrent aperture longer than excurrent and supra-anal apertures; excurrent aperture often slightly longer than supra-anal aperture.

Incurrent aperture length approximately 10% of shell length; creamy white within, pale tan basal to papillae; papillae in 2 rows, inner row larger, mostly simple, with few bifid, may have very few trifid papillae; papillae creamy white, few in outer row with dark gray edges basally.

Excurrent aperture length approximately 5% of shell length; creamy white within, marginal color band rusty tan with irregular dark brown lines more or less perpendicular to margin; margin smooth.

Supra-anal aperture length approximately 5% of shell length; creamy white within, without marginal coloration; margin smooth; mantle bridge separating supra-anal and excurrent apertures imperforate, length approximately 400% of supra-anal length.

Specimen examined: Coosa River (n = 1) (specimen previously preserved).

Glochidium Description

Length 323–328 μm; height 320–328 μm; with styliform hooks; outline subtriangular; dorsal margin straight, ventral margin pointed, anterior margin broadly convex, posterior margin slightly convex (Surber, 1915; Utterback, 1916a; Hoggarth, 1999).

Similar Species

An undescribed *Anodonta* in oxbows and backwaters of the Escambia River drainage and Mobile Basin resembles *A. suborbiculata*. However, the undescribed species is more elongate and inflated, with umbos elevated slightly above the hinge line.

General Distribution

Anodonta suborbiculata occurs in much of the Mississippi Basin from Minnesota (Graf, 1997) and Wisconsin (Mathiak, 1979) south to Louisiana (Vidrine, 1993), and from the Ohio River drainage in Ohio (Cummings and Mayer, 1992) west to Kansas (Murray and Leonard, 1962) and South Dakota (Backlund, 2000). *Anodonta suborbiculata* is found in lower reaches of the Cumberland River, Kentucky and Tennessee (Cicerello et al., 1991; Parmalee and Bogan, 1998). It occurs throughout the Tennessee River and is found in lower reaches of some large tributaries but is absent from headwater rivers (Parmalee and Bogan, 1998). *Anodonta suborbiculata* occurs in some Gulf Coast drainages from the Escambia River drainage in Alabama and Florida (Williams and Butler, 1994) west to the Brazos River in Texas (Howells et al., 1996).

Alabama and Mobile Basin Distribution

Anodonta suborbiculata is known from the Tennessee River drainage, Mobile Basin and Escambia River drainage. There are no known records of *A. suborbiculata* from Alabama and the Mobile Basin prior to 1946. This suggests that it has colonized or been introduced into the state since impoundment of the major rivers.

Anodonta suborbiculata remains widespread and may be locally common. Its range may continue to expand.

Ecology and Biology

Anodonta suborbiculata typically inhabits waters with little or no current, such as floodplain lakes, sloughs, oxbows and reservoirs associated with large creeks and rivers. It occurs in overbank habitat of reservoirs. Bates (1962) reported *A. suborbiculata* to be one of the first colonizers of newly created overbank habitat in Kentucky Reservoir of the Tennessee River. The preferred substrate of *A. suborbiculata* is mud or muddy sand.

Anodonta suborbiculata is a long-term brooder. In Missouri spawning occurs in September, and glochidia mature between October and December and are brooded until March (Utterback, 1915; Barnhart and Roberts, 1997). *Anodonta suborbiculata* sperm are released in the form of spermatozeugmata, which are approximately 57 μm in diameter and contain approximately 3,600 spermatozoa (Barnhart and Roberts, 1997). Spermatozoa migrate to one side of the spermatozeugmata to make directional propulsion possible. *Anodonta suborbiculata* spermatozeugmata were described by Utterback (1915) as "cysts rolling through the water like colonial Protozoa". Fishes reported to serve as glochidial hosts of *A. suborbiculata* in laboratory trials include *Lepomis cyanellus* (Green Sunfish), *Lepomis gulosus* (Warmouth), *Lepomis megalotis* (Longear Sunfish), *Micropterus salmoides* (Largemouth Bass) and *Pomoxis annularis* (White Crappie) (Centrarchidae); *Notemigonus crysoleucas* (Golden Shiner) (Cyprinidae); and *Ictalurus punctatus* (Channel Catfish) (Ictaluridae) (Barnhart and Roberts, 1996, 1997; Howells, 1997).

Current Conservation Status and Protection

Anodonta suborbiculata was considered stable throughout its range by Williams et al. (1993) and in Alabama by Lydeard et al. (1999). Garner et al. (2004) designated *A. suborbiculata* a species of low conservation concern in the state.

Remarks

The distribution of *Anodonta suborbiculata* has increased in recent years, and it appears to be a recent invader to Alabama. Bates (1962) was first to report *A. suborbiculata* from the Tennessee River, where it was found in soft sediments of overbank habitat in Kentucky Reservoir. Stansbery (1964) was first to report it from Alabama. Garner and McGregor (2001) reported *A. suborbiculata* to be a common species in Pickwick and Wheeler reservoirs. There is no reason to believe that its occurrence in the Tennessee River is not due to a natural upstream expansion as habitat became more favorable with impoundment of the river.

There are no records of *Anodonta suborbiculata* in the Mobile Basin until 1976, suggesting that the species is adventive in that drainage as well. Since the upper Coosa River is isolated from the remainder of the Mobile Basin by dams without locks, the presence of *A. suborbiculata* there is most likely due to introductions of fish infested with glochidia.

In the Escambia River drainage the distributional history of *Anodonta suborbiculata* is complicated by the presence of a similar, but distinct, undescribed species of *Anodonta*. The distinctness of the undescribed *Anodonta* was not recognized until recently. There have been intermittent reports of *A. suborbiculata* from the Escambia River drainage since 1917. *Anodonta suborbiculata* from the Alabama portion of Escambia drainage have come from Gantt and Point A reservoirs, Conecuh River, Covington County, Alabama. This population is likely the result of unintentional introduction.

Utterback (1915) stated that the meat of *Anodonta suborbiculata* "has been tested through Domestic Science to be of great food value", but provided no further information.

Synonymy

Anodonta suborbiculata Say, 1831 (January 29). Say, 1831a:[no pagination], pl. 11; Say, 1831b:[no pagination], pl. 11

Type locality: Ponds near the Wabash River, Indiana. Type specimen not found. Length not given in original description.

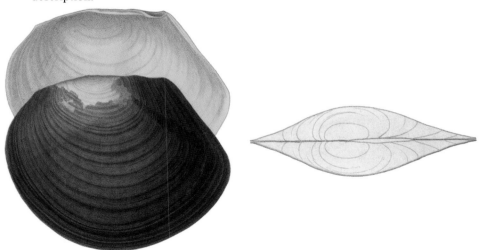

Anodonta (Nayadina) venusta De Gregorio, 1914. De Gregorio, 1914:65, pl. 12, fig. 2

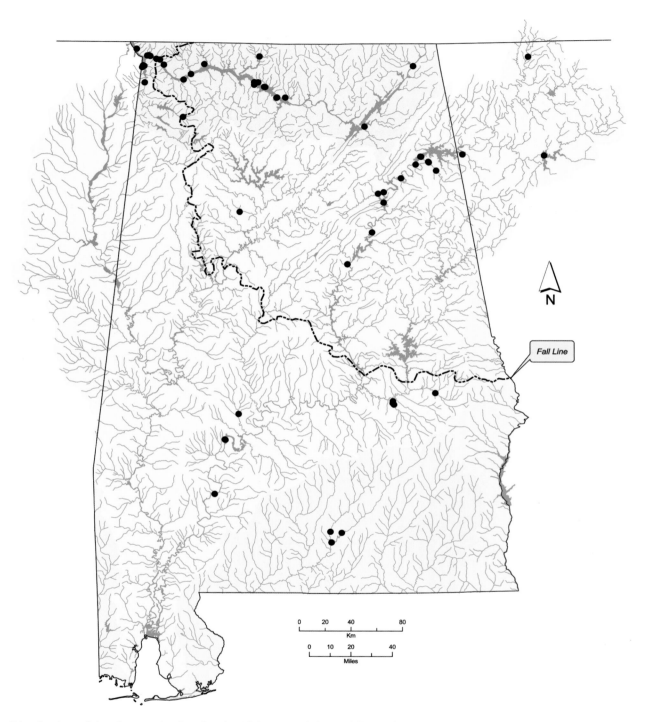

Distribution of *Anodonta suborbiculata* in Alabama and the Mobile Basin.

Anodonta sp.
Cypress Floater

Anodonta sp. – Upper figure: length 114 mm, UF 358657. Slough and gravel pits adjacent to Escambia River, Mystic Springs boat ramp, 1 mile southeast of McDavid, Escambia County, Florida, 20 September 1999. Lower figure: length 112 mm, UF 375595. Fish Lake, oxbow off Pascagoula River, 1 air mile southeast of Highway 614 bridge, southwest of Wade, Jackson County, Mississippi, 27 October 2000. © Richard T. Bryant.

Shell Description

Length to 115 mm; thin; moderately inflated; outline oval; posterior margin narrowly rounded to bluntly pointed; anterior margin broadly rounded; dorsal margin straight; ventral margin convex; posterior ridge low, rounded; posterior slope moderately steep, slightly concave, occasionally with very low dorsal wing; umbo broad, moderately inflated, barely elevated above hinge line, umbo sculpture nodulous ridges; periostracum tawny to olive or brown, may have very thin, olive brown rays.

Pseudocardinal and lateral teeth absent; umbo cavity wide, shallow; nacre white, sometimes with salmon tint in umbo cavity.

Soft Anatomy Description

Mantle creamy white to tan or golden tan, may be dull orange outside pallial line, outside of apertures dull orange to grayish brown; visceral mass creamy white, may become dull orange adjacent to foot; foot dull orange.

Gills gold to light brown; dorsal margin sinuous, ventral margin convex; gill length 55–65% of shell length; gill height 40–55% of gill length; outer gill height 85–95% of inner gill height; inner lamellae of inner gills only connected to visceral mass anteriorly.

Outer gills marsupial; glochidia held across gill length; well padded when gravid; light brown to brownish orange.

Labial palps tan, may have golden cast; straight to concave dorsally, convex ventrally, bluntly pointed distally; palp length 20–35% of gill length; palp height 45–70% of palp length; distal 30–60% of palps bifurcate.

Incurrent aperture longer than excurrent and supra-anal apertures; supra-anal aperture may be longer than excurrent aperture.

Incurrent aperture length 10–15% of shell length; creamy white to dull orange within, sometimes with grayish brown basal to papillae; papillae in 2–3 rows, inner row usually larger, simple, short, thick; papillae tan to dull orange, larger papillae often with black edges.

Excurrent aperture length 5–10% of shell length; creamy white to dull orange within, marginal color band dull orange with black lines in irregular reticulated pattern; margin smooth, may undulate.

Supra-anal aperture length 5–10% of shell length; creamy white to tan within, without marginal coloration; margin smooth; mantle bridge separating supra-anal and excurrent apertures usually imperforate, mantle bridge length 90–215% of supra-anal length.

Specimens examined: Conecuh River (n = 3).

Glochidium Description

Length 290–315 μm; height 310–325 μm; with styliform hooks; outline subtriangular; dorsal margin straight, ventral margin pointed, anterior margin broadly convex, posterior margin slightly convex.

Similar Species

The shell of *Anodonta* sp. resembles that of *Anodonta suborbiculata* but is less round and more inflated, with a more inflated umbo elevated slightly above the hinge line. It may also resemble *Pyganodon grandis*, but that species has a much more inflated umbo that is considerably more elevated above the hinge line. *Anodonta* sp. may vaguely resemble *Utterbackia imbecillis* and *Utterbackia peggyae*, but those species are more elongate and their umbos are not elevated above the hinge line.

General Distribution

The distribution of this species of *Anodonta* is unclear because its backwater habitats are often ignored during mussel surveys. It is known from the Escambia River drainage in Florida and Alabama and the Tombigbee and Pascagoula River drainages in Mississippi.

Alabama and Mobile Basin Distribution

The distribution of this species of *Anodonta* is poorly known. The only confirmed records from

Synonymy

There are no names available for this undescribed species.

Alabama are from Gantt and Point A reservoirs on the Conecuh River and Old Faulkner Lake, an oxbow lake on the Conecuh River floodplain in Escambia County. There are only two known specimens of this species from the Mobile Basin, collected from a floodplain lake along the Tombigbee River near Columbus, Mississippi. Backwater habitats and reservoir overbanks are infrequently sampled, thus the species is poorly represented in museum collections.

Anodonta sp. is extant in the Conecuh River drainage and probably the Mobile Basin.

Ecology and Biology

Anodonta sp. typically occurs in water with little or no current, such as reservoirs, oxbow lakes and sloughs. Substrates in these habitats are typically composed of soft mud or muddy sand, often with detritus.

This species of *Anodonta* is a long-term brooder, presumably gravid from late summer or autumn to the following summer. Gravid individuals brooding mature glochidia have been observed in early November in Gantt Reservoir, Conecuh River. Glochidial hosts of this species are unknown.

Current Conservation Status and Protection

The fact that this species of *Anodonta* has not been previously recognized has precluded its inclusion in conservation status reviews. Until a systematic survey of its habitat is carried out, its status will remain uncertain. However, its recent discovery in Gantt and Point A reservoirs, where it is locally common, suggests that its status is secure at this time.

Remarks

The habitats of this species are often overlooked during mussel surveys. Also, its natural oxbow and slough habitats have been greatly reduced with channelization and impoundment of large rivers. These factors probably contributed to the dearth of records and museum material.

A systematic survey of Gulf Coast and Mobile Basin floodplain lakes is required to determine the status of *Anodonta* sp. Comparative analyses of soft anatomy and genetics are needed to shed light on its taxonomic relationships.

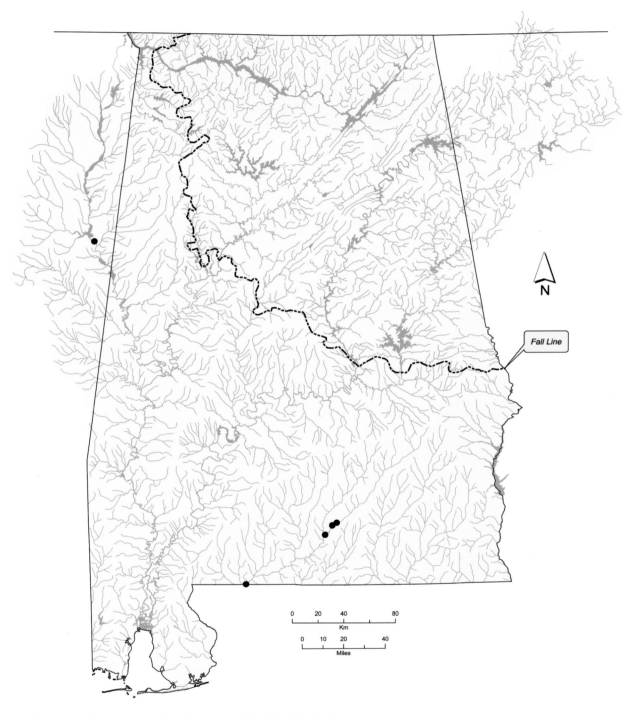

Distribution of *Anodonta* sp. in Alabama and the Mobile Basin.

Genus *Anodontoides*

Anodontoides Simpson *in* Baker, 1898b, is found in the Hudson Bay, Great Lakes, Mississippi and Susquehanna basins and in Gulf Coast drainages from western Florida to Louisiana. There are two species of *Anodontoides* (Turgeon et al., 1998), one of which occurs in Alabama. Cicerello and Schuster (2003) recognized an additional species, *Anodontoides denigrata* (Lea, 1852), which occurs in the Cumberland River above Cumberland Falls in Kentucky and Tennessee.

Anodontoides Simpson *in* Baker, 1898 (Baker, 1989b), is a replacement name for *Anodontopsis* Simpson *in* Baker, 1898, *non* M'Coy, 1851 (Baker, 1898a).

Type Species
Anodonta ferussacianus Lea, 1834

Diagnosis
Shell thin, smooth; elliptical; inflated; posterior ridge weak; umbo elevated above hinge line; umbo sculpture subparallel, concentric, curved ridges; hinge line slightly curved, edentulous or with rudimentary teeth.

Inner lamellae of inner gills only connected to visceral mass anteriorly; excurrent aperture with well-developed papillae; outer gills marsupial; marsupium not extended ventrally beyond original edge of gill when gravid; glochidium with styliform hooks (Ortmann, 1912a; Simpson, 1914; Baker, 1928a; Haas, 1969a).

Synonymy
Anodontopsis Simpson *in* Baker, 1898, *non* M'Coy, 1851 (Mollusca: Bivalvia)

Anodontoides radiatus (Conrad, 1834)
Rayed Creekshell

Anodontoides radiatus – Upper figure: length 56 mm, UF 358658. Big Sandy Creek, County Road 14, 12 air miles southeast of Union Springs, Bullock County, Alabama, 10 May 2000. Lower figure: length 44 mm, UF 388522. Pigeon Creek, County Road 62, 8 air miles east-northeast of Greenville, Butler County, Alabama, 10 May 2000. © Richard T. Bryant.

Shell Description

Length to 75 mm; thin; moderately inflated; outline oval to elliptical; posterior margin bluntly pointed to narrowly rounded; anterior margin rounded; dorsal margin straight; ventral margin convex; posterior ridge low, rounded; posterior slope low, flat; umbo broad, low, elevated slightly above hinge line, umbo sculpture single loops; periostracum yellowish green to dark olive, darkening to brownish green or black with age, with variable green rays.

Both valves with 1 small, lamellar pseudocardinal tooth; lateral teeth absent, but dorsal margin slightly thickened; umbo cavity wide, very shallow; nacre white.

Soft Anatomy Description

Mantle creamy white to tan or brown, often with golden cast, may be dull orange outside of pallial line, grayish tan to rusty tan or gold outside of apertures, often with small dark brown, rusty brown or black specks, sometimes with grayish brown bands perpendicular to margin; visceral mass creamy white or pearly white to tan or gold; foot creamy white to tan, gold or orange.

Gills creamy white to tan, rusty tan, gold or dull orange; dorsal margin straight to slightly sinuous, ventral margin almost straight to convex; gill length 50–65% of shell length; gill height 45–60% of gill length; outer gill height 55–90% of inner gill height; inner lamellae of inner gills connected to visceral mass only anteriorly.

Outer gills marsupial; glochidia held across entire gill; marsupium outline oval to somewhat reniform, padded when gravid; creamy white to gold or light brown when gravid, may have slight rusty cast.

Labial palps creamy white to tan, gold or brownish orange; broadly connected basally, straight to slightly concave dorsally, convex ventrally, bluntly pointed to narrowly rounded distally; palp length 20–35% of gill length; palp height 50–80% of palp length; distal 20–40% of palps bifurcate, occasionally greater.

Lengths of incurrent, excurrent and supra-anal apertures variable relative to one another.

Incurrent aperture length 10–20% of shell length; usually creamy white within, occasionally tan, usually with band of tan, gray, grayish brown, rusty brown or dark brown basal to papillae; papillae usually in 2–3 rows, long, simple, rarely with a few bifid, slender to thick; papillae usually some combination of creamy white, tan, rusty tan, rusty orange, brown, golden brown, dark brown black, often change color distally, often with dark edges basally.

Excurrent aperture length 10–15% of shell length; usually creamy white within, may be tan or light gray, marginal color band some combination of gold, tan, rusty tan, rusty orange, brown, gray or black, often with spots, rosettes or lines that may be perpendicular to margin or in open, reticulated pattern; margin crenulate or with small, simple papillae; papillae tan to rusty orange.

Supra-anal aperture length usually 5–20% of shell length, occasionally less; creamy white to tan or gold within, sometimes with sparse, irregular, black or brown marginally; margin smooth; mantle bridge separating supra-anal and excurrent apertures occasionally with 1 or more perforations, length usually 50–90% of supra-anal length, occasionally less.

Specimens examined: Bull Mountain Creek, Marion County (n = 4); Conecuh River (n = 2); Trussells Creek, Greene County (n = 1); Uchee Creek, Russell County (n = 2).

Glochidium Description

Length 220–300 μm; height 270–280 μm; with styliform hooks; outline subtriangular; dorsal margin straight to slightly convex, ventral margin bluntly pointed, anterior and posterior margins slightly convex.

Similar Species

Superficially, *Anodontoides radiatus* can resemble *Strophitus connasaugaensis* and *Strophitus subvexus*. The shells of both *Strophitus* species are considerably thicker than those of *A. radiatus*, especially along the hinge line. They are also usually more inflated, with a more broadly rounded posterior margin.

Shells of *Villosa lienosa* and *Villosa vibex* can be distinguished from those of *Anodontoides radiatus* by the presence of well-developed pseudocardinal and lateral teeth.

General Distribution

Anodontoides radiatus occurs in most Gulf Coast drainages from the Apalachicola Basin, Alabama, Florida and Georgia (Brim Box and Williams, 2000), west to the Amite River system, Lake Pontchartrain drainage, Louisiana (Vidrine, 1993), including the Mobile Basin. A disjunct population of *A. radiatus* is found in upper reaches of the Yazoo River drainage of the Mississippi Basin in Mississippi (Haag et al., 2002).

Alabama and Mobile Basin Distribution

Anodontoides radiatus is widespread in the Mobile Basin below the Fall Line. There are a few records from above the Fall Line. East of Mobile Basin it is found in the Escambia, Choctawhatchee, Chipola and Chattahoochee River drainages.

Anodontoides radiatus is extant in widespread isolated populations in much of its historical range.

Ecology and Biology

Anodontoides radiatus inhabits small to medium creeks in substrates composed of some combination of mud, sand and gravel. It generally prefers slight to moderate current. It has been found in large rivers along banks in areas of slack water.

Anodontoides radiatus is a long-term brooder. Gravid females have been collected from the Apalachicola Basin in August and September, from the Conecuh River drainage in December (Brim Box and Williams, 2000) and from the Noxubee River system in November (W.R. Haag, personal communication). Glochidial hosts of *A. radiatus* are unknown.

Current Conservation Status and Protection

Anodontoides radiatus was listed as a species of special concern throughout its range by Williams et al. (1993) and in Alabama by Lydeard et al. (1999). Garner et al. (2004) designated *A. radiatus* a species of high conservation concern in the state.

Remarks

In the original description of *Anodontoides radiatus* Conrad (1834) reported the type locality as "small streams in south Alabama." Based on historical accounts (Wheeler, 1935) the type material may have come from near the former community of Erie, Greene County, Alabama, which lies in the Black Warrior River drainage.

The systematic placement of *Anodontoides radiatus* has been a matter of conjecture. It has been placed in six genera: *Alasmidonta*, *Anodon*, *Anodonta*, *Anodontoides*, *Margaritana* and *Strophitus*.

Synonymy

Alasmodonta radiata Conrad, 1834. Conrad, 1834a:341, pl. 1, fig. 10
Type locality: Small streams in south Alabama. Holotype, ANSP 41147, length 64 mm.

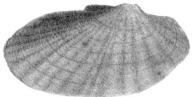

Margaritana elliottii Lea, 1858. Lea, 1858b:138; Lea, 1859f:226, pl. 31, fig. 108; Lea, 1859g:44, pl. 31, fig. 108
Type locality: Chattahoochee River, [below Uchee Bar,] near Columbus, [Muscogee County,] Georgia, Bishop
    Elliott. Lectotype, USNM 86257, length 45 mm, designated by Johnson (1967a).

Margaritana elliptica Lea, 1859. Lea, 1859a:113; Lea, 1862b:106, pl. 18, fig. 254; Lea, 1862d:110, pl. 18, fig. 254
Type locality: Tombigbee River, Columbus, [Lowndes County,] Mississippi, W. Spillman, M.D. Lectotype, USNM
    86258, length 51 mm, designated by Johnson (1967a).

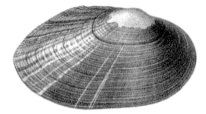  

Anodonta showalterii Lea, 1860. Lea, 1860c:307; Lea, 1862c:215, pl. 33, fig. 284; Lea, 1863a:37, pl. 33, fig. 284
Type locality: Coosa River, Watumpka [Wetumpka], [Elmore County,] Alabama, E.R. Showalter, M.D. Lectotype,
    USNM 86487, length 72 mm, designated by Johnson (1967a).

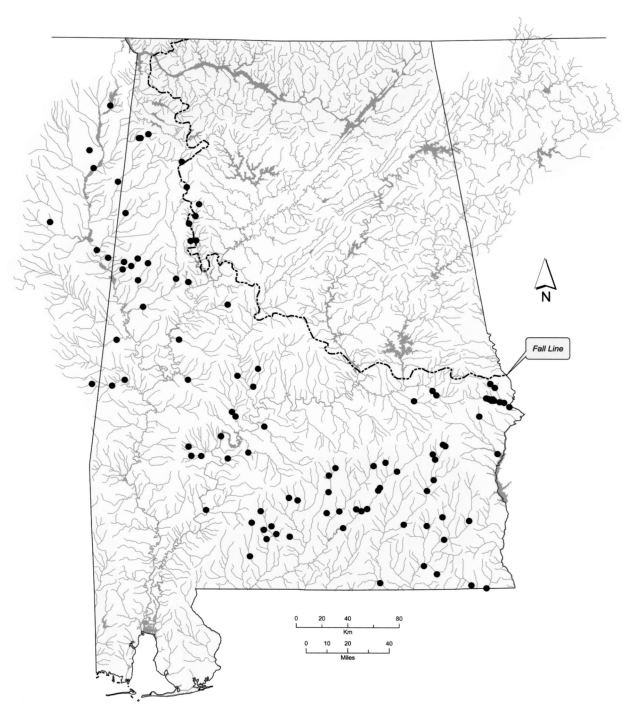

Distribution of *Anodontoides radiatus* in Alabama and the Mobile Basin.

Genus *Arcidens*

Arcidens Simpson, 1900, occurs in the Mississippi Basin and Gulf Coast drainages. In Alabama it is found in the Tennessee River drainage and Mobile Basin. *Arcidens* is a monotypic genus (Turgeon et al., 1998).

In a monograph of the tribe Alasmidontini, Clarke (1981b) treated *Arkansia* Ortmann and Walker, 1912, as a subgenus under *Arcidens*, but Turgeon et al. (1998) continued to recognize *Arkansia* as a genus.

Type Species
Alasmidonta confragosa Say, 1829

Diagnosis
Shell somewhat thin; subrhomboidal; inflated; umbo inflated, umbo sculpture strong, nodulous, double-looped corrugations, nodules form 2 radiating rows that continue across umbo; most of disk covered with fine wrinkles, central and posterior portion of shell with oblique, parallel ridges; interdental projection well-developed, usually fused to posterior pseudocardinal tooth in left valve, corresponding indentation in right valve interrupts interdentum.

Inner lamellae of inner gills only connected to visceral mass anteriorly; excurrent aperture with small papillae; outer gills marsupial; marsupium not extended ventrally beyond original edge of gill when gravid; gravid gills with peculiar granular texture due to visibility of large glochidia beneath thin marsupium wall; glochidium with styliform hooks (Simpson, 1900b; Ortmann, 1912a; Haas, 1969b; Clarke, 1981b).

Synonymy
None recognized.

Arcidens confragosus (Say, 1829)
Rock Pocketbook

 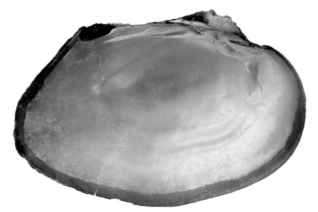

Arcidens confragosus – Length 133 mm, UF 244017. Alabama River, below State Highway 28 bridge, Millers Ferry [Dannelly] Reservoir, 12 miles northwest of Camden, Wilcox County, Alabama, 17 June 1995. © Richard T. Bryant.

Shell Description

Length to 153 mm; moderately thin; inflated; outline oval to rhomboidal; posterior margin broadly rounded to obliquely truncate, forming very blunt point ventrally; anterior margin rounded; dorsal margin straight to slightly convex; ventral margin convex to almost straight; posterior ridge high, rounded; posterior slope very steep dorsally, flat to slightly concave, adorned with corrugations, sometimes also with fine wrinkles; posterior 30% of shell with prominent, oblique, parallel ridges, smaller subradial wrinkles cover most of disk, occasional specimens with no wrinkles on anterior 30% of shell; umbo inflated, elevated well above hinge line, turned slightly anteriad, umbo sculpture strong, nodulous, double-looped corrugations, nodules form 2 radiating rows that continue across umbo; periostracum olive to brown in young individuals, darkening to very dark green or black with age, young individuals may have subtle green rays that usually become obscure with age.

Pseudocardinal teeth compressed, elongate, 2 low teeth in left valve, posterior tooth fused with interdental projection, curved upward and turned posteriorly, extending along hinge line for short distance, 1 large, erect, triangular tooth in right valve; lateral teeth poorly developed or absent; interdentum short in right valve, very short to absent in left valve; umbo cavity moderately deep; nacre white.

Soft Anatomy Description

Mantle creamy white to tan, may be partially gold, especially outside of pallial line, gray to grayish brown outside of apertures; visceral mass creamy white to light tan; foot tan, may be darker distally.

Gills tan to brown, may have slight golden cast, outer gill may be lighter than inner gill; dorsal margin sinuous, ventral margin usually slightly convex, occasionally straight; gill length 50–70% of shell length; gill height usually 40–50% of gill length; outer gill height 70–95% of inner gill height; inner lamellae of inner gills connected to visceral mass only anteriorly.

Outer gills marsupial; glochidia held across entire gill; marsupium well-padded when gravid; tan. Simpson (1900b) described marsupia as having "a peculiar, granular texture". Indeed, the marsupia can appear granular due to visibility of the large glochidia beneath the thin tissues of the marsupium wall. Utterback (1915) reported gills to be dark brown when brooding mature glochidia.

Labial palps tan to brown, often with irregular creamy white or tan area dorsally; elongate, straight to slightly concave dorsally, convex ventrally, bluntly pointed distally; palp length 30–45% of gill length; palp height 40–55% of palp length; distal 25–40% of palps bifurcate.

Incurrent and supra-anal apertures usually of similar length, supra-anal sometimes longer, both longer than excurrent aperture.

Incurrent aperture length 15–20% of shell length; creamy white to light tan within, often gray or grayish brown basal to papillae; papillae in 2–3 rows, simple, long or short, often wide basally, quickly tapering to sharp point; papillae usually some combination of tan, brown, gray and black, larger papillae with darker edges basally, often with white tips.

Excurrent aperture length 10–15% of shell length; light tan to creamy white within, marginal color band usually some combination of gray, tan, brown and black, often with bands perpendicular to margin; margin usually crenulate, occasionally with very small, simple papillae; papillae white to tan.

Supra-anal aperture length 15–30% of shell length; creamy white to light tan within, without marginal coloration, but may be grayish brown outside of margin; margin smooth; mantle bridge separating supra-anal and excurrent apertures occasionally with 1 or more perforations, length usually 10–25% of supra-anal length, occasionally greater, occasionally absent.

Specimens examined: Tennessee River (n = 4); Tombigbee River (n = 4).

Glochidium Description

Length 352–363 μm; height 350–360 μm; with prominent styliform hooks; outline pyriform; dorsal margin straight to undulate, ventral margin narrowly rounded, anterior and posterior margins broadly rounded dorsally, slightly concave just above ventral terminus, posterior margin more deeply rounded than anterior margin (Surber, 1912; Utterback, 1915; Clarke, 1981b; Hoggarth, 1999).

Similar Species

Arcidens confragosus may superficially resemble *Amblema plicata* and *Megalonaias nervosa*. However, *A. confragosus* usually has shallower ridges on the shell disk and posterior slope and a thinner, more inflated shell. Lateral teeth are poorly developed in *A. confragosus*, but well-developed in *A. plicata* and *M. nervosa*.

General Distribution

Arcidens confragosus occurs in most of the Mississippi Basin from Minnesota and Wisconsin (Dawley, 1947) south to Louisiana (Vidrine, 1993), and from the Ohio River drainage in southern Ohio (Cummings and Mayer, 1992) west to eastern Kansas (Murray and Leonard, 1962). *Arcidens confragosus* occurs in lower reaches of the Cumberland River and in the Tennessee River drainage from the vicinity of Muscle Shoals downstream to the mouth of the Tennessee River (Parmalee and Bogan, 1998). Along the Gulf Coast *A. confragosus* occurs from the Mobile Basin, Alabama, west to the Guadalupe River, eastern Texas (Howells et al., 1996).

Alabama and Mobile Basin Distribution

Arcidens confragosus occurs in the Tennessee River in northwestern Alabama and in a few tributaries. In the Mobile Basin it is found only on the Coastal Plain.

Arcidens confragosus remains in the Tennessee River, lower Bear Creek, Colbert County, and the Elk River. It has disappeared from much of the Mobile Basin but still occurs in several reaches of large rivers.

Ecology and Biology

Arcidens confragosus occurs in large creeks and rivers in gravel, sand or mud substrates. It may be found in riverine habitats as well as some overbank areas of reservoirs.

Arcidens confragosus is a long-term brooder, but its brooding period appears to be shorter than those of most other long-term brooders. Tennessee River populations become gravid during September and remain gravid until at least the following February (T.M. Haggerty, personal communication). In Missouri it has been reported gravid with mature glochidia in January and late embryos in March, which is unexpected for a long-term brooder (Utterback, 1915, 1916a).

Ictalurus punctatus (Channel Catfish) (Ictaluridae) has been reported to serve as a glochidial host of *Arcidens confragosus* in laboratory trials (Howells, 1997). Wilson (1916) reported laboratory trials in which encystment occurred on *Anguilla rostrata* (American Eel) (Anguillidae); *Ambloplites rupestris* (Rock Bass) and *Pomoxis annularis* (White Crappie) (Centrarchidae); *Dorosoma cepedianum* (Gizzard Shad) (Clupeidae); and *Aplodinotus grunniens* (Freshwater Drum) (Sciaenidae). However, Wilson (1916) was not clear as to whether transformation occurred.

Current Conservation Status and Protection

Arcidens confragosus is expanding its range upstream in the Tennessee River but has suffered serious declines in the Mobile Basin. It was listed as currently stable throughout its range by Williams et al. (1993) and in Alabama by Lydeard et al. (1999). Garner et al. (2004) designated *A. confragosus* a species of moderate conservation concern in the state.

Remarks

Arcidens confragosus appears to have colonized Alabama reaches of the Tennessee River since its impoundment (Garner and McGregor, 2001). The earliest known museum specimen was collected in 1966 from Wilson Dam tailwaters. Changes in habitat with impoundment of the river may have made conditions more favorable for this species. *Arcidens confragosus* also appears to have colonized Bear Creek, Colbert County, Alabama, since impoundment of the Tennessee River (McGregor and Garner, 2004).

There is one spurious *Arcidens confragosus* record from the Choctawhatchee River drainage in southeastern Alabama. This record is based on a single specimen labeled "Enterprise, Alabama." Clarke

(1981b) added without notation, "Double Bridge Creek" and "Coffee Co." in the geographical records section of his monograph. This was apparently an extrapolation since the stream nearest Enterprise is Double Bridges Creek and it is in Coffee County. However, there are several problems with this record. Double Bridges Creek near Enterprise is small and is not suitable habitat for *A. confragosus*. This is the only record of the species from drainages east of the Mobile Basin, even though systematic surveys of those drainages have been performed during the past decade (Blalock-Herod et al., 2005). There are at least three other Alabama communities named Enterprise, all within the Mobile Basin. Therefore, it appears that the record in question came from the Mobile Basin.

Synonymy

Alasmidonta confragosa Say, 1829. Say, 1829:339; Say, 1831:[no pagination], pl. 21

Type locality: Bayou Teche, St. Mary Parish, Louisiana. Type specimen not found. Length not given in original description.

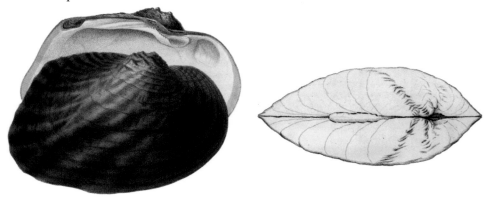

Arcidens confragosa jacintoensis Strecker, 1931. Strecker, 1931:13

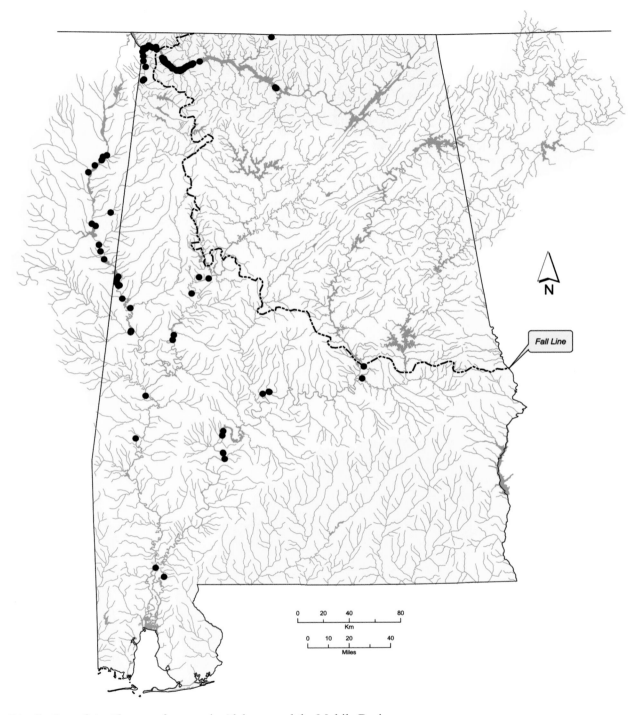

Distribution of *Arcidens confragosus* in Alabama and the Mobile Basin.

Genus *Cyclonaias*

Cyclonaias Pilsbry *in* Ortmann and Walker, 1922a, occurs in the Mississippi and Great Lakes basins. In Alabama it is restricted to the Tennessee River drainage. *Cyclonaias* is a monotypic genus (Turgeon et al., 1998).

This species had been placed erroneously in the genus *Rotundaria* Rafinesque, 1820 (e.g., Ortmann, 1919), but Ortmann and Walker (1922a) pointed out that *Rotundaria* belongs in the synonymy of *Obovaria* Rafinesque, 1820. This species has occasionally been aligned with *Quadrula* Rafinesque, 1820 (e.g., Simpson, 1914; Thiele, 1935).

Type Species
Obliquaria (*Rotundaria*) *tuberculata* Rafinesque, 1820

Diagnosis
Shell round to quadrate; somewhat thick; umbo sculpture broken and irregular concentric ridges; posterior ridge low; posterior portion of shell surface sculptured with pustules; umbo cavity deep, compressed; nacre purple.

Mantle bridge separating excurrent and supra-anal apertures consistently absent; inner lamellae of inner gills only connected to visceral mass anteriorly; outer gills marsupial; marsupium only slightly padded when gravid, not extended ventrally beyond original gill edge when gravid; glochidium without styliform hooks (Utterback, 1915; Ortmann and Walker, 1922a; Haas, 1969b).

Synonymy
Rotundaria Agassiz, 1852, *non* Rafinesque, 1820
(Mollusca: Unionidae)

Cyclonaias tuberculata (Rafinesque, 1820)
Purple Wartyback

Cyclonaias tuberculata – Upper figure: length 69 mm, UF 20568. Lower figure: length 45 mm, UF 20568. Tennessee River, U.S. Highway 231, south of Huntsville, Madison County, Alabama, December 1956. © Richard T. Bryant.

Shell Description

Length to 130 mm; thick; usually compressed, but individuals from large rivers may be slightly inflated; outline round to subquadrate, individuals from small rivers becoming more elongate than those from large rivers; posterior margin truncate to convex, may be slightly concave at junction of dorsal wing; anterior margin rounded; dorsal margin straight to slightly convex, often with low wing-like projection posteriorly; ventral margin broadly convex, may be almost straight in some individuals, especially in those from small rivers; posterior ridge low, rounded; posterior slope broad, flat to concave, often projected to form dorsal wing, some individuals with slight radial sulcus separating posterior ridge from dorsal wing, especially in those from small rivers; all but anterior 30% of surface, including posterior slope, covered with round or elongate tubercles that may form short, irregular, broken ridges distally and on posterior slope, individuals from small rivers may have more numerous tubercles than those from large rivers; umbo moderately inflated in individuals from large rivers, but may be compressed in those from small rivers, elevated above hinge line, umbo sculpture fine, irregular, broken or wavy ridges; periostracum dull, yellowish brown to dark brown, young individuals may have faint green rays.

Pseudocardinal teeth thick, deeply serrate, triangular in outline, 2 divergent teeth in left valve, disconnected dorsally, 1 large tooth in right valve, with accessory denticle anteriorly and posteriorly; lateral teeth short, thick, slightly curved to straight, 2 in left valve, 1 in right valve; interdentum moderately long, wide, flat; umbo cavity deep, compressed; nacre dark to light purple.

Soft Anatomy Description

Mantle tan to light brown, often brown outside of apertures; visceral mass usually creamy white or tan, becoming pearly white adjacent to foot, occasionally entirely pearly white; foot creamy white to tan, often darker distally.

Gills tan; dorsal margin straight to slightly sinuous, ventral margin straight to slightly curved, inner gill usually more convex than outer gill; gill length 65–75% of shell length; gill height 40–50% of gill length; outer gill height 30–70% of inner gill height; inner lamellae of inner gills connected to visceral mass only anteriorly. Utterback (1915) described gill color as dark brown and black in *Cyclonaias tuberculata* from Missouri.

Outer gills marsupial; glochidia held across entire gill except extreme anterior and posterior ends; marsupium slightly padded when gravid; creamy white to tan.

Labial palps tan, often with irregular white patch dorsally; broadly connected to mantle basally, straight to slightly concave dorsally, convex ventrally, bluntly pointed distally; palp length 15–30% of gill length; palp height usually 50–80% of palp length, sometimes greater; distal 45–60% of palps bifurcate.

Incurrent aperture longer than excurrent aperture; supra-anal aperture may be longer, shorter or equal to incurrent and excurrent apertures.

Incurrent aperture length 25–35% of shell length; creamy white to light tan within, usually with thin brown band basal to papillae; papillae in 1 row, arborescent, sometimes deeply branched; papillae of variable shades of brown or tan, often darker along edges. Utterback (1915) reported incurrent aperture papillae to be short and simple.

Excurrent aperture length 20–30% of shell length; creamy white to tan or brown within, marginal color band of variable shades of brown; margin smooth or with very fine crenulations.

Supra-anal aperture length 15–30% of shell length; creamy white to light brown within, with or without thin brown marginal band; margin smooth; mantle bridge separating supra-anal and excurrent apertures absent. Ortmann (1912a) and Utterback (1915) described the condition of continuous supra-anal and excurrent apertures as an absence of a supra-anal aperture. However, in specimens examined for the description herein, there was a distinct demarcation between the wide brown marginal color band of the excurrent aperture and a narrow or absent marginal color band of the supra-anal aperture. This demarcation was used to delineate excurrent and supra-anal apertures.

Specimens examined: Bear Creek, Colbert County (n = 2); Tennessee River (n = 4).

Glochidium Description

Length 267–294 μm; height 325–355 μm; without styliform hooks; outline subelliptical; dorsal margin straight, ventral margin deeply convex, anterior and posterior margins convex (Surber, 1912; Utterback, 1915, 1916a; Jirka and Neves, 1992). Ortmann (1919a) reported glochidia from the Black River, Arkansas, to be considerably smaller (length 250 μm, height 280 μm). Utterback (1915) and Jirka and Neves (1992) provided glochidial measurements that fall within the range given here but the longer measurement was given as length and the shorter measurement as "width".

Similar Species

Cyclonaias tuberculata most closely resembles *Plethobasus cooperianus* but has purple nacre instead of white, though *P. cooperianus* may be pale pink inside the pallial line. *Cyclonaias tuberculata* often has the posterior slope expanded into a moderately developed dorsal wing, which is absent or very poorly developed in *P. cooperianus*. Both *C. tuberculata* and *P. cooperianus* are heavily sculptured with tubercles, but *C. tuberculata* generally has a number of tubercles forming irregular, short, broken ridges, especially on the posterior slope. The periostracum of *P. cooperianus* often has small wrinkles associated with the tubercles, but *C. tuberculata* generally has few wrinkled tubercles. Also, the foot of *P. cooperianus* is orange, whereas that of *C. tuberculata* is white or tan.

Cyclonaias tuberculata may also resemble *Quadrula pustulosa*, but that species typically has a wide green ray on the umbo and shell disk, while *C. tuberculata* is rayless, or only weakly rayed in young individuals. *Quadrula pustulosa* also lacks the dorsal wing found on many *C. tuberculata* shells. Shell nacre of *Q. pustulosa* is always white, compared to the purple nacre of *C. tuberculata*.

General Distribution

Cyclonaias tuberculata is found in parts of the Great Lakes Basin, including lakes Erie and St. Clair (La Rocque and Oughton, 1937). It occurs in the Mississippi Basin from southern Minnesota (Graf, 1997) south to Arkansas (Harris and Gordon, 1990), and from headwaters of the Ohio River drainage in western Pennsylvania (Ortmann, 1909a) west to eastern Oklahoma (Oesch, 1995). A disjunct population occurs in the upper Mississippi Basin of eastern Minnesota and northwestern Wisconsin (Mathiak, 1979). *Cyclonaias tuberculata* is widespread in the Cumberland River drainage downstream of Cumberland Falls (Cicerello et al., 1991; Parmalee and Bogan, 1998) and found throughout the Tennessee River drainage, Alabama, Kentucky, Tennessee and Virginia (Ahlstedt, 1992a, 1992b; Parmalee and Bogan, 1998).

Alabama and Mobile Basin Distribution

Cyclonaias tuberculata is confined to the Tennessee River drainage, where it was historically widespread in the Tennessee River and tributaries.

In the Tennessee River *Cyclonaias tuberculata* is now found primarily in riverine reaches downstream of dams and in upper reaches of reservoirs. The only tributaries in which it is known to be extant are the Elk and Paint Rock rivers and Bear Creek in Colbert County.

Ecology and Biology

Cyclonaias tuberculata inhabits bodies of water ranging from large creeks to large rivers. It frequently occurs in moderate to swift current, in gravel substrate, but it may also occur in high densities in pools of unimpounded rivers and is found in some parts of the Great Lakes. In the Tennessee River *C. tuberculata* occurs primarily in tailwaters of dams at depths that may exceed 6 m. *Cyclonaias tuberculata* is found in some overbank habitats of Tennessee River reservoirs, particularly Pickwick Reservoir. However, they are usually in areas with firm sand and mud substrates where silt deposition is kept to a minimum by wave action (e.g., tops of submerged ridges and shelves along islands). Ahlstedt and McDonough (1993) documented the disappearance of *C. tuberculata* from a 13 km reach of Wheeler Reservoir, where the population of this species was estimated to be 3,440,800 in 1956–1957 (Scruggs, 1960).

Cyclonaias tuberculata is dioecious, but occasional hermaphroditic individuals may be encountered. Only a single specimen of 354 examined from the Tennessee River was found to be hermaphroditic (Haggerty et al., 1995). The youngest reported sexually mature specimen of *C. tuberculata* from New River, Virginia and West Virginia, was six years old (Jirka and Neves, 1992).

Cyclonaias tuberculata is a short-term brooder. In Pickwick Dam tailwaters of the Tennessee River spermatogenesis begins in August or September and continues through July of the following summer, with spawning beginning sometime between January and March and continuing through July (Haggerty et al., 1995). Oogenesis exhibits a similar pattern, with the number of oocytes per follicle and size of oocytes being lowest in August and peaking in April or May (Haggerty et al., 1995). The gametogenic cycle of a population of *C. tuberculata* in New River, Virginia and West Virginia, was reported to parallel that of the Tennessee River population, but spawning ceased in June (Jirka and Neves, 1992). *Cyclonaias tuberculata* is gravid from April through August, but glochidia do not mature until at least late May (Utterback, 1915, 1916a; Ortmann, 1919; Jirka and Neves, 1992; Haggerty et al., 1995). *Cyclonaias tuberculata* glochidia were reported from stream drift in New River only during May (Jirka and Neves, 1992). *Cyclonaias tuberculata* conglutinates are lanceolate and yellowish brown to pale orange (Ortmann, 1919).

Fishes reported to serve as hosts for *Cyclonaias tuberculata* glochidia based on laboratory trials include *Ameiurus melas* (Black Bullhead), *Ameiurus natalis* (Yellow Bullhead), *Ictalurus punctatus* (Channel Catfish) and *Pylodictis olivaris* (Flathead Catfish) (Ictaluridae) (Hove, 1997; Hove et al., 1997).

Current Conservation Status and Protection

Cyclonaias tuberculata was listed as a species of special concern throughout its range by Williams et al. (1993). It is declining in the northern portion of its range (Cummings and Mayer, 1992) but remains one of the more common species in riverine reaches of the Tennessee River in Alabama, so was assigned a status of currently stable in the state by Lydeard et al. (1999). Garner et al. (2004) designated *C. tuberculata* a species of lowest conservation concern in Alabama.

Remarks

Cyclonaias tuberculata can show marked ecophenotypic variation in shell morphology and has often been used as an example in studies regarding this topic (Wilson and Clark, 1914; Ortmann, 1920; Ball, 1922). In headwater rivers and tributary streams *C. tuberculata* is generally compressed and more quadrate in outline, often growing larger than individuals in large rivers (where it was generally referred to as *Cyclonaias tuberculata granifera* in older literature). There has been discussion about number and size of tubercles differing between the two forms, but these differences do not appear to hold true among all populations.

Cyclonaias tuberculata has been reported from prehistoric shell middens along the Tennessee River across north Alabama, but its abundance was variable

(Morrison, 1942; Warren, 1975; Hughes and Parmalee, 1999). However, Morrison (1942) reported it to be "one of the major fractions of the mussel fauna that was used for food" at Muscle Shoals.

Synonymy

Obliquaria (*Rotundaria*) *tuberculata* Rafinesque, 1820. Rafinesque, 1820:103
Type locality: Ohio River and its tributaries. Lectotype, ANSP 20215, designated by Johnson and Baker (1973).
Unio verrucosus Barnes, 1823. Barnes, 1823:123, pl. 5, fig. 6
Unio verrucosus purpureus Hildreth, 1828. Hildreth, 1828:281
Unio graniferus Lea, 1838. Lea, 1838a:69, pl. 19, fig. 60; Lea, 1838c:69, pl. 19, fig. 60
Quadrula (*Rotundaria*) *granifera* var. *pusilla* Simpson, 1900. Simpson, 1900a:795
Quadrula (*Cyclonaias*) *tuberculata utterbackiana* Frierson, 1927. Frierson, 1927:52

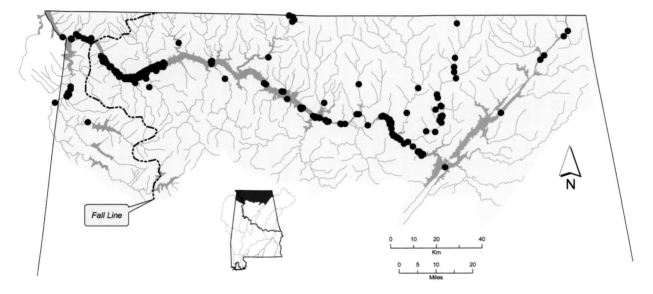

Distribution of *Cyclonaias tuberculata* in the Tennessee River drainage of Alabama.

Genus *Cyprogenia*

Cyprogenia Agassiz, 1852, is endemic to the Mississippi Basin. There are two species of *Cyprogenia* (Turgeon et al., 1998), one of which is found in Alabama, where it is confined to the Tennessee River drainage.

Species of *Cyprogenia* have a unique lateral outgrowth of the gill, which serves as a marsupium, first illustrated by Lea (1828) (Figure 13.4). Conglutinates produced by these species are worm-like: long, slender and curved. There is genetic evidence that *Cyprogenia* may include additional cryptic species in the Ozark region of Arkansas and Missouri (J.M. Serb, personal communication).

Type Species

Unio irroratus Lea, 1828 (= *Obovaria stegaria* Rafinesque, 1820)

Diagnosis

Shell thick; inflated; round; hinge teeth well-developed; shell surface finely sculptured with pustules or tubercles, rather faint in some individuals.

Excurrent aperture crenulate; mantle bridge separating excurrent and supra-anal apertures very short; inner lamellae of inner gills only connected to visceral mass anteriorly; females with finely crenulate, weak mantle fold just ventral to incurrent aperture; marsupium formed by lateral outgrowth of central portion of outer gill, coils posteriorly and inward; conglutinates reddish, fading as glochidia mature, very long, cylindrical to flattened; glochidium without styliform hooks (Agassiz, 1852; Simpson, 1900b, 1914;

Ortmann, 1912a; Chamberlain, 1934; Jones and Neves, 2002b).

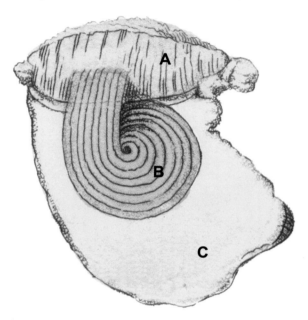

Figure 13.4. *Cyprogenia stegaria* gravid female. Outer gill (A), with specialized marsupium (B) and foot (C). From Lea (1828).

Synonymy

None recognized.

Cyprogenia stegaria (Rafinesque, 1820)
Fanshell

Cyprogenia stegaria – Length 52 mm, UF 64754. Tennessee River, Alabama. © Richard T. Bryant.

Shell Description

Length to 70 mm; thick; inflated; outline round; posterior margin broadly rounded to somewhat truncate; anterior margin broadly rounded; dorsal margin straight to slightly convex; ventral margin broadly rounded; posterior ridge well-developed, sharply angled dorsally, becoming broad and rounded distally; posterior slope flattened to slightly convex, many individuals with shallow radial sulcus on posterior slope and/or anterior to posterior ridge; all but anterior 25% of surface with numerous rounded and lachrymous tubercles, tubercles on center of valve often arranged in radial rows; umbo inflated, elevated well above hinge line, umbo sculpture indistinct double-looped ridges; periostracum greenish yellow to brown, usually with broad rays composed of interconnected dark green blotches.

Pseudocardinal teeth thick, low, roughened, triangular, 2 divergent teeth in left valve, 1 large tooth in right valve, sometimes with accessory denticle anteriorly and/or posteriorly; lateral teeth short, thick, slightly curved, 2 in left valve, 1 in right valve, right valve occasionally with weak secondary lateral tooth ventral to primary tooth; interdentum moderately short, moderately wide, flat; umbo cavity open, shallow; nacre white.

Soft Anatomy Description

No material was available for soft anatomy description. Lea (1863d) provided a brief description based on material from the Ohio River and White River, Indiana. The mantle is "thin, with a broad, thin margin, colored at the edges" and the "mass" is "whitish". The gills are large and rounded ventrally, "very oblique". Outer gills are marsupial, with glochidia brooded in the posterior part. The "ovisacks" were described as "volutes of different lengths". Inner lamellae of inner gills are connected to the visceral mass only anteriorly. Labial palps are small and subtriangular. The incurrent aperture is large, with numerous "minute papillae, slightly colored". The excurrent aperture is large and slightly crenulate. The supra-anal aperture is large, with colored margins, separated from the excurrent aperture by a short mantle bridge. Ortmann (1912a) provided further details based on material from the Ohio and Cumberland rivers. The visceral mass and mantle are "suffused with black" and mantle edges are brown with black spots, the "mottling extending all around". Gills are sinuous dorsally, with inner gill height considerably greater than outer gill height. When gravid the gills are greatly extended, coiled posteriorly and inward, red to white.

Glochidium Description

Length 180–210 µm; height 150–185 µm; without styliform hooks; outline elongate oval; dorsal margin straight and long, ventral margin broadly rounded, anterior and posterior margins broadly rounded (Sterki, 1898a; Ortmann, 1912a, 1919; Surber, 1912; Hoggarth, 1999).

Similar Species

The shell of *Cyprogenia stegaria* resembles that of *Dromus dromas* but differs in shell sculpture. *Cyprogenia stegaria* is adorned with numerous round and lachrymous pustules on the posterior 75 percent of its shell with the pustules often arranged in radial rows. *Dromus dromas* is without numerous well-defined pustules but may have a large, concentric ridge positioned centrally on the shell disk and a radial row of low, irregular knobs down the middle of the shell. *Cyprogenia stegaria* has a shallow umbo cavity, while that of *D. dromas* is deep and compressed. The large ridge of *D. dromas* is generally more prominent on individuals from large rivers than those from tributaries.

General Distribution

Cyprogenia stegaria historically occurred throughout much of the Ohio, Cumberland and Tennessee River drainages (Burch, 1975a). In the Ohio River drainage it is known from headwaters in Pennsylvania (Ortmann, 1909a) downstream to the mouth of the Ohio River, including the Wabash River in Indiana and Illinois, and the Green and Licking rivers in Kentucky (Cicerello et al., 1991; Cummings and Mayer, 1992). *Cyprogenia stegaria* was widespread in the Cumberland River downstream of Cumberland Falls, Kentucky and Tennessee (Cicerello et al., 1991; Parmalee and Bogan, 1998) and historically occurred throughout the Tennessee River drainage (Ahlstedt, 1992a, 1992b; Parmalee and Bogan, 1998).

Alabama and Mobile Basin Distribution

Cyprogenia stegaria is known from the Tennessee River proper, where it historically occurred across the state, and the Elk River.

Cyprogenia stegaria is known to be extant only in Wilson Dam tailwaters, but the population may not be viable.

Ecology and Biology

Cyprogenia stegaria occurs in riverine habitat of medium to large rivers at depths of less than 1 m to more than 6 m. Its preferred substrates are stable, coarse sand and gravel swept free of silt by current.

Cyprogenia stegaria is a long-term brooder, gravid from September or October to late May (Surber, 1912; Ortmann, 1919). It produces conglutinates that resemble oligochaetes. They are 20 mm to 80 mm long and brick-red when containing eggs but fade as glochidia develop (Sterki, 1898a; Jones and Neves, 2002b). *Cyprogenia stegaria* conglutinates protrude from the centers of outer gill demibranchs. Each mature female produces 6 to 14 conglutinates per reproductive season and discharges them from March until May. Conglutinates are released through the excurrent aperture via the suprabranchial chamber. Mean fecundity of four Clinch River females was 43,494 glochidia per female with a positive fecundity to size relationship. Conglutinate contents were estimated to be composed of 30 percent to 50 percent unfertilized eggs or embryos that may play a functional role in maintaining conglutinate structural integrity (Jones and Neves, 2002b).

Fishes found to serve as *Cyprogenia stegaria* glochidial hosts in laboratory trials include *Cottus bairdii* (Mottled Sculpin) and *Cottus carolinae* (Banded Sculpin) (Cottidae); and *Etheostoma blennioides* (Greenside Darter), *Etheostoma simoterum* (Snubnose Darter), *Etheostoma zonale* (Banded Darter), *Percina aurantiaca* (Tangerine Darter), *Percina burtoni* (Blotchside Logperch) and *Percina caprodes* (Logperch) (Percidae) (Jones and Neves, 2002b). One allopatric species, *Percina roanoka* (Roanoke Darter) (Percidae), was reported to serve as a glochidial host under laboratory conditions (Jones and Neves, 2002b).

Current Conservation Status and Protection

Cyprogenia stegaria was considered endangered throughout its range by Williams et al. (1993) and in Alabama by Stansbery (1976). Lydeard et al. (1999) listed *C. stegaria* as extirpated from the state, but it was subsequently rediscovered in Wilson Dam tailwaters, Tennessee River. Garner et al. (2004) designated *C. stegaria* a species of highest conservation concern in Alabama. It was listed as endangered under the federal Endangered Species Act in 1990.

Remarks

Cyprogenia stegaria has been reported from prehistoric shell middens along the Tennessee River across northern Alabama but was uncommon at most sites (Morrison, 1942; Warren, 1975; Hughes and Parmalee, 1999). It is extant at Muscle Shoals in Wilson Dam tailwaters but is very rare. The only recent *C. stegaria* record from Alabama is a single live individual collected adjacent to Sevenmile Island in 2001.

Synonymy

Obovaria stegaria Rafinesque, 1820. Rafinesque, 1820:312, pl. 82, figs. 4, 5
Obovaria stegaria var. *tuberculata* Rafinesque, 1820. Rafinesque, 1820:312
Type locality: Ohio River. Lectotype, ANSP 20241, length 47 mm, designated by Johnson and Baker (1973).

Unio irroratus Lea, 1828. Lea, 1828:269, pl. 5, fig. 5; Lea, 1834b:11, pl. 5, fig. 5
Unio verrucosus albus Hildreth, 1828. Hildreth, 1828:281
Cyprogenia irrorata var. *pusilla* Simpson, 1900. Simpson, 1900a:610

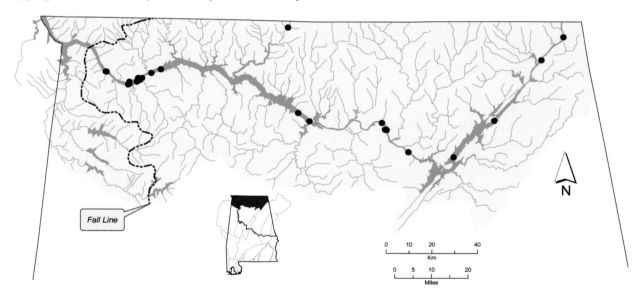

Distribution of *Cyprogenia stegaria* in the Tennessee River drainage of Alabama.

Genus *Dromus*

Dromus Simpson, 1900, is endemic to the Tennessee and Cumberland River drainages. In Alabama it is restricted to the Tennessee River drainage. *Dromus* is a monotypic genus (Turgeon et al., 1998).

Haas (1930) considered *Dromus* Selby, 1840, to have priority over *Dromus* Simpson, 1900, and coined the replacement name *Conchodromus* Haas, 1930. Baker (1964b) pointed out that *Dromus* Selby, 1840, is an "incorrect subsequent spelling" of a genus of birds, so is not a homonym. Thus, *Conchodromus* is an unnecessary replacement name and a junior synonym of *Dromus* Simpson, 1900.

Type Species
Unio dromas Lea, 1834

Diagnosis
Shell round to triangular; often with row of flattened tubercles down shell midline; shell disk often with large, concentric ridge positioned dorsally; hinge teeth well-developed.

Inner lamellae of inner gills only connected to visceral mass anteriorly; without mantle modification ventral to incurrent aperture; incurrent aperture very large, excurrent aperture crenulate; outer gills marsupial; glochidia held across gill except anterior portion; marsupium extended ventrally beyond original gill edge when gravid; glochidia discharged within elongate, flattened, red or white conglutinates; glochidium fabelliform, without styliform hooks (Simpson, 1900b, 1914; Ortmann, 1912a; Haas, 1969b; Jones et al., 2004).

Synonymy
Conchodromus Haas, 1930

Dromus dromas (Lea, 1834)
Dromedary Pearlymussel

Dromus dromas – Length 65 mm, CMNH 61.1108. Tennessee River, Tuscumbia, Colbert County, Alabama. © Richard T. Bryant.

Shell Description

Length to 100 mm; moderately thick; compressed to moderately inflated; outline round to oval or subtriangular; posterior margin rounded; anterior margin truncate but becoming rounded ventrally; dorsal margin convex; ventral margin broadly rounded; posterior ridge weak, disappearing posterioventrally; posterior slope convex; most individuals with large, concentric ridge high on disk, more prominent in individuals from large rivers, some individuals also with medial row of small, irregular tubercles extending from umbo to ventral margin; umbo broad, elevated above hinge line, directed anteriad, umbo sculpture fine concentric ridges; periostracum shiny, background tawny, covered with fine green rays and secondary streaks comprised of more closely set rays and/or broader blotches.

Pseudocardinal teeth low, rough, triangular, 2 divergent teeth in left valve, posterior tooth larger, 1 tooth in right valve, often with accessory denticle anteriorly and/or posteriorly; lateral teeth short, thick, slightly curved, 2 in left valve, 1 in right valve, right valve may have rudimentary lateral tooth ventrally; interdentum moderately long, wide, flat; umbo cavity deep, compressed; nacre white, occasionally with shades of pink or salmon.

Soft Anatomy Description

Mantle pale gray, creamy white outside of pallial line, pale gray outside of apertures with pale rusty orange in irregular bands perpendicular to margin; visceral mass creamy white with grayish cast; foot creamy white.

Mantle margin just ventral to incurrent aperture with weak fold; length of fold approximately 25% of shell length; margin mostly smooth but with small undulations and some minute crenulations; mantle surface adjacent to fold gray but not in distinct color band.

Gills creamy white with grayish cast; dorsal margin slightly sinuous, ventral margin elongate convex; gill length approximately 65% of shell length; gill height approximately 50% of gill length; outer gill height approximately 65% of inner gill height; inner lamellae of inner gills completely connected to visceral mass. Ortmann (1912a) reported the inner lamellae of inner gills to be free from the visceral mass except anteriorly.

No gravid females were available for marsupium description. Jones et al. (2004) reported outer gills to be marsupial with conglutinates typically held in all water tubes. Ortmann (1912a) and Heard and Guckert (1971) reported glochidia to be brooded only in the distal part of the posterior portion of outer gills. Jones et al. (2004) suggested that information used by Heard and Guckert (1970) was based on observations of specimens collected during early autumn before marsupia were fully gravid. Ortmann (1912a) described the marsupial gills as being irregularly folded, red or white when gravid.

Labial palps creamy white with faint gray wash distally and ventrally; straight dorsally, convex ventrally, bluntly pointed distally; palp length approximately 20% of gill length; palp height approximately 75% of palp length; distal 40% of palps bifurcate.

Supra-anal aperture longer than incurrent and excurrent apertures; incurrent aperture longer than excurrent aperture.

Incurrent aperture length approximately 25% of shell length; gray within, darker toward margin, rusty tan basal to papillae, with fine black mottling; papillae in 2 rows, inner row larger, simple, short, thick, blunt; papillae rusty tan with fine black mottling basally, creamy white distally. Ortmann (1912a) reported papillae of the incurrent aperture to gradually change to crenulations anteriorly.

Excurrent aperture length approximately 20% of shell length; dark gray within, marginal color band rusty tan with grayish brown mottling oriented perpendicular to margin, thin creamy white band located proximal to marginal band; margin with minute, irregular crenulations, disappearing dorsally.

Supra-anal aperture length measured from dorsal end of creamy white band and narrowing of marginal band approximately 30% of shell length; gray within, marginal color band rusty tan with grayish brown mottling; margin smooth; mantle bridge separating supra-anal and excurrent apertures absent. Ortmann (1912a) reported the mantle bridge to sometimes be present.

Specimen examined: Clinch River, Tennessee (n = 1).

Glochidium Description

Length 180–230 μm; height 90–120 μm; without styliform hooks; outline fabelliform; dorsal margin straight, ventral margin very broadly rounded, anterior and posterior margins narrowly rounded and equal (Lefevre and Curtis, 1912; Surber, 1912; Ortmann, 1919, 1921; Hoggarth, 1999).

Similar Species

The shell of *Dromus dromas* resembles that of *Cyprogenia stegaria* but differs in shell sculpture. *Dromus dromas* may have a large, concentric ridge located centrally on the shell disk, especially on individuals from large rivers. *Dromus dromas* also usually has a radial row of smaller, low, irregular knobs down the midline of the shell. *Cyprogenia stegaria* is adorned with numerous round and lachrymous pustules on the posterior 75 percent of its shell, the pustules often arranged in radial rows. The umbo cavity of *D. dromas* is deep and compressed, while that of *C. stegaria* is shallow.

General Distribution

Dromus dromas is endemic to the Cumberland and Tennessee River drainages. It was historically widespread in the Cumberland River drainage downstream of Cumberland Falls (Cicerello et al., 1991; Parmalee and Bogan, 1998). *Dromus dromas* also occurred throughout the Tennessee River drainage in Alabama, Kentucky, Tennessee and Virginia (Neves, 1991; Parmalee and Bogan, 1998).

Dromus dromas was found downstream of Muscle Shoals to the mouth of the Tennessee River based on archaeological remains (Casey, 1986; Parmalee and Bogan, 1998; Hughes and Parmalee, 1999). Casey (1986) examined mussels from archaeological sites along the lower Ohio River but did not find any *D. dromas*. However, a single valve of *D. dromas* was recovered from the Wicliffe Mounds located on the bluffs of the Mississippi River, Ballard County, Kentucky, about 3 miles downstream of the mouth of the Ohio River (Wesler, 2001).

Alabama and Mobile Basin Distribution

Dromus dromas occurred across the state in the Tennessee River and some tributaries. Ortmann (1925) reported it from the Elk River in Alabama, but no museum specimens were found.

Dromus dromas was extirpated from Alabama during the early 1900s. However, a reintroduction program is underway in tailwaters of Wilson Dam, Tennessee River.

Ecology and Biology

Dromus dromas is a species of shoal habitat in medium to large rivers. Its preferred substrate is a stable mixture of gravel and sand swept clean of silt by current. *Dromus dromas* usually remains well-buried in the substrate (Neves, 1991).

Dromus dromas is a long-term brooder, gravid from September until the following spring or early summer (Coker et al., 1921; Ortmann, 1921; Parmalee and Bogan, 1998; Jones et al., 2004). Glochidia are held in conglutinates that are elongate and flattened, in the shape of leeches, 20 mm to 50 mm long and 4 mm to 7 mm wide. Eggs or partially developed embryos comprise 50 percent to 70 percent of conglutinate content, primarily located in the interior, which are thought to help maintain conglutinate integrity. Some conglutinates are white, but others are red when containing immature glochidia, grading to pink as glochidia mature (Jones et al., 2004). Ortmann (1912a) suggested that glochidia are located on the exterior of the conglutinate to be as near as possible to the marsupium wall and thus closer to "fresh water" being flushed over the outer surface and described it as "one of the little special devices for the proper aeration of the glochidia". Ortmann (1912a) stated that glochidia appear to be connected to conglutinates by "fine threads, possibly embryonal threads". Conglutinates are discharged via the suprabranchial chamber and excurrent aperture (Jones et al., 2004). Most other members of the Lampsilinae (long-term brooders) discharge glochidia individually or in loose or irregular masses (Ortmann, 1910b). In the Clinch River, Tennessee and Virginia, *D. dromas* was reported to produce 33 to 152 conglutinates and approximately 55,000 to 253,000 glochidia per female per year (Jones et al., 2004). Most conglutinate discharge occurs between March and late April (Jones et al., 2004).

Fishes reported to serve as primary glochidial hosts in laboratory trials include *Etheostoma blennioides* (Greenside Darter), *Etheostoma flabellare* (Fantail Darter), *Percina aurantiaca* (Tangerine Darter), *Percina burtoni* (Blotchside Logperch) and *Percina evides* (Gilt Darter) (Percidae) (Jones et al., 2004). Secondary hosts include *Etheostoma simoterum* (Snubnose Darter), *Percina caprodes* (Logperch) and *Percina copelandi* (Channel Darter) (Percidae), as well as *Cottus baileyi* (Black Sculpin) (Cottidae) (Jones et al., 2004). One allopatric species, *Percina roanoka* (Roanoke Darter) (Percidae), was also reported to serve as a secondary glochidial host for *Dromus dromas* in laboratory trials (Jones et al., 2004).

Growth estimates for *Dromus dromas* in the Clinch River were reported to average 5 mm per year through age 10, then decreased to 1.8 mm per year (Jones et al., 2004).

Current Conservation Status and Protection

Dromus dromas was listed as endangered throughout its range by Stansbery (1970a) and Williams et al. (1993). Stansbery (1976) listed *D. dromas* as endangered but suggested that it was extirpated from Alabama. Lydeard et al. (1999) also listed it as extirpated from the state. Garner et al. (2004) designated *D. dromas* a species of highest conservation concern in Alabama. *Dromus dromas* was listed as endangered under the federal Endangered Species Act in 1976.

Dromus dromas was included on a list of species approved for a Nonessential Experimental Population in Wilson Dam tailwaters in 2001. A trial transplantation of 80 individuals was carried out in 2003. Recovery of almost 25 percent of transplanted individuals in 2005 suggests that conditions are favorable for the species.

Remarks

Dromus dromas exhibits clinal variation in shell morphology with specimens from headwaters and tributaries being more compressed, with thinner shells and a reduced concentric ridge or hump near the umbo. However, the two forms clearly intergrade (Wilson and Clark, 1914; Ortmann, 1920). Headwater specimens may have several low knobs down the middle of the shell instead of a dorsal hump. In some early literature the headwater form is referred to as *Dromus dromas caperatus* (Lea, 1845).

Dromus dromas was apparently one of the more abundant species in the Tennessee River prehistorically. It has been one of the major components of material recovered from middens along the Tennessee River across northern Alabama (Morrison, 1942; Warren, 1975; Hughes and Parmalee, 1999).

Wilson and Clark (1914) stated that "the size, shape and solidity of the shell of this species make it suitable for the manufacture of buttons, but unfortunately it is too brittle and hard".

Synonymy

Unio dromas Lea, 1834. Lea, 1834a:70, pl. 10, fig. 29; Lea, 1834b:182, pl. 10, fig. 29
Type locality: Harpeth River; Cumberland River, near Nashville, [Davidson County,] Tennessee. Lectotype, USNM
    25852, length 48 mm, designated by Johnson (1974), is from Harpeth River.

 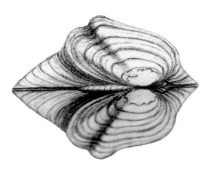 

Unio caperatus Lea, 1845. Lea, 1845:164; Lea, 1848a:75, pl. 5, fig. 14; Lea, 1848b:75, pl. 5, fig. 14
Unio abacoides Haldeman, 1846. Haldeman, 1846:75

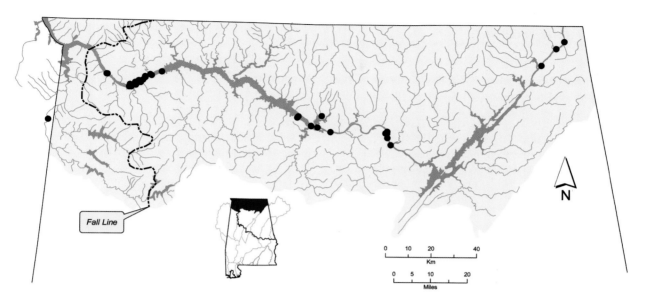

Distribution of *Dromus dromas* in the Tennessee River drainage of Alabama.

Genus *Ellipsaria*

Ellipsaria Rafinesque, 1820, is widespread in the Mississippi Basin and in Gulf Coast drainages from Alabama to Louisiana. In Alabama it occurs in the Tennessee River drainage and Mobile Basin. *Ellipsaria* is a monotypic genus (Turgeon et al., 1998).

The genus name *Plagiola* Rafinesque, 1820, was long used incorrectly based on a false assumption of the type species (Bogan, 1997). Frierson (1914) used *Unio fasciolaris* Rafinesque, 1820, as the type of *Ellipsaria* by elimination, but this is a subsequent type species designation and is invalid. Ortmann and Walker (1922a) pointed out that *Obliquaria ellipsaria* Rafinesque, 1820, is the type species of *Ellipsaria* by absolute tautonomy. The genus *Plagiolopsis* Thiele, 1934, was erected to replace the confusion of *Plagiola*. Baker (1964a) subsequently noted the priority of *Obliquaria lineolata* Rafinesque, 1820, over *O. ellipsaria* Rafinesque, 1820, based on Conrad (1834b) as the first reviser of the species.

Type Species
Obliquaria (*Ellipsaria*) *ellipsaria* Rafinesque, 1820
    (= *Obliquaria* (*Ellipsaria*) *lineolata* Rafinesque, 1820)

Diagnosis
Shell subtriangular; disk smooth; umbo elevated above hinge line; male shell compressed, posterior ridge sharp; female shell more inflated, with more rounded posterior ridge; umbo sculpture faint, double-looped ridges; hinge teeth well-developed.

Excurrent aperture smooth or crenulate; mantle bridge separating excurrent and supra-anal apertures very short; inner lamellae of inner gills only connected to visceral mass anteriorly; mantle margin thickened just ventral to incurrent aperture, with some fine dentations but no papillae, better developed in females; outer gills marsupial, glochidia held in posterior part of gill but not extending completely to posterior end; marsupium outline reniform, well-defined when gravid, rounded ventrally, extended ventrally beyond original gill edge when gravid; conglutinates lanceolate; glochidium without styliform hooks (Ortmann, 1912a; Simpson, 1914; Baker, 1928a; Haas, 1969b).

Synonymy
Plagiolopsis Thiele, 1934
Crenodonta Schlüter, 1838

Ellipsaria lineolata (Rafinesque, 1820)
Butterfly

Ellipsaria lineolata – Upper figure: female, length 63 mm, UF 370412. Tombigbee River, Memphis Landing, river mile 324.4, Pickens County, Alabama. Lower figure: male, length 54 mm, UF 64981. Coosa River, Three Island Shoals near Wilsonville, Shelby County, Alabama, November 1911. © Richard T. Bryant.

Shell Description

Length to 130 mm; thick; males compressed, females moderately to highly inflated; outline triangular; posterior margin bluntly pointed to narrowly rounded, females more rounded than males; anterior margin rounded; dorsal margin straight; ventral margin broadly rounded; valves gape anteriorly and posteriorly; posterior ridge sharp, distinct, located near and parallel to dorsal margin; posterior slope very steep; umbo very broad, inflated in females, compressed in males, oriented anteriad, umbo sculpture fine, irregular, slightly double-looped ridges; periostracum dull to slightly shiny, tawny to greenish yellow, with variable, thin green or greenish brown rays that may be comprised of chevrons or blotches, rays may be more prominent on males than females, periostracum may become dark brown with age, obscuring rays.

Pseudocardinal teeth massive, erect, triangular, deeply serrate, 2 divergent teeth in left valve, anterior tooth smaller than posterior tooth, 1 large tooth in right valve, may have small accessory denticle anteriorly and/or posteriorly; lateral teeth short in females, longer in males, straight to slightly curved, 2 in left valve, 1–2 in right valve; interdentum short, wide; umbo cavity shallow; nacre white.

Soft Anatomy Description

Mantle creamy white to light brown; visceral mass creamy white to pearly white; foot creamy white to tan, may be slightly darker distally.

Females with narrow, fleshy fold along mantle margin just ventral to incurrent aperture; fold length usually 40–60% of shell length, occasionally less; usually with narrow gray or black band on adjacent margin; fold margin often with variously developed crenulations or widely spaced simple papillae; males with rudimentary fold.

Gills creamy white to light brown; dorsal margin slightly sinuous to concave, curved around posterior adductor muscle, ventral margin convex; gill length 50–70% of shell length; gill height 40–55% of gill length, occasionally greater; outer gill height 65–85% of inner gill height, but greater in marsupial area of females (may exceed 100%); inner lamellae of inner gills connected to visceral mass only anteriorly.

Outer gills marsupial; glochidia held in posterior 50–75% of gill; marsupium oblong, reniform in outline; creamy white to pale tan. Lea (1863d) reported the ventral margins of marsupia to be "blackish".

Labial palps creamy white to light brown; short, straight to slightly concave dorsally, convex ventrally, round to bluntly pointed distally, broadly connected to visceral mass; palp length 15–30% of gill length; palp height 60–90% of palp length, occasionally greater; distal 30–70% of palps bifurcate.

Incurrent aperture longer than excurrent aperture; supra-anal aperture longer than incurrent aperture.

Incurrent aperture length 10–20% of shell length; creamy white within, may have sparse black pigmentation basal to papillae; papillae in 2 rows, simple, usually short, thick and blunt, occasionally long and slender, may be slightly pointed distally; papillae color variable, ranging from creamy white to pale orange or rusty brown, black pigment may be present between papillae, some specimens with black extending down onto aperture wall.

Excurrent aperture length 5–10% of shell length; creamy white to pale tan or light golden brown within, marginal color band pale orange to dark brown, occasionally creamy white, with or without regular black markings, tan spots or irregular black blotches; margin papillate, papillae very small, same color as marginal band, occasionally replaced by crenulations.

Supra-anal aperture length 20–30% of shell length; creamy white within, with or without a thin, sparsely pigmented black marginal band; margin smooth; mantle bridge separating supra-anal and excurrent apertures occasionally perforate, length 5–15% of supra-anal length.

Specimens examined: Alabama River (n = 3); Bear Creek, Colbert County (n = 1); Coosa River (n = 3); Locust Fork (n = 3); Sipsey River (n = 1); Tennessee River (n = 5); Tombigbee River (n = 4).

Glochidium Description

Length 229–260 μm, height 310–350 μm; without styliform hooks; outline subligulate; dorsal margin slightly convex, ventral margin broadly rounded, anterior and posterior margins slightly concave dorsally, becoming straight ventrally (Ortmann, 1912a; Surber, 1912; Utterback, 1916b; Hoggarth, 1999).

Similar Species

Some large female *Ellipsaria lineolata* may resemble large male *Lampsilis abrupta* but differ in having a sharper posterior ridge and often a more elevated umbo. *Ellipsaria lineolata* typically has rays that are comprised of chevrons or blotches, though they may be obscure in old, dark shells. *Lampsilis abrupta* is frequently without rays, but when present they are not comprised of chevrons or blotches, though they may be interrupted.

General Distribution

In the Mississippi Basin *Ellipsaria lineolata* is known from Minnesota (Dawley, 1947) south to Louisiana and southwestern Mississippi (Vidrine, 1993), and from headwaters of the Ohio River drainage in Pennsylvania (Ortmann, 1909a) west to eastern Kansas (Murray and Leonard, 1962). *Ellipsaria lineolata* is widespread in the Cumberland River downstream of Cumberland Falls (Cicerello et al., 1991; Parmalee and Bogan, 1998). It occurs throughout the Tennessee River proper and some large tributaries but is absent from headwater rivers (Parmalee and Bogan, 1998). On the Gulf Coast *E. lineolata* is known only from the Mobile Basin in Alabama, Georgia and Mississippi.

Alabama and Mobile Basin Distribution

In the Tennessee River *Ellipsaria lineolata* occurs across northern Alabama and is also in several tributaries. Ortmann (1925) reported *E. lineolata* from Limestone Creek, Limestone County, but no museum records were found. In the Mobile Basin *E. lineolata* occurs in the Tombigbee, Black Warrior, Alabama, Cahaba and Coosa rivers and some tributaries.

Ellipsaria lineolata is extant in some riverine reaches of large rivers, as well as a few tributaries. It is fairly common in the Tennessee River in Wilson and Guntersville Dam tailwaters but is generally less common in the Mobile Basin.

Ecology and Biology

Ellipsaria lineolata occurs in rivers and lower reaches of large creeks, typically in flowing water, including shoals and tailwaters of dams. Preferred substrates are sand and gravel with minimal silt. Depths range from less than 1 m in unimpounded reaches to more than 7 m in tailwaters of dams on large rivers. *Ellipsaria lineolata* is occasionally found in overbank habitats in Tennessee River reservoirs. However, they are usually in areas with firm sand and mud substrates where silt deposition is kept to a minimum by wave action (e.g., tops of submerged ridges and shelves along islands).

Ellipsaria lineolata is a long-term brooder and may be found gravid during any month of the year (Coker et al., 1921), but primarily September through June. *Aplodinotus grunniens* (Freshwater Drum) (Sciaenidae) is the only fish reported to serve as host for *E. lineolata* glochidia based on laboratory trials (Howard, 1914c; Coker et al., 1921). Species on which natural infestations of *E. lineolata* glochidia have been reported include *Lepomis cyanellus* (Green Sunfish) (Centrarchidae) and *Sander canadensis* (Sauger) (Percidae) (Surber, 1913; Coker et al., 1921). However, Coker et al. (1921) questioned the importance of *L. cyanellus* as a host for *E. lineolata*.

Current Conservation Status and Protection

Ellipsaria lineolata was listed as a species of special concern throughout its range by Williams et al. (1993), but Lydeard et al. (1999) listed it as currently stable in Alabama. Garner et al (2004) designated *E. lineolata* a species of low conservation concern in the state.

Ellipsaria lineolata was removed from a list of legally harvestable mussel species in 2004 to reduce the risk of incidental take of *Lampsilis abrupta*, which it may resemble in shell morphology.

Remarks

Ellipsaria lineolata appears to attain a larger size in the Mobile Basin than in the Tennessee River drainage. Parmalee and Bogan (1998) reported a maximum length of 110 mm for males and 70 mm for females. However, males more than 120 mm in length and females more than 100 mm are routinely encountered in some parts of the Mobile Basin (e.g., Locust Fork and Coosa River).

Ellipsaria lineolata is rare in prehistoric middens along the Tennessee River in Alabama (Morrison, 1942; Warren, 1975; Hughes and Parmalee, 1999). If its scarcity in these middens reflects its relative abundance prehistorically, it may be adapting better than many other species to modern habitat perturbations.

Synonymy

Obliquaria (Plagiola) depressa Rafinesque, 1820. Rafinesque, 1820:302, pl. 81, figs. 5–7
Type locality: Ohio River. Neotype, ANSP 20207, length 50 mm, designated by Johnson and Baker (1973).
Comment: Conrad (1834b), as first reviser, chose *lineolata* as the senior taxon and placed the other two Rafinesque taxa, *Obliquaria depressa* and *Obliquaria ellipsaria*, in synonymy. Baker (1964a) pointed out *Ellipsaria* Rafinesque, 1820, is the correct generic placement of *Obliquaria lineolata*, with *O. ellipsaria* the type species, by absolute tautonymy.

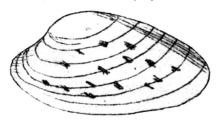

Obliquaria (Ellipsaria) ellipsaria Rafinesque, 1820. Rafinesque, 1820:303
Type locality: Ohio River. Lectotype, ANSP 20233, length 58.5 mm, designated by Johnson and Baker (1973).
Obliquaria (Plagiola) lineolata Rafinesque, 1820. Rafinesque, 1820:303
Type locality: Ohio River. Lectotype, ANSP 20242, length 71 mm, designated by Johnson and Baker (1973).
Unio securis Lea, 1829. Lea, 1829:437, pl. 11, fig. 17; Lea, 1834b:51, pl. 11, fig. 17
Unio compressus Rafinesque, 1831. Rafinesque, 1831:4 [unnecessary replacement name for *Obliquaria depressa* Rafinesque, 1820]

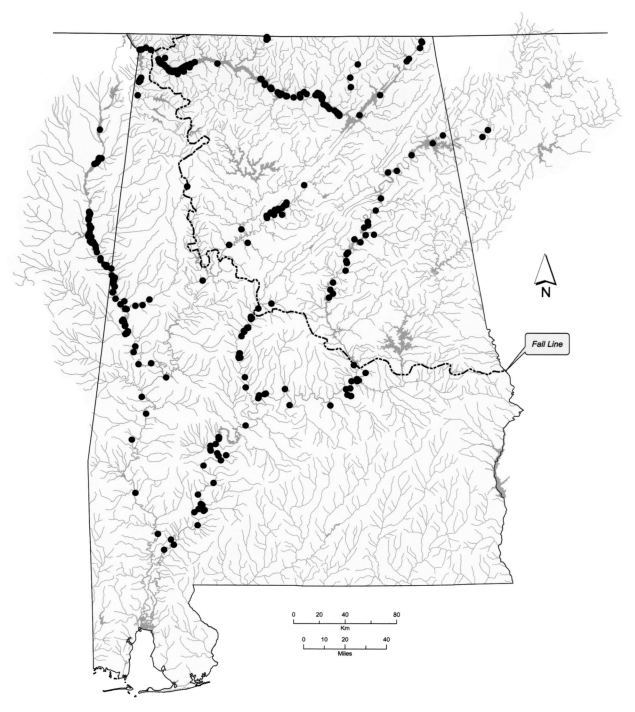

Distribution of *Ellipsaria lineolata* in Alabama and the Mobile Basin.

Genus *Elliptio*

Elliptio Rafinesque, 1819, occurs in the Hudson Bay, Great Lakes and Mississippi basins, Atlantic Coast drainages from Nova Scotia to peninsular Florida, and Gulf Coast drainages from peninsular Florida to Louisiana. As currently understood *Elliptio* is the most diverse genus of North American unionoids. Thirty-six species were recognized by Turgeon et al. (1998). Eleven *Elliptio* species are herein recognized from Alabama, of which three, *E. fumata* (Lea, 1857), *E. pullata* (Lea, 1856) and *E. purpurella* (Lea, 1857), were not included in Turgeon et al. (1998). *Elliptio fumata* and *E. pullata* are herein recognized based on shell morphology and zoogeographic patterns. *Elliptio purpurella* was previously elevated from synonymy by Brim Box and Williams (2000) based on shell morphology and electrophoretic data.

Agassiz (1852) and Simpson (1900b) used the genus name *Eurynia* Rafinesque, 1819, with the type species as *Unio rectus* Lamarck, 1819, but Herrmannsen (1847) had designated *Unio dilatatus* Rafinesque, 1820, as the type species of *Eurynia*, making *Eurynia* a junior synonym of *Elliptio*. Elaboration on the gender of *Elliptio*, a feminine noun, was provided by Baker (1964d).

Two different types of *Elliptio* glochidia were described by O'Brien et al. (2003). This suggests that the genus may be polyphyletic.

Type Species
Unio (*Elliptio*) *nigra* Rafinesque, 1820 (= *Unio crassidens* Lamarck, 1819)

Diagnosis
Shell elongate, varying from lanceolate to rectangular; thin to thick; shell surface smooth, except for spines or weak corrugations on a few species; umbo low, located near anterior end, umbo sculpture low concentric ridges; hinge plate narrow, but pseudocardinal and lateral teeth well-developed.

Excurrent aperture with simple papillae; mantle bridge separating excurrent and supra-anal apertures short; inner lamellae of inner gills only connected to visceral mass anteriorly; outer gills marsupial; glochidia held across entire gill; marsupium smooth and padded when gravid, not extended ventrally beyond original gill edge; glochidium without styliform hooks, may have weak, triangular hooks with microstylets (Ortmann, 1912a; Haas, 1969a; Oesch, 1995; O'Brien et al., 2003).

Synonymy
Eurynia Rafinesque, 1819
Cunicula Swainson, 1840
Canthyria Swainson, 1840
Ensinaia Starobogatov, 1970

Elliptio arca (Conrad, 1834)
Alabama Spike

Elliptio arca – Length 79 mm, UF 64872. Chattooga River, Lyerly, Chattooga County, Georgia, August 1915. © Richard T. Bryant.

Shell Description

Length to 90 mm; moderately thick; moderately inflated; outline elongate elliptical; posterior margin bluntly pointed to narrowly rounded or narrowly truncate; anterior margin rounded; dorsal margin convex; ventral margin straight to slightly convex or slightly concave; posterior ridge variable, well-defined and sharp to low and rounded, may be doubled posterioventrally; posterior slope typically steep, flat; umbo broad, flat, only slightly elevated above hinge line, umbo sculpture longitudinal bars, sometimes double-looped; periostracum green to brown or almost black, may have variable dark green rays.

Pseudocardinal teeth low, variable, 2 teeth in left valve, triangular and divergent to compressed and almost parallel, 1 well-developed, triangular or compressed tooth in right valve, sometimes with rudimentary accessory denticle anteriorly; lateral teeth short, straight, 2 in left valve, 1 in right valve; interdentum long, very narrow; umbo cavity wide, shallow; nacre variable, ranging from pink or salmon to dark purple, rarely white.

Soft Anatomy Description

Mantle tan to light brown; visceral mass creamy white; foot creamy white to light tan, may be slightly darker distally.

Gills tan to light brown; dorsal margin slightly sinuous, ventral margin convex to almost straight; gill length 50–60% of shell length; gill height 35–45% of gill length; outer gill height 65–90% of inner gill height; inner lamellae of inner gills connected to visceral mass only anteriorly.

Outer gills marsupial; glochidia held across entire gill except extreme anterior and posterior ends; marsupium slightly padded when gravid; light tan.

Labial palps creamy white to tan; straight to slightly concave dorsally, convex ventrally, bluntly pointed to rounded distally; palp length 20–30% of gill length; palp height 40–70% of palp length; distal 25–40% of palps bifurcate.

Incurrent aperture longer than or equal to excurrent aperture; supra-anal aperture longer than or equal to incurrent aperture.

Incurrent aperture length 10–15% of shell length; creamy white within, may be gray basal to papillae; papillae in 1–2 rows, long, slender, mostly simple, occasionally with some bifid; papillae white or gray, may have reddish brown tips, with black pigmentation between papillae. Lea (1863d) described incurrent aperture papillae as numerous, small and "brownish".

Excurrent aperture length 10–15% of shell length; creamy white within, marginal color band light brown to black; margin papillate, papillae small, simple, gray to grayish brown, with black pigment between papillae. Lea (1863d) described excurrent aperture papillae as "blackish".

Supra-anal aperture length 10–20% of shell length; creamy white within, with thin, brown or gray marginal band; margin smooth; mantle bridge separating supra-anal and excurrent apertures present or absent, when present imperforate, length 10–15% of supra-anal length.

Specimens examined: Coosa River (n = 2); Sipsey River (n = 3); Yellow Creek, Lowndes County, Mississippi (n = 3).

Glochidium Description

Length 225–238 μm; height 225–238 μm; without styliform hooks; outline depressed subelliptical; dorsal margin straight, ventral margin rounded, anterior and posterior margins convex.

Similar Species

Elliptio arca may resemble some elongate *Elliptio crassidens* but has lower, less massive pseudocardinal teeth. The interdentum of *E. arca* is usually longer and often narrower than that of *E. crassidens*. Also, the interdentum of *E. arca* is narrowest adjacent to the pseudocardinal teeth, but those of *E. crassidens* may be widest near the pseudocardinals. The ventral margin of the shell is usually straight, or nearly so, in *E. arca* but is usually convex in *E. crassidens*. The periostracum of *E. crassidens* is typically darker brown than that of *E. arca* and is usually without rays or with only weak rays whereas *E. arca* has variable dark green rays that may be prominent. *Elliptio arca* does not grow as large as *E. crassidens*.

Elliptio arca may also resemble *Elliptio arctata*, but the former has a much thicker, less arcuate shell. Pseudocardinal teeth of *E. arca* are larger and lateral teeth heavier than those of *E. arctata*. Lateral teeth of *E. arca* are slightly widened posteriorly, but those of *E. arctata* are not. The nacre of *E. arca* is variable in color but seldom white, whereas the nacre of *E. arctata* is usually white but may have a purplish cast. *Elliptio arca* grows larger than *E. arctata*.

Elliptio arca may superficially resemble *Ptychobranchus foremanianus* or *Ptychobranchus greenii* but is more elongate and less triangular in outline. The angle formed by the pseudocardinal teeth and interdentum is more obtuse in *E. arca* than in *Ptychobranchus*, which are often near 90°. The shell cavity is much deeper in *Ptychobranchus* than in *E. arca*.

General Distribution

Elliptio arca is known from several Gulf Coast drainages including the Mobile Basin, the Pascagoula and Pearl River drainages in Mississippi and the Amite River, a tributary to Lake Pontchartrain in Louisiana (Vidrine, 1993).

Alabama and Mobile Basin Distribution

Elliptio arca is confined to the Mobile Basin, where it is known from all major river systems, including many tributaries, above and below the Fall Line.

Elliptio arca is currently known to be extant in several disjunct populations, though it is uncommon everywhere except the Sipsey River and Yellow Creek, of the upper Tombigbee River drainage.

Ecology and Biology

Elliptio arca occurs in medium creeks to large rivers, usually in sand and gravel substrates. Though often found in water less than 1 m deep, it may also occur at depths of more than 6 m in large rivers.

Elliptio arca reaches sexual maturity at two years of age (Haag and Staton, 2003). However, in the Sipsey River, Alabama, trematode infestation has been reported to cause sterility in a small percentage of individuals (3.3 percent) (Haag and Staton, 2003). Trematodes were found only in older individuals. *Elliptio arca* is a short-term brooder, gravid from late May through late July, with glochidia becoming mature in late June (Haag and Warren, 2003). Haag and Staton (2003) reported 93 percent of mature females to be gravid during peak brooding period, with a low percentage (1.5 percent) of undeveloped eggs. Immature glochidia may be aborted as conglutinates when a gravid female is disturbed, but mature glochidia are released individually (Haag and Staton, 2003; Haag and Warren, 2003). Average fecundity of *E. arca* was reported as 136,227 glochidia per year, but ranged from 19,300 to 206,875 per individual (Haag and Staton, 2003).

Fishes found to serve as hosts for *Elliptio arca* glochidia in laboratory trials are *Etheostoma artesiae* (Redspot Darter) and *Percina nigrofasciata* (Blackbanded Darter) (Percidae). A secondary host for *E. arca* is *Ammocrypta meridiana* (Southern Sand Darter) (Percidae) (Haag and Warren, 2003).

Current Conservation Status and Protection

Elliptio arca was listed as threatened throughout its range by Williams et al. (1993). In Alabama it was designated as endangered by Stansbery (1976), imperiled by Lydeard et al. (1999) and a species of highest conservation concern by Garner et al. (2004).

Remarks

In some literature prior to the 1970s, *Elliptio arca* was variously treated as a synonym (e.g., van der Schalie, 1939b), variety (e.g., Simpson, 1914) or subspecies (e.g., Frierson, 1927) of *Elliptio dilatata*, which is confined to the Mississippi and Great Lakes basins. Stansbery (1976) recognized *E. arca* as a distinct species.

Synonymy

Unio arcus Conrad, 1834. Conrad, 1834a:340, pl. l, fig. 8

Type locality: Alabama River, [Alabama]. Lectotype, ANSP 56784a, length 51 mm, designated by Johnson and Baker (1973).

Unio luridus Lea, 1852, *nomen nudum*. Lea, 1852a:251

Unio luridus Lea, 1852. Lea, 1852b:273, pl. 20, fig. 29; Lea, 1852c:29, pl. 20, fig. 29

Type locality: Coosawattee River, Murray County, Georgia, Dr. Boykin. Lectotype, USNM 85253, length 49 mm, designated by Johnson (1974).

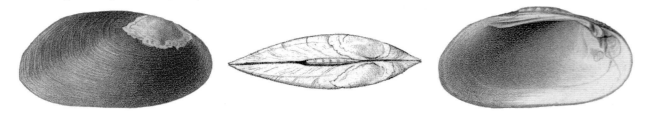

Unio subgibbosus Lea, 1857. Lea, 1857g:169; Lea, 1858e:53, pl. 6, fig. 36; Lea, 1858h:53, pl. 6, fig. 36

Type locality: Oostenaula [Oostanaula] River, Floyd County, and Etowah River, Georgia, Reverend G. White. Lectotype, USNM 86101, length 43 mm, designated by Johnson (1974), is from Oostanaula River, Floyd County, Georgia.

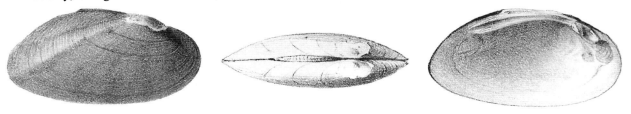

Unio rufus Lea, 1857. Lea, 1857g:171; Lea, 1858e:85, pl. 17, fig. 65; Lea, 1858f:85, pl. 17, fig. 65

Type locality: Etowah River, Cass [now Bartow] County, Georgia, Bishop Elliott. Lectotype, USNM 86070, length 41 mm, designated by Johnson (1974).

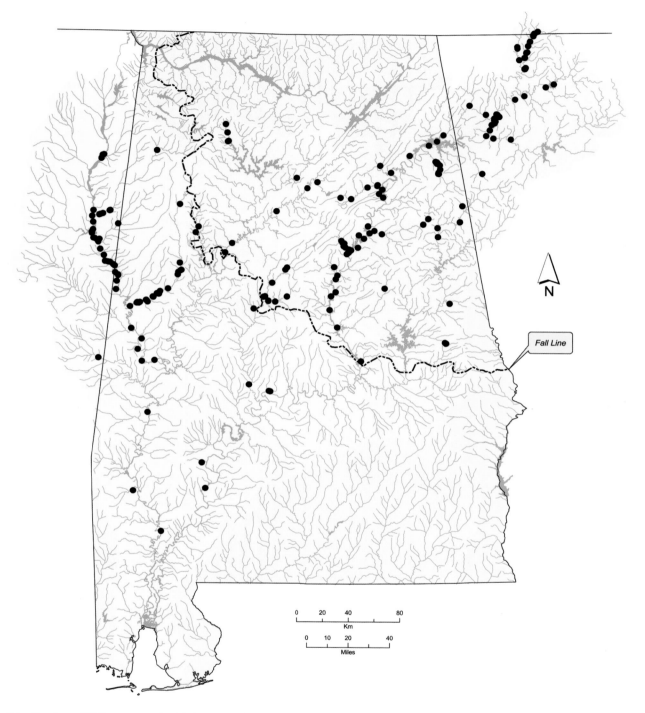

Distribution of *Elliptio arca* in Alabama and the Mobile Basin.

Elliptio arctata (Conrad, 1834)
Delicate Spike

Elliptio arctata – Upper figure: length 77 mm, UF 2670. Middle figure: length 48 mm, UF 2670. Jefferson County, Alabama, September 1911. Lower figure: length 60 mm, UF 370415. Sepulga River, Flat Rock Ford, Conecuh County, Alabama, 30 August 2000. © Richard T. Bryant.

Shell Description

Length to 90 mm; thin; compressed to slightly inflated; outline elongate elliptical, older individuals often arcuate; posterior margin truncate to rounded or bluntly pointed; anterior margin rounded; dorsal margin straight to slightly convex; ventral margin straight to slightly concave; posterior ridge low, rounded, sometimes doubled posterioventrally; posterior slope low, flat to slightly concave; umbo low, broad, not elevated above hinge line, umbo sculpture unknown; periostracum dark olive to brown or black, occasionally with variable dark green rays.

Pseudocardinal teeth small, low, triangular, 2 widely separated teeth in left valve, 1 tooth in right valve; lateral teeth long, thin, straight to slightly curved, 2 in left valve, 1 in right valve; interdentum moderately long, very narrow; umbo cavity wide, very shallow; nacre typically bluish white, occasionally purplish, nacre often discolored.

Soft Anatomy Description

Mantle tan, outside of apertures gray, grayish tan, grayish brown or rusty tan; visceral mass pearly white to creamy white or tan; foot creamy white to light tan, often darker distally.

Gills tan; dorsal margin slightly sinuous, ventral margin elongate convex to straight; gill length 45–65% of shell length; gill height usually 30–40% of gill length, occasionally greater; outer gill height 60–90% of inner gill height; inner lamellae of inner gills connected to visceral mass only anteriorly.

No gravid females were available for marsupium description; probably similar to *Elliptio arca*, with outer gills marsupial, marsupium slightly padded when gravid. In material from the Flint and Chattahoochee rivers, Georgia, Lea (1863d) reported glochidia to be brooded "the whole length of the outer branchiae, but not the upper portion".

Labial palps tan, sometimes with irregular creamy white area dorsally; straight to slightly convex or slightly concave dorsally, convex ventrally, bluntly pointed to narrowly rounded distally; palp length 15–30% of gill length; palp height 50–65% of palp length; distal 25–40% of palps bifurcate, bifurcation occasionally greater.

Incurrent and excurrent apertures usually of similar length; length of supra-anal aperture often less than incurrent and excurrent aperture lengths, but may be equal or slightly greater.

Incurrent aperture length 10–20% of shell length; tan within, usually with grayish brown, dark brown, reddish brown or black basal to papillae; papillae usually in 2 rows, moderately long, slender, simple; papillae usually some combination of tan, grayish brown, reddish brown, dark brown and black, usually changing color distally.

Excurrent aperture length 10–20% of shell length; tan within, marginal color band grayish brown, dark brown or black, often with small tan or light brown areas basal to some papillae; margin papillate, papillae simple, small but often well-developed; papillae creamy white, tan, grayish tan or light brown. Lea (1863d) reported excurrent aperture papillae to be dark brown.

Supra-anal aperture length 10–15% of shell length; creamy white within, often with thin, sparse, irregular gray or brown band marginally; margin smooth; mantle bridge separating supra-anal and excurrent apertures, imperforate, length 10–25% of supra-anal length, occasionally absent.

Specimens examined: Cahaba River (n = 5); Sepulga River (n = 1); Sipsey River (n = 1); West Fork Choctawhatchee River (n = 1).

Glochidium Description

Length 187–200 μm; height 200–237 μm; without styliform hooks; outline subelliptical; dorsal margin straight to slightly convex, ventral margin rounded, anterior and posterior margins convex. Lea (1863d) described the glochidium as having a "pouch-shape".

Similar Species

Elliptio arctata may resemble *Elliptio arca* but has a much thinner, more arcuate shell. *Elliptio arctata* has pseudocardinal teeth that are smaller and lateral teeth that are thinner relative to shell size than *E. arca*. Lateral teeth are typically slightly widened posteriorly in *E. arca* but not *E. arctata*. The nacre of *E. arctata* is usually bluish white or occasionally pale purple, whereas *E. arca* nacre is highly variable, ranging from dark purple to salmon, only occasionally white. *Elliptio arca* grows larger than *E. arctata*.

General Distribution

Elliptio arctata is known from most eastern Gulf Coast drainages from the Apalachicola Basin in Georgia and Florida (Brim Box and Williams, 2000), west to the Pearl River drainage, Mississippi, including the Mobile Basin (Jones et al., 2005).

Alabama and Mobile Basin Distribution

Elliptio arctata occurs in all major river systems within the Mobile Basin, both above and below the Fall Line. On the Gulf Coast it is known from the Apalachicola Basin, including Chipola River headwaters, as well as the Choctawhatchee and Escambia River drainages.

Elliptio arctata is widespread but uncommon in the Mobile Basin and populations are highly fragmented. The largest remaining populations known are in the Cahaba River and Alabama River just downstream of Claiborne Lock and Dam. It is rare in headwaters of the Conecuh, Choctawhatchee and Chipola River drainages, and possibly a few tributaries of the Chattahoochee River.

Ecology and Biology

Elliptio arctata inhabits creeks and rivers in moderate current, most often in crevices and under large rocks, where silt has been deposited. *Elliptio arctata* may also be found among roots in beds of macrophytes.

Elliptio arctata is dioecious, but occasional hermaphroditic individuals may be encountered. Heard (1979a) reported 4 hermaphroditic individuals among 126 from a stream in the Apalachicola Basin. *Elliptio arctata* is presumably a short-term brooder, gravid in spring and summer. Its glochidial hosts are unknown.

Current Conservation Status and Protection

Elliptio arctata was listed as a species of special concern throughout its range by Williams et al. (1993) and in Alabama by Lydeard et al. (1999). Garner et al. (2004) designated it a species of highest conservation concern in the state.

Remarks

Elliptio arctata has been reported from Atlantic Coast drainages from the Cape Fear River drainage in North Carolina to the Savannah River drainage in Georgia (Johnson, 1970). However, there is no evidence to support the presence of this species, which was described from the Mobile Basin, in Atlantic Coast drainages.

Systematic morphological and genetic analyses of *Elliptio arctata* among drainages may reveal sibling species. Conchological characters are not sufficiently distinctive among populations to recognize more than one species.

Synonymy

Unio arctatus Conrad, 1834. Conrad, 1834a:340, pl. 1, fig. 9
Type locality: Black Warrior and Alabama rivers, [Alabama]. Lectotype, ANSP 41356, length 55 mm, designated by Johnson (1970), is from Alabama River, Alabama; not found.

Unio strigosus Lea, 1840. Lea, 1840:287; Lea, 1842b:198, pl. 9, fig. 9; Lea, 1842c:36, pl. 9, fig. 9
Type locality: Chattahoochee River, Columbus, [Muscogee County,] Georgia, Dr. Boykin. Lectotype, USNM 85890, length 56 mm, designated by Johnson (1970).

Unio tortivus Lea, 1840. Lea, 1840:287; Lea, 1842b:204, pl. 12, fig. 17; Lea, 1842c:42, pl. 12, fig. 17
Type locality: Chattahoochee River, Columbus, [Muscogee County,] Georgia, Dr. Boykin. A specimen labeled "type", USNM 85674, length 49 mm, from "Lee County, Georgia", is not the specimen figured below. The specimen from Lee County is not mentioned in the original description or in Lea's subsequent publications.

Unio viridans Lea, 1859. Lea, 1859d:170; Lea, 1860d:337, pl. 54, fig. 162; Lea, 1862d:19, pl. 54, fig. 162
Type locality: Near Columbus, [Muscogee County,] Georgia. Lectotype, USNM 85579, length 52 mm, designated by Johnson (1970).

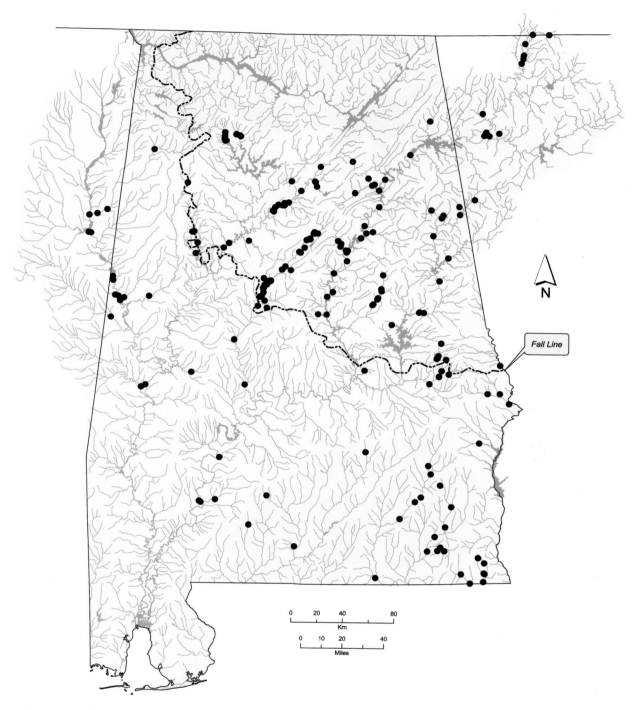

Distribution of *Elliptio arctata* in Alabama and the Mobile Basin.

Elliptio chipolaensis (Walker, 1905)
Chipola Slabshell

Elliptio chipolaensis – Length 53 mm, UMMZ 138409. Chipola River, Peacock Bridge, near Sink Creek, Jackson County, Florida, 1918. © Richard T. Bryant.

Shell Description

Length to 85 mm; moderately thin; moderately inflated; outline elongate elliptical; posterior margin bluntly pointed to narrowly truncate; anterior margin rounded; dorsal margin straight to slightly convex; ventral margin straight to slightly convex; posterior ridge high, rounded, may be doubled posterioventrally; posterior slope moderately steep, flat to slightly concave; umbo broad, inflated, elevated slightly above hinge line, umbo sculpture concentric ridges; periostracum light to dark brown, without rays, but typically with dark concentric bands.

Pseudocardinal teeth triangular, compressed, 2 teeth in left valve, long axis of teeth almost parallel to shell margin, 1 tooth in right valve, parallel to shell margin; lateral teeth long, straight, 2 in left valve, 1 in right valve; interdentum moderately long, very narrow; umbo cavity wide, shallow; nacre white to bluish white, sometimes with salmon tint.

Soft Anatomy Description

No material was available for soft anatomy description, but Brim Box and Williams (2000) provided some details. Incurrent and excurrent apertures are roughly equal in length. Papillae of the incurrent aperture are long and simple, occurring in two rows with the inner row larger and sparser than those of the outer row. Incurrent and excurrent apertures have darkly pigmented marginal coloration which is absent in the supra-anal aperture. The outer gills are marsupial, with glochidia filling the entire gills when gravid.

Glochidium Description

Glochidium unknown.

Similar Species

Elliptio chipolaensis can be distinguished from other Apalachicola Basin *Elliptio* species by its chestnut-colored periostracum and the presence of dark concentric bands on the shell. It most closely resembles *Elliptio fumata*, but is broader posteriorly and narrower anteriorly.

General Distribution

Elliptio chipolaensis is known only from the Chipola River system in Alabama and Florida, and a single tributary to the Chattahoochee River in southeastern Alabama (Brim Box and Williams, 2000).

Alabama and Mobile Basin Distribution

Elliptio chipolaensis is known from headwaters of the Chipola River and Howard's Mill Creek, a tributary of lower Chattahoochee River, all in Houston County.

Elliptio chipolaensis is believed to be extant in Big and Cowarts creeks, which are Chipola River headwaters in Houston County.

Ecology and Biology

Elliptio chipolaensis inhabits combinations of silt, clay, sand and occasionally gravel in moderate current, often along stream margins.

Elliptio chipolaensis is presumably a short-term brooder, gravid in spring and summer. Brim Box and Williams (2000) reported gravid *E. chipolaensis* in late June. Its glochidial hosts are unknown.

Current Conservation Status and Protection

Elliptio chipolaensis was considered threatened throughout its range by Williams et al. (1993) and in Alabama by Lydeard et al. (1999). Garner et al. (2004) listed it as extirpated from the state but it was subsequently rediscovered in Chipola River headwaters of Houston County. *Elliptio chipolaensis* was listed as threatened under the federal Endangered Species Act in 1998. Heard (1975b) reported it to be locally abundant in the early 1970s at some sites in the Chipola River in Florida.

Remarks

The first record of *Elliptio chipolaensis* outside of the Chipola River system was reported by Brim Box and Williams (2000). It is represented by a single museum specimen collected from Howard's Mill Creek, a tributary of the Chattahoochee River in Houston County, Alabama, in 1968.

Elliptio chipolaensis was named for the Chipola River. The word chipola is from the Native American Chatot dialect and means "sweet river" or "sweet water" (Simpson, 1956).

Synonymy

Unio chipolaensis Walker, 1905. Walker, 1905a:135, pl. 9, figs. 6, 7 [original figure not herein reproduced]
Type locality: Listed originally as Chipola River. Holotype, UMMZ 96363, length 57 mm. Clench and Turner (1956) restricted the type locality to Chipola River, 1 mile north of Marianna, Jackson County, Florida.

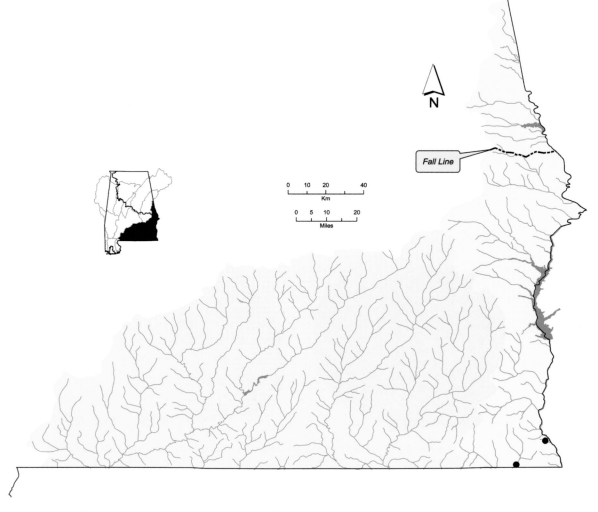

Distribution of *Elliptio chipolaensis* in the eastern Gulf Coast drainages of Alabama.

Elliptio crassidens (Lamarck, 1819)
Elephantear

Elliptio crassidens – Upper figure: length 111 mm, AUM 2045. Tennessee River, Buck Island Chute, river mile 249.8, Colbert and Lauderdale counties, Alabama, 14 June 2001. Middle figure: length 87 mm, UF 371. Escambia River, 3 miles southeast of Century, Escambia County, Florida, 27 August 1954. Lower figure: length 72 mm, UF 243942. Chipola River, 1.3 miles above State Highway 71 bridge, 12 miles southwest of Blountstown, Calhoun County, Florida, 23 August 1980. © Richard T. Bryant.

Shell Description

Length to 150 mm; thick; moderately inflated; outline highly variable, ranging from rhomboid to elliptical or subtriangular, often becoming arcuate with age; posterior margin obliquely truncate to narrowly rounded or bluntly pointed; anterior margin rounded;

dorsal margin straight to convex; ventral margin convex or straight, often becoming concave with age; posterior ridge well-defined, angular, becoming flattened and often doubled posterioventrally; posterior slope steep dorsally, flat to slightly concave, sometimes with small corrugations (particularly in Gulf Coast populations); umbo broad, flat to slightly inflated, elevated above hinge line, umbo sculpture coarse loops; periostracum dark brown to black, sometimes with weak dark green rays, especially in young individuals.

Pseudocardinal teeth large, triangular, erect, 2 divergent teeth in left valve, may be separated dorsally, anterior tooth smaller, 1 large tooth in right valve, sometimes with accessory denticle anteriorly; lateral teeth thick, short to moderately long, straight to slightly curved, 2 in left valve, 1 in right valve; interdentum variable, short to long, narrow to wide; umbo cavity wide, shallow; nacre variable, usually some shade of purple but varies from white to salmon or pink.

Soft Anatomy Description

Mantle tan to light brown, often creamy white or pale orange outside of pallial line, dark grayish brown outside of apertures; visceral mass pearly white or creamy white, occasionally tan or brown; foot creamy white to tan, often darker distally. Ortmann (1912a) reported the foot to be pale gray or brownish gray.

Gills tan to brown, occasionally gold; dorsal margin usually sinuous, sometimes straight, ventral margin convex to elongate convex or almost straight, outer gill ventral margin often more irregular than inner gill ventral margin; gill length 45–65% of shell length; gill height usually 35–55% of gill length, occasionally as high as 75%; outer gill height 60–90% of inner gill height, inner and outer gills occasionally of equal height; inner lamellae of inner gills connected to visceral mass only anteriorly.

Outer gills marsupial; glochidia held across entire gill; marsupium slightly padded when gravid; creamy white. Sterki (1895) included *Elliptio crassidens* in a group of species in which all four gills are marsupial. Ortmann (1912a) described gravid gills of *E. crassidens* as "moderately swollen".

Labial palps creamy white to tan, often with irregular pale area dorsally; straight to slightly concave dorsally, convex ventrally, bluntly pointed to rounded distally; palp length 20–35% of gill length; palp height usually 50–75% of palp length, occasionally higher, palps rarely higher than long; distal 25–60% of palps bifurcate.

Incurrent aperture usually longer than excurrent aperture, occasionally equal in length; supra-anal aperture longer, shorter or equal to incurrent and excurrent apertures.

Incurrent aperture length 10–20% of shell length; creamy white to tan within, occasionally rusty, usually with rusty orange, brown, gray or black basal to papillae; papillae in 1–2 rows, simple, bifid and trifid with occasional arborescent; papillae some combination of creamy white, tan, rusty brown, gray and black.

Excurrent aperture length 5–15% of shell length; creamy white to tan within, occasionally rusty, marginal color band dull orange or reddish brown to gray or black, occasionally with creamy white band proximally; margin papillate, papillae short, simple, often blunt; papillae creamy white, tan, reddish brown, gray or black.

Supra-anal aperture length 10–20% of shell length; usually creamy white within, occasionally tan or brown, may have sparse gray or black pigmentation marginally; margin smooth; mantle bridge separating supra-anal and excurrent apertures imperforate, length 10–20% of supra-anal length, occasionally absent.

Specimens examined: Alabama River (n = 3); Bear Creek, Colbert County (n = 3); Conecuh River (n = 3); Coosa River (n = 3); Flat Creek, Geneva County (n = 4); Sipsey River (n = 3); Tennessee River (n = 3); Tombigbee River (n = 2).

Glochidium Description

Length 130–150 μm; height 141–160 μm; with weak triangular hooks covered with microstylets; outline subtriangular; dorsal margin straight, ventral margin narrowly rounded, slightly produced, anterior and posterior margins broadly convex, anterior margin slightly more convex than posterior margin (Surber, 1915; O'Brien et al., 2003).

Similar Species

Elongate *Elliptio crassidens* may resemble *Elliptio arca*. However, the outline of *E. crassidens* is typically more triangular or rhomboidal than that of *E. arca*. *Elliptio crassidens* has more erect and massive pseudocardinal teeth and an interdentum that is usually shorter and often wider than those of *E. arca*. The interdentum of *E. crassidens* is often widest adjacent to the pseudocardinal teeth, but the interdentum of *E. arca* is typically narrowest near the pseudocardinals. The ventral margin of *E. crassidens* is variable and may be convex, straight or concave, while the ventral margin of *E. arca* is almost always straight. The periostracum of *E. crassidens* is often darker than that of *E. arca*. Rays of *E. crassidens*, when present, are usually less pronounced than those typically seen on *E. arca*.

General Distribution

Elliptio crassidens occurs in the Mississippi Basin from Minnesota (Dawley, 1947) south to Louisiana (Vidrine, 1993), and from headwaters of the Ohio River drainage in western Pennsylvania (Ortmann, 1909a) west to Missouri (Oesch, 1995). It is widespread in the Cumberland River drainage downstream of Cumberland Falls, Kentucky and Tennessee (Cicerello et al., 1991; Parmalee and Bogan, 1998) and occurs in most of the Tennessee River drainage (Neves, 1991; Parmalee and Bogan, 1998). *Elliptio crassidens* occurs in Gulf Coast drainages from the Ochlockonee River, Florida, west to the Amite River, Lake Ponchartrain drainage, Louisiana (Vidrine, 1993).

Alabama and Mobile Basin Distribution

Elliptio crassidens is found in the Tennessee River across northern Alabama and in several tributaries. It also occurs in much of the Mobile Basin, with the exception of the Tallapoosa River drainage above the Fall Line. On the Gulf Coast *E. crassidens* is found in the Escambia and Yellow River drainages and in the Chipola and Chattahoochee drainages of the Apalachicola Basin.

Elliptio crassidens remains a common species in tailwaters of Tennessee River dams and riverine reaches of some large Mobile Basin rivers. It is abundant in some large rivers but has disappeared from many tributaries. *Elliptio crassidens* is still common in some areas of Gulf Coast drainages as well.

Ecology and Biology

Elliptio crassidens inhabits creeks and rivers in a variety of stable substrates including combinations of mud, sand, gravel and cobble. It may be found in slow to moderately swift current, at depths from 1 m to more than 6 m. In large rivers *E. crassidens* is found primarily in riverine habitat such as tailwaters of dams. It is occasionally found in overbank habitats in Tennessee River reservoirs. However, they are most abundant in areas with firm sand and mud substrate where silt deposition is kept to a minimum by wave action (e.g., tops of submerged ridges and shelves along islands).

Elliptio crassidens is a short-term brooder with a gravid period that appears to vary among regions within its range. It has been reported gravid from April to June in the Mississippi River (Coker et al., 1921) and from April to August in the Apalachicola Basin (Brim Box and Williams, 2000; O'Brien et al., 2003). The smallest gravid female reported from the Apalachicola Basin was 51 mm long (Brim Box and Williams, 2000). Ortmann (1912a) and Utterback (1915) described *E. crassidens* conglutinates as white, well-developed and leaf-shaped. The only reported glochidial host of *E. crassidens* is *Alosa chrysochloris* (Skipjack Herring) (Clupeidae), which was based on an observation of a natural infestation (Howard, 1914d).

Current Conservation Status and Protection

Elliptio crassidens was listed as currently stable throughout its range by Williams et al. (1993) and in Alabama by Lydeard et al. (1999). Garner et al. (2004) designated it a species of lowest conservation concern in the state.

Commercial harvest of *Elliptio crassidens* is legal, but demand for its shell is usually low due to its colored nacre. Current regulations limit where and when harvest may occur, and harvest is limited to individuals greater than $2^5/_8$ inches (67 mm) in shell height.

Remarks

Elliptio crassidens in Gulf of Mexico drainages was previously recognized as a southeastern subspecies, *Elliptio crassidens incrassatus* (Clench and Turner, 1956). However, more recent publications (Johnson, 1970; Burch, 1975a; Heard, 1979b; Turgeon et al., 1998) have not recognized *incrassatus* as valid.

Elliptio crassidens has been reported as uncommon in prehistoric shell middens along the Tennessee River across northern Alabama (Morrison, 1942; Warren, 1975; Hughes and Parmalee, 1999). However, it is now the most abundant species in Guntersville Dam tailwaters, where it comprised 69 percent of all mussels collected during catch per unit effort sampling in 1997 (J.T. Garner, unpublished data). Ahlstedt and McDonough (1993) reported *E. crassidens* to be the most abundant species overall in Wheeler Reservoir in 1991, with an estimated population of 115,740,000. Thus, the population of *E. crassidens* appears to have increased since prehistoric times.

Elliptio crassidens was historically common in the Coosa River and some of its large tributaries. However, it has become rare since impoundment of the river. Coosa River dams have no fish passage or navigation locks. The only reported glochidial host of *E. crassidens* is *Alosa chrysochloris*, an anadromous fish.

Wilson and Clark (1914) reported that in beds where it was "exceedingly abundant" it was a nuisance, forming a large part of the catch and requiring considerable effort to separate from more valuable species. *Elliptio crassidens* was so abundant in one reach of the Cumberland River that commercial harvest was unprofitable and that "the problem of making this stretch a valuable clamming ground consists as much in the reduction of this species as in the increase in valuable kinds" (Wilson and Clark, 1914).

Synonymy

Unio crassidens Lamarck, 1819. Lamarck, 1819:71

Type locality: Mississippi and other rivers and lakes, restricted to Ohio River, Cincinnati, Ohio (Johnson, 1969b). Lectotype, MNHN specimen referred to by Lamarck as "var. b", length 66 mm, designated by Johnson (1969b).

Unio (Elliptio) nigra Rafinesque, 1820. Rafinesque, 1820:291, pl. 80, figs. 1–4

Unio (Elliptio) nigra var. *fusca* Rafinesque, 1820. Rafinesque, 1820:291

Unio (Elliptio) nigra var. *maculata* Rafinesque, 1820. Rafinesque, 1820:291

Unio cuneatus Barnes, 1823. Barnes, 1823:263

Obliquaria (Aximedia) venus Rafinesque, 1831. Rafinesque, 1831:3

Unio incrassatus Lea, 1840. Lea, 1840:286; Lea, 1842b:217, pl. 16, fig. 34; Lea, 1842c:55, pl. 16, fig. 34

Type locality: Chattahoochee River near Columbus, [Muscogee County,] Georgia, Dr. Boykin. Lectotype, USNM 84537, length 55 mm, designated by Johnson (1970).

Unio danielsii B.H. Wright, 1899. B.H. Wright, 1899a:31; Johnson, 1967b:5, pl. 6, fig. 1

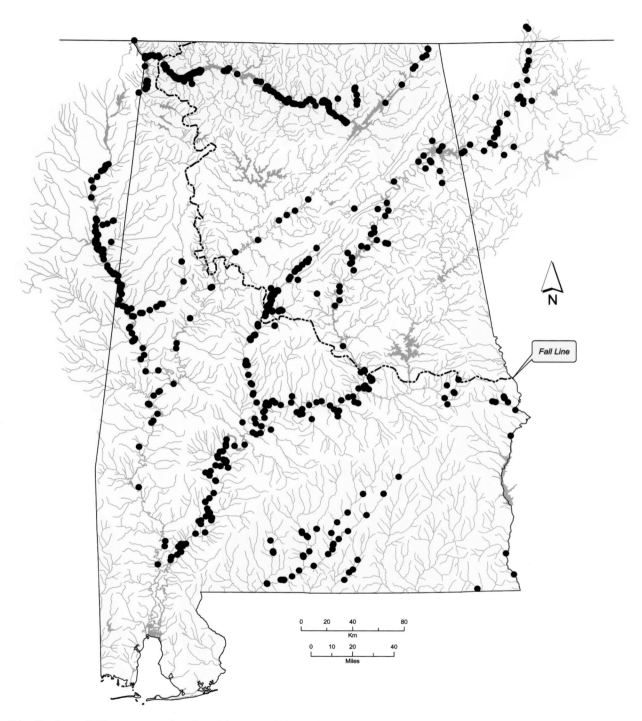

Distribution of *Elliptio crassidens* in Alabama and the Mobile Basin.

Elliptio dilatata (Rafinesque, 1820)
Spike

Elliptio dilatata – Upper figure: length 69 mm, UMMZ 94795. Lower figure: length 53 mm, UMMZ 94795. Tennessee River, Muscle Shoals, Lauderdale County, Alabama. © Richard T. Bryant.

Shell Description

Length to 136 mm; thick; compressed when young, becoming moderately inflated with age, individuals in large rivers become more inflated than those from small streams; outline elongate elliptical, old specimens may become slightly arcuate; posterior margin bluntly pointed, old individuals may be narrowly truncate; anterior margin rounded; dorsal margin straight to slightly convex; ventral margin straight to slightly convex or slightly concave; posterior ridge low, rounded, may be more angular in old individuals from large rivers; posterior slope usually not steep but may be steeper in old individuals from large rivers, typically flat; umbo broad, flat, elevated slightly above hinge line, umbo sculpture heavy parallel bars; periostracum usually smooth in young individuals but typically becomes roughened with age, young individuals light brown, tawny or greenish brown, often with faint rays, older shells dark reddish or greenish brown to black with obscure rays.

Pseudocardinal teeth thick, triangular, roughened, 2 divergent teeth in left valve, may be separated dorsally, 1 tooth in right valve; lateral teeth long, straight, thick, 2 in left valve, 1 in right valve; interdentum typically long, occasionally short, narrow, may become moderately wide with age; umbo cavity wide, very shallow; nacre usually purple but may be white or salmon.

Soft Anatomy Description

Mantle tan, may be lighter outside of pallial line, rusty tan to grayish brown outside of apertures, may have mottling in crude bands perpendicular to margin; visceral mass creamy white, pearly white adjacent to foot; foot creamy white to pale tan.

Gills creamy white to tan; dorsal margin straight to slightly sinuous, outer gill ventral margin straight to slightly convex, inner gill ventral margin elongate convex; gill length 60–65% of shell length; gill height 30–35% of gill length; outer gill height 65–80% of inner gill height; inner lamellae of inner gills connected to visceral mass only anteriorly.

Outer gills marsupial; glochidia held across entire gill; marsupium slightly padded when gravid; creamy white. Sterki (1895) included *Elliptio dilatata* in a group of species in which all four gills are marsupial. However, Lea (1863d) reported the marsupia to occupy the entire length of the outer gills.

Labial palps creamy white to tan; straight dorsally, convex ventrally, bluntly pointed to narrowly rounded distally; palp length 15–20% of gill length;

palp height 45–70% of palp length; distal 30–45% of palps bifurcate.

Incurrent, excurrent and supra-anal apertures of similar length, any may be slightly longer than the others.

Incurrent aperture length 10–15% of shell length; creamy white to pale tan within, usually with some combination of rusty tan, rusty orange, grayish brown and gray basal to papillae; papillae in 1 row, long, slender, simple; papillae some combination of creamy white, rusty orange, brown and grayish brown.

Excurrent aperture length 10–20% of shell length; creamy white to pale tan within, marginal color band rusty orange, rusty brown, dark brown, grayish brown or gray; margin with small, simple papillae, may be little more than crenulations; papillae creamy white, tan, rusty orange, rusty brown or gray. Lea (1863d) reported excurrent aperture papillae to be black.

Supra-anal aperture length 10–20% of shell length; creamy white to brown within, may have thin marginal band of gray or brown; margin smooth; mantle bridge separating supra-anal and excurrent apertures usually imperforate, length 5–10% of supra-anal length, occasionally greater.

Specimens examined: Duck River, Tennessee (n = 3); Tennessee River (n = 4).

Glochidium Description

Length 190–219 µm; height 215–225 µm; without styliform hooks; outline subelliptical; dorsal margin straight, ventral margin broadly rounded, anterior and posterior margins convex and almost equal; larval thread present (Lefevre and Curtis, 1910b; Surber, 1912; Utterback, 1915; Coker et al., 1921; Hoggarth, 1999). Jirka and Neves (1992) reported somewhat larger glochidia, length 235 µm and height 249 µm, but may have reported length as height and height as "width".

Similar Species

Elliptio dilatata resembles *Ptychobranchus fasciolaris*. It differs in usually having at least some purple nacre, while that of *P. fasciolaris* is always white. The interdental plate of *E. dilatata* is much narrower and less angular than that of *P. fasciolaris*. Also, the lateral teeth are proportionally longer in *E. dilatata* than in *P. fasciolaris*. The periostracum of *E. dilatata* is typically darker brown than that of *P. fasciolaris*, which is more yellowish brown. Though shells of young *E. dilatata* may be faintly rayed, adults are usually without rays, whereas *P. fasciolaris* typically has wide, broken, green rays throughout life.

Elliptio dilatata superficially resembles *Cumberlandia monodonta* but has a thicker, less arcuate shell, with a more pointed posterior end. Also, *E. dilatata* is less inflated along the posterior ridge than *C.*

monodonta. The pseudocardinal teeth of *E. dilatata* are well-developed, while those of *C. monodonta* are reduced, consisting of a single peg-like tooth in the right valve that fits into a depression in the left valve in adults. Also, *C. monodonta* has lateral mantle attachment scars that are only found in margaritiferids.

Hemistena lata may superficially resemble *Elliptio dilatata* but differs in having a much thinner shell with a periostracum that is typically lighter and usually has broken green rays. Pseudocardinal and lateral teeth of *E. dilatata* are well-developed, while pseudocardinal teeth of *H. lata* are represented by single raised knobs in each valve and lateral teeth by simple thickened hinge lines.

Elliptio dilatata may resemble *Ligumia recta*, but *E. dilatata* typically has thicker, triangular, divergent pseudocardinal teeth and shorter lateral teeth. Also, *E. dilatata* usually has purple nacre, whereas that of *L. recta* is typically purple only in or near the umbo cavity. *Ligumia recta* has a smoother, shinier periostracum than *E. dilatata* and is generally more heavily rayed.

General Distribution

Elliptio dilatata occurs in much of the Great Lakes Basin in southern Canada and north-central and northeastern U.S. (Goodrich and van der Schalie, 1932; La Rocque and Oughton, 1937; Mathiak, 1979). It is found in most of the Mississippi Basin from Minnesota (Dawley, 1947) south to Louisiana (Vidrine, 1993), and from western New York (Strayer et al., 1992) to eastern Kansas (Murray and Leonard, 1962). *Elliptio dilatata* is widespread in the Cumberland River drainage downstream of Cumberland Falls, Kentucky and Tennessee (Cicerello et al., 1991; Parmalee and Bogan, 1998). It occurs in much of the Tennessee River drainage (Ahlstedt, 1992a, 1992b; Parmalee and Bogan, 1998). There is one spurious record of *E. dilatata* from the San Marcos River, Texas (Howells et al., 1996). Daniels (1909) reported *E. dilatata* from the Hudson Bay Basin, but Graf (1997) considered that record questionable.

Alabama and Mobile Basin Distribution

Elliptio dilatata occurred in the Tennessee River across northern Alabama and was historically found in some of the larger tributaries.

Elliptio dilatata is extant only in Guntersville and Wilson Dam tailwaters, where it is rare, and possibly the Paint Rock River.

Ecology and Biology

Elliptio dilatata occurs in medium streams to large rivers. In occurs primarily in shoal habitat of unimpounded streams and rivers but can occasionally be found in tailwaters of Tennessee River dams in water from 4 m to 8 m deep. *Elliptio dilatata* can also be found in lakes under some conditions. Coker et al. (1921) included it on a list of species "characteristic of mussels in lakes of upper central states", where it occurs in areas subject to wave action. They reported it to comprise 13 percent of the mussel fauna in some areas of Lake Pepin. The preferred substrate of *E. dilatata* is a firm mixture of sand and gravel kept free of silt by current or wave action.

Elliptio dilatata has been reported to reach sexual maturity as young as four years of age (Collins, 1971; Jirka and Neves, 1992). It is dioecious, but occasional hermaphroditic individuals have been reported (van der Schalie, 1970; Heard, 1979b; Jirka and Neves, 1992). Gametogenesis begins during autumn in both sexes. Gonads of males fill with spermatozoa and females fill with mature ova by late autumn. Both sexes overwinter in that condition and spawning begins during spring and continues into mid-summer (Collins, 1971; Jirka and Neves, 1992).

Elliptio dilatata is a short-term brooder (Sterki, 1895). Gravid *E. dilatata* have been reported from May to August in various parts of its range (Ortmann, 1909a, 1912a, 1919; Surber, 1912; Utterback, 1915, 1916a; Coker et al., 1921; Jirka and Neves, 1992). Collins (1971) reported a slight thickening in the distal margin of the posterior part of the outer gills prior to spawning. Glochidial conglutinates are narrowly lanceolate and white (Utterback, 1915). *Elliptio dilatata* glochidia have been reported from stream drift in New River, Virginia and West Virginia, from late April to September, with a peak in June and July (Jirka and Neves, 1992).

Fishes reported to serve as glochidial hosts for *Elliptio dilatata* in laboratory trials include *Ambloplites rupestris* (Rock Bass) (Centrarchidae); *Etheostoma caeruleum* (Rainbow Darter) (Percidae); and *Cottus carolinae* (Banded Sculpin) (Cottidae) (Luo, 1993). Fishes reported to serve as glochidial hosts of *E. dilatata* based on observations of natural infestations include *Pylodictis olivaris* (Flathead Catfish) (Ictaluridae) and *Sander canadensis* (Sauger) (Percidae). Fishes reported to serve as glochidial hosts with no discussion of methods include *Pomoxis annularis* (White Crappie) and *Pomoxis nigromaculatus* (Black Crappie) (Centrarchidae); *Dorosoma cepedianum* (Gizzard Shad) (Clupeidae); and *Perca flavescens* (Yellow Perch) (Percidae) (Wilson, 1916; Clarke, 1981a).

Current Conservation Status and Protection

Elliptio dilatata was listed as currently stable throughout its range by Williams et al. (1993) and in Alabama by Lydeard et al. (1999). However, *E. dilatata* appears to be declining in the state, and Garner et al. (2004) designated it a species of highest conservation concern.

Remarks

Elliptio dilatata is rare in tailwaters of Wilson and Guntersville dams though it is abundant in prehistoric shell middens along the Tennessee River across northern Alabama (Morrison, 1942; Warren, 1975; Hughes and Parmalee, 1999). Very little recruitment has been observed recently, and conditions in the Tennessee River do not appear favorable for long-term survival of this species.

Synonymy

Unio nasuta Lamarck, 1819, *non* Say, 1817. Lamarck, 1819:75
Unio (*Eurynia*) *dilatata* Rafinesque, 1820. Rafinesque, 1820:297
Type locality: Kentucky River. Lectotype, ANSP 20248a, length 77 mm, designated by Johnson and Baker (1973).
Unio gibbosus Barnes, 1823. Barnes, 1823:262, pl. 11, fig. 12
Unio mucronatus Barnes, 1823. Barnes, 1823:266, pl. 13, fig. 13
Obliquaria violacea Rafinesque, 1831. Rafinesque, 1831:3 [unnecessary replacement name for *Unio dilatata* Rafinesque, 1820]
Unio bicolor Rafinesque, 1831. Rafinesque, 1831:3
Unio (*Eurynia*) *fulvus* Rafinesque, 1831. Rafinesque, 1831:3
Margarita (*Unio*) *arctior* Lea, 1836, *nomen nudum*. Lea, 1836:39
Unio arctior Lea, 1838. Lea, 1838a:10, pl. 4, fig. 10; Lea, 1838c:10, pl. 4, fig. 10
Unio stonensis Lea, 1840. Lea, 1840:286; Lea, 1842b:195, pl. 8, fig. 5; Lea, 1842c:33, pl. 8, fig. 5
Unio gibbosus var. *armathwaitensis* B.H. Wright, 1898. B.H. Wright, 1898b:123; Johnson, 1967b:5, pl. 3, fig. 1
Unio (*Elliptio*) *gibbosus* var. *delicatus* Simpson, 1900, *non* Lea, 1863. Simpson, 1900b:704
Unio propeverutus De Gregorio, 1914. De Gregorio, 1914:38, pl. 3, figs. 1a–c
Elliptio dilatatus var. *sterkii* Grier, 1918. Grier, 1918:9–10

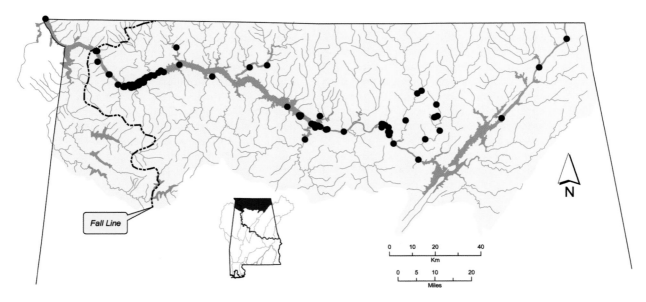

Distribution of *Elliptio dilatata* in the Tennessee River drainage of Alabama.

Elliptio fraterna (Lea, 1852)
Brother Spike

Elliptio fraterna – Upper figure: length 77 mm, UF 269381. Chattahoochee River near Eufaula, Barbour County, Alabama. Lower figure: length 55 mm, UF 388523. Apalachicola River, river mile 33, Liberty County, Florida, 16 June 2004. © Richard T. Bryant.

Shell Description

Length to 80 mm; moderately thin; moderately compressed; outline elliptical to subrhomboidal; posterior margin narrowly rounded to narrowly truncate; anterior margin rounded; dorsal margin straight to convex; ventral margin straight to slightly convex; posterior slope moderately to well-developed, becoming doubled posterioventrally, area between ridges flattened; posterior slope low, flat, adorned with weak, parallel corrugations radiating from posterior ridge to dorsal margin, corrugations typically disappear distally; umbo broad, flat, elevated slightly above hinge line, umbo sculpture unknown; periostracum brown, darkening to black with age, may have weak green rays.

Pseudocardinal teeth triangular, moderately thick, finely striate, 2 divergent teeth in left valve, separated dorsally, 1 tooth in right valve, may have accessory denticle anteriorly; lateral teeth moderately short, straight to slightly curved, 2 in left valve, 1 in right valve; interdentum long, narrow or very narrow; umbo cavity wide, shallow; nacre white to purple.

Soft Anatomy Description

Mantle tan with slight rusty cast, grayish brown outside of apertures; visceral mass creamy white and pale tan; foot tan, becoming pale orange distally.

Gills tan; dorsal margin straight, ventral margin elongate convex; gill length approximately 55% of shell length; gill height approximately 45% of gill length; outer gill height approximately 90% of inner gill height; inner lamellae of inner gills connected to visceral mass only anteriorly.

No gravid females were available for marsupium description; probably similar to other *Elliptio* species, with outer gills marsupial, glochidia held across most of gill, marsupium slightly padded when gravid.

Labial palps tan, with irregular creamy white area dorsally; straight dorsally, curved ventrally, bluntly pointed distally; palp length approximately 20% of gill length; palp height approximately 70% of palp length; palps bifurcate distally.

Supra-anal aperture longer than incurrent aperture; incurrent aperture longer than excurrent aperture.

Incurrent aperture length approximately 15% of shell length; tan within, with irregular rusty tan and black mottling basal to papillae; papillae in 2 rows, long, slender, mostly simple, few bifid; papillae tan to rusty tan, with dark brown edges basally, becoming grayish tan distally. Lea (1863d) reported incurrent aperture papillae to be small and dark brown.

Excurrent aperture length approximately 10% of shell length; tan within, marginal band rusty tan with irregular dark brown pattern; margin with small, simple papillae; papillae tan.

Supra-anal aperture length approximately 20% of shell length; creamy white within, without marginal coloration; margin smooth; mantle bridge separating supra-anal and excurrent apertures may be perforate; length approximately 25% of supra-anal length. Lea (1863d) reported the supra-anal aperture to be "not united below", suggesting an absence of the mantle bridge.

Specimen examined: Apalachicola River, Florida (n = 1).

Glochidium Description

Length approximately 200 μm; height 212–238 μm; without styliform hooks; outline depressed subelliptical; dorsal margin straight, ventral margin rounded, anterior and posterior margins convex.

Similar Species

Elliptio fraterna resembles some *Elliptio crassidens* in having corrugations on the posterior slope but has a more compressed, elongate outline than most *E. crassidens*. *Elliptio crassidens* also has a thicker shell than *E. fraterna*, with heavier lateral and pseudocardinal teeth. *Elliptio crassidens* attains a much greater size than *E. fraterna*.

Elliptio fraterna differs from other small, elongate *Elliptio* of the Apalachicola Basin—including *E. chipolaensis*, *E. fumata*, *E. nigella*, *E. pullata* and *E. purpurella*—in having corrugations on the posterior slope.

General Distribution

Elliptio fraterna is known from two disjunct populations, the Apalachicola Basin in Alabama, Florida and Georgia, and the Savannah Basin in Georgia and South Carolina. It is known from above and below the Fall Line in both basins.

Alabama and Mobile Basin Distribution

Elliptio fraterna is known only from the Chattahoochee River proper.

There are only two historical records of *Elliptio fraterna* from Alabama, and it is believed to be extirpated from the state. The only dated specimen was collected in 1852.

Ecology and Biology

There is very little information available on the ecology of *Elliptio fraterna*. It appears to have inhabited large rivers in sand substrates (Britton and Fuller, 1980; Brim Box and Williams, 2000).

Elliptio fraterna is presumably a short-term brooder, gravid in spring and summer. Its glochidial hosts are unknown.

Current Conservation Status and Protection

Elliptio fraterna was listed as endangered throughout its range by Williams et al. (1993). Garner et al. (2004) listed it extirpated from Alabama.

Remarks

There is no information on the historical abundance of *Elliptio fraterna*. The paucity of material in museum collections suggests that it was always rare or its populations were reduced by the mid-1800s. The most recent published record of *E. fraterna* was from the Savannah River, collected in 1972 (Britton and Fuller, 1980). However, two live individuals were collected from the Apalachicola River, Liberty County, Florida, in 2004 (J.D. Williams, personal observation). These represent the first record of *E. fraterna* from Florida.

Lea (1852b) based his original description of *Elliptio fraterna* on specimens from the Chattahoochee River, Columbus, Georgia (paralectotype), and Abbeville District, Savannah River drainage, South Carolina (lectotype) (Johnson, 1970, 1974). Johnson (1970) designated the Abbeville District specimen as lectotype, restricting the type locality.

The taxonomic relationship between the Apalachicola Basin and Savannah River drainage populations of putative *Elliptio fraterna* should be investigated. Johnson (1970) placed *Elliptio mcmichaeli* of the Choctawhatchee River drainage in the synonymy of *E. fraterna*. However, Fuller and Bereza (1974) recognized both species based on incurrent aperture papilla morphology. The systematic relationship between these species should also be addressed.

Synonymy

Unio fraternus Lea, 1852. Lea, 1852b:263, pl. l6, fig. 15; Lea, 1852c:19, pl. 16, fig. 15

Type locality: Columbus, [Muscogee County,] Georgia, Dr. Boykin; Abbeville District, South Carolina, J.P. Barratt, M.D. Lectotype, USNM 85396, length 61 mm, designated by Johnson (1970), is from Abbeville District, South Carolina.

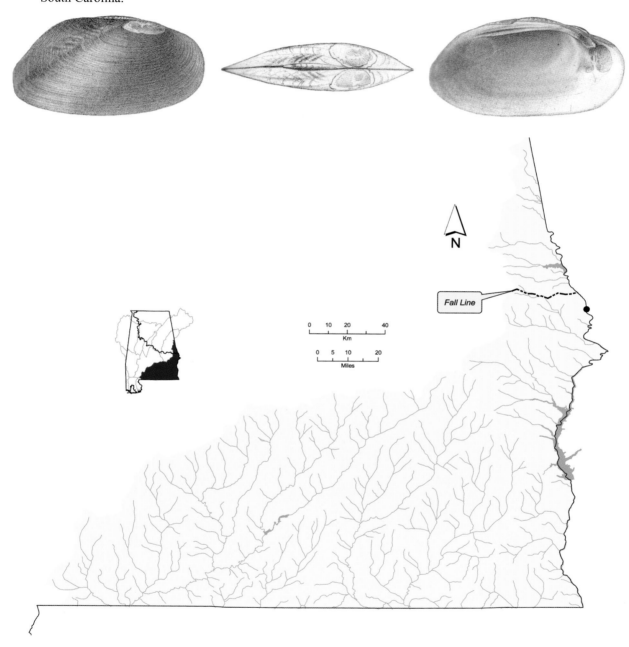

Distribution of *Elliptio fraterna* in the eastern Gulf Coast drainages of Alabama.

Elliptio fumata (Lea, 1857)
Gulf Slabshell

Elliptio fumata – Length 59 mm, USNM 85552 (lectotype). Chattahoochee River near Columbus, Muscogee County, Georgia. © Richard T. Bryant.

Shell Description

Length to 145 mm; moderately thin; moderately inflated; outline elliptical to subrhomboidal; posterior margin bluntly pointed to narrowly rounded or truncate; anterior margin rounded; dorsal margin straight to convex; ventral margin straight, may be slightly concave posteriorly; posterior ridge low, broadly rounded, may be doubled posterioventrally; posterior slope wide, flat to slightly convex; umbo broad, flat, elevated slightly above hinge line or not at all, umbo sculpture low, concentric loops; periostracum smooth to cloth-like, tawny or greenish brown to dark brown or black, may have variable rays, usually becoming obscure with age.

Pseudocardinal teeth triangular, 2 teeth in left valve, anterior tooth usually smaller and often pointed ventrally or anterioventrally, 1 tooth in right valve, sometimes with accessory denticle anteriorly; lateral teeth long, straight to slightly curved, 2 in left valve, 1 in right valve; interdentum long, very narrow; umbo cavity wide, very shallow; nacre white or various shades of purple or salmon.

Soft Anatomy Description

Mantle creamy white to pale tan, may have slight rusty cast in some areas, rusty tan or rusty brown to dark brown outside of apertures; visceral mass creamy white; foot creamy white.

Gills creamy white to pale tan, inner gills may be more golden than outer gills, both gills may have gray or brown areas posteriorly; dorsal margin straight to slightly sinuous, ventral margin elongate convex; gill length 50–60% of shell length; gill height 50–55% of gill length; outer gill height 65–75% of inner gill height; inner lamellae of inner gills connected to visceral mass only anteriorly.

Outer gills marsupial; glochidia held across most of gill; marsupium slightly padded when gravid; creamy white.

Labial palps creamy white; straight dorsally, convex ventrally, bluntly pointed to narrowly rounded distally; palp length 20–25% of gill length; palp height 55–70% of palp length; distal 35–50% of palps bifurcate.

Supra-anal aperture longer than incurrent and excurrent apertures; incurrent aperture longer than excurrent aperture.

Incurrent aperture length 15–20% of shell length; creamy white within, may have grayish cast, with rusty brown to grayish brown basal to papillae; papillae in 1–2 rows, short, may be slender or thick, mostly simple, may have few bifid; papillae tan to rusty tan or gray to grayish brown, may have dark brown edges basally.

Excurrent aperture length 10–15% of shell length; creamy white to pale gray within, marginal color band rusty brown to dark brown, may have irregular tan areas, some individuals with creamy white band proximal to marginal band; margin with small, simple papillae; papillae tan or rusty tan to gray. Lea (1859f) reported excurrent aperture papillae to be brown.

Supra-anal aperture length approximately 20% of shell length; creamy white within, may have narrow gray or grayish brown marginal band; margin smooth; mantle bridge separating supra-anal and excurrent apertures imperforate; length 15–25% of supra-anal length. Lea (1860d) reported the mantle bridge to be absent.

Specimens examined: Spring Creek, Jackson County, Florida (n = 3).

Glochidium Description

Glochidium unknown.

Similar Species

Elliptio fumata resembles several other Apalachicola Basin *Elliptio* species, including *E. arctata*, *E. chipolaensis*, *E. fraterna*, *E. nigella*, *E. pullata* and *E. purpurella*. It most closely resembles *E. pullata*, but *E. fumata* is typically higher in relation to length. *Elliptio arctata* is more elongate and arcuate than *E. fumata*. *Elliptio purpurella* is more inflated and elongate than *E. fumata*. Shells of *E. chipolaensis* and *E. nigella* are narrower anteriorly and broader posteriorly than *E. fumata*. *Elliptio fraterna* differs from *E. fumata* in having corrugations on the posterior slope.

General Distribution

Elliptio fumata appears to be endemic to the Apalachicola Basin in Alabama, Florida and Georgia.

Alabama and Mobile Basin Distribution

Elliptio fumata is known from the Chattahoochee River and tributaries and headwaters of the Chipola River.

Elliptio fumata is extant in some tributaries of the Chattahoochee River and headwaters of the Chipola River.

Ecology and Biology

Elliptio fumata inhabits small creeks to large rivers in slow to swift current. It occurs in various combinations of mud, sand and gravel as well as in crevices in bedrock.

Elliptio fumata is dioecious, but occasional hermaphroditic individuals may be encountered. Heard (1975b) reported 4 hermaphroditic individuals among 266 from a tributary of the Chipola River. *Elliptio fumata* is a short-term brooder, gravid in summer and possibly early spring. Brim Box and Williams (2000) reported gravid females from June to August for *Elliptio fumata* (as *Elliptio complanata*). Glochidial hosts of *E. fumata* are unknown.

Current Conservation Status and Protection

Elliptio fumata has not been recognized for more than a century and was included with *Elliptio complanata* in recent conservation assessments. Lydeard et al. (1999) listed *E. fumata* (as *E. complanata*) as currently stable in Alabama, and Garner et al. (2004) designated it a species of lowest conservation concern in the state.

Remarks

Elliptio complanata is one of the most widespread species on the Atlantic coast, and its morphological variability has long been recognized. Recent systematic work has shown it to be a complex of species (A.E. Bogan, personal observation). The oldest available name for the Apalachicola Basin homologue of *E. complanata* is *Elliptio fumata*.

Synonymy

Unio fumatus Lea, 1857. Lea, 1857g:171; Lea, 1858e:88, pl. 18, fig. 68; Lea, 1858h:88, pl. 18, fig. 68
Type locality: Chattahoochee River, near Columbus, [Muscogee County,] Georgia, Bishop Elliott; Hospaliga [Hospilika] Creek, [Lee and Russell counties,] Alabama; Chattahoochee River, Bishop Elliott. Lectotype, USNM 85552, length 59 mm, designated by Johnson (1970), is from Chattahoochee River, near Columbus, Muscogee County, Georgia.

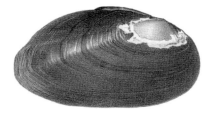  

Unio subniger Lea, 1857. Lea, 1857g:172; Lea, 1859f:196, pl. 22, fig. 79; Lea, 1859g:14, pl. 22, fig. 79
Unio hallenbeckii Lea, 1859. Lea, 1859d:170; Lea, 1860d:328, pl. 51, fig. 154; Lea, 1860g:10, pl. 51, fig. 154
Unio quadratus Lea, 1859. Lea, 1859d:172; Lea, 1860d:338, pl. 54, fig. 163; Lea, 1860g:20, pl. 54, fig. 163
Unio basalis Lea, 1872. Lea, 1872b:161; Lea, 1874c:48, pl. 16, fig. 46; Lea, 1874e:52, pl. 16, fig. 46

Unio gesnerii Lea, 1874. Lea, 1874b:424; Lea, 1874d:65, pl. 22, fig. 61; Lea, 1874e:69, pl. 22, fig. 61
Type locality: Uchee River [Creek, Lee and Russell counties, Alabama,] near Columbus, [Muscogee County,]
Georgia, Dr. J. Lewis. Lectotype, USNM 85670, length 82 mm, designated by Johnson (1970).

Unio invenustus Lea, 1874. Lea, 1874b:424; Lea, 1874d:66, pl. 22, fig. 62; Lea, 1874e:70, pl. 22, fig. 62

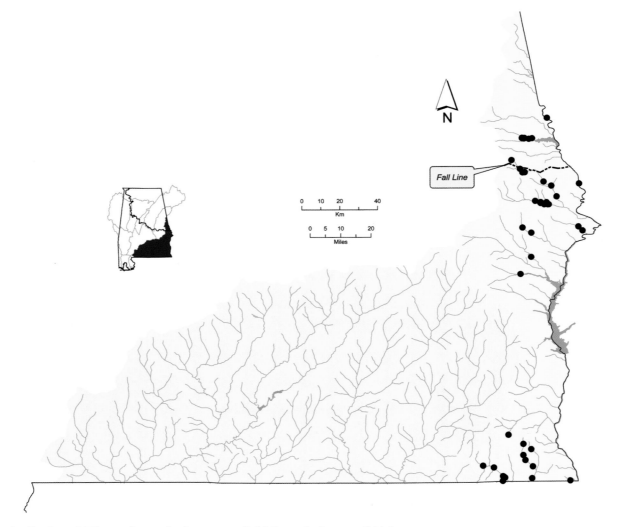

Distribution of *Elliptio fumata* in the eastern Gulf Coast drainages of Alabama.

Elliptio mcmichaeli Clench and Turner, 1956
Fluted Elephantear

Elliptio mcmichaeli – Upper figure: length 72 mm, USNM 710723 (paratype). Choctawhatchee River, 8 miles west of Miller Crossroads, State Route 2, Holmes County, Florida. Lower figure: juvenile, length 35 mm, UMMZ 163341. Pea River, Andrew's Fish Trap, Barbour County, Alabama, November 1915. © Richard T. Bryant.

Shell Description

Length to 130 mm; moderately thin; moderately inflated; outline subrhomboidal to elliptical, occasionally arcuate; posterior margin bluntly pointed to narrowly truncate; anterior margin rounded; dorsal margin slightly convex; ventral margin straight to convex or slightly concave; posterior ridge prominent dorsally, low and rounded posterioventrally, often doubled posterioventrally; posterior slope broad, flat, with small, radial plications near umbo, often eroded in old individuals; umbo broad, somewhat inflated, elevated slightly above hinge line, umbo sculpture unknown; periostracum smooth in young individuals, usually becoming roughened with age, young individuals often greenish brown, with rays, becoming dark brown to black with age, rays becoming obscure.

Pseudocardinal teeth low, triangular, 2 divergent teeth in left valve, separated dorsally, 1 tooth in right valve, may have accessory denticle anteriorly; lateral teeth long, straight to slightly curved, 2 in left valve, 1 in right valve; interdentum long, narrow to very narrow; umbo cavity wide, shallow; nacre bluish white to purplish, often with salmon tint.

Soft Anatomy Description

Mantle tan to light brown, may be lighter outside of pallial line, usually brown, gray or black outside of apertures, often mottled; visceral mass usually pearly white, occasionally creamy white, often with tan areas dorsally; foot creamy white, often darker distally.

Gills tan to light brown, occasionally gold; dorsal margin usually sinuous, occasionally straight, ventral margin convex; gill length 45–65% of shell length; gill height 45–55% of gill length; outer gill height 75–95% of inner gill height; inner lamellae of inner gills connected to visceral mass only anteriorly.

No gravid females were available for marsupium description; probably similar to other *Elliptio* species, with outer gills marsupial, glochidia held across entire gill except extreme anterior and posterior ends, marsupium slightly padded when gravid.

Labial palps creamy white to tan, occasionally with golden cast; straight to slightly concave dorsally, convex ventrally, bluntly pointed to rounded distally; palp length 25–35% of gill length; palp height 55–85% of palp length; distal 20–35% of palps bifurcate.

Incurrent, excurrent and supra-anal aperture lengths variable relative to one another.

Incurrent aperture length 15–25% of shell length; creamy white to tan within, usually with black, tan, gray or variable brown basal to papillae; papillae in 1–2 rows, variable, some combination of simple, bifid, trifid and arborescent, often lacking arborescent papillae, occasional individuals with arborescent papillae only; papillae some combination of creamy white, tan, dull orange, gray, black and various shades of brown, often changing color distally.

Excurrent aperture length 5–15% of shell length; creamy white to tan within, marginal color band tan, gray, brown or black; margin with small, simple papillae, rarely with few scattered bifid or trifid papillae; papillae creamy white, tan, dull orange or gray.

Supra-anal aperture length 10–20% of shell length; creamy white to tan within, often with thin marginal band of brown, gray or black; margin smooth; mantle bridge separating supra-anal and excurrent apertures usually imperforate, length 5–15% of supra-anal length, sometimes absent.

Specimens examined: Flat Creek, Geneva County (n = 4); Murder Creek, Conecuh County (n = 3); Sepulga River (n = 2); West Fork Choctawhatchee River (n = 3).

Glochidium Description

Length 130–157 μm, height 149–161 μm; with weak triangular hooks covered with microstylets; outline subtriangular; dorsal margin straight, ventral margin narrowly rounded, slightly produced, anterior and posterior margins broadly convex, anterior margin slightly more convex than posterior margin (O'Brien et al., 2003).

Similar Species

Elliptio mcmichaeli most closely resembles some *Elliptio crassidens* but has a thinner shell and lower, less angular posterior ridge. The outline of *E. mcmichaeli* is always elongate. The outline of *E. crassidens* is highly variable, with some individuals approaching the shape of *E. mcmichaeli*, but *E. crassidens* are typically more triangular. The two are sympatric only in the Escambia River drainage.

Elliptio mcmichaeli subadults could be confused with *Fusconaia burkei*, a Choctawhatchee River drainage endemic, as they both have a distinct posterior ridge and corrugations on the posterior slope. However, the pseudocardinal and lateral teeth are thicker in *E. mcmichaeli* than in *F. burkei*.

General Distribution

Elliptio mcmichaeli is endemic to the Choctawhatchee River drainage, Alabama and Florida, and the Conecuh River system, Alabama.

Alabama and Mobile Basin Distribution

Elliptio mcmichaeli is widespread in the Choctawhatchee River system. It is known from a few Conecuh River system localities.

Elliptio mcmichaeli is extant in much of the Choctawhatchee River drainage, with the exception of the Pea River system upstream of Elba Dam (Blalock-Herod et al., 2005). It is also extant in the Conecuh River system.

Ecology and Biology

Elliptio mcmichaeli occurs in large creeks to rivers, usually in areas with some current. Sand and sandy clay are its preferred substrates. It may be found at depths of less than 1 m to more than 5 m.

Elliptio mcmichaeli is dioecious. Heard (1979a) reported no hermaphroditic individuals among 36 examined histologically. However, occasional hermaphrodites are known for many species of *Elliptio*. *Elliptio mcmichaeli* is a short-term brooder, gravid in spring and possibly early summer. Mature glochidia have been reported in May from the Choctawhatchee River drainage (O'Brien et al., 2003).

Glochidial hosts of *Elliptio mcmichaeli* are unknown, but disappearance of this species upstream of Elba Dam, which has been in place since the early 1900s, on the Pea River suggests that it may utilize an anadromous fish (Hall and Hall, 1916). Anadromous fishes known to inhabit the Pea River include *Alosa alabamae* (Alabama Shad) and *Alosa chrysochloris* (Skipjack Herring) (Clupeidae); and *Acipenser oxyrinchus desotoi* (Gulf Sturgeon) (Acipenseridae).

Current Conservation Status and Protection

Elliptio mcmichaeli was listed as a species of special concern throughout its range by Williams et al. (1993) and in Alabama by Lydeard et al. (1999). Garner et al. (2004) designated *E. mcmichaeli* a species of highest conservation concern in the state.

Remarks

Clench and Turner (1956) described *Elliptio mcmichaeli* from the Choctawhatchee River drainage. Johnson (1970) placed *E. mcmichaeli* in the synonymy of *Elliptio fraterna* (Lea, 1852), a morphologically similar species of the Apalachicola and Savannah River drainages. Fuller and Bereza (1973, 1974) resurrected *E. mcmichaeli* based on pigmentation of the mantle margin and the presence of arborescent papillae on the incurrent aperture. However, the utility of papilla structure in separating many taxa at the species level is questionable, due to variation observed among individuals. *Elliptio mcmichaeli* is herein recognized, pending further taxonomic studies.

Specimens of *Elliptio mcmichaeli* from the Conecuh River system were identified based on conchological characters. This species was formerly believed to be endemic to the Choctawhatchee River drainage. Additional research is needed to resolve the relationship between Conecuh and Choctawhatchee populations.

Elliptio mcmichaeli was named in honor of Donald F. McMichael, an esteemed student and colleague of William J. Clench and Ruth D. Turner, who assisted them in eastern Gulf Coast drainage surveys in the 1950s (Clench and Turner, 1956).

Synonymy

Elliptio mcmichaeli Clench and Turner, 1956. Clench and Turner, 1956:170–171, pl. 7, figs. 1, 2 [original figure not herein reproduced]

Type locality: Choctawhatchee River, 8 miles west of Miller Crossroads, State Highway 2, Holmes County, Florida. Holotype, MCZ 191922, length 91 mm.

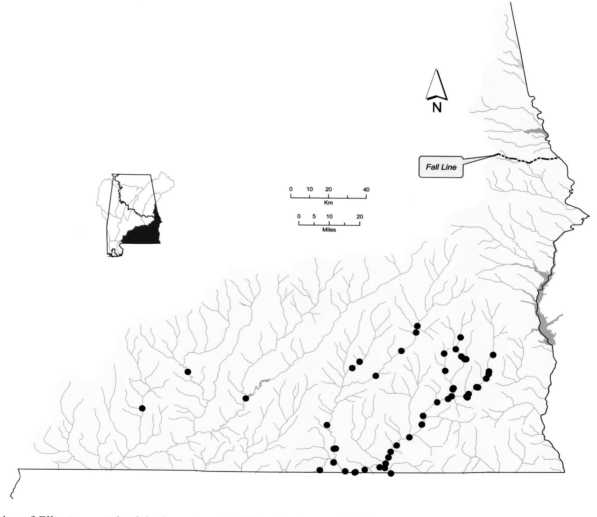

Distribution of *Elliptio mcmichaeli* in the eastern Gulf Coast drainages of Alabama.

Elliptio nigella (Lea, 1852)
Winged Spike

Elliptio nigella – Upper figure: length 55 mm, USNM 85566. Flint River near Albany, Dougherty County, Georgia. Lower figure: length 35 mm, USNM 85568 (lectotype). [Chattahoochee River] near Columbus, Muscogee County, Georgia. © Richard T. Bryant.

Shell Description

Length to 61 mm; moderately thin; moderately inflated, especially along posterior ridge; outline elliptical to subrhomboidal, greatest height midway between umbo and posterior end; posterior margin broadly rounded, may be slightly biangulate; anterior margin rounded; dorsal margin convex, with slight dorsal wing posteriorly; ventral margin straight to slightly convex; posterior ridge well-developed, often doubled posterioventrally; posterior slope flat to slightly convex, not steep; umbo low, wide, slightly elevated above hinge line, umbo sculpture unknown; periostracum dark olive brown in young individuals, becoming dark brown or black with age, young shells with dark green rays that usually become obscure with age.

Pseudocardinal teeth small, compressed, triangular, 2 teeth in left valve, oriented almost parallel to shell margin, 1 tooth in right valve; lateral teeth moderately thick, straight, 2 in left valve, 1 in right valve; interdentum moderately long, very narrow; umbo cavity wide, shallow; nacre white.

Soft Anatomy Description

No material was available for soft anatomy description, but Lea (1859g, 1863d) provided brief notes. The visceral mass is white. Gills are large and rounded ventrally, with inner lamellae of inner gills connected to the visceral mass only anteriorly. Outer gills are marsupial. Labial palps are small and rounded to bluntly pointed distally. The incurrent aperture is large with numerous brown papillae. The excurrent aperture is large with numerous very small brown papillae. The supra-anal aperture is small with a short mantle bridge separating it from the excurrent aperture.

Glochidium Description

Without styliform hooks; "pouch-shaped" (Lea, 1863d).

Similar Species

The shell of *Elliptio nigella* resembles those of several Apalachicola Basin *Elliptio* species including *E. chipolaensis, E. fraterna, E. pullata, E. purpurella* and small *E. fumata*. It most closely resembles *E. chipolaensis* but has more compressed pseudocardinal teeth that lay along the dorsal margin and a darker periostracum. The shell of *E. nigella* differs from those of *E. fraterna*, *E. pullata*, *E. purpurella* and small *E. fumata* in being much higher posteriorly and having smaller, more compressed pseudocardinal teeth. Also, *E. fraterna* has radial plications on the posterior slope, which are not found on *E. nigella*.

General Distribution

Elliptio nigella is endemic to the Apalachicola Basin, known to occur in the Chattahoochee River drainage in Alabama and Georgia and the Flint River drainage in Georgia. Simpson (1914) reported this species from "Chattahoochee River; south into Florida," but there are no known specimens from that state (Brim Box and Williams, 2000).

Alabama and Mobile Basin Distribution

Elliptio nigella is known from three sites, all in the Chattahoochee River. Two sites are near the Fall Line and the other is well downstream of it.

None of the *Elliptio nigella* material from Alabama is dated but were presumably collected during the mid- to late 1800s. It has not been collected anywhere since 1958 and is believed to be extinct.

Ecology and Biology

Almost nothing is known about the habitat of *Elliptio nigella*. Based on locality information of museum records, it apparently occurred in large rivers and large tributaries. Its scarcity in museum collections suggests that it was never common historically. Johnson (1968) reported only three *E. nigella* from "thousands" of mussels collected by Clench, Turner and McMichael during their Gulf Coast survey in 1954.

Elliptio nigella was presumably a short-term brooder, gravid in spring and summer. Its glochidial hosts are unknown.

Current Conservation Status and Protection

Elliptio nigella was reported as endangered throughout its range by Athearn (1970), Stansbery (1971) and Williams et al. (1993). Lydeard et al. (1999) considered it imperiled in Alabama and possibly extinct. Garner et al. (2004) listed *E. nigella* as extinct. The most recent collection of this species was from Coolewahee Creek near its junction with the Flint River, Baker County, Georgia, by H.D. Athearn in 1958.

Remarks

The type locality of *Elliptio nigella* has been mistakenly published as Columbus, Georgia (Johnson, 1968). This mistake appears to have resulted from a transcription error, as the USNM label reads "Columbus" instead of Columbia. Both the original description and a note written in the shell of the holotype read "Columbia". Columbia, Alabama, is located on the Chattahoochee River downstream of Columbus, Georgia.

Frierson (1927) and Johnson (1968) placed *Elliptio purpurella* (Lea, 1857) in the synonymy of *Elliptio nigella* but gave no justification. Brim Box and Williams (2000) recognized *E. purpurella* as a valid species.

Synonymy

Unio nigellus Lea, 1852, *nomen nudum*. Lea, 1852a:251
Unio nigellus Lea, 1852. Lea, 1852b:283, pl. 24, fig. 42; Lea, 1852c:39, pl. 24, fig. 42
Type locality: Chattahoochee River near Columbia, Georgia [Houston County, Alabama], Dr. Boykin. Holotype, USNM 85567, length 39 mm.

Unio denigratus Lea, 1857. Lea, 1857g:171; Lea, 1859f:200, pl. 23, fig. 83; Lea, 1859g:18, pl. 23, fig. 83
Type locality: Streams near Columbus, Georgia, Bishop Elliott. Lectotype, USNM 85568, length 39 mm, designated
    by Johnson (1968).

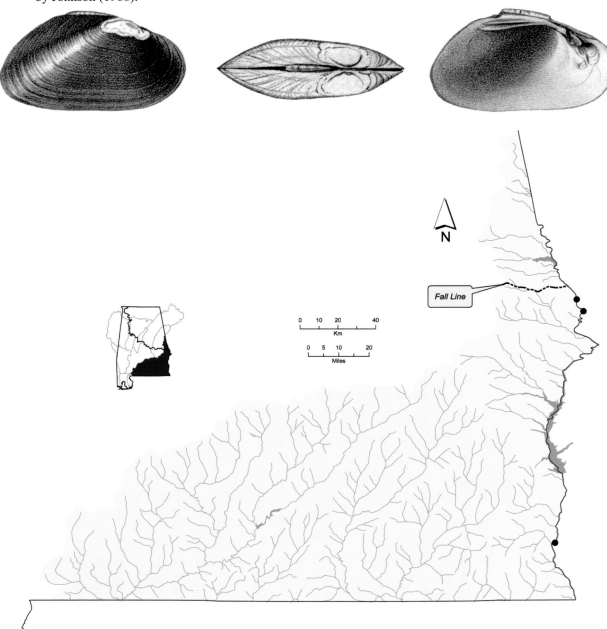

Distribution of *Elliptio nigella* in the eastern Gulf Coast drainages of Alabama.

Elliptio pullata (Lea, 1856)
Gulf Spike

Elliptio pullata – Upper figure: length 59 mm, UF 64842. Sepulga River above Herbert, Conecuh County, Alabama. Middle figure: length 58 mm, UF 64833. Conecuh River, Conecuh County, Alabama. Lower figure: length 58 mm, UF 375380. A canal tributary to Smokehouse Lake, Choctawhatchee River drainage, Walton County, Florida, 22 March 2000. © Richard T. Bryant.

Shell Description

Length to 95 mm; thin when young, becoming moderately thick with age; somewhat compressed to moderately inflated; outline elongate elliptical, sometimes arcuate; posterior margin rounded or bluntly pointed to truncate; anterior margin rounded; dorsal margin straight to slightly convex; ventral margin straight to slightly concave; posterior ridge low, rounded, usually doubled posterioventrally; posterior slope low, moderately wide, flat to slightly convex; umbo broad, flat to slightly inflated, not elevated above hinge line, umbo sculpture well-developed loops; periostracum greenish brown in young individuals, darkening to brown or black with age, occasionally faintly rayed.

Pseudocardinal teeth triangular, low to elevated, 2 teeth in left valve, anterior tooth often smaller, 1 tooth in right valve, occasionally with small accessory denticle anteriorly; lateral teeth long, straight to slightly curved, 2 in left valve, 1 in right valve; interdentum variable, short to moderately long, very narrow; umbo cavity wide, very shallow; nacre typically purplish, may have salmon tint, rarely white.

Soft Anatomy Description

Mantle opaque tan, often some shade of gray with darker blotches outside of apertures; visceral mass pearly white, often with tan patches, especially posteriorly and dorsally; foot creamy white.

Gills ranging from creamy white to light brown, often with golden cast; dorsal margin straight to slightly sinuous, ventral margin elongate convex; gill length 50–60% of shell length; gill height 35–55% of gill length; outer gill height 75–95% of inner gill height; inner lamellae of inner gills connected to visceral mass only anteriorly.

Outer gills marsupial; glochidia held across entire gill except extreme posterior end; marsupium slightly padded when gravid; creamy white.

Labial palps creamy white to tan; straight to slightly concave dorsally, convex ventrally, bluntly pointed distally; palp length 15–35% of gill length; palp height 50–80% of palp length, occasionally greater; distal 20–45% of palps bifurcate.

Incurrent aperture length usually slightly less than or equal to excurrent aperture length; supra-anal aperture usually longer than incurrent and excurrent apertures.

Incurrent aperture length 10–20% of shell length; creamy white or tan within, with black, gray or brown pigment basal to papillae; papillae in 1–2 rows, long, slender, simple; papillae creamy white to tan or rusty brown, often darker distally.

Excurrent aperture length 10–20% of shell length; creamy white to tan within, marginal color band dark brown, gray or black, with regular tan blotches proximal to papillae; margin papillate, papillae small but well-developed, simple; papillae tan to light brown.

Supra-anal aperture length 15–20% of shell length; creamy white to tan within, often with thin tan, sparse brown or gray marginal band; margin smooth; mantle bridge separating supra-anal and excurrent apertures occasionally perforate, length usually 10–30% of supra-anal length, occasionally greater.

Specimens examined: Conecuh River (n = 2); Eightmile Creek, Walton County, Florida (n = 6); Little Cedar Creek, Conecuh County (n = 3); Pauls Creek, Barbour County (n = 3); Pea River (n = 5); Sepulga River (n = 2); West Fork Choctawhatchee River (n = 1).

Glochidium Description

Length 195–215 µm; height 195–234 µm; without styliform hooks; outline depressed subelliptical; dorsal margin straight, ventral margin broadly rounded, anterior and posterior margins almost equally rounded (O'Brien et al., 2003).

Similar Species

The shell of *Elliptio pullata* resembles those of several other species of Gulf Coast *Elliptio* including *E.* *arctata*, *E. chipolaensis*, *E. fumata*, *E. nigella* and *E. purpurella*. The Apalachicola Basin is the only area where these species occur sympatrically in Alabama with the exception of *E. arctata*, with which it also occurs in the Choctawhatchee and Conecuh River drainages. *Elliptio pullata* most closely resembles young *E. fumata* but is more elliptical in outline and has more triangular and divergent pseudocardinal teeth. Although pseudocardinal teeth of *E. fumata* become more triangular with age, most remain less divergent than those of *E. pullata*. *Elliptio arctata* typically has a thinner shell and slightly concave ventral margin. *Elliptio purpurella* is more inflated and has darker purple nacre than *E. pullata*. Shells of *E. chipolaensis* and *E. nigella* are much higher posteriorly than *E. pullata* and also have smaller, more compressed pseudocardinal teeth.

General Distribution

Elliptio pullata occurs in Gulf Coast drainages from the Suwannee River, Georgia and Florida, west to the Escambia River, Alabama and Florida. It is found above and below the Fall Line.

Alabama and Mobile Basin Distribution

Elliptio pullata occurs in the Escambia, Yellow, Choctawhatchee, Chipola and Chattahoochee River drainages.

Elliptio pullata remains widespread and locally common in much of its historical range.

Ecology and Biology

Elliptio pullata occurs in creeks, rivers and floodplain lakes, most commonly in creeks with moderate current. It may be found in a variety of substrates including various mixtures of mud, sand and gravel. Jenkinson (1973) reported *E. pullata* from silt deposits between rocks. *Elliptio pullata* is the most abundant species in many areas but has been eliminated from some Chattahoochee River tributaries (Brim Box and Williams, 2000). *Elliptio pullata* remains locally common in the Escambia, Yellow and Choctawhatchee River drainages in Alabama and Florida.

Elliptio pullata is a dioecious species (Kotrla, 1988), but occasional hermaphroditic individuals may be encountered. Heard (1979a) reported 4 hermaphroditic individuals among 126 from a stream in the Apalachicola Basin. *Elliptio pullata* is a short-term brooder. It has been reported gravid in June and July in the Apalachicola Basin and late May in the Suwannee River (Brim Box and Williams, 2000). Glochidial hosts of *E. pullata*, determined using laboratory trials, include *Lepomis macrochirus* (Bluegill) and *Micropterus salmoides* (Largemouth Bass) (Centrarchidae) (Keller and Ruessler, 1997).

Current Conservation Status and Protection

Elliptio pullata (as *Elliptio icterina* in part) was listed as currently stable throughout its range by Williams et al. (1993) and in Alabama by Lydeard et al. (1999). Garner et al. (2004) designated *E. pullata* (as *E. icterina*) a species of lowest conservation concern in the state.

Remarks

Elliptio pullata is one of the most widespread, and often most abundant, species along the Gulf Coast, from the Escambia River drainage to the Suwannee River drainage. It is highly variable in shell morphology, both among and within populations. This,

along with its resemblance to some forms in Atlantic Coast drainages, has lead to confusion over its taxonomic status. Clench and Turner (1956) synonymized a number of currently recognized morphologically similar *Elliptio* species under the name *E. strigosa*, including *E. pullata*, *E. fraterna*, *E. fumata*, *E. nigella* and *E. purpurella*. However, *E. strigosa* is a synonym of *Elliptio arctata*, which is present but rare in Gulf Coast drainages east of the Mobile Basin. Johnson (1970) placed *E. pullata* in the synonymy of *Elliptio icterina* (Conrad, 1834), but *E. icterina* is a species of the southern Atlantic Coast and differs considerably from *E. pullata* in shell morphology.

Synonymy

Unio pullatis Lea, 1856. Lea, 1856d:262 [name corrected to *Unio pullatus* Lea, 1858]
Type locality: Creeks near Columbus, [Muscogee County,] Georgia. Lectotype, USNM 86020, length 84 mm, designated by Johnson (1970).
Unio sublatus Lea, 1857. Lea, 1857g:169; Lea, 1858e:82, pl. 16, fig. 62; Lea, 1858h:82, pl. 16, fig. 62
Type locality: Uchee Bar, [Chattahoochee River, mouth of Uchee Creek,] below Columbus, Georgia, Bishop Elliott. Lectotype, USNM 85897, length 50 mm, designated by Johnson (1970).

Unio tetricus Lea, 1857. Lea, 1857g:170; Lea, 1859f:195, pl. 22, fig. 78; Lea, 1859g:13, pl. 22, fig. 78
Unio aquilus Lea, 1857. Lea, 1857g:172; Lea, 1858e:92, pl. 20, fig. 72; Lea, 1858h:92, pl. 20, fig. 72
Unio roswellensis Lea, 1858. Lea, 1858c:165; Lea, 1859f:205, pl. 24, fig. 87; Lea, 1859g:23, pl. 24, fig. 87
Unio pullatus Lea, 1858. Lea, 1858e:57, pl. 8, fig. 39; Lea, 1858h:57, pl. 8, fig. 39
Comment: Lea (1858) proposed this name as a spelling correction for *Unio pullatis* (Lea, 1856).

Unio viridiradiatus Lea, 1859. Lea, 1859c:154; Lea, 1860d:336, pl. 53, fig. 161; Lea, 1860g:18, pl. 53, fig. 161
Type locality: Big Uchee [Creek, Lee and Russell counties,] Alabama, near Columbus, [Muscogee County,] Georgia, G. Hallenbeck. Lectotype, USNM 86018, length 66 mm, designated by Johnson (1970).

Unio salebrosus Lea, 1859. Lea, 1859d:170; Lea, 1860d:332, pl. 52, fig. 157; Lea, 1860g:14, pl. 52, fig. 157

Unio viridans Lea, 1859. Lea, 1859d:170; Lea, 1860d:337, pl. 54, fig. 162; Lea, 1860g:19, pl. 54, fig. 162
Unio verutus Lea, 1859. Lea, 1859d:171; Lea, 1860d:335, pl. 53, fig. 160; Lea, 1860g:17, pl. 53, fig. 160
Unio merceri Lea, 1862. Lea, 1862a:169; Lea, 1862c:209, pl. 31, fig. 278; Lea, 1863a:31, pl. 31, fig. 278
Unio corneus Lea, 1874. Lea, 1874a:423; Lea, 1874d:59, pl. 20, fig. 58; Lea, 1874e:63, pl. 20, fig. 58
Unio dooleyensis Lea, 1874. Lea, 1874b:424; Lea, 1874d:64, pl. 22, fig. 60; Lea, 1874e:68, pl. 22, fig. 60
Unio singularis B.H. Wright, 1899. B.H. Wright, 1899:75; Johnson, 1967b:8, pl. 5, fig. 7

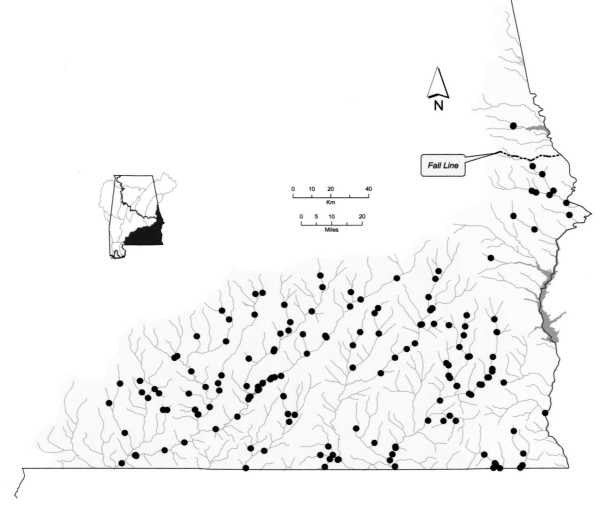

Distribution of *Elliptio pullata* in the eastern Gulf Coast drainages of Alabama.

Elliptio purpurella (Lea, 1857)
Inflated Spike

Elliptio purpurella – Length 49 mm, UF 388492. Muckalee Creek, State Highway 195, 3.5 air miles northeast of Leesburg, Lee County, Georgia, 11 August 1992. © Richard T. Bryant.

Shell Description

Length to 65 mm; moderately thin; somewhat inflated; outline elongate elliptical, may be somewhat arcuate in older individuals; posterior margin narrowly truncate; anterior margin rounded; dorsal margin straight to convex; ventral margin straight to slightly concave; posterior ridge usually low, rounded, may be doubled posteroventrally; posterior slope flat; umbo broad, low, elevated slightly above hinge line, usually eroded, umbo sculpture unknown; periostracum greenish brown to reddish brown, often with variable dark green rays.

Pseudocardinal teeth moderately thick, somewhat triangular, 2 divergent teeth in left valve, 1 tooth in right valve, may have accessory denticle anteriorly; lateral teeth short, thin, curved, 2 in left valve, 1 in right valve; interdentum long, very narrow; umbo cavity wide, shallow; nacre purple, may have salmon tint in umbo.

Soft Anatomy Description

Mantle tan, often with sparse rusty tan outside of apertures; visceral mass tan; foot tan.

Gills tan, may have slight golden cast; dorsal margin straight to slightly sinuous, ventral margin elongate convex; gill length 55–65% of shell length; gill height 35–45% of gill length; outer gill height 70–85% of inner gill height; inner lamellae of inner gills connected to visceral mass only anteriorly.

No gravid females were available for marsupium description. Lea (1859g) reported outer gills to be marsupial, with glochidia held across the entire gill.

Labial palps tan; straight dorsally, convex ventrally, bluntly pointed distally; palp length 20–25% of gill length; palp height 45–65% of palp length; distal 30–50% of palps bifurcate.

Incurrent aperture usually longer than excurrent aperture; incurrent and excurrent apertures may be equal to, longer or shorter than supra-anal aperture.

Incurrent aperture length 15–20% of shell length; tan within, often with slight golden cast, may have sparse, irregular rusty tan or light brown basal to papillae; papillae in 2 rows, moderately long, slender to thick, mostly simple, usually with few bifid; papillae tan to rusty tan. Lea (1859g) reported incurrent aperture papillae to be dark brown.

Excurrent aperture length 15–20% of shell length; tan within, often with slight golden cast, may have a pale tan, rusty tan or light brown marginal color band, sometimes irregular; margin crenulate or with very small, simple papillae, papillae and crenulations often diminish with a dorsal progression; papillae pale tan to rusty tan.

Supra-anal aperture length 10–25% of shell length; tan within, sometimes with sparse, irregular rusty tan or light brown marginally; margin smooth; mantle bridge separating supra-anal and excurrent apertures imperforate, length 15–40% of supra-anal length.

Specimens examined: Chickasawhatchee Creek, Terrell County, Georgia (n = 3); unnamed tributary to Abram's Creek, Worth County, Georgia (n = 1) (specimens previously preserved).

Glochidium Description

Glochidium unknown.

Similar Species

Elliptio purpurella resembles several Apalachicola Basin *Elliptio* species including *E. arctata*, *E. chipolaensis*, *E. nigella*, *E. pullata* and small *E. fumata*. It most closely resembles *Elliptio pullata* but is more inflated, with a thinner shell and darker purple nacre. Its shell is not as high posteriorly as those of *E. chipolaensis*, *E. nigella* and small *E. fumata*. Also, the pseudocardinal teeth of *E. purpurella* are more triangular than those of *E. chipolaensis* and *E. nigella*. Nacre of *E. purpurella* is typically purple, whereas that of *E. arctata* is typically bluish white, though nacre color may overlap.

General Distribution

Elliptio purpurella is endemic to the Apalachicola Basin in Alabama, Florida and Georgia, below the Fall Line. It is known from the Chattahoochee, Chipola and Flint River drainages, but there are no records from the Apalachicola River proper.

Alabama and Mobile Basin Distribution

Elliptio purpurella is known from the Chattahoochee River and headwaters of the Chipola River.

The only known extant population of *Elliptio purpurella* in Alabama is in headwaters of the Chipola River.

Ecology and Biology

Elliptio purpurella inhabits medium to large creeks and small rivers below the Fall Line. It is usually found in some combination of clay, sand, gravel and limestone in moderate current (Brim Box and Williams, 2000).

Elliptio purpurella is presumably a short-term brooder, gravid in spring. Brim Box and Williams (2000) reported no gravid females among 369 specimens checked between May and September, suggesting that its gravid period may be earlier than those of other species of *Elliptio*. Its glochidial hosts are unknown.

Current Conservation Status and Protection

Elliptio purpurella was not recognized in conservation status reviews prior to Garner et al. (2004), in which it was designated a species of highest conservation concern in Alabama.

Remarks

Elliptio purpurella has historically been placed in the synonymy of several species, including *Elliptio arctata* (Simpson, 1914), *Elliptio nigella* (Frierson, 1927; Johnson, 1968) and *Elliptio strigosa* (Clench and Turner, 1956). It was recently recognized as a valid species by Brim Box and Williams (2000), based on shell morphology and electrophoresis data.

Synonymy

Unio purpurellus Lea, 1857. Lea, 1857g:171; Lea, 1859f:198, pl. 23, fig. 81; Lea, 1859g:16, pl. 23, fig. 81
Type locality: Flint River near Albany, [Dougherty County,] Georgia. Lectotype, USNM 85675, length 36 mm, designated by Johnson (1974).

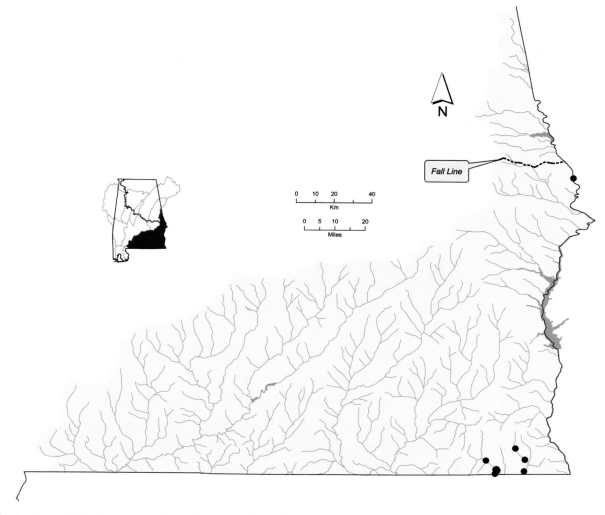

Distribution of *Elliptio purpurella* in the eastern Gulf Coast drainages of Alabama.

Genus *Elliptoideus*

Elliptoideus Frierson, 1927, is endemic to the Apalachicola and Ochlockonee basins. In Alabama it is known only from the Chattahoochee River. *Elliptoideus* is a monotypic genus (Turgeon et al., 1998).

Conrad (1853) placed *Unio sloatianus* Lea, 1840, into *Plectomerus* Conrad, 1853, along with nine other plicate species but failed to designate a type species. Frierson (1927) included *U. sloatianus* in *Elliptio* Rafinesque, 1819, and erected the subgenus *Elliptoideus* for the species. Modell (1949, 1964) treated *Elliptoideus* as a subjective synonym of *Plectomerus*. Phylogenetic trees of Serb et al. (2003) placed *Elliptoideus* and *Plectomerus* as sister taxa but with only weak support. In a more extensive study of North American unionids, *Elliptoideus* and *Plectomerus* appear in widely separated clades (Campbell et al., 2005).

Type species

Unio sloatianus Lea, 1840

Diagnosis

Shell rhomboidal; inflated; thick; umbo moderately inflated, elevated above hinge line; posterior ridge well-developed, with a few strong, irregular corrugations; shell disk sculptured with oblique ridges posteriorly; nacre purplish; periostracum brown to black; umbo cavity open, deep.

Incurrent aperture papillae arborescent; inner lamellae of inner gills usually connected to visceral mass only anteriorly; all 4 gills marsupial; marsupium not extended ventrally below original gill edge when gravid; glochidium without styliform hooks (Simpson, 1914; Frierson, 1927; Haas, 1969a, 1969b).

Synonymy

None recognized.

Elliptoideus sloatianus (Lea, 1840)
Purple Bankclimber

Elliptoideus sloatianus – Upper figure: length 117 mm, UF 388494. Flint River, 2 air miles upstream of State Highway 345 boat ramp, Decatur County, Georgia, 24 September 1992. Lower figure: archaeological specimen, length 79 mm, Columbus Museum, Columbus, Georgia, uncataloged. Mouth of Omusee Creek, B.W. Andrews Reservoir, Omusee Park (area 1 northwest slope), Chattahoochee River Drive, Houston County, Alabama. © Richard T. Bryant.

Shell Description

Length to 205 mm; thick; moderately inflated; outline quadrate to rhomboidal; posterior margin obliquely truncate; anterior margin rounded; dorsal margin straight to slightly convex; ventral margin straight to slightly concave; posterior ridge prominent, typically doubled posterioventrally; posterior slope moderately steep, but forming slight dorsal wing; posterior ridge and slope adorned with oblique plications from crest of posterior slope to dorsal and posterior margins, plications may become weak distally, posterior 75% of shell with variable, subradial wrinkles; umbo low, broad, elevated slightly above hinge line, umbo sculpture unknown; periostracum dull, may be somewhat shiny in young individuals, dark brown to black, without rays.

Pseudocardinal teeth triangular, rough, 2 divergent teeth in left valve, separated dorsally, 1 large tooth in right valve, often with accessory denticle anteriorly; lateral teeth thick, moderately long, straight to slightly curved, 2 in left valve, 1 in right valve; interdentum moderately long, narrow; umbo cavity wide, shallow; nacre various shades of purple, often darker outside of pallial line.

Soft Anatomy Description

Mantle tan to light brown, often with rusty brown outside of apertures; visceral mass tan to light brown; foot tan.

Gills tan; straight to slightly sinuous dorsally, elongate convex ventrally; gill length usually 60–70% of shell length; gill height 40–45% of gill length; outer gill height 75–90% of inner gill height; inner lamellae of inner gills usually connected to visceral mass only anteriorly, connection occasionally complete.

No gravid females were available for marsupium description. Frierson (1927) reported all four gills to be marsupial.

Labial palps tan; slightly convex dorsally, convex ventrally, bluntly pointed distally, occasionally narrowly rounded; palp length 15–25% of gill length; palp height 50–80% of palp length; distal 30–45% of palps bifurcate.

Incurrent, excurrent and supra-anal aperture lengths variable, any may be longer than the others.

Incurrent aperture length 15–25% of shell length; tan within, sometimes with rusty brown or dark brown mottling basal to papillae; papillae arborescent, interspersed with small, blunt, simple and bifid papillae; papillae tan to dark brown, often brown basally and tan distally.

Excurrent aperture length 15–30% of shell length; tan within, usually with rusty or dark brown marginal color band, band usually widest adjacent to incurrent aperture, tapering dorsally; margin minutely crenulate, crenulations may be better developed adjacent to incurrent aperture, sometimes disappearing dorsally. Lea (1863d) reported the excurrent aperture to have "numerous minute dark-brown papillae".

Supra-anal aperture length 10–25% of shell length; tan within, without marginal coloration; margin smooth; mantle bridge separating supra-anal and excurrent apertures absent.

Specimens examined: Flint River, Georgia (n = 4); Ochlockonee River, Florida (n = 3) (specimens previously preserved).

Glochidium Description

Length 120–170 μm; height 100–150 μm; without styliform hooks; outline elongate oval; dorsal margin straight, ventral margin broadly rounded, anterior and posterior margins convex (O'Brien and Williams, 2002).

Similar Species

Elliptoideus sloatianus may resemble *Megalonaias nervosa* but is more obliquely truncate posteriorly and has a more prominent posterior ridge. *Elliptoideus sloatianus* also has a shallow umbo cavity and purple nacre, whereas *M. nervosa* has a deep umbo cavity and white nacre. Also, the anterior adductor muscle scar of *E. sloatianus* is deep and moderately smooth, but that of *M. nervosa* is shallow and considerably roughened.

General Distribution

Elliptoideus sloatianus is endemic to the Apalachicola Basin in Alabama, Georgia and Florida and the Ochlockonee River drainage in Florida (Brim Box and Williams, 2000). A report of this species from the Escambia River, Florida, appears to be erroneous (Heard, 1979b).

Alabama and Mobile Basin Distribution

Elliptoideus sloatianus is restricted to the Chattahoochee River. There are few historical records of this species from the state.

Elliptoideus sloatianus appears to be extant only in the tailwaters of Bartletts Ferry Dam, Chattahoochee River, Lee County, Alabama, and Harris County, Georgia (R.C. Stringfellow, personal communication).

Ecology and Biology

Elliptoideus sloatianus occurs primarily in medium to large rivers in substrates composed of sand, muddy sand or fine gravel, often near limestone outcrops. This species often occurs in water more than 3 m deep. Though a few old individuals have been found in reservoirs, it does not appear to reproduce in impounded waters.

Elliptoideus sloatianus is presumably a short-term brooder, gravid during winter and spring, possibly summer. A population in the Ochlockonee River was monitored from September through April and females were found to brood mature glochidia from at least late February through April (O'Brien and Williams, 2002). The end of the brooding period is unclear. *Elliptoideus sloatianus* glochidia are not discharged in conglutinates but in loose, easily separated clumps (O'Brien and Williams, 2002).

The primary glochidial host of *Elliptoideus sloatianus* is unknown. Fishes shown to serve as secondary hosts during laboratory trials include *Percina nigrofasciata* (Blackbanded Darter) (Percidae) and *Gambusia holbrooki* (Eastern Mosquitofish) (Poeciliidae), as well as one nonindigenous species, *Poecilia reticulata* (Guppy) (Poeciliidae) (O'Brien and Williams, 2002). *Moxostoma lachneri* (Greater Jumprock) (Catostomidae) has also been found to serve as a secondary host for *E. sloatianus* glochidia (P.D. Johnson, personal communication).

Current Conservation Status and Protection

Elliptoideus sloatianus was considered endangered throughout its range by Athearn (1970) and Stansbery (1971) and threatened throughout its range by Williams et al. (1993). Lydeard et al. (1999) listed *E. sloatianus* as extirpated from Alabama. It was subsequently rediscovered in the Chattahoochee River, and Garner et al. (2004) designated it a species of highest conservation concern in the state.

Remarks

Elliptoideus sloatianus was historically treated as a species of *Elliptio*. Frierson (1927) erected the subgenus *Elliptoideus*, with *sloatianus* the only species recognized, based on the use of all four gills as marsupia instead of only the outer two. Heard and Guckert (1970) elevated *Elliptoideus* to generic level.

Elliptoideus sloatianus is a common element of archaeological material encountered along the Chattahoochee River in Alabama and Georgia, suggesting that it was common in that river prehistorically. *Elliptoideus sloatianus* is one of the few southeastern unionids known from fossil records. Two records of Pleistocene material exist, one from the Hillsborough River drainage (Bogan and Portell, 1995), the other from the Suwannee River drainage (J.D.

Williams, personal observation), suggesting that the species was formerly more widespread.

A report of one shell of *Elliptoideus sloatianus* from a boat ramp on the Escambia River in Florida (Heard, 1979b) appears to be based on a casual introduction and may be a misidentification of *Plectomerus dombeyanus* (Williams and Butler, 1994). There is no evidence of a reproducing population of this species in the Escambia River.

Synonymy

Unio sloatianus Lea, 1840. Lea, 1840:287; Lea, 1842b:217, pl. 16, fig. 33; Lea, 1842c:55, pl. 16, fig. 33
Type locality: Chattahoochee River, Georgia, L.W. Sloat. Holotype, AMNH 56104, length 107 mm. Clench and
  Turner (1956) restricted the type locality to [Chattahoochee River,] Columbus, [Muscogee County,] Georgia.

Unio atromarginatus Lea, 1840. Lea, 1840:288; Lea, 1842b:207, pl. 13, fig. 21; Lea, 1842c:45, pl. 13, fig. 21
Type locality: Chattahoochee River, Columbus, [Muscogee County,] Georgia, Dr. Boykin. Lectotype, USNM
  83977, length 47 mm, designated by Johnson (1974).

Unio plectrophorus Conrad, 1849. Conrad, 1849:154
Unio aratus Conrad, 1849. Conrad, 1849:302
Unio plectophorus Conrad, 1850. Conrad, 1850:277, pl. 38, fig. 7 [provided a figure of a syntype and corrected
  spelling of *Unio plectrophorus* to *Unio plectophorus*]

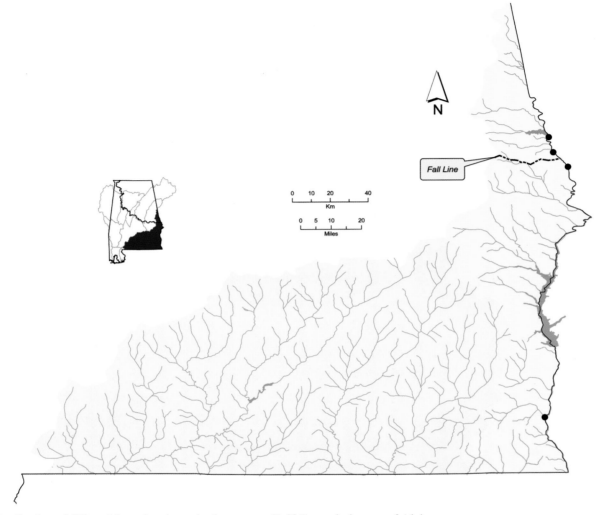

Distribution of *Elliptoideus sloatianus* in the eastern Gulf Coast drainages of Alabama.

Genus *Epioblasma*

Epioblasma Rafinesque, 1831, occurs in the Great Lakes, Mississippi and Mobile basins. Twenty species were recognized by Turgeon et al. (1998). Three species—*Epioblasma florentina* (Lea, 1857), *Epioblasma obliquata* (Rafinesque, 1820) and *Epioblasma torulosa* (Rafinesque, 1820)—were recognized by Turgeon et al. (1998) as having three, two and three subspecies, respectively. Twenty-one species are herein recognized as occurring in Alabama, including *Epioblasma cincinnatiensis* (Lea, 1840), which is herein elevated from synonymy and not included in Turgeon et al. (1998). *Epioblasma* has the largest number of extinct species (14) among North American unionid genera (Turgeon et al., 1998).

The name *Dysnomia* Agassiz, 1852, was long used for this group of species. Bogan (1997a) established the earliest name for this group as *Epioblasma* Rafinesque, 1831. The type species, *Epioblasma biloba* Rafinesque, 1831, was identified as a female *Epioblasma torulosa rangiana* (Lea, 1838). Thus *Dysnomia*, with type species *Unio foliatus* Hildreth, 1828, is a junior synonym of *Epioblasma*. Johnson (1978) used *Plagiola* Rafinesque, 1820, as the genus for this species group based on a misidentification of the type species *Plagiola interrupta* Rafinesque, 1820, which is in fact *Ptychobranchus fasciolaris* (Rafinesque, 1820) (see *Ptychobranchus* genus account).

Species of *Epioblasma* were formerly included in *Truncilla* Rafinesque, 1819 (e.g., Walker, 1910b). Ortmann and Walker (1922a) clarified the type species of *Truncilla* as *T. truncata* Rafinesque, 1820, and moved *Epioblasma* species to *Dysnomia*.

Epioblasma females develop an expanded extrapallial swelling posteriorly, which has serrated margins in some species (Figure 13.5). The extrapallial swelling is often erroneously referred to as a marsupial swelling. Some authors (e.g., Simpson, 1900b) suggested that marsupia are housed in the swelling. However, the extrapallial swelling is positioned outside the pallial line and is unrelated to the marsupia. The extrapallial swelling is filled with spongy mantle tissue and is often brightly pigmented in life. The mantle margin within the swelling typically has some form of modification believed to serve as a glochidial host lure (e.g., microlure, papillae and caruncle-like structures).

Type Species

Epioblasma biloba Rafinesque, 1831 (= *Unio rangiana* Lea, 1838)

Diagnosis

Shell rounded, ovate or triangular; thin to thick; moderately to greatly inflated; shell surface smooth or with low tubercles; posterior ridge well-developed; some species with radial sulcus; umbo sculpture double-looped ridges; sexual dimorphism typically exaggerated, with females having an extrapallial swelling.

Female mantle margin generally modified just ventral to incurrent aperture; inner lamellae of inner gills usually completely connected to visceral mass; outer gills marsupial; glochidia held in posterior part of gill; marsupium outline reniform when gravid, distended, extended ventrally beyond original gill edge when gravid; glochidium without styliform hooks, some species with supernumerary hooks (Ortmann, 1912a; Simpson, 1914; Johnson, 1978).

Figure 13.5. *Epioblasma haysiana* female (ventral view). Length 37 mm, USNM 84614. Florence, [Lauderdale County,] Alabama. The common name of this species, Acornshell, is derived from its appearance when viewed from the ventral aspect, as above. © Richard T. Bryant.

Synonymy

Dysnomia Agassiz, 1852
Pilea Simpson, 1900
Scalenilla Ortmann and Walker, 1922
Truncillopsis Ortmann and Walker, 1922
Capsaeformis Frierson, 1927
Obliquata Frierson, 1927
Penita Frierson, 1927
Torulosa Frierson, 1927

Epioblasma arcaeformis (Lea, 1831)
Sugarspoon

Epioblasma arcaeformis – Upper figure: female, length 40 mm, UF 4107. Tennessee River, Florence, Lauderdale County, Alabama. Lower figure: male, length 48 mm, UF 64222. Holston River, Sullivan County, Tennessee, [late 1800s]. © Richard T. Bryant.

Shell Description

Length to 70 mm; moderately thick; very inflated, females more inflated than males; outline oval to quadrate or trapezoidal, females more elongate than males; posterior margin broadly rounded, often with 1–2 small emarginations; anterior margin rounded; dorsal margin slightly convex; ventral margin straight to slightly convex or slightly concave; posterior ridge high, sharp, often extends slightly below ventral margin; extrapallial swelling of females formed by expanded and swollen posterior ridge, distal margin of extrapallial swelling weakly serrate; posterior slope steep, with radial undulations giving posterior ridge appearance of being doubled or tripled; umbo inflated, elevated well above hinge line, umbo sculpture weak, undulating ridges; periostracum tawny to yellowish green, with numerous narrow green rays.

Pseudocardinal teeth rough, triangular, erect, 2 divergent teeth in left valve, 1 large tooth in right valve, often with accessory denticle anteriorly and/or posteriorly; lateral teeth short, straight to slightly curved, 2 in left valve, 1 in right valve, right valve may also have rudimentary lateral tooth ventrally; interdentum short, narrow to moderately wide; umbo cavity shallow; nacre white.

Soft Anatomy Description

Mantle tan; visceral mass tan; foot tan, slightly darker distally.

Females with thick, spongy pad filling extrapallial swelling; mantle margin in extrapallial swelling papillate; papillae simple; small, fleshy, bulbous structure (microlure) near posterior end of swelling, thick basally, produced and acute distally, with row of small, papilla-like structures extending up 2 opposing sides (Figure 13.6).

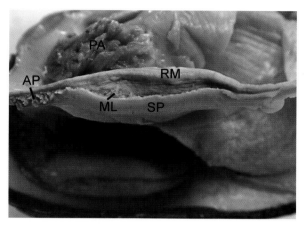

Figure 13.6. *Epioblasma arcaeformis* female modified mantle margin. Arborescent papillae (AP), right mantle (RM), microlure (ML), posterior adductor muscle (PA), spongy pad (SP). RMNH 2009. Tennessee, collected mid-1800s. Photograph by E. Neubert.

Gills tan; dorsal margin straight, outer gill ventral margin straight, curved up at ends, inner gill ventral margin convex, deepest anteriorly; gill length approximately 50% of shell length; gill height approximately 60% of gill length; outer gill height approximately 55% of inner gill height; inner lamellae of inner gills completely connected to visceral mass.

No gravid females were available for marsupium description; probably similar to other *Epioblasma* species, with glochidia held in posterior part of outer gill, marsupium padded when gravid.

Labial palps tan; curved ventrally, rounded distally; palp length approximately 15% of gill length; palp height 80% of palp length; distal 60% of palps bifurcate.

Incurrent and supra-anal apertures of similar length, longer than excurrent aperture.

Incurrent aperture length approximately 20% of shell length; tan within, without coloration basal to papillae; papillae in 2 rows, arborescent, inner row larger; papillae tan.

Excurrent aperture length approximately 15% of shell length; tan within, without marginal color band; margin with very small, simple papillae; papillae tan.

Supra-anal aperture length approximately 20% of shell length; tan within, without marginal coloration; margin smooth; mantle bridge separating supra-anal and excurrent apertures imperforate, length approxi-

mately 10% of supra-anal length; small secondary mantle bridge present.

Specimen examined: "Tennessee" (n = 1) (specimen previously preserved).

Glochidium Description
Glochidium unknown.

Similar Species
Some *Epioblasma arcaeformis* may resemble very large *Epioblasma triquetra*. The posterior slope of *E. arcaeformis* is less steep than that of *E. triquetra*. The green rays on the periostracum of *E. arcaeformis* are almost always thin, while those of *E. triquetra* are usually wide and broken into squares, triangles or chevrons.

General Distribution
Epioblasma arcaeformis is endemic to the Cumberland and Tennessee River drainages. It is known historically from the Cumberland River between Cumberland Falls and Nashville (Cicerello et al., 1991; Parmalee and Bogan, 1998) and from Tennessee River headwaters in eastern Tennessee downstream to Muscle Shoals. However, archaeological evidence suggests that it prehistorically occurred in both rivers downstream to their mouths (Parmalee and Bogan, 1998).

Epioblasma arcaeformis occurred downstream of Muscle Shoals to the mouth of the Tennessee River based on archaeological remains (Casey, 1986; Parmalee and Bogan, 1998; Hughes and Parmalee, 1999). Casey (1986) examined mussels from archaeological sites along the lower Ohio River but did not find any *E. arcaeformis*. However, six valves of *E. arcaeformis* were recovered from the Wicliffe Mounds located on the bluffs of the Mississippi River, Ballard County, Kentucky, about 3 miles downstream of the mouth of the Ohio River (Wesler, 2001).

Alabama and Mobile Basin Distribution
Epioblasma arcaeformis historically occurred in the Tennessee River across northern Alabama.

Epioblasma arcaeformis is believed to be extinct. There are no dated museum specimens collected from Alabama, but it probably disappeared soon after impoundment of the Tennessee River.

Ecology and Biology
Epioblasma arcaeformis occurred in medium to large rivers in shoal habitat (Ortmann, 1918; Parmalee and Bogan, 1998).

Epioblasma arcaeformis was presumably a long-term brooder, gravid from late summer or autumn to the following summer. Its glochidial hosts are unknown. Some species of *Epioblasma* are known to use darters (Percidae) and sculpins (Cottidae) as glochidial hosts.

Current Conservation Status and Protection

Epioblasma arcaeformis has been presumed extinct since 1970 (Stansbery, 1970a; Williams et al., 1993; Turgeon et al., 1998; Lydeard et al., 1999; Garner et al., 2004).

Remarks

There were apparently few museum specimens of *Epioblasma arcaeformis* from below Walden Gorge available for examination by Ortmann (1925), who questioned the validity of reports of the species from the lower Tennessee River. Though several lots of material are extant in museum collections, this species may have been uncommon historically. *Epioblasma arcaeformis* has been reported from prehistoric shell middens across northern Alabama but was not common at any of the sites studied (Morrison, 1942; Warren, 1975; Hughes and Parmalee, 1999).

The only known specimen of *Epioblasma arcaeformis* with soft tissues (Figure 13.6) is housed in the National Museum of Natural History, Leiden, The Netherlands. The specimen was collected by Gerard Troost, Tennessee State Geologist, during the mid-1800s.

Synonymy

Unio arcaeformis Lea, 1831. Lea, 1831:116, pl. 17, fig. 44; Lea, 1834b:126, pl. 17, fig. 44
Type locality: Tennessee River. Syntype not found, also not found by Johnson (1974). Length of figured shell in
    original description reported as 63 mm.

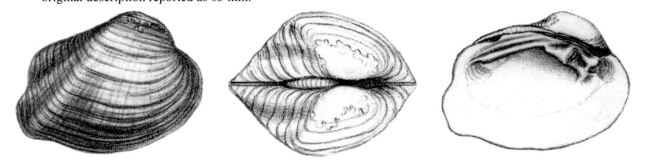

Unio nexus Say, 1831. Say, 1831c:527; Say, 1834:[no pagination], pl. 6, pl. 51

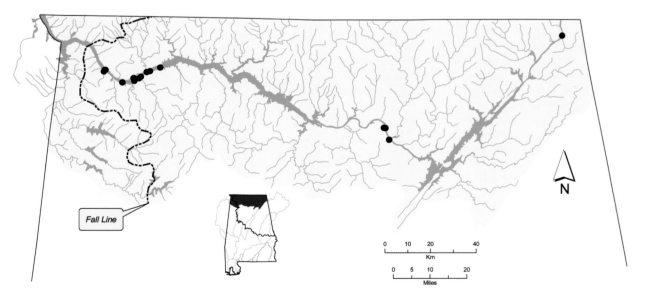

Distribution of *Epioblasma arcaeformis* in the Tennessee River drainage of Alabama.

Epioblasma biemarginata (Lea, 1857)
Angled Riffleshell

Epioblasma biemarginata – Upper figure: female, length 38 mm, UF 64224. Lower figure: male, length 39 mm, UF 64224. Tennessee River, Muscle Shoals, Lauderdale County, Alabama, November 1909. © Richard T. Bryant.

Shell Description

Length to 55 mm; moderately thick; slightly to moderately inflated; outline triangular to trapezoidal or irregularly ovate; male posterior margin truncate, often with slight oblique emargination, female posterior margin evenly rounded, often with slight emargination; anterior margin evenly rounded; dorsal margin straight to slightly convex; ventral margin rounded with an emargination posteriorly, ventral emargination of females may be obscured by extrapallial swelling; posterior ridge doubled distally, ridges flat or slightly rounded, straight in males, slightly curved in females; posterior slope steep, slightly concave; wide, shallow sulcus located anterior of posterior ridge, in females sulcus obscured distally by extrapallial swelling; umbo

moderately full, elevated above hinge line, umbo sculpture unknown; periostracum yellowish green to greenish brown, with numerous variable green rays.

Pseudocardinal teeth erect, triangular, 2 divergent teeth in left valve, 1 tooth in right valve, may have accessory denticle anteriorly and/or posteriorly; lateral teeth short, thick, straight, 2 in left valve, 1 in right valve; interdentum short, very narrow; umbo cavity wide, shallow; nacre white.

Soft Anatomy Description

Mantle tan, outside of apertures rusty tan to rusty brown; visceral mass tan; foot tan.

Females with papillate mantle margin within extrapallial swelling; papillae long, slender, widely separated; small, fleshy, bulbous structure (microlure) near posterior end of swelling; bulbous structure thick basally, produced and acute distally, with row of small papilla-like structures extending up 2 opposing sides; bulbous structure mottled with brown or rusty brown; mantle within extrapallial swelling with small dark brown pits in irregular rows parallel to margin, many pits oblong or reniform in outline. Males with rudimentary papillae just ventral to incurrent aperture, may have small, flattened structure homologous to bulbous structure of females.

Gills tan; dorsal margin slightly sinuous, ventral margin convex, may be scalloped; gill length 45–60% of shell length; gill height 60–80% of gill length; outer gill height 80–95% of inner gill height; inner lamellae of inner gills completely connected to visceral mass.

Outer gills marsupial; glochidia held in posterior 75% of gill; marsupium elongate, well-padded when gravid; creamy white.

Labial palps tan; straight to slightly convex dorsally, convex ventrally, narrowly rounded to bluntly pointed distally; palp length 20–35% of gill length; palp height 60–80% of palp length; distal 40–60% of palps bifurcate.

Incurrent and supra-anal apertures of similar length, longer than excurrent aperture.

Incurrent aperture length 15–20% of shell length; pale tan within, sometimes with thin band of rusty brown basal to papillae; papillae simple, bifid, trifid and small arborescent, long, thick or slender, some individuals with only simple papillae; papillae rusty tan, often with rusty brown between papillae.

Excurrent aperture length 10–15% of shell length; tan within, marginal color band rusty brown and tan or rusty tan, without distinctive pattern; margin with small, simple papillae, may be well-developed or little more than crenulations; papillae tan to rusty tan.

Supra-anal aperture length 15–20% of shell length; tan within, often with thin, irregular rusty tan or rusty brown marginal band; margin smooth or with minute crenulations; mantle bridge separating supra-anal and excurrent apertures imperforate, length 15–25% of supra-anal length.

Specimens examined: Elk River (n = 4) (specimens previously preserved).

Glochidium Description

Length 250–275 μm; height 237–262 μm; without styliform hooks; outline subrotund; dorsal margin straight, ventral margin rounded, anterior and posterior margins convex.

Similar Species

Female *Epioblasma biemarginata* superficially resemble female *Epioblasma capsaeformis*, *Epioblasma florentina*, *Epioblasma turgidula* and *Epioblasma* sp.

cf. *capsaeformis*. *Epioblasma biemarginata* differs from *E. turgidula* in being less elongate and having a less prominent posterior ridge. *Epioblasma capsaeformis*, *E. florentina* and *Epioblasma* sp. have more pronounced extrapallial swellings than *E. biemarginata*, which gives them a more bilobed outline.

Male *Epioblasma biemarginata* resemble male *Epioblasma stewardsonii* but lack a medial ridge.

Male *Epioblasma biemarginata* may also resemble small *Quadrula quadrula* in outline, but *E. biemarginata* lack pustules.

General Distribution

Epioblasma biemarginata is endemic to the Tennessee and Cumberland River drainages. It was found downstream of Cumberland Falls in the Cumberland River drainage (Cicerello et al., 1991) and in the Tennessee River drainage from headwaters in eastern Tennessee downstream to Muscle Shoals, Alabama, and some tributaries of the middle reaches (Parmalee and Bogan, 1998).

Alabama and Mobile Basin Distribution

Epioblasma biemarginata historically occurred in the Tennessee River across northern Alabama and some of its larger tributaries. No records of *E. biemarginata* from Alabama reaches of the Elk River exist, but historical records from Tennessee suggest that it occurred there (Parmalee and Bogan, 1998).

Epioblasma biemarginata is believed to be extinct. A specimen was collected from the Tennessee River near Florence in 1970, so it apparently remained extant in this reach of the river longer than any other *Epioblasma* species.

Ecology and Biology

Epioblasma biemarginata occurred in shoal habitat of medium to large rivers (Parmalee and Bogan, 1998).

Epioblasma biemarginata was presumably a long-term brooder, gravid from late summer or autumn to the following summer. Its glochidial hosts are unknown. Some species of *Epioblasma* have been found to use darters (Percidae) and sculpins (Cottidae) as glochidial hosts.

Current Conservation Status and Protection

Epioblasma biemarginata has been presumed extinct since at least 1970 (Stansbery, 1970a; Williams et al., 1993; Turgeon et al., 1998; Lydeard et al., 1999; Garner et al., 2004).

Remarks

Ortmann (1925) suggested that *Epioblasma biemarginata* is replaced by *Epioblasma turgidula* in upper reaches of the Tennessee River. However,

Parmalee and Bogan (1998) reported records of *E. biemarginata* from as far upstream as the Clinch and Holston rivers. Stansbery (1970a) listed *turgidula* as a headwater subspecies of *E. biemarginata*. Since the two forms occurred sympatrically from Muscle Shoals upstream to the head of the Tennessee River proper, they apparently warrant specific status.

Ortmann (1925) found *Epioblasma biemarginata* to be abundant at Muscle Shoals prior to impoundment of the Tennessee River, and it is well-represented in museum collections. However, it was rare in archaeological material recovered from Muscle Shoals and near Bridgeport, and not reported from middens in the intervening reach (Morrison, 1942; Warren, 1975; Hughes and Parmalee, 1999). Warren (1975) erroneously stated that the archaeological material from near Bridgeport represented the farthest downstream record of *E. biemarginata* but provided no further discussion.

Synonymy

Unio biemarginatus Lea, 1857. Lea, 1857b:83; Lea, 1866:47, pl. 16, fig. 45; Lea, 1867b:51, pl. 16, fig. 45
Type locality: [Tennessee River,] Florence, [Lauderdale County,] Alabama. Lectotype, USNM 84608, length 37 mm (female), designated by Johnson (1974).

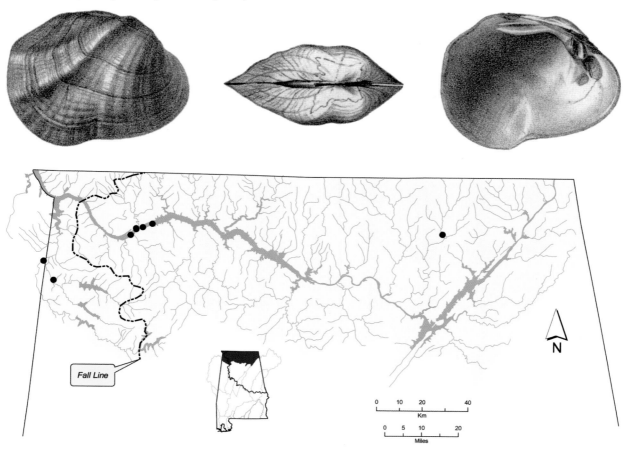

Distribution of *Epioblasma biemarginata* in the Tennessee River drainage of Alabama.

Epioblasma brevidens (Lea, 1831)
Cumberlandian Combshell

Epioblasma brevidens – Upper figure: female, length 47 mm, UF 64228. Tennessee River, Lauderdale County, Alabama. Lower figure: male, length 55 mm, UNA 351.7. Cedar Creek, creek mile 10, Franklin County, Alabama, 1966. © Richard T. Bryant.

Shell Description

Length to 80 mm; moderately thick; males compressed, females usually inflated posteriorly, older females often greatly inflated; male outline subquadrate to subtriangular, female outline rhomboidal to trapezoidal; male posterior margin narrowly rounded, female posterior margin broadly rounded, usually emarginate just posterior of extrapallial swelling; anterior margin rounded; dorsal margin convex; ventral margin usually gently convex, may be straight in females, females may have slight concavity just anterior to extrapallial swelling; females with narrow, radially striate extrapallial swelling along posterior ridge, usually separated from remainder of shell by slight sulci anteriorly and posteriorly, ventral margin of swelling serrate; posterior ridge rounded, curved; posterior slope steep, convex; umbo broad, moderately inflated,

elevated slightly above hinge line, umbo sculpture weak, double-looped ridges; periostracum smooth or cloth-like, yellowish to tawny, with narrow, broken green rays, rays usually more interrupted posteriorly.

Pseudocardinal teeth large, ragged, erect, triangular, 2 divergent teeth in left valve, separated dorsally, 1 tooth in right valve, may have accessory denticle anteriorly and/or posteriorly; lateral teeth short, thick, straight, 2 in left valve, 1 in right valve; interdentum short, narrow to moderately wide; umbo cavity shallow; nacre white.

Soft Anatomy Description

Mantle creamy white to tan, usually rusty brown with grayish brown mottling outside of apertures, mottling typically oriented perpendicular to margin; visceral mass creamy white to tan; foot tan.

Females with thick, spongy pad filling extrapallial swelling, pad grayish tan mottled with darker gray; mantle margin in extrapallial swelling papillate; with 2 specialized structures on each side, exposed in live, displaying individuals (Figure 13.7), posterior structure resembles flattened papilla folded over to the outside, with small bumps and wrinkles under the fold, anterior structure oblong, caruncle-like, both structures rusty tan (live, displaying females also have at least 1 tentacle-like papilla just anterior to caruncle-like structures, which were not visible in unrelaxed, preserved specimens examined); papillae posterior of structures long, thin, papillae anterior of structures thick, blunt; adjacent mantle may be dark brown or tan.

Gills creamy white to tan; dorsal margin straight to slightly sinuous, ventral margin convex to elongate convex, outer gill may be slightly bilobed in females; gill length 55–65% of shell length; gill height 40–65% of gill length; outer gill height 75–90% of inner gill height; inner lamellae of inner gills completely connected to visceral mass.

Outer gills marsupial; glochidia held in posterior 50–65% of gill; marsupium somewhat oval, well-padded when gravid; creamy white.

Labial palps creamy white to tan; straight to slightly concave dorsally, convex ventrally, bluntly pointed distally; palp length 20–30% of gill length; palp height 45–80% of palp length; distal 40–65% of palps bifurcate.

Incurrent aperture usually slightly longer than supra-anal aperture, occasionally equal in length; supra-anal aperture usually slightly longer than excurrent aperture, occasionally equal in length; occasional individuals with all 3 apertures of equal length.

Incurrent aperture length 15–25% of shell length; creamy white to tan within, sometimes with grayish or brownish cast, creamy white to light brown or grayish brown basal to papillae, may be mottled; papillae usually arborescent with scattered short, blunt, simple papillae, occasional individuals with simple papillae only, arborescent papillae folded, with branches oriented to the outside; papillae creamy white to rusty brown or grayish brown, often changing color distally, may be mottled basally.

Excurrent aperture length approximately 15% of shell length; creamy white to tan within, may have brownish cast, marginal color band grayish brown to rusty brown, may be mottled; margin crenulate or with very small, simple papillae; papillae tan to rusty brown.

Supra-anal aperture length 15–20% of shell length; creamy white to tan within, may have thin rusty tan, rusty brown or grayish brown marginal band; margin smooth to minutely crenulate; mantle bridge separating supra-anal and excurrent apertures imperforate, length 5–30% of supra-anal length.

Specimens examined: Bear Creek, Colbert County (n = 1); Big South Fork Cumberland River, Tennessee (n = 1); Clinch River, Tennessee (n = 2); "Hancock County, Tennessee" (n = 1) (specimens previously preserved).

Figure 13.7. *Epioblasma brevidens* female caruncle-like mantle modification. Photograph by W.N. Roston.

Glochidium Description

Length 213–220 μm; height 205–214 μm; with triangular to lanceolate supernumerary hooks; outline depressed subelliptical; dorsal margin straight, ventral margin broadly rounded, anterior and posterior margins convex (Hoggarth, 1999).

Similar Species

Male *Epioblasma brevidens* may resemble small *Actinonaias ligamentina*, *Ellipsaria lineolata* or *Ptychobranchus fasciolaris*. However, those species do not have the cloth-like periostracum with thin, close-set, broken rays of *E. brevidens*. Rays of *E. lineolata* may be broken but are usually widely separated.

General Distribution

Epioblasma brevidens is endemic to the Tennessee and Cumberland River drainages. In the Cumberland River drainage it is confined to that part of the river downstream of Cumberland Falls (Cicerello et al., 1991). It is known from Tennessee River headwaters in eastern Tennessee and southwestern Virginia downstream to near the mouth of the Duck River (Ahlstedt, 1992a, 1992b; Parmalee and Bogan, 1998).

Alabama and Mobile Basin Distribution

Epioblasma brevidens historically occurred in the Tennessee River across northern Alabama and in some large tributaries. There are no museum records of *E. brevidens* from Alabama reaches of the Elk River, but it is known from Tennessee reaches of that system.

The only known extant population of *Epioblasma brevidens* in Alabama is located in a short (approximately 6 km) reach of Bear Creek, in the vicinity of Natchez Trace Parkway, Colbert County (McGregor and Garner, 2004).

Ecology and Biology

Epioblasma brevidens occurs in shoal habitat of small to large rivers and large creeks. It occupies silt-free gravel, cobble and sand substrates, where it remains buried until spring and early summer, when it may be found completely exposed (Ahlstedt, 1992a). This exposure is presumably for glochidial discharge in females, but the behavior in males remains unexplained since spawning generally occurs in autumn in long-term brooders.

Epioblasma brevidens is a long-term brooder, gravid from late summer or autumn until the following summer. Female *E. brevidens* display a modified mantle margin at certain times, which is believed to serve as a glochidial host lure (Figure 13.7). The display includes two rusty tan caruncle-like structures. Just ventral to the caruncle-like structures are two long, thick, simple papillae, grayish brown in color.

Fishes demonstrated to serve as glochidial hosts of *Epioblasma brevidens* in laboratory trials include *Cottus baileyi* (Black Sculpin) and *Cottus carolinae* (Banded Sculpin) (Cottidae); and *Etheostoma blennioides* (Greenside Darter), *Etheostoma maculatum* (Spotted Darter), *Etheostoma rufilineatum* (Redline Darter), *Etheostoma simoterum* (Snubnose Darter) and *Percina caprodes* (Logperch) (Percidae) (Yeager and Saylor, 1995; Jones and Neves, 2002a). Limited success in transforming *E. brevidens* glochidia on *Etheostoma flabellare* (Fantail Darter) and *Etheostoma vulneratum* (Wounded Darter) (Percidae) suggests that they are secondary hosts (Yeager and Saylor, 1995; Jones and Neves, 2002a). Glochidia of *E. brevidens* have been successfully transformed on *Percina roanoka* (Roanoke Darter) (Percidae) under laboratory conditions, but they do not occur sympatrically (Jones and Neves, 2002a).

Current Conservation Status and Protection

Epioblasma brevidens was listed as endangered throughout its range by Williams et al. (1993). In Alabama it was considered threatened by Stansbery (1976) and endangered by Lydeard et al. (1999). Garner et al. (2004) designated *E. brevidens* a species of highest conservation concern in the state. In 1997 it was listed as endangered under the federal Endangered Species Act.

In 2001 *Epioblasma brevidens* was included on a list of species approved for a Nonessential Experimental Population in tailwaters of Wilson Dam on the Tennessee River. However, no reintroductions had taken place as of 2007.

Remarks

Epioblasma brevidens has been reported to be rare in prehistoric shell middens across northern Alabama (Morrison, 1942; Warren, 1975; Hughes and Parmalee, 1999). Morrison (1942) suggested that Muscle Shoals represented the downstream extent of the species in the Tennessee River. However, Parmalee and Bogan (1998) reported *E. brevidens* from archaeological studies downstream of Muscle Shoals near the mouth of the Duck River.

Epioblasma brevidens probably disappeared from the Tennessee River proper soon after it was impounded but is extant in scattered tributary populations. All known populations of *E. brevidens* appear to be declining, with the exceptions of those in the Clinch River and Big South Fork Cumberland rivers (D.W. Hubbs and S.A. Ahlstedt, personal communication). However, the Clinch River population appears to be under imminent threat from sedimentation resulting from mountaintop removal coal mining within the watershed in Virginia. In the only remaining Alabama population, Bear Creek in Colbert County, it is confined to a very short reach of stream and is uncommon there (McGregor and Garner, 2004). The Bear Creek population is potentially an important source of genetic diversity for the species, since it is more than 700 river km removed from the Clinch River population.

Synonymy

Unio brevidens Lea, 1831. Lea, 1831:75, pl. 6, fig. 6; Lea, 1834b:85, pl. 6, fig. 6

Type locality: Ohio [changed to Cumberland River in errata sheet]. Type specimen not found in USNM, also not found by Johnson (1974). Length of figured shell in original description reported as 43 mm.

Comment: *Obliquaria interrupta* was considered a synonym of *Epioblasma brevidens* until Bogan (1997) determined the type shell to be a specimen of *Ptychobranchus fasciolaris*.

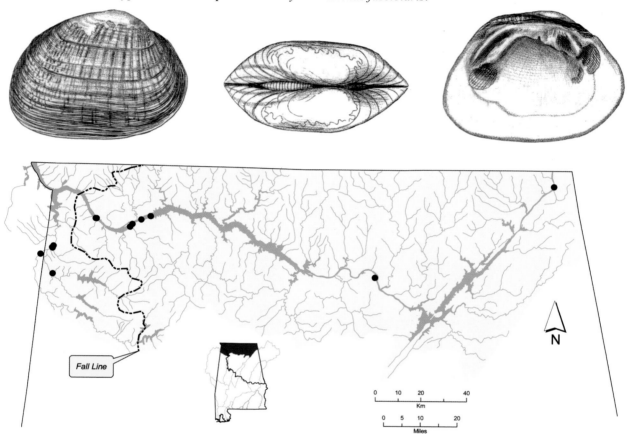

Distribution of *Epioblasma brevidens* in the Tennessee River drainage of Alabama.

Epioblasma capsaeformis (Lea, 1834)
Oyster Mussel

Epioblasma capsaeformis – Upper figure: female, length 37 mm, CMNH 61.7055. Lower figure: male, length 43 mm, CMNH 61.7055. Paint Rock River, Princeton, Jackson County, Alabama. © Richard T. Bryant.

Shell Description

Length to 70 mm; moderately thick anteriorly, thin posteriorly; moderately compressed; male outline irregularly oval to subelliptical, female outline oval with large extrapallial swelling posterioventrally; male posterior margin narrowly rounded, female posterior margin broadly rounded; anterior margin evenly rounded, may be slightly more narrowly rounded in females; dorsal margin straight; ventral margin convex, females with emargination just anterior of extrapallial swelling; extrapallial swelling of females large, rounded, thin, slightly inflated, extending well below ventral margin, may be separated from remainder of shell by shallow sulcus anteriorly, distal margin of swelling may be serrate; posterior ridge low, often doubled in males, obscured by extrapallial swelling in females; male posterior slope moderately steep, slightly concave, female posterior slope gently rounded; umbo moderately inflated, elevated above hinge line, umbo sculpture weak, parallel loops; periostracum smooth, often shiny, yellowish green, with thin green rays, extrapallial swelling of females green, corresponding area of males often green.

Pseudocardinal teeth small, triangular, 2 somewhat compressed teeth in left valve, with oblique gap between them, 1 triangular tooth in right valve, may be compressed, may have accessory denticle anteriorly and/or posteriorly; lateral teeth short, straight to slightly curved, 2 in left valve, 1 in right valve; interdentum short, very narrow; umbo cavity shallow; nacre white.

Soft Anatomy Description

Mantle creamy white, some rusty tan and grayish brown outside of apertures; visceral mass creamy white; foot creamy white.

Females with thick, spongy, creamy white pad filling extrapallial swelling; with slightly enlarged papilla (microlure) just ventral to incurrent aperture. Jones (2004) reported the pad within the extrapallial swelling of live females to be smooth and bluish white, with a black distal margin. Ortmann (1912a) reported tissue surfaces within the extrapallial swelling to have "low granules", that are finer than those found in *Epioblasma florentina*. Jones (2004) reported the microlure to be bluish to light gray with black fringes near the tips and to mimic insect cerci.

Gills pale tan; dorsal margin slightly sinuous, ventral margin convex, fold in outer gill gives appearance of being bilobed; gill length approximately 70% of shell length; gill height approximately 50% of gill length; outer gill height approximately 85% of inner gill height; inner lamellae of inner gills completely connected to visceral mass.

Outer gills marsupial; glochidia held distally, in posterior 50% of gill; marsupium somewhat oval, well padded when gravid; creamy white.

Labial palps creamy white; slightly concave dorsally, convex ventrally, bluntly pointed distally; palp length approximately 20% of gill length; palp height approximately 85% of palp length; distal 50% of palps bifurcate.

Incurrent aperture slightly longer than excurrent and supra-anal apertures; supra-anal aperture slightly longer than excurrent aperture.

Incurrent aperture length approximately 20% of shell length; creamy white within, marginal band basal to papillae rusty brown with a few dark brown lines perpendicular to margin; papillae simple, bifid, trifid and small arborescent; papillae rusty brown and creamy white.

Excurrent aperture length approximately 10% of shell length; creamy white, marginal color band irregular rusty brown; margin minutely crenulate.

Supra-anal aperture length approximately 15% of shell length; rusty brown within, creamy white marginally; margin smooth; mantle bridge separating supra-anal and excurrent apertures imperforate, length approximately 25% of supra-anal length.

Specimen examined: [Clinch River,] Hancock County, Tennessee (n = 1) (specimen previously preserved).

Glochidium Description

Length 252–256 μm; height 226–238 μm; with attenuate supernumerary hooks; outline depressed subelliptical; dorsal margin straight and long, ventral margin broadly rounded, anterior and posterior margins convex (Hoggarth, 1999; Jones, 2004).

Similar Species

Epioblasma capsaeformis closely resembles an undescribed *Epioblasma* species from the lower Tennessee River drainage. Differences in shell morphology are subtle. The extrapallial swelling of female *Epioblasma* sp. cf. *capsaeformis* is slightly larger and more darkly pigmented than that of female *E. capsaeformis*. Tissues within the extrapallial swelling of *E. capsaeformis* are bluish white in life, with a bluish to light gray microlure displayed from the posteriodorsal side of each mantle pad. Tissues within the extrapallial swelling of *Epioblasma* sp. are dark purple to slate gray with a spongy texture. A single microlure is displayed from an invagination near the posteriodorsal side of the mantle pad in *Epioblasma* sp. (Jones, 2004).

Epioblasma capsaeformis also closely resembles *Epioblasma florentina* but differs in periostracum color. The periostracum of both species is yellowish green to brownish green, with thin green rays over the entire shell, but *E. capsaeformis* tends to have more green posteriorly. Tissues within the extrapallial swelling of living *E. capsaeformis* are smooth and bluish white, whereas extrapallial swelling tissues of *E. florentina* (the headwater *walkeri* form) are pustulose and gray with black mottling. *Epioblasma capsaeformis* displays two microlures, but *E. florentina* displays only one (Jones, 2004).

Epioblasma capsaeformis bears a superficial resemblance to *Epioblasma biemarginata* and *Epioblasma turgidula*. However, females of those species have much reduced extrapallial swellings compared to *E. capsaeformis* females. Male *E. biemarginata* have a well-developed sulcus, whereas male *E. capsaeformis* usually do not, though they may have a slight radial depression just anterior of the posterior ridge.

General Distribution

Epioblasma capsaeformis is endemic to the Tennessee and Cumberland River drainages. It historically occurred in the Cumberland River drainage downstream of Cumberland Falls (Cicerello et al., 1991; Parmalee and Bogan, 1998). In the Tennessee River drainage *E. capsaeformis* historically occurred from headwaters in southwestern Virginia downstream to Muscle Shoals (Ahlstedt, 1992a, 1992b; Parmalee and Bogan, 1998; Jones, 2004).

Alabama and Mobile Basin Distribution

Epioblasma capsaeformis historically occurred in the Tennessee River across northern Alabama and in some tributaries.

Epioblasma capsaeformis is extirpated from the state. The most recent dated record was from the Estill Fork of the Paint Rock River in 1980 (Ahlstedt, 1992b).

Ecology and Biology

Epioblasma capsaeformis is a species of shoal habitat in small to large rivers, where it occurs in silt-free gravel and sand substrates.

Epioblasma capsaeformis is a long-term brooder, gravid from late summer or autumn until the following summer. Fecundity of a Clinch River population of *E. capsaeformis* has been reported to average 13,008 glochidia per female per year (n = 10) and to range from 7,780 to 16,876 glochidia per individual (Jones, 2004).

Male and female *Epioblasma capsaeformis* have been reported to emerge from the substrate during May and June, when females display paired microlures against bluish white pads within their extrapallial swellings (Figure 13.8) (Ahlstedt, 1992a). This behavior in females is presumably to attract glochidial hosts. However, that does not explain the behavior in males, since long-term brooders generally spawn during autumn. The display has been described as rhythmic movements, with the microlure of the left mantle pad rotating in a clockwise, circular manner and that of the right mantle pad rotating counterclockwise (Jones, 2004). Female *E. capsaeformis* have been observed to snap their valves together on darters that were investigating the lure and trap them between the valves (Jones, 2004). This behavior may facilitate infestation.

Figure 13.8. *Epioblasma capsaeformis* female with bluish white pads within extrapallial swellings displayed. Note the two small microlures near center left margin of pads. Clinch River. Photograph by W.N. Roston.

Fishes reported to serve as glochidial hosts for *Epioblasma capsaeformis* in laboratory trials include *Etheostoma blennioides* (Greenside Darter), *Etheostoma camurum* (Bluebreast Darter), *Etheostoma flabellare* (Fantail Darter), *Etheostoma rufilineatum* (Redline Darter), *Etheostoma simoterum* (Snubnose Darter), *Etheostoma vulneratum* (Wounded Darter) and *Percina sciera* (Dusky Darter) (Percidae); and *Cottus baileyi* (Black Sculpin), *Cottus bairdii* (Mottled Sculpin) and *Cottus carolinae* (Banded Sculpin) (Cottidae) (Yeager and Saylor, 1995; Jones and Neves, 2000; Jones, 2004).

Current Conservation Status and Protection

Epioblasma capsaeformis was reported as endangered throughout its range by Williams et al. (1993) and in Alabama by Lydeard et al. (1999). Garner et al. (2004) designated *E. capsaeformis* a species of highest conservation concern in Alabama. It was listed as endangered under the federal Endangered Species Act in 1997.

In 2001 *Epioblasma capsaeformis* was included on a list of species approved for a Nonessential Experimental Population in Wilson Dam tailwaters on the Tennessee River. No reintroductions had taken place as of 2006.

Remarks

Based on a recent systematic study, *Epioblasma capsaeformis* was found to be comprised of two species (Jones, 2004). *Epioblasma capsaeformis* occurs in upper reaches of the Tennessee River drainage, downstream to Muscle Shoals and in the Cumberland River drainage. The other species, presently unde-scribed (*Epioblasma* sp. cf. *capsaeformis*), is confined to the Duck River and Tennessee River upstream to Muscle Shoals, where the two historically occurred sympatrically (Jones, 2004).

The abundance of *Epioblasma capsaeformis* in prehistoric middens is difficult to assess due to its thin, brittle shell. It has been reported to be present but rare in most middens studied along the Tennessee River in Alabama (Morrison, 1942; Warren, 1975; Hughes and Parmalee, 1999). Relative abundance of *E. capsaeformis* and *Epioblasma* sp. in prehistoric middens at Muscle Shoals is unknown.

Synonymy

Unio capsaeformis Lea, 1834. Lea, 1834a:31, pl. 2, fig. 4; Lea, 1834b:143, pl. 2, fig. 4
Type locality: Cumberland River, [Tennessee]. Lectotype, MCZ 178570, length 48 mm, designated by Johnson (1956).

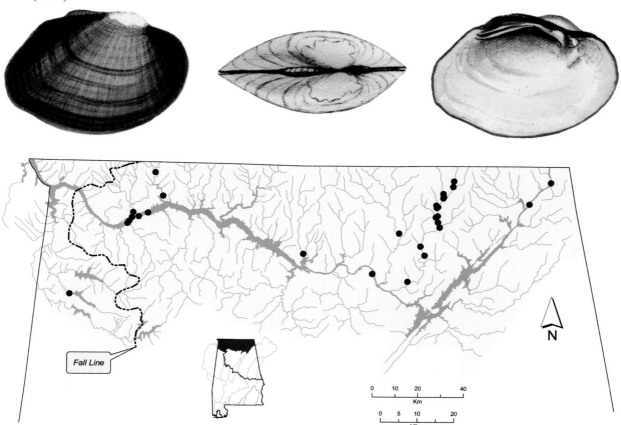

Distribution of *Epioblasma capsaeformis* in the Tennessee River drainage of Alabama.

Epioblasma cincinnatiensis (Lea, 1840)
Ohio Riffleshell

Epioblasma cincinnatiensis – Upper figure: female, length 35 mm, ANSP 126469. Lower figure: male, length 35 mm, ANSP 126469. Tennessee River, Jackson County, Alabama. © Richard T. Bryant.

Shell Description

Length to 48 mm; moderately thick; moderately to highly inflated; male outline irregularly quadrate to somewhat pentagonal, female outline irregularly ovate to somewhat pentagonal; posterior margin bluntly pointed, with slight emarginations dorsally and posterioventrally; anterior margin narrowly to broadly rounded; dorsal margin straight, usually with weak dorsal wing; ventral margin convex to bluntly pointed; posterior ridge narrow and distinct, usually adorned with low tubercles; low medial ridge lies anterior of posterior ridge, with well-developed tubercles; narrow, shallow sulcus separates posterior and medial ridges, sulcus becomes obscure distally by extrapallial swelling in females; extrapallial swelling may be adorned with irregular subradial wrinkles; posterior slope moderately steep, with very small tubercles sometimes arranged in radial rows; umbo moderately inflated, elevated well above hinge line, umbo sculpture unknown; periostracum shiny, yellowish green to light olive, with thin green rays.

Pseudocardinal teeth erect, triangular, 2 divergent teeth in left valve, separated dorsally, 1 tooth in right valve, may have accessory denticle anteriorly; lateral teeth thick, straight, 2 in left valve, 1 in right valve; interdentum short, narrow; umbo cavity shallow; nacre white.

Soft Anatomy Description

Soft anatomy unknown.

Glochidium Description

Glochidium unknown.

Similar Species

Epioblasma cincinnatiensis vaguely resembles nominal *Epioblasma torulosa* but has small, well-developed tubercles instead of low knobs on the medial and posterior ridges. It also slightly resembles *Epioblasma propinqua*. However, the medial and posterior ridges, as well as the sulcus, are curved anteriorly in *E. propinqua* but are almost straight in *E. cincinnatiensis*. When *E. propinqua* has sculpture on the posterior and medial ridges, they are little more than raised transverse ridges instead of the well-developed tubercles of *E. cincinnatiensis*.

General Distribution

Museum records of *Epioblasma cincinnatiensis* are uncommon, so its historical distribution is unclear. It was described from the Ohio River and is also known from the Green, Cumberland and Tennessee rivers (Patch, 1976, 2005; Cicerello and Schuster, 2003).

Alabama and Mobile Basin Distribution

Epioblasma cincinnatiensis historically occurred in the Tennessee River across northern Alabama.

Epioblasma cincinnatiensis is believed to be extinct. None of the material collected from Alabama is dated, but this species has presumably not been collected from the state since the mid- to late 1800s.

Ecology and Biology

Epioblasma cincinnatiensis occurred in large rivers, presumably in shoal habitat.

Epioblasma cincinnatiensis was presumably a long-term brooder, gravid from late summer or autumn to the following summer. Its glochidial hosts are unknown. Some species of *Epioblasma* are known to use darters (Percidae) and sculpins (Cottidae) as glochidial hosts.

Current Conservation Status and Protection

Epioblasma cincinnatiensis has not been recognized recently and appears on no endangered or extinct species lists. There are no museum specimens that were collected since modern perturbations of large river habitats, and the species appears to be extinct.

Remarks

Epioblasma cincinnatiensis has been included in the synonymy of *Unio phillipsii* Conrad, 1835, by some workers. However, *U. phillipsii* appears to be a misshapen *Obliquaria reflexa* with deformed knobs on the medial ridge, as pointed out by Simpson (1900b).

Epioblasma cincinnatiensis (as *Epioblasma torulosa cincinnatiensis*) was reported by Morrison (1942) to be rare in prehistoric shell middens at Muscle Shoals. He suggested that the form deserves species status but was apprehensive about recognizing it without examining more recent material. Warren (1975) reported *E. cincinnatiensis* from prehistoric shell middens near Bridgeport and found no intergradation between it and nominal *E. torulosa*.

Synonymy

Unio cincinnatiensis Lea, 1840. Lea, 1840:285; Lea, 1842b:194, pl. 8, fig. 4; Lea, 1842c:32, pl. 8, fig. 4
Type locality: Ohio River, Cincinnati, [Hamilton County, Ohio]. Lectotype, USNM 84199, length 48 mm (female), designated by Johnson (1974).

 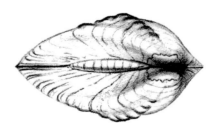 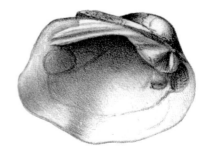

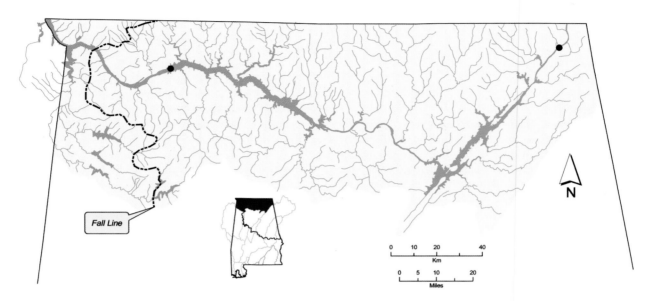

Distribution of *Epioblasma cincinnatiensis* in the Tennessee River drainage of Alabama.

Epioblasma flexuosa (Rafinesque, 1820)
Leafshell

Epioblasma flexuosa – Upper figure: female, length 64 mm, USNM 25803. Lower figure: male, length 54 mm, USNM 25803. Ohio River. © Richard T. Bryant.

Shell Description

Length to 80 mm; thick; inflated; male outline quadrate to irregularly oval, female outline irregularly quadrate to triangular depending on degree of extrapallial swelling; male posterior margin obliquely truncate to broadly rounded, female posterior margin extended in pronounced, narrow lobe dorsally, with emargination ventrally; anterior margin rounded; male dorsal margin slightly convex, female dorsal margin almost straight; male ventral margin broadly rounded anteriorly, with wide, moderately deep emargination at distal end of sulcus, female ventral margin slightly curved anteriorly, expanded into narrow, elongate extrapallial swelling posteriorly; extrapallial swelling a ventral extension of medial ridge; posterior ridge well-developed, rounded; posterior slope steep, slightly concave; wide, shallow sulcus separates posterior and medial ridges; umbo broad, elevated above hinge line, umbo sculpture faint corrugations; periostracum dull, yellowish green to tawny, some individuals with narrow green rays.

Pseudocardinal teeth thick, erect, 2 triangular, compressed, divergent teeth in left valve, 1 large triangular tooth in right valve, may have accessory denticle anteriorly; lateral teeth short, slightly curved in males, straight in females, 2 in left valve, 1 in right valve; interdentum short, narrow to moderately wide; umbo cavity wide, shallow; nacre white.

Soft Anatomy Description

No material was available for soft anatomy description. Lea (1863d) provided a brief description based on "an imperfect dried specimen" from the Ohio River. The mantle was described as having a thickened margin, "with elongate cilae on the inner edge". There is "an extended flap" just ventral to the incurrent aperture. Outer gills are marsupial, with glochidia held across the entire gill (Lea, 1863d).

Glochidium Description

Glochidium unknown. Lea (1863d) described the shape of *Unio foliatus* (= *Epioblasma flexuosa*) glochidia as subtriangular and without styliform hooks. However, this was from "an imperfect dried specimen". No other *Epioblasma* has subtriangular glochidia, so this description is questionable.

Similar Species

Shells of *Epioblasma flexuosa* resemble those of *Epioblasma lewisii* and *Epioblasma stewardsonii*. Both male and female *E. flexuosa* are larger and considerably thicker than *E. lewisii*. The periostracum of *E. flexuosa* is uniform brown with thin green rays, but *E. lewisii* females have a dark green extrapallial swelling. *Epioblasma flexuosa* and *E. stewardsonii* both have thick shells, but *E. flexuosa* attains a much larger size. In females, the sulcus of *E. flexuosa* lies posterior to the extrapallial swelling, while the extrapallial swelling of *E. stewardsonii* originates in the sulcus. Also, the ventral terminus of the extrapallial swelling of *E. flexuosa* is more elongate and narrow than that of *E. stewardsonii*.

General Distribution

Epioblasma flexuosa historically occurred in the Ohio River and some of its larger tributaries, including the Tennessee, Cumberland and Wabash rivers (Parmalee and Bogan, 1998).

Alabama and Mobile Basin Distribution

Epioblasma flexuosa is historically known only from the Tennessee River at Muscle Shoals. However, there are archaeological records in northeastern Alabama and southeastern Tennessee (Warren, 1975; Parmalee et al., 1982).

Epioblasma flexuosa is believed to be extinct. There are no dated museum specimens, but it appears to have disappeared around 1900.

Ecology and Biology

Epioblasma flexuosa was a species of large rivers. Call (1900) suggested that it occurred in muddy substrate in deep water, but Stansbery (1970a) surmised that it inhabited shoals.

Epioblasma flexuosa was presumably a long-term brooder, gravid from late summer or autumn to the following summer. Its glochidial hosts are unknown. Some species of *Epioblasma* are known to use darters (Percidae) and sculpins (Cottidae) as glochidial hosts.

Current Conservation Status and Protection

Epioblasma flexuosa has been presumed extinct since about 1900 (Stansbery, 1970a; Williams et al., 1993; Turgeon et al., 1998; Lydeard et al., 1999; Garner et al., 2004).

Remarks

The relationship between *Epioblasma flexuosa* and *Epioblasma lewisii* has been the matter of some conjecture. Some authors (e.g., Walker, 1910a; Stansbery, 1970a; Burch, 1975a; Parmalee and Bogan, 1998) recognize both species. However, Morrison (1942) listed *lewisii* as a subspecies of *E. flexuosa*, and Johnson (1978) synonymized *E. lewisii* under *E. flexuosa* as an "ecophenotypic variant". Warren (1975) reported nominal *E. flexuosa* and *E. lewisii* to be consistently separable in archaeological material but also reported "a considerable degree of variation and intergrading" and suggested that they were merely ecological variants.

Epioblasma flexuosa has been reported from prehistoric shell middens along the Tennessee River across northern Alabama but was not common at any of the sites (Warren, 1975; Hughes and Parmalee, 1999).

Synonymy

Obliquaria (*Quadrula*) *flexuosa* Rafinesque, 1820. Rafinesque, 1820:306
Type locality: Kentucky, Salt and Green rivers. Lectotype, ANSP 20249, length 57 mm (male), designated by Johnson and Baker (1973), is from Kentucky River.
Unio foliatus Hildreth, 1828. Hildreth, 1828:284, fig. 16

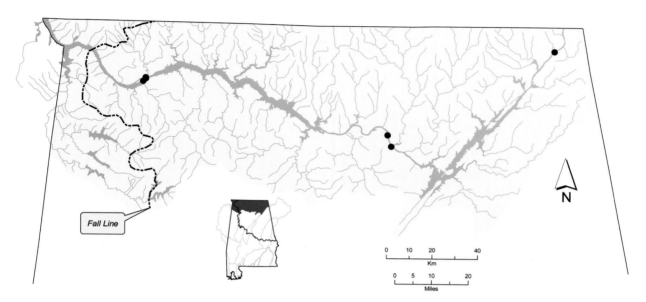

Distribution of *Epioblasma flexuosa* in the Tennessee River drainage of Alabama.

Epioblasma florentina (Lea, 1857)
Yellow Blossom

Epioblasma florentina – Upper figure: female, length 33 mm, CMNH 61.6304. Tennessee River, Tuscumbia, Colbert County, Alabama. Middle figure: female, *walkeri* form, length 38 mm, CMNH 61.7057. Lower figure: male, *walkeri* form, length 37 mm, CMNH 61.7057. Flint River, Gurley, Madison County, Alabama. © Richard T. Bryant.

Shell Description

Length to 60 mm; moderately thick anteriorly, thin posteriorly; slightly inflated; male outline irregularly oval to subelliptical, female outline oval with large extrapallial swelling posterioventrally; male posterior margin narrowly rounded, female posterior margin broadly rounded, often with slight emargination just dorsal to extrapallial swelling; anterior margin evenly rounded; dorsal margin short, straight to slightly convex; ventral margin gently rounded anteriorly, males with straight or slightly concave area at distal end of sulcus, females with posterior part of ventral margin expanded as part of extrapallial swelling; extrapallial swelling of females thin, round, in form of an exaggerated lobe posterioventrally; posterior ridge well-developed, rounded, may be doubled in males, female posterior ridge obscured by extrapallial swelling; male posterior slope moderately steep, flat to slightly concave, female posterior slope gently rounded; males with wide flattened area suggestive of shallow radial sulcus anterior of posterior ridge; umbo moderately inflated, elevated above hinge line, umbo sculpture weak, double-looped ridges; periostracum brownish green or yellowish green, with numerous thin green rays scattered over entire surface.

Pseudocardinal teeth small, erect, triangular to slightly compressed, 2 teeth in left valve, with oblique gap between them, 1 tooth in right valve, usually with accessory denticle anteriorly; lateral teeth short, slightly curved, 2 in left valve, 1 in right valve; interdentum short, very narrow; umbo cavity wide, shallow; nacre white.

Soft Anatomy Description

Mantle tan, outside of apertures slightly rusty and grayish brown, coloration extends along entire ventral margin but with only a very thin band anterior of mantle fold; visceral mass tan; foot tan.

No females were available for description of mantle modification. Males with very weak mantle fold just ventral to incurrent aperture; fold length approximately 40% of shell length; margin smooth, becoming crenulate anteriorly; with narrow, dark marginal color band, tapering and disappearing anterioventrally; mantle outside of fold slightly rusty and grayish brown. Ortmann (1912a) reported tissues within the extrapallial swelling of females of the nominal form to be "deep black" and spongy, covered with "low granules" (i.e., pustules). Jones (2004) reported live females of the *walkeri* form to display a microlure against a pustulose mantle pad within the extrapallial swelling. The surface of the pad is gray with black mottling in a Clinch River drainage population and brown with tan mottling in a Cumberland River drainage population. Pustules of the Cumberland River population are finer and more pointed than those of the Clinch River population (Jones, 2004).

Gills tan; dorsal margin slightly sinuous, ventral margin convex; gill length approximately 65% of shell length; gill height approximately 55% of gill length; outer gill height approximately 90% of inner gill height; inner lamellae of inner gills completely connected to visceral mass.

No gravid females were available for marsupium description. Outer gills are marsupial, well-padded when gravid (J.W. Jones, personal communication).

Labial palps tan; slightly convex dorsally, convex ventrally, narrowly rounded distally; palp length approximately 15% of gill length; palp height approximately 75% of palp length; distal 75% of palps bifurcate.

Incurrent and supra-anal apertures of similar length; excurrent aperture slightly shorter than incurrent and supra-anal apertures.

Incurrent aperture length approximately 20% of shell length; pale tan within, with irregular dark brown basal to papillae; papillae arborescent; papillae dark grayish brown basally, rusty brown distally.

Excurrent aperture length approximately 15% of shell length; pale within, with dark brown marginal color band; margin minutely crenulate.

Supra-anal aperture length approximately 20% of shell length; creamy white within, with thin dark brown marginal band; margin smooth; mantle bridge separating supra-anal and excurrent apertures imperforate, length approximately 15% of supra-anal length.

Specimen examined: Big South Fork Cumberland River [Tennessee?] (n = 1) (specimen of *walkeri* form, previously preserved).

Glochidium Description

Length 230 μm; height 220 μm; without styliform hooks; outline depressed subelliptical (Ortmann, 1912a). Jones (2004) reported the length of the *walkeri* form to be as great as 272 μm.

Similar Species

Epioblasma florentina closely resembles *Epioblasma capsaeformis*. Shells of the two can be separated primarily by periostracum color. The periostracum of both species is yellowish green to brownish green with thin green rays over the entire shell, but *E. capsaeformis* tends to have more green posteriorly. Tissues within the extrapallial swelling of the headwater *walkeri* form of *E. florentina* are pustulose and gray with black mottling. The soft anatomy of nominal *E. florentina* has never been described. Tissues within the extrapallial swelling of *E. capsaeformis* are smooth and bluish white. *Epioblasma florentina* displays only one microlure, while *E. capsaeformis* displays two (Jones, 2004).

Epioblasma florentina also closely resembles the undescribed species, *Epioblasma* sp. cf. *capsaeformis*, from the lower Tennessee River drainage but can be separated by periostracum color. The periostracum of both species is yellowish green to brownish green with thin green rays over the entire shell, but female *Epioblasma* sp. typically have an extrapallial swelling that is dark green to black, with the darker coloration continuing dorsally toward the umbo. Though not as obvious as in females, male *Epioblasma* sp. also have a wide, dark green band just anterior of the posterior ridge. Tissues within the extrapallial swelling of *E. florentina* (headwater *walkeri* form) are pustulose and gray with black mottling, whereas those of *Epioblasma* sp. are spongy and dark purple to slate gray. Both species display only one microlure, but that of *E. florentina* is dark brown to black and that of *Epioblasma* sp. is tan (Jones, 2004).

Epioblasma florentina bears a superficial resemblance to *Epioblasma biemarginata* and *Epioblasma turgidula*. However, females of those species have much reduced extrapallial swellings compared to those of *E. florentina* females. Male *E. biemarginata* have a well-developed sulcus, whereas male *E. florentina* usually do not, though they may have a slight radial depression just anterior of the posterior ridge.

General Distribution

Epioblasma florentina is endemic to the Tennessee and Cumberland River drainages as well as the White River drainage in northeastern Arkansas and southeastern Missouri (Harris and Gordon, 1990; Parmalee and Bogan, 1998). It was widespread in the Cumberland River drainage downstream of Cumberland Falls, Kentucky and Tennessee (Cicerello et al., 1991; Parmalee and Bogan, 1998). In the Tennessee River drainage *E. florentina* is known from headwaters in southwestern Virginia downstream to Muscle Shoals, Alabama, including many tributaries (Neves, 1991; Parmalee and Bogan, 1998). However, *E. florentina* records from the Tennessee River main stem are rare, with the exception of those from Muscle Shoals (Jones, 2004).

Alabama and Mobile Basin Distribution

Epioblasma florentina presumably occurred historically in the Tennessee River and large tributaries across northern Alabama. However, all Tennessee River proper records are from Muscle Shoals.

Epioblasma florentina is extirpated from Alabama. The most recent documented record of this species from the state was collected from the Paint Rock River in 1918.

Ecology and Biology

Epioblasma florentina occurs in bodies of water ranging from medium creeks to large rivers in shoal habitat (Parmalee and Bogan, 1998).

Nominal *Epioblasma florentina* was driven to extinction prior to study of its biology, so all descriptive studies have dealt with the headwater *walkeri* form. However, biology of the two forms is presumably similar. *Epioblasma florentina* is a long-term brooder, spawning during August and September, when adults, especially males, were found well-exposed above the substrate (Rogers et al., 2001). By November both sexes burrow into the substrate, where they remain until late winter or spring (Rogers et al., 2001).

Adult *Epioblasma florentina*, especially females, begin to emerge during February, and many are exposed by May, at which time glochidial discharge occurs (Rogers et al., 2001). In the process of attracting hosts and discharging glochidia, females often protrude well out of the substrate with gaped valves, displaying a modified mantle margin. A single brown microlure is displayed against a mantle pad that is gray with black mottling in a Clinch River drainage population and tan with brown mottling in a Cumberland River drainage population (Jones, 2004). The microlure emerges from an invagination at the dorsal edge of the mantle pad and is moved in a side-to-side motion (Jones, 2004).

An annual fecundity estimate of 20,000 glochidia was reported for a single female *Epioblasma florentina* (*walkeri* form) by Rogers et al. (2001). However, Jones (2004) reported fecundity to average 7,602 per female per year (n = 7) in a Clinch River drainage population and 9,606 per female per year (n = 6) in a Cumberland River drainage population. Annual fecundity in the Clinch River drainage population ranged from 3,261 to 12,558 glochidia per individual, and in the Cumberland River drainage population ranged from 1,828 to 16,921 glochidia per individual (Jones, 2004). *Epioblasma florentina* glochidia are discharged individually, not packaged as conglutinates (Rogers et al., 2001).

In laboratory trials glochidial hosts of *Epioblasma florentina* (*walkeri* form) from the Clinch River drainage population were found to include *Etheostoma blennioides* (Greenside Darter), *Etheostoma flabellare* (Fantail Darter), *Etheostoma rufilineatum* (Redline Darter) and *Etheostoma simoterum* (Snubnose Darter) (Percidae); and *Cottus baileyi* (Black Sculpin) (Cottidae) (Jones and Neves, 2001; Rogers et al., 2001). *Cottus bairdii* (Mottled Sculpin) and/or *Cottus carolinae* (Banded Sculpin) (Cottidae) also serve as glochidial hosts for the *walkeri* form of *E. florentina*.

Current Conservation Status and Protection

Nominal *Epioblasma florentina* has been considered extinct since at least 1970 (Stansbery, 1970a, 1976; Williams et al., 1993; Turgeon, et al.,

1998; Lydeard et al., 1999; Garner et al., 2004). The *walkeri* form was considered endangered throughout its range by Stansbery (1970a) and Williams et al. (1993) and extirpated from Alabama by Stansbery (1976). The USFWS recognized three subspecies of *E. florentina*, listing *Epioblasma florentina florentina* and *Epioblasma florentina curtisii* as endangered under the federal Endangered Species Act in 1976 and *Epioblasma florentina walkeri* as endangered in 1977.

In 2001 nominal *Epioblasma florentina* was included on a list of species approved for a Nonessential Experimental Population in Wilson Dam tailwaters on the Tennessee River. Should a population of nominal *E. florentina* be located and successfully propagated, the progeny of such a program could be reintroduced there.

Remarks

Two forms of *Epioblasma florentina* have long been recognized from the Tennessee and Cumberland River drainages. Nominal *E. florentina* is a more inflated large river form, and *Epioblasma florentina walkeri* is a compressed headwater form. A third form, *Epioblasma florentina curtisii*, is geographically confined to the Ozark region of Arkansas and Missouri (Utterback, 1916b). Ortmann (1918, 1924b) expounded on the relationship between nominal *E. florentina* and

the *walkeri* form and concluded that they are separable using ratios of height, length and width but intergrade in lower reaches of the Holston River. Though Ortmann (1918, 1924b) seemed to imply that *walkeri* was just a headwater form of *E. florentina*, and even went as far as stating that it is "practically a large, compressed *E. florentina*", he was ambiguous in his conclusions. Later, Ortmann (1925) confounded matters by referring to it as *E. f. walkeri*. This combination has been used to date with the exception of Johnson (1978), who placed *walkeri* in the synonymy of *E. florentina*.

Variation of shell characters over the course of a river is well-documented in the Unionidae, with compression of shells with an upstream progression the rule (Ortmann, 1918, 1920; Grier, 1920; Ball, 1922). This, along with the intergradation of nominal *florentina* and *walkeri* in the lower Holston River, suggest that *walkeri* does not warrant subspecific status. With no preserved soft anatomy of the nominal form and genetic material unavailable for comparison (nominal *florentina* is apparently extinct), resolution of the problem will have to be accomplished using shell morphology.

Epioblasma florentina has been reported from prehistoric shell middens at Muscle Shoals (Morrison, 1942; Hughes and Parmalee, 1999) but nowhere else in Alabama.

Synonymy

Unio florentinus Lea, 1857. Lea, 1857b:83; Lea, 1862b:64, pl. 5, fig. 213; Lea, 1862e:68, pl. 5, fig. 213
Type locality: [Tennessee River,] Florence, [Lauderdale County,] Alabama, Reverend G. White; Cumberland River, Tennessee, Drs. Troost and Edgar, and T.C. Downie, Esq., St. Simons Island, Georgia. Lectotype, USNM 84948, length 33 mm (female), and allotype, length 36 mm (male) (not figured), designated by Johnson (1974), are from Florence, Alabama.

 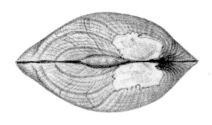 

Unio saccatus Küster, 1861. Küster, 1861:263, pl. 39, fig. 2
Unio sacculus Anthony *in* Reeve, 1864. Anthony *in* Reeve, 1864:[no pagination], pl. 15, fig. 67
Truncilla walkeri Wilson and Clark, 1914. Wilson and Clark, 1914:46, pl. 1, fig. 1
Truncilla curtisii Frierson and Utterback *in* Utterback, 1916. Utterback, 1916b:453–455, pl. 6, fig. 14a–d, pl. 28, fig. 109a–d

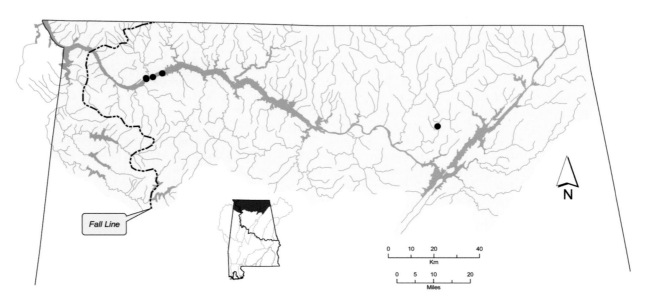

Distribution of *Epioblasma florentina* in the Tennessee River drainage of Alabama.

Epioblasma haysiana (Lea, 1834)
Acornshell

Epioblasma haysiana – Upper figure: female, length 37 mm, USNM 84614. Florence, [Lauderdale County,] Alabama. Lower figure: male, length 30 mm, USNM 30372. Tuscumbia, [Colbert County,] Alabama. © Richard T. Bryant.

Shell Description

Length to 45 mm; thick; moderately inflated; outline triangular to round or oval; posterior margin broadly rounded, males may be more elongate posteriorly than females; anterior margin rounded; dorsal margin rounded; ventral margin rounded to slightly concave, extrapallial swelling of females often extending slightly below ventral margin; posterior ridge weak in males, inflated into extrapallial swelling in females, distal margin of posterior ridge serrate, especially in females; male posterior slope steep,

usually concave, female posterior slope flat or slightly convex; posterior ridge and slope adorned with fine radial striae; umbo broad, slightly inflated, elevated above hinge line, umbo sculpture unknown; periostracum tawny to reddish brown, may have weak, narrow green rays.

Pseudocardinal teeth thick, triangular, 2 divergent teeth in left valve, 1 tooth in right valve, may have accessory denticle anteriorly and/or posteriorly; lateral teeth short, straight to slightly curved, 2 in left valve, 1 in right valve; interdentum moderately short, moderately wide; umbo cavity open, shallow; nacre white to dark purple.

Soft Anatomy Description

No material was available for soft anatomy description. Ortmann (1912a) provided brief details based on specimens from the Cumberland River, reporting that the soft anatomy "agrees in every particular" with that of *Epioblasma triquetra* except the specialized mantle margin located just ventral to the incurrent aperture. Female *Epioblasma haysiana* have a margin with four to six brown papillae and a white, rounded, caruncle-like structure located between the papillae and incurrent aperture. The mantle outside of the papillate margin is spongy, "black-brown" and lanceolate or ovate in outline. A black marginal band is located on the mantle proximal to the papillate margin. Males lack a spongy mantle area, papillae are present but reduced and the white caruncle-like structure is replaced by a white papilla. Ortmann (1912a) also reported the marsupial area of *E. haysiana* to be "more regularly kidney-shaped" than that of *E. triquetra*.

Glochidium Description

Length 240 μm; height 230 μm; outline depressed subelliptical (Ortmann, 1912a).

Similar Species

The shell of *Epioblasma haysiana* most resembles that of *Epioblasma obliquata*. The extrapallial swelling of female *E. haysiana* is less pronounced than that of female *E. obliquata*, presenting a more circular or oval outline. Male *E. haysiana* are also more oval in outline than male *E. obliquata*, which are more elongate and pointed posteriorly.

General Distribution

Epioblasma haysiana is endemic to the Tennessee and Cumberland River drainages. It historically occurred in most of the Cumberland River drainage downstream of Cumberland Falls (Cicerello et al., 1991; Parmalee and Bogan, 1998). In the Tennessee River drainage it occurred from headwaters in southwestern Virginia downstream to Muscle Shoals, Alabama. However, archaeological evidence suggests that *E. haysiana* prehistorically occurred downstream almost to the mouth of the Duck River (Parmalee and Bogan, 1998).

Alabama and Mobile Basin Distribution

Epioblasma haysiana apparently occurred in the Tennessee River across the state prehistorically, but all historical records were collected from Muscle Shoals. It also occurred in upper Elk River, Tennessee, so was presumably found in Alabama reaches of that river as well.

Epioblasma haysiana is believed to be extinct. The material collected from Alabama is not dated but was probably collected during the 1800s and early 1900s.

Ecology and Biology

Epioblasma haysiana occurred in shoal habitat of medium to large rivers (Parmalee and Bogan, 1998).

Epioblasma haysiana was presumably a long-term brooder, gravid from late summer or autumn to the following summer. Its glochidial hosts are unknown. Some species of *Epioblasma* are known to use darters (Percidae) and sculpins (Cottidae) as glochidial hosts.

Current Conservation Status and Protection

Stansbery (1970a) reported *Epioblasma haysiana* to be extant in a short reach of the Clinch River in Virginia, but it is considered extinct today (Turgeon et al., 1988; Williams et al., 1993; Lydeard et al., 1999; Garner et al., 2004).

Remarks

Ortmann (1925) reported *Epioblasma haysiana* to be rare in Alabama reaches of the Tennessee River historically. It has also been found to be rare, but widespread, in prehistoric middens across the state (Morrison, 1942; Warren, 1975; Hughes and Parmalee, 1999).

Wilson and Clark (1914) collected few *Epioblasma haysiana* during a survey of the Cumberland River and speculated that it was more common but "too small to bite the crowfoot [brail] hook and is easily overlooked". Wilson and Clark (1914) described *E. haysiana* as "one of the handsomest of the *Truncilla* [= *Epioblasma*] on account of its beautifully polished epidermis".

Synonymy

Unio haysianus Lea, 1834. Lea, 1834a:35, pl. 3, fig. 7; Lea, 1834b:147, pl. 3, fig. 7

Type locality: Tennessee. Lectotype, designated by Johnson (1974), not found in USNM, but length of figured shell in original description reported as 25 mm (female).

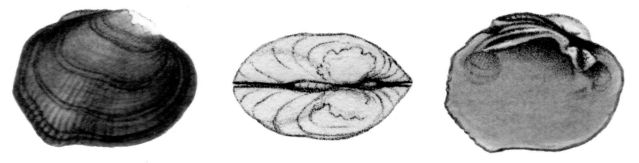

Unio sowerbianus Lea, 1834. Lea, 1834a:68, pl. 10, fig. 28; Lea, 1834b:180, pl. 10, fig. 28

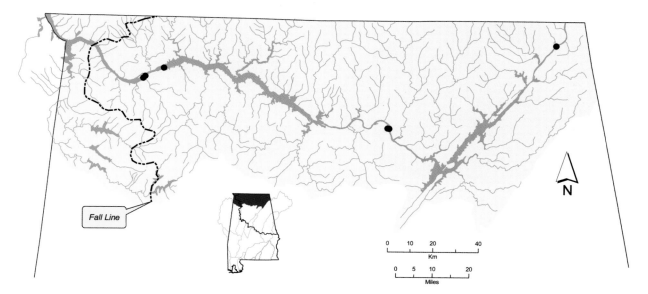

Distribution of *Epioblasma haysiana* in the Tennessee River drainage of Alabama.

Epioblasma lenior (Lea, 1842)
Narrow Catspaw

Epioblasma lenior – Upper figure: female, length 24 mm, USNM 150060. [Paint Rock River,] Woodville, Jackson County, Alabama. Lower figure: male, length 29 mm, CMNH 61.7048. Paint Rock River, Paint Rock, Jackson County, Alabama. © Richard T. Bryant.

Shell Description

Length to 35 mm; thin; somewhat compressed to moderately inflated; male outline elliptical, female outline oval; male posterior margin narrowly rounded, female posterior margin obliquely truncate to narrowly rounded, with narrow emargination just dorsal to extrapallial swelling; anterior margin rounded; dorsal margin straight to slightly convex; ventral margin straight to slightly convex; male posterior ridge low, rounded, female posterior ridge inflated into narrow, radially striate extrapallial swelling, serrate distally, with shallow sulcus anterior and posterior of swelling; posterior slope moderately steep, becoming flatter posterioventrally, may be adorned with thin, radial striae; umbo slightly inflated, barely elevated above hinge line, umbo sculpture irregular, double-looped ridges; periostracum pale yellowish green to yellowish brown, with numerous thin green rays.

Pseudocardinal teeth thin, compressed, 2 almost parallel teeth in left valve, 1 tooth in right valve, usually with parallel lamellar accessory denticle that ranges from weakly developed to almost as large as primary tooth; lateral teeth short, straight to slightly curved, 2 in left valve, 1 in right valve; interdentum short to moderately long, very narrow; umbo cavity wide, shallow; nacre white, may be tan.

Soft Anatomy Description

Soft anatomy unknown.

Glochidium Description

Glochidium unknown.

Similar Species

Epioblasma lenior is distinct from all other *Epioblasma* species. However, male *E. lenior* resemble *Lasmigona holstonia*. *Epioblasma lenior* is pale yellowish green with thin green rays posteriorly, and *L. holstonia* is typically some shade of brown without rays, with older individuals darkening to black. Lateral teeth of *E. lenior* are well-developed, but those of *L. holstonia* appear as simple thickened hinge plates.

Lasmigona holstonia attains a slightly larger size than *E. lenior*.

General Distribution

Epioblasma lenior is endemic to the Cumberland and Tennessee River drainages. The only known population of *E. lenior* in the Cumberland River drainage occurred in the Stones River (Parmalee and Bogan, 1998). In the Tennessee River drainage it historically occurred in headwaters of northeastern Tennessee and southwestern Virginia, with disjunct populations in the Duck and Paint Rock rivers (Parmalee and Bogan, 1998).

Alabama and Mobile Basin Distribution

Epioblasma lenior is known only from the Paint Rock River system.

There are no dated museum specimens of *Epioblasma lenior* from Alabama. The most recent record from the state was reportedly collected in 1918 (Stansbery, 1976).

Ecology and Biology

Epioblasma lenior occurred in clear, swift, medium to large rivers (Parmalee and Bogan, 1998). Ortmann (1924b) suggested it was common in the Paint Rock River system. Hickman (1937) reported collecting it from sand and gravel substrate in less than 1 m of water in the Clinch River, Tennessee.

Epioblasma lenior was presumably a long-term brooder, gravid from late summer or autumn to the following summer. Its glochidial hosts are unknown. Some species of *Epioblasma* use darters (Percidae) and sculpins (Cottidae) as glochidial hosts.

Current Conservation Status and Protection

Epioblasma lenior has been presumed extinct since at least 1970 (Stansbery, 1970a; Williams et al., 1993; Turgeon et al., 1998; Lydeard et al., 1999; Garner et al., 2004).

Remarks

Lea originally described *Epioblasma lenior* as *Unio lenis* in 1840 but subsequently changed it because the name *lenis* was preoccupied by a Conrad taxon that apparently represented a young *Lampsilis cardium* (Simpson, 1914).

The last known population of *Epioblasma lenior* was in the Stones River, a tributary of the Cumberland River in Tennessee. This species was driven to extinction with impoundment of the river behind J. Percy Priest Dam, which was built by the USACE in the 1960s (Stansbery, 1970a).

Synonymy

Unio lenis Lea, 1840, *non* Conrad, 1840. Lea, 1840:286 [changed to *Unio lenior*]
Type locality: Stones River, Tennessee.
Unio lenior Lea, 1842. Lea, 1842b:204, pl. 12, fig. 18; Lea, 1842c:42, pl. 12, fig. 18
Type locality: Stones River, Tennessee. Lectotype, USNM 86130, length 25 mm (female), designated by Johnson (1970).

 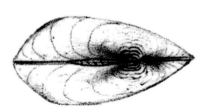 

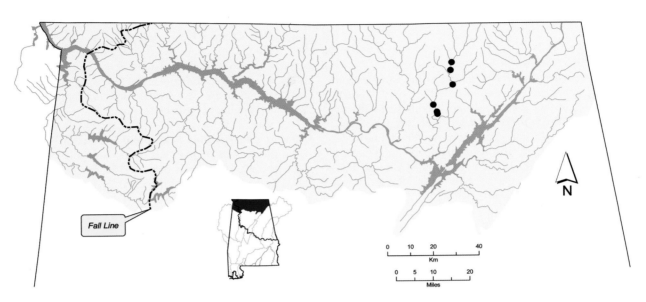

Distribution of *Epioblasma lenior* in the Tennessee River drainage of Alabama.

Epioblasma lewisii (Walker, 1910)
Forkshell

Epioblasma lewisii – Upper figure: female, length 62 mm, USNM 84051. Tennessee. Lower figure: male, length 42 mm, UMMZ 91458. Cumberland River, Burnside, Pulaski County, Kentucky. © Richard T. Bryant.

Shell Description

Length to 65 mm; moderately thick, females thinner than males; compressed; male outline quadrate to irregularly ovate, female outline irregularly quadrate to triangular depending on degree of extrapallial swelling; male posterior margin obliquely truncate to broadly rounded, female posterior margin extended into pronounced, narrow lobe dorsally, with emargination ventrally; anterior margin rounded; dorsal margin straight to slightly convex; male ventral margin broadly rounded with emargination at distal end of sulcus, female ventral margin oblique, slightly convex, greatly expanded into narrow, elongate extrapallial swelling posteriorly, may have slight emargination just anterior of extrapallial swelling; posterior ridge rounded, well-developed toward umbo, becoming indistinct posterioventrally; extrapallial swelling of females in form of medial ridge that extends below ventral margin; wide, shallow sulcus located between medial and posterior ridges; posterior slope flat to slightly convex; umbo compressed, barely elevated above hinge line, umbo sculpture faint corrugations; periostracum light greenish yellow to greenish brown, with faint, narrow green rays, extrapallial swelling of females almost always dark green.

Left valve with 2 pseudocardinal teeth, anterior tooth lamellar, oriented obliquely anteriad, posterior tooth triangular, 1 elongate triangular tooth in right valve, with accessory denticle anteriorly; lateral teeth straight in females, slightly curved in males, 2 teeth in left valve, 1 tooth in right valve; interdentum moderately long, narrow to moderately wide; umbo cavity shallow; nacre white.

Soft Anatomy Description

Mantle tan, slightly darker marginally; visceral mass tan; foot tan.

Females with papillate mantle margin just ventral to incurrent aperture; adjacent mantle thick, fleshy; papillae long, widely spaced, simple; enlarged, flattened, wrinkled papilla-like structure (microlure) located at posterior end of extrapallial swelling, adjacent to incurrent aperture (Figure 13.9).

Gills tan; dorsal margin straight, ventral margin of inner gill convex, higher anteriorly, ventral margin of outer gill gently curved; gill length approximately 45% of shell length; gill height approximately 55% of gill length; outer gill height approximately 70% of inner gill height; inner lamellae of inner gills incompletely connected to visceral mass.

No gravid females were available for marsupium description; probably similar to other *Epioblasma* species, with glochidia held in posterior part of outer gills, marsupium well-padded when gravid.

Labial palps white; straight dorsally, curved ventrally, bluntly pointed distally; palp length approximately 20% of gill length; palp height approximately 50% of palp length; distal 25% of palps bifurcate.

Incurrent aperture longer than excurrent aperture; supra-anal aperture longer than incurrent aperture.

Incurrent aperture length approximately 15% of shell length; tan within, without coloration basal to papillae; papillae in 2 rows, simple, inner row larger; papillae tan (Figure 13.9).

Excurrent aperture length approximately 10% of shell length; tan within, without marginal color band; margin with short simple papillae; papillae tan.

Supra-anal aperture length 18% of shell length; tan within, without marginal coloration; margin smooth; mantle bridge separating supra-anal and excurrent apertures imperforate, length approximately 15% of supra-anal length.

Specimen examined: "Tennessee" (n = 1) (specimen previously preserved).

Glochidium Description

Glochidium unknown.

Similar Species

Epioblasma lewisii resembles *Epioblasma flexuosa* and *Epioblasma stewardsonii*. *Epioblasma lewisii* is smaller and has a thinner shell than *E. flexuosa*. The periostracum of *E. lewisii* and *E. flexuosa* are uniform brown with thin green rays, but females of *E. lewisii* have a dark green extrapallial swelling. *Epioblasma lewisii* and *E. stewardsonii* are similar in size, but the shell of *E. stewardsonii* is thicker. In females, the sulcus of *E. lewisii* lies posterior to the extrapallial swelling, while the extrapallial swelling of *E. stewardsonii* originates in the sulcus. Also, the ventral terminus of the extrapallial swelling of *E. lewisii* is narrower than that of *E. stewardsonii*.

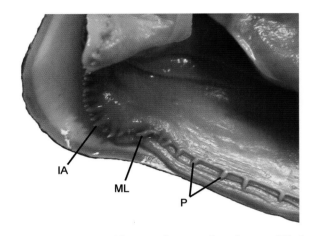

Figure 13.9. *Epioblasma lewisii* female modified mantle margin. Incurrent aperture (IA), microlure (ML), papillae (P). RMNH uncataloged. Tennessee, collected mid-1800s. Photograph by E. Neubert.

General Distribution

Epioblasma lewisii is endemic to the Tennessee and Cumberland River drainages (Cicerello et al., 1991; Parmalee and Bogan, 1998). In the Tennessee River drainage *E. lewisii* occurred from headwaters in eastern Tennessee downstream to Muscle Shoals, Alabama. No historical records exist downstream of Muscle Shoals, but *E. lewisii* has been recovered from prehistoric shell middens along the Tennessee River in Decatur County, Tennessee, and the Duck River, suggesting that it was formerly more widespread in the drainage.

Alabama and Mobile Basin Distribution

No historical specimens of *Epioblasma lewisii* from Alabama were found in museum collections. However, it has been collected from prehistoric middens at Muscle Shoals and near Bridgeport (Morrison, 1942; Warren, 1975), suggesting that it previously occurred across the state.

Epioblasma lewisii is believed to be extinct. It has apparently not been collected since at least the mid-1900s.

Ecology and Biology

Epioblasma lewisii was apparently a species of shoal habitat in medium to large rivers.

Epioblasma lewisii was presumably a long-term brooder, gravid from late summer or autumn to the following summer. Its glochidial hosts are unknown.

Some species of *Epioblasma* use darters (Percidae) and sculpins (Cottidae) as glochidial hosts.

Current Conservation Status and Protection

Epioblasma lewisii has been presumed extinct since the 1950s (Stansbery, 1970a; Williams et al., 1993; Turgeon et al., 1998; Lydeard et al., 1999; Garner et al., 2004).

Remarks

The relationship between *Epioblasma lewisii* and *Epioblasma flexuosa* has been the matter of some conjecture. Some authors (e.g., Stansbery, 1970a; Burch, 1975a; Parmalee and Bogan, 1998) recognize both species. However, Morrison (1942) listed *lewisii* as a subspecies of *E. flexuosa*, and Johnson (1978)

synonymized *E. lewisii* under *E. flexuosa*. Warren (1975) reported *E. flexuosa* and *E. lewisii* to be consistently separable in archaeological material but did report considerable variation and intergradation, suggesting that they are ecological variants.

Epioblasma lewisii has been reported from prehistoric shell middens along the Tennessee River at Muscle Shoals and near Bridgeport but not at intervening sites (Morrison, 1942; Warren, 1975; Hughes and Parmalee, 1999).

The only known specimen of *Epioblasma lewisii* with soft tissues (Figure 13.9) is housed in the National Museum of Natural History, Leiden, The Netherlands. The specimen was collected during the mid-1800s by Gerard Troost, Tennessee State Geologist.

Synonymy

Truncilla lewisii Walker, 1910. Walker, 1910a:42, pl. 3, figs. 3–5 [original figure not herein reproduced]
Type locality: Holston River, Tennessee. Lectotype, UMMZ 91456, length 51 mm (female), designated by Johnson (1978).

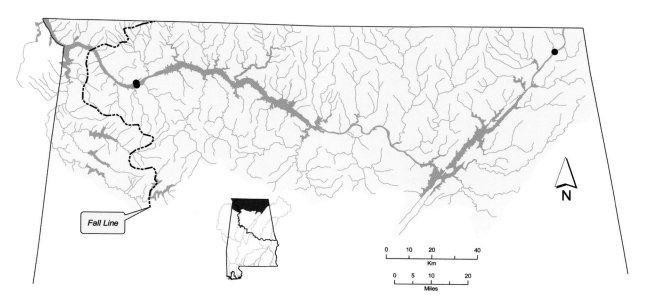

Distribution of *Epioblasma lewisii* in the Tennessee River drainage of Alabama.

Epioblasma metastriata (Conrad, 1838)
Upland Combshell

Epioblasma metastriata – Upper figure: female, length 40 mm, UF 64265. Lower figure: male, length 47 mm, UF 64265. Coosa River, Three Island Shoals near Wilsonville, Shelby County, Alabama, November 1911. © Richard T. Bryant.

Shell Description

Length to 60 mm; moderately thick; moderately inflated; outline subtriangular to trapezoidal; male posterior margin rounded to obliquely truncate, female posterior margin broadly rounded, usually becoming straight ventrally; anterior margin rounded; dorsal margin slightly convex; male ventral margin broadly convex, may become straight posteriorly, female ventral margin slightly convex with shallow emargination just anterior of extrapallial swelling; male posterior ridge well-developed, moderately angular, female posterior ridge obscured by extrapallial swelling; extrapallial swelling of females extends below ventral margin, adorned with faint radial striations, finely serrate distally, slight sulcus may separate extrapallial swelling from disk; umbo broad, elevated above hinge line, umbo sculpture unknown;

periostracum smooth, yellowish to tawny, often with thin green rays.

Pseudocardinal teeth triangular, somewhat compressed, 2 slightly divergent teeth in left valve, 1 tooth in right valve, may have accessory denticle anteriorly; lateral teeth short, straight to slightly curved, 2 in left valve, 1 in right valve; interdentum moderately long, narrow; umbo cavity wide, moderately shallow; nacre white.

Soft Anatomy Description

Mantle tan, usually with faint rusty brown outside of apertures; visceral mass tan; foot tan.

Females with papillate mantle margin within extrapallial swelling, with small fleshy fold midway down length of extrapallial swelling; fold forms small mound, open on interior side; surface of mound with very small, simple papilla-like structures; papillae posterior to fold small, simple, papillae anterior of fold wide basally, tapering, some with smaller, simple papillae joining them basally, papillae anterior of fold less numerous than those posteriorly.

Gills tan; dorsal margin slightly sinuous, ventral margin elongate convex; gill length 60–65% of shell length; gill height 35–55% of gill length; outer gill height 65–100% of inner gill height, inner gill height may be greater posteriorly; inner lamellae of inner gills completely connected to visceral mass. Lea (1863d) reported inner lamellae of inner gills to be connected to the visceral mass only anteriorly.

Outer gills marsupial; glochidia held in posterior 75% of gill; marsupium well-padded when gravid; creamy white, brighter white distally.

Labial palps tan; straight dorsally, convex ventrally, narrowly rounded to bluntly pointed distally; palp length 20–30% of gill length; palp height 50–85% of palp length; distal 40–65% of palps bifurcate.

Incurrent and supra-anal apertures longer than excurrent aperture; incurrent aperture slightly longer, slightly shorter or equal to supra-anal aperture.

Incurrent aperture length 15–20% of shell length; tan to rusty tan within, sometimes with thin or irregular rusty tan or rusty brown band basal to papillae; papillae arborescent; papillae rusty brown, may grade to rusty tan distally.

Excurrent aperture length 10–15% of shell length; tan within, marginal color band rusty tan, often with rusty brown lines perpendicular to margin or in open reticulated pattern; margin usually with small, simple papillae, occasionally crenulate, papillae may be well-developed or little more than crenulations; papillae tan.

Supra-anal aperture length 15–25% of shell length; tan within, may have thin rusty brown marginal band; margin smooth; mantle bridge separating supra-anal and excurrent apertures imperforate, length approximately 30% of supra-anal length.

Specimens examined: Cahaba River (n = 3); Coosa River (n = 1) (specimens previously preserved).

Glochidium Description

Length 250–262 µm; height 237–262 µm; without styliform hooks; outline subrotund; dorsal margin straight, ventral margin rounded, anterior and posterior margins convex.

Similar Species

Epioblasma metastriata closely resembles *Epioblasma penita*. Females can be distinguished by outline of the extrapallial swelling. *Epioblasma metastriata* typically has a ventrally curved extrapallial swelling, while that of *E. penita* is straight. The extrapallial swelling of *E. metastriata* is wider distally and continues up the posterior margin farther than that of *E. penita*, which is indicated by striations and distal serrations. *Epioblasma penita* attains greater size than *E. metastriata*.

General Distribution

Epioblasma metastriata is endemic to the Mobile Basin of Alabama, Georgia and Tennessee.

Alabama and Mobile Basin Distribution

Epioblasma metastriata is known from the Black Warrior, Cahaba and Coosa River drainages primarily above the Fall Line. There are also a few records from steep gradient reaches of those rivers a short distance downstream of the Fall Line.

Epioblasma metastriata has not been collected since 1973. It was last observed in the Little Cahaba River.

Ecology and Biology

Epioblasma metastriata is a species of shoal habitat in medium to large rivers.

Epioblasma metastriata is presumably a long-term brooder, gravid from late summer or autumn to the following summer. Its glochidial hosts are unknown. Some species of *Epioblasma* use darters (Percidae) and sculpins (Cottidae) as glochidial hosts.

Current Conservation Status and Protection

Epioblasma metastriata was recognized as endangered throughout its range by Athearn (1970) and Williams et al. (1993). In Alabama it was considered threatened by Stansbery (1976) and endangered by Lydeard et al. (1999). Garner et al. (2004) listed *E. metastriata* as extirpated from the state.

Remarks

Johnson (1978) synonymized *Epioblasma metastriata* with *Epioblasma penita*. However, most workers have recognized both species (e.g., Hurd, 1974; Stansbery, 1976).

Synonymy

Unio metastriatus Conrad, 1838. Conrad, 1838:Part 11, back cover; Conrad, 1840:104, pl. 57, fig. 2

Type locality: Black Warrior River, [near Blount Springs, Blount County,] Alabama. Lectotype, designated by
     Johnson and Baker (1973), not found; Conrad (1838) gave no shell measurement in original description.

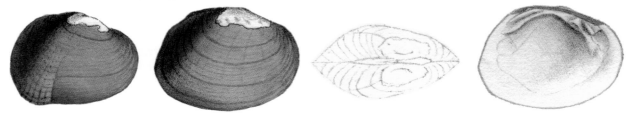

Unio compactus Lea, 1859. Lea, 1859b:154; Lea, 1859f:218, pl. 28, fig. 98; Lea, 1859g:36, pl. 28, fig. 98

Type locality: Etowah River, Georgia, Bishop Elliott and Reverend G. White. Lectotype, USNM 84447, length 33
     mm (male), designated by Johnson (1974).

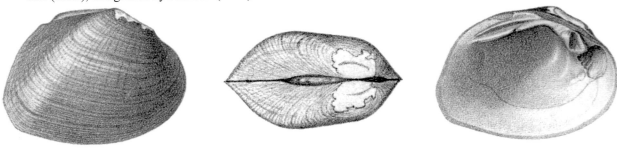

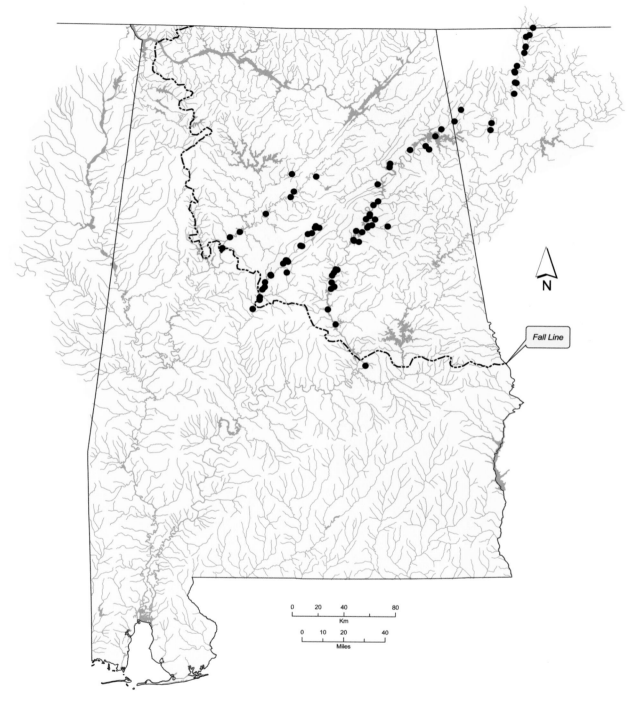

Distribution of *Epioblasma metastriata* in Alabama and the Mobile Basin.

Epioblasma obliquata (Rafinesque, 1820)
Catspaw

Epioblasma obliquata – Upper figure: female, length 35 mm, UF 269100. Lower figure: male, length 45 mm, UF 269102. Tennessee River, Florence, Lauderdale County, Alabama. © Richard T. Bryant.

Shell Description

Length of males to 70 mm, females to 50 mm; thick; moderately inflated; outline subquadrate to oval or triangular; male posterior margin bluntly pointed, often with weak emargination at distal end of sulcus, female posterior margin bilobed, upper lobe truncate to slightly convex, lower lobe rounded; male anterior margin truncate to slightly convex, female anterior margin rounded; dorsal margin convex, males may be more convex than females; ventral margin convex; posterior ridge low, rounded; females with narrow, radial extrapallial swelling just anterior of posterior ridge, separated from posterior ridge by deep, narrow sulcus, extrapallial swelling adorned with thin radial striae; posterior slope steep, flat to slightly concave or slightly convex, may have thin radial striae, especially females, males with wide, shallow sulcus just anterior of posterior ridge; umbo moderately inflated, elevated above hinge line, umbo sculpture weak corrugations, occasionally double-looped; periostracum usually smooth, shiny, often satiny in young individuals, tawny to yellowish green or reddish brown, often with numerous thin green rays, may darken with age, obscuring rays.

Pseudocardinal teeth erect, triangular, 2 teeth in left valve, become divergent with age, 1 tooth in right valve, often with accessory denticle anteriorly and/or posteriorly; lateral teeth short, straight to slightly curved, 2 in left valve, 1 in right valve, right valve may also have rudimentary lateral tooth ventrally;

interdentum moderately short, narrow to wide; umbo cavity wide, shallow; nacre usually some shade of purple, but may be white.

Soft Anatomy Description

Mantle tan, more pale outside of pallial line, with irregular rusty brown outside of apertures; visceral mass pale tan; foot pale tan.

No females were available for description of mantle modification. Lea (1863d) reported the mantle just ventral to the incurrent aperture to be "enlarged", very black and with "a few papillae".

Gills tan; dorsal margin straight, ventral margin convex; gill length approximately 65% of shell length; gill height approximately 40% of gill length; outer gill height approximately 80% of inner gill height; inner lamellae of inner gills completely connected to visceral mass.

No gravid females were available for marsupium description. Lea (1863d) reported outer gills to be marsupial, with glochidia held across "nearly the whole length".

Labial palps tan; straight dorsally, convex ventrally, bluntly pointed distally; palp length approximately 25% of gill length; palp height approximately 40% of palp length; distal 20% of palps bifurcate.

Incurrent aperture longer than excurrent aperture; supra-anal aperture longer than incurrent aperture.

Incurrent aperture length approximately 15% of shell length; tan within, irregular rusty brown basal to papillae; papillae arborescent; papillae pale tan and rusty brown. Lea (1863d) reported the incurrent aperture to be "blackish, with very small, very dark papillae".

Excurrent aperture length approximately 10% of shell length; tan within, marginal color band rusty brown; margin crenulate.

Supra-anal aperture length approximately 20% of shell length; tan within, with thin marginal band of rusty brown, disappearing dorsally; margin smooth; mantle bridge separating supra-anal and excurrent apertures imperforate, length approximately 15% of supra-anal length.

Specimen examined: Cumberland River, [Tennessee or Kentucky] (n = 1) (specimen previously preserved).

Glochidium Description

Length 200 μm; height 205 μm; without styliform hooks; outline subrotund; dorsal margin straight and long, ventral margin broadly rounded, anterior and posterior margins slightly convex (Surber, 1912).

Similar Species

The shell of *Epioblasma obliquata* resembles that of *Epioblasma haysiana*. In females, the extrapallial swelling of *E. obliquata* is better defined than that of *E. haysiana*. This gives the posterior margin of *E. obliquata* a bilobed outline and *E. haysiana* a convex outline. Male *E. obliquata* are more elongate and pointed posteriorly than male *E. haysiana*.

General Distribution

Nominal *Epioblasma obliquata* historically occurred in tributaries of lakes St. Clair and Erie of the Great Lakes Basin (Johnson, 1978). It was found in much of the Ohio River drainage from eastern Ohio to the mouth of the Ohio River (Cummings and Mayer, 1992). It was historically widespread in the Cumberland River drainage downstream of Cumberland Falls, Kentucky and Tennessee (Cicerello et al., 1991; Parmalee and Bogan, 1998). It occurred in middle and lower reaches of the Tennessee River (Johnson, 1970; Parmalee and Bogan, 1998).

Alabama and Mobile Basin Distribution

All historical records of *Epioblasma obliquata* are from the Tennessee River in the vicinity of Muscle Shoals. However, an archaeological specimen from the Tennessee River in eastern Tennessee suggests that it occurred across the state, at least prehistorically.

Epioblasma obliquata is extirpated from Alabama. Morrison (1942) reported collecting specimens from commercial mussel cull piles near the lower end of Muscle Shoals. Those piles were presumably created prior to impoundment of the river. The most recent dated historical material from Alabama was collected in 1924.

Ecology and Biology

Nominal *Epioblasma obliquata* was primarily a species of medium to large rivers (Parmalee and Bogan, 1998). Morrison (1942) speculated that it was a deep water species, resulting in its rarity in prehistoric shell middens at Muscle Shoals. However, *E. obliquata* is extant in a small Ohio stream (Hoggarth, 1998).

Epioblasma obliquata is a long-term brooder. Gravid females have been reported in September and October (Surber, 1912), presumably remaining gravid to the following summer. Fishes found to serve as glochidial hosts of *E. obliquata* during laboratory trials include *Ambloplites rupestris* (Rock Bass) (Centrarchidae); *Cottus bairdii* (Mottled Sculpin) (Cottidae); *Noturus flavus* (Stonecat) (Ictaluridae); and *Etheostoma blennioides* (Greenside Darter), *Percina caprodes* (Logperch) and *Percina maculata* (Blackside Darter) (Percidae) (G.T. Watters, personal communication).

Current Conservation Status and Protection

Epioblasma obliquata was listed as possibly extinct by Stansbery (1970a) and Turgeon et al. (1998)

but endangered throughout its range by Williams et al. (1993). It was reported as extirpated from Alabama by Lydeard et al. (1999) and Garner et al. (2004). Nominal *E. obliquata* is known to be extant only in Killbuck Creek, Ohio (Hoggarth, 1998). It was listed as endangered under the federal Endangered Species Act in 1990.

In 2001 *Epioblasma obliquata* was approved for a Nonessential Experimental Population in tailwaters of Wilson Dam, but no reintroductions had been made as of 2007.

Remarks

Nominal *Epioblasma obliquata* was collected alive from the Cumberland River during the late 1970s (Isom et al., 1979). However, it has not been found during recent sampling (D.W. Hubbs, personal communication). *Epioblasma obliquata perobliqua* is known to be extant in a single population in Fish Creek, a Lake Erie tributary in northeastern Indiana and northwestern Ohio (Keller et al., 1998).

The only prehistoric record of nominal *Epioblasma obliquata* was a single specimen from a shell midden at Muscle Shoals (Morrison, 1942). This species was not reported in Alabama reaches of the Tennessee River upstream of Muscle Shoals, but one archaeological specimen was reported farther upstream, from Chickamauga Reservoir in Tennessee (Parmalee et al., 1982; Hughes and Parmalee, 1999). This suggests that *E. obliquata* was rare in Alabama even prior to modern perturbations to the Tennessee River.

Synonymy

Obliquaria obliquata Rafinesque, 1820. Rafinesque, 1820:309
Type locality: Kentucky River. Lectotype, ANSP 20226, length 59 mm (male), designated by Johnson and Baker (1973).
Unio ridibundus Say, 1829. Say, 1829:308; Say, 1830a:[no pagination], pl. 5
Unio perplexus Say, 1829, *non* Lea, 1831. Say, 1829:309; Say, 1830a:[no pagination], pl. 5
Unio sulcatus Lea, 1829. Lea, 1829:430, pl. 8, fig. 12; Lea, 1834b:44, pl. 8, fig. 12
Unio flagellatus Say, 1830. Say, 1830a:[no pagination], pl. 5
Unio gibbosus perobliquus Conrad, 1836. Conrad, 1836:51, pl. 27, fig. 2
Unio pectitis Conrad, 1853, *nomen nudum*. Conrad, 1853:225
Unio pectitis Conrad, 1854. Conrad, 1854:297, pl. 27, fig. 4
Truncilla (Scalenaria) sulcata var. *delicata* Simpson, 1900. Simpson, 1900b:520
Unio propesulcatus De Gregorio, 1914. De Gregorio, 1914:60, pl. 10, fig. 2

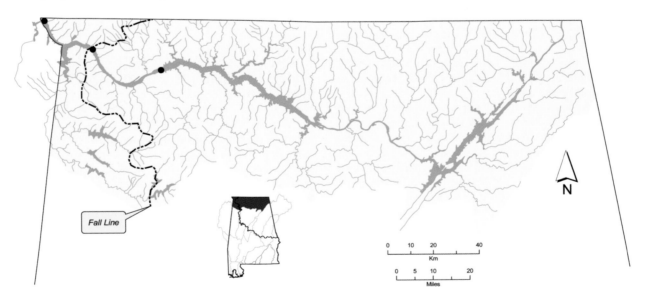

Distribution of *Epioblasma obliquata* in the Tennessee River drainage of Alabama.

Epioblasma othcaloogensis (Lea, 1857)
Southern Acornshell

Epioblasma othcaloogensis – Upper figure: female, length 23 mm, UF 64307. Lower figure: male, length 29 mm, UF 64307. Conasauga River, Upper King's Bridge, Murray County, Georgia, 14 October 1916. © Richard T. Bryant.

Shell Description

Length to 32 mm, males attain larger size than females; moderately thin; usually moderately inflated, occasionally becoming very inflated with age; male outline subtriangular, female outline oval; male posterior margin obliquely truncate, female posterior margin rounded; anterior margin rounded; dorsal margin straight to slightly convex; ventral margin straight to slightly convex, extrapallial swelling of females may extend slightly below ventral margin; posterior ridge rounded; extrapallial swelling of females poorly defined, typically with a few short radial striations; posterior slope flat to slightly convex in males, usually convex in females; umbo moderately inflated, elevated above hinge line, umbo sculpture unknown; periostracum smooth, shiny, yellowish to tawny, typically without rays.

Pseudocardinal teeth thin, triangular, 2 divergent teeth in left valve, may be less divergent in females than males, anterior tooth usually larger than posterior tooth, 1 tooth in right valve, usually with accessory denticle anteriorly, may have accessory denticle posteriorly; lateral teeth short, straight, 2 in left valve, 1 in right valve; interdentum moderately long, narrow; umbo cavity shallow; nacre white.

Soft Anatomy Description

Mantle pale tan, often with rusty cast outside of apertures and in extrapallial swelling; visceral mass tan; foot tan.

Females with papillate mantle margin within extrapallial swelling, with elongate, thickened flap or fleshy swelling midway down length of extrapallial swelling; papillae simple, papillae dorsal to flap slender, papillae ventral to flap wide basally, tapering to acute point, few rudimentary papillae may be present on outer margin of flap.

Gills pale tan; dorsal margin slightly sinuous, ventral margin usually convex, outer gills of some females slightly bilobed; gill length 60–65% of shell length; gill height 45–60% of gill length; outer gill height usually 75–85% of inner gill height; inner lamellae of inner gills completely connected to visceral mass.

Outer gills marsupial; glochidia held in posterior 65% of gill; marsupium distended when gravid; pale tan.

Labial palps tan; straight to slightly concave dorsally, convex ventrally, bluntly pointed distally; palp length 25–35% of gill length; palp height 60–70% of palp length; distal 30–40% of palps bifurcate.

Incurrent aperture longer than excurrent aperture; incurrent aperture slightly longer, shorter or equal to supra-anal aperture.

Incurrent aperture length 15–20% of shell length; pale tan within, sometimes with thin or irregular rusty tan band basal to papillae; papillae short, thick, blunt, simple, bifid and small arborescent; papillae tan and rusty tan.

Excurrent aperture length 10–15% of shell length, occasionally greater; pale tan within, marginal color band rusty tan, may be absent; margin with small, simple papillae; papillae pale tan.

Supra-anal aperture length 15–20% of shell length, occasionally less; pale tan within, without marginal coloration; margin smooth or with minute crenulations; mantle bridge separating supra-anal and excurrent apertures imperforate, length 15–50% of supra-anal length, occasionally absent.

Specimens examined: Conasauga River, Georgia (n = 5) (specimens previously preserved).

Glochidium Description

Length 237–275 μm; height 237–250 μm; without styliform hooks; outline subrotund; dorsal margin straight, ventral margin rounded, anterior and posterior margins convex.

Similar Species

Male *Epioblasma othcaloogensis* resemble small *Ptychobranchus foremanianus*. However, when *P. foremanianus* is small, it is much more compressed than *E. othcaloogensis* of comparable size. Also, *P. foremanianus* typically has well-defined green rays, whereas *E. othcaloogensis* is typically without rays.

General Distribution

Epioblasma othcaloogensis is endemic to the Coosa River drainage of the Mobile Basin in Alabama, Georgia and Tennessee.

Alabama and Mobile Basin Distribution

Epioblasma othcaloogensis is known from the Coosa River drainage in Alabama, Georgia and Tennessee. Only one record of *E. othcaloogensis* outside of the Coosa River drainage is known (UMMZ 91261). It was labeled "Cahaba River, Alabama", with no accompanying data. This record is questionable, since the Cahaba River was well-sampled during the early 1900s and no other specimens are known. Therefore this species is not considered to be part of the Cahaba River fauna.

The most recent documented museum collections of *Epioblasma othcaloogensis* were taken in 1973 from the Conasauga River, Georgia, and Little Canoe Creek, on the Etowah and St. Clair county line, Alabama. This species may be extinct.

Ecology and Biology

Epioblasma othcaloogensis appears to have been more common in tributaries than large rivers. It presumably occurred in shoals and riffles.

Epioblasma othcaloogensis is a long-term brooder, presumably gravid from late summer or autumn to the following summer. Preserved gravid females collected during September are housed at OSUM. Its glochidial hosts are unknown. Some species of *Epioblasma* use darters (Percidae) and sculpins (Cottidae) as glochidial hosts.

Current Conservation Status and Protection

Epioblasma othcaloogensis was recognized as endangered throughout its range by Athearn (1970) and Williams et al. (1993) and in Alabama by Stansbery (1976) and Lydeard et al. (1999). Garner et al. (2004) reported *E. othcaloogensis* as extirpated from the state. It was listed as endangered under the federal Endangered Species Act in 1993. Based on their failure to find the species during recent surveys, Johnson and Evans (2000) and Gangloff (2003) suggested that *E. othcaloogensis* may be extinct.

Remarks

There has been little question as to the taxonomic status of *Epioblasma othcaloogensis*. Johnson (1978) considered it a small headwater form of *Epioblasma penita* but provided no justification.

Epioblasma othcaloogensis is named for Othcalooga [Oothkalooga] Creek in Gordon County, Georgia. Othcalooga is derived from the Cherokee language and means beaver or where there are beaver dams (Krakow, 1975).

Synonymy

Unio othcaloogensis Lea, 1857. Lea, 1857a:32; Lea, 1858e:74, pl. 14, fig. 54; Lea, 1858h:74, pl. 14, fig. 54
Type locality: Othcalooga Creek, Gordon County, Georgia, Bishop Elliott. Lectotype, USNM 84615, length 22 mm (female), designated by Johnson (1974).

Unio modicellus Lea, 1859. Lea, 1859d:171; Lea, 1860d:347, pl. 57, fig. 172; Lea, 1860e:29, pl. 57, fig. 172
Type locality: Connasauga [Conasauga] River, Bishop Elliott; Chattanooga [Chattooga] River, Georgia, T. Stewardson, M.D. Lectotype, USNM 84841, length 28 mm (male), designated by Johnson (1974), is from Conasauga River, Georgia.

 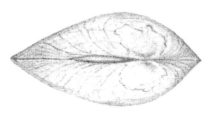 

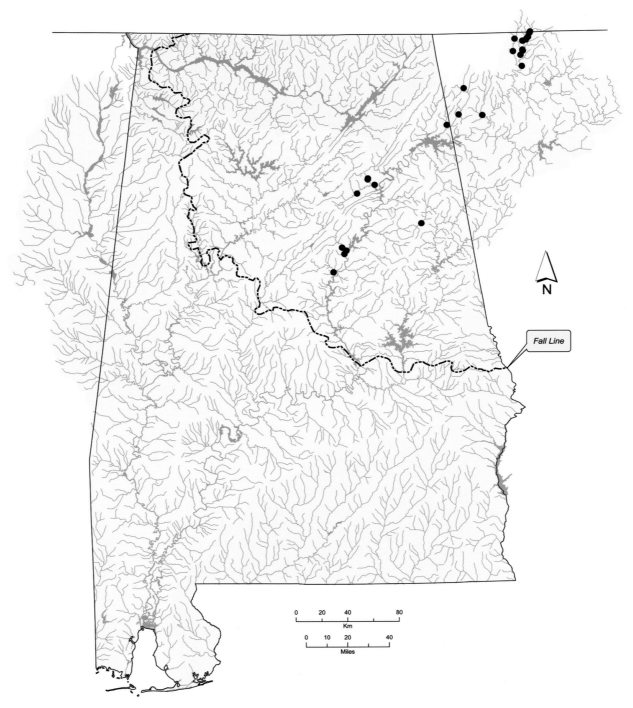

Distribution of *Epioblasma othcaloogensis* in Alabama and the Mobile Basin.

Epioblasma penita (Conrad, 1834)
Southern Combshell

Epioblasma penita – Upper figure: female, length 69 mm, USNM 809735. Tombigbee River, Memphis Landing, river mile 324.4, Pickens County, Alabama, 24 October 1976. Lower figure: male, length 64 mm, USNM 745457. Tombigbee River, about 2 miles west of Gainesville, Sumter County, Alabama, 14 September 1973. © Richard T. Bryant.

Shell Description

Length to 73 mm; thick; moderately inflated, becoming highly inflated with age; male outline triangular, female outline irregularly elliptical; male posterior margin narrowly rounded to bluntly pointed, female posterior margin bluntly pointed to obliquely truncate; anterior margin rounded; dorsal margin slightly convex; male ventral margin broadly convex, female ventral margin almost straight, extrapallial swelling extends below ventral margin with slight emargination just anterior of swelling; posterior ridge well-defined; in females extrapallial swelling occupies posterior 25% of shell, obscuring posterior ridge, swelling becomes wider posterioventrally, with thin radial striations, developing deeper striations with age, distal margin of swelling

serrate, slight sulcus separates extrapallial swelling from disk; posterior slope steep; umbo broad, moderately inflated, elevated well above hinge line, umbo sculpture unknown; periostracum tawny to greenish brown, may become dark brown with age, usually with thin, green to dark brown rays.

Pseudocardinal teeth triangular, erect, 2 divergent teeth in left valve, anterior tooth thinner and more elongate than posterior tooth, oriented ventrally or anterioventrally, 1 tooth in right valve, usually with accessory denticle anteriorly and posteriorly; lateral teeth thick, straight, 2 in left valve, 1 in right valve, lateral teeth of adult females relatively shorter than those of males; interdentum short, narrow; umbo cavity shallow; nacre white.

Soft Anatomy Description

Mantle tan, outside of apertures faint to moderately dark rusty brown; visceral mass tan; foot tan. Lea (1863d) reported the area outside of the apertures to be spotted.

Females with papillate mantle margin within extrapallial swelling, with small, fleshy fold midway down length of extrapallial swelling; fold forms small mound, open on interior side; surface of mound with very small, simple papillae, small papillae extend onto mantle wall as a small dense bed; marginal papillae posterior of fold small, simple, papillae anterior of fold wide basally, tapering, some with smaller, simple papillae joining them basally, some approaching arborescent in form, papillae anterior of fold less numerous than those posteriorly. Males with rudimentary papillae and fold. Lea (1863d) described the small, fleshy fold as a "subsigmoid fleshy process" and compared it to the caruncle of *Toxolasma*.

Gills tan; dorsal margin straight to slightly sinuous; ventral margin convex to elongate convex, outer gills may be slightly bilobed; gill length 50–65% of shell length; gill height 45–60% of gill length, occasionally greater; outer gill height 70–85% of inner gill height; inner lamellae of inner gills completely connected to visceral mass.

No gravid females were available for marsupium description. Outer gills are marsupial, padded when gravid, with patchy dark coloration on ventral margin (P.D. Johnson, personal communication).

Labial palps tan; straight dorsally, convex ventrally, narrowly rounded to bluntly pointed distally; palp length 15–30% of gill length; palp height 40–60% of palp length, occasionally greater; distal 30–35% of palps bifurcate, occasionally greater.

Incurrent and supra-anal apertures longer than excurrent aperture; incurrent aperture may be longer, shorter or equal to supra-anal aperture.

Incurrent aperture length 15–20% of shell length; tan within, may have rusty tan or rusty brown basal to papillae; papillae arborescent, often with scattered simple papillae; papillae rusty tan to rusty brown, color may change distally.

Excurrent aperture length usually 10–15% of shell length, occasionally less; tan within, marginal color band usually some combination of tan, rusty tan and rusty brown, often with lines forming an open reticulated pattern or bands perpendicular or oblique to margin; margin crenulate.

Supra-anal aperture length 15–20% of shell length; tan within, often with thin reddish brown marginal band, marginal band may disappear dorsally; margin mostly smooth, but may have minute crenulations adjacent to mantle bridge; mantle bridge separating supra-anal and excurrent apertures imperforate, length 20–30% of supra-anal length.

Specimens examined: Buttahatchee River (n = 1); Tombigbee River (n = 4) (specimens previously preserved).

Glochidium Description

Glochidium unknown.

Similar Species

Epioblasma penita closely resembles *Epioblasma metastriata*. Females can be distinguished by outline of the extrapallial swelling. *Epioblasma penita* typically has a straight extrapallial swelling, while that of *E. metastriata* is curved ventrally. The extrapallial swelling of *E. metastriata* is wider distally and continues up the posterior margin farther than that of *E. penita*, which is indicated by striations and distal denticulations. This gives the extrapallial swelling of *E. penita* the appearance of being oriented more ventrally. *Epioblasma penita* attains greater size than *E. metastriata*.

Large, inflated male *Epioblasma penita* can resemble female *Ellipsaria lineolata*. However, male *E. penita* are usually more elongate posteriorly, where they have weak radial striations. Rays on the shell disk of *E. lineolata* are usually thinner and more widely spaced than those of *E. penita*. Internally, the interdentum of *E. penita* is much narrower than that of *E. lineolata*.

General Distribution

Epioblasma penita is endemic to the Mobile Basin of Alabama, Georgia and Mississippi.

Alabama and Mobile Basin Distribution

Epioblasma penita was historically widespread in the Mobile Basin with most historic records from below the Fall Line. Above the Fall Line it is known from the Cahaba and Coosa River drainages. There are no records from the lower reaches of Tombigbee River, though it probably occurred there historically.

Epioblasma penita appears to have been extirpated from its entire range, with the exception of lower Buttahatchee River in Mississippi and possibly Alabama.

Ecology and Biology

Epioblasma penita occurs in small to large rivers, in shoals up to 1.5 m deep, with moderate to swift current and predominantly gravel and sand substrates. In the Tombigbee River it was most abundant in the reach between Gainesville and Pickensville in Greene, Pickens and Sumter counties, Alabama (Williams, 1982). *Epioblasma penita* is extant in the Buttahatchee River, a tributary of the upper Tombigbee River.

Epioblasma penita is a long-term brooder, gravid from late summer or autumn to the following spring or summer. Fishes known to serve as glochidial hosts, based on laboratory trials, are *Percina kathae* (Mobile Logperch) and *Percina nigrofasciata* (Blackbanded Darter) (Percidae) (P.D. Johnson, personal communication).

Current Conservation Status and Protection

Epioblasma penita was considered endangered throughout its range by Athearn (1970) and Williams et al. (1993), and in Alabama by Stansbery (1976). Though Lydeard et al. (1999) listed *E. penita* as possibly extirpated from the state, Garner et al. (2004) designated it a species of highest conservation concern in Alabama. *Epioblasma penita* was listed as endangered under the federal Endangered Species Act in 1987.

An effort to reestablish *Epioblasma penita* into tailwaters of Jordan Dam on the Coosa River was initiated in 2005. Juveniles were cultured at TNARI using brood stock from the Buttahatchee River.

Remarks

Unlike many species of *Epioblasma, E. penita* continued to thrive well into the second half of the twentieth century. However, construction of the Tennessee Tombigbee Waterway by the USACE pushed it to the brink of extinction.

Synonymy

Unio penitus Conrad, 1834. Conrad, 1834b:33, pl. 5, fig. 1
Type locality: Alabama River near Claiborne, [Monroe County,] Alabama. Lectotype, ANSP 59860, length 51 mm (female), designated by Johnson and Baker (1973).

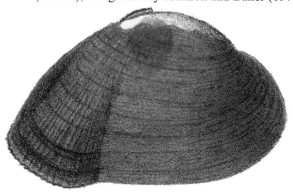

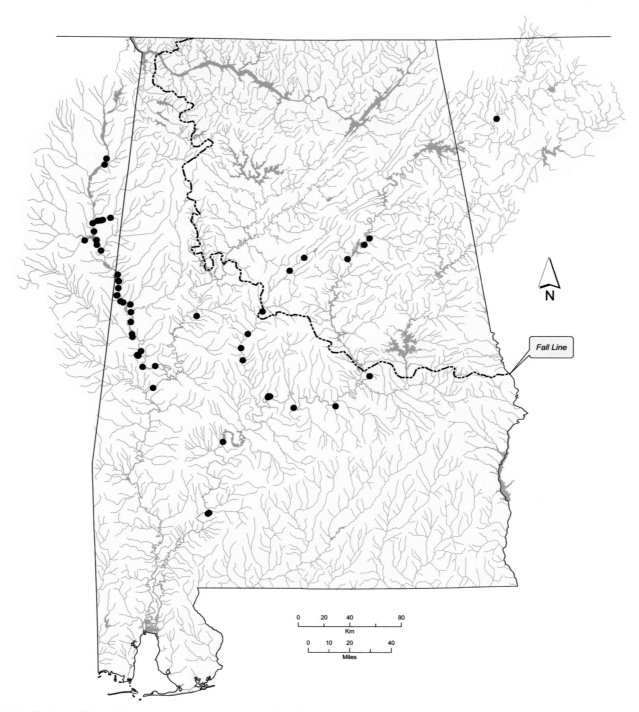

Distribution of *Epioblasma penita* in Alabama and the Mobile Basin.

Epioblasma personata (Say, 1829)
Round Combshell

Epioblasma personata – Upper figure: female, length 43 mm, USNM 84601. Lower figure: male, length 45 mm, USNM 84601. Tennessee. © Richard T. Bryant.

Shell Description

Length to 60 mm; thick; inflated; male outline round to subtriangular, female outline round; male posterior margin evenly rounded to obliquely truncate, female posterior margin rounded dorsally, with slight emargination above the extrapallial swelling; anterior margin rounded; dorsal margin straight to convex; ventral margin broadly rounded; male posterior ridge well-developed, slightly doubled, flattened, female posterior ridge low, rounded; extrapallial swelling of females wide, extending slightly below ventral margin, with weak radial striations and serrations distally; posterior slope steep, flatter posterioventrally in females; males with wide, shallow sulcus anterior of posterior ridge, sulcus obscured by extrapallial swelling in females; umbo inflated, elevated well above hinge line, turned slightly anteriad, umbo sculpture unknown; periostracum satiny, greenish yellow to brown, darkening with age, usually with narrow, wavy green rays.

Pseudocardinal teeth triangular, erect, 2 slightly divergent teeth in left valve, 1 tooth in right valve, often with accessory denticle anteriorly and/or posteriorly; lateral teeth short, straight to slightly curved, 2 in left valve, 1 in right valve; interdentum short, narrow to moderately wide; umbo cavity shallow; nacre white, occasionally tan.

Soft Anatomy Description

No material was available for soft anatomy description. Lea (1863d) offered an "imperfect description" of *Epioblasma personata*, which stated that marsupia occupy the entire length of the outer gills, but no glochidia were found proximally. This observation was made from a 25-year-old, dried specimen.

Glochidium Description

Without styliform hooks; outline subrotund (Lea, 1863d).

Similar Species

Male *Epioblasma personata* resemble female *Epioblasma propinqua*. The umbo of *E. personata* is positioned less anteriorly than that of *E. propinqua*. Also, the sulcus is complete distally in male *E. personata* but disappears with a weak extrapallial swelling in female *E. propinqua*. The round outline and shallow posterior slope of female *E. personata* make them distinct from any other *Epioblasma* species.

General Distribution

Epioblasma personata historically occurred in the Ohio, Tennessee, Cumberland and Wabash rivers (Parmalee and Bogan, 1998). In the Cumberland River it ranged upstream to near the mouth of the Stones River (Parmalee and Bogan, 1998). *Epioblasma personata* occurred in most of the Tennessee River

drainage, from headwaters in eastern Tennessee to the mouth of the Tennessee River (Parmalee and Bogan, 1998). However, a paucity of museum material from the Tennessee River suggests that it was rare there.

Alabama and Mobile Basin Distribution

All museum records of *Epioblasma personata* were collected at Muscle Shoals, but the species probably ranged across northern Alabama.

Epioblasma personata is believed to be extinct. The most recent documented museum material collected from Alabama was taken at Muscle Shoals in 1924.

Ecology and Biology

Epioblasma personata was a species of shoal habitat, primarily in large rivers (Parmalee and Bogan, 1998).

Epioblasma personata was presumably a long-term brooder, gravid from late summer or autumn to the following spring or summer. Its glochidial hosts are unknown. Some species of *Epioblasma* use darters (Percidae) and sculpins (Cottidae) as glochidial hosts.

Current Conservation Status and Protection

Epioblasma personata has been presumed extinct since at least 1970 (Stansbery, 1970a; Williams et al., 1993; Turgeon et al., 1998; Lydeard et al., 1999; Garner et al., 2004). It probably disappeared soon after the large rivers were impounded.

Remarks

The only reported prehistoric records of *Epioblasma personata* came from shell middens at Muscle Shoals, where it was rare (Morrison, 1942). Morrison (1942) suggested the reason for its scarcity may have been a preference for "deeper-water habitat" and not relative abundance, though no discussion of this supposition was given. Ortmann (1925) referred to *E. personata* as "one of the rarest naiades".

Synonymy

Unio personata Say, 1829. Say, 1829:309
Type locality: Wabash River. Neotype, MCZ 25763, length 54 mm (female), designated by Johnson and Baker
    (1973), is from Cumberland River, Tennessee.
Unio pileus Lea, 1831. Lea, 1831:119, pl. 18, fig. 47; Lea, 1834b:129, pl. 18, fig. 47
Unio capillaris Lea, 1834. Lea, 1834a:29, pl. 2, fig. 2; Lea, 1834b:141, pl. 2, fig. 2

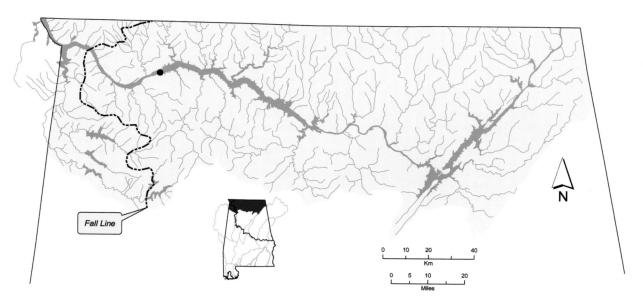

Distribution of *Epioblasma personata* in the Tennessee River drainage of Alabama.

Epioblasma propinqua (Lea, 1857)
Tennessee Rifleshell

Epioblasma propinqua – Upper figure: female, length 45 mm, UMMZ 90644. Holston River, Tennessee. Lower figure: male, length 35 mm, CMNH 61.644. Tennessee River, Tuscumbia, Colbert County, Alabama. © Richard T. Bryant.

Shell Description

Length to 55 mm; thick; inflated; male outline subtriangular, female outline ovate; male posterior margin narrowly rounded to obliquely truncate, female posterior margin broadly rounded; anterior margin rounded; dorsal margin straight to convex; male ventral margin broadly rounded anteriorly, with emargination posteriorly at distal end of sulcus, female ventral margin broadly rounded; females with weak extrapallial swelling posterioventrally; posterior ridge narrow, rounded, usually with low concentric ridges along annuli; low medial ridge also adorned with low concentric ridges; posterior and medial ridges separated by shallow, ventrally curved sulcus, sulcus obscured distally by extrapallial swelling in females; umbo inflated, elevated above hinge line, directed anteriad, umbo sculpture weak corrugations; periostracum dull to silky, yellowish green to brown, often with thin green rays.

Pseudocardinal teeth thick, triangular, 2 divergent teeth in left valve, 1 tooth in right valve, usually with accessory denticle anteriorly and/or posteriorly; lateral teeth thick, slightly curved, longer in males than females, 2 in left valve, 1 in right valve, right valve may also have rudimentary lateral tooth ventrally; interdentum short, moderately wide; umbo cavity shallow; nacre white, occasionally with pinkish tint.

Soft Anatomy Description

Mantle tan; visceral mass tan; foot tan, slightly darker distally.

Females with thick, spongy pad filling extrapallial swelling; mantle margin within extrapallial swelling papillate; papillae short, widely spaced, simple (Figure 13.10).

Gills tan; dorsal margin straight, inner gill ventral margin convex, higher anteriorly, outer gill ventral margin gently curved, with obvious convex marsupial protuberance just posterior of center in females; gill length approximately 55% of shell length; gill height approximately 55% of gill length; outer gill height similar to inner gill height in marsupial area, remainder considerably less; inner lamellae of inner gills completely connected to visceral mass.

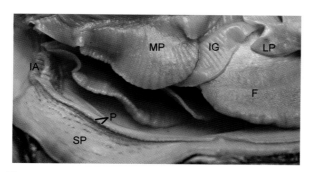

Figure 13.10. *Epioblasma propinqua* female soft anatomy. Foot (F), incurrent aperture (IA), inner gill (IG), labial palp (LP), marsupium (MP), papillae (P), spongy pad (SP). RMNH uncataloged. Tennessee, collected mid-1800s. Photograph by E. Neubert.

Outer gills marsupial; glochidia held just posterior of center, not extending completely to posterior end; marsupium outline semicircular, padded when gravid; tan (Figure 13.10).

Labial palps tan; straight dorsally, curved ventrally, pointed distally; palp length approximately 20% of gill length; palp height approximately 65% of palp length; distal 50% of palps bifurcate.

Incurrent aperture longer than excurrent aperture; supra-anal aperture longer than incurrent aperture.

Incurrent aperture length approximately 10% of shell length; tan within, without coloration basal to papillae; papillae in 2 rows, inner row simple, bifid, trifid and arborescent, outer row simple; papillae tan.

Excurrent aperture length approximately 7% of shell length; tan within, without marginal color band; margin with short, simple papillae; papillae white.

Supra-anal aperture length approximately 15% of shell length; tan within, without marginal coloration; margin smooth; mantle bridge separating supra-anal and excurrent apertures imperforate, length 25% of supra-anal length.

Specimen examined: "Tennessee" (n = 1) (specimen previously preserved).

Glochidium Description

Glochidium unknown.

Similar Species

Epioblasma propinqua resembles nominal *Epioblasma torulosa* but differs in being more inflated and having a steeper posterior slope. Also, the knobs on *E. torulosa* are usually more pronounced than the low concentric ridges of *E. propinqua*, which may be completely absent. Some female *E. propinqua* bear a resemblance to male *Epioblasma personata*. The umbo of *E. propinqua* is positioned more anteriorly than that of *E. personata*. Also, the sulcus is complete distally in male *E. personata* but disappears with a weak extrapallial swelling in female *E. propinqua*.

Male *Epioblasma propinqua* could be confused with small *Pleurobema cordatum* or *Pleurobema rubrum*. The periostracum of *E. propinqua* often has thin green rays over the shell disk. Rays of *P. cordatum* and *P. rubrum* are limited to umbos of subadults. Also the umbo cavity of *P. cordatum* is deep and that of *E. propinqua* is shallow. Both *P. cordatum* and *P. rubrum* reach considerably greater size than *E. propinqua*.

General Distribution

Epioblasma propinqua historically occurred in the Ohio and lower Wabash rivers as well as the Cumberland and Tennessee rivers. In the Cumberland River it is known historically only from the vicinity of Nashville, but archaeological material suggests that it was more widespread in that river prehistorically (Parmalee and Bogan, 1998). *Epioblasma propinqua* probably occurred in much of the Tennessee River drainage in Alabama, Kentucky and Tennessee. However, there are no historical records of the species downstream of Muscle Shoals, Alabama (Parmalee and Bogan, 1998).

Alabama and Mobile Basin Distribution

Epioblasma propinqua historically occurred in the Tennessee River across northern Alabama.

Epioblasma propinqua apparently disappeared soon after impoundment of the Tennessee River and is believed to be extinct. The most recent dated museum specimens from Alabama were collected at Muscle Shoals in 1901.

Ecology and Biology

Epioblasma propinqua occurred in shoal habitat of large rivers (Parmalee and Bogan, 1998).

Epioblasma propinqua was presumably a long-term brooder, gravid from late summer or autumn to the following summer. Its glochidial hosts are unknown. Some species of *Epioblasma* use darters (Percidae) and sculpins (Cottidae) as glochidial hosts.

Current Conservation Status and Protection

Epioblasma propinqua has been presumed extinct since at least 1970 (Stansbery, 1970a; Williams et al., 1993; Turgeon et al., 1998; Lydeard et al., 1999; Garner et al., 2004).

Remarks

The relationship between *Epioblasma propinqua* and the *Epioblasma torulosa* complex has been a matter of debate. Ortmann (1925) considered *propinqua* to be a subspecies of *E. torulosa*. However, most authors have considered them distinct (e.g., Simpson, 1914; Ball, 1922; Johnson, 1978; Parmalee and Bogan, 1998). Indeed, Morrison (1942) and Warren (1975) noted no intergrades between the two after examining large numbers of *E. torulosa* and *E. propinqua* from archaeological middens at Muscle Shoals and near Bridgeport, respectively.

Epioblasma propinqua has been found to be present, but uncommon, in prehistoric shell middens across northern Alabama (Morrison, 1942; Warren, 1975; Hughes and Parmalee, 1999). Morrison (1942) suggested that *E. propinqua* may have begun its decline prehistorically, based on diminishing numbers in archaeological middens over time.

The only known specimen of *Epioblasma propinqua* with soft tissues is housed in the National Museum of Natural History, Leiden, The Netherlands. The specimen was collected during the mid-1800s by Gerard Troost, Tennessee State Geologist.

Synonymy

Unio propinquus Lea, 1857. Lea, 1857b:83; Lea, 1862c:63, pl. 5, fig. 212; Lea, 1862d:67, pl. 5, fig. 212

Type locality: [Tennessee River,] Florence, [Lauderdale County,] Alabama, Reverend G. White; Tuscumbia, [Colbert County,] Alabama, L.B. Thornton, Esq. Lectotype, USNM 84332, length 53 mm (male), designated by Johnson (1974), is from Tennessee River, Florence, Lauderdale County, Alabama.

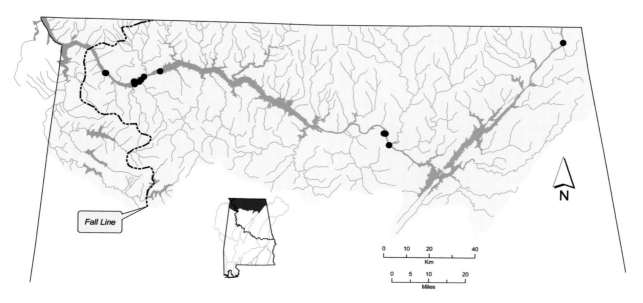

Distribution of *Epioblasma propinqua* in the Tennessee River drainage of Alabama.

Epioblasma stewardsonii (Lea, 1852)
Cumberland Leafshell

Epioblasma stewardsonii – Upper figure: female, length 33 mm, UF 269095. Tennessee River, Florence, Lauderdale County, Alabama. Lower figure: male, length 48 mm, CMNH 61.632. Tennessee River, Florence, Lauderdale County, Alabama. © Richard T. Bryant.

Shell Description

Length to 50 mm; moderately thick; slightly inflated; male outline subtriangular, female outline irregularly trapezoidal; male posterior margin narrowly rounded, female posterior margin rounded dorsally, with emargination and prominent lobe ventrally; anterior margin rounded, more narrowly rounded in females than males; dorsal margin convex; ventral margin convex, males with flattened or slightly concave area at distal end of sulcus, females with slight emargination anterior of extrapallial swelling; extrapallial swelling of females in form of a prominent,

rounded posterioventral lobe originating in the sulcus between posterior and medial ridges; posterior and medial ridges well-developed, rounded, ridges separated by wide, shallow sulcus, extrapallial swelling of females obscures sulcus distally; posterior slope steep, slightly concave; umbo broad, elevated above hinge line, umbo sculpture unknown; periostracum tawny to greenish yellow, often with faint greenish or brownish rays, extrapallial swelling of females usually darker than remainder of shell.

Pseudocardinal teeth erect, triangular, 2 slightly divergent teeth in left valve, 1 tooth in right valve, usually with accessory denticle anteriorly; lateral teeth short, longer in males than females, thick, straight, 2 in left valve, 1 in right valve, right valve also with rudimentary lateral tooth ventrally; interdentum moderately long, narrow to moderately wide; umbo cavity shallow; nacre white, occasionally with pinkish orange tint.

Soft Anatomy Description
Soft anatomy unknown.

Glochidium Description
Glochidium unknown.

Similar Species
The shell of *Epioblasma stewardsonii* resembles those of *Epioblasma flexuosa* and *Epioblasma lewisii*. Female *E. stewardsonii* differ from females of both species in having an extrapallial swelling that originates in the sulcus. The extrapallial swellings of *E. flexuosa* and *E. lewisii* lie anterior to the sulcus. Also, the extrapallial swelling of *E. stewardsonii* is blunter than those of *E. flexuosa* and *E. lewisii*. The extrapallial swelling of *E. stewardsonii* is often dark green, as it is in *E. lewisii*, but not in *E. flexuosa*. Adult *E. stewardsonii* are smaller than *E. flexuosa*, but both have thicker shells than *E. lewisii*.

General Distribution
Epioblasma stewardsonii is endemic to the Tennessee and Cumberland River drainages. It historically occurred in the Cumberland River drainage downstream of Cumberland Falls, Kentucky and Tennessee (Cicerello et al., 1991; Parmalee and Bogan, 1998). In the Tennessee River drainage it historically occurred from the headwaters in eastern Tennessee downstream to Muscle Shoals, Alabama (Parmalee and Bogan, 1998).

Alabama and Mobile Basin Distribution
Epioblasma stewardsonii occurred in the Tennessee River across northern Alabama.

Synonymy
Unio stewardsonii Lea, 1852. Lea, 1852a:252 *nomen nudum*

Epioblasma stewardsonii is believed to be extinct. The most recent dated museum record of this species from Alabama was collected in 1909.

Ecology and Biology
Epioblasma stewardsonii occurred in medium to large rivers in shoal habitat (Parmalee and Bogan, 1998).

Epioblasma stewardsonii was presumably a long-term brooder, gravid from late summer or autumn to the following summer. Its glochidial hosts are unknown. Some species of *Epioblasma* use darters (Percidae) and sculpins (Cottidae) as glochidial hosts.

Current Conservation Status and Protection
Epioblasma stewardsonii has been presumed extinct since the early 1900s (Stansbery, 1970a; Williams et al., 1993; Turgeon et al., 1998; Lydeard et al., 1999; Garner et al., 2004).

Remarks
Ortmann (1925) had access to only one specimen of *Epioblasma stewardsonii* from Alabama reaches of the Tennessee River and questioned its validity. However, the species is fairly well-represented in museum collections with material from Alabama, primarily from Muscle Shoals. *Epioblasma stewardsonii* has been reported to be present, but uncommon, in archaeological shell middens at Muscle Shoals and near Bridgeport but not at intervening sites (Morrison, 1942; Warren, 1975; Hughes and Parmalee, 1999).

The type locality of *Epioblasma stewardsonii* given by Lea was "Chattanooga River", which was the name he often applied to the Chattooga River of the upper Coosa River drainage in the Mobile Basin (Ortmann, 1918). However, neither *E. stewardsonii* nor any similar species occur anywhere in the Mobile Basin. The correct type locality may have been the Tennessee River at Chattanooga, Tennessee (Parmalee and Bogan, 1998).

Unio stewardsonii Lea, 1852. Lea, 1852b:278, pl. 23, fig. 36; Lea, 1852c:34, pl. 23, fig. 36
Type locality: Chattanooga River [Tennessee River, Chattanooga, Hamilton County,] Tennessee. Holotype, ANSP
      56572, length 31 mm (female).

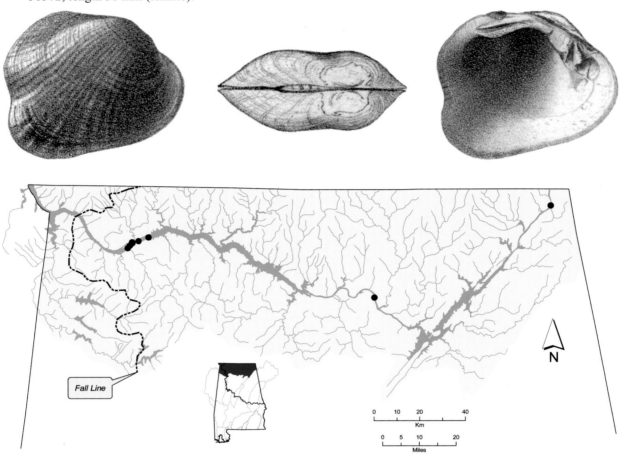

Distribution of *Epioblasma stewardsonii* in the Tennessee River drainage of Alabama.

Epioblasma torulosa (Rafinesque, 1820)
Tubercled Blossom

Epioblasma torulosa – Upper figure: female, length 57 mm, UF 64348. Holston River, Hodges, Jefferson County, Tennessee, 25 May 1914. Lower figure: male, length 50 mm, UF 64353. Tennessee River, Alabama. © Richard T. Bryant.

Shell Description

Length to 85 mm; thick anteriorly, thinner posteriorly; moderately inflated; outline elliptical to oval or trapezoidal; male posterior margin obliquely truncate to bluntly pointed, female posterior margin broadly rounded, sometimes with 1–2 shallow emarginations; anterior margin rounded; dorsal margin straight to slightly convex; male ventral margin slightly convex, with slight convex protrusion at distal end of medial ridge, female ventral margin slightly convex anteriorly, sometimes with slight emargination just anterior of extrapallial swelling; extrapallial swelling of females thin, broad, rounded distally, extending below ventral margin; posterior ridge low, wide, rounded, may have variable knobs; posterior slope moderately steep;

medial ridge narrower than posterior ridge, ventrally curved, usually with irregular, raised knobs; sulcus separating posterior and medial ridges wide, shallow, ending in emargination distally in males, in females sulcus obscured distally by extrapallial swelling; umbo slightly inflated, elevated above hinge line, turned slightly anteriad, umbo sculpture weak corrugations; periostracum smooth, often shiny, tawny to yellowish green or dark greenish brown, with numerous variable green rays.

Pseudocardinal teeth erect, triangular, 2 divergent teeth in left valve, 1 tooth in right valve, often with accessory denticle anteriorly and/or posteriorly; lateral teeth moderately long, straight, 2 in left valve, 1 in right valve, right valve may also have rudimentary lateral tooth ventrally; interdentum short, narrow to moderately wide; umbo cavity open, shallow; nacre usually white, may be salmon.

Soft Anatomy Description

The following is based on the headwater *gubernaculum* form of *Epioblasma torulosa*.

Mantle tan, outside of apertures rusty brown, more pale outside of pallial line; visceral mass tan; foot tan.

No females were available for description of mantle modification. Lea (1863d) reported the mantle margin of the nominal form, just ventral to the incurrent aperture, to be "extended into a flap offset on the posterior inferior margin, blackish on the border, the inner double edges having numerous small papillae". Lea described the mantle outside of the margin adjacent to the flap as brown with black spots. The sex of the individual on which Lea's description was based is unclear, but he later discussed the "fleshy flap" being "always found in the female of this species". Information inscribed on the inner surface of female shells included in a lot from the Holston River, collected by A.E. Ortmann in 1914, describes their "pads" as "black-gray" or "gray".

Gills tan; dorsal margin straight, ventral margin elongate convex; gill length approximately 60% of shell length; gill height approximately 30% of gill length; outer gill height approximately 80% of inner gill height; inner lamellae of inner gills completely connected to visceral mass. Lea (1863d) reported gills of nominal *Epioblasma torulosa* to be "light liver-brown, rather thick" and completely connected to the visceral mass dorsally.

No gravid females were available for marsupium description. Lea (1863d) reported outer gills of nominal *Epioblasma torulosa* to be marsupial, with glochidia held in the posterior part.

Labial palps tan, darker marginally; straight dorsally, convex ventrally, bluntly pointed distally; palp length approximately 25% of gill length; palp height approximately 60% of palp length; distal 40% of palps bifurcate.

Incurrent and excurrent apertures approximately equal in length; supra-anal aperture longer than incurrent and excurrent apertures.

Incurrent aperture length approximately 15% of shell length; tan within, rusty brown basal to papillae; papillae mostly long, slender, simple, occasional bifid and small arborescent; papillae rusty brown.

Excurrent aperture length approximately 15% of shell length; tan within, marginal color band rusty brown; margin with small, simple papillae, little more than crenulations; papillae tan.

Supra-anal aperture length approximately 20% of shell length; tan within, with thin rusty brown marginal band; margin minutely crenulate adjacent to mantle bridge, becoming smooth dorsally; mantle bridge separating supra-anal and excurrent apertures imperforate, length approximately 20% of supra-anal length.

Specimen examined: Clinch River, Tennessee (n = 1) (specimen previously preserved).

Glochidium Description

No nominal *Epioblasma torulosa* glochidia were available for description. The following is a description of the *rangiana* form of *E. torulosa* from Ohio.

Length 238–258 µm; height 210–238 µm; with triangular supernumerary hooks; outline depressed subelliptical; dorsal margin straight, ventral margin broadly rounded, anterior and posterior margins broadly rounded (Lea, 1858d; Hoggarth, 1999).

Similar Species

Nominal *Epioblasma torulosa* resembles *Epioblasma cincinnatiensis* and *Epioblasma propinqua*. It differs from *E. propinqua* in being more elongate and less inflated, with a shallower posterior slope and typically better-developed knobs on the posterior and medial ridges. Nominal *E. torulosa* differs from *E. cincinnatiensis* in having large, low knobs on the posterior and medial ridges, instead of small, well-defined tubercles.

General Distribution

Epioblasma torulosa is known from some tributaries of lakes Erie, Huron and Michigan in Indiana, Ohio and Michigan, as well as the Canadian province of Ontario (Johnson, 1978; Clarke, 1981a; Parmalee and Bogan, 1998). It also historically occurred in much of the Ohio River drainage from headwaters in Pennsylvania to the mouth of the Ohio River (Ortmann, 1909a; Cummings and Mayer, 1992). *Epioblasma torulosa* is also known from Tennessee reaches of the Cumberland River (Parmalee and Bogan, 1998), and occurred throughout the Tennessee River drainage (Parmalee and Bogan, 1998).

Three forms of *Epioblasma torulosa* are often recognized. The *rangiana* form is known from the Great Lakes drainage and some parts of the Ohio River drainage (Watters, 1993a). The nominal form of *E. torulosa* is the large river representative of the species. The *gubernaculum* form occurred in headwaters of the Tennessee River drainage (Ortmann, 1918). Neither the nominal form of *E. torulosa* nor the *gubernaculum* form are known to be extant. The *rangiana* form is extant in isolated Ohio River tributary systems.

Alabama and Mobile Basin Distribution

Nominal *Epioblasma torulosa* historically occurred in the Tennessee River across northern Alabama.

Epioblasma torulosa appears to have been eliminated from Alabama with impoundment of the Tennessee River. The most recent dated museum record of this species from the state was collected at Muscle Shoals in 1904.

Ecology and Biology

Nominal *Epioblasma torulosa* occurred in shoal habitats of large rivers (Parmalee and Bogan, 1998).

Nothing is known about the biology of nominal *Epioblasma torulosa*. Females of the *rangiana* form of *E. torulosa* have smooth, white pads within their extrapallial swellings but apparently do not have true microlures, as do some other *Epioblasma* species. The microlure of the *rangiana* form is replaced by a vestigial "nub" of tissue (Jones, 2004).

Epioblasma torulosa is presumably a long-term brooder, gravid from late summer or autumn to the following summer. Ortmann (1919) reported the *rangiana* form to become gravid by September. Glochidial hosts of nominal *E. torulosa* are unknown, but *Cottus bairdii* (Mottled Sculpin) (Cottidae); and *Etheostoma camurum* (Bluebreast Darter) and *Etheostoma zonale* (Banded Darter) (Percidae) have been reported to serve as hosts of the *rangiana* form in laboratory trials (Watters, 1996a). *Salmo trutta* (Brown Trout) (Salmonidae), a nonindigenous fish, was also found to serve as a host under laboratory conditions (Watters, 1996a).

Current Conservation Status and Protection

Nominal *Epioblasma torulosa* was reported to be endangered throughout its range by Stansbery (1970a)

but is now presumed extinct (Williams et al., 1993; Turgeon et al., 1998; Lydeard et al., 1999; Garner et al., 2004). Nominal *E. torulosa* and *Epioblasma torulosa gubernaculum* were listed as endangered under the federal Endangered Species Act in 1976. *Epioblasma torulosa rangiana* was listed as endangered in 1993.

In 2001 nominal *Epioblasma torulosa* was included on a list of species approved for a Nonessential Experimental Population in Wilson Dam tailwaters on the Tennessee River. Should a population of this form be located and successfully propagated, progeny of such a program could be reintroduced there.

Remarks

Epioblasma torulosa has been reported from archaeological shell middens along the Tennessee River across northern Alabama, where it was abundant at some sites (Morrison, 1942; Warren, 1975; Hughes and Parmalee, 1999). Morrison (1942) suggested that it was one of the most important species in the food supply of prehistoric Native Americans at Muscle Shoals. Ortmann (1918) reported *E. torulosa* to be abundant historically at Muscle Shoals, disappearing soon after impoundment of the river.

The high degree of variation in shell morphology observed in *Epioblasma torulosa* has resulted in descriptions of several "varieties" and shuffling of species and subspecies. The three forms—*rangiana*, *gubernaculum* and nominal *torulosa*—are often referred to as subspecies (e.g., Cicerello and Schuster, 2003) or species (e.g., Cummings and Mayer, 1992). Bogan (1997) established the identity of *Epioblasma biloba* as the *rangiana* form of *E. torulosa*. The name *biloba* Rafinesque, 1831, has priority over *rangiana* Lea, 1838. Since subspecies are not herein recognized, the most frequently used name of *rangiana* was retained for this form.

Synonymy

Amblema torulosa Rafinesque, 1820. Rafinesque, 1820:314, pl. 82, figs. 11, 12
Type locality: Ohio and Kentucky rivers. Lectotype, ANSP 20218, length 65 mm (female), designated by Johnson and Baker (1973).

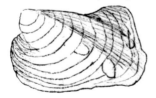 

Amblema gibbosa Rafinesque, 1820. Rafinesque, 1820:315
Amblema gibbosa var. *difformis* Rafinesque, 1820. Rafinesque, 1820:315
Amblema gibbosa var. *olivacea* Rafinesque, 1820. Rafinesque, 1820:315
Amblema gibbosa var. *radiata* Rafinesque, 1820. Rafinesque, 1820:315
Amblema torulosa var. *angulata* Rafinesque, 1820. Rafinesque, 1820:315

Epioblasma biloba Rafinesque, 1831. Rafinesque, 1831:2

Comment: Bogan (1997) has established the identity of *Epioblasma biloba* as *Epioblasma torulosa rangiana*. *Epioblasma biloba* Rafinesque, 1831, has priority over *rangiana* Lea, 1838, but *Amblema gibbosa* Rafinesque, 1820, may be the earliest name for the subspecies in the Ohio Basin.

Unio perplexus Lea, 1831, *non* Say, 1829. Lea, 1831:112, pl. 17, fig. 42; Lea, 1834b:122, pl. 17, fig. 42

Unio rangianus Lea, 1838. Lea, 1838a:95, pl. 18, fig. 56; Lea, 1838c:95, pl. 18, fig. 56

Unio gubernaculum Reeve, 1865. Reeve, 1865:[no pagination], species 146, pl. 28

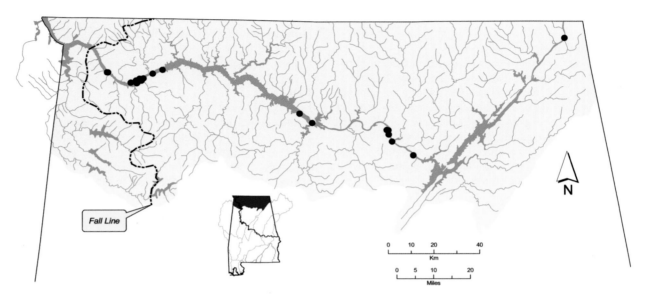

Distribution of *Epioblasma torulosa* in the Tennessee River drainage of Alabama.

Epioblasma triquetra (Rafinesque, 1820)
Snuffbox

Epioblasma triquetra – Upper figure: female, length 43 mm, UNA 350.11. Middle figure: male, length 47 mm, UNA 350.11. Cedar Creek, creek mile 10, Franklin County, Alabama, 1966. Lower figure: male, length 39 mm, UF 64367. Tennessee River, Muscle Shoals, Lauderdale County, Alabama. © Richard T. Bryant.

Shell Description

Length of males to 70 mm, length of females to 45 mm; moderately thick; inflated; outline triangular; posterior margin bluntly pointed, point often more acute in females; anterior margin narrowly rounded; dorsal margin convex; male ventral margin convex, female ventral margin almost straight; females with narrow, inflated, radially striate extrapallial swelling along posterior ridge, swelling often projects slightly below ventral margin, margin usually serrate, often with slight sulcus just anterior of swelling; posterior ridge well-defined, acute; posterior slope flat, steep, adorned with radial striations; umbo inflated, elevated above hinge line, turned slightly anteriad, umbo sculpture faint, somewhat double-looped ridges; periostracum smooth, shiny, yellow to yellowish green, with dark green rays usually broken into triangles, chevrons or squares.

Pseudocardinal teeth elevated, compressed, 2 obliquely angled teeth in left valve, anterior tooth often larger than posterior tooth, 1 large tooth in right valve, with well-developed accessory denticle anteriorly; lateral teeth short, straight, 2 in left valve, 1 in right valve; interdentum moderately long, narrow to very narrow; umbo cavity wide, deep; nacre white, often with silvery luster, may have grayish tint in umbo cavity.

Soft Anatomy Description

Mantle pale tan, outside of apertures creamy white to tan, with regular, short, dark brown bands along margin, giving appearance of a dashed line; visceral mass pale tan to creamy white or pearly white; foot pale orange.

Females with spongy, rusty tan pad filling extrapallial swelling, pad heavily mottled with dark brown and black; mantle margin inside pallial swelling with weak fold just ventral to incurrent aperture; length of fold approximately 25% of shell length; margin papillate, papillae simple, well-developed, widely spaced; mantle proximal to fold black. Lea (1863d) described this area as a "fleshy, brownish black enlargement" with "numerous rather large papillae" and later commented that "the fleshy enlargement on the flap is hard". Ortmann (1912a) reported males to have a few small papillae just ventral to the incurrent aperture.

Gills tan; dorsal margin slightly sinuous, steeply sloped from umbo to near incurrent aperture, ventral margin convex, may be slightly scalloped; gill length 50–55% of shell length; gill height 55–65% of gill length; outer gill height 75–90% of inner gill height; inner lamellae of inner gills attached to visceral mass only anteriorly. Lea (1863d) and Ortmann (1912a) reported dorsal margin of gills to be completely connected to the visceral mass. Lea (1863d) reported the gills to be white, with a crenulate ventral margin.

Outer gills marsupial; glochidia held in posterior 50% of gill; marsupium ovate, slightly narrower posteriorly, distended when gravid; creamy white, distal margin opaque white. Ortmann (1912a) reported the marsupial area to be reniform in outline.

Labial palps creamy white to pale tan; straight to slightly concave dorsally, convex ventrally, narrowly rounded distally; palp length 20–35% of gill length; palp height 65–80% of palp length; distal 40–60% of palps bifurcate.

Incurrent aperture longer than excurrent aperture; supra-anal aperture longer than incurrent aperture.

Incurrent aperture length 15–20% of shell length; creamy white within, sometimes with grayish or grayish brown cast, without coloration basal to papillae; papillae irregularly arborescent, branches variable; papillae rusty tan to rusty orange, sometimes with sparse, dark brown or black specks, papillae tips white.

Excurrent aperture length 10–15% of shell length; creamy white with grayish or brownish cast within, marginal color band rusty tan to rusty orange with dark brown specks or crude bands perpendicular to margin; margin papillate, papillae simple, small but well-developed; papillae rusty tan to rusty orange. Lea (1863d) and Ortmann (1912a) reported the excurrent aperture to be crenulate.

Supra-anal aperture length 20–25% of shell length; creamy white within, very thin marginal color band of gray or black mottling; margin smooth; mantle bridge separating supra-anal and excurrent apertures may have small perforation, length 10–20% of supra-anal length.

Specimens examined: Clinch River, Tennessee (n = 3).

Glochidium Description

Length 208–217 μm; height 205–214 μm; with triangular supernumerary hooks; outline depressed subelliptical; dorsal margin straight, ventral margin broadly rounded, anterior and posterior margins convex (Hoggarth, 1999).

Similar Species

Some large, inflated *Epioblasma triquetra* may resemble *Epioblasma arcaeformis*. The posterior slope of *E. triquetra* is steeper than that of *E. arcaeformis*. The green rays on the periostracum of *E. triquetra* are wider and broken into squares, triangles or chevrons, while those of *E. arcaeformis* are thin and unbroken.

Epioblasma triquetra, especially individuals from large rivers, may resemble male *Truncilla truncata* but are usually more elongate and less triangular in outline with less deeply rounded ventral margins. Also, *E. triquetra* is often more inflated posteriorly than *T. truncata*.

Large, inflated *Epioblasma triquetra* may also resemble *Alasmidonta marginata* but have a relatively thicker shell, are less elongate and have lateral teeth as well as better-developed pseudocardinal teeth.

General Distribution

Epioblasma triquetra is the most widespread species of *Epioblasma*. It is known from tributaries of lakes Erie, Huron, Michigan and St. Clair in Indiana, Michigan, New York, Ohio, Pennsylvania and Wisconsin (Johnson, 1978). In the Mississippi Basin, its historical range extended from Minnesota (Graf, 1997) and southern Wisconsin (Cummings and Mayer,

1992) south to Missouri (Oesch, 1995), and from headwaters of the Ohio River drainage in western Pennsylvania (Ortmann, 1909a) west to eastern Kansas (Murray and Leonard, 1962). *Epioblasma triquetra* was historically widespread in the Cumberland River drainage downstream of Cumberland Falls (Cicerello et al., 1991; Parmalee and Bogan, 1998). It also occurred in most of the Tennessee River drainage (Ahlstedt, 1992a, 1992b; Parmalee and Bogan, 1998). A disjunct population of *E. triquetra* is known from upper reaches of the White River drainage in Arkansas and Missouri (Johnson, 1978; Harris and Gordon, 1990; Oesch, 1995).

Alabama and Mobile Basin Distribution

Epioblasma triquetra historically occurred in the Tennessee River across northern Alabama and in some large tributaries. Records from Tennessee reaches of the Elk River suggest that it occurred in Alabama reaches as well, but there are no museum records.

The only known extant Alabama population of *Epioblasma triquetra* is in the Paint Rock River, where the species is rare.

Ecology and Biology

Epioblasma triquetra occurs in shoal habitat of small to large rivers. It is usually found buried in gravel or sand substrate with only the apertures exposed (Ortmann, 1919).

Epioblasma triquetra is a long-term brooder, gravid from September until May (Ortmann, 1919). Fishes found to serve as hosts for *E. triquetra* glochidia in laboratory trials include *Cottus baileyi* (Black Sculpin), *Cottus bairdii* (Mottled Sculpin) and *Cottus carolinae* (Banded Sculpin) (Cottidae); *Fundulus olivaceus* (Blackspotted Topminnow) (Fundulidae); and *Percina caprodes* (Logperch) and *Percina maculata* (Blackside Darter) (Percidae) (Yeager and Saylor, 1995; Hillegass and Hove, 1997; Barnhart et al., 1998; Jones and Neves, 2000). Glochidia of *E. triquetra* have been successfully transformed on *Percina roanoka* (Roanoke Darter) (Percidae) under laboratory conditions, but the two species do not occur sympatrically (Jones and Neves, 2000). Jones and Neves (2000) reported a more extended excystment period for *E. triquetra* juveniles transformed from glochidia collected during autumn or early winter than for those collected during spring, suggesting that they may not be fully mature until spring.

Current Conservation Status and Protection

Epioblasma triquetra was listed as endangered throughout its range by Williams et al. (1993) and imperiled in Alabama by Lydeard et al. (1999). Garner et al. (2004) designated *E. triquetra* a species of highest conservation concern in the state.

Remarks

The only reports of *Epioblasma triquetra* from archaeological studies have been from near Bridgeport in northeastern Alabama, where it was very rare (Warren, 1975). Morrison (1942) speculated that its absence in middens at Muscle Shoals is due to its preference for "deeper-water habitat" but provided no evidence or discussion.

Synonymy

Truncilla triqueter Rafinesque, 1820. Rafinesque, 1820:300, pl. 81, figs. 1–4
Type locality: Falls of the Ohio [River near Louisville, Jefferson County, Kentucky]. Lectotype, ANSP 20231, length 55 mm (male), designated by Johnson and Baker (1973).

 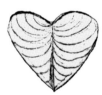  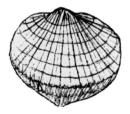

Unio cuneatus Swainson, 1823. Swainson, 1823:112
Unio triangularis Barnes, 1823. Barnes, 1823:272, pl. 13, figs. 17a, b
Unio formosus Lea, 1831. Lea, 1831:111, pl. 16, fig. 41; Lea, 1834b:121, pl. 16, fig. 41
Unio triangularis var. *longisculus* De Gregorio, 1914. De Gregorio, 1914:40, pl. 4, fig. 5
Unio triangularis var. *pergibosus* De Gregorio, 1914. De Gregorio, 1914:40, pl. 4, fig. 4

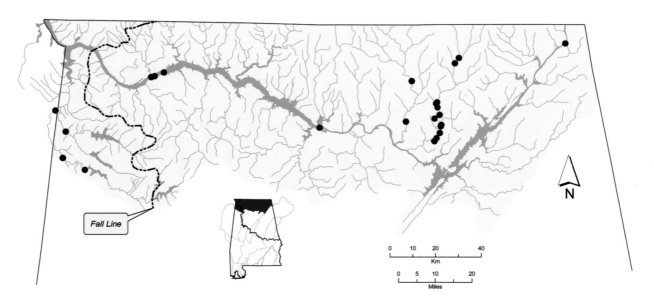

Distribution of *Epioblasma triquetra* in the Tennessee River drainage of Alabama.

Epioblasma turgidula (Lea, 1858)
Turgid Blossom

Epioblasma turgidula – Upper figure: female, length 39 mm, USNM 84944. Florence, [Lauderdale County,] Alabama. Lower figure: male, length 42 mm, CMNH 61.6750. Holston River, Holston Station, Grainger County, Tennessee, 15 September 1913. © Richard T. Bryant.

Shell Description

Length to 45 mm; moderately thin; moderately compressed; outline oval to irregularly elliptical; male posterior margin narrowly rounded, female posterior margin broadly rounded; anterior margin evenly rounded, females may be more narrowly rounded anteriorly; dorsal margin straight; male ventral margin broadly rounded, female ventral margin straight, sometimes with slight emargination just anterior of extrapallial swelling; females with large, thin, rounded, slightly inflated extrapallial swelling posterioventrally, may be separated from remainder of shell by shallow sulcus, ventral margin of swelling may be serrate; posterior ridge low, often doubled in males, usually obscured by extrapallial swelling in females; male posterior slope moderately steep, slightly concave, female posterior slope less steep; umbo moderately inflated, broad, elevated above

hinge line, umbo sculpture unknown; periostracum smooth, often shiny, yellowish green, with thin green rays.

Pseudocardinal teeth small, triangular, 2 slightly compressed teeth in left valve, 1 triangular tooth in right valve, may be slightly compressed, often with accessory denticle anteriorly; lateral teeth short, straight to slightly curved, 2 in left valve, 1 in right valve; interdentum short, very narrow; umbo cavity shallow; nacre white.

Soft Anatomy Description

Mantle tan, with faint rusty tan or rusty brown outside of apertures; visceral mass tan; foot tan.

No females were available for description of mantle modification. Many, possibly all, *Epioblasma* females have some form of host attraction lure.

Gills pale tan; dorsal margin slightly sinuous, ventral margin convex; gill length 50–60% of shell length; gill height 65–85% of gill length; outer gill height 75–85% of inner gill height; inner lamellae of inner gills completely connected to visceral mass.

No gravid females were available for marsupium description; probably similar to other *Epioblasma* species, with glochidia held in posterior part of outer gills, marsupium padded when gravid.

Labial palps tan; straight dorsally, curved ventrally, narrowly rounded distally; palp length 20–30% of gill length; palp height approximately 50% of palp length; distal 35% of palps bifurcate.

Incurrent aperture longer than excurrent aperture; incurrent aperture longer than or equal to supra-anal aperture; supra-anal aperture usually longer than excurrent aperture.

Incurrent aperture length 15–20% of shell length; tan within, may have golden tan basal to papillae; papillae in 2 rows, inner row larger, long, slender, simple; papillae golden tan.

Excurrent aperture length 10–15% of shell length; pale tan within, marginal color band some combination of golden tan, rusty tan and rusty brown, without distinct pattern; margin with small, simple papillae, may be well-developed or little more than crenulations; papillae golden tan.

Supra-anal aperture length 10–20% of shell length; pale tan with very narrow rusty brown marginal band; margin smooth, but may have small undulations; mantle bridge separating supra-anal and excurrent apertures imperforate, length 30–40% of supra-anal length.

Specimens examined: Duck River, Tennessee (n = 2) (specimens previously preserved).

Glochidium Description
Glochidium unknown.

Similar Species
The shell of *Epioblasma turgidula* most closely resembles that of *Epioblasma biemarginata*, but female *E. turgidula* have a more expansive and wrinkled extrapallial swelling. Male *E. turgidula* are more elongate posteriorly than male *E. biemarginata*. *Epioblasma turgidula* also resembles *Epioblasma capsaeformis*, *Epioblasma florentina* and *Epioblasma* sp. cf. *capsaeformis*. Female *E. turgidula* are more inflated posteriorly, with the extrapallial swelling obscuring the posterior ridge. The posterior margin of female *E. turgidula* is less broadly rounded than those of female *E. capsaeformis*, *E. florentina* and *Epioblasma* sp.

General Distribution
Epioblasma turgidula is endemic to the Tennessee and Cumberland River drainages in Alabama and Tennessee. One specimen of Isaac Lea's labeled "Cumberland River" is the only evidence that *E. turgidula* occurred in that drainage. In the Tennessee River it historically occurred from headwaters in northeastern Tennessee to Muscle Shoals, Alabama. Johnson (1978) included a record in upper St. Francis River, Arkansas, on the distributional map of *E. turgidula*. However, no specific locality data were provided.

Alabama and Mobile Basin Distribution
All Alabama records of *Epioblasma turgidula* from the Tennessee River proper are from Muscle Shoals. However, its presence in eastern Tennessee suggests that it occurred across northern Alabama. *Epioblasma turgidula* was also known from a few tributaries. There are records of *E. turgidula* from Tennessee reaches of the Elk River in Tennessee, so it likely occurred in Alabama reaches of the river as well.

Epioblasma turgidula is believed to be extinct. None of the museum specimens from Alabama have collection dates, but all appear to have been collected during the late 1800s or early 1900s.

Ecology and Biology
Epioblasma turgidula occurred in large creeks and small to large rivers in shoal habitat (Parmalee and Bogan, 1998).

Epioblasma turgidula was presumably a long-term brooder, gravid from late summer or autumn to the following spring or summer. Its glochidial hosts are unknown. Some species of *Epioblasma* use darters (Percidae) and sculpins (Cottidae) for glochidial hosts.

Current Conservation Status and Protection
Epioblasma turgidula was considered endangered throughout its range and possibly extinct by Stansbery (1970a). It was listed as extinct by Williams et al. (1993), Turgeon et al. (1998), Lydeard et al. (1999) and Garner et al. (2004). The most recent reports of this species were from the Elk River, Tennessee, during the late 1960s and upper reaches of the Duck River in 1972 (Stansbery, 1970a, 1976). *Epioblasma turgidula* was listed as endangered under the federal Endangered Species Act in 1976.

In 2001 *Epioblasma turgidula* was included on a list of species approved for a Nonessential Experimental Population in tailwaters of Wilson Dam on the Tennessee River. Should a population of *E. turgidula* be located and successfully propagated, progeny of such a program could be reintroduced there.

Remarks

Simpson (1914) synonymized *Epioblasma turgidula* under *Epioblasma florentina*. However, Ortmann (1918) considered it a close relative of *Epioblasma biemarginata*. Stansbery (1970a) considered *turgidula* to be a subspecies of *E. biemarginata* but later afforded it specific status (1976), stating that intermediate specimens between the two are lacking though they occurred sympatrically at Muscle Shoals.

Epioblasma turgidula has not been reported from archaeological excavations in Alabama (Morrison, 1942; Warren, 1975; Hughes and Parmalee, 1999), suggesting that it was rare or difficult to harvest prehistorically.

Synonymy

Unio turgidulus Lea, 1858. Lea, 1858a:40; Lea, 1862b:62, pl. 5, fig. 211; Lea, 1862d:66, pl. 5, fig. 211
Type locality: Cumberland River, Tennessee, Dr. Troost and T.C. Downie; [Tennessee River,] Florence, [Lauderdale County,] Alabama. Lectotype, USNM 84946, length 42 mm (male), designated by Johnson (1974), is from Cumberland River, Tennessee.

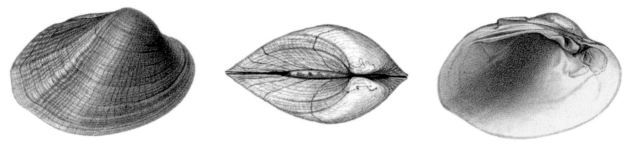

Unio deviatus Reeve, 1864. Reeve, 1864:[no pagination], species 61, pl. 15
Truncilla lefevrei Utterback, 1916. Utterback, 1916b:455, pl. 6, figs. 13a–d, pl. 28, figs. 108a–d

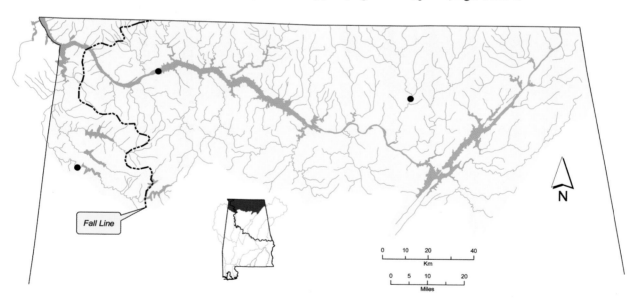

Distribution of *Epioblasma turgidula* in the Tennessee River drainage of Alabama.

Epioblasma sp. cf. *capsaeformis*
Duck River Oyster Mussel

Epioblasma sp. cf. *capsaeformis* – Upper figure: female, length 54 mm, UF 226003. Lower figure: male, length 53 mm, UF 226003. Duck River, Hickman County, Tennessee. © Richard T. Bryant.

Shell Description

Length to 60 mm; moderately thick anteriorly, thin posteriorly; moderately compressed; male outline irregularly oval to subelliptical, female outline oval, with large, rounded extrapallial swelling posterioventrally; male posterior margin narrowly rounded, female posterior margin broadly rounded with emargination just dorsal to extrapallial swelling; anterior margin evenly rounded; dorsal margin straight; ventral margin convex, females with emargination just anterior of extrapallial swelling, distal margin of extrapallial swelling extending well below ventral margin, usually serrate; posterior ridge low, often doubled in males, obscured by extrapallial swelling in females; shallow radial depression usually present just anterior of posterior ridge in males, sulcus obscured by extrapallial swelling in females, females may have slight sulcus just posterior of extrapallial swelling; male posterior slope moderately steep, flat to slightly concave, female posterior slope gently rounded; umbo moderately inflated, elevated above hinge line, umbo sculpture weak, parallel loops; periostracum smooth, somewhat shiny, tawny to olive brown or dark brown, with numerous thin green rays, extrapallial swelling of females very dark greenish brown to almost black.

Pseudocardinal teeth small, triangular, 2 somewhat compressed teeth in left valve, with oblique gap between them, 1 triangular tooth in right valve, may be compressed, may have accessory denticle anteriorly; lateral teeth short, straight to slightly curved, 2 in left valve, 1 in right valve; interdentum short, very narrow; umbo cavity shallow; nacre white to bluish white.

Soft Anatomy Description

The soft anatomy of *Epioblasma* sp. has never been described in detail. Jones (2004) reported the mantle display of females to consist of a pad within the extrapallial swelling, against which a microlure is displayed. The pad is spongy and dark purple to slate gray. The microlure is tan and protrudes from a small invagination just ventral to the incurrent aperture.

Glochidium Description

Length 248 μm; height 234 μm; outline depressed subelliptical; dorsal margin straight and long, ventral margin broadly rounded, anterior and posterior margins convex, posterior margin slightly more convex than anterior margin (J.W. Jones, personal communication).

Similar Species

Epioblasma sp. closely resembles *Epioblasma capsaeformis*. Differences in shell morphology are subtle. The extrapallial swelling of female *Epioblasma* sp. is slightly larger and more darkly pigmented than those of *E. capsaeformis* females. Tissues within the extrapallial swelling of *Epioblasma* sp. are dark purple to slate gray in life, with a spongy texture. A single microlure is displayed from an invagination near the posteriodorsal side of the mantle pad. Tissues within the extrapallial swelling of *E. capsaeformis* are bluish white in life, with a bluish to light gray microlure displayed from the posteriodorsal side of each mantle pad (Jones, 2004).

Epioblasma sp. also closely resembles *Epioblasma florentina*. The shell can be separated from that of *E. florentina* primarily by periostracum color. The periostracum of both species is yellowish green to brownish green, with thin green rays over the entire shell. *Epioblasma florentina* is usually uniform in color, but *Epioblasma* sp. females typically have an extrapallial swelling that is dark green to almost black, with the darker coloration continuing dorsally toward the umbo. Though not as obvious as in females, male *Epioblasma* sp. also have a wide, dark green band just anterior to the posterior ridge. Tissues within the extrapallial swelling of *Epioblasma* sp. are spongy and dark purple to slate gray, whereas extrapallial swelling tissues of *E. florentina* (the headwater *walkeri* form) are pustulose and gray with black mottling. Both species display only one microlure, but that of *Epioblasma* sp. is tan and that of *E. florentina* is dark brown to black (Jones, 2004).

Epioblasma sp. bears a superficial resemblance to *Epioblasma biemarginata* and *Epioblasma turgidula*. However, females of those species have relatively smaller extrapallial swellings than *Epioblasma* sp. females. Male *E. biemarginata* have a well-defined sulcus, whereas male *Epioblasma* sp. usually do not, though they usually have a slight radial depression just anterior of the posterior ridge.

General Distribution

Epioblasma sp. is endemic to the lower Tennessee River drainage. It historically occurred from the Duck and Buffalo rivers and the Tennessee River upstream to Muscle Shoals (Jones, 2004).

Epioblasma sp. is known to be extant only in the Duck River, Tennessee (Jones, 2004).

Alabama and Mobile Basin Distribution

Epioblasma sp. historically occurred in the Tennessee River at Muscle Shoals and Shoal Creek, Lauderdale County.

Epioblasma sp. is extirpated from Alabama. However, a reintroduction program is underway in tailwaters of Wilson Dam.

Ecology and Biology

Epioblasma sp. is a species of shoal habitat in small to large rivers, where it occurs on silt-free gravel and sand substrates.

Epioblasma sp. is a long-term brooder, gravid from late summer or autumn to the following spring or summer. Annual fecundity of *Epioblasma* sp. in the Duck River, Tennessee, was reported to average 18,757 glochidia per female (n = 6) and range from 6,668 to 38,716 glochidia per female (Jones, 2004). However, the largest females were not used for fecundity estimates, and Jones (2004) suggested that maximum fecundity may exceed 50,000 glochidia per individual. Females attract potential glochidial hosts with display of a tan microlure against dark purple to slate gray pads in the extrapallial swelling. Only one microlure is prominently displayed, protruding from the invagination and moved in a side-to-side manner, resembling a caddisfly larva (Jones, 2004).

Fishes reported to serve as glochidial hosts of *Epioblasma* sp. in laboratory trials include *Etheostoma blennioides* (Greenside Darter), *Etheostoma flabellare* (Fantail Darter) and *Etheostoma rufilineatum* (Redline Darter) (Percidae). Greatest numbers of juveniles were recovered from *E. flabellare* (Jones, 2004).

Current Conservation Status and Protection

Epioblasma sp. was recently split from *Epioblasma capsaeformis* but is not yet formally described. Therefore, it has not appeared in previous conservation status reviews. *Epioblasma capsaeformis* (which included *Epioblasma* sp.) was reported as endangered throughout its range by Williams et al. (1993) and in Alabama by Lydeard et al. (1999). Garner et al. (2004) designated *E. capsaeformis* a species of highest conservation concern in Alabama. *Epioblasma capsaeformis* was listed as endangered under the federal Endangered Species Act in 1997.

In 2001 *Epioblasma capsaeformis* (= *Epioblasma* sp. in part) was included on a list of species approved for a Nonessential Experimental Population in Wilson Dam tailwaters on the Tennessee River. A trial transplantation of 80 *Epioblasma* sp. individuals was carried out in 2003. Survival of 16 percent of transplanted individuals was observed in a follow-up evaluation in 2005 (J.T. Garner, unpublished data).

Remarks

The abundance of *Epioblasma* sp. in prehistoric middens is difficult to assess due to its thin, brittle shell and conchological similarity to *Epioblasma capsaeformis*. Reference to "*E. capsaeformis*" in archaeological studies at Muscle Shoals probably included both *Epioblasma* sp. and *E. capsaeformis*. It has been reported to be present, but rare, in most prehistoric middens studied along the Tennessee River in Alabama (Morrison, 1942; Warren, 1975; Hughes and Parmalee, 1999). There are no historically collected specimens of *Epioblasma* sp. from upstream of Muscle Shoals (Jones, 2004). Therefore, prehistoric material from upstream of Muscle Shoals may have been comprised exclusively of *E. capsaeformis*.

Synonymy

There are no names available for this undescribed species.

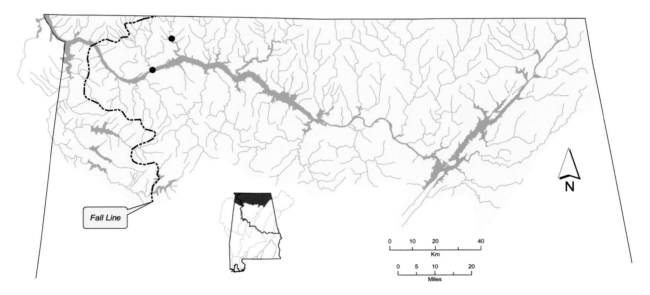

Distribution of *Epioblasma* sp. cf. *capsaeformis* in the Tennessee River drainage of Alabama.

Genus *Fusconaia*

Fusconaia Simpson, 1900, occurs in the Hudson Bay, Great Lakes and Mississippi basins, Atlantic Coast drainages from Virginia to Georgia, and Gulf Coast drainages from northern Florida to Texas. Turgeon et al. (1998) listed 13 species in *Fusconaia*. One species, *Fusconaia apalachicola* Williams and Fradkin, 1999, has been described since Turgeon et al. (1998). Two species, *Obovaria rotulata* (Wright, 1899), and *Quincuncina burkei* Walker *in* Ortmann and Walker, 1922, were herein moved into *Fusconaia* based on shell morphology and/or genetic analysis. Two species were removed from *Fusconaia*: *F. barnesiana* (Lea, 1838), to *Pleuronaia* Frierson, 1927, and *F. succissa* (Lea, 1852), to *Quadrula* Rafinesque, 1820. Nine species are herein reported from Alabama. There are two specimens of what appear to be an undescribed species of *Fusconaia* from the Chattahoochee River (Figure 13.11).

Obovaria Rafinesque, 1820, actually has priority over *Fusconaia* Simpson, 1900, for this species group with the type species *O. obovalis* Rafinesque, 1820 (= *Unio ebenus* Lea, 1831). A petition for conservation of modern use of *Fusconaia* and *Obovaria* has been submitted to the ICZN (Case 3353) (see *Obovaria* genus account) (Bogan et al., 2006).

Fusconaia barnesiana along with *Lexingtonia dolabelloides* (Lea, 1840), and *Pleurobema gibberum* (Lea, 1838) were recently recognized as a distinct clade (Campbell et al., 2005). These taxa are now recognized as belonging to *Pleuronaia* Frierson, 1927.

Based on DNA analyses, *Unio subplana* Conrad, 1837, was found to be genetically indistinguishable from *Fusconaia masoni* (Conrad, 1834) (Bogan et al., unpublished data). Since *U. subplana* is the type species of *Lexingtonia*, this genus is a junior synonym of *Fusconaia* (see *Pleuronaia* genus account).

The genus *Quincuncina* Ortmann *in* Ortmann and Walker, 1922, was diagnosed based on shell sculpture. However, Ortmann noted that soft anatomy was almost identical to that of *Fusconaia*. Lydeard et al. (2000) presented phylogenetic analyses supporting inclusion of *Q. burkei* Walker *in* Ortmann and Walker, 1922, the type species of *Quincuncina*, in a clade with *Fusconaia flava* (Rafinesque, 1820), type species of that genus. Thus, *Quincuncina* is a junior synonym of *Fusconaia*. This leaves *Quincuncina mitchelli* (Simpson, 1896), a Texas endemic, in need of modern generic placement.

Type Species
Unio trigonus Lea, 1831 (= *Unio (Quadrula) flava* Rafinesque, 1820)

Diagnosis
Shell quadrate to triangular; shell surface typically smooth; umbo elevated above hinge line; lateral and pseudocardinal teeth well-developed; umbo cavity deep; nacre white, occasionally pink.

Inner lamellae of inner gills only connected to visceral mass anteriorly; excurrent aperture smooth; mantle bridge separating excurrent and supra-anal apertures very short or absent; all 4 gills marsupial; glochidia held across entire gill; marsupium not extended ventrally beyond original gill edge when gravid; ova and embryos pink or purplish, coloring marsupia when brooding, becoming white or tan as glochidia mature; glochidium without styliform hooks (Simpson, 1900b, 1914; Ortmann, 1912a; Baker, 1928a; Haas, 1969b).

Synonymy
Lexingtonia Ortmann, 1914
Quincuncina Ortmann *in* Ortmann and Walker, 1922

Figure 13.11. *Fusconaia* sp. Length 53 mm, OSUM 20860.2. Chattahoochee River, Georgia. © Richard T. Bryant.

Fusconaia apalachicola Williams and Fradkin, 1999
Apalachicola Ebonyshell

Fusconaia apalachicola – Length 34 mm, UF 358659. Mouth of Omusee Creek, [Chattahoochee River,] Omusee Park, Houston County, Alabama. © Richard T. Bryant.

Shell Description

Length to 45 mm; moderately thick; compressed; outline round; posterior, anterior and ventral margins broadly rounded, dorsal margin slightly convex; posterior ridge low, rounded; posterior slope flat to slightly convex; umbo moderately inflated, elevated above hinge line, umbo sculpture unknown; periostracum unknown.

Pseudocardinal teeth thick, triangular, 2 divergent teeth in left valve, 1 tooth in right valve; lateral teeth short, straight to slightly curved, 2 in left valve, 1 in right valve; interdentum short, wide; umbo cavity deep, compressed; nacre appears to have been white.

Soft Anatomy Description

Soft anatomy unknown.

Glochidium Description

Glochidium unknown.

Similar Species

Fusconaia apalachicola may superficially resemble young *Glebula rotundata*. However, *G. rotundata* has distinctive pseudocardinal teeth that are comprised of small, radially arranged serrate ridges. Pseudocardinal teeth of *F. apalachicola* are triangular and divergent. Also, *G. rotundata* lacks a wide interdentum and deep, compressed umbo cavity.

Some *Quadrula infucata* lack shell sculpture and superficially resemble *Fusconaia apalachicola*.

However, *Q. infucata* are typically quadrate to subtriangular in outline instead of round. Pseudocardinal teeth of *Q. infucata* are not as massive, the interdentum not as wide and the umbo cavity not as deep and compressed as those of *F. apalachicola*.

General Distribution

Fusconaia apalachicola was probably an Apalachicola Basin endemic. Archaeological specimens have been found at sites along the Apalachicola River in Florida and the Chattahoochee River in southeastern Alabama and southwestern Georgia. There have been no collections of *F. apalachicola* from above the Fall Line.

Alabama and Mobile Basin Distribution

There are no historical records of *Fusconaia apalachicola*. The only known record from Alabama is a single valve from Omusee Creek where it joins the Chattahoochee River in Houston County.

Ecology and Biology

All records of *Fusconaia apalachicola* are from archaeological excavations along medium to large Coastal Plain rivers. These areas typically have moderately flowing water and soft sediments.

Fusconaia apalachicola was presumably a short-term brooder, gravid during spring and summer. Its glochidial hosts are unknown.

Current Conservation Status and Protection

Fusconaia apalachicola was not addressed in earlier status assessments. It appears to be extinct.

Remarks

Fusconaia apalachicola was discovered in archaeological samples. It was assigned to the genus *Fusconaia* based on conchological characters and appears most closely related to *Fusconaia rotulata*, an Escambia River drainage endemic.

The causal factors and time of extinction of *Fusconaia apalachicola* are not known. The presence of this mussel in archaeological remains dated to AD 1350 (Schnell et al., 1981) suggests that it was extant when Europeans arrived in North America.

Fusconaia apalachicola was named for the Apalachicola River. The word apalachicola has several interpretations. One is from the Native American Hitchiti words *apalahchi*, which means "on the other side," and *okli*, which means "people" (Read, 1937). Another interpretation is from the Choctaw words *apelichi*, which means "ruling place", and *okla*, which means "people" (Simpson, 1956).

Synonymy

Fusconaia apalachicola Williams and Fradkin, 1999. Williams and Fradkin, 1999:51–62, fig. 2

Type locality: Archaeological Site 8LI76, 500 m east of Apalachicola River (T1N; R8W; SE ¼ Sec. 1) near river mile 88, about 5 miles north of Bristol, Liberty County, Florida. Holotype, UF 5260690.1 (right valve, left figure below); paratype, UF 5260528.9 (left valve, right figure below); length 40 mm.

 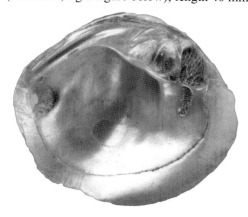

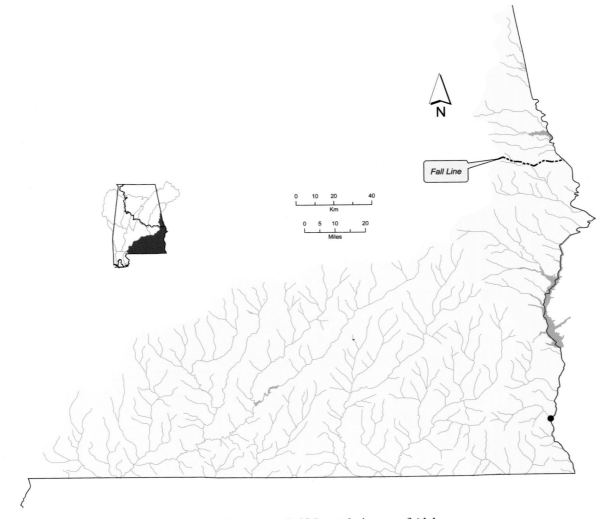

Distribution of *Fusconaia apalachicola* in the eastern Gulf Coast drainages of Alabama.

Fusconaia burkei (Walker *in* Ortmann and Walker, 1922)
Tapered Pigtoe

Fusconaia burkei – Length 55 mm, UF 64972. Holmes Creek, Jackson County, Florida. © Richard T. Bryant.

Shell Description

Length to 75 mm; moderately thick; moderately inflated, occasionally compressed; outline elliptical to subtriangular; posterior margin obliquely truncate to bluntly pointed; anterior margin rounded; dorsal margin straight to convex; ventral margin convex; posterior ridge well-defined, somewhat angular, low and rounded in juveniles; posterior slope wide, slightly concave, with parallel plications originating on posterior ridge; disk usually with short, raised chevrons, older individuals often with subtle sculpture, young individuals may have pronounced sculpture over entire shell; umbo low, broad, elevated slightly above hinge line, umbo sculpture concentric ridges; periostracum smooth on disk, usually roughened on posterior slope, greenish brown to yellowish brown, occasionally with faint dark green rays, becoming dark brown to black with age.

Pseudocardinal teeth small, triangular, 2 divergent teeth in left valve, 1 tooth in right valve; lateral teeth thin, straight to slightly curved, 2 in left valve, 1 in right valve; interdentum short, narrow; umbo cavity wide, shallow; nacre bluish white.

Soft Anatomy Description

Mantle creamy white to tan or light brown, may be dull orange outside of pallial line, gray or brown outside of apertures; visceral mass pearly white to creamy white, often with tan areas dorsally; foot creamy white to tan or pale orange, may be darker distally.

Gills creamy white to tan or gold; dorsal margin straight to slightly sinuous, ventral margin convex; gill length 50–60% of shell length; gill height 45–55% of gill length; outer gill height usually 80–90% of inner gill height, occasionally greater; inner lamellae of inner gills connected to visceral mass only anteriorly.

No gravid females were available for marsupium description. Ortmann and Walker (1922b) and Pilarczyk et al. (2005) reported all four gills to be marsupial and inflated only slightly when gravid. Pilarczyk et al. (2005) reported marsupial gills to be pinkish when gravid.

Labial palps creamy white to tan or gold, often with irregular lighter area dorsally; straight to slightly concave dorsally, convex ventrally, bluntly pointed distally; palp length 25–35% of gill length; palp height 50–70% of palp length; distal 30–50% of palps bifurcate.

Incurrent, excurrent and supra-anal aperture lengths variable relative to one another.

Incurrent aperture length 15–20% of shell length; tan within, usually without coloration basal to papillae, occasionally with sparse brown, may have dark brown or gray from edges of papillae extending onto aperture wall; papillae variable, in 1–2 rows, mostly simple, may have scattered bifid and trifid, trifid papillae approach form of arborescent papillae when contracted; papillae tan, often with rusty or golden cast, may have dark brown or gray edges basally. Ortmann and Walker (1922b) described incurrent aperture papillae as "subfalciform, their posterior margins connected for about one-half of their length".

Excurrent aperture length 15–20% of shell length; tan within, may have rusty brown cast, marginal color band some combination of tan, gray and brown, usually with irregular spots or lines perpendicular to margin; margin with simple papillae, small but usually well-developed; papillae tan. Ortmann and Walker (1922b) reported the excurrent aperture margin to be finely crenulate.

Supra-anal aperture length 10–20% of shell length; tan within, may have slight golden cast, often with thin brown or gray band marginally; margin smooth; mantle bridge separating supra-anal and excurrent aperture occasionally perforate, length variable, 15–90% of supra-anal length, occasionally absent.

Specimens examined: Eightmile Creek, Walton County, Florida (n = 9); Flat Creek, Geneva County (n = 2).

Glochidium Description

Length 167 µm; height 160 µm; without styliform hooks; outline subelliptical (Pilarczyk et al., 2005).

Similar Species

Some *Fusconaia burkei* have little shell sculpture and can resemble *Pleurobema strodeanum*. However, close examination usually reveals at least traces of sculpture on the posterior ridge and posterior slope of *F. burkei*. The pseudocardinal teeth of *P. strodeanum* are larger than those of *F. burkei*.

General Distribution

Fusconaia burkei is endemic to the Choctaw-hatchee River drainage of Alabama and Florida.

Alabama and Mobile Basin Distribution

Fusconaia burkei is found in the Choctawhatchee River drainage, including the Pea River system.

Fusconaia burkei remains in isolated tributaries of the Pea River and headwaters of the Choctawhatchee River.

Ecology and Biology

Fusconaia burkei inhabits creeks and rivers. It is found in stable substrates ranging from silty sand to sandy gravel in slow to moderate current. *Fusconaia burkei* can occasionally be found in floodplain lakes.

The biology of *Fusconaia burkei* in Eightmile Creek, Walton County, Florida, was studied by Pilarczyk et al. (2005). It is a short-term brooder, gravid from mid-March through May, possibly June. *Fusconaia burkei* conglutinates are pinkish and cylindrical, tapered at both ends. The pinkish color is derived from undeveloped eggs incorporated in the conglutinate, even those which also contain mature glochidia (M.M. Pilarczyk, personal communication). Fecundity was measured in 4 females and ranged from 3,880 to 10,395 per year, with an average of 6,058 glochidia per individual. *Cyprinella venusta* (Blacktail

Shiner) (Cyprinidae) was found to serve as a glochidial host of *F. burkei* in a preliminary host fish study.

Current Conservation Status and Protection

Fusconaia burkei was considered endangered throughout its range by Athearn (1970) and threatened throughout its range by Williams et al. (1993). In Alabama it was listed as threatened by Lydeard et al. (1999), and Garner et al. (2004) designated it a species of high conservation concern in the state. In 2003 *F. burkei* (as *Quincuncina burkei*) was elevated to a candidate for protection under the federal Endangered Species Act.

Remarks

The original description of *Fusconaia burkei* is interesting in that it was published jointly by Ortmann and Walker (1922). It was described as a new genus and species, *Quincuncina burkei*, with Ortmann credited with the genus description and Walker with the species description.

The phylogenetic relationships of *Fusconaia burkei* have been the subject of speculation since its description. Ortmann pointed out the similarity of soft anatomy between *Quincuncina* and *Fusconaia*, stating that the former "resembles only one genus, *Fusconaia*" (Ortmann and Walker, 1922b). However, he concluded that the presence of sculpturing on the shell prevented its inclusion in *Fusconaia*, which was his basis for creating the genus *Quincuncina*. Frierson (1927), without comment, treated *Quincuncina* as a subgenus of *Quadrula*, but Clench and Turner (1956) and subsequent workers continued to recognize *Quincuncina*. In a phylogenetic analysis based on genetic sequence data, *Quincuncina burkei* was determined to be sister to *Fusconaia escambia* (Lydeard, et al., 2000; Serb et al., 2003). Based on similarity of soft anatomy and results of genetic analyses, *burkei* is herein recognized as belonging to the genus *Fusconaia*.

A report of *Fusconaia burkei* from Dade County, Georgia, by Webb (1942) is erroneous. Dade County is entirely within the Tennessee River drainage.

Fusconaia burkei was named for Joseph B. Burke, who was employed by H.H. Smith to collect mussels from southern Alabama. Burke discovered the species while sampling in the Pea River and Smith requested that it be named in his honor (Ortmann and Walker, 1922b).

Synonymy

Quincuncina burkei Walker, 1922. Walker *in* Ortmann and Walker, 1922b:3, pl. 1, figs. 1, 4

Type locality: Sikes Creek, tributary of Choctawhatchee River, Barbour County, Alabama. Holotype, UMMZ
94496, length 51 mm.

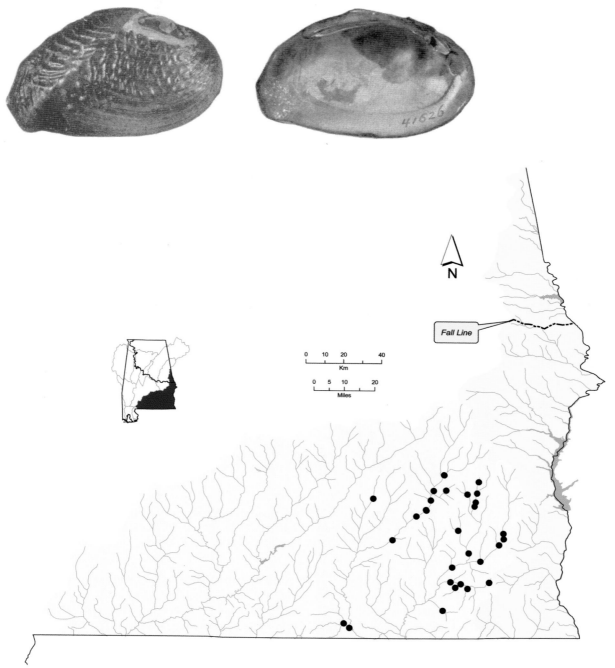

Distribution of *Fusconaia burkei* in the eastern Gulf Coast drainages of Alabama.

Fusconaia cerina (Conrad, 1838)
Gulf Pigtoe

Fusconaia cerina – Upper figure: length 53 mm, UF 271974. Tombigbee River, east bank, lower end of Big Creek cutoff, Tennessee Tombigbee Waterway, Pickens County, Alabama, 22 October 1980. Lower figure: length 69 mm, UMMZ 54026. Big Prairie Creek, Marengo [Hale] County, Alabama. © Richard T. Bryant.

Shell Description

Length to 80 mm; individuals from large rivers thick, individuals from small streams moderately thin; individuals from large rivers inflated, individuals from small streams compressed; outline triangular to quadrate, individuals from large rivers typically more triangular than those from small streams; posterior margin obliquely truncate to bluntly pointed; anterior margin rounded; dorsal margin convex; ventral margin straight to gently rounded; posterior ridge well-developed, angular; posterior slope steep, flat to slightly concave; most individuals with wide, very shallow sulcus just anterior of posterior ridge; umbo elevated above hinge line, umbo of individuals from large rivers more inflated and elevated higher above hinge line than those from small streams, umbo sculpture concentric ridges, may be double-looped; periostracum satiny, chestnut brown to black, occasionally with faint green rays.

Pseudocardinal teeth thick, rough, triangular, 2 divergent teeth in left valve, 1 tooth in right valve, may have very small accessory denticle anteriorly and/or posteriorly; lateral teeth short, thick, straight when young, becoming curved with age, 2 in left valve, 1 in right valve; interdentum short to moderately long, narrow to wide; umbo cavity deep to moderately shallow; nacre white to pink.

Soft Anatomy Description

Mantle tan to light brown, often dull orange to reddish brown outside of pallial line; visceral mass creamy white to tan, becoming orange toward foot in some individuals; foot creamy white to orange, may be darker distally.

Gills tan to brown; dorsal margin slightly sinuous, ventral margin convex to elongate convex, may be slightly bilobed in some females; gill length 40–65% of shell length; gill height 40–65% of gill length; outer gill height 75–90% of inner gill height, occasionally less; inner lamellae of inner gills connected to visceral mass only anteriorly.

All 4 gills marsupial; glochidia held across entire gill except anterior and posterior ends; marsupium slightly padded when gravid; pink when brooding embryos, becoming tan as glochidia mature.

Labial palps tan to pale orange; straight to slightly concave dorsally, curved ventrally, bluntly pointed distally; palp length 25–40% of gill length; palp height 30–80% of palp length; distal 25–45% of palps bifurcate, occasionally greater.

Incurrent and supra-anal apertures longer than excurrent aperture; incurrent aperture longer, shorter or equal to supra-anal aperture.

Incurrent aperture length 15–20% of shell length; creamy white to tan within, occasionally with faint orange tint, may have black to rusty brown basal to papillae; papillae variable, 1–2 rows, long, some combination of simple, bifid and arborescent; papillae creamy white to tan or rusty brown, black pigment usually between and at base of papillae.

Excurrent aperture length 10–20% of shell length; creamy white within, usually with black marginal color band, band may be rusty brown or absent; margin papillate, papillae short, slender, simple, creamy white to tan or reddish brown, often with black pigment between papillae.

Supra-anal aperture length 15–25% of shell length; creamy white to pale orange within, sometimes with thin gray or black marginal band; margin smooth; mantle bridge separating supra-anal and excurrent apertures occasionally perforate, length variable, 5–65% of supra-anal length, sometimes absent.

Specimens examined: Alabama River (n = 3); Bogue Chitto, Dallas County (n = 1); Bull Mountain Creek, Marion County (n = 3); Cahaba River (n = 3); Sipsey River (n = 3); Tombigbee River (n = 3); Trussells Creek, Greene County (n = 3); Yellow Creek, Lowndes County, Mississippi (n = 3).

Glochidium Description

Length 162–175 μm; height 175–187 μm; without styliform hooks; outline depressed subelliptical; dorsal margin slightly concave to slightly undulating, ventral margin rounded, anterior and posterior margins convex.

Similar Species

Fusconaia cerina resembles *Pleurobema marshalli* and *Pleurobema taitianum*, but *F. cerina* has a sharper posterior ridge. The posterior margin of *F. cerina* is more pointed than that of *P. marshalli*. Also, the umbo cavity of *F. cerina* is deep and somewhat compressed, whereas those of *P. taitianum* and *P. marshalli* are shallow. The nacre of *P. marshalli* is always white, whereas that of *F. cerina* is often pink.

General Distribution

Fusconaia cerina is found in Gulf Coast drainages from the Mobile Basin west to the Amite River system, Lake Pontchartrain drainage, Louisiana (Vidrine, 1993).

Alabama and Mobile Basin Distribution

Fusconaia cerina is known from all major drainages of the Mobile Basin. It is not known from above the Fall Line in the Tallapoosa River drainage. *Fusconaia cerina* is widespread in tributaries.

Fusconaia cerina remains widespread in the Mobile Basin but is common in only a few tributaries.

Ecology and Biology

Fusconaia cerina occurs in a variety of habitats, from medium creeks to large rivers. It can be found in various combinations of mud, sand and gravel, usually in water with at least some current. In creeks *F. cerina* often occurs in water less than 1 m deep but in large rivers it can be found at depths of more than 6 m.

Fusconaia cerina is a short-term brooder, gravid during late spring and summer. It was reported gravid in the Buttahatchee River, Mississippi, from late May to late July, but glochidia were not mature until early June (Haag and Warren, 2003). In a Sipsey River, Alabama, population of *F. cerina*, 100 percent of adult females were reported to be gravid during the peak of the brooding period (Haag and Staton, 2003). Glochidia are held in lanceolate conglutinates, thicker along the margin than medially. *Fusconaia cerina* conglutinates contain an average of 48 percent undeveloped eggs, which are believed to function to maintain conglutinate integrity (Haag and Staton, 2003; Haag and Warren, 2003). However, Haag and Staton (2003) reported an individual *F. cerina* in which 100 percent of eggs in its conglutinates were undergoing development. Fecundity of *F. cerina* has been reported to average 23,890 glochidia per year (Haag and Staton, 2003). In the Sipsey River, Alabama, 6.6 percent of *F. cerina* specimens examined were infested with trematodes to the point of sterility (Haag and Staton, 2003). Trematode infestation was not limited to large individuals.

Fishes reported to serve as glochidial hosts of *Fusconaia cerina* in laboratory trials include *Cyprinella callistia* (Alabama Shiner), *Cyprinella venusta* (Blacktail Shiner), *Hybopsis winchelli* (Clear Chub), *Luxilus chrysocephalus* (Striped Shiner), *Lythrurus bellus* (Pretty Shiner) and *Notemigonus crysoleucas* (Golden Shiner) (Cyprinidae) (Haag and Warren, 2003). Fishes with inconsistent transformation, suggesting that they are secondary hosts, include *Campostoma oligolepis* (Largescale Stoneroller), *Nocomis leptocephalus* (Bluehead Chub), *Notropis ammophilus* (Orangefin Shiner), *Notropis atherinoides* (Emerald Shiner), *Notropis stilbius* (Silverstripe Shiner) and *Pimephales notatus* (Bluntnose Minnow) (Cyprinidae) (Haag and Warren, 2003).

Current Conservation Status and Protection

Fusconaia cerina was listed as currently stable throughout its range by Williams et al. (1993) and in Alabama by Lydeard et al. (1999). Garner et al. (2004) designated *F. cerina* a species of lowest conservation concern in the state.

Remarks

Fusconaia cerina has one of the most conchologically variable shells among Mobile Basin unionids. Specimens from creeks and rivers often differ in shell thickness, inflatedness, shape and periostracum texture. Also, some creek populations appear to have a higher percentage of individuals with pink nacre than populations in large rivers.

Synonymy

Unio cerinus Conrad, 1838. Conrad, 1838:95–96, pl. 52
Type locality: Waters of Louisiana, not far from New Orleans. Lectotype, ANSP 41644, length 63 mm, designated by Johnson and Baker (1973).

Unio glandaceus Lea, 1861. Lea, 1861c:59; Lea, 1862b:77, pl. 9, fig. 226; Lea, 1862d:81, pl. 9, fig. 226
Type locality: Cahawba [Cahaba] River, Alabama, E.R. Showalter, M.D. Lectotype, USNM 84577, length 51 mm, designated by Johnson (1974).

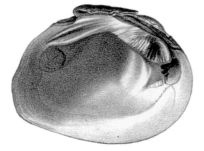

Unio negatus Lea, 1861. Lea, 1861c:59; Lea, 1862b:76, pl. 9, fig. 225; Lea, 1862d:80, pl. 9, fig. 225
Type locality: Big Prairie Creek, [tributary of Black Warrior River,] Alabama, E.R. Showalter, M.D.; Columbus, [Lowndes County,] Mississippi. Lectotype, USNM 84382, length 69 mm, designated by Johnson (1974), is from Big Prairie Creek, Alabama.

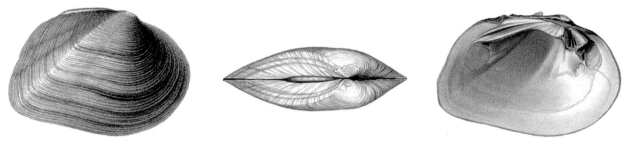

Unio rubidus Lea, 1861. Lea, 1861a:40; Lea, 1862b:95, pl. 14, fig. 244; Lea, 1862d:99, pl. 14, fig. 244
Type locality: Tombigbee River, Mississippi, W. Spillman, M.D; Coosa River and Big Prairie Creek, [tributary of Black Warrior River,] Alabama, E.R. Showalter, M.D. Lectotype, USNM 84381, length 57 mm, designated by Johnson (1974), is from Tombigbee River, Mississippi.

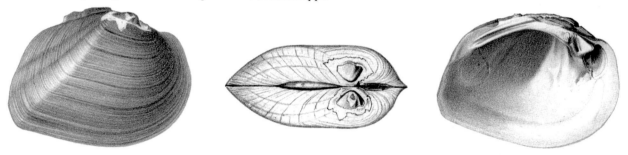

Quadrula rubidula Frierson, 1905. Frierson, 1905:14, pl. 1, figs. 3, 4
Type locality: Mulberry [Fork] River; Black Warrior River; North River, [Tyner, Tuscaloosa County,] Alabama. Lectotype, UMMZ 113535, length 36 mm, designated by Johnson (1979), is from North River, [Tyner, Tuscaloosa County,] Alabama.

Quadrula castanea Simpson, 1914. Simpson, 1914:877
Type locality: Tombigbee River, Moscow, [Sumter County,] Alabama. Lectotype, USNM 159959, length 64 mm, designated by Johnson (1975).

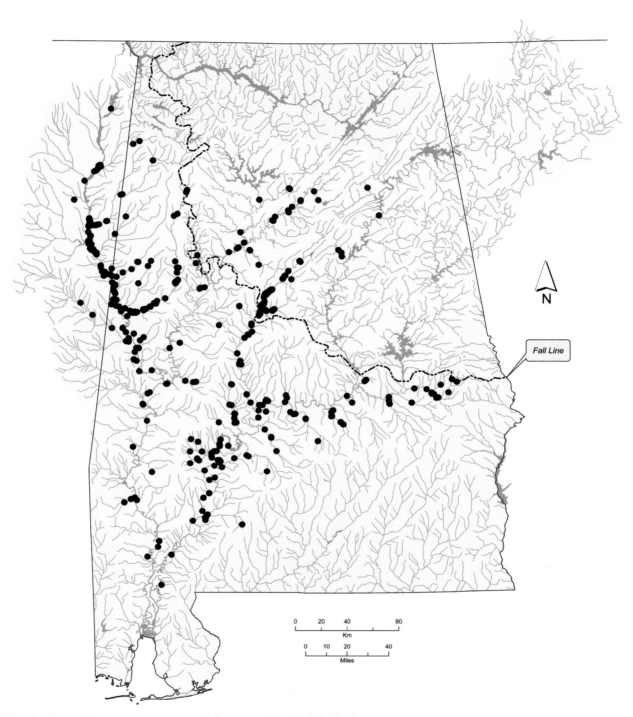

Distribution of *Fusconaia cerina* in Alabama and the Mobile Basin.

Fusconaia cor (Conrad, 1834)
Shiny Pigtoe

Fusconaia cor – Length 36 mm, UF 65052. Tennessee River, Alabama. © Richard T. Bryant.

Shell Description

Length to 80 mm; moderately thick; somewhat compressed; outline subtriangular, becoming elongate posteriorly with age; posterior margin narrowly rounded to obliquely truncate; anterior margin rounded; dorsal margin slightly convex; ventral margin rounded anteriorly, becoming straight at an oblique angle posteriorly; posterior ridge well-developed, angular near umbo, sometimes doubled posterioventrally; wide, shallow sulcus located anterior of posterior ridge, disk swollen anterior of sulcus; posterior slope steep dorsally, becoming shallow and slightly convex ventrally; umbo broad, elevated above hinge line, umbo sculpture broken, knobbed ridges; periostracum smooth, shiny, may be dull distally, greenish yellow to brown, may darken to almost black with age, with numerous variable dark green rays that may become obscure with age.

Pseudocardinal teeth thick, triangular, 2 divergent teeth in left valve, anterior tooth smaller, crest of both teeth approximately equal in height; 1 tooth in right valve, may have very small accessory denticle anteriorly and/or posteriorly; lateral teeth thick, slightly curved, 2 in left valve, 1 in right valve; interdentum moderately short, wide; umbo cavity deep, moderately compressed; nacre white.

Soft Anatomy Description

No material was available for soft anatomy description. Ortmann (1921) provided limited information, stating that "soft parts [are] identical to those of *Fusconaia cuneolus*, and with those of the *flava*-group in general". Coloration of the soft anatomy is usually deep orange, especially the foot and adductor muscles, rarely pale orange. Gonads, ova and conglutinates are usually crimson, occasionally pink.

Glochidium Description

Length 133–155 μm; height 156–185 μm; shape not reported (Kitchel, 1985).

Similar Species

Fusconaia cor resembles several species, including *Fusconaia cuneolus*, some young *Fusconaia subrotunda*, *Pleuronaia barnesiana*, *Pleuronaia dolabelloides* and *Pleurobema oviforme*. It can be separated from all of these species by its shiny periostracum, compared to the satiny or dull periostracum of the other species. The periostracum of *F. cor* becomes dull distally as the shell grows, but the upper portion of the shell disk typically remains shiny for life.

Fusconaia cor most closely resembles *Fusconaia cuneolus*. In addition to the distinction in periostracum texture, there are subtle differences in their pseudocardinal teeth. In the left valve of adult *F. cuneolus* individuals the crest of the anterior pseudocardinal tooth is often higher than the posterior tooth, whereas the anterior tooth of *F. cor* is about equal to the posterior tooth in height.

The posterior ridge of *Fusconaia cor* is much higher and the posterior margin usually more oblique than those of *Fusconaia subrotunda*.

Fusconaia cor has a deeper, more compressed umbo cavity and heavier, less compressed pseudocardinal teeth than *Pleuronaia barnesiana*. The pseudocardinal teeth in the left valve of *F. cor* are more divergent than those of *P. barnesiana*, giving them an overall less perpendicular appearance relative to the lateral teeth.

The posterior ridge of *Fusconaia cor* is straight, whereas that of *Pleuronaia dolabelloides* curves ventrally. Also, the umbo cavity of *F. cor* is deeper and more compressed than that of *P. dolabelloides*.

The deep, compressed umbo cavity of *Fusconaia cor* distinguishes it from *Pleurobema oviforme*. The shells of *P. oviforme* are often more elongate than those of *F. cor*.

General Distribution

Fusconaia cor is endemic to the Tennessee River drainage where it historically occurred from the headwaters in southwestern Virginia and eastern Tennessee, downstream to Muscle Shoals, Alabama (Ahlstedt, 1992a, 1992b; Parmalee and Bogan, 1998). It is also known from medium to large Tennessee River tributaries (Parmalee and Bogan, 1998).

Alabama and Mobile Basin Distribution

Fusconaia cor historically occurred in the Tennessee River across northern Alabama and some tributaries.

Fusconaia cor is known to be extant only in the Paint Rock River system.

Ecology and Biology

Fusconaia cor occurs in shoal habitats of small to large rivers with stable gravel substrates. It is frequently well-buried. *Fusconaia cor* is extirpated from the Tennessee River proper, but remained extant downstream of Wilson Dam until at least 1968. In headwater rivers it is generally found in shallow shoals, but below Wilson Dam it occurred in water approximately 5 m deep.

Fusconaia cor is a short-term brooder. Ortmann (1921) reported gravid females from mid-May to mid-July. However, he stated that "none happened to have glochidia." Presumably they were gravid with eggs or developing embryos. Females gravid with mature glochidia have been reported in July and August (Jones and Neves, 2001) and glochidia have been reported in stream drift in the North Fork Holston River, southwestern Virginia, during the same period (Kitchel, 1985).

Cyprinella galactura (Whitetail Shiner) (Cyprinidae) was found to serve as a glochidial host of *Fusconaia cor* in laboratory trials (Neves, 1991). Fishes found to be marginal hosts in laboratory trials include *Luxilus chrysocephalus* (Striped Shiner) and *Luxilus coccogenis* (Warpaint Shiner) (Cyprinidae); and *Etheostoma rufilineatum* (Redline Darter) (Percidae) (Jones and Neves, 2001). Fishes in the North Fork Holston River for which observations of natural infestation with *F. cor* glochidia have been reported include *Luxilus cornutus* (Common Shiner) and *Notropis telescopus* (Telescope Shiner) (Cyprinidae) (Kitchel, 1985).

Current Conservation Status and Protection

Fusconaia cor was considered endangered throughout its range by Stansbery (1970a) and Williams et al. (1993) and in Alabama by Lydeard et al. (1999). Garner et al. (2004) designated *F. cor* a species of highest conservation concern in the state. It was listed as endangered under the federal Endangered Species Act in 1976.

In 2001 *Fusconaia cor* was included on a list of species approved for a Nonessential Experimental Population in Wilson Dam tailwaters on the Tennessee River. However, no reintroductions had taken place as of 2007.

Remarks

Fusconaia cor has been reported from prehistoric shell middens along the Tennessee River at Muscle Shoals and near Bridgeport but not from intervening sites (Morrison, 1942; Warren, 1975; Hughes and Parmalee, 1999). It was uncommon at sites where it occurred.

Prior to the mid-1980s the name *Fusconaia edgariana* (Lea, 1840) was used for *Fusconaia cor*. However, *edgariana* is a junior synonym of *F. cor*. The confusion appears to have arisen because Conrad's illustration accompanying the description represented a specimen of a different species (Frierson, 1916). The written description is adequate to identify the species as *F. cor* (Ortmann and Walker, 1922a). Frierson (1916) designated an ANSP specimen as the lectotype and provided three figures. Unfortunately, Johnson (1956) and Johnson and Baker (1973) subsequently designated the specimen figured by Conrad as the lectotype. Therefore, their designations are invalid.

Synonymy

Unio cor Conrad, 1834. Conrad, 1834b:28, pl. 3, fig. 3
Type locality: Elk and Flint rivers, Alabama.

Comment: The history surrounding the type specimen of *Unio cor* is convoluted and the subject of debate (Frierson, 1916, 1927; Johnson, 1956; Johnson and Baker, 1973). It appears that Conrad (1834b) mistakenly inserted a figure of *Fusconaia ebena* (figure above), MCZ 178792, in place of *Unio cor* (= *Fusconaia cor*) in the original description. Frierson (1916) designated and figured one of Conrad's syntypes (figure below), ANSP 20113a, length 36 mm, as the "type"—thus this specimen becomes the lectotype of *Unio cor*. Johnson's (1956) selection of Conrad's original figured shell (MCZ 178792) as the lectotype is invalid.

Unio edgarianus Lea, 1840. Lea, 1840:288; Lea, 1842b:214, pl. 15, fig. 30; Lea, 1842c:52, pl. 15, fig. 30
Unio obuncus Lea, 1871. Lea, 1871:192; Lea, 1874c:9, pl. 2, fig. 5; Lea, 1874e:13, pl. 2, fig. 5
Unio andersonensis Lea, 1872. Lea, 1872b:155; Lea, 1874c:36, pl. 12, fig. 33; Lea, 1874e:40, pl. 12, fig. 33
Fusconaia edgariana var. *analoga* Ortmann, 1918. Ortmann, 1918:533

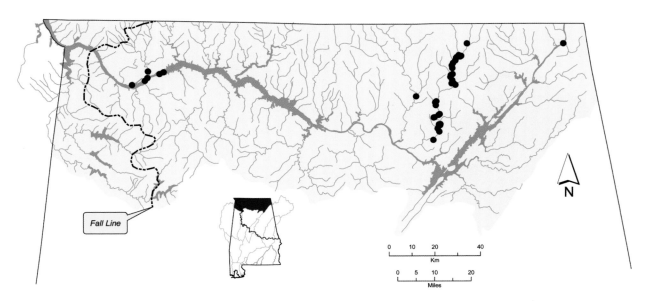

Distribution of *Fusconaia cor* in the Tennessee River drainage of Alabama.

Fusconaia cuneolus (Lea, 1840)
Finerayed Pigtoe

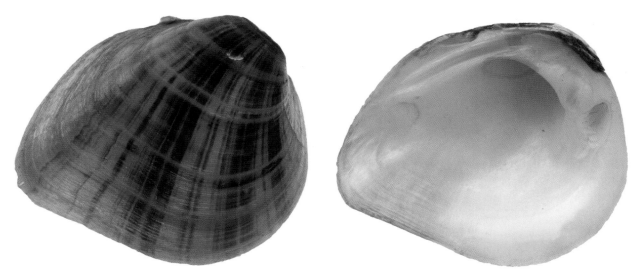

Fusconaia cuneolus – Length 29 mm, UF 229593. The Points, [Limestone Creek,] Mooresville, Limestone County, Alabama. © Richard T. Bryant.

Shell Description

Length to 80 mm; moderately thick; moderately inflated; outline subtriangular; posterior margin narrowly rounded to obliquely truncate; anterior margin straight to slightly convex; dorsal margin convex to almost straight; ventral margin broadly rounded, becoming straight at an oblique angle posteriorly; posterior ridge well-developed, narrow adjacent to umbo, wider and lower posterioventrally; wide, shallow sulcus located anterior of posterior ridge; posterior slope steep dorsally, becoming flatter posterioventrally; umbo moderately inflated, elevated slightly above hinge line, umbo sculpture broken, knobbed ridges; periostracum satiny, dull tawny to olive or yellowish green, with variable green rays that may become indistinct with age.

Pseudocardinal teeth moderately thick, triangular, 2 divergent teeth in left valve, posterior tooth wider, but anterior tooth with higher crest, 1 tooth in right valve, sometimes with accessory denticle anteriorly and/or posteriorly; lateral teeth straight to slightly curved, 2 in left valve, 1 in right valve; interdentum short, moderately wide; umbo cavity moderately deep; nacre white.

Soft Anatomy Description

No material was available for soft anatomy description, but Ortmann (1921) provided limited information. The mantle bridge separating excurrent and supra-anal apertures is short, may be absent in some individuals. Soft parts are orange, often pale, with intensity of coloration usually greater in the foot, mantle margin and adductor muscles. All four gills are marsupial, but in young individuals the marsupial area of the inner gill is often restricted to the central part. Gonads are red, and ova and conglutinates are pink to pinkish orange or crimson.

Glochidium Description

Length 160 µm; height 160 µm; outline "subelliptical, nearly semielliptical" Ortmann (1921).

Similar Species

Fusconaia cuneolus most closely resembles *Fusconaia cor*. They can be distinguished by periostracum texture, with *F. cuneolus* being satiny and *F. cor* shiny. The periostracum on the upper part of the shell disk of *F. cor* remains shiny throughout life but often becomes dull distally. Eroded individuals may be very difficult to distinguish from *F. cuneolus*. In addition to the distinction in periostracum texture, there are subtle differences in their pseudocardinal teeth. In the left valve of adult *F. cuneolus* the crest of the anterior pseudocardinal tooth is usually higher than that of the posterior tooth, whereas the anterior and posterior teeth of *F. cor* are usually about equal in height.

Fusconaia cuneolus resembles *Fusconaia subrotunda*. The posterior ridge of *F. cuneolus* is much higher and the posterior margin usually more oblique than those of *F. subrotunda*.

Fusconaia cuneolus resembles *Pleuronaia barnesiana* but has a deeper, more compressed umbo cavity and heavier, less compressed pseudocardinal teeth. Pseudocardinal teeth in the left valve of *F. cuneolus* are much more divergent than those of *P. barnesiana*, giving them an overall less perpendicular appearance relative to the lateral teeth.

Fusconaia cuneolus resembles *Pleuronaia dolabelloides*. The posterior ridge of *F. cuneolus* is straight, whereas that of *P. dolabelloides* is ventrally curved. Also, the umbo cavity of *F. cuneolus* is deeper and more compressed of that of *P. dolabelloides*.

Fusconaia cuneolus resembles *Pleurobema oviforme* but has a deep, compressed umbo cavity instead of a shallow umbo cavity. Shells of *P. oviforme* are often more elongate than those of *F. cuneolus*.

General Distribution

Fusconaia cuneolus is endemic to the Tennessee River drainage, where it historically occurred from headwaters in southwestern Virginia and eastern Tennessee downstream to Muscle Shoals, Alabama (Neves, 1991; Parmalee and Bogan, 1998).

Alabama and Mobile Basin Distribution

Fusconaia cuneolus historically occurred in the Tennessee River across northern Alabama and some tributaries.

The only known extant population of *Fusconaia cuneolus* is in the Paint Rock River system.

Ecology and Biology

Fusconaia cuneolus occurs in shoal habitats of creeks and rivers. Historical distribution data suggest that *F. cuneolus* occurred in smaller streams than many other species of *Fusconaia*. It prefers stable gravel substrates in moderate current.

Bruenderman and Neves (1993) detailed the life history of *Fusconaia cuneolus* in the Clinch River, southwestern Virginia. It is a short-term brooder, apparently spawning in early May, with females gravid from mid-May through early August. Developing embryos are bound in conglutinates and change color from pink to orange to light peach as they mature.

However, mature glochidia are discharged in a loose, gelatinous matrix instead of well-defined conglutinates. Fecundity was assessed in one gravid female, which contained approximately 113,000 embryos. Glochidia were recovered from stream drift as early as late May and as late as mid-August. Ortmann (1921) reported a similar gravid period for *F. cuneolus*, mid-May through mid-July.

Fishes on which natural infestations of *Fusconaia cuneolus* glochidia have been observed and confirmed in a laboratory include *Cyprinella galactura* (Whitetail Shiner), *Nocomis micropogon* (River Chub) and *Notropis leuciodus* (Tennessee Shiner) (Cyprinidae) (Bruenderman and Neves, 1993). Additional species found to serve as glochidial hosts in laboratory trials are *Cottus bairdii* (Mottled Sculpin) (Cottidae); and *Campostoma anomalum* (Central Stoneroller), *Luxilus albeolus* (White Shiner), *Notropis telescopus* (Telescope Shiner) and *Pimephales notatus* (Bluntnose Minnow) (Cyprinidae) (Bruenderman and Neves, 1993).

Based on examination of thin sections of shell, *Fusconaia cuneolus* have been found to live at least 32 years (Bruenderman and Neves, 1993).

Current Conservation Status and Protection

Fusconaia cuneolus was considered endangered throughout its range by Stansbery (1970a) and Williams et al. (1993) and in Alabama by Lydeard et al. (1999). Garner et al. (2004) designated it a species of highest conservation concern in the state. *Fusconaia cuneolus* was listed as endangered under the federal Endangered Species Act in 1976.

In 2001 *Fusconaia cuneolus* was included on a list of species approved for a Nonessential Experimental Population in Wilson Dam tailwaters on the Tennessee River. However, no reintroductions had taken place as of 2007.

Remarks

Fusconaia cuneolus has been reported to be present, but rare, in prehistoric shell middens along the Tennessee River at Muscle Shoals and near Bridgeport (Morrison, 1942; Warren, 1975; Hughes and Parmalee, 1999). It was not reported from intervening sites (Hughes and Parmalee, 1999).

Synonymy

Unio cuneolus Lea, 1840. Lea, 1840:286; Lea, 1842b:193, pl. 7, fig. 3; Lea, 1842c:31, pl. 7, fig. 3
Type locality: Holston River, Tennessee. Lectotype, USNM 84591, length 35 mm, designated by Johnson (1974).

Unio appressus Lea, 1871. Lea, 1871:189; Lea, 1874c:12, pl. 3, fig. 8; Lea, 1874e:16, pl. 3, fig. 8
Type locality: [Tennessee River,] Tuscumbia, [Colbert County,] Alabama, B. Pybas and J.G. Anthony; Holston River, Miss Law and C.M. Wheatley. Lectotype, USNM 84483, length 43 mm, designated by Johnson (1974), is from Tennessee River, Tuscumbia, Colbert County, Alabama.

Unio tuscumbiensis Lea, 1871. Lea, 1871:190; Lea, 1874c:11, pl. 3, fig. 7; Lea, 1874e:15, pl. 3, fig. 7
Type locality: Tuscumbia, [Colbert County,] Alabama, L.B. Thornton, Esq; Holston River, east Tennessee, Dr. Edgar. Lectotype, USNM 84593, length 39 mm, designated by Johnson (1974), is from Tuscumbia, Colbert County, Alabama.

Unio flavidus Lea, 1872. Lea, 1872b:156; Lea, 1874c:28, pl. 9, fig. 25; Lea, 1874e:32, pl. 9, fig. 25

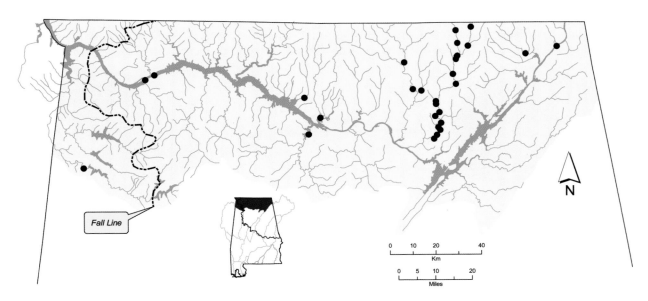

Distribution of *Fusconaia cuneolus* in the Tennessee River drainage of Alabama.

Fusconaia ebena (Lea, 1831)
Ebonyshell

Fusconaia ebena – Upper figure: length 90 mm, UF 175059. Decatur, Morgan County, Alabama. Lower figure: length 47 mm, UF 65081. Alabama River, Alabama. © Richard T. Bryant.

Shell Description

Length to 110 mm; thick; inflated; outline round, becoming oval in some old individuals; posterior margin broadly to narrowly rounded; anterior margin rounded; dorsal margin straight to convex; ventral margin convex; posterior ridge low, rounded, may be absent; posterior slope flat, moderately steep; disk surface broadly rounded; umbo inflated, turned anteriad, located near anterior margin, elevated well above hinge line, umbo sculpture a few weak ridges; periostracum reddish brown to black, without rays, very young specimens often with yellowish green umbo.

Pseudocardinal teeth massive, rough, triangular, 2 somewhat divergent teeth in left valve, arranged dorsal and ventral to each other, crest of dorsal tooth parallel to lateral teeth, teeth joined anteriorly with the junction elevated well above hinge plate, pseudocardinal teeth in left valve of young individuals arranged anterior and posterior to each other, 1 tooth in right valve, arising from a depression that accommodates elevated anterior junction of pseudocardinal teeth in left valve, adult individuals with 1–3 ridge-like accessory denticles located dorsally; lateral teeth thick, long, extend anteriorly along dorsal margin of interdentum, 2 in left valve, 1 in right valve; interdentum short, wide; umbo cavity deep, compressed; nacre white.

Soft Anatomy Description

Mantle tan to light brown, rarely pale orange outside of pallial line; visceral mass creamy white to tan; foot creamy white to tan, occasionally pale orange, often darker ventrally.

Gills tan to light brown; dorsal margin slightly sinuous, ventral margin straight to slightly curved; gill length 60–70% of shell length; gill height 40–50% of gill length; outer gill height 60–90% of inner gill height; inner lamellae of inner gills connected to visceral mass only anteriorly.

All 4 gills marsupial; glochidia held across entire gill; marsupium slightly padded when gravid; pink when holding embryos, creamy white to tan with mature glochidia.

Labial palps creamy white to tan; straight to slightly concave dorsally, curved ventrally, bluntly pointed distally; palp length 20–40% of gill length; palp height 40–75% of palp length; distal 25–45% of palps bifurcate.

Incurrent aperture longer than excurrent and supra-anal apertures; supra-anal aperture usually longer than excurrent aperture, occasionally of equal length.

Incurrent aperture length 20–35% of shell length; creamy white to tan within, may have brown to golden brown basal to papillae; papillae arborescent, occasionally a secondary row of simple papillae outside of arborescent row; papillae tan to brown or rusty brown, often gray or black distally. Utterback (1915) described incurrent aperture papillae simply as "branched". Whether he was referring to arborescent or bifid is unclear.

Excurrent aperture length 10–20% of shell length; creamy white within, marginal color band brown to reddish brown or gray; margin may have small, simple papillae or crenulations; papillae creamy white to dark brown or reddish brown. Utterback (1915) indicated that excurrent aperture margins may be smooth.

Supra-anal aperture length 15–25% of shell length; creamy white to golden brown within, usually with thin brown or reddish brown marginal band; margin smooth; mantle bridge separating supra-anal and excurrent apertures imperforate, length 5–20% of supra-anal length, sometimes absent.

Specimens examined: Alabama River (n = 3); Bear Creek, Colbert County (n = 1); Sipsey River (n = 2); Tennessee River (n = 9); Tombigbee River (n = 3).

Glochidium Description

Length 140–160 μm; height 148–150 μm; without styliform hooks; outline depressed subelliptical; dorsal margin straight, ventral margin broadly rounded, anterior and posterior margins convex (Lefevre and Curtis, 1910b; Surber, 1912; Howard, 1914b).

Similar Species

Fusconaia ebena resembles several other *Fusconaia* and *Pleurobema* species. Pseudocardinal teeth in the left valve, which are oriented more or less parallel to the lateral teeth, separate *F. ebena* from all similar species.

Fusconaia subrotunda from large rivers may be identical to *F. ebena* in all respects except pseudocardinal teeth but may be adorned with subtle, dark green rays. *Fusconaia cerina* and *Fusconaia flava* differ from *F. ebena* in having a flattened area on the central portion of the shell disk, more angular posterior ridges and more truncate or pointed posterior margins. Umbo cavities of *F. cerina* and *F. flava* are deep, as are those of *F. ebena*, but the cavities of the former two species are not as compressed as that of *F. ebena*.

Fusconaia ebena closely resembles *Pleurobema marshalli* externally. Some individuals can be distinguished by the slightly flattened area just anterior of the posterior ridge in *P. marshalli*, an area that is usually rounded in *F. ebena*. Internally, the two are easily distinguished since *P. marshalli* has a shallow umbo cavity and that of *F. ebena* is deep and compressed.

Fusconaia ebena resembles several other species of *Pleurobema*, including *P. cordatum*, *P. plenum*, *P. rubrum*, *P. sintoxia* and *P. taitianum*. All of those species differ from *F. ebena* in having a sulcus or at least a flattened area on the central portion of the disk. Unlike *F. ebena*, the umbo cavity of *P. rubrum*, *P. sintoxia* and *P. taitianum* is shallow. The umbo cavity of *P. cordatum* and *P. plenum* is moderately deep but is not as compressed as that of *F. ebena*.

Fusconaia ebena may also resemble *Obovaria olivaria*, but *F. ebena* has a much deeper, more compressed umbo cavity. Young *O. olivaria* are often distinctly rayed, whereas *F. ebena* typically lacks rays.

General Distribution

Fusconaia ebena occurs in the Mississippi Basin from southeastern Minnesota (Dawley, 1947) south to Louisiana (Vidrine, 1993), and from the Ohio River in southern Ohio (Cummings and Mayer, 1992) west to Missouri (Oesch, 1995) and possibly Kansas (Scammon, 1906). *Fusconaia ebena* is found in the Cumberland River drainage downstream of Cumberland Falls (Cicerello et al., 1991; Parmalee and Bogan, 1998). In the Tennessee River drainage it occurs from northeastern Alabama to the mouth of the Tennessee River (Isom, 1972; Parmalee and Bogan, 1998; Garner and McGregor, 2001). *Fusconaia ebena* is found in Gulf Coast drainages from the Mobile Basin of Alabama and Mississippi west to the Pearl River in Louisiana and Mississippi (Vidrine, 1993).

Alabama and Mobile Basin Distribution

Fusconaia ebena occurs in the Tennessee River drainage and Mobile Basin. In the Tennessee River, Muscle Shoals was its upstream limit historically, but it has expanded its range upstream since impoundment of the river (Ortmann, 1925; Garner and McGregor, 2001). It also occurs in some tributaries, which are thought to be the result of recent colonizations as well (McGregor and Garner, 2004). In the Mobile Basin, *F. ebena* is confined almost entirely to large rivers and mouths of some large tributaries on the Coastal Plain. However, it historically occurred above the Fall Line in the Coosa River drainage. There is one recent record of the species from the Conecuh River, Escambia River drainage, near the Alabama and Florida state line (D.N. Shelton, personal communication) that represents an introduction. There is no evidence of *F. ebena* having established a reproducing population in the Conecuh River.

Fusconaia ebena remains widespread and locally abundant in Coastal Plain reaches of large Mobile Basin rivers, where it is often the dominant species. It is the most common species in the Tennessee River in tailwaters of Wilson Dam but remains uncommon upstream of Muscle Shoals.

Ecology and Biology

Fusconaia ebena typically inhabits large rivers, primarily in free-flowing reaches and tailwaters of dams. The preferred substrate is sandy gravel kept clear of heavy silt accumulations by moderate to swift current. *Fusconaia ebena* is rarely found in the silty substrates of most overbank habitat. However, in some overbank areas in which silt accumulations are reduced (e.g., ridge tops, point bars and island shelves that are subject to wave action or occasional current) *F. ebena* can be fairly common. In large rivers it is usually found at depths of 4 m to 7 m, but may be found at depths of more than 20 m. In unimpounded tributaries *F. ebena* occurs in runs and shoals.

Fusconaia ebena is currently one of the most abundant mussels in Alabama. It is the dominant species in many areas of the Alabama and Tombigbee rivers. It is also dominant in Wilson Dam tailwaters on the Tennessee River, where densities of this species may exceed 30 per m^2 (J.T. Garner, unpublished data).

Fusconaia ebena is a short-term brooder, gravid from April or May to August or September (Surber, 1912; Howard, 1914b; Wilson and Clark, 1914; Coker et al., 1921). When brooding eggs or embryos, the gills are pink but become creamy white as glochidia mature.

Alosa chrysochloris (Skipjack Herring) (Clupeidae) is a glochidial host of *Fusconaia ebena*, which was determined by observations of natural infestations (Surber, 1913; Howard, 1914b). Observations of limited natural infestations of *F. ebena*

glochidia on "Black Bass" (= Largemouth Bass, *Micropterus salmoides*, and/or Spotted Bass, *Micropterus punctulatus*), *Pomoxis annularis* (White Crappie) and *Pomoxis nigromaculatus* (Black Crappie) (Centrarchidae) have been reported but could not be confirmed with laboratory trials (Howard, 1914b). Surber (1913) reported heavy infestations of *F. ebena* glochidia on *A. chrysochloris* from June through September in the upper Mississippi River.

All *Fusconaia ebena* life history work has taken place in the Mississippi Basin, so glochidial hosts of the Mobile Basin population remain unclear. However, *F. ebena* disappeared from the Coosa River drainage upstream of a series of dams that are barriers to upstream fish migration. This suggests that Mobile Basin populations also use anadromous fish such as *Alosa chrysochloris*.

Current Conservation Status and Protection

Fusconaia ebena was considered currently stable throughout its range by Williams et al. (1993) and in Alabama by Lydeard et al. (1999). Garner et al. (2004) designated *F. ebena* a species of lowest conservation concern in the state.

Fusconaia ebena is a legally harvestable commercial species in Alabama and is protected by state regulations, with the most important being a minimum size limit of $2^3/_8$ inches (60 mm) in shell height.

Remarks

In the Tennessee River *Fusconaia ebena* historically reached its upstream limit at Muscle Shoals, where the river crosses the Fall Line (Ortmann, 1925). It has not been reported from prehistoric middens along the Tennessee River in Alabama (Hughes and Parmalee, 1999). However, with impoundment of the river *F. ebena* has expanded its range upstream as far as Chickamauga Dam in southeastern Tennessee (Garner and McGregor, 2001). In another interesting range expansion *F. ebena* has apparently colonized Bear Creek in Colbert County, Alabama, which is a tributary of the Tennessee River with its mouth near the downstream end of Muscle Shoals as well as lower reaches of the Flint and Paint Rock rivers (McGregor and Garner, 2004).

The expansion of *Fusconaia ebena* in the Tennessee River drainage contrasts greatly to its situation in the Coosa River. There the species has disappeared since the river was impounded. Etnier and Starnes (1993) reported that *Alosa chrysochloris*, the presumed host fish of *F. ebena*, remains common in some main channel reservoirs but has disappeared from many storage reservoirs. Coosa River dams are for storage and hydroelectric power production and do not have locks or any other form of fish passage.

Fusconaia ebena was described by Wilson and Clark (1914) as being "generally regarded as the producer of the most valuable shell for the manufacture of buttons". It is also one of the most important shells for use as nuclei in the cultured pearl industry and some years it makes up the majority of Alabama's harvest.

There is uncertainty surrounding the generic placement of *Fusconaia ebena*. In electrophoretic studies and genetic analyses it does not cluster with *Fusconaia flava* or other species currently recognized as *Fusconaia* (Davis and Fuller, 1981; Campbell et al., 2005). *Fusconaia ebena* does appear to be related to *Fusconaia rotulata*, which it resembles morphologically (Lydeard et al., 2000; Campbell et al., 2005).

Synonymy

Obovaria obovalis Rafinesque, 1820. Rafinesque, 1820:311

Type locality: Ohio River. Lectotype, ANSP 20224, length 41 mm, designated by Johnson and Baker (1973).

Comment: Senior synonym of *Fusconaia ebena*. Petition to suppress *Obovaria obovalis* is being considered by the ICZN (Case 3353).

Obovaria pachostea Rafinesque, 1820. Rafinesque, 1820:312

Amblema antrosa Rafinesque, 1820. Rafinesque, 1820:322

Comment: Rafinesque created this replacement name for *Obovaria pachostea* Rafinesque, 1820. The species was renamed because he moved it to a new genus.

Unio ebenus Lea, 1831. Lea, 1831:84, pl. 9, fig. 14; Lea, 1834b:94, pl. 9, fig. 14

Type locality: Ohio River. Lectotype, USNM 85792, length 40 mm, designated by Johnson (1974).

 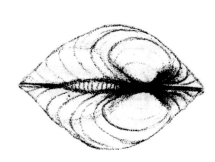 

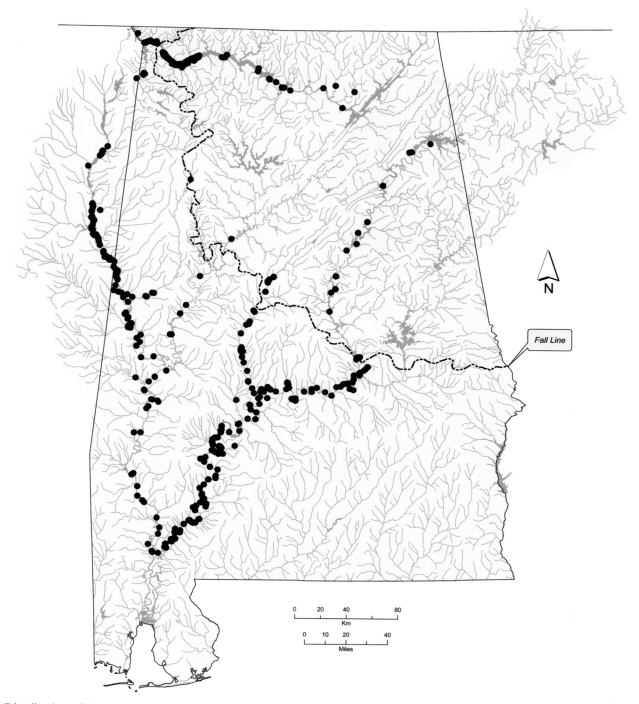

Distribution of *Fusconaia ebena* in Alabama and the Mobile Basin.

Fusconaia escambia Clench and Turner, 1956
Narrow Pigtoe

Fusconaia escambia – Length 40 mm, UF 4997. Escambia River, 3 miles southeast of Century, Escambia County, Florida, 27 August 1954. © Richard T. Bryant.

Shell Description

Length to 75 mm; moderately thick; moderately compressed; outline subtriangular to quadrate; posterior margin obliquely truncate to bluntly pointed; anterior margin rounded; dorsal margin convex; ventral margin straight to broadly rounded; posterior ridge sharp; some individuals with very shallow, wide radial depression just anterior of posterior ridge; posterior slope moderately steep, flat to slightly concave; umbo broad, slightly inflated, elevated above hinge line, umbo sculpture unknown; periostracum reddish brown to black, without rays.

Pseudocardinal teeth triangular, 2 divergent teeth in left valve, anterior tooth somewhat compressed, 1 heavily serrate tooth in right valve; lateral teeth short, straight to slightly curved, 2 in left valve, 1 in right valve; interdentum short, narrow to moderately wide; umbo cavity moderately deep, somewhat compressed; nacre white, pinkish orange to purple.

Soft Anatomy Description

Mantle pale orange to tan or light brown, often with rusty brown or grayish brown mottling outside of apertures, mottling may be oriented perpendicular to margin; visceral mass pearly white to creamy white or tan, may have orange cast; foot orange to tan.

Gills tan to gold or pale orange; dorsal margin straight to slightly sinuous, gills steeply sloped from near umbo to vicinity of apertures, ventral margin convex to elongate convex; gill length 55–70% of shell length; gill height 40–60% of gill length; outer gill height 70–85% of inner gill height; inner laminae of inner gills only connected to visceral mass anteriorly.

All four gills marsupial; glochidia held across gill length; slightly padded when gravid; tan.

Labial palps pale orange to tan; straight to slightly concave or convex dorsally, convex ventrally, bluntly pointed distally; palp length 20–45% of gill length; palp height 50–75% of palp length; distal 25–40% of palps bifurcate.

Incurrent and supra-anal apertures longer than excurrent aperture; incurrent aperture longer or shorter than supra-anal aperture.

Incurrent aperture length 15–25% of shell length; creamy white to tan or pale orange within, sometimes with irregular rusty brown or grayish brown basal to papillae; papillae in 1–3 rows, inner row largest, simple, bifid and trifid, occasionally with a few arborescent, some individuals with only simple; papillae pale orange to tan or rusty brown.

Excurrent aperture length 10–20% of shell length; creamy white to tan within, marginal color band rusty brown to grayish brown with pale orange or tan lines or oblong blotches perpendicular to margin; margin with very short simple papillae, often little more than crenulations; papillae pale orange to tan.

Supra-anal aperture length 15–20% of shell length; pale orange to tan within, may have sparse gray marginal coloration; margin smooth; mantle bridge separating supra-anal and excurrent apertures sometimes perforate, mantle bridge length variable, 20–90% of supra-anal length.

Specimens examined: Conecuh River (n = 4); Escambia River (n = 3); Patsaliga Creek, Crenshaw County (n = 4) (some specimens previously preserved).

Glochidium Description

Glochidium unknown.

Similar Species

Fusconaia escambia resembles *Quadrula succissa* but has a sharper posterior ridge, more truncate posterior margin, less inflated umbo and less compressed umbo cavity. The anterior pseudocardinal tooth in the left valve of *F. escambia* is oriented more anteriorly than that of *Q. succissa*, in which the tooth lies posterior to the anterior adductor muscle scar. Nacre of *Q. succissa* is usually pale purple, whereas that of *F. escambia* is often white to pinkish orange, but may also be purple.

Fusconaia escambia may also resemble *Pleurobema strodeanum* but has a deeper, more compressed umbo cavity. *Pleurobema strodeanum* is usually more elongate and pointed posteriorly. The posterior ridge of *F. escambia* is typically sharper and oriented more ventrally than those of most *P. strodeanum*. Nacre of *P. strodeanum* is always white, but that of *F. escambia* is often pinkish orange or purple.

General Distribution

Fusconaia escambia is known only from the Escambia River drainage in Alabama and Florida and the Yellow River in Florida. There are only two records of *F. escambia* from the Yellow River, both from just south of the Alabama and Florida state line in Okaloosa County, Florida.

Alabama and Mobile Basin Distribution

Fusconaia escambia occurs in the Conecuh River from the Alabama and Florida state line upstream to the vicinity of Troy, Pike County. It also occurs in some Conecuh River tributaries. There are no known records from Alabama reaches of the Yellow River, but it likely occurred there based on the proximity of Florida records.

Fusconaia escambia is extant but rare in most reaches of the Conecuh River. However, it was recently found to be locally common in some portions of Gantt and Point A reservoirs.

Ecology and Biology

Fusconaia escambia occurs in creeks and small to medium rivers with slow to moderate current. It is also locally common in Gantt and Point A reservoirs. *Fusconaia escambia* is found in substrates consisting of sand and gravel or silty sand.

Fusconaia escambia is presumably a short-term brooder, gravid during spring and summer. It was found gravid in the Escambia River drainage in late June, brooding eggs and immature glochidia, which were reddish (J.D. Williams, personal observation). Glochidial hosts of *F. escambia* are unknown.

Current Conservation Status and Protection

Fusconaia escambia was considered threatened throughout its range by Williams et al. (1993). In Alabama it was considered a species of special concern by Stansbery (1976), imperiled by Lydeard et al. (1999) and a species of highest conservation concern by Garner et al. (2004). In 2003 *F. escambia* was elevated to a candidate for protection under the federal Endangered Species Act.

Remarks

The distribution pattern of *Fusconaia escambia* is unusual. Several unionids range from the Escambia River drainage east to the Choctawhatchee River drainage (e.g., *Hamiota australis*, *Pleurobema strodeanum*, *Quadrula succissa* and *Villosa choctawensis*), but *F. escambia* is unique in being restricted to the Escambia and Yellow River drainages.

Fusconaia escambia was named for the Escambia River, where it was originally discovered. The word escambia is from the Native American Choctaw words *oski*, meaning "cane", and *ambeha*, meaning "therein", which suggests the presence of extensive cane thickets along the river historically (Foscue, 1989).

Synonymy

Fusconaia escambia Clench and Turner, 1956. Clench and Turner, 1956:152–153, pl. 7, figs. 3, 4

Type locality: Escambia River, 3 miles southeast of Century, Escambia County, Florida. Holotype, MCZ 191470, length 46 mm.

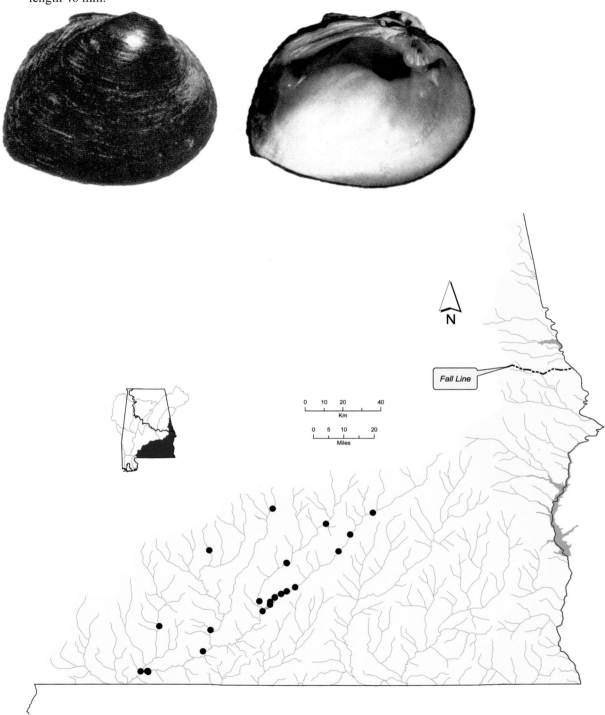

Distribution of *Fusconaia escambia* in the eastern Gulf Coast drainages of Alabama.

Fusconaia rotulata (Wright, 1899)
Round Ebonyshell

Fusconaia rotulata – Upper figure: length 61 mm, Doug Shelton personal collection, catalog number 4997.1. Conecuh River, 1 mile upstream of State Highway 41, south of East Brewton, Escambia County, Alabama, 17 August 1996. Lower figure: length 50 mm, UF 358660. Escambia River, boat ramp 0.8 miles southeast of Bluff Springs, Escambia County, Florida, 29 June 1995. © Richard T. Bryant.

Shell Description

Length to 70 mm; thick; moderately inflated; outline round to oval; posterior margin rounded; anterior margin rounded; ventral margin broadly rounded; dorsal margin convex; posterior ridge very low and rounded to absent; posterior slope broad, flat; umbo inflated, elevated above hinge line, umbo sculpture unknown; periostracum smooth, dark brown to black, without rays.

Pseudocardinal teeth triangular, thick, low, 2 divergent teeth in left valve, anterior tooth smaller, 1 tooth in right valve; lateral teeth short, straight to slightly curved, 2 in left valve, 1 in right valve; interdentum very short, wide; umbo cavity deep, compressed; nacre white.

Soft Anatomy Description

Mantle tan to light brown, often creamy white outside of pallial line, gray to grayish brown outside of apertures; visceral mass creamy white to tan or light brown; foot creamy white to tan.

Gills creamy white to tan or gold; dorsal margin straight to slightly sinuous, ventral margin convex, may be somewhat acute; gill length 50–65% of shell length; gill height 55–75% of gill length; outer gill height 60–

80% of inner gill height; inner lamellae of inner gills connected to visceral mass only anteriorly.

No gravid females were available for marsupium description; probably similar to other *Fusconaia* species, with all four gills marsupial and glochidia held across most of gill, marsupium slightly padded when gravid.

Labial palps creamy white to tan; straight to slightly convex dorsally, convex ventrally, bluntly pointed to rounded distally; palp length 20–40% of gill length; palp height 55–70% of palp length; distal 20–45% of palps bifurcate.

Excurrent and supra-anal apertures continuous, without mantle bridge; some individuals with marginal color band ending anterioventrally, suggestive of demarcation between excurrent and supra-anal apertures; continuous excurrent and supra-anal apertures longer than incurrent aperture.

Incurrent aperture length 25–30% of shell length; tan to light brown within, may have rusty cast in some areas; papillae in 1 row, arborescent; papillae some combination of creamy white, light brown, reddish brown and grayish brown, often changing color distally.

Excurrent and supra-anal apertures continuous, some individuals with demarcation between the two in mantle margin coloration; length 30–40% of shell length; creamy white to tan within, may have slight brownish cast, often with narrow rusty brown or grayish brown marginal band; margin with crenulations or small simple papillae, papillae may gradually taper to crenulations or smooth margin dorsally; papillae creamy white.

Specimens examined: Conecuh River (n = 4) (some specimens previously preserved).

Glochidium Description
Glochidium unknown.

Similar Species
Fusconaia rotulata may superficially resemble young *Glebula rotundata*. However, *G. rotundata* has distinctive pseudocardinal teeth that are comprised of small radially arranged serrate ridges. Also, *G. rotundata* lacks a wide interdentum and deep, compressed umbo cavity, which are present in *F. rotulata*.

Some *Quadrula succissa* from lower reaches of the Escambia River superficially resemble *Fusconaia rotulata*. However, *Q. succissa* nacre typically has a purplish tint, and the shell is more compressed with a better developed posterior ridge and cloth-like periostracum.

General Distribution
Fusconaia rotulata is endemic to the Escambia River drainage in Alabama and Florida.

Alabama and Mobile Basin Distribution
Fusconaia rotulata occurs in the Conecuh River upstream past the Covington and Conecuh County line.

Fusconaia rotulata is extant but extremely rare in the Conecuh River.

Ecology and Biology
Fusconaia rotulata occurs in small to medium rivers. It is usually found in slow to moderate current, typically in sand and/or small gravel substrates, occasionally in sandy mud.

Fusconaia rotulata is presumably a short-term brooder, gravid in spring and summer. Glochidial hosts of *F. rotulata* are unknown.

Current Conservation Status and Protection
Fusconaia rotulata was considered endangered throughout its range by Williams et al. (1993). In Alabama it was reported as imperiled by Lydeard et al. (1999) and designated a species of highest conservation concern by Garner et al. (2004). In 2003 *F. rotulata* was elevated to a candidate for protection under the federal Endangered Species Act.

Remarks
Fusconaia rotulata is listed in Turgeon et al., 1988, and Turgeon et al., 1998, as *Obovaria rotulata*. Williams and Butler (1994) placed this species in the genus *Fusconaia* based on conchological characters. In a subsequent genetic analysis, *F. rotulata* was found to be most closely related to *Fusconaia ebena* (Lydeard et al., 2000).

Fusconaia rotulata is the only freshwater mussel species known to be endemic to the Escambia River drainage.

Synonymy
Unio rotulatus Wright, 1899. Wright, 1899:22–23; Simpson, 1900b:78, pl. 4, fig. 2
Type locality: Escambia River, Escambia County, Florida. Holotype, USNM 159969, length 49 mm.

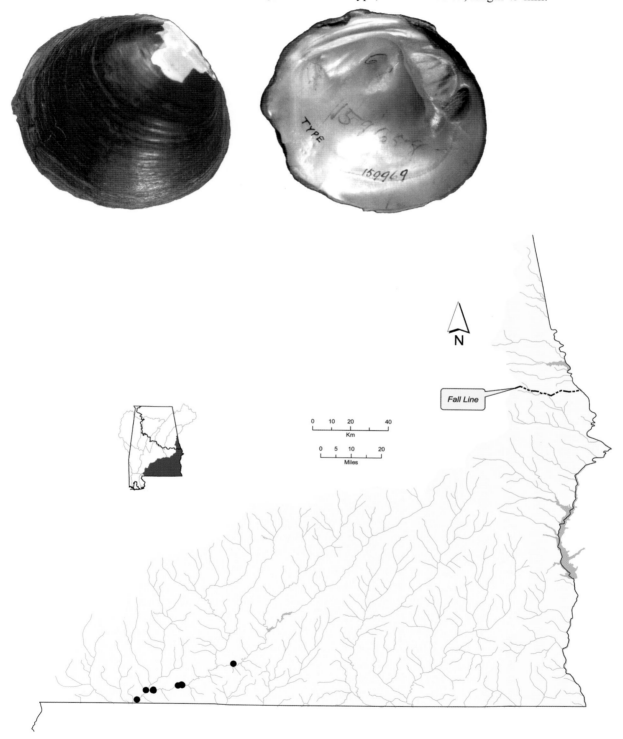

Distribution of *Fusconaia rotulata* in the eastern Gulf Coast drainages of Alabama.

Fusconaia subrotunda (Lea, 1831)
Longsolid

Fusconaia subrotunda – Upper figure: length 71 mm, UF 65239. Tennessee River, Muscle Shoals, Lauderdale County, Alabama, November 1909. Lower figure: length 56 mm, UF 388519. Clinch River at Brooks Island, Hancock County, Tennessee, 30 October 1998. © Richard T. Bryant.

Shell Description

Length to 125 mm; thick, individuals from large rivers thicker than those from small rivers; individuals from large rivers inflated, individuals from small rivers more compressed; outline oval to broadly elliptical; posterior margin broadly to narrowly rounded; anterior margin rounded, may be somewhat straight dorsally; dorsal margin straight to convex; ventral margin broadly rounded; posterior ridge very low, rounded, may be weakly doubled posterioventrally; posterior slope flat to slightly convex, steep dorsally; umbo inflated, elevated well above hinge line, turned slightly anteriad, umbo sculpture a few knobbed ridges; periostracum dull, may have satiny texture, tawny to greenish brown, darkening to almost black with age, subtle green rays sometimes present, primarily on umbo.

Pseudocardinal teeth triangular, 2 divergent teeth in left valve, anterior tooth smaller, 1 tooth in right valve, may have accessory denticle anteriorly and/or posteriorly; lateral teeth moderately long, thick, straight to slightly curved, 2 in left valve, 1 in right valve; interdentum short to moderately long, wide; umbo cavity deep, compressed; nacre white.

Soft Anatomy Description

Mantle tan, paler outside of pallial line, dark brown outside of apertures; visceral mass tan; foot tan (note accompanying specimen stated that foot and adductors were white prior to preservation). Lea (1863d) and Ortmann (1912a) reported the foot and adductor muscles to be white to orange or reddish orange.

Gills tan; dorsal margin slightly sinuous, ventral margin convex; gill length approximately 60% of shell length; gill height approximately 60% of gill length; outer gill height approximately 85% of inner gill height; inner lamellae of inner gills connected to visceral mass only anteriorly.

No gravid females were available for marsupium description. Lea (1863d) and Sterki (1898) reported all four gills to be marsupial. Lea (1863d) described the gills as red when holding "eggs". Ortmann (1912a) reported "eggs" to be generally red, but pale pink or white "in rare instances". Since fertilization occurs in the suprabranchial chamber as eggs are passed into the marsupia, Lea and Ortmann were likely observing early embryos, possibly mixed with some unfertilized eggs.

Labial palps tan; slightly concave dorsally, convex ventrally, bluntly pointed distally; palp length approximately 20% of gill length; palp height approximately 55% of palp length; distal 35% of palps bifurcate.

Incurrent aperture slightly longer than excurrent aperture; supra-anal aperture slightly longer than incurrent aperture.

Incurrent aperture length approximately 20% of shell length; tan within, dark brown basal to papillae; papillae arborescent; papillae dark brown.

Excurrent aperture length approximately 15% of shell length; grayish brown within, marginal color band dark brown, with tan band proximal to marginal band; margin with very small simple papillae, little more than crenulations; papillae dark brown. Ortmann (1912a) reported the excurrent aperture margin to have fine crenulations.

Supra-anal aperture length approximately 20% of shell length; grayish brown within, without marginal coloration; margin minutely crenulate; mantle bridge separating supra-anal and excurrent apertures absent.

Specimen examined: Clinch River, Tennessee (n = 1) (specimen previously preserved).

Glochidium Description

Length 130 μm; height 150 μm; without styliform hooks; outline subelliptical (Ortmann, 1912a, 1921).

Similar Species

In large rivers, *Fusconaia subrotunda* becomes inflated and closely resembles *Fusconaia ebena*. Rays are often present on umbos of *F. subrotunda*, but those of *F. ebena* are rarely rayed. Adult shells can be distinguished by pseudocardinal teeth. Those of *F. subrotunda* are triangular and more divergent, while those of *F. ebena* are somewhat compressed, joined anteriorly and more or less parallel to the lateral teeth. Pseudocardinal teeth in the left valve of adult *F. subrotunda* are separated to slightly connected anteriorly. In *F. ebena* the two pseudocardinals are always connected anteriorly, and the junction of the two teeth is elevated well above the hinge plate. The pseudocardinal tooth of the right valve arises from a depression in *F. ebena* but not in *F. subrotunda*.

In headwater and tributary rivers *Fusconaia subrotunda* is typically more compressed, and young individuals may resemble *Fusconaia cor* and *Fusconaia cuneolus*. The posterior ridge of *F. subrotunda* is much lower and more rounded than those of *F. cor* and *F. cuneolus*. Young *F. subrotunda* usually have a more rounded posterior margin than *F. cor* and *F. cuneolus*. The satiny periostracum texture of *F. subrotunda* can also be used to distinguish it from *F. cor*, which has a shiny periostracum.

Fusconaia subrotunda resembles *Pleuronaia dolabelloides* and *Pleuronaia barnesiana*. The posterior ridge of *F. subrotunda* is straight, whereas that of *P. dolabelloides* is ventrally curved. The foot of *F. subrotunda* may be orange but is often dull and may be creamy white or tan, whereas the foot of *P. dolabelloides* is always bright orange. *Fusconaia subrotunda* has a deeper, more compressed umbo cavity and heavier, less compressed pseudocardinal teeth than *P. barnesiana*. Pseudocardinal teeth in the left valve of *F. subrotunda* are much more divergent than those of *P. barnesiana*, giving them an overall appearance of being less perpendicular to the lateral teeth.

Fusconaia subrotunda may resemble *Pleurobema sintoxia*, but *F. subrotunda* has a deeper umbo cavity. *Pleurobema sintoxia* is often more triangular in outline than *F. subrotunda* from large river habitats. Soft anatomy of *F. subrotunda* may have some shade of orange, but that of *P. sintoxia* is always white.

General Distribution

Fusconaia subrotunda is known from the Detroit River and Lake Erie (Goodrich and van der Schalie, 1932). It occurs in the Ohio River drainage from headwaters in Pennsylvania (Ortmann, 1909a) downstream to the mouth of the Ohio River in Illinois and Kentucky (Cummings and Mayer, 1992). *Fusconaia subrotunda* is widespread in the Cumberland River drainage downstream of Cumberland Falls (Cicerello et al., 1991) and occurs throughout the Tennessee River drainage (Ahlstedt, 1992a, 1992b; Parmalee and Bogan, 1998).

Alabama and Mobile Basin Distribution

Fusconaia subrotunda historically occurred in the Tennessee River across the state and in some tributaries.

Fusconaia subrotunda is apparently extant only in tailwaters of Wilson and possibly Guntersville dams on the Tennessee River, and the Paint Rock River. It is rare in all extant populations.

Ecology and Biology

Fusconaia subrotunda occurs in small to large rivers. In headwater rivers it occurs in shoal habitats. In the Tennessee River today it can be found only in tailwaters of dams, where it is rare. There it may occur at depths of more than 6 m. The preferred substrate of *F. subrotunda* is gravel or a mixture of gravel and sand, kept free of silt by current.

Fusconaia subrotunda is a short-term brooder, gravid from early May through July (Sterki, 1898; Ortmann, 1909a, 1913b, 1919, 1921). Mature glochidia have been reported as early as mid-May (Ortmann, 1921), but they may mature later in some populations (Ortmann, 1913b). Glochidial conglutinates of *F. subrotunda* are subcylindrical and red, pink or white (Ortmann, 1912a, 1918, 1921). Conglutinates of *F. subrotunda* may be discharged intact or in pieces (Ortmann, 1910c, 1913b). Glochidial hosts of *F. subrotunda* are unknown.

Current Conservation Status and Protection

Fusconaia subrotunda was listed as a species of special concern throughout its range by Williams et al. (1993). In Alabama it was considered endangered by Stansbery (1976) and a species of special concern by Lydeard et al. (1999). Garner et al. (2004) designated *F. subrotunda* a species of highest conservation concern in the state.

Remarks

Clinal variation in shell morphology is well-documented in *Fusconaia subrotunda*. Ortmann (1918, 1920) reported such a phenomenon in the upper Ohio and Tennessee River drainages. With an upstream progression, shells become smaller and more compressed. This is partially responsible for the considerable synonymy of this species.

Fusconaia subrotunda has been reported from prehistoric shell middens along the Tennessee River across northern Alabama but was uncommon at all sites (Morrison, 1942; Warren, 1975; Hughes and Parmalee, 1999). Morrison (1942) speculated that its rarity may have been due to its "relatively deep-water habitat". Material from some shell piles, apparently historical preimpoundment cull piles from commercial mussel harvesters, contain a considerable proportion of *F. subrotunda*. The likelihood that these shells were collected from deeper water using brail adds credence to Morrison's speculation.

Synonymy

Obliquaria (*Rotundaria*) *subrotunda* var. *maculata* Rafinesque, 1820. Rafinesque, 1820:308
Comment: The variety *maculata* was not figured, and no types are known (Johnson and Baker, 1973). Lea (1836, 1838b, 1852a, 1870) and Simpson (1900b, 1914) considered *subrotunda* a *nomen dubium*. Stansbery (1976) was the first to recognize this variety as a species, *Fusconaia maculata maculata* (Rafinesque, 1820). See Article 45.6 (ICZN, 1999). However, since there is no recognized type and no original figure, this taxon is considered a *nomen dubium*.
Unio subrotundus Lea, 1831. Lea, 1831:117, pl. 18, fig. 45; Lea, 1834b:127, pl. 18, fig. 45
Type locality: Ohio. Lectotype, USNM 85832, length 66 mm, designated by Johnson (1974).

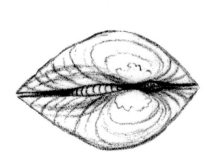
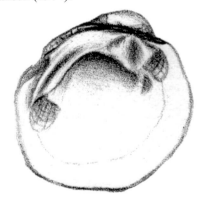

Unio kirtlandianus Lea, 1834. Lea, 1834a:98, pl. 14, fig. 41; Lea, 1834b:210, pl. 14, fig. 41
Unio personatus Conrad, 1834, *non* Say, 1829. Conrad, 1834b:71
Unio pilaris Lea, 1840. Lea, 1840:285; Lea, 1842b:209, pl. 14, fig. 24; Lea, 1842c:47, pl. 14, fig. 24
Unio lesueurianus Lea, 1840. Lea, 1840:286; Lea, 1842b:195, pl. 8, fig. 6; Lea, 1842c:33, pl. 8, fig. 6

Unio globatus Lea, 1871. Lea, 1871:191; Lea, 1874c:5, pl. 1, fig. 1; Lea, 1874e:9, pl. 1, fig. 1
Unio bursa-pastoris B.H. Wright, 1896. B.H. Wright, 1896:133, pl. 3
Quadrula flexuosa Simpson, 1900. Simpson, 1900a:83, pl. 2, fig. 8
Quadrula (*Fusconaia*) *kirtlandiana* var. *minor* Simpson, 1900. Simpson, 1900b:791
Quadrula andrewsii Marsh, 1902. Marsh, 1902a:115
Quadrula andrewsae Marsh, 1902. Marsh 1902b:8, pl. 1 figure caption, upper two figs. [unjustified emendation]
Quadrula beauchampii Marsh, 1902. Marsh, 1902b:7, pl. 1, lower two figs.
Fusconaja subrotunda var. *leucogona* Ortmann, 1913. Ortmann, 1913b:89–90

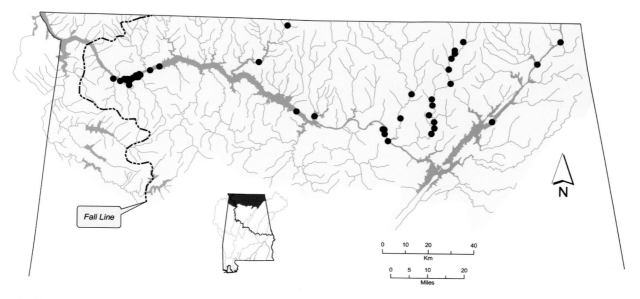

Distribution of *Fusconaia subrotunda* in the Tennessee River drainage of Alabama.

Genus *Glebula*

Glebula Conrad, 1853, is restricted to Gulf Coast drainages from northern Florida to Texas, including the Mobile and Mississippi basins. In Alabama it occurs in lower reaches of the Mobile Basin and Conecuh River. *Glebula* is a monotypic genus (Turgeon et al., 1998).

Type Species
Unio rotundatus Lamarck, 1819

Diagnosis
Shell elliptical; moderately thick; posterior ridge low; umbo flattened; periostracum cloth-like; female shells swollen posterioventrally; pseudocardinal teeth divided into small radially arranged serrate ridges.

Inner lamellae of inner gills completely connected to visceral mass; outer gills marsupial; glochidia held in posterior portion of gill; marsupium reniform in outline when gravid, extended ventrally beyond original gill edge when gravid; glochidium without styliform hooks (Conrad, 1853; Simpson, 1900b, 1914; Haas, 1969b).

Synonymy
None recognized.

Glebula rotundata (Lamarck, 1819)
Round Pearlshell

Glebula rotundata – Upper figure: female, length 103 mm, UF 375985. Lower figure: male, length 77 mm, UF 375985. Slough off Tensaw Lake, 1 air mile southwest of Hubbard Fish Camp and Landing, Baldwin County, Alabama, 18 September 1999. © Richard T. Bryant.

Shell Description

Length to 150 mm; moderately thick; moderately to greatly inflated; outline oval; posterior margin rounded to obliquely truncate; anterior margin rounded, may be somewhat obliquely truncate; dorsal margin convex; ventral margin broadly rounded; posterior ridge rounded to somewhat angular; posterior slope steep, somewhat convex; sexual dimorphism subtle, females somewhat swollen posteriorly, slightly less elongate than males; umbo broad, inflated, even with hinge line in young individuals, elevated slightly above hinge line in older individuals, umbo sculpture irregular, very weak, almost nonexistent; periostracum dull, with cloth-like texture, brown in young individuals, underlying purple nacre may give very young individuals pinkish cast, old individuals usually black, without rays.

Pseudocardinal teeth of both valves small, radially arranged serrate ridges, individual lamellae serrate, juveniles with 2 thin, blade-like teeth in each valve, almost parallel to margin; lateral teeth short, strong, straight to slightly curved, 2 in left valve, 1 in right valve; interdentum long, very narrow; umbo cavity broad, moderately deep; nacre purplish pink to white, often paler ventrally.

Soft Anatomy Description

Mantle tan to light brown; visceral mass white to light tan; foot tan, usually darker distally.

Gills light brown; slightly sinuous dorsally, convex ventrally; gill length 50–60% of shell length; gill height 30–45% of gill length; outer gill height variable compared to inner gill height (greater, lesser or equal); inner lamellae of inner gills completely connected to visceral mass.

Outer gills marsupial; glochidia held in posterior 65% of gills; marsupia padded when gravid; creamy white.

Labial palps creamy white to light brown; straight to slightly concave dorsally, curved ventrally, rounded to bluntly pointed distally; palp length 30–45% of gill length; palp height 25–30% of palp length; distal 25–30% of palps bifurcate.

Incurrent and supra-anal apertures longer than excurrent aperture; incurrent and supra-anal apertures of variable length, either may be longer than the other.

Incurrent aperture length 10–15% of shell length; creamy white to tan within, usually without coloration basal to papillae; papillae in 2 rows, long, slender, simple and bifid; papillae creamy white to tan or pale orange.

Excurrent aperture length 5–10% of shell length; creamy white to tan within, with or without black or brown marginal color band; margin papillate, papillae simple, short, blunt, sometimes triangular; papillae pale orange, may have black mottling.

Supra-anal aperture length 10–20% of shell length; creamy white within, without marginal coloration; margin smooth; mantle bridge separating supra-anal and excurrent apertures often with multiple perforations, length usually at least 50% of supra-anal length.

Specimens examined: Tombigbee River (n = 3).

Glochidium Description

Length 165–170 µm; height 320–325 µm; without styliform hooks; outline depressed subelliptical; dorsal margin straight, ventral, anterior and posterior margins evenly curved (Howells et al., 1996).

Similar Species

Glebula rotundata superficially resembles *Elliptio crassidens* and *Lampsilis ornata*. It can be distinguished from those and all other species by its unique pseudocardinal teeth comprised of radially arranged serrate ridges. Additionally, *E. crassidens* is usually less inflated and more elongate. *Lampsilis ornata* has a much higher umbo and a lighter, yellowish periostracum.

General Distribution

Glebula rotundata occurs in Gulf Coast drainages from the Ochlockonee River in Florida west to the Guadalupe River in eastern Texas (Howells et al., 1996). It is found northward in the Mississippi Basin to Arkansas and northeastern Oklahoma. Cicerello et al. (1991) reported a disjunct record in the Ohio River at Louisville, Kentucky.

Alabama and Mobile Basin Distribution

Glebula rotundata occurs throughout the Mobile Delta and upstream into lower reaches of the Alabama and Tombigbee rivers. There is one museum record labeled "Black Warrior" with no specific locality data. *G. rotundata* is known from most Gulf Coast drainages, but the only record from Alabama reaches of those systems is from the Conecuh River at East Brewton, Escambia County.

Glebula rotundata remains locally abundant in parts of the Mobile Delta. It is rare in the Conecuh River.

Ecology and Biology

Glebula rotundata is most abundant in lower reaches of rivers, near the coast. It is tolerant of brackish water and is syntopic with euryhaline bivalves *Mytilopsis leucophaeata* and *Rangia cuneata*. *Glebula rotundata* inhabits creeks to large rivers with slow to moderate current, as well as sloughs, oxbows and other backwaters with no current. The substrate in which it is found is typically a mixture of mud, silt and sand, often with detritus. At a site in Louisiana with a density of 14 *G. rotundata* per m^2, substrate was characterized as 73 percent clay, 12.5 percent sand, 8.4 percent silt and 6.1 percent organic material (Parker et al., 1984). *Glebula rotundata* is widespread but is absent from many areas with apparently suitable habitat. It is locally abundant in the Mobile Delta, but its abundance decreases upstream in the Alabama and Tombigbee rivers.

Glebula rotundata has generally been considered to belong to the Lampsilinae, which are primarily long-term brooders. However, its brooding period differs from those of most other long-term brooders, which are typically gravid from late summer or autumn into the following summer. In a detailed study in Louisiana *G. rotundata* was reported to be gravid from March to October (Parker et al., 1984). Gravid females have been reported from June to August in the Apalachicola River, Florida (Brim Box and Williams, 2000), and in June and July in Texas (Howells, 2000). Parker et al. (1984) suggested that female *G. rotundata* have at least three periods of glochidial discharge per year. Marsupial content of 5 individuals averaged 531,000 glochidia, but overall fecundity depends on the number of broods per year (Parker et al., 1984).

Lepomis cyanellus (Green Sunfish) and *Lepomis macrochirus* (Bluegill) (Centrarchidae) have been found to serve as glochidial hosts of *Glebula rotundata* in laboratory trials (Parker et al., 1984). In Louisiana natural infestations of *G. rotundata* glochidia have been reported for *Trinectes maculatus* (Hogchoker) (Achiridae); *Anchoa mitchilli* (Bay Anchovy) (Engraulidae); *Lepisosteus oculatus* (Spotted Gar) (Lepisosteidae); and *Morone chrysops* (White Bass) (Moronidae), as well as the nonindigenous *Cyprinus carpio* (Common Carp) (Cyprinidae) (Parker et al., 1984). All of these fishes occur in lower reaches of the Mobile Basin.

Current Conservation Status and Protection

Glebula rotundata was considered currently stable throughout its range by Williams et al. (1993) and in Alabama by Lydeard et al. (1999). Garner et al. (2004) designated *G. rotundata* a species of moderate conservation concern in the state.

Remarks

Unio grandensis Conrad, 1855, was described from the Rio Grande, Texas, and has been included as a synonym of *Glebula rotundata* by some authors. However, its identity is difficult to confirm, since the type specimen is lost (Johnson and Baker, 1973). Howells et al. (1996) suggested that *U. grandensis* is a synonym of *Cyrtonaias tampicoensis* because the accepted range of *G. rotundata* reaches its western limit in the Guadalupe River system in Texas.

Synonymy

Unio rotundatus Lamarck, 1819. Lamarck, 1819:75

Type locality: Unknown. Holotype, USNM 85760, length 80 mm. Clench and Turner (1956) restricted the type locality to Bayou Teche, St. Mary Parish, Louisiana.

Comment: Lea was presented what is believed to be the type specimen of *Unio rotundatus* by Ferussac, which is reported to be the individual cited by Lamarck. Ferussac pointed out that Lamarck based his description of *Unio rotundata* and *Unio suborbiculata* on the same specimen in the same paper (Johnson, 1969b, 1975). The holotype was figured by Johnson (1969b).

Unio suborbiculata Lamarck, 1819. Lamarck, 1819:81

Unio glebulus Say, 1831. Say, 1831c:526

Unio subglobosus Lea, 1834. Lea, 1834a:89, pl. 2, fig. 3; Lea, 1834b:142, pl. 2, fig. 3

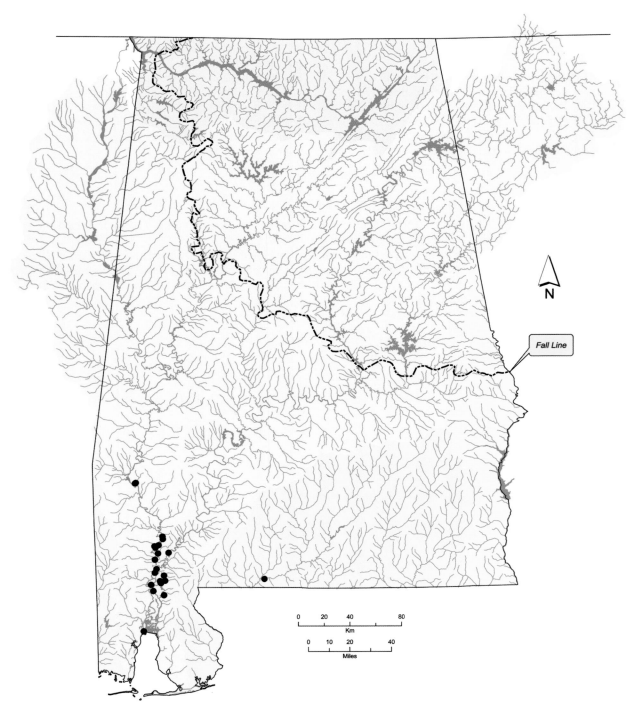

Distribution of *Glebula rotundata* in Alabama and the Mobile Basin.

Genus *Hamiota*

Hamiota Roe and Hartfield, 2005, is restricted to Gulf Coast drainages from the Apalachicola Basin to the Mobile Basin. Four species are included in *Hamiota*, all of which occur in Alabama.

These four species—*Hamiota altilis* (Conrad, 1834), *Hamiota australis* (Simpson, 1900), *Hamiota perovalis* (Conrad, 1834) and *Hamiota subangulata* (Lea, 1840)—were previously included in *Lampsilis* Rafinesque, 1820, and/or *Villosa* Frierson, 1927. Recent genetic analyses indicate that the group is a monophyletic clade separate from *Lampsilis* and *Villosa* (Roe et al., 2001), and *Hamiota* was erected to accommodate them (Roe and Hartfield, 2005). Species of *Hamiota* produce superconglutinates (Figures 13.12, 13.13), a character believed to be unique among the Unionidae (Haag et al., 1995).

Type Species
Unio subangulatus Lea, 1840

Diagnosis
Shell oval to elliptical; thin to moderately thick; somewhat compressed to inflated; females often slightly more inflated posterioventrally than males; disk surface smooth; periostracum often shiny.

Females usually with mantle margin just ventral to incurrent aperture variously developed into flap or papillate fold, variously pigmented; flap or fold rudimentary in males. Inner lamellae of inner gills completely connected or only connected to visceral mass anteriorly; outer gills marsupial; glochidia held in posterior portion of gill; marsupium broadest anteriorly, tapering posteriorly, extended ventrally below original gill edge when gravid, ventral margin darkly pigmented; glochidia discharged as superconglutinates; glochidium without styliform hooks (Haag et al., 1995; O'Brien and Brim Box, 1999; Roe and Hartfield, 2005).

Synonymy
None recognized.

Figure 13.12. *Hamiota perovalis*. Early stages of superconglutinate discharge. Black Warrior River drainage, Alabama. Photograph by W.N. Roston.

Figure 13.13. *Hamiota subangulata* in an artificial stream, discharging a superconglutinate attached to paired mucus tubes. Individual from Flint River drainage, Georgia, May 1995. Photograph by N.M. Burkhead.

Hamiota altilis (Conrad, 1834)
Finelined Pocketbook

Hamiota altilis – Upper figure: female, length 53 mm, UF 3255. Coosa River, Weduska Shoals, Shelby County, Alabama, August 1913. Lower figure: male, length 87 mm, UF 358661. Uphapee Creek, State Highway 81, 2.5 miles north of Tuskegee, Macon County, Alabama, 24 June 1973. © Richard T. Bryant.

Shell Description

Length to 117 mm; thin when young, becoming moderately thick with age; moderately inflated; outline oval; posterior margin narrowly rounded; anterior margin rounded; dorsal margin convex; ventral margin broadly rounded; sexual dimorphism very subtle, females slightly more inflated posteriorly than males; posterior ridge, low, rounded; posterior slope moderately steep, flat; umbo broad, moderately inflated, elevated above hinge line, umbo sculpture unknown; periostracum tawny to brown, usually with thin dark green or brown rays.

Pseudocardinal teeth triangular, compressed, 2 slightly divergent teeth in left valve, 1 tooth in right valve, usually with accessory denticle anteriorly; lateral teeth short, moderately thick, straight to slightly curved, 2 in left valve, 1 in right valve; interdentum long, very narrow; umbo cavity shallow; nacre white to bluish white.

Soft Anatomy Description

Mantle tan, heavy rusty tan mottling outside of apertures; visceral mass tan; foot tan.

Females with well-developed mantle flap just ventral to incurrent aperture; flap length approximately 40% of shell length; mantle adjacent to flap with heavy but irregular rusty brown mottling; margin with well-developed, widely separated papillae; papillae thick basally, tapering distally, most simple but a few suggestive of bifid or trifid, papillae dark rusty brown, rudimentary papillae may continue for some distance anterior of mantle flap.

Gills tan; dorsal margin slightly sinuous, ventral margin convex, outer gill ventral margin may be slightly bilobed in females; gill length approximately 55% of shell length; gill height approximately 60% of gill length; outer gill height approximately 65% of inner gill height; inner lamellae of inner gills completely connected to visceral mass.

Outer gills marsupial; glochidia held in posterior 50% of gill (posterior lobe when bilobed); marsupium distended when gravid; creamy white with black or gray distal margin.

Labial palps tan; straight dorsally, convex ventrally, bluntly pointed distally; palp length approximately 20% of gill length; palp height approximately 65% of palp length; distal 45% of palps bifurcate.

Incurrent and supra-anal apertures of similar length; excurrent aperture shorter than incurrent and supra-anal apertures.

Incurrent aperture length approximately 15% of shell length; tan within, rusty tan with rusty brown mottling basal to papillae; papillae in 2–3 rows, long, slender, mostly simple, a few scattered bifid; papillae rusty tan mottled with rusty brown.

Excurrent aperture length approximately 10% of shell length; tan within, marginal color band rusty brown, with faint rusty tan lines distally; margin crenulate.

Supra-anal aperture length approximately 15% of shell length; tan within, with sparse, irregular rusty brown marginally; margin smooth; mantle bridge separating supra-anal and excurrent apertures imperforate, length approximately 40% of supra-anal length.

Specimen examined: Little River, Cherokee County (n = 1) (specimen previously preserved).

Glochidium Description

Length 250–275 μm; height 300–325 μm; without styliform hooks; outline subelliptical; dorsal margin straight to slightly concave, ventral margin rounded, anterior and posterior margins convex.

Similar Species

Hamiota altilis closely resembles *Hamiota perovalis* but typically has a thinner shell and yellowish periostracum. The periostracum of *H. perovalis* is usually brown to black. Nacre of *H. altilis* is almost always white, whereas that of *H. perovalis* is often pale pink or pinkish orange.

Hamiota altilis may also resemble *Lampsilis straminea*, but is less elongate and less inflated. The periostracum of *L. straminea* is typically without rays, whereas *H. altilis* usually has rays. Sexual dimorphism is more pronounced in *L. straminea*, in which females often have greatly exaggerated marsupial swellings.

General Distribution

Hamiota altilis is endemic to the Mobile Basin in Alabama, Georgia and Tennessee.

Alabama and Mobile Basin Distribution

Hamiota altilis occurs in the eastern portion of the Mobile Basin from headwaters of the Coosa River drainage downstream to Claiborne, Monroe County, on the Alabama River, and in the Cahaba and Tallapoosa rivers. Some specimens from the Black Warrior and Tombigbee River drainages are very similar to *H. altilis*, but their identity remains unresolved.

Hamiota altilis is extant in widespread, isolated populations.

Ecology and Biology

Hamiota altilis occurs in small creeks to large rivers. It is typically found in areas with at least some current, though usually not in swift current. Substrates in which *H. altilis* usually occur include sand and mixtures of sand and gravel without heavy silt accumulations.

Hamiota altilis is a long-term brooder, gravid from late summer or autumn to the following spring. Ortmann (1924a) reported one female *H. altilis* gravid with eggs during mid-May, which he noted was "entirely abnormal". *Hamiota altilis* is one of only four freshwater mussel species known to produce superconglutinates. The superconglutinates of *H. altilis* are fusiform in shape, larger at one end than the other, creamy white with a dusky stripe dorsally and no eyespot (Haag and Warren, 1999). It is discharged into paired mucus tubes that remain attached to the female parent, at least for a period. Female *H. altilis* also display mantle flaps, which mimic fish (Figure 13.14). While displaying, the female either positions itself with the posterior one-third to one-half of its shell above the substrate or lies completely out of the substrate on its dorsum. Each displayed flap has numerous papillae marginally. The flap is wider and papillae larger toward the anterior end (Haag et al., 1999).

Figure 13.14. *Hamiota altilis* gravid female with fish-mimicking mantle flaps displayed (dorsal view). Charged gills are visible between the flaps. Conasauga River, Whitfield County, Georgia, May 2004. Photograph by P.D. Johnson.

Fishes found to serve as glochidial hosts of *Hamiota altilis* in laboratory trials include *Lepomis cyanellus* (Green Sunfish), *Micropterus coosae* (Redeye

Bass), *Micropterus punctulatus* (Spotted Bass) and *Micropterus salmoides* (Largemouth Bass) (Centrarchidae) (Haag et al., 1999). Some of the trials with *L. cyanellus* successfully produced juvenile *H. altilis*, but others failed.

Current Conservation Status and Protection

Hamiota altilis was considered endangered throughout its range by Athearn (1970) and threatened throughout its range by Williams et al. (1993). In Alabama Lydeard et al. (1999) reported it as threatened, and Garner et al. (2004) designated it a species of high conservation concern. *Hamiota altilis* (as *Lampsilis altilis*) was listed as threatened under the federal Endangered Species Act in 1993.

Remarks

This species has historically been placed in *Lampsilis*. However, using genetic analyses, it was found to belong to a monophyletic clade of four superconglutinate producers (Roe et al., 2001). Roe and Hartfield (2005) erected the genus *Hamiota* for the group.

Exact distributions of *Hamiota altilis* and *Hamiota perovalis* are uncertain due to overlapping conchological characters. Some individuals from the Black Warrior and Tombigbee River drainages closely resemble typical *H. altilis*. Likewise, some individuals in the Coosa and Tallapoosa River drainages closely resemble *H. perovalis*. Roe et al. (2001), using genetic analysis, validated the distinctness of the two taxa, but small sample sizes precluded a clear resolution of their ranges. Since the exact distributions of these species remain unresolved, *H. altilis* is herein tentatively considered restricted to eastern and southern reaches of the Mobile Basin and *H. perovalis* restricted to western and southern reaches of the basin.

Hamiota altilis was described from the Alabama River at Claiborne, Alabama, along with *Hamiota perovalis*. The original descriptions included colored plates (Conrad, 1834b), but the current interpretation of these two species does not exactly coincide with Conrad's descriptions and plates. A detailed study of these species, including shell morphology, soft anatomy, life history traits and genetics, is needed to resolve the problem.

Synonymy

Unio altilis Conrad, 1834. Conrad, 1834b:43, pl. 2, fig. 1
Type locality: Alabama River, near Claiborne, [Monroe County, Alabama]. Lectotype, ANSP 56419, length 54 mm, designated by Johnson and Baker (1973).
Comment: Johnson and Baker (1973) designated ANSP 56419 as the "figured holotype", which is the valid lectotype selection. Roe and Hartfield (2005) unnecessarily redesignated the same specimen as the lectotype.

Unio clarkianus Lea, 1852, *nomen nudum*. Lea, 1852a:251
Unio clarkianus Lea, 1852. Lea, 1852b:273, pl. 21, fig. 30; Lea, 1852c:29, pl. 21, fig. 30
Type locality: Williamsport, Tennessee, J. Clark; Georgia or Alabama. Lectotype not found, also not found by Johnson (1974). Length of figured shell in original description reported as about 56 mm. This specimen appears to be from the Coosa River drainage of Alabama or Georgia.

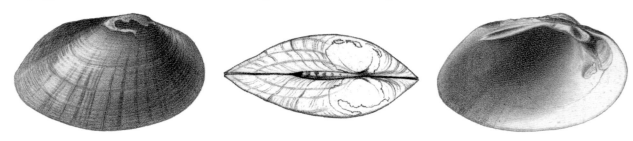

Unio gerhardtii Lea, 1862. Lea, 1862a:168; Lea, 1862c:208, pl. 31, fig. 277; Lea, 1863a:30, pl. 31, fig. 277
Type locality: Chattanooga, [Chattooga River,] Georgia, Alexander Gerhardt. Holotype, USNM 25711, length 70
   mm.

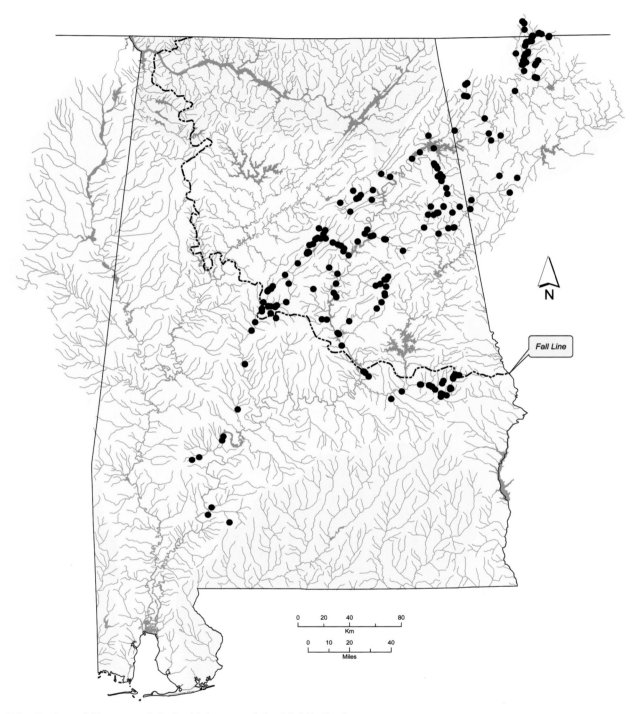

Distribution of *Hamiota altilis* in Alabama and the Mobile Basin.

Hamiota australis (Simpson, 1900)
Southern Sandshell

Hamiota australis – Upper figure: female, length 67 mm, UF 65312. Conecuh River near Searight, Crenshaw County, Alabama, June 1915. Lower figure: male, length 56 mm, UMMZ 54108. Little Choctawhatchie [Choctawhatchee] River, southeastern Alabama. © Richard T. Bryant.

Shell Description

Length to 83 mm; moderately thin; moderately inflated; outline elliptical; posterior margin bluntly pointed; anterior margin rounded; dorsal margin straight to slightly convex; ventral margin broadly rounded to straight; sexual dimorphism subtle, females slightly more inflated posterioventrally; posterior ridge rounded; posterior slope moderately steep, flat to slightly concave; umbo broad, elevated above hinge line, umbo sculpture unknown; periostracum shiny, greenish when young, becoming dark greenish brown to black with age, usually with many variable green rays.

Pseudocardinal teeth triangular, compressed, 2 slightly divergent teeth in left valve, 1 tooth in right valve, often with accessory denticle anteriorly; lateral teeth thin, straight, 2 in left valve, 1 in right valve; interdentum moderately long, very narrow; umbo cavity shallow; nacre white.

Soft Anatomy Description

Mantle creamy white to tan or light brown, outside of apertures creamy white to dull orange with black specks or gray to brown mottling, usually in form of crude bands perpendicular to margin; visceral mass creamy white to tan; foot creamy white to tan.

No mantle modification ventral to incurrent aperture.

Gills tan to light brown, sometimes with slight golden cast; dorsal margin straight to slightly sinuous, ventral margin convex; gill length 50–65% of shell length; gill height 40–65% of gill length; outer gill height usually 65–95% of inner gill height, sometimes greater; inner lamellae of inner gills connected to visceral mass only anteriorly.

Outer gills marsupial; glochidia held across entire gill except extreme anterior and posterior ends; marsupium elongate, concave dorsally, convex ventrally, bluntly pointed posteriorly, rounded anteriorly; gray with thin black band and white specks on distal margin, nongravid marsupia reddish brown with thin black band and white specks distally.

Labial palps creamy white to tan; straight to slightly concave dorsally, convex ventrally, rounded to bluntly pointed distally; palp length 25–35% of gill length; palp height 55–75% of palp length; distal 25–40% of palps bifurcate.

Incurrent aperture slightly longer or equal to excurrent aperture; supra-anal aperture slightly longer or equal to incurrent aperture.

Incurrent aperture length 10–15% of shell length; creamy white within, usually with some combination of gray, black and various shades of brown basal to papillae, coloration often irregular; papillae in 1–3 rows, inner row usually largest, long, slender, simple, occasionally with very few bifid; papillae some combination of creamy white, tan, pale orange, brown and black, may be mottled, often with darker edges basally, may change color distally.

Excurrent aperture length 5–10% of shell length, creamy white to light brown within, may be mottled with various shades of brown, occasionally with slight golden cast, marginal color band some combination of creamy white, tan, black and various shades of brown, often with dark background and irregular lighter bands perpendicular to margin, some bands may converge proximally, creamy white or tan band often present just proximal to marginal band; margin with small simple papillae; papillae tan, pale orange, brown or gray.

Supra-anal aperture length 10–20% of shell length; creamy white within, may have thin gray or black marginal band; margin smooth; mantle bridge separating supra-anal and excurrent apertures imperforate, length 30–60% of supra-anal length.

Specimens examined: Eightmile Creek, Walton County, Florida (n = 1); Flat Creek, Geneva County (n = 2); West Fork Choctawhatchee River (n = 4).

Glochidium Description

Length 225 µm; height 294 µm; without styliform hooks; outline subspatulate; dorsal margin straight, ventral margin broadly rounded, anterior and posterior margins convex (Blalock-Herod et al., 2002).

Similar Species

Hamiota australis most closely resembles *Ptychobranchus jonesi* but has a more regular, pointed posterior margin. The posterior ridge of *H. australis* is lower and rounder than that of *P. jonesi*, which is typically doubled distally and has a shallow radial depression just dorsal to the posterior ridge, ending in a weak emargination posteriorly. Also, the ventral margin of *H. australis* is usually more broadly rounded than that of *P. jonesi*. When gravid, *P. jonesi* has folded outer gills, which is characteristic of the genus.

Hamiota australis resembles *Villosa villosa*, but the periostracum of that species is cloth-like, unlike that of *H. australis*, which is always shiny. The posterior margin of *H. australis* is typically more pointed than that of *V. villosa*.

Hamiota australis may superficially resemble *Villosa vibex*, but the posterior margin of *V. vibex* is much more rounded. Pseudocardinal teeth of *V. vibex* are much thinner and more blade-like than those of *H. australis*.

Hamiota australis may also superficially resemble male *Villosa lienosa* but is more elongate, has a shinier periostracum and bluish white nacre. The nacre of *V. lienosa* is typically purple to pinkish orange, only rarely white.

General Distribution

Hamiota australis is endemic to the Yellow and Choctawhatchee River drainages in Alabama and Florida and the Escambia River drainage in Alabama.

Alabama and Mobile Basin Distribution

Hamiota australis occurs in the Conecuh, Yellow and Choctawhatchee River drainages.

Hamiota australis remains widespread in the Choctawhatchee River and many of its tributaries, but its range is fragmented and it appears to be declining in many areas. There are few recent records of *H. australis* from the Escambia and Yellow River drainages (Pilarczyk et al., 2006).

Ecology and Biology

Hamiota australis inhabits small creeks to rivers in sand or mixtures of sand and fine gravel substrates and slow to moderate current.

Blalock-Herod et al. (2002) provided a partial description of *Hamiota australis* reproductive biology. It is a long-term brooder, gravid from late summer or autumn to the following spring. It produces a superconglutinate, which is discharged in mid-spring. The superconglutinate from a 62 mm-long female was described as lanceolate, resembling a small fish, composed of 36 pairs of oval conglutinates arranged in oblique rows. The superconglutinate was creamy white, but black dorsally and reddish brown ventrally, with a black eyespot on the proximal end. The superconglutinate was tethered to the female by paired mucus tubes 45 mm long but may have been limited by its confinement in an aquarium. Fecundity was estimated at 61,200 to 122,400 glochidia.

Glochidial hosts of *Hamiota australis* are unknown. However, other *Hamiota* species utilize predatory centrarchids as glochidial hosts (Haag et al., 1995; O'Brien and Brim Box, 1999).

Current Conservation Status and Protection

Hamiota australis was considered endangered throughout its range by Athearn (1970) and Stansbery (1971). Williams et al. (1993) reported it as threatened throughout its range. *Hamiota australis* was considered imperiled in Alabama by Lydeard et al. (1999) and a species of highest conservation concern by Garner et al. (2004). In 2003 it was elevated to a candidate for protection under the federal Endangered Species Act.

Remarks

Hamiota australis has been variously treated as a species of *Lampsilis* or *Villosa*. However, the absence of a mantle flap or papillate mantle fold made this species difficult to place in either genus with certainty. *Hamiota australis* is one of four superconglutinate-producing species that form a monophyletic clade. Roe and Hartfield (2005) erected the genus *Hamiota* for the group.

Genetic analyses provided by Roe et al. (2001) suggest that *Hamiota australis* consists of two clades. One clade is confined to the Choctawhatchee River drainage and the other occurs in the Escambia and Yellow River drainages. These two clades should be considered as evolutionary significant units for conservation purposes.

Synonymy

Lampsilis australis Simpson, 1900. Simpson, 1900a:75, pl. 2, fig. 2 [original figure not herein reproduced]
Type locality: Little Patsaliga Creek, [Crenshaw County,] southeastern Alabama, Dr. R. Kirkland. Holotype, USNM 150473, length 52 mm.

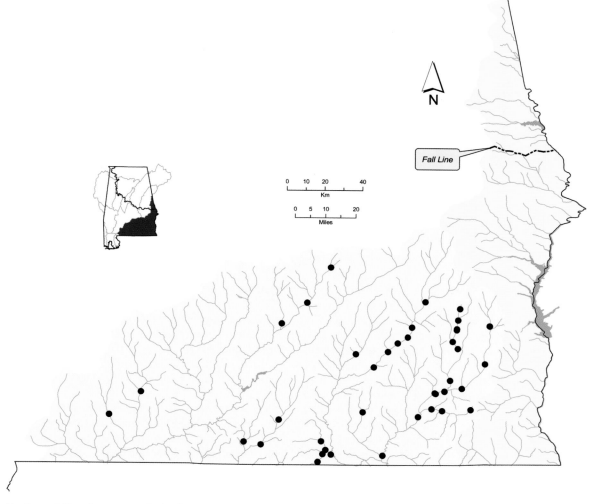

Distribution of *Hamiota australis* in the eastern Gulf Coast drainages of Alabama.

Hamiota perovalis (Conrad, 1834)
Orangenacre Mucket

Hamiota perovalis – Upper figure: female, length 71 mm, UF 65304. Lower figure: male, length 64 mm, UF 65304. Valley Creek near Toadvine, [Black] Warrior River drainage, Jefferson County, Alabama, June 1913. © Richard T. Bryant.

Shell Description

Length to 89 mm; thin when young, becoming moderately thick with age; moderately inflated; outline oval to elliptical; posterior margin narrowly rounded; anterior margin rounded; dorsal margin convex; ventral margin convex, may be almost straight; sexual dimorphism very subtle, females slightly more inflated posterioventrally; posterior ridge low, rounded; posterior slope moderately steep, slightly convex; umbo broad, moderately inflated, elevated above hinge line, umbo sculpture strong, slightly double-looped ridges; periostracum typically brown to black, rays weak to prominent, sometimes obscure.

Pseudocardinal teeth triangular, 2 divergent teeth in left valve, 1 tooth in right valve, usually with accessory denticle anteriorly; lateral teeth short, moderately thick, straight to slightly curved, 2 in left valve, 1 in right valve; interdentum long, narrow; umbo cavity shallow; nacre white, often with pink or pinkish orange tint.

Soft Anatomy Description

Mantle tan, outside of apertures rusty brown; visceral mass tan; foot tan. The foot is often pale orange in live individuals (P.D. Johnson, personal communication).

Females with weak to moderately well-developed, pigmented mantle fold just ventral to incurrent aperture (Figure 13.15); fold length 30–35% of shell length; adjacent margin tan to rusty brown or grayish brown, not in distinct band; fold with small, simple, irregularly spaced papillae, most originating inside fold instead of on margin, papillae may be better developed anteriorly, anterior-most papillae may be larger, flattened, occasionally with 2 fused to form bifid papillae, anterior-most papillae originate from margin instead of within fold. Males with more rudimentary mantle fold, shorter relative to shell length, with fewer papillae.

Gills tan; dorsal margin slightly sinuous, ventral margin convex, outer gills of females may be slightly bilobed, with marsupial area in posterior lobe; gill length 55–65% of shell length; gill height 45–55% of gill length; outer gill height 75–90% of inner gill height; inner lamellae of inner gills completely connected to visceral mass.

Outer gills marsupial; glochidia held in posterior 50–65% of gill; marsupium distended when gravid; tan with gray, grayish tan or grayish brown distal margin (Figure 13.16).

Labial palps tan; straight dorsally, curved ventrally, bluntly pointed distally; palp length 15–20% of gill length; palp height 55–85% of palp length; distal 45–50% of palps bifurcate.

Incurrent aperture longer than excurrent aperture, usually longer than supra-anal aperture, occasionally equal; supra-anal aperture usually longer than excurrent aperture, occasionally equal or shorter.

Incurrent aperture length 15–20% of shell length; tan within, with irregular rusty brown basal to papillae; papillae in 2–3 rows, inner row largest, simple, short to moderately long, thick, may become slender distally; papillae tan to rusty tan or rusty brown.

Excurrent aperture length 10–15% of shell length; tan within, with rusty brown marginal color band, may have paler band proximally; margin minutely crenulate or with very small, simple papillae, little more than crenulations; papillae tan to rusty brown.

Supra-anal aperture length 10–15% of shell length; tan within, some individuals with sparse, irregular rusty tan marginally; margin smooth; mantle bridge separating supra-anal and excurrent apertures imperforate, length usually 40–50% of supra-anal length, occasionally less.

Specimens examined: Brushy Creek, Winston County (n = 2); Sipsey Fork, Winston County (n = 2) (specimens previously preserved).

Glochidium Description

Length 225–262 μm; height 275–325 μm; without styliform hooks; outline subelliptical; dorsal margin straight, ventral margin rounded, anterior and posterior margins convex.

Similar Species

Hamiota perovalis closely resembles *Hamiota altilis* but has a thicker shell and usually a darker periostracum. *Hamiota altilis* is almost always adorned with thin rays, but the dark periostracum of *H. perovalis* often obscures rays in that species, if they are present at all. Nacre of *H. perovalis* is often pale pink or pinkish orange, whereas that of *H. altilis* is almost always white.

Hamiota perovalis may also resemble *Lampsilis straminea* but is less elongate and less inflated. The periostracum of *H. perovalis* is typically darker brown than that of *L. straminea*, which is usually yellowish. *Hamiota perovalis* often has rays, though they may be obscured by the dark periostracum, whereas *L. straminea* is typically without rays.

Small *Hamiota perovalis* may superficially resemble *Villosa lienosa* and *Villosa umbrans*. However, those species have more compressed pseudocardinal teeth and usually have purple nacre.

Figure 13.15. *Hamiota perovalis* gravid female with pigmented mantle flaps exposed. Black Warrior River drainage, Alabama. Photograph by W.N. Roston.

General Distribution

Hamiota perovalis is endemic to the Mobile Basin in Alabama and Mississippi.

Alabama and Mobile Basin Distribution

Hamiota perovalis occurs in the Alabama, Black Warrior and Tombigbee River drainages. It is found above the Fall Line in the Black Warrior River drainage. Some specimens from the Cahaba, Coosa and Tallapoosa River drainages are very similar to *H. perovalis*, but their identity remains unresolved.

Hamiota perovalis is extant in widespread, isolated populations.

Ecology and Biology

Hamiota perovalis is a species of medium creeks to large rivers. However, it is extirpated from most large rivers. *Hamiota perovalis* is usually found in slow to moderate current in sand and gravel substrates.

Hamiota perovalis is a long-term brooder, gravid from late summer or autumn to the following spring. It is one of four superconglutinate-producing species. The superconglutinates are 37 mm to 50 mm in length, 10 mm to 12 mm in height and 5 mm to 7 mm in width, rounded proximally and gradually tapering to a point distally (Haag et al., 1995). Superconglutinates are discharged into paired mucus tubes that remain attached to the excurrent aperture of the female parent, at least for a period. The tube may be up to 2.5 m long (Haag et al., 1995). Superconglutinate release is usually completed in eight hours or less (Hartfield and Butler, 1997). When displayed, the superconglutinate is creamy white ventrally, with a broad black stripe medially and pinkish cast dorsally, and a distinct eyespot at the proximal end (Haag et al., 1995). However, conglutinates from the two demibranchs may be discharged offset to each other, giving the superconglutinate a checkered appearance (Hartfield and Butler, 1997). Displays of *H. perovalis* superconglutinates have been reported to elicit attack by predatory fish under laboratory conditions (Haag and Warren, 1999).

Glochidial hosts of *Hamiota perovalis*, determined by laboratory trials, include *Micropterus coosae* (Redeye Bass), *Micropterus punctulatus* (Spotted Bass) and *Micropterus salmoides* (Largemouth Bass) (Centrarchidae) (Haag and Warren, 1997).

Current Conservation Status and Protection

Hamiota perovalis was considered endangered throughout its range by Athearn (1970) and threatened throughout its range by Williams et al. (1993). In Alabama it was considered endangered by Stansbery (1976), threatened by Lydeard et al. (1999) and a species of high conservation concern by Garner et al. (2004). *Hamiota perovalis* (as *Lampsilis perovalis*) was listed as threatened under the federal Endangered Species Act in 1993.

Remarks

This species has historically been placed in *Lampsilis*. However, using genetic analyses it was found to belong to a monophyletic clade of four superconglutinate producers (Roe et al., 2001). Roe and Hartfield (2005) erected the genus *Hamiota* for the group.

Hamiota perovalis was described from the Alabama River at Claiborne, Monroe County, Alabama, along with *Hamiota altilis*. The original descriptions included colored plates (Conrad, 1834b), but the current interpretation of these two species does not exactly coincide with Conrad's descriptions and plates. A detailed study of these species, including shell morphology, soft anatomy, life history traits and genetics, is needed to resolve the problem.

Exact distributions of *Hamiota perovalis* and *Hamiota altilis* are uncertain due to overlapping conchological characters. Some individuals from the

Cahaba, Coosa and Tallapoosa River drainages closely resemble typical *H. perovalis*. Likewise, some individuals in the Black Warrior and Tombigbee River drainages closely resemble *H. altilis*. Roe et al. (2001), using genetic data, validated the distinctness of the two taxa, but small sample sizes precluded a clear resolution of their ranges. Since the exact distributions of these species remain unresolved, *H. perovalis* is herein tentatively considered restricted to western and southern reaches of the Mobile Basin and *H. altilis* restricted to eastern and southern reaches of the basin.

Figure 13.16. *Hamiota perovalis* gravid females from Bankhead National Forest, Winston County, Alabama, illustrating variation in marsupium coloration and shape. Upper specimen: length 71 mm, UF uncataloged. Sipsey Fork, collected 20 May 1985. Lower specimen: length 70 mm, UF uncataloged. Brushy Creek, collected 26 May 1985. Photographs by J.D. Williams.

Synonymy

Unio perovalis Conrad, 1834. Conrad, 1834b:43, pl. 2, fig. 2
Type locality: Alabama River, Claiborne, [Monroe County,] Alabama. Lectotype, ANSP 56416a, length 51 mm,
    designated by Johnson and Baker (1973).

Unio placitus Lea, 1852, *nomen nudum*. Lea, 1852a:252
Unio placitus Lea, 1852. Lea, 1852b:279, pl. 23, fig. 38; Lea, 1852c:35, pl. 23, fig. 38
Type locality: Alabama, C.M. Wheatley. Holotype, USNM 85152, length 34 mm (female).

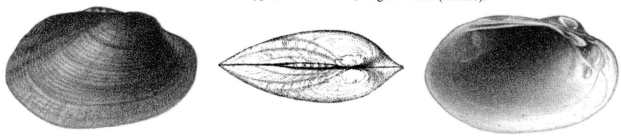

Unio spillmanii Lea, 1861. Lea, 1861a:39; Lea, 1862b:98, pl. 15, fig. 246; Lea, 1862d:102, pl. 15, fig. 246
Type locality: Luxpalila [Luxapallila] Creek, near Columbus, [Lowndes County,] Mississippi, W. Spillman, M.D.
    Lectotype, USNM 84925, length 75 mm (male), designated by Johnson (1974).

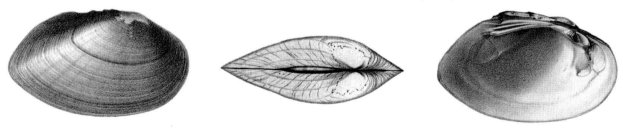

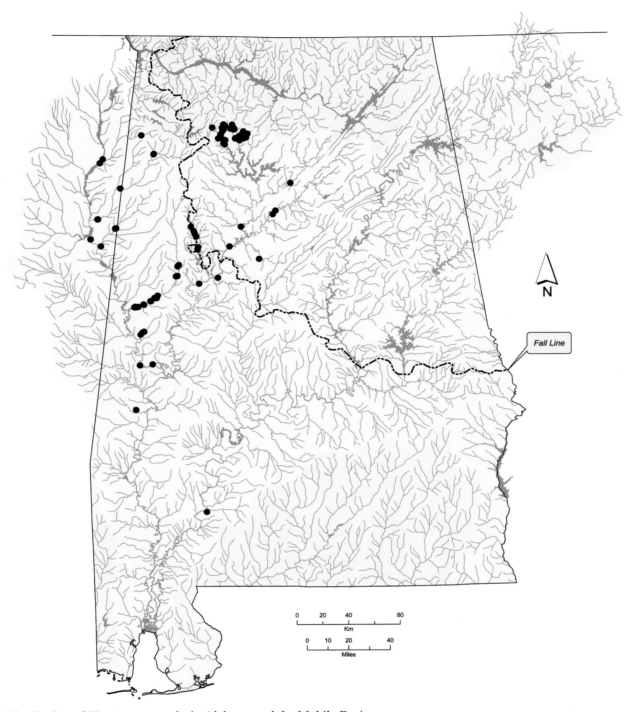

Distribution of *Hamiota perovalis* in Alabama and the Mobile Basin.

Hamiota subangulata (Lea, 1840)
Shinyrayed Pocketbook

Hamiota subangulata – Upper figure: female, length 48 mm, USNM 85081 (lectotype). Chattahoochee River, Georgia. Photo by J.D. Williams. Lower figure: male, length 61 mm, UF 388509. Kinchafoonee Creek, State Highway 45, 5.3 air miles southwest of Plains, Webster County, Georgia, 10 August 1992. © Richard T. Bryant.

Shell Description

Length to 85 mm; moderately thin; somewhat inflated; outline elongate elliptical; posterior margin bluntly pointed to narrowly rounded; anterior margin rounded; dorsal margin straight to slightly convex; ventral margin broadly rounded; sexual dimorphism expressed as slight swelling posterioventrally in females; posterior ridge well-developed near umbo but becomes low and rounded posterioventrally; posterior slope moderately steep, flat to slightly concave; umbo broad, somewhat inflated, elevated slightly above hinge line, umbo sculpture unknown; periostracum yellow to dark brown, with variable dark green rays, occasional specimens almost rayless.

Pseudocardinal teeth compressed, 2 teeth in left valve, anterior tooth larger, 1 tooth in right valve, with very small accessory denticle anteriorly; lateral teeth short, straight to slightly curved, 2 in left valve, 1 in right valve; interdentum long, narrow to very narrow; umbo cavity broad, shallow; nacre white to bluish white, occasionally with pinkish orange tint in umbo cavity.

Soft Anatomy Description

Mantle tan to light brown, may have rusty cast inside pallial line, tan with dark gray or brown mottling outside of apertures; visceral mass tan, may have rusty brown mottling; foot tan to pale pinkish orange. Lea (1863d) reported the "color of the mass" to be "light salmon".

Females with rudimentary mantle flap just ventral to incurrent aperture; flap length approximately 40–55% of shell length, very narrow; with narrow dark brown and/or rusty brown band marginally; margin crenulate, crenulations may be minute, occasionally with very small papillae.

Gills tan to rusty brown; dorsal margin straight to slightly sinuous, ventral margin elongate convex, outer gills may be more convex than inner gills; gill length 60–70% of shell length; gill height 40–55% of gill length; outer gill height 75–85% of inner gill height; inner lamellae of inner gills completely connected to visceral mass.

Outer gills marsupial; glochidia held in posterior 50% of gill, but not extending completely to posterior end; marsupium padded when gravid; tan, distal margin rusty brown or grayish brown.

Labial palps tan, may have rusty cast; straight to slightly convex or undulating dorsally, convex ventrally, bluntly pointed distally, occasionally rounded; palp length 20–35% of gill length; palp height 55–75% of palp length; distal 35–50% of palps bifurcate.

Incurrent, excurrent and supra-anal aperture lengths variable in relation to each other.

Incurrent aperture length 10–20% of shell length; tan within, often with irregular gray or rusty brown basal to papillae, brown of papillae edges may extend onto aperture wall as irregular lines; papillae in 2 rows, inner row larger, usually long, slender, simple, rarely with occasional bifid papillae on inner row; papillae creamy white to tan, often with brown edges basally. Lea (1863d) reported incurrent aperture papillae to be brown.

Excurrent aperture length approximately 10% of shell length; creamy white to light brown within, marginal color band usually tan with irregular brown bands perpendicular to margin, may have pale band proximal to marginal band; margin usually crenulate, occasionally with very small, simple papillae, margin may become smooth dorsally; papillae tan. Lea (1863d) reported excurrent aperture papillae to be brown.

Supra-anal aperture length 10–15% of shell length; tan within, without marginal coloration; margin smooth; mantle bridge separating supra-anal and excurrent apertures imperforate, length 30–90% of supra-anal length.

Specimens examined: Big Creek, Houston County (n = 2); Chipola River, Florida (n = 1); Kinchafoonee Creek, Terrell County, Georgia (n = 2); Spring Creek, Decatur County, Georgia (n = 4) (specimens previously preserved).

Glochidium Description

Length 189–211 μm; height 245–271 μm; without styliform hooks; outline subspatulate; dorsal margin straight, ventral margin rounded, anterior and posterior margins convex (O'Brien and Brim Box, 1999).

Similar Species

Hamiota subangulata resembles some *Lampsilis floridensis*, but typically has well-developed green rays over much of the shell, while *L. floridensis* has rays restricted to the posterior ridge and slope when present. The posterior terminus of the *H. subangulata* shell is usually positioned higher on the shell than in species of *Lampsilis*. Also, the periostracum of *L. floridensis* is almost always yellow, whereas that of *H. subangulata* is frequently dark brown. *Lampsilis floridensis* attains a greater size than *H. subangulata* and is highly variable

in shell morphology, often being more inflated, with a more rounded posterior margin than *H. subangulata*.

Hamiota subangulata resembles two species of *Villosa*. *Villosa vibex* differs in having a much thinner shell and hinge teeth. *Villosa lienosa* also differs in typically having purple nacre. Rays of *H. subangulata* are usually better developed and wider than those of *V. lienosa*. Adult female *V. lienosa* have a more pronounced marsupial swelling posteriorly than female *H. subangulata*.

General Distribution

Hamiota subangulata is endemic to the Apalachicola Basin in Alabama, Florida and Georgia and the Ochlockonee River drainage in Florida and Georgia. It was erroneously reported from the Choctawhatchee River drainage by Clench and Turner (1956) and Burch (1975a). Most records of *H. subangulata* are from below the Fall Line, but Brim Box and Williams (2000) reported a few records from above the Fall Line.

Alabama and Mobile Basin Distribution

Hamiota subangulata is known from the Chattahoochee River and some of its larger tributaries as well as headwaters of the Chipola River system.

The only known extant populations of *Hamiota subangulata* are in the Uchee Creek system, Lee and Russell counties, and Big and Cowarts creeks, Houston County.

Ecology and Biology

Hamiota subangulata is known from medium creeks to large rivers in slow to moderate current. However, it has apparently been extirpated from large rivers. It has been found in a range of substrates, including various combinations of clay, sand and gravel. Individuals in a population in Cowarts Creek, Houston County, Alabama, were found primarily in small, silt-filled scour holes in a clay bottom.

Hamiota subangulata is a long-term brooder. O'Brien and Brim Box (1999) reported gravid females from December to August, but examined no females between August and December. *Hamiota subangulata* produces superconglutinates, which attract glochidial hosts (Figure 13.17). Superconglutinates are released from mid-May to mid-July (O'Brien and Brim Box, 1999). They range in length from 30 mm to 50 mm and are tethered to the female parent by paired mucus tubes up to 2.3 m long (O'Brien and Brim Box, 1999).

Fishes reported to serve as primary glochidial hosts of *Hamiota subangulata* in laboratory trials are *Micropterus punctulatus* (Spotted Bass) and *Micropterus salmoides* (Largemouth Bass) (Centrarchidae) (O'Brien and Brim Box, 1999). Fishes on which limited numbers of juveniles transformed were *Lepomis macrochirus* (Bluegill) (Centrarchidae); and *Gambusia holbrooki* (Eastern Mosquitofish) and the nonindigenous *Poecilia reticulata* (Guppy) (Poeciliidae) (O'Brien and Brim Box, 1999).

Figure 13.17. *Hamiota subangulata* superconglutinate "fishing" on end of mucus strand. Flint River drainage, Georgia. Photograph by J. Brim Box.

Conservation Status

Hamiota subangulata was reported as threatened throughout its range by Williams et al. (1993) and in Alabama by Lydeard et al. (1999). Garner et al. (2004) designated *H. subangulata* a species of highest conservation concern in the state. It was listed as endangered under the federal Endangered Species Act in 1998.

Remarks

Hamiota subangulata has been variously treated as a species of *Lampsilis* or *Villosa*. However, the rudimentary nature of the mantle modification made this species difficult to place in either genus with certainty. *Hamiota subangulata* is one of four superconglutinate-producing species that form a monophyletic clade. The genus *Hamiota* was erected for this group by Roe and Hartfield (2005). *Unio subangulatus* is the type species of the genus.

Synonymy

Unio subangulatus Lea, 1840. Lea, 1840:287; Lea, 1842b:209, pl. 13, fig. 23; Lea, 1842c:47, pl. 13, fig. 23
Type locality: Chattahoochee River, Columbus, [Muscogee County,] Georgia, Dr. Boykin. Lectotype, USNM 85081, length 48 mm (female), designated by Clench and Turner (1956). This specimen does not appear to be the individual figured by Lea (1842b).

Unio kirklandianus S.H. Wright, 1897, *non* Lea, 1834. S.H. Wright, 1897:136–137; Johnson, 1967b:7, pl. 5, fig. 3

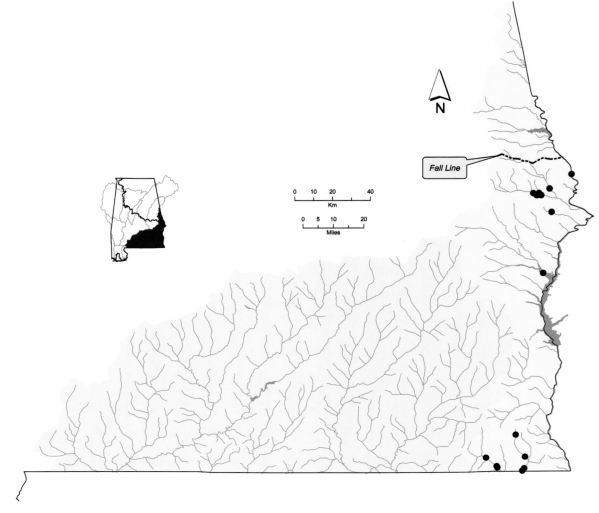

Distribution of *Hamiota subangulata* in the eastern Gulf Coast drainages of Alabama.

Genus *Hemistena*

Hemistena Rafinesque, 1820, is restricted to the Ohio, Cumberland and Tennessee River drainages. In Alabama it is restricted to the Tennessee River drainage. *Hemistena* is a monotypic genus (Turgeon et al., 1998).

This species has previously been assigned to *Lastena* Rafinesque, 1820. However, the type species of *Lastena* is *Anodonta* (*Lastena*) *ohiensis* Rafinesque, 1820 (= *Potamilus ohiensis*). Thus, *Lastena* is a junior synonym of *Potamilus*.

Type Species
Anodonta (*Lastena*) *lata* Rafinesque, 1820

Diagnosis
Shell elongate; compressed; thin; without posterior ridge; umbo low; periostracum shiny, with green rays; pseudocardinal teeth reduced; lateral teeth rudimentary; umbo sculpture irregular concentric bars.

Inner lamellae of inner gills only connected to visceral mass anteriorly; foot long, club-like; excurrent aperture large, margin crenulate; mantle bridge separating excurrent and supra-anal apertures short; outer gills marsupial; glochidium without styliform hooks (Rafinesque, 1820; Ortmann, 1912a, 1915; Haas, 1969b).

Synonymy
Odatelia Rafinesque, 1832
Stenelasma Herrmannsen, 1849
Hemilastena Agassiz, 1852
Sayunio De Gregorio, 1914

Hemistena lata (Rafinesque, 1820)
Cracking Pearlymussel

Hemistena lata – Upper figure: length 68 mm, UT 179.8. Clinch River, [Hancock County, Tennessee,] 17 November 1978. Lower figure: length 46 mm, NCSM 30912. Elk River, upstream of State Highway 127, Limestone County, Alabama, 22 October 1999. © Richard T. Bryant.

Shell Description

Length to 108 mm; thin; compressed to moderately inflated; outline elongate elliptical to elongate rhomboidal; posterior margin bluntly pointed to obliquely truncate; anterior margin rounded; dorsal margin straight to slightly convex; ventral margin straight to slightly convex; valves gape anteriorly and posteriorly; posterior ridge low, rounded; posterior slope not steep, flat, roughened; umbo flat, broad, may be elevated slightly above hinge line, umbo sculpture a few strong ridges; periostracum dull yellow to brownish green or brown, usually with irregular, broken, dark green rays.

Pseudocardinal teeth rudimentary, represented by 1 raised knob or ridge in each valve; lateral teeth appear as thickened hinge lines in both valves; interdentum absent; umbo cavity very shallow or absent; nacre pale bluish white, purple in deepest portion of shell.

Soft Anatomy Description

Mantle grayish tan, very dark brown outside of apertures; visceral mass creamy white and tan; foot dark tan. The foot is very elongate and swollen distally when extended in living individuals. Lea (1863d) reported the foot to be "slightly tinted with salmon".

Mantle fold just ventral to incurrent aperture rudimentary; fold length approximately 45% of shell length; unremarkable in appearance, very shallow, margin smooth, without marginal coloration.

Gills tan; dorsal margin straight, ventral margin elongate convex; gill length approximately 60% of shell length; gill height approximately 40% of gill length; outer gill height approximately 80% of inner gill height; inner lamellae of inner gills connected to visceral mass only anteriorly. Lea (1863d) reported gills to be "light liver brown".

No gravid females were available for marsupium description. Ortmann (1912a) reported outer gills to be marsupial.

Labial palps tan; slightly convex dorsally, convex ventrally, narrowly rounded distally; palp length approximately 10% of gill length; palp height approximately 80% of palp length; distal 60% of palps bifurcate.

Supra-anal aperture slightly longer than incurrent and excurrent apertures; incurrent aperture slightly longer than excurrent aperture.

Incurrent aperture length approximately 10% of shell length; tan within, dark brown basal to papillae; papillae arborescent, with scattered short, simple papillae; papillae dark brown, almost black.

Excurrent aperture length approximately 10% of shell length; tan within, marginal color band dark brown, almost black; margin minutely crenulate, becoming smooth dorsally.

Supra-anal aperture length approximately 15% of shell length; creamy white within, with thin, dark brown marginal band; margin smooth; mantle bridge separating supra-anal and excurrent apertures imperforate, length approximately 35% of supra-anal length.

Specimen examined: Clinch River, Tennessee (n = 1) (specimen previously preserved).

Glochidium Description

Length 159 μm; height 190μ; without styliform hooks; outline subrotund; ventral margin broadly rounded; anterior and posterior margins convex (Ortmann, 1915).

Similar Species

The thin, elongate elliptical to elongate rhomboidal shell, with rudimentary pseudocardinal and lateral teeth, distinguish *Hemistena lata* from all other freshwater mussels.

General Distribution

The historical distribution of *Hemistena lata* includes much of the Ohio, Tennessee and Cumberland River drainages. In the Ohio River drainage it is known from western Pennsylvania downstream to eastern Illinois (Cummings and Mayer, 1992). In the Cumberland River drainage *H. lata* was historically widespread downstream of Cumberland Falls (Cicerello et al., 1991; Parmalee and Bogan, 1998). *Hemistena lata* is known from much of the Tennessee River drainage, from headwaters in eastern Tennessee and southwestern Virginia to the mouth of the Tennessee River (Neves, 1991; Parmalee and Bogan, 1998).

Alabama and Mobile Basin Distribution

Hemistena lata presumably occurred in the Tennessee River across northern Alabama, though all museum records are from the vicinity Muscle Shoals. It also occurred in some large tributaries.

Hemistena lata appears to be extirpated from the Tennessee River, most recently collected there in 1966. It is extant in Alabama only in a short reach of free-flowing Elk River in Limestone County, near the Alabama and Tennessee state line.

Ecology and Biology

Hemistena lata occurs in shoal habitat of medium to large rivers. However, it survived in riverine habitat of the Tennessee River in tailwaters of Pickwick and Wilson dams, where depths exceed 5 m in many areas, until the late 1960s or early 1970s (Yokley, 1972). It typically remains deeply buried in substrates of sand and gravel or under flat rocks. *Hemistena lata* anchors itself in the substrate using its long, club-shaped foot.

There have been perplexing reports of the brooding strategy of *Hemistena lata*. Ortmann (1915) reported it to be a short-term brooder, with some females brooding eggs and some brooding mature glochidia in May in the Clinch River, Tennessee; females examined in September were not gravid. However, Ortmann (1921) reported gravid females from early September through May, which is the long-term brooding period. Jones and Neves (2000) reported collecting individuals with mature glochidia in early June.

Laboratory host fish trials involving *Hemistena lata* glochidia have documented marginal success with *Cottus carolinae* (Banded Sculpin) (Cottidae); *Campostoma anomalum* (Central Stoneroller), *Cyprinella galactura* (Whitetail Shiner) and *Erimystax dissimilis* (Streamline Chub) (Cyprinidae); and *Etheostoma flabellare* (Fantail Darter) (Percidae) (Jones and Neves, 2000; Jones et al., 2003).

Current Conservation Status and Protection

Hemistena lata was considered endangered throughout its range by Stansbery (1970a) and Williams et al. (1993) and in Alabama by Stansbery (1976) and Lydeard et al. (1999). It was listed as an endangered species by the USFWS in 1989. Garner et al. (2004) designated *H. lata* a species of highest conservation concern in the state.

In 2001 *Hemistena lata* was included on a list of species approved for a Nonessential Experimental Population in Wilson Dam tailwaters on the Tennessee River. However, no reintroductions had taken place as of 2007.

Remarks

Ortmann (1915) surmised that *Hemistena lata* is related to *Elliptio* based on shell shape and umbo sculpture but suggested that it may warrant a separate subfamily, should the Unioninae (Ambleminae in part) be elevated to family. Characters separating *H. lata* from other unionids include its weakened hinge plate, restriction of the marsupium to the middle portion of outer gills and extreme development of the foot. Ortmann (1915) also commented on the somewhat unusual shape of *H. lata* glochidia, specifically their "obliquity".

Hemistena lata has not been reported from prehistoric middens along the Tennessee River in Alabama. Morrison (1942) suggested that the burrowing habits of the species may have limited its availability or desirability as a food resource for Native Americans. To this Warren (1975) added rarity and its fragile shell as attributes to explain the absence of *H. lata* in the middens.

Synonymy

Anodonta (*Lastena*) *lata* Rafinesque, 1820. Rafinesque 1820:317, pl. 82, figs. 17, 18
Type locality: Kentucky River. Lectotype, ANSP 20227, length 62 mm, designated by Johnson and Baker (1973).

Unio dehiscens Say, 1829. Say, 1829:308; Say, 1830a:[no pagination], pl. 24
Unio oriens Lea, 1831. Lea, 1831:68, 73, pl. 6, fig. 5; Lea, 1834b:83, pl. 6, fig. 5
Odatelia radiata Rafinesque, 1832. Rafinesque, 1832:154
Unio hildrethi Delessert, 1841. Delessert, 1841:[no pagination], pl. 19, figs. 4a, b
Unio (*Sayunio*) *dehiscens orienopsis* De Gregorio, 1914. De Gregorio, 1914:39, pl. 7, figs. 2a, b

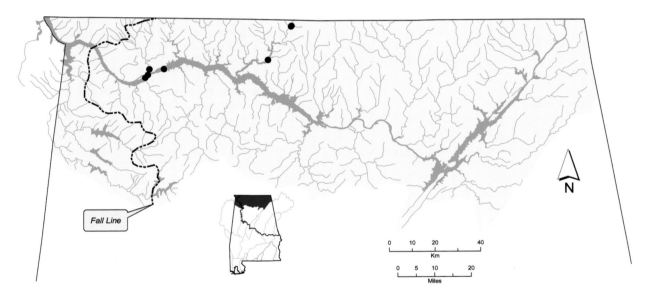

Distribution of *Hemistena lata* in the Tennessee River drainage of Alabama.

Genus *Lampsilis*

Lampsilis Rafinesque, 1820, is a very diverse genus found in the Hudson Bay, Great Lakes and Mississippi basins, Atlantic Coast drainages from Nova Scotia to Georgia, and Gulf Coast drainages from peninsular Florida to the Rio Grande. Turgeon et al. (1998) listed 28 species of *Lampsilis*. Three of those— *Lampsilis radiata* (Gmelin, 1791), *Lampsilis reeviana* (Lea, 1852) and *Lampsilis straminea* (Conrad, 1834)— were recognized by Turgeon et al. (1998) as having subspecies. Nine *Lampsilis* species are known from Alabama, including one species, *L. floridensis* (Lea, 1852), herein elevated from the synonymy of *L. teres* (Rafinesque, 1820) based on shell morphology and genetic analysis. Four species—*Lampsilis altilis* (Conrad, 1834), *Lampsilis australis* (Simpson, 1900), *Lampsilis perovalis* (Conrad, 1834) and *Lampsilis subangulata* (Lea, 1840)—were recently moved to a new genus, *Hamiota* Roe and Hartfield, 2005. One species, *Lampsilis haddletoni* Athearn, 1964, included in Turgeon et al. (1998), is herein recognized as a species of *Obovaria* Rafinesque, 1819. Soft anatomy of *Obovaria haddletoni* (Athearn, 1964) is unknown. It was moved from *Lampsilis* to *Obovaria* based on its circular shape, shallow umbo cavity and triangular, divergent, striated pseudocardinal teeth. There is also zoogeographical precedent for species being shared between Gulf Coast drainages and the Mobile Basin.

Type Species
Unio ovatus Say, 1817

Diagnosis
Shell oval to elliptical; disk smooth; umbo sculpture sinuous or double-looped; sexual dimorphism distinct, females inflated posterioventrally; hinge teeth well-developed, often compressed.

Females usually with well-developed mantle flap just ventral to incurrent aperture (Figures 13.18, 13.19, 13.20); flap usually rudimentary in males. Inner lamellae of inner gills mostly or completely connected to visceral mass, may have small opening at posterior end; outer gills marsupial; glochidium held in posterior portion of gill; marsupium outline reniform, ventral margin blunt, may be beaded, often pigmented, extended ventrally below original gill edge when gravid; glochidium without styliform hooks (Simpson, 1900b; Ortmann, 1912a; Haas, 1969b; Kraemer, 1970).

Synonymy
Aeglia Swainson, 1840
Simpsonunio Starobogatov, 1970

Figure 13.18. *Lampsilis ovata* gravid female with mantle flaps and charged gills displayed. Clinch River, Virginia, June, 1990s. Photograph by R.J. Neves.

Figure 13.19. *Lampsilis floridensis* gravid female with mantle flaps and charged gills displayed. Santa Fe River, Alachua and Bradford counties, Florida, 20 April 1996. Photograph by D.S. Ruessler.

Figure 13.20. *Lampsilis virescens* gravid female with mantle flaps displayed. Charged gills are visible inside incurrent aperture. Estill Fork, Jackson County, Alabama, May 2004. Photograph by P.D. Johnson.

Lampsilis abrupta (Say, 1831)
Pink Mucket

Lampsilis abrupta – Upper figure: female, length 87 mm, CMNH 61.7320. Lower figure: male, length 82 mm, CMNH 61.7320. Tennessee River, Florence, Lauderdale County, Alabama. © Richard T. Bryant.

Shell Description

Length to 120 mm; thick; inflated, females often greatly expanded posterioventrally; male outline usually oval, female outline oval to subquadrate; male posterior margin broadly to narrowly rounded, female posterior margin slightly rounded to truncate; anterior margin rounded; dorsal margin convex; ventral margin convex to straight; valves gape anteriorly; posterior ridge rounded, may be less distinct in females; posterior slope steep; umbo inflated, elevated above hinge line, umbo sculpture faint, looped ridges; periostracum tawny to olive brown or dark brown, juveniles may be light yellow, some individuals with thin, faint green rays.

Pseudocardinal teeth large, erect, triangular, 2 divergent teeth in left valve, 1 tooth in right valve, often with accessory denticle anteriorly and/or posteriorly; lateral teeth short, thick, slightly curved, 2 in left valve, 1 in right valve; interdentum short, moderately wide; umbo cavity wide, moderately deep; nacre white to pink or pinkish orange.

Soft Anatomy Description

No material was available for soft anatomy description, but Lea (1863d) provided a brief description. The mantle is "rather thin, double along the inferior edges" and the "color of the mass" is "whitish inclining to salmon". Gills are "very much rounded" ventrally, and inner lamellae of inner gills are completely connected to the visceral mass. Labial palps are "large, thick, rather oblique, suboval". The incurrent aperture has "small, dark papillae," and the excurrent aperture has "numerous, very small, brownish papillae". A mantle bridge separates the supra-anal and excurrent apertures. Ortmann (1912a) described the mantle flap just ventral to the incurrent aperture as having numerous papillae and projecting as a lobe on the anterior end. The inside of the flap is pigmented with black. No eyespot was observed, but it may have been obscured by the "contracted condition" of the specimens. The outer gills are marsupial, greatly distended when fully gravid, with a black distal margin that fades as the gill swells (D.W. Hubbs, personal communication).

Glochidium Description

Length 207–214 μm; height 251–259 μm; without styliform hooks; outline subspatulate; dorsal margin straight, ventral margin broadly rounded, anterior and posterior margins convex (Hoggarth, 1999). Ortmann (1912a) reported two sizes of glochidia, with no intergrades (190 μm x 210 μm, and more rarely 220 μm x 250 μm), but did not rule out the possibility of cross contamination of the sample with glochidia from another species.

Similar Species

Large male *Lampsilis abrupta* may resemble large female *Ellipsaria lineolata* but differ in having a more rounded posterior ridge and usually a less elevated umbo. *Ellipsaria lineolata* typically has rays comprised of chevrons or blotches, though they may be obscured in old, dark shells. *Lampsilis abrupta* frequently lacks rays, but when present they are not comprised of chevrons or blotches, though they may be interrupted.

Male and young, uninflated female *Lampsilis abrupta* resemble *Actinonaias ligamentina*, but the posterior slope of *L. abrupta* is usually steeper. The periostracum of *A. ligamentina* is typically greener and more heavily rayed than that of *L. abrupta*.

Male *Lampsilis abrupta* may superficially resemble *Obovaria olivaria*, but *L. abrupta* has a broader, less inflated umbo. The umbo cavity of *L. abrupta* is deeper than that of *O. olivaria*.

Female *Lampsilis abrupta* are distinctive, being more swollen posteriorly than *Ellipsaria lineolata*, *Actinonaias ligamentina* and *Obovaria olivaria*. They are also often quadrate, as opposed to oval or triangular in outline.

General Distribution

Lampsilis abrupta occurs in middle and lower reaches of the Mississippi Basin from Missouri and Illinois downstream to Louisiana (Cummings and Mayer, 1992; Vidrine, 1993; Oesch, 1995). It is known from much of the Ohio River drainage from headwaters in Pennsylvania (Ortmann, 1909a) to the mouth of the Ohio River in Illinois and Kentucky (Cummings and Mayer, 1992). There are *L. abrupta* records from the Cumberland River upstream to the Obey River, Tennessee (Parmalee and Bogan, 1998). *Lampsilis abrupta* occurs in much of the Tennessee River drainage from headwaters in southwestern Virginia to the mouth of the Tennessee River (Ortmann, 1909a; Neves, 1991; Parmalee and Bogan, 1998).

Alabama and Mobile Basin Distribution

Lampsilis abrupta is known from the Tennessee River across northern Alabama and a few tributaries.

The only known viable populations of *Lampsilis abrupta* in Alabama are located in Guntersville and Wilson Dam tailwaters. A single gravid female was encountered in Bear Creek, Colbert County, in 1999 (McGregor and Garner, 2004).

Ecology and Biology

Lampsilis abrupta typically occurs in free-flowing reaches of large rivers, though it is occasionally reported from large creeks and small rivers. Its preferred substrates appear to be gravel with interstitial sand, kept free of silt by current. However, it occurs in overbank habitat of reservoirs under some conditions (D.W. Hubbs, personal communication).

Lampsilis abrupta is a long-term brooder, gravid from late summer or autumn to the following summer. It has been reported gravid with eggs in August and with glochidia in early September in the Ohio River drainage of Pennsylvania (Ortmann, 1912a). Fishes shown to serve as glochidial hosts of *L. abrupta* in laboratory trials include *Micropterus dolomieu* (Smallmouth Bass), *Micropterus punctulatus* (Spotted Bass), *Micropterus salmoides* (Largemouth Bass) and *Pomoxis annularis* (White Crappie) (Centrarchidae); and *Sander canadensis* (Sauger) and *Sander vitreus* (Walleye) (Percidae) (Barnhart et al., 1997; J.B. Layzer and L.M. Madison, personal communication). Fuller (1974) reported *Aplodinotus grunniens* (Freshwater Drum) (Sciaenidae) as a host for *L. abrupta*, citing Coker et al. (1921) and Wilson (1916). However, those citations included *A. grunniens* as host for *Lampsilis higginsi* (Lea, 1857) not *L. abrupta*.

Current Conservation Status and Protection

Lampsilis abrupta was reported as endangered throughout its range by Stansbery (1970a) and Williams et al. (1993). In Alabama it was considered a species of special concern by Stansbery (1976) and endangered by Lydeard et al. (1999). Garner et al. (2004) designated *L. abrupta* a species of highest conservation concern in the state. It was listed as endangered under the federal Endangered Species Act in 1976.

Remarks

    Lampsilis abrupta is rare in prehistoric shell middens along the Tennessee River in Alabama. Specimens have been reported only from the mouth of the Elk River (Hughes and Parmalee, 1999).

    The taxonomic history of *Lampsilis abrupta* has been the subject of some confusion with regard to the name *Lampsilis orbiculata* (Hildreth, 1828). Although

Frierson (1924, 1927) clearly resolved the issue, the name *L. orbiculata* was used incorrectly for this species for more than a half century. The name *orbiculata* is actually a junior synonym of *Obovaria subrotunda*.

    Wilson and Clark (1914) stated that *Lampsilis abrupta* shell was "a very good button species, but so uncommon that it is not much of an item in the trade."

Synonymy

Unio abruptus Say, 1831. Say, 1831(II):[no pagination], pl. 17
Type locality: Muskingum River, Ohio. Type specimen not found. Length not given in original description.

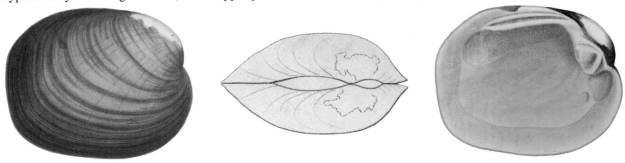

Toxolasma cyclips Rafinesque, 1831. Rafinesque, 1831:2

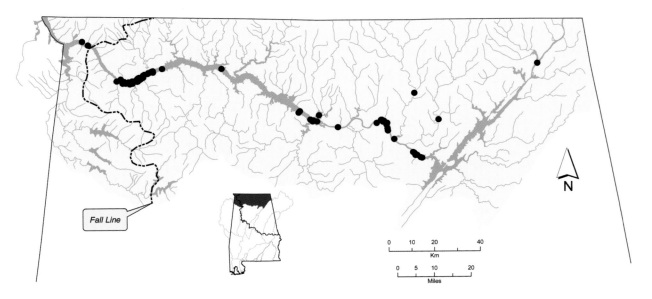

Distribution of *Lampsilis abrupta* in the Tennessee River drainage of Alabama.

Lampsilis binominata Simpson, 1900
Lined Pocketbook

Lampsilis binominata – Upper figure: female, length 41 mm, USNM 84883. Middle figure: male, length 46 mm, USNM 84883. Lower figure: male, length 37 mm, USNM 84883. [Chattahoochee River,] Columbus, Muscogee County, Georgia. © Richard T. Bryant.

Shell Description

Length to 60 mm; thin; moderately inflated; outline oval; male posterior margin narrowly rounded, female posterior margin broadly rounded; anterior margin rounded; dorsal margin convex; ventral margin convex; valves gape anteriorly and posteriorly; females more swollen posteriorly than males; posterior ridge well-defined near umbo, broadly rounded posterioventrally; posterior slope flat to slightly concave, often with a few fine, radial wrinkles; umbo broad, inflated, elevated above hinge line, umbo sculpture slightly double-looped ridges; periostracum yellowish, with thin, widely spaced, dark green rays.

Pseudocardinal teeth compressed, 2 teeth in left valve, roughly parallel to dorsal margin, 1 tooth in right valve, usually with well-developed lamellar accessory denticle anteriorly, roughly parallel to dorsal margin; lateral teeth thin, moderately long, straight, 2 in left valve, 1 in right valve; interdentum moderately long, narrow; umbo cavity wide, moderately deep; nacre bluish white.

Soft Anatomy Description

Soft anatomy unknown.

Glochidium Description

Glochidium unknown.

Similar Species

Lampsilis binominata may resemble subadults of other species of *Lampsilis* but can be distinguished by its thin shell, shiny yellow periostracum with thin green rays and compressed lateral teeth.

General Distribution

Lampsilis binominata is endemic to the Chattahoochee and Flint rivers of the Apalachicola Basin, where it appears to have been restricted to areas near and above the Fall Line.

Alabama and Mobile Basin Distribution

Lampsilis binominata is known only from the Chattahoochee River. There are no records of its occurrence in tributaries of that river.

Lampsilis binominata was most recently collected in the Chattahoochee River in 1942. It is believed to be extinct.

Ecology and Biology

Based on locality data in museum collections, *Lampsilis binominata* occurred in large river habitats. There is a single tributary record, from Line Creek, one of the headwater forks of the Flint River in Georgia. Habitat requirements for *L. binominata* are unknown except for notes of H.D. Athearn (personal communication) who reported finding the species in stable sand and gravel in moderate to swift current.

Lampsilis binominata was presumably a long-term brooder, gravid from late summer or autumn to the following spring or summer. Its glochidial hosts are unknown, but other species of *Lampsilis* primarily use members of the Centrarchidae. Host specificity appears to be a highly conserved trait at the genus level (Haag and Warren, 2003).

Conservation Status

Lampsilis binominata was reported as endangered throughout its range by Athearn (1970), Stansbery (1971) and Williams et al. (1993). In Alabama it was considered endangered by Stansbery (1976). *Lampsilis binominata* was listed as extinct by Turgeon et al. (1998), Lydeard et al. (1999) and Garner et al. (2004).

Remarks

Based on museum records, *Lampsilis binominata* appears not to have been common historically. A note accompanying two specimens collected from the Chattahoochee River, Georgia, in the 1800s stated that the collector was a "lucky fellow to get a pair." The largest known collection contained eight specimens, also collected in the 1800s.

There appears to be only 13 historical collections of *Lampsilis binominata* from 11 sites in the Apalachicola Basin (Brim Box and Williams, 2000). The most recent known occurrence in the Chattahoochee River was at West Point, Georgia, in 1942. Habitat at this site was destroyed by construction of the West Point Dam by the USACE in 1975. The most recent observation of live *L. binominata* was in October 1967 by H.D. Athearn, who collected four individuals at two localities on the Flint River, Pike and Meriwether counties, Georgia. A dead shell of *L. binominata* (35 mm long, approximately 3 years of age) was collected from the Flint River near Warm Springs, Meriwether County, Georgia, in April 1976.

Lampsilis binominata has been treated as a synonym of *Lampsilis ornata*, which occurs in Gulf Coast drainages from Louisiana east to the Escambia River drainage of Alabama and Florida (Frierson, 1927; Clench and Turner, 1956). Johnson (1967a) recognized *L. binominata* as a valid species and confirmed the type locality as the Chattahoochee River, Columbus, Georgia.

The original species name given to *Lampsilis binominata* was *Unio lineatus* Lea, 1840. That name was preoccupied, necessitating a replacement name. Simpson (1900b) coined the name *binominatus*, which means "second name".

Synonymy

Unio lineatus Lea, 1840, *non* Valenciennes, 1827. Lea, 1840:287; Lea, 1842a:44, pl. 12, fig. 20

Type locality: Chattahoochee River, Columbus, [Muscogee County,] Georgia, Dr. Boykin. Lectotype, USNM 84884, length 33 mm, designated by Johnson (1974).

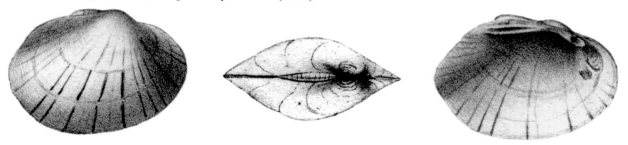

Lampsilis binominatus Simpson, 1900. Simpson, 1900b:528

Comment: *Lampsilis binominatus* is a replacement name for *Unio lineatus* Lea, 1840 (Simpson, 1900b).

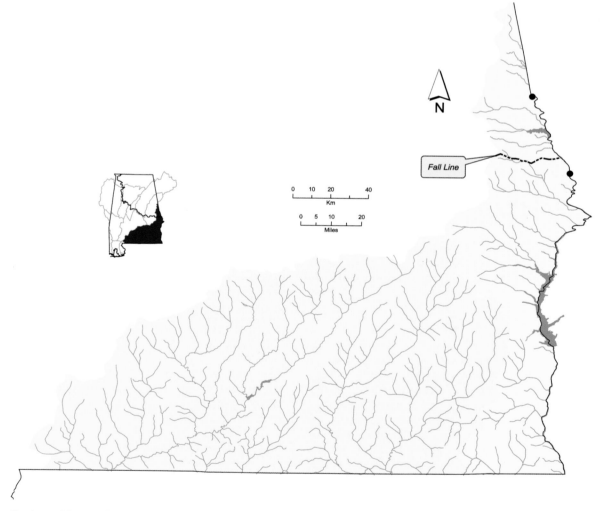

Distribution of *Lampsilis binominata* in the eastern Gulf Coast drainages of Alabama.

Lampsilis fasciola Rafinesque, 1820
Wavyrayed Lampmussel

Lampsilis fasciola – Upper figure: female, length 59 mm, UF 65510. Hurricane Creek (branch of Flint River) near Gurley, Madison County, Alabama, July 1910. Lower figure: male, length 61 mm, CMNH 61.6989. Paint Rock River, Paint Rock, Jackson County, Alabama. © Richard T. Bryant.

Shell Description

Length to 100 mm; thin when young, becoming moderately thick with age; inflated; male outline oval to elliptical, female outline oval; male posterior margin narrowly rounded, female posterior margin broadly rounded; anterior margin rounded; dorsal margin straight to slightly convex; ventral margin straight to slightly convex; sexual dimorphism expressed as greater inflation posteriorly in females, partially obscuring posterior ridge; posterior ridge broadly rounded; posterior slope moderately steep, flat to slightly concave; umbo broad, somewhat inflated, elevated above hinge line, umbo sculpture indistinct wavy ridges; periostracum light yellow to greenish yellow, with numerous green rays of variable width, rays usually wavy distally.

Pseudocardinal teeth erect, moderately thick, 2 divergent triangular teeth in left valve, oriented at obtuse angle to lateral teeth, 1 blunt tooth in right valve, usually with small compressed accessory denticle anteriorly; lateral teeth short, thick, slightly curved, 2 in left valve, 1 in right valve; interdentum long, narrow to very narrow; umbo cavity open, moderately deep; nacre white to bluish white.

Soft Anatomy Description

Mantle tan to light brown, usually pale orange with black or gray mottling outside of apertures; visceral mass creamy white to pale tan; foot tan to pinkish orange, often darker distally.

Females with well-developed mantle flap just ventral to incurrent aperture; flap length approximately 25% of shell length; flap coloration variable within and among populations, with 2–4 marginal color bands in some combination of tan, dull orange, reddish brown, black and gray, eyespot may be present at posterior end; margin papillate, papillae simple, fused into wide "tail" anteriorly. Males with rudimentary mantle flap.

Gills tan to light brown; straight to slightly sinuous or slightly convex dorsally, convex ventrally, occasionally bilobed in females; gill length 45–65% of shell length; gill height 55–75% of gill length; outer gill height usually 80–95% of inner gill height, female outer gills in marsupial area may be higher than inner gills; inner lamellae of inner gills usually completely connected to visceral mass, occasionally incomplete.

Outer gills marsupial; glochidia held in posterior 35% of gill; marsupium greatly distended when gravid; creamy white with black distal margin. Lea (1863d) reported the posterior 50% of outer gills to be marsupial.

Labial palps creamy white to pale tan; straight to slightly convex or slightly concave dorsally, convex ventrally, bluntly pointed distally; palp length 30–40% of gill length; palp height 50–80% of palp length; distal 30–45% of palps bifurcate.

Incurrent aperture longer than or equal to excurrent aperture; supra-anal aperture longer than or equal to incurrent aperture.

Incurrent aperture length 10–15% of shell length; creamy white within, usually pale orange with black mottling basal to papillae, black from some papillae edges may extend onto aperture wall as irregular lines; papillae in 2–3 rows, inner row largest, long, slender, simple, occasionally with a few bifid or trifid; papillae usually pale orange, may have black mottling, often with black edges basally.

Excurrent aperture length 5–15% of shell length; creamy white within, marginal color band usually pale orange with irregular black spots or lines, some lines may be interconnected; margin crenulate or with very short, blunt, simple papillae; papillae pale orange, occasionally with black mottling.

Supra-anal aperture length 10–20% of shell length; creamy white within, usually with thin orange or brown marginal band; margin smooth; mantle bridge separating supra-anal and excurrent apertures imperforate, length 15–45% of supra-anal length.

Specimens examined: Bear Creek, Colbert County (n = 1); Duck River, Tennessee (n = 2); Paint Rock River (n = 6).

Glochidium Description

Length 218–266 μm; height 266–318 μm; without styliform hooks; outline subelliptical; dorsal margin straight, ventral margin rounded, anterior and posterior margins convex (Ortmann, 1912a, 1919; Surber, 1915; Zale and Neves, 1982b). Jirka and Neves (1992) included similar measurements, but gave the longer measurement (316 μm) as length and the shorter measurement (262 μm) as "width".

Similar Species

Large *Lampsilis fasciola* may superficially resemble *Lampsilis ovata* but have lower, less inflated umbos and a rounder posterior ridge. Pseudocardinal teeth of *L. fasciola* are thicker but lower than those of *L. ovata*. The periostracum of *L. fasciola* is typically adorned with numerous wavy rays, but *L. ovata* usually has only a few thin, straight rays or none at all. *Lampsilis ovata* attains a much greater size than *L. fasciola*.

General Distribution

In the Great Lakes Basin *Lampsilis fasciola* occurs in lakes Erie, Huron and St. Clair as well as some of their tributaries (Goodrich and van der Schalie, 1932; La Rocque and Oughton, 1937). It occurs in the Ohio River drainage from headwaters in Pennsylvania (Ortmann, 1909a) downstream to Illinois and Kentucky (Cummings and Mayer, 1992). *Lampsilis fasciola* is widespread in the Cumberland River drainage downstream of Cumberland Falls, Kentucky and Tennessee (Cicerello et al., 1991). It historically occurred throughout the Tennessee River drainage, from southwestern Virginia, western North Carolina and eastern Tennessee downstream to the mouth of the Tennessee River (Ahlstedt, 1992a, 1992b; Parmalee and Bogan, 1998).

Alabama and Mobile Basin Distribution

Lampsilis fasciola is known from the Tennessee River proper across northern Alabama and many tributaries.

Lampsilis fasciola is extant in the Elk River and Paint Rock River system as well as Bear Creek, Colbert County, and Cypress and Shoal creeks, Lauderdale County. It has apparently not been collected from the Tennessee River proper since 1966.

Ecology and Biology

Lampsilis fasciola is primarily a species of large creeks and small to medium rivers. However, it was extant in the Tennessee River proper at Muscle Shoals until at least 1966. It occurs primarily in stable sand and gravel substrates in flowing water. Ortmann (1919) reported *L. fasciola* to be most common in riffles with abundant *Justicia americana* (Waterwillow) and other aquatic macrophytes. *Lampsilis fasciola* also occurs in some northern lakes. In most areas where it is extant, *L. fasciola* occurs in less than 1 m of water. However, individuals collected from Tennessee River proper following its impoundment came from water more than 5 m deep.

Lampsilis fasciola is a long-term brooder. Spawning occurs in late summer, probably August (Zale and Neves, 1982b), with females becoming gravid in August or September and brooding into the following July (Ortmann, 1912a, 1919; Zale and Neves, 1982b). However, Ortmann (1909a, 1919) reported a possible overlap in brooding period in August. *Lampsilis fasciola* females display an elaborate mantle flap, presumably to attract glochidial hosts (Figure 13.21). Kraemer (1970) described the flapping behavior of individuals in the River Raisin, Michigan. Glochidia have been reported to be discharged in loose, irregular masses from the marsupia via the distal gill margin, instead of through the suprabranchial chamber (Ortmann, 1910a). However, if glochidia are discharged in such a manner, the mechanism by which they are transported from the mantle cavity to the excurrent aperture remains unexplained. *Lampsilis fasciola* glochidia were present, but rare, in stream drift of Big Moccasin Creek, Virginia, from May to August (Zale and Neves, 1982b). Glochidia of several long-term brooders have been found in stream drift over longer periods (e.g., *Medionidus conradicus* and *Villosa* spp.) (Zale and Neves, 1982b).

Fishes reported to serve as *Lampsilis fasciola* glochidial hosts in laboratory trials include *Micropterus dolomieu* (Smallmouth Bass) and *Micropterus salmoides* (Largemouth Bass) (Centrarchidae) (Zale and Neves, 1982a; O'Beirn et al., 1998; Parmalee and Bogan, 1998).

Current Conservation Status and Protection

Williams et al. (1993) reported *Lampsilis fasciola* as currently stable throughout its range. In Alabama Lydeard et al. (1999) considered it a species of special concern and Garner et al. (2004) designated it a species of moderate conservation concern.

Remarks

Ortmann and Walker (1922a) discussed confusion over the identity of *Lampsilis fasciola*. Some early workers placed this name in the synonymy of *Lampsilis*

siliquoidea. However, Ortmann and Walker recognized Rafinesque's description based primarily on the wavy rays of its periostracum.

Lampsilis fasciola demonstrates a wide range of variation in mantle flap morphology and pigmentation across its range (Figure 13.21) (S.A. Ahlstedt, personal communication). Two distinct forms are extant in the Paint Rock River. Based on this variability, it is unclear whether *L. fasciola* represents one highly variable species, or a complex of sibling species. Comprehensive taxonomic studies are needed.

Figure 13.21. *Lampsilis fasciola* females with dissimilar mantle flaps displayed. North Fork Holston River, Washington County, Virginia. 1995. Photographs by W.F. Henley.

Lampsilis fasciola has been reported to be rare in prehistoric shell middens along the Tennessee River in northern Alabama (Warren, 1975; Hughes and Parmalee, 1999).

Synonymy

Lampsilis fasciola Rafinesque, 1820. Rafinesque, 1820:299
Type locality: Kentucky River. Syntype in the MNHN, Paris (Johnson and Baker, 1973), was not found.
Comment: The "type" mentioned by Vanatta (1915), ANSP 20203, length 66 mm, is *Lampsilis siliquoidea* (Barnes, 1823).
Unio multiradiatus Lea, 1829. Lea, 1829:434, pl. 9, fig. 15; Lea, 1834b:48, pl. 9, fig. 15

Unio perradiatus Lea, 1858. Lea, 1858a:40; Lea, 1862b:66, pl. 6, fig. 215; Lea, 1862d:70, pl. 6, fig. 215
Type locality: Florence, [Lauderdale County,] Alabama, Reverend G. White. Holotype, USNM 84501, length 60
    mm (male).

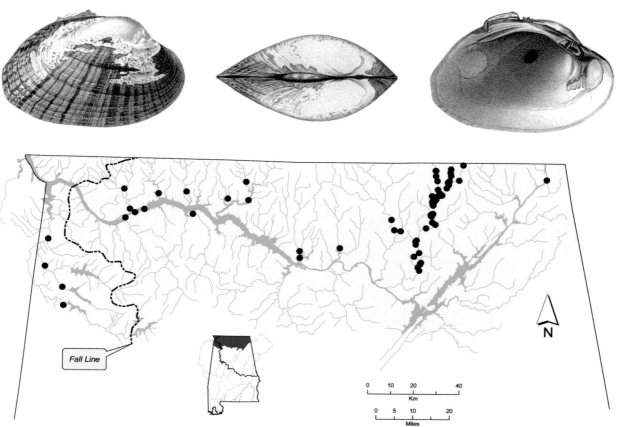

Distribution of *Lampsilis fasciola* in the Tennessee River drainage of Alabama.

Lampsilis floridensis (Lea, 1852)
Florida Sandshell

Lampsilis floridensis – Upper figure: female, length 85 mm, UF 370781. Lower figure: male, length 92 mm, UF 370781. Choctawhatchee River slough, about 1 mile south of Interstate Highway 10 bridge, Holmes and Washington counties, Florida, 3 June 1998. © Richard T. Bryant.

Shell Description

Length to 131 mm; moderately thin; moderately compressed; outline elliptical; posterior margin bluntly pointed to narrowly rounded, females more rounded posteriorly than males; anterior margin rounded; dorsal margin straight to slightly convex; ventral margin straight to convex; posterior ridge low, rounded; posterior slope broad, flat; umbo broad, low, elevated slightly above hinge line, umbo sculpture unknown; periostracum shiny, yellow to tawny, typically with 1 or more green rays restricted to posterior ridge and posterior slope.

Pseudocardinal teeth small, triangular, may be compressed, 2 in left valve, 1 in right valve, may have accessory denticle anteriorly; lateral teeth thin, straight to slightly curved, 2 in left valve, 1 in right valve, tooth in right valve longer than those in left valve; interdentum moderately long, very narrow; umbo shallow, open; nacre white.

Soft Anatomy Description

Mantle tan, usually more pale outside of pallial line, pale orange to rusty orange or tan outside of apertures, with gray and/or black mottling; visceral mass creamy white to pale tan; foot pale orange to pale tan.

Females with mantle flap just ventral to incurrent aperture (Figure 13.19); flap length 20–30% of shell length; usually with 3 color bands adjacent to flap, proximal and distal bands gray to grayish brown, medial band rusty orange, some individuals with only gray to grayish brown band; flap margin crenulate; without distinctive eyespot or "tail" structure. Males with rudimentary mantle flap in form of shallow fold; marginal color band reduced or absent.

Gills tan; dorsal margin straight to slightly sinuous, ventral margin convex to elongate convex, often slightly sinuous in female outer gills; gill length 40–55% of shell length; gill height 45–60% of gill length; outer gill height 75–85% of inner gill height, height of marsupial area of females almost equal to inner gill height; inner lamellae of inner gills completely connected to visceral mass.

Outer gills marsupial; glochidia held in posterior 35% of gill; marsupium distended when gravid; creamy white, distal margin black.

Labial palps tan; straight to slightly concave dorsally, convex ventrally, narrowly rounded to bluntly pointed distally; palp length 25–40% of gill length; palp height 65–75% of palp length; distal 35–45% of palps bifurcate.

Incurrent aperture longer than excurrent aperture; supra-anal aperture longer or shorter than incurrent aperture; excurrent and supra-anal apertures of similar length, supra-anal may be longer.

Incurrent aperture length approximately 10% of shell length; creamy white within, may be rusty brown basal to papillae, brown lines extend from edges of papillae onto aperture wall, some converging proximally; papillae in 1–3 rows, inner row largest, moderately long, simple; papillae creamy white to rusty tan, with brown edges basally.

Excurrent aperture length 5–10% of shell length; creamy white within, marginal color band rusty tan, with dark brown lines in open reticulated pattern distally; margin papillate, papillae very small, simple, may be little more than crenulations; papillae gray to tan or rusty brown.

Supra-anal aperture length 5–15% of shell length; creamy white within, may have sparse, irregular rusty tan marginally; margin smooth; mantle bridge separating supra-anal and excurrent apertures may have one or more small perforations, length 50–180% of supra-anal length.

Specimens examined: Apalachicola River, Florida (n = 4).

Glochidium Description

Without styliform hooks; outline "pouch-shaped" (Lea, 1863a).

Similar Species

Lampsilis floridensis may superficially resemble *Lampsilis straminea*. However, *L. floridensis* is more elongate and males are more pointed posteriorly than *L. straminea*.

Some *Lampsilis floridensis* resemble *Hamiota subangulata*, but that species typically has a brown periostracum with prominent green rays, and the periostracum of *L. floridensis* is typically shiny, yellow and rayless with the exception of a few rays on the posterior slope. *Lampsilis floridensis* attains a greater size than *H. subangulata* and is highly variable in shell morphology, often being more inflated with a more rounded posterior margin than *H. subangulata*.

Synonymy

Unio floridensis Lea, 1852, *nomen nudum*. Lea, 1852a:251

General Distribution

Lampsilis floridensis occurs in large Gulf Coast drainages from the Tampa Bay drainages in Florida west to the Escambia River drainage in Alabama and Florida.

Alabama and Mobile Basin Distribution

Lampsilis floridensis occurs in the Apalachicola Basin, as well as the Escambia and Choctawhatchee River drainages.

Lampsilis floridensis is extant in all Alabama drainages within its range.

Ecology and Biology

Lampsilis floridensis occurs in slow to moderate current of large creeks and rivers in sand and sandy mud substrates. It also may be found in reservoirs and floodplain lakes.

Lampsilis floridensis is a long-term brooder, gravid from late summer or autumn to the following spring or summer. Brim Box and Williams (2000) reported mantle flap displays to occur nocturnally in the Apalachicola and Suwannee rivers. Its glochidial hosts are unknown. Many species of *Lampsilis* use members of the Centrarchidae as hosts.

Fishes shown to serve as glochidial hosts for *Lampsilis floridensis* in laboratory trials include *Lepisosteus osseus* (Longnose Gar) and *Lepisosteus platyrhincus* (Florida Gar) (Lepisosteidae); and *Micropterus salmoides* (Largemouth Bass) (Centrarchidae) (Keller and Ruessler, 1997).

Current Conservation Status and Protection

Lampsilis floridensis has not been previously recognized so was not included in past conservation status assessments. It does not appear to be imperiled.

Remarks

Lampsilis floridensis has been considered a subspecies (e.g., Clench and Turner, 1956) or synonym (e.g., Burch, 1975a) of *Lampsilis teres* by most authors. However, recent genetic analyses, along with subtle differences in shell morphology, suggest that it deserves species status.

Unio floridensis Lea, 1852. Lea, 1852b:274, pl. 21, fig. 31; Lea, 1852c:30, pl. 21, fig. 31
Type locality: Chácktaháchi [Choctawhatchee] River, west Florida. Holotype, ANSP 42081, length 76 mm. Clench
       and Turner (1956) restricted type locality to Choctawhatchee River, 1 mile west of Caryville, Holmes County,
       Florida.

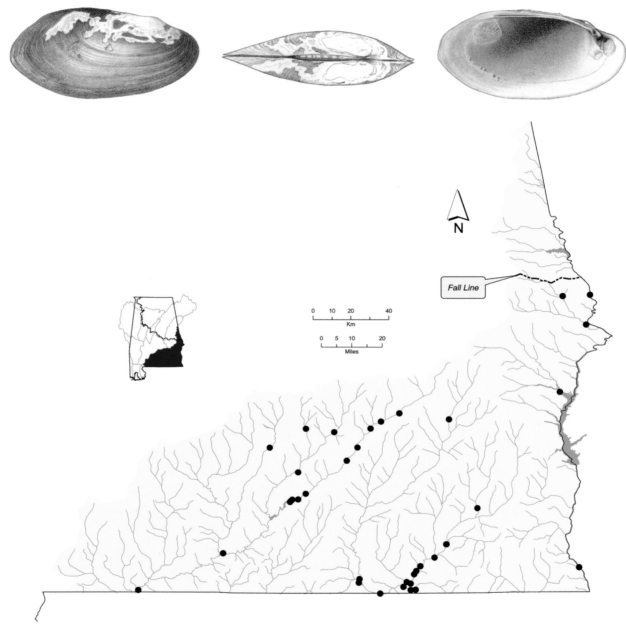

Distribution of *Lampsilis floridensis* in the eastern Gulf Coast drainages of Alabama and the Mobile Basin.

Lampsilis ornata (Conrad, 1835)
Southern Pocketbook

Lampsilis ornata – Upper figure: female, length 78 mm, UF 20740. Lower figure: male, length 68 mm, UF 20740. Escambia River, 3 miles southeast of Century, Escambia County, Florida, 27 August 1954. © Richard T. Bryant.

Shell Description

Length to 115 mm; thin when young, becoming moderately thick with age; inflated; outline oval to subtriangular; posterior margin narrowly rounded to bluntly pointed; anterior margin rounded; dorsal margin straight to convex; ventral margin rounded; valves gape anteriorly and posteriorly; sexual dimorphism often subtle, females more inflated posteriorly, with more rounded posterior margin; posterior ridge well-defined, sharp; posterior slope steep, flat to slightly concave; umbo broad, greatly inflated, elevated well above hinge line, turned anteriad, umbo sculpture unknown; periostracum smooth, shiny, yellowish, may darken to brown with age, typically with variable narrow dark green rays.

Pseudocardinal teeth well-developed, erect, compressed, 2 teeth in left valve, posterior tooth smaller, usually partially fused with anterior tooth, 1 tooth in right valve, with well-developed accessory denticle anteriorly; lateral teeth short, straight to slightly curved, 2 in left valve, 1 in right valve; interdentum long, narrow; umbo cavity open, deep; nacre white, often with pinkish or salmon tint.

Soft Anatomy Description

Mantle pearly or creamy white to tan or light brown, outside of apertures often fleshy and usually dull orange with black or gray mottling, may be in form of crude lines perpendicular to margin; visceral mass usually creamy white, may be pearly white or light tan; foot creamy white or pale orange to tan or light brown, often darker distally.

Females with well-developed mantle flap just ventral to incurrent aperture; flap length 25–40% of shell length; typically with 3 color bands in various combinations of creamy white, tan, rusty brown, rusty orange, gray and black; usually with eyespot at posterior end, eyespot occasionally with white ring; flap margin may be papillate, crenulate or smooth and undulating; well-developed "tail" located at anterior end, composed of enlarged, partially fused papillae or folds. Males usually with rudimentary flap, reduced in size and coloration, without well-developed "tail." Ortmann (1924a) described the "tail" of the mantle flap as "a lacerated lobe".

Gills creamy white to light brown; dorsal margin concave to slightly sinuous, ventral margin convex, female outer gills often folded anterior of marsupia; gill length 40–50% of shell length; gill height 50–80% of gill length, occasional females with gills as high as long; outer gill height 70–90% of inner gill height, but height of marsupial area in females may be greater than inner gill height; inner lamellae of inner gills usually completely connected to visceral mass, occasionally incomplete.

Outer gills marsupial; glochidia held in posterior 30–50% of gill; marsupium outline ovate to reniform, greatly distended when gravid; creamy white to light tan, distal margin gray or black.

Labial palps creamy white to tan or light brown; straight dorsally, slightly convex ventrally, usually bluntly pointed distally, occasionally narrowly rounded; palp length 30–55% of gill length; palp height 50–75% of palp length; distal 30–45% of palps bifurcate.

Incurrent aperture usually considerably longer than excurrent aperture; supra-anal aperture usually considerably longer than incurrent aperture.

Figure 13.22. *Lampsilis ornata* apertures (left to right): supra-anal (partially shown), excurrent, incurrent. Bogue Chitto, Dallas County, Alabama, July 2004. Photograph by A.D. Huryn.

Incurrent aperture length 10–15% of shell length; creamy white within, with gray, rusty brown or pale orange basal to papillae, often mottled with gray or dark brown, gray or dark brown from papillae edges often extending onto aperture wall as dark lines, some lines may converge proximally; papillae usually in 2–3 rows, long, simple, rarely with a few bifid, may be slender or thick, inner row largest; papillae usually some combination of creamy white, tan, orange, brown and gray, edges of papillae often dark brown, dark gray or black basally.

Excurrent aperture length 5–10% of shell length; creamy white within, marginal color band variable, usually some combination of dull orange, tan, brown, gray and black, often with irregular mottling or bands perpendicular to margin, bands may converge proximally or be connected to form an open reticulated pattern; margin usually with small simple papillae, some individuals with crenulate margin; papillae orange, brown or gray, some with black between papillae (Figure 13.22).

Supra-anal aperture length 15–25% of shell length; creamy white within, thin marginal color band dull orange or gray; margin smooth; mantle bridge separating supra-anal and excurrent apertures imperforate, length 15–50% of supra-anal length.

Specimens examined: Conecuh River (n = 1); Coosa River (n = 2); Locust Fork (n = 5); Sipsey River (n = 4).

Glochidium Description

Length 198–225 μm; height 240–287 μm; without styliform hooks; outline subelliptical; dorsal margin straight, ventral margin rounded, anterior and posterior margins convex (Ortmann, 1923b, 1924a; Hoggarth, 1999).

Similar Species

Lampsilis ornata superficially resembles *Hamiota altilis* and *Hamiota perovalis* but has a much more inflated, more elevated umbo and attains a greater size. *Hamiota altilis* and *H. perovalis* are usually more elongate in outline than *L. ornata*.

General Distribution

Lampsilis ornata occurs in Gulf Coast drainages from the Escambia River drainage, Alabama and Florida, west to the Amite River, Lake Pontchartrain drainage, Louisiana (Vidrine, 1993). Vidrine (1993) reported a disjunct population in the Ouachita River drainage, Arkansas. However, Harris and Gordon (1990) did not report *L. ornata* from Arkansas.

Alabama and Mobile Basin Distribution

Lampsilis ornata is widespread in the Mobile Basin and occurs in all drainages, but there are no records from the Tallapoosa River above the Fall Line. It also occurs in lower reaches of the Conecuh River system.

Lampsilis ornata is extant in widespread, isolated localities in the Mobile Basin, where it may be locally common. It is uncommon in the Conecuh River system.

Ecology and Biology

Lampsilis ornata is found in medium creeks to large rivers. It may occur in flowing water or pools but is generally not found in reservoirs. *Lampsilis ornata* may occur in sand or gravel substrates, in which it is usually deeply buried.

Lampsilis ornata is a dioecious species, but a single hermaphroditic individual was reported from Little Tallahatchie River, Mississippi (Haag and Staton, 2003). It is a long-term brooder, gravid from August to June of the following year (Haag and Warren, 2003). In the Sipsey River, Alabama, sexual maturity was reported in all individuals two years of age, but some individuals were reported to be sexually mature during their first year (Haag and Staton, 2003). In the same Sipsey River population, 94 percent of sexually mature females were reported to be gravid, with a low percentage of unfertilized eggs (1.5 percent) (Haag and Staton, 2003). *Lampsilis ornata* displays an elaborate mantle flap, presumably to attract glochidial hosts. Fecundity of *L. ornata* was reported to average 281,776 glochidia per year, but ranged from 48,625 to 739,600 per individual (Haag and Staton, 2003). Glochidia are discharged singly, not bound within conglutinates (Haag and Staton, 2003). *Micropterus salmoides* (Largemouth Bass) (Centrarchidae) was found to serve as glochidial host of *L. ornata* in laboratory trials (Haag and Warren, 2003).

Current Conservation Status and Protection

Lampsilis ornata was reported as a species of special concern throughout its range by Williams et al. (1993) and in Alabama by Lydeard et al. (1999). Garner et al. (2004) designated *L. ornata* a species of low conservation concern in the state.

Remarks

For many years the binomen *Lampsilis excavata* (Lea, 1857) was used for *Lampsilis ornata*. During the 1970s, D.H. Stansbery (personal communication) began using *L. ornata* after noting a brief statement in Conrad's monography (1835b) that designated this form, "common in the rivers of south Alabama", as *Unio ornatus*, a variety of *Unio ovatus*. Cvancara (1963) reported *L. ornata* to possibly be a subspecies of *L. ovata*. However, the study was based only on shell measurement ratios.

The form of *Lampsilis ornata* that historically occurred in some Coosa River drainage headwater streams differed considerably from the typical form found in most of its range. The shell of this form, described as *Unio doliaris* Lea, 1865, is less inflated, with a much less prominent umbo and lacks a sharp posterior ridge (Figure 13.23). This form is conchologically similar to *Lampsilis binominata* of the Apalachicola Basin, but both are poorly known and appear to be extinct. Taxonomic placement of the *doliaris* form remains unresolved, but it is tentatively included as a synonym of *L. ornata*.

Figure 13.23. *Lampsilis ornata*, "*doliaris* form". Female, length 53 mm, UF 269891. Etowah River, Georgia. © Richard T. Bryant.

Synonymy

Unio ovatus var. *ornatus* Conrad, 1835. Conrad, 1835b:4
Type locality: Rivers of south Alabama. Type specimen not found.

Unio excavatus Lea, 1857. Lea, 1857a:32; Lea, 1858e:71, pl. 13, fig. 52; Lea, 1858h:71, pl. 13, fig. 52
Type locality: Othcalooga Creek, Gordon County, Georgia, Bishop Elliott; Etowah River, Georgia, Reverend George White; Alabama River, Claiborne, [Monroe County,] Alabama, Judge Tait. Lectotype, USNM 84498, length 101 mm (male), designated by Johnson (1974), is from Othcalooga Creek, Gordon County, Georgia.

Unio doliaris Lea, 1865. Lea, 1865:88; Lea, 1868b:260, pl. 32, fig. 75; Lea, 1869a:20, pl. 32, fig. 75
Type locality: Etowah River, Georgia, Reverend G. White. Holotype, USNM 84936, length 60 mm.

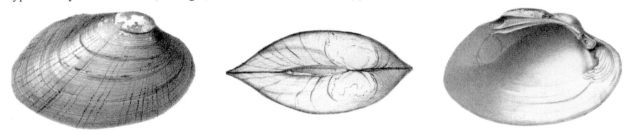

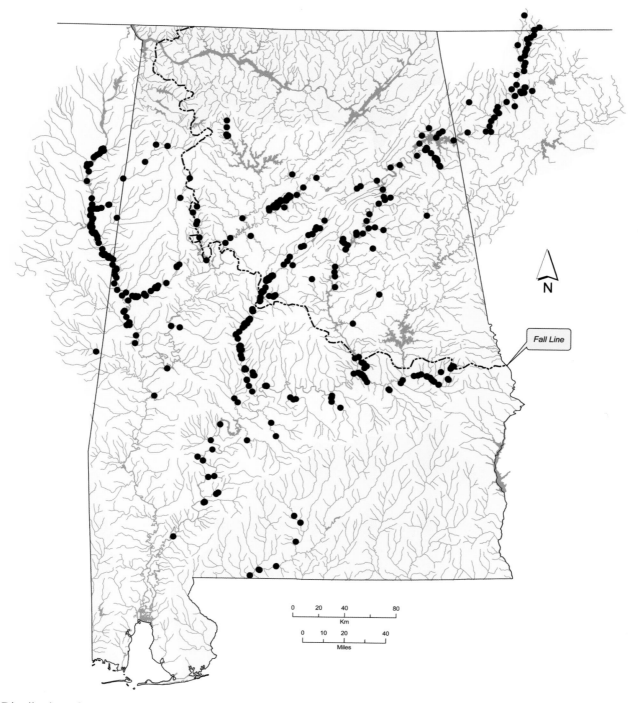

Distribution of *Lampsilis ornata* in Alabama and the Mobile Basin.

Lampsilis ovata (Say, 1817)
Pocketbook

Lampsilis ovata – Upper figure: female, length 110 mm, UF 65580. Tennessee River, Florence, Lauderdale County, Alabama. Lower figure: male, length 55 mm, UF 65594. Shoal Creek, Lauderdale County, Alabama, October 1909. © Richard T. Bryant.

Shell Description

Length to 170 mm; thin when young, becoming moderately thick with age; inflated; outline usually oval to subtriangular; posterior margin bluntly pointed to rounded or obliquely truncate, males usually more pointed than females; anterior margin narrowly rounded; dorsal margin convex; ventral margin broadly rounded; valves gape anteriorly and posteriorly; sexual dimorphism subtle, females more inflated posteriorly; posterior ridge sharp, usually less sharp in individuals from creeks; posterior slope steep, flat to slightly concave; umbo greatly inflated, elevated well above hinge line, posterior ridge remains sharp onto umbo, making it appear to be oriented anteriad, umbo sculpture coarse ridges, may be slightly double-looped ventrally; periostracum pale yellow in juveniles, yellowish green to brown in adults, some individuals with narrow, dark green rays.

Pseudocardinal teeth well-developed, erect, somewhat compressed, 2 slightly divergent triangular teeth in left valve, 1 tooth with flattened crest in right valve, usually with low thin accessory denticle anteriorly; lateral teeth short, straight to slightly curved, 2 in left valve, 1 in right valve, tooth in right valve truncate posteriorly; interdentum long, narrow; umbo cavity deep, open; nacre white.

Soft Anatomy Description

Mantle creamy white to light tan, outside of apertures fleshy, usually dull orange with black spots and gray mottling in irregular bands perpendicular to margin; visceral mass white to light tan; foot creamy white to tan, often darker distally.

Females with well-developed mantle flap just ventral to incurrent aperture (Figure 13.18); flap length 15–25% of shell length; with 3–4 variable color bands in combinations of creamy white, tan, brown, pale orange, gray and black; eyespot at posterior end of mantle flap round or oblong; flap margin usually papillate, often with only a few simple papillae, papillae usually long; well-developed "tail" at anterior end of flap, comprised of enlarged, flattened, partially fused papillae, some individuals with papillae almost entirely fused, giving "tail" a dentate margin. Ortmann (1912a) reported the eyespot to be "indistinct in the contracted condition".

Gills creamy white to tan; dorsal margin straight to slightly sinuous, ventral margin convex, female outer gills may be folded anterior of marsupia; gill length 35–50% of shell length; gill height 55–80% of gill length; outer gill height 75–90% of inner gill height; inner lamellae of inner gills completely connected to visceral mass. Ortmann (1912a) reported some individuals to have a small hole in the dorsal gill connection, near the posterior end of the foot.

Outer gills marsupial; glochidia held in posterior 25–50% of gill; marsupium usually somewhat reniform in outline, greatly distended when gravid; creamy white with grayish cast, distal margin dark gray or black.

Labial palps creamy white to tan; straight to slightly concave dorsally, slightly convex ventrally, bluntly pointed to rounded distally; palp length 25–35% of gill length; palp height 50–90% of palp length; distal 35–65% of palps bifurcate.

Incurrent aperture usually longer than excurrent aperture, occasionally equal in length; supra-anal aperture longer than incurrent and excurrent apertures.

Incurrent aperture length 5–15% of shell length; creamy white within, often with dull orange or rusty orange basal to papillae, with black or gray lines extending from papillae edges onto aperture wall, some lines converge proximally; papillae in 2 rows, long, simple, occasionally with some bifid, may be slender or thick, inner row larger; papillae creamy white to dull orange, often with gray or black mottling, edges of papillae usually black basally.

Excurrent aperture length 5–10% of shell length; usually creamy white within, occasionally dull orange, marginal color band dull orange with black mottling or black lines forming reticulated pattern; margin crenulate or with very small, simple papillae; papillae dull orange or black.

Supra-anal aperture length 15–20% of shell length; creamy white within, usually with thin band of sparse black or brown marginally; margin smooth; mantle bridge separating supra-anal and excurrent apertures often with 1 or more small perforations, length 15–50% of supra-anal length.

Specimens examined: Duck River, Tennessee (n = 2); Tennessee River (n = 6).

Glochidium Description

Length 200–240 μm; height 240–280 μm; without styliform hooks; outline subspatulate; dorsal margin straight, ventral margin broadly rounded, anterior and posterior margins convex, anterior margin slightly more deeply convex than posterior margin (Ortmann, 1912a, 1924a; Hoggarth, 1999). Jirka and Neves (1992) gave length as 309 μm and "width" as 264 μm.

Similar Species

Lampsilis ovata may superficially resemble *Lampsilis fasciola* but has a higher, more inflated umbo. *Lampsilis ovata* often has a much sharper posterior ridge than *L. fasciola*, especially individuals from large rivers. *Lampsilis ovata* attains a much greater size than *L. fasciola*. Also, the periostracum of *L. fasciola* is typically adorned with numerous wavy rays, whereas *L. ovata* usually has only a few thin, straight rays or none at all.

General Distribution

Lampsilis ovata occurs in most of the Ohio River drainage from western New York (Strayer and Jirka, 1997) to the mouth of the Ohio River (Cummings and Mayer, 1992). It is widespread in the Cumberland River drainage upstream and downstream of Cumberland Falls, though it was reportedly historically introduced above the falls (Wilson and Clark, 1914). In the Tennessee River drainage *L. ovata* occurs from the headwaters in eastern Tennessee and Virginia downstream to the mouth of the Tennessee River (Ahlstedt, 1992a, 1992b; Parmalee and Bogan, 1998).

Alabama and Mobile Basin Distribution

Lampsilis ovata occurs in the Tennessee River and some of its tributaries.

In the Tennessee River *Lampsilis ovata* is now confined to tailwaters of dams. Tributaries in which it is extant include the Elk and Paint Rock rivers and Bear Creek in Colbert County.

Ecology and Biology

Lampsilis ovata occurs in large creeks to large rivers, including tailwaters of dams. It may be found in flowing water or pools but is generally not found in overbank habitats of reservoirs. Ortmann (1919) stated that *L. ovata* inhabits "the roughest parts, riffles with strong current, the bottoms consisting of large stones, loosely piled over each other, with little finer material packing them together". In tailwaters of Tennessee River dams *L. ovata* usually occurs in mixtures of sand and gravel, in which it is usually deeply buried. Females emerge somewhat to display a mantle flap.

However, Ortmann (1919) observed it "moving along in coarse gravel, and pushing aside stones, which were much larger than itself".

Lampsilis ovata is a long-term brooder, gravid from August to the following spring or summer (Ortmann, 1909a, 1919). Females display elaborate mantle flaps to lure potential glochidial hosts (Figure 13.18). Fishes found to serve as glochidial hosts in laboratory trials include *Micropterus dolomieu* (Smallmouth Bass), *Micropterus punctulatus* (Spotted Bass) and *Micropterus salmoides* (Largemouth Bass) (Centrarchidae) (J.B. Layzer, personal communication).

Current Conservation Status and Protection

Lampsilis ovata was considered a species of special concern throughout its range by Williams et al. (1993). In Alabama it was considered endangered by Stansbery (1976) and a species of special concern by Lydeard et al. (1999). However, Garner et al. (2004) considered it a species of low conservation concern in the state.

Remarks

Lampsilis ovata has been found to be uncommon in prehistoric shell middens along the Tennessee River (Morrison, 1942; Warren, 1975; Hughes and Parmalee, 1999).

Lampsilis ovata and *Lampsilis cardium* have long been recognized as valid species. However, distributional limits and possible convergence in shell morphologies of these species are matters of conjecture. Typical large river *L. ovata* are characterized by well-developed, angular posterior ridges. In the Tennessee River drainage, this feature becomes less prominent in headwaters and tributary streams, resulting in individuals that resemble *L. cardium* (Figure 13.24). This change in shell morphology follows a clear progression from the large river to the headwater form. Therefore, this appears to simply be an expression of ecophenotypic variation, and *L. cardium* is not herein recognized as occurring in Alabama. Detailed anatomical and genetic studies are needed to clarify the matter. Some authors have reported *L. cardium* from Alabama.

Figure 13.24. *Lampsilis ovata*, headwater form. Length 119 mm, OSUM 59071. Bear Creek downstream of Factory Falls, Marion County, Alabama, 26 June 1996. Photograph by J.D. Williams.

Synonymy

Unio ovatus Say, 1817. Say, 1817:[no pagination], pl. 2, fig. 7

Type locality: Ohio River and its tributary streams. Haas (1930) designated a neotype, SMF 4338, length 130 mm, from the U.S.

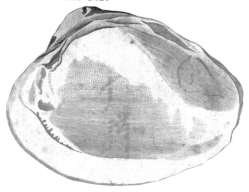

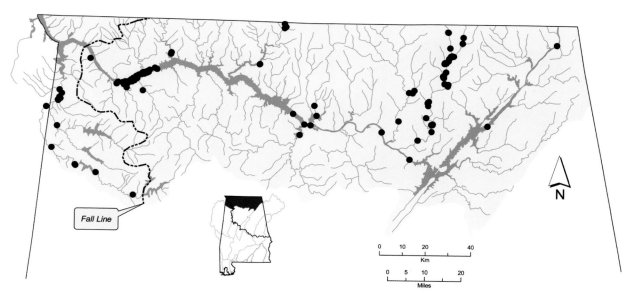

Distribution of *Lampsilis ovata* in the Tennessee River drainage of Alabama.

Lampsilis straminea (Conrad, 1834)
Southern Fatmucket

Lampsilis straminea – Upper figure: female, length 57 mm, USNM 84403. Lower figure: male, length 60 mm, USNM 84403. Columbus, [Lowndes County,] Mississippi. © Richard T. Bryant.

Shell Description

Length to 100 mm; moderately thick; inflated; outline elliptical; posterior margin broadly to narrowly rounded; anterior margin rounded; dorsal margin straight to convex; ventral margin straight to broadly curved; sexual dimorphism expressed as greater inflation posteriorly and more broadly rounded posterior margin in females; posterior ridge low, rounded; posterior slope flat to slightly convex; umbo broad, elevated slightly above hinge line, umbo sculpture slightly double-looped, parallel ridges; periostracum yellowish, may darken to brown with age, typically rayless.

Pseudocardinal teeth triangular to compressed, erect, 2 teeth in left valve, crests may be parallel, oriented at angle oblique to dorsal margin, 1 tooth in right valve, with accessory denticle anteriorly; lateral teeth long, thin, slightly curved, 2 in left valve, 1 in right valve; interdentum long, narrow to very narrow;

umbo cavity open, shallow; nacre white, sometimes with pinkish tint.

Soft Anatomy Description

Mantle creamy white to tan or light brown, often with golden cast, creamy white to pale orange outside of apertures, with black, gray or brown mottling in irregular bands perpendicular to margin; visceral mass creamy white to tan, occasionally pearly white; foot creamy white to tan or gold, may be slightly darker distally.

Females with well-developed mantle flap just ventral to incurrent aperture; flap length 15–30% of shell length; usually with 1–3 color bands, bands may be creamy white, dull orange or various shades of brown, often mottled; usually without distinct eyespot; flap margin usually papillate, but papillae may be irregular, papillae at anterior end of flap enlarged, flattened, wrinkled, several may be fused to form enlarged palmate "tail" structure, enlarged anterior papillae often colored differently than other papillae. Males may have small but well-developed or rudimentary mantle flap. Lea (1859f) described the flap papillae as "thickly set, perfectly white and look like rows of fishes' teeth".

Gills creamy white to tan or light brown, often with golden cast; dorsal margin straight to sinuous, ventral margin convex, female outer gills usually with several folds anterior of marsupia, outer gill often ragged distally; gill length 40–60% of shell length; gill height 40–75% of gill length; outer gill height 75–90% of inner gill height, outer gill height occasionally equal to inner gill height; inner lamellae of inner gills usually connected to visceral mass only anteriorly, occasionally completely connected.

Outer gills marsupial; glochidia held in posterior 25–50% of gill; marsupium oblong and curved to oval, distended when gravid; creamy white to light brown, distal margin gray or black.

Labial palps creamy white to tan; straight to slightly concave dorsally, convex ventrally, usually bluntly pointed distally, occasionally narrowly rounded; palp length 30–45% of gill length; palp height 50–85% of palp length; distal 30–45% of palps bifurcate, occasionally less.

Incurrent and supra-anal apertures longer than excurrent aperture; incurrent aperture usually longer than supra-anal aperture, occasionally equal in length.

Incurrent aperture length 10–20% of shell length; creamy white to tan within, usually with some combination of dull orange, gray, black and various shades of brown basal to papillae, usually with mottling, occasionally in form of crude bands perpendicular to margin; papillae in 1–3 rows, inner row usually largest, papillae long, slender, simple, rarely with few bifid; papillae usually some combination of creamy white, gold, tan, dull orange, gray, brown and black, often mottled, may change color distally, often with brown or black edges basally.

Excurrent aperture length 5–15% of shell length; creamy white to tan within, marginal color band usually dull orange to light brown, with markings in various shades of brown or black, markings variable, in form of irregular mottling, oblong spots, lines oriented perpendicular to margin or lines in reticulated pattern, often with irregular mottling proximally and more regular pattern distally; margin papillate, papillae small, simple; papillae dull orange, tan, brown or black.

Supra-anal aperture length 5–15% of shell length; creamy white to tan within, may have slight golden cast, rarely with thin marginal color band of tan, brown, gray, black or dull orange; margin smooth; mantle bridge separating supra-anal and excurrent apertures occasionally perforate, length highly variable, 20–100% of supra-anal length.

Specimens examined: Bull Mountain Creek, Marion County (n = 2); Burnt Corn Creek, Conecuh County (n = 1); Conecuh River (n = 3); Oakmulgee Creek, Dallas County (n = 2); Pea River (n = 1); Red Creek, Washington County (n = 1); Sepulga River (n = 4); Sipsey River (n = 2); Tombigbee River (n = 1); Trussells Creek, Greene County (n = 1); West Fork Choctawhatchee River (n = 2).

Glochidium Description

Length 170–210 μm; height 220–270 μm; without styliform hooks; outline subelliptical; dorsal margin straight, ventral margin rounded, anterior and posterior margins slightly convex (Lea, 1858d; Ortmann, 1924a).

Similar Species

Lampsilis straminea may superficially resemble *Lampsilis floridensis* and *Lampsilis teres*, but *L. straminea* is less elongate. Male *L. teres* and *L. floridensis* are usually more pointed posteriorly than *L. straminea*.

Lampsilis straminea may resemble *Hamiota altilis* and *Hamiota perovalis*, but both *Hamiota* species are less elongate and inflated. Additionally, *H. altilis* and *H. perovalis* have rays, though they are often obscure in *H. perovalis*. The periostracum of *L. straminea* is typically yellowish and without rays, whereas that of *H. perovalis* is typically much darker brown.

General Distribution

Lampsilis straminea occurs in Gulf Coast drainages from the Suwannee River in Florida and Georgia west to the Amite River, Lake Pontchartrain drainage, Louisiana (Vidrine, 1993).

Alabama and Mobile Basin Distribution

Lampsilis straminea is widespread in the Mobile Basin but is more common below the Fall Line. In Gulf Coast drainages it occurs from the Chattahoochee River drainage west to the Conecuh River system.

Lampsilis straminea remains widespread in the Mobile Basin as well as Gulf Coast drainages. It may be locally common, but has disappeared from some areas, especially large rivers.

Ecology and Biology

Lampsilis straminea occurs in large creeks to large rivers. It may be found in slow to moderate current, including pools, but it is generally absent from reservoirs. *Lampsilis straminea* is found in sand, sandy mud and gravel substrates.

Lampsilis straminea is a long-term brooder, gravid from late summer or autumn to the following summer. Female *L. straminea* as small as 48 mm in length have been found gravid. *Lampsilis straminea* glochidial hosts, determined by laboratory trials, include *Lepomis macrochirus* (Bluegill) and *Micropterus salmoides* (Largemouth Bass) (Centrarchidae); *Notemigonus crysoleucas* (Golden Shiner) and *Notropis texanus* (Weed Shiner) (Cyprinidae); *Ictalurus punctatus* (Channel Catfish) (Ictaluridae); and *Gambusia affinis* (Western Mosquitofish) (Poeciliidae) (Keller and Ruessler, 1997; Brim Box and Williams, 2000). The *Gambusia* stock used by Keller and Ruessler (1997) may have been *Gambusia holbrooki* (Eastern Mosquitofish). Highest transformation rates were obtained using *L. macrochirus* and *M. salmoides*.

Current Conservation Status and Protection

Lampsilis straminea was listed as currently stable throughout its range by Williams et al. (1993) and in Alabama by Lydeard et al. (1999). Garner et al. (2004) designated it a species of low conservation concern in the state.

Remarks

Two forms of *Lampsilis straminea*, the nominal form and *claibornensis* form, have historically been treated as distinct species (e.g., Simpson, 1914; Clench and Turner, 1956) or subspecies (e.g., Frierson, 1927; Turgeon et al., 1998). However, the two forms appear to represent a single taxon. Nominal *L. straminea*

differs from *claibornensis* in having numerous fine concentric ridges on its shell, whereas *claibornensis* is smooth. The ridged form is primarily confined to the Black Belt region of the Gulf Coastal Plain. The ridges appear to be an ecophenotypic character, as some syntopic species (e.g., *Fusconaia cerina* and *Villosa vibex*) also frequently possess such sculpture, which is rare outside of the Black Belt.

The common name of the *claibornensis* form, Southern Fatmucket, is herein retained since it is the more widespread and commonly recognized form. The common name applied to nominal *Lampsilis straminea*, Rough Fatmucket (Turgeon et al., 1998), is descriptive of individuals in only a small portion of its range.

A figure of *Lampsilis straminea*, a color illustration, appeared in Conrad (1834b) without a description. It was identified as *L. stramineus* by Frierson (1916).

The *Lampsilis straminea* population in headwaters of the Coosa River differs conchologically from typical *L. straminea*. It often has thin rays and variably colored nacre, ranging from white to purple. This form was originally described as *Unio perpastus* Lea, 1861, but has rarely been recognized by subsequent workers (Figure 13.25). It may be a distinct species but has not been found live in the Coosa River drainage since the early 1900s and appears to be extinct.

Figure 13.25. *Lampsilis straminea*, "*perpasta* form". Length 51 mm, USNM 84960. Coosa River, Alabama. © Richard T. Bryant.

Synonymy

Unio stramineus Conrad, 1834. Conrad, 1834a:339, pl. 1, fig. 6
Type locality: Small creeks of south Alabama. Lectotype designated by Johnson and Baker (1973), ANSP 20420.
Specimen not found, also not found by Johnson and Baker (1973). Length of shell in original description reported as about 64 mm.

Unio claibornensis Lea, 1838. Lea, 1838a:105, pl. 24, fig. 115; Lea, 1838c:105, pl. 24, fig. 115
Type locality: Alabama River, near Claiborne, [Monroe County, Alabama,] Judge Tait. Lectotype, USNM 85020, length 45 mm (male), designated by Johnson (1974).

Unio obtusus Lea, 1840. Lea, 1840:287; Lea, 1842b:201, pl. 11, fig. 13; Lea, 1842c:39, pl. 11, fig. 13
Type locality: Chattahoochee River, Columbus, [Muscogee County,] Georgia, Dr. Boykin. Lectotype, USNM 86142, length 55 mm, designated by Johnson (1974).

Unio pallescens Lea, 1845. Lea, 1845:164; Lea, 1848a:79, pl. 7, fig. 20; Lea, 1848b:79, pl. 7, fig. 20
Type locality: [Black Warrior River,] Tuscaloosa, [Tuscaloosa County,] Alabama, B.W. Budd, M.D. Holotype, USNM 84958, length 78 mm.

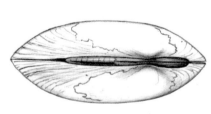

Unio contrarius Conrad, 1849. Conrad, 1849:153, pl. 37, fig. 7
Unio perpastus Lea, 1861. Lea, 1861c:60; Lea, 1862b:69, pl. 7, fig. 219; Lea, 1862d:73, pl. 7, fig. 219
Type locality: Coosa River, Alabama, E.R. Showalter, M.D. Holotype, USNM 84960, length 51 mm.

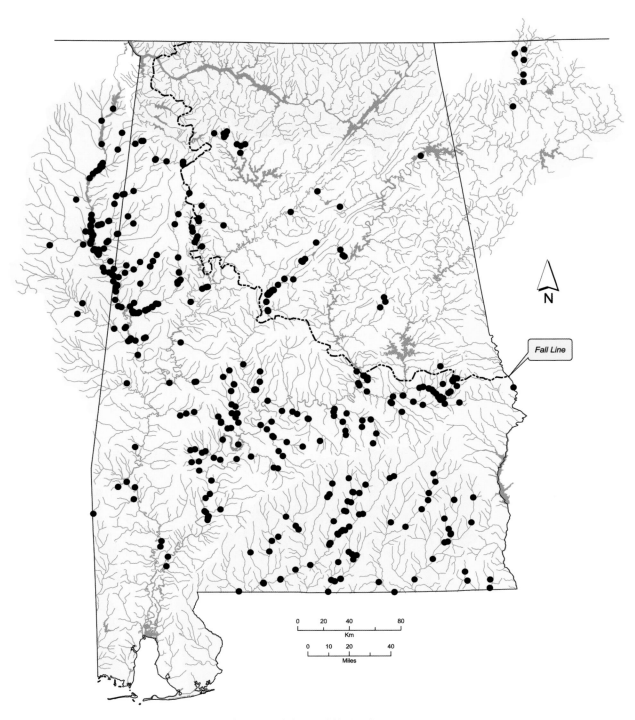

Distribution of *Lampsilis straminea* in Alabama and the Mobile Basin.

Lampsilis teres (Rafinesque, 1820)
Yellow Sandshell

Lampsilis teres – Upper figure: female, length 91 mm, UF 243998. Lower figure: male, length 107 mm, UF 243998. Alabama River, mouth of creek across from Isaac Creek Park, 500 m north of Claiborne Dam, 13 miles northwest of Monroeville, Monroe County, Alabama, 25 September 1988. © Richard T. Bryant.

Shell Description

Length to 185 mm; moderately thick; moderately to highly inflated; outline elongate elliptical; posterior margin bluntly pointed, females less pointed posteriorly than males; anterior margin rounded; dorsal margin straight to slightly convex; ventral margin straight to slightly convex or slightly concave; sexual dimorphism expressed as posterioventral swelling in females; posterior ridge low, rounded; posterior slope moderately steep; umbo broad, slightly inflated, elevated slightly above hinge line, umbo sculpture a few indistinct ridges; periostracum shiny, yellowish to tawny, typically without rays.

Pseudocardinal teeth triangular, usually compressed, blade-like, 2 divergent teeth in left valve, 1 tooth in right valve, usually with accessory denticle anteriorly; lateral teeth long, slender, straight to slightly curved, 2 in left valve, 1 in right valve; interdentum long, very narrow; umbo cavity wide, shallow; nacre white, sometimes with salmon or pink tint in umbo cavity.

Soft Anatomy Description

Mantle tan to light brown; visceral mass creamy white; foot creamy white to tan, usually darker distally.

Females with mantle flap just ventral to incurrent aperture; flap length 15–30% of shell length; margin

may have very short, simple papillae, crenulations or be smooth and undulating; no eyespot or enlarged "tail"; usually with 1–3 thin, dark brown and/or rusty brown bands, may be mottled. There appears to be variation in mantle flap coloration among populations. The mantle flap in females of some Alabama River populations is uniform creamy white with a pinkish tint and no marginal color bands. Ortmann (1912a) reported the anterior end of the flap to be truncate or to form a small lobe.

Gills creamy white to tan; dorsal margin straight, may be slightly curved around posterior adductor muscle, ventral margin straight to slightly convex; gill length 50–60% of shell length; gill height 25–50% of gill length; outer gill height 70–100% of inner gill height; inner lamellae of inner gills usually completely connected to visceral mass, occasionally connected only anteriorly.

Outer gills marsupial; glochidia held in posterior 35–50% of gill; marsupium, distended when gravid; creamy white with gray or black distal margin. Ortmann (1912a) and Utterback (1916b) reported gravid marsupia to be reniform in outline.

Labial palps creamy white to tan; straight dorsally, convex ventrally, usually bluntly pointed distally, may be more acute; palp length 30–40% of gill length; palp height 50–80% of palp length; distal 20–50% of palps bifurcate.

Incurrent and supra-anal apertures usually longer than excurrent aperture; incurrent aperture slightly longer, shorter or equal to supra-anal aperture.

Incurrent aperture length approximately 10% of shell length; creamy white within, may have black to pale rusty brown basal to papillae; papillae in 2 rows, simple and bifid, occasionally with simple papillae only, long, slender; papillae creamy white to golden brown or dull orange, may change color distally, some with black pigment between and along papillae edges.

Excurrent aperture length 5–10% of shell length; creamy white to golden brown within, usually with marginal color band, creamy white to pale orange with black markings; margin with small, mostly simple papillae, occasional bifid papillae in some individuals; papillae pale orange or golden brown. Ortmann (1912a) and Utterback (1916b) reported a crenulate excurrent aperture.

Supra-anal aperture length 5–10% of shell length; creamy white to golden brown within, usually without marginal coloration; margin smooth; mantle bridge separating supra-anal and excurrent apertures usually with 1 or more perforations, length 50–160% of supra-anal length.

Specimens examined: Alabama River (n = 4); Bogue Chitto, Dallas County (n = 1); Locust Fork (n = 1); Second Creek, Lauderdale County (n = 1); Sipsey River (n = 1); Tennessee River (n = 1); Tombigbee River (n = 2); Yellow Creek, Lowndes County, Mississippi (n = 1).

Glochidium Description

Length 180–207 μm, height 200–260 μm; without styliform hooks, outline subspatulate; dorsal margin straight, ventral margin rounded, anterior and posterior margins slightly curved (Ortmann, 1912a; Surber, 1912; Utterback, 1916b; Merrick, 1930; Hoggarth, 1999).

Similar Species

Lampsilis teres resembles *Lampsilis straminea* and *Lampsilis virescens*, but both of those species are less elongate and less pointed posteriorly. *Lampsilis virescens* is less inflated than *L. teres*.

Lampsilis teres resembles *Ligumia recta* in shape, but the periostracum of that species is always dark olive to black, never yellowish.

General Distribution

Lampsilis teres is known from the Niagara River, New York, in the Great Lakes and St. Lawrence Basin. It is widespread in the Mississippi Basin from southern Minnesota (Dawley, 1947) south to Louisiana (Vidrine, 1993), and from the Ohio River drainage in Ohio (Cummings and Mayer, 1992) west to Kansas (Murray

and Leonard, 1962). In the Cumberland River drainage *L. teres* is found downstream of the Obey River (Parmalee and Bogan, 1998). It occurs in the Tennessee River drainage downstream of the Paint Rock River. *Lampsilis teres* is widespread in Gulf Coast drainages, ranging from the Mobile Basin west to the Rio Grande in Mexico and Texas (Howells et al., 1996; Johnson, 1999).

Alabama and Mobile Basin Distribution

Lampsilis teres is found in most of the Tennessee River drainage and Mobile Basin. It is unknown from the Tennessee River upstream of the Paint Rock River and from the Tallapoosa River drainage upstream of the Fall Line.

Lampsilis teres remains widespread in all drainages from which it was historically known. It may be locally common.

Ecology and Biology

Lampsilis teres occurs in a variety of habitats from medium creeks to large rivers, but is more common in large water bodies. It is found in mud, sandy mud, sand and gravel substrates, often in slow to moderate current, but also may be found in swift current. *Lampsilis teres* is most frequently encountered along shore and channel slopes and overbanks of some reservoirs. Wilson and Clark (1914) stated that "it is one of the most active of the mussels, responding quickly to changes in environment by moving about."

Lampsilis teres has been reported to reach sexual maturity at age three in the upper Mississippi River, Iowa, and White River, Arkansas, and at age two in the Rio Grande, Texas, determined based on decreases in shell growth during those years (Chamberlain, 1931). It is a long-term brooder, and gravid individuals may be found most of the year (Surber, 1912; Utterback, 1916a; Coker et al., 1921; Howells, 2000). Conglutinates are white and "sole-shaped", not very solid (Utterback, 1916b). Females display a mantle flap, presumably to attract glochidial hosts.

Fishes shown to serve as glochidial hosts for *Lampsilis teres* in laboratory trials include *Atractosteus spatula* (Alligator Gar) and *Lepisosteus platostomus* (Shortnose Gar) (Lepisosteidae) (Howard, 1914c; Coker et al., 1921). Prentice (1994) reported encystment of *L. teres* glochidia on *Lepomis auritus* (Redbreast Sunfish) and *Lepomis macrochirus* (Bluegill) (Centrarchidae); and *Etheostoma lepidum* (Greenthroat Darter) (Percidae) in laboratory trials, but did not follow the experiment through to transformation. Natural infestations of *L. teres* glochidia have been reported for *Scaphirhynchus platorynchus* (Shovelnose Sturgeon) (Acipenseridae); and *Lepomis cyanellus* (Green Sunfish), *Lepomis humilis* (Orangespotted Sunfish), *Micropterus salmoides* (Largemouth Bass), *Pomoxis annularis* (White Crappie) and *Pomoxis nigromaculatus* (Black Crappie) (Centrarchidae) (Surber, 1913). Wilson (1916) included *Lepomis gulosus* (Warmouth) (Centrarchidae) as a host of *L. teres* but gave no indication as to whether this was determined under laboratory conditions or observed as a natural infestation.

Current Conservation Status and Protection

Lampsilis teres was listed as currently stable throughout its range by Williams et al. (1993) and in Alabama by Lydeard et al. (1999). Garner et al. (2004) designated it a species of lowest conservation concern in the state.

Remarks

Lampsilis teres is highly variable in shell morphology throughout its range, which has resulted in several synonyms. Three subspecies—*Lampsilis teres teres* (Rafinesque, 1820), *Lampsilis teres anodontoides* (Lea, 1834) and *Lampsilis teres fallaciosa* (Smith, 1899)—have been recognized in past literature. Two of these, *L. t. teres* and *L. t. anodontoides*, were reported from Alabama. However, conchological differences exhibited by these forms appear to represent ecophenotypic variation, and the subspecies were not recognized herein or by Turgeon et al. (1998).

Coker (1919) called *Lampsilis teres* "the most highly prized of all commercial shells" during the early 1900s. Though harvested along with shells used in the pearl button industry, *L. teres* were exported separately for use in making knife handles and other novelties. In addition to its pearly nacre and smooth texture, its shell dimensions and uniform thickness made it ideal for such uses (Wilson and Clark, 1914; Coker, 1919).

It is interesting to note that *Lampsilis teres* has not been reported from any prehistoric shell middens across the Tennessee River drainage (Hughes and Parmalee, 1999). It was present prior to impoundment of the river, reported from Muscle Shoals and Flint Creek, Morgan County, by Ortmann (1925).

Synonymy

Elliptio teres Rafinesque, 1820. Rafinesque, 1820:321

Type locality: La Rivière Wabash[, Indiana]. Type specimen not found, also not found by Johnson and Baker (1973).

Comment: Figure below is an illustration (93 mm) of the type in the Charles Poulson collection [ANSP], published in Conrad (1836).

Unio anodontoides Lea, 1834. Lea, 1834a:81, pl. 8, fig. 11; Lea, 1834b:91, pl. 8, fig. 11

Type locality: Mississippi River, T.W. Robeson; Alabama River, [Alabama,] Judge Tait; Ohio River, T.H. Taylor.
    Syntype not found, but length of figured shell in original description reported as about 104 mm.

Lampsilis fallaciosus Smith, 1899. Smith, 1899:291, pl. 79

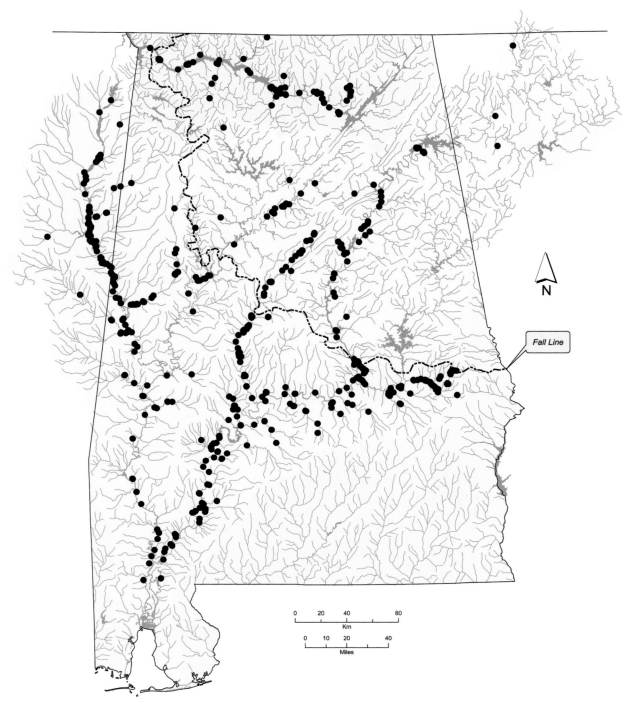

Distribution of *Lampsilis teres* in Alabama and the Mobile Basin.

Lampsilis virescens (Lea, 1858)
Alabama Lampmussel

Lampsilis virescens – Upper figure: female, length 50 mm, UF 65389. Hurricane Creek (branch of Flint River) near Gurley, Madison County, Alabama. Lower figure: male, length 73 mm, UF 65843. Paint Rock River, Alabama. © Richard T. Bryant.

Shell Description

Length to 75 mm; usually thin, old individuals may be moderately thick; moderately inflated; outline oval to elliptical; male posterior margin narrowly rounded, female posterior margin broadly rounded; anterior margin rounded; dorsal margin slightly convex; ventral margin straight to slightly convex; sexual dimorphism expressed as slightly more inflation posterioventrally and broader posterior margin in females; posterior ridge low, rounded; posterior slope not steep, slightly concave to slightly convex; umbo broad, slightly inflated, elevated slightly above hinge line, umbo sculpture delicate, double-looped ridges; periostracum shiny, tawny to greenish yellow, often pale, may have faint green rays, especially on posterior slope.

Pseudocardinal teeth compressed, triangular, 2 teeth in left valve, 1 tooth in right valve, with thin accessory denticle anteriorly; lateral teeth moderately long, slightly curved, 2 in left valve, 1 in right valve; interdentum long, very narrow; umbo cavity wide, moderately deep; nacre white.

Soft Anatomy Description

Mantle creamy white to tan; visceral mass creamy white to tan; foot creamy white to tan, may be slightly darker distally.

Females with mantle flap just ventral to incurrent aperture (Figure 13.20); flap length approximately 20% of shell length; with irregular black spots marginally, both inside and outside of flap; margin papillate, papillae small, simple; papillae tan.

Gills creamy white to tan; dorsal margin straight to slightly sinuous, ventral margin convex to elongate convex, outer gill may be almost straight ventrally; gill length approximately 60% of shell length; gill height 40–65% of gill length; outer gill height 70–90% of inner gill height; inner lamellae of inner gills completely connected to visceral mass.

Outer gills marsupial; glochidia held in posterior portion of gill; marsupium distended when gravid; creamy white, distal margin reddish brown.

Labial palps creamy white to tan; straight dorsally, straight to curved ventrally, bluntly pointed to rounded distally; palp length 20–25% of gill length;

palp height approximately 50% of palp length; distal 20–45% of palps bifurcate.

Incurrent and excurrent aperture length variable, either may be longer than the other; supra-anal aperture shorter than incurrent and excurrent apertures.

Incurrent aperture length approximately 15% of shell length; creamy white to tan within, may have black to reddish brown spots basal to papillae, spots somewhat square in outline, aligned perpendicular to margin; papillae in 2–3 rows, inner row largest, simple; papillae creamy white to dark tan or reddish brown, may be darker distally, may have black spots on edges basally.

Excurrent aperture length approximately 15% of shell length; creamy white to tan within, marginal color band with reddish brown to black markings in form of spots or squares; margin crenulate or with short, simple papillae; papillae creamy white to brown.

Supra-anal aperture length approximately 10% of shell length; creamy white to tan within, may have sparse black spots marginally; margin smooth; mantle bridge separating supra-anal and excurrent apertures imperforate, length 55–65% of supra-anal length.

Specimens examined: Estill Fork (n = 2); Paint Rock River (n = 1) (specimens previously preserved; note that specimens preserved for long periods are darker than those more recently preserved).

Glochidium Description

Length 240–260 μm; height 180–190 μm; without styliform hooks; outline subelliptical; dorsal margin straight, ventral margin broadly rounded, anterior and posterior margins slightly convex.

Similar Species

Lampsilis virescens resembles *Lampsilis teres* but is less elongate and less pointed posteriorly. *Lampsilis virescens* is also usually less inflated.

General Distribution

Lampsilis virescens is endemic to the Tennessee River drainage. It historically occurred from headwaters in eastern Tennessee downstream to Muscle Shoals in northwestern Alabama, including some large tributaries (Parmalee and Bogan, 1998). It appears to have been eliminated from its entire range, with the exception of the Paint Rock River in Alabama.

Alabama and Mobile Basin Distribution

Lampsilis virescens occurred in the Tennessee River across northern Alabama and in some tributaries.

Lampsilis virescens is extant only in upper reaches of the Paint Rock River system, Jackson County, where it is extremely rare.

Ecology and Biology

Lampsilis virescens historically occurred in streams ranging from small creeks to large rivers. However, it no longer occurs in large rivers. In small streams it is found in areas of slow to moderate current with sand and gravel substrates. It may be associated with beds of *Justicia americana* (Waterwillow).

Lampsilis virescens is a long-term brooder, gravid from late summer or autumn to the following summer. Female *L. virescens* display a modified mantle margin to attract glochidial hosts (P.D. Johnson, personal communication). In laboratory trials *L. virescens* glochidia have been found to utilize *Ambloplites rupestris* (Rock Bass), *Lepomis cyanellus* (Green Sunfish) and *Micropterus salmoides* (Largemouth Bass) (Centrarchidae) as glochidial hosts. *Lepomis macrochirus* (Bluegill) (Centrarchidae) and *Cottus carolinae* (Banded Sculpin) (Cottidae) appear to be marginal hosts (P.D. Johnson, personal communication).

Current Conservation Status and Protection

Lampsilis virescens was considered endangered throughout its range by Stansbery (1970a) and Williams et al. (1993) and in Alabama by Stansbery (1976) and Lydeard et al. (1999). Garner et al. (2004) designated *L. virescens* a species of highest conservation concern in the state. This species was listed as endangered under the federal Endangered Species Act in 1976.

In 2001 *Lampsilis virescens* was included on a list of species approved for a Nonessential Experimental Population in Wilson Dam tailwaters on the Tennessee River. However, no reintroductions had taken place as of 2007.

Remarks

Lampsilis virescens is often considered to be one of the most imperiled freshwater mussel species in North America. It, along with *Toxolasma cylindrellus*, is now restricted to upper reaches of the Paint Rock River system in Jackson County, Alabama, and possibly Franklin County, Tennessee, where it is very rare.

Lampsilis virescens has been reported to be uncommon in prehistoric shell middens at Muscle Shoals (Morrison, 1942), but has not been found in middens elsewhere in Alabama (Hughes and Parmalee, 1999).

Synonymy

Unio virescens Lea, 1858. Lea, 1858a:40; Lea, 1860d:341, pl. 55, fig. 166; Lea, 1860e:23, pl. 55, fig. 166
Type locality: Tennessee River, at Florence, [Lauderdale County,] Alabama, B. Pybas. Lectotype, USNM 84927,
    length 55 mm (male), designated by Johnson (1974).

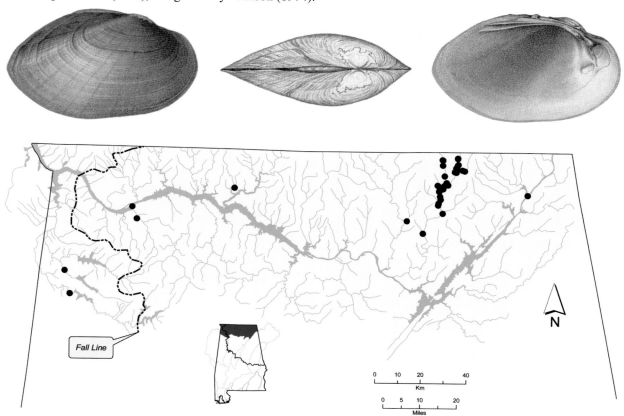

Distribution of *Lampsilis virescens* in the Tennessee River drainage of Alabama.

Genus *Lasmigona*

Lasmigona Rafinesque, 1831, occurs in the Hudson Bay, Great Lakes and Mississippi basins, Atlantic Coast drainages from New York to Georgia, and Gulf Coast drainages from the Apalachicola Basin to the Mississippi Basin. Turgeon et al. (1998) listed six species of *Lasmigona*. One species, *Lasmigona complanata* (Barnes, 1823), was recognized as being comprised of two subspecies, *L. c. complanata* (Barnes, 1823) and *L. c. alabamensis* Clarke, 1985. *Lasmigona c. alabamensis* is herein elevated to species status based on shell morphology and preliminary genetic analysis. An additional species, *Lasmigona etowaensis* (Conrad, 1849), is herein resurrected from synonymy based on preliminary genetic analysis. This brings the number of *Lasmigona* species that occur in Alabama to six.

Lasmigona was included in a monograph of the tribe Alasmidontini by Clarke (1985).

Type Species
Alasmidonta costata Rafinesque, 1820

Diagnosis
Shell elliptical or rhomboidal to ovate; compressed; disk typically smooth, posterior slope sometimes with ridges or corrugations; umbo low, not inflated, umbo sculpture ranging from a few simple bars to distinctly sinuous or double-looped; pseudocardinal teeth always present; lateral teeth may be reduced or absent; all species with moderate to prominent interdental projection in left valve.

Soft tissues usually some shade of orange; inner lamellae of inner gills connected to visceral mass only anteriorly; excurrent aperture smooth or crenulate; mantle bridge separating excurrent and supra-anal apertures short; outer gills marsupial; glochidia held across entire gill (Figure 13.26); secondary water tubes present when gravid; marsupium thickly padded when gravid; glochidium with styliform hooks (Rafinesque, 1831; Ortmann, 1912a; Simpson, 1914; Haas, 1969b; Clarke, 1985).

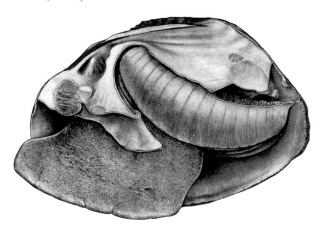

Figure 13.26. *Lasmigona complanata* gravid female soft anatomy. From Lefevre and Curtis (1912).

Synonymy
Amblasmodon Rafinesque, 1831
Pterosyna Rafinesque, 1831
Sulcularia Rafinesque, 1831
Complanaria Swainson, 1840
Megadomus Swainson, 1840 [placement uncertain]
Alasminota Ortmann, 1914
Platynaias Walker, 1918

Lasmigona alabamensis Clarke, 1985
Alabama Heelsplitter

Lasmigona alabamensis – Upper figure: length 143 mm, UF 65813. Lower figure: juvenile, length 88 mm, UF 65813. Chattooga River near Lyerly, Chattooga County, Georgia, 1 August 1915. © Richard T. Bryant.

Shell Description

Length to 150 mm; thin when young, becoming moderately thick with age; compressed; outline triangular, alate, may be oval if dorsal wing is eroded or broken; posterior margin obliquely truncate to broadly rounded, may have slight emargination just ventral to dorsal wing; anterior margin rounded; dorsal margin with triangular wing, often convex in eroded individuals; ventral margin straight to convex; posterior ridge very low, doubled posterioventrally; posterior slope wide, flat to slightly concave; posterior 65% of shell typically sculptured with subradial wrinkles, posterior slope and dorsal wing with wrinkles formed into more regular, shallow plications;

umbo broad, compressed, elevated slightly above hinge line if at all, umbo sculpture double-looped, nodulous ridges; periostracum greenish brown when young, darkening to brown or black with age, typically without rays.

Pseudocardinal teeth triangular, 2 divergent teeth in left valve, 1 tooth in right valve; lateral teeth short, poorly developed, may consist of low rudimentary ridges, tapering posteriorly, 2 in left valve, 1 in right valve; interdentum short, usually moderately wide, interdental projection in left valve compressed, elevated, may be confluent with posterior pseudocardinal tooth and lateral teeth, interdental portion of hinge plate may be absent in right valve, obscured by depression that accommodates interdental projection of opposing valve; umbo cavity wide, shallow, slightly deeper in left valve; nacre white.

Soft Anatomy Description

Mantle tan, brown, gold or orange, often with dark brown area dorsally, usually darker orange or brown outside of pallial line; visceral mass creamy white to tan, orange or pink adjacent to foot; foot orange, may be darker distally.

Gills light brown to reddish brown or orange, often with golden or grayish cast; dorsal margin straight to sinuous, ventral margin elongate convex; gill length 50–60% of shell length; gill height 30–45% of gill length; outer gill height 60–90% of inner gill height; inner lamellae of inner gills connected to visceral mass only anteriorly.

Outer gills marsupial; glochidia held across entire gill; marsupium padded when gravid; brown.

Labial palps brown or brownish orange, typically with tan area dorsally; straight to slightly concave dorsally, slightly convex ventrally, bluntly pointed distally; palp length 20–35% of gill length; palp height 40–60% of palp length; distal 20–40% of palps bifurcate, occasionally greater.

Incurrent and supra-anal apertures longer than excurrent aperture; incurrent aperture longer, shorter or equal to supra-anal aperture.

Incurrent aperture length 20–25% of shell length, occasionally less; creamy white to light tan within, occasionally with grayish cast, gray or grayish brown to rusty brown or rusty orange basal to papillae; papillae in 2–3 rows, inner row largest, long, simple, slender, sometimes wide basally, tapering to a point; papillae some combination of creamy white, tan, rusty orange, brown and gray, edges of papillae sometimes dark brown basally.

Excurrent aperture length 10–15% of shell length; creamy white to tan within, marginal color band some combination of creamy white, tan, gray and brown, may have open reticulated pattern, oblong spots or bands perpendicular to margin, some individuals without distinct pattern; margin usually crenulate, but may have small, simple papillae; papillae creamy white to grayish tan.

Supra-anal aperture length 15–25% of shell length, occasionally greater; creamy white or gold to orange or brownish orange within, often with thin brown or brownish orange band marginally; margin smooth; mantle bridge separating supra-anal and excurrent apertures often perforate, length highly variable, 10–120% of supra-anal length.

Specimens examined: Bogue Chitto, Dallas County (n = 2); Coosa River (n = 3); Locust Fork (n = 4); Tombigbee River (n = 3).

Glochidium Description

Length 325–350 μm; height 300–325 μm; with styliform hooks; outline pyriform; dorsal margin straight, ventral margin bluntly pointed, anterior and posterior margins rounded, anterior margin slightly more broadly rounded than posterior margin.

Similar Species

Small *Lasmigona alabamensis* may superficially resemble small *Megalonaias nervosa* due to their wrinkled sculpture. However, the high dorsal wing of *L. alabamensis* gives it a more triangular outline than the quadrate or oval *M. nervosa*.

General Distribution

Lasmigona alabamensis is endemic to the Mobile Basin of Alabama, Georgia and Mississippi.

Alabama and Mobile Basin Distribution

Lasmigona alabamensis occurs in much of the Mobile Basin but is absent from the Tallapoosa River drainage above the Fall Line.

Lasmigona alabamensis is extant in widespread localities in large rivers and some tributaries.

Ecology and Biology

Lasmigona alabamensis occurs primarily in medium to large rivers, though there are a few records from large creeks. It may be found in still or flowing water in various combinations of mud, sand and gravel substrates. *Lasmigona alabamensis* occurs in overbank habitats and was considered by Hurd (1974) to be "preadapted to the reservoir environment". However, it is uncommon in such habitat.

Lasmigona alabamensis is a long-term brooder, and is presumed to be gravid from late summer or autumn to the following spring or summer. Its glochidial hosts are unknown.

Current Conservation Status and Protection

Lasmigona alabamensis (as *Lasmigona complanata alabamensis*) was listed as a taxon of special concern throughout its range by Williams et al. (1993) and in Alabama by Lydeard et al. (1999). Garner et al. (2004) designated *L. alabamensis* a species of moderate conservation concern in the state.

Remarks

Lasmigona alabamensis was originally described as a subspecies of *Lasmigona complanata*, which is widespread in the Mississippi Basin and Pearl River drainage (Clarke, 1985; Vidrine, 1993). However, *L. alabamensis* has much heavier sculpture over a larger portion of the shell than nominal *L. complanata*. The two are also separable genetically (A.E. Bogan, unpublished data). *Lasmigona alabamensis* is the

Mobile Basin homologue of *L. complanata*. Clarke (1985) reported specimens from the Pearl and Pascagoula River drainages that could be identified as neither nominal *L. complanata* nor *L. alabamensis* but included them on the map for *L. alabamensis*. However, examination of several large lots of material from the Pearl and Pascagoula rivers revealed only nominal *L. complanata*.

Synonymy

Lasmigona (Lasmigona) complanata alabamensis Clarke, 1985. Clarke, 1985:36–40, figs. 10, 11

Type locality: H. Neely Henry Lake [Reservoir], Coosa River, Calhoun County, Georgia [Alabama]. Holotype, USNM 809996, length 148 mm.

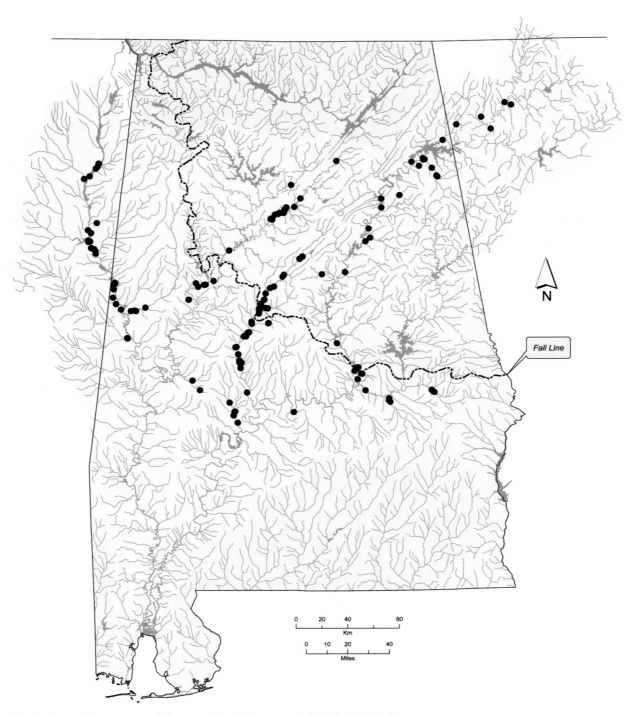

Distribution of *Lasmigona alabamensis* in Alabama and the Mobile Basin.

Lasmigona complanata (Barnes, 1823)
White Heelsplitter

Lasmigona complanata – Upper figure: length 143 mm, CMNH 61.11429. Duck River, 4 miles above Columbia, Maury County, Tennessee, 5 September 1922. Lower figure: length 50 mm, MMNS 6354. Pascagoula River, first shoal below State Highway 614, Jackson County, Mississippi, 23 June 2000. © Richard T. Bryant.

Shell Description

Length to 200 mm; thin when young, becoming moderately thick with age; compressed to moderately inflated; outline triangular, alate, may become oval if dorsal wing is eroded or broken; posterior margin obliquely truncate to rounded; anterior margin rounded; dorsal margin with triangular wing, often convex in eroded individuals; ventral margin straight to convex; posterior ridge very low, doubled posterioventrally; posterior slope wide, flat to slightly concave, typically with regular shallow plications radiating from posterior ridge onto dorsal wing; umbo broad, compressed, elevated slightly above hinge line if at all, umbo sculpture heavy ridges, becoming double-looped distally, posterior loop more angular than anterior loop; periostracum greenish brown, darkening to brown or black with age, juveniles sometimes with green rays.

Pseudocardinal teeth triangular, 2 divergent teeth in left valve, 1 tooth in right valve, occasionally with accessory denticle posteriorly; lateral teeth short, poorly developed, may consist of low rudimentary ridges, tapering posteriorly, 2 in left valve, 1 in right valve; interdentum short, typically moderately wide, interdental projection in left valve compressed, elevated, may be confluent with posterior pseudocardinal tooth and lateral teeth, interdental portion of hinge plate may be absent in right valve, replaced by depression that accommodates interdental projection of opposing valve; umbo cavity shallow, slightly deeper in left valve; nacre white.

Soft Anatomy Description

Mantle dull orange, brighter orange outside pallial line, grayish brown to tan outside of apertures; visceral mass pale orange; foot orange. Clarke (1985) described the mantle as pale grayish brown with a narrow band of pale orange and "dark purple-brown" located along the margin posteriorly and posterioventrally. Lea (1863d) reported the mantle margin to have small crenulations just ventral to the incurrent aperture.

Gills dull orange, may have grayish brown or creamy white areas; dorsal margin straight to sinuous, ventral margin elongate convex; gill length 55–60% of shell length; gill height 35–45% of gill length; outer gill height approximately 75% of inner gill height; inner laminae of inner gills only connected to visceral mass anteriorly.

Outer gills marsupial; glochidia held throughout gill. Clarke (1985) reported gravid gills to be thick and padded, Lea (1863d) described them as "enormously extended".

Labial palps tan to brown, may have creamy white area dorsally; concave dorsally, convex ventrally, bluntly pointed distally; palp length 25–40% of gill length; palp height 50–55% of palp length; distal 20–30% of palps bifurcate.

Incurrent and supra-anal apertures longer than excurrent aperture; incurrent aperture longer or shorter than supra-anal aperture.

Incurrent aperture length 15–20% of shell length; creamy white within, gray to grayish brown basal to papillae; papillae in 2 rows, inner row larger, simple, slender; papillae tan to brown or grayish brown, some with white tips. Clarke (1985) reported papillae to occur in three to four rows.

Excurrent aperture length 5–15% of shell length; creamy white within, marginal color band brown to dull orange, may have irregular grayish brown bands perpendicular to margin; margin papillate, papillae simple, short, may be wide, margin may be partially crenulate; papillae creamy white to tan or brown.

Supra-anal aperture length 15–25% of shell length; pale orange within, without marginal color band; margin smooth; mantle bridge separating supra-anal and excurrent apertures often perforate, length 45–55% of supra-anal length. Clarke (1985) reported the mantle bridge separating excurrent and supra-anal apertures to be only 4% of the supra-anal aperture length.

Specimens examined: Duck River, Tennessee (n = 1); Tennessee River (n = 1).

Glochidium Description

Length 289–310 μm; height 290–320 μm; with styliform hooks; outline pyriform; dorsal margin straight, ventral margin bluntly pointed, anterior and posterior margins rounded, anterior margin slightly more broadly rounded than posterior margin (Lefevre and Curtis, 1910b; Surber, 1912; Hoggarth, 1999). Ortmann (1912a) gave slightly larger measurements, length and height 340 μm.

Similar Species

Lasmigona complanata superficially resembles *Potamilus alatus* but typically has a less inflated umbo and plications on the dorsal wing. Also, *L. complanata* has poorly developed lateral teeth, whereas *Potamilus alatus* has well-developed lateral teeth. The shell nacre of *L. complanata* is white, but that of *P. alatus* is typically pink or purple.

General Distribution

Lasmigona complanata occurs in the southern Hudson Bay Basin, Canada and Minnesota, as well as western and southern portions of the Great Lakes and St. Lawrence Basin east to western New York (Clarke, 1973, 1985; Graf, 1997; Strayer and Jirka, 1997). In the Mississippi Basin it occurs from Minnesota (Dawley, 1947) south to Louisiana (Vidrine, 1993), and from headwaters of the Ohio River drainage in Pennsylvania (Ortmann, 1909a) west to North Dakota (Cvancara, 1983). Gangloff and Gustafson (2000) reported *L. complanata* from the Missouri River in Montana, where it appears to have been recently introduced. In the Cumberland River drainage *L. complanata* occurs from near the mouth of Caney Fork downstream to the mouth of the Cumberland River (Parmalee and Bogan, 1998). It is found in the Tennessee River drainage from headwaters in eastern Tennessee downstream to the mouth of the Tennessee River (Ahlstedt, 1992a; Parmalee and Bogan, 1998). *Lasmigona complanata* occurs in Gulf Coast tributaries from the Pascagoula River drainage west to the Mississippi River. Howells et al. (1996) reported a single specimen from the San Marcos River, Texas, a western Gulf Coast tributary.

Alabama and Mobile Basin Distribution

Lasmigona complanata occurs in the Tennessee River and some large tributaries.

Lasmigona complanata is extant in Wheeler Reservoir of the Tennessee River, in the Elk and Paint Rock rivers and in Bear Creek, Colbert County.

Ecology and Biology

Lasmigona complanata occurs in medium creeks to large rivers. It frequently inhabits unimpounded streams but is almost always found in pools and slow runs. Most of the recent records of *L. complanata* from the Tennessee River proper have come from overbank habitat of Wheeler Reservoir, though the species is occasionally found in riverine habitat in upper reaches of the reservoir. It may be found in mud, sand or gravel substrates. *Lasmigona complanata* may occur in water less than 1 m deep in unimpounded streams, but in the Tennessee River it can be found at depths exceeding 6 m.

Spermatozeugmata are discharged by male *Lasmigona complanata*, which were described by Utterback (1931) as "hollow globular masses of sperm revolving in the water by means of flagella thrust through a matrix from hundreds of individual sperm-cells". *Lasmigona complanata* is a long-term brooder, gravid with embryos in August or September and fully developed glochidia from late September through June (Lefevre and Curtis, 1912; Utterback, 1916a; Ortmann, 1919; Coker et al., 1921; Clarke, 1985; Howells, 2000). The hooked glochidia of *L. complanata* typically attach to the fins and tail of host fishes (Lefevre and Curtis, 1910b; Young, 1911).

Fishes proven to serve as glochidial hosts for *Lasmigona complanata* in laboratory trials include *Lepomis cyanellus* (Green Sunfish) and *Pomoxis annularis* (White Crappie) (Centrarchidae) (Lefevre and Curtis, 1912). In a study on glochidium implantation, Young (1911) exposed *L. complanata* glochidia to *Lepomis humilis* (Orangespotted Sunfish) and *Micropterus salmoides* (Largemouth Bass) (Centrarchidae); and *Fundulus diaphanus* (Banded Killifish) (Fundulidae), but the report was ambiguous as to whether transformation occurred. Observations of natural infestations of *L. complanata* glochidia have been reported for *Moxostoma carinatum* (River Redhorse) (Catostomidae); *Dorosoma cepedianum*

(Gizzard Shad) (Clupeidae); *Lepisosteus osseus* (Longnose Gar) (Lepisosteidae); and *Sander canadensis* (Sauger) (Percidae) (Weiss and Layzer, 1995). Glochidia of *L. complanata* have been reported to infest fins of the nonindigenous *Cyprinus carpio* (Common Carp) (Cyprinidae) in laboratory trials, but there have been no reports of confirmed transformation on the species (Lefevre and Curtis, 1910b).

Current Conservation Status and Protection

Lasmigona complanata was reported as currently stable throughout its range by Williams et al. (1993). Lydeard et al. (1999) considered it a species of special concern in Alabama. Garner et al. (2004) designated *L. complanata* a species of moderate conservation concern in the state.

Remarks

Lasmigona complanata appears to be a recent colonizer of the Tennessee River (Garner and McGregor, 2001). The earliest report of this species is from the late 1970s (Gooch et al., 1979). No museum records collected from Alabama prior to the mid-1990s were found. This species has not been reported from prehistoric shell middens along the Tennessee River in Alabama (Hughes and Parmalee, 1999). *Lasmigona complanata* has also been reported to be a recent colonizer of the Missouri River drainage in Nebraska (Hoke, 2000) and Montana (Gangloff and Gustafson, 2000).

Synonymy

Alasmodonta complanata Barnes, 1823. Barnes, 1823:278, pl. 13, fig. 21
Type locality: Fox River, Wisconsin. Type specimen lost, likely in fire with remainder of Barnes collection (Johnson, 2006). Length of figured shell in original description reported as 127 mm.

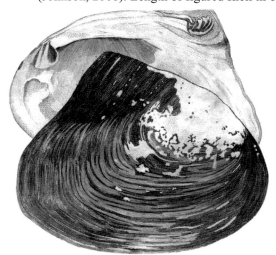

Unio gigas Swainson, 1824. Swainson, 1824:15–17 [identification uncertain]
Unio katherinae Lea, 1838. Lea, 1838a:143; Lea, 1838c:143

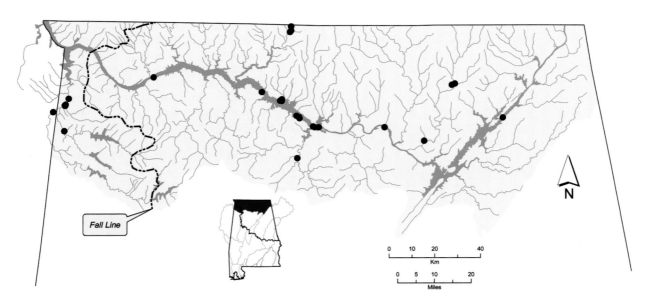

Distribution of *Lasmigona complanata* in the Tennessee River drainage of Alabama.

Lasmigona costata (Rafinesque, 1820)
Flutedshell

Lasmigona costata – Length 103 mm, UF 65840. Shoal Creek, Lauderdale County, Alabama. © Richard T. Bryant.

Shell Description

Length to 190 mm; thin when young, becoming moderately thick with age; moderately inflated; outline rectangular to rhomboidal; posterior margin obliquely truncate to rounded; anterior margin rounded; dorsal margin straight to slightly convex; ventral margin straight, may become slightly concave with age; posterior ridge low, rounded; posterior slope broad, typically with well-developed plications radiating from posterior ridge to posterior margin, often with small secondary wrinkles; posterior 25–50% of disk may be wrinkled; umbo broad, compressed, elevated slightly above hinge line, umbo sculpture well-developed ridges, parallel to hinge line, becoming slightly double-looped ventrally; periostracum tawny to greenish yellow, usually darkening to brown or black with age, young may have variable green rays.

Pseudocardinal teeth variable, 2 low teeth in left valve, sometimes fused into 1 tooth, may be thick and triangular or compressed, 1–2 triangular or compressed teeth in right valve; lateral teeth poorly developed, represented by low rudimentary ridges in both valves; interdentum short to moderately long, very narrow to moderately wide, interdental projection well-developed, erect, may be fused with posterior pseudocardinal tooth; umbo cavity shallow; nacre white.

Soft Anatomy Description

Mantle creamy white to tan or golden brown, may be dull orange outside of pallial line; visceral mass creamy white; foot orange.

Gills reddish orange to light brown; dorsal margin straight to slightly sinuous, ventral margin convex to elongate convex; gill length 50–60% of shell length; gill height 35–45% of gill length; outer gill height 75–80% of inner gill height; inner lamellae of inner gills completely connected to visceral mass or connected only anteriorly.

No gravid females were available for marsupium description. Lea (1863d) reported outer gills to be marsupial, with glochidia held across the entire gill and the marsupium to be brownish, forming "a large massive lobe which extends below the margin". Ortmann (1912a) reported gravid gills of to vary from yellowish to brown and Utterback (1915) reported the gills to be "rich and brown when charged".

Labial palps brown, with irregular creamy white area dorsally; straight to slightly concave dorsally, convex ventrally, rounded distally; palp length 20–30% of gill length; palp height 55–60% of palp length, occasionally greater; distal 25–35% of palps bifurcate. Utterback (1915) reported triangular labial palps.

Incurrent aperture slightly longer than excurrent aperture; supra-anal aperture shorter or equal to excurrent aperture.

Incurrent aperture length 15–25% of shell length; creamy white within, grayish brown to black basal to papillae; papillae in 2 rows, long, slender, simple; papillae golden brown to grayish brown, may have dark brown or black edges basally.

Excurrent aperture length 15–20% of shell length; creamy white within, marginal band golden brown to grayish brown or dark brown; margin smooth or with very short, stubby papillae, little more than crenulations; papillae brown to golden brown.

Supra-anal aperture length approximately 15% of shell length; creamy white to golden brown within, may have small brown flecks, sometimes with thin dark brown marginal band; margin smooth; mantle bridge separating supra-anal and excurrent apertures imperforate, length 5–15% of supra-anal length.

Specimens examined: Paint Rock River (n = 3).

Glochidium Description

Length 340–385 μm; height 363–390 μm; with styliform hooks; outline pyriform; dorsal margin straight to undulate, ventral margin narrowly rounded, anterior and posterior margins convex above, slightly incurved below, anterior margin slightly more convex than posterior margin (Lefevre and Curtis, 1910b; Ortmann, 1912a; Surber, 1912; Utterback, 1915; Hoggarth, 1999).

Similar Species

Lasmigona costata may superficially resemble *Ptychobranchus subtentum*, but the shell sculpture of *P. subtentum* is restricted to the posterior ridge and slope, whereas *L. costata* typically has sculpture over the posterior 25–50% of the shell. *Lasmigona costata* lateral teeth are rudimentary, but those of *P. subtentum* are well-developed. Rays on the periostracum of young *L. costata* are generally thinner than those of *P. subtentum* and often disappear with age. The offset rays often found near the umbo of *P. subtentum* are not present on *L. costata*.

General Distribution

Lasmigona costata occurs in the southern Hudson Bay Basin and western and southern portions of the Great Lakes and St. Lawrence Basin east to Vermont (Clarke, 1985). In the Mississippi Basin it occurs from Minnesota south to Arkansas, and from headwaters of the Ohio River in western New York west to eastern Kansas (Clarke, 1985). *Lasmigona costata* occurs in the Cumberland River downstream of Cumberland Falls and most of the Tennessee River drainage (Cicerello et al., 1991; Ahlstedt, 1992a, 1992b; Parmalee and Bogan, 1998).

Alabama and Mobile Basin Distribution

Lasmigona costata historically occurred in the Tennessee River across northern Alabama and in many tributaries.

Lasmigona costata is extant only in the Elk and Paint Rock rivers and Bear Creek in Colbert County.

Ecology and Biology

Lasmigona costata occurs in large creeks to medium rivers, though historically it was found in large rivers. It is primarily a species of shoal habitats, where it occurs in mixed sand and gravel substrates, but may also be found in pools. *Lasmigona costata* is also known from the Great Lakes.

Lasmigona costata is a long-term brooder. It becomes gravid as early as August and glochidia mature as early as September (Ortmann, 1919; Baker, 1928a; Clarke, 1985). Glochidia are brooded until the following spring (Ortmann, 1912a, 1919; Utterback,

1915; Baker, 1928a). Ortmann (1909a) reported late May as the latest date that gravid females were observed. Ortmann (1910a) reported *L. costata* glochidia to be discharged in "irregular masses, which do not stick together, so as to preserve the shape of the placentae" (i.e., they are not discharged as well-formed conglutinates).

Lasmigona costata has been found to utilize a variety of glochidial hosts in laboratory trials, but results varied among species. Fishes that produced at least one juvenile include *Amia calva* (Bowfin) (Amiidae); *Hypentelium nigricans* (Northern Hog Sucker) (Catostomidae); *Ambloplites rupestris* (Rock Bass), *Lepomis cyanellus* (Green Sunfish), *Lepomis macrochirus* (Bluegill), *Lepomis megalotis* (Longear Sunfish), *Micropterus dolomieu* (Smallmouth Bass) and *Micropterus salmoides* (Largemouth Bass) (Centrarchidae); *Cottus carolinae* (Banded Sculpin) (Cottidae); *Campostoma anomalum* (Central Stoneroller), *Rhinichthys cataractae* (Longnose Dace) and *Semotilus atromaculatus* (Creek Chub) (Cyprinidae); *Esox lucius* (Northern Pike) (Esocidae); *Fundulus catenatus* (Northern Studfish) (Fundulidae); *Ameiurus nebulosus* (Brown Bullhead) (Ictaluridae); and *Etheostoma caeruleum* (Rainbow Darter), *Etheostoma flabellare* (Fantail Darter), *Etheostoma virgatum* (Striped Darter), *Etheostoma zonale* (Banded Darter), *Perca flavescens* (Yellow Perch) and *Sander vitreus* (Walleye) (Percidae) (Luo, 1993; Hove et al., 1994). The nonindigenous *Carassius auratus* (Goldfish) and *Cyprinus carpio* (Common Carp) (Cyprinidae) have also been reported to serve as glochidial hosts in laboratory trials (Lefevre and Curtis, 1910c; Watters et al., 2005). Natural infestations of *Dorosoma cepedianum* (Gizzard Shad) (Clupeidae) and *Moxostoma carinatum* (River Redhorse) (Catostomidae) with *L. costata* glochidia have been reported (Weiss and Layzer, 1995). Two types of *L. costata* glochidia have been reported, and Watters et al. (1998) suggested that one of them develops to the juvenile stage without parasitism.

Current Conservation Status and Protection

Lasmigona costata was listed as currently stable throughout its range by Williams et al. (1993) and in Alabama by Lydeard et al. (1999). However, Garner et al. (2004) designated it a species of high conservation concern in the state.

Remarks

The only prehistoric records of *Lasmigona costata* from Alabama are two specimens from shell middens at Muscle Shoals on the Tennessee River (Morrison, 1942; Hughes and Parmalee, 1999).

Synonymy

Alasmidonta costata Rafinesque, 1820. Rafinesque, 1820:318, pl. 82, figs. 15, 16
Type locality: Kentucky River. Type specimen not found.

Alasmodonta rugosa Barnes, 1823. Barnes, 1823:278, pl. 13, fig. 21
Alasmodon rugosum Rafinesque, 1831. Rafinesque, 1831:5
Amblasmodon hians Rafinesque, 1831. Rafinesque, 1831:5
Lasmigona costata var. *ereganensis* Grier, 1918. Grier, 1918:10
Lasmigona costata pepinensis F.C. Baker, 1928. F.C. Baker, 1928a:144
Lasmigona costata nuda F.C. Baker, 1928. F.C. Baker, 1928a:145

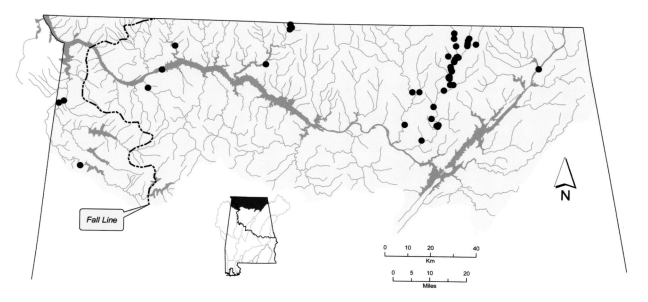

Distribution of *Lasmigona costata* in the Tennessee River drainage of Alabama.

Lasmigona etowaensis (Conrad, 1849)
Etowah Heelsplitter

Lasmigona etowaensis – Length 50 mm, UF 65874. Shoal Creek, St. Clair County, Alabama, 20 October 1914. © Richard T. Bryant.

Shell Description

Length to 70 mm; thin; moderately inflated; outline elliptical; posterior margin rounded to obliquely truncate; anterior margin rounded; dorsal margin convex; ventral margin straight to slightly convex; posterior ridge broadly rounded, may be slightly doubled; posterior slope moderately steep, slightly concave; umbo slightly inflated, barely elevated above hinge line, umbo sculpture thick, weakly double-looped ridges; periostracum tawny to greenish brown, becoming dark brown with age, usually without rays.

Pseudocardinal teeth low, compressed, triangular, 2 teeth in left valve, 1 tooth in right valve, teeth in both valves oriented almost parallel to dorsal margin; lateral teeth rudimentary, represented by little more than thickened hinge line in both valves; interdentum short, very narrow, interdental projection small, compressed, may be confluent with posterior pseudocardinal tooth; umbo cavity shallow, open; nacre bluish white, may have pale salmon tint in umbo cavity.

Soft Anatomy Description

Mantle tan, outside of apertures thick, fleshy, dark brown with small black spots; visceral mass creamy white with slight golden cast; foot pale gold.

Gills tan with slight golden cast; dorsal margin slightly sinuous, ventral margin of inner gill long, slightly curved, ventral margin of outer gill convex; gill length approximately 50% of shell length; gill height approximately 70% of gill length; outer gill height approximately 85% of inner gill height; inner lamellae of inner gills connected to visceral mass only anteriorly.

No gravid females were available for marsupium description, probably similar to other *Lasmigona* species, with outer gills marsupial, marsupium padded when gravid.

Labial palps tan; slightly concave dorsally, convex ventrally, bluntly pointed distally; palp length approximately 25% of gill length; palp height approximately 70% of palp length; distal 40% of palps bifurcate.

Incurrent aperture longer than excurrent and supra-anal apertures; excurrent aperture slightly longer than supra-anal aperture.

Incurrent aperture length approximately 25% of shell length; tan within, dark brown basal to papillae; papillae in 2–3 rows, long, slender, simple; papillae tan.

Excurrent aperture length approximately 20% of shell length; grayish brown within, marginal color band dark brown; margin smooth but with small undulations.

Supra-anal aperture length approximately 15% of shell length; creamy white within, thin dark brown marginal band; margin smooth; mantle bridge separating supra-anal and excurrent apertures imperforate, length approximately 40% of supra-anal length.

Specimen examined: Spring Creek, Cherokee County (n = 1).

Glochidium Description

Length 325–350 μm; height 300–337 μm; with styliform hooks; outline pyriform; dorsal margin straight to undulate or slightly convex, ventral margin narrowly rounded, anterior and posterior margins convex above, slightly incurved below, anterior margin slightly more convex than posterior margin.

Similar Species

Lasmigona etowaensis may resemble *Villosa lienosa*, *Villosa nebulosa* and *Villosa umbrans*. However, those species have well-developed lateral teeth, thicker shells and lack an interdental projection in the left valve, which is present in *L. etowaensis*. Also, the nacre of *V. lienosa* and *V. umbrans* is usually purple, whereas that of *L. etowaensis* is white. *Villosa nebulosa* usually has rays on the periostracum, which are absent on *L. etowaensis*.

Lasmigona etowaensis may also superficially resemble *Anodontoides radiatus* or subadult *Strophitus* (*S. connasaugaensis* or *S. subvexus*), but those species have only rudimentary pseudocardinal teeth and lack an interdental projection in the left valve. Also, *A. radiatus* has prominent rays, which are lacking in *L. etowaensis*.

General Distribution

Lasmigona etowaensis occurs in the Coosa, Cahaba and Black Warrior River drainages above the Fall Line in Alabama, Georgia and Tennessee.

Alabama and Mobile Basin Distribution

Lasmigona etowaensis is widespread in tributaries of the Coosa River and has been found at a few localities in the Cahaba River system and in eastern tributaries of the Black Warrior River drainage, all above the Fall Line.

Lasmigona etowaensis is extant in scattered isolated tributaries. Its current distribution is poorly known due to a dearth of surveys in small headwater streams.

Ecology and Biology

Lasmigona etowaensis is a species of small to medium creeks, often found in very small streams where no other mussels occur. It is usually encountered in areas with at least some current, often in sandy substrates, though may also be found in mud or gravel.

Lasmigona etowaensis is a long-term brooder, gravid from late summer or autumn to the following summer. Fishes found to serve as glochidial hosts of *E. etowaensis* in laboratory trials include *Ambloplites rupestris* (Rock Bass), *Lepomis macrochirus* (Bluegill) and *Micropterus coosae* (Redeye Bass) (Centrarchidae); *Cottus carolinae* (Banded Sculpin) (Cottidae); *Campostoma oligolepis* (Largescale Stoneroller), *Cyprinella trichroistia* (Tricolor Shiner) and *Notropis xaenocephalus* (Coosa Shiner) (Cyprinidae); *Ictalurus punctatus* (Channel Catfish) (Ictaluridae); and *Etheostoma coosae* (Coosa Darter) and *Percina nigrofasciata* (Blackbanded Darter) (Percidae) (P.D. Johnson, personal communication).

Current Conservation Status and Protection

Most conservation assessments have included *Lasmigona etowaensis* as part of *Lasmigona holstonia*, which was listed as a species of special concern throughout its range by Williams et al. (1993) and in Alabama by Lydeard et al. (1999). Stansbery (1976) recognized *L. etowaensis* (as *Lasmigona georgiana* Lea, 1859) and listed it as a species of uncertain status in the state. *Lasmigona holstonia* (including *L. etowaensis*) was designated a species of high conservation concern in Alabama by Garner et al. (2004).

Remarks

Lasmigona etowaensis is herein recognized as distinct from *Lasmigona holstonia* based on genetic analyses (A.E. Bogan, unpublished data). These two species cannot reliably be distinguished using conchological characters.

The range of *Lasmigona etowaensis* crosses several drainages of the Mobile Basin. However, it appears to be confined to streams of the Valley and Ridge physiographic province.

Synonymy

Margaritana etowahensis Conrad, 1849. Conrad, 1849:154; Conrad, 1849:302
Type locality: Etowah River, [Georgia]. Type specimen not found.
Margaritana etowaensis Conrad, 1849. Conrad, 1849:154; Conrad, 1849:302
Type locality: Etowah River, [Georgia]. Type specimen not found, also not found by Johnson and Baker (1973).
Comment: The specific name, *etowaensis,* was in reference to the Etowah River and has often been taken to represent an incorrect original spelling. However, subsequently Conrad (1853) specifically pointed out that the spelling in the original description, *etowaensis,* was correct.

Margaritana etowahensis Lea, 1858, *non* Conrad 1849. Lea, 1858b:138; Lea, 1859f:227, pl. 31, fig. 110; Lea, 1859g:45, pl. 31, fig. 110

Type locality: Tennessee, Dr. Troost; Etowah River, Georgia, Reverend G. White. Lectotype, USNM 86259, length 52 mm, designated by Johnson (1974).

Margaritana georgiana Lea, 1859. Lea, 1859e:280 [new name for *Margaritana etowahensis* Lea, 1858]

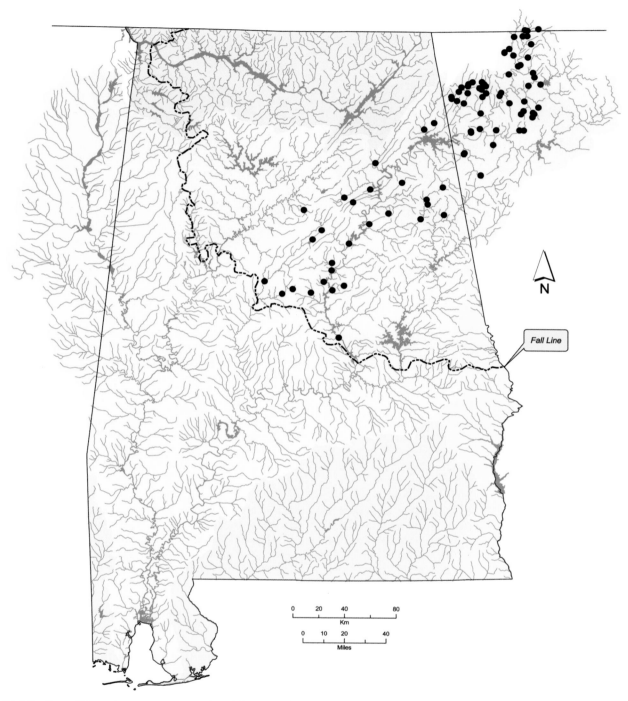

Distribution of *Lasmigona etowaensis* in Alabama and the Mobile Basin.

Lasmigona holstonia (Lea, 1838)
Tennessee Heelsplitter

Lasmigona holstonia – Length 47 mm, NCSM 5709. Tiger Creek, 2.9 km southwest of Cohutta, Whitfield County, Georgia, 22 April 1969. © Richard T. Bryant.

Shell Description

Length to 75 mm; thin; moderately inflated; outline elliptical to rhomboidal; posterior margin narrowly rounded to obliquely truncate; anterior margin rounded; dorsal margin convex; ventral margin straight to slightly convex; posterior ridge broadly rounded, may be slightly doubled; posterior slope moderately steep, slightly concave; umbo slightly inflated, barely elevated above hinge line, umbo sculpture thick, weakly double-looped ridges; periostracum tawny to greenish brown, becoming dark brown with age, typically without rays.

Pseudocardinal teeth low, compressed, triangular, 2 teeth in left valve, 1 tooth in right valve, teeth in both valves oriented almost parallel to dorsal margin; lateral teeth rudimentary, represented by little more than thickened hinge line in both valves; interdentum short, very narrow, interdental projection high, compressed, triangular, may be confluent with posterior pseudocardinal tooth; umbo cavity shallow, open; nacre bluish white, may have pale salmon tint in umbo cavity.

Soft Anatomy Description

Mantle pale tan anteriorly, creamy white posteriorly, outside of apertures dark brown to black; visceral mass creamy white with small grayish brown area posteriorly; foot pale tan to pale orange. Clarke (1985) reported the mantle margin to have a narrow "purplish brown" pigmented area beginning mid-ventrally and extending posteriorly, becoming broader at the apertures. Clarke (1985) also described the outer surface of the mantle as having a broken brownish band in the region of the apertures. The foot was reported by Clarke (1985) as being white in a previously preserved specimen.

Gills tan with grayish brown cast posteriorly; dorsal margin straight to slightly sinuous, ventral margin convex; gill length 45–50% of shell length; gill height 55–60% of gill length; outer gill height 80–90% of inner gill height; inner lamellae of inner gills connected to visceral mass only anteriorly. Clarke (1985) described longer, narrower gills in an individual from the Holston River drainage, with gill length being 69% of shell length and gill height 40% of gill length.

No gravid females were available for marsupium description. Clarke (1985) reported outer gills to be marsupial, when gravid "engorged" across the entire gill except the posterior 25%. Ortmann (1914) described the ventral edges of gravid gills as "more or less distended".

Labial palps pale tan, may be creamy white distally; straight dorsally, convex ventrally, bluntly pointed distally; palp length approximately 20% of gill length; palp height 50–70% of palp length; distal 35–40% of palps bifurcate.

Incurrent and excurrent apertures of similar length; incurrent and excurrent apertures longer than supra-anal aperture.

Incurrent aperture length approximately 20–25% of shell length; solid dark brown or creamy white with brownish cast within; papillae in 2 rows, long, slender, simple; papillae dark brown and tan. Clarke (1985) reported the incurrent aperture (32% of shell length) to be longer than the excurrent aperture (21% of shell length).

Excurrent aperture length approximately 20–25% of shell length; solid dark brown or creamy white with brownish cast within, may have dark brown marginal color band; margin smooth but with small irregular undulations.

Supra-anal aperture length approximately 10–15% of shell length; creamy white within, without marginal coloration; margin smooth; mantle bridge separating supra-anal and excurrent apertures imperforate, length approximately 40–45% of supra-anal length. The specimen described by Clarke (1985) lacked a mantle bridge, but excurrent and supra-anal apertures could be distinguished marginally. Clarke noted that the individual was "apparently anomalous in the absence of a mantle connection".

Specimens examined: Hurricane Creek, Jackson County (n = 2).

Glochidium Description

Length 281–303 µm; height 275–313 µm; with styliform hooks; outline subtriangular; dorsal margin straight, ventral margin narrowly rounded, anterior and posterior margins convex, anterior margin slightly more convex than posterior margin (Clarke, 1985; Hoggarth, 1999). Ortmann (1914) reported glochidia that differed in size from those described elsewhere. Length was near the range given above at 320 µm, but height was considerably greater at 380 µm.

Similar Species

Lasmigona holstonia may resemble *Villosa iris* and *Villosa vanuxemensis*, but those species have well-developed lateral teeth and thicker shells and also lack an interdental projection in the left valve, which is present in *L. holstonia*. *Villosa iris* has rays on the periostracum, which are absent on *L. holstonia*. Also, the nacre of *V. vanuxemensis* is usually purple whereas that of *L. holstonia* is white.

Lasmigona holstonia may also superficially resemble subadult *Strophitus undulatus*, but that species has only rudimentary pseudocardinal teeth and lacks an interdental projection in the left valve.

General Distribution

Lasmigona holstonia occurs in the Ohio and Tennessee River drainages. In the Ohio River drainage it has been reported only from the New River system in Virginia and West Virginia (Pinder et al., 2003). In the Tennessee River drainage *L. holstonia* is known from headwaters in southwestern Virginia, western North Carolina and eastern Tennessee downstream to the Paint Rock River system in northeastern Alabama (Clarke, 1985). There are also a few records from headwaters of the Duck River, central Tennessee (Parmalee and Bogan, 1998). In the Cumberland River drainage *L. holstonia* is known from headwaters of the Caney Fork (Parmalee and Bogan, 1998).

Alabama and Mobile Basin Distribution

Lasmigona holstonia is known only from the Paint Rock River system and western tributaries of the Tennessee River in Jackson County.

Lasmigona holstonia is extant in the Paint Rock River system, but its current distribution is poorly known due to a lack of surveys in small headwater streams.

Ecology and Biology

Lasmigona holstonia is a species of small to medium creeks, often found in very small streams where no other mussels occur. Ortmann (1918) reported *L. holstonia* from larger rivers (i.e., Holston and Hiwassee) but only from small side channels. It is usually found in areas with at least some current, usually in sandy substrates, though may also be found in mud or gravel.

Lasmigona holstonia is a long-term brooder. In the upper Tennessee River drainage gravid females have been reported with embryos in August and mature glochidia in September and May of the following year (Ortmann, 1921; Clarke, 1985). Fishes reported to serve as glochidial hosts of *L. holstonia* in laboratory trials include *Ambloplites rupestris* (Rock Bass) (Centrarchidae); *Cottus carolinae* (Banded Sculpin) (Cottidae); *Campostoma anomalum* (Central Stoneroller), *Luxilus chrysocephalus* (Striped Shiner), *Pimephales notatus* (Bluntnose Minnow), *Semotilus atromaculatus* (Creek Chub) and an unidentified species of *Notropis* (Cyprinidae); and *Etheostoma caeruleum* (Rainbow Darter), *Etheostoma rufilineatum* (Redline Darter) and *Etheostoma simoterum* (Snubnose Darter) (Percidae) (Gordon, 1993; Steg, 1998).

Current Conservation Status and Protection

Most conservation assessments of *Lasmigona holstonia* have included *Lasmigona etowaensis*, a Mobile Basin endemic with which it was considered synonymous until recent genetic analyses suggested otherwise. It was listed as a species of special concern throughout its range by Williams et al. (1993) and in Alabama by Lydeard et al. (1999). However, Stansbery (1976) considered it to be endangered in the state. Garner et al. (2004) designated *L. holstonia* a species of high conservation concern in Alabama.

Remarks

Until recently *Lasmigona etowaensis* was considered conspecific with *Lasmigona holstonia*, but the two were distinguished using genetic analyses (A.E. Bogan, unpublished data). These two species cannot be reliably separated using conchological characters.

Lasmigona holstonia appears to demonstrate a case of distributional limitations based on physiographic province. Records are almost exclusively confined to the Valley and Ridge physiographic province, regardless of river system.

Synonymy

Alasmodon (*Sulcularia*) *badium* Rafinesque 1831, *nomen dubium*. Rafinesque 1831:5

Comment: This name has traditionally been placed in the synonymy of *Lasmigona holstonia* (Ortmann and Walker, 1922a; Clarke, 1985) or recognized as a senior synonym of *L. holstonia* (Morrison, 1969). The legitimate but inadequate original description and lack of a type specimen or figure preclude positive placement of this name. The type locality, "small streams of the Knobs [Kentucky]", is outside of the known range of *L. holstonia*, which is confined to the Valley and Ridge physiographic province. Thus, *Alasmodon badium* is herein considered a *nomen dubium*.

Margarita (*Margaritana*) *holstonia* Lea, 1836, *nomen nudum*. Lea, 1836:46

Margaritana holstonia Lea, 1838. Lea, 1838a:42, pl. 13, fig. 37; Lea, 1838c:42, 136, 148, pl. 13, fig. 37

Type locality: Holston River[, Tennessee or Virginia]. Lectotype, USNM 86320, length 61 mm, designated by Johnson (1974).

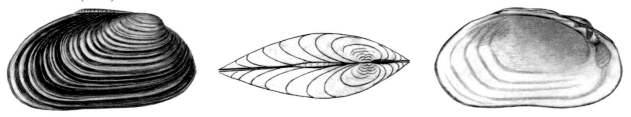

Alasmodon impressa Anthony, 1865. Anthony, 1865:157, pl. 76, fig. 4

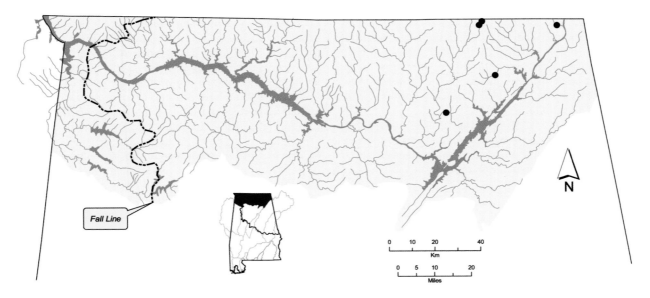

Distribution of *Lasmigona holstonia* in the Tennessee River drainage of Alabama.

Lasmigona subviridis (Conrad, 1835)
Green Floater

Lasmigona subviridis – Length 48 mm, UMMZ 23324. Chattahoochee River, Columbus, Muscogee County, Georgia. © Richard T. Bryant.

Shell Description

Length to 63 mm; thin; compressed; outline elliptical; posterior margin rounded, may be slightly truncate; anterior margin rounded; dorsal margin straight to slightly convex; ventral margin straight to slightly convex; posterior ridge rounded, may be weakly doubled posterioventrally; posterior slope moderately steep, flat to slightly concave; umbo broad, elevated slightly above hinge line, umbo sculpture thick ridges, becoming double-looped ventrally; periostracum green to tawny or brown, usually with variable green rays.

Pseudocardinal teeth rudimentary, compressed, almost parallel to dorsal margin, typically 2 teeth in left valve, 1 tooth in right valve; lateral teeth thin, straight, 1 in each valve, tooth in left valve divided posteriorly; interdentum moderately long, very narrow, interdental projection in left valve poorly developed, confluent with pseudocardinal tooth, interdentum of right valve may be divided by depression that accommodates interdental projection of opposing valve; umbo cavity shallow; nacre white. Clarke (1985) examined specimens from Atlantic Coast and Ohio River drainages and reported two lateral teeth in the left valve. However, Apalachicola Basin specimens have only one lateral tooth in each valve, with the tooth in the left valve divided toward its posterior end.

Soft Anatomy Description

No material was available for a soft anatomy description. The following is based on a description by Clarke (1985) of a single specimen from the Neuse River system, Wake County, North Carolina.

Mantle yellowish white, translucent centrally; visceral mass yellowish white; foot orange.

Gills pale orange; dorsal margin straight, ventral margin uneven; gill length approximately 50% of shell length; gill height approximately 45% of gill length; outer gill height approximately 65% of inner gill height; inner lamellae of inner gills connected to visceral mass only anteriorly.

Marsupia not examined. Ortmann (1912a) reported marsupia to be creamy white, pale orange or brown.

Labial palps straight dorsally, curved ventrally, bluntly pointed distally, slightly turned up.

Incurrent aperture length approximately 20% of shell length; papillae in 2 rows, flattened, pyriform; papillae with pigmented edges basally.

Excurrent aperture length approximately 10% of shell length; more darkly pigmented than incurrent aperture; margin with small, simple papillae.

Supra-anal aperture length approximately 20% of shell length; without marginal coloration; margin smooth; mantle bridge separating supra-anal and excurrent apertures short.

Glochidium Description

Length 368–383 µm; height 285–318 µm; with styliform hooks; outline depressed pyriform; dorsal margin straight, ventral margin with very slight, nipple-like projection, anterior and posterior margins convex, anterior margin distinctly more convex than posterior margin (Ortmann, 1912a; Clarke, 1985; Hoggarth, 1999).

Similar Species

Lasmigona subviridis resembles *Anodontoides radiatus*, but *A. radiatus* lacks a lateral tooth and has more rudimentary pseudocardinal teeth. The shell of *L. subviridis* is thicker than that of *A. radiatus*.

Lasmigona subviridis may also resemble *Villosa vibex*, but *V. vibex* has well-developed pseudocardinal and lateral teeth, including two lateral teeth in the left valve.

General Distribution

Lasmigona subviridis occurs in Atlantic Coast drainages from the St. Lawrence and Hudson River drainages in New York south to the Cape Fear River drainage in North Carolina. A disjunct Atlantic Coast population is known from the Savannah River drainage (Fuller, 1971). Disjunct populations also occur in the Kanawha River system of the Ohio River drainage in North Carolina, Virginia and West Virginia (Clarke, 1985; Pinder et al., 2003); the Tennessee River drainage, eastern Tennessee and western North Carolina (Parmalee and Bogan, 1998; R.S. Butler, personal communication); and the Apalachicola Basin, Alabama and Georgia (Brim Box and Williams, 2000).

Alabama and Mobile Basin Distribution

There are only three records of *Lasmigona subviridis* from the Apalachicola Basin, including one from Alabama. The Alabama record is from the Chattahoochee River at Columbus, between Muscogee County, Georgia, and Russell County, Alabama. The specimen was apparently collected during the 1800s. The type of *Unio neglectus* Lea, 1843, is from north Alabama and represents the only record of *L. subviridis* from the Tennessee River drainage (locality data were imprecise, so the distribution map does not include the Tennessee River drainage).

Lasmigona subviridis appears to be extirpated from the Apalachicola Basin and the Tennessee River drainage of Alabama.

Ecology and Biology

Lasmigona subviridis occurs primarily in creeks to small rivers and occasionally in large rivers. The only Apalachicola Basin records are from medium and large rivers near and above the Fall Line. Clarke (1985) reported that when *L. subviridis* is found in large rivers,

it is generally in small side channels. *Lasmigona subviridis* is generally not found in swift current, usually occurring in pools, eddies or along stream margins with mud, sand or gravel substrates (Clarke, 1985; Adams et al., 1990; R.S. Butler, personal communication). *Lasmigona subviridis* has also been reported from canals (Ortmann, 1919; Johnson, 1970).

Lasmigona subviridis is typically hermaphroditic (Ortmann, 1912a, 1919; van der Schalie, 1970) and a long-term brooder. In Pennsylvania it becomes gravid in mid-August, with glochidia maturing by September, and remains gravid until the following June (Ortmann, 1912a, 1919). *Lasmigona subviridis* glochidia have been reported to undergo direct development, bypassing the parasitic stage, in two widely separated Atlantic Coast drainages (Barfield and Watters, 1998; Lellis and King, 1998). Development takes place inside the marsupia. It is unknown whether *L. subviridis* glochidia are capable of facultative parasitism, as no glochidial hosts have been identified.

Current Conservation Status and Protection

Williams et al. (1993) considered *Lasmigona subviridis* threatened throughout its range. It was listed as extirpated from Alabama by Lydeard et al. (1999) and Garner et al. (2004).

Remarks

The distribution of *Lasmigona subviridis* has historically been reported to encompass the Atlantic Coast south to the Cape Fear River drainage, North Carolina (Clarke, 1985). It has also been recovered from archaeological material in the Altamaha River drainage in Georgia (A.E. Bogan, personal observation). Brim Box and Williams (2000) reported this species from the Apalachicola Basin in Georgia and Alabama, based on a total of three records. The Apalachicola specimens are slightly different conchologically from those of the southern Atlantic Coast, with the primary difference being the lateral teeth. These specimens could represent an undescribed species, but it appears to be extinct, so the question may never be resolved.

Synonymy

Unio subviridis Conrad, 1835. Conrad, 1835a:4 (appendix), pl. 9, fig. 1

Type locality: Schuylkill River; Juniata River, [Blair County, Pennsylvania]; creeks in Lancaster County, Penn[sylvania]. Lectotype, ANSP 2105, length not given, designated by Johnson and Baker (1973), is from Juniata River[, Blair County, Pennsylvania], not found.

Comment: Clarke (1985) erroneously listed the type locality as "creeks in Lancaster Co., Penn."

Margarita (*Unio*) *tappanianus* Lea, 1836, *nomen nudum*. Lea, 1836:39

Unio tappanianus Lea, 1838. Lea, 1838a:62, pl. 17, fig. 55 [replacement name for *Unio viridis* Rafinesque *sensu* Conrad, 1836]; Lea, 1838c:62, pl. 17, fig. 55

Unio neglectus Lea, 1843. Lea, 1843:[1]; Lea, 1846:280, pl. 42, fig. 10

Type locality: North Alabama.

Comment: The figured holotype, USNM 85701, was sent to Isaac Lea by Dr. Budd and was reported to be from "North Alabama". The shell appears to be *Lasmigona subviridis*. If the locality information, north Alabama, is correct, then this would represent the only known record of this species from the Tennessee drainage of Alabama. The species is known from headwaters of the Tennessee River (Parmalee and Bogan, 1998).

Unio hyalinus Lea, 1845. Lea, 1845:164; Lea, 1848a:69, pl. 2, fig. 4; Lea, 1848b:69, pl. 2, fig. 4

Margaritana quadrata Lea, 1861. Lea, 1861a:41; Lea, 1862c:210, pl. 32, fig. 279; Lea, 1863a:32, pl. 32, fig. 279

Unio pertenuis Lea, 1863. Lea, 1863c:193; Lea, 1866:8, pl. 2, fig. 4; Lea, 1867b:12, pl. 2, fig. 4

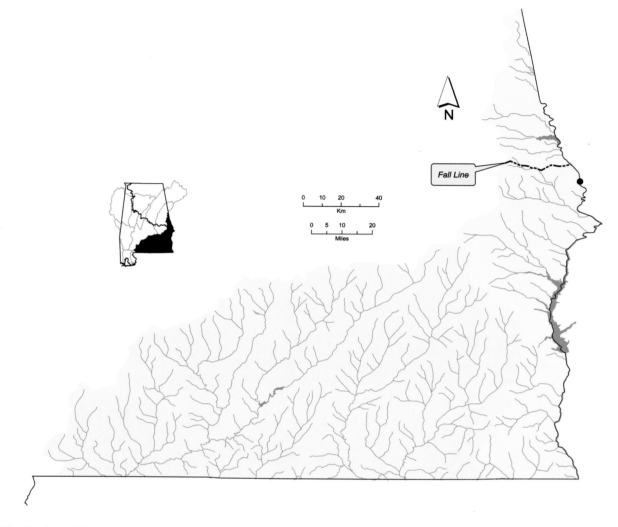

Distribution of *Lasmigona subviridis* in the eastern Gulf Coast drainages of Alabama

Genus *Lemiox*

Lemiox Rafinesque, 1831, is endemic to the Cumberland and Tennessee River drainages. *Lemiox* is a monotypic genus (Turgeon et al., 1998). Female *L. rimosus* display a modified mantle margin, presumably to attract glochidial hosts. The display is in the form of a small, grayish brown, bulbous knob, formed by a merging of mantle margins of the two valves (D.W. Hubbs, personal communication).

Frierson (1914) listed the type species *Unio rimosus* Rafinesque, 1831, by monotypy and noted it was synonymous with *Unio caelatus* Conrad, 1834. Ortmann (1921) erected *Conradilla* Ortmann, 1921, to separate *U. caelatus* into a distinct genus. Ortmann and Walker (1922a) discussed the problems of identification of Rafinesque's species and decided it was not identifiable. *Conradilla* Ortmann, 1921, is a junior synonym of *Lemiox* Rafinesque, 1831. Use of *Lemiox* over *Conradilla* has been gradual, with no single source documenting its priority.

Type Species

Unio rimosus Rafinesque, 1831 (= *Unio caelatus* Conrad, 1834)

Diagnosis

Shell thick; subtriangular to oval; slightly to moderately inflated, females more inflated than males; umbo high, turned anteriad, umbo sculpture distinct double-looped bars; posterior ridge well-developed, rounded; posterior half of shell strongly sculptured with irregular subradial corrugations; hinge teeth well-developed.

Excurrent aperture finely papillate; mantle bridge separating excurrent and supra-anal apertures moderately long; inner lamellae of inner gills completely attached to visceral mass except for small slit posteriorly; mantle margin just ventral to incurrent aperture crenulate or papillate; outer gills marsupial; glochidia held in posterior portion of gill, encompassing less than half of gill; marsupium outline reniform, extended ventrally below original gill edge when gravid; glochidium without styliform hooks (Rafinesque, 1831; Ortmann, 1916a; Haas, 1969b).

Synonymy

Conradilla Ortmann, 1921

Lemiox rimosus (Rafinesque, 1831)
Birdwing Pearlymussel

Lemiox rimosus – Upper figure: female, length 36 mm, UF 370405. Duck River, Lillard Mill, Marshall County, Tennessee. Lower figure: male, length 34 mm, UF 66111. Tennessee River, Alabama. © Richard T. Bryant.

Shell Description

Length to 50 mm; thick; slightly compressed; outline triangular to oval; posterior margin obliquely truncate to narrowly rounded; anterior margin rounded; dorsal margin straight to convex; ventral margin broadly convex, almost straight in some individuals; posterior ridge rounded but distinct; females with moderate marsupial swelling posteriorly, encompassing posterior ridge; posterior slope steep, flat to slightly concave, sculptured with regular, curved, parallel ridges extending from posterior slope to posteriodorsal margin; often with wide, very shallow sulcus just anterior of posterior ridge; posterior 30–65% of disk sculptured with irregular subradial corrugations; umbo elevated above hinge line, turned slightly anteriad, umbo sculpture double-looped ridges; periostracum greenish brown, becoming dark brown or black with

age, young individuals may have weak green rays that become obscure with age.

Pseudocardinal teeth thick, triangular, 2 divergent teeth in left valve, 1 tooth in right valve, often with accessory denticle anteriorly and/or posteriorly; lateral teeth short, thick, slightly curved, 2 in left valve, 1 in right valve, some individuals with rudimentary tooth ventrally in right valve; interdentum moderately long, moderately wide; umbo cavity shallow; nacre white.

Soft Anatomy Description

Mantle tan to light brown, may have rusty cast outside of apertures; visceral mass tan; foot tan.

Females with small, elongate, fleshy flap just ventral to incurrent aperture; flap length approximately 10% of shell length; tan; margin somewhat crenulate.

Gills tan; dorsal margin straight, ventral margin convex to elongate convex, female outer gills may be slightly bilobed; gill length 45–60% of shell length; gill height 45–55% of gill length; outer gill height 75–85% of inner gill height; inner lamellae of inner gills connected to visceral mass only anteriorly.

Outer gills marsupial; glochidia held in posterior 50% of gill; marsupium outline reniform, greatly distended when gravid; tan.

Labial palps tan; straight dorsally, convex ventrally, bluntly pointed distally; palp length 10–20% of gill length; palp height 50–65% of palp length; distal 35–65% of palps bifurcate.

Incurrent and supra-anal apertures longer than excurrent aperture; incurrent aperture shorter or equal to supra-anal aperture.

Incurrent aperture length 15–20% of shell length; creamy white to tan within, tan to dark tan basal to papillae; papillae tan. Specimens at hand were poorly preserved, and papilla morphology was difficult to establish, but they appeared to possibly be small arborescent. Ortmann (1916a) simply described incurrent aperture papillae as "large".

Excurrent aperture length 10–15% of shell length; creamy white to tan within, may have tan marginal color band; margin crenulate.

Supra-anal aperture length approximately 20% of shell length; tan within, without marginal coloration; margin smooth; mantle bridge separating supra-anal and excurrent apertures imperforate, length 15–40% of supra-anal length.

Specimens examined: Duck River, Tennessee (n =2) (specimens previously preserved; preservation quality poor).

Glochidium Description

Length 210 µm; height 260 µm; without styliform hooks, outline subrotund (Ortmann, 1916a).

Similar Species

With its distinctive, sculptured shell, *Lemiox rimosus* is easily distinguished from other sympatric species.

General Distribution

Lemiox rimosus is endemic to the Tennessee and Cumberland River drainages. It was described from material collected from the Cumberland River (Rafinesque, 1831). However, there are no subsequent reports or museum material from that drainage. All other records of *L. rimosus* are from the Tennessee River drainage, southwestern Virginia downstream to Muscle Shoals, Alabama (Ahlstedt, 1992a, 1992b; Parmalee and Bogan, 1998).

Alabama and Mobile Basin Distribution

Lemiox rimosus occurred in the Tennessee River across northern Alabama. The presence of this species in the Tennessee reach of the Elk River suggests that it also occurred in Alabama reaches of the river (Ahlstedt, 1983; Parmalee and Bogan, 1998). Type material for *Unio caelatus* Conrad, 1834, was probably collected from an Alabama reach of the Elk River.

Lemiox rimosus was extirpated from Alabama, but a reintroduction program is underway in Wilson Dam tailwaters of the Tennessee River.

Ecology and Biology

Lemiox rimosus is a species of shoal habitats in small to large rivers but is extirpated from large rivers. It generally occurs in gravel substrates, usually with some interstitial sand.

Lemiox rimosus is a long-term brooder, with mature glochidia by mid-September, which it broods until the following spring or summer (Ortmann, 1916a). In laboratory trials poor infestation rates were obtained using glochidia removed from conglutinates during May and early June. Infestation rates improved during late June and July, after conglutinates disintegrated, suggesting that conglutinates do not serve as a host-attracting mechanism for *L. rimosus* and that glochidia may not be completely mature until summer (TVA, 1986). Female *L. rimosus* have been observed to display a modified mantle margin, presumably as a glochidial host attractant. During the display the female is positioned with its ventral margin oriented upward, slightly elevated above the surrounding substrate. The mantle is protruded beyond the shell margin on the posterior half of the animal. Just posterior of the midpoint, mantle tissue from the two sides come together to give the appearance of a small, grayish brown, bulbous knob. Movement is slight, and it is unclear as to whether this is due to muscular movement by the animal or water current (D.W. Hubbs, personal communication).

Fishes reported to serve as glochidial hosts for *Lemiox rimosus* in laboratory trials include *Etheostoma simoterum* (Snubnose Darter) and *Etheostoma zonale* (Banded Darter) (Percidae) (TVA, 1986; Watson and Neves, 1998). A possible third host is *Etheostoma blennioides* (Greenside Darter) (Percidae), but results from laboratory trials were not conclusive (TVA, 1986).

Current Conservation Status and Protection

Lemiox rimosus was reported as endangered throughout its range by Stansbery (1970a) and Williams et al. (1993). It was listed as extirpated from Alabama by Stansbery (1976) and Lydeard et al. (1999). *Lemiox rimosus* (as *Conradilla caelata*) was listed as endangered under the federal Endangered Species Act in 1976.

Lemiox rimosus was included on a list of species approved for a Nonessential Experimental Population in Wilson Dam tailwaters in 2001. A trial transplantation of 80 individuals from the Duck River, Tennessee, was carried out in 2003. Recovery of 56% of transplanted individuals in 2005 suggests that conditions are favorable for the species. Garner et al. (2004) included L. rimosus as an extirpated species for which conservation action is underway in Alabama.

Remarks

The nomenclature of Lemiox rimosus has been a matter of debate (Walker, 1918). Most literature prior to

1980 used the binomen *Conradilla caelata* (Conrad, 1834). The use of *L. rimosus* has been gradual, with no single source documenting its priority (Morrison, 1969). Most authors currently recognize *L. rimosus* and it was included in both editions of Turgeon et al. (1988, 1998). However, *C. caelata* is still occasionally used. Indeed, the USFWS continues to use *C. caelata* on the federal endangered species list.

Lemiox rimosus has been reported as uncommon in prehistoric shell middens along the Tennessee River across northern Alabama (Morrison, 1942; Warren, 1975; Hughes and Parmalee, 1999).

Synonymy

Unio (Lemiox) rimosus Rafinesque, 1831. Rafinesque, 1831:3
Type locality: Cumberland River. Type specimen not found.
Unio coelatus Conrad, 1834. Conrad, 1834a:338, pl. 1, fig. 2 [original figure not herein reproduced]
Type locality: Tennessee, Elk and Flint rivers, [probably Alabama]. Type specimen not found, but length of figured shell in original description reported as about 46 mm.
Comment: Type material most likely came from Alabama since Conrad sampled these rivers only in Alabama, with the exception of a short reach of the Elk River just above the Alabama state line. An early map of Conrad's travels in Alabama indicates collections from the Tennessee River at Muscle Shoals and the Elk River near its mouth, and the Flint River. The collection from the Flint River was actually in what is today known as Flint Creek in Morgan County, Alabama (Wheeler, 1935).
Unio caelatus Conrad, 1834. Conrad, 1834b:29, pl. 3, fig. 4 [corrected spelling]

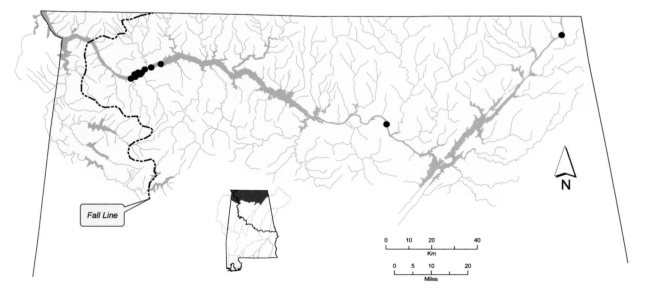

Distribution of *Lemiox rimosus* in the Tennessee River drainage of Alabama.

Genus *Leptodea*

Leptodea Rafinesque, 1820, occurs in the Great Lakes and Mississippi basins, Atlantic Coast drainages from Nova Scotia to Georgia, and Gulf Coast drainages from the Mobile Basin to Texas. Turgeon et al. (1998) listed three *Leptodea* species, two of which are known from Alabama.

The type species of *Leptodea* was designated twice. Rafinesque (1820) described three species of *Leptodea* but did not designate a type species. Rafinesque subsequently described *Lasmonos* Rafinesque, 1831, with a single species, *Lasmonos fragilis* Rafinesque, 1831, *non* Rafinesque, 1820, as type species by monotypy. Herrmannsen (1847) designated *Unio fragilis* Rafinesque, 1820, as the type species for *Leptodea*. Ortmann (1911a) erected the genus *Paraptera* using *Unio gracilis* Barnes, 1823, as the type species, which is a junior synonym of *Leptodea fragilis*. This makes *Paraptera* a junior synonym of *Leptodea*. Subsequently Frierson (1914) used *Unio leptodon* Rafinesque, 1820, as the type species of *Leptodea*. Utterback (1916b) used *Lasmonos* and assumed the type species to be *Unio fragilis* Rafinesque, 1820. Ortmann (1919) noted that the correct identity of *Lasmonos fragilis* is *U. leptodon*. Thus *Lasmonos* Rafinesque, 1831, is a junior synonym of *Leptodea* Rafinesque, 1820. Ortmann and Walker (1922a) ended this confusion by recognizing the genus *Leptodea*.

Type Species
Unio (*Leptodea*) *fragilis* Rafinesque, 1820

Diagnosis
Shell thin; elliptical; somewhat compressed; weak posteriodorsal wing often present; umbo low; disk without sculpture; pseudocardinal teeth compressed, poorly developed; females usually more inflated posteriorly than males.

Mantle edge just ventral to excurrent aperture slightly lamellar, weakly crenulate; mantle bridge separating excurrent and supra-anal apertures well-developed; inner lamellae of inner gills partially or completely connected to visceral mass; outer gills marsupial; glochidia held in posterior portion of gill; marsupium outline reniform, extended ventrally below original gill edge when gravid; glochidium without styliform hooks (Rafinesque, 1831; Simpson, 1900b, 1914; Ortmann, 1912a; Baker, 1928a; Haas, 1969b).

Synonymy
Lasmonos Rafinesque, 1831
Monelasmus Agassiz, 1846
Paraptera Ortmann, 1911

Leptodea fragilis (Rafinesque, 1820)
Fragile Papershell

Leptodea fragilis – Upper figure: female, length 115 mm, UF 388518. Tennessee River, river mile 93, Benton County, Tennessee, 2 June 1972. Middle figure: female?, length 115 mm, UF 374072. Coosawattee River, 0.3 miles below U.S. Highway 411, Gordon and Murray counties, Georgia, 5 September 1997. Lower figure: male, length 69 mm, UF 388521. Cedar Creek, Colbert County, Alabama, 13 January 1967. © Richard T. Bryant.

Shell Description

Length to 150 mm; thin; moderately inflated, females often more inflated posteriorly than males; outline elliptical to oval; posterior margin rounded, females often more broadly rounded than males; anterior margin rounded; dorsal margin straight, may be projected into low wing posterior of umbo, often eroded or broken, some individuals with smaller wing anterior of umbo; ventral margin straight to broadly convex; posterior ridge rounded; posterior slope steep dorsally, lower posterioventrally, flat to slightly concave; umbo broad, elevated slightly above hinge line, umbo sculpture weak ridges, becoming double-looped ventrally; periostracum greenish yellow to tawny, sometimes with a few thin, variable green rays.

Pseudocardinal teeth poorly developed, compressed, parallel to hinge line, 1–2 in left valve, 1 in right valve, some individuals with accessory denticle in right valve; lateral teeth thin, short, curved, 2 in left valve, 1 in right valve; interdentum long, very narrow; umbo cavity wide, shallow; nacre white to pink.

Soft Anatomy Description

Mantle creamy white to tan, outside of apertures dull orange or tan with dark gray or brown mottling, sometimes arranged as broken lines parallel to mantle margin, usually distinct; visceral mass creamy white to pearly white; foot creamy white to tan, often darker distally. Utterback (1916b) described the foot as "large, powerful, very extensile".

Both males and females with rudimentary mantle fold just ventral to incurrent aperture, thick and fleshy; fold length 15–45% of shell length; usually with heavily pigmented black marginal band, occasionally gray; margin smooth and undulating or crenulate.

Gills creamy white to tan, may have grayish cast; dorsal margin usually slightly sinuous, occasionally straight or slightly concave, ventral margin convex, outer gills may be more convex than inner gills or slightly bilobed, female gills often folded anterior of marsupia, outer gill ventral margins often ragged; gill length 40–60% of shell length; gill height 40–55% of gill length, occasionally greater; outer gill height 60–95% of inner gill height, inner and outer gills occasionally of equal height; inner lamellae of inner gills completely connected to visceral mass or connected only anteriorly. Utterback (1916b) described the gills as pointed posteriorly.

Outer gills marsupial; glochidia held in posterior 50% of gills; marsupium outline oval to reniform, distended when gravid; creamy white to light tan. Lea (1863d) reported glochidia to be brooded in only the posterior 25% of outer gills and reported marsupia ventral margins to have "three rows of crenulations".

Labial palps creamy white to tan, often with translucent, horizontal, jagged line; straight to slightly concave dorsally, convex ventrally, bluntly pointed distally, basal connection may extend dorsally to bifurcation; palp length 25–40% of gill length; palp height 50–90% of palp length; distal 20–50% of palps bifurcate.

Incurrent and supra-anal apertures longer than excurrent aperture; incurrent aperture shorter or equal to supra-anal aperture.

Incurrent aperture length 10–15% of shell length; creamy white to tan within, usually tan to dull orange basal to papillae, with black or brown bands perpendicular to margin, bands originate on papillae edges and extend onto aperture wall, often with some lines converging proximally; papillae in 1–3 rows, inner row largest, simple, long, slender; papillae some combination of creamy white, dull orange, tan, brown and black, usually with dark brown or black edges basally, often changing color distally.

Excurrent aperture length 5–10% of shell length; creamy white to tan within, marginal color band dull orange or tan with black or brown lines, lines may be in reticulated pattern or as simple, irregular bands perpendicular to margin, often with some bands converging proximally; margin papillate, papillae very small, simple, some specimens with crenulate margin; papillae white to tan or dull orange.

Supra-anal aperture length 10–20% of shell length; creamy white to light tan within, usually with thin brown, gray or black marginal band; margin smooth; mantle bridge separating supra-anal and excurrent apertures usually with 1 or more perforations, length variable, 30–130% of supra-anal length.

Specimens examined: Alabama River (n = 2); Coosa River (n = 3); Sipsey River (n = 1); Tennessee River (n = 7); Tombigbee River (n = 2).

Glochidium Description

Length 70–80 μm; height 80–95 μm; without styliform hooks; outline subelliptical; dorsal margin straight, ventral margin rounded, anterior and posterior margins convex (Lefevre and Curtis, 1910b; Ortmann, 1912a; Surber, 1912; Hoggarth, 1993, 1999).

Similar Species

Leptodea fragilis resembles *Leptodea leptodon* but is less elongate, less pointed posteriorly and has better developed lateral teeth. Female *L. leptodon* have a wrinkled, uncalcified, extrapallial extension of periostracum posteriorly, which is absent in *L. fragilis*.

Leptodea fragilis also resembles *Potamilus ohiensis* but has a more elongate shell. The posteriodorsal wing of *P. ohiensis* is generally better developed than that of *L. fragilis*. The periostracum of both species may be yellow in subadults, but the yellow color often persists into adulthood in *L. fragilis*, whereas the periostracum of *P. ohiensis* generally turns dark brown. Subadult *L. fragilis* may have faint green rays, but *P. ohiensis* do not.

General Distribution

Leptodea fragilis occurs in the Great Lakes and St. Lawrence Basin from the tidewater of the St. Lawrence River west to Lake Huron (Clarke, 1981a). It is widespread in the Mississippi Basin, ranging from Minnesota (Dawley, 1947) downstream to Louisiana (Vidrine, 1993) and from headwaters of the Ohio River in western Pennsylvania (Ortmann, 1919) west to eastern Nebraska and South Dakota (Backlund, 2000; Hoke, 2000). *Leptodea fragilis* occurs in the Cumberland River drainage downstream of Cumberland Falls (Cicerello et al., 1991) and in the Tennessee River drainage from headwaters in southwestern Virginia to the mouth of the Tennessee River (Ahlstedt, 1992a, 1992b; Parmalee and Bogan, 1998). *Leptodea fragilis* occurs in Gulf Coast drainages from the Mobile Basin west to the Colorado River in south-central Texas (Howells et al., 1996).

Alabama and Mobile Basin Distribution

Leptodea fragilis is widespread in the Tennessee River drainage, occurring in the Tennessee River and medium to large tributaries. It is found in all major drainages of the Mobile Basin, but is absent from the Tallapoosa River above the Fall Line. *Leptodea fragilis* does not occur in Gulf Coast drainages east of the Mobile Basin.

Leptodea fragilis remains widespread in the Tennessee River drainage and Mobile Basin.

Ecology and Biology

Leptodea fragilis usually occurs in large creeks and rivers, though it can occasionally be found in small creeks. It also inhabits some northern lakes (Ortmann, 1919). *Leptodea fragilis* may occur in flowing water but is more often encountered in areas such as pools and overbanks of reservoirs. *Leptodea fragilis* was one of the first colonizers of newly created overbank habitat in Kentucky Reservoir, Tennessee River (Bates, 1962).

It may occur in highly disturbed areas. It can be found in almost any substrate, including combinations of mud, sand and gravel, as well as under large rocks. *Leptodea fragilis* is often well-buried in the substrate, but Ortmann (1919) called it "a lively shell, crawling around frequently, with a speed unusual in other shells".

Leptodea fragilis is a long-term brooder, gravid from late August to the following July or August (Surber, 1912; Utterback, 1916a; Ortmann, 1919; Howells, 2000). Utterback (1916b) described conglutinates of *L. fragilis* as "white, elongate, leaf-shaped, not very solid, usually surrounded by brick-red matter". Natural infestations of *L. fragilis* glochidia on *Aplodinotus grunniens* (Freshwater Drum) (Sciaenidae) have been reported (Howard and Anson, 1922).

Current Conservation Status and Protection

Leptodea fragilis was considered currently stable throughout its range by Williams et al. (1993) and in Alabama by Lydeard et al. (1999). Garner et al. (2004) designated it a species of lowest conservation concern in the state.

Remarks

Medionidus mcglameriae van der Schalie, 1939, was described from the Tombigbee River, Sumter County, Alabama. This species, known only from the type material (holotype and one paratype), has been recognized in subsequent literature. However, these two specimens appear to be subadult *Leptodea fragilis* (Figure 13.27). They are characterized by a thin shell, blade-like pseudocardinal teeth, a slight dorsal wing and absence of corrugations on the posterior slope.

The only prehistoric record of *Leptodea fragilis* from the Tennessee River drainage of Alabama is a single specimen collected from an aboriginal shell midden near Bridgeport (Warren, 1975).

Figure 13.27. *Leptodea fragilis.* Subadult, length 20 mm, USNM 198984. Dalton, [Whitfield County,] Georgia. © Richard T. Bryant.

Synonymy

Unio (Leptodea) fragilis Rafinesque, 1820. Rafinesque, 1820:295

Type locality: Creeks in Kentucky. Lectotype, ANSP 20209, length 97 mm, designated by Johnson and Baker (1973).

Unio plantus Barnes, 1823. Barnes, 1823:272, fig. 16

Unio gracilis Barnes, 1823. Barnes, 1823:274

Lasmonos fragilis Rafinesque, 1831. Rafinesque, 1831:5

Unio (Naia) atrata Swainson, 1841, *non* Sowerby, 1839. Swainson, 1841:[no pagination], pl. 171

Unio permiscens Lea, 1859. Lea, 1859a:112; Lea, 1862b:102, pl. 17, fig. 251; Lea, 1862d:106, pl. 17, fig. 251

Type locality: Tombigbee River, Columbus, [Lowndes County,] Mississippi, W. Spillman, M.D. Lectotype, USNM 86118, length 55 mm, designated by Johnson (1974).

Lampsilis simpsoni Ferriss, 1900. Ferriss, 1900:38

Paraptera gracilis lacustris F.C. Baker, 1922. F.C. Baker, 1922:131

Medionidus mcglameriae van der Schalie, 1939. van der Schalie, 1939a:1–6, pl. 1

Type locality: Tombigbee River, Epes, Sumter County, Alabama. Holotype, UMMZ 130460, length 24 mm.

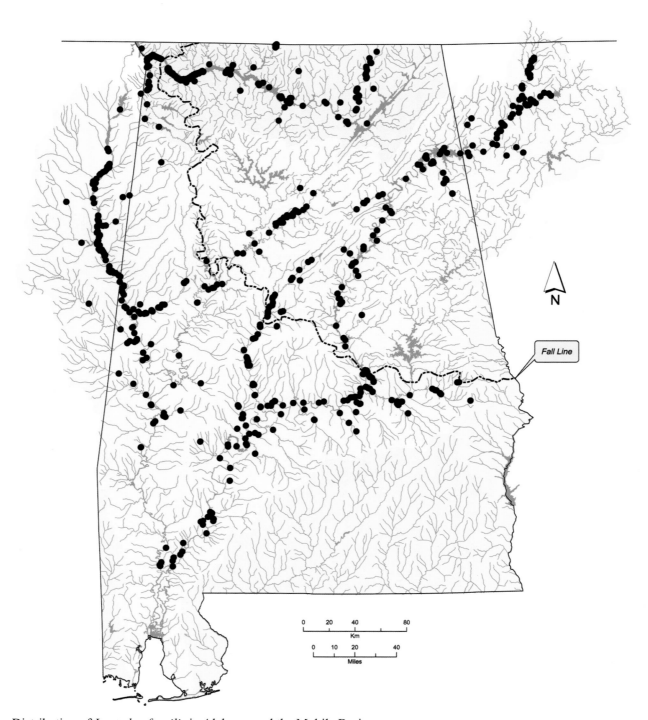

Distribution of *Leptodea fragilis* in Alabama and the Mobile Basin.

Leptodea leptodon (Rafinesque, 1820)
Scaleshell

Leptodea leptodon – Male, length 89 mm, USNM 150158. [Clinch River,] Scott County, Virginia. © Richard T. Bryant.

Shell Description

Length to 120 mm, males larger than females; thin; compressed; outline elliptical to rhomboidal; male posterior margin bluntly pointed, female posterior margin with wrinkled, uncalcified extrapallial extension of periostracum; anterior margin rounded; dorsal margin straight to slightly convex; ventral margin convex to nearly straight; posterior ridge low, rounded; posterior slope not steep, often roughened; umbo compressed, barely elevated above hinge line, positioned anteriorly, umbo sculpture faint double-looped ridges; periostracum yellowish to olive or brown, often with faint green rays.

Pseudocardinal teeth rudimentary, represented as weak swellings in both valves; lateral teeth weak, indistinct, moderately long, 2 in left valve, 1 in right valve, tooth in right valve somewhat better developed than those in left valve; interdentum long, very narrow; umbo cavity very shallow or absent; nacre pink or purplish to salmon.

Soft Anatomy Description

Mantle creamy white, some grayish brown outside of apertures; visceral mass creamy white; foot creamy white.

Mantle margin just ventral to incurrent aperture with very weak mantle fold; fold length approximately 25% of shell length; fold unremarkable in appearance, margin smooth, with color band; mantle wall adjacent to fold rusty brown.

Gills creamy white; dorsal margin straight, ventral margin elongate convex; gill length approximately 65% of shell length; gill height approximately 45% of gill length; outer gill height approximately 85% of inner gill height; inner lamellae of inner gills completely connected to visceral mass.

No gravid females were available for marsupium description. However, USFWS (2004) reported that glochidia are held in the posterior portion of the outer gill, with the marsupium extending beyond the original gill edge when gravid.

Labial palps creamy white; straight dorsally, convex ventrally, bluntly pointed distally; palp length approximately 25% of gill length; palp height approximately 65% of palp length; distal 45% of palps bifurcate.

Incurrent and supra-anal apertures considerably longer than excurrent aperture; incurrent and supra-anal apertures approximately equal in length.

Incurrent aperture length approximately 20% of shell length; creamy white within, rusty brown basal to papillae, with a few dark brown lines perpendicular to margin; papillae in 2 rows, short, thick, simple; papillae creamy white and rusty brown, with dark brown edges basally.

Excurrent aperture length approximately 10% of shell length; creamy white within, marginal color band rusty tan with dark brown lines in reticulated pattern; margin with very small, simple papillae, little more than crenulations; papillae creamy white.

Supra-anal aperture length approximately 20% of shell length; creamy white within, with a very thin dark brown marginal band; margin smooth; mantle bridge separating supra-anal and excurrent aperture imperforate, length approximately 25% of supra-anal length.

Specimen examined: Meramec River (n = 1) (specimen previously preserved; shell not available, and percentages based on approximation of shell length).

Glochidium Description

Length 68 µm; height 81 µm; without styliform hooks; outline subelliptical, dorsal margin straight to slightly convex, ventral margin rounded, anterior and posterior margins convex (USFWS, 2004).

Similar Species

Leptodea leptodon resembles *Leptodea fragilis* but is more elongate, often with a more pointed posterior margin, and has more rudimentary lateral teeth. Female *L. leptodon* have an uncalcified extrapallial expansion (Figure 13.28), which is absent in *L. fragilis*.

Figure 13.28. *Leptodea leptodon* live female, length 50 mm, with broad, uncalcified extension of the shell posteriorly. Photograph by Chris Barnhart.

General Distribution

Leptodea leptodon is known from the Mississippi Basin in Iowa and northern Illinois (Cummings and Mayer, 1992) south to Arkansas (Harris and Gordon, 1990) and from the Ohio River drainage in southwestern Ohio (Cummings and Mayer, 1992) west to the Missouri River drainage in Nebraska (Hoke, 2000) and South Dakota (Backlund, 2000). Its range in the Ohio River drainage includes the Wabash River system in Illinois and Indiana (Cummings and Mayer, 1992). *Leptodea leptodon* historically occurred in the Cumberland River downstream of Cumberland Falls (Cicerello et al., 1991) and most of the Tennessee River drainage (Parmalee and Bogan, 1998).

Alabama and Mobile Basin Distribution

Leptodea leptodon probably occurred in the Tennessee River across northern Alabama, but all records are from Muscle Shoals.

Leptodea leptodon appears to be extirpated from Alabama. It is believed to have disappeared from the state during the early 1900s (Stansbery, 1976).

Ecology and Biology

Leptodea leptodon is a species of medium to large rivers with low to medium gradient. Call (1900) reported it to inhabit muddy bottoms in the Ohio and Wabash rivers. However, it has also been reported to primarily occur in riffles with clear water and good current, where it often remains well-buried in the substrate (Oesch, 1995; USFWS, 2004). Substrates from which *L. leptodon* have been reported include gravel, cobble and boulders as well as sand and mud (USFWS, 2004).

Leptodea leptodon is a long-term brooder, gravid from August through June of the following summer (USFWS, 2004). At least some populations of *L. leptodon* appear to have sex ratios skewed toward males. Indeed, female specimens are exceedingly rare in museum collections. Fecundity was estimated as 419,000 glochidia in a 44 mm-long female from the Gasconade River, Missouri. The only known glochidial host of *L. leptodon* is *Aplodinotus grunniens* (Freshwater Drum) (Sciaenidae). *Leptodea leptodon* glochidia grow during the encystment period and may increase more than four times in size. Longevity of *L. leptodon* is estimated to be less than ten years (USFWS, 2004).

Current Conservation Status and Protection

Leptodea leptodon was considered endangered throughout its range by Stansbery (1970a) and Williams et al. (1993). It was listed as extirpated from Alabama by Stansbery (1976), Lydeard et al. (1999) and Garner et al. (2004). *Leptodea leptodon* was listed as an endangered species under the federal Endangered Species Act in 2001.

Remarks

Leptodea leptodon appears to have always been rare in the Tennessee River. There is a paucity of museum material of this species from Alabama, and it has not been reported from archaeological studies across northern Alabama (Hughes and Parmalee, 1999).

Synonymy

Unio (*Leptodea*) *leptodon* Rafinesque, 1820. Rafinesque, 1820:295, pl. 80, figs. 5–7
Type locality: Lower Ohio River. Neotype, ANSP 20214, length 50 mm (male), designated by Johnson and Baker
  (1973), is from Kentucky River[, Kentucky].

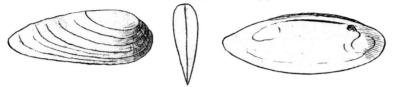

Anodon purpurascens Swainson, 1823. Swainson, 1823:[no pagination], pl. 160
Symphynota tenuissima Lea, 1829. Lea, 1829:453, pl. 11, fig. 21; Lea, 1834b:67, pl. 11, fig. 21
Unio velum Say, 1829. Say, 1829:293
Lampsilis blatchleyi Daniels, 1902. Daniels, 1902:13, pl. 2

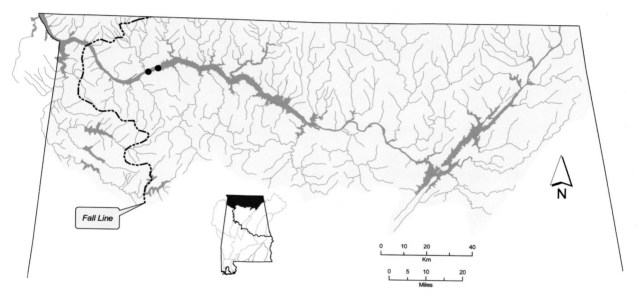

Distribution of *Leptodea leptodon* in the Tennessee River drainage of Alabama.

Genus *Ligumia*

Ligumia Swainson, 1840, occurs in the Hudson Bay, Great Lakes and Mississippi basins, Atlantic Coast drainages from Canada to Georgia, and Gulf Coast drainages from the Mobile Basin to Texas. Turgeon et al. (1998) listed three species, two of which occur in Alabama. Females display pigmented, papillate mantle folds that are more extensive anteriorly than other genera with papillate mantle folds.

Agassiz (1852) and Simpson (1900b) used the genus name *Eurynia* Rafinesque, 1819, with the type species as *Unio rectus* Lamarck, 1819, but Herrmannsen (1847) had designated *Unio dilatatus* Rafinesque, 1820, as the type species of *Eurynia*, making *Eurynia* a junior synonym of *Elliptio* Rafinesque, 1820. Davis and Fuller (1981) reported that *Ligumia* is not monophyletic based on immunological analysis.

Type Species
Unio recta Lamarck, 1819

Diagnosis
Shell elongate; smooth; umbo sculpture fine, double-looped ridges; females usually swollen posterioventrally.

Papillate mantle margin just ventral to incurrent aperture; supra-anal opening long; inner lamellae of inner gills usually completely attached to visceral mass; outer gills marsupial, glochidia held in posterior portion of gill; marsupium outline reniform, extended ventrally below original gill edge when gravid (Figure 13.29);

glochidium without styliform hooks (Utterback, 1916a; Haas, 1969b; Oesch, 1995).

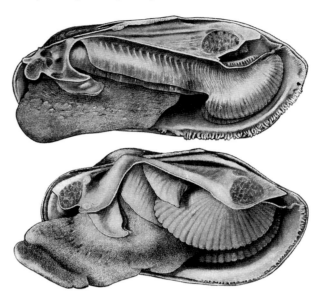

Figure 13.29. *Ligumia recta* (above) and *Ligumia subrostrata* (below) gravid female soft anatomy. Note distended marsupia in posterior portion of outer gills. From Lefevre and Curtis (1912).

Synonymy
None recognized.

Ligumia recta (Lamarck, 1819)
Black Sandshell

Ligumia recta – Upper figure: female, length 142 mm, AUM 2129. Middle figure: male, length 149 mm, AUM 2129. Tennessee River, Buck Island Chute, river mile 249, Colbert and Lauderdale counties, Alabama, 20 May 2001. Lower figure: male, length 124 mm, UF 64605. Alabama River, Selma, Dallas County, Alabama. © Richard T. Bryant.

Shell Description

Length to 195 mm; thick anteriorly, somewhat thin posteriorly; moderately inflated; outline elongate elliptical; male posterior margin bluntly pointed, female posterior margin obliquely truncate; anterior margin rounded; dorsal margin straight to slightly convex; ventral margin straight to slightly convex, may be slightly concave in females; females with moderate marsupial swelling posterioventrally; posterior ridge well-developed, rounded; posterior slope very steep adjacent to umbo, less steep posteriorly; umbo very broad, moderately inflated, elevated slightly above hinge line, umbo sculpture indistinct, double-looped ridges; periostracum dark green to brown, usually darkening to black with age, often with variable dark green rays.

Pseudocardinal teeth large, erect, triangular, 2 divergent teeth in left valve, 1 tooth in right valve, usually with low lamellar accessory denticle anteriorly; lateral teeth long, straight, 2 in left valve, 1 in right valve; interdentum long, very narrow; umbo cavity broad, shallow; nacre white, umbo cavity usually with some purple or pink, rare individuals entirely pink or purple.

Soft Anatomy Description

Mantle creamy white to tan, may have thin dark brown or black band ventrally, outside of apertures creamy white with numerous small black specks, specks may fuse to form bands; visceral mass creamy white to pearly white; foot tan or pinkish orange, often darker distally.

Both males and females with papillate mantle fold ventral to incurrent aperture; fold length 35–45% of shell length; papillae in 1 row, simple, short, numerous, evenly spaced, may extend anterior of fold; marginal color band some combination of brown, gray and black, usually mottled; papillae some combination of white, gray and brown, white almost always included, usually distally; papillae may be reduced in males. Ortmann

(1912a) reported the papillae to increase in size with an anterior progression.

Gills creamy white to tan; dorsal margin sinuous, ventral margin slightly curved; gill length 50–60% of shell length; gill height 25–40% of gill length; outer gill height 75–90% of inner gill height; inner lamellae of inner gills completely connected to visceral mass or connected only anteriorly.

Outer gills marsupial; glochidia held in posterior 30–50% of gill, but marsupium does not extend completely to posterior end; marsupium elongate, reniform, distended when gravid; creamy white to tan, with bright white distal margin.

Labial palps creamy white to tan; usually short and blunt, straight dorsally, convex ventrally, bluntly pointed to rounded distally; palp length 15–20% of gill length; palp height 55–80% of palp length, occasionally greater; distal 20–55% of palps bifurcate.

Incurrent and supra-anal apertures longer than excurrent aperture; incurrent aperture shorter, longer or equal to supra-anal aperture.

Incurrent aperture length 10–15% of shell length; creamy white within, some combination of tan, brown, gray and black basal to papillae, usually with dark brown or black lines originating on basal edges of larger papillae and extending onto aperture wall, where some converged proximally; papillae in 2 rows, inner row larger, long, slender, simple; papillae some combination of creamy white, tan, gray and brown, often with darker edges basally, usually changing color distally.

Excurrent aperture length 5–10% of shell length; creamy white within, marginal color band some combination of creamy white or tan and dark brown or black, with markings in form of irregular spots, regular squares, elongate blocks or lines perpendicular to margin, lines often with crossbars to form reticulated pattern; margin usually with short, simple papillae, occasionally crenulate; papillae white, gray, brown or black. Lea (1863d) described the excurrent aperture as having "granulations on the inner edges".

Supra-anal aperture length 10–15% of shell length; creamy white within, often with thin, sparse brown or black band marginally; margin smooth; mantle bridge separating supra-anal and excurrent apertures often with 1 or more perforations, length 20–70% of supra-anal length.

Specimens examined: Tennessee River (n = 7).

Glochidium Description

Length 200–220 μm; height 240–280 μm; without styliform hooks, but with lanceolate micropoints; outline subelliptical; dorsal margin straight to slightly convex, ventral margin rounded, anterior and posterior margins convex (Lefevre and Curtis, 1910b; Surber, 1912; Utterback, 1916b; Hoggarth, 1999).

Similar Species

Ligumia recta superficially resembles *Elliptio dilatata*, but *E. dilatata* typically has thicker pseudocardinal teeth and shorter lateral teeth. Also, *L. recta* typically has mostly white nacre with purple only in the umbo cavity, whereas nacre of *E. dilatata* is usually purple but ranges from white to salmon. *Ligumia recta* has a smoother, shinier periostracum than *E. dilatata* and is typically more heavily rayed.

Ligumia recta resembles *Lampsilis teres* in shape, but the periostracum of that species is always yellowish, never dark olive brown or black.

General Distribution

Ligumia recta occurs in extreme southern Hudson Bay Basin, southern Canada, Minnesota and North Dakota (Dawley, 1947; Clarke, 1973; Cvancara, 1983). It also occurs in most of the Great Lakes basin including the St. Lawrence River (La Rocque and Oughton, 1937; Dawley, 1947; Strayer and Jirka, 1997). *Ligumia recta* is widespread in the Mississippi Basin, ranging from Minnesota (Dawley, 1947) downstream to Louisiana (Vidrine, 1993), and from headwaters of the Ohio River in New York (Strayer and Jirka, 1997) west to South Dakota (Backlund, 2000). Gangloff and Gustafson (2000) reported *L. recta* from Montana, where it appears to have been recently introduced. *Ligumia recta* is found in the Cumberland River drainage downstream of Cumberland Falls (Cicerello et al., 1991) and in the Tennessee River drainage from headwaters in southwestern Virginia to the mouth of the Tennessee River (Neves, 1991; Parmalee and Bogan, 1998). On the Gulf Coast *L. recta* occurs in the Mobile Basin of Alabama and Mississippi and the Pearl River drainage in Mississippi and Louisiana (Jones et al., 2005).

Alabama and Mobile Basin Distribution

Ligumia recta is known from the Tennessee River across the state and in some large tributaries. In the Mobile Basin it is known from all major drainages, but there are no records from the Tallapoosa River above the Fall Line.

In the Tennessee River drainage *Ligumia recta* is now confined to tailwaters of dams. It has declined drastically in the Mobile Basin and is very rare. The only recent record of *L. recta* from the Mobile Basin is from the Sipsey River (McCullagh et al., 2002).

Ecology and Biology

Ligumia recta inhabits medium creeks to large rivers and some northern lakes. It usually occurs in areas with at least moderate current. Its preferred substrates are mixtures of sand and gravel. In unimpounded creeks and rivers *L. recta* may be found in water less than 1 m deep, but in tailwaters of large river dams it may be found at depths exceeding 6 m. It does not occur in overbank habitat of reservoirs.

Ligumia recta is a long-term brooder, gravid from August to the following July (Conner, 1909; Surber, 1912; Ortmann, 1919; Coker et al., 1921). Utterback (1916b) reported white conglutinates in *L. recta* but gave no further details. Female *L. recta* display a papillate mantle fold during certain periods, presumably to attract glochidial hosts. When displaying, the female lays prone on the substrate, with the papillae pulsing in undulating waves. Males usually remain deeply buried in the substrate, as do females when not displaying. Corey et al. (2006) reported female *L. recta* to display at night.

Fishes reported to serve as primary and secondary glochidial hosts of *Ligumia recta* in laboratory trials include *Ambloplites rupestris* (Rock Bass), *Lepomis auritus* (Redbreast Sunfish), *Lepomis cyanellus* (Green Sunfish), *Lepomis gibbosus* (Pumpkinseed), *Lepomis macrochirus* (Bluegill), *Lepomis megalotis* (Longear Sunfish), *Micropterus salmoides* (Largemouth Bass) and *Pomoxis annularis* (White Crappie) (Centrarchidae); *Campostoma anomalum* (Central Stoneroller), *Lythrurus umbratilis* (Redfin Shiner) and *Notropis rubellus* (Rosyface Shiner) (Cyprinidae); *Morone americana* (White Perch) (Moronidae); *Perca flavescens* (Yellow Perch) and *Sander canadensis* (Sauger) (Percidae); and two nonindigenous taxa, *Cichlasoma nigrofasciatum* (Convict Cichlid) (Cichlidae); and *Xiphophorus maculatus* (Southern Platyfish) (Poeciliidae) (Steg and Neves, 1997; Watters et al., 1999; Khym and Layzer, 2000). A study of glochidium implantation by Young (1911) included glochidia of *L. recta* implanted on *Lepomis humilis* (Orangespotted Sunfish) (Centrarchidae) and *Fundulus diaphanus* (Banded Killifish) (Fundulidae). Glochidia of *L. recta* encysted on those species, but no details of whether transformation occurred were reported. Coker et al. (1921) reported a natural infestation of *Anguilla rostrata* (American Eel) (Anguillidae) with *L. recta* glochidia, but questioned its significance as a host. Gut contents of *Ligumia recta* from the Wisconsin River, Wisconsin, were examined (Bisbee, 1984). Diatoms were most prevalent, but green and blue-green algae were also common.

Current Conservation Status and Protection

Ligumia recta was considered a species of special concern throughout its range by Williams et al. (1993) and in Alabama by Lydeard et al. (1999). Garner et al. (2004) designated it a species of high conservation concern in the state.

Remarks

Ligumia recta has been reported to be rare in prehistoric shell middens along the Tennessee River across northern Alabama (Warren, 1975; Hughes and Parmalee, 1999).

Synonymy

Unio recta Lamarck, 1819. Lamarck, 1819:74
Type locality: Lake Erie. Holotype, MNHN uncataloged, length 100 mm. Original label has additional data: "de la [vicin] ité de Niaga" [Niagara Falls, New York] (Johnson, 1969b).
Unio (*Eurynia*) *latissima* Rafinesque, 1819, *nomen nudum*. Rafinesque, 1819:426
Unio (*Eurynia*) *latissima* Rafinesque, 1820. Rafinesque, 1820:297, pl. 80, figs. 14, 15
Unio praelongus Barnes, 1823. Barnes, 1823:261, pl. 13, fig. 11
Unio sageri Conrad, 1836. Conrad, 1836:53, pl. 29, fig. 1
Unio arquatus Conrad, 1854. Conrad, 1854:297, pl. 26, fig. 8
Unio leprosus Miles, 1861. Miles, 1861:240

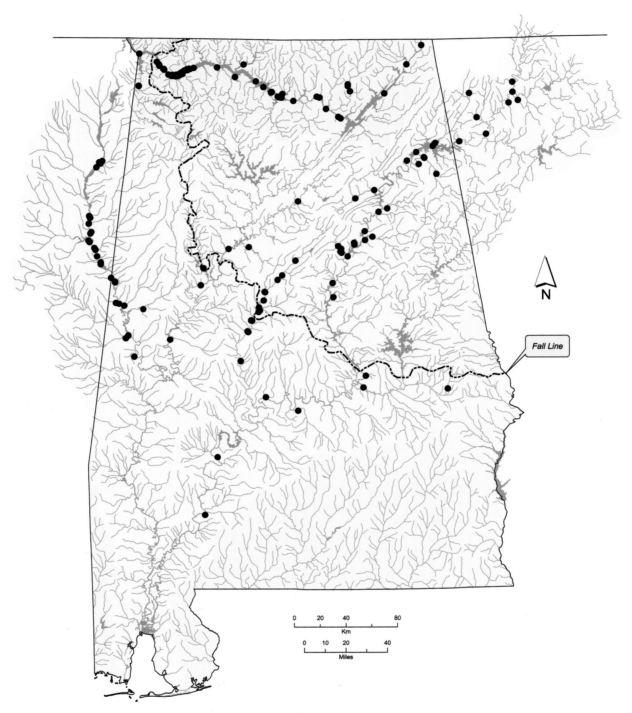

Distribution of *Ligumia recta* in Alabama and the Mobile Basin.

Ligumia subrostrata (Say, 1831)
Pondmussel

Ligumia subrostrata – Upper figure: female, length 69 mm, USNM 85982. Lower figure: male, length 79 mm, USNM 85982. Coffee Creek near Uniontown, [Perry County,] Alabama. © Richard T. Bryant.

Shell Description

Length to 81 mm; moderately thin; moderately inflated; outline elongate elliptical; male posterior margin bluntly pointed, female posterior margin obliquely truncate; anterior margin rounded; dorsal margin straight to slightly convex; ventral margin straight to slightly convex, females often becoming slightly concave with age; females with a moderately to greatly expanded marsupial swelling posterioventrally; posterior ridge moderately developed, rounded; posterior slope flat to slightly concave; umbo broad, flat, elevated slightly above hinge line, umbo sculpture fine, weakly double-looped ridges; periostracum shiny to cloth-like, yellowish green to brown or black, rays usually weak.

Pseudocardinal teeth compressed, almost parallel to margin, 2 teeth in left valve, anterior tooth larger, 1 tooth in right valve, may have lamellar accessory denticle anteriorly; lateral teeth thin, straight, 2 in left valve, 1 in right valve; interdentum moderately long, very narrow; umbo cavity wide, shallow; nacre white.

Soft Anatomy Description

Mantle creamy white to tan, usually dull orange outside of apertures, mottled with various shades of brown; visceral mass creamy white to tan; foot creamy white to tan or pale orange.

Both males and females with papillate mantle fold just ventral to incurrent aperture; fold length 20–40% of shell length; papillae simple, long, thick, blunt, evenly spaced; usually with brown marginal band. The fold may be reduced in males, as reported by Ortmann (1916b).

Gills creamy white to tan or light brown, may have golden cast; dorsal margin straight to slightly sinuous or slightly concave, ventral margin convex; gill length 55–65% of shell length, occasionally less; gill height 35–45% of gill length; outer gill height 70–100% of inner gill height; inner lamellae of inner gills connected to visceral mass only anteriorly. Lea (1863d) and Utterback (1916b) reported gills to be completely connected to the visceral mass.

Outer gills marsupial; glochidia held in posterior 33% of gill; marsupium distended when gravid; tan with slight golden cast; ventral margin white. Lea (1863d) reported glochidia to be held in the posterior 50% of the outer gills. Utterback (1916b) reported gravid gills to be reniform or fan-shaped, distended when gravid. Ortmann (1916b) reported marsupial gills to have "blackish-brown pigment" distally.

Labial palps creamy white to tan; straight to slightly concave or slightly convex dorsally, convex ventrally, bluntly pointed to rounded distally; palp length 25–40% of gill length; palp height 45–70% of palp length; distal 25–45% of palps bifurcate.

Incurrent aperture usually longer than excurrent and supra-anal apertures, occasionally equal to excurrent aperture; excurrent aperture longer, shorter or equal to supra-anal aperture.

Incurrent aperture length 10–15% of shell length; creamy white within, often with slight golden cast, usually with some combination of rusty orange, brown, gray and black mottling basal to papillae; papillae usually in 1–2 rows, long, slender, simple, with occasional bifid; papillae usually black or dark brown with dull orange, gray or creamy white mottling.

Excurrent aperture length 5–10% of shell length; creamy white within, often with golden cast, marginal color band black with creamy white or dull orange mottling; margin with small, simple papillae, may be little more than crenulations, portions of margin may be crenulate; papillae black or dull orange.

Supra-anal aperture length 5–10% of shell length; creamy white or gold within, often with thin marginal band of brown, black, gray or dull orange; margin smooth; mantle bridge separating supra-anal and excurrent apertures usually with multiple perforations, length 80–180% of supra-anal length.

Specimens examined: Threemile Creek, Mobile County (n = 4).

Glochidium Description

Length 210–270 μm; height 260–330 μm; without styliform hooks; outline subelliptical; dorsal margin slightly convex, ventral margin rounded, anterior and posterior margins convex (Lefevre and Curtis, 1912; Surber, 1912; Ortmann, 1916b; Utterback, 1916b).

Similar Species

Ligumia subrostrata resembles *Villosa lienosa*, especially sexually mature females. However, *V. lienosa* is less elongate, has a thicker shell and thicker hinge teeth, and its nacre is often purple instead of white. Mature female *L. subrostrata* are more pointed posteriorly than female *V. lienosa*, with the terminal point located higher on the posterior margin.

Ligumia subrostrata also resembles *Villosa vibex*, but has a more pointed posterior margin, giving it a more elliptical outline. Also, the marsupial swelling of *V. vibex* does not become as exaggerated as that of female *L. subrostrata*. Rays on the periostracum of *V. vibex* are typically more pronounced than those of *L. subrostrata*.

General Distribution

Ligumia subrostrata occurs in the Mississippi Basin from Iowa and Illinois (Cummings and Mayer, 1992) south to Louisiana (Vidrine, 1993) and from the Ohio River drainage and Wabash River system in Indiana west to Nebraska (Hoke, 2000) and South Dakota (Backlund, 2000). There are a few records from the lower Cumberland and Tennessee River drainages, western Kentucky and Tennessee (Cicerello et al., 1991; Parmalee and Bogan, 1998). *Ligumia subrostrata* occurs on the Gulf Coast from the Mobile Basin west to the Nueces River, Texas (Vidrine, 1993; Howells et al., 1996).

Alabama and Mobile Basin Distribution

Ligumia subrostrata is known only from the Mobile Basin and some direct tributaries of Mobile Bay. Most records are from the Tombigbee River drainage, but one old record is from Bogue Chitto in the Alabama River drainage. A recent record of *L. subrostrata* from a commercial fish producer in the Black Warrior River drainage may represent an introduction. The population in a direct tributary to Mobile Bay is in an impoundment in a municipal park and may also represent a recent introduction.

Ligumia subrostrata occurs in several disjunct populations. It has potential for expanding its range in commercial and private fish ponds and associated drainages.

Ecology and Biology

Ligumia subrostrata typically occurs in areas with little or no current, such as pools and sloughs of creeks and rivers, floodplain lakes, impoundments and man-made ponds, in mud or sand substrates. It may become extremely abundant in artificial ponds (Lefevre and Curtis, 1912). Densities of *L. subrostrata* were reported to average 200 per m^2 in a pond on a golf course in Louisiana (Stern and Felder, 1978).

The diet of *Ligumia subrostrata* was studied in Lake Maxinkuckee, Indiana (Evermann and Clark, 1920). Stomach contents included gray or green flocculent material as well as black mud, suggestive of the substrate of Maxinkuckee, which has a significant component of gray marl. Identified ingested organic material included primarily green algae and diatoms.

Ligumia subrostrata is a long-term brooder, gravid from late summer or autumn to the following summer (Lefevre and Curtis, 1912). In Texas it becomes gravid in July (Howells, 2000). Utterback (1916b) reported conglutinates of *L. subrostrata* to be club-shaped, with "ovisacs" beaded distally and bearing a bluish pigment. Corey et al. (2006) reported conglutinates to by creamy white. Female *L. subrostrata* have been reported to display their papillate mantle folds, which was detailed by Corey et al. (2006). The display includes white, feathery papillae along the posterior portion of the fold and darkly pigmented eyespots ventral to the incurrent aperture. It is presented while the mussel is positioned in an upright position with valves agape. The display was described as fluttering activity, alternating with a rest period of several seconds. Display frequency was shown to increase with light intensity and no displays were observed in dark. A fish was observed to attack the mantle edge during a display, at which time glochidia were released. Fishes shown to serve as glochidial hosts for *L. subrostrata* in laboratory trials include *Lepomis cyanellus* (Green Sunfish), *Lepomis humilis* (Orangespotted Sunfish) and *Micropterus salmoides* (Largemouth Bass) (Centrarchidae) (Lefevre and Curtis, 1912). Stern and Felder (1978) added *Lepomis gulosus* (Warmouth) and *Lepomis macrochirus* (Bluegill) (Centrarchidae) to the list of *L. subrostrata* hosts based on observation of natural infestations.

Current Conservation Status and Protection

Ligumia subrostrata was reported as currently stable throughout its range by Williams et al. (1993) and in Alabama by Lydeard et al. (1999). Garner et al. (2004) designated it a species of moderate conservation concern in the state.

Remarks

Ligumia subrostrata appears to be readily introduced into artificial ponds, including commercial aquaculture operations, presumably as glochidia on stocked fish. This provides an opportunity for range expansion. Indeed, *L. subrostrata* has appeared in at least one fish pond outside of its historical range, in the Apalachicola Basin, Harris County, Georgia (C. Blanchard and R.C. Stringfellow, personal communication). The pond was stocked using fish from a commercial aquaculture operation in Alabama that is known to harbor a population of *L. subrostrata*.

Synonymy

Unio subrostratus Say, 1831. Say, 1831b:[no pagination]
Type locality: Wabash River. Neotype, SMF 4348, length 57 mm, designated by Haas (1930) is from the U.S.
Unio nashvillianus Lea, 1834. Lea, 1834a:100, pl. 14, fig. 43; Lea, 1834b:212, pl. 14, fig. 43
Margaron (Unio) nashvilliensis (Lea, 1834). Lea, 1870:45 [emendation]
Unio mississippiensis Conrad, 1850. Conrad, 1850:277, pl. 38, fig. 11
Unio rutersvillensis Lea, 1859. Lea, 1859b:155; Lea, 1860d:356, pl. 60, fig. 181; Lea, 1860g:38, pl. 60, fig. 181
Unio cocoduensis Reeve, 1865. Reeve, 1865:[no pagination], pl. 24, fig. 117
Unio topekaensis Lea, 1868. Lea, 1868a:144; Lea, 1868c:313, pl. 49, fig. 126; Lea, 1869a:73, pl. 49, fig. 126
Lampsilis subrostrata var. *furva* Simpson, 1914. Simpson, 1914:100

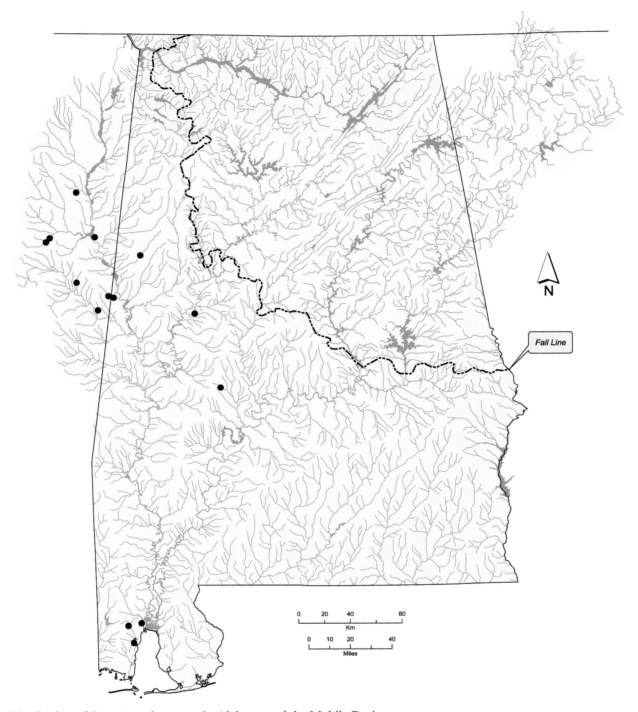

Distribution of *Ligumia subrostrata* in Alabama and the Mobile Basin.

Genus *Medionidus*

Medionidus Simpson, 1900, occurs in the Cumberland and Tennessee River drainages and eastern Gulf Coast drainages from northern Florida to the Mobile Basin. Turgeon et al. (1998) listed seven species in the genus. One species, *Medionidus mcglameriae* van der Schalie, 1939, is herein placed in the synonymy of *Leptodea fragilis* (Rafinesque, 1820), based on its thin shell, slight dorsal wing, blade-like pseudocardinal teeth and absence of corrugations on the posterior slope. Four species of *Medionidus* occur in Alabama.

Females of some *Medionidus* species have been observed to display modified mantle margins during the spring (Brim Box and Williams, 2000; Haag and Warren, 2001).

Type Species
Unio conradicus Lea, 1834

Diagnosis
Shell elongate; inflated; posterior slope sculptured with fine wrinkles or plications; pseudocardinal teeth small, well-developed; lateral teeth short; females somewhat swollen just posterior to center of ventral margin; nacre often distinctive gray or blue-green.

Inner lamellae of inner gills completely or partially connected to visceral mass; mantle edge slightly thickened ventral to incurrent aperture, with finely crenulate, pigmented margin; outer gills marsupial; glochidia usually held in central portion of gill; marsupium extended ventrally below original gill edge when gravid; glochidium without styliform hooks (Simpson, 1900b, 1914; Ortmann, 1912a; Haas, 1969b; Johnson, 1977).

Synonymy
None recognized.

Medionidus acutissimus (Lea, 1831)
Alabama Moccasinshell

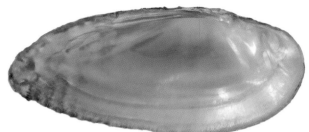

Medionidus acutissimus – Length 42 mm, UF 269733. [Tombigbee River,] Columbus, Lowndes County, Mississippi, [late 1800s].

Shell Description

Length to 55 mm, thin; moderately inflated; outline elliptical, may become arcuate with age; posterior margin obliquely truncate to bluntly pointed; anterior margin rounded; dorsal margin straight to slightly convex; ventral margin straight to slightly concave or slightly convex; posterior ridge moderately sharp, well-defined; posterior slope moderately steep, with corrugations extending from posterior slope to margin, corrugations extend anterior of posterior slope in some individuals; umbo broad, low, barely elevated above hinge line, umbo sculpture weak, undulating ridges; periostracum yellowish to tawny or greenish brown, covered with chevrons or zigzag lines that may form rays.

Pseudocardinal teeth small, triangular, 2 slightly compressed teeth in left valve, become thicker and more divergent with age, 1 tooth in right valve; lateral teeth short to moderately long, thin, straight to slightly curved, 2 in left valve, 1 in right valve; interdentum moderately long, very narrow; umbo cavity wide, shallow; nacre white, may have salmon tint, occasionally greenish in populations above the Fall Line.

Soft Anatomy Description

Mantle tan, outside of apertures rusty tan to grayish brown; visceral mass creamy white to pale tan; foot tan, may have slight orange tint.

Females with well-developed mantle fold just ventral to incurrent aperture; fold length 35–50% of shell length; margin rusty tan to rusty brown or grayish brown, may not form distinct band; margin with widely spaced papillae anteriorly, anterior-most papillae larger, margin smooth but undulating posteriorly. Males with weak mantle fold, without papillae or with very few at anterior end. Haag and Warren (2001) described the mantle fold of live individuals as matte black, with a small white patch.

Gills tan; dorsal margin straight to sinuous, ventral margin elongate convex; gill length 55–65% of shell length; gill height 35–50% of gill length; outer gill height 70–100% of inner gill height; inner lamellae of inner gills usually completely connected to visceral mass.

Outer gills marsupial; glochidia held across entire gill except anterior and posterior ends; marsupium outline somewhat reniform, greatly distended when gravid; pale tan.

Labial palps tan; straight to slightly convex dorsally, convex ventrally, bluntly pointed distally; palp length 10–15% of gill length; palp height approximately 75% of palp length; distal 50% of palps bifurcate.

Incurrent aperture usually considerably longer than excurrent and supra-anal apertures, occasionally of similar length; excurrent aperture may be shorter, longer or equal to supra-anal aperture.

Incurrent aperture length 15–35% of shell length; creamy white to tan within, may have slight rusty cast, may have rusty tan or rusty brown basal to papillae; papillae in 1–2 rows, inner row larger, simple, short, thick, blunt; papillae rusty tan, may be grayer distally. Lea (1863d) reported incurrent aperture papillae to be brown.

Excurrent aperture length 10–20% of shell length; tan within, may have slight rusty cast, marginal color band rusty tan to rusty brown; margin crenulate, crenulations may be irregular. Lea (1863d) reported excurrent aperture papillae that were "very minute" and brownish.

Supra-anal aperture length 15–20% of shell length; tan to creamy white within, may have very thin or irregular rusty brown or gray marginal band; margin smooth; mantle bridge separating supra-anal and excurrent apertures imperforate, length 50–60% of supra-anal length.

Specimens examined: Brushy Creek, Winston County (n = 4) (specimens previously preserved).

Glochidium Description

Length 175–213 μm; height 225–275 μm; without styliform hooks; outline subelliptical; dorsal margin straight, ventral margin rounded, anterior and posterior margins convex. Ortmann (1924a) gave glochidia measurements that were considerably larger: length 260–270 μm, height 320–330 μm. Ortmann was ambiguous as to the collection locality, but in a list of material examined the only mention of specimens with soft anatomy, including a gravid female, were from Chattooga Creek, Chattooga County, Georgia.

Similar Species

Medionidus acutissimus may be very similar to *Medionidus parvulus* in shell morphology. Occasional specimens cannot be positively identified. However, several subtle characters may be helpful in distinguishing the two. The posterior ridge is better developed in *M. acutissimus* and the posterior slope is steeper and narrower than that of *M. parvulus*. Also, the posterior end of *M. acutissimus* tends to be more pointed than that of *M. parvulus*.

General Distribution

Medionidus acutissimus is widespread in the Mobile Basin. Populations in the Escambia, Yellow and Choctawhatchee River drainages on the Gulf Coast are believed to represent this species.

Alabama and Mobile Basin Distribution

Medionidus acutissimus occurs in most of the Mobile Basin, with the exception of the Tallapoosa River above the Fall Line. Populations in the Escambia, Yellow and Choctawhatchee River drainages along the Gulf Coast are herein tentatively assigned to this species.

Medionidus acutissimus is extant in isolated and widely separated localities in the Mobile Basin. It has not been collected from Gulf Coast drainages since the 1960s.

Biology and Ecology

Medionidus acutissimus inhabits sand and gravel substrates in medium creeks to rivers. It may occur in slow to swift current. Lea (1863d) reported one of five *Medionidus acutissimus* examined to have a byssal thread.

Medionidus acutissimus is a long-term brooder, gravid from late summer or autumn into the following summer. Females brooding mature glochidia have been reported from late February to mid-March in the Black Warrior River drainage (Haag and Warren, 1997). Female *M. acutissimus* have been reported to emerge

completely from the substrate and display a modified mantle margin. The display consists of a small white patch, which is flickered against a black background (Haag and Warren, 2001).

Fishes reported to serve as glochidial hosts of *Medionidus acutissimus* in laboratory trials are *Fundulus olivaceus* (Blackspotted Topminnow) (Fundulidae); and *Ammocrypta beanii* (Naked Sand Darter), *Ammocrypta meridiana* (Southern Sand Darter), *Etheostoma artesiae* (Redspot Darter), *Etheostoma douglasi* (Tuskaloosa Darter), *Etheostoma nigrum* (Johnny Darter), *Etheostoma stigmaeum* (Speckled Darter), *Etheostoma swaini* (Gulf Darter), *Etheostoma whipplei* (Redfin Darter), *Percina kathae* (Mobile Logperch), *Percina nigrofasciata* (Blackbanded Darter) and *Percina vigil* (Saddleback Darter) (Percidae) (Haag and Warren, 1997, 2001). A marginal host is *Etheostoma rupestre* (Rock Darter) (Percidae) (Haag and Warren, 2001).

Current Conservation Status and Protection

Medionidus acutissimus was considered threatened throughout its range by Williams et al. (1993) and in Alabama by Lydeard et al. (1999). Garner et al. (2004) designated it a species of high conservation concern in the state. In 1993 *M. acutissimus* was listed as a threatened species under the federal Endangered Species Act.

Remarks

The relationship of *Medionidus acutissimus* to *Medionidus parvulus* is unclear. The two overlap in shell morphology in parts of their ranges. Detailed comparative studies, including life history, soft anatomy and genetic analyses, are needed to determine if they represent two valid species.

Further taxonomic questions exist with regard to the relationship between Mobile Basin *Medionidus acutissimus* and what appears to be that species in the Choctawhatchee, Yellow and Escambia River drainages (Figure 13.30). Those populations are represented in museum collections by small numbers of specimens. Some individuals appear conchologically different from those of Mobile Basin, but differences are slight and the amount of material available is not sufficient for a thorough comparison. Future soft anatomy comparisons and genetic analyses are dependent on discovery of an extant population.

Medionidus acutissimus from the Escambia, Yellow and Choctawhatchee River drainages are also similar morphologically to *Medionidus penicillatus*. Some authors have included those populations in the range of *M. penicillatus* (Johnson, 1977; Butler, 1990; Williams and Butler, 1994). However, those populations are herein included in the range of *M. acutissimus* because faunal distribution patterns suggest that faunas in the Choctawhatchee, Yellow and Escambia River drainages are more closely related to those of the Mobile Basin than those of the Apalachicola Basin.

Figure 13.30. *Medionidus acutissimus*. Upper figure: subadult, length 23 mm, UF 66220. Sepulga River, Flat Rock Shoals, Conecuh County, Alabama. Lower figure: length 41 mm, UF 89877. Little Choctawhatchee River, 5 miles south of Pinchard, Dale County, Alabama, April 1916. © Richard T. Bryant.

Synonymy

Unio acutissimus Lea, 1831. Lea, 1831:89, pl. 10, fig. 18; Lea, 1834b:99, pl. 10, fig. 18
Type locality: Alabama River, [Alabama,] Judge Tait. Type specimen not found, also not found by Johnson and
    Baker (1973). Length of figured shell in original description reported as about 28 mm.

Unio rubellinus Lea, 1857. Lea, 1857a:32; Lea, 1858e:70, pl. 13, fig. 51; Lea, 1858h:51, 70, pl. 13, fig. 51
Type locality: Othcalooga Creek, Gordon County, Georgia, Bishop Elliott. Lectotype, USNM 84136, length 39 mm,
    designated by Johnson (1974).

Unio semiplicatus Küster, 1862. Küster, 1862:279, pl. 94, fig. 4
Type locality: New Holland. Location of types unknown. Length given as 10 lines (1 line = approximately 2.1 mm),
    or about 21 mm.
Comment: Simpson (1900b) placed *Unio semiplicatus* in the synonymy of *Medionidus acutissimus* and pointed out
    that Küster (1862) erroneously listed "New Holland" as the type locality.

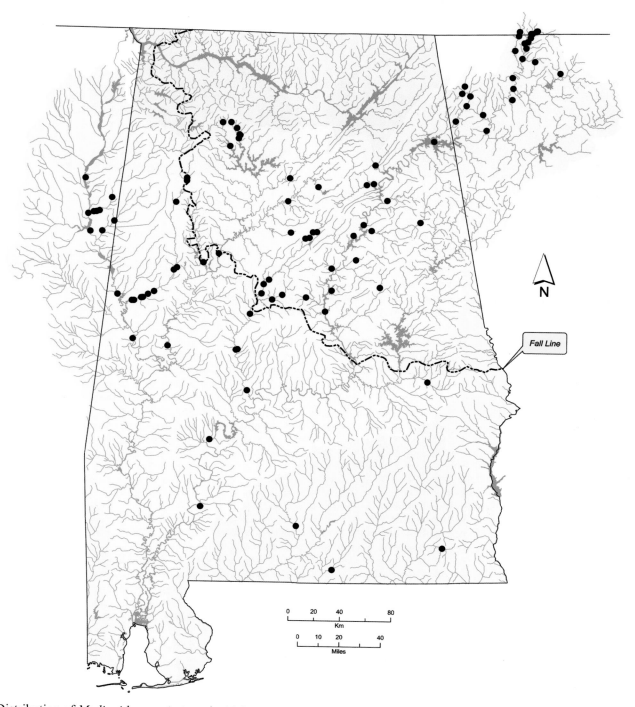

Distribution of *Medionidus acutissimus* in Alabama and the Mobile Basin.

Medionidus conradicus (Lea, 1834)
Cumberland Moccasinshell

Medionidus conradicus – Length 37 mm, CMNH 61.7296. Tennessee River, Florence, Lauderdale County, Alabama. © Richard T. Bryant.

Shell Description

Length to 60 mm; moderately thin, compressed when young, becoming moderately inflated with age; outline elongate elliptical, old individuals often arcuate, especially males; posterior margin narrowly rounded; anterior margin rounded; dorsal margin straight to convex; ventral margin straight to slightly concave; females with slight marsupial swelling posterioventrally, slight sulcus often located just anterior of marsupial swelling; posterior ridge low, rounded; posterior slope low, flat to slightly convex; posterior ridge and slope may be sculptured with fine subradial wrinkles; umbo broad, flat, elevated slightly above hinge line, umbo sculpture fine ridges, may be double-looped; periostracum somewhat shiny, tawny, with numerous wavy green rays that are often interrupted.

Pseudocardinal teeth low, triangular, 2 divergent teeth in left valve, 1 erect peg-like tooth in right valve; lateral teeth straight to slightly curved, 2 in left valve, 1 in right valve; interdentum moderately long, very narrow; umbo cavity wide, very shallow; nacre white to gray or bluish green.

Soft Anatomy Description

Mantle creamy white on outside surface, inside surface often black, giving overall gray appearance, outside of apertures white or pale orange to dark brown, often with dark brown or black mottling; visceral mass creamy white or pearly white to gray or grayish brown, often darker dorsally and/or posteriorly; foot creamy white to tan.

Both males and females with weakly developed mantle fold just anterior to incurrent aperture; fold length approximately 30% of shell length; margin crenulate and/or papillate; papillae simple, may be long; fold and adjacent mantle black. Ortmann (1912a) reported papillae of the mantle fold of males to be considerably shorter than those on folds of females.

Gills grayish tan to dark brown, may be pale tan ventrally; dorsal margin straight to slightly sinuous, ventral margin straight to elongate convex; gill length 55–70% of shell length; gill height 35–45% of gill length; outer gill height 75–100% of inner gill height; inner lamellae of inner gills connected to visceral mass only anteriorly.

Outer gills marsupial; glochidia held in approximately 35% of gill, just posterior of center of gill; marsupium padded when gravid; creamy white. Ortmann (1915) reported the gravid marsupial area to be reniform in outline.

Labial palps creamy white to tan, often with irregular pale spot dorsally; straight to slightly convex dorsally, convex ventrally, bluntly pointed distally; palp length 15–20% of gill length; palp height 55–80% of palp length, distal 45–65% of palps bifurcate.

Incurrent aperture usually longer than excurrent and supra-anal apertures, occasionally equal; excurrent aperture usually longer than supra-anal aperture, occasionally equal or shorter.

Incurrent aperture length 15–20% of shell length; gray to grayish brown within; papillae in 2 rows, inner row larger, long, slender, simple; papillae some combination of pale orange, tan, grayish tan, rusty tan, grayish brown, dark brown and black, often changing color distally.

Excurrent aperture length 10–20% of shell length; creamy white to grayish brown within, with 1–3 marginal color bands, distal band usually grayish brown to dark brown, some individuals with creamy white color band proximal to marginal band, some individuals with creamy white band medially between two dark bands; margin crenulate or with short, simple papillae; papillae rusty orange to brown.

Supra-anal aperture length 10–20% of shell length, occasionally less; gray to grayish brown within, some individuals with thin dark brown band marginally; margin smooth; mantle bridge separating supra-anal and excurrent apertures imperforate, length 15–40% of supra-anal length, occasionally absent.

Specimens examined: Clinch River, Tennessee (n = 4); Estill Fork (n = 2).

Glochidium Description

Length 200–237 μm; height 237–300 μm; without styliform hooks; outline subelliptical; dorsal margin straight, ventral margin rounded, anterior and posterior margins convex (Zale and Neves, 1982b).

Similar Species

Medionidus conradicus may resemble some elongate male *Villosa iris*. However, the posterior ridge and slope of *M. conradicus* is sculptured with fine wrinkles, though they may be weak or obscure. The posterior terminus of male *V. iris* is higher on the shell than that of *M. conradicus*. Some old individuals of *M. conradicus* are arcuate, whereas *V. iris* is rarely arcuate.

General Distribution

Medionidus conradicus is endemic to the Cumberland and Tennessee River drainages. In the Cumberland River drainage it is widespread downstream of Cumberland Falls, Kentucky and Tennessee (Cicerello et al., 1991; Parmalee and Bogan, 1998). It is widespread in the Tennessee River drainage from southwestern Virginia, western North Carolina and eastern Tennessee downstream to Muscle Shoals (Ahlstedt, 1992a, 1992b; Parmalee and Bogan, 1998).

Alabama and Mobile Basin Distribution

Medionidus conradicus probably occurred in the Tennessee River across northern Alabama, but all records are from Muscle Shoals. It is also known from several Tennessee River tributaries.

Medionidus conradicus is known to be extant only in the Paint Rock River system and Foxtrap Creek, a tributary of Spring Creek in Colbert County.

Ecology and Biology

Medionidus conradicus occurs in small creeks to large rivers. It is restricted to shoal and run habitats, where it is found in substrates composed of mixtures of sand and gravel, often with cobble and boulders. *Medionidus conradicus* frequently occurs under large, flat rocks. They are often attached to the substrate by byssal threads.

Medionidus conradicus is a long-term brooder. Both sexes mature by age three (Zale and Neves,

1982b). Gametogenesis takes place during the four months just prior to spawning, which occurs in mid-July, somewhat earlier than in many other species of long-term brooders (Zale and Neves, 1982b). Glochidia mature by early September, at which time they have been observed to appear in stream drift in Big Moccasin Creek, Virginia. Glochidia were sporadic in stream drift from September through December but were abundant from January through June (Zale and Neves, 1982a, 1982b). Ortmann (1921) reported the brooding period of *M. conradicus* to last from September to May. A single hermaphroditic individual was encountered among 158 histologically examined by Zale and Neves (1982b). It was functionally female and gravid, with approximately 5 percent male tissues in the gonad.

Fishes reported to serve as glochidial hosts of *Medionidus conradicus* in laboratory trials, include *Etheostoma flabellare* (Fantail Darter) and *Etheostoma rufilineatum* (Redline Darter) (Percidae) (Zale and Neves, 1982a). Natural infestations of *M. conradicus* glochidia were observed on *E. flabellare* in Big Moccasin Creek, Virginia, during every month except August (Zale and Neves, 1982a). *Lepomis gulosus* (Warmouth) (Centrarchidae) was erroneously reported as a glochidial host of *M. conradicus* by Watters (1994a) and Parmalee and Bogan (1998).

Trematode-infested *Medionidus conradicus* were frequently encountered in Big Moccasin Creek, Virginia, by Zale and Neves (1982b). Older individuals were more frequently and more heavily parasitized than younger individuals.

Current Conservation Status and Protection

Medionidus conradicus was considered a species of special concern throughout its range by Williams et al. (1993). In Alabama it was considered endangered by Stansbery (1976) and a species of special concern by Lydeard et al. (1999). Garner et al. (2004) designated *M. conradicus* a species of highest conservation concern in the state.

Remarks

Medionidus conradicus is generally considered a small stream species and appears to have always been uncommon in the Tennessee River. Historical records from Alabama reaches of the Tennessee River are restricted to Muscle Shoals. Only three specimens of this species have been reported from prehistoric shell middens along the Tennessee River in Alabama, all from Muscle Shoals (Hughes and Parmalee, 1999).

Synonymy

Unio plateolus Rafinesque, 1831. Rafinesque, 1831:3 [identification uncertain]

Unio conradicus Lea, 1834. Lea, 1834a:63, pl. 9, fig. 23; Lea, 1834b:175, pl. 9, fig. 23
Type locality: No locality given in original description, but Johnson (1977) lists it as "no locality [Caney Fork of the
    Cumberland River, Tennessee]," without explanation. Lectotype, USNM 84134, length 47 mm, designated by
    Johnson (1974).

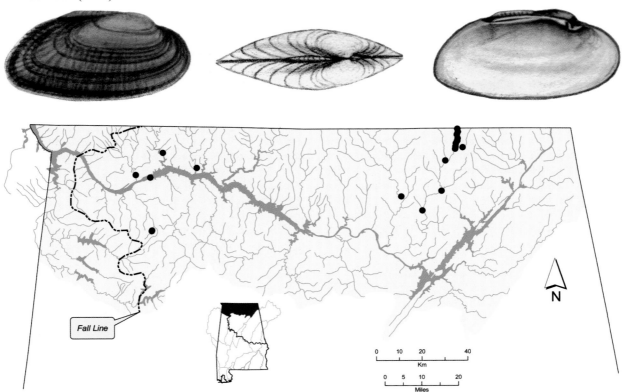

Distribution of *Medionidus conradicus* in the Tennessee River drainage of Alabama.

Medionidus parvulus (Lea, 1860)
Coosa Moccasinshell

Medionidus parvulus – Length 48 mm, UF 66216. Conasauga River, Campbell's Mill, Murray County, Georgia, 17 October 1916. © Richard T. Bryant.

Shell Description

Length to 58 mm; moderately thin; moderately compressed; outline elliptical, becoming arcuate with age; posterior margin obliquely truncate to bluntly pointed; anterior margin rounded; dorsal margin convex; ventral margin straight, becoming concave with age; posterior ridge low, rounded; posterior slope typically low, usually weakly to moderately sculptured with corrugations extending from posterior slope to margin, weak corrugations rarely extend anterior of posterior slope; umbo low, broad, barely elevated above hinge line, umbo sculpture unknown; periostracum yellowish green to green or greenish brown, with numerous weak zigzag lines that may be grouped together to form rays.

Pseudocardinal teeth small, triangular, 2 slightly compressed teeth in left valve, become thicker and more divergent with age, 1 tooth in right valve; lateral teeth short to moderately long, thin, straight to slightly curved, 2 in left valve, 1 in right valve; interdentum short, very narrow; umbo cavity wide, shallow; nacre gray to greenish gray.

Soft Anatomy Description

Mantle dark gray, creamy white outside of pallial line; visceral mass creamy white with grayish brown cast; foot creamy white.

No females were available for mantle modification description. Lea (1866) reported the mantle to be "much thickened on the posterior half of the margin, which is there crenulate and very dark colored".

Gills grayish brown; dorsal margin straight, ventral margin elongate convex; gill length approximately 60% of shell length; gill height approximately 45% of gill length; outer gill height approximately 90% of inner gill height; inner lamellae of inner gills completely connected to visceral mass. Lea (1866) reported the dorsal gill margin to be connected to the visceral mass for less than half its length.

No gravid females were available for marsupium description. Lea (1866) reported outer gills to be marsupial, with glochidia held in approximately 33% of the central portion of the gill.

Labial palps creamy white; straight dorsally, curved ventrally, bluntly pointed distally; palp length approximately 10% of gill length; palp height approximately 50% of palp length; distal 50% of palps bifurcate.

Excurrent and supra-anal apertures continuous, with no mantle bridge or distinction in marginal coloration or morphology; combined excurrent and supra-anal apertures longer than incurrent aperture.

Incurrent aperture length approximately 15% of shell length; gray within, without coloration basal to papillae; papillae short, simple; papillae creamy white with grayish cast. Lea (1866) reported incurrent aperture papillae to be light brown.

Combined excurrent and supra-anal aperture length approximately 20% of shell length; creamy white within, with thin gray marginal band; margin minutely crenulate. Lea (1866) reported a mantle bridge separating the supra-anal and excurrent apertures and the excurrent aperture to have "very small light brown papillae".

Specimen examined: Conasauga River, Georgia (n = 1) (specimen previously preserved).

Glochidium Description

Length 225–237 µm; height 287–312 µm; without styliform hooks; outline subelliptical; dorsal margin straight, ventral margin rounded, anterior and posterior margins convex.

Similar Species

Medionidus parvulus may be very similar to *Medionidus acutissimus* in shell morphology, and occasional specimens cannot be positively identified. However, several subtle characters may be helpful in distinguishing the two. The posterior ridge is sharper and the posterior slope is steeper and narrower in *M. acutissimus* than *M. parvulus*. Also, the posterior end of *M. acutissimus* tends to be more pointed than that of *M. parvulus*. Lea (1866) reported the marsupia of *M. parvulus* to occupy "not more than one-third" of the outer gill, whereas the marsupia of *M. acutissimus* were reported to "extend nearly the whole breadth" of the outer gill. However, as was often the case, Lea based his observations on small sample sizes (in this case two *M. parvulus* and one *M. acutissimus*).

General Distribution

Medionidus parvulus is endemic to the Mobile Basin of Alabama, Georgia and Tennessee.

Alabama and Mobile Basin Distribution

Medionidus parvulus is known from streams above the Fall Line in the Black Warrior, Cahaba and Coosa River drainages, primarily in tributaries.

Medionidus parvulus is extant in at least one Coosa River tributary in northwestern Georgia and the Conasauga River in Tennessee (P.D. Johnson, personal communication).

Ecology and Biology

Medionidus parvulus occurs in shoal areas of medium creeks to rivers with sand and gravel substrates.

Medionidus parvulus is believed to be a long-term brooder. It has been observed gravid in April and May in tributaries of the upper Coosa River (P.D. Johnson, personal communication). Glochidial hosts of *M. parvulus* are unknown.

Current Conservation Status and Protection

Medionidus parvulus was considered endangered throughout its range by Williams et al. (1993) and in Alabama by Lydeard et al. (1999). Garner et al. (2004) listed it as extirpated from the state. *Medionidus parvulus* was listed as an endangered species under the federal Endangered Species Act in 1993.

Remarks

The relationship of *Medionidus parvulus* to *Medionidus acutissimus* is unclear. The two overlap in shell morphology in parts of their ranges. Johnson (1977) placed *M. parvulus* in the synonymy of *M. acutissimus*, but most recent authors have recognized both. Detailed comparative studies, including life history, soft anatomy and genetic analyses, are needed to determine if they represent two valid species. Future studies will be hampered by the extreme rarity of *M. parvulus*, which appears to be on the verge of extinction.

Synonymy

Unio parvulus Lea, 1860. Lea, 1860c:307; Lea, 1866:45, pl. 16, fig. 43; Lea, 1867b:49, pl. 16, fig. 43

Type locality: Coosa River, Alabama, E.R. Showalter, M.D. Lectotype, USNM 84139, length 42 mm, designated by Johnson (1974), is from Chattooga River, Georgia. This lectotype designation is invalid, however, as the figured specimen is not from the stated type locality "Coosa River, Alabama".

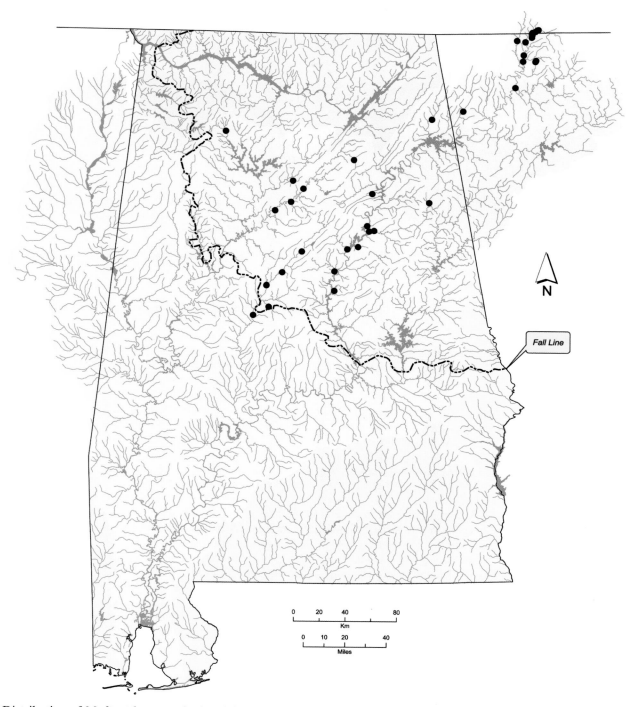

Distribution of *Medionidus parvulus* in Alabama and the Mobile Basin.

Medionidus penicillatus (Lea, 1857)
Gulf Moccasinshell

Medionidus penicillatus – Length 29 mm, UMMZ 139210. Reedie [Reedy] Creek near Madrid, Houston County, Alabama, August 1916. © Richard T. Bryant.

Shell Description

Length to 55 mm; moderately thin; moderately inflated; outline elongate oval; posterior margin narrowly rounded to bluntly pointed; anterior margin rounded; dorsal margin slightly convex; ventral margin straight to slightly convex; posterior ridge rounded; posterior slope moderately steep, sculptured with fine, radial corrugations extending from posterior ridge to margin, corrugations sometimes weak; umbo broad, low, elevated slightly above hinge line, umbo sculpture slightly double-looped ridges, becoming single loops ventrally; periostracum greenish yellow or tawny to dark brown, usually with variable interrupted green rays often composed of chevrons or blotches.

Pseudocardinal teeth small, compressed, oriented almost parallel to margin, 2 teeth in left valve, aligned almost end to end, 1 tooth in right valve; lateral teeth thin, moderately short, straight to slightly curved, 2 in left valve, 1 in right valve; interdentum short, narrow to very narrow; umbo cavity wide, shallow; nacre white to bluish or gray.

Soft Anatomy Description

Mantle tan, outside of apertures dark grayish brown; visceral mass creamy white and tan, often with gray areas; foot creamy white, may be darker distally.

Weakly developed mantle fold just ventral to incurrent aperture (sex of individuals examined uncertain); fold narrow; margin crenulate, may be incompletely crenulate; fold and adjacent mantle dark brown.

Gills light tan; dorsal margin straight to slightly sinuous, ventral margin elongate convex; gill length 55–65% of shell length; gill height 40–50% of gill length; outer gill height 75–100% of inner gill height;

inner lamellae of inner gills connected to visceral mass only anteriorly.

No gravid females were available for marsupium description. Lea (1859f) reported "about two-thirds" of the outer gills to serve as marsupia.

Labial palps light tan with small rusty brown specks; slightly concave dorsally, convex ventrally, bluntly pointed distally; palp length approximately 25% of gill length; palp height 50–60% of palp length; distal 50–70% of palps bifurcate.

Incurrent aperture longer than excurrent and supra-anal apertures; excurrent aperture usually longer than supra-anal aperture.

Incurrent aperture length approximately 20% of shell length; tan with grayish cast or grayish brown mottling within, coloration darker or mottling heavier basal to papillae; papillae in 2 rows, inner row larger, long, slender, simple; papillae tan or brown basally, may be mottled, gray distally.

Excurrent aperture length 10–15% of shell length; gray within, with 2 marginal color bands, bands creamy white, pale gray or dark brown; margin papillate, papillae small, simple, may be well-developed or little more than crenulations; papillae grayish brown.

Supra-anal aperture length 10–15% of shell length; grayish brown or gray with grayish brown mottling within, mottling may be heavier marginally; margin smooth; mantle bridge separating supra-anal and excurrent apertures imperforate, length approximately 35% of supra-anal length, occasionally absent.

Specimens examined: Big Creek, Houston County (n = 2); Swift Creek, Crisp County, Georgia (n = 1) (some specimens previously preserved).

Glochidium Description

Length 218–241 µm; height 280–310 µm; without styliform hooks; outline subspatulate; dorsal margin straight, ventral margin rounded, anterior and posterior margins convex (O'Brien and Williams, 2002).

Similar Species

Most *Medionidus penicillatus* can easily be distinguished from other sympatric species by the corrugations on the posterior slope. Some *M. penicillatus* with weak sculpture may superficially resemble small *Elliptio* (e.g., *E. pullata* or *E. purpurella*), *Hamiota subangulata* or *Villosa* (e.g., *V. lienosa*, *V. vibex* or *V. villosa*), but close examination almost always reveals at least some suggestion of sculpture on the posterior slope of *M. penicillatus*.

General Distribution

Medionidus penicillatus is endemic to the Apalachicola Basin in Alabama, Florida and Georgia and the Econfina Creek system, a direct tributary to St. Andrews Bay, immediately to the west in Florida.

Alabama and Mobile Basin Distribution

Medionidus penicillatus is restricted to the Chattahoochee River drainage and the headwaters of the Chipola River.

Medionidus penicillatus is known to be extant in Alabama only in headwaters of the Chipola River.

Ecology and Biology

Medionidus penicillatus is known from small creeks to large rivers but has been extirpated from large rivers. It occurs in water with at least moderate current. *Medionidus penicillatus* may be found in a variety of substrates, including combinations of sand and small gravel as well as rocky shoals with bedrock and cobble. It may occasionally be found in sandy mud.

Male *Medionidus penicillatus* have been observed to produce spermatozeugmata during late summer and autumn (C.A. O'Brien, personal communication). *Medionidus penicillatus* is a long-term brooder, becoming gravid in late summer or autumn and brooding until the following spring or summer. A female has been observed lying exposed on the surface of the substrate, waving its mantle margin in March, suggestive of glochidial host attraction behavior (Brim Box and Williams, 2000). In laboratory trials glochidial hosts of *M. penicillatus* were found to include *Etheostoma edwini* (Brown Darter) and *Percina nigrofasciata* (Blackbanded Darter) (Percidae) (O'Brien and Williams, 2002). *Gambusia holbrooki* (Eastern Mosquitofish) (Poeciliidae) was found to serve as a secondary host (O'Brien and Williams, 2002). The nonindigenous *Poecilia reticulata* (Guppy) (Poeciliidae) was also found to serve as a marginal host in laboratory trials (O'Brien and Williams, 2002).

Current Conservation Status and Protection

Medionidus penicillatus was considered endangered throughout its range by Athearn (1970) and Williams et al. (1993) and in Alabama by Lydeard et al. (1999). Garner et al. (2004) designated it a species of highest conservation concern in the state. *Medionidus penicillatus* was listed as endangered under the federal Endangered Species Act in 1998.

Remarks

Lea (1857g) included three localities with the original description: "Chattahoochee, near Columbus, Georgia"; "near Atlanta"; and "Flint River, near Albany, Georgia". Clench and Turner (1956) erroneously restricted the type locality to the Chattahoochee River, Columbus, Georgia. However, Johnson (1974) designated the specimen figured by Lea (1859f) as the lectotype. It was from the Flint River, near Albany, Georgia.

Species of *Medionidus* from the Apalachicola Basin eastward have been variously treated as one to four species: *M. penicillatus*, *M. kingii* (Wright, 1900), *M. simpsonianus* Walker, 1905, and *M. walkeri* (Wright, 1897). Currently, three species are recognized, with *M. simpsonianus* in the Ochlockonee River drainage, *M. walkeri* in the Suwannee River drainage and *M. penicillatus* in the Apalachicola Basin.

Shell morphology of *Medionidus* in Gulf Coast drainages exhibits considerable variation, often with overlap in characters among populations. Populations of *Medionidus* in the Choctawhatchee, Yellow and Escambia River drainages have been included in the distribution of *M. penicillatus* by some authors (e.g., Johnson, 1977; Butler, 1990; Williams and Butler, 1994). However, *M. penicillatus* is herein tentatively restricted to the Apalachicola Basin based on zoogeographic patterns observed with other freshwater mussels and fishes. Few species occur in both the Apalachicola Basin and drainages to the west, suggesting a well-defined barrier to distribution. The *Medionidus* in Gulf Coast drainages west of the Apalachicola Basin are herein tentatively assigned to *Medionidus acutissimus*.

Synonymy

Unio penicillatus Lea, 1857. Lea, 1857g:171; Lea, 1859f:203, pl. 23, fig. 85; Lea, 1859g:21, pl. 23, fig. 85

Type locality: Chattahoochee River, near Columbus, [Muscogee County, Georgia,] Dr. Boykin; [Chattahoochee River] near Atlanta, [Fulton County, Georgia,] Bishop Elliott; Flint River, near Albany, [Dougherty County,] Georgia, Reverend G. White. Lectotype, USNM 84142, length 34 mm, designated by Johnson (1974), is from Flint River, near Albany, Dougherty County, Georgia.

Unio kingii B.H. Wright, 1900. B.H. Wright, 1900:138; Johnson, 1967b:7, pl. 5, fig. 6

Distribution of *Medionidus penicillatus* in the eastern Gulf Coast drainages of Alabama.

Genus *Megalonaias*

Megalonaias Utterback, 1915, occurs in the Mississippi Basin and Gulf Coast drainages from the Ochlockonee River drainage in Florida to Central America. It is known from the fossil record in Gulf Coast drainages of peninsular Florida (Bogan and Portell, 1995). Two *Megalonaias* species are currently recognized, one of which occurs south of the Rio Grande (Turgeon et al., 1998). *Megalonaias nervosa* (Rafinesque, 1820) is the largest freshwater mussel in North America, attaining lengths of 280 mm. This species occurs in Alabama.

The ICZN (1988: Opinion 1487) conserved the use of *Megalonaias* Utterback, 1915, over the unused senior synonym *Magnonaias* Utterback, 1915 (Bogan and Williams, 1986).

Type Species
Unio heros Say, 1829 (= *Unio gigantea* Barnes, 1823 = *Unio nervosa* Rafinesque, 1820)

Diagnosis
Shell large; thick; ovate to trapezoidal; low posteriodorsal wing often present; umbo sculpture coarse, double-looped corrugations, remainder of shell covered with various folds, plications and wrinkles; umbo cavity deep, compressed; pseudocardinal teeth massive; lateral teeth long, straight; anterior adductor muscle scar moderately shallow, considerably roughened.

Excurrent aperture almost smooth; mantle bridge separating excurrent and supra-anal apertures short; inner lamellae of inner gills usually completely connected to visceral mass; all 4 gills marsupial; glochidia held throughout gill; gravid marsupium inflated, padded, not extended ventrally below original gill edge; glochidium without styliform hooks (Utterback, 1915; Haas, 1969b).

Synonymy
Magnonaias Utterback, 1915

Megalonaias nervosa (Rafinesque, 1820)
Washboard

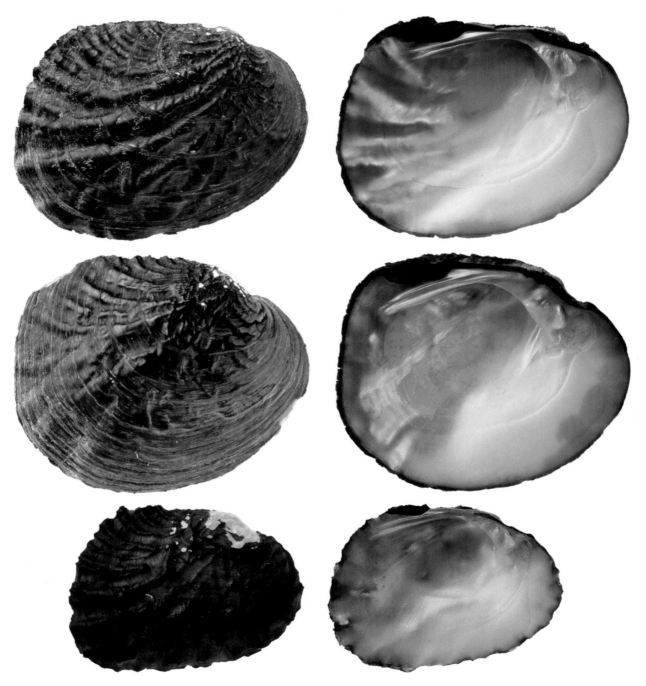

Megalonaias nervosa – Upper figure: length 120 mm, AUM 2134. Tennessee River, Kogers Island, river mile 239, Colbert and Lauderdale counties, Alabama, 29 May 2001. Middle figure: length 129 mm, UMMZ 50971. Chattooga River, 1 mile north of Cedar Bluff, Cherokee County, Alabama. Lower figure: juvenile, length 44 mm, UF 271975. Sipsey River, State Highway 14, Greene County, Alabama, 12 September 1984. © Richard T. Bryant.

Shell Description

Length to 280 mm; thick; somewhat compressed; outline trapezoidal to ovate; posterior margin obliquely truncate to rounded; anterior margin rounded, may be somewhat truncate; dorsal margin slightly convex to straight, often obliquely angled as part of poorly developed dorsal wing; ventral margin straight to slightly convex; posterior ridge poorly developed; posterior slope low, extending into low dorsal wing, usually adorned with regular, parallel, arcuate plications radiating from posterior slope to dorsal margin, often with smaller, irregular wrinkles; disk usually with 3–5 large, oblique plications that do not extend onto the anterior 30% of shell, umbo with corrugations that may extend onto disk, irregular wrinkles may be present across entire disk; umbo broad, elevated slightly above hinge line, umbo sculpture double-looped, zigzag ridges; periostracum dark brown to black, without rays, young specimens may be greenish brown.

Pseudocardinal teeth massive, jagged, triangular, 2 divergent teeth in left valve, 1 tooth in right valve, usually with accessory denticle anteriorly, sometimes with a suggestion of an accessory denticle posteriorly; lateral teeth thick, raised, moderately long, straight to slightly curved, 2 in left valve, 1 in right valve; interdentum short to long, narrow to moderately wide; umbo cavity deep; nacre white, rarely pale pink.

Soft Anatomy Description

Mantle creamy white to light brown; visceral mass creamy white or pearly white to light brown, may be unevenly colored, often pearly white adjacent to foot; foot creamy white to tan or light brown, may be darker distally.

Gills tan to light brown; dorsal margin straight to slightly sinuous, ventral margin convex to elongate convex; gill length 50–70% of shell length; gill height 30–45% of gill length, occasionally greater; outer gill height 55–80% of inner gill height; inner lamellae of inner gills usually completely connected to visceral mass. Utterback (1915) suggested that the dorsal gill connection was often incomplete. Ortmann (1923a) reported "short holes" in the connection of gills to visceral mass in a few specimens that he examined from the Coosa River. The holes were located at the posterior end, near the foot.

All 4 gills marsupial; glochidia held across entire gill except posterior 10–30%; marsupium, well-padded when gravid; creamy white to light tan. Utterback (1915) described gravid gills as "enormous purplish pads".

Labial palps creamy white to tan; elongate, straight dorsally, curved ventrally, bluntly pointed to rounded distally; palp length 30–55% of gill length; palp height 25–50% of palp length, occasionally greater; distal 10–30% of palps bifurcate.

Excurrent and supra-anal apertures usually not separated by mantle bridge, but distinguished by marginal crenulations and color band; incurrent and supra-anal apertures usually considerably longer than excurrent aperture; incurrent aperture longer, shorter or equal to supra-anal aperture.

Incurrent aperture length 20–30% of shell length; creamy white to tan within, often with brown or black basal to papillae; papillae variable, typically some mixture of simple, bifid and trifid with occasional arborescent, occasional specimens with numerous arborescent papillae; papillae usually dark brown or black, occasionally creamy white.

Excurrent aperture length usually 10–25% of shell length, occasionally greater; creamy white to tan or light gray within, usually with black or dark brown marginal color band; margin smooth to crenulate.

Supra-anal aperture length 10–30% of shell length, occasionally greater; creamy white within, sometimes with thin dark marginal band; margin smooth; mantle bridge separating supra-anal and excurrent apertures absent. Utterback (1915) stated that the supra-anal and excurrent apertures are "slightly but distinctly separated" by a mantle bridge.

Specimens examined: Bear Creek, Colbert County (n = 3); Conecuh River (n = 1); Coosa River (n = 3); Sepulga River (n = 2); Sipsey River (n = 3); Tennessee River (n = 6); Tombigbee River (n = 3).

Glochidium Description

Length 254–280 μm, height 340–380 μm; without styliform hooks; outline subelliptical; dorsal margin straight, ventral margin rounded, anterior and posterior margins convex (Surber, 1912, 1915; Utterback, 1915; Hoggarth, 1999). Surber (1915) stated that variation in glochidial size is "remarkable".

Similar Species

Megalonaias nervosa may closely resemble *Amblema elliottii* and *Amblema plicata*, but those species lack corrugations or wrinkles on the umbo. Also, *A. elliotti* and *A. plicata* from large rivers often lack plications on the posterior slope.

Megalonaias nervosa may resemble *Arcidens confragosus* but is less inflated, has a much thicker shell and has well-developed lateral teeth.

Plectomerus dombeyanus and *Elliptoideus sloatianus* also may resemble *Megalonaias nervosa*. However, the posterior ridge of those species is much better developed and usually ends in a point posterioventrally. They also have shallow umbo cavities and purple nacre.

General Distribution

Megalonaias nervosa is widespread in the Mississippi Basin from southern Minnesota and

Wisconsin (Dawley, 1947; Mathiak, 1979) south to Louisiana (Vidrine, 1993) and from the upper Ohio River drainage in eastern Ohio and West Virginia (Cummings and Mayer, 1992) west to eastern Kansas (Murray and Leonard, 1962). *Megalonaias nervosa* occurs in the Cumberland River drainage downstream of Cumberland Falls and in the Tennessee River drainage from eastern Tennessee downstream to the mouth of the Tennessee River (Cicerello et al., 1991; Parmalee and Bogan, 1998). On the Gulf Coast it occurs from the Ochlockonee River drainage, Florida and Georgia, west to the Rio Grande Basin, Texas, and south into Nuevo Leon, Mexico (Contreras Arquieta, 1995; Howells et al., 1996; Johnson, 1999).

Alabama and Mobile Basin Distribution

Megalonaias nervosa occurs across northern Alabama in the Tennessee River and in some tributaries. In the Mobile Basin it is widespread in rivers and large creeks above and below the Fall Line. *Megalonaias nervosa* also occurs in the Chattahoochee and Escambia River drainages.

Megalonaias nervosa is extant in the Tennessee River drainage and Mobile Basin, where it is widespread and may be locally abundant. It is also extant in small, isolated populations in the Conecuh and Chattahoochee rivers.

Ecology and Biology

Megalonaias nervosa inhabits bodies of water ranging from large creeks to large rivers and reservoirs. It is found in areas with sand and gravel substrates, swept clean of silt by swift current, as well as muddy substrates of pools and reservoir overbanks. *Megalonaias nervosa* may be common on slopes of impounded river channels in reservoirs, sometimes at depths approaching 20 m. In some reservoirs where *M. nervosa* has colonized overbank habitat it is the dominant species.

Megalonaias nervosa has been reported to reach sexual maturity by age eight in upper reaches of the Mississippi River (Woody and Holland-Bartels, 1993). The gametogenic cycle of *M. nervosa* differs from that of other short-term brooders, being similar to those of many species of long-term brooders. Gamete production occurs over a short period just prior to spawning. In an upper Mississippi River population most gonadal activity was reported to occur in July, followed by spawning in August (Woody and Holland-Bartels, 1993). In a Tennessee River population most gonadal activity was reported to occur between mid-July and mid-September, followed by spawning that was completed in a two-week period in late September (Haggerty et al., 2005).

Megalonaias nervosa glochidia were reported to mature in one month in the Tennessee River population (Haggerty et al., 2005). Although *M. nervosa* is a short-term brooder, its brooding period occurs in autumn and winter, unlike most other short-term brooders, which are gravid during spring and summer (Utterback, 1915, 1916a; Woody and Holland-Bartels, 1993; Howells, 2000; Haggerty et al., 2005). Brooding periods were reported to differ between populations in the upper Mississippi River (August through October) and Tennessee River (October to late January) (Woody and Holland-Bartels, 1993; Haggerty et al., 2005). Brood size of *M. nervosa* (100 mm to 150 mm in length) in the Tennessee River averaged 750,000 glochidia (n = 15) (Haggerty et al., 2005). Utterback (1915) reported *M. nervosa* to discharge conglutinates, describing them as "sole-shaped", thick and brown. However, no conglutinates were reported by Haggerty et al. (2005) or Woody and Holland-Bartels (1993), and Ortmann (1910a) stated that its glochidia are discharged as "irregular masses, that do not stick together, so as to preserve the shape of the placentae". *Megalonaias nervosa* from the Apalachicola Basin have been observed to discharge glochidia embedded in a thin, web-like, mucous mass (C.A. O'Brien, personal communication).

Populations of freshwater mussels in tailwaters of some dams, especially those with releases from the hypolimnion, suffer from poor reproductive success. Low water temperatures have been suspected as a causal factor. A population of *Megalonaias nervosa* affected by hypolimnetic releases in the Cumberland River, Tennessee, was reproductively inactive, but individuals began to produce gametes and became gravid two years after being moved to warmer water in Kentucky Reservoir, Tennessee River (Heinricher and Layzer, 1999). Though *M. nervosa* is dioecious, almost half of the transplanted individuals developed into hermaphrodites (Heinricher and Layzer, 1999).

Megalonaias nervosa is a generalist with regard to glochidial hosts. Hosts determined based on laboratory trials include *Lepomis cyanellus* (Green Sunfish), *Lepomis macrochirus* (Bluegill), *Lepomis megalotis* (Longear Sunfish), *Micropterus salmoides* (Largemouth Bass), *Pomoxis annularis* (White Crappie) and *Pomoxis nigromaculatus* (Black Crappie) (Centrarchidae); *Campostoma anomalum* (Central Stoneroller) (Cyprinidae); *Ameiurus melas* (Black Bullhead), *Ameiurus nebulosus* (Brown Bullhead), *Ictalurus punctatus* (Channel Catfish) and *Pylodictis olivaris* (Flathead Catfish) (Ictaluridae); *Lepisosteus osseus* (Longnose Gar) (Lepisosteidae); *Perca flavescens* (Yellow Perch), *Percina caprodes* (Logperch) and *Percina phoxocephala* (Slenderhead Darter) (Percidae); and *Aplodinotus grunniens* (Freshwater Drum) (Sciaenidae) (Howard, 1914b; Coker et al., 1921; Woody and Holland-Bartels, 1993; Keller and Ruessler, 1997; O'Dee and Watters, 2000). Howard (1914b) reported *M. nervosa* glochidia to encyst on *Carpiodes velifer* (Highfin Carpsucker) (Catostomidae), but the fish died before transformation was complete. Fishes added to the list of suspected *M. nervosa* glochidial hosts based on observations of natural infestations include *Amia calva* (Bowfin) (Amiidae); *Anguilla rostrata* (American Eel) (Anguillidae); *Lepomis gulosus* (Warmouth) and *Micropterus punctulatus* (Spotted Bass) (Centrarchidae); *Dorosoma cepedianum* (Gizzard Shad) (Clupeidae); *Noturus gyrinus* (Tadpole Madtom) (Ictaluridae); and *Morone chrysops* (White Bass) (Moronidae) (Coker et al., 1921; Weiss and Layzer, 1993). Coker et al. (1921) also reported a natural infestation of *Alosa chrysochloris* (Skipjack Herring) (Clupeidae) with glochidia of *M. nervosa* but deemed it of doubtful significance. Howard (1914b) reported glochidia of *M. nervosa* attached to the amphibian *Necturus maculosus* (Mudpuppy) (Proteidae), but the glochidia never became encysted, even though they remained attached for several days. Surber (1915) reported a natural infestation of a single *M. nervosa* glochidium on *N. maculosa*, but did not provide details as to development of the glochidium.

Glochidia of *Megalonaias nervosa* have been reported to infest both gills and fins of hosts (Howard, 1914b; Weiss and Layzer, 1995). Arey (1924) reported *M. nervosa* to be the only species with hookless glochidia to sometimes infest fins.

Current Conservation Status and Protection

Megalonaias nervosa was reported as currently stable throughout its range by Williams et al. (1993) and in Alabama by Lydeard et al. (1999). Garner et al. (2004) designated it a species of lowest conservation concern in the state.

Megalonaias nervosa is currently one of the most significant commercial species being exported from Alabama for pearl culture. It is protected by state regulations, with the most important being a minimum size limit of 4 inches (102 mm) in shell height.

Remarks

The genus *Megalonaias* was erected by Utterback (1915) based on its "uniqueness of breeding season" as well as differences in its soft anatomy and shell. Utterback originally coined the name *Magnonaias*, which is a mixture of Latin and Greek, but subsequently changed it to *Megalonaias* to correct the inconsistency. In a ruling by the ICZN (1988), *Magnonaias* was suppressed and placed into the synonymy of *Megalonaias* (Bogan and Williams, 1986).

Systematics of the genus *Megalonaias* in some reaches of Gulf Coast drainages remains in question. A distinctive form, *Megalonaias boykiniana* (Lea, 1840), occurs in the Apalachicola and Ochlockonee River drainages. Based on a preliminary genetic analysis with very small sample sizes, Mulvey et al. (1997) placed it in the synonymy of *Megalonaias nervosa*. This classification was herein followed, but the need for further taxonomic work is recognized.

Megalonaias triumphans (Wright, 1898), another distinctive form, is found in the Coosa River drainage. It has long been considered a synonym of *Megalonaias nervosa*. However, adult *M. triumphans* are distinctive from *M. nervosa* conchologically and may represent a valid species.

Muscle Shoals appears to have been near the upstream limit of *Megalonaias nervosa* in the Tennessee River historically (Ortmann, 1925). It has not been reported from prehistoric shell middens along Alabama reaches of the Tennessee River (Hughes and Parmalee, 1999). Its range expansion upstream of Muscle Shoals followed impoundment of the river (Garner and McGregor, 2001). This could have been the result of habitat alterations or changes in the assemblage of potential glochidial hosts. *Megalonaias nervosa* was reported to be the second most abundant mussel species in Wheeler Reservoir in 1991, with an estimated population of 87,660,000 (Ahlstedt and McDonough, 1993).

Synonymy

Unio (*Leptodea*) *nervosa* Rafinesque, 1820. Rafinesque, 1820:296, pl. 80, figs. 8–10
Type locality: "rapides de l'Ohio". Type specimen not found.

Unio crassus var. *giganteus* Barnes, 1823. Barnes, 1823:119
Unio undulatus Barnes, 1823. Barnes, 1823:120, figs. 2a, b
Unio heros Say, 1829. Say, 1829:291
Unio multiplicatus Lea, 1831. Lea, 1831:70, pl. 4, fig. 2; Lea, 1834b:80, pl. 4, fig. 2

Unio boykinianus Lea, 1840. Lea, 1840:288; Lea, 1842b:208, pl. 13, fig. 22; Lea, 1842c:46, pl. 13, fig. 22
Type locality: Chattahoochee River, Columbus, [Muscogee County,] Georgia, Dr. Boykin. Type specimen not
     found, also not found by Johnson and Baker (1973). Length of figured shell in original description reported as
     about 71 mm.

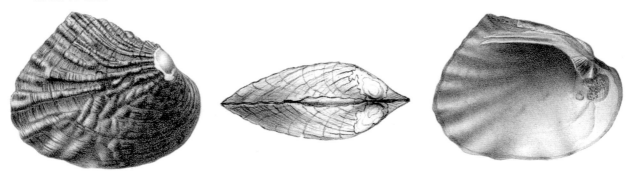

Unio eightsii Lea, 1860. Lea, 1860c:306; Lea, 1860d:367, pl. 64, fig. 192; Lea, 1860g:49, pl. 64, fig. 192
Unio triumphans B.H. Wright, 1898. B.H. Wright, 1898a:101
Type locality: Coosa River, St. Clair County, Alabama. Lectotype, USNM 150554, length 105 mm, designated by
     Johnson (1967b).

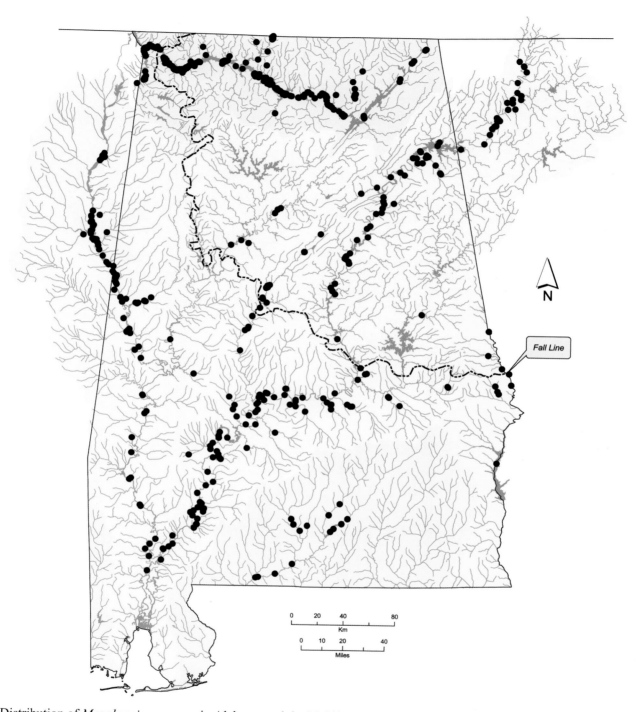

Distribution of *Megalonaias nervosa* in Alabama and the Mobile Basin.

Genus *Obliquaria*

Obliquaria Rafinesque, 1820, occurs in the Great Lakes and Mississippi basins and Gulf Coast drainages from the Mobile Basin to Texas. In Alabama it is found in the Tennessee River drainage and Mobile Basin. *Obliquaria* is a monotypic genus (Turgeon et al., 1998).

Type Species
Obliquaria (*Quadrula*) *reflexa* Rafinesque, 1820

Diagnosis
Shell ovate; inflated; disk marked by row of prominent knobs arranged in alternating pattern between valves.

Inner lamellae of inner gills connected to visceral mass only anteriorly; excurrent aperture crenulate; mantle bridge separating excurrent and supra-anal apertures short; outer gills marsupial; glochidia held in only a few water tubes just posterior to center of gill; marsupium extended ventrally below original gill edge when gravid (Figure 13.31); conglutinates subcylindrical, curved, solid; glochidium without styliform hooks (Rafinesque, 1820; Simpson, 1900b, 1914; Ortmann, 1912a; Haas, 1969b).

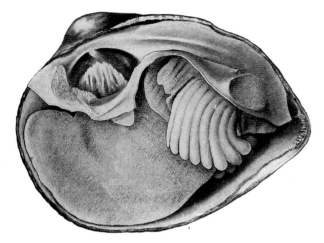

Figure 13.31. *Obliquaria reflexa* gravid female soft anatomy. Note distended marsupium of outer gill. From Lefevre and Curtis (1912).

Synonymy
None recognized.

Obliquaria reflexa Rafinesque, 1820
Threehorn Wartyback

Obliquaria reflexa – Upper figure: length 63 mm, UF 243994. Old Alabama River channel, Millers Ferry [Dannelly] Reservoir, 500 m north of State Highway 28 bridge, 9 miles northwest of Camden, Wilcox County, Alabama, 24 September 1988. Lower figure: length 44 mm, UF 66312. Cahaba River, Pratts Ferry, Bibb County, Alabama, June 1912. © Richard T. Bryant.

Shell Description

Length to 80 mm; thick; moderately inflated, occasionally highly inflated; outline subtriangular to ovate; posterior margin obliquely truncate to bluntly pointed; anterior margin rounded; dorsal margin slightly convex; ventral margin rounded; posterior ridge well-defined, usually angular, occasionally rounded; posterior slope steep, flat to slightly concave, usually adorned with small, regular, arcuate, parallel plications extending from posterior ridge to dorsal margin; disk usually with row of 2–5 large, conspicuous, dorsoventrally compressed knobs, aligned from umbo to center of ventral margin, arranged in alternating pattern between valves, knobs rarely absent; disk and posterior slope may also be adorned with small, irregular wrinkles or poorly developed pustules; broad, shallow sulcus may separate posterior ridge and row of knobs; umbo broad, moderately inflated, elevated above hinge line, turned slightly anteriad, umbo sculpture low, parallel ridges, with indistinct tubercles on posterior ridge; periostracum dull, smooth, yellowish tan to olive or brown, darkening with age, may have thin dark green rays or chevrons, individuals with pale periostracum and pink nacre may appear reddish brown to black.

Pseudocardinal teeth thick, triangular, 2 divergent teeth in left valve, 1 tooth in right valve, often with accessory denticle anteriorly and/or posteriorly; lateral teeth short, straight to slightly curved, 2 in left valve, 1 in right valve; interdentum short, moderately wide; umbo cavity broad, moderately deep; nacre typically white, but ranges from white to pink or reddish purple in Mobile Basin.

Soft Anatomy Description

Mantle tan to golden brown, may have grayish cast, often with translucent line running parallel to dorsal margin of gills, connected to gill margin by thinner, branching lines; visceral mass usually pearly white, occasionally light brown or tan, may have grayish cast; foot creamy white to tan, may be darker distally.

Gills creamy white to golden brown, may have grayish cast; dorsal margin slightly sinuous, ventral margin convex; gill length 45–60% of shell length; gill height 45–55% of gill length; outer gill height 60–90% of inner gill height; inner lamellae of inner gills connected to visceral mass only anteriorly.

Outer gills marsupial; glochidia held in portion of gill just posterior of center, extending completely to posterior end, confined to distal half of gill; marsupium distended when gravid; creamy white.

Labial palps creamy white to tan, occasionally light golden brown; straight dorsally, convex ventrally, rounded distally; palp length 25–30% of gill length; palp height 50–80% of palp length; distal 30–55% of palps bifurcate.

Supra-anal aperture longer than incurrent and excurrent apertures; incurrent aperture longer than excurrent aperture.

Incurrent aperture length 10–20% of shell length; usually creamy white to tan or golden brown within, occasionally black, sometimes with sparse black coloration basal to papillae; papillae in 2 rows, short, thick, mostly simple, occasionally with few bifid; papillae creamy white to tan, sometimes with sparse black specks, giving overall gray cast, some with dark brown edges, brown occasionally extending onto aperture wall between larger papillae.

Excurrent aperture length 10–15% of shell length; creamy white to golden brown within, occasionally black; margin crenulate or with small, simple papillae, margin occasionally smooth; papillae golden brown to gray, darker papillae often tipped with gray or white, crenulations may also be tipped in white.

Supra-anal aperture length 20–25% of shell length; creamy white within, usually without marginal coloration, may have thin, brown marginal band; margin smooth; mantle bridge separating supra-anal and excurrent apertures imperforate, length 5–15% of supra-anal length.

Specimens examined: Alabama River (n = 3); Sipsey River (n = 3); Tennessee River (n = 2); Tombigbee River (n = 3).

Glochidium Description

Length 213–225 μm; height 206–235 μm; without styliform hooks; outline subrotund; dorsal margin slightly convex, ventral rounded, anterior and posterior margins convex (Ortmann, 1912a; Surber, 1912; Utterback, 1915; Hoggarth, 1999).

Similar Species

Obliquaria reflexa may superficially resemble several species of *Quadrula*, including *Q. metanevra, Q. quadrula* and *Q. stapes*. However, no species of *Quadrula* has a row of large, horizontally compressed knobs down the center of the shell disk, alternating between valves.

General Distribution

Obliquaria reflexa occurs in Lake Erie and its tributaries in southeastern Ontario, Canada, and eastern Michigan (La Rocque and Oughton, 1937). It is widespread in the Mississippi Basin from Minnesota (Dawley, 1947) south to Louisiana (Vidrine, 1993) and in headwaters of the Ohio River drainage in western Pennsylvania (Ortmann, 1919) west to South Dakota (Backlund, 2000) and Arkansas (Harris and Gordon, 1990). It is found in the Cumberland River drainage downstream of Cumberland Falls (Cicerello et al., 1991) and in the Tennessee River drainage from eastern Tennessee to the mouth of the Tennessee River (Parmalee and Bogan, 1998). On the Gulf Coast *O. reflexa* occurs from the Mobile Basin west to the Trinity River in Texas (Howells et al., 1996). Dawley (1947) reported *O. reflexa* from the southern Hudson Bay Basin in Minnesota, but this appears to have been based on a misidentification (Cvancara, 1970; Clarke, 1981a; Graf, 1997).

Alabama and Mobile Basin Distribution

Obliquaria reflexa is found throughout the Tennessee River drainage and Mobile Basin in Alabama, Georgia and Mississippi, with the exception of the Tallapoosa River drainage above the Fall Line. There is one record of *O. reflexa* from the Conecuh River near the Alabama and Florida state line that is apparently the result of a recent introduction (D.N. Shelton, personal communication). There is no evidence of *O. reflexa* having established a reproducing population in the Conecuh River.

Obliquaria reflexa remains widespread and locally common in the Tennessee River drainage and Mobile Basin.

Ecology and Biology

Obliquaria reflexa is typically found in large rivers, reservoirs and medium to large tributaries. It has colonized overbanks of reservoirs. *Obliquaria reflexa* occurs in a variety of substrates, ranging from gravel to sand and mud.

Obliquaria reflexa is a short-term brooder, gravid from May to August (Surber, 1912; Utterback, 1915, 1916a; Ortmann, 1919). Haag and Staton (2003) reported 97 percent of mature females to be gravid during the peak of the brooding period in a study of Sipsey River, Alabama, and Little Tallahatchie River, Mississippi, populations. Glochidia are packaged into well-formed conglutinates that are creamy white, long, cylindrical and curved. *Obliquaria reflexa* conglutinates are bound by membranes, unlike conglutinates of many other species. In the Sipsey and Little Tallahatchie River populations *O. reflexa* conglutinates were found to average 0.6 percent undeveloped eggs in composition (Haag and Staton, 2003). *Obliquaria reflexa* fecundity appears to be highly variable, averaging 25,767 glochidia per year in the Sipsey River and 40,975 per year in the Little Tallahatchie River. The latter population ranged from 447 to 135,750 glochidia per individual (Haag and Staton, 2003).

Fishes shown to serve as glochidial hosts of *Obliquaria reflexa* in laboratory trials include *Luxilus cornutus* (Common Shiner), *Notropis buccatus* (Silverjaw Minnow) and *Rhinichthys cataractae* (Longnose Dace) (Cyprinidae) (Watters et al., 1998).

Lefevre and Curtis (1912) and Utterback (1915) suggested that *Obliquaria reflexa* does not require a glochidial host, based on the fact that glochidia remain incorporated in the conglutinate after it is discharged. However, they did not address the possibility of conglutinates being ruptured when bitten by fish. There appears to be little evidence supporting the suggestion that glochidia of *O. reflexa* are not obligate parasites.

Current Conservation Status and Protection

Obliquaria reflexa was considered currently stable throughout its range by Williams et al. (1993) and in Alabama by Lydeard et al. (1999). Garner et al. (2004) designated *O. reflexa* a species of lowest conservation concern in the state.

Though seldom harvested, *Obliquaria reflexa* is a legally harvestable shell. It is protected by state regulations, with the most important being a minimum size limit of 1¾ inches (44 mm) in shell height.

Remarks

The systematic position of *Obliquaria reflexa* has been a subject of debate. It has generally been considered to belong to the Lampsilinae (e.g., Utterback, 1915; Heard and Guckert, 1970). However, it has a short-term brooding period, which is atypical of lampsilines. Recent genetic analyses have been inconclusive as to the systematic position of *O. reflexa* (Lydeard et al., 1996; Serb et al., 2003).

Frierson (1927) described a new subspecies, *Obliquaria reflexa conradi*, from the Alabama River. The pink to reddish purple nacre was presented as the diagnostic character. However, this character was subsequently recognized as intraspecific variation within the species. Preliminary genetic analysis suggests that populations (Tennessee River drainage versus Mobile Basin) and individuals with differing nacre color (white versus pink) are conspecific (A.E. Bogan, unpublished data).

Obliquaria reflexa is rare in prehistoric shell middens along Alabama reaches of the Tennessee River, where it has been reported from Muscle Shoals and near Bridgeport (Warren, 1975; Hughes and Parmalee, 1999). However, *O. reflexa* is currently one of the most commonly encountered species. In 1991 it was found to be the fourth most abundant species in Wheeler Reservoir, with an estimated population of 44,590,000 (Ahlstedt and McDonough, 1993).

Synonymy

Obliquaria (*Quadrula*) *reflexa* Rafinesque, 1820. Rafinesque, 1820:306
Type locality: Kentucky River and Rapids of Letart, [Falls of the Ohio River, Meigs County, Ohio]. Lectotype, ANSP 20206, length 54 mm, designated by Johnson and Baker (1973), is from Rapids of Letart, Falls of the Ohio River, Meigs County, Ohio.
Unio cornutus Barnes, 1823. Barnes, 1823:122, pl. 4, figs. 5a–c
Unio phillipsii Conrad, 1835. Conrad 1835b:9, pl. 5, fig. 1
Obliquaria reflexa var. *conradi* Frierson, 1927. Frierson, 1927:65
Type locality: Alabama River, [Alabama]. Lectotype, ANSP 56695a, length 32 mm, designated by Johnson (1972a).

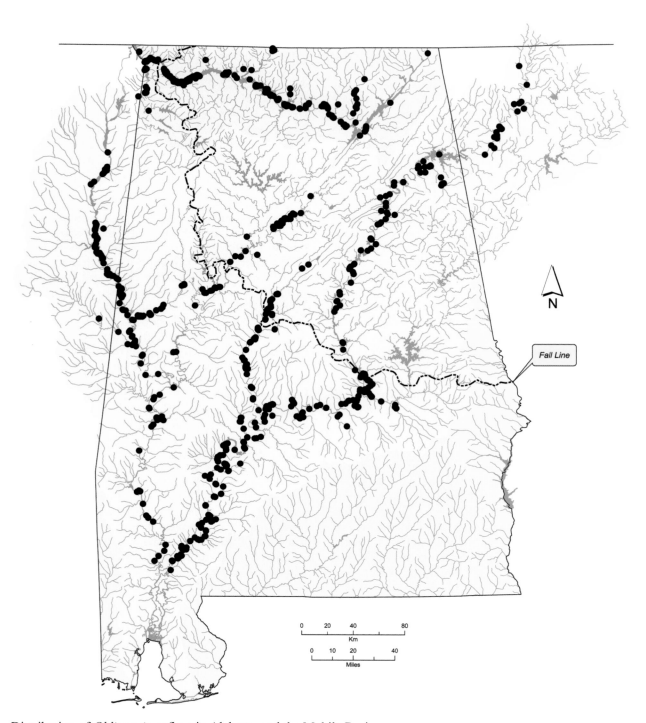

Distribution of *Obliquaria reflexa* in Alabama and the Mobile Basin.

Genus *Obovaria*

Obovaria Rafinesque, 1819, is found in the Great Lakes and Mississippi basins and Gulf Coast drainages from the Choctawhatchee River drainage to eastern Texas. Turgeon et al. (1998) listed six *Obovaria* species, five of which occur in Alabama. An additional species found in Alabama, *Lampsilis haddletoni* Athearn, 1964, is herein moved to *Obovaria*. Soft anatomy of *Obovaria haddletoni* is unknown. It was moved from *Lampsilis* to *Obovaria* based on its circular shape, shallow umbo cavity and triangular, divergent, striated pseudocardinal teeth.

Ortmann and Walker (1922a) identified the specimen of *Obovaria obovalis* Rafinesque, 1820, as *Fusconaia ebena* (Lea, 1831). Herrmannsen (1847) designated *O. obovalis* as the type species for *Obovaria*. However, Agassiz (1852) was unable to recognize the species and chose to use *Unio retusa* Lamarck, 1819, as the type species of *Obovaria*. Fischer (1886) and Simpson (1900b, 1914) also assumed *U. retusa* to be the type species of *Obovaria*, apparently independent of Agassiz (1852). If *O. obovalis* was recognized as a species belonging to *Fusconaia* Simpson, 1900, *Obovaria* would be the senior synonym of *Fusconaia*. A petition to suppress all type species designations for *Obovaria* and to designate *U. retusa* Lamarck, 1819, as the type species of *Obovaria* is currently before the ICZN (Bogan et al. 2006: Case 3353). This proposed action would conserve the modern usage of *Obovaria* and *Fusconaia*.

Type Species

Obovaria obovalis Rafinesque, 1820, *fide* Herrmannsen (1847). *Obovaria torsa* Rafinesque, 1820 (= *Unio retusa* Lamarck, 1819), is used as the type species today *fide* Agassiz (1852). These type species designations are proposed to be suppressed and *Unio retusa* Lamarck, 1819 designated as the type species of *Obovaria* (Bogan et al., 2006: Case 3353).

Diagnosis

Shell round to oval or subtriangular; inflated; posterior ridge weak or absent; periostracum dull or cloth-like, rarely with rays; sexual dimorphism subtle, females somewhat inflated posterioventrally.

Inner lamellae of inner gills completely attached to visceral mass; mantle margin just ventral to incurrent aperture slightly thickened in females, finely crenulate and pigmented; outer gills marsupial; glochidia held in posterior portion of gill; marsupium outline reniform, extended ventrally beyond original gill edge when gravid; glochidium without styliform hooks (Rafinesque, 1819, 1820; Simpson, 1900b, 1914; Ortmann, 1912a; Haas, 1969b).

Synonymy

Striata Frierson, 1927, *non* Boettger, 1878 (Mollusca)
Luteacarnea Frierson, 1927, *nomen novum* for *Striata* Frierson, 1927
Pseudoon Simpson, 1900

Obovaria haddletoni (Athearn, 1964)
Haddleton Lampmussel

Obovaria haddletoni – Upper figure: length 31 mm, MFM 6705 (paratype). Lower figure: length 30 mm, CMNML 20095 (holotype). Choctawhatchee River, West Fork, 7 miles southeast of Ozark, Dale County, Alabama, 23 September 1956. © Richard T. Bryant.

Shell Description

Length to 31 mm; moderately thin; moderately inflated; outline round to oval; posterior margin rounded; anterior margin rounded; dorsal margin convex; ventral margin convex; posterior ridge broad, rounded; posterior slope flat, not steep; umbo low, somewhat compressed, not elevated above hinge line, umbo sculpture unknown; periostracum yellowish brown with thin dark green rays posteriorly.

Pseudocardinal teeth moderately thick, triangular, 2 divergent teeth in left valve, anterior tooth somewhat compressed, roughly parallel to dorsal margin, directed toward middle of anterior adductor muscle scar, posterior tooth smaller, somewhat knobby, 1 tooth in right valve, with accessory denticle anteriorly and posteriorly; lateral teeth thin, slightly curved, 2 in left valve, 1 in right valve; interdentum moderately long, very narrow; umbo cavity shallow; nacre white, stained yellow near umbo cavity.

Soft Anatomy Description

Soft anatomy unknown.

Glochidium Description

Glochidium unknown.

Similar Species

Obovaria haddletoni most closely resembles *Villosa choctawensis,* but *V. choctawensis* typically is more elongate and usually has more prominent rays.

General Distribution

Obovaria haddletoni is known only from the Choctawhatchee River drainage in southeastern Alabama.

Alabama and Mobile Basin Distribution

Obovaria haddletoni is known only from its type locality in the West Fork Choctawhatchee River, Dale County, Alabama.

Obovaria haddletoni has not been collected since the type lot was taken and may be extinct.

Ecology and Biology

Obovaria haddletoni was collected from a shoal in the West Fork Choctawhatchee River, which is a small river. There is no other information on its habitat.

Obovaria haddletoni is presumably a long-term brooder, gravid from late summer or autumn to the following summer. Its glochidial hosts are unknown.

Current Conservation Status and Protection

Obovaria haddletoni (as *Lampsilis haddletoni*) was considered endangered throughout its range by Athearn (1970), Stansbery (1971) and Williams et al. (1993) and in Alabama by Lydeard et al. (1999). Garner et al. (2004) listed it as extinct.

Remarks

Obovaria haddletoni was originally described as a species of *Lampsilis*. In the original description it was compared to *Lampsilis ochracea* (= *Leptodea ochracea*) and *Villosa choctawensis* Athearn, 1964. There was no justification or discussion by Athearn (1964) regarding placement of *haddletoni* in the genus *Lampsilis*. However, based on its circular shape, shallow umbo cavity and triangular, divergent, striated pseudocardinal teeth, it is herein placed in *Obovaria*.

Obovaria haddletoni was named for Arthur Haddleton Clarke, Jr., an esteemed malacologist and colleague of H.D. Athearn.

Synonymy

Lampsilis haddletoni Athearn, 1964. Athearn, 1964:135–136, pl. 9, figs. g, h [figured at beginning of account]

Type locality: Choctawhatchee River, West Fork, 7 miles southwest [southeast] of Ozark, Dale County, Alabama, 23 September 1956. Holotype, CMNML 20095, length 30 mm.

Comment: There was an error in the locality information given in the original description. Athearn (1964) reported the collection from the West Fork Choctawhatchee River, 7 miles southwest of Ozark, but it should read southeast of Ozark.

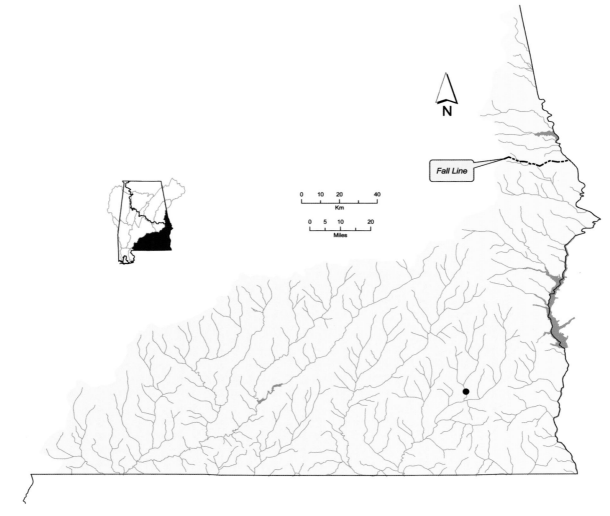

Distribution of *Obovaria haddletoni* in the eastern Gulf Coast drainages of Alabama.

Obovaria jacksoniana (Frierson, 1912)
Southern Hickorynut

Obovaria jacksoniana – Upper figure: female, length 52 mm, UF 358663. Tombigbee River above Pickensville boat ramp, about 1 mile northwest of Pickensville, Pickens County, Alabama, 4 June 1972. Lower figure: male, length 41 mm, UF 358662. Buttahatchee River, U.S. Highway 45, 12 miles north of Columbus, Lowndes County, Mississippi, 13 September 1984. © Richard T. Bryant.

Shell Description

Length to 55 mm; moderately thick; moderately inflated, females may be more inflated posterioventrally; outline oval to subtriangular; posterior margin narrowly rounded; anterior margin rounded to somewhat truncate; dorsal margin straight to convex; ventral margin convex; posterior ridge low, rounded; posterior slope steep; umbo broad, inflated, elevated above hinge line, umbo sculpture unknown; periostracum dark greenish brown to black, may have green rays, often without rays.

Pseudocardinal teeth triangular, erect, 2 divergent teeth in left valve, 1 tooth in right valve, may have accessory denticle anteriorly and/or posteriorly; lateral teeth short, straight, 2 in left valve, 1 in right valve;

interdentum short to moderately long, narrow to moderately wide; umbo cavity moderately deep, not compressed; nacre white.

Soft Anatomy Description

Mantle tan, slight rusty cast outside of apertures; visceral mass tan; foot tan.

Females with weak mantle fold just ventral to incurrent aperture, fold length 20–25% of shell length; margin with widely spaced simple papillae anteriorly and widely spaced crenulations posteriorly; adjacent margin with rusty tan posteriorly.

Gills tan; dorsal margin straight, ventral margin convex, outer gill may be slightly bilobed; gill length 50–55% of shell length; gill height approximately 55%

of gill length; outer gill height 65–85% of inner gill height; inner lamellae of inner gills completely connected to visceral mass.

No gravid females were available for marsupium description; probably similar to other *Obovaria* species, with outer gills marsupial, marsupium well-padded when gravid.

Labial palps tan; slightly convex dorsally, convex ventrally, narrowly rounded to bluntly pointed distally; palp length approximately 30% of gill length; palp height approximately 65% of palp length; distal 40–45% of palps bifurcate.

Incurrent and supra-anal apertures approximately equal in length, incurrent aperture may be slightly longer; incurrent and supra-anal apertures considerably longer than excurrent aperture.

Incurrent aperture length 15–20% of shell length; tan within, without coloration basal to papillae; papillae in 2 rows, long, slender, mostly simple, with occasional bifid; papillae tan, with rusty cast distally.

Excurrent aperture length approximately 10% of shell length; tan within, with slight rusty cast marginally; margin crenulate or with small, simple papillae; papillae tan with slight rusty cast.

Supra-anal aperture length approximately 15% of shell length; tan within, without marginal coloration; margin smooth; mantle bridge separating supra-anal and excurrent apertures imperforate, length 30–45% of supra-anal length.

Specimens examined: Lubbub Creek, Pickens County (n = 2) (specimens previously preserved).

Glochidium Description

Length 175–187 μm; height 230–243 μm; without styliform hooks; outline subelliptical; dorsal margin slightly convex, ventral margin rounded, anterior and posterior margins convex (Hoggarth, 1999).

Similar Species

It can be very difficult to distinguish *Obovaria jacksoniana* from some *Obovaria unicolor* because shell characters of these species overlap. *Obovaria unicolor* has an overall rounder appearance, while *O. jacksoniana* is more elongate and has a more narrowly rounded posterior margin and a more anteriorly positioned umbo. Shell nacre of *O. unicolor* is occasionally pink, whereas that of *O. jacksoniana* is always white.

More elongate *Obovaria jacksoniana* may resemble *Pleurobema perovatum*. However, the umbo of *P. perovatum* is more centrally positioned. The umbo cavity of *O. jacksoniana* is deeper than that of *P. perovatum*.

Elongate *Obovaria jacksoniana* may also resemble *Pleurobema curtum* but the umbo of *P. curtum* is generally more anterior than that of *O. jacksoniana*.

Pleurobema curtum usually has a shallow sulcus just anterior to the posterior ridge, which is absent in *O. jacksoniana*. Also, *O. jacksoniana* has a deeper umbo cavity than that of *P. curtum*. The crest of the anterior pseudocardinal tooth of *O. jacksoniana* is oriented more anteriorly than that of *P. curtum*.

General Distribution

Obovaria jacksoniana occurs in the lower Mississippi Basin from southeastern Missouri (Oesch, 1995) and western Tennessee south to the Gulf Coast. In Gulf Coast drainages it occurs from the Mobile Basin west to the Neches River drainage in eastern Texas (Howells et al., 1996).

Alabama and Mobile Basin Distribution

Obovaria jacksoniana occurs in the Mobile Basin primarily below the Fall Line. It is known from the Alabama, Cahaba and Tombigbee River drainages.

Obovaria jacksoniana is known to be extant in several tributaries of the upper Tombigbee River, including Lubbub Creek, Pickens County, Alabama, and Yellow Creek, Lowndes County, Mississippi, as well as Buttahatchee, Sipsey and East Fork Tombigbee rivers (McGregor and Haag, 2004; R.L. Jones, personal communication).

Ecology and Biology

Obovaria jacksoniana inhabits rivers and large creeks with sand and gravel substrates in slow to moderate current.

Obovaria jacksoniana is presumably a long-term brooder, gravid from late summer or autumn to the following summer. Its glochidial hosts are unknown.

Current Conservation Status and Protection

Obovaria jacksoniana was considered a species of special concern throughout its range by Williams et al. (1993). In Alabama Stansbery (1976) considered it endangered, but Lydeard et al. (1999) listed it as a species of special concern. Garner et al. (2004) designated *O. jacksoniana* a species of moderate conservation concern in the state.

Remarks

The conchological similarity of *Obovaria jacksoniana* and *Obovaria unicolor* has prompted questions of the validity of the two species. However, a detailed conchological and electrophoretic comparison suggested that the two are distinct (Hoggarth, 1980).

Obovaria jacksoniana appears to be named for the city of Jackson, Mississippi, located along the Pearl River, one of the localities from which type material of the species was collected.

Synonymy

Unio castaneus Lea, 1831 (December), *non* Rafinesque, 1831. Lea, 1831:91, pl. 11, fig. 21; Lea, 1834b:101, pl. 11, fig. 21

Type locality: Alabama River, [Alabama,] Judge Tait. Lectotype, USNM 84824, length 25 mm, designated by Johnson (1974).

Unio (*Obovaria*) *jacksonianus* Frierson, 1912. Frierson, 1912:23, pl. 3, figs. 1–3

Type locality: Pearl River, [Jackson,] and Yalabusha River, Mississippi. Lectotype, ANSP 106063a, length 44 mm, designated by Johnson (1972a), is from Pearl River, Jackson, Mississippi.

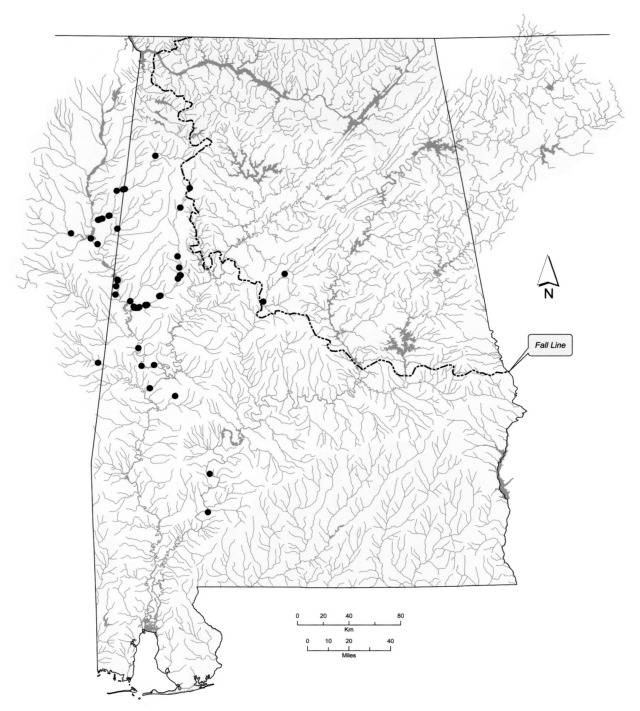

Distribution of *Obovaria jacksoniana* in Alabama and the Mobile Basin.

Obovaria olivaria (Rafinesque, 1820)
Hickorynut

Obovaria olivaria – Upper figure: length 60 mm, USNM 84820. Cumberland River, Tennessee. Lower figure: length 55 mm, UMMZ 66441. Tennessee River, 6 miles east of Decatur, Morgan County, Alabama. © Richard T. Bryant.

Shell Description

Length to 100 mm; thick; inflated; outline oval to elliptical, sex is difficult to distinguish; posterior margin bluntly pointed to broadly rounded; anterior margin broadly rounded; dorsal margin convex; ventral margin broadly rounded; posterior ridge weak, rounded or absent; posterior slope steep, flat; umbo inflated, elevated well above hinge line, oriented anteriad, located near anterior margin, umbo sculpture weak, double-looped ridges, disappearing on posterior slope; periostracum olive to yellowish brown, with numerous fine green rays, periostracum becomes dark brown with age, rays often become obscured.

Pseudocardinal teeth large, triangular, 2 teeth in left valve, divergent in young individuals, becoming less divergent with age, 1 tooth in right valve, usually with small lamellar accessory denticle anteriorly, teeth in both valves become almost parallel to lateral teeth with age; lateral teeth moderately long, curved, 2 in left valve, 1 in right valve; interdentum short, moderately wide; umbo cavity shallow; nacre white, sometimes with pink or cream tint in umbo cavity.

Soft Anatomy Description

Mantle tan, may have thin dark brown band marginally; visceral mass tan; foot tan, may have slight orange cast. Lea (1863c) reported the mantle to be thickened marginally. Ortmann (1919) reported mantle margins to be "inclining to blackish", primarily toward the incurrent and excurrent apertures, with the coloration being "more intense in the male sex".

Females with weak mantle fold just ventral to incurrent aperture; fold length 20–55% of shell length; margin crenulate or with very small, simple papillae, papillae may change to crenulations anteriorly; mantle adjacent to fold may be brown.

Gills tan; dorsal margin straight, ventral margin broadly convex; gill length 50–60% of shell length; gill height 35–45% of gill length; outer gill height 70–95% of inner gill height; inner lamellae of inner gills completely connected to visceral mass.

Outer gills marsupial; glochidia held across entire gill; marsupium distended when gravid; white to tan. Ortmann (1919) reported glochidia to be held only in the posterior half of the gills and the marsupial area to be reniform in outline, with a "purplish gray" distal margin.

Labial palps tan; straight dorsally, straight to convex ventrally, bluntly pointed distally; palp length 30–35% of gill length; palp height 55–65% of palp length; distal 25–35% of palps bifurcate.

Incurrent and supra-anal apertures longer than excurrent aperture; incurrent and supra-anal aperture lengths variable relative to each other.

Incurrent aperture length 15–20% of shell length; tan within, may have white or brown basal to papillae; papillae in 2 rows, mostly simple, may have few bifid; papillae tan to white. Lea (1863c) reported incurrent aperture papillae to be "very small" and "brownish".

Excurrent aperture length 5–10% of shell length; tan within, without marginal color band; margin with small, simple papillae; papillae tan to white. Lea (1863c) reported excurrent aperture margins to have "very minute crenulations".

Supra-anal aperture length 15–20% of shell length; tan within, without marginal color band; margin crenulate, may have very small papillae adjacent to mantle bridge; mantle bridge separating supra-anal and excurrent apertures imperforate, length 15–20% of supra-anal length.

Specimens examined: Mississippi River, Illinois and Iowa (n = 2); Wisconsin River, Wisconsin (n = 1) (specimens previously preserved).

Glochidium Description

Length 190–210 μm; height 220–265 μm; without styliform hooks; outline subelliptical; dorsal margin slightly convex, ventral margin rounded, anterior margin convex, posterior margin obliquely straight above, convex below (Ortmann, 1912a; Surber, 1912; Hoggarth, 1999).

Similar Species

Obovaria olivaria superficially resembles *Obovaria subrotunda* but is more elliptical in outline. The crest of the posterior pseudocardinal tooth in the left valve of adult *O. olivaria* is oriented more anterior to posterior than that of *O. subrotunda*, which is oriented more dorsal to ventral.

Obovaria olivaria may resemble *Fusconaia ebena* but can be distinguished by its lighter, more green or yellow periostracum, which often has faint green rays. Mature *F. ebena* are typically dark brown to black, without rays. Also, the pseudocardinal teeth of *O. olivaria* are divergent and separated anteriorly, whereas those of *F. ebena* are more acutely angled and joined anteriorly. The umbo cavity of *O. olivaria* is shallow, and that of *F. ebena* is deep and compressed.

Obovaria olivaria may also resemble *Fusconaia subrotunda* from large rivers, but *O. olivaria* has a broader, more inflated umbo and shallow umbo cavity. The periostracum of *O. olivaria* is usually more yellow or green than that of *F. subrotunda*, which is typically dark brown to black. Both may have faint green rays, but those of *F. subrotunda* are usually confined to the umbo, whereas those of *O. olivaria* extend to the ventral margin.

Obovaria olivaria may superficially resemble *Lampsilis fasciola* but has a much thicker shell and higher, more inflated umbo. Both may have green rays, but those of *L. fasciola* are thinner and wavy.

General Distribution

Obovaria olivaria is known from some parts of the Great Lakes Basin, southern Canada and north-central U.S., including drainages of lakes Erie, Huron, Ontario and St. Clair, as well as the St. Lawrence River system (Goodrich and van der Schalie, 1932; La Rocque and Oughton, 1937; Burch, 1975a; Clarke, 1981a; Strayer et al., 1992). It is widespread in the Mississippi Basin, ranging from Minnesota and Wisconsin (Dawley, 1947) south to Louisiana (Vidrine, 1993), and in headwaters of the Ohio River drainage in Pennsylvania (Ortmann, 1919) west to eastern Kansas (Murray and Leonard, 1962). *Obovaria olivaria* was historically widespread in the Cumberland River drainage downstream of Cumberland Falls, Kentucky and Tennessee (Cicerello et al., 1991; Parmalee and Bogan, 1998). In the Tennessee River drainage *O. olivaria* occurred from northeastern Alabama downstream to the mouth of the Tennessee River (Parmalee and Bogan, 1998).

Alabama and Mobile Basin Distribution

All Alabama records of *Obovaria olivaria* are from the Tennessee River, Guntersville downstream to Muscle Shoals.

Obovaria olivaria appears to be extirpated from Alabama. The most recent known collection of this species was from Muscle Shoals in 1966.

Ecology and Biology

Obovaria olivaria is generally a species of large to medium rivers, but can also be found in some northern lakes. It occurs primarily in sand and gravel substrates at depths usually exceeding 2 m (Ortmann, 1919; Parmalee and Bogan, 1998).

Obovaria olivaria is a long-term brooder, gravid from August to June of the following summer (Surber, 1912; Ortmann, 1919). *Acipenser fulvescens* (Lake Sturgeon) and *Scaphirhynchus platorynchus* (Shovel-nose Sturgeon) (Acipenseridae) have been shown to serve as glochidial hosts of *O. olivaria* in laboratory trials (Coker et al., 1921; Brady et al., 2004).

Current Conservation Status and Protection

Obovaria olivaria was considered currently stable throughout its range by Williams et al. (1993). In Alabama Stansbery (1976) designated it as endangered, but Lydeard et al. (1999) erroneously listed it as currently stable in the state. Garner et al. (2004) listed it as extirpated from the state.

Remarks

The only known glochidial hosts of *Obovaria olivaria* are *Acipenser fulvescens* and *Scaphirhynchus platorynchus*, which are extirpated from the Tennessee River. The disappearance of these migratory species from the state appears to be related to construction of Pickwick Dam in 1938 by TVA. The most recent reports of these species from Alabama were from near Decatur, Morgan County, *A. fulvescens* in 1938 (Boschung and Mayden, 2004) and *S. platorynchus* in 1940 (Bailey and Cross, 1954). *Obovaria olivaria* is not likely to return to the state unless the problem of fish passage at Pickwick Dam is addressed.

Synonymy

Amblema olivaria Rafinesque, 1820. Rafinesque, 1820:314
Type locality: Kentucky River. Lectotype, ANSP 20251a, length 58 mm, designated by Johnson and Baker (1973).
Unio ellipsis Lea, 1828. Lea, 1828:268, pl. 4, fig. 4; Lea, 1834b:10, pl. 4, fig. 4
Unio pealei Lea, 1871. Lea, 1871:191; Lea, 1874c:26, pl. 8, fig. 23; Lea, 1874e:30, pl. 8, fig. 23

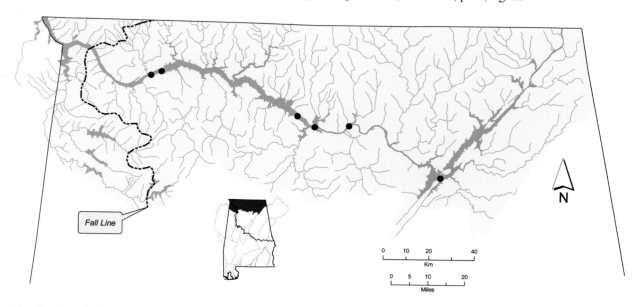

Distribution of *Obovaria olivaria* in the Tennessee River drainage of Alabama.

Obovaria retusa (Lamarck, 1819)
Ring Pink

Obovaria retusa – Upper figure: female, length 67 mm, UF 370409. Tennessee River near Diamond Island, above Savannah, Tennessee, 24 June 1964. Lower figure: male, length 48 mm, USNM 894514. Florence, [Lauderdale County,] Alabama. © Richard T. Bryant.

Shell Description

Length to 95 mm; very thick; inflated; outline oval to quadrate or triangular; male posterior margin broadly rounded, female posterior margin emarginate; anterior margin rounded; dorsal margin convex; ventral margin broadly rounded; posterior ridge rounded, made distinct in females by narrow sulcus at posterior base of ridge; male posterior slope steep, flat to slightly convex, female posterior slope less steep, flat to broadly convex; umbo greatly inflated, elevated well above hinge line, turned anteriad, located near anterior margin, umbo sculpture weak, double-looped ridges;

periostracum often shiny, yellowish green to reddish brown, without rays, becoming dark brown or black with age.

Pseudocardinal teeth massive, triangular, deeply serrate, 2 divergent teeth in left valve, 1 tooth in right valve, often with accessory denticle anteriorly and/or posteriorly; lateral teeth short, thick, slightly curved, 2 in left valve, 1 in right valve; interdentum short, wide; umbo cavity deep, compressed; nacre purple, pink or salmon inside of pallial line, white outside of pallial line.

Soft Anatomy Description

Mantle tan; visceral mass tan; foot tan, may have slight orange cast. Lea (1863c) described the mantle as thin, "with a broad margin, thickened and colored on the edges" and the color of "the mass" as "dilute salmon".

Females with weak mantle fold just ventral to incurrent aperture; fold length 20–35% of shell length; margin crenulate; mantle adjacent to fold brown posteriorly. Ortmann (1912a) reported the mantle margin to have a "black streak" adjacent to the incurrent aperture and males to have a much weaker fold.

Gills tan; dorsal margin straight, ventral margin deeply convex, gill filaments curved posteriorly; gill length 55–60% of shell length; gill height 55–65% of gill length; outer gill height 70–80% of inner gill height; inner lamellae of inner gills completely connected to visceral mass.

No gravid females were available for marsupium description. Lea (1863c) and Ortmann (1912a) reported outer gills to be marsupial, with glochidia held in the posterior portion. Ortmann (1912a) added that the marsupial area is white and reniform in outline.

Labial palps tan; straight dorsally, straight to broadly convex ventrally, bluntly pointed distally; palp length approximately 25% of gill length; palp height 50–70% of palp length; distal 40–55% of palps bifurcate.

Incurrent aperture longer than excurrent and supra-anal apertures; supra-anal aperture longer or equal to excurrent aperture.

Incurrent aperture length 20–25% of shell length; tan within, may have white basal to papillae; papillae in 2 rows, inner row larger, mostly simple, may have few bifid; papillae brown basally, white distally, occasionally entirely white.

Excurrent aperture length 15–20% of shell length; tan within, marginal color band brown; margin with small, simple papillae, may be partially crenulate; papillae brown to tan, may have white tips.

Supra-anal aperture length 15–20% of shell length; tan within, may have thin brown band marginally; margin smooth, may be crenulate adjacent to mantle bridge; mantle bridge separating supra-anal and excurrent apertures imperforate, length approximately 40% of supra-anal length.

Specimens examined: Green River, Kentucky (n = 1); Tennessee River, Tennessee (n = 2) (specimens previously preserved).

Glochidium Description

Length 218–240 µm; height 270–295 µm; without styliform hooks; outline subelliptical; dorsal margin slightly convex, ventral margin rounded, anterior margin straight above, convex below, posterior margin slightly convex, straight dorsally (Ortmann, 1912a; Surber, 1912; Hoggarth, 1993, 1999).

Similar Species

Obovaria retusa may superficially resemble some *Obovaria subrotunda* but the umbo is much more inflated and oriented more anteriad. Pseudocardinal teeth of *O. retusa* are massive and divergent, whereas those of *O. subrotunda* are thinner and less divergent.

General Distribution

Obovaria retusa was historically widespread in the Ohio River from western Pennsylvania (Ortmann, 1919) downstream to the mouth of the Ohio River in Illinois and Kentucky (Cummings and Mayer, 1992). It occurred in the Cumberland River drainage downstream of Cumberland Falls, Kentucky and Tennessee (Cicerello et al., 1991; Parmalee and Bogan, 1998). In the Tennessee River drainage *O. retusa* was historically found from headwaters in eastern Tennessee downstream to the mouth of the Tennessee River (Parmalee and Bogan, 1998).

Alabama and Mobile Basin Distribution

All Alabama records of *Obovaria retusa* are from the Tennessee River, where it historically occurred across the state.

Obovaria retusa may be extirpated from Alabama. The most recent confirmed record of this species was from Muscle Shoals in 1962. However, an unconfirmed report from Muscle Shoals in the 1990s offers hope that it remains in the state (Garner and McGregor, 2001).

Ecology and Biology

Obovaria retusa is primarily a species of medium to large rivers. It occurs in sand and gravel substrates in flowing water.

Obovaria retusa is a long-term brooder, gravid from late August to the following summer (Ortmann, 1909b, 1919). Its glochidial hosts are unknown.

Current Conservation Status and Protection

Obovaria retusa was considered endangered throughout its range by Stansbery (1970a) and Williams et al. (1993) and in Alabama by Stansbery (1976) and Lydeard et al. (1999). Garner et al. (2004) designated *O. retusa* a species of highest conservation concern in the state. It was listed as endangered under the federal Endangered Species Act in 1989.

Remarks

Obovaria retusa has been reported from all archaeological studies of prehistoric shell middens along the Tennessee River in Alabama (Morrison, 1942; Warren, 1975; Hughes and Parmalee, 1999). Morrison (1942) reported it to be more abundant in preimpoundment cull piles left by commercial harvesters than in prehistoric middens.

Synonymy

Unio retusa Lamarck, 1819. Lamarck, 1819:72

Type locality: Lamarck (1819) erroneously gave the type locality as Nova Scotia. Johnson (1969b) corrected this and designated the type locality as the Ohio River at Cincinnati, Ohio. Johnson (1969b) reported the length of the lectotype as 44 mm. The lectotype is in MNHN, Paris, and is uncataloged.

Obovaria torsa Rafinesque, 1820. Rafinesque, 1820:311, pl. 82, figs. 1–3

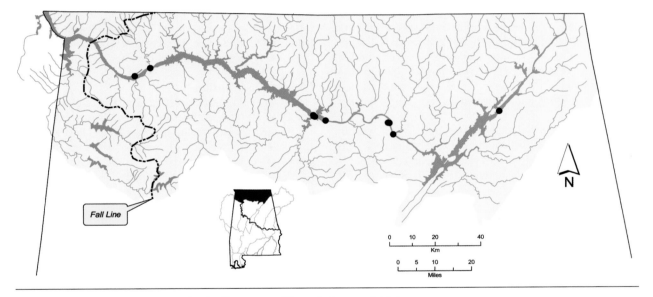

Distribution of *Obovaria retusa* in the Tennessee River drainage of Alabama.

Obovaria subrotunda (Rafinesque, 1820)
Round Hickorynut

Obovaria subrotunda – Length 37 mm, UF 4126. Paint Rock River near Huntsville, Madison County, Alabama. © Richard T. Bryant.

Shell Description

Length to 60 mm; moderately thick; compressed when young, becoming inflated with age; outline round to oval; posterior margin rounded, may be more narrowly rounded in females; anterior margin truncate to rounded; dorsal margin convex; ventral margin broadly rounded; posterior ridge weak, rounded, may be absent; posterior slope not steep, flat to slightly concave; umbo moderately inflated, elevated above hinge line, umbo sculpture weak bars, slightly sinuous, disappearing on posterior slope; periostracum usually olive to dark brown, may have yellowish band dorsally, occasional specimens mostly yellowish.

Pseudocardinal teeth erect, triangular, 2 divergent teeth in left valve, oriented approximately perpendicular to lateral teeth in adults, 1 tooth in right valve, may have small accessory denticle anteriorly and posteriorly; lateral teeth short, slightly curved, 2 in left valve, 1 in right valve, sometimes with incomplete second lateral tooth ventrally in right valve; interdentum short, moderately wide to narrow; umbo cavity moderately deep; nacre white, some individuals with pink tint inside of pallial line.

Soft Anatomy Description

Mantle tan, outside of apertures rusty tan to pale rusty orange, with grayish brown to black mottling that may form crude bands perpendicular to margin; visceral mass creamy white, pearly white adjacent to foot; foot pale tan to pale pinkish orange.

Females with weak mantle fold just ventral to incurrent aperture; fold length variable, 10–40% of shell length; margin with short, simple papillae, may grade to crenulations and disappear anteriorly; mantle adjacent to fold with rusty tan color band distally, dark brown or black color band proximally, coloration may be present only on posterior portion of fold. Ortmann (1912a) described the mantle margin just ventral to the incurrent aperture as "slightly lamellar and crenulated", with some brown coloration in females.

Gills tan; dorsal margin straight to slightly sinuous, steeply sloped from near umbo to apertures, ventral margin convex; gill length 35–55% of shell length; gill height 60–85% of gill length; outer gill height 70–100% of inner gill height; inner lamellae of inner gills completely connected to visceral mass.

Outer gills marsupial; glochidia held in posterior 50–90% of gill; marsupium distended when gravid; creamy white. Ortmann (1912a) reported gravid marsupia to be "essentially the same as *Obovaria retusa*", which he described as being restricted to a small section in the posterior half of the outer gills, but not extending completely to the posterior end, swollen and reniform.

Labial palps creamy white, may be tan distally; straight to slightly convex dorsally, convex ventrally, bluntly pointed to rounded distally; palp length 25–40% of gill length; palp height 50–100% of palp length; distal 40–60% of palps bifurcate.

Incurrent aperture longer than excurrent aperture; supra-anal aperture longer or equal to incurrent aperture.

Incurrent aperture length 15–20% of shell length; creamy white within, may have irregular dark brown lines basal to papillae; papillae in 2 rows, inner row larger, long, slender, simple, may have few bifid; papillae rusty tan to rusty orange, may have dark brown edges basally. Lea (1863d) reported incurrent aperture papillae to be brown.

Excurrent aperture length 10–15% of shell length; creamy white within, marginal color band rusty orange with wavy, dark brown lines perpendicular to margin or in reticulated pattern; margin with simple papillae, small but well-developed; papillae pale rusty tan to rusty orange.

Supra-anal aperture length approximately 15–20% of shell length; creamy white to pale tan within, may have thin band of rusty tan and grayish brown marginally; margin smooth; mantle bridge separating supra-anal and excurrent apertures imperforate, length variable, 15–60% of supra-anal length.

Specimens examined: Duck River, Tennessee (n = 4).

Glochidium Description

Length 170–180 μm; height 197–215 μm; without styliform hooks; outline subelliptical; dorsal margin slightly convex, ventral margin rounded, anterior margin convex, posterior margin obtusely straight above, convex below (Surber, 1915; Hoggarth, 1999).

Similar Species

Obovaria subrotunda superficially resembles *Obovaria olivaria* but is rounder in outline. The crest of the posterior pseudocardinal tooth in the left valve of adult *O. subrotunda* is oriented more dorsal to ventral than that of *O. olivaria*, which is oriented more anterior to posterior.

Obovaria subrotunda may also superficially resemble *Obovaria retusa* but the umbo is less inflated and oriented less anteriad. Pseudocardinal teeth of *O. subrotunda* are much less massive and divergent than those of *O. retusa*.

Obovaria subrotunda may superficially resemble *Fusconaia ebena* but is typically rounder in outline with a broader, less anteriorly oriented umbo. Pseudocardinal teeth of *O. subrotunda* are aligned approximately dorsal to ventral, and those of *F. ebena* are oriented more anterior to posterior. The umbo cavity of *F. ebena* is much more compressed than that of *O. subrotunda*.

General Distribution

Obovaria subrotunda is known from parts of the Great Lakes Basin, including lakes Erie and St. Clair (Goodrich and van der Schalie, 1932; La Rocque and Oughton, 1937). In the Ohio River drainage *O. subrotunda* ranges from headwaters in western Pennsylvania (Ortmann, 1909a) downstream to the mouth of the Ohio River in Illinois and Kentucky (Cummings and Mayer, 1992). It is widespread in the Cumberland River drainage downstream of Cumberland Falls, Kentucky and Tennessee (Cicerello et al., 1991; Parmalee and Bogan, 1998). In the Tennessee River drainage *O. subrotunda* is known from headwaters in eastern Tennessee downstream to the mouth of the Tennessee River (Parmalee and Bogan, 1998).

Alabama and Mobile Basin Distribution

Obovaria subrotunda historically occurred across northern Alabama in the Tennessee River and some tributaries.

Obovaria subrotunda has been extirpated from Alabama, with the exceptions of Bear Creek in Colbert County and the Paint Rock River system.

Ecology and Biology

Obovaria subrotunda occurs in medium to large rivers as well as some large lakes (e.g., lakes Erie and St. Clair), usually in sand and gravel substrates. In rivers it is often found in shoals.

Obovaria subrotunda is a long-term brooder, gravid from September through at least May of the following summer (Ortmann, 1919). A female observed discharging glochidia in early August was described by Ortmann (1919) as "exceptionally belated". Glochidial hosts of *O. subrotunda* are unknown.

Current Conservation Status and Protection

Obovaria subrotunda was considered a species of special concern throughout its range by Williams et al. (1993). In Alabama it was considered endangered by Stansbery (1976) and of special concern by Lydeard et al. (1999). Garner et al. (2004) designated *O. subrotunda* a species of highest conservation concern in the state.

Remarks

Obovaria subrotunda has been reported from prehistoric shell middens along the Tennessee River in northern Alabama (Morrison, 1942; Warren, 1975; Hughes and Parmalee, 1999).

Clinal variation in *Obovaria subrotunda* shell morphology has resulted in description of several synonyms. It tends to be more inflated in large rivers than in headwaters and tributaries.

Synonymy

Obliquaria (*Rotundaria*) *subrotunda* Rafinesque, 1820. Rafinesque, 1820:308, pl. 81, figs. 21–23 [original figure not herein reproduced]

Type locality: The Ohio and its tributary rivers. Neotype, ANSP 20254, length 27 mm, designated by Johnson and Baker (1973), is from Kentucky River.

Unio (*Aximedia*) *levigata* Rafinesque, 1820. Rafinesque, 1820:296, pl. 80, figs. 11–13

Obovaria striata Rafinesque, 1820. Rafinesque, 1820:311

Unio orbiculatus Hildreth, 1828. Hildreth, 1828:284, fig. 15

Unio circulus Lea, 1829. Lea, 1829:433, pl. 9, fig 14; Lea, 1834b:47, pl. 9, fig. 14

Unio lens Lea, 1831. Lea, 1831:80, pl. 8, fig. 10; Lea, 1834b:90, pl. 8, fig. 10

Unio leibii Lea, 1862. Lea, 1862a:168; Lea, 1866:44, pl. 15, fig. 42; Lea, 1867b:48, pl. 15, fig. 42

Unio depygis Conrad, 1866. Conrad, 1866a:107, pl. 10, fig. 1

Obovaria lens var. *parva* Simpson, 1914. Simpson, 1914:294

Obovaria lens var. *elongata* Simpson, 1914. Simpson, 1914:294–295

Obovaria subrotunda globula Morrison, 1942. Morrison, 1942:360; Johnson, 1975:29, pl. 1, fig. 7

Type locality: [Tennessee River,] Tuscumbia, [Colbert County,] Alabama. Holotype, USNM 85789, length 27 mm.

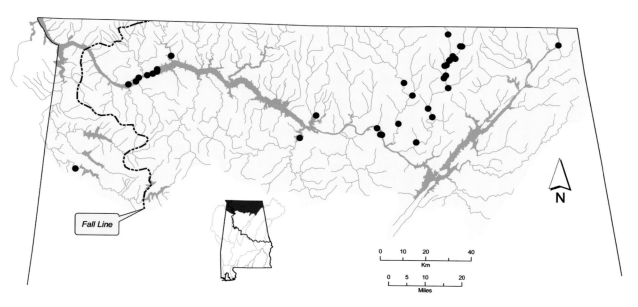

Distribution of *Obovaria subrotunda* in the Tennessee River drainage of Alabama.

Obovaria unicolor (Lea, 1845)
Alabama Hickorynut

Obovaria unicolor – Length 44 mm, UF 197609. Sipsey River, 3.5 to 4.5 miles downstream of County Road 2, Greene and Pickens counties, Alabama, 27 September 1991. © Richard T. Bryant.

Shell Description

Length to 70 mm; moderately thick; moderately inflated; outline round to oval; posterior margin narrowly to broadly rounded; anterior margin rounded; dorsal margin convex; ventral margin convex; posterior ridge low, rounded; posterior slope steep; umbo broad, inflated, elevated well above hinge line, umbo sculpture weak, imperfectly looped ridges; periostracum tawny to dark greenish brown or black, typically without rays.

Pseudocardinal teeth triangular, erect, 2 divergent teeth in left valve, 1 tooth in right valve, may have accessory denticle anteriorly and/or posteriorly; lateral teeth short, straight, 2 in left valve, 1 in right valve; interdentum short, usually narrow; umbo cavity moderately deep, not compressed; nacre white, occasionally pale pink.

Soft Anatomy Description

Mantle creamy white to tan, outside of apertures rusty tan or pale orange with grayish brown mottling; visceral mass pale tan; foot creamy white to tan.

Females with weak mantle fold just ventral to incurrent aperture, fold length 20–40% of shell length; margin with widely spaced, simple papillae; adjacent margin with rusty tan posteriorly, sometimes with black specks.

Gills tan; dorsal margin straight, ventral margin convex, outer gill may be slightly bilobed; gill length 55–70% of shell length; gill height 45–65% of gill length; outer gill height 60–80% of inner gill height; inner laminae of inner gills completely connected to visceral mass.

Outer gills marsupial; glochidia held in posterior 50% of gill, marsupium outline reniform, greatly distended when gravid; pale tan.

Labial palps tan; straight to slightly convex or slightly concave dorsally, convex ventrally, narrowly rounded to bluntly pointed distally; palp length 15–30% of gill length; palp height 60–70% of palp length; distal 40–50% of palps bifurcate.

Incurrent aperture slightly longer or equal to supra-anal aperture; incurrent and supra-anal apertures longer than excurrent aperture.

Incurrent aperture length 15–20% of shell length; creamy white to tan within, may have brown specks basal to papillae; papillae in 2 rows, long, mostly simple, may have occasional bifid papillae; papillae tan, may have dark brown specks or rusty cast distally.

Excurrent aperture length approximately 10% of shell length; creamy white to tan within, marginal color band rusty orange with irregular dark brown or rusty brown bands perpendicular to margin; margin with small, simple papillae; papillae tan to rusty brown.

Supra-anal aperture length 15–20% of shell length; creamy white to tan within, may have narrow band of rusty tan marginally; margin smooth; mantle bridge separating supra-anal and excurrent apertures imperforate, length 50–60% of supra-anal length.

Specimens examined: Lubbub Creek, Pickens County (n = 2); Sipsey River (n = 1) (some specimens previously preserved).

Glochidium Description

Length 160–180 μm; height 210–234 μm; without styliform hooks; outline subelliptical; dorsal margin slightly convex, ventral margin rounded, anterior and posterior margins convex (Hoggarth, 1999).

Similar Species

Obovaria unicolor and some *Obovaria jacksoniana* can be very difficult to distinguish because they overlap in shell morphology. *Obovaria unicolor* has an overall rounder appearance, whereas *O. jacksoniana* is usually more elongate and has a more narrowly rounded posterior margin. Nacre of *O. unicolor* is occasionally pink, but that of *O. jacksoniana* is always white.

General Distribution

Obovaria unicolor occurs in eastern Gulf Coast drainages from the Mobile Basin west to the Lake Pontchartrain drainage of Louisiana and Mississippi (Vidrine, 1993; Jones et al., 2005).

Alabama and Mobile Basin Distribution

Obovaria unicolor occurs in the Alabama, Coosa, Cahaba, Tombigbee and Black Warrior River drainages in Alabama, Georgia and Mississippi. It is found primarily below the Fall Line.

The distribution of *Obovaria unicolor* has declined since the mid-1900s. It appears to be extant in only a few tributaries of the upper Tombigbee River.

Ecology and Biology

Obovaria unicolor inhabits streams ranging in size from large creeks to rivers with sand and gravel substrates in slow to moderate current.

Obovaria unicolor is a long-term brooder, gravid from August to June of the following summer, with glochidia fully developed by November (Haag and Warren, 2003). Glochidial discharge has been reported to occur from April to June (Haag and Warren, 2001). Glochidial hosts of *O. unicolor*, determined by laboratory trials, include *Ammocrypta beanii* (Naked Sand Darter), *Ammocrypta meridiana* (Southern Sand Darter) and *Etheostoma artesiae* (Redspot Darter) (Percidae) as primary hosts and *Etheostoma nigrum* (Johnny Darter), *Etheostoma swaini* (Gulf Darter), *Percina nigrofasciata* (Blackbanded Darter) and *Percina sciera* (Dusky Darter) (Percidae) as secondary hosts (Haag and Warren, 2003).

Current Conservation Status and Protection

Obovaria unicolor was listed as a species of special concern throughout its range by Williams et al. (1993). In Alabama Stansbery (1976) considered it endangered, but Lydeard et al. (1999) listed it as a species of special concern. Garner et al. (2004) designated *O. unicolor* a species of highest conservation concern in the state.

Remarks

The conchological similarity of *Obovaria unicolor* and *Obovaria jacksoniana* has prompted questions of the validity of the two species. However, a detailed conchological and electrophoretic comparison found the two to be distinct (Hoggarth, 1980).

Synonymy

Unio unicolor Lea, 1845. Lea, 1845:163; Lea, 1848a:74, pl. 4, fig. 12; Lea, 1848b:48, pl. 4, fig. 12
Type locality: [Black Warrior River,] Tuscaloosa, [Tuscaloosa County,] Alabama, B.W. Budd, M.D. Holotype, USNM 85750, length 27 mm.

Unio uber Conrad, 1866. Conrad, 1866b:279, pl. 15, fig. 16
Type locality: Alabama River [, Alabama]. Type specimen not found, also not found by Johnson and Baker (1973). Conrad (1866b) gave no measurement of the figured shell in the original description.

Unio tinkeri B.H. Wright, 1899. B.H. Wright, 1899:7; Simpson 1900a:78, pl. 4, fig. 3
Type locality: Tombigbee River, Alabama, B.H. Wright. Lectotype, USNM 159193, length 48 mm, designated by Johnson (1967b).

Obovaria nux Simpson, 1914. Simpson, 1914:297; Johnson, 1975:16, pl. 1, fig. 11
Type locality: Tombigbee River, Moscow, [Sumter County,] Alabama; Cannisaria Lake, [De Soto Parish,] Louisiana. Holotype, USNM 152968, length 39 mm, is from Tombigbee River, Moscow, Sumter County, Alabama.

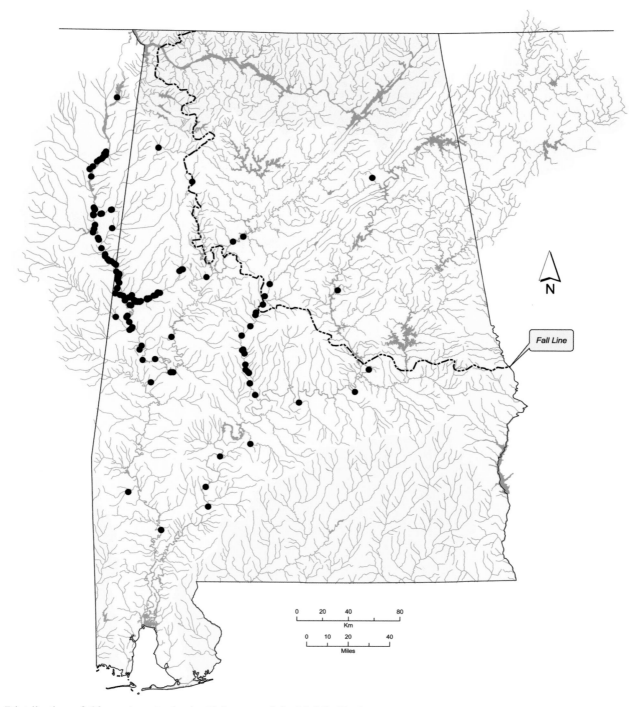

Distribution of *Obovaria unicolor* in Alabama and the Mobile Basin.

Genus *Pegias*

Pegias Simpson, 1900, is endemic to the Cumberland and Tennessee River drainages. In Alabama it is restricted to the Tennessee River drainage. *Pegias* is a monotypic genus (Turgeon et al., 1998).

This genus was included in a monograph of the tribe Alasmidontini by Clarke (1981b). It is one of the smallest species of North American unionids, seldom exceeding 35 mm in length.

Type Species
Margaritana fabula Lea, 1838

Diagnosis
Shell thickened anteriorly; posterior ridge sharp, may be adorned with irregular, subconcentric knobs; wide radial sulcus just anterior of posterior ridge; umbo sculpture subconcentric corrugations; disk without sculpture; periostracum typically decorticated in adults; sexual dimorphism expressed as slightly wider posterior slope in females than males, resulting in more truncate or emarginate posterior margin; well-developed pseudocardinal teeth present, lateral teeth absent.

Inner lamellae of inner gills connected to visceral mass only anteriorly; mantle margin marked with squarish blotches on exterior surface; excurrent aperture smooth; mantle bridge separating excurrent and supra-anal apertures variable in length; outer gills marsupial; glochidium with styliform hooks (Simpson, 1900b; Ortmann, 1914; Clarke, 1981b).

Synonymy
None recognized.

Pegias fabula (Lea, 1838)
Littlewing Pearlymussel

Pegias fabula – Upper figure: length 28 mm, UMMZ 105469. Elk River, [Limestone County,] Alabama. Middle figure: length 30 mm, UF 63732. Tennessee. Lower figure: juvenile, length 13 mm, UMMZ 105468. Elk River, Estill Springs, Franklin County, Tennessee. © Richard T. Bryant.

Shell Description

Length to 36 mm; thin, thicker anteriorly; moderately compressed; outline irregularly oval to subquadrate; posterior margin rounded to obliquely truncate or bilobed; anterior margin rounded; dorsal margin slightly convex; ventral margin straight to slightly convex, often slightly concave posteriorly; posterior ridge well-defined, often adorned with irregular subconcentric knobs that persist from umbo sculpture, separated from disk by shallow sulcus; posterior slope moderately steep, often with low secondary radial ridge and shallow intermediate sulcus; sexual dimorphism expressed as slightly wider posterior slope in females, forming more truncate or emarginate posterior margin; umbo low to moderately inflated, elevated above hinge line, umbo sculpture prominent, subconcentric to slightly double-looped ridges, often forming knobs on posterior ridge; periostracum tawny to brown, often with variable green rays, periostracum of adults often completely decorticated.

Pseudocardinal teeth irregular, triangular, 1 well-developed tooth in left valve, large relative to shell, with accessory denticle anteriorly, 1 large triangular tooth in right valve; lateral teeth absent, some individuals with short irregular ridge in each valve; umbo cavity variable, usually deep; nacre white, may be tan or gold posteriorly and in umbo cavity.

Soft Anatomy Description

Mantle tan, with series of irregular, somewhat rectangular, dark brown blotches on posterior 50% of outside surface; visceral mass pale tan; foot creamy white.

Gills pale tan; dorsal margin straight, ventral margin elongate convex; gill length approximately 65% of shell length; gill height approximately 50% of gill length; outer gill height approximately 70% of inner gill height; inner lamellae of inner gills connected to visceral mass only anteriorly.

No gravid females were available for marsupium description. Ortmann (1914) reported outer gills to be marsupial, having secondary water tubes and being distended when gravid. Glochidia fill the marsupia in a mass but do not form distinct conglutinates.

Labial palps pale tan; straight dorsally, curved ventrally, bluntly pointed distally; palp length approximately 20% of gill length; palp height approximately 75% of palp length; distal 50% of palps bifurcate.

Incurrent and excurrent apertures approximately equal in length; supra-anal aperture shorter than incurrent and excurrent apertures. Clarke (1981b) reported the supra-anal aperture to be approximately twice as long as incurrent and excurrent apertures.

Incurrent aperture length approximately 20% of shell length; pale tan within, without coloration basal to papillae; papillae in 2 rows, inner row larger, long, slender, simple; papillae reddish tan.

Excurrent aperture length approximately 20% of shell length; pale tan within, marginal color band reddish tan with a few irregular pale tan blotches; margin smooth but undulating. Clarke (1981b) reported the excurrent aperture to be 15% of shell length, based on a single specimen from the North Fork Holston River, Virginia. Ortmann (1914) reported the excurrent

aperture of North Fork Holston *Pegias fabula* to be "entirely black" inside and to have a crenulate margin.

Supra-anal aperture length approximately 15% of shell length; pale tan within, with thin reddish tan marginal band; margin smooth; mantle bridge separating supra-anal and excurrent apertures absent. Ortmann (1914) reported North Fork Holston River *Pegias fabula* to have a well-developed but short mantle bridge and the supra-anal aperture to be "entirely black" inside.

Specimen examined: Big South Fork Cumberland River, Tennessee (n = 1) (specimen previously preserved).

Glochidium Description

Length 380–400 μm; height 319–360 μm; with styliform hooks; outline irregularly oval; dorsal margin straight, ventral margin irregular, with slight flattened area just anterior to its rounded ventral terminus, anterior and posterior margins convex, anterior margin distinctly more deeply convex than posterior margin (Ortmann, 1914; Clarke, 1981b; Hoggarth, 1993, 1999).

Similar Species

Pegias fabula may superficially resemble *Alasmidonta viridis*. However, *P. fabula* has a biangulate posterior ridge, usually adorned with irregular knobs, whereas that of *A. viridis* is not biangulate or knobbed.

General Distribution

Pegias fabula is endemic to the Cumberland and Tennessee River drainages. It is known from the Cumberland River drainage downstream of Cumberland Falls, Kentucky and Tennessee (Cicerello et al., 1991; Parmalee and Bogan, 1998; Ahlstedt et al., 2005). In the Tennessee River drainage *P. fabula* is known from headwaters downstream to a tributary at the upstream end of Muscle Shoals, northwestern Alabama (Ahlstedt and Saylor, 1998; Parmalee and Bogan, 1998). It has also been recovered from archaeological excavations along the Elk and Duck rivers, central Tennessee (Bogan, 1990).

Alabama and Mobile Basin Distribution

There are few records of *Pegias fabula* from Alabama, all from Bluewater Creek, Lauderdale County, and the Elk River.

The most recent record of *Pegias fabula* from Alabama was in the early 1900s. It appears to be extirpated from the state.

Ecology and Biology

Pegias fabula is a species of creeks and small to medium rivers. Parmalee and Bogan (1998) give "clear, cool, high-gradient streams" as its habitat. However,

Bluewater Creek, Lauderdale County, Alabama, from which most of the Alabama records came, has only moderate gradient. *Pegias fabula* remains well-buried in the substrate or under large rocks for most of the year (Ahlstedt and Saylor, 1998).

Pegias fabula is a long-term brooder, becoming gravid in September and presumably remaining gravid until the following spring or summer (Ortmann, 1914, 1921). It has been observed lying exposed on the substrate surface in late September and early October in Cumberland and Tennessee River tributaries (Blankenship, 1971; Starnes and Starnes, 1980; Ahlstedt and Saylor, 1998). This behavior is thought to be related to reproductive activities. Those reported by Ahlstedt and Saylor (1998) were gravid females, but Ahlstedt et al. (2005) reported nongravid *P. fabula* lying in a similar posture during March.

Fishes reported to serve as glochidial hosts of *Pegias fabula* include *Cottus baileyi* (Black Sculpin) (Cottidae); and *Etheostoma baileyi* (Emerald Darter) and *Etheostoma blennioides* (Greenside Darter) (Percidae) (Layzer and Anderson, 1992; Jones et al., 2003).

Current Conservation Status and Protection

Pegias fabula was recognized as endangered throughout its range by Stansbery (1970a) and Williams et al. (1993). In Alabama it was considered endangered by Stansbery (1976) and listed as extirpated from the state by Lydeard et al. (1999) and Garner et al. (2004). In 1988 *P. fabula* was listed as endangered under the federal Endangered Species Act.

Remarks

Pegias fabula is one of the smallest mussel species. This, along with its habit of remaining well-buried in the substrate for most of the year, may account for the paucity of museum material. However, current rarity of this species is not debatable, as extirpation of many populations is well-documented.

Synonymy

Margarita (*Margaritana*) *fabula* Lea, 1836, *nomen nudum*. Lea, 1836:46
Margaritana fabula Lea, 1838. Lea, 1838a:44, pl. 13, fig. 39; Lea, 1838c:44, pl. 13, fig. 39
Type locality: Cumberland River, Tennessee. Lectotype, USNM 86325, length 20 mm, designated by Johnson (1974).

Margaritana curreyiana Lea, 1840. Lea, 1840:288; Lea, 1842b:223, pl. 18, fig. 40; Lea, 1842c:61, pl. 18, fig. 40
Unio propecaelatus De Gregorio, 1914. De Gregorio, 1914:30, pl. 8, figs. 1a–d

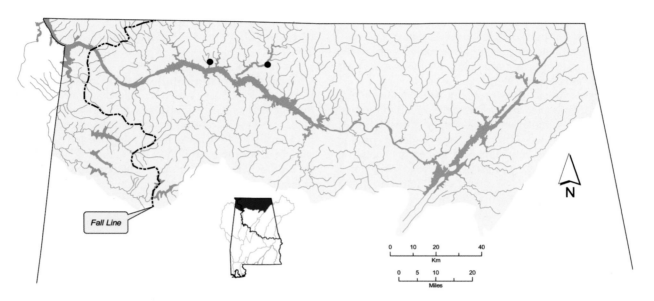

Distribution of *Pegias fabula* in the Tennessee River drainage of Alabama.

Genus *Plectomerus*

Plectomerus Conrad, 1853, is restricted to Gulf Coast drainages from the Mobile Basin to eastern Texas, the Mississippi Basin south of the Ohio River, and lower reaches of the Ohio and Tennessee rivers. *Plectomerus* is a monotypic genus (Turgeon et al., 1998). In Alabama it is restricted to the Mobile Basin.

Conrad (1853) erected *Plectomerus* with ten plicate species but failed to designate a type species for the genus. Subsequently, Ortmann and Walker (1922a) designated *Unio trapezoides* Lea, 1831 (= *P. dombeyanus* (Valenciennes, 1827)), as the type species of *Plectomerus*. Thiele (1934) considered *Unio dombeyana* Valenciennes, 1827, to belong to *Amblema* Rafinesque, 1820. Davis and Fuller (1981) noted that *Plectomerus, Megalonaias* Utterback, 1915, and *Amblema* are very closely related and should be combined into a single genus. Phylogenetic trees of Serb et al. (2003) placed *Plectomerus* and *Elliptoideus* Frierson, 1927, as sister taxa but with only weak support. In a more extensive study of North American unionids *Plectomerus* and *Elliptoideus* appear in widely separated clades (Campbell et al., 2005).

Type Species

Unio trapezoides Lea, 1831 (= *Unio dombeyana* Valenciennes, 1827)

Diagnosis

Shell rectangular to rhomboidal; inflated; thick; umbo moderately inflated, elevated slightly above hinge line; posterior ridge well-developed, with a few strong, irregular corrugations; disk sculptured with oblique plications posteriorly; periostracum brown to black; umbo cavity wide, deep; nacre purplish.

Incurrent aperture papillae simple; inner lamellae of inner gills usually connected to visceral mass only anteriorly; all 4 gills marsupial; marsupium not extended ventrally below original gill edge when gravid; glochidium without styliform hooks (Simpson, 1914; Frierson, 1927; Haas, 1969a, 1969b).

Synonymy

None recognized.

Plectomerus dombeyanus (Valenciennes, 1827)
Bankclimber

Plectomerus dombeyanus – Length 115 mm, UF 375969. Isaac Creek near entrance into Claiborne Reservoir, [Alabama River,] 500 m north of Claiborne Dam, Monroe County, Alabama, 25 September 1988. © Richard T. Bryant.

Shell Description

Length to 150 mm; thick; moderately inflated; outline rectangular to rhomboidal; posterior margin obliquely truncate, ending in point posterioventrally; anterior margin rounded; dorsal margin straight to slightly convex, may have very low dorsal wing; ventral margin straight to slightly convex; posterior ridge high, sharp; posterior slope moderately steep, flat to slightly concave; posterior ridge and posterior slope sculptured with parallel plications, originating on crest of posterior ridge, extending to dorsal and posterior margin, disk with shallow oblique ridges that may be obscure in old individuals, posterior 75% of shell with variable subradial wrinkles; umbo broad, may be slightly elevated above hinge line, umbo sculpture irregular, double-looped ridges; periostracum greenish brown to brown, usually darkening to black with age.

Pseudocardinal teeth thick, erect, triangular, 2 divergent teeth in left valve, may be separate dorsally, 1 tooth in right valve, usually with thin accessory denticle anteriorly, sometimes with accessory denticle posteriorly; lateral teeth long, slightly curved, 2 in left valve, 1 in right valve; interdentum moderately long, narrow; umbo cavity wide, shallow; nacre purple, may be coppery.

Soft Anatomy Description

Mantle tan to light brown; visceral mass pearly white to creamy white or tan; foot tan, may be creamy white proximally.

Gills tan to light brown; dorsal margin straight to slightly sinuous, ventral margin long, straight to slightly convex; gill length 55–65% of shell length; gill height 30–40% of gill length; outer gill height 65–70% of inner gill height; inner lamellae of inner gills connected to visceral mass only anteriorly.

All 4 gills marsupial; glochidia held across entire gill; marsupium padded when gravid; creamy white. Frierson (*in* Ortmann, 1912a) reported that sometimes only two gills are used as marsupia, but Ortmann (1912a) described all four gills as being "built to receive eggs and serve as marsupia", suggesting that in the inner gills glochidia are brooded primarily in the distal half.

Labial palps creamy white to light golden brown; broadly connected to mantle, straight dorsally, convex ventrally, rounded distally; palp length 25–30% of gill length; palp height 45–65% of palp length; distal 15–35% of palps bifurcate.

Incurrent aperture longer than excurrent aperture; supra-anal aperture longer than incurrent aperture.

Incurrent aperture length 15–20% of shell length; creamy white within, without coloration basal to papillae; papillae in 2 rows, inner row long, thick, simple, outer row short, simple; papillae creamy white, occasionally with sparse black pigment between papillae. Lea (1863d) reported incurrent aperture papillae to be brown.

Excurrent aperture length 10–15% of shell length; creamy white within, marginal color band tan to rusty orange with irregular rusty brown or dark brown bands perpendicular to margin; margin crenulate, may have very small papillae on portions of aperture margin. Lea (1863d) reported the excurrent aperture to have "very minute colored papillae".

Supra-anal aperture length 20–30% of shell length; creamy white within, without marginal coloration; margin smooth; mantle bridge separating supra-anal and excurrent apertures imperforate, length less than 5% of supra-anal length, sometimes absent. Ortmann (1912a) reported the mantle bridge to be moderately long.

Specimens examined: Tombigbee River (n = 3).

Glochidium Description

Length 223–231 µm; height 238–259 µm; without styliform hooks; outline subelliptical; dorsal margin straight, ventral margin rounded, anterior and posterior margins convex, anterior margin slightly more convex than posterior margin (Hoggarth, 1999).

Similar Species

Plectomerus dombeyanus may resemble *Megalonaias nervosa* but is more obliquely truncate posteriorly, with a more pronounced posterior ridge that ends in a point posterioventrally. *Plectomerus dombeyanus* also has a shallow umbo cavity and purple nacre compared to a deep, compressed umbo cavity and white nacre in *M. nervosa*.

Some small to medium *Plectomerus dombeyanus* may resemble *Quadrula verrucosa*. However, the posterior ridge of *Q. verrucosa* is usually not as high as that of *P. dombeyanus* and often extends below the ventral margin of the shell, giving it a slightly arcuate outline. The posterior slope of *Q. verrucosa* is often lower and flatter than that of *P. dombeyanus*. The umbo cavity of *Q. verrucosa* is deeper and more compressed than that of *P. dombeyanus*. Shell nacre of *Q. verrucosa* is typically white, though it is occasionally purple in some Mobile Basin populations whereas nacre of *P. dombeyanus* is generally some shade of purple.

General Distribution

Plectomerus dombeyanus occurs in the lower Mississippi Basin from near the mouth of the Ohio River, Kentucky and Missouri (Oesch, 1995), south to Louisiana (Vidrine, 1993). It appears to have expanded its range into the lower Tennessee River in Kentucky and Tennessee since its impoundment (Pharris et al., 1982; Parmalee and Bogan, 1998). *Plectomerus dombeyanus* occurs in Gulf Coast drainages from the Mobile Basin of Alabama west to the San Jacinto River in eastern Texas (Howells et al., 1996).

Alabama and Mobile Basin Distribution

Plectomerus dombeyanus is found in the Mobile Basin where almost all records are from the Coastal Plain. A recent discovery of a shell of this species in the Escambia River is most likely an introduction as there is no evidence of an established population.

Plectomerus dombeyanus remains widespread in the Alabama and Tombigbee rivers, where it may be locally common.

Ecology and Biology

Plectomerus dombeyanus inhabits medium to large rivers and oxbow lakes. It occurs in areas with no current as well as in swift current in tailwaters of dams. *Plectomerus dombeyanus* is often most common along channel margins and may be found embedded in steep clay or mud slopes, or among cobble and boulders, a considerable distance up from the channel bottom. It also occurs in sand and gravel substrates.

Plectomerus dombeyanus is a short-term brooder. Gravid females have been reported from May to September (Frierson, 1904; Howells, 2000). In Texas gravid females were reported from July through September, but mature glochidia were observed only in July (Howells, 2000). Glochidial hosts of *P. dombeyanus* are unknown.

Current Conservation Status and Protection

Plectomerus dombeyanus was considered currently stable throughout its range by Williams et al. (1993) and in Alabama by Lydeard et al. (1999). Garner et al. (2004) designated it a species of lowest conservation concern in the state.

Plectomerus dombeyanus may be legally harvested in Alabama. It is protected by state regulations, with the most important being a minimum size limit of 3 inches (76 mm) in shell height.

Remarks

The presence of *Plectomerus dombeyanus* in lower reaches of the Coosa River above the Fall Line appears to be the result of a recent colonization. That river was thoroughly sampled prior to its impoundment, but there are no historical records of *P. dombeyanus*. A single individual was encountered in tailwaters of H. Neely Henry Dam in 1996 (J.T. Garner, personal observation). A healthy population, including gravid females, was discovered in tailwaters of Lay Dam in 2004 (J.T. Garner, personal observation). The means by which *P. dombeyanus* colonized the Coosa River upstream of the Fall Line is unclear, since none of the dams on the river are equipped with locks or mechanisms for fish passage.

All records of *Plectomerus dombeyanus* from the Escambia River drainage appear to be recent introductions. Two records are represented by single valves. The third record was of live individuals found near a public boat ramp, but they were in the company of three other Mobile Basin species and had apparently been released (D.N. Shelton, personal communication). There is no evidence of a reproducing population of *P. dombeyanus* in the Escambia River drainage.

Synonymy

Unio crassidens var. a Lamarck, 1819. Lamarck, 1819:71

Unio dombeyana Valenciennes, 1827. Valenciennes, 1827:227, pl. 53, figs. 1a, b

Type locality: Peru. Type specimen not found. The reported type locality is erroneous. Measurement of figured holotype not given in original description.

Unio interruptus Say, 1831. Say, 1831a:525

Unio trapezoides Lea, 1831. Lea, 1831:69, pl. 3, fig. 1; Lea, 1834b:79, pl. 3, fig. 1

Quadrula trapezoides var. *pentagonoides* Frierson, 1902. Frierson, 1902:39

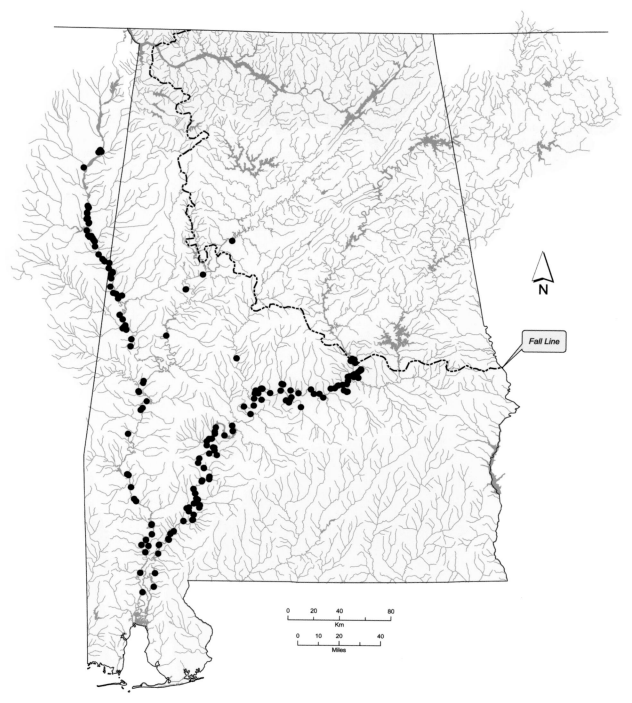

Distribution of *Plectomerus dombeyanus* in Alabama and the Mobile Basin

Genus *Plethobasus*

Plethobasus Simpson, 1900, is restricted to the Mississippi Basin, primarily in the Ohio, Cumberland and Tennessee River drainages but also in a few western tributaries of the central Mississippi Basin. It contains three species (Turgeon et al., 1998), all of which are known from the Tennessee River in Alabama.

Type Species

Unio aesopus Green, 1827 (= *Obliquaria* (*Quadrula*) *cyphya* Rafinesque, 1820)

Diagnosis

Shell ovate; inflated; umbo moderately elevated above hinge line; disk sculptured with elongate nodules, lachrymous tubercles or pustules; hinge plate well-developed; umbo cavity shallow to moderately deep.

Excurrent aperture smooth to crenulate; inner lamellae of inner gills connected to visceral mass only anteriorly; mantle bridge separating excurrent and supra-anal apertures short; outer gills marsupial; glochidia held across entire gill; marsupium slightly padded, not extended ventrally below original gill edge when gravid; glochidium without styliform hooks (Simpson, 1900b; Ortmann, 1912a; Haas 1969b).

Synonymy

None recognized.

Plethobasus cicatricosus (Say, 1829)
White Wartyback

Plethobasus cicatricosus – Upper figure: length 80 mm, UNA 168.1. Lower figure: length 61 mm, UNA 168.2. Tennessee River, river mile 249–251, Lauderdale County, Alabama, 13 May 1965. © Richard T. Bryant.

Shell Description

Length to 109 mm; thick; moderately inflated; outline oval to subtriangular; posterior margin broadly to narrowly rounded; anterior margin rounded; dorsal margin straight to convex; ventral margin convex; posterior ridge high and sharp near umbo, disappearing posterioventrally; posterior slope moderately steep; oblique medial swelling comprised of variable low, lachrymous tubercles and concentric nodules, sculpture may be weak in young individuals; some individuals with shallow sulcus just posterior of medial swelling; umbo narrow, moderately inflated, elevated above hinge line, turned slightly anteriad, umbo sculpture unknown; periostracum cloth-like, tawny to brown, without rays.

Pseudocardinal teeth thick, triangular, elevated, 2 divergent teeth in left valve, 1 tooth in right valve, may have accessory denticle anteriorly and/or posteriorly; lateral teeth thick, slightly curved, 2 in left valve, 1 in right valve; interdentum short, wide; umbo cavity moderately shallow, may be deeper in left valve; nacre white.

Soft Anatomy Description

Soft anatomy unknown.

Glochidium Description

Glochidium unknown.

Similar Species

Plethobasus cicatricosus most closely resembles *Plethobasus cyphyus*, but the nodules are smaller and tubercles more numerous in *P. cicatricosus*. *Plethobasus cicatricosus* is generally less elongate than *P. cyphyus*, with a more broadly rounded ventral margin. The posterior margin of *P. cyphyus* is often bluntly pointed, whereas that of *P. cicatricosus* is usually rounded. The periostracum of *P. cicatricosus* is cloth-like, whereas that of *P. cyphyus* is typically smooth and shiny. Subadult *P. cyphyus* are usually more yellow than subadult *P. cicatricosus*.

Plethobasus cicatricosus may superficially resemble *Fusconaia ebena*, *Fusconaia subrotunda* or *Pleurobema sintoxia* in overall shape, but none of those species are sculptured. The periostracum of *P. cicatricosus* is usually not as dark as those of *F. ebena*, *F. subrotunda* and *P. sintoxia*.

General Distribution

Plethobasus cicatricosus is known only from the Ohio, Cumberland and Tennessee River drainages. In the Ohio River it is known from southwestern Ohio downstream to near the mouth of the Cumberland River, northwestern Kentucky (Cicerello et al., 1991; Cummings and Mayer, 1992). It was also found in the Wabash River system, Illinois and Indiana. In the Cumberland River drainage *P. cicatricosus* ranged as far upstream as the Caney Fork in Tennessee (Parmalee and Bogan, 1998). In the Tennessee River drainage it historically occurred from the Clinch and Holston rivers in eastern Tennessee to the mouth of the Tennessee River (Parmalee and Bogan, 1998).

Alabama and Mobile Basin Distribution

Plethobasus cicatricosus historically occurred in the Tennessee River across northern Alabama.

Plethobasus cicatricosus appears to have been extirpated from its entire range with the exception of the Tennessee River in Wilson Dam tailwaters.

Ecology and Biology

Plethobasus cicatricosus is a species of flowing water in medium to large rivers. It is generally found in substrates composed of a mixture of sand and gravel, where current keeps silt accumulation to a minimum.

Plethobasus cicatricosus is presumably a short-term brooder, gravid during spring and summer. Its glochidial hosts are unknown.

Current Conservation Status and Protection

Plethobasus cicatricosus was considered endangered throughout its range by Stansbery (1970a) and Williams et al. (1993) and in Alabama by Stansbery (1976) and Lydeard et al. (1999). Garner et al. (2004) designated *P. cicatricosus* a species of highest conservation concern in the state. This species was listed as endangered under the federal Endangered Species Act in 1976.

Remarks

Plethobasus cicatricosus has apparently been eliminated from its entire range with the exception of approximately 18 km of the Tennessee River downstream of Wilson Dam (Garner and McGregor, 2001). Though evidence of recent recruitment has been observed, *P. cicatricosus* remains very rare. Its existence is tenuous and dependent on the quality and quantity of water released from Wilson Dam, which is controlled by TVA. In 2007 TVA altered the hydropower generation schedule at Wilson Dam to provide more consistent flows and better water quality to mussels and other species in the tailwaters.

Plethobasus cicatricosus has been reported to be rare in prehistoric shell middens along the Tennessee River in northern Alabama (Morrison, 1942; Warren, 1975; Hughes and Parmalee, 1999).

Synonymy

Unio cicatricosus Say, 1829. Say, 1829:292
Type locality: Wabash River. Neotype, SMF 4320, length 109 mm, designated by Haas (1930) is from the U.S.
Unio varicosus Lea, 1829, *non* Lamarck, 1819. Lea, 1829:424; Lea, 1831:90, pl. 11, fig. 20; Lea, 1834b:100, pl. 11, fig. 20
Unio detectus Frierson, 1911. Frierson, 1911:52, pl. 2, lower fig., pl. 3, upper fig.
Unio cicatricoides Frierson, 1911. Frierson, 1911:53, pl. 2, upper fig.

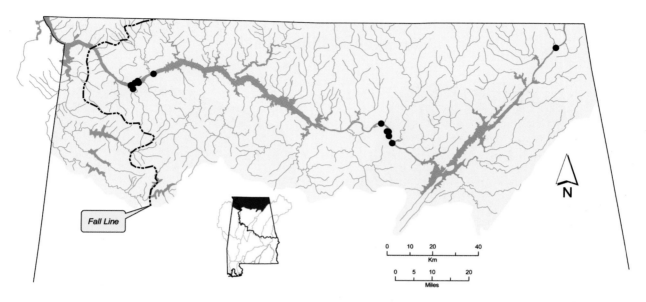

Distribution of *Plethobasus cicatricosus* in the Tennessee River drainage of Alabama.

Plethobasus cooperianus (Lea, 1834)
Orangefoot Pimpleback

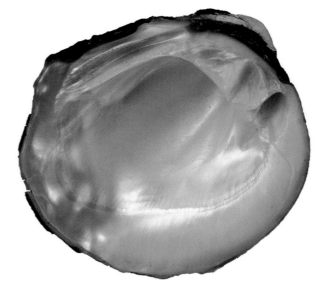

Plethobasus cooperianus – Length 69 mm, UF 229727. Tennessee River, Decatur, Morgan County, Alabama. © Richard T. Bryant.

Shell Description

Length to 90 mm; thick; moderately inflated; outline round to subtriangular; posterior margin broadly to narrowly rounded; anterior margin rounded, may be straight; dorsal margin convex; ventral margin convex; posterior ridge low, rounded; posterior slope not steep, wide, flat; posterior 75%, including posterior slope, sculptured with numerous round, oblong and lachrymous tubercles; umbo narrow, moderately inflated, elevated above hinge line, directed anteriad, umbo sculpture unknown; periostracum often with fine wrinkles associated with tubercles, reddish brown to brown, darkening with age, subadults often with green rays that usually become obscure with age.

Pseudocardinal teeth thick, triangular, low, 2 divergent teeth in left valve, 1 tooth in right valve, may have accessory denticle anteriorly and/or posteriorly; lateral teeth thick, straight to slightly curved, 2 in left valve, 1 in right valve; interdentum short, wide; umbo cavity deep, compressed; nacre white to pale pink.

Soft Anatomy Description

No material was available for soft anatomy description. Ortmann (1912a) stated that the soft anatomy and color of *Plethobasus cooperianus* are "practically identical with that of *Plethobasus aesopus*" (= *Plethobasus cyphyus*). Wilson and Clark (1914) and Ortmann (1919) reported outer gills to be marsupial.

Glochidium Description

Glochidium unknown.

Similar Species

Plethobasus cooperianus most closely resembles *Cyclonaias tuberculata*, but *C. tuberculata* always has purple nacre, whereas that of *P. cooperianus* is typically white, or if colored, only pale pink. The posterior slope of *C. tuberculata* is often expanded into a moderately developed dorsal wing, which is absent or very poorly developed in *P. cooperianus*. Both *P. cooperianus* and *C. tuberculata* are heavily sculptured with tubercles, but *C. tuberculata* generally has a number of tubercles forming irregular, short, broken ridges, especially on the dorsal wing. Fine wrinkles are associated with some tubercles of *P. cooperianus* but are absent in *C. tuberculata*. Also, the foot of *P. cooperianus* is orange, whereas that of *C. tuberculata* is white or tan.

Plethobasus cooperianus may superficially resemble *Quadrula pustulosa*. However, *Q. pustulosa* almost always has a dark green umbo that expands into a broad green ray. *Plethobasus cooperianus* often has smaller and more numerous tubercles than *Q. pustulosa*. Fine wrinkles are associated with some tubercles of *P. cooperianus* but are absent in *Q. pustulosa*. Also, the foot of *P. cooperianus* is orange, whereas that of *Q. pustulosa* is white or tan.

General Distribution

Plethobasus cooperianus is known only from the Ohio, Cumberland and Tennessee River drainages. In the Ohio River it occurred from western Pennsylvania downstream to its mouth in Illinois and Kentucky (Ortmann, 1919; Cummings and Mayer, 1992). *Plethobasus cooperianus* occurred in the Cumberland River downstream of Cumberland Falls (Cicerello et al., 1991; Parmalee and Bogan, 1998). In the Tennessee River drainage *P. cooperianus* ranged from headwater tributaries in eastern Tennessee downstream to the mouth of the Tennessee River (Parmalee and Bogan, 1998).

Alabama and Mobile Basin Distribution

Plethobasus cooperianus historically occurred in the Tennessee River across northern Alabama and a few tributaries.

Plethobasus cooperianus may be extirpated from Alabama. However, it is possibly extant in tailwaters of Wilson and/or Guntersville dams.

Ecology and Biology

Plethobasus cooperianus occurs in flowing water of medium to large rivers. In the Tennessee River it is found primarily in substrates composed of a mixture of gravel and sand.

Plethobasus cooperianus is a short-term brooder. Gravid females have been reported in early June in the Cumberland River (Wilson and Clark, 1914). Its glochidial hosts are unknown.

Current Conservation Status and Protection

Plethobasus cooperianus was considered endangered throughout its range by Stansbery (1970a) and Williams et al. (1993) and in Alabama by Stansbery (1976) and Lydeard et al. (1999). Garner et al. (2004) designated *P. cooperianus* a species of highest conservation concern in the state. It was listed as endangered under the federal Endangered Species Act in 1976.

Remarks

Wilson and Clark (1914) reported *Plethobasus cooperianus* to be "not rare" in the Cumberland River during the early 1900s. However, its range has been greatly reduced with impoundment of the major rivers.

Plethobasus cooperianus has been found to be rare in prehistoric shell middens along the Tennessee River in northern Alabama (Morrison, 1942; Warren, 1975; Hughes and Parmalee, 1999).

Plethobasus cooperianus was named for William Cooper, Esq., "as a light acknowledgement of the many favours received in the way of communications, and the loan of specimens" (Lea, 1834a).

Synonymy

Unio cooperianus Lea, 1834. Lea, 1834a:61, pl. 8, fig. 21; Lea, 1834b:173, pl. 8, fig. 21

Type locality: Ohio River. Type specimen not found, also not found by Johnson (1974). Length of figured shell in original description reported as about 81 mm.

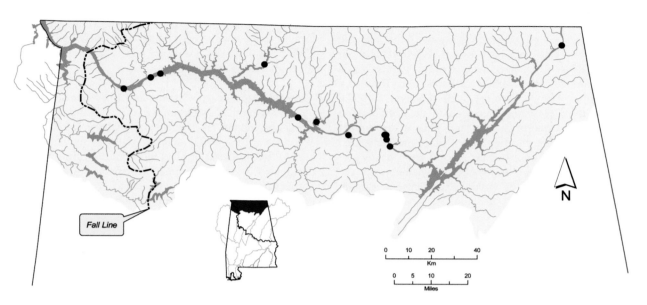

Distribution of *Plethobasus cooperianus* in the Tennessee River drainage of Alabama.

Plethobasus cyphyus (Rafinesque, 1820)
Sheepnose

Plethobasus cyphyus – Length 55 mm, UF 66454. Tennessee River, Florence, Lauderdale County, Alabama. © Richard T. Bryant.

Shell Description

Length to 122 mm; thick; moderately inflated; outline irregularly oval to subtriangular; posterior margin narrowly rounded to bluntly pointed; anterior margin rounded; dorsal margin convex; ventral margin convex, often forming an oblique angle; posterior ridge well-developed, becoming low and rounded posterioventrally; posterior slope moderately steep, flat; oblique medial ridge extends from umbo to ventral margin, comprised of variable low, elongate, concentric nodules and a few low irregular tubercles; most individuals with shallow sulcus between posterior and medial ridges; umbo narrow, moderately inflated, elevated above hinge line, umbo sculpture a few concentric ridges; periostracum smooth, shiny, yellow to tawny or brown, without rays.

Pseudocardinal teeth thick, triangular, 2 divergent teeth in left valve, 1 tooth in right valve, may have accessory denticle anteriorly; lateral teeth short, thick, slightly curved, 2 in left valve, 1 in right valve, may also have rudimentary lateral tooth ventrally in right valve; interdentum moderately long, wide; umbo cavity moderately shallow, deeper in left valve than right; nacre white.

Soft Anatomy Description

Mantle tan to gold or orange, usually bright orange outside of pallial line, outside of apertures some combination of orange, gold, tan and various shades of brown, usually mottled, sometimes in crude bands perpendicular to margin; visceral mass creamy white to tan, may have pinkish cast, sometimes becoming pearly white or orange adjacent to foot; foot usually bright orange, occasionally tan; adductor muscles may be bright orange.

Gills tan to gold or orange, may have brownish cast in places; dorsal margin slightly sinuous to slightly concave, ventral margin convex, sometimes deeply convex; gill length 55–65% of shell length; gill height usually 30–45% of gill length, occasionally as much as 70% of gill length; outer gill height usually 75–90% of inner gill height, occasionally less; inner lamellae of inner gills connected to visceral mass only anteriorly.

No gravid females were available for marsupium description. Ortmann (1912a) and Utterback (1915) reported outer gills to be marsupial and to swell moderately when gravid. However, Sterki (1895) included *Plethobasus cyphyus* in a short list of examples of species that use all four gills as marsupia. Ortmann (1909a) reported gills to be red when brooding eggs but described them as being more of a "lilac" hue than other species of "*Quadrula*" (= *Fusconaia* in part).

Labial palps tan to rusty brown or orange, often with lighter area dorsally; straight dorsally, convex ventrally, bluntly pointed distally; palp length 20–35% of gill length; palp height 50–70% of palp length, occasionally greater; distal 30–55% of palps bifurcate.

Incurrent aperture longer or equal to excurrent aperture; supra-anal aperture longer, shorter or equal to incurrent aperture.

Incurrent aperture length 15–30% of shell length; creamy white to tan within, often some shade of orange, occasionally gray, basal to papillae; papillae arborescent, often intermingled with simple, bifid and trifid papillae; papillae some combination of orange, brown and gray, color often changes distally.

Excurrent aperture length 15–20% of shell length; creamy white to light tan within, marginal color band some combination of dull orange, gray and brown, may have regular bands and spots or reticulated pattern; margin papillate, papillae simple, small, may be little more than crenulations; papillae orange to rusty tan, may have grayish cast.

Supra-anal aperture length 15–20% of shell length; usually orange within, occasionally creamy white, may have thin gray or black marginal band; margin smooth; mantle bridge separating supra-anal and excurrent apertures imperforate, length 5–15% of supra-anal length. Ortmann (1912a) reported the mantle bridge to be occasionally absent.

Specimens examined: Clinch River, Tennessee (n = 2); Tennessee River (n = 4).

Glochidium Description

Length 220 μm; height 200 μm; without styliform hooks; outline subrotund; dorsal margin straight, ventral margin rounded, anterior and posterior margins convex, anterior margin more convex than posterior margin (Surber, 1912).

Similar Species

Plethobasus cyphyus most closely resembles *Plethobasus cicatricosus*, but the knobs are larger and tubercles less numerous in *P. cyphyus*. *Plethobasus cyphyus* is generally more elongate than *P. cicatricosus*, with a more angular ventral margin. The posterior margin of *P. cyphyus* is often bluntly pointed, whereas that of *P. cicatricosus* is usually more rounded. The periostracum of *P. cyphyus* is typically smooth and shiny, whereas that of *P. cicatricosus* has a cloth-like texture. Subadult *P. cyphyus* are usually more yellow than subadult *P. cicatricosus*.

Plethobasus cyphyus may superficially resemble *Fusconaia ebena*, *Fusconaia subrotunda* or *Pleurobema sintoxia* in shape, but none of those species are sculptured. The periostracum of *P. cyphyus* is generally not as dark as those of *F. ebena*, *F. subrotunda* and *P. sintoxia*.

General Distribution

Plethobasus cyphyus occurs in the Mississippi Basin from Minnesota and Wisconsin (Dawley, 1947) downstream to northern Mississippi (Jones et al., 2005) and in the Ohio River drainage from headwaters in Pennsylvania (Ortmann, 1919; Williams and Schuster, 1989) west to Missouri (Oesch, 1995). *Plethobasus cyphyus* is known from the Cumberland River downstream of Cumberland Falls, Kentucky and

Tennessee, and from headwaters of the Tennessee River drainage in southwestern Virginia downstream to the mouth of the Tennessee River (Cicerello et al., 1991; Ahlstedt, 1992a, 1992b; Parmalee and Bogan, 1998).

Alabama and Mobile Basin Distribution

Plethobasus cyphyus historically occurred in the Tennessee River across northern Alabama.

Plethobasus cyphyus is apparently now restricted to tailwaters of Guntersville and Wilson dams.

Ecology and Biology

Plethobasus cyphyus occurs in flowing water of medium to large rivers. Its preferred substrate is a mixture of sand and gravel. In tailwaters of dams on large rivers *P. cyphyus* may be found at depths exceeding 6 m.

Plethobasus cyphyus is a short-term brooder, gravid from May to July (Ortmann, 1909a, 1919; Surber, 1912). However, a single pink conglutinate was observed discharged from an individual while being measured on the Tennessee River, Alabama, during August (J.T. Garner, personal observation). Conglutinates are discharged intact (Ortmann, 1910a). *Campostoma anomalum* (Central Stoneroller) (Cyprinidae) has been found to serve as a glochidial host for *Plethobasus cyphyus* in laboratory trials (Watters et al., 2005). *Sander canadensis* (Sauger) (Percidae) has been reported as a host based on observations of natural infestations. In one of the observations the encysted glochidium had almost completed transformation to the juvenile stage (Surber, 1913).

Current Conservation Status and Protection

Plethobasus cyphyus was considered threatened throughout its range by Williams et al. (1993). In Alabama Stansbery (1976) listed it as a species of special concern, and Lydeard et al. (1999) considered it threatened. Garner et al. (2004) designated *P. cyphyus* a species of highest conservation concern in the state. It was elevated to a candidate for protection under the federal Endangered Species Act in 2002.

Remarks

Plethobasus cyphyus has been reported to be rare in prehistoric shell middens along the Tennessee River in northern Alabama (Morrison, 1942; Warren, 1975; Hughes and Parmalee, 1999). It was the rarest of the three *Plethobasus* species in those studies.

Because the shell of *Plethobasus cyphyus* is hard and brittle, it was not desirable to pearl button producers. The shell is reportedly hard enough to damage the tubular saws used in cutting button blanks (Wilson and Clark, 1914; Coker, 1919). Some commercial mussel harvesters referred to *P. cyphyus* as "clear profit", as they could be included to add weight to a load, profiting the harvester at a loss to the button producer (Wilson and Clark, 1914; Coker, 1919).

A record of *Plethobasus cyphyus* reported by Murray and Leonard (1962) from southeastern Kansas appears to be based on specimens of *Quadrula quadrula* (R.S. Butler, personal communication).

Synonymy

Obliquaria (*Quadrula*) *cyphya* Rafinesque, 1820. Rafinesque, 1820:305
Type locality: Ohio River. Lectotype, ANSP 20239, length 83 mm, designated by Johnson and Baker (1973).
Unio aesopus Green, 1827. Green, 1827:46, fig. 3
Unio compertus Frierson, 1911. Frierson, 1911:53, pl. 3, middle and lower figs.

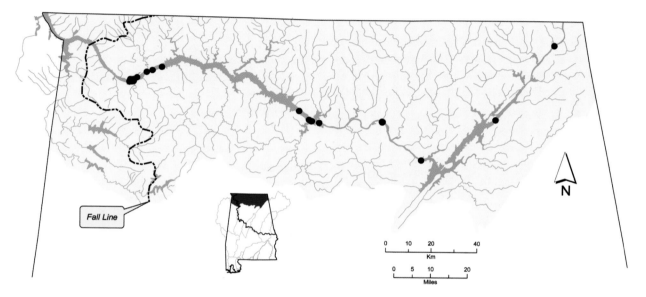

Distribution of *Plethobasus cyphyus* in the Tennessee River drainage of Alabama.

Genus *Pleurobema*

Pleurobema Rafinesque, 1819, occurs in the Mississippi and Great Lakes basins and Gulf Coast drainages from northern Florida to Texas. One species purported to be a *Pleurobema*, *P. collina*, occurs in Atlantic Coast drainages of North Carolina and Virginia. Preliminary genetic data do not support the placement of this spined species in the clade with the type species of *Pleurobema* (Bogan et al., unpublished data.).

Twenty-one species of *Pleurobema* are herein recognized as having occurred in Alabama. Of 32 species recognized in Turgeon et al. (1998), 17 known from Alabama are herein considered valid. Extensive comparative study of type material of all Mobile Basin *Pleurobema* resulted in a revised list of valid taxa (Table 13.1 and Table 13.2). As a result, some previously unrecognized older taxa take priority, and some long-used taxa are herein placed into synonymy. Ten species described from the Mobile Basin that were included in Turgeon et al. (1998) are herein placed into synonymy. Three species not recognized in Turgeon et al. (1998), *Pleurobema fibuloides*, *Pleurobema hartmanianum* and *Pleurobema stabilis*, are herein elevated from synonymy.

There has been one species, *Pleurobema athearni* Gangloff et al., 2006, described since publication of Turgeon et al. (1998). One species that has historically been placed in *Pleurobema*, *Unio altus* Conrad, 1849, was determined to be a *nomen dubium* based on an inadequate description, ambiguous type locality and lack of extant type material (see *Pleurobema fibuloides* account). After examination of type material, another species frequently assigned to *Pleurobema*, *Unio brumbyanus* Lea, 1841, was determined to be *Ptychobranchus greenii* (Conrad, 1834).

The genus *Pleurobema* is the most perplexing group of unionids in the Mobile Basin. In letters to Bryant Walker in December 1908, H.H. Smith wrote:

"Unless I am much mistaken you will find the Pleurobemas a terror."

"If you don't swear at *Pleurobema* before you get through with that box, you are a better Christian than I think you are."

Type Species
Pleurobema mytiloides Rafinesque, 1820 (= *Unio clava* Lamarck, 1819)

Diagnosis
Shell subquadrate to oval or elongate oval; umbo often positioned near anterior end; posterior ridge low, rounded; periostracum yellow to olive, rays absent or broken into squarish blocks; hinge plate well-developed; umbo cavity shallow to moderately deep.

Inner lamellae of inner gills connected to visceral mass only anteriorly; outer gills marsupial, occasional individuals with all 4 marsupial; marsupium not extended ventrally below original gill edge when gravid; glochidium without styliform hooks (Rafinesque, 1820; Simpson, 1900b; Ortmann, 1912a; Haas, 1969b).

Synonymy
Aximedia Rafinesque, 1820
Scalenaria Rafinesque, 1820
Sintoxia Rafinesque, 1820

Table 13.1. Treatment of the species assigned to the genus *Pleurobema* from the Mobile Basin by various authors during the past century. Species names followed by an asterisk were recognized in Turgeon et al. (1998). The 13 species in bold are those herein recognized as valid.

Species	Simpson (1914)	Frierson (1927)	Hurd (1974)	Stansbery (1976)	Herein Recognized
aldrichi Frierson, 1927	-----	aldrichi	aldrichianum	-----	taitianum
aldrichianum Goodrich, 1931	-----	-----	aldrichianum	-----	hanleyianum
altus* Conrad, 1849	altum	altum	altum	altum	nomen dubium
anaticulus Lea, 1861	decisum	decisum	decisum	decisum	curtum
athearni Gangloff, Williams and Feminella, 2006	-----	-----	-----	-----	athearni
avellana* Simpson, 1900	avellana	avellans	-----	-----	rubellum
chattanoogaensis* Lea, 1858	chattanoogaense	chattanoogaensis	hanleyanum	-----	decisum
cinnamomicus Lea, 1861	nux	perovatum	-----	perovatum	perovatum
concolor Lea, 1861	brumbyanum	concolor	-----	-----	perovatum
consanguineus Lea, 1861	decisum	decissus	decisum	decisum	decisum
crapulus Lea, 1861	cor	lewisi	-----	-----	fibuloides
crebrivittatus Lea, 1861	decisum	decissus	decisum	decisum	decisum
curtus* Lea, 1859	curtum	curtum	-----	curtum	curtum
decisus* Lea, 1831	decisum	decissus	decisum	decisum	decisum
favosus Lea, 1856	favosum	georgiana	georgianum	-----	georgianum
fibuloides Lea, 1859	altum	fibuloides	altum	altum	fibuloides
fictum Frierson, 1927	-----	fictum	-----	-----	rubellum
flavidulus* Lea, 1861	flavidulum	flavidulum	johannis	-----	perovatum
furvus* Conrad, 1834	furvum	furvum	-----	-----	rubellum
georgianus* Lea, 1841	georgianum	georgianum	georgianum	-----	georgianum
hagleri* Frierson, 1900	hagleri	hagleri	georgianum	-----	rubellum
hanleyianus* Lea, 1852	hanleyanum [sic]	hanleyanum [sic]	hanleyanum [sic]	-----	hanleyianum
hartmanianus Lea, 1860	-----	hartmanianum	hartmanianum	hartmanianum	hartmanianum
instructus Lea, 1861	instructum	instructum	-----	-----	verum
interventus Lea, 1861	interventum	interventum	georgianum	-----	perovatum
irrasus Lea, 1861	irrasum	showalteri	rubellum	rubellum ?	georgianum
johannis* Lea, 1859	johannis	johannis	johannis	-----	perovatum
lewisii Lea, 1861	cor	lewisi	lewisi	nucleopsis	fibuloides
litus Lea, 1871	litum	striatum	georgianum	-----	perovatum
marshalli* Frierson, 1927	-----	marshalli	-----	marshalli	marshalli
medius Lea, 1861	stabile	nucleopsis	georgianum	nucleopsis	georgianum
murrayensis* Lea, 1868	murrayense	murrayensis	stabile	-----	stabilis
nucleopsis* Conrad, 1849	nucleopsis	nucleopsis	rubellum	nucleopsis	georgianum
nux Lea, 1852	nux	perovatum	nux	perovatum	perovatum

Species	Simpson (1914)	Frierson (1927)	Hurd (1974)	Stansbery (1976)	Herein Recognized
pallidofulvus Lea, 1861	*interventum*	*interventum*	---------	---------	*perovatum*
perovatus* Conrad, 1834	*perovatum*	*perovatum*	*perovatum*	*perovatum*	*perovatum*
pinkstoni S.H. Wright, 1897	*pinkstoni*	*concolor*	---------	---------	*perovatum*
pulvinulus Lea, 1845	*rubellum*	*pulvinulum*	*rubellum*	*rubellum*	*rubellum*
rubellus* Conrad, 1834	*rubellum*	*rubellum*	*rubellum*	*rubellum*	*rubellum*
showalterii Lea, 1860	*showalterii*	*showalteri*	*showalteri*	*showalteri*	*hartmanianum*
simulans Lea, 1871	*simulans*	*simulans*	---------	---------	*perovatum*
stabilis Lea, 1861	*stabile*	*rubellum*	*stabile*	*hartmanianum*	*stabilis*
taitianus* Lea, 1834	*taitianum*	*taitianum*	*taitianum*	*taitianum*	*taitianum*
tombigbeanum Frierson, 1908	*taitianum*	*taitianum*	---------	*taitianum*	*taitianum*
*troschelianus*** Lea, 1852	*troschelianum*	*pulvinulum*	*troschelianum*	---------	*georgianum*
verus* Lea, 1861	*verum*	*instructum*	*rubellum*	*rubellum ?*	*verum*

Table 13.2. Geographic arrangement by drainage of the type locality of the 45 nominal species of *Pleurobema* described from the Mobile Basin. The taxa are divided into eastern and western Mobile Basin drainages. The eastern Mobile Basin includes the Coosa, Alabama, Cahaba and Tallapoosa River drainages. The western Mobile Basin includes the Black Warrior and Tombigbee River drainages. Species names followed by an asterisk were recognized in Turgeon et al. (1998). The 13 species in bold are those herein recognized as valid. One species included in Turgeon et al. (1998), *Pleurobema altum*, is herein considered a *nomen dubium* and not included on this table.

EASTERN RIVER DRAINAGES OF THE MOBILE BASIN

Coosa River Drainage

Species	Type Locality	Synonymy
aldrichianum Goodrich, 1931	Conasauga River near Conasauga, Tennessee	= *hanleyianum*
athearni Gangloff, Williams and Feminella, 2006	Big Canoe Creek, Alabama	= *athearni*
*chattanoogaensis** Lea, 1858	Chattanooga, Tennessee [Chattooga, Georgia]	= *decisum*
crapulus Lea, 1861	Etowah River, Georgia	= *fibuloides*
crebrivittatus Lea, 1861	Coosawattee River, Alabama [Georgia]	= *decisum*
favosus Lea, 1856	Othcalooga Creek, Georgia	= *georgianum*
fibuloides Lea, 1859	Connasauga [Conasauga] River, Georgia	= *fibuloides*
georgianus * Lea, 1841	Stump [Stamp] Creek, Georgia	= *georgianum*
hanleyianus * Lea, 1852	Coosawattee River, Georgia	= *hanleyianum*
hartmanianus Lea, 1860	Coosa River, Watumpka [Wetumpka], Alabama	= *hartmanianum*
irrasus Lea, 1861	Etowah River, Georgia	= *georgianum*
lewisii Lea, 1861	Coosa River, Alabama	= *fibuloides*
medius Lea, 1861	Near Coosa River, Alabama	= *georgianum*
*murrayensis** Lea, 1868	Conasauga Creek [River], Georgia	= *stabilis*
*nucleopsis** Conrad, 1849	Etowah River, Georgia	= *georgianum*
showalterii Lea, 1860	Coosa River, Wetumpka, Alabama	= *hartmanianum*
stabilis Lea, 1861	Coosa River, Alabama	= *stabilis*
*troschelianus** Lea, 1852	Coosawattee River, Georgia	= *georgianum*

Alabama, Cahaba and Tallapoosa River Drainages

Species	Type Locality	Synonymy
aldrichi Frierson, 1927	[Alabama River,] Selma, Alabama	= *taitianum*
*avellana** Simpson, 1900	Cahaba River, Alabama	= *rubellum*
consanguineus Lea, 1861	Cahaba River, Alabama	= *decisum*
decisus * Lea, 1831	Alabama River, [Alabama]	= *decisum*
fictum Frierson, 1927	Cahaba River, Alabama	= *rubellum*
instructus Lea, 1861	Cahaba River, Alabama	= *verum*
interventus Lea, 1861	Cahaba River, Alabama	= *perovatum*
*johannis** Lea, 1859	Alabama River, [Alabama]	= *perovatum*

litus Lea, 1871	Cahaba River, Alabama	= *perovatum*
nux Lea, 1852	Alabama River, Alabama	= *perovatum*
pallidofulvus Lea, 1861	Cahaba River, Alabama	= *perovatum*
pinkstoni S.H. Wright, 1897	Tuscaloosa [Tallapoosa] River, Alabama	= *perovatum*
simulans Lea, 1871	Cahaba River, Alabama	= *perovatum*
taitianus* Lea, 1834	Alabama River, [Claiborne, Alabama]	= *taitianum*
verus* Lea, 1861	Cahawba [Cahaba] River, Alabama	= *verum*

WESTERN RIVER DRAINAGES OF THE MOBILE BASIN

Black Warrior River Drainage

Species	**Type Locality**	**Synonymy**
concolor Lea, 1861	Big Prairie Creek, Alabama	= *perovatum*
furvus* Conrad, 1834	Black Warrior River, [Alabama]	= *rubellum*
hagleri* Frierson, 1900	North River near Tynes, Alabama	= *rubellum*
perovatus* Conrad, 1834	[Big] Prairie Creek, Alabama	= *perovatum*
pulvinulus Lea, 1845	[Black Warrior River,] Tuscaloosa, Alabama	= *rubellum*
rubellus* Conrad, 1834	Black Warrior River, Alabama	= *rubellum*

Tombigbee River Drainage

Species	**Type Locality**	**Synonymy**
anaticulus Lea, 1861	Near Columbus, Mississippi	= *curtum*
cinnamomicus Lea, 1861	Tombigbee River, Columbus, Mississippi	= *perovatum*
curtus* Lea, 1859	Tombigbee River, Columbus, Mississippi	= *curtum*
flavidulus* Lea, 1861	Stream near Columbus, Mississippi	= *perovatum*
marshalli* Frierson, 1927	Tombigbee River, Boligee, Alabama	= *marshalli*
tombigbeanum Frierson, 1908	[Tombigbee River,] Columbus, Mississippi	= *taitianum*

Pleurobema athearni Gangloff, Williams and Feminella, 2006
Canoe Creek Clubshell

Pleurobema athearni – Length 84 mm, USNM 1078388 (holotype). Big Canoe Creek, approximately 1 km downstream of St. Clair County Road 36, near mouth of Mukleroy Creek, St. Clair County, Alabama, 23 September 2001. © Richard T. Bryant.

Shell Description

Length to 93 mm; moderately thick; compressed; outline oval; posterior margin narrowly rounded; anterior margin rounded; dorsal margin convex; ventral margin convex; posterior ridge very low, rounded; posterior slope low, flat; disk smooth or with irregular subradial wrinkles; umbo broad, compressed, not elevated above hinge line, umbo sculpture unknown; periostracum tawny to brown, without rays.

Pseudocardinal teeth low, triangular, 2 divergent teeth in left valve, 1 tooth in right valve, may have accessory denticle; lateral teeth short, thick, slightly curved, 2 in left valve, 1 in right valve; interdentum moderately long, wide; umbo cavity moderately deep; nacre white.

Soft Anatomy Description

Mantle pale tan, some rusty tan and grayish brown outside of apertures; visceral mass pale tan; foot pale tan.

Mantle margin just ventral to, but slightly removed from, incurrent aperture, with slight fold; tissues within fold grayish brown; very narrow brown marginal band on inner surface of mantle; fold margin smooth.

Gills tan; dorsal margin slightly sinuous, ventral margin convex; gill length approximately 60% of shell length; gill height approximately 50% of gill length; outer gill height approximately 75% of inner gill height; inner lamellae of inner gills connected to visceral mass only anteriorly.

No gravid females were available for marsupium description. Outer gills are marsupial (P.D. Johnson, personal communication).

Labial palps pale tan; straight dorsally, convex ventrally, bluntly pointed distally; palp length approximately 20% of gill length; palp height approximately 55% of palp length; distal 25% of palps bifurcate.

Supra-anal aperture considerably longer than incurrent and excurrent apertures; incurrent and excurrent apertures of similar length.

Incurrent aperture length approximately 10% of shell length; pale tan within, with irregular gray cast toward margin, dark grayish brown basal to papillae; papillae in 2 rows, inner row very sparse, comprised of small arborescent papillae, outer row short, slender, blunt, simple; papillae rusty tan.

Excurrent aperture length approximately 10% of shell length; pale tan within, marginal color band dark grayish brown; margin with well-developed, short, mostly simple papillae, may have few scattered bifid papillae; papillae rusty tan.

Supra-anal aperture length approximately 20% of shell length; pale tan within, pale rusty tan marginally; margin smooth; mantle bridge separating supra-anal and excurrent apertures imperforate, length approximately 6% of supra-anal length.

Specimen examined: Big Canoe Creek, St. Clair County (n = 1, holotype, previously preserved).

Glochidium Description

Glochidium unknown.

Similar Species

Pleurobema athearni superficially resembles *Pleurobema georgianum*. However, the umbo cavity of *P. athearni* is deeper than that of *P. georgianum*, and *P. georgianum* often has green rays or blotches on upper part of disk or along posterior ridge, which are absent in *P. athearni*. The disk of *P. georgianum* lacks subradial wrinkles that are often present on *P. athearni*.

Pleurobema athearni resembles *Fusconaia cerina*, but *F. cerina* has a strong posterior ridge and darker periostracum.

General Distribution

Pleurobema athearni is endemic to the Coosa River drainage of Alabama.

Alabama and Mobile Basin Distribution

Pleurobema athearni is known only from Big Canoe Creek, a western tributary of the Coosa River in St. Clair County, Alabama.

Pleurobema athearni is extant but extremely rare in Big Canoe Creek, it appears to be extirpated from the Coosa River proper.

Ecology and Biology

Pleurobema athearni occurs in shoal habitat in a medium to large Coosa River tributary. Its preferred substrate is gravel.

Pleurobema athearni is believed to be a short-term brooder, gravid in spring and summer. Its glochidial hosts are unknown, but some *Pleurobema* species utilize members of the Cyprinidae (Haag and Warren, 2003).

Current Conservation Status and Protection

Pleurobema athearni is a newly described species and has not received conservation evaluation. However, considering its restricted distribution and rarity, it should be considered a species of highest conservation concern.

Remarks

Historically, *Pleurobema athearni* has been misidentified as *Fusconaia cerina*, based on its deep umbo cavity. However, based on genetic analyses, it is clearly a *Pleurobema* (Campbell et al., 2005). *Pleurobema athearni* is the only species of the genus in the Mobile Basin with a deep umbo cavity.

Pleurobema athearni was named for Herbert D. Athearn of Cleveland, Tennessee, who first collected the species.

Synonymy

Pleurobema athearni Gangloff, Williams and Feminella, 2006. Gangloff, Williams and Feminella, 2006:43–56, fig. 1 [figured at beginning of account]

Type locality: Big Canoe Creek, approximately 1 km downstream of St. Clair County Road 36, near mouth of Mukleroy Creek, St. Clair County, Alabama, 23 September 2001. Holotype, USNM 1078388, length 84 mm.

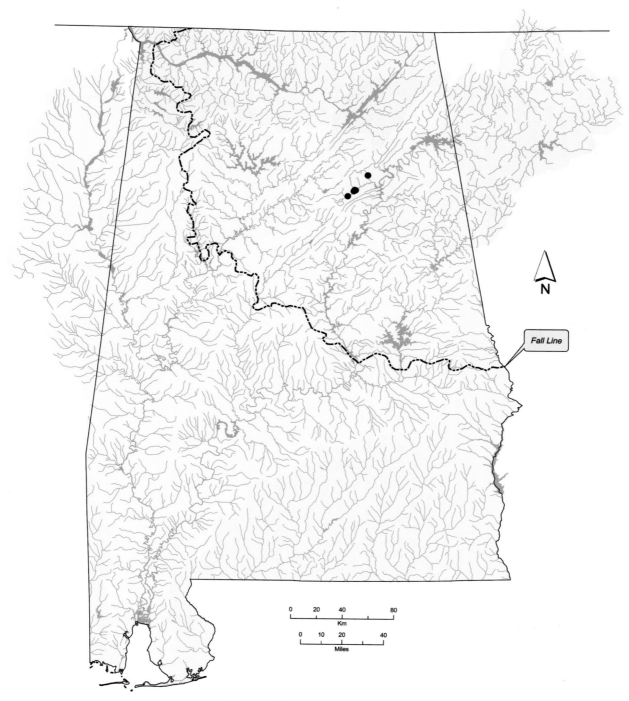

Distribution of *Pleurobema athearni* in Alabama and the Mobile Basin.

Pleurobema clava (Lamarck, 1819)
Clubshell

Pleurobema clava – Length 55 mm, USNM 30381. [Tennessee River,] Tuscumbia, [Colbert County,] Alabama.
© Richard T. Bryant.

Shell Description

Length to 78 mm; thick; inflated anteriorly; outline usually elongate triangular (scalene triangle), occasionally elliptical; posterior margin bluntly pointed to narrowly rounded; anterior margin obliquely truncate to rounded; dorsal margin convex; ventral margin straight to slightly convex; posterior ridge angular dorsally, becoming low and rounded posterioventrally; posterior slope usually steep, flat to slightly convex; broad, shallow, ventrally curved sulcus often just anterior of posterior ridge, originating just posterior of umbo; umbo narrow, inflated, positioned anteriorly, elevated above hinge line, turned anteriad, umbo sculpture rudimentary subconcentric ridges; periostracum tawny to brown, usually with variable, often wide, broken green rays, rays may not extend to ventral margin.

Pseudocardinal teeth thick, triangular, 2 teeth in left valve, often joined anteriodorsally, 1 tooth in right valve, arising from depression; lateral teeth long, thick, curved, 2 in left valve, 1 in right valve, may also have rudimentary lateral tooth ventrally in right valve; interdentum short, variable, usually wide; umbo cavity very shallow; nacre white.

Soft Anatomy Description

Mantle tan; visceral mass tan; foot tan. Lea (1863c) reported the "color of the mass" to be "whitish". Ortmann (1912a) reported the foot to be pale gray to pale orange and mantle margins to sometimes be pale orange and black posteriorly.

Gills tan; dorsal margin straight, ventral margin broadly convex; gill length 55–66% of shell length; gill height 35–40% of gill length; outer gill height 70–80% of inner gill height; inner lamellae of inner gills connected to visceral mass only anteriorly. Ortmann (1912a) reported gills to be grayish to grayish brown.

No gravid females were available for marsupium description. Ortmann (1912a) reported outer gills to be marsupial.

Labial palps tan; straight dorsally, convex ventrally, bluntly pointed distally; palp length 15–25% of gill length; palp height 50–55% of palp length; distal 35–40% of palps bifurcate.

Supra-anal aperture usually longer than incurrent and excurrent apertures; incurrent and excurrent aperture lengths variable relative to each other.

Incurrent aperture length approximately 15% of shell length; tan within, may have brown basal to papillae; papillae in 2 rows, outer row simple, inner row with arborescent and bifid; papillae dark tan to dark brown, may be white distally.

Excurrent aperture length 10–15% of shell length; tan within, with brown marginal color band; margin with small simple papillae; papillae white.

Supra-anal aperture length 15–20% of shell length; tan within, with thin dark brown marginal band; margin smooth; mantle bridge separating supra-anal and excurrent apertures imperforate, length usually 5–15% of supra-anal length.

Specimens examined: Elk River, West Virginia (n = 1); Mohican River, Ohio (n = 1) (specimens previously preserved).

Glochidium Description

Length 160 μm; height 160 μm; without styliform hooks; outline subelliptical (Ortmann, 1912a).

Similar Species

Pleurobema clava is distinctive from almost all sympatric taxa. Elongate *Pleurobema oviforme* may resemble *P. clava*. However, the umbo of *P. clava* is at or overhanging the anterior end of the shell, whereas the umbo of *P. oviforme* is positioned less anteriorly. Also, the sulcus is usually better developed in *P. clava* than *P. oviforme*.

General Distribution

Simpson (1914) gave Lake Erie as the type locality of *Pleurobema clava*. However, it has appeared in few accounts of the Great Lakes fauna. Winslow (1918) reported *P. clava* from a small stream of the Maumee River system, tributary of Lake Erie. In the Ohio River drainage *P. clava* is known from headwaters in Pennsylvania (Ortmann, 1909a) downstream to the mouth of the Ohio River in Illinois and Kentucky (Cummings and Mayer, 1992). Parmalee and Bogan (1998) reported *P. clava* to be widespread in Tennessee reaches of the Cumberland River, but Cicerello et al. (1991) did not report it from Kentucky reaches of the drainage. Ortmann (1925) questioned the presence of *P. clava* in the Cumberland River drainage. *Pleurobema clava* is known from most of the Tennessee River drainage, Alabama, Kentucky and Tennessee, with the exceptions of headwaters in Virginia and North Carolina (Parmalee and Bogan, 1998).

Alabama and Mobile Basin Distribution

Pleurobema clava occurred in the Tennessee River across northern Alabama.

Pleurobema clava appears to be extirpated from the Tennessee River. The most recent known collections of live individuals were in tailwaters of Wilson Dam in 1966.

Ecology and Biology

Pleurobema clava is a species of small to large rivers, where it occurs in flowing water, usually in sand and/or gravel substrates without heavy silt deposits. *Pleurobema clava* often remains deeply buried in the substrate (Ortmann, 1919).

Pleurobema clava is a short-term brooder. Gravid females have been reported from May to July in Pennsylvania, with glochidia mature in mid-June (Ortmann, 1912a, 1919). Fishes reported to serve as glochidial hosts of *P. clava* in laboratory trials include *Campostoma anomalum* (Central Stoneroller) and *Luxilus chrysocephalus* (Striped Shiner) (Cyprinidae); and *Percina caprodes* (Logperch) and *Percina maculata* (Blackside Darter) (Percidae) (O'Dee and Watters, 2000).

Current Conservation Status and Protection

Pleurobema clava was considered endangered throughout its range by Stansbery (1970a) and Williams et al. (1993). It was listed as extirpated from Alabama by Lydeard et al. (1999) and Garner et al. (2004). *Pleurobema clava* was listed as endangered under the federal Endangered Species Act in 1993.

In 2001 *Pleurobema clava* was included on a list of species approved for a Nonessential Experimental Population in the tailwaters of Wilson Dam on the Tennessee River. However, no reintroductions had taken place as of 2007.

Remarks

Pleurobema clava is reported to be uncommon in prehistoric shell middens along the Tennessee River in northern Alabama (Morrison, 1942; Warren, 1975; Hughes and Parmalee, 1999). Morrison (1942) suggested that it disappeared from the Muscle Shoals area prior to recent times. However, there are historical museum specimens, and *P. clava* is occasionally found in commercial mussel cull piles, presumably left prior to impoundment of the river (piles submerged but located on the edges of historical river channels).

The relationship between *Pleurobema clava* and *Pleurobema oviforme* has been a matter of conjecture. *Pleurobema oviforme* replaces *P. clava* in headwaters of the Tennessee River, and it has been suggested that the two may be "only varieties of one species" (Ortmann, 1925). However, Warren (1975) reported no intergrades between the two species in prehistoric midden material from near Bridgeport.

Synonymy

Unio clava Lamarck, 1819. Lamarck, 1819:74
Type locality: Reported as "Habite dans le lac Erié" [Lake Erie], but this is likely erroneous. Holotype, MNHN uncataloged, length 75 mm (Johnson, 1969b).
Unio (Aximedia) elliptica Rafinesque, 1820. Rafinesque, 1820:296
Obliquaria (Scalenaria) scalenia Rafinesque, 1820. Rafinesque, 1820:309, pl. 81, figs. 24, 25
Pleurobema cuneata Rafinesque, 1820. Rafinesque, 1820:313
Pleurobema mytiloides Rafinesque, 1820. Rafinesque, 1820:313, pl. 82, figs. 8–10
Unio patulus Lea, 1829. Lea, 1829:441, pl. 12, fig. 20; Lea, 1834b:55, pl. 12, fig. 20
Unio bournianus Lea, 1840. Lea, 1840:288; Lea, 1842b:213, pl.15, fig. 28; Lea, 1842c:51, pl.15, fig. 28
Unio consanguineus De Gregorio, 1914, *non* Lea, 1863. De Gregorio, 1914:46

Unio anaticulus var. *ohiensis* De Gregorio, 1914. De Gregorio, 1914:51

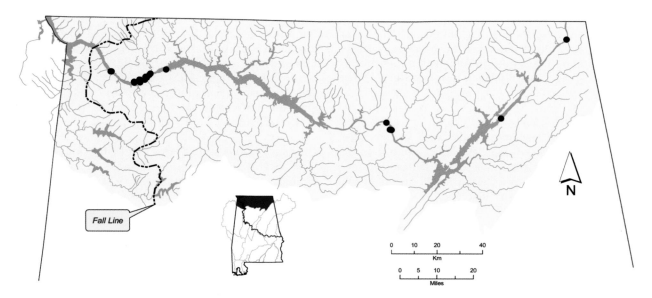

Distribution of *Pleurobema clava* in the Tennessee River drainage of Alabama.

Pleurobema cordatum (Rafinesque, 1820)
Ohio Pigtoe

Pleurobema cordatum – Length 72 mm, UF 66600. [Tennessee River,] Muscle Shoals, Lauderdale County, Alabama, November 1909. © Richard T. Bryant.

Shell Description

Length to 109 mm; thick; inflated; outline subtriangular; posterior margin bluntly pointed to narrowly rounded; anterior margin usually straight; dorsal margin straight to convex; ventral margin rounded; posterior ridge low, well-defined, often slightly curved; posterior slope steep to moderately steep; wide shallow sulcus anterior of posterior ridge, ventrally curved; umbo inflated, projecting anteriorly, turned inward, elevated well above hinge line, umbo sculpture a few irregular ridges; periostracum tawny to dark brown, darkening with age, green rays may be present on umbo of subadults.

Pseudocardinal teeth rough, triangular, 2 divergent teeth in left valve, anterior tooth usually thinner than posterior tooth, 1 tooth in right valve, may have accessory denticle anteriorly and/or posteriorly; lateral teeth thick, straight to slightly curved, 2 in left valve, 1 in right valve, may also have rudimentary to moderately developed lateral tooth ventrally in right valve; interdentum short, wide; umbo cavity deep, not compressed; nacre white, rarely pale pink.

Soft Anatomy Description

Mantle creamy white to light brown; visceral mass pearly white to creamy white; foot tan.

Gills tan to light brown; dorsal margin straight to slightly sinuous, steeply sloped from umbo to near incurrent aperture, outer gill ventral margin straight to slightly convex, inner gill ventral margin slightly convex to deeply convex; gill length 50–65% of shell length; gill height 40–60% of gill length; outer gill height 65–85% of inner gill height; inner lamellae of inner gills connected to visceral mass only anteriorly.

Outer gills marsupial, some gravid individuals with a few glochidia in inner gills; glochidia held across entire gill except extreme anterior and posterior ends; marsupium slightly padded when gravid; creamy white. Yokley (1972) reported only outer gills to be marsupial. Wilson and Clark (1914) reported "three gills, the two outer and one of the inner" to be used as marsupia. Ortmann (1919) also reported inner gills to occasionally be partially gravid.

Labial palps creamy white to tan; straight to slightly curved dorsally, convex ventrally, rounded to bluntly pointed distally; palp length 15–30% of gill length; palp height 55–90% of palp length; distal 30–50% of palps bifurcate.

Incurrent aperture usually slightly longer than excurrent aperture; supra-anal aperture usually slightly longer than incurrent aperture.

Incurrent aperture length 15–20% of shell length, occasionally greater; creamy white to tan within, without coloration basal to papillae; papillae in 2 rows, inner row may be larger, short, mostly simple with varying percentage of bifid, bifid papillae absent in some individuals; papillae tan to pale orange, may have black pigment basally or along edges.

Excurrent aperture length 15–20% of shell length, occasionally less; creamy white to tan within, may have thin pale orange and black marginal color band; margin papillate, papillae very short, mostly simple, occasional individuals with few bifid; papillae tan or dull orange, some with black basally. Lea (1863d) reported excurrent aperture papillae to be brown.

Supra-anal aperture length 15–25% of shell length; creamy white within, may have sparse black pigment flecks, some individuals with thin gray marginal band; margin smooth; mantle bridge separating supra-anal and excurrent apertures imperforate, length 5–15% of supra-anal length.

Specimens examined: Tennessee River (n = 8).

Glochidium Description

Length 140–160 μm; height 150–175 μm; without styliform hooks; outline depressed subelliptical; dorsal margin straight, ventral margin rounded, anterior and posterior margins slightly convex (Surber, 1915; Ortmann, 1919; Yokley, 1972).

Similar Species

Pleurobema cordatum resembles *Pleurobema plenum*. However, *P. cordatum* is much more elongate posteriorly, with a more pointed posterior margin. The truncate posterior margin of *P. plenum* gives the rounded anterioventral margin an exaggerated, rounded appearance compared to that of *P. cordatum*.

Pleurobema cordatum resembles *Pleurobema rubrum*, but the outline of *P. cordatum* approaches an equilateral triangle, whereas that of *P. rubrum* more closely resembles a scalene triangle. The difference is most evident in the anterior margin, which is oriented ventrally to anterioventrally in *P. cordatum* and ventrally to posterioventrally in *P. rubrum* when the shell is held with the hinge ligament horizontal. The sulcus of *P. rubrum* is often more poorly developed than that of *P. cordatum*. Also, the umbo cavity of *P. cordatum* is usually deeper than that of *P. rubrum*.

Pleurobema cordatum resembles *Pleurobema sintoxia*. However, *P. cordatum* is typically more triangular and has a better-developed sulcus. *Pleurobema sintoxia* is sometimes triangular, but may be round or ovate, and is usually without a sulcus. The umbo cavity of *P. cordatum* is deep, whereas that of *P. sintoxia* is shallow.

Pleurobema cordatum also resembles *Fusconaia flava*, but has a rounder, more curved posterior ridge and narrower umbo. *Pleurobema cordatum* has a well-defined sulcus, whereas *F. flava* has a flattened area on the shell disk.

General Distribution

The northern extent of *Pleurobema cordatum* distribution is perplexing since some previous authors treated *Pleurobema sintoxia*, which has a more northerly distribution, as conspecific with *P. cordatum*.

It is unclear whether *P. cordatum* occurred in the upper Mississippi River. In the Ohio River drainage *P. cordatum* is known from headwaters in western Pennsylvania (Ortmann, 1909a) to the mouth of the Ohio River in Illinois and Kentucky (Williams and Schuster, 1989; Cummings and Mayer, 1992). *Pleurobema cordatum* is widely distributed in the Cumberland River drainage downstream of Cumberland Falls, Kentucky and Tennessee (Cicerello et al., 1991; Parmalee and Bogan, 1998). It occurs in most of the Tennessee River drainage from headwaters in southwestern Virginia to the mouth of the Tennessee River (Ahlstedt, 1992a; Parmalee and Bogan, 1998).

Alabama and Mobile Basin Distribution

Pleurobema cordatum historically occurred in the Tennessee River across northern Alabama and in a few tributaries.

Pleurobema cordatum is now restricted to the Tennessee River in tailwaters of Guntersville and Wilson dams.

Ecology and Biology

Pleurobema cordatum is found in medium to large rivers, where it occurs in flowing water with substrates composed of mixtures of sand and gravel. Ortmann (1919) described it as "the shell, which largely contributes in forming the shell-banks, in rather deep, steadily flowing water". In large impounded rivers it is now limited to tailwaters of dams, where it may occur at depths exceeding 5 m.

The life history of Tennessee River *Pleurobema cordatum* was described by Yokley (1972). It is dioecious and reaches sexual maturity by four years of age. Gametogenesis begins in autumn and continues through the following spring. Spawning begins in early spring and continues into summer. *Pleurobema cordatum* is a short-term brooder, gravid from late April to mid-July. Wilson and Clark (1914) reported gravid *P. cordatum* until late August in the Cumberland River. Conglutinates of *P. cordatum* are white, oblong and flattened, with rounded ends and a linear row of holes down the center. When viewed from a lateral angle, the conglutinates have a zigzag outline (J.T. Garner, personal observation). Two or more conglutinates may be connected at one end. Wilson and Clark (1914) reported *P. cordatum* conglutinates to resemble "the seed of the green cucumber".

Fishes proven to serve as glochidial host of *Pleurobema cordatum* in laboratory trials include *Lythrurus fasciolaris* (Scarlet Shiner) and *Semotilus atromaculatus* (Creek Chub) (Cyprinidae); and *Culaea inconstans* (Brook Stickleback) (Gasterosteidae) (Yokley, 1972; Watters and Kuehnl, 2004). One nonindigenous fish, *Poecilia reticulata* (Guppy) (Poeciliidae), was also reported to serve as a glochidial host in laboratory trials (Watters and Kuehnl, 2004). Fuller (1974) and Parmalee and Bogan (1998) listed *Lepomis macrochirus* (Bluegill) (Centrarchidae) as a host of *P. cordatum*. However, the original references cited in those publications listed *L. macrochirus* as host for *Pleurobema sintoxia*, which was considered a synonym of *P. cordatum* by some authors.

Current Conservation Status and Protection

Pleurobema cordatum was listed as a species of special concern throughout its range by Williams et al. (1993) and in Alabama by Lydeard et al. (1999). Garner et al. (2004) designated it a species of moderate conservation concern in the state.

Pleurobema cordatum was formerly an important commercial species. However, due to its decline in Alabama, it was removed from the list of legally harvestable mussels in 2004.

Remarks

Scruggs (1960) estimated that *Pleurobema cordatum* comprised almost 53 percent of the mussel fauna in Guntersville Dam tailwaters, Tennessee River. Relative abundance in prehistoric middens compared to the findings of Scruggs (1960) suggests that *P. cordatum* populations increased during historical times (Morrison, 1942; Warren, 1975; Hughes and Parmalee, 1999). This species has not adapted well to impoundment of the river. In 1997 it comprised only 6% of the fauna, determined with catch per unit effort sampling in Guntersville Dam tailwaters (J.T. Garner, unpublished data). Scruggs (1960) estimated the *P. cordatum* population in an 8-mile reach of lower Guntersville Dam tailwaters to be 20,566,000 in 1956–1957. However, Ahlstedt and McDonough (1993) reported finding no *P. cordatum* in that reach in 1991. Reasons for the decline of *P. cordatum* are unclear but may be the result of depletion or elimination of its glochidial hosts, *Lythrurus fasciolaris* and *Semotilus atromaculatus*, which typically do not occur in deep water habitats of large rivers. The presence of occasional subadult *P. cordatum* in Wilson and Guntersville Dam tailwaters suggests that primary hosts have not been completely eliminated or that there is at least one secondary host. Unless causal factors of the decline of *P. cordatum* are determined and mitigated, this species may disappear from the Tennessee River.

Pleurobema cordatum was the most important commercial mussel species harvested from the Tennessee River during the mid-1900s. Scruggs (1960) reported it to comprise more than 80 percent of the harvest from upper reaches of Wheeler Reservoir in 1956 and 1957. The precipitous decline of *P. cordatum* following impoundment of the river suggests that the abundance of this species found by harvesters during that period were remnants of the preimpoundment population.

The resemblance of *Pleurobema cordatum* to *Pleurobema plenum*, *Pleurobema rubrum* and *Pleurobema sintoxia* has been the source of considerable confusion. The four appear to represent valid species but the latter three have been recognized as subspecies of *P. cordatum* by some authors.

Use of all four gills as marsupia in *Pleurobema cordatum* is perplexing. One of the defining characteristics of the genus *Pleurobema* is use of only the outer gills as marsupia (Simpson, 1900b; Ortmann, 1912a). Ortmann (1919) reported only outer gills to serve as marsupia after examination of "nearly a thousand individuals". However, Wilson and Clark (1914) reported some *P. cordatum* to brood glochidia only in the outer gills and a few to brood in both outer and one inner gill. Based on these findings, Wilson and Clark (1914) suggested that removal of this species from *Quadrula* and placing it in *Pleurobema* was "hardly advisable". Of six gravid *P. cordatum* recently examined from the Alabama reach of the Tennessee River, two were gravid in all four gills and four were gravid in outer gills only (J.T. Garner, personal observation).

Synonymy

Unio obliqua Lamarck, 1819. Lamarck, 1819:72
Comment: There is confusion surrounding the identification of the type (holotype lost, Johnson, 1969b) of *Unio obliqua* Lamarck, 1819 (see Lea, 1834a; Ortmann and Walker, 1922a). Ortmann and Walker (1922a) suggested the next available name, *Obovaria cordata,* should be used for this taxon.
Obovaria cordata Rafinesque, 1820. Rafinesque, 1820:312
Type locality: Ohio River. Lectotype, ANSP 20221, length 61 mm, designated by Johnson and Baker (1973).
Obovaria cordata var. *rosea* Rafinesque, 1820. Rafinesque, 1820:312

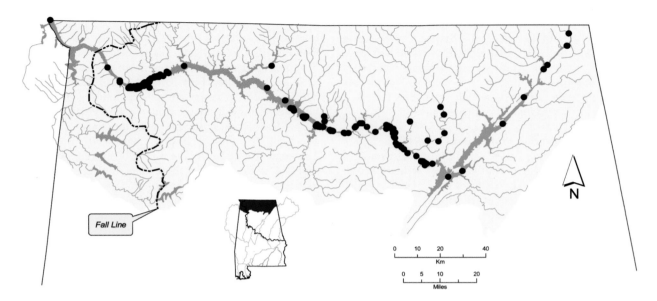

Distribution of *Pleurobema cordatum* in the Tennessee River drainage of Alabama.

Pleurobema curtum (Lea, 1859)
Black Clubshell

Pleurobema curtum – Length 56 mm, UF 358664. Tombigbee River, Pickensville boat ramp, about 1 mile northwest of Pickensville, Pickens County, Alabama, 4 June 1972. © Richard T. Bryant.

Shell Description

Length to 63 mm; thick; inflated, appears swollen anteriorly; outline elongate triangular (scalene triangle); posterior margin bluntly pointed; anterior margin straight to slightly convex; dorsal margin convex; ventral margin convex; posterior ridge pronounced, rounded; posterior slope narrow, steep; wide, very shallow sulcus just anterior of posterior ridge, distal end of sulcus often straight or slightly concave; umbo broad, inflated, turned inward, located near anterior terminus of shell, elevated well above hinge line, umbo sculpture unknown; periostracum smooth, reddish brown to olive brown or black, rarely with very weak green rays.

Pseudocardinal teeth thick, triangular, elevated, 2 divergent teeth in left valve, united dorsally to form cusp, 1 tooth in right valve, arises from pit-like depression, often with low ridge-like accessory denticle anteriorly and/or posteriorly; lateral teeth long, thick, slightly curved, 2 in left valve, 1 in right valve; interdentum very short, narrow; umbo cavity very shallow; nacre white.

Soft Anatomy Description

Mantle tan, may be pale rusty tan outside of apertures; visceral mass tan; foot tan.

Gills tan; dorsal margin straight to slightly sinuous, ventral margin convex; gill length 45–55% of shell length; gill height approximately 55% of gill length; outer gill height 75–85% of inner gill height; inner lamellae of inner gills connected to visceral mass only anteriorly.

No gravid females were available for marsupium description; probably similar to most other species of *Pleurobema*, with outer gills marsupial, padded when gravid.

Labial palps tan; straight dorsally, convex ventrally, bluntly pointed distally; palp length approximately 30% of gill length; palp height 35–40% of palp length; distal 30–35% of palps bifurcate.

Incurrent and supra-anal apertures longer than excurrent aperture; incurrent aperture longer or shorter than supra-anal aperture.

Incurrent aperture length 10–20% of shell length; tan within; papillae in 2 rows, inner row larger, long, simple, blunt; papillae tan to rusty tan.

Excurrent aperture length 5–10% of shell length; tan within, margin may be rusty tan, not in distinct color band; margin with simple, small but well-developed papillae; papillae tan.

Supra-anal aperture length 15–25% of shell length; tan within, without marginal color band but tan may be slightly darker marginally; margin smooth; mantle bridge separating supra-anal and excurrent apertures imperforate, length 15–30% of supra-anal length.

Specimens examined: Tombigbee River (n = 2) (specimens previously preserved).

Glochidium Description

Glochidium unknown.

Similar Species

Pleurobema curtum resembles *Pleurobema decisum* but its ventral margin has a more convex outline. *Pleurobema curtum* also has a better developed sulcus, giving an impression that the posterior terminus is more dorsally positioned than in *P. decisum*.

Pleurobema curtum resembles some *Obovaria jacksoniana*, but the umbo of *P. curtum* is generally located more anteriorly. *Pleurobema curtum* usually has a shallow sulcus just anterior of the posterior ridge, which is absent in *O. jacksoniana*. The umbo cavity of *O. jacksoniana* is deeper than that of *P. curtum*. The crest of the anterior pseudocardinal tooth of the left valve of *O. jacksoniana* is oriented more anterioventrally than that of *P. curtum*.

General Distribution

Pleurobema curtum is endemic to the Tombigbee River drainage of Alabama and Mississippi.

Alabama and Mobile Basin Distribution

Pleurobema curtum is known only from the Tombigbee River from Pickens County, Alabama, upstream to the East Fork Tombigbee River, Monroe and Itawamba counties, Mississippi.

Pleurobema curtum is extirpated from Alabama and may be extinct. It may be extant in the East Fork Tombigbee River but has not been collected since 1992 (P.D. Hartfield, personal communication).

Ecology and Biology

Pleurobema curtum is known only from flowing water in medium to large rivers. Preferred substrate of *P. curtum* is a mixture of sand and gravel or pure sand. It may be found in water less than 1 m deep.

Pleurobema curtum is probably a short-term brooder, gravid in spring and summer. Its glochidial hosts are unknown, but it may utilize members of the Cyprinidae (Haag and Warren, 2003).

Current Conservation Status and Protection

Pleurobema curtum was recognized as endangered throughout its range by Williams et al. (1993) and in Alabama by Stansbery (1976). Turgeon et al. (1998) included it on a list of possibly extinct species. Lydeard et al. (1999) and Garner et al. (2004) listed *P. curtum* as extirpated from Alabama. It was listed as endangered under the federal Endangered Species Act in 1976.

Remarks

Hinkley (1906) reported *Pleurobema curtum* from the Big Black River, a tributary of the Mississippi River, in west-central Mississippi. The specimens on which this record is based have not been located. Hinkley may have made an error in identification, possibly confusing *P. curtum* with *Obovaria jacksoniana*, which occurs in the Big Black River and superficially resembles *P. curtum*. Recent collections from the Big Black River have not yielded *P. curtum* (Jones et al., 2005).

The disappearance of *Pleurobema curtum* from Alabama and its possible extinction are a direct result of habitat destruction resulting from construction of the Tennessee Tombigbee Waterway by the USACE.

Synonymy

Unio curtus Lea, 1859. Lea, 1859a:112; Lea, 1862b:103, pl. 17, fig. 253; Lea, 1862d:107, pl. 17, fig. 253
Type locality: Tombigbee River, Columbus, [Lowndes County,] Mississippi, W. Spillman, M.D. Lectotype, USNM
     84737, length 40 mm, designated by Johnson (1974).

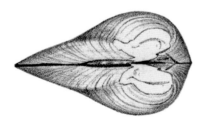

Unio anaticulus Lea, 1861. Lea, 1861a:40; Lea, 1862b:92, pl. 13, fig. 240; Lea, 1862d:96, pl. 13, fig. 240
Type locality: Near Columbus, [Lowndes County,] Mississippi, W. Spillman, M.D. Lectotype, USNM 84735, length
  35 mm, designated by Johnson (1974).

 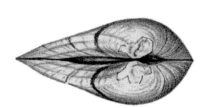 

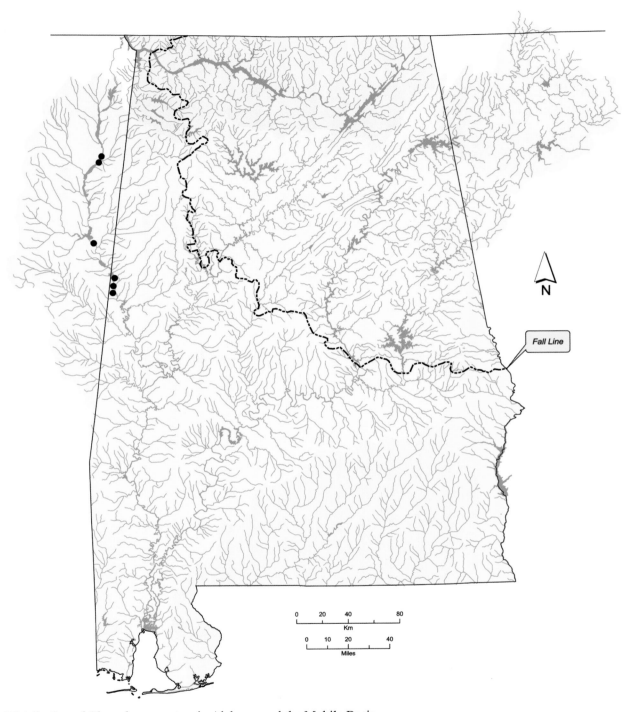

Distribution of *Pleurobema curtum* in Alabama and the Mobile Basin.

Pleurobema decisum (Lea, 1831)
Southern Clubshell

Pleurobema decisum – Upper figure: length 53 mm, UF 358665. Middle figure: length 39 mm, UF 358665. Buttahatchee River, U.S. Highway 45, 12 miles north of Columbus, Lowndes County, Mississippi, 13 September 1984. Lower figure: length 50 mm, UF 66501. Coosa River, Fort Williams Shoals, Talladega County, Alabama, 12 July 1913. © Richard T. Bryant.

Shell Description

Length to 93 mm; moderately thick; inflated anteriorly; outline elongate triangular to elliptical; posterior margin bluntly pointed to narrowly rounded; anterior margin truncate to rounded; dorsal margin straight to convex; ventral margin straight to slightly convex; posterior ridge angular, becoming low and rounded posterioventrally; posterior slope flat, usually steep dorsally; old individuals may have very shallow sulcus just anterior of posterior ridge; umbo narrow to moderately broad, inflated, positioned anteriorly, may be turned slightly anteriad, umbo sculpture unknown; periostracum tawny to dark brown, often with variable, broken green rays or concentric bands.

Pseudocardinal teeth thick, triangular, 2 divergent teeth in left valve, 1 tooth in right valve, may have accessory denticle anteriorly and/or posteriorly; lateral teeth moderately long, thick, straight to slightly curved, 2 in left valve, 1 in right valve, may also have rudimentary lateral tooth ventrally in right valve; interdentum short, variable, usually moderately wide; umbo cavity shallow; nacre white.

Soft Anatomy Description

Mantle tan, often rusty brown outside of apertures; visceral mass tan; foot tan.

Gills tan; dorsal margin straight to slightly sinuous, ventral margin elongate convex; gill length 60–70% of shell length; gill height 40–55% of gill length; outer gill height 65–76% of inner gill height; inner lamellae of inner gills connected to visceral mass only anteriorly.

Outer gills marsupial; glochidia held across entire gill except extreme posterior end; marsupium padded when gravid; tan.

Labial palps tan; straight or undulating to slightly convex dorsally, convex ventrally, bluntly pointed distally; palp length 15–25% of gill length; palp height 50–70% of palp length; distal 25–40% of palps bifurcate. Lea (1863d) described the labial palps as ovate.

Incurrent and supra-anal apertures usually longer than excurrent aperture; incurrent aperture may be shorter, longer or equal to supra-anal aperture.

Incurrent aperture length 10–20% of shell length; tan within, sometimes with rusty brown basal to papillae; papillae in 1–2 rows, short, blunt, simple and bifid, with occasional trifid, larger trifid papillae may approach arborescent in form; papillae tan to rusty tan or rusty brown.

Excurrent aperture length 10–15% of shell length; tan within, usually with rusty brown marginal color band, may have tan bands perpendicular to margin; margin papillate, papillae small, simple, blunt, decreasing to crenulations dorsally; papillae tan.

Supra-anal aperture length 15–20% of shell length; tan within, sometimes with thin rusty brown marginal band; margin smooth; mantle bridge separating supra-anal and excurrent apertures usually imperforate, length 25–35% of supra-anal length, occasionally absent.

Specimens examined: Conasauga River, Georgia (n = 1); Sipsey River (n = 5) (specimens previously preserved).

Glochidium Description

Length 175–200 µm; height 187–212 µm; without styliform hooks; outline subrotund; dorsal margin slightly concave, ventral margin rounded, anterior and posterior margins convex.

Similar Species

Pleurobema decisum resembles *Pleurobema curtum* but has an almost straight ventral margin, whereas the ventral margin of *P. curtum* is well rounded. *Pleurobema curtum* also has a better-developed sulcus, giving an impression that the posterior terminus is more dorsally positioned than in *P. decisum*.

Some small *Pleurobema decisum* resemble *Pleurobema perovatum*. However, the umbo of *P. perovatum* is positioned more centrally than that of *P. decisum*.

General Distribution

Pleurobema decisum is endemic to the Mobile Basin of Alabama, Georgia, Mississippi and Tennessee.

Alabama and Mobile Basin Distribution

Pleurobema decisum is known from most of the Mobile Basin, with the exception of the Tallapoosa River drainage above the Fall Line.

Pleurobema decisum is extant in scattered, isolated localities in the Alabama, Coosa, Tallapoosa and Tombigbee River drainages. Most remaining populations are in tributaries.

Ecology and Biology

Pleurobema decisum is a species of flowing water in large creeks and rivers. It is usually found in substrates composed of gravel with interstitial sand.

Pleurobema decisum has been reported to reach sexual maturity when as small as 26.3 mm (Haag and Staton, 2003). It is a short-term brooder, gravid from late May to late July, with glochidia mature by the second week of June (Haag and Warren, 2003). Haag and Staton (2003) estimated 94 percent of females in a population to be gravid during the peak of the brooding period. Glochidia are released in conglutinates that are ovate in outline, thin and orange or white (Haag and Staton, 2003; Haag and Warren, 2003). In addition to glochidia, *P. decisum* conglutinates are composed of undeveloped eggs, which are believed to help maintain the conglutinate integrity. Percentage of undeveloped eggs in *P. decisum* conglutinates has been reported to average 47 percent (Haag and Staton, 2003). Annual fecundity was found to be variable in two Sipsey River,

Alabama, populations, averaging 29,433 glochidia per female in one and 40,887 per female in the other (Haag and Staton, 2003). One primary glochidial host, *Cyprinella venusta* (Blacktail Shiner) (Cyprinidae), and one secondary host, *Luxilus chrysocephalus* (Striped Shiner) (Cyprinidae), were reported based on laboratory trials (Haag and Warren, 2003).

Current Conservation Status and Protection

Pleurobema decisum was considered endangered throughout its range by Athearn (1970) and Williams et al. (1993) and in Alabama by Stansbery (1976) and Lydeard et al. (1999). Garner et al. (2004) designated *P. decisum* a species of high conservation concern in the

state. It was listed as endangered under the federal Endangered Species Act in 1993.

Remarks

Some *Pleurobema decisum* have a more rounded posterior margin, dark concentric rings on the periostracum and an umbo that does not extend to the anterior margin. This form was described as *Unio chattanoogaensis* Lea, 1858. The taxonomic status of this form has been a matter of debate. It occurs sympatrically with typical *P. decisum*, and shell characters overlap. Preliminary genetic analyses including these two forms were inconclusive. *Unio chattanoogaensis* was herein tentatively placed into the synonymy of *P. decisum*.

Synonymy

Unio decisus Lea, 1831. Lea, 1831:92, pl. 12, fig. 23; Lea, 1834b:102, pl. 12, fig. 23
Type locality: Alabama River, [Alabama,] Judge Tait. Lectotype, USNM 84723, length 79 mm, designated by Johnson (1974).

Unio chattanoogaensis Lea, 1858. Lea, 1858c:166; Lea 1859f:209, pl. 25, fig. 90; Lea, 1859g:27, pl. 25, fig. 90
Type locality: Chattanooga [Chattooga], Tennessee [Georgia], T. Stewardson, M.D.; Etowah River, [Georgia,] Reverend G. White; Coosawattee and Oostenaula [Oostanaula] rivers, Georgia, Bishop Elliott. Lectotype, USNM 84729, length 55 mm, designated by Johnson (1974), is from Chattooga, Georgia.

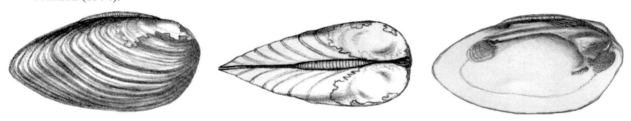

Unio consanguineus Lea, 1861. Lea, 1861c:60; Lea, 1862b:67, pl. 7, fig. 217; Lea, 1862d:71, pl. 7, fig. 217
Type locality: Etowah River, Reverend G. White; Oostenaula [Oostanaula] River, Georgia, Bishop Elliott; Cahawba [Cahaba] River, Alabama, E.R. Showalter. Lectotype, USNM 84726, length 56 mm, designated by Johnson (1974), is from Cahaba River, Alabama.

Unio crebrivittatus Lea, 1861. Lea, 1861c:60; Lea, 1866:43, pl. 15, fig. 41; Lea, 1867b:47, pl. 15, fig. 41
Type locality: Coosawattee River, Alabama [Georgia], Bishop Elliott. Lectotype, USNM 84725, length 53 mm,
    designated by Johnson (1974).

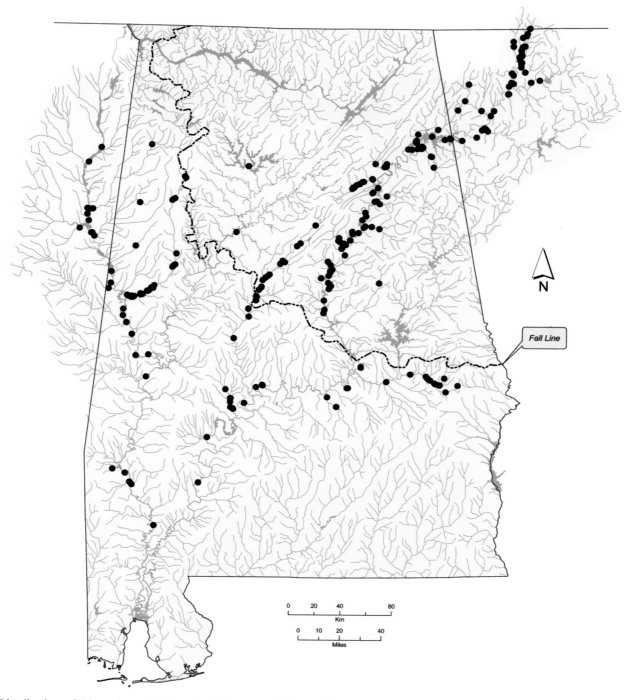

Distribution of *Pleurobema decisum* in Alabama and the Mobile Basin.

Pleurobema fibuloides (Lea, 1859)
Kusha Pigtoe

Pleurobema fibuloides – Length 34 mm, USNM 85809 (holotype). Conasauga River, Georgia. © Richard T. Bryant.

Shell Description

Length to 50 mm; thick; moderately inflated; outline round to oval or subtriangular; posterior margin rounded to bluntly pointed; anterior margin truncate; dorsal margin convex; ventral margin broadly rounded; posterior ridge rounded; posterior slope steep; umbo narrow, inflated, positioned anteriorly, turned slightly anteriad, elevated well above hinge line, umbo sculpture unknown; periostracum tawny to brown, may have a few small green blotches on umbo or disk.

Pseudocardinal teeth thick, triangular, 2 divergent teeth in left valve, 1 tooth in right valve, may have accessory denticle anteriorly and/or posteriorly; lateral teeth short, slightly curved, 2 in left valve, 1 in right valve, with rudimentary lateral tooth ventrally in right valve; interdentum short, wide; umbo cavity shallow; nacre white.

Soft Anatomy Description

No material was available for soft anatomy description, but Lea (1863d) provided a brief description based on material from the Coosa River: Gills are rounded ventrally and inner lamellae of inner gills are connected to the visceral mass only anteriorly. Labial palps are small and subovate in outline. Incurrent and excurrent apertures are small, with "very small" and "minute" brownish papillae, respectively. Supra-anal and excurrent apertures are separated by a mantle bridge.

Glochidium Description

Glochidium unknown.

Similar Species

Pleurobema fibuloides most closely resembles *Pleurobema hartmanianum* but is more oval in outline, compared to the more triangular *P. hartmanianum*.

Pleurobema fibuloides may resemble some *Pleurobema georgianum*, but *P. fibuloides* usually has a thicker, more inflated shell. The umbo of *P. fibuloides* is typically oriented more anteriorly. Also, the interdentum is much broader in *P. fibuloides* than in *P. georgianum*.

Pleurobema fibuloides superficially resembles *Pleurobema stabilis*, but *P. stabilis* may have a slight sulcus and is usually more elongate. The shell of *P. stabilis* is typically not as thick as that of *P. fibuloides*.

Pleurobema fibuloides may resemble some *Quadrula kieneriana*. Many *Q. kieneriana* have at least some pustules, which are never present on *P. fibuloides*. The umbo cavity of *P. fibuloides* is shallow, whereas that of *Q. kieneriana* is deep and compressed.

General Distribution

Pleurobema fibuloides is endemic to the Coosa River drainage of Alabama and Georgia.

Alabama and Mobile Basin Distribution

Pleurobema fibuloides historically occurred from headwaters of the Coosa River drainage downstream through its middle reaches.

Pleurobema fibuloides is believed to be extinct. The most recent dated museum material was collected by H.D. Athearn from the Conasauga River in 1958.

Ecology and Biology

Pleurobema fibuloides appears to have been restricted to shoal habitats in the Coosa River and major headwater tributaries, based on historical collection data.

Pleurobema fibuloides is believed to have been a short-term brooder, gravid during spring and summer. Its glochidial hosts are unknown, but some *Pleurobema* species utilize members of the Cyprinidae (Haag and Warren, 2003).

Current Conservation Status and Protection

Pleurobema fibuloides was listed as endangered by Athearn (1970) but was not recognized by subsequent workers. It is now believed to be extinct.

Remarks

The species to which the name *Pleurobema altum* has previously been applied is *Pleurobema fibuloides*.

Simpson (1900b, 1914) included *P. fibuloides* in the synonymy of *P. altum*. Conrad (1854) described *Unio altum*, giving the type locality as "one of the western states, probably Tennessee". However, the brief description was incomplete, omitting details about the umbo cavity, an important characteristic in the identification of Mobile Basin *Pleurobema* (for which a shallow umbo cavity is characteristic in all but one species). Unfortunately, the type specimen of *U. altum* has been lost, which precludes identification of the species to which this name has been applied. Without comment or discussion, Simpson (1900b) gave the distribution of *P. altum* as Conasauga River, Georgia. Simpson (1914) later gave the type locality of *P. altum* as "Tennessee?" Most subsequent authors have recognized *P. altum* as a Coosa River drainage endemic (Frierson, 1927; Hurd, 1974; Stansbery, 1976). However, with an incomplete description, vague type locality and absence of a type specimen, recognition of *U. altum* is unjustified and is herein considered a *nomen dubium*.

Synonymy

Unio fibuloides Lea, 1859. Lea, 1859b:154; Lea, 1859f:219, pl. 28, fig. 100; Lea, 1859g:37, pl. 28, fig. 100
Type locality: Connasauga [Conasauga] River, Georgia, Bishop Elliott. Lectotype, USNM 85809, length 34 mm, designated by Johnson (1974).

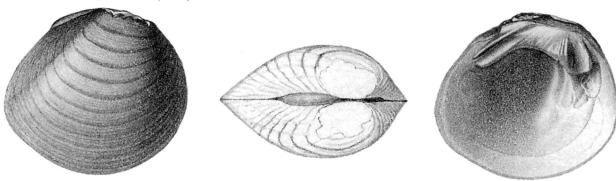

Unio crapulus Lea, 1861. Lea, 1861a:39; Lea, 1866:42, pl. 15, fig. 40; Lea, 1867b:46, pl. 15, fig. 40
Type locality: Etowah River, Georgia, Reverend G. White. Lectotype, USNM 84821, length 38 mm, designated by Johnson (1974).

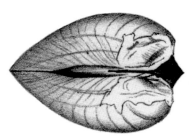
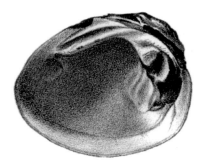

Unio lewisii Lea, 1861. Lea, 1861a:40; Lea, 1862b:71, pl. 8, fig. 220; Lea, 1862d:75, pl. 8, fig. 220
Type locality: Coosa River, Alabama, E.R. Showalter, M.D. Lectotype, USNM 85806, length 34 mm, designated by
    Johnson (1974).

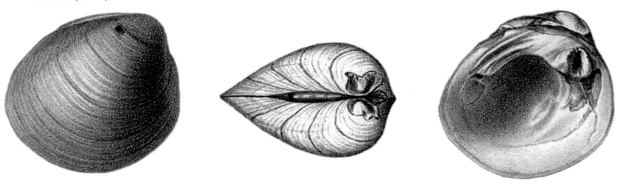

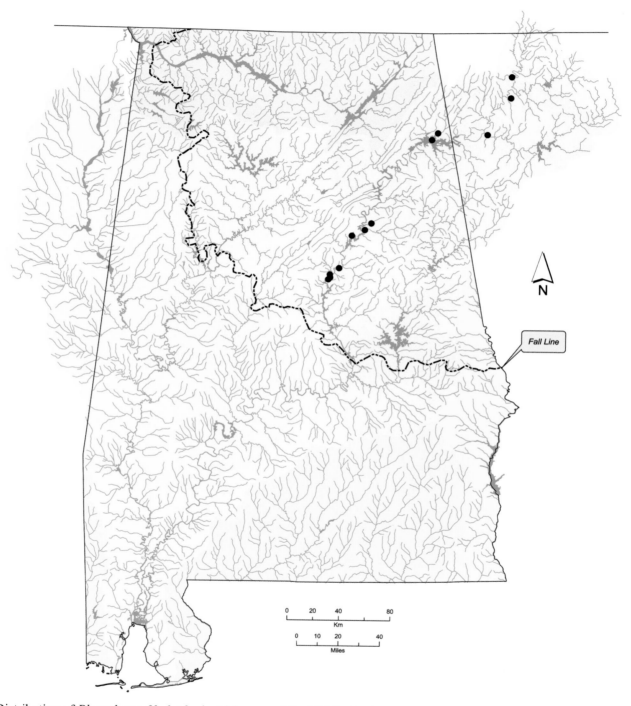

Distribution of *Pleurobema fibuloides* in Alabama and the Mobile Basin.

Pleurobema georgianum (Lea, 1841)
Southern Pigtoe

Pleurobema georgianum – Upper figure: length 58 mm, UF 66700. Shoal Creek, St. Clair County, Alabama, October 1914. Lower figure: length 32 mm, UF 358667. Conasauga River, 0.8 miles above Sumac Creek, Murray County, Georgia, 10 June 1997. © Richard T. Bryant.

Shell Description

Length to 65 mm; moderately thick; moderately compressed to moderately inflated; outline oval to subtriangular; posterior margin narrowly to broadly rounded; anterior margin rounded; dorsal margin convex; ventral margin straight to broadly rounded; posterior ridge rounded; posterior slope moderately steep, flat to slightly convex; umbo broad, slightly inflated, elevated slightly above hinge line, umbo sculpture unknown; periostracum yellowish to brown, often with interrupted green ray or radially arranged green blotches, usually disappearing ventrally.

Pseudocardinal teeth triangular, may be slightly compressed, 2 teeth in left valve, may be divergent, 1 tooth in right valve, sometimes with accessory denticle anteriorly and/or posteriorly; lateral teeth short to moderately long, straight to slightly curved, 2 in left

valve, 1 in right valve, may also have rudimentary lateral tooth ventrally in right valve; interdentum short to moderately long; narrow to moderately wide; umbo cavity shallow; nacre white.

Soft Anatomy Description

Mantle tan, paler outside of pallial line, rusty tan outside of apertures; visceral mass tan; foot tan. Ortmann (1923a) reported soft tissues to be "whitish or pale orange". The soft anatomy of live individuals may also be bright orange (P.D. Johnson, personal communication).

Gills tan; dorsal margin slightly sinuous, ventral margin convex; gill length approximately 65% of shell length; gill height approximately 55% of gill length; outer gill height approximately 80% of inner gill height;

inner lamellae of inner gills connected to visceral mass only anteriorly.

No gravid females were available for marsupium description. Outer gills are marsupial (P.D. Johnson, personal communication), creamy white to pale orange when gravid (Ortmann, 1923a).

Labial palps tan; straight dorsally, convex ventrally, bluntly pointed distally; palp length approximately 20% of gill length; palp height approximately 65% of palp length; distal 35% of palps bifurcate.

Incurrent aperture longer than excurrent and supra-anal apertures; excurrent and supra-anal apertures of approximately equal length.

Incurrent aperture length approximately 20% of shell length; rusty tan within, rusty brown basal to papillae; papillae arborescent; papillae rusty brown basally, rusty tan distally.

Excurrent aperture length approximately 15% of shell length; rusty tan within, marginal color band rusty brown; margin with small simple papillae, little more than crenulations; papillae rusty tan.

Supra-anal aperture length approximately 15% of shell length; pale tan within, without marginal coloration; margin smooth; mantle bridge separating supra-anal and excurrent apertures imperforate, length approximately 25% of supra-anal length.

Specimen examined: Conasauga River (n = 1) (specimen previously preserved).

Glochidium Description

Length 130–160 μm; height 150–160 μm; without styliform hooks; outline subelliptical; dorsal margin straight, ventral margin broadly rounded, anterior and posterior margins convex (Ortmann, 1923a).

Similar Species

Pleurobema georgianum resembles *Pleurobema hanleyianum*, but *P. hanleyianum* is more elongate. Typically the anterior pseudocardinal tooth in the left valve of *P. hanleyianum* is oriented anterioventrally, whereas that of *P. georgianum* is oriented more ventrally.

Some *Pleurobema georgianum* resemble *Pleurobema fibuloides*, but *P. georgianum* has a thinner, less inflated shell. The umbo of *P. fibuloides* is typically positioned more anteriorly than that of *P. georgianum*. The interdentum of *P. fibuloides* is much broader than that of *P. georgianum*.

Some *Pleurobema georgianum* resemble *Pleurobema hartmanianum*, but *P. georgianum* has a thinner shell, typically with green rectangular blotches on the periostracum, which are absent in *P. hartmanianum*. The interdentum of *P. hartmanianum* is much wider than that of *P. georgianum*.

Pleurobema georgianum may resemble small compressed *Ptychobranchus foremanianus* from tributary populations. However, *P. foremanianus* is more elongate and triangular than *P. georgianum*, with a broader umbo and wider interdentum.

Some *Pleurobema georgianum* resemble some *Quadrula kieneriana*. However, *Q. kieneriana* has a deep umbo cavity, heavier pseudocardinal teeth and wider interdentum. *Quadrula kieneriana* frequently has at least a few pustules, which are absent on *P. georgianum*.

General Distribution

Pleurobema georgianum is endemic to the Coosa River drainage of the Mobile Basin in Alabama, Georgia and Tennessee.

Alabama and Mobile Basin Distribution

Pleurobema georgianum occurs above the Fall Line in the Coosa River drainage.

Pleurobema georgianum is known to be extant in only a few Coosa River tributaries and the Conasauga River.

Ecology and Biology

Pleurobema georgianum occurs in riffles, runs and shoals of medium creeks to large rivers, typically in sand and gravel substrates.

Pleurobema georgianum is a short-term brooder, gravid during spring and early summer. Conglutinates are lanceolate, compressed and pink (P.D. Johnson, personal communication). Preliminary laboratory trials suggest that *Cyprinella callistia* (Alabama Shiner), *Cyprinella trichroistia* (Tricolor Shiner) and *Cyprinella venusta* (Blacktail Shiner) (Cyprinidae) are glochidial hosts of *P. georgianum* (P.D. Johnson, personal communication).

Current Conservation Status and Protection

Pleurobema georgianum was considered endangered throughout its range by Williams et al. (1993) and in Alabama by Lydeard et al. (1999). It was designated a species of highest conservation concern in the state by Garner et al. (2004). In 1993 *P. georgianum* was listed as endangered under the federal Endangered Species Act.

Remarks

Large collections of *Pleurobema georgianum* taken during the early 1900s suggest that it was historically one of the more common and widespread species of *Pleurobema* in the Coosa River drainage. However, it is now very rare and occurs in only a few isolated populations.

Synonymy

Unio georgianus Lea, 1841. Lea, 1841a:31; Lea, 1842b:235, pl. 21, fig. 49; Lea, 1842c:73, pl. 21, fig. 49
Type locality: Stump [Stamp] Creek, [Bartow County,] Georgia, T.R. Dutton. Holotype, USNM 84928, length 41 mm.

 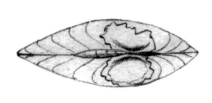 

Unio nucleopsis Conrad, 1849. Conrad, 1849:154; Conrad, 1850:276, pl. 37, fig. 8
Type locality: Etowah River, Georgia. Holotype, ANSP 41421, length 32 mm.

Unio troschelianus Lea, 1852. Lea, 1852b:280, pl. 23, fig. 39; Lea, 1852c:36, pl. 23, fig. 39
Type locality: Coosawattee River, Murray County, Georgia, Dr. Boykin. Lectotype, USNM 84790, length 32 mm, designated by Johnson (1974).

Unio favosus Lea, 1856. Lea, 1856d:262; Lea, 1858e:58, pl. 8, fig. 40; Lea, 1858h:58, pl. 8, fig. 40
Type locality: Othcalooga Creek, Gordon County, Georgia, Bishop Elliott. Lectotype, USNM 84585, length 51 mm, designated by Johnson (1974).

Unio irrasus Lea, 1861. Lea, 1861a:38; Lea, 1862b:91, pl. 13, fig. 239; Lea, 1862d:95, pl. 13, fig. 239
Type locality: Etowah River, Georgia, Reverend G. White. Lectotype, USNM 84637, length 36 mm, designated by Johnson (1974).

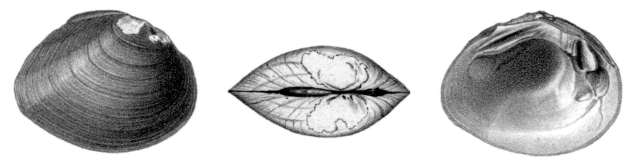

Unio medius Lea, 1861. Lea, 1861a:40; Lea, 1862b:78, pl. 10, fig. 227; Lea, 1862d:82, pl. 10, fig. 227
Type locality: Near Coosa River, Alabama, E.R. Showalter, M.D. Lectotype, USNM 84794, length 33 mm, designated by Johnson (1974).

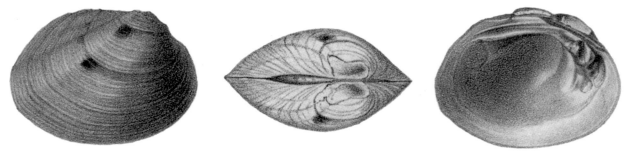

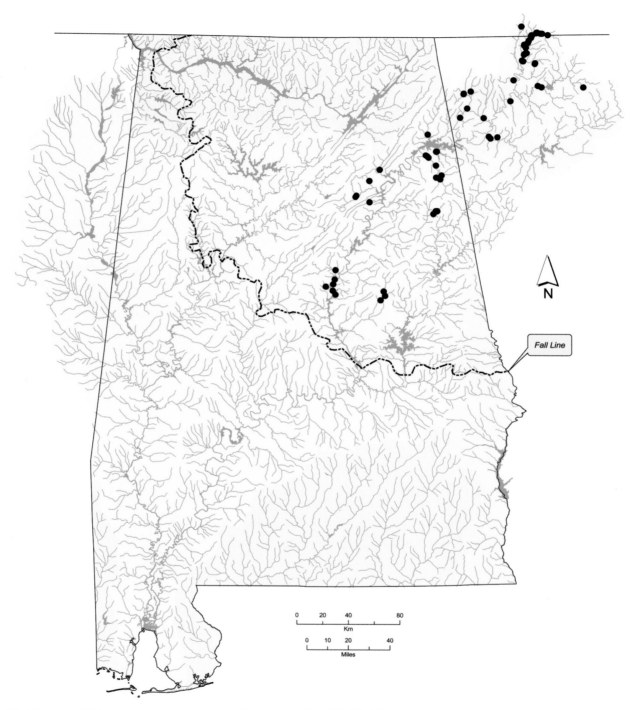

Distribution of *Pleurobema georgianum* in Alabama and the Mobile Basin.

Pleurobema hanleyianum (Lea, 1852)
Georgia Pigtoe

Pleurobema hanleyianum – Length 37 mm, USNM 84717 (holotype). Coosawattee River, Georgia. © Richard T. Bryant.

Shell Description

Length to 50 mm; moderately thin; moderately inflated; outline oval; posterior margin narrowly rounded; anterior margin rounded; dorsal margin convex; ventral margin straight to slightly convex; posterior ridge low, rounded; posterior slope moderately wide, flat; umbo broad, may be slightly inflated, elevated slightly above hinge line, umbo sculpture unknown; periostracum yellowish to dark brown, may have concentric green rings that usually become thinner and less distinct ventrally.

Pseudocardinal teeth triangular, 2 divergent teeth in left valve, anterior tooth more compressed than posterior tooth, crest of anterior tooth approximately parallel to margin, oriented anterioventrally, 1 tooth in right valve, may have accessory denticle anteriorly and/or posteriorly; lateral teeth moderately long, thin to moderately thick, straight to slightly curved, 2 in left valve, 1 in right valve; interdentum short, narrow; umbo cavity shallow; nacre white.

Soft Anatomy Description

Mantle tan or grayish tan to brown, often grayish brown outside of apertures, may be paler outside of pallial line; visceral mass tan; foot tan.

Gills tan; dorsal margin straight to slightly sinuous, ventral margin elongate convex; gill length 55–70% of shell length; gill height 50–60% of gill length, occasionally less; outer gill height 70–85% of inner gill height; inner lamellae of inner gills connected to visceral mass only anteriorly.

Outer gills marsupial; glochidia held across entire gill; marsupium padded when gravid; pale tan.

Labial palps tan, sometimes with brown areas; straight dorsally, convex ventrally, bluntly pointed distally; palp length 15–25% of gill length; palp height 45–50% of palp length; distal 25–35% of palps bifurcate.

Incurrent aperture longer than excurrent and supra-anal apertures; supra-anal aperture usually slightly longer than excurrent aperture, occasionally of equal length.

Incurrent aperture length 15–20% of shell length; tan to rusty tan within, often with rusty brown or dark brown basal to papillae; papillae arborescent, often with occasional simple and bifid; papillae tan to rusty brown or dark brown, often lighter distally, areas between papillae may be darker.

Excurrent aperture length 10–15% of shell length; tan or rusty tan within, often with brown or rusty brown marginal color band, some individuals with 1–2 additional tan or brown bands proximally; margin papillate, papillae small, may be well-developed or little more than crenulations; papillae tan or rusty tan to brown.

Supra-anal aperture length 10–15% of shell length; creamy white to pale tan or pale rusty tan within, margin often with sparse rusty tan or rusty brown; margin smooth; mantle bridge separating supra-anal and excurrent apertures imperforate, length 15–25% of supra-anal length, occasionally absent.

Specimens examined: Conasauga River, Georgia (n = 4) (specimens previously preserved).

Glochidium Description

Glochidium unknown.

Similar Species

Pleurobema hanleyianum most closely resembles *Pleurobema georgianum* but is more elongate. Some *P. georgianum* are more triangular than *P. hanleyianum*. The anterior pseudocardinal tooth in the left valve is oriented anterioventrally in *P. hanleyianum*, whereas that of *P. georgianum* is typically oriented more ventrally.

Pleurobema hanleyianum may also superficially resemble *Pleurobema stabilis* but is less inflated, more elongate and typically has a thinner shell. *Pleurobema stabilis* may have a very shallow sulcus, which is absent in *P. hanleyianum*. Also, the hinge plate and teeth of *P. hanleyianum* are not as thick as those of *P. stabilis*.

General Distribution

Pleurobema hanleyianum is endemic to the Coosa River drainage in Alabama, Georgia and Tennessee.

Alabama and Mobile Basin Distribution

Pleurobema hanleyianum is known from the Coosa River and some tributaries and headwater rivers.

Pleurobema hanleyianum is believed to be extant in the Conasauga River, Georgia, but is extremely rare.

Ecology and Biology

Pleurobema hanleyianum inhabits shoals of large creeks and small to large rivers.

Pleurobema hanleyianum is believed to be a short-term brooder, gravid in spring and summer. Its glochidial hosts are unknown, but some *Pleurobema* species use members of the Cyprinidae (Haag and Warren, 2003).

Current Conservation Status and Protection

Pleurobema hanleyianum was reported as endangered throughout its range by Athearn (1970) and Williams et al. (1993) and in Alabama by Lydeard et al. (1999). Garner et al. (2004) erroneously listed it as extinct. *Pleurobema hanleyianum* was elevated to a candidate for protection under the federal Endangered Species Act in 2003.

Remarks

The species name of *Pleurobema hanleyianum* has been variously spelled, with and without the "i". The incorrect spelling, "*hanleyanum*", has been used by many authors.

Pleurobema hanleyianum was named for Sylvanus Hanley, Esq., one of the authors of *History of British Mollusca* and several other major conchological volumes.

Synonymy

Unio hanleyianus Lea, 1852, *nomen nudum*. Lea, 1852a:26, 61, 71
Unio hanleyianus Lea, 1852. Lea, 1852b:279, pl. 23, fig. 37; Lea, 1852c:35, pl. 23, fig. 37
Type locality: Coosawattee River, Murray County, Georgia, Dr. Boykin. Lectotype, USNM 84717, length 37 mm, designated by Johnson (1974).

Pleurobema aldrichianum Goodrich, 1931. Goodrich, 1931:1–4, pl. 1
Type locality: Conasauga River, near Conasauga, Polk County, Tennessee, H.H. Smith, 20 October 1916. Holotype originally deposited in the Alabama Museum of Natural History, number 83, but was subsequently transferred to the Florida Museum of Natural History, UF 66466, length 46 mm.

 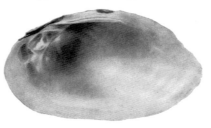 

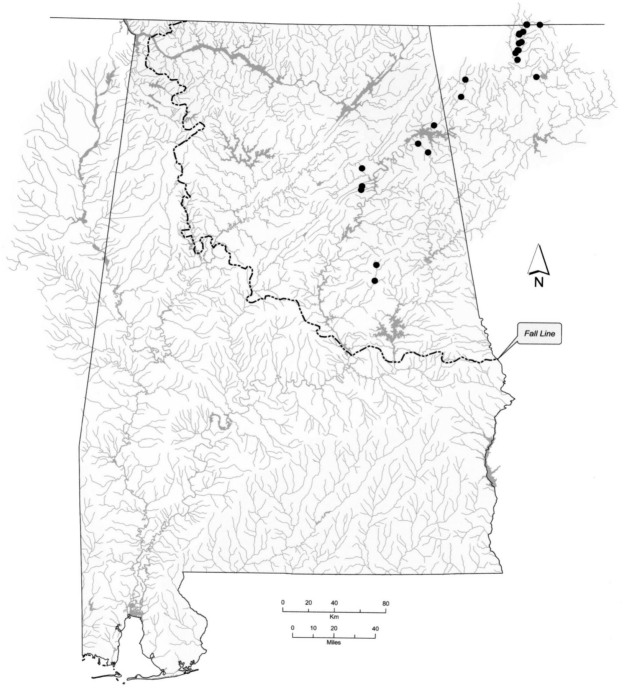

Distribution of *Pleurobema hanleyianum* in Alabama and the Mobile Basin.

Pleurobema hartmanianum (Lea, 1860)
Cherokee Pigtoe

 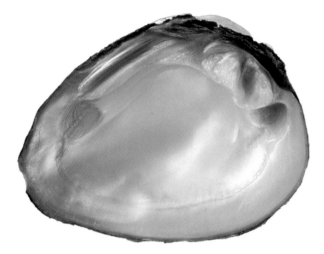

Pleurobema hartmanianum – Length 43 mm, UMMZ 81017. Coosa River, Weduska Shoals, Shelby County, Alabama, August 1913. © Richard T. Bryant.

Shell Description

Length to 55 mm; thick; moderately inflated; outline round in young individuals, becoming subtriangular with age; posterior margin narrowly rounded; anterior margin straight to convex; dorsal margin convex; ventral margin straight to broadly rounded; posterior ridge low, rounded; posterior slope steep to moderately steep; umbo moderately broad, inflated, elevated above hinge line, umbo sculpture unknown; periostracum satiny, olive brown to brown, occasionally with dark green rays.

Pseudocardinal teeth thick, triangular, 2 slightly divergent teeth in left valve, 1 tooth in right valve, usually with accessory denticle posteriorly, occasionally anteriorly; lateral teeth moderately short, thick, straight to slightly curved, 2 in left valve, 1 in right valve, may also have rudimentary lateral tooth ventrally in right valve; interdentum short, moderately wide; umbo cavity very shallow; nacre white.

Soft Anatomy Description

Soft anatomy unknown.

Glochidium Description

Glochidium unknown.

Similar Species

Pleurobema hartmanianum most closely resembles *Pleurobema fibuloides* but is more triangular in outline. *Pleurobema fibuloides* is oval in outline.

Pleurobema hartmanianum also resembles some *Pleurobema georgianum* but has a thicker shell. The periostracum of *P. georgianum* often has an interrupted green ray or rectangular blotches, which are absent in *P. hartmanianum*.

Pleurobema hartmanianum also superficially resembles *Pleurobema stabilis* but typically has a much steeper posterior slope, giving it a more triangular outline. The periostracum of *P. hartmanianum* is generally more greenish than that of *P. stabilis*, which is yellowish to tawny.

General Distribution

Pleurobema hartmanianum is endemic to the Coosa River drainage in Alabama and Georgia.

Alabama and Mobile Basin Distribution

Pleurobema hartmanianum occurred from Coosa headwaters downstream to Wetumpka, near its mouth.

Pleurobema hartmanianum was believed to be extinct for most of the 1900s, but a single live individual was collected from the Coosawattee River, Georgia, in 1997 (Figure 13.32).

Ecology and Biology

Pleurobema hartmanianum appears to have been restricted to shoal habitats based on historical collection data.

Pleurobema hartmanianum is believed to be a short-term brooder, gravid in spring and summer. Its glochidial hosts are unknown, but some other

Pleurobema species utilize members of the Cyprinidae (Haag and Warren, 2003).

Current Conservation Status and Protection

Pleurobema hartmanianum was recognized as endangered throughout its range by Athearn (1970) and in Alabama by Stansbery (1976). It was listed as extinct by the USFWS (2000). *Pleurobema hartmanianum* was not recognized by Lydeard et al. (1999) or Garner et al. (2004). A single live individual was collected in the lower reach of the Coosawattee River, Gordon County, Georgia, in September 1997 (Figure 13.32).

Remarks

Pleurobema hartmanianum was recognized by some authors (e.g., Frierson, 1927; Stansbery, 1976), but was not included in Turgeon et al. (1988, 1998).

Pleurobema hartmanianum was named for W.D. Hartman, M.D., a conchologist who provided Lea with large numbers of mussel specimens.

Figure 13.32. *Pleurobema hartmanianum*. Length 64 mm, UF 374239. Coosawattee River, 2.6 miles above Georgia State Highway 25, Gordon County, Georgia, 25 September 1997. © Richard T. Bryant.

Synonymy

Unio hartmanianus Lea, 1860. Lea, 1860c:307; Lea, 1862b:73, pl. 8, fig. 222; Lea, 1862d:77, pl. 8, fig. 222
Type locality: Coosa River, Watumpka [Wetumpka], [Elmore County,] Alabama, E.R. Showalter, M.D. Lectotype, USNM 84652, length 39 mm, designated by Johnson (1974).

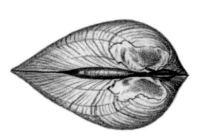
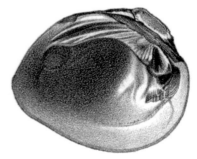

Unio showalterii Lea, 1860. Lea, 1860c:307; Lea, 1862b:73, pl. 8, fig. 223; Lea, 1862d:77, pl. 8, fig. 223
Type locality: Coosa River, Watumpka [Wetumpka], [Elmore County,] Alabama, E.R. Showalter, M.D. Lectotype, USNM 85748, length 30 mm, designated by Johnson (1974).

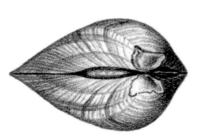
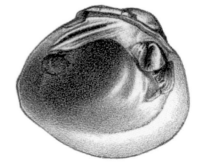

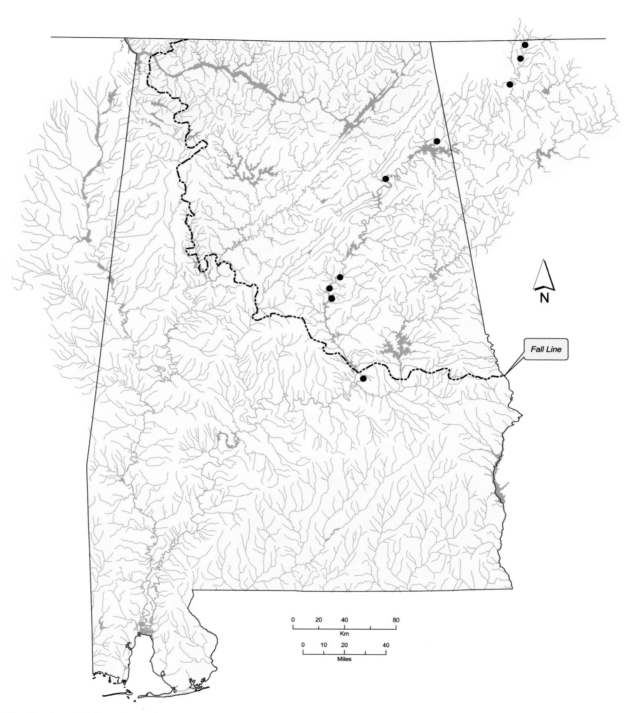

Distribution of *Pleurobema hartmanianum* in Alabama and the Mobile Basin.

Pleurobema marshalli Frierson, 1927
Flat Pigtoe

Pleurobema marshalli – Length 50 mm, UF 358666. Tombigbee River, Memphis Landing, river mile 324.4, Pickens County, Alabama, 24 October 1976. © Richard T. Bryant.

Shell Description

Length to 65 mm; thick, thinner posteriorly; moderately inflated; outline oval; posterior margin narrowly rounded; anterior margin obliquely truncate to convex; dorsal margin convex; ventral margin broadly rounded; posterior ridge low, rounded; posterior slope moderately steep; disk usually with flattened area just anterior of posterior ridge; umbo broad, inflated, directed anteriad, elevated well above hinge line, umbo sculpture unknown; periostracum tawny to dark brown or olive brown, occasionally with weak rays.

Pseudocardinal teeth moderately thick, triangular, 2 divergent teeth in left valve, 1 tooth in right valve, may have accessory denticle anteriorly; lateral teeth moderately long, straight to slightly curved, 2 in left valve, 1 in right valve; interdentum short, narrow to moderately wide; umbo cavity shallow; nacre white to creamy white.

Soft Anatomy Description

Mantle tan, may have rusty tan mottling outside of pallial line; visceral mass tan; foot tan.

Gills tan; dorsal margin slightly sinuous, ventral margin straight to elongate convex; gill length 55–65% of shell length; gill height 45–60% of gill length; outer gill height 85–90% of inner gill height; inner lamellae of inner gills connected to visceral mass only anteriorly.

Outer gills marsupial; marsupium slightly padded when gravid; pale tan.

Labial palps tan; straight to slightly convex dorsally, convex ventrally, bluntly pointed distally; palp length 20–30% of gill length; palp height 40–50% of palp length; distal 30–40% of palps bifurcate.

Incurrent aperture longer than excurrent aperture; supra-anal aperture usually longer than incurrent aperture, occasionally equal in length.

Incurrent aperture length 15–20% of shell length; tan within, may have rusty brown basal to papillae; papillae arborescent, usually with occasional simple and bifid papillae; papillae rusty brown.

Excurrent aperture length 10–15% of shell length; tan within, marginal color band rusty brown, may have irregular dark brown spots; margin with small, simple papillae, may be little more than crenulations; papillae tan to rusty brown.

Supra-anal aperture length 15–25% of shell length; tan within, with narrow rusty brown marginal band; margin smooth, sometimes undulating; mantle bridge separating supra-anal and excurrent apertures imperforate, length 5–15% of supra-anal length.

Specimens examined: Tombigbee River (n = 3) (specimens previously preserved).

Glochidium Description

Glochidium unknown.

Similar Species

Pleurobema marshalli closely resembles *Fusconaia ebena* externally. Some *P. marshalli* can be distinguished by the slightly flattened area just anterior of the posterior ridge. That area is usually rounded in *F. ebena*. Also, the periostracum of *P. marshalli* tends to be more yellowish brown than that of *F. ebena*. Internally the two are easily distinguished. *Pleurobema marshalli* has a shallow umbo cavity, whereas that of *F. ebena* is deep and compressed.

Pleurobema marshalli superficially resembles *Pleurobema taitianum* but is less triangular in outline. *P. taitianum* has a shallow sulcus just anterior to the posterior ridge, where *P. marshalli* is typically flattened.

Pleurobema marshalli may superficially resemble *Obovaria unicolor* but the umbo of *P. marshalli* is placed much farther anteriorly. Also, pseudocardinal teeth of *P. marshalli* are heavier than those of *O. unicolor*.

General Distribution

Pleurobema marshalli is endemic to the Tombigbee River drainage in Alabama and Mississippi.

Alabama and Mobile Basin Distribution

Pleurobema marshalli is known historically from the Tombigbee River from Epes, Sumter County, Alabama, upstream to the vicinity of Columbus, Lowndes County, Mississippi. An archaeological record from the lower Tombigbee River suggests that it had a wider distribution in the drainage prehistorically.

Pleurobema marshalli is extirpated from Alabama and may be extinct. No live individuals have been collected since the late 1970s.

Ecology and Biology

Pleurobema marshalli is a species of large river shoals with moderate to swift current. Its preferred substrates are sand and gravel.

Pleurobema marshalli is presumably a short-term brooder, gravid in spring and summer. Its conglutinates are creamy white, lanceolate and compressed. Glochidial hosts of *P. marshalli* are unknown, but some species of *Pleurobema* utilize members of the Cyprinidae (Haag and Warren, 2003).

Current Conservation Status and Protection

Pleurobema marshalli was considered endangered throughout its range by Williams et al. (1993) and in Alabama by Stansbery (1976) and Lydeard et al. (1999). Garner et al. (2004) listed *P. marshalli* as extirpated from the state. This species was listed as endangered under the federal Endangered Species Act in 1987.

Remarks

In describing *Pleurobema marshalli*, Frierson (1927) noted that Conrad (1834a) may have illustrated this species under the name *Unio mytiloides* (Rafinesque, 1820). However, the specimen figured by Conrad (1834a) has a deep umbo cavity, broad hinge plate and pink nacre, which are not characteristic of *P. marshalli*.

The extinction of *Pleurobema marshalli* was a direct result of destruction of riverine habitat during construction of the Tennessee Tombigbee Waterway by the USACE. This species has not been collected since completion of the project.

Pleurobema marshalli was named for William B. Marshall of the USNM, a malacologist who included unionids among his many molluscan interests.

Synonymy

Pleurobema marshalli Frierson, 1927. Frierson, 1927:43; Frierson 1928:139, pl. 3, fig. 3
Type locality: Tombigbee River, Boligee, [Greene County,] Alabama. Holotype, UMMZ 80859, length 55 mm.

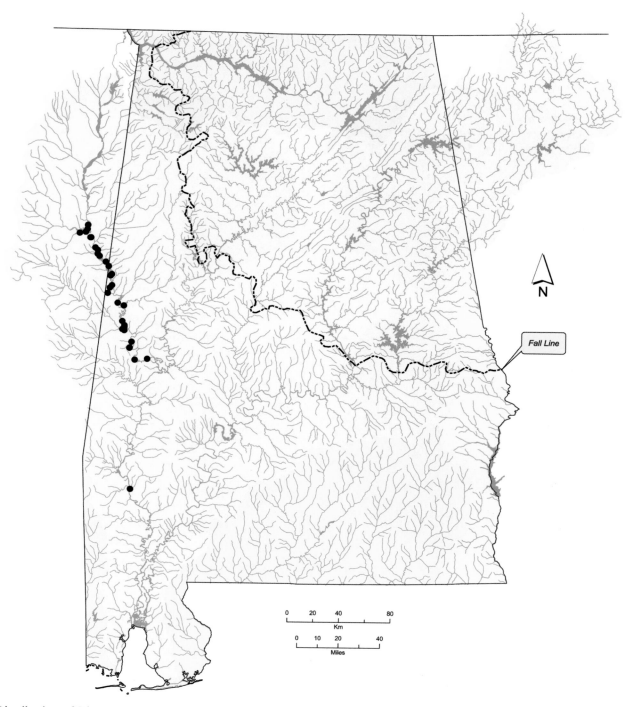

Distribution of *Pleurobema marshalli* in Alabama and the Mobile Basin.

Pleurobema oviforme (Conrad, 1834)
Tennessee Clubshell

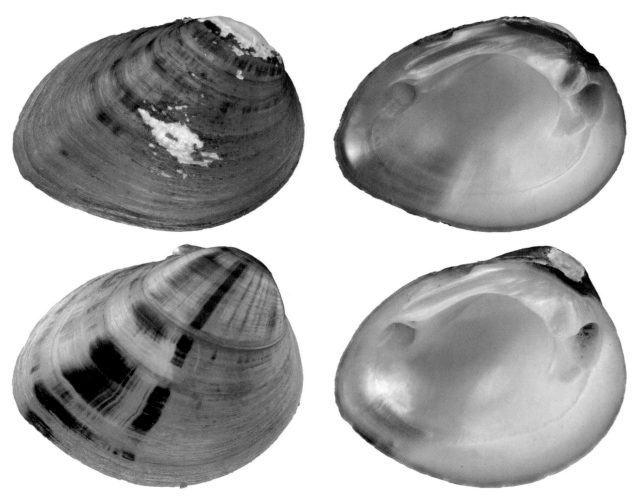

Pleurobema oviforme – Upper figure: length 40 mm, USNM 26071. Lower figure: length 28 mm, USNM 26071. Holston River, Tennessee. © Richard T. Bryant.

Shell Description

Length to 90 mm; moderately thin to moderately thick, individuals from large rivers thicker than those from small streams; moderately compressed to moderately inflated, individuals from large rivers more inflated than those from small streams; outline oval to triangular, usually more triangular in individuals from large rivers; posterior margin broadly to narrowly rounded; anterior margin truncate to rounded; dorsal margin convex; ventral margin straight to convex; posterior ridge low, rounded; posterior slope flat, moderately steep; umbo broad, may be slightly inflated, elevated above hinge line, umbo sculpture unknown; periostracum tawny to brown, usually with variable, often wide, broken green rays, rays may not extend to ventral margin, rays often absent in individuals from small streams.

Pseudocardinal teeth triangular, often compressed, usually thicker in individuals from large rivers, 2 somewhat divergent teeth in left valve, may be more divergent in individuals from large rivers, 1 tooth in right valve, may have accessory denticle anteriorly; lateral teeth usually moderately long, occasionally short, slightly curved, 2 in left valve, 1 in right valve, may also have rudimentary lateral tooth ventrally in right valve; interdentum short to moderately long, narrow to moderately wide; umbo cavity shallow; nacre white.

Soft Anatomy Description

Mantle creamy white to tan, outside of apertures rusty tan to grayish brown; visceral mass creamy white to pale tan, may have pink cast from underlying gonadal tissues; foot white to very pale orange.

Gills tan; dorsal margin straight to slightly sinuous, ventral margin convex; gill length 60–65% of shell length; gill height 45–60% of gill length; outer gill height 75–90% of inner gill height; inner lamellae of inner gills connected to visceral mass only anteriorly.

Outer gills marsupial; glochidia held across entire gill except extreme anterior and posterior ends; marsupium slightly padded when gravid; pink when holding embryos, becoming creamy white or tan as glochidia mature.

Labial palps tan to golden tan, often with irregular creamy white area dorsally; straight to slightly concave dorsally, convex ventrally, bluntly pointed distally; palp length 15–25% of gill length; palp height 40–75% of palp length, occasionally greater; distal 35–50% of palps bifurcate.

Incurrent, excurrent and supra-anal apertures of variable relative lengths, any of the three may be longer than the others.

Incurrent aperture length 15–20% of shell length, occasionally less; creamy white to tan within, often with rusty tan or grayish brown basal to papillae; papillae in 1–2 rows, simple, bifid, trifid and small arborescent, thick, blunt; papillae tan or rusty tan to grayish brown or dark brown, may have darker edges basally.

Excurrent aperture length 15–25% of shell length; creamy white to tan within, marginal color band rusty tan to grayish brown; margin usually with small, simple papillae, may be well-developed, occasional papillae slightly bifid, some individuals with crenulate margin; papillae rusty tan to grayish tan or grayish brown.

Supra-anal aperture length 15–25% of shell length; creamy white within, usually without marginal coloration, irregular grayish brown when present; margin smooth; mantle bridge separating supra-anal and excurrent apertures imperforate, length 5–15% of supra-anal length, occasionally absent.

Specimens examined: Paint Rock River (n = 5).

Glochidium Description

Length 150–179 μm; height 155–170 μm; without styliform hooks; outline subelliptical (Ortmann, 1921; Kitchel, 1985; Weaver et al., 1991).

Similar Species

Elongate individuals of *Pleurobema oviforme* resemble *Pleurobema clava*. However, *P. clava* has the umbo at or overhanging the anterior end of the shell. *Pleurobema oviforme* generally lacks a sulcus, whereas *P. clava* usually has one.

Pleurobema oviforme resembles *Fusconaia cor*, *Fusconaia cuneolus* and some *Fusconaia subrotunda* but can be separated from those species by its shallow umbo cavity. The umbo cavity of the *Fusconaia* species is deep and compressed.

Pleurobema oviforme also resembles *Pleuronaia barnesiana* but has more triangular, divergent pseudo-cardinal teeth.

Pleurobema oviforme resembles *Pleuronaia dolabelloides*. However, *P. dolabelloides* has a posterior ridge that is ventrally curved, whereas that of *P. oviforme* is straight. Also, the foot of *P. dolabelloides* is orange compared to the white or tan foot of *P. oviforme*.

General Distribution

Pleurobema oviforme is endemic to the Cumberland and Tennessee River drainages. It occurs only downstream of Cumberland Falls in the Cumberland River drainage, Kentucky and Tennessee (Cicerello et al., 1991; Parmalee and Bogan, 1998). In the Tennessee River drainage *P. oviforme* is known from headwaters in southwestern Virginia, western North Carolina and eastern Tennessee downstream to Muscle Shoals, northwestern Alabama (Ahlstedt, 1992a, 1992b; Parmalee and Bogan, 1998). A disjunct population exists in the Duck River system in central Tennessee.

Alabama and Mobile Basin Distribution

Pleurobema oviforme is known from the Tennessee River and many tributaries across northern Alabama.

Pleurobema oviforme is known to be extant only in the Paint Rock River system.

Ecology and Biology

Pleurobema oviforme is a species of small to large rivers, where it occurs in shoal habitats. Its preferred substrate is a mixture of gravel and sand.

Pleurobema oviforme is a dioecious species and reaches sexual maturity by age four or five. It is a short-term brooder, spawning from late March into May, with glochidia fully developed by late June (Weaver et al., 1991). Females remain gravid as late as mid-July (Ortmann, 1921). *Pleurobema oviforme* conglutinates are lanceolate, flattened proximally, cylindrical distally, and often occur in clumps of two or three joined proximally. The conglutinates are pink when composed of embryos. Numbers of *P. oviforme* glochidia in stream drift have been reported to peak in mid-July (Kitchel, 1985).

Glochidial hosts of *Pleurobema oviforme* determined using laboratory trials include *Campostoma anomalum* (Central Stoneroller), *Cyprinella galactura* (Whitetail Shiner) and *Nocomis micropogon* (River

Chub) (Cyprinidae); and *Etheostoma flabellare* (Fantail Darter) (Percidae) (Weaver et al., 1991). Weaver et al. (1991) observed transformation of *P. oviforme* glochidia on an additional species, but the report is unclear as to its identity, listing *Luxilus cornutus* (Common Shiner) (Cyprinidae) in a table but mentioning *Luxilus chrysocephalus* (Striped Shiner) (Cyprinidae) in the results.

Current Conservation Status and Protection

Pleurobema oviforme was considered a species of special concern throughout its range by Williams et al. (1993). In Alabama it was reported as endangered by Stansbery (1976) and a species of special concern by Lydeard et al. (1999). Garner et al. (2004) designated *P. oviforme* a species of highest conservation concern in the state.

Remarks

Pleurobema oviforme demonstrates clinal variation in degree of shell inflation, with headwater forms more compressed than those from large rivers. This has resulted in considerable confusion and a sizable synonymy (Ortmann, 1920).

Pleurobema oviforme is reported to be rare in prehistoric shell middens along the Tennessee River at Muscle Shoals (Morrison, 1942; Hughes and Parmalee, 1999) and near Bridgeport (Warren, 1975).

The relationship between *Pleurobema oviforme* and *Pleurobema clava* has been a matter of debate. *Pleurobema oviforme* replaces *P. clava* in headwaters of the Tennessee River, and it has been suggested by some authors that they are conspecific. Ortmann (1925) stated that *P. oviforme* and *P. clava* may be "only varieties of one species". However, Warren (1975) reported no intergrades between the two species in prehistoric midden material from near Bridgeport.

Synonymy

Unio oviformis Conrad, 1834. Conrad, 1834b:46, 70, pl. 3, fig. 6
Type locality: Rivers in Tennessee. Type specimen not found, also not found by Johnson and Baker (1973). Length not given in original description.

Unio ravenelianus Lea, 1834. Lea, 1834a:32, pl. 3, fig. 5; Lea, 1834b:144, pl. 3, fig. 5
Unio rudis Conrad, 1837. Conrad, 1837:76, pl. 43, fig. 1 [unnecessary replacement name for *Unio ravenelianus* Lea, 1834]
Unio holstonensis Lea, 1840. Lea, 1840:288; Lea, 1842b:212, pl. 15, fig. 27; Lea, 1842c:50, pl. 15, fig. 27
Unio argenteus Lea, 1841. Lea, 1841b:82; Lea, 1842b:242, pl. 25, fig. 57; Lea, 1842c:80, pl. 25, fig. 57
Unio mundus Lea, 1857. Lea, 1857b:83; Lea, 1866:40, pl. 14, fig. 38; Lea, 1867b:44, pl. 14, fig. 38
Type locality: [Tennessee River,] Tuscumbia, [Colbert County,] Alabama, L.B. Thornton, Esq. Lectotype, USNM 84767, length 38 mm, designated by Johnson (1974).

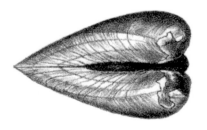

Unio lesleyi Lea, 1860. Lea, 1860c:306; Lea, 1860d:352, pl. 58, fig. 177; Lea, 1860g:34, pl. 58, fig. 177
Unio tesserulae Lea, 1861. Lea, 1861d:392; Lea, 1866:40, pl. 15, fig. 39; Lea, 1867b:44, pl. 15, fig. 39

Unio striatissimus Anthony, 1865. Anthony, 1865:156, pl. 12, fig. 1

Unio clinchensis Lea, 1867. Lea, 1867a:81; Lea, 1868b:278, pl. 37, fig. 91; Lea, 1869a:38, pl. 37, fig. 91

Unio planior Lea, 1868. Lea, 1868a:145; Lea, 1868c:316, pl. 1, fig. 129; Lea, 1869a:76, pl. 50, fig. 129

Unio lawii Lea, 1871. Lea, 1871:189; Lea, 1874c:8, pl. 2, fig. 4; Lea, 1874e:12, pl. 2, fig. 4

Unio acuens Lea, 1871. Lea, 1871:190; Lea, 1874c:27, pl. 8, fig. 24; Lea, 1874e:35, pl. 8, fig. 24

Unio pattinoides Lea, 1871. Lea, 1871:193; Lea, 1874c:16, pl. 4, fig. 12; Lea, 1874e:20, pl. 4, fig. 12

Unio conasaugaensis Lea, 1872. Lea, 1872b:155; Lea, 1874c:33, pl. 10, fig. 30; Lea, 1874e:37, pl. 10, fig. 30

Comment: The type locality for this species is Conasauga Creek, Monroe County, Tennessee, a tributary of the Hiwassee River, Tennessee River drainage, not to be confused with the Conasauga River of the Mobile Basin.

Unio brevis Lea, 1872. Lea, 1872b:157; Lea, 1874c:35, pl. 12, fig. 32; Lea, 1874e:39, pl. 12, fig. 32

Unio bellulus Lea, 1872. Lea, 1872b:161; Lea, 1874c:50, pl. 17, fig. 48; Lea, 1874e:54, pl. 17, fig. 48

Type locality: Holston River, Dr. Edgar; Tennessee River, Rev. G. White; Mussel [Muscle] Shoals, Tennessee River, [Colbert County,] Alabama, C.M. Wheatley. Lectotype, USNM 84772, length 37 mm, designated by Johnson (1974), is from Muscle Shoals, Tennessee River, Colbert County, Alabama.

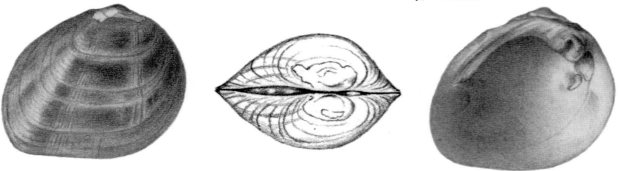

Unio swordianus S.H. Wright, 1897. S.H. Wright, 1897:4; Johnson, 1967b:9, pl. 3, fig. 3

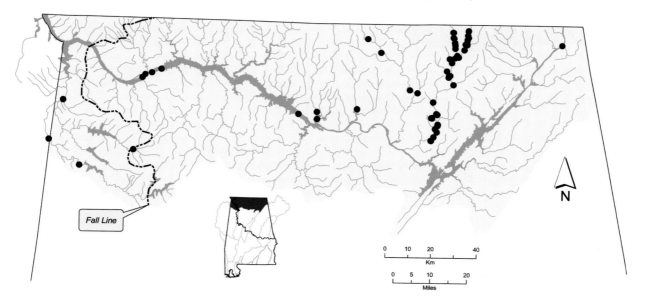

Distribution of *Pleurobema oviforme* in the Tennessee River drainage of Alabama.

Pleurobema perovatum (Conrad, 1834)
Ovate Clubshell

Pleurobema perovatum – Upper figure: length 56 mm, UF 358668. Tombigbee River, Memphis Landing, river mile 324.4, Pickens County, Alabama, 24 October 1976. Lower figure: length 35 mm, UF 358669. McCalls Creek, County Road 56, 3.8 miles northwest of State Highway 41 at Hybart, Wilcox County, Alabama, 19 September 1999. © Richard T. Bryant.

Shell Description

Length 60 mm; moderately thick; moderately inflated; outline oval to subtriangular; posterior margin narrowly rounded to obliquely truncate; anterior margin rounded to somewhat truncate; dorsal margin slightly convex; ventral margin convex; posterior ridge rounded; posterior slope moderately steep, less steep posphalteroventrally; umbo narrow, inflated, elevated above hinge line, turned inward, umbo sculpture unknown; periostracum tawny, greenish brown, brown or black, usually without rays.

Pseudocardinal teeth triangular, compressed, 2 slightly divergent teeth in left valve, anterior tooth usually larger, its crest oriented approximately parallel to margin, 1 tooth in right valve; lateral teeth short, usually straight, may be slightly curved, 2 in left valve, 1 in right valve, may also have rudimentary lateral tooth ventrally in right

valve; interdentum short, narrow; umbo cavity wide, shallow; nacre white to bluish white.

Soft Anatomy Description

Mantle tan, outside of apertures rusty tan; visceral mass pearly white to tan; foot tan to pale orange.

Gills tan, with slight golden cast; dorsal margin slightly sinuous, ventral margin convex; gill length 50–60% of shell length; gill height 50–55% of gill length; outer gill height 75–80% of inner gill height; inner lamellae of inner gills connected to visceral mass only anteriorly.

No gravid females were available for marsupium description; probably similar to other *Pleurobema* species, with outer gills marsupial, padded when gravid.

Labial palps tan, may have a paler area dorsally; straight to slightly concave dorsally, convex ventrally, bluntly pointed distally; palp length 25–40% of gill length; palp height 55–65% of palp length; distal 25–35% of palps bifurcate.

Incurrent, excurrent and supra-anal apertures of similar length.

Incurrent aperture length approximately 15% of shell length; light tan within, with sparse grayish brown or rusty brown basal to papillae; papillae in 1–2 rows, long, mostly simple, sometimes with a few bifid or small arborescent; papillae some combination of tan, light brown and rusty brown, may change color distally, may have dark brown edges basally.

Excurrent aperture length approximately 15% of shell length; light tan within, rusty brown or dark brown marginal color band may have regular tan markings basal to alternating papillae or oblong blotches perpendicular to margin; margin papillate, papillae simple, small but well-developed; papillae tan to rusty brown.

Supra-anal aperture length approximately 15% of shell length; light tan with slight grayish cast within, may have very thin dark brown marginal band; margin smooth; mantle bridge separating supra-anal and excurrent apertures imperforate, length 15–25% of supra-anal length.

Specimens examined: Chewacla Creek, Macon County (n = 1); Yellow Creek, Lowndes County, Mississippi (n = 1) (specimens previously preserved).

Glochidium Description

Glochidium unknown.

Similar Species

Some *Pleurobema perovatum* resemble *Pleurobema rubellum* but are more elongate and often more narrowly rounded posteriorly. Also, the umbo of *P. perovatum* is often narrower and more inflated than that of *P. rubellum*. The crest of the anterior pseudocardinal tooth in the left valve of *P. perovatum* is generally oriented almost parallel to the shell margin, whereas pseudocardinal teeth of *P. rubellum* are oriented more ventrally.

Some *Pleurobema perovatum* resemble small *Pleurobema decisum*. However, the umbo of *P. perovatum* is positioned more posteriorly than that of *P.*

decisum, in which the umbo is typically at the anterior margin.

General Distribution

Pleurobema perovatum is endemic to the Mobile Basin of Alabama and Mississippi.

Alabama and Mobile Basin Distribution

Pleurobema perovatum is widespread in the Mobile Basin with the exception of the Coosa and Tallapoosa River drainages above the Fall Line.

Pleurobema perovatum is extant in localized populations in tributaries of the Tombigbee and Alabama rivers, including the Cahaba River.

Ecology and Biology

Pleurobema perovatum occurs in riffles, runs and shoals of small creeks to large rivers. It is usually found in sand and gravel substrates.

Pleurobema perovatum is presumably a short-term brooder, gravid during late spring and early summer. Glochidial hosts of *P. perovatum* are unknown, but some species of *Pleurobema* use members of the Cyprinidae (Haag and Warren, 2003).

Current Conservation Status and Protection

Pleurobema perovatum was listed as endangered throughout its range by Athearn (1970) and Williams et al. (1993) and in Alabama by Lydeard et al. (1999). Garner et al. (2004) designated *P. perovatum* a species of highest conservation concern in the state. In 1993 it was listed as endangered under the federal Endangered Species Act.

Remarks

According to Conrad (1834a) the type material of *Pleurobema perovatum* came from "small streams in Greene County, Alabama". However, the label associated with the type specimen suggests that it came from Big Prairie Creek, Marengo County, Alabama. Big Prairie Creek is now in Hale County, Alabama, which was formed from parts of Marengo, Greene and Perry counties in 1867 (Foscue, 1989).

Synonymy

Unio perovatus Conrad, 1834. Conrad, 1834a:338, pl. 1, fig. 3

Type locality: [Big] Prairie Creek, Marengo County, Alabama. Lectotype, ANSP 129864, length 48 mm, designated by Johnson and Baker (1973).

Unio nux Lea, 1852, *nomen nudum*. Lea, 1852a:252

Unio nux Lea, 1852. Lea, 1852b:283, pl. 24, fig. 43; Lea, 1852c:39, pl. 24, fig. 43

Type locality: Alabama River, [Alabama,] C.M. Wheatley. Lectotype, USNM 85316, length 29 mm, designated by Johnson (1974).

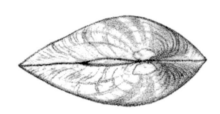

Unio johannis Lea, 1859. Lea, 1859d:171; Lea, 1860d:343, pl. 55, fig. 168; Lea, 1860g:25, pl. 55, fig. 168

Type locality: Connasauga [Conasauga] River, Bishop Elliott; Etowah River, Georgia, Reverend G. White; Alabama River, Dr. Budd. Reported lectotype, USNM 84830, length 26 mm, designated by Johnson (1974), is from Tuscaloosa, Alabama, not one of the reported type localities, and is invalid.

Comment: Two of the three localities given for *Unio johannis* are outside the range of *Pleurobema perovatum*, but the third locality, Alabama River, is not. The written shell description more closely matches *P. perovatum* and the specimen subsequently illustrated by Lea (1860d) was from [the Black Warrior River,] Tuscaloosa, Alabama.

Unio cinnamomicus Lea, 1861. Lea, 1861a:39; Lea, 1862b:100, pl. 16, fig. 248; Lea, 1862d:104, pl. 16, fig. 248
Type locality: Tombigbee River, Columbus, [Lowndes County,] Mississippi, W. Spillman, M.D. Lectotype, USNM
   85317, length 36 mm, designated by Johnson (1974).

Unio flavidulus Lea, 1861. Lea, 1861a:39; Lea, 1862b:97, pl. 15, fig. 245; Lea, 1862d:101, pl. 15, fig. 245
Type locality: Stream near Columbus, [Lowndes County,] Mississippi, W. Spillman, M.D. Lectotype, USNM
   84719, length 37 mm, designated by Johnson (1974).

Unio concolor Lea, 1861. Lea, 1861a:40; Lea, 1862b:89, pl. 12, fig. 237; Lea, 1862d:93, pl. 12, fig. 237
Type locality: Big Prairie Creek, Alabama, E.R. Showalter, M.D. Lectotype, USNM 85319, length 52 mm,
   designated by Johnson (1974).

Unio interventus Lea, 1861. Lea, 1861c:60; Lea, 1862b:84, pl. 11, fig. 233; Lea, 1862d:88, pl. 11, fig. 233
Type locality: Cahawba [Cahaba] River, Alabama, E.R. Showalter, M.D. Lectotype, USNM 84722, length 38 mm,
   designated by Johnson (1974).

Unio pallidofulvus Lea, 1861. Lea, 1861c:60; Lea, 1862b:83, pl. 11, fig. 232; Lea, 1862d:87, pl. 11, fig. 232
Type locality: Cahawba [Cahaba] River, Alabama, E.R. Showalter, M.D. Lectotype, USNM 84721, length 36 mm, designated by Johnson (1974).

Unio litus Lea, 1871. Lea, 1871:189; Lea, 1874c:17, pl. 5, fig. 13; Lea, 1874e:21, pl. 5, fig. 13
Type locality: Cahaba River, Shelby County, Alabama, E.R. Schowalter [Showalter], M.D. Lectotype, USNM 85142, length 40 mm, designated by Johnson (1974).

Unio simulans Lea, 1871. Lea, 1871:190; Lea, 1874c:18, pl. 5, fig. 15; Lea, 1874e:22, pl. 5, fig. 15
Type locality: Cahaba River, Shelby County, Alabama, E.R. Schowalter [Showalter], M.D. Lectotype, USNM 84840, length 39 mm, designated by Johnson (1974).

Unio pinkstoni S.H. Wright, 1897. S.H. Wright, 1897:136; Simpson, 1900a:81, pl. 1, fig. 8
Type locality: Tuscaloosa [Tallapoosa] River, Macon County, Alabama. Lectotype, USNM 149649, length 46 mm, designated by Johnson (1967b). This specimen was also figured by Simpson (1900a).

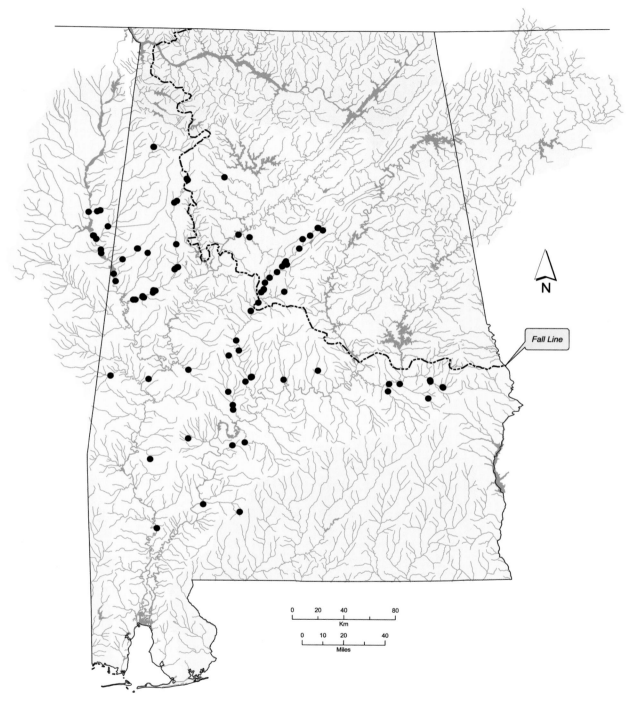

Distribution of *Pleurobema perovatum* in Alabama and the Mobile Basin.

Pleurobema plenum (Lea, 1840)
Rough Pigtoe

Pleurobema plenum – Length 46 mm, UMMZ 255264. Tennessee River, mouth of Flint River [Creek], Morgan County, Alabama. © Richard T. Bryant.

Shell Description

Length to 80 mm; thick; inflated; outline subtriangular; posterior margin broadly rounded; anterior margin straight; dorsal margin convex; ventral margin rounded; posterior ridge low, rounded, curved; posterior slope steep; medial ridge high, rounded, curved; narrow, moderately developed sulcus separating posterior and medial ridges; umbo inflated, elevated well above hinge line, projected anteriad, umbo sculpture a few irregular nodulous ridges; periostracum satiny, tawny to reddish brown or black, occasionally with faint green rays.

Pseudocardinal teeth rough, triangular, 2 divergent teeth in left valve, 1 tooth in right valve, often with accessory denticle anteriorly and/or posteriorly; lateral teeth long, curved, 2 in left valve, 1 in right valve, may also have rudimentary to moderately developed lateral tooth ventrally in right valve; interdentum short, wide; umbo cavity moderately deep, open; nacre white, rarely pink.

Soft Anatomy Description

Mantle tan, may be pale, often rusty brown to grayish brown or dark brown outside of apertures; visceral mass tan; foot tan.

Gills tan; dorsal margin straight, gills steeply sloped from near umbo to near apertures, ventral margin convex to elongate convex; gill length 55–70% of shell length; gill height 45–55% of gill length; outer gill height 70–90% of inner gill height; inner lamellae of inner gills connected to visceral mass only anteriorly.

No gravid females were available for marsupium description. Outer gills are marsupial and white when gravid (J.W. Jones, personal communication).

Labial palps tan; straight dorsally, convex ventrally, bluntly pointed distally; palp length 15–25% of gill length; palp height 40–75% of palp length; distal 20–55% of palps bifurcate.

Incurrent and supra-anal apertures longer than excurrent aperture; incurrent aperture usually shorter than supra-anal aperture, occasionally of equal length.

Incurrent aperture length 15–25% of shell length; tan within, may have rusty brown or dark brown basal to papillae; papillae usually in 1 row, variable, often simple, bifid and trifid with a few small arborescent, sometimes completely arborescent; papillae tan to rusty brown or dark brown.

Excurrent aperture length 10–15% of shell length; tan within, usually with rusty brown marginal color band; margin with simple papillae, small but well-developed, occasionally with few bifid; papillae tan, rusty brown or dark brown.

Supra-anal aperture length 20–35% of shell length; tan within, often with sparse rusty brown or dark brown marginally; margin mostly smooth, may have minute crenulations adjacent to mantle bridge; mantle bridge separating supra-anal and excurrent apertures imperforate, length 5–10% of supra-anal length.

Specimens examined: Tennessee River (n = 5) (specimens previously preserved).

Glochidium Description

Glochidium unknown.

Similar Species

Pleurobema plenum resembles *Pleurobema cordatum*. However, *P. plenum* is less elongate and more truncate posteriorly, giving the anterioventral margin of *P. plenum* a more exaggerated rounded appearance than that of *P. cordatum*.

Pleurobema plenum resembles *Pleurobema rubrum* but is less elongate posteriorly. The outline of *P. plenum* approaches an equilateral triangle, whereas that of *P. rubrum* more closely resembles a scalene triangle. The difference is most evident in the anterior margin, which is angled ventrally to slightly anterioventrally in *P. plenum* and ventrally to posterioventrally in *P. rubrum* when the shell is held with the hinge ligament horizontal. Also, the umbo cavity of *P. plenum* is usually deeper than that of *P. rubrum*.

Pleurobema plenum resembles large river *Pleurobema sintoxia* but is less elongate posteriorly. The outline of *P. plenum* is triangular. The outline of *P. sintoxia* is sometimes triangular, but may also be round or ovate. The sulcus of *P. plenum* is usually moderately developed, whereas the sulcus of *P. sintoxia* is often weak and may be absent. The umbo cavity of *P. plenum* is moderatly deep and that of *P. sintoxia* is shallow.

Pleurobema plenum superficially resembles *Obovaria retusa*. The umbo of *P. plenum* is less inflated than that of *O. retusa*, and its anterior projection is not as pronounced. The nacre of *P. plenum* is usually white, occasionally pale pink, whereas that of *O. retusa* is always pink or purple inside the pallial line. When the nacre of *P. plenum* is pink, it is uniform across the shell.

General Distribution

Pleurobema plenum is found in the Ohio, Cumberland and Tennessee River drainages. In the Ohio River drainage it is known from western Pennsylvania downstream to the mouth of the Ohio River in Illinois and Kentucky, including the Wabash River drainage of Illinois and Indiana. Parmalee and Bogan (1998) report *P. plenum* to be widespread in lower and middle reaches of the Cumberland River drainage in Tennessee, but Cicerello et al. (1991) did not report it from Kentucky reaches of that drainage. *Pleurobema plenum* occurs in most of the Tennessee River drainage (Ahlstedt, 1992a; Parmalee and Bogan, 1998).

Alabama and Mobile Basin Distribution

Pleurobema plenum is known from the Tennessee River across northern Alabama and lower reaches of the Elk and Paint Rock rivers.

Pleurobema plenum is extant only in tailwaters of Wilson and possibly Guntersville dams on the Tennessee River.

Ecology and Biology

Pleurobema plenum occurs in medium to large rivers, usually in flowing water with clean sand and gravel substrates.

Pleurobema plenum is a short-term brooder, presumably gravid in spring and summer. Ortmann (1919) reported gravid females in late May in upper reaches of the Tennessee River drainage. Glochidial hosts of *P. plenum* are unknown, but some *Pleurobema* species utilize members of the Cyprinidae (Haag and Warren, 2003).

Current Conservation Status and Protection

Pleurobema plenum was considered endangered throughout its range by Stansbery (1971) and Williams et al. (1993) and in Alabama by Stansbery (1976) and Lydeard et al. (1999). Garner et al. (2004) designated it a species of highest conservation concern in the state. *Pleurobema plenum* was listed as endangered under the federal Endangered Species Act in 1976.

Remarks

Pleurobema plenum is rare in prehistoric shell middens along the Tennessee River across northern Alabama (Morrison, 1942; Hughes and Parmalee, 1999).

Some authors have considered *Pleurobema plenum* a subspecies of *Pleurobema cordatum*.

Synonymy

Unio plenus Lea, 1840. Lea, 1840:286; Lea, 1842b:211, pl. 14, fig. 26; Lea, 1842c:49, pl. 14, fig. 26

Type locality: Ohio River, Cincinnati, Ohio. Lectotype, USNM 84677, length 41 mm, designated by Johnson (1974).

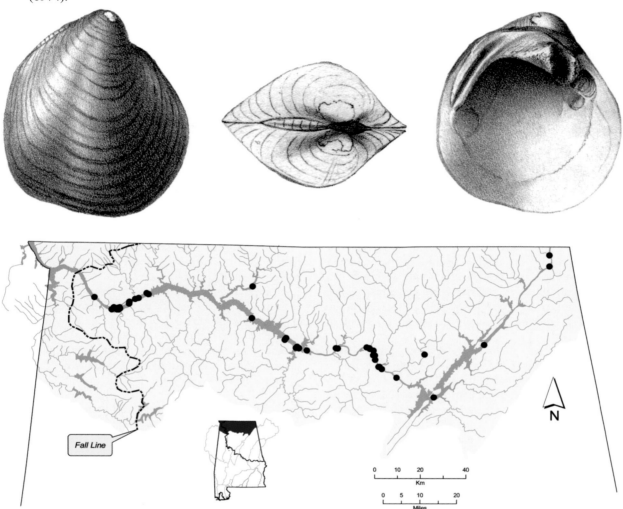

Distribution of *Pleurobema plenum* in the Tennessee River drainage of Alabama.

Pleurobema pyriforme (Lea, 1857)
Oval Pigtoe

Pleurobema pyriforme – Length 38 mm, AUM 1238. Little Uchee Creek, 11.5 miles southeast of Opelika, Lee County, Alabama, 6 October 1972. © Richard T. Bryant.

Shell Description

Length to 60 mm; moderately thin; moderately inflated; outline oval; posterior margin narrowly rounded; anterior margin rounded to somewhat truncate; dorsal margin convex; ventral margin straight to slightly convex; posterior ridge rounded; posterior slope moderately steep, flat; umbo broad, elevated slightly above hinge line, umbo sculpture subconcentric undulations; periostracum shiny, tawny to dark brown, without rays.

Pseudocardinal teeth low, triangular, compressed, 2 in left valve, 1 in right valve; lateral teeth of variable length, short to moderately long, straight to slightly curved, 2 in left valve, 1 in right valve; interdentum short, narrow; umbo cavity wide, shallow; nacre white, often with golden cast.

Soft Anatomy Description

Mantle creamy white to tan, outside of apertures some combination of tan, brown and gray, may be mottled; visceral mass pearly white to tan; foot creamy white to pale orange or tan. Lea (1863d) described the "color of the mass" as "whitish or light-salmon".

Gills creamy white to tan, may have slight rusty cast; dorsal margin usually slightly sinuous, occasionally straight, ventral margin straight to slightly convex; gill length 55–75% of shell length; gill height 45–60% of gill length, occasionally less; outer gill height 75–95% of inner gill height; inner lamellae of inner gills connected to visceral mass only anteriorly.

Outer gills marsupial; glochidia held across entire gill; marsupium slightly padded when gravid; creamy white to pale orange or tan.

Labial palps creamy white to light tan; straight to slightly convex or concave dorsally, convex ventrally, bluntly pointed to narrowly rounded distally; palp length 15–30% of gill length; palp height 50–65% of palp length, occasionally greater; distal 35–60% of palps bifurcate.

Incurrent and supra-anal apertures usually longer than excurrent aperture; incurrent and supra-anal apertures of variable length relative to each other.

Incurrent aperture length 15–20% of shell length; creamy white to tan within, may have irregular grayish cast, usually with irregular gray or brown basal to papillae; papillae in 1–2 rows, mostly long, slender, simple, with occasional bifid and trifid, some large trifid approach the form of arborescent when contracted; papillae some combination of creamy white, tan, brown and gray, color often changing distally, edges may be darker basally. Lea (1863d) reported incurrent aperture papillae to be "very small".

Excurrent aperture length 10–15% of shell length, occasionally greater; creamy white to tan within, usually with marginal color band in some combination of creamy white, tan, gray and various shades of brown, may have lightly colored band proximal to marginal band; margin with small simple papillae, may be little more than crenulations; papillae creamy white to tan or light brown.

Supra-anal aperture length 15–20% of shell length; creamy white to tan or light brown within, usually without marginal coloration; margin smooth; mantle bridge separating supra-anal and excurrent apertures imperforate, length 10–30% of supra-anal length, occasionally absent.

Specimens examined: Big Creek, Houston County (n = 5); Chipola River, Florida (n = 3) (some specimens previously preserved).

Glochidium Description

Length 155–185 μm; height 152–180 μm; without styliform hooks; outline depressed subelliptical; dorsal margin straight, ventral margin rounded, anterior and posterior margins convex (O'Brien and Williams, 2002).

Similar Species

Pleurobema pyriforme may resemble male *Villosa villosa*. However, the bluntly pointed posterior terminus of the *P. pyriforme* shell is positioned more ventrally than that of *V. villosa*, which is located near the midline of the posterior margin. The periostracum of *P. pyriforme* is typically without rays, but that of *V. villosa* is usually rayed.

General Distribution

Pleurobema pyriforme is endemic to the Apalachicola Basin of Alabama, Florida and Georgia.

Alabama and Mobile Basin Distribution

Pleurobema pyriforme is confined to the Chattahoochee River drainage, above and below the Fall Line, and headwaters of the Chipola River.

Pleurobema pyriforme is apparently extant only in Big and Cowarts creeks of the Chipola River drainage, Houston County.

Ecology and Biology

Pleurobema pyriforme is a species of creeks and small to large rivers. It generally occurs in slow to moderate current in pool, run and riffle habitats. *Pleurobema pyriforme* may be found in various combinations of clay, sand and gravel substrates.

Pleurobema pyriforme is a short-term brooder, gravid from March to July (Ortmann, 1909a; O'Brien and Williams, 2002). Glochidial hosts determined in laboratory trials include *Pteronotropis hypselopterus* (Sailfin Shiner) (Cyprinidae) and *Gambusia holbrooki* (Eastern Mosquitofish) (Poeciliidae). One nonindigenous fish, *Poecilia reticulata* (Guppy) (Poeciliidae), was reported to serve as a marginal host of *P. pyriforme* in laboratory trials (O'Brien and Williams, 2002).

Current Conservation Status and Protection

Pleurobema pyriforme was listed as endangered throughout its range by Athearn (1970) and Williams et al. (1993). In Alabama it was listed as threatened by Stansbery (1976) and endangered by Lydeard et al. (1999). Garner et al. (2004) listed *P. pyriforme* as a species of highest conservation concern in the state. It was listed as endangered under the federal Endangered Species Act in 1994.

Remarks

There has been some confusion regarding the taxonomy and distribution of *Pleurobema pyriforme*. Clench and Turner (1956), Johnson (1970) and Burch (1975a) gave the range of *P. pyriforme* as the Suwannee River drainage west to the Apalachicola Basin. Williams and Butler (1994) considered *P. pyriforme* to be an Apalachicola Basin endemic, distinct from *Pleurobema reclusum* (B.H. Wright, 1898) of the Ochlockonee and Suwannee River drainages. Kandl et al. (2001) determined that the *Pleurobema* in the Apalachicola Basin and Suwannee River drainage are not conspecific.

Synonymy

Unio striatus Lea, 1840, *non* Rafinesque, 1820. Lea, 1840:287; Lea, 1842b:203, pl. 12, fig. 16; Lea, 1842c:41, pl. 12, fig. 16

Type locality: Chattahoochee River, Columbus, [Muscogee County,] Georgia, Dr. Boykin. Lectotype, USNM 84797, length 32 mm, designated by Johnson (1974).

Unio pyriformis Lea, 1857. Lea, 1857a:31; Lea, 1858e:69, pl. 12, fig. 50; Lea, 1858h:69, pl. 12, fig. 50
Type locality: Near Columbus, [Muscogee County,] Georgia, Bishop Elliott. Lectotype, USNM 84781, length 54
    mm, designated by Johnson (1974).

Unio modicus Lea, 1857. Lea, 1857g:171; Lea, 1859f:204, pl. 24, fig. 86; Lea, 1859g:22, pl. 24, fig. 86
Type locality: Chattahoochee River, near Columbus, [Muscogee County,] Georgia, Bishop Elliott. Lectotype,
    USNM 84787, length 35 mm, designated by Johnson (1974).

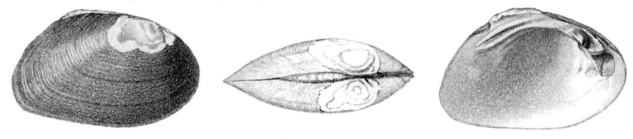

Pleurobema simpsoni Vanatta, 1915. Vanatta, 1915:559 [new name proposed for *Unio striatus* Lea, 1840, which
    was preoccupied]

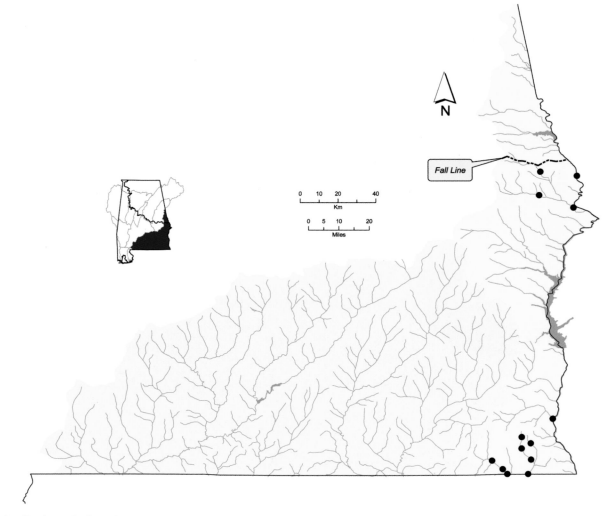

Distribution of *Pleurobema pyriforme* in the eastern Gulf Coast drainages of Alabama.

Pleurobema rubellum (Conrad, 1834)
Warrior Pigtoe

Pleurobema rubellum – Upper figure: length 31 mm, INHS 30419. Black Warrior River, Jefferson County, Alabama. 1903. Lower figure: length 34 mm, UF 358670. Brushy Creek, Bankhead National Forest, Forest Service Road 3159, Alabama, 18 June 2002. © Richard T. Bryant.

Shell Description

Length to 55 mm; moderately thick; moderately compressed to moderately inflated; outline round or oval to subtriangular; posterior margin narrowly rounded; anterior margin rounded; dorsal margin convex; ventral margin convex; posterior ridge low, rounded, may be weakly doubled distally; posterior slope moderately steep; umbo broad, inflated, elevated above hinge line, umbo sculpture unknown; periostracum tawny to dark olive or dark brown, without rays.

Pseudocardinal teeth moderately thick, triangular, slightly compressed, 2 slightly divergent teeth in left valve, 1 tooth in right valve, may have accessory denticle anteriorly and/or posteriorly; lateral teeth thick, short to moderately long, straight to slightly curved, 2 in left valve, 1 in right valve; interdentum short, moderately

wide; umbo cavity wide, shallow; nacre usually white, may be pink.

Soft Anatomy Description

Mantle pale tan, outside of apertures rusty tan to rusty brown; visceral mass tan; foot tan.

Gills tan; dorsal margin slightly sinuous, ventral margin convex; gill length 55–60% of shell length; gill height 65–80% of gill length; outer gill height approximately 85% of inner gill height; inner lamellae of inner gills connected to visceral mass only anteriorly.

No gravid females were available for marsupium description; probably similar to other *Pleurobema* species, with outer gills marsupial, padded when gravid.

Labial palps tan; straight dorsally, convex ventrally, bluntly pointed distally; palp length approximately 15% of gill length; palp height 50% of palp length; distal 25–35% of palps bifurcate.

Incurrent aperture longer than excurrent aperture; supra-anal aperture slightly longer than incurrent aperture.

Incurrent aperture length 15–20% of shell length; creamy white to tan within, may have rusty brown basal to papillae; papillae in 1 row, arborescent, may have occasional simple and bifid papillae; papillae rusty brown, some may be paler distally.

Excurrent aperture length 10–15% of shell length; creamy white to tan within, with 3 marginal color bands, rusty brown bands separated by creamy white or tan band; margin with small but well-developed papillae, mostly simple, with occasional bifid papillae; papillae rusty tan.

Supra-anal aperture length 15–20% of shell length; creamy white to rusty tan within, may have sparse irregular rusty tan marginally; margin smooth; mantle bridge separating supra-anal and excurrent apertures imperforate, length approximately 20% of supra-anal length.

Specimens examined: West Fork Sipsey River (n = 2) (specimens previously preserved).

Glochidium Description

Length 187–200 μm; height approximately 150 μm; without styliform hooks; outline subelliptical; dorsal margin slightly concave or slightly undulating, ventral margin rounded, anterior and posterior margins almost straight.

Similar Species

Some *Pleurobema rubellum* resemble *Pleurobema perovatum*, but *P. rubellum* is less elongate and more broadly rounded posteriorly. Also, the umbo of *P. perovatum* is often narrower and more inflated than that of *P. rubellum*. The crest of the anterior pseudocardinal tooth in the left valve of *P. perovatum* is generally oriented almost parallel to the shell margin, whereas pseudocardinal teeth of *P. rubellum* are oriented more ventrally.

General Distribution

Pleurobema rubellum is endemic to the Black Warrior River drainage and Cahaba River system in Alabama.

Alabama and Mobile Basin Distribution

Pleurobema rubellum historically occurred throughout the Black Warrior River drainage, but most records are from above the Fall Line. In the Cahaba River it is only known from a few individuals from above the Fall Line.

Pleurobema rubellum is now limited to headwaters of the Sipsey Fork in Bankhead National Forest and the North River upstream of Lake Tuscaloosa.

Ecology and Biology

Pleurobema rubellum is a species of creeks and medium to large rivers. It generally occurs in sandy gravel of shoal habitats.

Pleurobema rubellum is a short-term brooder and has been reported gravid with mature glochidia in June (Haag and Warren, 1997). Its conglutinates are subcylindrical and flattened, pink or peach in color (Haag and Warren, 1997). Glochidial hosts of *P. rubellum* determined in laboratory trials include *Campostoma oligolepis* (Largescale Stoneroller), *Cyprinella callistia* (Alabama Shiner), *Cyprinella venusta* (Blacktail Shiner) and *Semotilus atromaculatus* (Creek Chub) (Cyprinidae); and *Fundulus olivaceus* (Blackspotted Topminnow) (Fundulidae). Results varied among hosts, but *C. callistia*, *C. venusta* and *S. atromaculatus* generally produced more juveniles than the others (Haag and Warren, 1997).

Current Conservation Status and Protection

Pleurobema rubellum was considered endangered by Athearn (1970), Stansbery (1976), Williams et al. (1993), Lydeard et al. (1999) and Garner et al. (2004).

Garner et al. (2004) recognized both *Pleurobema rubellum* and *Pleurobema furvum* as distinct species and listed *P. rubellum* as extinct and *P. furvum* as a species of highest conservation concern. The USFWS (2000) also recognized both species, *P. furvum* as endangered and *P. rubellum* as extinct. *Pleurobema hagleri* was also recognized and included on a list of extinct species (USFWS, 2000). *Pleurobema furvum* is currently included on the federal endangered species list. A formal reconciliation of the list should replace *P. furvum* with *P. rubellum*.

Remarks

Conrad (1834b) described two species, *Pleurobema furvum* and *Pleurobema rubellum*, from headwaters of the Black Warrior River. Though *P. rubellum* has generally been recognized, *P. furvum* has been recognized by only some authors. Based on conchological characters and preliminary genetic analyses (D.C. Campbell, personal communication), they appear to represent a single, highly variable species. Since both descriptions appeared in the same publication (Conrad, 1834b), they have equal nomenclatural status. The type specimen of *P. rubellum* is housed in the Academy of Natural Sciences of Philadelphia, but the type of *P. furvum* could not be found. Therefore, *P. furvum* was herein placed into the synonymy of *P. rubellum*.

Pleurobema hagleri (Frierson, 1900) was described from the North River, a tributary of the Black

Warrior River. It appears to represent an extreme in the range of variation of *Pleurobema rubellum* so was also herein treated as a synonym of *P. rubellum*. However, more comprehensive genetic and morphological analyses may prove this to be a valid species.

Synonymy

Unio rubellus Conrad, 1834. Conrad, 1834a:38, pl. 6, fig. 2
Type locality: Black Warrior River, near its source, among the mountains of Alabama. Holotype, ANSP 41433, length 33 mm.

Unio furvus Conrad, 1834. Conrad, 1834a:39, pl. 6, fig. 3
Type locality: Black Warrior River, [Alabama]. Figured type specimen not found in ANSP, also not found by Johnson and Baker (1973). Measurement of figured shell not given in original description.

Unio pulvinulus Lea, 1845. Lea, 1845:164; Lea, 1848a:81, pl. 8, fig. 24; Lea, 1848b:55, pl. 8, fig. 24
Type locality: [Black Warrior River,] Tuscaloosa, [Tuscaloosa County,] Alabama, B.W. Budd, M.D. Lectotype, USNM 84838, length 29 mm, designated by Johnson (1974).

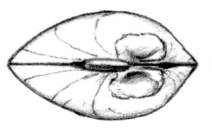

Pleurobema avellana Simpson, 1900. Simpson, 1900a:81, pl. 2, figs. 6, 7
Type locality: Catawba [Cahaba] River, Alabama. Lectotype, UMMZ 79653, length 30 mm, designated by Johnson
(1975).

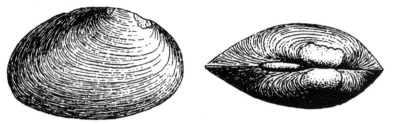

Unio (*Pleurobema*) *hagleri* Frierson, 1900. Frierson, 1900:109, pl. 2
Type locality: North River, near Tynes [Tyner, Tuscaloosa County,] Alabama. Lectotype, ANSP 77902a, length 52
mm, designated by Johnson (1972a), is the specimen figured by Frierson (1900).

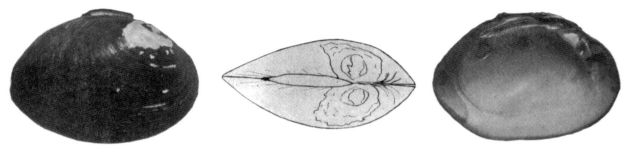

Pleurobema fictum Frierson, 1927. Frierson, 1927:43; Frierson, 1928:139, pl. 3, figs. 2, 2a
Type locality: Cahaba River, Alabama. Holotype, UMMZ 87575, length 28 mm.

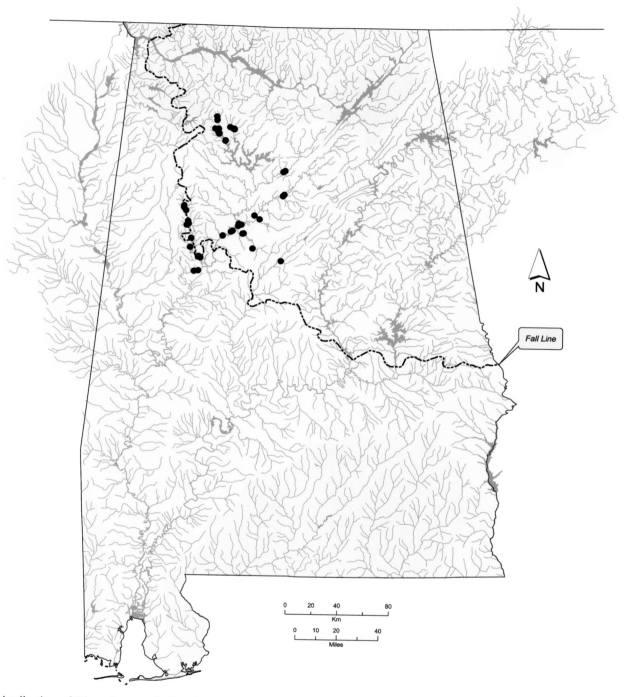

Distribution of *Pleurobema rubellum* in Alabama and the Mobile Basin.

Pleurobema rubrum (Rafinesque, 1820)
Pyramid Pigtoe

Pleurobema rubrum – Upper figure: length 69 mm, UMMZ 80958. Lower figure: length 57 mm, UMMZ 80958. Tennessee River, Bridgeport, Jackson County, Alabama. © Richard T. Bryant.

Shell Description

Length to 91 mm; thick; moderately inflated; outline triangular; posterior margin narrowly rounded; anterior margin straight to slightly convex; dorsal margin convex; ventral margin broadly rounded; posterior ridge low, rounded; posterior slope flat, steep; narrow, shallow sulcus separates posterior ridge from broad, elevated medial swelling; umbo narrow, inflated, elevated above hinge line, turned anteriad, umbo sculpture unknown; periostracum greenish brown to reddish brown, darkening with age, subadults often with weak green rays that become obscure with age.

Pseudocardinal teeth thick, low, triangular, 2 divergent teeth in left valve, 1 tooth in right valve, often with accessory denticle anteriorly and/or posteriorly; lateral teeth short to moderately long, moderately thick, straight to slightly curved, 2 in left valve, 1 in right valve, may also have rudimentary lateral tooth ventrally in right valve; interdentum short, moderately wide; umbo cavity usually shallow; nacre white to pink.

Soft Anatomy Description

Mantle tan to light brown, may have slight golden cast, outside of apertures some combination of tan, dull orange, gray and various shades of brown, usually mottled, sometimes with bands perpendicular to margin; visceral mass creamy white to tan; foot tan to pinkish orange.

Gills tan; dorsal margin straight to slightly sinuous, ventral margin convex; gill length 60–70% of shell length; gill height 40–50% of gill length, occasionally slightly greater; outer gill height 70–85% of inner gill height; inner lamellae of inner gills connected to visceral mass only anteriorly. Lea (1863d) reported ventral margins of gills to be "irregular and very minutely crenulate".

No gravid females were available for marsupium description; probably similar to other *Pleurobema* species, with outer gills marsupial, padded when gravid.

Labial palps creamy white to tan, usually with irregular paler area dorsally; straight to slightly concave dorsally, convex ventrally, usually bluntly pointed distally, occasionally rounded; palp length 20–30% of gill length; palp height usually 60–70% of palp length, occasionally slightly more or less; distal 25–45% of palps bifurcate.

Incurrent, excurrent and supra-anal aperture lengths variable relative to one another.

Incurrent aperture length 15–25% of shell length; creamy white to tan within, usually with rusty orange to brown or gray basal to papillae, coloration may be irregular or sparse; papillae variable, in 1–2 rows, some combination of simple, bifid, trifid and arborescent, arborescent papillae often lacking, occasionally with arborescent papillae only; papillae some combination of tan, rusty orange, gray and various shades of brown, often changing color distally.

Excurrent aperture length 10–20% of shell length; creamy white to tan within, usually with marginal color band in various combinations of tan, rusty orange, brown and gray, often sparsely mottled, sometimes with short bands perpendicular to margin; margin papillate, papillae small, simple, may be little more than crenulations; papillae tan to brown or gray.

Supra-anal aperture length 15–30% of shell length; creamy white within, often with thin gray or brown marginal color band, may be sparse and irregular; margin smooth; mantle bridge separating supra-anal and excurrent apertures occasionally with 1 or more small perforations, length 5–30% of supra-anal length.

Specimens examined: Clinch River, Tennessee (n = 1); Duck River, Tennessee (n = 1); Tennessee River (n = 7).

Glochidium Description

Glochidium unknown.

Similar Species

Pleurobema rubrum resembles *Pleurobema cordatum*, differing primarily in shell outline. The outline of *P. cordatum* approaches an equilateral triangle, whereas that of *P. rubrum* more closely resembles a scalene triangle. The difference is most evident in the anterior margin when the shell is held with the hinge ligament horizontal. The anterior margin of *P. rubrum* is angled ventrally to posterioventrally and it is angled ventrally to anterioventrally in *P. cordatum*. Also, the umbo cavity of *P. rubrum* is usually shallower than that of *P. cordatum*.

Pleurobema rubrum resembles *Pleurobema plenum* but is more elongate posteriorly. The outline of *P. rubrum* approaches a scalene triangle, whereas that of *P. plenum* approaches an equilateral triangle. The difference is most evident in the anterior margin, which is angled ventrally to posterioventrally in *P. rubrum* and ventrally to anterioventrally in *P. plenum* when the shell is held with the hinge ligament horizontal. Also, the umbo cavity of *P. rubrum* is usually shallower than that of *P. plenum*.

Pleurobema rubrum may resemble some *Pleurobema sintoxia*. *Pleurobema sintoxia* is highly variable in shell morphology, and many individuals are round to oval in outline, easily distinguished from the triangular outline of *Pleurobema rubrum*. Some *P. sintoxia* are triangular in outline but are more equilateral than *P. rubrum*, which typically has the outline of a scalene triangle. The umbo cavity of *P. rubrum* is slightly deeper than that of *P. sintoxia*.

General Distribution

Pleurobema rubrum is widespread in the Mississippi Basin from southwestern Wisconsin (Havlik and Stansbery, 1977) south to Louisiana (Vidrine, 1993), and from Ohio River headwaters in western Pennsylvania (Ortmann, 1909a) west to eastern Kansas (Murray and Leonard, 1962). It is known from the Cumberland River drainage downstream of Cumberland Falls, Kentucky and Tennessee (Cicerello et al., 1991; Parmalee and Bogan, 1998). In the Tennessee River drainage *P. rubrum* occurs from headwaters in southwestern Virginia downstream to the mouth of the Tennessee River, Kentucky (Ahlstedt, 1992a; Parmalee and Bogan, 1998).

Alabama and Mobile Basin Distribution

Pleurobema rubrum historically occurred in the Tennessee River across northern Alabama. It is also known from the Paint Rock River and extreme lower Limestone Creek, Limestone County.

Pleurobema rubrum is extant in tailwaters of Guntersville and Wilson dams.

Ecology and Biology

Pleurobema rubrum is a species of shoal habitats in medium to large rivers, usually in sand and gravel substrates. It occurs in tailwaters of dams at depths exceeding 6 m.

Pleurobema rubrum is a short-term brooder, reported gravid from late May to late July (Ortmann, 1919). It is dioecious, but Sterki (1898a) reported 2 hermaphroditic individuals among 120 from the Ohio River.

Fishes found to serve as glochidial hosts of *Pleurobema rubrum* in laboratory trials include *Cyprinella spiloptera* (Spotfin Shiner), *Erimystax dissimilis* (Streamline Chub), *Lythrurus fasciolaris* (Scarlet Shiner) and *Notropis photogenis* (Silver Shiner) (Cyprinidae) (Culp et al., 2006).

Current Conservation Status and Protection

Pleurobema rubrum (as *Pleurobema pyramidatum*) was listed as endangered throughout its range by Stansbery (1970a) and threatened throughout its range by Williams et al. (1993). In Alabama it was listed as extirpated by Stansbery (1976) and considered imperiled by Lydeard et al. (1999). Garner et al. (2004) designated *P. rubrum* a species of highest conservation concern in the state.

Remarks

Pleurobema rubrum is rare in prehistoric shell middens along the Tennessee River in northern Alabama (Morrison, 1942; Hughes and Parmalee, 1999).

Some authors have considered *Pleurobema rubrum* a subspecies of *Pleurobema cordatum*.

Synonymy

Obliquaria rubra Rafinesque, 1820. Rafinesque, 1820:314
Type locality: Kentucky. Lectotype, ANSP 20237, length 87 mm, designated by Johnson and Baker (1973).
Unio cardiacea Guérin-Méneville, 1828. Guérin-Méneville, 1828:[no pagination], pl. 28, fig. 7
Unio pyramidatus Lea, 1834. Lea, 1834a:109; pl. 16, fig. 39; Lea, 1834b:119, pl. 16, fig. 39

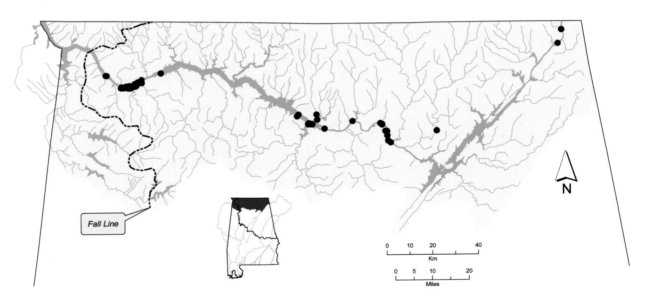

Distribution of *Pleurobema rubrum* in the Tennessee River drainage of Alabama.

Pleurobema sintoxia (Rafinesque, 1820)
Round Pigtoe

Pleurobema sintoxia – Length 75 mm, UNA 834.1. Cumberland River, Tennessee. © Richard T. Bryant.

Shell Description

Length to 120 mm; thick; moderately inflated; outline subtriangular to oval; posterior margin rounded; anterior margin rounded; dorsal margin straight; ventral margin broadly rounded; posterior ridge rounded; posterior slope moderately steep; may have very slight sulcus just anterior of posterior ridge; umbo somewhat inflated, elevated well above hinge line, umbo sculpture coarse irregular ridges; periostracum reddish brown to black, juveniles may have green rays.

Pseudocardinal teeth thick, triangular, 2 divergent teeth in left valve, 1 tooth in right valve, may have small accessory denticle anteriorly and/or posteriorly; lateral teeth thick, straight, 2 in left valve, 1 in right valve; interdentum moderately long, wide; umbo cavity shallow; nacre white, occasionally pink.

Soft Anatomy Description

Mantle creamy white to light tan, outside of apertures grayish to rusty brown; visceral mass creamy white to tan, may be pearly white adjacent to foot; foot light tan to pale pinkish orange.

Gills tan; dorsal margin straight to slightly sinuous, ventral margin convex; gill length 60–70% of shell length; gill height 40–45% of gill length; outer gill height 75–85% of inner gill height; inner lamellae of inner gills connected to visceral mass only anteriorly.

Outer gills marsupial; glochidia held across entire gill except extreme anterior and posterior ends; marsupium padded when gravid; creamy white. Utterback (1915) reported gravid gills to sometimes be brownish.

Labial palps creamy white to tan, often with an irregular pale area dorsally; straight dorsally, convex ventrally, bluntly pointed distally; palp length 20–30% of gill length; palp height 60–70% of palp length; distal 25–45% of palps bifurcate.

Incurrent aperture longer than excurrent aperture; supra-anal aperture length variable with regard to incurrent and excurrent aperture lengths.

Incurrent aperture length 20–25% of shell length; creamy white within, brown or gray basal to papillae; papillae variable, in 1–2 rows, usually mostly simple with a few arborescent, arborescent papillae may be absent; papillae in various shades of brown, may change color distally, may have darker edges basally.

Excurrent aperture length 15–20% of shell length; creamy white within, marginal color band usually some combination of tan, brown and gray, may be mottled or have regular markings; margin with small but well-developed papillae; papillae creamy white to tan, may have rusty cast. The excurrent aperture margin may also be smooth or crenulate (Lea, 1863d; Utterback, 1915; Ortmann, 1919).

Supra-anal aperture length 15–25% of shell length; creamy white within, may have thin pale gray marginal band; margin smooth; mantle bridge separating supra-anal and excurrent apertures sometimes perforate, length 10–30% of supra-anal length. Ortmann (1912a) reported the mantle bridge to be absent in some individuals.

Specimens examined: Tennessee River (n = 3).

Glochidium Description

Length 150–170 μm; height 150–180 μm; without styliform hooks; outline depressed subelliptical; dorsal margin straight, ventral margin rounded, anterior and posterior margins convex, posterior margin slightly more convex than anterior margin (Ortmann, 1912a; Surber, 1912, 1915; Utterback, 1915, 1916a).

Similar Species

Pleurobema sintoxia resembles *Pleurobema cordatum*, but *P. sintoxia* is typically less triangular, often round or oval in outline. The sulcus of *P. sintoxia* is generally less developed than that of *P. cordatum* and may be absent. The umbo cavity of *P. cordatum* is deep, whereas that of *P. sintoxia* is shallow.

Pleurobema sintoxia resembles *Pleurobema plenum* but is more elongate posteriorly. The outline of *P. sintoxia* ranges from triangular to round or ovate, whereas the outline of *P. plenum* is always triangular. The sulcus of *P. sintoxia* is less developed than that of *P. plenum* and may be absent. The umbo cavity of *P. sintoxia* is shallow and that of *P. plenum* is moderately deep.

Pleurobema sintoxia resembles *Pleurobema rubrum* but is typically less triangular, often round or oval in outline. Individual *P. sintoxia* that are triangular are generally equilateral, whereas the outline of *P. rubrum* approaches a scalene triangle. The sulcus of *P. sintoxia* is generally less developed than that of *P. rubrum* and may be absent.

Pleurobema sintoxia also resembles *Fusconaia ebena* and *Fusconaia subrotunda*. However, *P. sintoxia* has a shallower umbo cavity. *Pleurobema sintoxia* is often more triangular in outline than *F. ebena* and *F. subrotunda* from large river habitats and may have a more anteriorly positioned umbo.

General Distribution

Pleurobema sintoxia is known from the eastern half of the Great Lakes Basin, including drainages of lakes Ontario, Erie and Huron (La Rocque and Oughton, 1937). It is widespread in the Mississippi Basin, occurring from Minnesota (Dawley, 1947) south to Arkansas (Ahlstedt and Jenkinson, 1993), and from the Ohio River drainage in western Pennsylvania (Ortmann, 1909a) west to eastern Kansas (Murray and Leonard, 1962), Nebraska (Hoke, 2000) and South Dakota (Backlund, 2000). *Pleurobema sintoxia* is also widespread in the Cumberland River drainage downstream of Cumberland Falls, Kentucky and Tennessee (Cicerello et al., 1991; Parmalee and Bogan, 1998). It is known from most of the Tennessee River drainage (Parmalee and Bogan, 1998). Wilson and

Danglade (1914) reported *P. sintoxia* from the lower Hudson Bay Basin, but Graf (1997) questioned the record.

Alabama and Mobile Basin Distribution

Pleurobema sintoxia is known from the Tennessee River across northern Alabama and the Paint Rock River.

Pleurobema sintoxia is currently confined to tailwaters of Guntersville and Wilson dams on the Tennessee River, where it is rare.

Ecology and Biology

Pleurobema sintoxia is found in riverine habitats of small to large rivers, including tailwaters of dams. It typically occurs in sand and gravel substrates with silt kept to a minimum by current. *Pleurobema sintoxia* is also known from some portions of the Great Lakes. Ortmann (1919) reported *P. sintoxia* to generally be deeply buried in the substrate.

Pleurobema sintoxia is a short-term brooder, gravid from May to late July (Ortmann, 1912a; 1919; Surber, 1912; Coker et al., 1921). Ortmann (1919) reported *P. sintoxia* to begin discharging glochidia during the second half of June. *Pleurobema sintoxia* conglutinates are white and lanceolate, with a zigzagged outline when viewed laterally. *Pleurobema sintoxia* conglutinates are discharged intact (Ortmann, 1910a, 1919).

Glochidial hosts determined using laboratory trials include *Campostoma anomalum* (Central Stoneroller), *Cyprinella spiloptera* (Spotfin Shiner), *Phoxinus eos* (Northern Redbelly Dace), *Phoxinus erythrogaster* (Southern Redbelly Dace), *Pimephales notatus* (Bluntnose Minnow) and *Semotilus atromaculatus* (Creek Chub) (Cyprinidae); and *Lepomis macrochirus* (Bluegill) (Centrarchidae) (Hove et al., 1997; Watters et al., 2005).

Current Conservation Status and Protection

Pleurobema sintoxia was considered currently stable throughout its range by Williams et al. (1993). Lydeard et al. (1999) listed it as imperiled in Alabama, and Garner et al. (2004) designated it a species of highest conservation concern in the state.

Remarks

Pleurobema sintoxia is rare in prehistoric shell middens along the Tennessee River across northern Alabama (Warren, 1975; Hughes and Parmalee, 1999).

Pleurobema sintoxia has been considered a subspecies of *Pleurobema cordatum* by some authors.

Synonymy

Obliquaria sintoxia Rafinesque, 1820. Rafinesque, 1820:310

Type locality: Ohio River. Lectotype, ANSP 20208, length 97 mm, designated by Johnson and Baker (1973).

Unio rubens Menke, 1828. Menke, 1828:90

Margarita (*Unio*) *solidus* Lea, 1836, *nomen nudum*. Lea, 1836:20

Unio coccineus Conrad, 1836. Conrad, 1836:29, pl. 13, fig. 1

Unio catillus Conrad, 1836. Conrad, 1836:30, pl. 13, fig. 2

Unio cuneus Conrad, 1838. Conrad, 1838:Part 11, back cover; Conrad, 1840:105, pl. 58, fig. 1

Unio solidus Lea, 1838. Lea, 1838a:13, pl. 5, fig. 13; Lea, 1838c:13, pl. 5, fig. 13

Unio gouldianus Jay, 1839, *nomen nudum*. Jay, 1839:24

Unio fulgidus Lea, 1845. Lea, 1845:164; Lea, 1848a:73, pl. 4, fig. 10; Lea, 1848b:73, pl. 4, fig. 10

Quadrula (*Fusconaia*) *coccinea* var. *paupercula* Simpson, 1900, *non* Lea, 1861. Simpson, 1900b:789

Pleurobema missouriensis Marsh, 1901. Marsh, 1901:74; Walker, 1915:140, pl. 5, figs. 1, 2

Quadrula (*Fusconaia*) *coccinea* var. *magnalacustris* Simpson, 1914. Simpson, 1914:884 [replacement name for *paupercula* Simpson, 1900, a secondary homonym]

Pleurobema coccineum var. *mississippiensis* F.C. Baker, 1928. F.C. Baker, 1928a:121, pl. 53, figs. 1–5

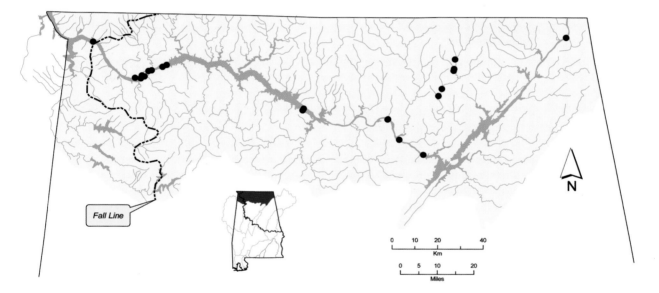

Distribution of *Pleurobema sintoxia* in the Tennessee River drainage of Alabama.

Pleurobema stabilis (Lea, 1861)
Coosa Pigtoe

Pleurobema stabilis – Length 43 mm, USNM 84639 (holotype). Coosa River, Alabama. © Richard T. Bryant.

Shell Description

Length to 45 mm; thick; inflated; outline subelliptical; posterior margin narrowly rounded; anterior margin rounded; dorsal margin straight to slightly convex; ventral margin straight to slightly convex; posterior ridge low, rounded, may be slightly doubled; posterior slope moderately steep, flat to slightly concave; may have slight sulcus just anterior of posterior ridge; umbo broad, inflated, elevated above hinge line, positioned anteriorly, umbo sculpture unknown; periostracum yellowish to tawny, without rays.

Pseudocardinal teeth moderately thick, triangular, 2 divergent teeth in left valve, 1 tooth in right valve, may have accessory denticle anteriorly and/or posteriorly; lateral teeth short, moderately thick, typically straight, 2 in left valve, 1 in right valve, may also have rudimentary lateral tooth ventrally in right valve; interdentum short, moderately wide; umbo cavity shallow; nacre white.

Soft Anatomy Description

Soft anatomy unknown.

Glochidium Description

Glochidium unknown.

Similar Species

Pleurobema stabilis resembles *Pleurobema fibuloides* but may have a slight sulcus, which is absent in *P. fibuloides*. The shell of *P. stabilis* is typically more oblong and thinner than that of *P. fibuloides*.

Pleurobema stabilis resembles *Pleurobema hartmanianum* but typically has a less steep posterior slope which gives it a less triangular outline. Also, the

periostracum of *P. hartmanianum* is generally more greenish than that of *P. stabilis*, which is yellowish to tawny.

Pleurobema stabilis resembles *Pleurobema hanleyianum* but is more inflated and less elongate. The shell of *P. stabilis* is typically thicker than that of *P. hanleyianum*. *Pleurobema stabilis* may have a very shallow sulcus, which is absent in *P. hanleyianum*. Also, the hinge plate and teeth of *P. stabilis* are thicker than those of *P. hanleyianum*.

General Distribution

Pleurobema stabilis is endemic to the Coosa River drainage in Alabama and Georgia.

Alabama and Mobile Basin Distribution

Pleurobema stabilis was apparently limited to the Coosa River.

Pleurobema stabilis appears to be extinct. The most recent dated museum material was collected by H.D. Athearn below Lock and Dam 4 on the Coosa River, St. Clair County, Alabama, in 1956.

Ecology and Biology

Pleurobema stabilis was probably restricted to shoal habitat in medium to large rivers.

Pleurobema stabilis was presumably a short-term brooder, gravid in spring and summer. Its glochidial hosts are unknown, but some species of *Pleurobema* utilize members of the Cyprinidae (Haag and Warren, 2003).

Current Conservation Status and Protection

Pleurobema stabilis appears to be extinct and was not recognized in most previous conservation assessments. A badly eroded specimen similar in shape to *P. stabilis* was collected from the Conasauga River, Georgia, in 2002, but the identity of the specimen remains uncertain.

Remarks

Pleurobema stabilis is uncommon in museum collections, which possibly contributed to its neglect by previous workers. It has one of the more distinctive shells among all of the Mobile Basin *Pleurobema*.

Synonymy

Unio stabilis Lea, 1861. Lea, 1861c:59; Lea, 1862b:71, pl. 8, fig. 221; Lea, 1862d:75, pl. 8, fig. 221
Type locality: Coosa River, Alabama, E.R. Showalter, M.D. Lectotype, USNM 84639, length 43 mm, designated by Johnson (1974).

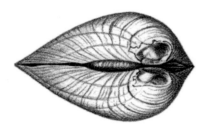

Unio murrayensis Lea, 1868. Lea, 1868a:143; Lea, 1868c:303, pl. 46, fig. 115; Lea, 1869a:62, pl. 46, fig. 115
Type locality: Connesauga [Conasauga] Creek [River], Whitfield County, Georgia; Etowah River, Georgia. Lectotype, USNM 84769, length 37 mm, designated by Johnson (1974), is from Conasauga River, Whitfield County, Georgia.

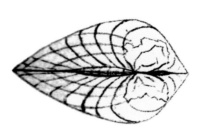

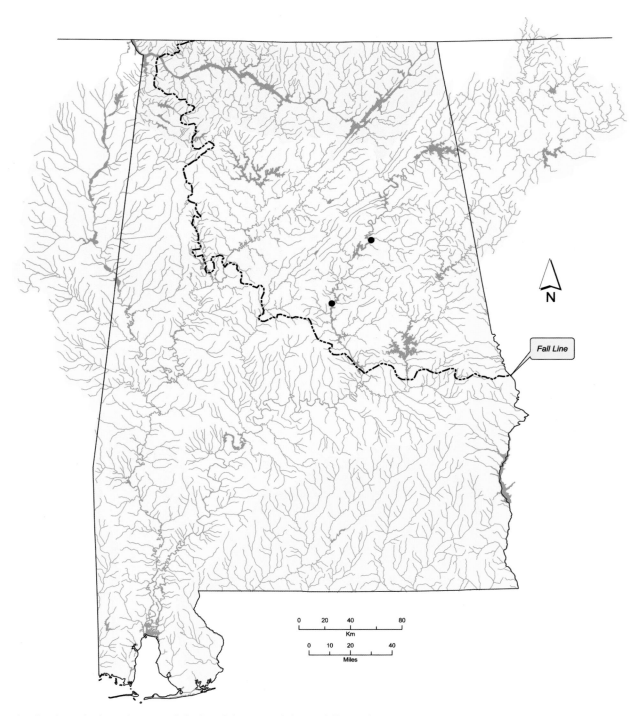

Distribution of *Pleurobema stabilis* in Alabama and the Mobile Basin.

Pleurobema strodeanum (Wright, 1898)
Fuzzy Pigtoe

Pleurobema strodeanum – Upper figure: length 55 mm, UF 358671. Patsaliga Creek, State Highway 106, 7 miles northwest of Dozier, Crenshaw County, Alabama, 27 July 1993. Lower figure: length 32 mm, UAUC 3239. West Fork Choctawhatchee River, State Highway 10, Barbour County, Alabama, 2 June 1999. © Richard T. Bryant.

Shell Description

Length to 75 mm; moderately thin; moderately inflated; outline oval to subtriangular; posterior margin narrowly rounded to bluntly pointed; anterior margin rounded; dorsal margin convex; ventral margin nearly straight to rounded; posterior ridge rounded to angled, often doubled; posterior slope moderately steep, flat to slightly concave; umbo broad, compressed to moderately inflated, elevated slightly above hinge line, umbo sculpture unknown; periostracum cloth-like, usually dark brown to black, occasionally greenish brown, usually without rays.

Pseudocardinal teeth triangular, 2 slightly divergent teeth in left valve, 1 tooth in right valve, may have accessory denticle posteriorly; lateral teeth moderately short, straight, 2 in left valve, 1 in right valve; interdentum short to moderately long, narrow; umbo cavity wide, shallow; nacre bluish white.

Soft Anatomy Description

Mantle tan to light brown, sometimes with golden cast, outside of apertures often pale orange with brown or gray mottling or evenly gray; visceral mass mostly pearly white, sometimes with creamy white or tan areas dorsally; foot creamy white to pale pinkish orange or tan, often slightly darker distally.

Gills tan, often with golden cast; dorsal margin straight to slightly sinuous, ventral margin elongate convex, outer gill ventral margin often ragged; gill length 45–60% of shell length; gill height 45–75% of gill length; outer gill height 85–100% of inner gill height, occasionally greater; inner lamellae of inner gills connected to visceral mass only anteriorly.

Outer gills marsupial, glochidia held across entire gill except extreme anterior and posterior ends; marsupium well-padded when gravid; creamy white.

Pilarczyk et al. (2005) reported gills to sometimes be orange when gravid.

Labial palps creamy white to tan, usually with irregular pale spot centrally or dorsally; straight to slightly concave dorsally, convex ventrally, bluntly pointed to rounded distally; palp length 30–40% of gill length; palp height 40–70% of palp length; distal 20–45% of palps bifurcate.

Incurrent, excurrent and supra-anal apertures of variable relative length, any of the three may be longer than the others.

Incurrent aperture length 15–20% of shell length; creamy white or tan within, may have irregular brown or black basal to papillae; papillae usually in 2 rows, inner row larger, thick or slender, usually mostly simple, often with a few bifid or trifid, occasionally with arborescent, sporadic individuals with mostly bifid and trifid; papillae tan or dull orange, often brown basally or on edges.

Excurrent aperture length 10–20% of shell length; creamy white or tan within, usually with brown marginal color band, often with regular oblong tan patches; margin with short simple papillae; papillae tan to dull orange.

Supra-anal aperture length 10–20% of shell length; creamy white within, occasionally with very thin gray marginal band; margin smooth; mantle bridge separating supra-anal and excurrent apertures rarely perforate, length 10–25% of supra-anal length, occasionally greater.

Specimens examined: Burnt Corn Creek, Conecuh County (n = 2); Conecuh River (n = 5); Eightmile Creek, Walton County, Florida (n = 6); Murder Creek, Conecuh County (n = 1); West Fork Choctawhatchee River (n = 6).

Glochidium Description

Length 176 μm; height 166 μm; without styliform hooks; outline subelliptical (Pilarczyk et al., 2005).

Similar Species

Pleurobema strodeanum may resemble *Fusconaia escambia*. However, *P. strodeanum* is usually more elongate, less truncate posteriorly and has a shallow umbo cavity. The posterior ridge of *F. escambia* is sharper and oriented more posterioventrally than most *P. strodeanum*. The nacre of *P. strodeanum* is always white, but that of *F. escambia* is often salmon to pink.

Pleurobema strodeanum superficially resembles some *Quadrula succissa*, but *P. strodeanum* is more elongate. The umbo cavity of *P. strodeanum* is shallow, whereas that of *Q. succissa* is moderately deep and compressed. Shell nacre of *P. strodeanum* is white, but that of *Q. succissa* is pale purple.

Small *Pleurobema strodeanum* may resemble *Villosa choctawensis*, especially when the periostracum

of the former is covered by mineral deposits. However, the posterior ridge of *P. strodeanum* is much better developed and terminates in a blunt point posterioventrally, giving it an overall more angular outline than that of *V. choctawensis*. The shell of *P. strodeanum* is thicker than that of *V. choctawensis*, and its pseudocardinal teeth are thicker. The periostracum of *V. choctawensis* is typically adorned with green rays, whereas *P. strodeanum* is seldom rayed.

General Distribution

Pleurobema strodeanum is endemic to the Choctawhatchee, Yellow and Escambia River drainages of western Florida and southern Alabama.

Alabama and Mobile Basin Distribution

Pleurobema strodeanum is found in the Choctawhatchee, Yellow and Conecuh rivers and some tributaries.

Pleurobema strodeanum is extant in all three drainages but is generally uncommon.

Ecology and Biology

Pleurobema strodeanum inhabits medium creeks to rivers, where it occurs in stable substrates of sand and silty sand in slow to moderate current.

The biology of *Pleurobema strodeanum* in Eightmile Creek, Walton County, Florida, was studied by Pilarczyk et al. (2005). It is a short-term brooder, gravid from mid-March through May, possibly June. *Pleurobema strodeanum* conglutinates are elongate, flattened and tapered on both ends, creamy white to pinkish orange. The pinkish orange conglutinate color is derived from undeveloped eggs and diminishes as glochidia mature. Fecundity was measured in 11 individuals and ranged from 330 to 20,800, with an average of 15,173 glochidia per individual. *Cyprinella venusta* (Blacktail Shiner) (Cyprinidae) was found to serve as a glochidial host of *P. strodeanum* in a preliminary host fish study (Pilarczyk et al., 2005).

Current Conservation Status and Protection

Pleurobema strodeanum was considered a species of special concern throughout its range by Williams et al. (1993) and in Alabama by Lydeard et al. (1999). Garner et al. (2004) listed *P. strodeanum* as a species of highest conservation concern in the state. In 2003 this species was elevated to a candidate for protection under the federal Endangered Species Act.

Remarks

In a phylogenetic analysis of *Pleurobema* in eastern Gulf drainages, *P. strodeanum* was found to be sister to a clade containing *P. pyriforme* and *P. reclusum* (B.H. Wright, 1898) (Kandl et al., 2001). Additional genetic analyses of the various shell forms of *P. strodeanum* revealed no differences (K.L. Kandl, personal communication).

Pleurobema strodeanum was named for conchologist W.S. Strode of Lewiston, Illinois.

Synonymy

Unio strodeanus B.H. Wright, 1898. B.H. Wright, 1898c:5; Simpson, 1900a:81, pl. 1, fig. 3
Type locality: Escambia River, west Florida. Lectotype, USNM 150498, length 38 mm, designated by Johnson (1974).

Pleurobema patsaligensis Simpson, 1900. Simpson, 1900a:82, pl. 2, fig. 1
Type locality: Little Patsaliga Creek, [Crenshaw County,] Alabama, Kirkland. Holotype, USNM 150475, length 44 mm.

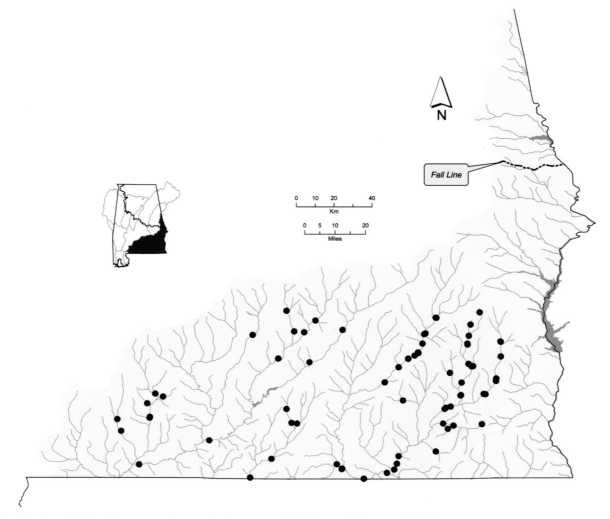

Distribution of *Pleurobema strodeanum* in the eastern Gulf Coast drainages of Alabama.

Pleurobema taitianum (Lea, 1834)
Heavy Pigtoe

Pleurobema taitianum – Length 45 mm, USNM 84653. Alabama River, Claiborne, Monroe County, Alabama. © Richard T. Bryant.

Shell Description

Length to 70 mm; thick; inflated anteriorly; outline triangular; posterior margin obliquely truncate to bluntly pointed; anterior margin straight to rounded; dorsal margin convex; ventral margin broadly rounded; posterior ridge low to well-developed, rounded; posterior slope steep; shallow sulcus just anterior of posterior ridge; umbo narrow, inflated, positioned anteriorly, turned anteriad, elevated well above hinge line, umbo sculpture unknown; periostracum brown or olive brown to black, usually without rays.

Pseudocardinal teeth large, erect, triangular, 2 divergent teeth in left valve, 1 tooth in right valve, often with accessory denticle posteriorly, sometimes anteriorly; lateral teeth long, thick, straight in young individuals, becoming slightly curved with age, 2 in left valve, 1 or 2 in right valve; interdentum short, wide; umbo cavity shallow; nacre white, occasionally pale pink.

Soft Anatomy Description

Mantle tan, may be more pale outside of pallial line, outside of apertures creamy white, tan or rusty tan, sometimes with gray mottling; visceral mass creamy white to tan; foot tan, often slightly darker distally.

Gills tan; dorsal margin straight to slightly sinuous, ventral margin convex; gill length 55–60% of shell length; gill height 45–60% of gill length; outer gill height 85–90% of inner gill height; inner lamellae of inner gills connected to visceral mass only anteriorly.

No gravid females were available for marsupium description. Outer gills are marsupial, slightly padded when gravid (P.D. Johnson, personal communication).

Labial palps tan, often with irregular opaque area centrally; straight dorsally, convex ventrally, bluntly pointed to narrowly rounded distally; palp length 25–35% of gill length; palp height 55–80% of palp length; distal 25–45% of palps bifurcate.

Incurrent and supra-anal apertures longer than excurrent aperture; incurrent and supra-anal aperture variable, either may be longer than the other.

Incurrent aperture length 15–25% of shell length; tan within, often creamy white basal to papillae; papillae in 1 row, simple, bifid and small arborescent, some individuals without arborescent; papillae some combination of creamy white, rusty tan and pale gray, spaces between papillae may be rusty brown.

Excurrent aperture length 10–15% of shell length; tan within, marginal color band creamy white to pale tan, may have rusty brown or grayish brown spots or bands perpendicular to margin; margin with small simple papillae, may be well-developed; papillae creamy white to pale tan.

Supra-anal aperture length 15–25% of shell length; creamy white to pale tan within, may have sparse grayish brown marginally; margin smooth; mantle bridge separating supra-anal and excurrent apertures imperforate, length 15–20% of supra-anal length, occasionally absent.

Specimens examined: Alabama River (n = 3); Tombigbee River (n = 1) (some specimens previously preserved).

Glochidium Description

Glochidium unknown.

Similar Species

Pleurobema taitianum most closely resembles *Fusconaia cerina*, but *P. taitianum* has a much less angular posterior ridge and the shell is less pointed posterioventrally. *Pleurobema taitianum* often has a weak sulcus, which is replaced by a flattened area on the shell disk in *F. cerina*. *Pleurobema taitianum* has a shallow umbo cavity, whereas that of *F. cerina* is deep and somewhat compressed.

Pleurobema taitianum resembles *Pleurobema verum*. However, *P. verum* lacks a sulcus and its pseudocardinal teeth are less massive than those of *P. taitianum*. The periostracum of *P. taitianum* is darker than that of *P. verum*, which is usually yellowish.

Pleurobema taitianum superficially resembles *Fusconaia ebena* but has a more triangular outline and shallow sulcus. The umbo cavity of *P. taitianum* is shallow, whereas that of *F. ebena* is deep and compressed. The pseudocardinal teeth of *P. taitianum* are more divergent than those of *F. ebena*.

General Distribution

Pleurobema taitianum is endemic to the Mobile Basin in Alabama and Mississippi.

Alabama and Mobile Basin Distribution

Pleurobema taitianum historically occurred throughout the Tombigbee and Alabama rivers and lower reaches of some large tributaries. There is one isolated record of this species from above the Fall Line in the Black Warrior River.

Pleurobema taitianum is currently known to be extant at one site in the Alabama River, Dallas County, and one site in the Tombigbee River, Choctaw and Marengo counties.

Ecology and Biology

Pleurobema taitianum is a species of rivers and large creeks. Williams (1982) reported it from gravel shoals. However, substrates where *P. taitianum* is extant in the Alabama River are composed of gravel with a large component of coarse sand in water exceeding 6 m in depth with variable current.

Pleurobema taitianum is a short-term brooder, gravid in spring and summer (P.D. Johnson, personal communication). Its glochidial hosts are unknown, but some species of *Pleurobema* use members of the Cyprinidae (Haag and Warren, 2003).

Current Conservation Status and Protection

Pleurobema taitianum was considered endangered throughout its range by Williams et al. (1993) and in Alabama by Stansbery (1976) and Lydeard et al. (1999). Garner et al. (2004) designated it a species of highest conservation concern in the state. *Pleurobema taitianum* was listed as endangered under the federal Endangered Species Act in 1997.

Remarks

Pleurobema taitianum disappeared from most of the Tombigbee River when its natural habitat was destroyed with construction of the Tennessee Tombigbee Waterway by the USACE. It declined in the Alabama River following impoundment during the 1960s. The absence of *P. taitianum* records from the lower Tombigbee River are apparently due to a lack of collections prior to its destruction.

Pleurobema taitianum was named for Judge Charles Tait, who resided at Claiborne, Alabama, and hosted naturalist Timothy A. Conrad during his explorations in Alabama during the 1830s.

Synonymy

Unio taitianus Lea, 1834. Lea, 1834a:39, pl. 4, fig. 11; Lea, 1834b:151, pl. 4, fig. 11

Type locality: Alabama River, [Claiborne, Monroe County, Alabama,] Judge Tait. Lectotype, USNM 84653, length 42 mm, designated by Johnson (1974).

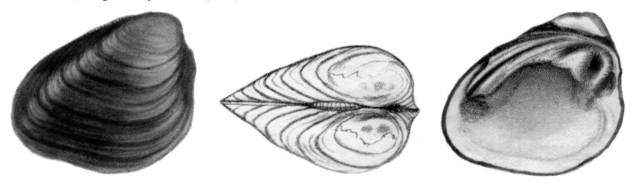

Pleurobema tombigbeanum Frierson, 1908. Frierson, 1908:27, pl. 3, figs. 3, 4

Type locality: [Tombigbee River,] Demopolis, Marengo County, Alabama; [Tombigbee River,] Columbus, [Lowndes County,] Mississippi. Lectotype, ANSP 173339a, length 38 mm, designated by Johnson (1972a), is from Tombigbee River, Columbus, Lowndes County, Mississippi.

Pleurobema aldrichi Frierson, 1927. Frierson, 1927:44; Frierson, 1928:139, pl. 3, fig. 1

Type locality: [Alabama River,] Selma, [Dallas County,] Alabama. Lectotype, ANSP 145246, length 50 mm, designated by Johnson (1972a), is not the specimen figured by Frierson (1928).

Comment: Johnson (1972a) designated and figured ANSP 145246 as the lectotype and stated that the type specimens in the original lot, including the one figured by Frierson (1928), could not be found in the UMMZ. However, the type lot is present in the UMMZ, making Johnson's lectotype designation invalid. The specimen below was figured by Frierson (1928).

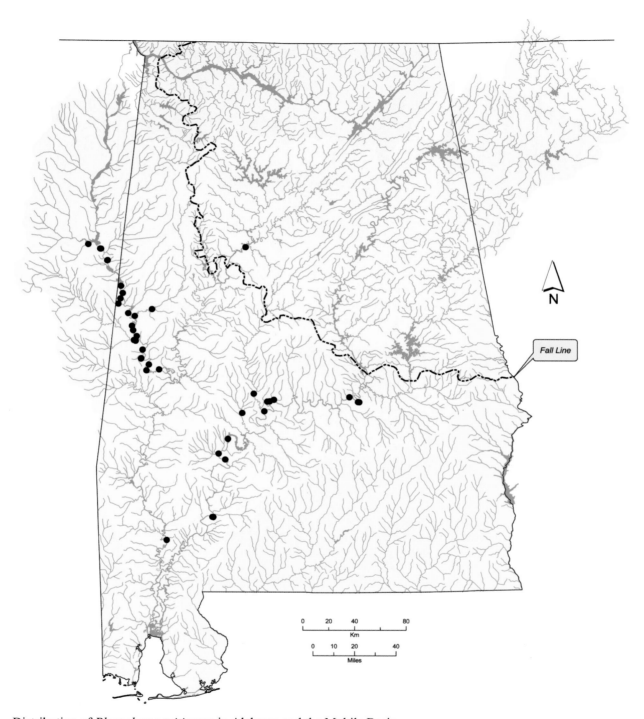

Distribution of *Pleurobema taitianum* in Alabama and the Mobile Basin.

Pleurobema verum (Lea, 1861)
True Pigtoe

Pleurobema verum – Length 60 mm, UMMZ 58363. Cahaba River, 10 miles west of Selma, Dallas County, Alabama. © Richard T. Bryant.

Shell Description

Length to 60 mm; thick, thinner posteriorly; moderately inflated; outline oval to subtriangular; posterior margin narrowly rounded to obliquely truncate; anterior margin rounded; dorsal margin convex; ventral margin rounded; posterior ridge low, rounded; posterior slope moderately steep; flattened area just anterior of posterior ridge; umbo broad, inflated, directed anteriad, elevated above hinge line, umbo sculpture unknown; periostracum yellowish to tawny.

Pseudocardinal teeth moderately thick, triangular, 2 divergent teeth in left valve, 1 tooth in right valve; lateral teeth moderately long, straight, 2 in left valve, 1 in right valve; interdentum short, moderately wide; umbo cavity shallow; nacre white.

Soft Anatomy Description

Soft anatomy unknown.

Glochidium Description

Glochidium unknown.

Similar Species

Pleurobema verum resembles *Pleurobema taitianum*. However, *P. verum* is less triangular in outline and lacks the shallow sulcus just anterior of the posterior ridge, though that portion of the shell disk is flattened. Pseudocardinal teeth of *P. verum* are less massive than those of *P. taitianum*. Also, the periostracum of *P. verum* is much more yellow than that of *P. taitianum*, which is brown to black.

General Distribution

Pleurobema verum is endemic to the Cahaba and Alabama River drainages of the Mobile Basin in Alabama.

Alabama and Mobile Basin Distribution

Pleurobema verum is known only from the Cahaba River below the Fall Line and Alabama River.

Pleurobema verum has not been collected since the 1930s and appears to be extinct.

Ecology and Biology

Locality data from historical collections suggest that *Pleurobema verum* inhabited shoal habitats.

Nothing is known about the biology of *Pleurobema verum*. However, other species of *Pleurobema* are short-term brooders and utilize members of the Cyprinidae as glochidial hosts (Haag and Warren, 2003).

Current Conservation Status and Protection

Pleurobema verum was listed as endangered by Williams et al. (1993) and imperiled by Lydeard et al. (1999). The USFWS (2000) and Garner et al. (2004) listed *P. verum* as extinct.

Remarks

There are three records of *Pleurobema verum* from the Cahaba River, including the type. In an annotated checklist of Cahaba River mussels, van der Schalie (1938a) reported one record under "*Pleurobema cordatum* and varieties" and commented on the perplexity of its identification. The specimen herein figured is likely from that collection.

Synonymy

Unio verus Lea, 1861. Lea, 1861a:40; Lea, 1862b:83, pl. 11, fig. 231; Lea, 1862d:87, pl. 11, fig. 231

Type locality: Cahawba [Cahaba] River, Perry County, Alabama, E.R. Showalter, M.D. Holotype, USNM 84589, length 32 mm.

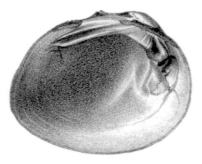

Unio instructus Lea, 1861. Lea, 1861c:59; Lea, 1862b:82, pl. 10, fig. 230; Lea, 1862d:86, pl. 10, fig. 230
Type locality: Cahawba [Cahaba] River, Alabama, E.R. Showalter, M.D. Holotype, USNM 84640, length 37 mm.

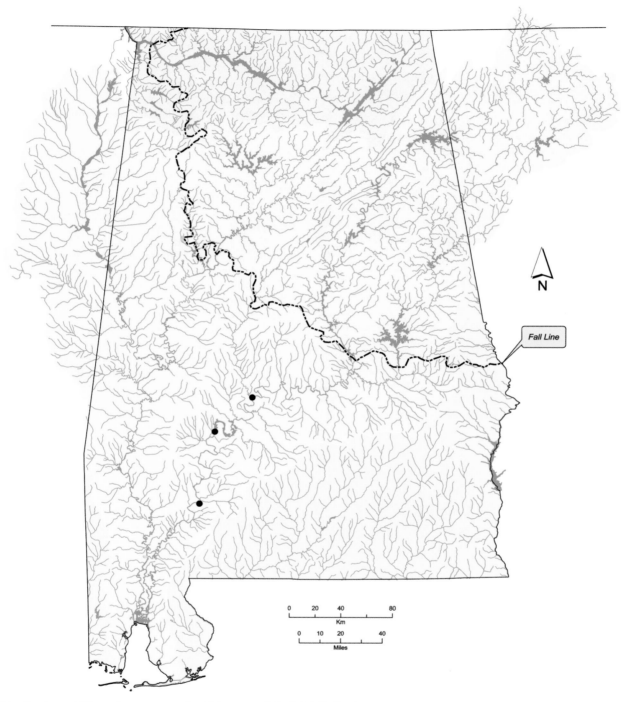

Distribution of *Pleurobema verum* in Alabama and the Mobile Basin.

Genus *Pleuronaia*

Pleuronaia Frierson, 1927, is composed of three species, all restricted to the Cumberland and/or Tennessee River drainages. Two of these occur in Alabama.

Pleuronaia is herein resurrected based on phylogenetic analyses of DNA sequence data (Campbell et al., 2005; Bogan et al., unpublished data) and shell morphology (i.e., shallow umbo cavity). *Pleuronaia* includes the taxa formerly recognized as *Fusconaia barnesiana* (Lea, 1838), *Lexingtonia dolabelloides* (Lea, 1840) and *Pleurobema gibberum* (Lea, 1838). This clade is distinct from *Fusconaia* Simpson, 1900, *Pleurobema* Rafinesque, 1819, and *Quadrula* Rafinesque, 1820 (Campbell et al., 2005). *Lexingtonia* Ortmann, 1914, is no longer an available name for this clade because the type species of the genus, *Unio subplana* Conrad, 1837, was found to be a synonym of *Fusconaia masoni* (Conrad, 1834) (Bogan et al., unpublished data). Thus, *Lexingtonia* Ortmann, 1914, becomes a junior synonym of *Fusconaia* Simpson, 1900. The available name for this group is *Quadrula* (*Pleuronaia*) Frierson, 1927. Frierson (1927) also included *Fusconaia cor* (Conrad, 1834) and *Fusconaia cuneolus* (Lea, 1840) in this group, but DNA analyses do not support their inclusion (Campbell et al., 2005).

Frierson (1927) observed in erecting this subgenus, "We propose this Sub-Genus for the accommodation of those exceedingly difficult *Naiades* whose headquarters are chiefly found in the Tennessee River drainage, typed by the '*Unio barnesianus* Lea'."

Type Species
Unio barnesianus Lea, 1838

Diagnosis
Shell oval or subquadrate to triangular; moderately thick; compressed to slightly inflated; posterior ridge sharp to broadly rounded; hinge plate well-developed; periostracum tawny to greenish brown or dark brown, with variable interrupted green lines; umbo cavity shallow.

Inner lamellae of inner gills only connected to visceral mass anteriorly; marsupium not extended ventrally below original gill edge when gravid; glochidium without styliform hooks (Frierson, 1927).

Synonymy
None recognized.

Pleuronaia barnesiana (Lea, 1838)
Tennessee Pigtoe

Pleuronaia barnesiana – Length 52 mm, UF 67019. Shoal Creek, Lauderdale County, Alabama, October 1904. © Richard T. Bryant.

Shell Description

Length to 95 mm; moderately thick; compressed, to moderately inflated; outline oval to subtriangular or subquadrate; posterior margin obliquely truncate to broadly rounded; anterior margin rounded; dorsal margin convex; ventral margin usually broadly rounded, may be straight or slightly concave in old individuals; posterior ridge low, rounded; posterior slope flat to slightly concave; umbo broad, slightly to moderately inflated, usually elevated slightly above hinge line, umbo sculpture fine, concentric ridges; periostracum satiny, may be roughened by prominent annuli, yellowish green to brown, darkening with age to dark brown or black, may have variable dark green rays.

Pseudocardinal teeth moderately thick, erect, slightly compressed, axis of pseudocardinal teeth usually oriented nearly perpendicular to axis of lateral teeth, 2 slightly divergent teeth in left valve, 1 tooth in right valve, may have accessory denticle anteriorly and/or posteriorly; lateral teeth moderately long, slightly curved, 2 in left valve, 1 in right valve; interdentum moderately long, narrow to moderately wide; umbo cavity moderately shallow; nacre white, may have salmon tint.

Soft Anatomy Description

Mantle tan to light brown, may be dull orange outside of pallial line anteriorly, grayish brown to rusty tan outside of apertures, may have gray bands perpendicular to margin; visceral mass creamy white to tan; foot creamy white to tan, occasionally pale orange.

Gills tan to light brown; dorsal margin straight to slightly sinuous, ventral margin elongate convex to deeply convex; gill length 45–60% of shell length; gill height 40–70% of gill length; outer gill height 65–80% of inner gill height; inner lamellae of inner gills connected to visceral mass only anteriorly, occasionally completely connected.

All 4 gills marsupial, occasional individuals gravid in outer gills only; glochidia held across entire gill; marsupium slightly padded when gravid; pink when holding embryos, becoming pale tan as glochidia mature.

Labial palps tan to dull orange, may have irregular white spot dorsally; straight to slightly concave dorsally, convex ventrally, usually bluntly pointed distally, occasionally rounded; palp length 10–30% of gill length; palp height 60–100% of palp length, occasionally less; distal 30–60% of palps bifurcate.

Incurrent, excurrent and supra-anal aperture lengths variable relative to each other; supra-anal aperture usually longer than incurrent and excurrent apertures.

Incurrent aperture length 15–25% of shell length; creamy white to gray or brown within, rusty tan to golden brown or grayish brown basal to papillae; papillae in 1–2 rows, long, slender, usually mostly simple, often with some bifid, occasionally with some trifid or small arborescent; papillae brown to golden brown or rusty brown, occasionally dull orange.

Excurrent aperture length 15–20% of shell length; creamy white or tan to grayish brown or rusty brown within, marginal color band rusty brown to grayish brown, often with white band proximally; margin usually with small, simple papillae, may be crenulate or smooth; papillae tan to rusty brown or dull orange.

Supra-anal aperture length 15–25% of shell length; creamy white or pale tan to golden brown or grayish brown within, often with thin grayish brown or rusty brown marginal band; margin smooth; mantle bridge separating supra-anal and excurrent apertures usually imperforate, length 5–10% of supra-anal length, occasionally greater.

Specimens examined: Duck River, Tennessee (n = 2); Estill Fork, Jackson County (n = 3); Paint Rock River (n = 5).

Glochidium Description

Length 150–185 μm; height 162–185 μm; without styliform hooks; outline subelliptical (Ortmann, 1917; Kitchel, 1985).

Similar Species

Pleuronaia barnesiana resembles *Pleuronaia dolabelloides*. However, the posterior ridge of *P. dolabelloides* is more angular and ventrally curved. The posterior slope of *P. dolabelloides* is usually steeper than that of *P. barnesiana*, except in *P. barnesiana* from large rivers. The pseudocardinal teeth of *P. barnesiana* are more compressed and somewhat less divergent than those of *P. dolabelloides*. Also, the umbo cavity is often slightly deeper in *P. dolabelloides* than in *P. barnesiana*. The foot of *P. barnesiana* is usually creamy white or tan, occasionally pale orange, whereas *P. dolabelloides* almost always has a bright orange foot.

Pleuronaia barnesiana also closely resembles *Fusconaia cor*, *Fusconaia cuneolus* and some young *Fusconaia subrotunda*. Those species usually have a deeper, more compressed umbo cavity and heavier, less compressed pseudocardinal teeth than *P. barnesiana*. The pseudocardinal teeth in the left valve of *P. barnesiana* are less divergent than those of *F. cor*, *F. cuneolus* and *F. subrotunda* and their axis is almost perpendicular to the lateral teeth. The periostracum of *F. cor* is almost always shiny, whereas that of *P. barnesiana* is satiny.

Pleuronaia barnesiana resembles *Pleurobema oviforme*. However, the pseudocardinal teeth of *P. oviforme* are more massive, less compressed and more divergent than those of *P. barnesiana*. Also, the axis of *P. barnesiana* pseudocardinal teeth is almost perpendicular to the lateral teeth, whereas that of *P. oviforme* is not. The shell of *P. oviforme* is often more elliptical than that of *P. barnesiana*.

General Distribution

Pleuronaia barnesiana is endemic to the Tennessee and possibly Cumberland River drainages. The reported type locality of *P. barnesiana* is the Cumberland River (Simpson, 1914), but there are no other reports of the species from outside the Tennessee River drainage, suggesting that the reported type locality may have been in error. In the Tennessee River drainage *P. barnesiana* historically occurred from headwaters in southwestern Virginia, western North Carolina and eastern Tennessee downstream to Muscle Shoals in northwestern Alabama (Ortmann, 1925; Neves, 1991; Parmalee and Bogan, 1998).

Alabama and Mobile Basin Distribution

Pleuronaia barnesiana historically occurred in the Tennessee River across northern Alabama as well as in many tributaries.

Pleuronaia barnesiana is extant in several Tennessee River tributaries. It has not been collected from the Tennessee River proper since 1966, when it was found in Wilson Dam tailwaters.

Ecology and Biology

Pleuronaia barnesiana occurs in a wide variety of habitats, ranging from small headwater streams to large rivers, generally in flowing water. The preferred substrate of *P. barnesiana* appears to be stable gravel with interstitial sand.

Pleuronaia barnesiana is a short-term brooder, spawning in spring and early summer and brooding glochidia from mid-May to mid-July (Ortmann, 1917). Glochidial hosts of *P. barnesiana* are unknown.

Current Conservation Status and Protection

Pleuronaia barnesiana was considered a species of special concern throughout its range by Williams et al. (1993). In Alabama it was listed as endangered by Stansbery (1976), a species of special concern by Lydeard et al. (1999) and a species of high conservation concern by Garner et al. (2004).

Remarks

Pleuronaia barnesiana has historically been treated as a member of the genus *Fusconaia*. However, it has been recognized as differing from other members of that genus since the early 1900s. Ortmann (1917, 1918) commented on conchological differences, specifically the shallow umbo cavity of *P. barnesiana*, which is atypical of *Fusconaia*. Preliminary genetic analyses suggest that *P. barnesiana* is more closely related to *Pleuronaia dolabelloides* than to species of *Fusconaia* (Campbell et al., 2005; Bogan et al., unpublished data). Bogan et al. (unpublished data) found that *Lexingtonia subplana* is synonymous with *Fusconaia masoni*. Since *L. subplana* is the type species of the genus, *Lexingtonia* becomes a synonym of *Fusconaia*. The next available genus name for *barnesiana* and *dolabelloides* is *Pleuronaia* Frierson, 1927. *Pleurobema gibberum*, a Cumberland River drainage endemic, was also found to belong to the *Pleuronaia* clade.

Pleuronaia barnesiana, like many other species, demonstrates clinal variation in shell morphology, principally becoming more compressed with an upstream progression. Ortmann (1920) discussed this variation in *P. barnesiana*, noting a clear gradation from the headwater form (*bigbyensis*) to the large river form (*tumescens*) with typical *P. barnesiana* falling in between. Ortmann (1917) reported Shoal Creek, Lauderdale County, Alabama, to have "a very peculiar mixture", containing "all three types side by side,

intergrading completely, but with typical *barnesiana* prevailing". The *tumescens* form was historically found in the Tennessee River proper in Alabama, but forms in tributaries range from typical *P. barnesiana* to the *bigbyensis* form.

Pleuronaia barnesiana has been reported from prehistoric shell middens along the Tennessee River at Muscle Shoals and near Bridgeport but not from intervening sites (Morrison, 1942; Warren, 1975; Hughes and Parmalee, 1999). It was rare at sites where it occurred.

Synonymy

Margarita (*Unio*) *barnesianus* Lea, 1836, *nomen nudum*. Lea, 1836:20
Unio barnesianus Lea, 1838. Lea, 1838a:31, pl. 10, fig. 26; Lea, 1838c:31, pl. 10, fig. 26
Type locality: Cumberland River, Tennessee. Lectotype, USNM 84586, length 37 mm, designated by Johnson (1974).

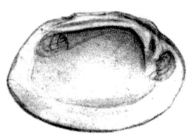

Unio bigbyensis Lea, 1841. Lea, 1841a:30; Lea, 1842b:237, pl. 22, fig. 51; Lea, 1842c:75, pl. 22, fig. 51
Unio estabrookianus Lea, 1845. Lea, 1845:164; Lea, 1848a:77, pl. 6, fig. 17; Lea, 1848b:77, pl. 6, fig. 17
Unio tumescens Lea, 1845. Lea, 1845:164; Lea, 1848a:71, pl. 3, fig. 7; Lea, 1848b:71, pl. 3, fig. 7
Unio meredithii Lea, 1858. Lea, 1858a:40; Lea, 1862b:65, pl. 6, fig. 214; Lea, 1862d:69, pl. 6, fig. 214
Type locality: Tennessee River, Florence, [Lauderdale County,] Alabama, L.B. Thornton, Esq. Lectotype, USNM 84393, length 45 mm, designated by Johnson (1974).

Unio pudicus Lea, 1860. Lea, 1860a:92; Lea, 1860d:346, pl. 56, fig. 171; Lea, 1860g:28, pl. 56, fig. 171
Type locality: North Alabama, Professor Tuomey; Florence, [Lauderdale County,] Alabama, L.B. Thornton, Esq.
    Lectotype, USNM 84553, length 33 mm, designated by Johnson (1974).

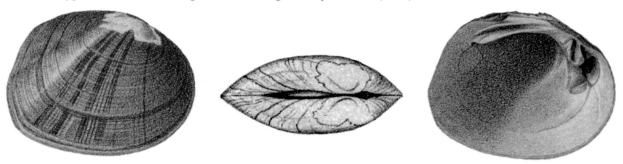

Unio ornatus Lea, 1861, *non* Conrad, 1835. Lea, 1861a:41; Lea, 1862b:85, pl. 11, fig. 234; Lea, 1862d:89, pl. 11,
    fig. 234
Unio lyonii Lea, 1865. Lea, 1865:89; Lea, 1868a:259, pl. 32, fig. 74; Lea, 1869a:19, pl.32, fig. 74
Unio striatissimus Anthony, 1865. Anthony, 1865:156, pl. 12, fig. 1
Unio fassinans Lea, 1868. Lea, 1868a:143; Lea, 1868c:305, pl. 47, fig. 118; Lea, 1869a:65, pl. 47, fig. 118
Unio crudus Lea, 1871. Lea, 1871:190; Lea, 1874c:14, pl. 4, fig. 10; Lea, 1874e:18, pl. 4, fig. 10
Unio radiosus Lea, 1871. Lea, 1871:192; Lea, 1874c:13, pl. 3, fig. 9; Lea, 1874e:17, pl. 3, fig. 9
Unio lenticularis Lea, 1872. Lea, 1872b:155; Lea, 1874c:30, pl. 9, fig. 27; Lea, 1874e:34, pl. 9, fig. 27
Unio tellicoensis Lea, 1872. Lea, 1872b:155; Lea, 1874c:31, pl. 10, fig. 28; Lea, 1874e:35, pl. 10, fig. 28
Pleurobema fassinans var. *rhomboidea* Simpson, 1900. Simpson, 1900b:762

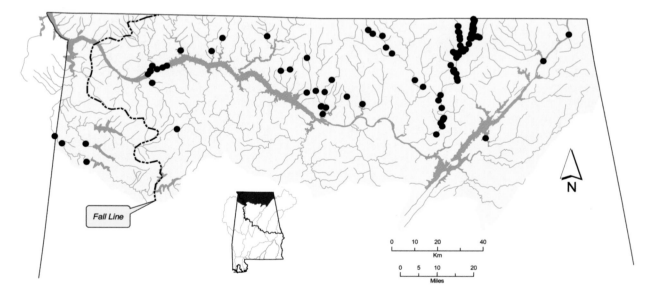

Distribution of *Pleuronaia barnesiana* in the Tennessee River drainage of Alabama.

Pleuronaia dolabelloides (Lea, 1840)
Slabside Pearlymussel

Pleuronaia dolabelloides – Upper figure: length 68 mm, UF 370417. [Tennessee River at mouth of] Pond Creek, below Wilson Dam, Alabama, 12 October 1966. Lower figure: length 61 mm, CMNH 61.8604. Paint Rock River, Princeton, Jackson County, Alabama. © Richard T. Bryant.

Shell Description

Length to 85 mm; thick; moderately compressed; outline triangular to subquadrate; posterior margin bluntly pointed to obliquely truncate; anterior margin rounded to truncate; dorsal margin convex; ventral margin convex; posterior ridge rounded but distinct, ventrally curved; posterior slope moderately steep, flat to slightly concave; disk wide, flat, may have wide, very shallow radial depression just anterior of posterior ridge; umbo broad, not inflated, elevated above hinge line, turned slightly anteriad, umbo sculpture fine, irregular ridges; periostracum tawny to brown, often greenish yellow in juveniles, usually with a few interrupted green rays of variable width, rays often do not extend to ventral margin.

Pseudocardinal teeth moderately thick, 2 divergent teeth in left valve, posterior tooth triangular, anterior tooth somewhat compressed, 1 narrowly triangular tooth in right valve, sometimes with accessory denticle anteriorly and/or posteriorly; lateral teeth slightly curved, 2 in left valve, usually 1 in right valve, occasionally with rudimentary lateral tooth ventrally in right valve; interdentum short, moderately wide; umbo cavity wide, moderately shallow, moderately deep in large individuals from large rivers; nacre white.

Soft Anatomy Description

Mantle tan to light brown or golden brown, sometimes partially orange, often orange outside of pallial line; visceral mass creamy white to tan, often with orange cast; foot bright orange. Ortmann (1921) reported soft tissues to rarely be pale orange, occasional white in young individuals.

Gills creamy white to light brown, gold or orange; dorsal margin straight to slightly sinuous, ventral margin convex; gill length 45–65% of shell length; gill height 40–60% of gill length, occasionally greater; outer gill height 65–90% of inner gill height; inner lamellae of inner gills connected to visceral mass only anteriorly.

Outer gills marsupial; glochidia held across entire gill; marsupium padded when gravid; pink when holding embryos, creamy white when holding glochidia.

Labial palps orange, often with irregular creamy white area dorsally; usually straight dorsally, sometimes slightly convex or concave, convex ventrally, bluntly pointed to round distally; palp length 20–30% of gill length, occasionally less; palp height usually 50–70% of palp length, occasionally greater; distal 25–50% of palps bifurcate.

Incurrent and excurrent apertures of similar length; supra anal aperture longer than incurrent and excurrent apertures.

Incurrent aperture length 15–20% of shell length; creamy white within, some combination of orange, gold, brown and gray basal to papillae, often with only one color; papillae variable, in 1–3 rows, simple, bifid and trifid, often with a few scattered arborescent; papillae some combination of dull orange, brown, golden brown and grayish brown, may have dark brown or black edges basally.

Excurrent aperture length 15–20% of shell length; creamy white within, marginal color band some combination of tan, golden brown, dark brown, grayish brown and pale orange, often mottled with no distinct pattern; margin with short but well developed, simple papillae, may have a few bifid; papillae tan to gold, orange, brown or gray.

Supra anal aperture length 20–30% of shell length; golden brown to pale orange within, sometimes with thin brown marginal band, often irregular; margin smooth; mantle bridge separating supra anal and excurrent apertures imperforate, length 5–10% of supra anal length.

Specimens examined: Bear Creek, Colbert County (n = 3); Duck River, Tennessee (n = 3); Paint Rock River (n = 2).

Glochidium Description

Length 130–159 μm; height 160–203 μm; without styliform hooks (Ortmann, 1921; Kitchel, 1985).

Similar Species

Pleuronaia dolabelloides resembles *Pleuronaia barnesiana* and *Pleurobema oviforme* as well as *Fusconaia cor*, *Fusconaia cuneolus* and some young *Fusconaia subrotunda*. *Pleuronaia dolabelloides* can be separated from all of those species by its ventrally curved posterior ridge and generally steeper posterior slope. The pseudocardinal teeth of *P. dolabelloides* are less compressed and more divergent than those of *P. barnesiana*. Also, the umbo cavity in *F. cor*, *F. cuneolus* and *F. subrotunda* is deeper and more compressed than that of *P. dolabelloides*. The foot of *P. dolabelloides* is almost always bright orange, whereas those of *F. cor*, *F. cuneolus* and *P. oviforme* are always tan or white. The foot of *P. barnesiana* and *F. subrotunda* may be pale orange but are often tan or white.

General Distribution

Pleuronaia dolabelloides is endemic to the Cumberland and Tennessee River drainages. Only a few records exist from the Cumberland River in Kentucky and Tennessee (Cicerello et al., 1991; Parmalee and Bogan, 1998). *Pleuronaia dolabelloides* is widespread in the Tennessee River drainage, where it historically occurred from headwaters in southwestern Virginia downstream at least to, and including, the Duck River, Tennessee (Neves, 1991; Parmalee and Bogan, 1998).

Alabama and Mobile Basin Distribution

Pleuronaia dolabelloides historically occurred in the Tennessee River across northern Alabama and in some tributaries.

Pleuronaia dolabelloides is extant in a short reach of Bear Creek, Colbert County, and the Paint Rock River system. Ahlstedt (1998) reported *P. dolabelloides* to be the most commonly encountered mussel species in a 1991 survey of the Paint Rock River.

Ecology and Biology

Pleuronaia dolabelloides occurs in shoal habitats of large creeks to large rivers. It was historically present in the Tennessee and Cumberland rivers but was extirpated with their impoundment. *Pleuronaia dolabelloides* is generally found in gravel substrates with some interstitial sand.

Pleuronaia dolabelloides is a short term brooder. Females brooding glochidia have been reported from mid May through early August (Ortmann, 1921; Neves, 1991). Glochidia have been reported in stream drift from mid June to mid or late August in the North Fork Holston River, southwestern Virginia (Kitchel, 1985). Observations of natural infestations with *P. dolabelloides* glochidia have been reported for *Notropis ariommus* (Popeye Shiner), *Notropis leuciodus*

(Tennessee Shiner), *Notropis photogenis* (Silver Shiner), *Notropis rubellus* (Rosyface Shiner), *Notropis rubricroceus* (Saffron Shiner) and *Notropis telescopus* (Telescope Shiner) (Cyprinidae) (Kitchel, 1985).

Current Conservation Status and Protection

Pleuronaia dolabelloides was reported as endangered throughout its range by Stansbery (1970a) and threatened throughout its range by Williams et al. (1993). In Alabama it was considered endangered by Stansbery (1976), imperiled by Lydeard et al. (1999) and a species of highest conservation concern by Garner et al. (2004).

Remarks

Pleuronaia dolabelloides has historically been treated as a member of the genus *Lexingtonia*.

Lexingtonia subplana is the type species of the genus, but recent genetic analyses suggest that it is synonymous with *Fusconaia masoni*. Therefore, the genus *Lexingtonia* falls into the synonymy of *Fusconaia*. The next available genus name for *dolabelloides* is *Pleuronaia* Frierson, 1927.

Clinal variation in shell morphology has long been recognized in *Pleuronaia dolabelloides*. The compressed small stream form was historically referred to by the trinomen *Pleuronaia dolabelloides conradi* (Vanatta, 1915). However, Ortmann (1920) determined that the two forms "pass gradually into each other".

Pleuronaia dolabelloides has been reported from prehistoric shell middens along the Tennessee River in Alabama in moderate abundance (Morrison, 1942; Warren, 1975; Hughes and Parmalee, 1999).

Synonymy

Unio maculatus Conrad, 1834, *non* Rafinesque, 1820. Conrad, 1934b:30, pl. 4, fig. 4
Type locality: Banks of Elk and Flint rivers, tributaries of the Tennessee [River, Alabama]. Holotype, ANSP 41413, length 53 mm.
Comment: *Unio maculatus* Conrad, 1834, is a junior primary homonym of *Unio nigra* var. *maculata* Rafinesque, 1820, recognized by Vanatta (1915) who erected the replacement name *Unio conradi* Vanatta, 1915.

Unio dolabelloides Lea, 1840. Lea, 1840:288; Lea, 1842b:215, pl. 15, fig. 31; Lea, 1842c:53, pl. 15, fig. 31
Type locality: Holston River, Tennessee. Holotype, USNM 85828, length 36 mm.

 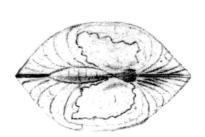 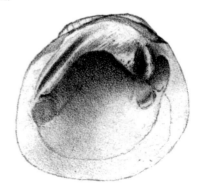

Unio mooresianus Lea, 1857. Lea, 1857b:83; Lea, 1866:39, pl. 14, fig. 37; Lea, 1867b:43, pl. 14, fig. 37
Type locality: [Tennessee River,] Tuscumbia, [Colbert County,] Alabama, H. Moores. Lectotype, USNM 84695,
	length 40 mm, designated by Johnson (1974).

 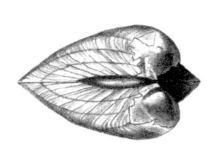 

Unio thorntonii Lea, 1857. Lea, 1857b:83; Lea, 1866:38, pl. 14, fig. 36; Lea, 1867b:42, pl. 14, fig. 36
Type locality: [Tennessee River,] Tuscumbia, [Colbert County,] Alabama, L.B. Thornton, Esq. Holotype, USNM
	85829, length 37 mm.

 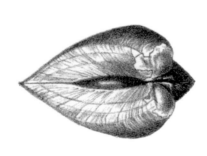 

Unio subglobatus Lea, 1871. Lea, 1871:191; Lea, 1874c:7, pl. 1, fig. 3; Lea, 1874e:11, pl. 1, fig. 3
Type locality: Florence, [Lauderdale County,] Alabama, B. Pybas; Nashville, Tennessee, Pres. J.B. Lindsley.
	Lectotype, USNM 85823, length 42 mm, designated by Johnson (1974).

 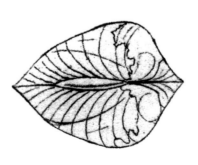 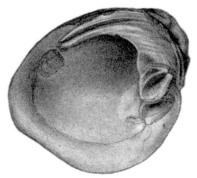

Unio circumactus Lea, 1871. Lea, 1871:192; Lea, 1874c:15, pl. 4, fig. 11; Lea, 1874e:19, pl. 4, fig. 11
Unio recurvatus Lea, 1871. Lea, 1871:192; Lea, 1874c:10, pl. 2, fig. 6; Lea, 1874e:14, pl. 2, fig. 6
Pleurobema conradi Vanatta, 1915. Vanatta, 1915:559 [replacement name for *Unio maculatus* Conrad, 1834]

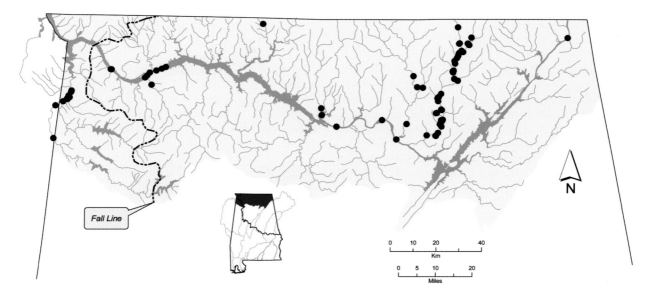

Distribution of *Pleuronaia dolabelloides* in the Tennessee River drainage of Alabama.

Genus *Potamilus*

Potamilus Rafinesque, 1818, occurs in the Hudson Bay, Great Lakes and Mississippi basins and Gulf Coast drainages from the Mobile Basin to the Rio Grande. Six *Potamilus* species are currently recognized (Turgeon et al., 1998), four of which occur in Alabama.

The genus name *Proptera* Rafinesque, 1819, has been used for this group of species over the earlier name *Potamilus* Rafinesque, 1818. Clarke (1986) provided a summary of the controversy surrounding usage of *Proptera* versus *Potamilus*. Subsequently, the ICZN was petitioned to conserve the use of *Proptera* over the earlier *Potamilus* (Bogan et al., 1990; Clarke, 1990; Gordon, 1990; Smith, 1991). However, the ICZN (1992: Opinion 1665) ruled on the basis of simple priority to support the use of *Potamilus*. Roe and Lydeard (1998) produced a molecular phylogeny of *Potamilus* that suggested it is not monophyletic. *Lastena* Rafinesque, 1820, is a synonym of *Potamilus*. Herrmannsen (1847) designated *Anodonta* (*Lastena*) *ohiensis* Rafinesque, 1820 (= *Potamilus ohiensis*) as the type species of *Lastena*.

Type Species
Unio alatus Say, 1817

Diagnosis
Shell oval to elliptical or subtriangular; thin to moderately thick; inflated to compressed; usually strongly alate posteriorly, often with smaller wing anterior of umbo; disk without sculpture; hinge plate variable; sexual dimorphism subtle, females more swollen posterioventrally; nacre usually pink to purple.

Inner lamellae of inner gills usually completely connected to visceral mass; outer gills marsupial; glochidia held in posterior portion of gill; marsupium outline reniform, swollen when gravid; glochidium ligulate, often with lateral hooks on ventral flange (Rafinesque, 1818, 1820; Ortmann, 1912a; Simpson, 1914; Haas, 1969b; Roe and Lydeard, 1998; Hoggarth, 1999).

Figure 13.33. *Potamilus purpuratus* apertures (left to right): supra-anal, excurrent, incurrent. Bogue Chitto, Dallas County, Alabama, July 2004. Photograph by A.D. Huryn.

Synonymy
Proptera Rafinesque, 1819
Lastena Rafinesque, 1820
Metaptera Rafinesque, 1820
Symphynota Lea, 1829
Lymnadia G.B. Sowerby II, 1839
Naidea Swainson, 1840

Potamilus alatus (Say, 1817)
Pink Heelsplitter

Potamilus alatus – Upper figure: female, length 113 mm, UF 370416. Tennessee River at Round Island Creek embayment, river mile 292, Limestone County, Alabama, 10 December 1996. Lower figure: male, length 135 mm, UF 67557. Tennessee River, Florence, Lauderdale County, Alabama. © Richard T. Bryant.

Shell Description

Length to 190 mm; thin when young, becoming moderately thick with age; compressed, females slightly more inflated posteriorly; outline triangular, alate, may become oval when dorsal wing eroded or broken; posterior margin rounded in males, broadly rounded to slightly truncate in females; anterior margin narrowly rounded; dorsal margin with triangular wing, convex in eroded individuals; ventral margin straight to slightly convex; posterior ridge absent; posterior slope flattened to slightly concave, merging with dorsal wing; umbo low, broad, elevated slightly above hinge line, umbo sculpture a few subconcentric bars, becoming double-looped ventrally; periostracum usually greenish brown in young individuals, occasionally tawny, becoming dark brown or black with age, young individuals may have green rays.

Pseudocardinal teeth variable, triangular to somewhat compressed, 2 triangular divergent teeth or 2 compressed, almost parallel teeth in left valve, 1 triangular or compressed tooth in right valve, usually with accessory denticle anteriorly; lateral teeth moderately short, curved, 2 in left valve, 1 in right valve; interdentum long, very narrow; umbo cavity wide, very shallow; nacre pink to purple.

Soft Anatomy Description

Mantle creamy white to tan or light brown; visceral mass pearly white to creamy white; foot creamy white to tan, often darker distally.

Rudimentary mantle fold or fleshy area just ventral to incurrent aperture; fold length variable, 10–30% of shell length; with gray, reddish brown or dark brown marginal band; margin may be irregularly crenulate.

Gills creamy white to tan or light brown; dorsal margin straight to sinuous, ventral margin gently convex; gill length 45–60% of shell length; gill height 35–50% of gill length; outer gill height 75–95% of inner gill height; inner lamellae of inner gills completely connected to visceral mass, occasionally connected only anteriorly.

Outer gills marsupial; glochidia held in posterior 30–50% of gill; marsupium reniform in outline, distended when gravid; tan to light brown. Utterback (1916b) reported gravid gills to have purplish blotches.

Labial palps creamy white to tan; straight to slightly concave dorsally, gently convex ventrally, bluntly pointed distally, rarely rounded; palp length 20–35% of gill length; palp height 55–85% of palp length; distal 30–55% of palps bifurcate.

Incurrent aperture longer than excurrent aperture; supra-anal aperture longer than incurrent aperture.

Incurrent aperture length 15–20% of shell length, occasionally less; creamy white to light tan within, may have some combination of tan, dull orange, golden brown and black basal to papillae, may be as irregular mottling, reticulated pattern or crude bands perpendicular to margin, some bands converge proximally; papillae in 2 rows, long, slender to thick, mostly simple, often with a few bifid, papillae of outer row shorter, more numerous; papillae some combination of white, tan, pale orange and golden brown, usually with black or dark brown edges basally.

Excurrent aperture length 5–10% of shell length; creamy white within, occasionally with grayish cast, marginal color band some combination of tan, orange, golden brown and black, may have irregular mottling or open reticulated pattern; margin papillate; papillae short, simple, margin occasionally crenulate; papillae creamy white, dull orange or golden brown, may have black between papillae.

Supra-anal aperture 20–25% of shell length; creamy white to pearly white within, often with very thin black or brown marginal band; margin smooth; mantle bridge separating supra-anal and excurrent apertures often with one or more small perforations, length 10–50% of supra-anal length.

Specimens examined: Bear Creek, Colbert County (n = 1); Paint Rock River (n = 2); Tennessee River (n = 7).

Glochidium Description

Length 206–230 µm; height 371–410 µm; with lanceolate hooks on lateral margins of ventral flange; outline ligulate; dorsal margin very short, straight, ventral margin convex, expanded laterally, anterior and posterior margins straight dorsally, becoming slightly concave ventrally, forming moderately sharp angle at junction with ventral margin; valves with lateral gape when closed (Lefevre and Curtis, 1910b; Ortmann, 1912a; Surber, 1912; Hoggarth, 1999).

Similar Species

Potamilus alatus resembles *Potamilus ohiensis*, but *P. alatus* never has a dorsal wing anterior of the umbo as does *P. ohiensis*, though they may be eroded or broken. Pseudocardinal teeth of *P. alatus* are much better developed than those of *P. ohiensis*, which are very compressed and delicate. *Potamilus alatus* has two pseudocardinal teeth in the left valve, whereas *P. ohiensis* has only one.

Potamilus alatus also resembles *Lasmigona complanata* but typically has a slightly more inflated umbo. The dorsal wing of *P. alatus* is unsculptured, whereas that of *L. complanata* is usually adorned with plications. Lateral teeth of *P. alatus* are well-developed, and those of *L. complanata* are rudimentary. Shell nacre of *P. alatus* is always pink or purple, and that of *L. complanata* is white.

General Distribution

Potamilus alatus is known from the Hudson Bay Basin of Minnesota, North Dakota and Canada (Dawley, 1947; Clarke, 1981a; Picha and Swenson, 2000). It occurs in eastern portions of the Great Lakes Basin, including drainages of lakes Ontario, Erie and Huron (La Rocque and Oughton, 1937). In the Mississippi Basin *P. alatus* occurs from southern Minnesota (Dawley, 1947) south to Arkansas (Harris and Gordon, 1990) and from headwaters of the Ohio River drainage in western Pennsylvania (Ortmann, 1909a) west to eastern Kansas (Murray and Leonard, 1962), Nebraska (Hoke, 2000) and South Dakota (Backlund, 2000). *Potamilus alatus* is widespread in the Cumberland River drainage downstream of Cumberland Falls, Kentucky and Tennessee (Cicerello et al., 1991; Parmalee and Bogan, 1998) and in the Tennessee River drainage from headwaters in southwestern Virginia (Ahlstedt, 1992a, 1992b) downstream to the mouth of the Tennessee River (Parmalee and Bogan, 1998).

Alabama and Mobile Basin Distribution

Potamilus alatus occurs in the Tennessee River across northern Alabama and is widespread in tributaries.

Potamilus alatus is extant in the Tennessee, Elk and Paint Rock rivers as well as Bear Creek, Colbert County, and Cypress Creek, Lauderdale County. It can also be found in impounded lower reaches of many tributaries.

Ecology and Biology

Potamilus alatus occurs primarily in slackwater habitats of creeks and rivers, including overbank areas of reservoirs. However, it is sometimes found in shoals and may be fairly common in tailwaters of dams. Ortmann (1919) stated that in Lake Erie the species "seems to prefer open shore". *Potamilus alatus* occurs in a variety of substrates, including gravel, sand and mud, and is tolerant of silty conditions. Ortmann (1919) observed that "it is a rather active species, crawling around a good deal."

Potamilus alatus is a dioecious, long-term brooder. In a Tennessee River population gametes are produced between July and August, and spawning occurs over a two- or three-week period in late August and September (Haggerty and Garner, 2000). Holland-Bartels and Kammer (1989) reported maturation of female gametes by mid-June and male gametes by early July in upper reaches of the Mississippi River. In the Tennessee River glochidia are fully developed by late October and brooded to the following spring, with discharge occurring in April and possibly May (Haggerty and Garner, 2000). In the upper Mississippi River females begin to become gravid in early August and are brooded to early July of the following year (Holland-Bartels and Kammer, 1989). *Potamilus alatus* has also been reported to brood until early July in Pennsylvania (Ortmann, 1919). Ortmann (1912a)

reported a female gravid with eggs in June but stated that it may be "an exceptional case". The only reported glochidial host for *P. alatus* is *Aplodinotus grunniens* (Freshwater Drum) (Sciaenidae), which was based on laboratory trials and observations of natural infestations (Howard, 1913; Brady et al., 2004).

Current Conservation Status and Protection

Potamilus alatus was considered currently stable throughout its range by Williams et al. (1993) and in Alabama by Lydeard et al. (1999). Garner et al. (2004) designated it a species of lowest conservation concern in the sate.

Potamilus alatus is commercially harvestable in Alabama. Market demand for its pink or purple shell is sporadic. *Potamilus alatus* is protected by state regulations, the most important being a minimum size limit of 4 inches (102 mm) in shell height.

Remarks

Potamilus alatus has been found to be rare in prehistoric shell middens along the Tennessee River at Muscle Shoals, represented by only a few fragments (Morrison, 1942). It has not been reported from other archaeological sites in Alabama (Hughes and Parmalee, 1999). *Potamilus alatus* is probably more abundant today than prehistorically. It is a species for which modern perturbations to habitat have resulted in an increase in abundance. Its ability to survive in muddy habitats has allowed it to expand its range into overbank areas of reservoirs. In 1991 it was reported as the third most abundant species in Wheeler Reservoir, with a population of 56,150,000 (Ahlstedt and McDonough, 1993).

Synonymy

Unio alatus Say, 1817. Say, 1817:[no pagination], pl. 4, fig. 2
Type locality: Ohio River. Figured type lost (Johnson and Baker, 1973). Haas (1930) designated a neotype, SMF 4349, length 140 mm, from the U.S.

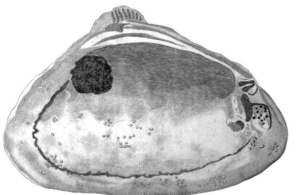

Metaptera megaptera Rafinesque, 1820. Rafinesque, 1820:300, pl. 80, figs. 20–22

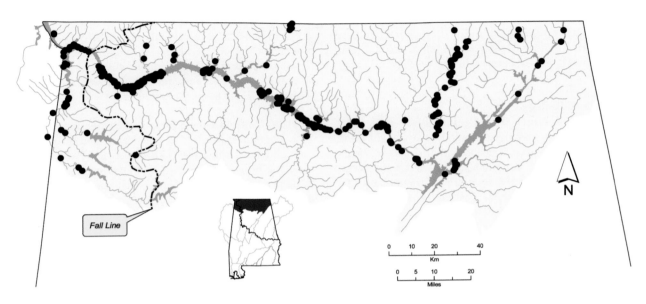

Distribution of *Potamilus alatus* in the Tennessee River drainage of Alabama.

Potamilus inflatus (Lea, 1831)
Inflated Heelsplitter

Potamilus inflatus – Length 110 mm, USNM 83908. Columbus, Mississippi. © Richard T. Bryant.

Shell Description

Length to 160 mm, mature males considerably larger than females; thin; moderately inflated; outline triangular, alate, may become oval when dorsal wings eroded or broken; posterior margin obliquely truncate to rounded, slightly more expanded in females; anterior margin narrowly rounded; dorsal margin with large rounded dorsal wing posterior to umbo and small triangular dorsal wing anterior to umbo, both wings often missing due to erosion or breakage; ventral margin convex, occasionally straight; posterior ridge low, rounded; posterior slope flattened to slightly concave, merging with posteriodorsal wing; umbo low, broad, slightly elevated above hinge line if at all, umbo sculpture irregular nodulous ridges; periostracum olive brown to almost black, usually without rays.

Pseudocardinal teeth elongate, extremely thin, delicate, 1 in each valve, right valve sometimes with accessory denticle anteriorly; lateral teeth short, thin, 2 in left valve, 1 in right valve; interdentum long, very narrow; umbo cavity wide, shallow; nacre bluish in females and young males, purple in large males.

Soft Anatomy Description

No material was available for soft anatomy description, but Lea (1863d) provided a brief description based on material from [Tombigbee River,] Columbus, Mississippi: Gills are rounded ventrally and inner lamellae of inner gills are completely connected to the visceral mass dorsally. The incurrent aperture is "large, with numerous small, brownish papillae", and the excurrent aperture is "small, with numerous small papillae". The supra-anal aperture is large, separated from the excurrent aperture by a mantle bridge. Outer gills are marsupial, distended when gravid, as they are in other species of *Potamilus*.

Glochidium Description

Length 81–126 μm; height 180–234 μm; with enlarged, lanceolate hooks on anterior and posterior ends of ventral flange, 5–7 smaller bifurcate hooks between enlarged marginal hooks; outline ligulate; dorsal margin straight, short, ventral margin convex, expanded laterally (Roe et al., 1997).

Similar Species

Potamilus inflatus with damaged or missing dorsal wings may superficially resemble *Leptodea fragilis*. However, the dorsal wings of *L. fragilis*, when present, are lower and more triangular than those of *P. inflatus*. Also, *L. fragilis* is more elongate and has a yellow to tawny periostracum.

Potamilus inflatus with broken or eroded dorsal wings may resemble some *Potamilus purpuratus*, but *P. inflatus* has a thinner shell that is broader anteriorly. *Potamilus purpuratus* is usually more inflated than *P. inflatus*. The posterior slope of *P. purpuratus* is usually adorned with two or three weak radial ridges, but the slope of *P. inflatus* is without ridges. Pseudocardinal teeth of *P. inflatus* are thin, compressed and delicate, whereas those of *P. purpuratus* are triangular.

General Distribution

Potamilus inflatus is endemic to the Mobile Basin in Alabama and Mississippi, primarily below the Fall Line.

Alabama and Mobile Basin Distribution

Potamilus inflatus is known from the Alabama, Black Warrior and Tombigbee rivers. This species is also known from lower reaches of some large tributaries, including the Cahaba River in the Alabama River drainage and the Sipsey and Noxubee rivers in the Tombigbee River drainage.

Potamilus inflatus is extant in the Alabama, Black Warrior and Tombigbee rivers. However, recent records of this species from the Alabama River drainage are limited to a single fresh dead shell collected in 1998 near its confluence with the Tombigbee River.

Ecology and Biology

Potamilus inflatus is primarily a species of large rivers, but a few records exist from small to medium rivers. It usually occurs in sandy and muddy substrates in areas with slow to moderate current. It may rarely be found in reservoirs.

Potamilus inflatus is a long-term brooder, gravid from autumn to the following summer. Roe et al. (1997) reported *P. inflatus* to discharge glochidia in June and July. The only reported glochidial host of *P. inflatus* is *Aplodinotus grunniens* (Freshwater Drum) (Sciaenidae), which was determined by observations of natural infestations (Roe et al., 1997). Other species of *Potamilus* also utilize this species.

Current Conservation Status and Protection

Potamilus inflatus was reported to be threatened throughout its range by Williams et al. (1993). In Alabama it was considered endangered by Stansbery (1976) and threatened by Lydeard et al. (1999). Garner et al. (2004) designated *P. inflatus* a species of high conservation concern in the state. In 1990 it was listed as threatened under the federal Endangered Species Act.

Remarks

Potamilus inflatus is considered conspecific with a similar form in the Amite River, Louisiana. However, recent genetic analyses have indicated significant differences, though the shells are virtually indistinguishable (Roe and Lydeard, 1998). A similar form of *Potamilus* also occurs in the Pearl River, Mississippi and Louisiana, but no live specimens have been collected from that drainage in recent years, so its genetic relationships with other populations remain unknown.

Potamilus inflatus has variously been referred to colloquially as Inflated Heelsplitter (Turgeon et al., 1988) and Alabama Heelsplitter (Turgeon et al., 1998). The name was changed in the second edition of Turgeon et al. (1998) from that in the first edition (Turgeon et al., 1988) without comment.

Synonymy

Symphynota inflata Lea, 1831. Lea, 1831:99, pl. 14, fig. 28; Lea, 1834b:109, pl. 14, fig. 28

Type locality: Alabama River, [Claiborne, Monroe County, Alabama,] Judge Tait. Lectotype, USNM 83909, length 112 mm, designated by Johnson (1974).

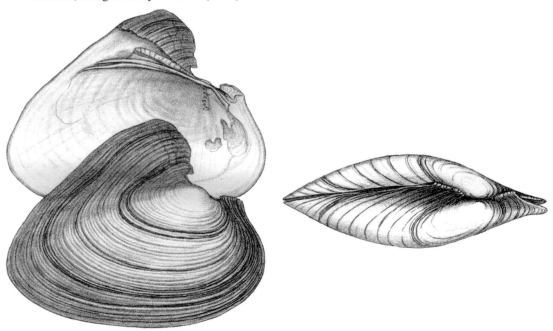

Unio alabamensis Conrad, 1834. Conrad, 1834b:67
Comment: Conrad (1834b) proposed this replacement name for *Symphynota inflata* Lea, 1831, when he moved this
    taxon to the genus *Unio*, creating a secondary junior homonym.

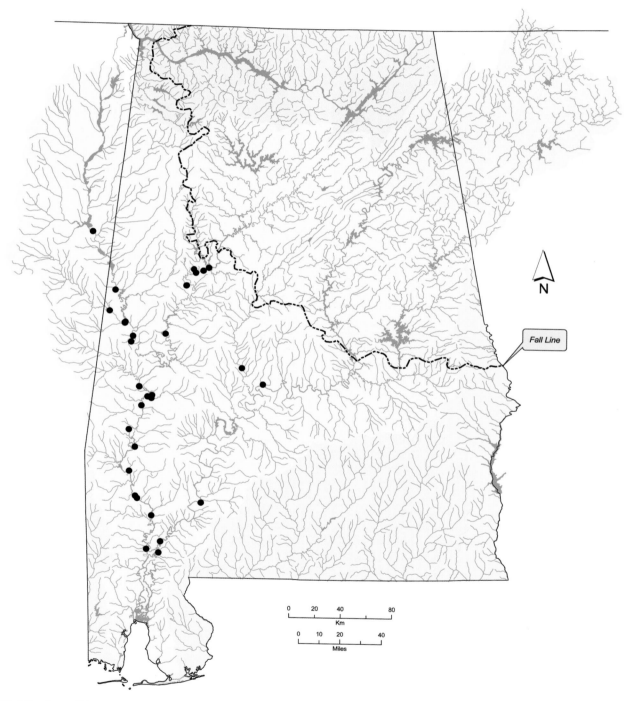

Distribution of *Potamilus inflatus* in Alabama and the Mobile Basin.

Potamilus ohiensis (Rafinesque, 1820)
Pink Papershell

Potamilus ohiensis – Length 107 mm, NCSM 6704. Tennessee River, Fort Loudon Reservoir, Blount County, Tennessee, 10 February 1977. © Richard T. Bryant.

Shell Description

Length to 140 mm; thin; compressed; outline triangular, alate, may be oval when dorsal wings eroded or broken; posterior margin rounded; anterior margin rounded; dorsal margin with well-developed dorsal wing posterior to umbo, often with small triangular dorsal wing anterior to umbo, both wings often missing due to erosion or breakage; ventral margin straight to convex; posterior ridge very low or absent; posterior slope flattened to slightly concave, merging with posteriodorsal wing; umbo low, broad, barely elevated above hinge line if at all, umbo sculpture a few weak nodulous ridges; periostracum shiny, greenish or yellowish, underlying pink nacre may give pinkish cast, becoming dark brown with age.

Pseudocardinal teeth very compressed, delicate, 1 tooth in each valve, right valve usually with accessory denticle; lateral teeth moderately long, slightly curved, 2 in left valve, 1 in right valve; interdentum moderately long, very narrow; umbo cavity wide, very shallow; nacre usually pink or purple, occasionally white to bluish white.

Soft Anatomy Description

Mantle pale tan, grayish brown outside of apertures with numerous thin wrinkles approximately parallel to margin; visceral mass pearly white; foot tan.

Females with thickened mantle margin just ventral to apertures; length of thickened area approximately 15% of shell length; margin smooth but with numerous small undulations, giving appearance of crenulations; with sparse, irregular gray marginally.

Gills creamy white with grayish cast; dorsal margin straight, ventral margin convex; gill length approximately 50% of shell length; gill height approximately 45% of gill length; outer gill height approximately 80% of inner gill height; inner lamellae of inner gills completely connected to visceral mass.

No gravid females were available for marsupium description. Lea (1863d) reported outer gills to be marsupial, with glochidia held in the posterior 33% of the gill.

Labial palps tan; slightly concave dorsally, elongate convex ventrally, bluntly pointed distally; palp length approximately 40% of gill length; palp height approximately 45% of palp length; distal 40% of palps bifurcate.

Supra-anal aperture much longer than incurrent and excurrent apertures; incurrent aperture longer than excurrent aperture.

Incurrent aperture length approximately 10% of shell length; creamy white with rusty cast within, dark brown of papillae edges extend onto aperture wall as irregular lines, some lines converge proximally; papillae in 2 rows, moderately long, simple, inner row larger, very sparse; papillae creamy white, with dark brown edges basally.

Excurrent aperture length approximately 5% of shell length; creamy white within, margin with dark brown lines in open reticulated pattern; margin with simple, very small but well-developed papillae; papillae creamy white with pale rusty cast. Lea (1863d) reported the excurrent aperture to be smooth.

Supra-anal aperture length approximately 20% of shell length; creamy white within, very sparse grayish brown marginally; margin smooth; mantle bridge separating supra-anal and excurrent apertures with multiple perforations, length approximately 60% of supra-anal length.

Specimen examined: Second Creek embayment, Lauderdale County (n = 1).

Glochidium Description

Length 100–126 µm; height 155–187 µm; without lateral hooks, with well-developed, compressed micropoints on ventral margin; outline ligulate; dorsal margin very short, straight, ventral margin rounded, expanded laterally, anterior and posterior margins straight dorsally, becoming deeply concave ventrally, forming blunt point at junction with ventral margin; valves with lateral gape when closed (Ortmann, 1912a; Surber, 1912; Hoggarth, 1999).

Similar Species

Potamilus ohiensis resembles *Potamilus alatus*. Both have a well-developed dorsal wing posterior to the umbo, but *P. ohiensis* often has a second dorsal wing anterior of the umbo. Pseudocardinal teeth of *P. ohiensis* are much more compressed and delicate than those of *P. alatus*, which are often triangular. *Potamilus ohiensis* has only one pseudocardinal tooth in the left valve, whereas *P. alatus* has two.

Potamilus ohiensis may also resemble *Leptodea fragilis* but has a less elongate shell. The periostracum of both species may be yellow in subadults, but the yellow color often persists into adulthood in *L. fragilis*, whereas the periostracum of *P. ohiensis* generally turns dark brown. Subadult *L. fragilis* may have faint green rays, but *P. ohiensis* does not.

General Distribution

Potamilus ohiensis occurs in most of the Mississippi Basin, ranging from Minnesota (Dawley, 1947) south to Louisiana (Vidrine, 1993) and Texas (Howells et al., 1996), and in headwaters of the Ohio River drainage in western New York (Simpson, 1914) west to North Dakota (Hoke, 2000). *Potamilus ohiensis* occurs in the Cumberland River drainage downstream of Cumberland Falls (Cicerello et al., 1991) and in the Tennessee River drainage from eastern Tennessee to the mouth of the Tennessee River (Parmalee and Bogan, 1998). Along the Gulf Coast *P. ohiensis* extends from the Mississippi River west to the Brazos River in eastern Texas (Howells et al., 1996).

Alabama and Mobile Basin Distribution

Potamilus ohiensis is restricted to the Tennessee River and some tributaries.

Potamilus ohiensis is extant in many reaches of Tennessee River reservoirs. It may be locally common.

Ecology and Biology

Potamilus ohiensis is generally found in water with little or no current and sand or mud substrates. It has colonized overbank habitat of reservoirs. Bates (1962) reported *P. ohiensis* to be one of the first mussels to colonize overbank habitat of Kentucky Reservoir on the Tennessee River, where it was the most commonly encountered species in qualitative sampling. Although Bates (1962) reported *P. ohiensis* from the lower Tennessee River proper for the first time, an early report of this species from Flint Creek, Morgan County, Alabama, which is farther upstream than Kentucky Reservoir, suggests that it occurred in low numbers in the lower Tennessee River prior to its impoundment (Ortmann, 1925).

Potamilus ohiensis is a long-term brooder. Gravid females have been reported in all months except October, December, February and May (Surber, 1912, 1913; Utterback, 1916a; Coker et al., 1921). It likely becomes gravid during the fall and broods glochidia until the following summer, but there appears to be overlap in brooding period among individuals. Utterback (1916a) reported a gravid female that was 39 mm in length, suggesting that the species reaches sexual maturity at a young age. *Potamilus ohiensis* exhibits very little sexual dimorphism, which is atypical for a long-term brooder.

Fishes reported to serve as glochidial hosts of *Potamilus ohiensis*, based on observations of natural infestations, are *Pomoxis annularis* (White Crappie) (Centrarchidae) and *Aplodinotus grunniens* (Freshwater Drum) (Sciaenidae) (Howard, 1913). *Potamilus ohiensis* glochidia increase in size considerably, and lose their ligulate shape, during infestation on gills of *A. grunniens* (Coker and Surber, 1911; Howard, 1913).

Current Conservation Status and Protection

Potamilus ohiensis was considered currently stable throughout its range by Williams et al. (1993) and in Alabama by Lydeard et al. (1999). It was designated a species of low conservation concern in the state by Garner et al. (2004).

Remarks

Potamilus ohiensis appears to be more abundant today than historically. There are few museum specimens from Alabama collected prior to impoundment of the Tennessee River, and the species has not been reported from archaeological studies (Hughes and Parmalee, 1999). The ability of *P. ohiensis* to occupy muddy habitats has allowed it to expand its range to overbank areas of reservoirs.

Potamilus ohiensis is often common in shallow water along reservoir margins but is rarely collected in large numbers during dive surveys in deeper water. It is most frequently encountered on exposed mud flats following reservoir draw down to winter pool levels.

Synonymy

Anodonta (*Lastena*) *ohiensis* Rafinesque, 1820. Rafinesque, 1820:316–317
Type locality: "l'Ohio et toutes les rivières adjacentes." Type specimen not found, also not found by Johnson and Baker (1973).
Symphynota laevissima Lea, 1829. Lea, 1829:444, pl. 13, fig. 23; Lea, 1834b:58, pl. 13, fig. 23

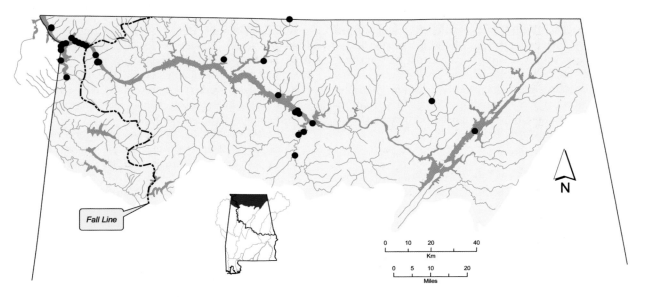

Distribution of *Potamilus ohiensis* in the Tennessee River drainage of Alabama.

Potamilus purpuratus (Lamarck, 1819)
Bleufer

Potamilus purpuratus – Upper figure: length 123 mm, UF 376686. Alabama River, mouth of Holly Creek, Montgomery Hill Landing, Baldwin County, Alabama, 20 October 1976. Lower figure: length 195 mm, specimen lost. Coosawattee River, 2.6 miles above State Highway 225, Gordon County, Georgia, 25 September 1997. © Richard T. Bryant.

Shell Description

Length to 200 mm; moderately thick; moderately inflated; outline elliptical to oval or subtriangular; posterior margin broadly rounded, often with straight area or shallow concavity dorsally; anterior margin rounded; dorsal margin with low wing posterior to umbo, often straight to convex in individuals with eroded dorsal wing; ventral margin convex; posterior ridge broadly rounded; posterior slope broad, with 2–3 weak radial ridges, posterior slope merging with dorsal wing when present; umbo broad, somewhat inflated, elevated slightly above hinge line, umbo sculpture faint corrugations; periostracum olive brown to black, subadults may have weak rays.

Pseudocardinal teeth triangular, somewhat compressed, 2 somewhat divergent teeth in left valve, 1 tooth in right valve, often with accessory denticle anteriorly; lateral teeth short, moderately thick, straight to slightly curved, 2 in left valve, 1 in right valve;

interdentum long, narrow to very narrow; umbo cavity wide, shallow; nacre purple to salmon or pink.

In populations above the Fall Line, shells are generally thinner and more compressed, with a higher dorsal wing and more compressed pseudocardinal teeth than those from below the Fall Line.

Soft Anatomy Description

Mantle creamy white to tan, outside of apertures pale orange or rusty tan with black or dark grayish brown mottling, some individuals solid dark brown outside of apertures; visceral mass creamy white to light tan; foot creamy white to tan, may be darker distally.

Rudimentary mantle fold just ventral to incurrent aperture; fold length 15–33% of shell length; with 1–2 gray, brown or black marginal bands; margin may be smooth, crenulate or dentate.

Gills tan to light brown; dorsal margin slightly sinuous to concave, ventral margin convex; gill length

45–65% of shell length; gill height 35–60% of gill length; outer gill height 75–100% of inner gill height; inner lamellae of inner gills completely connected to visceral mass.

Outer gills marsupial; glochidia held in posterior 25–50% of gill; marsupium distended when gravid; light tan.

Labial palps creamy white to tan; straight, convex or concave dorsally, convex ventrally, bluntly pointed distally; palp length 30–40% of gill length; palp height 50–70% of palp length; distal 30–50% of palps bifurcate.

Incurrent, excurrent and supra-anal aperture lengths variable in relation to each other.

Incurrent aperture length 10–20% of shell length; creamy white to light tan within, some combination of tan, brown, gray and black basal to papillae, often irregularly mottled, occasionally with crude bands perpendicular to margin; papillae usually in 2 rows, long, slender or thick, simple, inner row larger; papillae some combination of creamy white, tan, brown and gray, often with rusty cast, usually with dark brown or black edges basally.

Excurrent aperture length 5–15% of shell length; creamy white to light tan within, marginal color band some combination of creamy white, tan, brown and black, usually with light background and darker markings in reticulated pattern, sometimes with lines or oblong markings perpendicular to margin; margin crenulate or with short simple papillae; papillae tan to light brown, often with rusty cast.

Supra-anal aperture length 10–15% of shell length, occasionally slightly greater; creamy white within, sometimes with thin gray, brown or black marginal band; margin smooth; mantle bridge separating supra-anal and excurrent apertures occasionally perforate, length 20–50% of supra-anal length, occasionally greater.

Specimens examined: Bogue Chitto, Dallas County (n = 3); Sipsey River (n = 2); Tombigbee River (n = 4).

Glochidium Description

Length 180–220 μm; height 347–378 μm; with lanceolate hooks located ventrally on anterior and posterior ends of ventral flange; outline ligulate; dorsal margin very short, straight, ventral margin convex, expanded laterally, anterior and posterior margins straight dorsally, becoming slightly concave ventrally, forming moderately sharp angle at junction with ventral margin; valves with lateral gape when closed (Surber, 1915; Roe et al., 1997; Hoggarth, 1999).

Similar Species

Potamilus purpuratus may resemble some *Potamilus inflatus* with broken or eroded dorsal wings,

but *P. purpuratus* has a thicker shell that is narrower anteriorly. *Potamilus purpuratus* is usually more inflated than *P. inflatus*. The posterior slope of *P. purpuratus* is typically adorned with two or three low radial ridges, but the posterior slope of *P. inflatus* is without ridges. Pseudocardinal teeth of *P. purpuratus* are well-developed and triangular, whereas those of *P. inflatus* are thin, compressed and delicate.

General Distribution

Potamilus purpuratus occurs in the Mississippi Basin from southern Illinois and southeastern Missouri (Cummings and Mayer, 1992) south to Louisiana (Vidrine, 1993) and from western Tennessee (Parmalee and Bogan, 1998) and western Mississippi (Jones et al., 2005) west to Kansas (Murray and Leonard, 1962). On the Gulf Coast *P. purpuratus* ranges from the Mobile Basin west to the Guadalupe River in Texas. An introduced population of *P. purpuratus* exists in Lake Corpus Christi, Nueces River, Texas (Howells et al., 1996).

Alabama and Mobile Basin Distribution

Potamilus purpuratus is confined to the Mobile Basin and is widespread above and below the Fall Line. However, there is only one record from the Tallapoosa River drainage above the Fall Line.

Potamilus purpuratus remains widespread in the Mobile Basin, where it may be locally common.

Ecology and Biology

Potamilus purpuratus occurs in creeks to large rivers, where it may be found in substrates composed of various combinations of mud, sand, gravel and cobble, as well as under rocks. It occurs in areas with slow to swift current and may also be found in some reservoirs and oxbows. In some riverine areas, especially those subject to swift current, *P. purpuratus* is most commonly encountered on and along the base of the channel slope.

Potamilus purpuratus is a long-term brooder, gravid from autumn to the following summer. In Texas gravid females were reported from August through May (Howells, 2000). Fishes reported to serve as glochidial hosts of *P. purpuratus* in laboratory trials are *Lepomis gulosus* (Warmouth) (Centrarchidae) and *Notemigonus crysoleucas* (Golden Shiner) (Cyprinidae) (Howells, 1995). *Aplodinotus grunniens* (Freshwater Drum) (Sciaenidae) has been reported as a host of *P. purpuratus* based on observations of natural infestations (Surber, 1913, 1915). Surber (1915) stated that the glochidium of *P. purpuratus* undergoes "wonderful changes" while encysted on the host and suggested a considerable size increase during the encystment period.

Current Conservation Status and Protection

Potamilus purpuratus was considered currently stable throughout its range by Williams et al. (1993) and in Alabama by Lydeard et al. (1999). Garner et al. (2004) designated it a species of lowest conservation concern in the state.

In 2004 *Potamilus purpuratus* was added to the list of species that can be legally harvested in Alabama. It is protected by state regulations, with the most important being a minimum size limit of 3 inches (76 mm) in shell height.

Remarks

Conrad (1834b) described *Unio poulsoni*, a compressed form with a well-developed dorsal wing, from upper reaches of the Black Warrior River and compared it to *Potamilus alatus*, which it superficially resembles. *Potamilus alatus* is widespread in the Mississippi Basin but is absent from the Mobile Basin. *Potamilus poulsoni*, which has generally not been recognized as a valid species, is considered a form of *Potamilus purpuratus* restricted to areas above the Fall Line in the Black Warrior, Cahaba and Coosa River drainages. It is herein placed in the synonymy of *P. purpuratus* but may prove to be a valid species once it is thoroughly studied.

Synonymy

Unio purpurata Lamarck, 1819. Lamarck, 1819:71
Type locality: "Africa." Spurious type locality. Type specimen not found.
Unio ater Lea, 1829, *non* Nilsson, 1822. Lea, 1829:426, pl. 7, fig. 9; Lea, 1834b:40, pl. 7, fig. 9
Unio atra Deshayes, 1830. Deshayes, 1830:582
Unio lugubris Say, 1832. Say, 1832:[no pagination], pl. 43, fig. 6 [replacement name for *Unio ater* Lea, 1829]
Unio poulsoni Conrad, 1834. Conrad, 1834b:25, pl. 1
Type locality: Black Warrior River, north Alabama. Type specimen not found, also not found by Johnson and Baker (1973). Measurement of figured shell not given in original description (Conrad, 1834b).

Unio coloradoensis Lea, 1856. Lea, 1856b:103; Lea, 1858f:314, pl. 31, fig. 29; Lea, 1858g:34, pl. 31, fig. 29
Unio dolosus Lea, 1860. Lea, 1860c:307; Lea, 1862b:75, pl. 9, fig. 224; Lea, 1862d:79, pl. 9, fig.224
Type locality: Alabama River, Claiborne, [Monroe County,] Alabama, Judge Tait; Coosa River, E.R. Showalter, M.D. Lectotype, USNM 86119, length 59 mm, designated by Johnson (1974), is from Alabama River, Claiborne, Monroe County, Alabama.
Comment: A small specimen in the type lot, USNM 86119, length 24 mm, has a note in the shell: "Coosa R., Dr. Showalter"; it appears to be *Leptodea fragilis*.

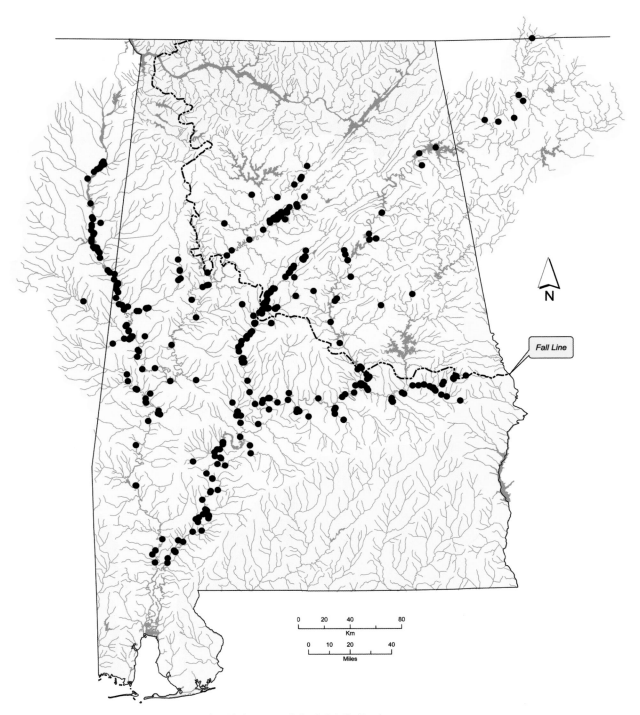

Distribution of *Potamilus purpuratus* in Alabama and the Mobile Basin.

Genus *Ptychobranchus*

Ptychobranchus Simpson, 1900, occurs in the Great Lakes and Mississippi basins and Gulf Coast drainages from western Florida to Louisiana. Turgeon et al. (1998) listed five species in *Ptychobranchus,* four of which occur in Alabama. An additional species also found in Alabama, *Ptychobranchus foremanianus* (Lea, 1842), is herein elevated from synonymy based on shell morphology and preliminary genetic analysis (K.J. Roe, personal communication).

Bogan (1997) determined the identity of *Plagiola interrupta* Rafinesque, 1820, the type species of *Plagiola* Rafinesque, 1819, as *Ptychobranchus fasciolaris* (Rafinesque, 1820), making *Plagiola* an unused senior synonym of *Ptychobranchus* Simpson, 1900. The genus name *Plagiola* has been used for various species groups in the past. Use of *Plagiola* for species currently included in *Ptychobranchus* would disrupt current stable taxonomy (ICZN, 1999: Article 23.9.1). The genus *Plagiola*, with *interrupta* as the type species identified as *Ptychobranchus fasciolaris*, has not been used since *Ptychobranchus* was described (ICZN, 1999: Article 23.9.1.1). *Ptychobranchus* has been solely used for this group of species since its description in 1900 and continues to be widely used in the literature (ICZN, 1999: Article 23.9.1.2). Both of the conditions of ICZN (1999) Article 23.9.1 are met. *Plagiola*, with the type species *interrupta*, is a senior synonym of *Ptychobranchus*, with the type species *fasciolaris*. The younger genus name is valid, and this action is taken in accordance with ICZN (1999) Article 23.9.2. Thus, *Ptychobranchus* Simpson, 1900, is a *nomen protectum*, and the invalid but older genus name *Plagiola* Rafinesque, 1819, is a *nomen oblitum*.

Ptychobranchus is known for its interesting mimetic conglutinates, which may resemble insect larvae, fish eggs or fish larvae (Hartfield and Hartfield, 1996; Watters, 1999). Single populations of some species have been observed to produce multiple types of conglutinates (Hartfield and Hartfield, 1996; Haag and Warren, 1997; Watters, 1999).

Type Species

Unio phaseolus Hildreth, 1828 (= *Obliquaria (Ellipsaria) fasciolaris* Rafinesque, 1820)

Diagnosis

Shell moderately thin to thick; elliptical to triangular; hinge teeth thick; umbo cavity very shallow; disk smooth; male and female shells very similar externally, female shell with oblique depression internally.

Inner lamellae of inner gills completely connected to visceral mass or connected only anteriorly; mantle margin only weakly modified ventral to incurrent aperture; mantle bridge separating supra-anal and excurrent apertures short; outer gills marsupial; glochidia held in ventral portion of gill; entire marsupium developed into short radial folds (Figure 13.34), extended slightly beyond original gill ventral edge when gravid; conglutinates variable, in form of insect larvae, fish eggs or fish larvae; glochidium without styliform hooks (Simpson, 1900b, 1914; Ortmann, 1912a; Haas, 1969b; Hartfield and Hartfield, 1996; Watters, 1999).

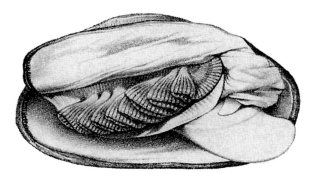

Figure 13.34. *Ptychobranchus foremanianus* gravid female soft anatomy. Note folded outer gill. From Lea (1859f).

Synonymy

Plagiola Rafinesque, 1819
Subtentus Frierson, 1927

Ptychobranchus fasciolaris (Rafinesque, 1820)
Kidneyshell

Ptychobranchus fasciolaris – Length 61 mm, CMNH 61.6956. Paint Rock River, Paint Rock, Jackson County, Alabama. © Richard T. Bryant.

Shell Description

Length to 150 mm; moderately thick, thinner posteriorly; compressed; outline elongate elliptical; posterior margin narrowly rounded to obliquely truncate; anterior margin rounded; dorsal margin convex; ventral margin straight to slightly convex; posterior ridge well-developed, rounded; posterior slope steep, flat to slightly convex; umbo broad, low, may be elevated slightly above hinge line, umbo sculpture fine wavy ridges, interrupted, may appear double-looped; periostracum yellow to yellowish green, darkening to brown with age, typically with prominent, wide, often interrupted green rays.

Pseudocardinal teeth low, triangular, 2 divergent teeth in left valve, 1 tooth in right valve, may have accessory denticle anteriorly and/or posteriorly; lateral teeth short, thick, straight, 2 in left valve, 1 in right valve; interdentum moderately long, usually wide, occasionally narrow; umbo cavity very shallow; females with shallow, oblique depression posterioventral to umbo cavity; nacre white.

Soft Anatomy Description

Mantle creamy white to tan, may have golden cast in some areas, may be more pale outside of pallial line, rusty brown or grayish brown outside of apertures; visceral mass pearly white to creamy white or tan; foot creamy white to tan, often darker distally.

Gills creamy white to tan; dorsal margin slightly sinuous, ventral margin straight to elongate convex; gill length 60–70% of shell length; gill height 30–45% of gill length; outer gill height 70–80% of inner gill height; inner lamellae of inner gills completely attached to visceral mass or connected only anteriorly.

Outer gills marsupial; glochidia held across distal 50% of gill with exception of anterior and posterior ends; marsupium with radial folds distally; light brown with conglutinate coloration visible giving distal margin pink, beaded appearance. Ortmann (1912a) and Lea (1859f) reported gravid gills to be purplish or blackish and the "beads at edge" (i.e., the distal ends of conglutinates) to be red or purple.

Labial palps creamy white to tan, may have lighter area dorsally; straight to slightly concave dorsally, convex ventrally, bluntly pointed distally; palp length 10–20% of gill length; palp height 60–80% of palp length; distal 40–45% of palps bifurcate.

Incurrent and supra-anal apertures longer than excurrent aperture; incurrent and supra-anal apertures approximately equal in length.

Incurrent aperture length 10–20% of shell length; creamy white to pale tan within, creamy white to dull orange basal to papillae, may be mottled with gray or grayish brown; papillae in 2 rows, long, slender, simple; papillae some combination of creamy white, tan, dull orange, gray, grayish brown and dark brown, may be mottled basally, may be of lighter color distally.

Excurrent aperture length 10–15% of shell length, occasionally less; creamy white to pale tan within, marginal color band dull orange to grayish tan or grayish brown, may have dark brown mottling or small gray specks; margin with small simple papillae; papillae rusty tan to grayish brown. Ortmann (1912a) reported the excurrent aperture margin to be finely crenulate.

Supra-anal aperture length 15–20% of shell length; creamy white within, with a thin sparse gray to brown marginal band; margin smooth; mantle bridge separating supra-anal and excurrent apertures occasionally perforate, length 5–15% of supra-anal length.

Specimens examined: Paint Rock River (n = 2); Tennessee River (n = 2).

Glochidium Description

Length 170–175 μm; height 182–195 μm; without styliform hooks; outline subelliptical; dorsal margin slightly convex, ventral margin rounded, anterior and posterior margins convex (Ortmann, 1912a; Hoggarth, 1999).

Similar Species

Ptychobranchus fasciolaris may resemble *Ptychobranchus subtentum* but lacks distinctive plications on the posterior ridge and slope. The rays of *P. subtentum* are often offset to form a zigzag pattern near the umbo, whereas those of *P. fasciolaris* are separated and form no such pattern.

Ptychobranchus fasciolaris may superficially resemble *Elliptio dilatata*, but *E. dilatata* generally has a darker brown periostracum, with rays weak or absent. The inner margin of the hinge plate of *P. fasciolaris* is usually angular, whereas the hinge plate of *E. dilatata* is more gently curved. Shell nacre of *P. fasciolaris* is white, but that of Tennessee River drainage populations of *E. dilatata* range from pink to purple or salmon, seldom white (some more northerly populations have a larger percentage of individuals with white nacre).

General Distribution

Ptychobranchus fasciolaris is known from the Great Lakes Basin, including western reaches of Lake Ontario, west to lakes Erie and St. Clair and tributaries (Ortmann, 1909a; La Rocque and Oughton, 1937; Clarke, 1981a). In the Ohio River drainage *P. fasciolaris* ranges from headwaters in western New York (Strayer et al., 1992) downstream to near the mouth of the Ohio River (Cummings and Mayer, 1992). *Ptychobranchus fasciolaris* is widely distributed in the Cumberland River, Kentucky and Tennessee, downstream of Cumberland Falls (Cicerello et al., 1991; Parmalee and Bogan, 1998). It is widespread in the Tennessee River drainage from headwaters in southwestern Virginia, western North Carolina and eastern Tennessee downstream to the mouth of the Tennessee River (Ahlstedt, 1992a, 1992b; Parmalee and Bogan, 1998).

Alabama and Mobile Basin Distribution

Ptychobranchus fasciolaris is limited to the Tennessee River drainage. It historically occurred across the state in the Tennessee River and several tributaries. There are no records of *P. fasciolaris* from the Alabama reaches of the Elk River, but it probably occurred there since records exist from several localities in Tennessee reaches.

Ptychobranchus fasciolaris is extant but rare in tailwaters of Guntersville and Wilson dams. It is more common in Bear Creek in Colbert County and the Paint Rock River system.

Ecology and Biology

Ptychobranchus fasciolaris occurs in small to large rivers, typically in habitats with flowing water. Combinations of sand and gravel appear to be its preferred substrate. *Ptychobranchus fasciolaris* can frequently be found along margins of *Justicia americana* (Waterwillow) beds. In tailwaters of Tennessee River dams *P. fasciolaris* occurs at depths exceeding 6 m.

Ptychobranchus fasciolaris is a long-term brooder, gravid from August to the following summer (Ortmann, 1919). Ortmann (1919) reported a female discharging glochidia in late August but deemed it "exceptionally belated". *Ptychobranchus fasciolaris* glochidia are discharged as conglutinates. Watters (1999) provided a description of *P. fasciolaris* conglutinates from Big Darby Creek, Ohio. Two forms are produced, major conglutinates and minor conglutinates, which differ in size but not gross morphology. Both forms occur within the same population, but individual mussels produce only a single type. The conglutinates are cylindrical, with a bulbous end that may taper to a point and a twisted, lamellar adhesive structure on the opposite end. They are light tan to yellowish overall but have brown pigmentation on each side that is irregularly banded, giving the appearance of segmentation. Near the bulbous end are two to five darkly pigmented oval areas that resemble eyes. The "eyes" are smaller and lighter in minor conglutinates and hold a red disk-shaped inclusion giving the overall appearance of an insect larva (e.g., Chironomidae or Simuliidae). Major conglutinates have darker pigmentation on the "eyes" and down the sides giving them the overall appearance of fish fry (Figure 13.35). The primary sites of glochidial release from the conglutinates are the "eyes", where the conglutinate wall is thinnest. The adhesive end adheres to a variety of hard substrates, including wood, stone, glass and metal, and retains its adhesive properties for days of repeated sticking and unsticking (Watters, 1999). Conglutinates are positioned inside the marsupia with the bulbous end oriented distally at the gill margin, which gives the gill edge a beaded appearance. *Ptychobranchus fasciolaris* conglutinates from Paint Rock River, Alabama, closely resemble those described by Watters (1999). Sterki (1898a) reported 283 conglutinates from one gill of "a large specimen". Ortmann (1910a, 1912a) reported *P. fasciolaris* conglutinates "repeatedly observed" being discharged through holes in the distal ends of the marsupia. However, the mechanism and pathway by which the conglutinates make their way from the mantle cavity to the excurrent aperature is unknown.

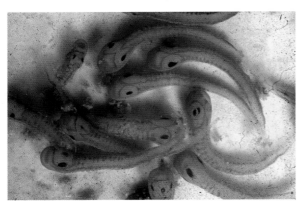

Figure 13.35. *Ptychobranchus fasciolaris*. Recently released conglutinates. Photograph by W.N. Roston.

Culaea inconstans (Brook Stickleback) (Gasterosteidae) has been reported to serve as a glochidial host for *Ptychobranchus fasciolaris* in laboratory trials (Watters et al., 2005). In French Creek,

Pennsylvania, *Etheostoma blennioides* (Greenside Darter), *Etheostoma flabellare* (Fantail Darter), *Etheostoma nigrum* (Johnny Darter) and *Etheostoma zonale* (Banded Darter) (Percidae) were listed as potential hosts (White et al., 1996). They were determined by comparison of genetic material of glochidia removed from the fish with genetic material of adult mussels.

Current Conservation Status and Protection

Ptychobranchus fasciolaris was reported as currently stable throughout its range by Williams et al. (1993). In Alabama it was listed as a species of special concern by Stansbery (1976) and Lydeard et al. (1999). Garner et al. (2004) designated *P. fasciolaris* a species of highest conservation concern in the state.

Remarks

Ptychobranchus fasciolaris has been reported to be uncommon in prehistoric shell middens along the Tennessee River in northern Alabama (Morrison, 1942; Warren, 1975; Hughes and Parmalee, 1999).

Synonymy

Potamilus fasciolaris Rafinesque, 1818, *nomen nudum*. Rafinesque, 1818:355

Unio (Plagiola) fasciolaris Rafinesque, 1819, *nomen nudum*. Rafinesque, 1819:426

Obliquaria (Plagiola) interrupta Rafinesque, 1820. Rafinesque, 1820:302

Comment: Bogan (1997) identified the syntype of *Obliquaria interrupta* as a specimen of *Ptychobranchus fasciolaris*. Based on the correct identification of the type species, *Plagiola* Rafinesque, 1819, is the senior synonym of *Ptychobranchus* Simpson, 1900.

Obliquaria (Ellipsaria) fasciolaris Rafinesque, 1820. Rafinesque, 1820:303

Type locality: Ohio, Wabash and Kentucky rivers. Lectotype, ANSP 20253, length 81 mm, designated by Johnson and Baker (1973), is from Kentucky River.

Obliquaria (Ellipsaria) fasciolaris var. *interrupta* Rafinesque, 1820. Rafinesque, 1820:303

Obliquaria (Ellipsaria) fasciolaris var. *fuscata* Rafinesque, 1820. Rafinesque, 1820:304

Obliquaria (Ellipsaria) fasciolaris var. *longa* Rafinesque, 1820. Rafinesque, 1820:304

Obliquaria (Ellipsaria) fasciolaris var. *obliterata* Rafinesque, 1820. Rafinesque, 1820:304

Unio phaseolus Hildreth, 1828. Hildreth, 1828:283

Unio planulatus Lea, 1829. Lea, 1829:431, pl. 9, fig. 13; Lea, 1834b:45, pl. 9, fig. 13

Unio camelus Lea, 1834. Lea, 1834a:102, pl. 15, fig. 45; Lea, 1834b:214, pl. 15, fig. 45

Unio compressissimus Lea, 1845. Lea, 1845:163; Lea, 1848a:81, pl. 8, fig. 23; Lea, 1848b:81, pl. 8, fig. 23

Unio imperitus De Gregorio, 1914. De Gregorio, 1914:45–46

Unio lanceolatus var. *blandus* De Gregorio, 1914. De Gregorio, 1914:52

Unio compressissimus var. *performosus* De Gregorio, 1914. De Gregorio, 1914:53–54

Ptychobranchus fasciolaris var. *lacustris* F.C. Baker, 1928. F.C. Baker, 1928b:52

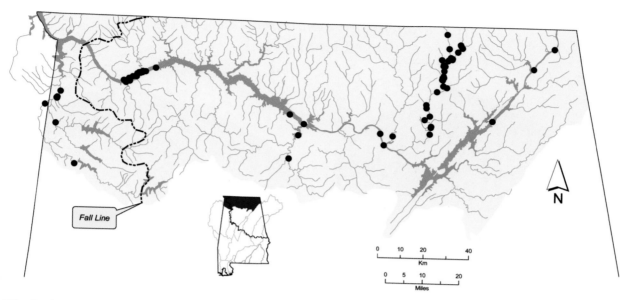

Distribution of *Ptychobranchus fasciolaris* in the Tennessee River drainage of Alabama.

Ptychobranchus foremanianus (Lea, 1842)
Rayed Kidneyshell

Ptychobranchus foremanianus – Upper figure: length 94 mm, UF 376055. Coosawattee River, 2.5 miles above State Highway 225, Gordon County, Georgia, 12 September 1997. Middle figure: length 61 mm, CMNH 61.8399. Lower figure: length 50 mm, CMNH 61.8399. Coosa River, Weduska Shoals, [Coosa and Shelby county line,] Alabama. © Richard T. Bryant.

Shell Description

Length to 100 mm; moderately thick; moderately inflated; outline triangular; posterior margin obliquely truncate to bluntly pointed or narrowly rounded; anterior margin rounded; dorsal margin convex; ventral margin straight to slightly convex; posterior ridge distinct, rounded; posterior slope typically steep, flat; umbo broad, low, becoming inflated with age in some individuals, umbo sculpture unknown; periostracum yellowish green to tawny or brown, usually with variable dark green rays, often wide and broken.

Pseudocardinal teeth triangular, somewhat compressed, 2 slightly divergent teeth in left valve, 1 tooth in right valve, may have accessory denticle anteriorly; lateral teeth thick, short, straight, 2 in left valve, 1 in right valve; interdentum moderately long, narrow to moderately wide; umbo cavity shallow; nacre white to bluish white.

Soft Anatomy Description

Mantle creamy white to tan, outside of apertures rusty orange to rusty brown, may have gray or grayish brown mottling oriented somewhat perpendicular to margin; visceral mass creamy white to tan; foot tan, may be slightly darker distally.

Very weak mantle fold just ventral to incurrent aperture; fold length 20–40% of shell length; margin smooth but with small irregular undulations; marginal color band thin, gray to grayish brown.

Gills tan; dorsal margin slightly sinuous, ventral margin convex to elongate convex; gill length 60–65% of shell length; gill height 40–50% of gill length; outer gill height 60–85% of inner gill height; inner lamellae of inner gills completely connected to visceral mass or connected only anteriorly.

Outer gills marsupial; glochidia held in posterior 50% of gill; marsupium radially folded, restricted to distal 35% of gill; tan, with thin rusty brown band distally. Note that the gravid female used in this description was probably brooding incompletely developed conglutinates (collected in late July). Lea (1859f) reported gravid marsupia to be black on their distal halves, with a white knob at the distal end of each water tube, giving it "the appearance of a string of pearls". These were presumably representative of completely developed conglutinates.

Labial palps tan; straight to slightly concave dorsally, convex ventrally, bluntly pointed distally; palp length 15–20% of gill length; palp height 55–80% of palp length; distal 45–80% of palps bifurcate.

Incurrent and supra-anal apertures of similar length, both longer than excurrent aperture; some individuals with excurrent and supra-anal apertures continuous.

Incurrent aperture length 15–20% of shell length; creamy white to tan within, may have slight rusty cast, may have rusty orange with sparse black mottling basal to papillae; papillae in 1–2 rows, short to moderately long, thick, blunt, simple; papillae rusty orange to rusty brown.

Excurrent aperture length 10–15% of shell length; creamy white to tan within, marginal color band rusty orange to rusty brown, may have sparse black or gray mottling; margin crenulate.

Supra-anal aperture length approximately 20% of shell length; creamy white to tan within, may have narrow rusty brown or grayish brown marginal band; margin smooth; mantle bridge separating supra-anal and excurrent apertures imperforate, length 15–20% of supra-anal length, sometimes absent.

Specimens examined: Cahaba River (n = 1); Conasauga River, Georgia (n = 2) (specimens previously preserved).

Glochidium Description

Length 140–175 µm; height 170–195 µm; without styliform hooks; outline subelliptical; dorsal margin slightly convex, ventral margin rounded, anterior and posterior margins convex (Hoggarth, 1999).

Similar Species

Ptychobranchus foremanianus from large rivers, which tend to be thick-shelled and inflated, are easily distinguished from all sympatric mussel species.

Small, compressed *Ptychobranchus foremanianus* from tributaries may resemble *Pleurobema georgianum* but have a broader umbo and are more elongate and triangular.

Ptychobranchus foremanianus may resemble some *Elliptio arca* but has a more inflated umbo and less pointed posterior margin. Internally the shells are easily distinguished. *Ptychobranchus foremanianus* always has white nacre, whereas that of *E. arca* is usually some shade of pink, purple or salmon, only occasionally white. Also, the shell cavity of *P. foremanianus* is deeper than that of *E. arca*.

Ptychobranchus foremanianus may also resemble *Pleurobema perovatum*, but *P. foremanianus* is usually more elongate, with a lower and more compressed umbo. The anterior pseudocardinal tooth of *P. perovatum* is typically compressed and aligned almost parallel to the shell margin, whereas that of *P. foremanianus* is usually more triangular and oriented more ventrally.

Ptychobranchus foremanianus usually has fine green rays, in contrast to *Ptychobranchus greenii*, which does not. The two are not believed to occur sympatrically.

General Distribution

Ptychobranchus foremanianus is endemic to the Alabama, Cahaba, Coosa and Tallapoosa River drainages of the Mobile Basin in Alabama, Georgia and Tennessee.

Alabama and Mobile Basin Distribution

Most records of *Ptychobranchus foremanianus* are from the Cahaba and Coosa River drainages near and above the Fall Line. There is a single record from "Claiborne, Alabama", located on the Alabama River.

Ptychobranchus foremanianus is extant in isolated reaches of the Coosa River drainage, where it is very rare. It also remains in the Cahaba River above the Fall Line.

Ecology and Biology

Ptychobranchus foremanianus is a species of flowing water habitats in medium to large rivers. It usually occurs in mixtures of sand and gravel in moderate to swift current.

Ptychobranchus foremanianus is a long-term brooder, gravid from late summer or autumn to the following spring or summer (Ortmann, 1923b). A gravid female used for the soft anatomy description in this account was collected in late July. Glochidia of *P. foremanianus* are packaged in conglutinates. Lea (1852c) provided a figure of a *P. foremanianus* conglutinate, which is oblong, rounded on one end and very acute on the other. A primary glochidial host of *P. foremanianus* is *Etheostoma jordani* (Greenbreast Darter) (Percidae) and a secondary host is *Cottus carolinae* (Banded Sculpin) (Cottidae) (P.D. Johnson, personal communication).

Current Conservation Status and Protection

Ptychobranchus foremanianus was considered endangered by Athearn (1970). In subsequent conservation status reviews *P. foremanianus* was considered conspecific with *Ptychobranchus greenii*, which was considered endangered throughout its range by Williams et al. (1993) and in Alabama by Lydeard et al. (1999). Garner et al. (2004) considered *P. greenii* a species of highest conservation concern in the state. *Ptychobranchus greenii* (including *P. foremanianus*)

was listed as endangered under the federal Endangered Species Act in 1993.

Remarks

Ptychobranchus foremanianus was placed in the synonymy of *Ptychobranchus greenii* in most previous works. It is herein recognized based on subtle differences in shell coloration, with *P. foremanianus* typically having well-defined dark green rays, which are usually absent on *P. greenii*, as noted by Ortmann (1923b). Preliminary genetic analyses suggest at least two species in the Mobile Basin (K.J. Roe, personal communication). Ortmann (1923b) aligned Cahaba River *Ptychobranchus* with *P. greenii* but noted that individuals from that population share characteristics of both *P. greenii* and *P. foremanianus*. The range of *P. foremanianus* herein includes the Cahaba River drainage, which is suggested by preliminary genetic analyses (K.J. Roe, personal communication).

Conglutinate morphology differs among individuals within populations of *Ptychobranchus foremanianus* (e.g., may resemble dipteran larvae or membrane-bound fish embryos). Detailed comparative studies, including soft anatomy, life history, conglutinate morphology and additional genetic analyses, may reveal cryptic species.

Ptychobranchus foremanianus was named after Dr. Foreman, a colleague of Isaac Lea's who provided specimens for the original description of the species.

Synonymy

Unio foremanianus Lea, 1842. Lea, 1842a:224; Lea, 1842b:247, pl. 27, fig. 64; Lea, 1842c:85, pl. 27, fig. 64
Type locality: Coosa River, Alabama, Dr. Brumby. Lectotype, USNM 84439, length 43 mm, designated by Johnson (1974).

Unio velatus Conrad, 1854. Conrad, 1854:298, pl. 27, fig. 6
Type locality: River St. Fois. Type specimen not found, also not found by Johnson and Baker (1973). Conrad (1854) gave no measurement of figured shell not given in original description (Conrad 1854), but length of illustration is 49 mm. It appears that Conrad (1854) made a mistake in reporting the type locality as "River St. Fois"; this shell appears to be *Ptychobranchus foremanianus* from the Mobile Basin.

  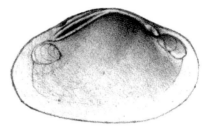

Unio woodwardius Lea, 1857. Lea, 1857g:170
Type locality: Etowah and Connasauga [Conasauga] rivers, Cass [now Bartow] County, Georgia, Bishop Elliott and Reverend G. White.
Unio woodwardianus Lea, 1859. Lea, 1859f:199, pl. 23, fig. 82; pl. 29, fig. 103; Lea, 1859g:17, pl. 23, fig. 82; pl. 29, fig. 103 [new name proposed for *Unio woodwardius* Lea, 1857]
Type locality: Etowah and Connasauga [Conasauga] rivers, Cass [now Bartow] County, Georgia, Bishop Elliott and Reverend G. White. Lectotype, USNM 84444, length 46 mm, designated by Johnson (1974), is from Conasauga River, Bartow County, Georgia.

Unio trinacrus Lea, 1861. Lea, 1861c:59; Lea, 1862b:86, pl. 12, fig. 235; Lea, 1862d:90, pl. 12, fig. 235
Type locality: Coosa River, Alabama, E.R. Showalter, M.D. Holotype, USNM 84445, length 49 mm.

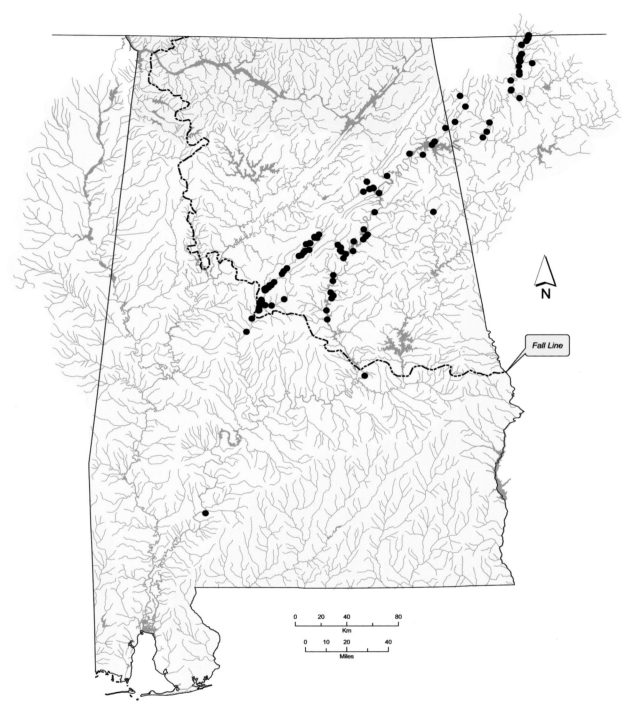

Distribution of *Ptychobranchus foremanianus* in Alabama and the Mobile Basin.

Ptychobranchus greenii (Conrad, 1834)
Triangular Kidneyshell

Ptychobranchus greenii – Length 57 mm, UF 67715. North River near Haglers Mill, Tuscaloosa County, Alabama, 27 August 1911. © Richard T. Bryant.

Shell Description

Length to 85 mm; moderately thick; moderately compressed to inflated; outline subtriangular to oval; posterior margin narrowly rounded; anterior margin rounded; dorsal margin convex; ventral margin straight to slightly convex; posterior ridge high, rounded; posterior slope usually moderately steep; umbo broad, moderately inflated, elevated above hinge line, umbo sculpture unknown; periostracum tawny to brown, usually without rays, when present thin, sparse, often confined to posterior slope.

Pseudocardinal teeth low, triangular, 2 divergent teeth in left valve, 1 tooth in right valve, may have accessory denticle anteriorly and/or posteriorly; lateral teeth short, thick, straight to slightly curved, 2 in left valve, 1 in right valve; interdentum long, narrow to moderately wide; umbo cavity shallow; shell cavity deep, typically with 2 parallel creases extending from umbo posterioventrally, becoming less pronounced distally; nacre white.

Soft Anatomy Description

Mantle tan, may be creamy white outside of pallial line, rusty tan outside of apertures, with grayish brown mottling oriented perpendicular to margin, more distinct posteriorly; visceral mass tan, may have grayish brown areas; foot pale tan to pale orange.

Weak mantle fold just ventral to incurrent aperture; fold length 30–35% of shell length; margin mostly smooth, may be crenulate or with sparse simple papillae near posterior end of fold (one individual with 1 short blunt papilla near anterior end of fold); adjacent mantle rusty brown to grayish brown, mantle outside of fold grayish brown. The specimens examined appeared to be male. Mantle folds of females may be better developed.

Gills tan; dorsal margin straight to slightly sinuous, ventral margin convex to elongate convex; gill length approximately 65% of shell length; gill height approximately 45% of gill length; outer gill height 75–90% of inner gill height; inner lamellae of inner gills completely connected to visceral mass.

No gravid females were available for marsupium description. Hartfield and Hartfield (1996) reported outer gills to be marsupial, with glochidia held in the ventral portion of the entire gill with the exception of the extreme posterior end. The marsupial area has strong radial folds. Coloration of the conglutinates is visible through the marsupium wall, with the red "head" of the conglutinate giving the ventral gill edge a beaded appearance, a narrow black band proximal to the beaded edge and the remainder of the marsupium yellowish brown.

Labial palps tan; straight dorsally, convex ventrally, bluntly pointed distally; palp length 10–15% of gill length; palp height 60–75% of palp length; distal 35–65% of palps bifurcate.

Incurrent and supra-anal apertures longer than excurrent aperture; incurrent aperture longer or equal to supra-anal aperture.

Incurrent aperture length approximately 20% of shell length; creamy white to tan within, with grayish brown or rusty brown basal to papillae; papillae in 1–2 rows, short, simple; papillae tan to rusty brown or grayish brown.

Excurrent aperture length approximately 15% of shell length; creamy white to tan within, may have grayish brown cast, marginal color band rusty tan to rusty brown, may have grayish brown mottling oriented perpendicular to margin; margin with small, simple papillae, may be little more than crenulations; papillae tan to rusty brown or grayish brown.

Supra-anal aperture length 15–20% of shell length; creamy white to tan within, may have grayish cast, with thin grayish brown or rusty brown marginal band; margin smooth to minutely crenulate; mantle bridge separating supra-anal and excurrent apertures imperforate, length 10–15% of supra-anal length.

Specimens examined: Brushy Creek, Winston County, (n = 1); Sipsey Fork (n = 1) (specimens previously preserved).

Glochidium Description

Length 150–200 μm; height 180–230 μm; without styliform hooks; outline subelliptical (Ortmann, 1923b). Ortmann gave a glochidium description based on specimens from both the Black Warrior River drainage (*Ptychobranchus greenii*) and the Coosa River drainage (*Ptychobranchus foremanianus*), but did not distinguish between the two.

Similar Species

Ptychobranchus greenii may resemble some *Elliptio arca* but has a more inflated umbo and less pointed posterior margin. Internally the shells are easily distinguished. *Ptychobranchus greenii* always has white nacre, whereas that of *E. arca* is usually some shade of pink, purple or salmon, only occasionally white. Also, the shell cavity of *P. greenii* is deeper than that of *E. arca*.

Ptychobranchus greenii may also resemble *Pleurobema perovatum*, but *P. greenii* is usually more elongate, with a lower and more compressed umbo. The anterior pseudocardinal tooth of *P. perovatum* is typically compressed and aligned almost parallel to the shell margin, whereas that of *P. greenii* is usually more triangular and oriented more ventrally.

Ptychobranchus greenii typically lacks rays, while *Ptychobranchus foremanianus* usually has fine green rays. When *P. greenii* has rays they are thin, sparse and often confined to the posterior slope. The two species are not believed to occur sympatrically.

General Distribution

Ptychobranchus greenii is endemic to the Black Warrior and Tombigbee River drainages of the Mobile Basin in Alabama.

Alabama and Mobile Basin Distribution

Most records of *Ptychobranchus greenii* are from the Black Warrior River drainage near and above the Fall Line. However, one record of this species exists from Coalfire Creek, a tributary of the Tombigbee River well below the Fall Line in Pickens County.

Ptychobranchus greenii is extant in isolated, localized populations.

Ecology and Biology

Ptychobranchus greenii occurs in shoal habitats in a variety of stream sizes, ranging from small creeks to large rivers. It is usually found in sand and gravel substrates.

Ptychobranchus greenii is a long-term brooder, gravid from autumn to the following spring or summer (Ortmann, 1923b; Haag and Warren, 1997). Glochidia of *P. greenii* are discharged bound in conglutinates. Two forms of conglutinates have been reported for this species. One form resembles a dipteran larva (Chironomidae) in size, shape and color, with regions corresponding to a reddish "head", black "thorax" and tapering yellowish brown "abdomen", which has narrow, dark lateral stripes and regular constrictions that give the appearance of segments (Hartfield and Hartfield, 1996). These conglutinates may attach to the substrate by means of a short, transparent filament or tube (Hartfield and Hartfield, 1996). The other conglutinate form is pearl colored, with two black spots, resembling membrane-bound larval fish (Haag and Warren, 1997).

Fishes reported to serve as glochidial hosts for *Ptychobranchus greenii*, based on laboratory trials, include *Etheostoma bellator* (Warrior Darter), *Etheostoma douglasi* (Tuskaloosa Darter), *Percina kathae* (Mobile Logperch) and *Percina nigrofasciata* (Blackbanded Darter) (Percidae) (Haag and Warren, 1997).

Current Conservation Status and Protection

Ptychobranchus greenii was considered endangered throughout its range by Athearn (1970) and Williams et al. (1993). In Alabama *P. greenii* was reported as threatened by Stansbery (1976) and endangered by Lydeard et al. (1999). Garner et al. (2004) designated *P. greenii* a species of highest conservation concern in the state. In 1993 it was listed as endangered under the federal Endangered Species Act. All previous conservation assessments of *P. greenii* included *Ptychobranchus foremanianus* as a conspecific.

Remarks

Ptychobranchus greenii has previously been recognized as occurring throughout the Mobile Basin, primarily above the Fall Line. However, it appears to be restricted to the Black Warrior and Tombigbee River drainages. The form in the remainder of the Mobile Basin is herein recognized as *Ptychobranchus foremanianus*. *Ptychobranchus greenii* can be distinguished from *P. foremanianus* by the absence or poor development of dark green rays on its periostracum, which are characteristic of *P. foremanianus*. This distinction was first reported by Ortmann (1923b). Also, preliminary genetic analyses suggest the presence of at least two species in the Mobile Basin (K.J. Roe, personal communication). Ortmann (1923b) aligned Cahaba River *Ptychobranchus* with *P. greenii* but noted that individuals from that population shared characteristics of both *P. greenii* and *P. foremanianus*. The range of *P. foremanianus* herein includes the Cahaba River drainage, which is suggested by preliminary genetic analyses (K.J. Roe, personal communication).

Conglutinate morphology differs among individuals within populations of *Ptychobranchus greenii* (e.g., may resemble dipteran larvae or membrane-bound fish embryos). Further detailed comparative studies, including soft anatomy, conglutinate morphology, life history and more extensive genetic analyses, may reveal cryptic species.

Conrad (1834b) named *Ptychobranchus greenii* for Jacob Green, M.D., professor of chemistry at Jefferson College. Conrad described Dr. Green as "a gentlemen well known as a contributor to the science of conchology".

Synonymy

Unio greenii Conrad, 1834. Conrad, 1834b:32, pl. 4, fig. 1
Type locality: Headwaters of the Black Warrior River, Alabama. Lectotype, ANSP 20413, not found, also not found by Johnson and Baker (1973). Measurement of figured shell not given in original description (Conrad, 1834b).

Unio brumleyanus Lea, 1841. Lea, 1841b:82
Type locality: Tuscaloosa, [Tuscaloosa County,] Alabama, Professor Brumley [Brumby]. Type specimen not found (see *Unio brumbyanus*).
Unio brumbyanus Lea, 1842. Lea, 1842b:245, pl. 26, fig. 62; Lea, 1842c:83, pl. 26, fig. 62
Comment: Lea (1842b) changed the spelling of the name from *brumleyanus* Lea, 1841, to *brumbyanus*. Length of figured shell in original description reported as about 30 mm. Holotype lost (Johnson, 1974). This taxon has erroneously been placed in the synonymy of *Pleurobema rubellum* by some authors (e.g., Frierson, 1927).

Unio flavescens Lea, 1845. Lea, 1845:163; Lea, 1848a:72, pl. 3, fig. 9; Lea, 1848b:72, pl. 3, fig. 9
Type locality: Black Warrior River, Alabama, B.W. Budd, M.D. Lectotype, USNM 84952, length 43 mm, designated by Johnson (1974).

Unio simplex Lea, 1845. Lea, 1845:163; Lea, 1848a:76, pl. 5, fig. 15; Lea, 1848b:50, pl. 5, fig. 15
Type locality: Black Warrior River, Alabama, B.W. Budd, M.D. Lectotype, USNM 84951, length 41 mm,
    designated by Johnson (1974).

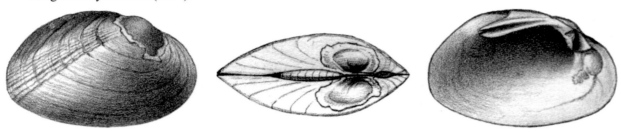

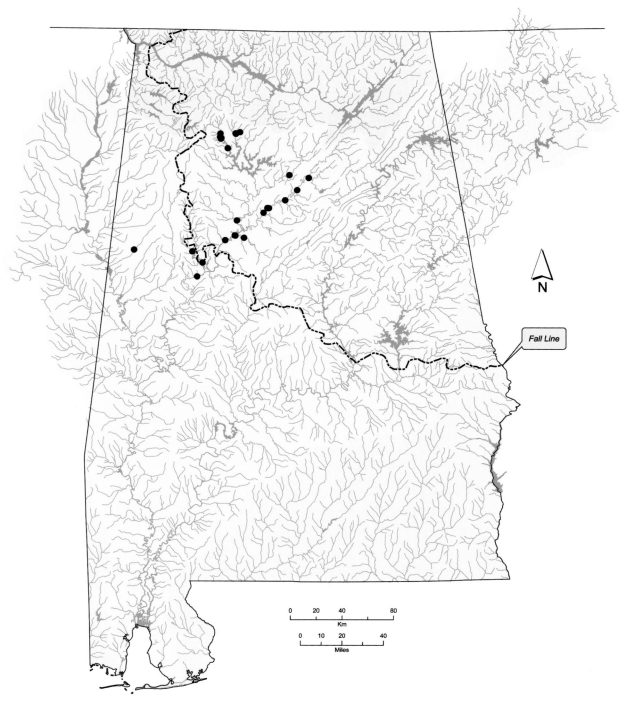

Distribution of *Ptychobranchus greenii* in Alabama and the Mobile Basin.

Ptychobranchus jonesi (van der Schalie, 1934)
Southern Kidneyshell

Ptychobranchus jonesi – Length 61 mm, UF 65567. Conecuh River, Bozeman's Landing near Crenshaw County line, Covington County, Alabama, 1 July 1915. © Richard T. Bryant.

Shell Description

Length to 72 mm; moderately thin; moderately inflated; outline elliptical; posterior margin narrowly truncate, sometimes obliquely; anterior margin rounded; dorsal margin straight to slightly convex; ventral margin straight to slightly convex; posterior ridge well-developed, doubled posterioventrally, slightly concave between ridges; posterior slope moderately steep, flat; umbo broad, elevated slightly above hinge line, umbo sculpture unknown; periostracum smooth, shiny, greenish yellow to dark brown or black, may be weakly rayed.

Pseudocardinal teeth low, triangular, 2 divergent teeth in left valve, 1 tooth in right valve, may have accessory denticle anteriorly; lateral teeth moderately long, straight, 2 in left valve, 1 in right valve; interdentum moderately long, very narrow; umbo cavity wide, shallow; nacre bluish white.

Soft Anatomy Description

Mantle tan, some grayish brown mottling outside of apertures; visceral mass tan; foot tan.

Gills tan; dorsal margin slightly sinuous, ventral margin elongate convex; gill length approximately 45% of shell length; gill height approximately 60% of gill length; outer gill height approximately 80% of inner gill height; inner lamellae of inner gills connected to visceral mass only anteriorly.

No gravid females were available for marsupium description. Fuller and Bereza (1973) reported glochidia to be held in the distal portions of the outer gills, which are folded when gravid.

Labial palps tan; straight dorsally, convex ventrally, bluntly pointed distally; palp length approximately 35% of gill length; palp height approximately 55% of palp length; distal 35% of palps bifurcate.

Incurrent, excurrent and supra-anal apertures of similar length.

Incurrent aperture length approximately 10% of shell length; creamy white within, tan with grayish brown cast basal to papillae; papillae in 1 row, short, simple; papillae creamy white and grayish brown.

Excurrent aperture length approximately 10% of shell length; pale tan within, marginal color band rusty brown; margin crenulate.

Supra-anal aperture length approximately 10% of shell length; pale tan within, with irregular, sparse grayish brown marginally; margin smooth; mantle bridge separating supra-anal and excurrent apertures absent.

Specimen examined: West Fork Choctawhatchee River (n = 1) (specimen previously preserved).

Glochidium Description

Glochidium unknown.

Similar Species

Ptychobranchus jonesi most closely resembles *Hamiota australis* but is more cylindrical with a rounder posterior margin. The posterior ridge of *P. jonesi* is typically doubled distally with a shallow radial concavity dorsal to the posterior ridge, which is absent in *H. australis*. Also, the ventral margin of *H. australis* is usually more broadly rounded than that of *P. jonesi*. *Ptychobranchus jonesi* has folded outer gills when gravid, which is characteristic of the genus. The marsupia of *Hamiota* are not folded.

General Distribution

Ptychobranchus jonesi is endemic to the Choctawhatchee, Yellow and Escambia River drainages of the eastern Gulf Coast, in Alabama and Florida.

Alabama and Mobile Basin Distribution

Ptychobranchus jonesi was historically widespread in the Choctawhatchee River drainage in

Alabama and Florida. Few records of this species exist from the Escambia and Yellow River drainages.

The only recent records of *Ptychobranchus jonesi* are from the lower portion of the Pea River (M.S. Gangloff, personal communication), West Fork Choctawhatchee River (Pilarczyk et al., 2006) and Conecuh River (D.N. Shelton, personal communication).

Ecology and Biology

Ptychobranchus jonesi occurs in medium creeks to medium rivers, usually in habitats with at least some current. Its preferred substrate appears to be firm sand.

Ptychobranchus jonesi is presumably a long-term brooder, gravid from autumn to the following spring or summer. The conglutinate of *P. jonesi* is small, being approximately 3 mm in length. It is oblong, bluntly pointed and slightly bulbous on the distal end and tapered on the proximal end. The distal half of the conglutinate is surrounded by a membrane that may have adhesive properties (Paul Johnson and Michael Buntin, personal communication). Glochidial hosts of *P. jonesi* are unknown, but other members of *Ptychobranchus* utilize species of Cottidae and Percidae.

Current Conservation Status and Protection

Ptychobranchus jonesi was listed as threatened throughout its range by Williams et al. (1993). In Alabama it was considered imperiled by Lydeard et al. (1999). *Ptychobranchus jonesi* was designated a species of highest conservation concern in the state by Garner et al. (2004). In 2003 this species was elevated to a candidate for protection under the federal Endangered Species Act.

Remarks

Ptychobranchus jonesi was originally described as a member of the genus *Lampsilis*, where it remained for approximately four decades. It was determined to belong to the genus *Ptychobranchus* following examination of gravid marsupia (Fuller and Bereza, 1973).

The decline of *Ptychobranchus jonesi* is perplexing. It appears to have been common historically. Athearn (1964) reported finding 98 individuals at one site on the West Fork Choctawhatchee River in Dale County, Alabama. Although habitat appears to have changed little over the past 40 years at some sites where several mussel species remain common, *P. jonesi* is now on the verge of extinction.

The reported type locality of *Ptychobranchus jonesi* is the Pea River, Pristons Mill, Dale County, Alabama (van der Schalie, 1934). However, the reference to Pea River in the type locality appears to be an error. Only about 4 km of the Pea River lies within Dale County, located in the extreme northwestern corner, but there is no Pristons Mill or other mill of a similar name known to have existed in that reach. However, there was a Prestons Mill located on the East Fork Choctawhatchee River in southeastern Dale County (Remington and Kallsen, 1999), and it is most likely the locality from which the type material was collected. The site of Prestons Mill on the East Fork Choctawhatchee River is near the present-day County Road 67 crossing, about 7 km north of Midland City.

Ptychobranchus jonesi was named for Dr. Walter Jones, who served as Alabama State Geologist and Director of the Alabama Museum of Natural History for many years.

Synonymy

Lampsilis jonesi van der Schalie, 1934. van der Schalie, 1934:125, pl. 15, figs. 1–3

Type locality: Pea River [East Fork Choctawhatchee River], Pristons [Prestons] Mill, [near County Road 67 crossing, about 7 km north of Midland City,] Dale County, Alabama, J.A. Burke, 15 November 1915.

Comment: The holotype figured by van der Schalie (1934) was originally sent to the Alabama Museum of Natural History but is now in the Florida Museum of Natural History, UF 65558 (male), length 46 mm. Paratypes were sent to the MCZ by van der Schalie. Johnson (1967a) invalidly designated a paratype as the lectotype (MCZ 98802).

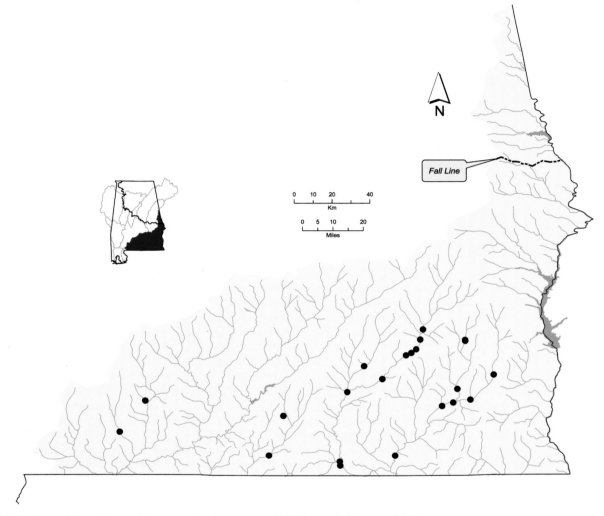

Distribution of *Ptychobranchus jonesi* in the eastern Gulf Coast drainages of Alabama.

Ptychobranchus subtentum (Say, 1825)
Fluted Kidneyshell

Ptychobranchus subtentum – Length 69 mm, UF 67706. Tennessee River, Alabama. © Richard T. Bryant.

Shell Description

Length to 120 mm; moderately thick; slightly inflated; outline elongate elliptical to subrhomboidal, may become somewhat arcuate with age; posterior margin narrowly rounded, females may be more broadly rounded posteriorly than males; anterior margin rounded; dorsal margin convex; ventral margin straight to slightly convex; posterior ridge well-developed, rounded; posterior slope moderately steep, with evenly spaced plications extending from posterior ridge to posteriodorsal margin; umbo broad, low, elevated slightly above hinge line, umbo sculpture low wavy ridges; periostracum tawny to greenish yellow, darkening to brown with age, typically with prominent wide green rays, usually interrupted into rectangular blotches, blotches may be offset, appearing as zigzag pattern near umbo.

Pseudocardinal teeth triangular, somewhat compressed, 2 slightly divergent teeth in left valve, may be almost parallel, 1 tooth in right valve, may have accessory denticle anteriorly; lateral teeth short, moderately thick, straight to slightly curved, 2 in left valve, 1 in right valve; interdentum moderately long, narrow to very narrow; umbo cavity wide, shallow; nacre white, often with salmon tint in umbo cavity.

Soft Anatomy Description

Mantle tan, outside of apertures tan or brown, often rusty, mottled with dark brown or grayish brown; visceral mass pearly white to creamy white, may have grayish cast; foot creamy white to tan.

Gills tan; dorsal margin straight to slightly sinuous, ventral margin elongate convex; gill length 60–70% of shell length; gill height 35–40% of gill length; outer gill height 75–80% of inner gill height; inner lamellae of inner gills connected to visceral mass only anteriorly.

Outer gills marsupial; glochidia held across entire gill except anterior and posterior ends; marsupium padded, arranged in short folds when gravid; dark grayish brown. Ortmann (1912a) reported gravid marsupia to be blackish purple and "pale along the beaded edge".

Labial palps tan; straight to slightly convex dorsally, convex ventrally, bluntly pointed distally; palp length 15–20% of gill length; palp height 45–70% of palp length; distal 45–75% of palps bifurcate.

Excurrent and supra-anal apertures continuous, without mantle bridge; continuous excurrent and supra-anal apertures approximately twice as long as incurrent aperture.

Incurrent aperture length approximately 15% of shell length; creamy white within, rusty orange and various shades of brown basal to papillae, usually mottled; papillae in 2 rows, long, thick, simple, blunt, inner row larger; papillae usually rusty orange, may have some gray or brown.

Excurrent and supra-anal apertures continuous, with no demarcation in either mantle margin morphology or coloration; length approximately 30% of shell length; gray or grayish brown within, 1–2 marginal color bands, some combination of rusty tan, rusty orange, reddish brown and dark brown, may have mottling; margin usually smooth and undulating, may be partially crenulate.

Specimens examined: Clinch River, Tennessee (n = 3).

Glochidium Description

Length 180-195 μm; height 220-251 μm; without styliform hooks; outline subelliptical; dorsal margin short, slightly convex, ventral margin rounded, anterior and posterior margins straight above, convex below (Ortmann, 1912a; Hoggarth, 1999).

Similar Species

Ptychobranchus subtentum may resemble *Ptychobranchus fasciolaris* but has distinctive plications on the posterior ridge and slope. The rays of

P. subtentum are often offset to form a zigzag pattern near the umbo, but rays of *P. fasciolaris* are separated and form no such pattern.

Ptychobranchus subtentum may superficially resemble *Lasmigona costata*. However, shell sculpture of *P. subtentum* is restricted to the posterior ridge and slope, whereas the sculpture of *L. costata* usually extends onto the shell disk. Rays on the periostracum of young *L. costata* are generally thinner than those of *P. subtentum* and often disappear with age. Rays near the umbo of *P. subtentum* are often in a zigzag pattern, whereas those of *L. costata* are not. Lateral teeth of *L. costata* are rudimentary, but those of *P. subtentum* are well-developed. The foot of *L. costata* is orange, whereas that of *P. subtentum* is white to tan.

General Distribution

Ptychobranchus subtentum is endemic to the Cumberland and Tennessee River drainages. There are a few widespread records of this species from the Cumberland River drainage downstream of Cumberland Falls (Cicerello et al., 1991; Parmalee and Bogan, 1998). In the Tennessee River drainage *P. subtentum* occurred historically from headwaters in southwestern Virginia downstream to Muscle Shoals, with disjunct populations in the Buffalo and Duck rivers in central Tennessee (Ahlstedt, 1992a, 1992b; Parmalee and Bogan, 1998). *Ptychobranchus subtentum* appears to have prehistorically occurred in the Tennessee River downstream of Muscle Shoals (Parmalee and Bogan, 1998).

Alabama and Mobile Basin Distribution

Ptychobranchus subtentum is confined to the Tennessee River drainage, where it appears to have historically occurred across the state in the Tennessee River and several tributaries.

Ortmann (1925) reported this species from Muscle Shoals, as well as several Tennessee River tributaries, but it has not been reported in subsequent literature.

Ecology and Biology

Ptychobranchus subtentum is a species of shoal habitats, primarily in small to large rivers. However, a few records exist from medium to large creeks. Its preferred substrate appears to be a mixture of sand and gravel. It can often be found under large, flat rocks.

Ptychobranchus subtentum is a long-term brooder, gravid from autumn to the following summer. Glochidia are discharged as conglutinates, which individually resemble pupae of blackflies (Simuliidae), complete with "eyes" and "antennae" (Figure 13.36).

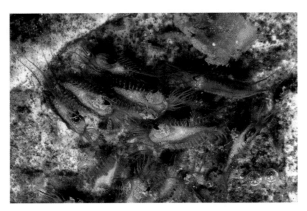

Figure 13.36. *Ptychobranchus subtentum*. Recently released conglutinates. Photograph by W.N. Roston

Fishes reported to serve as glochidial hosts of *Ptychobranchus subtentum* in laboratory trials include *Etheostoma caeruleum* (Rainbow Darter), *Etheostoma flabellare* (Fantail Darter), *Etheostoma obeyense* (Barcheek Darter) and *Etheostoma rufilineatum* (Redline Darter) (Percidae); and *Cottus carolinae* (Banded Sculpin) (Cottidae) (Luo, 1993).

Current Conservation Status and Protection

Ptychobranchus subtentum was reported as endangered throughout its range by Stansbery (1970a) and as a species of special concern by Williams et al. (1993). In Alabama it was considered endangered by Stansbery (1976) and imperiled by Lydeard et al. (1999). *Ptychobranchus subtentum* was listed as extirpated from the state by Garner et al. (2004).

Remarks

Ptychobranchus subtentum has been reported to be uncommon in prehistoric shell middens along the Tennessee River in northern Alabama (Morrison, 1942; Warren, 1975; Hughes and Parmalee, 1999).

Synonymy

Unio subtentus Say, 1825. Say, 1825:130; Say, 1831a:[no pagination], pl. 15
Type locality: North Fork Holston River, Tennessee. Type specimen not found.

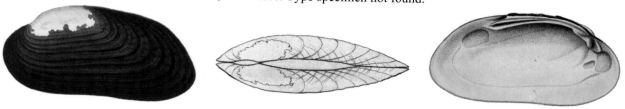

Unio subteritus purcheornatus De Gregorio, 1914. De Gregorio, 1914:31, pl. 9, fig. 2

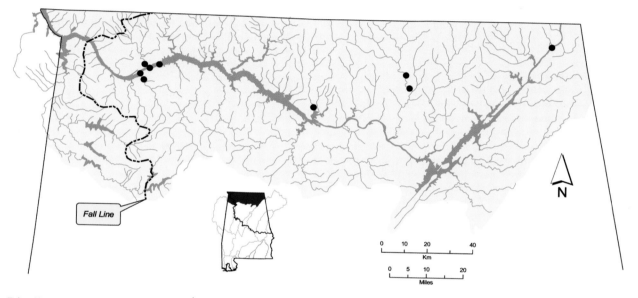

Distribution of *Ptychobranchus subtentum* in the Tennessee River drainage of Alabama.

Genus *Pyganodon*

Pyganodon Crosse and Fischer *in* Fischer and Crosse, 1894, occurs in the Hudson Bay, Great Lakes and Mississippi basins, Atlantic Coast drainages from Newfoundland to Georgia, and Gulf Coast drainages from the Apalachicola Basin west into Mexico. Turgeon et al. (1998) listed five species of *Pyganodon*, two of which occur in Alabama.

Pyganodon is based on a type species from Mexico. Using morphological and allozymic data, Hoeh (1990) provided a phylogeny that divided North American *Anodonta* Lamarck, 1799—along with the type species, *Anodonta cygnea* (Linnaeus, 1758)—into three clades. Based on these phylogenetic analyses, Hoeh (1990) elevated *Pyganodon* and *Utterbackia* Baker, 1928a, from subgeneric to generic level. However, he did not include the type species of *Pyganodon*, *Anodonta globosa* Lea, 1841, which is native to Mexico.

Type Species
Anodonta globosa Lea, 1841

Diagnosis
Shell inflated; thin to moderately thick; elliptical to oval; without hinge teeth; umbo sculpture double-looped ridges; umbo inflated, elevated above hinge line.

Inner lamellae of inner gills connected to visceral mass only anteriorly; outer gills marsupial; marsupium well-padded across entire gill, not extended beyond original gill ventral edge when gravid; glochidium with styliform hooks (Fischer and Crosse, 1894; Frierson, 1927; Haas, 1969a, 1969b; Hoeh, 1990; Vidrine, 1993).

Synonymy
None recognized (see discussion of *Flexiplis* Rafinesque, 1831, in Genera *Incertae Sedis*).

Pyganodon cataracta (Say, 1817)
Eastern Floater

Pyganodon cataracta – Length 92 mm, UF 366865. Chipola River, east bank at southern tip of cutoff island, 0.1 nautical mile above Apalachicola River, Gulf County, Florida, 1 August 1991. © Richard T. Bryant.

Shell Description

Length to 170 mm; thin, moderately thick in some older individuals; inflated; outline elliptical; posterior margin rounded; anterior margin rounded; dorsal margin straight to slightly convex; ventral margin straight to slightly convex; posterior ridge rounded; posterior slope moderately steep, slightly concave; umbo broad, inflated, elevated above hinge line, umbo sculpture low, double-looped ridges, loops uniform; periostracum olive or tawny to dark brown, may have weak green rays.

Pseudocardinal and lateral teeth absent; umbo cavity moderately deep, open; nacre white to bluish white.

Soft Anatomy Description

Mantle pale tan, translucent, outside of apertures tan with numerous small black spots or specks; visceral mass creamy white to pearly white; foot pale orange.

Gills tan; dorsal margin straight to slightly sinuous, ventral margin convex; gill length 50–60% of shell length; gill height 50–55% of gill length; outer gill height 80–95% of inner gill height; inner lamellae of inner gills connected to visceral mass only anteriorly.

No gravid females were available for marsupium description. Tankersley and Dimock (1992, 1993a) reported outer gills to be marsupial and to be greatly padded when gravid, with a tripartite system of water tubes that is not present in nonmarsupial gills.

Labial palps tan; straight to slightly concave or slightly convex dorsally, convex ventrally, narrowly rounded to bluntly pointed distally; palp length 40–45% of gill length; palp height 45–55% of palp length; distal 30–45% of palps bifurcate.

Incurrent aperture longer than excurrent and supra-anal apertures; excurrent and supra-anal apertures of similar length.

Incurrent aperture length approximately 15% of shell length; creamy white within, some individuals with rusty orange, rusty tan or dark brown basal to papillae, may have dark brown lines perpendicular to margin; papillae in 2–3 rows, short to long, slender to thick, mostly simple, may have some bifids, inner row larger; papillae some combination of creamy white, tan, rusty tan and rusty orange, may have dark brown edges basally.

Excurrent aperture length 5–10% of shell length; creamy white within, marginal color band rusty tan to rusty orange, with dark brown lines perpendicular to margin, some lines converge proximally; margin smooth but undulating, may be crenulate adjacent to incurrent aperture.

Supra-anal aperture length 5–10% of shell length; creamy white within, may have thin tan marginal color band; margin smooth; mantle bridge separating supra-anal and excurrent apertures may be perforate, length 140–225% of supra-anal length.

Specimens examined: Uchee Creek, Russell County (n = 3).

Glochidium Description

Length 366–382 µm; height 351–383 µm; with styliform hooks; outline subtriangular; dorsal margin straight, ventral margin bluntly pointed, anterior and posterior margins convex, anterior margin slightly more convex than posterior margin (Simpson, 1884; Lefevre and Curtis, 1910b; Wiles, 1975; Hoggarth, 1999).

Similar Species

Pyganodon cataracta closely resembles *Pyganodon grandis* but its umbo is positioned more anteriorly and the ventral margin is usually straighter. Umbo sculpture of both *P. cataracta* and *P. grandis* is double-looped. However, ventral margins of the loops in *P. cataracta* are without nodules, unlike those of *P. grandis*.

Pyganodon cataracta may also resemble *Anodonta heardi* (a species of hypothetical occurrence), but the umbo of *A. heardi* is barely elevated above the hinge line, and its ventral margin is more rounded.

General Distribution

Pyganodon cataracta occurs in Atlantic Coast drainages from the St. Lawrence River drainage, Canada, south to the Altamaha River drainage, Georgia (Johnson, 1970). A disjunct population occurs in the Apalachicola Basin of the eastern Gulf Coast in Alabama, Florida and Georgia (Brim Box and Williams, 2000).

Alabama and Mobile Basin Distribution

Pyganodon cataracta is restricted to the Chattahoochee River drainage of the Apalachicola Basin and the upper Tallapoosa River drainage.

The status of *Pyganodon cataracta* in Alabama is unclear. It was not encountered in the state during a recent survey of the Apalachicola Basin (Brim Box and Williams, 2000). It has since been collected from Uchee Creek, Russell County. Also, genetic analyses of individuals from Lake Martin, Tallapoosa River, suggest that they belong to *P. cataracta*. The origin of the Lake Martin population is unknown; however, it may be the result of an introduction. The paucity of records from Alabama may be a reflection of the less intensive sampling of reservoir and backwater habitats.

Ecology and Biology

Pyganodon cataracta occurs in a variety of habitats, including rivers, creeks, ponds, lakes and reservoirs. It is typically found in sand or mud substrates in areas with little or no current. However, *P. cataracta* is occasionally encountered in areas with gravel substrates and moderate current. Ortmann (1919) stated that *P. cataracta* occurs "under special

Synonymy

Anodonta fluviatilis Gmelin, 1791 [of authors]

conditions in the estuary of the Delaware [River]", suggesting that it has some tolerance for salinity.

Pyganodon cataracta is a long-term brooder, but its brooding period appears to vary geographically. Wiles (1975) reported gravid females from September to the following May in Nova Scotia. Conner (1905, 1909) reported gravid females during the same period at unspecified localities. However, Tankersley and Dimock (1993b) made observations on "postbrooding" *P. cataracta* in February and March in North Carolina. Ortmann (1919) reported *P. cataracta* to become gravid in August and brood through April in Pennsylvania. Ortmann (1919) reported a single gravid female in July, but referred to it as "an exceptional case".

Fishes reported to serve as glochidial hosts of *Pyganodon cataracta* in laboratory trials include *Catostomus commersonii* (White Sucker) (Catostomidae); and *Ambloplites rupestris* (Rock Bass) and *Lepomis gibbosus* (Pumpkinseed) (Centrarchidae) (van Snik Gray et al., 1999). Wiles (1975) reported *Apeltes quadracus* (Fourspine Stickleback), *Gasterosteus aculeatus* (Threespine Stickleback) and *Pungitius pungitius* (Ninespine Stickleback) (Gasterosteidae) to be infested with anodontine glochidia, which may have been *P. cataracta*. Lefevre and Curtis (1910b) reported observations of *P. cataracta* glochidia implanting on fins and gills of *Cyprinus carpio* (Common Carp) (Cyprinidae). However, it is unclear whether those infestations were carried through to transformation.

Current Conservation Status and Protection

Pyganodon cataracta was considered currently stable throughout its range by Williams et al. (1993) and in Alabama by Lydeard et al. (1999). Garner et al. (2004) designated *P. cataracta* a species of moderate conservation concern in the state.

Remarks

Pyganodon cataracta is possibly the most widespread freshwater mussel species on the Atlantic Coast. Johnson (1970) reported the range of *P. cataracta* to extend west along the Gulf Coast to the Mobile Basin. However, its range does not appear to extend west of the Apalachicola Basin (Brim Box and Williams, 2000).

Anodonta cataracta Say, 1817. Say, 1817:[no pagination], pl. 3, fig. 4

Type locality: No specific geographic locality stated in original description, deep part of a mill dam. Johnson (1970) restricted type locality to presumably near Philadelphia, Philadelphia County, Pennsylvania. Type specimen not found in ANSP (Johnson, 1970). Length of figured shell in original description reported as about 114 mm.

Anodonta teres Conrad, 1834. Conrad, 1834b:47, pl. 7, fig. 2

Anodon excurvata De Kay, 1843. De Kay, 1843:202, pl. 17, fig. 233

Anodonta virgulata Lea, 1857. Lea, 1857d:86; Lea, 1862c:213, pl. 33, fig. 282; Lea, 1863e:35, pl. 33, fig. 282

Anodonta dariensis Lea, 1858. Lea, 1858b:139; Lea, 1859f:230, pl. 28, fig. 99; Lea, 1860e:48, pl. 28, fig. 99

Anodonta williamsii Lea, 1862. Lea, 1862a:169; Lea, 1866:27, pl. 10, fig. 26; Lea, 1867b:31, pl. 10, fig. 26

Anodonta tryonii Lea, 1862. Lea, 1862a:169; Lea, 1866:28, pl. 10, fig. 27; Lea, 1867b:32, pl. 10, fig. 27

Anodonta dolearis Lea, 1863. Lea, 1863c:193

Anodonta doliaris Lea, 1863. Lea, 1866:24, pl. 8, fig. 23 [corrected spelling of *Anodonta dolearis* Lea, 1863]; Lea, 1867b:28, pl. 8, fig. 23

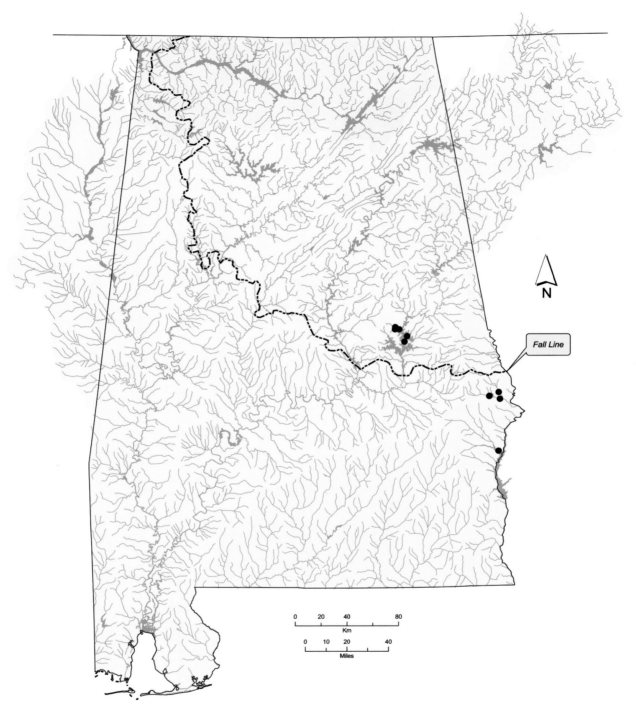

Distribution of *Pyganodon cataracta* in Alabama and the Mobile Basin.

Pyganodon grandis (Say, 1829)
Giant Floater

Pyganodon grandis – Length 81 mm, UF 8379. Pigeon Creek, Luverne-Greenville Highway [Highway 10, east of Greenville], Butler County, Alabama, July 1937. © Richard T. Bryant.

Shell Description

Length to 254 mm; thin, moderately thick in some older individuals; inflated; outline oval to triangular; posterior margin rounded to bluntly pointed; anterior margin rounded; dorsal margin straight to slightly convex; ventral margin rounded; posterior ridge rounded; posterior slope moderately steep, somewhat concave; umbo broad, inflated, elevated well above hinge line, umbo sculpture ridged, double-looped distally, anterior loop rounded, posterior loop angular, nodulous; periostracum shiny, tawny to olive brown or black, may be faintly rayed.

Pseudocardinal and lateral teeth absent; hinge may be thickened; umbo cavity deep, open; nacre white, may have pink or salmon tint.

Soft Anatomy Description

Mantle creamy white to tan or gold, occasionally dull orange outside of pallial line, rusty tan, brown or gray outside of apertures, usually with numerous small black specks; visceral mass pearly white or creamy white to tan, may have golden cast; foot tan to pale orange.

Gills tan to light brown; dorsal margin straight to slightly sinuous or slightly concave, ventral margin elongate convex; gill length 40–60% of shell length; gill height 40–60% of gill length; outer gill height 75–100% of inner gill height; inner lamellae of inner gills connected to visceral mass only anteriorly.

Outer gills marsupial; glochidia held across entire gill; marsupium well-padded when gravid; light brown. Ortmann (1912a) reported gravid gills to be yellowish when brooding eggs and "liver-brown" when gravid with glochidia.

Labial palps light tan to light brown; straight to slightly concave dorsally, curved ventrally, bluntly pointed to narrowly rounded distally; palp length 25–45% of gill length; palp height 45–60% of palp length; distal 30–55% of palps bifurcate.

Incurrent aperture longer than excurrent and supra-anal apertures; excurrent aperture slightly longer, slightly shorter or equal to supra-anal aperture.

Incurrent aperture length 10–15% of shell length; creamy white within, may have slight golden cast, brown from edges of papillae may extend onto aperture wall as irregular lines, some lines converge proximally; papillae in 2–4 rows, long, slender, simple, inner row larger; papillae white, tan or golden brown, often with dark brown or black edges basally. Lea (1863d) stated that incurrent aperture papillae are "somewhat branched".

Excurrent aperture length 5–10% of shell length; creamy white within, marginal color band some combination of tan, orange, rusty brown, dark brown and black, with lighter background and dark lines in reticulated pattern, bands perpendicular to margin or irregularly mottled; margin smooth and undulating or crenulate.

Supra-anal aperture length 5–10% of shell length; creamy white to gold within, occasionally with thin marginal band of gray or rusty brown; margin smooth; mantle bridge separating supra-anal and excurrent apertures usually imperforate, length 95–230% of supra-anal length.

Specimens examined: Bear Creek, Colbert County (n = 1); Conecuh River (n = 2); unnamed Flint River tributary, Madison County (n = 1); Tennessee River (n = 5); Tombigbee River (n = 1).

Glochidium Description

Length 350–410 µm; height 343–420 µm; with styliform hooks; outline subtriangular; dorsal margin straight, ventral margin bluntly pointed, anterior and posterior margins broadly convex, anterior margin slightly more convex than posterior margin (Ortmann, 1912a; Surber, 1912; Utterback, 1916b; Tucker, 1928; Hoggarth, 1999).

Similar Species

Pyganodon grandis closely resembles *Pyganodon cataracta*, but *P. grandis* has a rounder ventral margin, and the umbo is positioned more centrally. Umbo sculpture of both *P. grandis* and *P. cataracta* is double-looped. However, ventral margins of the loops in *P. grandis* are nodulous, unlike those of *P. cataracta*.

Pyganodon grandis may also resemble *Anodonta heardi*. However, the umbo of *A. heardi* (a species of hypothetical occurrence) is barely elevated above the hinge line and is less inflated than that of *P. grandis*.

Pyganodon grandis resembles *Strophitus undulatus* but is more inflated and has a higher umbo. The ventral margin of *P. grandis* is generally more rounded than that of *S. undulatus*. Umbo sculpture of *P. grandis* is double-looped, but that of *S. undulatus* consists of single loops. *Strophitus undulatus* typically has some suggestion of rudimentary pseudocardinal teeth and a thickened hinge plate that is somewhat suggestive of lateral teeth, whereas *P. grandis* is without any trace of teeth.

General Distribution

Pyganodon grandis is the most widespread freshwater mussel in North America. It occurs in the Hudson Bay Basin of southern Canada (Clarke, 1981a), Minnesota (Dawley, 1947) and North Dakota (Dyke, 2000) as well as throughout the Great Lakes drainage from the tidewater of the St. Lawrence River to western Lake Superior (Goodrich and van der Schalie, 1932; Clarke, 1981a; Graf, 1997). *Pyganodon grandis* is also found throughout the Mississippi Basin from Minnesota (Dawley, 1947) south to Louisiana (Vidrine, 1993) and from New York (Strayer et al., 1992) west to Montana (Gangloff and Gustafson, 2000). *Pyganodon grandis* is found in the Cumberland River drainage downstream of Cumberland Falls (Cicerello et al., 1991) and in the Tennessee River drainage from eastern Tennessee downstream to the mouth of the Tennessee River (Parmalee and Bogan, 1998). Along the Gulf Coast *P. grandis* occurs from the Apalachicola Basin (Brim Box and Williams, 2000) west to the lower Rio Grande (Howells et al., 1996). *Pyganodon grandis* may have been introduced outside of its native range, including upper reaches of the Rio Grande in Arizona (Bequaert and Miller, 1973) and New Mexico (Johnson, 1999).

Alabama and Mobile Basin Distribution

Pyganodon grandis occurs throughout Alabama, with the exception of the Yellow, Blackwater and Perdido River drainages.

Pyganodon grandis remains widespread and may be locally abundant. It is frequently found in ponds that have been stocked with fish.

Ecology and Biology

Pyganodon grandis occurs in a variety of habitats, including ponds, lakes and reservoirs as well as pools and backwater areas of creeks and rivers. It is usually found in sand or mud substrates in areas with little or no current. Bates (1962) reported *P. grandis* to be one of the first colonizers of newly formed overbank habitat following the completion of Kentucky Dam on the Tennessee River. It is frequently found in ponds that have been stocked with fish. This may have resulted in expansion of its range. Under laboratory conditions, Gatenby et al. (1996) reported juvenile *P. grandis* to survive and grow better in cultures to which silt had been added compared to those without silt.

Pyganodon grandis is generally considered to be dioecious, though occasional hermaphroditic individuals are encountered (van der Schalie and Locke, 1941). However, Jansen and Hanson (1991) reported a population of *P. grandis* in Alberta, Canada, in which more than 90 percent of adults brooded glochidia, suggesting that the population is hermaphroditic. *Pyganodon grandis* sperm are released as motile spermatozeugmata, 40–50 µm in diameter, composed of a gelatinous sphere in which sperm are embedded, with their flagella to the exterior (Lynn, 1994).

Pyganodon grandis is most often reported to be a long-term brooder, but this appears to be variable among populations across its range. Its gravid period has been reported to extend from August or September to the following April or May in Alabama, Kansas, Louisiana and Pennsylvania (Ortmann, 1912a) as well as Wisconsin (Baker, 1928a) and Texas (Howells, 2000). A longer brooding period, July to the following June, was reported for a population in an Alberta, Canada, lake (Jansen and Hanson, 1991). Shorter brooding periods have been observed: December to March in Missouri (Utterback, 1915, 1916a); September to November in Louisiana (Penn, 1939) and July to September in Montreal, Canada, near the northern limits of its range (Lewis, 1985).

Fecundity of *Pyganodon grandis* was estimated as 235,210 glochidia in a female 125 mm long (USFWS, 2004). Conglutinates are not formed and glochidia are discharged loose, in irregular masses (Ortmann, 1912a). Richard et al. (1991) suggested that glochidia of *P. grandis* are discharged from the ventral margin of the gill. However, Tankersley and Dimock (1993b) questioned this, based on observations of glochidial discharge via the suprabranchial chamber in closely related *Pyganodon cataracta*. If glochidia are discharged via the ventral gill margin, the mechanism and pathway by which they get to the excurrent aperture is unknown.

Pyganodon grandis is a generalist with regard to glochidial hosts. Fishes reported to serve as glochidial hosts based on laboratory trials include *Labidesthes sicculus* (Brook Silverside) (Atherinopsidae); *Ambloplites rupestris* (Rock Bass), *Lepomis cyanellus* (Green Sunfish), *Lepomis gibbosus* (Pumpkinseed), *Lepomis macrochirus* (Bluegill), *Micropterus salmoides* (Largemouth Bass), *Pomoxis annularis* (White Crappie) and *Pomoxis nigromaculatus* (Black Crappie) (Centrarchidae); *Cichlasoma cyanoguttatum* (Rio Grande Cichlid) (Cichlidae); *Campostoma anomalum* (Central Stoneroller), *Luxilus cornutus* (Common Shiner), *Lythrurus umbratilis* (Redfin Shiner), *Notemigonus crysoleucas* (Golden Shiner), *Notropis heterodon* (Blackchin Shiner), *Notropis heterolepis* (Blacknose Shiner), *Pimephales notatus* (Bluntnose Minnow), *Rhinichthys obtusus* (Western Blacknose Dace) and *Semotilus atromaculatus* (Creek Chub) (Cyprinidae); *Fundulus diaphanus* (Banded Killifish) (Fundulidae); *Culaea inconstans* (Brook Stickleback) (Gasterosteidae); *Lepisosteus osseus* (Longnose Gar) (Lepisosteidae); and *Etheostoma caeruleum* (Rainbow Darter), *Etheostoma exile* (Iowa Darter), *Etheostoma nigrum* (Johnny Darter) and *Perca flavescens* (Yellow Perch) (Percidae) (Tucker, 1928; Trdan and Hoeh, 1982; Howells, 1997). Nonindigenous fishes shown to serve as hosts of *P. grandis* in laboratory trials include *Carassius auratus* (Goldfish) (Cyprinidae); *Neogobius melanostomus* (Round Goby) (Gobiidae); and *Poecilia reticulata* (Guppy) (Poeciliidae) (Watters et al., 2005). Species reported to serve as glochidial hosts of *P. grandis* based on observations of natural infestations include *Lepomis megalotis* (Longear Sunfish) (Centrarchidae); *Alosa chrysochloris* (Skipjack Herring) (Clupeidae); *Margariscus margarita* (Pearl Dace) (Cyprinidae); *Fundulus chrysotus* (Golden Topminnow) (Fundulidae); *Ameiurus nebulosus* (Brown Bullhead) (Ictaluridae); and *Gambusia affinis* (Western Mosquitofish) (Poeciliidae) (Lefevre and Curtis, 1910b; Penn, 1939; Trdan and Hoeh, 1982). Penn (1939) observed excystment of transformed *P. grandis* juveniles from naturally infested *L. macrochirus*, *L. megalotis* and *M. salmoides* in aquaria. Natural infestations on the nonindigenous *Cyprinus carpio* (Common Carp) (Cyprinidae) have been reported (Lefevre and Curtis, 1910b). In a study of its parasitic period, Blystad (1923) reported observations of *P. grandis* glochidia on *Lepomis humilis* (Orangespotted Sunfish) (Centrarchidae) until "the end of the parasitic period", but did not indicate whether transformation occurred. Wilson (1916) included *Dorosoma cepedianum* (Gizzard Shad) (Clupeidae); *Ameiurus natalis* (Yellow Bullhead) (Ictaluridae); *Morone chrysops* (White Bass) (Moronidae); and *Aplodinotus grunniens* (Freshwater Drum) (Sciaenidae) as glochidial hosts of *P. grandis* but did not indicate the methods by which these observations were made. Clarke and Berg (1959) included *Carpiodes carpio* (River Carpsucker) (Catostomidae) as a glochidial host of *P. grandis*, citing J.P.E. Morrison, but gave no details as to how this observation was made.

Current Conservation Status and Protection
Pyganodon grandis was considered currently stable throughout its range by Williams et al. (1993) and in Alabama by Lydeard et al. (1999). It was designated a species of lowest conservation concern in the state by Garner et al. (2004).

Remarks
Shell morphology of *Pyganodon grandis* is highly variable (Jass and Glenn, 1999), which is reflected in its synonymy.

Pyganodon grandis has increased in abundance in the Tennessee River during the past century. It is present, but rare, in prehistoric middens at Muscle Shoals and near Bridgeport (Morrison, 1942; Warren, 1975; Hughes and Parmalee, 1999), but is now locally common in some reservoirs.

Synonymy
Anodonta grandis Say, 1829. Say, 1829:341
Type locality: Fox River of the Wabash River, Indiana. Neotype, SMF 4300, length 170 mm, designated by Haas (1930), is from Ohio River.
Anodonta lugubris Say, 1829. Say, 1829:340
Anodonta corpulenta Cooper, 1834. Cooper, 1834:154

Anodonta declivis Conrad, 1834. Conrad, 1834a:341, pl. 1, fig. 11
Type locality: Flint River, Morgan County, Alabama. Type specimen lost (Johnson and Baker, 1973). Length of figured shell in original description reported as 84 mm.

Anodonta stewartiana Lea, 1834. Lea, 1834a:47, pl. 6, fig. 17; Lea, 1834b:159, pl. 6, fig. 17
Anodonta palna Lea, 1834. Lea, 1834a:48, pl. 7, fig. 18; Lea, 1834b:160, pl. 7, fig. 18 [error for *plana*]
Anodonta plana Lea, 1834. Lea, 1834a:64
Symphynota benedictensis Lea, 1834. Lea, 1834a:104, pl. 16, fig. 48; Lea, 1834b:216, pl. 16, fig. 48
Anodonta gigantea Lea, 1836, *nomen nudum*. Lea, 1836:52
Margarita (*Anodonta*) *decora* Lea, 1836, *nomen nudum*. Lea, 1836:52
Margarita (*Anodonta*) *salmonia* Lea, 1836, *nomen nudum*. Lea, 1836:51
Margarita (*Anodonta*) *ovata* Lea, 1836, *nomen nudum*. Lea, 1836:52
Anodonta gigantea Lea, 1838. Lea, 1838a:1, pl. 1, fig. 1; Lea, 1838c:1, pl. 1, fig. 1
Anodonta ovata Lea, 1838. Lea, 1838a:2, pl. 2, fig. 2; Lea, 1838c:2, pl. 2, fig. 2
Anodonta salmonia Lea, 1838. Lea, 1838a:45, pl. 14, fig. 41; Lea, 1838c:45, pl. 14, fig. 41
Anodonta decora Lea, 1838. Lea, 1838a:64, pl. 20, fig. 63; Lea, 1838c:64, pl. 20, fig. 63
Anodonta pepinianus Lea, 1838. Lea, 1838a:96, pl. 16, fig. 51; Lea, 1838c:96, pl. 16, fig. 51
Anodonta footiana Lea, 1840. Lea, 1840:289; Lea, 1842b:225, pl. 20, fig. 44; Lea, 1842c:63, pl. 20, fig. 44
Anodonta harpethensis Lea, 1840. Lea, 1840:289; Lea, 1842b:224, pl. 19, fig. 42; Lea, 1842c:62, pl. 19, fig. 42
Anodonta maryattiana Lea, 1840. Lea, 1840:289
Anodonta maryattana Lea, 1842. Lea, 1842b:226, pl. 20, fig. 45 [change in spelling]; Lea, 1842c:64, pl. 20, fig. 45
Anodonta opaca Lea, 1852, *nomen nudum*. Lea, 1852a:252
Anodonta linnaeana Lea, 1852, *nomen nudum*. Lea, 1852a:252
Anodonta virens Lea, 1852, *nomen nudum*. Lea, 1852a:252
Anodonta opaca Lea, 1852. Lea, 1852b: 285, pl. 25, fig. 46; Lea, 1852c:41, pl. 25, fig. 46
Anodonta linnaeana Lea, 1852. Lea, 1852b:289, pl. 27, fig. 51; Lea, 1852c:45, pl. 27, fig. 51
Anodonta virens Lea, 1852. Lea, 1852b:290, pl. 28, fig. 53; Lea, 1852c:46, pl. 28, fig. 53
Anodonta nilssonii Küster, 1853. Küster, 1853:61, pl. 17, figs. 3, 4
Anodonta lewisii Lea, 1857. Lea, 1857c:84; Lea, 1860d:362, pl. 62, fig. 187; Lea, 1860g:44, pl. 62, fig. 187
Anodonta danielsii Lea, 1858. Lea, 1858b:139; Lea, 1860d:365, pl. 63, fig. 190; Lea, 1860g:47, pl. 63, fig. 190
Anodonta gesnerii Lea, 1858. Lea, 1858b:139; Lea, 1859f:231, pl. 31, fig. 109; Lea, 1859g:49, pl. 31, fig. 109
Type locality: Uphaupee [Uphapee] Creek, Macon County, Alabama, W. Gesner. Lectotype, USNM 86427, length 118 mm, designated by Johnson (1974).

Anodonta texasensis Lea, 1859. Lea, 1859a:113; Lea, 1860d:366, pl. 63, fig. 191; Lea, 1860g:48, pl. 63, fig. 191
Anodonta simpsoniana Lea, 1861. Lea, 1861b:56; Lea, 1862c:212, pl. 32, fig. 281; Lea, 1863a:34, pl. 32, fig.281
Anodonta leonensis Lea, 1862. Lea, 1862a:169; Lea, 1866:25, pl. 9, fig. 24; Lea, 1867b:29, pl. 9, fig. 24
Anodonta dallasiana Lea, 1863. Lea, 1863b:190; Lea, 1866:29, pl. 11, fig. 28; Lea, 1867b:33, pl. 11, fig. 28
Anodonta bealei Lea, 1863. Lea, 1863c:194; Lea, 1866:26, pl. 9, fig. 25; Lea, 1867b:30, pl. 9, fig. 25
Anodon subangulata Anthony, 1865. Anthony, 1865:158, pl. 13, fig. 1

Anodon imbricata Anthony, 1865. Anthony, 1865:159, pl. 14, fig. 1
Anodon opalina Anthony, 1865. Anthony, 1865:159, pl. 14, fig. 2
Anodon subinflata Anthony, 1865. Anthony, 1865:160, pl. 15, fig. 1
Anodon micans Anthony, 1865. Anthony, 1865:162, pl. 16, fig. 1
Anodon mcnielii Anthony, 1866. Anthony, 1866:144, pl. 6, fig. 1
Anodon subgibbosa Anthony, 1866. Anthony, 1866:144, pl. 6, fig. 2
Anodon inornata Anthony, 1866. Anthony, 1866:145
Anodonta houghtonensis Currier, 1868, *nomen nudum*. Currier, 1868:11
Margaron (*Anodonta*) *benedictii* Lea, 1870. Lea, 1870:75 [unjustified emendation]
Anodonta sulcata Küster, 1873. Küster, 1873:62, pl. 18, fig. 1
Anodonta hockingensis "Moores" Call, 1880, *nomen nudum*. Call, 1880:530
Anodonta somersii "Moores" Call, 1880, *nomen nudum*. Call, 1880:530
Anodonta houghtonensis Currier *in* DeCamp, 1881. DeCamp, 1881:14, pl. 1, fig. 2
Anodonta dakota Frierson, 1910. Frierson, 1910:113, pl. 10

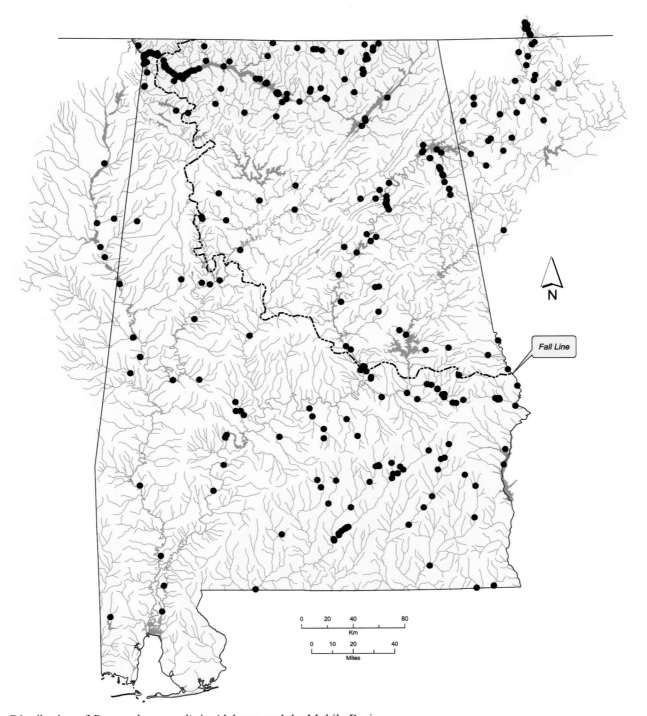

Distribution of *Pyganodon grandis* in Alabama and the Mobile Basin.

Genus *Quadrula*

Quadrula Rafinesque, 1820, occurs in the Hudson Bay, Great Lakes and Mississippi basins and Gulf Coast drainages from northern Florida to the Rio Grande. Turgeon et al. (1998) included 18 species of *Quadrula*. Two of those, *Quadrula cylindrica* (Say, 1817) and *Quadrula pustulosa* (Lea, 1831), were each recognized as being composed of two subspecies. One species, *Quadrula nobilis* (Conrad, 1853), was recently elevated from synonymy based on shell morphology and electrophoretic analysis (Howells et al., 1996) but was not recognized in Turgeon et al. (1998). *Quadrula nobilis* occurs in Alabama and is herein included. An additional species, *Quadrula kieneriana* (Lea, 1852), is herein elevated from synonymy based on shell morphology. Three additional species—*Fusconaia succissa* (Lea, 1852), *Quincuncina infucata* (Conrad, 1834) and *Tritogonia verrucosa* (Rafinesque, 1820)— were herein moved to *Quadrula* based on various combinations of genetic analyses (Lydeard et al., 2000; Serb et al., 2003; Campbell et al., 2005), shell morphology and soft anatomy. The total number of *Quadrula* species known from Alabama is 15.

Quadrula verrucosa (Rafinesque, 1820) has been variously recognized in the genera *Quadrula* and *Tritogonia* Agassiz, 1852 (e.g., Simpson, 1900b; Ortmann, 1912a). Davis and Fuller (1981) noted that *Tritogonia* was immunologically closest to two groups within *Quadrula*. Serb et al. (2003) provided the first molecular phylogenetic analysis of *Quadrula* and recommended that *Tritogonia* should be moved to that genus. This action makes *Tritogonia* Agassiz, 1852, a junior synonym of *Quadrula* Rafinesque, 1820.

Type Species
Obliquaria (*Quadrula*) *quadrula* Rafinesque, 1820

Diagnosis
Shell moderately thick; rectangular to square, triangular or round; umbos elevated above hinge line; surface usually sculptured, may be smooth; posterior ridge usually well-developed; hinge plate well-developed; umbo cavity usually deep.

Supra-anal aperture usually long; mantle bridge separating supra-anal and excurrent apertures short, sometimes absent; excurrent aperture smooth; inner lamellae of inner gills connected to visceral mass only anteriorly; all 4 gills marsupial; glochidia held across entire gill; marsupium slightly padded when gravid, not extended beyond original gill ventral edge; glochidium without styliform hooks (Rafinesque, 1820; Simpson, 1900b, 1914; Ortmann, 1912a; Haas, 1969b).

Synonymy
Rotundaria Rafinesque, 1820
Theliderma Swainson, 1840
Orthonymus Agassiz, 1852
Tritogonia Agassiz, 1852
Amphinaias Crosse and Fischer *in* Fischer and Crosse, 1894
Bullata Frierson, 1927, *non* Jousseaume, 1875
Pustulosa Frierson, 1927, *nomen novum* for *Bullata* Frierson, 1927

Quadrula apiculata (Say, 1829)
Southern Mapleleaf

Quadrula apiculata – Upper figure: length 101 mm, UF 372437. Tombigbee River, river mile 289, between Lubbub Creek and Cochran Cutoff, Pickens County, Alabama, 8 July 2003. Lower figure: length 75 mm, UF 243905. Alabama River, west bank across from Isaac Creek Park, 500 m north of Claiborne Dam, Monroe County, Alabama, 26 August 1990. © Richard T. Bryant.

Shell Description

Length to 129 mm; moderately thick; moderately inflated; outline subquadrate to trapezoidal; posterior margin obliquely truncate, may be slightly concave; anterior margin rounded; dorsal margin convex; ventral margin broadly rounded anteriorly, concave posteriorly; posterior ridge variable, low to high, sharply angled; posterior slope low to steep, flat, sculptured with small pustules, often elongate and arranged in broken radial plications, extending from posterior ridge to margin; usually with narrow to wide sulcus just anterior of posterior ridge; disk covered with variable pustules, may

be regularly arranged or random, pustules usually present in sulcus, almost absent in occasional individuals; umbo narrow to broad, elevated above hinge line, umbo sculpture nodulous ridges; periostracum tawny to greenish brown, reddish brown or black.

Pseudocardinal teeth large, triangular, 2 divergent teeth in left valve, 1 tooth in right valve, may have accessory denticle anteriorly and/or posteriorly; lateral teeth short to moderately long, thick, straight to slightly curved, 2 in left valve, 1 in right valve; interdentum short, moderately wide; umbo cavity deep; nacre white.

Soft Anatomy Description

Mantle creamy white to tan or light brown, outside of apertures grayish brown to grayish tan; visceral mass creamy white or pearly white to tan; foot tan, often darker ventrally.

Gills creamy white to tan or light brown; dorsal margin straight to slightly sinuous, ventral margin elongate convex; gill length 45–65% of shell length; gill height 35–60% of gill length; outer gill height 50–75% of inner gill height; inner lamellae of inner gills connected to visceral mass only anteriorly.

Outer gills marsupial; glochidia held across entire gill; marsupium slightly padded when gravid; pale tan. This observation was based on a field observation of a single gravid individual late in the brooding period. Species of *Quadrula* typically use all four gills as marsupia.

Labial palps creamy white to tan; straight to slightly concave dorsally, convex ventrally, bluntly pointed distally; palp length 35–60% of gill length; palp height 40–50% of palp length, occasionally greater; distal 20–35% of palps bifurcate, occasionally less.

Incurrent and supra-anal apertures longer than excurrent aperture; supra-anal aperture usually longer than incurrent aperture, occasionally shorter or equal in length.

Incurrent aperture length 20–25% of shell length, occasionally less; creamy white to light tan within, may have sparse gray basal to papillae; papillae arborescent, may be deeply branched; papillae some combination of creamy white, tan, brown, gray and black, usually mottled, often with white tips.

Excurrent aperture length 10–15% of shell length; creamy white to tan within, marginal color band gray, grayish brown or black; margin crenulate, crenulations may be minute.

Supra-anal aperture length 15–35% of shell length; creamy white within, occasionally with sparse gray or black marginally; margin smooth; mantle bridge separating supra-anal and excurrent apertures imperforate, length 5–30% of supra-anal length, occasionally greater.

Specimens examined: Bogue Chitto, Dallas County (n = 3); Coosa River (n = 3); Tennessee River (n = 4); Tombigbee River (n = 3).

Glochidium Description

Length 65–80 μm; height 77–80 μm; without styliform hooks; outline subrotund; dorsal margin straight, ventral margin broadly rounded, anterior and posterior margins convex (Howells et al., 1996).

Similar Species

Some *Quadrula apiculata* resemble *Quadrula nobilis*. The sulcus of *Q. nobilis* generally has few pustules, if any, whereas the sulcus of *Q. apiculata* is moderately to highly pustulose, though occasional individuals have very few. Pustules on either side of the sulcus in *Q. nobilis* are usually in a radial row and are dorsoventrally compressed. The anterior pseudocardinal tooth in the left valve of *Q. nobilis* is often shorter than that of *Q. apiculata*.

Some *Quadrula apiculata* resemble *Quadrula rumphiana*. However, *Q. rumphiana* lacks pustules on the posterior ridge, which is typically pustulose on *Q. apiculata*. The posterior ridge of *Q. rumphiana* is often more prominent and the sulcus narrower than those of *Q. apiculata*.

Quadrula apiculata may also resemble *Quadrula metanevra* but lacks a narrow, elevated, knobby posterior ridge. Also, *Q. apiculata* has a wide, shallow sulcus down the shell disk, which is absent in *Q. metanevra*.

Quadrula apiculata may superficially resemble *Quadrula stapes*. However, the posterior ridge of *Q. stapes* is much sharper and the posterior slope much steeper than those of *Q. apiculata*. Also, the periostracum of *Q. stapes* is usually adorned with small dark green chevrons and triangles, which are lacking on the periostracum of *Q. apiculata*.

General Distribution

Quadrula apiculata occurs in Gulf Coast drainages from the Mobile Basin west to the Rio Grande drainage in Texas and Mexico (Howells et al., 1996). Its range extends north in the Mississippi Basin to southern Arkansas and Oklahoma (Vidrine, 1993). *Quadrula apiculata* was introduced to the lower Tennessee River in the 1980s (Parmalee and Bogan, 1998).

Alabama and Mobile Basin Distribution

Quadrula apiculata occurs in most of the Mobile Basin. However, the majority of records are from below the Fall Line. It occurs in the Tennessee River in northwestern Alabama where it was introduced.

Quadrula apiculata remains widespread and locally abundant in the Mobile Basin and Pickwick Reservoir of the Tennessee River.

Ecology and Biology

Quadrula apiculata occurs in a variety of habitats, from medium creeks to large rivers as well as floodplain lakes, sloughs and reservoirs. It may be found in swift to sluggish water in substrates ranging from mud to sand or gravel. *Quadrula apiculata* is often abundant in coastal waters of the Mobile Delta, where it occurs sympatrically with *Mytilopsis leucophaeata* (Dreissenidae) and *Rangia cuneata* (Mactridae), which are brackish water species.

Quadrula apiculata is a short-term brooder, gravid in spring and summer. In Texas it has been reported gravid from May to July and from June to

August, apparently dependent on weather conditions (Howells et al., 1996; Howells, 2000). Glochidial hosts of *Q. apiculata* are unknown.

Current Conservation Status and Protection

Quadrula apiculata was considered currently stable throughout its range by Williams et al. (1993). In Alabama Stansbery (1976) considered it a species of special concern, but Lydeard et al. (1999) reported it to be currently stable, and Garner et al. (2004) designated it a species of lowest conservation concern.

Quadrula apiculata is on the list of species that can be legally harvested in Alabama. It is protected by state regulations, the most important of which is a minimum size limit of $2^5/_8$ inches (67 mm) in shell height.

Remarks

Quadrula apiculata is one of the most conchologically variable unionids. Neel (1941)

compared it with *Quadrula quadrula* and *Quadrula rumphiana*. Shells that appear to be intermediate in form and sculpture between those two species and *Q. apiculata* are often encountered. Two "phases" of *Q. apiculata* were discussed by Neel (1941), the *apiculata* phase and the *aspera* phase. Within many Mobile Basin populations, individuals match both phases. *Quadrula apiculata* also closely resembles *Q. nobilis*, with which it occurs sympatrically the Mobile Basin. The taxonomic relationships of *Q. apiculata* to *Q. quadrula*, *Q. rumphiana* and *Q. nobilis* are unclear. Serb et al. (2003) reported all four of these taxa to be closely related, but the study did not include *Q. apiculata* from the Mobile Basin.

Most records of *Quadrula apiculata* are from below the Fall Line. However, it is currently one of the more common species in some riverine and overbank reaches of the Coosa River, suggesting that its numbers have increased since impoundment of the river.

Synonymy

Unio apiculatus Say, 1829. Say, 1829:309; Say, 1834:[no pagination], pl. 52
Type locality: New Orleans, Louisiana. Type specimen not found. Length not given in original description.

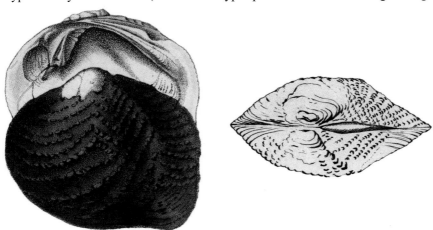

Unio asper Lea, 1831. Lea, 1831:85, pl. 9, fig. 15; Lea, 1834b:95, pl. 9, fig. 15
Type locality: Alabama River, [Alabama,] Judge Tait. Syntype, USNM 84212, length 45 mm.

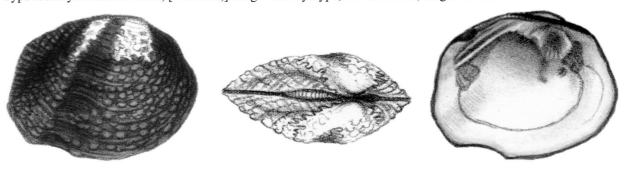

Unio forsheyi Lea, 1859. Lea, 1859c:155; Lea, 1860d:357, pl. 60, fig. 182; Lea, 1860g:39, pl. 60, fig. 182

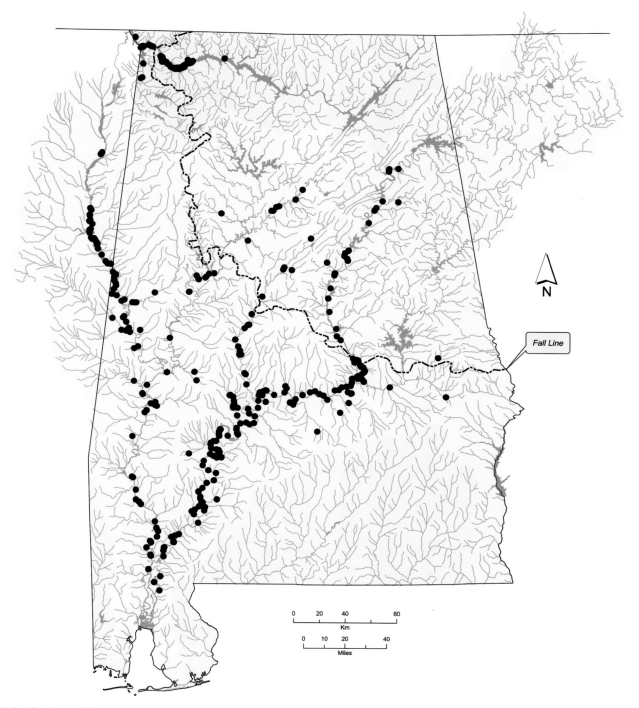

Distribution of *Quadrula apiculata* in Alabama and the Mobile Basin.

Quadrula asperata (Lea, 1861)
Alabama Orb

Quadrula asperata – Upper figure: length 38 mm, UF 68816. Lower figure: juvenile, length 20 mm, UF 68816. Conasauga River, Whitfield County, Georgia, 1 August 1915. © Richard T. Bryant.

Shell Description

Length to 90 mm; moderately thin to very thick; moderately inflated; outline round to subtriangular or subquadrate; posterior margin rounded to obliquely truncate; anterior margin rounded to truncate; dorsal margin convex; ventral margin rounded to almost straight; posterior ridge low, rounded; posterior slope moderately steep, flat to convex, usually sculptured with small corrugations or elongate pustules extending from posterior ridge to posteriodorsal margin; disk typically with variable pustules, often elongate, anterior 30% of disk usually without pustules, occasional individuals with pustules almost absent; umbo narrow to broad, compressed to inflated, turned slightly anteriad, elevated above hinge line, umbo sculpture nodulous ridges;

periostracum tawny to brown, may have concentric green bands on umbo.

Pseudocardinal teeth thick, triangular, 2 divergent teeth in left valve, anterior tooth larger and higher, 1 tooth in right valve, often with accessory denticle anteriorly and/or posteriorly; lateral teeth short, thick, straight to slightly curved, 2 in left valve, 1 in right valve; interdentum short, wide; umbo cavity deep, compressed; nacre white.

Soft Anatomy Description

Mantle tan to light brown, often pale orange outside of apertures; visceral mass pearly white or creamy white, occasionally tan; foot creamy white or tan, often darker ventrally.

Gills tan or light brown; dorsal margin slightly sinuous, occasionally straight, steeply sloped from umbo to near incurrent aperture, ventral margin deeply to slightly convex; gill length 45–60% of shell length; gill height 45–70% of gill length; outer gill height 60–80% of inner gill height; inner lamellae of inner gills connected to visceral mass only anteriorly.

All 4 gills marsupial; glochidia held across entire gill; marsupium slightly padded when gravid; creamy white when brooding embryos, becoming tan as glochidia mature.

Labial palps tan, often with irregular creamy white spot dorsally; straight to slightly concave dorsally, slightly to deeply convex ventrally, bluntly pointed distally; palp length 20–50% of gill length; palp height 45–80% of palp length; distal 30–40% of palps bifurcate.

Incurrent and supra-anal apertures longer than excurrent aperture; incurrent aperture longer, shorter or equal to supra-anal aperture (Figure 13.37).

Incurrent aperture length 20–30% of shell length; creamy white to tan within, occasionally with brown basal to papillae; papillae arborescent, short, some individuals with row of very short, simple papillae outside of arborescent papillae; papillae creamy white to rusty brown or dull orange, may be darker distally.

Excurrent aperture length 10–15% of shell length; creamy white to tan within, marginal color band reddish brown to black; margin usually crenulate or smooth and undulating, occasional individuals with small simple papillae; papillae same color as marginal color band.

Supra-anal aperture length 25–30% of shell length; creamy white within, may have thin black marginal band; margin smooth; mantle bridge separating supra-anal and excurrent apertures usually imperforate, length 5–15% of supra-anal length, occasionally greater.

Specimens examined: Alabama River (n = 3); Sipsey River (n = 3); Tombigbee River (n = 3).

Figure 13.37. *Quadrula asperata* apertures (left to right): supra-anal, excurrent, incurrent. Bogue Chitto, Dallas County, Alabama, July 2004. Photograph by A.D. Huryn.

Glochidium Description

Length 200–260 μm; height 180–200 μm; without styliform hooks; outline subelliptical; dorsal margin straight, ventral margin broadly rounded, anterior and posterior margins convex.

Similar Species

Quadrula asperata may be confused with *Quadrula kieneriana* in the Coosa River, where they occur sympatrically. *Quadrula asperata* is usually more pustulose than *Q. kieneriana* and lacks a sulcus or flattened area just anterior to the posterior ridge. However, some *Q. kieneriana* also lack these features.

Other species of *Quadrula* in the Mobile Basin (i.e., *Q. apiculata, Q. metanevra, Q. rumphiana, Q. stapes* and *Q. verrucosa*) are easily distinguishable from *Q. asperata* by the presence of a well-defined posterior ridge. *Quadrula apiculata, Q. rumphiana* and *Q. stapes* have a sulcus anterior of the posterior ridge, which is absent in *Q. asperata*.

General Distribution

Quadrula asperata is endemic to the Mobile Basin of Alabama, Georgia, Mississippi and Tennessee.

Alabama and Mobile Basin Distribution

Quadrula asperata occurs in all major drainages of the Mobile Basin, above and below the Fall Line. There is one record of this species from the Conecuh River near the Alabama and Florida state line, apparently the result of a recent introduction (D.N. Shelton, personal communication). There is no evidence of *Q. asperata* having established a reproducing population in the Conecuh River.

Quadrula asperata remains widespread in the Mobile Basin and may be locally abundant.

Ecology and Biology

Quadrula asperata occurs in large creeks to rivers, usually in areas with at least some current. It appears to be rare or absent in overbank habitats of reservoirs. *Quadrula asperata* is found in substrates composed of varying proportions of mud, sand and gravel.

Quadrula asperata was found to be variable among individuals within populations with regard to age at sexual maturity, ranging from three to nine years of age (Haag and Staton, 2003). A small percentage (1.2 percent) of individuals in Buttahatchee River, Mississippi, and Sipsey River, Alabama, populations was found to be infested with trematodes to the point of sterility (Haag and Staton, 2003). *Quadrula asperata* is a short-term brooder, gravid from mid-April to late July. Haag and Staton (2003) reported a high percentage (92 percent) of adult females to be gravid during the peak of the brooding period and a low percentage of undeveloped eggs in the marsupia (2.8 percent). Annual fecundity of *Q. asperata* was found to be highly variable, with an average of 9,647 glochidia per year, but ranged from 250 to 28,250 per individual (Haag and Staton, 2003). Glochidia are discharged freely, not as conglutinates, but are often suspended in mucus (Haag and Staton, 2003; Haag and Warren, 2003).

In laboratory trials *Ictalurus punctatus* (Channel Catfish) (Ictaluridae) was found to serve as a glochidial host of *Q. asperata*. *Noturus leptacanthus* (Speckled Madtom) (Ictaluridae) was found to be a marginal host (Haag and Warren, 2003).

Current Conservation Status and Protection

Quadrula asperata was considered a species of special concern throughout its range by Williams et al. (1993). Lydeard et al. (1999) considered it currently stable in Alabama, and Garner et al. (2004) designated it a species of lowest conservation concern in the state.

Remarks

Recognition of *Quadrula asperata* as distinctive, as opposed to being a synonym of *Quadrula pustulosa*, has been a matter of debate. Serb et al. (2003) reported *Q. asperata* and *Quadrula kieneriana* to form a distinct clade within the *Q. pustulosa* group.

A distinctive form of *Quadrula asperata*, described as *Quadrula archeri* Frierson, 1905, is restricted to the Tallapoosa River above the Fall Line (Figure 13.38). The taxonomic status (species or subspecies) of this form remains unclear. Its rarity has precluded genetic analyses. At a minimum it warrants subspecific status, as it is morphologically distinct, occurs in a geographically defined area and was isolated by the falls at Tallassee, Elmore County, Alabama, prior to impoundment of the Tallapoosa River. Evaluation of *Quadrula asperata archeri* is especially critical because additional impoundments in the upper Tallapoosa River drainage in Georgia could drive it to extinction. In a recent Tallapoosa River drainage mussel survey no live individuals or shells of *Q. asperata archeri* were found (Johnson and DeVries, 2002).

Figure 13.38. *Quadrula asperata archeri*, Tallapoosa Orb. Length 52 mm, UF 358673. Tallapoosa River, Alabama State Highway 49, Horseshoe Bend National Military Park, Tallapoosa County, Alabama, 10 October 1972. © Richard T. Bryant.

Synonymy

Unio asperatus Lea, 1861. Lea, 1861a:41; Lea, 1862b:68, pl. 7, fig. 218; Lea, 1862d:72, pl. 7, fig. 218
Type locality: Alabama River, Claiborne, [Monroe County,] Alabama, Judge Tait. Lectotype, USNM 84272, length
43 mm, designated by Johnson (1974).

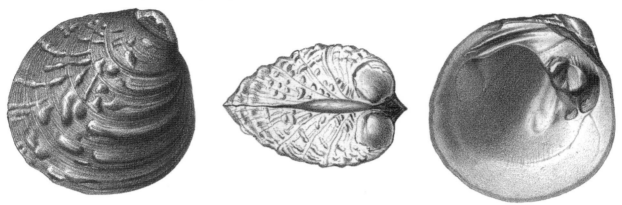

Unio pauperculus Lea, 1861. Lea, 1861a:39; Lea, 1862b:99, pl. 15, fig. 247; Lea, 1862d:103, pl. 15, fig. 247
Type locality: Stream near Columbus, [Lowndes County,] Mississippi, W. Spillman, M.D. Lectotype, USNM
85763, length 30 mm, designated by Johnson (1974).

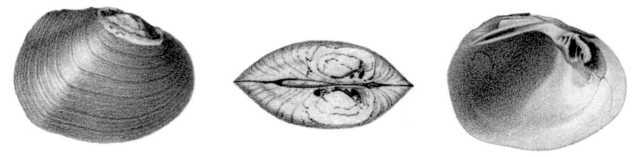

Unio vallatus Lea, 1868. Lea, 1868a:145; Lea, 1868c:315, pl. 50, fig. 128; Lea, 1869a:75, pl. 50, fig. 128
Type locality: Alabama River, [Alabama,] Dr. Showalter. Holotype, USNM 84275, length 60 mm.

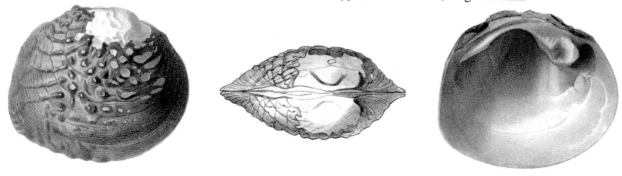

Unio cahabensis Lea, 1871. Lea, 1871:190; Lea, 1874c:17, pl. 5, fig. 14: Lea, 1874e:21, pl. 5, fig. 14
Type locality: Cahaba River, Shelby County, Alabama, E.R. Schowalter [Showalter], M.D. Lectotype, USNM
     84204, length 46 mm, designated by Johnson (1974).

Quadrula archeri Frierson, 1905. Frierson, 1905:13, pl. 1, figs. 1, 2
Type locality: Tallapoosa River, Tallassee, [Elmore County,] Alabama. Holotype, UMMZ 113538, length 34 mm.

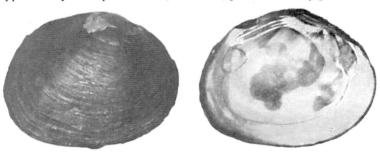

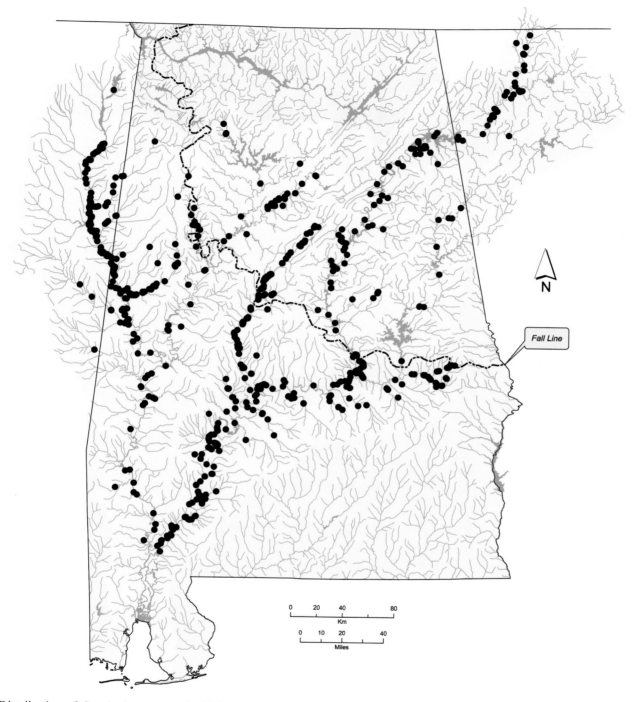

Distribution of *Quadrula asperata* in Alabama and the Mobile Basin.

Quadrula cylindrica (Say, 1817)
Rabbitsfoot

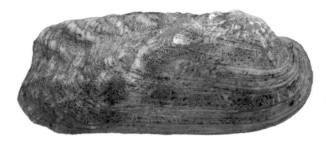 

Quadrula cylindrica – Length 92 mm, UF 68708. Tennessee River, Alabama. © Richard T. Bryant.

Shell Description

Length to 123 mm; thick; highly inflated, juveniles somewhat compressed; outline rectangular; posterior margin truncate to concave; anterior margin rounded; dorsal margin straight to slightly convex; ventral margin straight; posterior ridge well-developed; posterior slope wide, sometimes with shallow sulcus basal to posterior ridge; sculpture variable, consisting of radial series of large knobs along posterior ridge, posterior slope with ridges and large wrinkles, disk with small lachrymous tubercles, individuals from large rivers may lack all sculpture except knobs on posterior ridge, headwater form is more compressed and highly sculptured; umbo broad, somewhat inflated, elevated above hinge line, umbo sculpture irregular double-looped ridges, posterior loop angular and nodulous on posterior ridge; periostracum smooth, yellowish green to brown, usually with dark olive chevrons and triangles.

Pseudocardinal teeth moderately thick, 2 radially striate teeth in left valve, 1 tooth in right valve, usually with accessory denticle anteriorly, sometimes posteriorly; lateral teeth long, slightly curved, 2 in left valve, 1 in right valve; interdentum long, narrow; umbo cavity deep; nacre pearly white.

Soft Anatomy Description

Mantle tan to brown, may be golden brown outside of pallial line, inner surface with scattered black pigment giving overall grayish brown cast; visceral mass grayish brown; foot dark brown proximally, grading to bright orange distally. Lea (1863d) described the mantle margin as "saffron-yellow". Ortmann (1912a) reported a grayish foot, grading to black proximally, but did not indicate whether he was describing fresh or previously preserved specimens. Utterback (1915) reported the foot to have an orange background with black stripes. Vidrine (1993) reported "black flesh" without further comment.

Gills golden tan to grayish brown; dorsal margin straight to slightly sinuous, ventral margin straight to elongate convex; gill length 55–75% of shell length; gill height 25–40% of gill length; outer gill height 75–95% of inner gill height; inner lamellae of inner gills connected to visceral mass only anteriorly.

All 4 gills marsupial; glochidia held across entire gill; marsupium slightly padded when gravid; gold. Yeager and Neves (1986) found some individuals to be gravid in outer gills only.

Labial palps golden tan to rusty orange or grayish brown; straight to slightly concave dorsally, convex ventrally, bluntly pointed distally; palp length 15–30% of gill length; palp height 45–90% of palp length; distal 30–45% of palps bifurcate.

Incurrent and supra-anal apertures longer than excurrent aperture; incurrent aperture shorter or equal to supra-anal aperture.

Incurrent aperture length approximately 15% of shell length; gray to grayish brown within, dark rusty orange to dark grayish brown or black basal to papillae; papillae arborescent; papillae dark brown to grayish brown, some with white tips. Utterback (1915) described incurrent aperture papillae as "brownish yellow tentacles".

Excurrent aperture length 10–15% of shell length; creamy white to gray or grayish brown within, marginal color band rusty orange to dark brown, often with creamy white band proximally; margin smooth to crenulate.

Supra-anal aperture length 15–20% of shell length; creamy white to dull orange, rusty tan or grayish brown within, may be darker marginally; margin smooth; mantle bridge separating supra-anal and excurrent apertures often perforate, length 10–30% of supra-anal length, sometimes absent.

Specimens examined: Bear Creek, Colbert County (n = 1); Duck River, Tennessee (n = 3); Paint Rock River (n = 2).

Glochidium Description

Length 220 μm; height 220 μm; without styliform hooks; outline subrotund (Ortmann, 1919; Yeager and Neves, 1986).

Similar Species

The distinctive shape and sculpturing of *Quadrula cylindrica* make it easily distinguishable from all other mussels.

General Distribution

Quadrula cylindrica is found throughout the Ohio River drainage from headwaters in Pennsylvania to the mouth of the Ohio River (Ortmann, 1919; Cummings and Mayer, 1992). It is widespread in the Cumberland River drainage downstream of Cumberland Falls (Cicerello et al., 1991; Parmalee and Bogan, 1998) and in the Tennessee River drainage from headwaters in southwestern Virginia downstream to the mouth of the Tennessee River (Ahlstedt, 1992a, 1992b; Parmalee and Bogan, 1998). *Quadrula cylindrica* occurs in some tributaries of the lower Mississippi River from southeastern Kansas (Murray and Leonard, 1962) and Missouri (Oesch, 1995) south to Arkansas (Harris and Gordon, 1990), northern Louisiana (Vidrine, 1993) and Mississippi (Jones et al., 2005).

Alabama and Mobile Basin Distribution

Quadrula cylindrica is confined to the Tennessee River drainage. It historically occurred in the Tennessee River across the state and in some large tributaries. There are no museum records of *Q. cylindrica* from Alabama reaches of the Elk River, but it is known from Tennessee reaches of the river, so likely historically occurred in Alabama reaches (Ahlstedt, 1983).

The current range of *Quadrula cylindrica* in Alabama is limited to Bear Creek, Colbert County, and the Paint Rock River system. Though it is extirpated from the Tennessee River in Alabama, *Q. cylindrica* still occurs in the riverine reach downstream of Pickwick Dam, a short distance north of the Alabama state line.

Ecology and Biology

Quadrula cylindrica occurs in large creeks to large rivers. It is often found along margins of shoals in gravel substrate in slow to moderate current. In Pickwick Dam tailwaters on the Tennessee River it is most often encountered in muddy sand substrates on the submerged shelf along the river margin, in water approximately 2 m deep.

Quadrula cylindrica is a short-term brooder, gravid in spring and summer. Gravid females have been reported from late May to early July, with mature glochidia observed over the entire period, suggesting that the brooding period begins somewhat earlier (Ortmann, 1919). The reproductive biology of *Quadrula cylindrica strigillata* in headwaters of the Tennessee River was described by Yeager and Neves (1986). The sex ratio was roughly equal. Spermatogenesis occurs from August to July and oogenesis from late July to May. Spawning takes place in May and June. *Quadrula c. strigillata* is gravid from May to July. Its conglutinates are lanceolate and whitish to reddish brown but disintegrated upon discharge. Ortmann (1919) reported conglutinates of nominal *Q. cylindrica* from Pennsylvania to be lanceolate and yellowish brown or pale orange. Fertilization rates for *Q. c. strigillata* are high, with only 5 percent of conglutinate contents consisting of unfertilized eggs in May and June. Fecundity was reported to be 114,246 glochidia per female annually (Yeager and Neves, 1986).

Fishes identified as glochidial hosts of *Quadrula cylindrica strigillata* in laboratory trials include *Cyprinella galactura* (Whitetail Shiner), *Cyprinella spiloptera* (Spotfin Shiner) and *Hybopsis amblops* (Bigeye Chub) (Cyprinidae) (Yeager and Neves, 1986).

Current Conservation Status and Protection

Quadrula cylindrica was considered threatened throughout its range by Williams et al. (1993). In Alabama it was reported as endangered by Stansbery (1976), threatened by Lydeard et al. (1999) and a species of highest conservation concern by Garner et al. (2004). The headwater form, *Quadrula cylindrica strigillata*, was listed as endangered under the federal Endangered Species Act in 1997.

Remarks

Two subspecies of *Quadrula cylindrica* are currently recognized (Turgeon et al., 1998). *Quadrula cylindrica strigillata* is confined to the headwaters of the Tennessee River in northeastern Tennessee and southwestern Virginia. *Quadrula cylindrica cylindrica* occurs throughout the remainder of its range. *Quadrula c. strigillata* is characterized by being much more heavily sculptured and more compressed. However, shell characters gradually grade into the nominal *Q. cylindrica* form with a downstream progression. Thus, the *strigillata* form may not represent a valid subspecies.

Quadrula cylindrica is rare in prehistoric shell middens along the Tennessee River in northern Alabama (Morrison, 1942; Warren, 1975; Hughes and Parmalee, 1999).

Synonymy

Unio cylindricus Say, 1817. Say, 1817:[no pagination], pl. 4, fig. 3

Type locality: Wabash River. Haas (1930) designated a neotype, SMF 4310, length 95 mm, from "No. [North] America".

Unio naviformis Lamarck, 1819. Lamarck, 1819:75
Unio cylindricus var. *strigillatus* B.H. Wright, 1898. B.H. Wright, 1898c:6; Johnson, 1967b:9, pl. 3, fig. 2

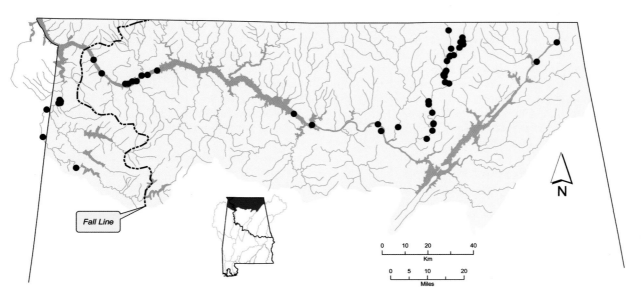

Distribution of *Quadrula cylindrica* in the Tennessee River drainage of Alabama.

Quadrula infucata (Conrad, 1834)
Sculptured Pigtoe

Quadrula infucata – Upper figure: length 50 mm, FMNH 22792. Kidson, near Hogansville, Troup County, Georgia. Lower figure: length 41 mm, UMMZ 163764. Cowikee Creek, near Batesville, Barbour County, Alabama, April 1917. © Richard T. Bryant.

Shell Description

Length to 63 mm; moderately thin to moderately thick; moderately compressed; outline subrhomboidal to subtriangular; posterior margin obliquely truncate; anterior margin rounded; dorsal margin convex; ventral margin straight to convex; posterior ridge rounded; posterior slope usually not steep, typically with fine interrupted plications extending from posterior ridge to margin; disk with variable sculpture, typically in form of pustules, often with small chevrons near umbo, sometimes with small irregular pustules scattered among other sculpture, sculpture may be absent; umbo broad, low, barely elevated above hinge line, umbo sculpture unknown; periostracum brown or greenish brown to black, without rays.

Pseudocardinal teeth moderately thick, triangular, 2 divergent teeth in left valve, 1 tooth in right valve, may have accessory denticle anteriorly and/or posteriorly; lateral teeth short, straight to slightly curved, 2 in left valve, 1 in right valve; interdentum short, narrow to moderately wide; umbo cavity shallow; nacre white.

Soft Anatomy Description

Mantle tan, outside of apertures gray, grayish tan or grayish brown; visceral mass tan, pearly white adjacent to foot; foot creamy white to pale pinkish orange, may be darker distally.

Gills tan with rusty or golden cast; dorsal margin slightly sinuous, ventral margin convex; gill length 50–60% of shell length; gill height 60–70% of gill length;

outer gill height 75–90% of inner gill height; inner lamellae of inner gills connected to visceral mass only anteriorly.

No gravid females were available for marsupium description; probably similar to other *Quadrula* species with all four gills marsupial, slightly padded when gravid.

Labial palps creamy white to light tan; straight to slightly concave dorsally, convex ventrally, bluntly pointed distally; palp length 25–45% of gill length; palp height 55–80% of palp length; distal 35–50% of palps bifurcate.

Incurrent aperture longer than excurrent and supra-anal apertures; supra-anal aperture usually longer than excurrent aperture.

Incurrent aperture length 20–30% of shell length; light tan within, may have sparse light brown or grayish brown basal to papillae; papillae arborescent; papillae light tan to brown, usually with grayish cast, edges may be dark brown basally.

Excurrent aperture length 15–20% of shell length; light tan within, marginal color band grayish brown; margin crenulate, crenulations may be weak. Lea (1863d) and Ortmann and Walker (1922a) described the excurrent aperture margin as smooth.

Supra-anal aperture length 20–25% of shell length; light tan to tan within, without marginal coloration; margin smooth; mantle bridge separating supra-anal and excurrent apertures imperforate, length 10–30% of supra-anal length, occasionally absent.

Specimens examined: Big Creek, Houston County (n = 4).

Glochidium Description

Length 234–242 μm; height 275–287 μm; without styliform hooks; outline subelliptical; dorsal margin straight, short, ventral margin rounded, anterior and posterior margins convex (Hoggarth, 1999).

Similar Species

With its distinctive subrhomboidal to subtriangular shell, which is typically adorned with pustules and small chevrons, *Quadrula infucata* is easily distinguishable from all other sympatric species.

General Distribution

Quadrula infucata is endemic to the Apalachicola Basin of Alabama, Florida and Georgia.

Alabama and Mobile Basin Distribution

Quadrula infucata is confined to the Chattahoochee River drainage and headwaters of the Chipola River in southeastern Alabama. In the Chattahoochee River drainage it occurs above and below the Fall Line.

Quadrula infucata is known to be extant only in headwaters of the Chipola River and Uchee Creek, a tributary of the Chattahoochee River.

Ecology and Biology

Quadrula infucata is often found in creeks with shallow sandy pools and rocky areas with moderate to swift current. It is also found in deep waters of rivers with sand, muddy sand and fine gravel substrates. Brim Box and Williams (2000) reported *Q. infucata* to occur primarily at sites with sand, limestone rock and detritus substrates. *Quadrula infucata* is rarely found in impounded waters (Brim Box and Williams, 2000).

Quadrula infucata is a short-term brooder, gravid in spring and summer (Brim Box and Williams, 2000). Its glochidial hosts are unknown.

Current Conservation Status and Protection

Quadrula infucata was considered a species of special concern throughout its range by Williams et al. (1993) and in Alabama by Lydeard et al. (1999). Garner et al. (2004) designated it a species of highest conservation concern in the state.

Remarks

Quadrula infucata appears to have been relatively common prehistorically and historically. At one archaeological site, a midden at Middle Weeden Island, Jackson County, Florida, on the Apalachicola River, it comprised more than 86 percent of the shells excavated (Percy, 1976). Heard (1964) also reported it to be present in considerable numbers in the Apalachicola River at Chattahoochee, Florida. Subsequently, Heard (1975b) noted that the population was drastically reduced in the Apalachicola River following a heavy invasion of *Corbicula fluminea*. Changes in substrate from sand to mostly *Corbicula* shells appear to have adversely affected the *Q. infucata* population in the Apalachicola River (R.S. Butler, personal communication).

Simpson (1900b, 1914) placed *infucata* in the genus *Quadrula*. Ortmann and Walker (1922a) suggested that the species belonged in the genus *Quincuncina*, where it remained until Serb et al. (2003) reported it to belong to *Quadrula*, based on genetic analyses.

Synonymy

Unio infucatus Conrad, 1834. Conrad, 1834b:45, pl. 3, fig. 2
Type locality: Flint River, Georgia. Figured type not found, also not found by Johnson and Baker (1973). Measurement of figured shell not given in original description.

Unio securiformis Conrad, 1849. Conrad, 1849:152; Conrad, 1850:275, pl. 37, fig. 1
Type locality: Flint River, Georgia. Figured type not found, also not found by Johnson and Baker (1973). Length of figured shell in original description reported as 38 mm.

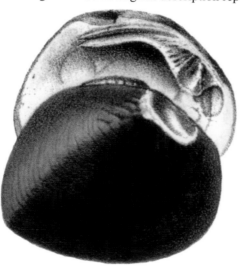

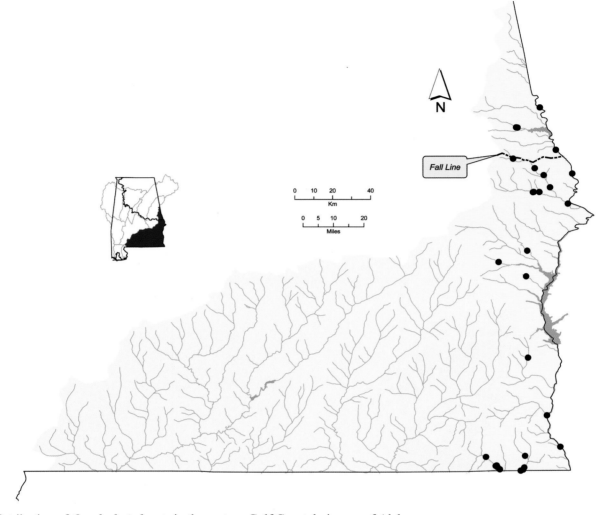

Distribution of *Quadrula infucata* in the eastern Gulf Coast drainages of Alabama.

Quadrula intermedia (Conrad, 1836)
Cumberland Monkeyface

Quadrula intermedia – Upper figure: length 42 mm, UF 270067. Tennessee River, Florence, Lauderdale County, Alabama. Lower figure: length 35 mm, UNA 722. Elk River, U.S. Highway 64, Lincoln County, Tennessee, 28 December 1970. © Richard T. Bryant.

Shell Description

Length to 80 mm; moderately thick; compressed; outline round to subquadrate; posterior margin emarginate; anterior margin rounded; dorsal margin convex; ventral margin convex; posterior ridge low, rounded; posterior slope wide, flat to slightly convex, with variable tubercles, often in form of short, interrupted, nodulous corrugations extending from posterior ridge to posteriodorsal margin, with narrow sulcus basal to posterior ridge; all but anterior 25% of disk sculptured with numerous lachrymous tubercles of variable size and

shape; umbo broad, low, elevated slightly above hinge line, umbo sculpture coarse, irregular, nodulous loops; periostracum yellowish green to brown, usually with dark olive chevrons and triangles.

Pseudocardinal teeth large, triangular, erect, 2 divergent teeth in left valve, 1 tooth in right valve, may have accessory denticle anteriorly and/or posteriorly; lateral teeth short, straight, 2 in left valve, 1 in right valve, may also have rudimentary lateral tooth ventrally in right valve; interdentum moderately short, moderately wide; umbo cavity deep, compressed; nacre white, sometimes with salmon tint.

Soft Anatomy Description

Mantle tan, with heavy grayish brown cast on inside surface, visible through translucent mantle tissue, grayish brown cast disappears outside of pallial line anteriorly; visceral mass grayish brown; foot tan, demarcation between grayish brown visceral mass and tan foot abrupt.

Gills tan; dorsal margin straight, ventral margin elongate convex; gill length approximately 70% of shell length; gill height approximately 50% of gill length; outer gill height approximately 75% of inner gill height; inner lamellae of inner gills connected to visceral mass only anteriorly.

No gravid females were available for marsupium description; probably similar to other *Quadrula* species, with all four gills marsupial, slightly padded when gravid.

Labial palps tan; straight dorsally, convex ventrally, bluntly pointed distally; palp length approximately 15% of gill length; palp height approximately 55% of palp length; distal 35% of palps bifurcate.

Supra-anal aperture much longer than incurrent and excurrent apertures; incurrent aperture much longer than excurrent aperture.

Incurrent aperture length approximately 20% of shell length; grayish brown within, slightly darker basal to papillae; papillae arborescent; papillae dark brown basally, rusty brown distally.

Excurrent aperture length approximately 10% of shell length; grayish brown within, marginal color band dark grayish brown and rusty brown, creamy white band proximal to marginal band slightly narrower than marginal band; margin smooth.

Supra-anal aperture length approximately 45% of shell length; grayish brown within, thin rusty brown marginal band; margin smooth; mantle bridge separating supra-anal and excurrent apertures absent.

Specimen examined: Powell River, Tennessee (n = 1) (specimen previously preserved).

Glochidium Description

Glochidium unknown.

Similar Species

Quadrula intermedia resembles *Quadrula sparsa* but has more numerous pustules. Also, *Q. sparsa* has a more prominent posterior ridge and is generally more elongate than *Q. intermedia*.

Quadrula intermedia may superficially resemble *Quadrula metanevra* but is much more compressed and lacks an elevated, knobby posterior ridge.

General Distribution

Quadrula intermedia is endemic to the Tennessee River drainage of Alabama, Tennessee and Virginia (Ahlstedt, 1992b; Parmalee and Bogan, 1998).

Alabama and Mobile Basin Distribution

Quadrula intermedia is known only from the Tennessee River across northern Alabama and the Elk River.

Quadrula intermedia is extirpated from Alabama and has not been reported from the state since the early 1900s.

Ecology and Biology

Quadrula intermedia is a species of flowing water in medium to large rivers. It generally occurs in substrates comprised of gravel with interstitial sand.

Quadrula intermedia is a short-term brooder, gravid from April through June (Yeager and Saylor, 1995). Individuals have been observed lying partially covered on the substrate from April to June but usually remain well-buried during the remainder of the year. This behavior is presumably related to reproductive activities (S.A. Ahlstedt, personal communication).

Fishes shown to serve as glochidial hosts of *Quadrula intermedia* in laboratory trials are *Erimystax dissimilis* (Streamline Chub) and *Erimystax insignis* (Blotched Chub) (Cyprinidae) (Yeager and Saylor, 1995).

Current Conservation Status and Protection

Quadrula intermedia was considered endangered throughout its range by Stansbery (1970a) and Williams et al. (1993). In Alabama it was reported as endangered by Stansbery (1976) but listed as extirpated by Lydeard et al. (1999) and Garner et al. (2004). *Quadrula intermedia* was listed as endangered under the federal Endangered Species Act in 1976.

In 2001 *Quadrula intermedia* was included on a list of species approved for a Nonessential Experimental Population in tailwaters of Wilson Dam on the Tennessee River. However, no reintroductions had taken place as of 2007.

Remarks

Quadrula intermedia is rare in prehistoric shell middens along the Tennessee River in northern Alabama (Morrison, 1942; Warren, 1975; Hughes and Parmalee, 1999). At Muscle Shoals Morrison (1942) reported a few typical *Q. intermedia*, which has a compressed shell, as well as a more common inflated form. Morrison (1942) described the inflated form as *Quadrula biangulata*. Shells similar to the *Q. biangulata* form were recovered from archaeological remains in the lower Clinch River, Tennessee, and identified as *Quadrula sparsa* (Parmalee and Bogan,

1986). Subsequently, Parmalee and Bogan (1998) placed *Q. biangulata* in the synonymy of *Quadrula quadrula*. Whether or not the *biangulata* form represents a valid species is unclear. It is conchologically different from typical *Q. intermedia* but appears to represent one extreme in range of variation of this species (i.e., inflated, with larger tubercles over the entire shell). It was apparently rare even prior to impoundment of the river and is not well-represented in museum collections. This paucity of historical material and extinction of the form make resolution of its specific status unlikely.

Synonymy

Unio intermedius Conrad, 1836. Conrad, 1836:63, pl. 35, fig. 2

Type locality: Nolichucky River, Tennessee. Lectotype, ANSP 41606, length 44 mm, designated by Johnson and Baker (1973). Measurement of figured shell not given in original description.

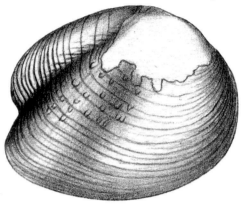

Unio tuberosus perlobatus De Gregorio, 1914. De Gregorio, 1914:9

Quadrula biangulata Morrison, 1942. Morrison, 1942:356

Type locality: [Tennessee River,] Tuscumbia, [Colbert County,] Alabama. Holotype, USNM 84221, length 33 mm.

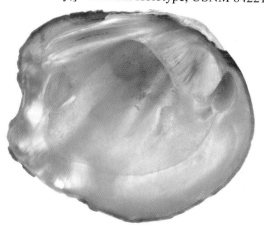

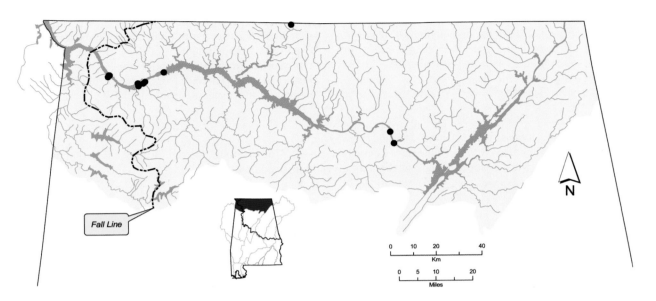

Distribution of *Quadrula intermedia* in the Tennessee River drainage of Alabama.

Quadrula kieneriana (Lea, 1852)
Coosa Orb

Quadrula kieneriana – Upper figure: length 46 mm, UF 66910. Coosa River, Three Island Shoals near Wilsonville, Shelby County, Alabama, November 1911. Lower figure: juvenile, length 25 mm, UF 68906. Holly Creek, Conasauga Springs, Murray County, Georgia, 8 August 1914. © Richard T. Bryant.

Shell Description

Length to 58 mm; thick; moderately to highly inflated; outline oval to subtriangular; posterior margin narrowly rounded to obliquely truncate; anterior margin rounded to truncate; dorsal margin convex; ventral margin convex; posterior ridge rounded; posterior slope moderately steep; most individuals with very shallow sulcus or flattened area just anterior of posterior ridge; pustules absent on disk, if present sparse, small and low; umbo broad, elevated slightly above hinge line, umbo sculpture irregular knobby loops; periostracum tawny to brown, often with concentric green blotches on umbo and disk, disappearing ventrally.

Pseudocardinal teeth thick, triangular, 2 divergent teeth in left valve, 1 tooth in right valve, may have accessory denticle anteriorly and/or posteriorly; lateral teeth short, thick, straight, 2 in left valve, 1 in right valve; interdentum short to moderately long, wide; umbo cavity deep, compressed; nacre white.

Soft Anatomy Description

Soft anatomy unknown.

Glochidium Description

Glochidium unknown.

Similar Species

Quadrula kieneriana resembles *Quadrula asperata* in the Coosa River, where they occur sympatrically. *Quadrula asperata* is usually more pustulose than *Q. kieneriana* and lacks a sulcus or flattened area, but some *Q. kieneriana* may also lack these features.

Quadrula kieneriana may closely resemble several species of *Pleurobema* (e.g., *P. fibuloides* and *P. hartmanianum*). However, the umbo cavity of *Q. kieneriana* is deep and compressed, whereas the umbo cavity of similar Mobile Basin *Pleurobema* is shallow. Also, those species never have pustules, which are often present on *Q. kieneriana*.

General Distribution

Quadrula kieneriana is endemic to the Coosa River drainage in Alabama and Georgia.

Alabama and Mobile Basin Distribution

Quadrula kieneriana occurs from the Conasauga and Etowah rivers, Georgia, downstream in the Coosa River to the Fall Line at Wetumpka.

Quadrula kieneriana is extant in the Conasauga and Oostanaula rivers in Georgia.

Ecology and Biology

Quadrula kieneriana occurs in medium to large rivers in moderate to swift current. Its preferred substrate is a mixture of sand and gravel.

Quadrula kieneriana is presumably a short-term brooder. Its glochidial hosts are unknown.

Current Conservation Status and Protection

Quadrula kieneriana has not been recognized in recent literature and has thus not been included in conservation status reviews. The range of *Q. kieneriana* is reduced and fragmented, but the species does not appear to be in imminent danger of extinction.

Remarks

Quadrula kieneriana has not been recognized since Simpson (1914), who considered it a variety of *Quadrula pustulosa*. It occurs sympatrically with *Quadrula asperata* in the Coosa River drainage and is herein recognized as distinct based on shell morphology. Serb et al. (2003) included one individual of *Q. kieneriana* in a genetic analysis of *Quadrula*, but that analysis did not resolve the relationship between *Q. kieneriana* and *Q. asperata*. More detailed examination of the relationship of *Q. kieneriana* and *Q. asperata* is needed to resolve the taxonomic status of these species. If they prove to be synonymous, *Q. kieneriana* (Lea, 1852) takes priority over *Q. asperata* (Lea, 1861).

Quadrula kieneriana was named for Louis Charles Kiener, author of "Spécies général et des coquilles vivantes comprenant la collection du Museum d'Histoire naturelle de Paris". Lea originally misspelled the specific name as *keineriana*, but changed it to *kieneriana* in the fourth edition of his synopsis (Lea, 1870). This was noted as *lapsus calami* by Simpson (1914).

Synonymy

Unio keinerianus Lea, 1852, *nomen nudum*. Lea, 1852a:251

Unio keineriana Lea, 1852. Lea, 1852b:281, pl. 23, fig. 40; Lea, 1852c:37, pl. 23, fig. 40

Type locality: Coosawattee River, Murray County, Georgia, Dr. Boykin. Lectotype, USNM 84273, length 32 mm, designated by Johnson (1974).

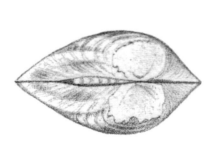
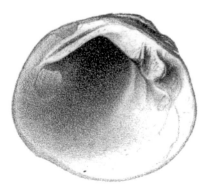

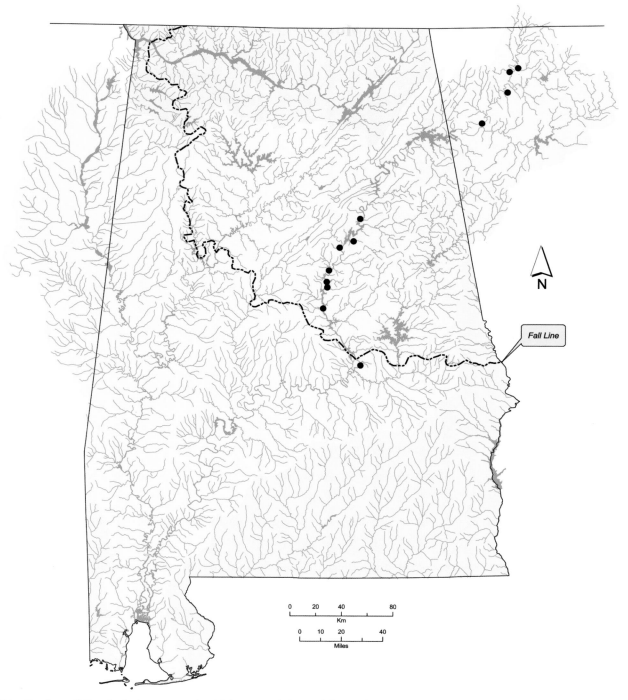

Distribution of *Quadrula kieneriana* in Alabama and the Mobile Basin.

Quadrula metanevra (Rafinesque, 1820)
Monkeyface

Quadrula metanevra – Length 73 mm, UF 68744. Tennessee River, Muscle Shoals, Lauderdale County, Alabama, November 1909. © Richard T. Bryant.

Shell Description

Length to 110 mm; thick; moderately inflated; outline quadrate to rhomboidal; posterior margin emarginate to obliquely truncate; anterior margin rounded; dorsal margin straight to slightly convex; ventral margin rounded; posterior ridge high, elevated above disk, narrow dorsally, broader posterioventrally, usually adorned with large knobs (Tennessee River drainage individuals tend to have better defined knobs than those from Mobile Basin); posterior slope steep, becoming flatter posterioventrally, often with narrow, shallow sulcus at base of posterior ridge, usually terminating in emargination on posterior margin; surface of posterior slope may be sculptured with small pustules or corrugations that extend from posterior ridge to margin; disk sculptured with variable pustules, often elongate; umbo narrow, inflated, elevated well above hinge line, umbo sculpture short ridges, nodulous posteriorly; periostracum tawny to greenish brown or dark brown, often with small green chevrons and triangles, especially in young individuals.

Pseudocardinal teeth thick, triangular, striate, 2 divergent teeth in left valve, 1 tooth in right valve, usually with well-developed accessory denticle anteriorly and posteriorly; lateral teeth short, thick, 2 in left valve, usually 1 in right valve, may also have rudimentary lateral tooth ventrally or dorsally in right valve; interdentum short, wide; umbo cavity deep; nacre white.

Soft Anatomy Description

Mantle creamy white to tan or light brown, outside of apertures dull orange with black mottling; visceral mass pearly white to creamy white or pale tan; foot creamy white to tan or pale pinkish orange, usually darker distally. Ortmann (1912a) reported the mantle edge to be "more or less blackish or brownish" posteriorly and the visceral mass to often be suffused with black.

Gills creamy white to tan or light brown; dorsal margin straight to slightly sinuous, steeply sloped from umbo to near apertures, ventral margin elongate convex; gill length 55–65% of shell length; gill height 30–50% of gill length; outer gill height 65–95% of inner gill height; inner lamellae of inner gills connected to visceral mass only anteriorly.

No gravid females were available for marsupium description. Ortmann (1912a) and Utterback (1915) reported all four gills to be marsupial, padded when gravid.

Labial palps creamy white to tan; straight to slightly convex dorsally, convex ventrally, rounded to bluntly pointed distally; palp length 20–30% of gill length; palp height 65–90% of palp length; distal 35–45% of palps bifurcate.

Incurrent aperture considerably longer than excurrent aperture; supra-anal aperture considerably longer than incurrent aperture.

Incurrent aperture length 20–25% of shell length; creamy white within, may have rusty orange basal to papillae; papillae arborescent; papillae tan to rusty orange, may have some creamy white areas, may be brown distally.

Excurrent aperture length 10–15% of shell length; creamy white within, marginal color band orange to rusty brown, may have some grayish brown; margin smooth to slightly crenulate.

Supra-anal aperture length 30–45% of shell length; creamy white to grayish brown within, may have thin rusty orange or dark brown marginal band; margin smooth; mantle bridge separating supra-anal and excurrent apertures imperforate, length 5–15% of supra-anal length, sometimes less.

Specimens examined: Alabama River (n = 1); Tennessee River (n = 3).

Glochidium Description

Length 170–180 μm; height 185–200 μm; without styliform hooks; outline subrotund; dorsal margin straight to slightly concave, ventral margin rounded, anterior and posterior margins convex (Lefevre and Curtis, 1910b; Surber, 1912; Howard, 1914b). Ortmann (1919) reported glochidia of individuals from headwaters (*wardii* form) to be slightly larger (length 200 μm, height 220 μm) than those of the nominal form.

Similar Species

In the Tennessee River drainage *Quadrula metanevra* may superficially resemble *Quadrula apiculata*, *Quadrula intermedia*, *Quadrula quadrula* and *Quadrula sparsa*. However, none of those species have a prominent, knobby posterior ridge. Also, *Q. intermedia* and *Q. sparsa* are much more compressed. *Quadrula apiculata* and *Q. quadrula* have a wide, shallow sulcus down the shell disk that is absent in *Q. metanevra*.

In the Mobile Basin *Quadrula metanevra* may superficially resemble *Quadrula apiculata*, *Quadrula rumphiana* and *Quadrula stapes*. All of those species lack a narrow, elevated, knobby posterior ridge. The posterior ridge of *Q. apiculata* is usually adorned with small pustules, and that of *Q. rumphiana* is usually smooth. Additionally, *Q. apiculata* and *Q. rumphiana* have a shallow sulcus down the shell disk that is absent in *Q. metanevra*. The posterior slope of *Q. stapes* is much shorter and steeper than that of *Q. metanevra*.

General Distribution

Quadrula metanevra occurs in the Mississippi Basin from southern Minnesota (Dawley, 1947) south to the Ouachita River in Louisiana (Vidrine, 1993) and from headwaters of the Ohio River drainage in western Pennsylvania (Ortmann, 1919) west to southeastern Kansas (Murray and Leonard, 1962). It is widespread in the Cumberland River drainage downstream of Cumberland Falls (Cicerello et al., 1991; Parmalee and Bogan, 1998). In the Tennessee River drainage *Q. metanevra* occurs from headwaters in eastern Tennessee downstream to the mouth of the Tennessee River (Parmalee and Bogan, 1998). *Quadrula metanevra* is also widespread in the Mobile Basin of Alabama, Georgia and Mississippi.

Alabama and Mobile Basin Distribution

Quadrula metanevra historically occurred in the Tennessee River across northern Alabama and some large tributaries. In the Mobile Basin most records of this species are from below the Fall Line, with the exception of the Coosa River where it is known to occur upstream to headwaters in northwestern Georgia.

The only remaining populations of *Quadrula metanevra* in the Tennessee River drainage are in tailwaters of Guntersville and Wilson dams and extreme upper reaches of Guntersville Reservoir, a short distance downstream of Nickajack Dam. It is also extant in a very short reach of the Elk River at the Alabama and Tennessee state line, and lower reaches of the Paint Rock River. In the Mobile Basin *Q. metanevra* is extant in some riverine reaches of the Alabama, Cahaba and Tombigbee rivers. However, there has been no evidence of robust recruitment in any Alabama population of *Q. metanevra* during the past decade.

Ecology and Biology

Quadrula metanevra occurs in medium to large rivers in flowing water. It can be found in substrates comprised of varying mixtures of sand and gravel.

In a Tennessee River population of *Quadrula metanevra*, Pickwick Dam tailwaters, Hardin County, Tennessee, the sex ratio was skewed toward females at 1.5 to 1. *Quadrula metanevra* is typically dioecious, but 2 percent of the Tennessee River population was hermaphroditic (Garner et al., 1999). In the Tennessee River, gamete production began during autumn in both sexes, but slowed during winter. Spawning began in early March, during which time gamete production increased, and gamete production and spawning continued through July (Garner et al., 1999). *Quadrula metanevra* is a short-term brooder, reported gravid from May through July (Surber, 1912; Howard, 1914b; Ortmann, 1919; Garner et al., 1999).

Fishes found to serve as glochidial hosts of *Quadrula metanevra* in laboratory trials include *Cyprinella spiloptera* (Spotfin Shiner), *Pimephales notatus* (Bluntnose Minnow), *Rhinicthys atratulus* (Eastern Blacknose Dace) and *Semitolus atromaculatus* (Creek Chub) (Cyprinidae) (Crownhart et al., 2006). Fishes reported to serve as glochidial hosts of *Q. metanevra* based on observations of natural infestations include *Lepomis cyanellus* (Green Sunfish) and *Lepomis macrochirus* (Bluegill) (Centrarchidae); and

Sander canadensis (Sauger) (Percidae) (Surber, 1913; Howard, 1914b; Coker et al., 1921).

Current Conservation Status and Protection

Quadrula metanevra was considered currently stable throughout its range by Williams et al. (1993) and in Alabama by Lydeard et al. (1999). Garner et al. (2004) designated it a species of moderate conservation concern in the state.

In 2004 *Quadrula metanevra* was removed from the list of legally harvestable mussels in Alabama due to its decline statewide.

Remarks

Populations of *Quadrula metanevra* in Alabama have declined considerably during the past century. However, a population in tailwaters of Pickwick Dam, Hardin County, Tennessee, on the Tennessee River appears to be stable.

Quadrula metanevra is uncommon in prehistoric shell middens along the Tennessee River in northern Alabama (Morrison, 1942; Warren, 1975; Hughes and Parmalee, 1999).

Synonymy

Obliquaria (Quadrula) metanevra Rafinesque, 1820. Rafinesque, 1820:305, pl. 81, figs. 15, 16

Type locality: Ohio River. Syntypes, ANSP 20238, lengths 31 and 87 mm, recognized by Vanatta (1915). Johnson and Baker (1973) designated a neotype, ANSP 20238a, but Johnson (1973) subsequently designated this same lot as a lectotype.

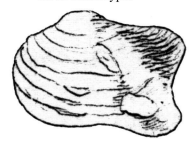 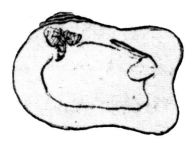

Unio nodosus Barnes, 1823. Barnes, 1823:124, pl. 6, figs. 7a, b
Unio tuberosus Lea, 1840. Lea, 1840:286; Lea, 1842b:210, pl. 14, fig. 25; Lea, 1842c:48, 81, pl. 14, fig. 25
Unio wardii Lea, 1861. Lea, 1861d:392; Lea, 1862c:187, pl. 24, fig. 257; Lea, 1863a:9, pl. 24, fig. 257

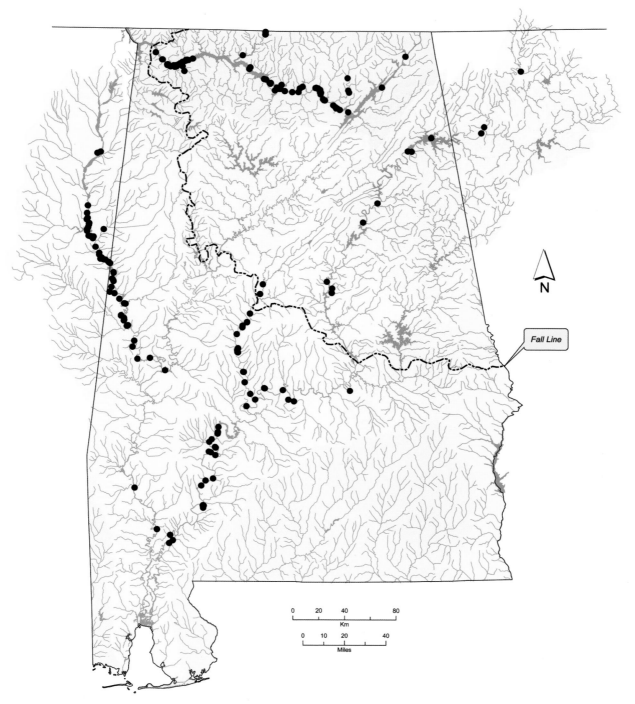

Distribution of *Quadrula metanevra* in Alabama and the Mobile Basin.

Quadrula nobilis (Conrad, 1853)
Gulf Mapleleaf

Quadrula nobilis – Upper figure: length 83 mm, UF 372438. Tombigbee River, river mile 289, between Lubbub Creek and Cochran Cutoff, Pickens County, Alabama, 8 July 2003. Lower figure: length 53 mm, UF 370414. Tombigbee River, 6 miles north of Gainesville, Green and Sumter counties, Alabama, 8 June 1972. © Richard T. Bryant.

Shell Description

Length to 130 mm; moderately thick; moderately inflated; outline subquadrate to trapezoidal; posterior margin obliquely truncate, may be slightly concave; anterior margin rounded; dorsal margin convex; ventral margin broadly rounded, concave at distal end of sulcus; posterior ridge variable, low to high, narrow dorsally, wider posterioventrally; posterior slope low to steep, slightly concave, sculptured with small pustules, often elongate and arranged in broken radial plications, extending from posterior ridge to margin; narrow to wide sulcus just anterior of posterior ridge, pustules in sulcus absent or sparse, usually restricted to ventral portion of sulcus when present, often with low medial ridge anterior of sulcus; disk with variable pustules, often present in rows on either side of sulcus, pustules adjacent to sulcus usually dorsoventrally compressed; umbo narrow to broad, elevated above hinge line, umbo sculpture nodulous ridges; periostracum brown or reddish brown to black.

Pseudocardinal teeth large, triangular, 2 divergent teeth in left valve, 1 tooth in right valve, may have accessory denticle anteriorly and/or posteriorly; lateral teeth short to moderately long, moderately thick, straight to slightly curved, 2 in left valve, 1 in right valve, may also have rudimentary lateral tooth ventrally in right valve; interdentum short, moderately wide; umbo cavity deep, not compressed; nacre white.

Soft Anatomy Description

Mantle creamy white to pale tan, outside of apertures light to dark grayish brown; visceral mass creamy white to pale tan; foot pale tan.

Gills creamy white to pale tan; dorsal margin slightly sinuous, ventral margin elongate convex, may be straight centrally; gill length 40–65% of shell length; gill height 40–55% of gill length; outer gill height 55–65% of inner gill height; inner lamellae of inner gills connected to visceral mass only anteriorly.

No gravid females were available for marsupium description; probably similar to other *Quadrula* species, with all four gills marsupial, slightly padded when gravid.

Labial palps creamy white; slightly concave dorsally, convex ventrally, bluntly pointed distally; palp length 40–50% of gill length; palp height 40–50% of palp length; distal 20–35% of palps bifurcate.

Supra-anal aperture longer than incurrent and excurrent apertures; incurrent aperture longer than excurrent aperture.

Incurrent aperture length 15–20% of shell length; creamy white within, without coloration basal to papillae; papillae arborescent; papillae grayish brown to dark gray.

Excurrent aperture length 10–15% of shell length; creamy white within, marginal color band grayish brown, extreme margin may be creamy white; margin usually crenulate, occasional individuals with very small simple papillae; papillae pale tan to grayish brown.

Supra-anal aperture length 20–45% of shell length; creamy white within, without marginal coloration; margin smooth; mantle bridge separating supra-anal and excurrent apertures imperforate, length 5–10% of supra-anal length.

Specimens examined: Alabama River (n = 4).

Glochidium Description

Glochidium unknown.

Similar Species

Quadrula nobilis resembles some *Quadrula apiculata*. The sulcus of *Q. nobilis* generally has few pustules, if any, whereas the sulcus of *Q. apiculata* is pustulose. *Quadrula apiculata* never has radial rows of dorsoventrally compressed pustules on either side of the sulcus, as does *Q. nobilis*. The anterior pseudocardinal tooth in the left valve of *Q. nobilis* is often shorter than that of *Q. apiculata*.

Some *Quadrula nobilis* resemble *Quadrula rumphiana*. However, *Q. rumphiana* lacks pustules on the posterior ridge, which is typically pustulose on *Q. nobilis*. The posterior ridge of *Q. rumphiana* is often more prominent and the sulcus narrower than those of *Q. nobilis*.

General Distribution

Quadrula nobilis was not recognized in most modern literature, so its distribution is poorly understood. It occurs in Gulf Coast drainages from the Mobile Basin west to the San Jacinto River drainage, Texas (Howells et al., 1996; Jones et al., 2005). In the Mississippi Basin its range extends north at least to the Ohio River, northwestern Kentucky (Serb et al., 2003). Mobile Basin *Quadrula apiculata* were introduced to the lower Tennessee River in the 1980s (Parmalee and Bogan, 1998). It is unclear whether *Q. nobilis* was included in the introduction, but some of the phenotypes now found in Pickwick Reservoir resemble *Q. nobilis*.

Alabama and Mobile Basin Distribution

Quadrula nobilis occurs in most of the Mobile Basin, primarily below the Fall Line. There are a few records of this species from above the Fall Line in the Coosa River and Black Warrior River drainages.

Quadrula nobilis remains widespread and locally common in the Mobile Basin.

Ecology and Biology

Quadrula nobilis occurs in large rivers. It may be found in swift to sluggish water in substrates ranging from mud to sand or gravel.

Quadrula nobilis is presumably a short-term brooder, gravid in spring and summer. However, Howells (2000) reported finding no gravid females in approximately 50 individuals examined, mostly May through September in Texas. Howells (2000) did find two gravid females in the Neches River in January, but other females in the same collection were not gravid and had oocytes in early stages of development.

Fishes reported to serve as glochidial hosts of *Quadrula nobilis* in laboratory trials include *Ictalurus punctatus* (Channel Catfish) and *Pylodictis olivaris* (Flathead Catfish) (Ictaluridae) (Howells, 1997).

Current Conservation Status and Protection

Quadrula nobilis has not been recognized as a valid species in modern conservation status assessments. However, it appears to be currently stable, at least in Alabama.

Remarks

Quadrula nobilis belongs to a group of mussels that includes some of the most conchologically variable unionids. This contributed to its neglect by malacologists since Frierson (1927) included it as a subspecies of *Q. quadrula*. Howells et al. (1996) appears to be the first recent publication to recognize *Q. nobilis* as a valid species. Though intermediate phenotypes between *Q. nobilis* and *Q. apiculata* occur in many populations, *Q. nobilis* was found to be genetically distinct (Serb et al., 2003; Campbell et al., 2005).

Synonymy

Unio nobilis Conrad, 1853. Conrad, 1853:297, pl. 27, figs. 2, 3

Type locality: Bayou Teche, Louisiana. Lectotype, ANSP 43019a, length 114 mm.

Comment: Conrad (1853) included figures of two specimens. The largest specimen (fig. 3 in Conrad, 1853, reproduced below) was designated by Johnson and Baker (1973) as the lectotype. The second specimen (fig. 2 in Conrad, 1853) is *Quadrula apiculata*.

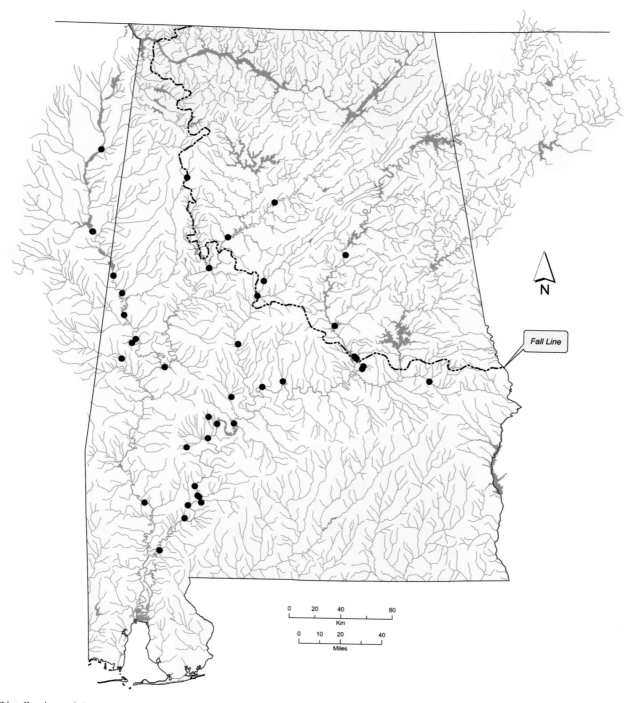

Distribution of *Quadrula nobilis* in Alabama and the Mobile Basin.

Quadrula pustulosa (Lea, 1831)
Pimpleback

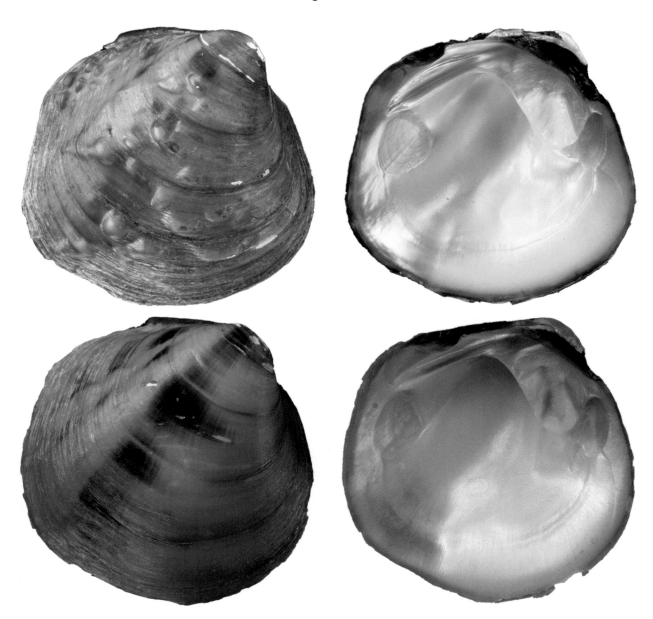

Quadrula pustulosa – Upper figure: length 52 mm, UMMZ 77228. Paint Rock River, between Paint Rock and New Hope, Jackson County, Alabama. Lower figure: juvenile, length 33 mm, UF 68864. Bear Creek, Burleson, Franklin County, Alabama, August 1909. © Richard T. Bryant.

Shell Description

Length to 80 mm; moderately thick; moderately inflated; outline usually round, occasionally subtriangular or quadrate; posterior margin rounded to obliquely truncate; anterior margin straight to convex; dorsal margin convex; ventral margin rounded; posterior ridge low, rounded; posterior slope broad, flat, often with round or oblong pustules; disk usually adorned with round or oblong pustules that range from very sparse to numerous, occasionally absent, anterior part of disk usually without sculpture; umbo somewhat narrow, moderately inflated, elevated above hinge line, umbo sculpture coarse ridges, somewhat nodulous on posterior ridge; periostracum usually satiny, tawny to reddish brown, typically with at least 1 broad, broken green ray, some individuals with ray confined to umbo.

Pseudocardinal teeth moderately thick, triangular, 2 divergent teeth in left valve, anterior tooth larger and higher than posterior tooth, 1 tooth in right valve, often with well-developed accessory denticle anteriorly and/or posteriorly; lateral teeth short, straight, moderately thick, 2 in left valve, 1 in right valve; interdentum short, wide; umbo cavity deep; nacre white.

Soft Anatomy Description

Mantle tan, occasionally dull orange outside of pallial line, grayish brown to golden brown outside of apertures; visceral mass creamy white; foot tan to pinkish orange.

Gills tan; dorsal margin straight to slightly concave or sinuous, ventral margin convex; gill length 60–75% of shell length; gill height 40–55% of gill length; outer gill height 60–80% of inner gill height; inner lamellae of inner gills connected to visceral mass only anteriorly.

All 4 gills marsupial; glochidia held across entire gill except extreme anterior and posterior ends; marsupium slightly padded when gravid; creamy white.

Labial palps tan; straight to slightly concave dorsally, convex ventrally, usually bluntly pointed distally, occasionally rounded; palp length 20–30% of gill length; palp height 60–85% of palp length; distal 40–55% of palps bifurcate.

Incurrent aperture considerably longer than excurrent aperture; supra-anal aperture usually considerably longer than incurrent aperture.

Incurrent aperture length 25–30% of shell length; creamy white within, may be tan basal to papillae; papillae arborescent, may be variable in size; papillae brown, may have rusty or grayish cast. Lea (1863d) described incurrent aperture papillae as "nearly white", and Utterback (1915) described them as "yellowish plumed tentacles".

Excurrent aperture length usually 10–20% of shell length; creamy white within, marginal band grayish brown; margin crenulate.

Supra-anal aperture length 30–35% of shell length, occasionally greater; creamy white within, may have very thin brown or grayish brown marginal color band; margin smooth; mantle bridge separating supra-anal and excurrent apertures imperforate, length 5–15% of supra-anal length.

Specimens examined: Tennessee River (n = 9).

Glochidium Description

Length 200–235 μm; height 250–320 μm; without styliform hooks; outline subelliptical; dorsal margin straight, ventral margin broadly rounded, anterior and posterior margins convex (Lefevre and Curtis, 1910b;

Ortmann, 1912a, 1919; Surber, 1912; Howard, 1914b; Utterback, 1915).

Similar Species

Quadrula pustulosa may resemble *Cyclonaias tuberculata*. *Quadrula pustulosa* typically has a wide green ray located centrally on the umbo and shell disk, while *C. tuberculata* is rayless or only weakly rayed in young individuals. *Quadrula pustulosa* lacks the dorsal wing found on many *C. tuberculata* shells. Shell nacre of *Q. pustulosa* is always white, whereas that of *C. tuberculata* is purple.

Quadrula pustulosa may also resemble *Plethobasus cooperianus*, but that species lacks a dark green ray across the umbo and shell disk. Also, the foot of *P. cooperianus* is orange, whereas the foot of *Q. pustulosa* is white or tan. *Plethobasus cooperianus* often has smaller and more numerous tubercles than *Q. pustulosa*. The *P. cooperianus* periostracum often has small wrinkles associated with the tubercles, but *Q. pustulosa* generally has few wrinkled tubercles.

General Distribution

Quadrula pustulosa occurs in eastern reaches of the Great Lakes Basin from Lake St. Clair to the Niagara River (La Rocque and Oughton, 1937; Clarke, 1981a). It is found through much of the Mississippi Basin from southern Minnesota (Dawley, 1947) south to Louisiana (Vidrine, 1993) and from Ohio River headwaters in western Pennsylvania (Ortmann, 1919) west to Nebraska (Hoke, 2000) and South Dakota (Backlund, 2000). Vidrine (1993) reported *Q. pustulosa* from the Mermentau and Calcasieu rivers just west of the Mississippi River in Louisiana. Howells et al. (1996) reported a form that may represent *Q. pustulosa* as far west as the Neches River, Texas. *Quadrula pustulosa* is widespread in the Cumberland River downstream of Cumberland Falls (Cicerello et al., 1991; Parmalee and Bogan, 1998). In the Tennessee River drainage it occurs from headwaters in southwestern Virginia, western North Carolina and eastern Tennessee, downstream to the mouth of the Tennessee River (Ahlstedt, 1992a, 1992b; Parmalee and Bogan, 1998). Coker and Southall (1915) reported *Q. pustulosa* from the southern Hudson Bay Basin, but this record was questioned by Graf (1997).

Alabama and Mobile Basin Distribution

Quadrula pustulosa occurs in the Tennessee River across northern Alabama. It was historically widespread in tributaries.

Quadrula pustulosa is extant and locally common in the Tennessee River across northern Alabama. Tributary populations are extant in the Elk and Paint Rock rivers, as well as Bear Creek in Colbert County. It can be found in impounded reaches of some tributaries.

Ecology and Biology

Quadrula pustulosa is a species of large creeks to large rivers. It occurs in shoals, runs and pools in substrates ranging from clean gravel to sand and gravel with a silt layer. *Quadrula pustulosa* also occurs in some parts of the Great Lakes and is found in overbank habitats of reservoirs, often in muddy substrates. It is one of the more common species in some overbank habitats of Tennessee River reservoirs as well as tailwaters of dams.

Quadrula pustulosa is a dioecious species, but a single hermaphroditic individual was reported from the Little Tallahatchie River, Mississippi (Haag and Staton, 2003). A low percentage of this species (2.7 percent) in the Little Tallahatchie River were infested with trematodes to the point of sterility (Haag and Staton, 2003). *Quadrula pustulosa* has been reported to reach sexual maturity as early as three years of age, but individuals may take as long as seven years to mature (Haag and Staton, 2003). It is a short-term brooder, gravid from May or June to July or August (Surber, 1912; Howard, 1914b; Utterback, 1916a; Ortmann, 1919). Haag and Staton (2003) reported 94 percent of mature females from the Little Tallahatchie River to be gravid during the peak of the brooding period, with only 3.4 percent of eggs undeveloped. Glochidia of *Q. pustulosa* are discharged singly, not packaged as conglutinates (Haag and Staton, 2003). *Quadrula pustulosa* fecundity has been reported to average 28,369 glochidia per year, but appears to be highly variable, ranging from 49 to 50,625 per individual (Haag and Staton, 2003).

Fishes reported to serve as *Quadrula pustulosa* glochidial hosts in laboratory trials include *Ameiurus melas* (Black Bullhead), *Ameiurus nebulosus* (Brown Bullhead), *Ictalurus punctatus* (Channel Catfish) and *Pylodictis olivaris* (Flathead Catfish) (Ictaluridae) (Howard, 1914b). Wilson (1916) reported *Pomoxis annularis* (White Crappie) (Centrarchidae) as a *Q. pustulosa* host but gave no details about its determination. Coker et al. (1921) reported a natural infestation on *Scaphirhynchus platorynchus* (Shovelnose Sturgeon) (Acipenseridae) but deemed it of doubtful significance, and Surber (1913) reported a natural infestation of *Pomoxis annularis* (White Crappie) (Centrarchidae) involving a single *Q. pustulosa* glochidium.

Current Conservation Status and Protection

Quadrula pustulosa was considered currently stable throughout its range by Williams et al. (1993) and in Alabama by Lydeard et al. (1999). Garner et al. (2004) designated *Q. pustulosa* a species of lowest conservation concern in the state.

Remarks

Quadrula pustulosa demonstrates a high degree of conchological variability. Even within populations, variation in outline, degree of inflation and density of pustules can be great. Some populations of *Q. pustulosa* west of the Mississippi River (e.g., Ouachita River, Arkansas) lack a green ray (J.L. Harris, personal communication).

Quadrula pustulosa is uncommon in prehistoric shell middens along the Tennessee River in northern Alabama (Morrison, 1942; Warren, 1975; Hughes and Parmalee, 1999).

Synonymy

Obliquaria (Quadrula) retusa Rafinesque, 1820. Rafinesque, 1820:306, pl. 81, figs. 19, 20
Comment: The original type locality was given as "Ohio and Kentucky rivers". Johnson and Baker (1973) invalidly designated ANSP 20220 as a neotype. This taxon is considered a *nomen dubium* and *nomen oblitum* (Ortmann and Walker, 1922a).

Obliquaria bullata Rafinesque, 1820. Rafinesque, 1820:307
Comment: Vanatta (1915) recognized this species as *Quadrula pustulosa pernodosa* (Lea, 1845). If *Obliquaria bullata* and *Quadrula pustulosa pernodosa* are the same, *Obliquaria bullata* would have priority over *Unio pustulosus* Lea, 1831, as reported by Morrison (1969). Johnson and Baker (1973) invalidly designated the type mentioned by Vanatta (1915) as a neotype (ICZN, 1999: Article 75). The long established name *pustulosa* Lea, 1831, is herein used in the interest of nomenclatorial stability.

Unio premorsus Rafinesque, 1831. Rafinesque, 1831:4 [unnecessary replacement name for *Obliquaria retusa* Rafinesque, 1820]

Unio pustulosus Lea, 1831. Lea, 1831:76, pl. 7, fig. 7; Lea, 1834b:86, pl. 7, fig. 7
Type locality: Ohio, T.G. Lea; Alabama River, [Alabama,] Judge Tait. Lectotype designated by Johnson (1974), not
    found in USNM, but length of figured shell in original description reported as about 53 mm. Syntypes, ANSP
    43058, are from Ohio.

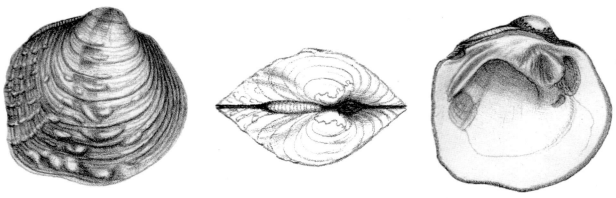

Unio nodulosus Say, 1834. Say, 1834:[no pagination]
Unio prasinus Conrad, 1834. Conrad, 1834b:44, 71, pl. 3, fig. 1
Unio schoolcraftensis Lea, 1834. Lea, 1834a:37, pl. 3, fig. 9; Lea, 1834b:149, pl. 3, fig. 9
Margarita (*Unio*) *turgidus* Lea, 1836, *nomen nudum*. Lea, 1836:16
Unio mortoni Conrad, 1836. Conrad, 1836:11, pl. 6, fig. 1
Unio turgidus Lea, 1838. Lea, 1838a:11, pl. 5, fig. 11; Lea, 1838c:11, pl. 5, fig. 11
Margarita (*Unio*) *dorfeuillianus* Lea, 1838, *nomen nudum*. Lea 1838b:15
Unio dorfeuillianus Lea, 1838. Lea, 1838a:73, pl. 17, fig. 54; Lea, 1838c:73, pl. 17, fig. 54
Unio pernodosus Lea, 1845. Lea, 1845:163; Lea, 1848a:71, pl. 3, fig. 8; Lea, 1848b:71, pl. 3, fig. 8
Margaron (*Unio*) *schoolcraftii* Lea, 1870. Lea, 1870:33 [unjustified emendation]

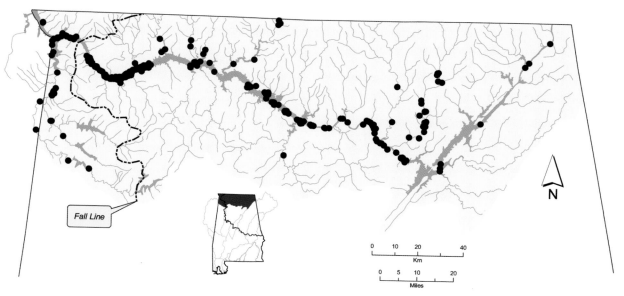

Distribution of *Quadrula pustulosa* in the Tennessee River drainage of Alabama.

Quadrula quadrula (Rafinesque, 1820)
Mapleleaf

Quadrula quadrula – Length 59 mm, UMMZ 76690. Tennessee River, Alabama. © Richard T. Bryant.

Shell Description

Length to 140 mm; thick; moderately inflated; outline usually quadrate, occasionally trapezoidal; posterior margin truncate, may be slightly concave; anterior margin rounded; dorsal margin straight to slightly convex; ventral margin rounded; posterior ridge narrow, adorned with small, round or elongate tubercles; posterior slope moderately steep, usually with small irregular pustules or low corrugations extending from posterior ridge to margin; wide shallow sulcus separates posterior ridge from low medial ridge, sulcus typically without pustules or tubercles, medial ridge with radial row of irregular tubercles, some individuals with additional variable tubercles anterior of medial ridge; umbo narrow, moderately inflated, elevated above hinge line, umbo sculpture 2 radial rows of nodules; periostracum greenish yellow to reddish brown, sometimes with weak, variable green rays in young individuals.

Pseudocardinal teeth thick, prominently striate, triangular, 2 divergent teeth in left valve, 1 tooth in right valve, may have accessory denticle anteriorly and/or posteriorly; lateral teeth long, moderately thick, slightly curved, 2 in left valve, 1 in right valve, may also have rudimentary lateral tooth ventrally in right valve; interdentum short, narrow to moderately wide; umbo cavity deep; nacre white.

Soft Anatomy Description

Mantle creamy white to light tan, may be darker outside of pallial line, grayish tan to grayish brown outside of apertures; visceral mass pearly white to creamy white or tan; foot creamy white to tan to pale pinkish orange, may be darker ventrally. Utterback (1915) reported a "crenulated flap on post-dorsal part of foot" and suggested that it is sexual in function since it was only observed in females, but no further elaboration on the structure was given. Nothing matching the description of this structure was observed in specimens examined for this work.

Gills tan; dorsal margin straight to slightly sinuous, steeply sloped from umbo to near incurrent aperture, ventral margin convex; gill length 50–65% of shell length; gill height 45–60% of gill length; outer gill height 65–75% of inner gill height; inner lamellae of inner gills connected to visceral mass only anteriorly.

No gravid females were available for marsupium description. Ortmann (1912a) and Utterback (1915) reported all four gills to be marsupial. Utterback (1915) reported gravid gills to be brown.

Labial palps creamy white to light tan; slightly concave dorsally, convex ventrally, usually bluntly pointed distally, occasionally rounded; palp length 30–55% of gill length; palp height 45–60% of palp length; distal 25–40% of palps bifurcate.

Incurrent aperture longer than excurrent aperture; supra-anal aperture longer than incurrent aperture.

Incurrent aperture length 15–25% of shell length; creamy white within, may have gray basal to papillae; papillae arborescent; papillae some combination of creamy white, brown and gray, often changing color distally.

Excurrent aperture length 10–15% of shell length; creamy white within, marginal color band gray, tan, brown or black; margin smooth, crenulate or with very small, simple papillae; papillae pale gray.

Supra-anal aperture length 25–30% of shell length; creamy white within, may have irregular gray or black marginally; margin smooth; mantle bridge separating supra-anal and excurrent apertures imperforate, length 10–30% of supra-anal length.

Specimens examined: Bear Creek, Colbert County (n = 2); Tennessee River (n = 5).

Glochidium Description

Length 78–85 μm; height 85–90 μm; without styliform hooks; outline subrotund; dorsal margin straight, ventral margin broadly rounded, anterior and posterior margins convex, slightly unequal in convexity (Howard, 1914b; Surber, 1915).

Similar Species

Quadrula quadrula resembles *Quadrula apiculata*. However, *Q. quadrula* generally lacks pustules in the sulcus, unlike *Q. apiculata*. *Quadrula apiculata* often grows proportionally longer with age than *Q. quadrula*, which generally retains a more quadrate outline.

Quadrula quadrula may superficially resemble *Quadrula metanevra*. However, *Q. metanevra* has a prominent, elevated, knobby posterior ridge and lacks a sulcus down the shell disk.

General Distribution

Quadrula quadrula is found in extreme southern Hudson Bay Basin (Dawley, 1947; Clarke, 1981a) and in the Great Lakes Basin from Lake Erie to Lake St. Clair (La Rocque and Oughton, 1937; Clarke, 1981a). It is widespread in the Mississippi Basin, occurring from southern Minnesota (Dawley, 1947) south to Texas (Howells et al., 1996) and Louisiana (Vidrine, 1993) and from headwaters of the Ohio River in Pennsylvania (Ortmann, 1919) west to Nebraska (Hoke, 2000), South Dakota (Backlund, 2000) and North Dakota (Picha and Swenson, 2000). *Quadrula quadrula* was common in the lower Cumberland River during the early 1900s (Wilson and Clark, 1914) and has expanded its range upstream since impoundment of the river, but it does not occur upstream of Cumberland Falls (Cicerello et al., 1991; Parmalee and Bogan, 1998). Flint Creek, Morgan County, Alabama, was apparently the upstream limit of *Q. quadrula* in the Tennessee River prior to its impoundment (Ortmann, 1925). However, it now occurs at least as far upstream as Guntersville Reservoir. Reports of *Q. quadrula* in drainages west of the Red River in Texas probably represent *Quadrula nobilis* (R.G. Howells, personal communication).

Alabama and Mobile Basin Distribution

Quadrula quadrula historically occurred only in the Tennessee River in northwestern Alabama. Since impoundment of the river, it has expanded its range upstream to Guntersville Reservoir in northeastern Alabama.

Quadrula quadrula is common in Pickwick Reservoir and Wilson Dam tailwaters but uncommon farther upstream.

Ecology and Biology

Quadrula quadrula is a species of medium to large rivers. It may occur in still or flowing water. Bates (1962) reported *Q. quadrula* to be one of the first colonizers of overbank habitat in Kentucky Reservoir, Tennessee River. It occurs in a variety of substrates ranging from mud and sand to gravel.

Quadrula quadrula is a short-term brooder, gravid from May to August (Howard, 1914a; Utterback, 1916a; Baker, 1928a). The only reported glochidial hosts of *Q. quadrula* are *Ictalurus punctatus* (Channel Catfish) (Ictaluridae), determined with laboratory trials (Schwebach et al., 2002), and *Pylodictis olivaris* (Flathead Catfish) (Ictaluridae), based on an observation of a natural infestation (Howard and Anson, 1922).

Current Conservation Status and Protection

Quadrula quadrula was considered currently stable throughout its range by Williams et al. (1993) and in Alabama by Lydeard et al. (1999). Garner et al. (2004) designated *Q. quadrula* a species of lowest conservation concern in the state.

Quadrula quadrula is an important commercial species being exported from Alabama for pearl culture. It is protected by state regulations, the most important being a minimum size limit of $2^5/_8$ inches (67 mm) in shell height.

Remarks

Some authors have reported *Quadrula fragosa* (Conrad, 1835) from the Tennessee River in Alabama. These appear to be the result of misidentification of *Quadrula quadrula* (Ortmann, 1925) or misuse of the binomen *Q. fragosa* for *Q. quadrula* (Scruggs, 1960). No museum records of *Q. fragosa* from Alabama are known.

Quadrula quadrula has never been reported from the archaeological record in Alabama. It was reported from Flint Creek, Morgan County, Alabama, and Muscle Shoals prior to impoundment of the river (Ortmann, 1925) but has since expanded its range upstream to at least Guntersville Dam.

Synonymy

Obliquaria (*Quadrula*) *quadrula* Rafinesque, 1820. Rafinesque, 1820:305

Type locality: Ohio River. Neotype, ANSP 20224, length 60 mm, designated by Johnson and Baker (1973), is from Salt River, [Kentucky].

Unio rugosus Barnes, 1823. Barnes, 1823:126, pl. 8, fig. 9

Unio lacrymosus Lea, 1828. Lea, 1828:272, pl. 6, fig. 8; Lea, 1834b:14, pl. 5, fig. 8

Unio asperrimus Lea, 1831. Lea, 1831:71, pl. 5, fig. 3; Lea, 1834b:81, pl. 5, fig. 3

Unio lunulatus Pratt, 1876. Pratt, 1876:167, pl. 31, fig. 1

Quadrula quadrula contraryensis Utterback, 1915. Utterback, 1915:138, pl. 18, figs. 47a, b

Quadrula (*Quadrula*) *quadrula* var. *bullocki* F.C. Baker, 1928. F.C. Baker, 1928a:87, pl. 46, figs. 1–3

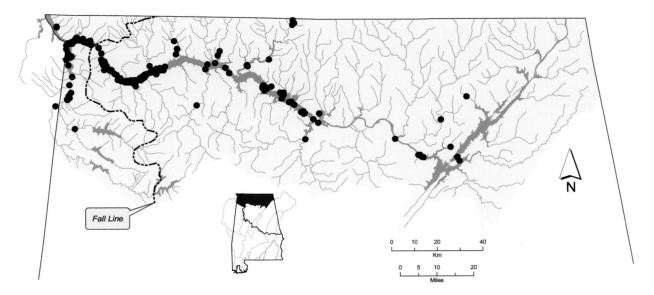

Distribution of *Quadrula quadrula* in the Tennessee River drainage of Alabama.

Quadrula rumphiana (Lea, 1852)
Ridged Mapleleaf

Quadrula rumphiana – Length 58 mm, UF 67877. Coosa River, Ten Island Shoals, Calhoun County, Alabama, October 1914. © Richard T. Bryant.

Shell Description

Length to 110 mm; moderately thick; moderately inflated; outline subquadrate; posterior margin truncate or concave; anterior margin rounded; dorsal margin convex; ventral margin broadly rounded anteriorly, with emargination at ventral terminus of sulcus; posterior ridge prominent, rounded, without sculpture, narrow dorsally, wider posterioventrally; posterior slope not steep, flat to slightly concave, sculptured with small pustules, may be in form of broken, elongate plications extending from base of posterior slope to margin; shallow sulcus just anterior of posterior ridge; disk with variable pustules; umbo narrow to broad, somewhat inflated, elevated well above hinge line, umbo sculpture knobby ridges; periostracum greenish brown to tawny, reddish brown or dark brown, without rays.

Pseudocardinal teeth large, triangular, 2 divergent teeth in left valve, 1 in right valve, may have accessory denticle anteriorly and/or posteriorly; lateral teeth short, straight to slightly curved, 2 in left valve, 1 in right valve; interdentum short, wide; umbo cavity deep, moderately compressed; nacre white.

Soft Anatomy Description

Mantle creamy white to tan, outside of apertures gray to grayish brown; visceral mass creamy white to pale tan; foot tan to pale pinkish orange, may be darker distally.

Gills creamy white to tan or light brown; dorsal margin straight to slightly sinuous, gills steeply sloped from umbo to near incurrent aperture, ventral margin convex, outer gill often less convex than inner gill; gill length 50–70% of shell length; gill height usually 40–60% of gill length, occasionally greater or less; outer gill height 50–85% of inner gill height, occasionally greater; inner lamellae of inner gills connected to visceral mass only anteriorly.

No gravid females were available for marsupium description; probably similar to other *Quadrula* species, with all four gills marsupial, slightly padded when gravid.

Labial palps creamy white to tan; straight to slightly concave dorsally, convex ventrally, bluntly pointed to rounded distally; palp length 25–60% of gill length; palp height 40–65% of palp length, occasionally greater; distal 20–35% of palps bifurcate.

Incurrent and supra-anal apertures longer than excurrent aperture; incurrent aperture may be shorter, longer or equal to supra-anal aperture in length.

Incurrent aperture length 20–30% of shell length; creamy white to tan within, may have gray or grayish tan basal to papillae; papillae arborescent, may be deeply branched, branches not uniform; papillae some combination of creamy white, tan, gray and black, often lighter distally.

Excurrent aperture length 10–20% of shell length, occasionally greater; creamy white to tan within, usually with gray or grayish brown marginal color band; margin smooth or crenulate.

Supra-anal aperture length 20–30% of shell length, occasionally greater; creamy white within, may have narrow gray or black marginal band; margin smooth; mantle bridge separating supra-anal and excurrent apertures imperforate, length 10–25% of supra-anal length, occasionally absent.

Specimens examined: Big Canoe Creek, St. Clair County (n = 2); Coosa River (n = 2); Mobile River (n = 2); Sipsey River (n = 3); Tombigbee River (n = 2).

Glochidium Description

Length 78 μm; height 39 μm; without styliform hooks; outline subrotund; dorsal margin straight, ventral margin broadly rounded, anterior and posterior margins convex.

Similar Species

Quadrula rumphiana resembles *Quadrula apiculata* and *Quadrula nobilis*, but those species haves pustules on the posterior ridge, which is typically smooth on *Q. rumphiana*. The posterior ridge of *Q. rumphiana* is often more prominent and the sulcus narrower than those of *Q. apiculata* and *Q. nobilis*.

Quadrula rumphiana resembles *Quadrula stapes*. However, the posterior ridge of *Q. stapes* is much sharper and the posterior slope much steeper than those of *Q. rumphiana*. The periostracum of *Q. stapes* is usually adorned with small, dark green chevrons and triangles, which are lacking on the periostracum of *Q. rumphiana*.

Quadrula rumphiana may also resemble *Quadrula metanevra*, but the posterior ridge of that species is typically higher and sculptured with large knobs or undulations. The posterior slope of *Q. metanevra* is steeper than that of *Q. rumphiana*. The periostracum of *Q. metanevra* is often marked by small, dark green chevrons and triangles, which are absent on *Q. rumphiana*.

General Distribution

Quadrula rumphiana is endemic to the Mobile Basin of Alabama, Georgia and Mississippi.

Alabama and Mobile Basin Distribution

Quadrula rumphiana is widespread in the Mobile Basin, where it occurs in all major river systems, above and below the Fall Line.

Quadrula rumphiana is extant in widespread, isolated populations.

Ecology and Biology

Quadrula rumphiana is a species of small to large rivers, where it occurs primarily in flowing water. Its preferred substrate is gravel or a mixture of sand and gravel.

Quadrula rumphiana is presumably a short-term brooder, gravid in spring and summer. Its glochidial hosts are unknown.

Current Conservation Status and Protection

Quadrula rumphiana was considered a species of special concern throughout its range by Williams et al. (1993) and in Alabama by Lydeard et al. (1999). Garner et al. (2004) designated *Q. rumphiana* a species of low conservation concern in the state.

Remarks

Lea (1852b) received the type specimen of *Quadrula rumphiana* from Dr. Budd, who indicated that "he thinks it came from the west of Georgia". The type of this distinctive species is undoubtedly from the Mobile Basin, probably the upper Coosa River drainage in Georgia.

Quadrula rumphiana was named in memory of George Eberhard Rumph (1620–1702), author of *D'Amboinsche Rariteitkamer* ("Ambionese Curiosity Cabinet"), an early text on invertebrates from the Molucca Islands.

Synonymy

Unio rumphianus Lea, 1852, *nomen nudum*. Lea, 1852a:252
Unio rumphianus Lea, 1852. Lea, 1852b:276, pl. 22, fig. 34; Lea, 1852c:32, pl. 22, fig. 34
Type locality: West Georgia [Coosa River drainage], Dr. Budd. Holotype, USNM 84189, length 50 mm.

 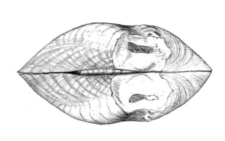 

Unio blandianus Lea, 1856. Lea, 1856d:263; Lea, 1858e:65, pl. 11, fig. 47; Lea, 1858h:65, pl. 11, fig. 47
Type locality: Othcalooga Creek, Gordon County, Georgia, Bishop Elliott. Lectotype, USNM 84190, length 87 mm,
  designated by Johnson (1974).

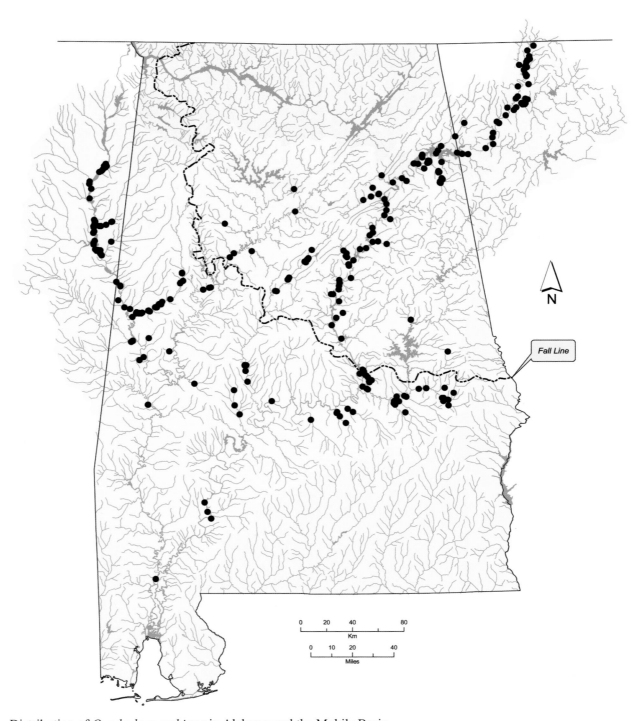

Distribution of *Quadrula rumphiana* in Alabama and the Mobile Basin.

Quadrula sparsa (Lea, 1841)
Appalachian Monkeyface

Quadrula sparsa – Upper figure: length 63 mm, NCSM 6334. Powell River, river mile 109 [106.6], McDowell Ford, Hancock County, Tennessee, 10 May 1980. Lower figure: length 46 mm, NCSM 30916. Powell River, McDowell Ford, river mile 106.6, Hancock County, Tennessee. 15 September 1977. © Richard T. Bryant.

Shell Description

Length to 80 mm; moderately thick; somewhat compressed; outline quadrate to irregularly rhomboidal; posterior margin obliquely truncate to concave; anterior margin rounded; dorsal margin straight; ventral margin broadly rounded anteriorly, with slight emargination at ventral terminus of sulcus; posterior ridge low, rounded; posterior slope wide, usually with corrugations or oblong pustules more or less perpendicular to margin; often with weak sulcus anterior and posterior of posterior ridge; disk with sparse round and lachrymous tubercles, absent from anterior 30% of shell; umbo broad, elevated slightly above hinge line, umbo sculpture unknown; periostracum dull, yellowish green to brown, often with small green chevrons or triangles.

Pseudocardinal teeth thick, triangular, 2 divergent teeth in left valve, 1 tooth in right valve, may have well-developed accessory denticle anteriorly and/or posteriorly; lateral teeth thick, short, straight, 2 in left valve, 1 in right valve; interdentum moderately long, wide; umbo cavity deep, compressed; nacre white, may have salmon tint posteriorly.

Soft Anatomy Description

No material was available for soft anatomy description. Ortmann (1912a) stated that it is "identical in every detail with *Quadrula metanevra*, to which it is closely allied by the shell. The agreement extends so far, that minor details are also identical, as the smooth edge of the anal, the shape of the palpi, and the black pigment of the posterior part of the abdominal sac."

Glochidium Description
Glochidium unknown.

Similar Species
Quadrula sparsa resembles *Quadrula intermedia*. It differs in having a well-developed posterior ridge that is not covered with pustules. The outline of *Q. sparsa* is generally more elongate than that of *Q. intermedia*.

Quadrula sparsa resembles *Quadrula metanevra* but is typically more compressed and has a less elevated posterior ridge that lacks large knobs.

General Distribution
Quadrula sparsa is endemic to the Cumberland and Tennessee River drainages of Tennessee and possibly Kentucky (Cicerello et al., 1991; Parmalee and Bogan, 1998). It occurred downstream into the lower Tennessee River at least prehistorically.

Alabama and Mobile Basin Distribution
The occurrence of *Quadrula sparsa* in Alabama is based on archaeological material from prehistoric middens along the Tennessee River (Morrison, 1942).

Ecology and Biology
Quadrula sparsa is a species of medium to large rivers where it occurs in shoal habitat with sand and gravel substrates.

Quadrula sparsa is presumably a short-term brooder, gravid during spring and summer. Its glochidial hosts are unknown.

Current Conservation Status and Protection
Quadrula sparsa was considered endangered throughout its range by Stansbery (1970a) and Williams et al. (1993). It has not been included in any assessments of the Alabama fauna. *Quadrula sparsa* was listed as endangered under the federal Endangered Species Act in 1976.

Remarks
Quadrula sparsa has been reported from prehistoric shell middens at Muscle Shoals and Hobbs Island, Madison County, where it was rare (Morrison, 1942). There are no historical records with specific locality for this species from Alabama. However, Lewis (1876) reported *Q. sparsa* from "Tennessee River" in a list of freshwater and land shells of Alabama.

Synonymy
Unio sparsus Lea, 1841. Lea, 1841b:82; Lea, 1842b:242, pl. 25, fig. 58; Lea 1842c:80, pl. 25, fig. 58
Type locality: Holston River, east Tennessee. Lectotype, USNM 84222, length 38 mm, designated by Johnson (1974).

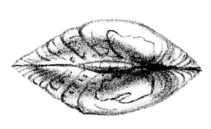

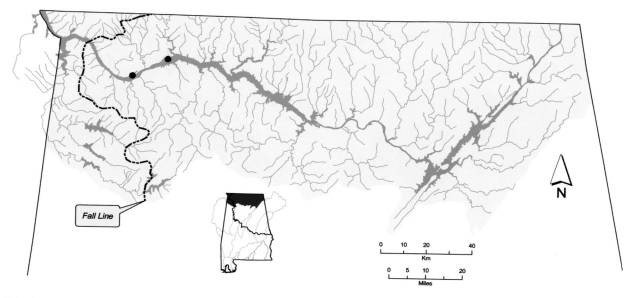

Distribution of *Quadrula sparsa* in the Tennessee River drainage of Alabama.

Quadrula stapes (Lea, 1831)
Stirrupshell

Quadrula stapes – Length 45 mm, UF 385674. Tombigbee River, Memphis Landing, river mile 324.4, Pickens County, Alabama, 24 October 1976. © Richard T. Bryant.

Shell Description

Length to 67 mm; moderately thick; somewhat compressed to moderately inflated; outline triangular; posterior margin bluntly pointed; anterior margin rounded; dorsal margin convex; ventral margin convex; posterior ridge well-developed, angular; posterior slope very steep, concave, with concentric corrugations extending from posterior ridge to posteriodorsal margin; shallow sulcus just anterior of posterior ridge; most of disk covered with variable lachrymous tubercles, tubercles on anterior 25% may be reduced or absent; umbo narrow, moderately inflated, elevated above hinge line, turned slightly anteriad, umbo sculpture unknown; periostracum yellowish green to olive or brown, usually with dark olive chevrons and triangles.

Pseudocardinal teeth thick, triangular, 2 divergent teeth in left valve, 1 tooth in right valve, usually with accessory denticle anteriorly, sometimes posteriorly; lateral teeth short, thick, straight, 2 in left valve, 1 in right valve; interdentum moderately long, wide; umbo cavity deep; nacre white.

Soft Anatomy Description

Mantle tan to brown, may be lighter outside of pallial line, brown to rusty brown outside of apertures; visceral mass tan, may have brown cast; foot tan.

Gills tan; dorsal margin straight to slightly sinuous, gills steeply sloped from umbo to near incurrent aperture, ventral margin convex; gill length 60–70% of shell length; gill height 50–55% of gill length; outer gill height 55–70% of inner gill height; inner lamellae of inner gills connected to visceral mass only anteriorly.

No gravid females were available for marsupium description; probably similar to other *Quadrula* species, with all four gills marsupial, slightly padded when gravid.

Labial palps tan; straight dorsally, curved ventrally, bluntly pointed distally; palp length 20–25% of gill length; palp height 50–65% of palp length; distal 35–50% of palps bifurcate.

Incurrent aperture considerably longer than excurrent aperture; supra-anal aperture considerably longer than incurrent aperture.

Incurrent aperture length 15–20% of shell length; tan within, may have rusty or brownish cast, often darker basal to papillae; papillae arborescent, small; papillae brown to rusty brown.

Excurrent aperture length 5–15% of shell length; tan within, marginal color band rusty brown; margin crenulate or smooth and undulating.

Supra-anal aperture length 20–40% of shell length; tan within, with thin rusty tan or rusty brown marginal band; margin smooth; mantle bridge separating supra-anal and excurrent apertures occasionally with 1 or more small perforations, length 5–30% of supra-anal length.

Specimens examined: Tombigbee River (n = 4) (specimens previously preserved).

Glochidium Description

Glochidium unknown.

Similar Species

Quadrula stapes may superficially resemble some *Quadrula apiculata*, but *Q. stapes* has a much sharper posterior ridge and steeper posterior slope. The sulcus of *Q. apiculata* is generally deeper than that of *Q. stapes*. Also, *Q. apiculata* lacks distinctive green chevrons and triangles on its periostracum.

Obliquaria reflexa that lack knobs will occasionally be encountered and may superficially resemble *Quadrula stapes* with reduced pustules. However, the posterior ridge of *Q. stapes* is steeper, and *O. reflexa* never has green chevrons and triangles on its periostracum. Also, the umbo cavity of *Q. stapes* is deeper than that of *O. reflexa*.

General Distribution

Quadrula stapes is endemic to the Mobile Basin below the Fall Line in Alabama and Mississippi.

Alabama and Mobile Basin Distribution

Quadrula stapes historically occurred in Coastal Plain reaches of the Tombigbee, Black Warrior and Alabama rivers. There are few tributary records of this species.

Quadrula stapes appears to be extinct. Its last stronghold in the upper Tombigbee River was destroyed during the 1970s when the Tennessee Tombigbee Waterway was constructed by the USACE.

Ecology and Biology

Quadrula stapes inhabited shoals of large rivers with moderate to swift current over clean, coarse gravel substrate. It could be found in water less than 1 m deep.

Quadrula stapes was probably a short-term brooder, gravid in spring and summer. Its glochidial hosts are unknown.

Current Conservation Status and Protection

Quadrula stapes was recognized as endangered throughout its range by Athearn (1970) and Williams et al. (1993) and in Alabama by Stansbery (1976) and Lydeard et al. (1999). Garner et al. (2004) listed *Q. stapes* as possibly extinct. In 1987 it was listed as endangered under the federal Endangered Species Act.

Remarks

Quadrula stapes was found to be common in a small prehistoric midden along upper reaches of the Alabama River, near Montgomery (J.T. Garner, unpublished data). A single specimen was recovered from a prehistoric midden on the Black Warrior River, southwest of Tuscaloosa, Alabama (Hanley, 1982).

Synonymy

Unio stapes Lea, 1831. Lea, 1831:77, pl. 7, fig. 8; Lea, 1834b:87, pl. 7, fig. 8
Type locality: Alabama River, [Alabama,] Judge Tait. Lectotype, USNM 84218, length 43 mm, designated by Johnson (1974).

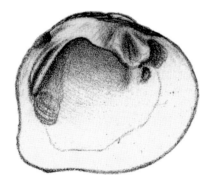

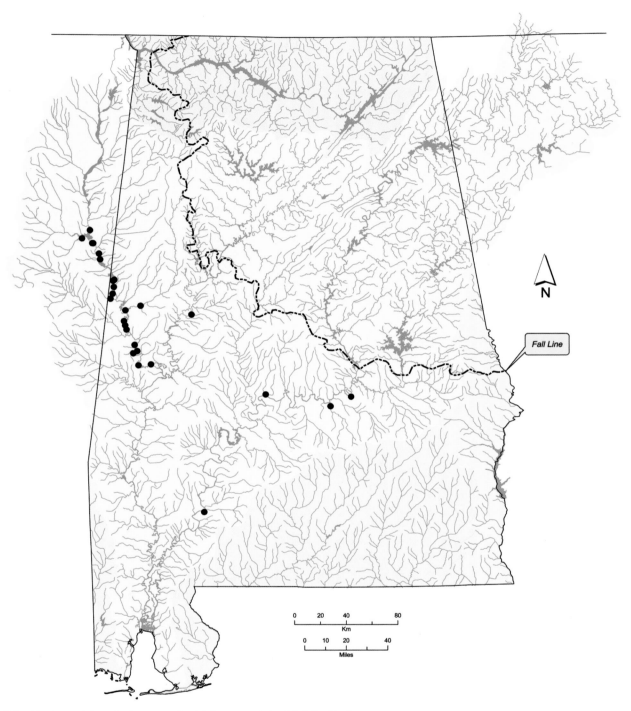

Distribution of *Quadrula stapes* in Alabama and the Mobile Basin.

Quadrula succissa (Lea, 1852)
Purple Pigtoe

Quadrula succissa – Length 45 mm, UF 65274. Conecuh River, Searight, Crenshaw County, Alabama, October 1915. © Richard T. Bryant.

Shell Description

Length to 65 mm; moderately thin to moderately thick; usually compressed, old individuals may be moderately inflated; outline subrhomboidal to subquadrate; posterior margin obliquely truncate; anterior margin rounded; dorsal margin convex; ventral margin convex to almost straight; posterior ridge low, rounded; posterior slope not steep, flat, may be steeper in old more inflated individuals; umbo broad, compressed, elevated slightly above hinge line, but almost always eroded, umbo sculpture unknown; periostracum dark brown to black, without rays.

Pseudocardinal teeth moderately thick, triangular, 2 divergent teeth in left valve, 1 tooth in right valve, with well-developed accessory denticle anteriorly, accessory denticle large in some individuals; lateral teeth short, straight to slightly curved, 2 in left valve, 1 in right valve; interdentum short, narrow to moderately wide; umbo cavity moderately deep; nacre pale purple.

Soft Anatomy Description

Mantle tan to light brown, outside of apertures usually gray or grayish brown; visceral mass pearly white, with varying degrees of tan; foot creamy white to tan, but may have pale golden cast.

Gills usually very dark brown, almost black but tan in occasional individuals, often with irregular opaque tan streaks oriented parallel to water tubes, paler distally, distal margins may be almost colorless, inner gills usually lighter in color than outer gills; dorsal margin straight to slightly sinuous, steeply sloped from umbo to vicinity of apertures, ventral margin convex; gill length 45–60% of shell length; gill height 50–75% of gill length; outer gill height 70–95% of inner gill height; inner lamellae of inner gills connected to visceral mass only anteriorly.

All 4 gills marsupial; glochidia held across entire gill; marsupium slightly padded when gravid; creamy white with grayish brown cast.

Labial palps creamy white to tan; straight to slightly concave dorsally, convex ventrally, bluntly pointed distally; palp length 30–50% of gill length; palp height 50–75% of palp length; distal 20–40% of palps bifurcate.

Incurrent and supra-anal apertures longer than excurrent aperture; incurrent aperture may be shorter, longer or equal to supra-anal aperture. However, Ortmann (1923a) reported the supra-anal aperture to be "slightly shorter" than the excurrent aperture.

Incurrent aperture length 20–25% of shell length; creamy white to tan within, may have brown or grayish brown basal to papillae; papillae arborescent, occasionally with a few simple or bifid near ends of aperture; papillae creamy white to grayish brown or grayish orange, coloration often changing distally. Ortmann (1923a) reported incurrent aperture papillae to be "of the normal, subfalciform shape" (i.e., sickle-shaped).

Excurrent aperture length 10–20% of shell length; creamy white to tan within, distal margin dark brown to grayish brown or rusty brown; margin usually crenulate, occasionally smooth.

Supra-anal aperture length 20–30% of shell length; creamy white to pale tan within, may have sparse tan, black or gray marginally; margin smooth; mantle bridge separating supra-anal and excurrent apertures imperforate, length 15–25% of supra-anal length.

Specimens examined: Conecuh River (n = 7); Murder Creek, Conecuh County (n = 4); West Fork Choctawhatchee River (n = 1).

Glochidium Description

Glochidium unknown.

Similar Species

Quadrula succissa may be confused with *Fusconaia escambia*. *Quadrula succissa* has a rounder posterior ridge, less steep posterior slope and less truncate posterior margin. The anterior pseudocardinal tooth in the left valve of *F. escambia* is oriented more anteriorly than that of *Q. succissa*.

Quadrula succissa superficially resembles some *Pleurobema strodeanum* but is less elongate with a rounder posterior ridge. *Quadrula succissa* has a deep, moderately compressed umbo cavity and purple nacre, whereas *P. strodeanum* has a shallow umbo cavity and white nacre.

General Distribution

Quadrula succissa is endemic to eastern Gulf Coast drainages from the Choctawhatchee River drainage west to the Escambia River drainage in Alabama and Florida (Clench and Turner, 1956).

Alabama and Mobile Basin Distribution

Quadrula succissa occurs in the Choctawhatchee, Yellow and Conecuh River drainages.

Quadrula succissa remains widespread in much of its historical range. Populations are fragmented, but it may be locally common.

Ecology and Biology

Quadrula succissa occurs in medium creeks to large rivers. It is usually found in areas with slow to moderate current, in sand or sandy mud substrates. *Quadrula succissa* is sometimes common along channel slopes in shallow water.

Quadrula succissa is presumably a short-term brooder, gravid in spring and summer. It has been observed brooding embryos in June in the Escambia River. Glochidial hosts of this species are unknown.

Current Conservation Status and Protection

Quadrula succissa was considered a species of special concern throughout its range by Williams et al. (1993) and in Alabama by Lydeard et al. (1999). Garner et al. (2004) designated *Q. succissa* a species of low conservation concern in the state.

Remarks

Simpson (1900b) described *Fusconaia* as a subgenus of *Quadrula*, and included *Unio successus*. It remained in *Fusconaia*, which was later elevated to generic level, for a century. However, Lydeard et al. (2000) reported *succissa* to be most closely related to *Quadrula infucata*, based on comparison of mitochondrial DNA (16S rRNA gene). Both *Quadrula succissa* and *Q. infucata* typically have darkly pigmented gills. Occasional *Q. infucata* are without shell sculpture and are very similar in appearance to *Quadrula succissa*.

Synonymy

Unio successus Lea, 1852, *nomen nudum*. Lea, 1852a:252
Unio successus Lea, 1852. Lea, 1852b:275, pl. 21, fig. 32; Lea, 1852c:31, pl. 21, fig. 32
Type locality: West Florida. Type locality subsequently restricted to Choctawhatchee River, Caryville, Holmes County, Florida, by Clench and Turner (1956). Lectotype, USNM 84574, length 43 mm, designated by Johnson (1974).

Unio cacao Lea, 1859. Lea, 1859c:154; Lea, 1860d:344, pl. 56, fig. 169; Lea, 1860g:94, pl. 56, fig. 169
Quadrula wrighti Simpson, 1914. Simpson, 1914:868

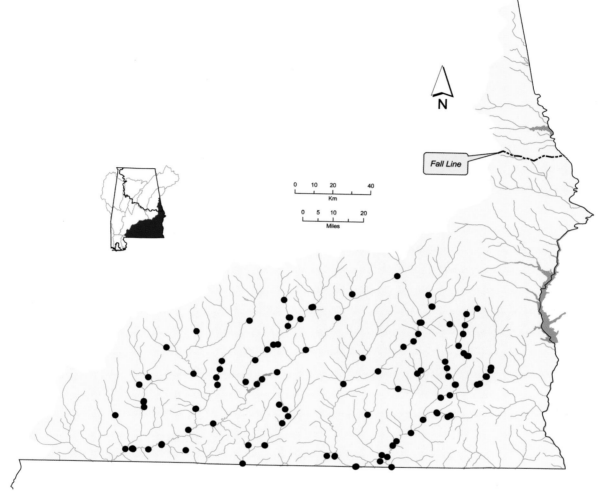

Distribution of *Quadrula succissa* in the eastern Gulf Coast drainages of Alabama.

Quadrula verrucosa (Rafinesque, 1820)
Pistolgrip

Quadrula verrucosa – Upper figure: female, length 144 mm, UF 574539. Oostanaula River, 1.3 miles below Interstate 75, Gordon County, Georgia, 3 September 1997. Lower figure: male, length 104 mm, UF 68074. Valley Creek near Toadvine, Jefferson County, Alabama, June 1913. © Richard T. Bryant.

Shell Description

Length to 162 mm; thick, thinner posteriorly; somewhat compressed, females more compressed than males; outline irregular, elongate rhomboidal to elliptical; posterior margin bluntly pointed to obliquely truncate or rounded, females with flattened expansion posteriorly; anterior margin rounded; dorsal margin straight to convex; ventral margin straight to convex, becoming concave posteriorly with age; posterior ridge well-defined, with irregular nodules; posterior slope wide, usually sculptured with ridges extending from posterior ridge to margin, often with smaller secondary wrinkles and/or pustules; most individuals with broad shallow sulcus just anterior of posterior ridge, terminating in emargination on ventral margin; disk sculptured with variable pustules and wrinkles, pustules becoming larger and lachrymous ventrally; umbo broad, flat, elevated slightly above hinge line, umbo sculpture nodulous ridges, may be somewhat double-looped distally; periostracum greenish brown to dark brown or black.

Pseudocardinal teeth thick, triangular, erect, 2 divergent teeth in left valve, 1 in right valve, often with well-developed accessory denticle anteriorly; lateral teeth long, usually straight, 2 in left valve, 1 in right valve; interdentum long, moderately wide; umbo cavity deep, may be compressed; nacre usually white, may be purple in individuals from Mobile Basin.

Soft Anatomy Description

Mantle creamy white to tan or light brown, outside of apertures gray, grayish tan or grayish brown, usually wrinkled; visceral mass pearly white to creamy white or light tan; foot tan, occasionally pale orange, often darker distally.

Some females with thickened pad of mantle tissue in posterior expansion of shell; thickened tissues with wrinkles parallel to margin; wrinkles with sparse, irregular, somewhat flattened, simple and bifid papilla-like structures. Simpson (1900b) reported females to have a "thickened flap" of mantle filling the posterior expansion of the shell.

Gills creamy white to tan or light brown, may have slight golden cast; dorsal margin straight to slightly sinuous, ventral margin slightly convex to straight; gill length 50–65% of shell length; gill height 35–50% of gill length; outer gill height 65–85% of inner gill height; inner lamellae of inner gills connected to visceral mass only anteriorly. Utterback (1915) stated that the inner lamellae of inner gills are "connected to visceral mass except for a short distance anteriorly".

All 4 gills marsupial; glochidia held across entire gill; marsupium slightly padded when gravid; pale tan.

Labial palps creamy white to light tan; straight to slightly concave dorsally, convex ventrally, bluntly pointed to rounded distally; palp length 20–40% of gill length; palp height 35–70% of palp length; distal 20–40% of palps bifurcate.

Incurrent aperture usually longer than excurrent aperture, occasionally equal; supra-anal aperture longer than incurrent and excurrent apertures.

Incurrent aperture length 15–25% of shell length; creamy white to tan within, may have grayish cast, may be grayish tan to brown or black basal to papillae; papillae arborescent, doubly or triply branched, occasionally with some simple and bifid; papillae gray to grayish brown.

Excurrent aperture length 10–20% of shell length, occasionally greater; creamy white to gray, grayish tan or grayish brown within, with 1–2 marginal color bands, proximal band creamy white, distal band gray, brown or black; margin smooth to crenulate or with small simple papillae; papillae gray to grayish brown. Lea (1863d) described the excurrent aperture as "enormously large".

Supra-anal aperture length 20–35% of shell length; creamy white within, may have thin gray, brown or black marginal band; margin smooth; mantle bridge separating supra-anal and excurrent apertures imperforate, length 5–15% of supra-anal length, occasionally greater, occasionally absent. Lea (1863d) described the supra-anal aperture as "enormously large".

Specimens examined: Big Canoe Creek, St. Clair County (n = 3); Chewacla Creek, Macon County (n = 1); Locust Fork (n = 1); Sipsey River (n = 4); Tennessee River (n = 2); Yellow Creek, Lowndes County, Mississippi (n = 3).

Glochidium Description

Length 85–94 µm; height 97–101 µm; without styliform hooks; outline subrotund; dorsal margin straight, ventral margin rounded, anterior and posterior margins convex (Surber, 1912; Hoggarth, 1999). Jirka and Neves (1992) reported slightly larger glochidia, but gave the longer measurement (122 µm) as length and the shorter measurement (109 µm) as "width".

Similar Species

Quadrula verrucosa superficially resembles *Margaritifera marrianae*, but is usually heavily tuberculate over most of the shell disk while *M. marrianae* is sculptured with plications that are typically confined to the posterior portion of the shell. The posterior ridge of *Q. verrucosa* is more prominent than that of *M. marrianae*.

Quadrula verrucosa may resemble some small to medium *Plectomerus dombeyanus*. However, *Q. verrucosa* has a lower posterior ridge and deeper, more compressed umbo cavity than *P. dombeyanus*. Shell nacre of *Q. verrucosa* is usually white, but may be purple in some Mobile Basin populations, whereas *P. dombeyanus* nacre is always some shade of purple.

General Distribution

Quadrula verrucosa is widespread in the Mississippi Basin from Minnesota (Dawley, 1947) south to Louisiana (Vidrine, 1993) and from headwaters of the Ohio River drainage in western Pennsylvania (Ortmann, 1919) west to South Dakota (Backlund, 2000) and Oklahoma (Oesch, 1995). It is widespread in the Cumberland River drainage downstream of Cumberland Falls (Cicerello et al., 1991; Parmalee and Bogan, 1998). *Quadrula verrucosa* is widespread in the Tennessee River drainage from eastern Tennessee to the mouth of the Tennessee River (Parmalee and Bogan, 1998). *Quadrula verrucosa* occurs in Gulf Coast drainages from the Mobile Basin west to the San Antonio River drainage of Texas (Howells et al., 1996).

Alabama and Mobile Basin Distribution

Quadrula verrucosa occurs in the Tennessee River drainage and Mobile Basin. In the Mobile Basin it is known from all major drainages with the exception of the Tallapoosa River drainage above the Fall Line.

Quadrula verrucosa remains in widespread populations in the Tennessee River drainage and Mobile Basin. However, it has disappeared from some large river reaches in the Mobile Basin.

Ecology and Biology

Quadrula verrucosa occurs in a variety of habitats from pools to runs and riffles. It may be found in small creeks to large rivers, but is generally more abundant in large creeks and small rivers with moderate current and sand and gravel substrates. *Quadrula verrucosa* is often found only partially buried in the substrate.

Quadrula verrucosa is dioecious, but van der Schalie (1970) reported the gonad in one of nine specimens from the Guadalupe River, Texas, to be hermaphroditic. In New River, Virginia and West Virginia, gametogenesis begins during autumn in both sexes. Gonads of males fill with spermatozoa and females fill with mature ova by late autumn. Both sexes overwinter in that condition, and spawning begins in March and continues through May (Jirka and Neves, 1992). *Quadrula verrucosa* is a short-term brooder, gravid from April to July or August (Utterback, 1915, 1916a; Ortmann, 1919; Jirka and Neves, 1992). A single gravid female was reported from Texas in January, but other females in that population were not gravid and had gonads with ova early in development, suggesting that the gravid individual was aberrant (Howells, 2000). Conglutinates are lanceolate, white and not very solid (Ortmann, 1919). During spring and possibly summer female *Q. verrucosa* can sometimes be found displaying the fleshy mantle pad located in the posterior expansion of the shell.

Fishes reported to serve as glochidial hosts of *Quadrula verrucosa* based on laboratory trials include *Ameiurus natalis* (Yellow Bullhead), *Ameiurus nebulosus* (Brown Bullhead) and *Pylodictis olivaris* (Flathead Catfish) (Ictaluridae) (Howells, 1997; Pepi and Hove, 1997).

Current Conservation Status and Protection

Quadrula verrucosa was considered currently stable throughout its range by Williams et al. (1993) and in Alabama by Lydeard et al. (1999). Garner et al. (2004) designated *Q. verrucosa* a species of low conservation concern in the state.

Commercial harvest of *Quadrula verrucosa* is legal, but demand for its shell is usually low due to the thinness of its posterior portion (making it uneconomical for pearl nucleous production). Currently regulations limit where and when harvest may occur, and harvest is limited to individuals greater than $2^5/_8$ inches (67 mm) in shell height.

Remarks

Quadrula verrucosa was recognized as belonging to the monotypic genus *Tritogonia* for most of the last century. Ortmann (1912a) stated that "in the structure of the soft parts, this species is essentially a *Quadrula*", and Utterback (1915) stated that the soft parts of *Q. verrucosa* "are so identical with those of typical *Quadrula* that there is no reason for its groupings with any other genus". Recent genetic analyses support inclusion of this species in *Quadrula* (Serb et al., 2003).

The nacre of *Quadrula verrucosa* is almost always white in Tennessee River populations. However, individuals with purple nacre are not uncommon in some populations of the Mobile Basin. In the Cahaba River, van der Schalie (1938) noted that purple-nacred individuals were more common in headwaters than in lower reaches of that system.

Quadrula verrucosa is one of the few short-term brooders reported to exhibit sexual dimorphism (Sterki, 1898a; Ortmann, 1919). Ortmann (1919) reported that, on average, females are more compressed than males and have a flattened extension of the shell posteriorly. This is contrary to sexually dimorphic long-term brooders (with the exception of the *Epioblasma*), in which the dimorphism is a result of marsupial swelling in females that makes room for greatly distended marsupia. Sterki (1898a) stated that sexual dimorphism in *Q. verrucosa* is "of another kind than in *Lampsilis*, a small portion of the branchiae, if any, finding room in the extended part". The flattened posterioventral shell extension holds the fleshy mantle pad. Thus, the dimorphism appears to possibly be an aspect of host-attracting adaptation rather than accommodation for gravid marsupia.

Quadrula verrucosa has not been reported from prehistoric shell middens in the Tennessee River drainage (Hughes and Parmalee, 1999). This suggests that it was uncommon or is a recent invader to the drainage. However, Ortmann (1925) reported it to occur in the Tennessee River across northern Alabama during the early 1900s prior to the river's impoundment.

Synonymy

Obliquaria (*Ellipsaria*) *verrucosa* Rafinesque, 1820. Rafinesque, 1820:304, pl. 81, figs. 10–12
Type locality: Ohio River. Lectotype, ANSP 20235a, length 100 mm, designated by Johnson and Baker (1973).

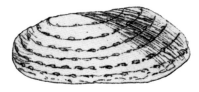 

Unio tuberculatus Barnes, 1823. Barnes, 1823:125, pl. 7, figs. 8a, b
Unio (*Theliderma*) *pustulata* Swainson, 1840, *non* Lea, 1831. Swainson, 1840:271, fig. 54d
Unio conjugans B.H. Wright, 1899. B.H. Wright, 1899:89
Quadrula tritogonia Ortmann, 1909. Ortmann, 1909b:101 [new name for *Tritogonia tuberculata* (Barnes, 1823)]
Quadrula parkeri Geiser, 1911. Geiser, 1911:15 [unnecessary replacement name for *Tritogonia tuberculata* (Barnes, 1823)]

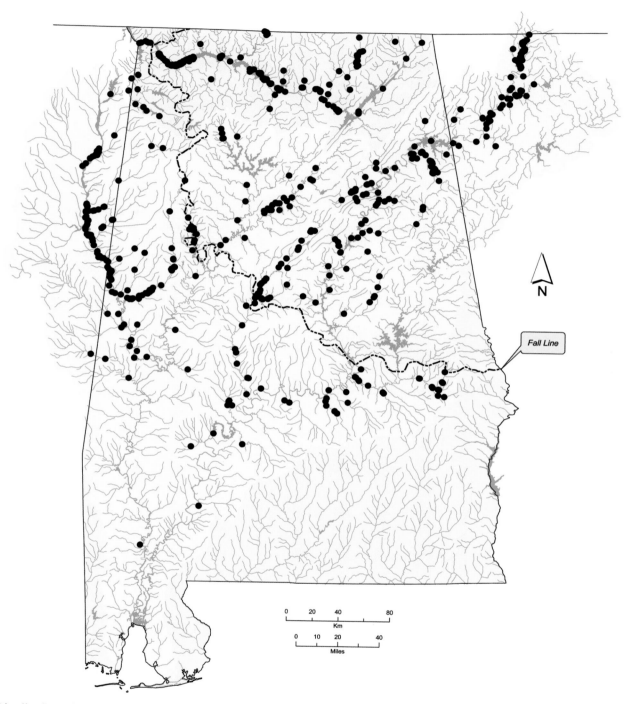

Distribution of *Quadrula verrucosa* in Alabama and the Mobile Basin.

Genus *Strophitus*

Strophitus Rafinesque, 1820, occurs in the Hudson Bay, Great Lakes and Mississippi basins. It is also found in Atlantic Coast drainages from Nova Scotia to Georgia and Gulf Coast drainages from Alabama to Texas. Turgeon et al. (1998) listed three *Strophitus* species, all of which occur in Alabama.

Preliminary molecular phylogenetic studies suggest the genus may not be monophyletic (Bogan, unpublished data). One species was reported to discharge glochidia bound in a matrix of mucus (Haag and Warren, 1997). This difference in life history traits is further suggestive of polyphyly in the genus.

At least one species of *Strophitus* discharges glochidia embedded in unusual rod-shaped gelatinous conglutinates (Figure 13.39). Glochidia emerge from the interior of the conglutinate after it is discharged (Sterki, 1898a; Watters, 2002).

Figure 13.39. *Strophitus undulatus* conglutinate. From Lefevre and Curtis (1912).

Type Species
Anodonta undulata Say, 1817

Diagnosis
Shell elliptical to rhomboidal; thin to moderately thick; inflated; hinge plate reduced, pseudocardinal teeth rudimentary, lateral teeth rarely present; posterior ridge broadly rounded; disk unsculptured; umbo inflated, umbo sculpture a few strong concentric ridges.

Outer surface of mantle margin with squarish blotches posteriorly; inner lamellae of inner gills usually connected to visceral mass only anteriorly; mantle bridge separating excurrent and supra-anal apertures short; excurrent aperture papillate or crenulate; outer gills marsupial; glochidia held across entire gill; marsupium distended when gravid, not extended beyond original gill ventral edge; secondary water tubes present when gravid; glochidium with styliform hooks (Rafinesque, 1820; Simpson, 1900b, 1914; Ortmann, 1912a; Haas, 1969b; Britton and Fuller, 1980).

Synonymy
Pseudodontoideus Frierson, 1927

Strophitus connasaugaensis (Lea, 1858)
Alabama Creekmussel

Strophitus connasaugaensis – Length 93 mm, UF 4244. Conasauga River, Polk County, Tennessee. © Richard T. Bryant.

Shell Description

Length to 120 mm; thin; moderately compressed to moderately inflated; outline elongate oval to rhomboidal, older individuals may be arcuate; posterior margin narrowly rounded to obliquely truncate; anterior margin rounded; dorsal margin straight to convex; ventral margin straight to slightly convex or slightly concave; posterior ridge low, rounded; posterior slope usually not steep; umbo broad, usually compressed, elevated slightly above hinge line, umbo sculpture concentric loops, wider on posterior ridge; periostracum yellowish green to tawny, may darken to brown or black with age, occasionally with sparse thin green rays posteriorly.

Pseudocardinal teeth poorly developed, compressed, may be erect, 1 in each valve; lateral teeth rudimentary, little more than thickened hinge lines; umbo cavity shallow, open; nacre white, often with gold blotches.

Soft Anatomy Description

Mantle tan, may become creamy white posteriorly, grayish brown outside of apertures; visceral mass creamy white to pale tan; foot orange.

Gills tan; dorsal margin slightly sinuous, ventral margin elongate convex; gill length 55–60% of shell length; gill height 50–55% of gill length; outer gill height 70–80% of inner gill height; inner lamellae of inner gills connected to visceral mass only anteriorly.

Outer gills marsupial; marsupium padded when gravid; brown (M.L. Buntin, personal communication).

Labial palps tan; straight to slightly concave dorsally, convex ventrally, narrowly rounded to bluntly pointed distally; palp length 15–20% of gill length; palp height 65–75% of palp length; distal 35–50% of palps bifurcate.

Incurrent aperture longer or equal to excurrent aperture; supra-anal aperture shorter than incurrent and excurrent apertures.

Incurrent aperture length approximately 20% of shell length; pale tan within, brown basal to papillae; papillae in 2 rows, inner row larger, thick, blunt, mostly simple, with occasional bifid papillae; papillae mostly brown, some may be tan.

Excurrent aperture length 15–20% of shell length; pale tan with grayish cast within, marginal color band brown, grayish cast usually absent from area just proximal to marginal band; margin smooth.

Supra-anal aperture length 10–15% of shell length; creamy white within, with grayish or rusty cast, no marginal color band; margin smooth; mantle bridge separating supra-anal and excurrent apertures imperforate, length 45–55% of supra-anal length.

Specimens examined: Shoal Creek, Cleburne County (n = 3).

Glochidium Description

Length 359–375 μm; height 375–391 μm; with styliform hooks; outline depressed pyriform; dorsal margin straight to slightly convex, ventral margin produced into small nipple-like projection at ventral terminus, anterior and posterior margins deeply convex, anterior margin slightly more convex than posterior margin (M.L. Buntin, personal communication).

Similar Species

Strophitus connasaugaensis resembles *Strophitus subvexus* but is usually more elongate, has a more compressed umbo and has a lower, rounder posterior ridge. The periostracum of *S. connasaugaensis* is often more yellowish than that of *S. subvexus*, which tends to be darker and browner. *Strophitus connasaugaensis* often becomes arcuate with age, but *S. subvexus* seldom does. Pseudocardinal teeth of *S. connasaugaensis* are poorly developed, but may be better developed than those of *S. subvexus*. Additional comparative morphological and genetic information is needed to determine if these species occur sympatrically.

Strophitus connasaugaensis also resembles *Anodontoides radiatus* but is often more rhomboidal in outline. The periostracum of *S. connasaugaensis* is usually more yellow than that of *A. radiatus*, which is usually dark greenish brown with green rays of varying width covering much of the shell disk. *Strophitus connasaugaensis* seldom has rays, but when present, they are weak and associated with the posterior ridge. Pseudocardinal teeth of *A. radiatus* are more weakly developed than those of *S. connasaugaensis*.

General Distribution

Strophitus connasaugaensis is endemic to the eastern portion of the Mobile Basin of Alabama, Georgia and Tennessee (but see Remarks).

Alabama and Mobile Basin Distribution

Strophitus connasaugaensis appears to be confined to the Alabama, Cahaba, Coosa and lower Tallapoosa River drainages (but see Remarks).

Strophitus connasaugaensis is extant in widespread, isolated populations. It is rare in most areas where extant.

Ecology and Biology

Strophitus connasaugaensis primarily inhabits creeks, but it occasionally occurred in large rivers historically. It is usually found in areas with moderate current in sand and gravel substrates.

Strophitus connasaugaensis is a short-term brooder, gravid from mid October to January. Mean annual glochidial production was estimated at 91,433 per female (M.L. Buntin, personal communication). *Strophitus connasaugaensis* is a glochidial host generalist. Species found to serve as hosts in laboratory trials include *Hypentelium etowanum* (Alabama Hog Sucker) (Catostomidae); *Lepomis cyanellus* (Green Sunfish), *Lepomis macrochirus* (Bluegill), *Lepomis megalotis* (Longear Sunfish), *Micropterus coosae* (Redeye Bass) and *Micropterus punctulatus* (Spotted Bass) (Centrarchidae); *Cottus carolinae* (Banded Sculpin) (Cottidae); *Cyprinella callistia* (Alabama Shiner), *Cyprinella trichroistia* (Tricolor Shiner), *Cyprinella venusta* (Blacktail Shiner), *Luxilus chrysocephalus* (Striped Shiner), *Notropis baileyi* (Rough Shiner) and *Notropis buccatus* (Silverjaw Minnow) (Cyprinidae); *Fundulus olivaceous* (Blackspotted Topminnow) (Fundulidae); *Ameirus natalis* (Yellow Bullhead) (Ictaluridae); and *Etheostoma jordani* (Greenbreast Darter), *Percina kathae* (Mobile Logperch), *Percina palmeris* (Bronze Darter) and *Percina nigrofasciata* (Blackbanded Darter) (Percidae) (M.L. Buntin, personal communication). *Cottus carolinae* appears to be the best host. *Ameirus natalis* appears to be a scondary host.

Current Conservation Status and Protection

Strophitus connasaugaensis was considered a species of special concern throughout its range by Williams et al. (1993) and in Alabama by Lydeard et al. (1999). Garner et al. (2004) designated it a species of high conservation concern in the state.

Remarks

There remains some uncertainty regarding the exact distribution of *Strophitus connasaugaensis* and *Strophitus subvexus*. The distribution of *S. connasaugaensis* was herein confined to eastern drainages of the Mobile Basin, and that of *S. subvexus* restricted to the Tombigbee and Black Warrior River drainages. However, individuals resembling *S. connasaugaensis* can occasionally be found in the Black Warrior and Tombigbee River drainages and individuals resembling *S. subvexus* can sometimes be found in eastern reaches of the Mobile Basin. Therefore, it is unclear if these two species occur sympatrically in one or more drainages, especially downstream of the Fall Line. Further work is needed to elucidate distributions of these taxa.

Strophitus connasaugaensis was named for the Conasauga River, from which the type specimens were collected. Historical spelling of the river's name was Connasauga, but modern usage is Conasauga. The word conasauga is from the Cherokee word *kahnasagah*, meaning "grass" (Krakow, 1975).

Synonymy

Margaritana connasaugaensis Lea, 1858. Lea, 1858b:138; Lea, 1859f:229, pl. 32, fig. 113; Lea, 1859g:47, pl. 32, fig. 113

Type locality: Connasauga [Conasauga] River, one of the headwaters of the Alabama River, Gilmer County, Georgia, Bishop Elliott. Lectotype, USNM 86277, length 63 mm, designated by Johnson (1974).

Margaritana gesnerii Lea, 1858. Lea, 1858b:138; Lea, 1862c:211, pl. 32, fig. 280; Lea, 1863a:33, pl. 32, fig. 280

Type locality: Uphaupee [Uphapee] Creek, Alabama; [Chattahoochee River] below Columbia, [Houston County,] Georgia [Alabama], W. Gesner and G. Hallenbeck. Lectotype, USNM 86212, length 92 mm, is from Uphapee Creek, Alabama.

Anodonta hallenbeckii Lea, 1858. Lea, 1858b:139; Lea, 1859f:232, pl. 32, fig. 112; Lea, 1859g:50, pl. 32, fig. 112

Type locality: Uphaupee [Uphapee] Creek, Macon [County], Georgia [Alabama]. Lectotype, USNM 86428, length 103 mm.

Margaritana alabamensis Lea, 1861. Lea, 1861a:41; Lea, 1862b:104, pl. 16, fig. 249; Lea, 1862d:108, pl. 16, fig. 249

Type locality: Talladega Creek, [Talladega County,] Alabama. Holotype, USNM 86262, length 105 mm.

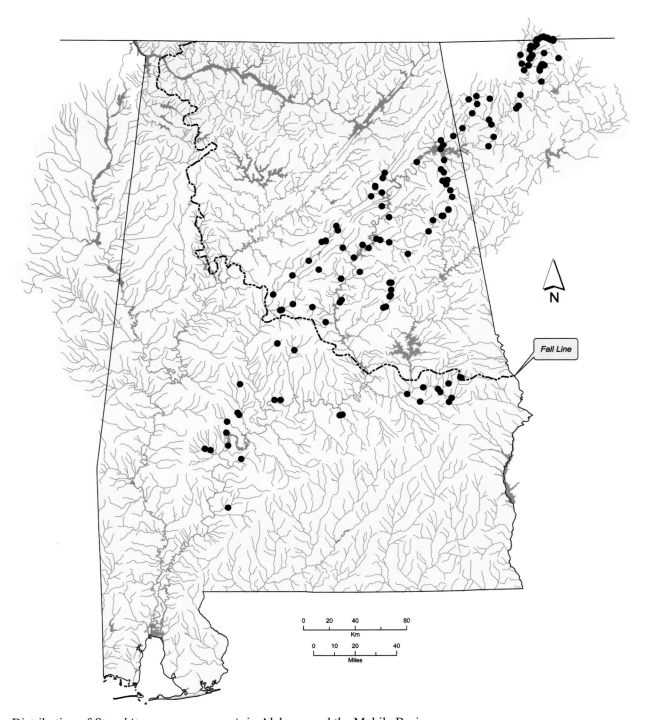

Distribution of *Strophitus connasaugaensis* in Alabama and the Mobile Basin.

Strophitus subvexus (Conrad, 1834)
Southern Creekmussel

Strophitus subvexus – Length 77 mm, UF 358675. Bull Mountain Creek, U.S. Highway 78 (new highway) near Tremont, Itawamba County, Mississippi, 19 September 2000. © Richard T. Bryant.

Shell Description

Length to 111 mm; thin; moderately compressed to moderately inflated; outline oval to rhomboidal; posterior margin rounded to bluntly pointed or obliquely truncate; anterior margin rounded; dorsal margin convex; ventral margin usually convex, may be straight or slightly concave; posterior ridge distinct, angled dorsally; posterior slope moderately steep; umbo broad, moderately inflated, elevated above hinge line, umbo sculpture thick concentric ridges, may be knobby on posterior ridge; periostracum greenish brown to black, often with weak green rays posteriorly.

Pseudocardinal teeth rudimentary, usually represented by thickened hinge line; lateral teeth absent but hinge line may be slightly thickened; umbo cavity shallow, open; nacre white, may have gold blotches.

Soft Anatomy Description

Mantle creamy white to tan, gold or light brown, occasionally pale orange outside of pallial line, usually dark grayish brown outside of apertures, occasionally pale orange with grayish mottling; visceral mass pearly white to tan; foot dull orange to pinkish orange.

Gills tan to gold or light brown; dorsal margin slightly sinuous, ventral margin elongate convex, occasionally almost straight; gill length 50–60% of shell length; gill height 40–50% of gill length; outer gill height 40–50% of inner gill height; inner lamellae of inner gills usually connected to visceral mass only anteriorly, occasionally completely connected.

Outer gills marsupial; glochidia held in posterior 30% of gill; marsupium padded when gravid; reddish brown.

Labial palps creamy white to tan or light brown; straight to slightly concave or convex dorsally, convex ventrally, bluntly pointed distally; palp length 15–35% of gill length; palp height 50–70% of palp length; distal 30–45% of palps bifurcate.

Incurrent aperture usually longer than excurrent and supra-anal apertures; excurrent aperture longer, shorter or equal to supra-anal aperture.

Incurrent aperture length 10–20% of shell length; creamy white to tan or rusty brown within, dark brown or grayish brown basal to papillae, may be dull orange with brown mottling; papillae in 2–3 rows, usually short, thick, blunt, occasionally long and slender, mostly simple, usually with a few bifid; papillae usually some combination of pale orange, tan, reddish brown and dark brown, often changing color distally.

Excurrent aperture length 10–15% of shell length; creamy white to pale tan within, marginal color band usually dark brown, often with tan, dull orange or reddish brown blotches that may be in regular pattern or as irregular mottling; margin smooth but undulating, slightly crenulate or with very small simple papillae; papillae rusty brown to dark brown.

Supra-anal aperture length 10–15% of shell length; creamy white to pale orange or gold within, may have thin gray or brown marginal band; margin smooth; mantle bridge separating supra-anal and excurrent apertures imperforate, length 40–65% of supra-anal length.

Specimens examined: Brushy Creek, Lawrence County (n = 4); Bull Mountain Creek, Marion County (n = 1); Yellow Creek, Lowndes County, Mississippi (n = 1).

Glochidium Description

Length 348–359 µm; height 288–292 µm; with styliform hooks; outline depressed pyriform; dorsal margin straight, ventral margin produced into small nipple-like projection at ventral terminus, anterior and posterior margins deeply convex, anterior margin slightly more convex than posterior margin (Hoggarth, 1999).

Similar Species

Strophitus subvexus resembles *Strophitus connasaugaensis* but is usually less elongate, has a more inflated umbo and has a more prominent posterior ridge. The periostracum of *S. connasaugaensis* is usually yellowish except in some old individuals, whereas that of *S. subvexus* is darker and greenish brown to black. *Strophitus subvexus* seldom becomes arcuate with age, as does *S. connasaugaensis*. Pseudocardinal teeth of *S. connasaugaensis* are poorly developed, but may be better developed than those of *S. subvexus*. Additional comparative morphological and genetic information is needed to determine if these species occur sympatrically.

Strophitus subvexus also resembles *Anodontoides radiatus* but is less elongate in outline. *Anodontoides radiatus* usually has distinct dark green rays of varying width covering the shell disk, whereas the rays of *S. subvexus* are absent or weak and usually associated with the posterior slope.

General Distribution

Strophitus subvexus has been reported from the Mobile Basin, the Tangipahoa River of the Lake Pontchartrain drainage and a tributary of the Mississippi River in extreme southwestern Mississippi (Vidrine, 1993; Jones et al., 2005). The taxonomic status of *Strophitus* in the Calcasieu and Sabine River drainages of western Louisiana is unclear (Vidrine, 1993; Howells et al., 1996).

Alabama and Mobile Basin Distribution

Strophitus subvexus appears to be confined to the Black Warrior and Tombigbee River drainages of Alabama and Mississippi (but see Remarks).

Strophitus subvexus is extant in widespread, isolated tributary populations.

Ecology and Biology

Strophitus subvexus occurs in a variety of habitats, ranging from small creeks to large rivers, where it may be found in pools or areas with moderate current. Substrates in which *S. subvexus* may occur include sandy mud, sand and sandy gravel.

Strophitus subvexus is a long-term brooder, gravid from late summer or autumn to the following spring or summer. Gravid females have been reported in February and March (Haag and Warren, 1997). Glochidia are discharged bound in a matrix of mucus, presumably to serve as an aid for infesting host fish by entanglement (Haag and Warren, 1997).

Glochidial hosts reported for *Strophitus subvexus*, based on laboratory trials, include *Hypentelium etowanum* (Alabama Hog Sucker) (Catostomidae); *Lepomis megalotis* (Longear Sunfish), *Micropterus coosae* (Redeye Bass) and *Micropterus salmoides* (Largemouth Bass) (Centrarchidae); *Campostoma oligolepis* (Largescale Stoneroller), *Cyprinella callistia* (Alabama Shiner) and *Semotilus atromaculatus* (Creek Chub) (Cyprinidae); *Fundulus olivaceus* (Blackspotted Topminnow) (Fundulidae); *Ictalurus punctatus* (Channel Catfish) (Ictaluridae); and *Etheostoma artesiae* (Redspot Darter), *Etheostoma douglasi* (Tuskaloosa Darter), *Etheostoma whipplei* (Redfin Darter) and *Percina nigrofasciata* (Blackbanded Darter) (Percidae) (Haag and Warren, 1997). However, the viability of juveniles produced on *I. punctatus* and *M. coosae* was questionable. More juveniles were recovered from *H. etowanum* and *E. douglasi* than from the other species (Haag and Warren, 1997).

Current Conservation Status and Protection

Strophitus subvexus was considered endangered throughout its range by Stansbery (1971). It was deemed a species of special concern throughout its range by Williams et al. (1993) and in Alabama by Lydeard et al. (1999). Garner et al. (2004) designated *S. subvexus* a species of moderate conservation concern in the state.

Remarks

There remains some uncertainty regarding the exact distributions of *Strophitus subvexus* and *Strophitus connasaugaensis*. The distribution of *S. subvexus* was herein confined to the Tombigbee and Black Warrior River drainages, and that of *S. connasaugaensis* restricted to eastern drainages of the Mobile Basin. However, individuals resembling *S. subvexus* can sometimes be found in eastern reaches of the Mobile Basin and individuals resembling *S. connasaugaensis* can occasionally be found in the Black Warrior and Tombigbee River drainages. Therefore, it is unclear if these two species occur sympatrically in one or more drainages, especially downstream of the Fall Line. Further work is needed to elucidate distributions of these taxa.

Strophitus subvexus has been reported from Gulf Coast drainages east of the Mobile Basin by some authors (e.g., van der Schalie, 1940; Johnson, 1967a; Brim Box and Williams, 2000). However, based on recent genetic analyses, individuals conchologically similar to *S. subvexus* in those drainages were determined to be *Anodontoides radiatus* (Hsiu-Ping Liu, personal communication).

The taxonomic status of *Strophitus* in western Gulf Coast drainages is unclear. Strecker (1931) reported *Strophitus subvexus* from eastern Texas, but

Howells et al. (1996) considered all individuals of the genus in that state to belong to *Strophitus undulatus*. Vidrine (1993) suggested that *Strophitus* from western Louisiana (i.e., those in the Calcasieu and Sabine rivers) represent an undescribed species similar to *S. subvexus*.

Synonymy

Anodonta subvexa Conrad, 1834. Conrad, 1834a:341, pl. 1, fig. 12
Type locality: Black Warrior River, [Alabama]. Type specimen not found, reported lost by Johnson and Baker (1973). Length of figured shell in original description reported as about 51 mm.

Margaritana spillmanii Lea, 1858. Lea, 1858b:138; Lea, 1862b:105, pl. 17, fig. 252; Lea, 1862d:109, pl. 17, fig. 252
Type locality: Tombecbee [Tombigbee] River, near Columbus, [Lowndes County,] Mississippi, W. Spillman, M.D. Lectotype, USNM 86278, length 93 mm, designated by Johnson (1974).

Margaritana tombecbeensis Lea, 1858. Lea, 1858b:138
Type locality: Tombecbee [Tombigbee] River, near Columbus, [Lowndes County,] Mississippi, W. Spillman, M.D. Lectotype, USNM 86253, length 77 mm, designated by Johnson (1974).
Comment: The name Tombecbee may be derived from a fort erected in 1736 by French traders, Fort Tombecbee, along the Tombigbee River in Sumter County, Alabama.
Margaritana tombigbeensis Lea, 1862. Lea, 1862b:107, pl. 18, fig. 255; Lea, 1862d:111, pl. 18, fig. 255
Comment: This appears to be a new name proposed for *Margaritana tombecbeensis* Lea, 1858, as the same figured shell is referred to as the type of both taxa. However, there is no statement to this effect in the description of *M. tombigbeensis*.

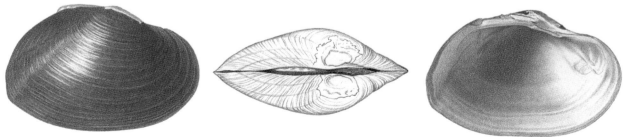

Margaritana columbensis Lea, 1867. Lea, 1867a:81
Type locality: Tombigbee River, near Columbus, [Lowndes County,] Mississippi, W. Spillman, M.D. Type specimen not found, reported lost by Johnson and Baker (1973).

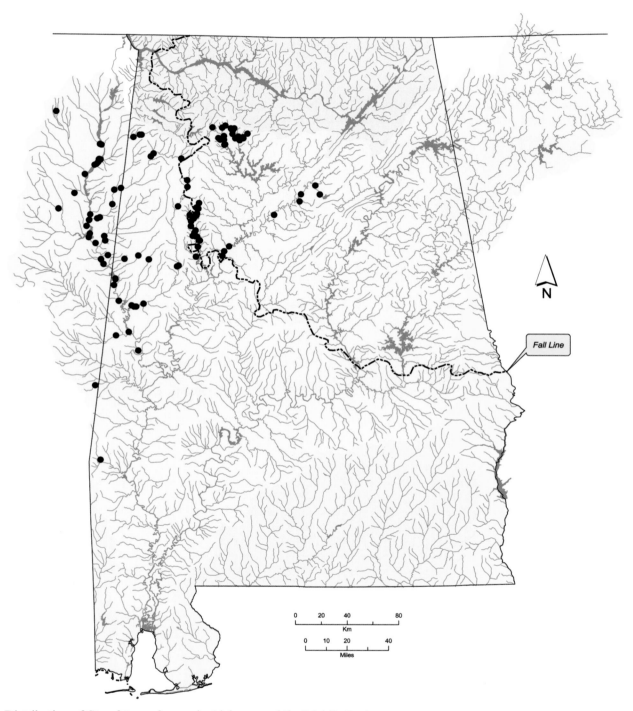

Distribution of *Strophitus subvexus* in Alabama and the Mobile Basin.

Strophitus undulatus (Say, 1817)
Creeper

Strophitus undulatus – Length 78 mm, UNA 118. Muskingum River, below dam, east bank, Lowell, Washington County, Ohio, 11 August 1962. © Richard T. Bryant.

Shell Description

Length to 115 mm; thin, may become moderately thick with age; moderately inflated; outline subrhomboidal to oval; posterior margin narrowly rounded to obliquely truncate; anterior margin narrowly to broadly rounded; dorsal margin sinuous; ventral margin straight to slightly convex; posterior ridge rounded; posterior slope moderately steep; umbo broad, flat, elevated above hinge line, umbo sculpture single-looped ridges, oblique to hinge line, somewhat nodulous at posterior ridge; periostracum yellowish to greenish in subadults, becoming brown or black with age, young individuals often with green rays that may be wavy.

Pseudocardinal teeth rudimentary, represented by thickened hinge line, slightly sinuous; lateral teeth absent but hinge line somewhat thickened; umbo cavity moderately deep, open; nacre white, may have salmon tint in shell cavity.

Soft Anatomy Description

No material was available for soft anatomy description, but Lea (1863d) provided a brief description based on material from the Ohio River and Fox River, Illinois, as well as Atlantic Coast drainages. The color of the "mass" is salmon, more intense on the mantle margin and foot. Gills are convex ventrally, with inner lamellae of inner gills completely connected to the visceral mass or connected only anteriorly. Inner gills are larger than outer gills. Outer gills are marsupial, with glochidia held across the entire gill "in transverse ovisacks" and are brownish when gravid. Labial palps are "light salmon color, transverse, subelliptical". Incurrent and excurrent apertures are large, with brown papillae, though those of the excurrent aperture are minute, and may be "few" and

"imperfect". The supra-anal aperture is "moderately large", separated from the excurrent aperture by a mantle bridge.

Glochidium Description

Length 350–369 μm; height 285–300 μm; with styliform hooks; outline depressed pyriform; dorsal margin straight, ventral margin produced into small nipple-like projection at ventral terminus, anterior and posterior margins deeply convex, anterior margin slightly more convex than posterior margin (Surber, 1912; Hoggarth, 1993, 1999).

Similar Species

Strophitus undulatus resembles *Pyganodon grandis* but is less inflated, with a lower umbo. The ventral margin of *S. undulatus* is generally straighter than that of *P. grandis*, which tends to be broadly rounded. *Pyganodon grandis* umbo sculpture is double-looped, but that of *S. undulatus* consists of single loops. *Strophitus undulatus* typically has rudimentary pseudocardinal teeth, whereas *P. grandis* is completely without teeth.

Small *Strophitus undulatus* may resemble *Lasmigona holstonia* externally. However, *L. holstonia* has small but distinctive pseudocardinal teeth, whereas those of *S. undulatus* are rudimentary.

General Distribution

Strophitus undulatus is known from the southern Hudson Bay Basin as well as parts of the Great Lakes Basin (Goodrich and van der Schalie, 1932; La Rocque and Oughton, 1937; Dawley, 1947). It occurs in Atlantic Coast drainages from Nova Scotia south to the Savannah River drainage in South Carolina. It is widespread in the Mississippi Basin from southern

Minnesota (Dawley, 1947) south to Louisiana (Vidrine, 1993) and from headwaters of the Ohio River drainage in New York (Strayer et al., 1992) west to South Dakota (Backlund, 2000) and Kansas (Murray and Leonard, 1962). *Strophitus undulatus* historically occurred in most of the Cumberland and Tennessee River drainages in Alabama, Kentucky, North Carolina, Tennessee and Virginia (Cicerello et al., 1991; Ahlstedt, 1992a; Parmalee and Bogan, 1998). It has been reported from some Gulf Coast drainages from the Mississippi River west to the Guadalupe River drainage in Texas (Howells et al., 1996).

Alabama and Mobile Basin Distribution

Strophitus undulatus is confined to the Tennessee River drainage, where it is known from the Tennessee River and a few tributaries.

Strophitus undulatus appears to be extirpated from Alabama with the exception of a short reach of Bear Creek, Colbert County, where it is very rare.

Ecology and Biology

Strophitus undulatus occurs in a variety of habitats across a number of physiographic areas including streams with steep gradient as well as coastal plain streams. *Strophitus undulatus* is occasionally found in shoals but typically inhabits areas with less current, such as pools, lakes and canals. It can be found in substrates of mud, sand or gravel and may occur in beds of *Justicia americana* (Waterwillow) (Ortmann, 1919). Ortmann (1919) reported it to occur in the tidewater of the Delaware River.

Strophitus undulatus is a long-term brooder, reported gravid from as early as July to as late as June of the following year (Ortmann, 1909a, 1919; Surber, 1912; Utterback, 1915, 1916a). However, brooding period appears to be somewhat variable among populations in different parts of its range (Conner, 1909; Baker, 1928a). *Strophitus undulatus* glochidia are discharged bound in conglutinates that are short, white, gelatinous, cylindrical rods resembling maggots or worms (Sterki, 1898a; Lefevre and Curtis, 1911; Watters, 2002). Upon their discharge from the marsupium the conglutinates swell and become more translucent (Sterki, 1898a). With the swelling, individual glochidia emerge from the interior of the conglutinate but remain attached to it by glochidial threads (Lea, 1863a; Sterki, 1898a; Watters, 2002). Sterki (1898a) suggested that emergence of the glochidia is in response to the swelling of the conglutinate, as each glochidium is "dislodged from its cavity, evidently expelled by swelling of the surrounding substance, and the exit facilitated by its softening". The hypothesis that emergence of glochidia from the conglutinate is due to osmotic pressure was confirmed with laboratory experiments (Watters, 2002).

Emergence of glochidia from *S. undulatus* conglutinates is accompanied by a writhing motion, which may help attract the attention of glochidial hosts (Watters, 2002). Ortmann (1910a) reported that sometimes glochidia become free of the conglutinate in the suprabranchial chamber prior to discharge.

Strophitus undulatus is a generalist with regard to glochidial hosts. Fishes shown to serve as hosts in laboratory trials include *Acipencer oxyrinchus* (Atlantic Sturgeon) (Acipenseridae); *Ambloplites rupestris* (Rock Bass), *Lepomis cyanellus* (Green Sunfish), *Lepomis gibbosus* (Pumpkinseed), *Lepomis macrochirus* (Bluegill), *Lepomis microlophus* (Redear Sunfish), *Micropterus dolomieu* (Smallmouth Bass), *Micropterus salmoides* (Largemouth Bass), *Pomoxis annularis* (White Crappie) and *Pomoxis nigromaculatus* (Black Crappie) (Centrarchidae); *Cottus cognatus* (Slimy Sculpin) (Cottidae); *Campostoma anomalum* (Central Stoneroller), *Cyprinella spiloptera* (Spotfin Shiner), *Luxilus cornutus* (Common Shiner), *Nocomis micropogon* (River Chub), *Notropis hudsonius* (Spottail Shiner), *Notropis stramineus* (Sand Shiner), *Phoxinus eos* (Northern Redbelly Dace), *Pimephales notatus* (Bluntnose Minnow), *Pimephales promelas* (Fathead Minnow), *Rhinichthys atratulus* (Eastern Blacknose Dace), *Rhinichthys cataractae* (Longnose Dace) and *Semotilus atromaculatus* (Creek Chub) (Cyprinidae); *Lota lota* (Burbot) (Gadidae); *Culaea inconstans* (Brook Stickleback) (Gasterosteidae); *Ameiurus melas* (Black Bullhead), *Ameiurus natalis* (Yellow Bullhead) and *Ictalurus punctatus* (Channel Catfish) (Ictaluridae); *Etheostoma caeruleum* (Rainbow Darter), *Etheostoma exile* (Iowa Darter), *Etheostoma flabellare* (Fantail Darter), *Etheostoma nigrum* (Johnny Darter), *Etheostoma olmstedi* (Tessellated Darter), *Etheostoma zonale* (Banded Darter), *Perca flavescens* (Yellow Perch), *Percina caprodes* (Logperch), *Percina maculata* (Blackside Darter), *Percina phoxocephala* (Slenderhead Darter) and *Sander vitreus* (Walleye) (Percidae); and *Umbra limi* (Central Mudminnow) (Umbridae) (Hove et al., 1997; Watters et al., 1998; Cliff et al., 2001; van Snik Gray et al., 2002; Watters et al., 2005). *Strophitus undulatus* glochidia have also been reported to transform on a fish nonindigenous to the Mississippi Basin, *Oncorhynchus mykiss* (Rainbow Trout) (Salmonidae), as well as on *Notophthalmus viridescens* (Red-spotted Newt) (van Snik Gray et al., 2002). Ellis and Keim (1918) reported *S. undulatus* glochidia attaching to excised gills of *Fundulus zebrinus* (Plains Killifish) (Fundulidae) but did not follow the experiment with infestation of living fish. Lefevre and Curtis (1910b, 1911) provided strong support that some *S. undulatus* undergo metamorphosis from glochidium to juvenile mussel inside the conglutinate, without benefit of a host. However, in a control treatment van Snik Gray et al. (2002) observed no metamorphosis without potential hosts.

Current Conservation Status and Protection

Strophitus undulatus was considered currently stable throughout its range by Williams et al. (1993). Lydeard et al. (1999) listed it as possibly extirpated from Alabama. Indeed, the viability of the only known remaining population of *S. undulatus* in Alabama is questionable. Garner et al. (2004) designated *S.*

undulatus a species of highest conservation concern in the state.

Remarks

Strophitus undulatus is represented in the archaeological record of Alabama by a single specimen collected from a prehistoric midden at Muscle Shoals on the Tennessee River (Morrison, 1942; Hughes and Parmalee, 1999).

Synonymy

Anodonta undulata Say, 1817. Say, 1817:[no pagination], pl. 3, fig. 6

Type locality: No locality given for *Anodonta undulata*. Johnson (1970) restricted the type locality to Schuylkill River, near Philadelphia, Philadelphia County, Pennsylvania. Type specimen not found in ANSP.

Anodonta pensylvanica Lamarck, 1819. Lamarck, 1819:86
Anodon rugosus Swainson, 1822. Swainson, 1822:[no pagination], pl. 96
Alasmodonta edentula Say, 1829. Say, 1829:340
Anodon areolatus Swainson, 1829. Swainson, 1829:[no pagination], pl. 18
Anodonta virgata Conrad, 1836. Conrad, 1836:Part 5, front cover
Margarita (Anodonta) wardiana Lea, 1836, *nomen nudum*. Lea, 1836:50
Anodonta wardiana Lea, 1838. Lea, 1838a:46, pl. 14, fig. 42; Lea, 1838c:46, pl. 14, fig. 42
Anodonta pavonia Lea, 1838. Lea, 1838a:78, pl. 21, fig. 65; Lea, 1838c:78, pl. 21, fig. 65
Anodon unadilla De Kay, 1843. De Kay, 1843:199, pl. 15, fig. 228
Anodonta tetragona Lea, 1845. Lea, 1845:165; Lea, 1848a:82, pl. 8, fig. 25; Lea, 1848b:82, pl. 8, fig. 25
Anodonta arkansasensis Lea, 1852, *nomen nudum*. Lea, 1852a:251–252
Anodonta shaefferiana Lea, 1852, *nomen nudum*. Lea, 1852a:252
Anodonta shaefferiana Lea, 1852. Lea, 1852b:288, pl. 26, fig. 50; Lea, 1852c:44, pl. 26, fig. 50
Anodonta arkansasensis Lea, 1852. Lea, 1852b:293, pl. 29, fig. 56; Lea, 1852c:49, pl. 29, fig. 56
Alasmodon rhombica Anthony, 1865. Anthony, 1865:158, pl. 12, fig. 5
Anodon papyraceus Anthony, 1865. Anthony, 1865:161
Anodon annulatus Sowerby, 1867. Sowerby, 1867:[no pagination], pl. 18, fig. 67
Anodon quadriplicatus Sowerby, 1867. Sowerby, 1867:[no pagination], pl. 28, fig. 110
Strophitus undulatus ovatus Frierson, 1927. Frierson, 1927:22
Strophitus undulatus tennesseensis Frierson, 1927. Frierson, 1927:22
Strophitus rugosus pepinensis F.C. Baker, 1928. F.C. Baker, 1928a:204, pl. 74, fig. 8
Strophitus rugosus winnebagoensis F.C. Baker, 1928. F.C. Baker, 1928a:205, pl. 74, figs. 1–6
Strophitus rugosus lacustris F.C. Baker, 1928. F.C. Baker, 1928a:207, pl. 75, figs. 6–8

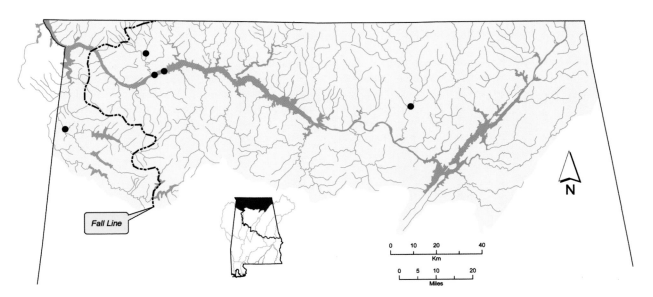

Distribution of *Strophitus undulatus* in the Tennessee River drainage of Alabama.

Genus *Toxolasma*

Toxolasma Rafinesque, 1831, is found in the Great Lakes and Mississippi basins, Atlantic Coast drainages from North Carolina to Florida, and Gulf Coast drainages from peninsular Florida to the Rio Grande. Turgeon et al. (1998) listed eight species in this genus, five of which occur in Alabama. Also included is a *Toxolasma* found in Gulf Coast drainages in Alabama and the Florida panhandle that is believed to represent an undescribed species. Species of *Toxolasma* are among the smallest of North American unionids, most seldom exceeding 40 mm in length.

Toxolasma was described as a new genus with four species. The first attempt at systematic revision of the genus was by Call (1896). However, his otherwise comprehensive review was limited by a dearth of material. Simpson *in* Baker, 1898b, erected the subgenus *Carunculina* in *Lampsilis* Rafinesque, 1820, and included in it one species, *L. parvus* (Barnes, 1823), which is the type species of *Carunculina* by monotypy (Baker, 1964c). Frierson (1914) designated the type species for *Toxolasma* by elimination as *Unio lividus* Rafinesque, 1831 (= *Unio glans* Lea, 1831), and suggested its elevation from a subgenus of *Lampsilis* to genus rank. Ortmann and Walker (1922a) considered *Toxolasma* Rafinesque, 1831, and *Unio lividus* Rafinesque, 1831, *nomina dubia*. Morrison (1969) identified *Toxolasma lividus* Rafinesque, November 1831, as the senior synonym of *Unio glans* Lea,

December 1831. Valentine and Stansbery (1971) first defined and used *Toxolasma* as the senior synonym of *Carunculina*.

When the genus name *Toxolasma* was proposed by Rafinesque (1831), he did not designate a gender nor was it clear from the species he included in the genus. A recent review of *Toxolasma* (Lee, 2006) determined that the gender is neuter. Three species, *Toxolasma corvunculus*, *Toxolasma cylindrellus* and *Toxolasma pullus*, require no change from the current usage as they are nouns in apposition. However, endings for four species are changed: *Toxolasma lividus* to *lividum*; *Toxolasma parvus* to *parvum*; *Toxolasma paulus* to *paulum*; and *Toxolasma texasensis* to *texasense*.

There are two species of *Toxolasma* from the Mobile Basin—described as *Unio germanus* Lea, 1861 (Figure 13.40), and *Unio granulatus* Lea, 1861 (Figure 13.41)—that have not been placed in the synonymy of recognized species. After examination of the types, plus additional lots of museum material, no reliable conchological characters were found to identify or distinguish these taxa with certainty. The status of these taxa will remain unresolved until phylogenetic relationships within the genus *Toxolasma* can be determined, especially regarding those forms in the Mobile Basin and Gulf Coast drainages. Preliminary phylogenetic data suggest that *Toxolasma* as currently used may not be a monophyletic genus.

Figure 13.40. *Unio germanus* Lea, 1861. Lea, 1861a:40; Lea, 1866:49, pl. 19, fig. 54; Lea, 1867b:53, pl. 19, fig. 54. Type locality: Coosa River, Alabama, E.R. Showalter, M.D. Holotype, USNM 85296, length 38 mm.

Figure 13.41. *Unio granulatus* Lea, 1861. Lea, 1861c:60; Lea, 1866:48, pl. 16, fig. 46; Lea, 1867b:52, pl. 16, fig. 46. Type locality: Big Prairie Creek, Alabama, E.R. Showalter, M.D. Holotype, USNM 85297, length 28 mm.

Type Species
Unio lividus Rafinesque, 1831

Diagnosis
Shell moderately thick; somewhat compressed to inflated; oval to elliptical; periostracum cloth-like, dark brown to black, without rays or only faintly rayed; umbo sculpture strong concentric ridges; females often more truncate posteriorly than males.

Mantle margin just ventral to incurrent aperture papillate, in females 1 papilla per valve enlarged into a caruncle (Figure 13.42); caruncle reduced or absent in males; inner lamellae of inner gills connected to visceral mass only anteriorly; outer gills marsupial; glochidia held in posterior portion of gill; marsupium distended when gravid, extended beyond original gill ventral edge; glochidium without styliform hooks (Ortmann, 1912a; Simpson, 1914; Haas, 1969b; Britton and Fuller, 1980; Bogan and Parmalee, 1983).

Synonymy
Carunculina Simpson *in* Baker, 1898

Figure 13.42. *Toxolasma cylindrellus* gravid female with pink caruncles displayed. Estill Fork, Jackson County, Alabama, May 2004. Photograph by P.D. Johnson.

Toxolasma corvunculus of Authors, *non* Lea, 1868
Southern Purple Lilliput

Toxolasma corvunculus – Upper figure: female, length 28 mm, CMNH 61.8469. Middle figure: male, length 23 mm, CMNH 61.8469. Beaver Creek, St. Clair County, Alabama. Lower figure: female, length 28 mm, CMNH 61.8474. Black Warrior River, Squaw Shoals, Jefferson County, Alabama. © Richard T. Bryant.

Shell Description

Length to 33 mm; thin to moderately thick; moderately inflated; outline oval; posterior margin narrowly to broadly rounded or obliquely truncate, females more truncate or broadly rounded than males; anterior margin narrowly rounded; dorsal margin convex; ventral margin straight to convex; posterior ridge low, rounded; posterior slope moderately steep; umbo broad, elevated slightly above hinge line, umbo sculpture single-looped ridges; periostracum tawny to olive or greenish brown, occasionally with weak green rays.

Pseudocardinal teeth moderately thick, triangular, 2 divergent teeth in left valve, 1 tooth in right valve, pseudocardinal teeth of subadults thinner, less divergent; lateral teeth short, straight, 2 in left valve, 1 in right valve, may also have rudimentary lateral tooth ventrally in right valve; interdentum short to moderately long, narrow; umbo cavity shallow; nacre purple, usually lighter distally.

Soft Anatomy Description

No material was available for soft anatomy description, but Lea (1863d) provided some details based on material from Othcalooga Creek, Georgia. The mantle is "thin, with a broad thick margin". A caruncle is located just ventral to the incurrent aperture (probably reduced or absent in males). Gills are "rather small" and convex ventrally, with inner gills larger than outer gills.

Glochidium Description

Without styliform hooks; outline subelliptical (Lea, 1863d).

Similar Species

Toxolasma corvunculus resembles *Toxolasma parvum* but has a yellowish or greenish periostracum and thicker shell. It also attains a larger size, is more strongly sexually dimorphic and has purple nacre instead of white.

Toxolasma corvunculus may resemble small *Villosa lienosa* or *Villosa umbrans*. However, *T. corvunculus* is more inflated than similar-sized individuals of those species. The periostracum of *T. corvunculus* is typically greener than that of *V. lienosa* and *V. umbrans*. Live *T. corvunculus* can be distinguished from either *Villosa* species by the presence of a caruncle, which is replaced by a papillate mantle fold in *Villosa*. These features are rudimentary in males of these species.

General Distribution

Toxolasma corvunculus is endemic to the Mobile Basin of Alabama and Georgia.

Alabama and Mobile Basin Distribution

The distribution of *Toxolasma corvunculus* is unclear. It is known from the Coosa, Cahaba and Black Warrior River drainages. Individuals which appear similar to *T. corvunculus* have also been found in the Alabama, Tallapoosa and Tombigbee River drainages.

Toxolasma corvunculus is believed to be extant but rare in isolated tributaries of the Coosa and Black Warrior rivers and possibly in tributaries of the Alabama, Cahaba and Tombigbee rivers.

Ecology and Biology

Toxolasma corvunculus occurs primarily in creeks but occasionally in medium to large rivers.

Toxolasma corvunculus is presumably a long-term brooder, gravid from late summer or autumn to the following summer. Ortmann (1924a) reported a gravid female from Gordon County, Georgia, collected in July. Glochidial hosts of *T. corvunculus* are unknown.

Current Conservation Status and Protection

The status of *Toxolasma corvunculus* was listed as unknown by Williams et al. (1993). Lydeard et al. (1999) considered it imperiled in Alabama. Garner et al. (2004) designated *T. corvunculus* a species of highest conservation concern in the state.

Remarks

Toxolasma corvunculus is one of the most poorly understood extant mussel species in Alabama. It is rare but an individual was collected from a Black Warrior River tributary in 2001 (Haag and Warren, 2003).

Synonymy

Unio corvunculus Lea, 1868. Lea, 1868a:144; Lea, 1868c:314, pl. 50, fig. 127; Lea, 1869a:74, pl. 50, fig. 127
Type locality: Flint River, Georgia, J.C. Plant and Dr. Neisler; Darien, [Altamaha River, Georgia,] J.H. Couper.
Comment: There was confusion in reporting the type locality for this species. Lea (1868a) gave the above information in the original description, but the specimen he subsequently figured (Lea, 1868c) was from Swamp Creek, Whitfield County, Georgia [Coosa River drainage], collected by Major T.C. Downie. Lea (1868c) did not mention the type localities—Flint River and Darien, Georgia—reported in the original description. Simpson (1914) also reported the type locality as Swamp Creek, Whitfield County, Georgia. The lectotype, USNM 85293, length 19 mm, designated by Johnson (1974), is from Swamp Creek, Whitfield County, Georgia. The lectotype designated does not appear to be the specimen figured below by Lea (1869a).

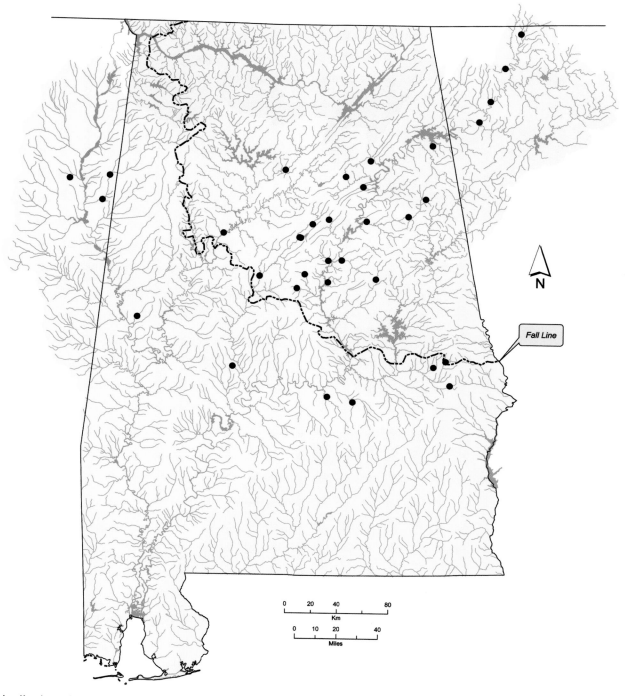

Distribution of *Toxolasma corvunculus* in Alabama and the Mobile Basin.

Toxolasma cylindrellus (Lea, 1868)
Pale Lilliput

Toxolasma cylindrellus – Upper figure: female, length 38 mm, USNM 783176. Lower figure: male, length 28 mm, USNM 783176. Larkin Fork Paint Rock River, ford 3.4 miles southeast of Francisco, Jackson County, Alabama, 28 September 1976. © Richard T. Bryant.

Shell Description

Length to 44 mm; moderately thin; somewhat compressed; outline oval; posterior margin narrowly rounded to obliquely truncate, females more truncate or broadly rounded posteriorly than males; anterior margin rounded; dorsal margin straight to slightly convex; ventral margin straight to slightly convex, marsupial swelling of females often expanded posterioventrally; posterior ridge low, rounded; posterior slope flat, not steep; umbo broad, somewhat compressed, elevated slightly above hinge line, umbo sculpture coarse ridges, somewhat nodulous at posterior ridge; periostracum tawny, without rays.

Pseudocardinal teeth triangular, 2 divergent teeth in left valve, 1 tooth in right valve, may have small blade-like accessory denticle along margin anteriorly; lateral teeth moderately long, straight to slightly curved, 2 in left valve, 1 in right valve; interdentum short to moderately long, narrow to very narrow; umbo cavity shallow, open; nacre light purple or coppery, typically paler outside of pallial line.

Soft Anatomy Description

Mantle light brown, outside of apertures dull orange with black mottling; visceral mass creamy white; foot white to pale pinkish orange.

Females with well-developed, low caruncle just ventral to incurrent aperture; outline somewhat oval; red and golden brown.

Gills tan; dorsal margin straight, ventral margin convex, may be ragged; gill length approximately 40% of shell length; gill height approximately 70% of gill length; outer gill height approximately 75% of inner gill height, inner lamellae of inner gills connected to visceral mass only anteriorly.

Outer gills marsupial; glochidia held in posterior 50% of gill; marsupium thick and moderately distended when gravid, distal margin posteriorly curved; creamy white, distal margin rusty brown.

Labial palps tan; straight dorsally, curved ventrally, bluntly pointed distally; palp length approximately 25% of gill length; palp height approximately 50% of palp length; distal 33% of palps bifurcate.

Incurrent aperture longer than excurrent and supra-anal apertures; supra-anal aperture longer than excurrent aperture.

Incurrent aperture length approximately 15% of shell length; creamy white within; papillae long, slender, simple, in females papillae adjacent to caruncle mostly deeply bifurcate, thick; papillae orange, some with sparse black along edges basally, in females papillae adjacent to caruncle orange with sparse black specks.

Excurrent aperture length approximately 5% of shell length; creamy white within, marginal color band orange; margin with short simple papillae; papillae golden brown.

Supra-anal aperture length approximately 10% of shell length; creamy white within, thin marginal color band golden brown with sparse black specks; margin smooth; mantle bridge separating supra-anal and excurrent apertures imperforate, length approximately 45% of supra-anal length.

Specimen examined: Estill Fork, Jackson County, (n = 1).

Glochidium Description

Glochidium unknown.

Similar Species

Toxolasma cylindrellus resembles *Toxolasma lividum* but is less inflated and has a lighter, often more yellowish, periostracum. Also, the purple nacre of *T. cylindrellus* often has a coppery sheen and is usually lighter in color outside of the pallial line unlike that of *T. lividum*.

Toxolasma cylindrellus may also resemble *Villosa vanuxemensis*. However, *V. vanuxemensis* attains a larger size than *T. cylindrellus*. The periostracum of *V. vanuxemensis* may have green rays, especially posteriorly, whereas that of *T. cylindrellus* is without rays and often more yellowish. Male *V. vanuxemensis* are often more pointed posteriorly than male *T. cylindrellus*. Female *T. cylindrellus* have modified mantle margins just ventral to the incurrent aperture that include a pair of caruncles, whereas *V. vanuxemensis* females have papillate mantle folds and no caruncles. These features are rudimentary in males of both species.

General Distribution

Toxolasma cylindrellus is endemic to middle reaches of the Tennessee River drainage in Alabama and Tennessee and the Duck River system in central Tennessee (Parmalee and Bogan, 1998).

Alabama and Mobile Basin Distribution

Toxolasma cylindrellus historically occurred in the Tennessee River drainage across northern Alabama. There are a few records from the Tennessee River proper, but most are from tributaries.

Toxolasma cylindrellus has apparently been extirpated from its entire range, with the exception of upper reaches of the Paint Rock River system.

Ecology and Biology

Toxolasma cylindrellus has been eliminated from all medium to large river habitats and now remains only in small to medium reaches of the Paint Rock River system. It occurs in moderate current, usually in gravel substrates.

Toxolasma cylindrellus is a long-term brooder, gravid from late summer or autumn to the following summer. *Toxolasma cylindrellus* glochidia have been found to utilize *Lepomis macrochirus* (Bluegill) (Centrarchidae) as a glochidial host in laboratory trials (P.D. Johnson, personal communication).

Current Conservation Status and Protection

Toxolasma cylindrellus was considered endangered throughout its range by Stansbery (1970a) and Williams et al. (1993) and in Alabama by Stansbery (1976) and Lydeard et al. (1999). Garner et al. (2004) designated *T. cylindrellus* a species of highest conservation concern in the state. In 1976 this species was listed as endangered under the federal Endangered Species Act.

Remarks

Toxolasma cylindrellus is often considered to be one of the most imperiled mussel species in North America. It is now restricted to the Paint Rock River system in Jackson County, Alabama, and adjacent Franklin County, Tennessee, where it is rare.

Synonymy

Unio cylindrellus Lea, 1868. Lea, 1868a:144; Lea, 1868c:308, pl. 48, fig. 121; Lea, 1869a:68, pl. 48, fig. 121

Type locality: Duck Creek, Tennessee; Swamp Creek, Murray County, Georgia, Major T.C. Downie; north Alabama, Professor Tuomey. Lectotype, USNM 85300, length 37 mm, designated by Johnson (1974), is from north Alabama.

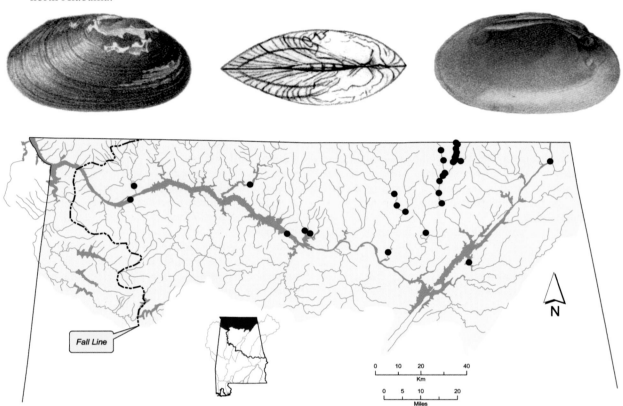

Distribution of *Toxolasma cylindrellus* in the Tennessee River drainage of Alabama.

Toxolasma lividum (Rafinesque, 1831)
Purple Lilliput

Toxolasma lividum – Upper figure: female, length 21 mm, NCSM 27350. Lower figure: male, length 27 mm, NCSM 27350. Paint Rock River, Butler Mill, Marshall County, Alabama, 10 July 2002. © Richard T. Bryant.

Shell Description

Length to 38 mm; moderately thick; moderately inflated; outline oval; posterior margin narrowly rounded, females more broadly rounded posteriorly than males; anterior margin rounded; dorsal margin straight to slightly convex; ventral margin straight to slightly convex; posterior ridge rounded; posterior slope moderately steep; umbo broad, somewhat inflated, elevated above hinge line, umbo sculpture irregular ridges, somewhat nodulous at posterior ridge; periostracum dark brown, without rays.

Pseudocardinal teeth triangular, may be slightly compressed, 2 divergent teeth in left valve, 1 tooth in right valve, often with small blade-like accessory denticle along margin anteriorly; lateral teeth moderately long, slightly curved, 2 in left valve, 1 in right valve; interdentum short, narrow to very narrow; umbo cavity shallow, open; nacre purple.

Soft Anatomy Description

Mantle tan, may have golden cast, outside of apertures with gray or black specks or mottling; visceral mass creamy white, may be pale orange adjacent to foot; foot creamy white to tan or pale orange.

Females with well-developed caruncle just ventral to incurrent aperture; outline round to oval; rounded to bluntly conical; white, usually with numerous small wrinkles on surface. Ortmann (1921) reported the caruncle to generally be brown, varying to white or "blackish".

Gills creamy white to tan, rusty tan or pale orange; dorsal margin slightly sinuous to slightly concave, ventral margin convex to elongate convex, outer gills may be slightly bilobed; gill length 35–50% of shell length; gill height 45–55% of gill length, occasionally greater; outer gill height 65–90% of inner gill height, marsupial area of outer gill in females higher than inner gill; inner lamellae of inner gills connected to visceral mass only anteriorly.

Outer gill marsupial; glochidia held in posterior 50–65% of gill; marsupium distended when gravid; creamy white to tan or pale orange, distal margin dark brown or black.

Labial palps creamy white to tan, may have slight rusty cast; straight to slightly concave dorsally, convex ventrally, bluntly pointed to rounded distally; palp length 20–35% of gill length; palp height 35–80% of palp length; distal 30–50% of palps bifurcate.

Incurrent aperture considerably longer than excurrent aperture; incurrent aperture usually slightly longer than supra-anal aperture; supra-anal aperture longer than excurrent aperture.

Incurrent aperture length 15–25% of shell length; creamy white to tan within, may have slight rusty cast, may have black specks or mottling basal to papillae, in females area basal to papillae adjacent to caruncle with heavy black specks on reddish brown background; papillae in 2–3 rows, long, slender, mostly simple, occasional bifid; papillae creamy white to tan, brown or grayish brown, often lighter distally, may be mottled basally, in females papillae adjacent to caruncle reddish brown.

Excurrent aperture length 5–10% of shell length; creamy white to tan within, marginal color band usually pale orange to rusty tan with irregular dark brown or black spots or lines perpendicular to margin; margin with crenulations or small simple papillae; papillae pale orange to tan or light brown.

Supra-anal aperture length 15–25% of shell length; creamy white to tan within, may have sparse tan or grayish brown marginally; margin smooth; mantle bridge separating supra-anal and excurrent apertures occasionally perforate, length 30–60% of supra-anal length.

Specimens examined: Second Creek, Lauderdale County (n = 4); Tennessee River (n = 1).

Glochidium Description

Length 170–180 μm; height 190–200 μm; without styliform hooks; outline subelliptical (Ortmann, 1921).

Similar Species

Toxolasma lividum superficially resembles *Toxolasma parvum* but attains greater size, is usually more inflated and has purple nacre rather than white. *Toxolasma lividum* expresses moderate sexual dimorphism, whereas at least some populations of *T. parvum* are hermaphroditic, expressing no sexual dimorphism.

Toxolasma lividum resembles *Toxolasma cylindrellus* but is more inflated and has a darker brown periostracum. Also, the purple nacre of *T. cylindrellus* often has a coppery sheen and is usually lighter in color outside of the pallial line, sometimes white. The nacre of *T. lividum* is uniform purple.

Toxolasma lividum may resemble small *Villosa vanuxemensis*, but *T. lividum* has a proportionally thicker shell and is more inflated overall, especially the umbo. Live *T. lividum* can be distinguished from *V. vanuxemensis* by the presence of a caruncle, which is replaced by a papillate mantle fold in *V. vanuxemensis*. These features are rudimentary in males of both species.

Toxolasma lividum may resemble *Villosa trabalis* but is less elongate and has a rounder posterior margin. The *V. trabalis* periostracum is typically rayed, whereas that of *T. lividum* is usually without rays. Shell nacre of *V. trabalis* is white, but that of *T. lividum* is purple.

Toxolasma lividum may resemble *Villosa fabalis*, but the periostracum of *V. fabalis* is typically rayed, whereas that of *T. lividum* is usually without rays. Male *V. fabalis* are usually less inflated than *T. lividum*. Shell nacre of *V. fabalis* is white, but that of *T. lividum* is purple.

General Distribution

Toxolasma lividum occurs in the lower Ohio River drainage from southwestern Ohio downstream to near the mouth of the Ohio River (Cummings and Mayer, 1992). It occurs west of the Mississippi River in southern Missouri (Oesch, 1995), northern Arkansas (Harris and Gordon, 1990) and possibly eastern Kansas. *Toxolasma lividum* is widespread, but sporadic, in the Cumberland River drainage downstream of Cumberland Falls (Cicerello et al., 1991; Parmalee and Bogan, 1998). It occurs in most of the Tennessee River drainage from southwestern Virginia, western North Carolina and eastern Tennessee downstream to the mouth of the Tennessee River (Neves, 1991; Parmalee and Bogan, 1998).

Alabama and Mobile Basin Distribution

Toxolasma lividum occurs across northern Alabama in the Tennessee River and some tributaries.

Toxolasma lividum is extant in several reaches of the Tennessee River and impounded lower reaches of tributaries. It also remains in some free-flowing tributary reaches.

Ecology and Biology

Toxolasma lividum occurs in a variety of habitats in bodies of water ranging from small streams to large rivers. In unimpounded reaches, it can be found in water with slow to swift current, usually in substrates comprised of sand and gravel. *Toxolasma lividum* also occurs in some overbank habitats of Tennessee River reservoirs with substrates of sand and mud. It may be locally common in embayments at the mouths of some Tennessee River tributaries.

Toxolasma lividum is a long-term brooder, gravid from late summer or autumn to the following summer (Ortmann, 1921). Females have caruncles just ventral to the incurrent aperture, which are presumably used for host-attracting displays. Wilson and Clark (1914) reported observing one in the Roaring River in which "the peculiar glands [*sic*] of the mantle, small white cylindrical objects on each side, were protruded and were undergoing spasmodic movements." Fishes reported to serve as glochidial hosts in laboratory trials include *Lepomis cyanellus* (Green Sunfish) and *Lepomis megalotis* (Longear Sunfish) (Centrarchidae) (Jenkinson, 1982; Hill, 1986).

Current Conservation Status and Protection

Stansbery (1970a) recognized two subspecies of *Toxolasma lividum* (see Remarks) and considered the form outside of the Tennessee and Cumberland River drainages to be endangered. However, Stansbery (1976) reported "nominal" *T. lividum* to be endangered in Alabama. *Toxolasma lividum* was considered a species of special concern throughout its range by Williams et al. (1993) and in Alabama by Lydeard et al. (1999). Garner et al. (2004) designated *T. lividum* a species of low conservation concern in the state.

Remarks

Two subspecies of *Toxolasma lividum* have been recognized by some authors, *T. lividum lividum* in the Cumberland and upper Tennessee River drainages and *T. lividum glans* in the remainder of its range (Oesch, 1995). However, there appears to be little evidence to support recognition of the two subspecies.

The status of *Toxolasma lividum* has been a matter of conjecture for some time. However, its ability to survive in some overbank habitats, as well as tributaries and tailwaters of dams, suggests that it may not be imminently imperiled.

Toxolasma lividum is reported to be rare in prehistoric shell middens along the Tennessee River in northern Alabama (Morrison, 1942; Warren, 1975; Hughes and Parmalee, 1999).

Synonymy

Unio lividus Rafinesque, 1831. Rafinesque, 1831:2
Type locality: Rockcastle River, [Kentucky]. Type specimen not found.
Unio glans Lea, 1831. Lea, 1831:82, pl. 8, fig. 12; Lea 1834b:92, pl. 8, fig. 12
Unio moestus Lea, 1841. Lea, 1841b:82; Lea, 1842b:244, pl. 26, fig. 60; Lea 1842c:82, pl. 26, fig. 60

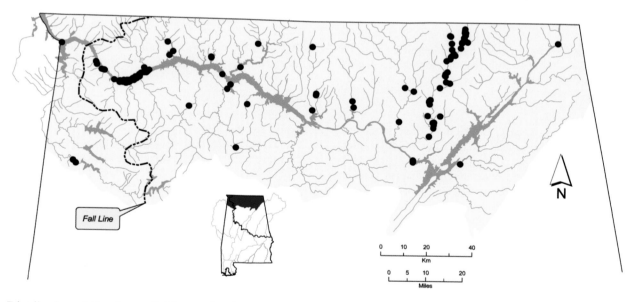

Distribution of *Toxolasma lividum* in the Tennessee River drainage of Alabama.

Toxolasma parvum (Barnes, 1823)
Lilliput

Toxolasma parvum – Upper figure: length 29 mm, UF 370407. Commercial sport and bait fish ponds near Elliotts Creek, about 2 miles southwest of Moundville, Hale County, Alabama, August 1997. Lower figure: length 23 mm, USNM 85288. Tennessee River. © Richard T. Bryant.

Shell Description

Length to 40 mm; moderately thin; moderately inflated; outline elliptical, height often slightly greater posteriorly; posterior margin rounded, becoming obliquely truncate with age; anterior margin rounded; dorsal margin straight to slightly convex; ventral margin straight to slightly convex; posterior ridge rounded; posterior slope moderately steep, steeper in larger individuals; umbo broad, low, somewhat inflated in larger individuals, umbo sculpture single-looped ridges, may be somewhat angular posteriorly; periostracum cloth-like, greenish brown to black, without rays.

Pseudocardinal teeth triangular, compressed, 2 almost parallel teeth in left valve, 1 tooth in right valve, often with accessory denticle anteriorly; lateral teeth straight to slightly curved, thin, 2 in left valve, 1 in right valve; interdentum short, narrow to very narrow; umbo cavity shallow; nacre white to bluish white.

Soft Anatomy Description

Mantle creamy white to tan, outside of apertures creamy white to tan, sometimes with black specks; visceral mass creamy white; foot creamy white, may become tan distally.

Females with well-developed caruncle just ventral to incurrent aperture; spherical when contracted, bulbous when relaxed; rusty orange to light brown. Utterback (1916b) reported the caruncle to be "cellular" at 87x magnification, with each cell hexagonal. Ortmann (1915) described the caruncle as "white to brownish (chestnut), of various shapes, cylindro-conical, or pyramidal, or semi-globular, sometimes somewhat divided". Lea (1859f) described the caruncle as "a black spongy looking mass". Ortmann (1915) reported males to lack a caruncle, but to have the coloration present in females.

Gills creamy white to light tan; dorsal margin straight with slight concavity around posterior adductor muscle, ventral margin elongate convex; gill length 45–55% of shell length; gill height 35–55% of gill length; outer gill height 50–85% of inner gill height; inner lamellae of inner gills connected to visceral mass only anteriorly.

Outer gills marsupial; glochidia held in posterior 50% of gill; marsupium distended when gravid, somewhat reniform in outline; creamy white to light tan, ventral edge darker.

Labial palps creamy white to light tan; straight to slightly concave dorsally, convex ventrally, bluntly pointed distally, may be broadly connected to mantle basally; palp length 30–35% of gill length; palp height 60–65% of palp length, occasionally less; distal 35–65% of palps bifurcate.

Incurrent aperture longer than excurrent and supra-anal apertures; supra-anal aperture longer than excurrent aperture.

Incurrent aperture length 15–20% of shell length; creamy white within, in females margin adjacent to caruncle dark brown or with heavy black speckling; papillae in 2–4 rows, long, slender, simple, in females papillae adjacent to caruncle fused into bifid or trifid papillae, some may approach arborescent; papillae white to tan, inner rows often lighter, may have dark brown edges basally, in females papillae adjacent to caruncle usually reddish brown.

Excurrent aperture length 5–10% of shell length; creamy white within, marginal color band tan to rusty orange, with dark brown lines perpendicular to margin or interconnected to form open reticulated pattern, occasional individuals with colors reversed (i.e., dark brown background with tan lines); margin usually with small simple papillae, occasionally crenulate; papillae tan to dark brown.

Supra-anal aperture length 10–15% of shell length; creamy white within, often with sparse, irregular gray or dark brown marginally; margin smooth; mantle bridge separating supra-anal and excurrent apertures occasionally perforate, length 100–175% of supra-anal length. Ortmann (1912a, 1915) reported the mantle bridge separating the excurrent and supra-anal apertures to sometimes be absent.

Specimens examined: Second Creek, Lauderdale County (n = 1); Tennessee River (n = 4).

Glochidium Description

Length 170–180 μm; height 200μm; without styliform hooks; outline subelliptical; dorsal margin straight, ventral margin rounded, anterior and posterior margins convex (Surber, 1915; Utterback, 1916a; Ortmann, 1919). Glochidium size given by Utterback (1916b) was 175 μm by 100 μm, which was deemed a misprint by Ortmann (1915, 1919).

Similar Species

Toxolasma parvum superficially resembles *Toxolasma lividum* but is smaller in size, is usually less inflated and has white nacre instead of purple. *Toxolasma lividum* expresses moderate sexual dimorphism, whereas at least some populations of *T. parvum* are hermaphroditic, expressing no sexual dimorphism.

Toxolasma parvum may resemble *Toxolasma corvunculus* but has a thinner shell and more compressed pseudocardinal teeth. The nacre of *T. parvum* is white, whereas that of *T. corvunculus* is often some shade of purple. The periostracum of *T. corvunculus* is often more yellowish than that *T. parvum*.

General Distribution

Toxolasma parvum occurs in portions of lakes Erie and Michigan in the Great Lakes drainage (Goodrich and van der Schalie, 1932; La Rocque and Oughton, 1937; Mathiak, 1979). It is widespread in the Mississippi Basin, ranging from southern Minnesota (Dawley, 1947) south to Louisiana (Vidrine, 1993) and from headwaters of the Ohio River in western Pennsylvania (Ortmann, 1919) west to the Missouri River drainage of South Dakota (Backlund, 2000) and Kansas (Murray and Leonard, 1962). *Toxolasma parvum* is present in the Cumberland River drainage above and below Cumberland Falls (Cicerello et al., 1991; Parmalee and Bogan, 1998). In the Tennessee River drainage it ranges from near the confluence of the French Broad and Holston rivers downstream to the mouth of the Tennessee River (Parmalee and Bogan, 1998). *Toxolasma parvum* also occurs in the Mobile Basin, but its distribution there is poorly understood. There are few Mobile Basin records of the species dated prior to the mid-1900s. It may have been recently introduced to the Mobile Basin via fish with glochidial infestations. Widespread incidental introductions of *T. parvum* have been noted, including interbasin transfers (e.g., Flint River drainage, Georgia; R.C. Stringfellow, personal communication). Along the Gulf Coast, *T. parvum* has been reported from most drainages in Louisiana and Texas (Vidrine, 1993; Howells et al., 1996).

Alabama and Mobile Basin Distribution

Toxolasma parvum occurs across northern Alabama in the Tennessee River and some tributaries. It occurs sporadically in the Mobile Basin, where its distribution is poorly understood.

Toxolasma parvum remains widespread in the Tennessee River drainage and Mobile Basin and is apparently being unintentionally introduced to additional drainages via fish stockings.

Ecology and Biology

Toxolasma parvum typically occurs in sluggish water of creeks, rivers, lakes and ponds. It is often found in disturbed habitats such as artificial ponds, borrow pits and reservoirs. *Toxolasma parvum* may be common in ponds used to culture fish. Bates (1962) reported it to be among the first species to colonize overbank habitat created by impoundment of the

Tennessee River by Kentucky Dam. It appears to favor soft substrates, such as mud and sand.

Toxolasma parvum has been reported to be a hermaphroditic species (Sterki, 1898a). Utterback (1916b) described a population in a lake in northwestern Missouri in which all of "hundreds" of individuals examined had marsupial gills, and there was no evidence of sexual dimorphism with regard to shell morphology. Hermaphroditic populations have also been reported from the Vermillion River in Illinois (Tepe, 1943) and the Tennessee River (van der Schalie, 1970), based on histological examinations of gonadal tissues. However, all populations of *T. parvum* may not be hermaphroditic. Populations exhibiting sexual dimorphism in shell morphology have been reported in Missouri (Utterback, 1916b), Pennsylvania (Ortmann, 1919) and Wisconsin (Baker, 1928a). Ortmann (1919) reported sexual dimorphism in Pennsylvania populations to be very slight, pronounced only in very old individuals.

The brooding strategy of *Toxolasma parvum* is perplexing. It is a long-term brooder, but there appears to be variation among populations in the timing of reproductive events. Female *T. parvum* have been reported with "eggs" in April in Texas (Howells, 2000), June in Pennsylvania and Arkansas, and August in Indiana and Wisconsin (Ortmann, 1915, 1919; Baker, 1928a). Most long-term brooders do not become gravid until at least September. Conglutinates of *T. parvum* are white and club-shaped (Utterback, 1916b).

Fishes reported to serve as glochidial hosts of *Toxolasma parvum* based on laboratory trials include *Lepomis cyanellus* (Green Sunfish) (Centrarchidae) and *Etheostoma nigrum* (Johnny Darter) (Percidae) (Hove, 1995; Watters et al., 2005). *Lepomis gulosus* (Warmouth), *Lepomis humilis* (Orangespotted Sunfish), *Lepomis macrochirus* (Bluegill) and *Pomoxis annularis* (White Crappie) (Centrarchidae) were reported as hosts by Mermilloid *in* Fuller (1974), but the method of their determination was not given.

Current Conservation Status and Protection

Toxolasma parvum was considered currently stable throughout its range by Williams et al. (1993) and in Alabama by Lydeard et al. (1999). Garner et al. (2004) designated *T. parvum* a species of moderate conservation concern in the state, based on a poor understanding of its systematics and taxonomy.

Remarks

The native range of *Toxolasma parvum* is unclear. This species appears to have increased its range during the past century. Its expansion was apparently associated with habitat alterations as large rivers were impounded, along with mass culture and widespread introduction of game fishes (e.g., Centrarchidae, which *T. parvum* uses as glochidial hosts). There are few preimpoundment records of *T. parvum* from middle and upper reaches of the Tennessee River (Ortmann, 1925), and the species has not appeared in the archaeological record from that area (Hughes and Parmalee, 1999), but it is locally common there today. Ortmann (1925) reported only one individual from the Tennessee River drainage and questioned its validity based on a lack of additional material from the drainage. Interbasin transfer of cultured fish may have resulted in introduction of *T. parvum* into drainages outside of its native range. It is unclear if the presence of *T. parvum* in the Mobile Basin is the result of such an introduction. There is evidence of *T. parvum* being transferred to the Apalachicola Basin of Georgia (R.C. Stringfellow, personal communication). Discovery of *T. parvum* anywhere in the southeastern U.S. would not be unexpected, since it appears to be easily introduced with stocked game fish.

The taxonomic status of *Toxolasma* in the Choctawhatchee, Yellow and Escambia River drainages is unclear. Some authors have considered them to be *Toxolasma parvum* (e.g., Heard, 1979b). However, individuals from those drainages differ from typical *T. parvum* in being strongly sexually dimorphic, with larger pseudocardinal teeth and thicker shells. *Toxolasma* in the Choctawhatchee, Yellow and Escambia River drainages likely represent an undescribed species.

In dioecious populations, sexual dimorphism is expressed as subtle differences in shell morphology. Females are more inflated than males and have a somewhat rounder posterior margin. However, the dimorphism is not pronounced and Ortmann (1919) stated that "only in the older shells are these differences unmistakable".

Synonymy

Unio parvus Barnes, 1823. Barnes, 1823, pl. 13, fig. 18
Type locality: Fox River, Wisconsin. Type specimen not found, likely lost in fire with remainder of Barnes collection (Johnson, 2006).

Carunculina parva cahni F.C. Baker, 1928. F.C. Baker, 1928a:253–254, pl. 105, figs. 14–18

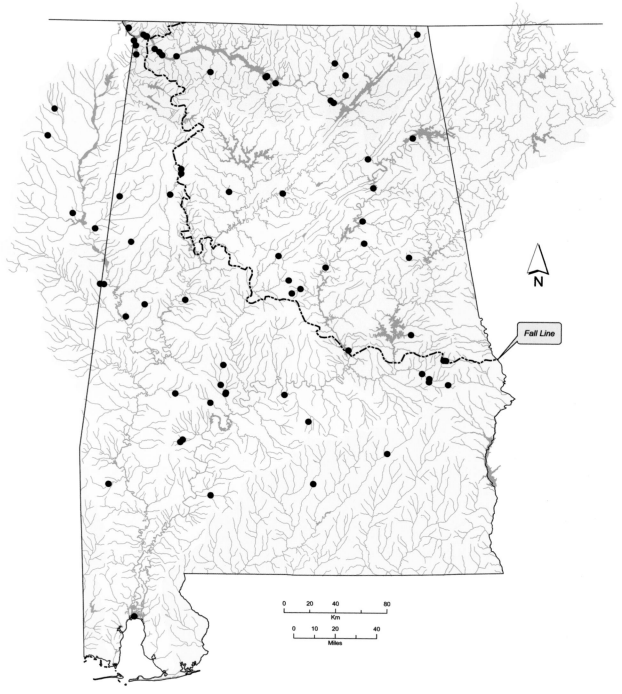

Distribution of *Toxolasma parvum* in Alabama and the Mobile Basin.

Toxolasma paulum (Lea, 1840)
Iridescent Lilliput

Toxolasma paulum – Upper figure: female, length 35 mm, UF 366786. Walter F. George Reservoir, fish attractor F9, Henry County, Alabama, 11 July 1991. Lower figure: male, length 34 mm, UF 373559. Spring Creek, County Road 282, 3.5 air miles southeast of Bluffton, Early County, Georgia, 3 September 1992. © Richard T. Bryant.

Shell Description

Length to 49 mm; moderately thin; moderately inflated; outline elliptical; male posterior margin narrowly rounded, female posterior margin obliquely truncate; anterior margin rounded; dorsal margin slightly convex; ventral margin straight to slightly convex; posterior ridge rounded; posterior slope moderately steep, steeper in large individuals; umbo broad, low, somewhat inflated in larger individuals, umbo sculpture single-looped ridges; periostracum cloth-like, greenish brown to black, without rays.

Pseudocardinal teeth triangular, somewhat compressed, elevated, 2 divergent teeth in left valve, 1 tooth in right valve, may have accessory denticle anteriorly; lateral teeth thin to moderately thick, usually straight, 2 in left valve, 1 in right valve; interdentum short, very narrow; umbo cavity shallow; nacre white to pale purple.

Soft Anatomy Description

Mantle tan, outside of apertures rusty tan, grayish brown or dark gray, may have black specks; visceral mass pearly white or creamy white to tan; foot creamy white to tan or pale orange.

Females with low caruncle just ventral to incurrent aperture; brown, adjacent mantle dark brown or grayish brown. Males may have rudimentary caruncle or thickened mantle area just ventral to incurrent aperture.

Gills tan, may have slight golden cast; dorsal margin straight to slightly sinuous, ventral margin convex to elongate convex; gill length 50–65% of shell length; gill height 40–50% of gill length; outer gill height 60–70% of inner gill height; inner lamellae of inner gills connected to visceral mass only anteriorly.

Outer gills marsupial; glochidia held in posterior 50% of gill; marsupium distended when gravid; tan, distal margin brown.

Labial palps tan; straight to slightly concave dorsally, convex ventrally, bluntly pointed distally; palp length 25–40% of gill length; palp height 45–65% of palp length; distal 35–55% of palps bifurcate.

Incurrent aperture longer than excurrent and supra-anal apertures; supra-anal aperture usually longer than excurrent aperture, occasionally slightly shorter.

Incurrent aperture length 20–25% of shell length; tan within, may have brownish cast, may be dark brown basal to papillae; papillae in 2–3 rows, long, slender, mostly simple, with occasional bifid; papillae tan to grayish tan, may become gray distally, in females papillae adjacent to caruncle reddish brown.

Excurrent aperture length approximately 10% of shell length; tan within, marginal color band rusty tan with dark brown lines perpendicular to margin, band may be solid dark brown proximally; margin with simple papillae, small but well-developed; papillae light brown to rusty tan.

Supra-anal aperture length 10–15% of shell length; creamy white to tan within, may have thin tan marginal band; margin smooth; mantle bridge separating supra-anal and excurrent apertures usually imperforate, length usually 40–100% of supra-anal length, occasionally greater.

Specimens examined: Big Creek, Houston County (n = 1); Uchee Creek, Russell County (n = 3).

Glochidium Description

Glochidium unknown.

Similar Species

Toxolasma paulum may resemble small *Villosa lienosa*. However, small *V. lienosa* are more compressed than similar-sized *T. paulum*. Pseudocardinal teeth of *T. paulum* are thicker, relative to shell size, than those of *V. lienosa*, which are often compressed. Also, shell nacre of *V. lienosa* is typically darker purple than that of *T. paulum*. Live *T. paulum* can be distinguished from *V. lienosa* by the presence of a caruncle, which is replaced by a papillate mantle fold in *V. lienosa*. These features are rudimentary in males of both species.

Toxolasma paulum may resemble *Toxolasma parvum*. There are no records of *T. parvum* from the Apalachicola Basin, but it is a widely introduced species and may be found in the drainage at some point. Most populations of *T. parvum* are not sexually dimorphic, but *T. paulum* females are much more inflated posteriorly than males. Also, *T. paulum* typically has at least some purple nacre, whereas *T. parvum* nacre is always white. Pseudocardinal teeth of *T. paulum* are heavier than those of *T. parvum*.

General Distribution

Toxolasma paulum appears to be endemic to an area encompassing the Apalachicola Basin and much of peninsular Florida (Brim Box and Williams, 2000).

Alabama and Mobile Basin Distribution

Toxolasma paulum is limited to the Chattahoochee River drainage and headwaters of the Chipola River.

Toxolasma paulum is known to be extant in Big and Cowarts creeks, headwater streams of the Chipola River in Houston County. In the Chattahoochee River drainage it has been recently reported from one site on the Chattahoochee River proper and in Halawakee Creek, Lee County, and Uchee Creek, Russell County.

Ecology and Biology

Toxolasma paulum occurs in small streams to large rivers, usually in areas with slight current. It may occasionally be found in reservoir overbank habitat. *Toxolasma paulum* appears to prefer sandy substrates but may be found in mud or gravel. This species is often encountered along margins of creeks in less than 15 cm of water. Clench and Turner (1956) noted that *T. paulum* migrates up and down stream slopes with changes in water level, remaining near the water's edge.

Toxolasma paulum is presumably a long-term brooder, gravid from late summer or autumn to the following summer. Brim Box and Williams (2000) reported gravid females from May to July but not in August or September. Glochidial hosts of *T. paulum* are unknown.

Current Conservation Status and Protection

Toxolasma paulum was considered currently stable throughout its range by Williams et al. (1993) and in Alabama by Lydeard et al. (1999). Garner et al. (2004) designated *T. paulum* a species of moderate conservation concern in the state, primarily due to a dearth of information on its status.

Remarks

The taxonomic status of *Toxolasma* species in eastern Gulf Coast and southern Atlantic Coast drainages remains poorly understood. *Toxolasma paulum* has been placed into the synonymy of *Toxolasma parvum* by some authors (e.g., Johnson, 1972b; Heard, 1979b). However, *T. paulum* is dioecious and strongly sexually dimorphic with regard to shell morphology, whereas *T. parvum* is monoecious, exhibiting a single shell form, at least in some populations. In dioecious populations of *T. parvum*, sexual dimorphism is very slight.

Some authors (e.g., Clench and Turner, 1956) consider the range of *Toxolasma paulum* to include the Choctawhatchee, Yellow and Escambia River drainages. However, the *Toxolasma* found in those drainages is herein considered to represent an underscribed species.

Synonymy

Unio paulus Lea, 1840. Lea, 1840:287; Lea, 1842b:213, pl. 15, fig. 29; Lea, 1842c:51, pl. 15, fig. 29

Type locality: Chattahoochee River, Columbus, [Muscogee County,] Georgia, Dr. Boykin. Lectotype, USNM 85274, length 24 mm, designated by Johnson (1974).

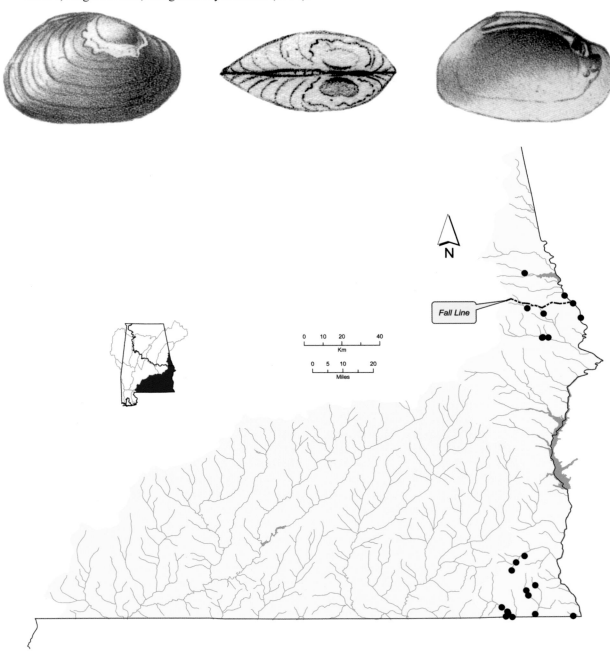

Distribution of *Toxolasma paulum* in the eastern Gulf Coast drainages of Alabama.

Toxolasma sp.
Gulf Lilliput

Toxolasma sp. – Upper figure: female, length 30 mm, UF 69099. Yellow River near Harmony and Andalusia, [County Route 70,] Covington County, Alabama. Middle figure: male, length 32 mm, CMNH 61.8465. Sandy Creek, Evergreen, Conecuh County, Alabama. Lower figure: male, length 28 mm, UF 69101. Conecuh River, Bozeman's Landing near Crenshaw County line, Covington County, Alabama, July 1915. © Richard T. Bryant.

Shell Description

Length to 40 mm; moderately thin; moderately inflated, females more inflated posteriorly than males; outline elliptical; posterior margin rounded, females more broadly rounded than males, may become obliquely truncate with age; anterior margin rounded; dorsal margin straight to slightly convex; ventral margin straight to slightly convex; posterior ridge rounded; posterior slope moderately steep; umbo broad, low, somewhat inflated in large individuals, umbo sculpture unknown; periostracum olive brown to dark brown, without rays.

Pseudocardinal teeth triangular, may be somewhat compressed, 2 teeth in left valve, 1 tooth in right valve, may have accessory denticle anteriorly; lateral teeth short, straight, 2 in left valve, 1 in right valve; interdentum long, narrow to very narrow; umbo cavity shallow; nacre salmon to bluish white.

Soft Anatomy Description

Mantle creamy white to tan, may have slight rusty or golden cast, outside of apertures often with black specks, sometimes grayish brown; visceral mass creamy white to pearly white, may have pinkish cast; foot creamy white to tan, may be slightly darker distally.

Females with well-developed caruncle ventral to incurrent aperture; outline round to oblong; high to somewhat flattened; surface may be wrinkled; creamy white to pink, pinkish brown or light rusty brown, often pink basally with white apex.

Gills creamy white to tan or gold, may have slight rusty cast; dorsal margin straight to slightly sinuous, ventral margin convex to elongate convex; gill length 50–55% of shell length; gill height 50–70% of gill length; outer gill height 65–80% of inner gill height; inner lamellae of inner gills connected to visceral mass only anteriorly.

Outer gills marsupial; glochidia held in posterior 35–50% of gill; marsupium distended when gravid; creamy white to light tan, ventral edge may be black.

Labial palps creamy white to tan; straight to slightly concave dorsally, convex ventrally, rounded to bluntly pointed distally; palp length 30–35% of gill length; palp height 50–70% of palp length; distal 25–45% of palps bifurcate.

Incurrent aperture longer than excurrent and supra-anal apertures; supra-anal aperture usually longer than excurrent aperture.

Incurrent aperture length 20–25% of shell length; creamy white to tan within, in females margin adjacent to caruncle brown to gray; papillae usually in 2–3 rows, long, simple, usually slender, occasionally thick, in females papillae adjacent to caruncle enlarged, flattened; papillae creamy white to tan, dark brown or gray, often changing color distally, may have dark brown edges basally, in females papillae adjacent to caruncle may be rusty brown.

Excurrent aperture length 10–15% of shell length; creamy white to tan within, marginal color band usually rusty tan to rusty brown, with dark brown lines perpendicular to margin or interconnected to form open reticulated pattern, occasional individuals with dark brown marginal color band with tan spots or bars perpendicular to margin; margin with small simple papillae; papillae creamy white to tan, brown or gray.

Supra-anal aperture length 15–20% of shell length; creamy white within, often tan marginally; margin smooth; mantle bridge separating supra-anal and excurrent apertures occasionally perforate, length 45–70% of supra-anal length.

Specimens examined: Conecuh River (n = 6); Eightmile Creek, Walton County, Florida (n = 2); West Fork Choctawhatchee River, Barbour County (n=1).

Glochidium Description

Glochidium unknown.

Similar Species

Toxolasma sp. resembles *Villosa choctawensis*. Rays are absent in *Toxolasma* sp., but are typically present in *V. choctawensis*, though they may become obscure with age or covered by mineral deposits. *Toxolasma* sp. is generally more elongate than *V. choctawensis*. Live *Toxolasma* sp. can be distinguished from *V. choctawensis* by the presence of a caruncle which is replaced by a papillate mantle fold in *V. choctawensis*. These features are rudimentary in males of both species.

Toxolasma sp. may resemble small *Villosa lienosa*. However, *V. lienosa* is generally more compressed than *Toxolasma* sp. of comparable size. Live *Toxolasma* sp. can be distinguished from *V. lienosa* by the presence of a caruncle, which is replaced by a papillate mantle fold in *V. lienosa*. These features are rudimentary in males of both species.

It is unclear if *Toxolasma* sp. occurs sympatrically with *Toxolasma parvum* in all Gulf Coast drainages. However, with the propensity for *T. parvum* to be introduced into new localities, it is reasonable to believe that it will be found in one or more Gulf Coast drainages. The two can be distinguished by the thicker shell and larger pseudocardinal teeth of *Toxolasma* sp. Nacre color of *Toxolasma* sp. is sometimes salmon, whereas the nacre of *T. parvum* is always white. Also, *Toxolasma* sp. is strongly sexually dimorphic, whereas *T. parvum* is only weakly dimorphic, and sexual dimorphism appears to be nonexistent in some *Toxolasma parvum* populations.

General Distribution

Toxolasma sp. occurs in Gulf Coast drainages from the Choctawhatchee to the Escambia River drainage.

Alabama and Mobile Basin Distribution

Toxolasma sp. occurs in the Choctawhatchee, Yellow and Escambia River drainages.

Toxolasma sp. is extant in all drainages where it was historically found. It may be locally common, but some populations are confined to isolated stream reaches.

Ecology and Biology

Toxolasma sp. typically occurs in slow to moderately swift water, usually in substrates of varying combinations of sand and mud. It is often more common along stream margins, on and just basal to channel slopes.

Nothing is known about the biology of *Toxolasma* sp. It is presumably a long-term brooder, gravid from late summer or autumn to the following spring or summer. Glochidial hosts of this species are unknown.

Current Conservation Status and Protection

Toxolasma sp. was included with *Toxolasma parvum* in Garner et al. (2004), in which it was designated a species of moderate conservation concern in Alabama, based on a poor understanding of its systematics and taxonomy. This undescribed species has not been addressed in other conservation status reviews.

Remarks

The *Toxolasma* sp. in the Choctawhatchee, Yellow and Escambia River drainages has previously been considered to belong to *Toxolasma paulum* (e.g., Clench and Turner, 1956) or *Toxolasma parvum* (e.g., Heard, 1979b; Garner et al., 2004). However, the species found from the Choctawhatchee to the Escambia River drainages differs considerably from both *T. paulum* and *T. parvum* and appears to represent an undescribed species.

Synonymy

There are no names available for this undescribed species.

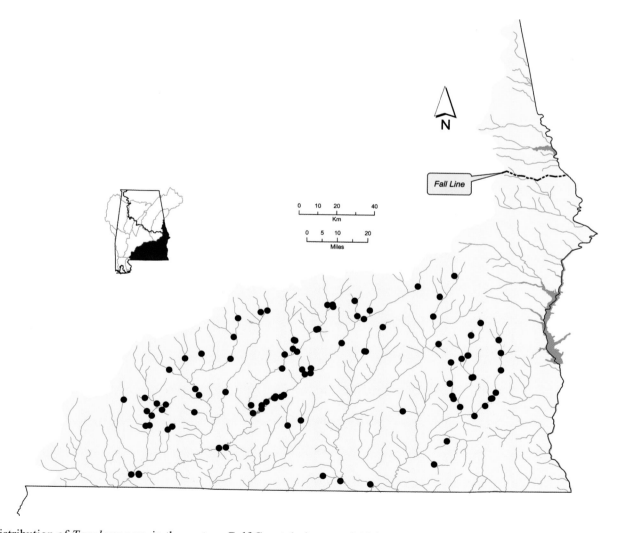

Distribution of *Toxolasma* sp. in the eastern Gulf Coast drainages of Alabama.

Genus *Truncilla*

Truncilla Rafinesque, 1819, occurs in the Great Lakes and Mississippi basins and Gulf Coast drainages from the Mobile Basin to the Rio Grande. Turgeon et al. (1998) listed four species in the genus, two of which occur in Alabama.

Species in this genus were placed in *Amygdalonaias* Crosse and Fischer *in* Fischer and Crosse, 1894, until Ortmann and Walker (1922a) clarified the identification of *Truncilla truncata* Rafinesque, 1820, and moved the *Epioblasma* Rafinesque, 1831, species from *Truncilla* to *Dysnomia* Agassiz, 1852 (see *Epioblasma* genus account).

Type Species

Truncilla truncata Rafinesque, 1820

Diagnosis

Shell oval to triangular; inflated; posterior ridge usually sharp; posterior slope steep; pseudocardinal teeth somewhat compressed; sexual dimorphism very subtle, females slightly more inflated than males posteriorly; periostracum yellowish green with broken rays that may be comprised of chevrons.

Excurrent aperture crenulate; mantle bridge separating excurrent and supra-anal apertures wide; inner lamellae of inner gills completely connected to visceral mass or connected only anteriorly; outer gills marsupial; glochidia held in posterior portion of gill; marsupium extended beyond original gill ventral edge when gravid; glochidium without styliform hooks (Rafinesque, 1819, 1820; Ortmann and Walker, 1922a; Haas, 1969b).

Synonymy

Amygdalonaias Crosse and Fischer *in* Fischer and Crosse, 1894

Truncilla donaciformis (Lea, 1828)
Fawnsfoot

Truncilla donaciformis – Upper figure: length 34 mm, UF 68144. Alabama River, Alabama. Lower figure: length 44 mm, UF 68159. Alabama River, Alabama. © Richard T. Bryant.

Shell Description

Length to 52 mm; moderately thin; moderately inflated; outline subelliptical; posterior margin bluntly pointed; anterior margin rounded; dorsal margin convex; ventral margin convex; posterior ridge prominent, angled; posterior slope steep, slightly concave; umbo broad, moderately inflated, elevated above hinge line, umbo sculpture fine double-looped ridges; periostracum yellowish to greenish, usually with dark green chevrons and green rays.

Pseudocardinal teeth thin, compressed, 2 slightly divergent teeth in left valve, 1 tooth in right valve, may have thin accessory denticle anteriorly; lateral teeth short, moderately thick, straight, 2 teeth in left valve, 1 tooth in right valve, may also have rudimentary lateral tooth ventrally in right valve; interdentum short, narrow; umbo cavity shallow; nacre white to bluish white.

Soft Anatomy Description

Mantle creamy white to tan or light brown; visceral mass creamy white to pearly white; foot creamy white. Utterback (1916b) described soft parts as "dirty white except for blackish mantle edge at siphonal openings".

Gills creamy white to tan, may have grayish cast; dorsal margin straight to slightly sinuous, ventral margin convex; gill length usually 35–60% of shell length, occasionally less; gill height usually 35–50% of gill length, occasionally greater; outer gill height 60–90% of inner gill height; inner lamellae of inner gills connected to visceral mass only anteriorly. Ortmann (1914) reported most of the dorsal margin of gills to be connected to the visceral mass, with only "a small hole open at posterior end of foot".

Outer gills marsupial; glochidia held in posterior 50% of gill; marsupium padded when gravid; creamy white.

Labial palps creamy white to tan or gray; straight dorsally, convex ventrally, bluntly pointed to rounded distally; palp length 20–30% of gill length; palp height 50–75% of palp length; distal 35–65% of palps bifurcate.

Incurrent aperture longer than excurrent aperture; supra-anal aperture and incurrent apertures usually of similar length.

Incurrent aperture length 10–15% of shell length; creamy white within, may be tan basal to papillae, with black lines extending from edges of papillae onto aperture wall, some lines converging proximally; papillae in 2 rows, long, slender, simple, occasionally with some bifid; papillae usually creamy white, tan or rusty orange, often darker distally, occasionally white distally, often with black or gray edges basally. Lea (1863d) reported incurrent aperture papillae to be brown.

Excurrent aperture length 5–10% of shell length; creamy white within, marginal color band pale orange to rusty tan, often with dark brown or black bands perpendicular to margin, sometimes with reticulated pattern; margin crenulate or with small simple papillae; papillae pale orange to rusty brown.

Supra-anal aperture length 10–20% of shell length; creamy white within, may have sparse black or brown marginally; margin smooth; mantle bridge separating supra-anal and excurrent apertures imperforate, length 35–50% of supra-anal length, occasionally less.

Specimens examined: Alabama River (n = 2); Bear Creek, Colbert County (n = 3); Coosa River (n = 2).

Glochidium Description

Length 50–60 μm; height 60–63 μm; without styliform hooks; outline subelliptical; dorsal margin straight, ventral margin rounded, anterior and posterior margins convex (Surber, 1912; Utterback, 1916a, 1916b; Ortmann, 1919).

Similar Species

Truncilla donaciformis resembles *Truncilla truncata* but does not grow as large and has a thinner shell. *Truncilla donaciformis* is less inflated, more elongate and less triangular or quadrate in outline than *T. truncata*. The posterior ridge of *T. donaciformis* is more rounded than that of *T. truncata*.

General Distribution

Truncilla donaciformis is known from the Great Lakes drainage in lakes Erie and St. Clair (Goodrich and van der Schalie, 1932; La Rocque and Oughton, 1937). It is widespread in the Mississippi Basin, ranging from southern Minnesota (Dawley, 1947) south to Louisiana (Vidrine, 1993) and from Ohio River drainage headwaters in western Pennsylvania (Ortmann, 1919) west to South Dakota (Backlund,

2000) and Kansas (Murray and Leonard, 1962). *Truncilla donaciformis* is known from the Cumberland River drainage downstream of Cumberland Falls (Cicerello et al., 1991; Parmalee and Bogan, 1998) and the Tennessee River drainage from eastern Tennessee downstream to the mouth of the Tennessee River (Parmalee and Bogan, 1998). *Truncilla donaciformis* also occurs in most Gulf Coast drainages from the Mobile Basin west to the San Jacinto River in Texas (Vidrine, 1993; Howells et al., 1996).

Alabama and Mobile Basin Distribution

In the Tennessee River drainage *Truncilla donaciformis* is known from across northern Alabama in the Tennessee River and some major tributaries. It has been collected from most of the large rivers and some tributary streams of the Mobile Basin, with the exception of the Tallapoosa River above the Fall Line.

In the Tennessee River *Truncilla donaciformis* is extant in tailwaters of all dams as well as some reservoir overbank areas. The only known extant tributary populations are in the Elk River and a short reach of Bear Creek in Colbert County. In the Mobile Basin *T. donaciformis* is known to be extant in some reaches of the Alabama, Cahaba, Coosa, Sipsey and Tombigbee rivers.

Ecology and Biology

Truncilla donaciformis is primarily a species of large rivers but may be found in some small rivers and creeks. It occurs most commonly in flowing water, including tailwaters of dams, where it may be found at depths exceeding 6 m. Substrates in those areas are composed primarily of gravel with interstitial sand. *Truncilla donaciformis* may also be found in overbank habitats where substrates are composed of firm sand, sandy mud or gravel. Bates (1962) reported it to be one of the early colonizers of overbank habitat following impoundment of the Tennessee River behind Kentucky Dam. *Truncilla donaciformis* spend much of their lives well-buried in the substrate.

Truncilla donaciformis is a long-term brooder, gravid from late summer or autumn to the following summer. It has been reported gravid with early embryos in July (Surber, 1912) and mature glochidia in August (Ortmann, 1919). Fishes reported to serve as glochidial hosts, based on observations of natural infestations, include *Sander canadensis* (Sauger) (Percidae) and *Aplodinotus grunniens* (Freshwater Drum) (Sciaenidae) (Surber, 1913). *Truncilla donaciformis* juveniles are often found attached to rocks by byssal threads.

Current Conservation Status and Protection

Truncilla donaciformis was considered currently stable throughout its range by Williams et al. (1993). Lydeard et al. (1999) listed it as imperiled in Alabama.

However, Garner et al. (2004) designated *T. donaciformis* a species of moderate conservation concern in the state.

Remarks

Juvenile *Truncilla donaciformis* are often found with byssal threads entangled on rocks or other submerged objects. Fuller (1980b, 1980c) reported *T. donaciformis* to be the second most abundant species in upper reaches of the Mississippi River, based on observations of numerous "immediately post metamorphic individuals" brought to the surface with byssal threads entangled on a crowfoot brail.

Truncilla donaciformis is absent from the archaeological record of the Tennessee River (Hughes and Parmalee, 1999). However, it was present in the Tennessee River prior to its impoundment (Ortmann, 1925).

Synonymy

Unio donaciformis Lea, 1828. Lea, 1828:267, pl. 4, fig. 3; Lea, 1834b:9, pl. 4, fig. 3
Type locality: Ohio. Lectotype, USNM 84457, length 45 mm (male), designated by Johnson (1974).

 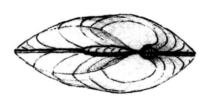 

Unio zigzag Lea, 1829. Lea, 1829:440, pl. 12, fig. 19; Lea 1834b:54, pl. 12, fig. 19

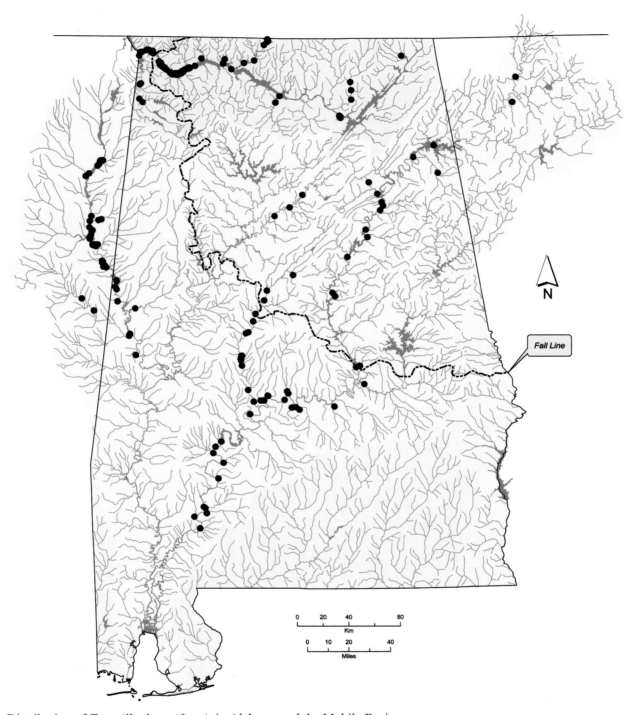

Distribution of *Truncilla donaciformis* in Alabama and the Mobile Basin.

Truncilla truncata Rafinesque, 1820
Deertoe

Truncilla truncata – Length 26 mm, UF 68171. Bear Creek, Burleson, Franklin County, Alabama, August 1909. © Richard T. Bryant.

Shell Description

Length to 83 mm; moderately thick; moderately inflated; outline triangular to subquadrate; posterior margin bluntly pointed; anterior margin rounded; dorsal margin convex; ventral margin convex; posterior ridge prominent, narrow; posterior slope typically steep; often with wide shallow sulcus anterior of posterior ridge, usually less distinct in females; umbo somewhat broad, flat, slightly inflated, elevated above hinge line, umbo sculpture fine ridges, becoming double-looped ventrally; periostracum tawny to olive brown, darkening with age, typically with variable green rays, usually composed of radially arranged rows of chevrons or blotches, occasional individuals without rays.

Pseudocardinal teeth moderately thick, erect, triangular, usually compressed, 2 slightly divergent teeth in left valve, 1 tooth in right valve, may have thin accessory denticle anteriorly; lateral teeth short, straight to slightly curved, 2 in left valve, 1 in right valve, may also have rudimentary lateral tooth ventrally in right valve; interdentum short, narrow; umbo cavity shallow; nacre white, occasionally pale pink or salmon.

Soft Anatomy Description

Mantle creamy white to tan, outside of apertures grayish tan to grayish brown; visceral mass creamy white to pearly white; foot creamy white to tan.

Gills creamy white to tan; dorsal margin sinuous, ventral margin convex, female outer gills may be bilobed; gill length 50–65% of shell length; gill height 55–60% of gill length; outer gill height 75–85% of inner gill height; inner lamellae of inner gills usually completely connected to visceral mass.

Outer gills marsupial; glochidia held in posterior 50–60% of gill; marsupium padded when gravid; creamy white to tan.

Labial palps creamy white to tan; straight to slightly concave or convex dorsally, convex ventrally, bluntly pointed to narrowly rounded distally; palp length 25–35% of gill length; palp height 60–90% of palp length; distal 35–45% of palps bifurcate.

Incurrent and supra-anal apertures longer than excurrent aperture; incurrent aperture slightly longer, shorter or equal to supra-anal aperture.

Incurrent aperture length 15–20% of shell length; creamy white within, usually with sparse gray or black basal to papillae, lines from edges of papillae may extend onto aperture wall, some lines converge proximally; papillae in 2 rows, short or long, slender or thick, simple and bifid, some individuals without bifids, inner row larger; papillae creamy white to tan or rusty brown, may be mottled, some lighter distally.

Excurrent aperture length 10–15% of shell length; creamy white within, marginal color band grayish brown or black with grayish tan bands perpendicular to margin, some individuals without bands; margin crenulate or with very small simple papillae; papillae rusty brown.

Supra-anal aperture length 15–20% of shell length; creamy white to light tan within, often with sparse irregular gray or grayish brown marginally; margin smooth; mantle bridge separating supra-anal and excurrent apertures imperforate, length 20–25% of supra anal length.

Specimens examined: Bear Creek, Colbert County (n = 1); Duck River, Tennessee (n = 3).

Glochidium Description

Length 60–75 μm; height 70–90 μm; without styliform hooks; outline depressed subelliptical; dorsal margin straight, ventral margin rounded, anterior and posterior margins convex (Lefevre and Curtis, 1910b; Surber, 1912; Utterback, 1916b; Ortmann, 1919).

Similar Species

Truncilla truncata resembles *Truncilla donaciformis* but attains a larger size and has a thicker shell. *Truncilla truncata* is more inflated, less elongate and more triangular or quadrate in outline than *T. donaciformis*. The posterior ridge of *T. truncata* is sharper than that of *T. donaciformis*.

Truncilla truncata may resemble male *Epioblasma triquetra*, especially individuals from large rivers. However, *T. truncata* is less elongate and more triangular or quadrate in outline, with a more deeply rounded ventral margin. *Epioblasma triquetra* is often more inflated posteriorly than *T. truncata*.

Truncilla truncata may superficially resemble *Fusconaia flava*, but the periostracum of *F. flava* typically lacks rays and never has the green chevrons that are often present on *T. truncata*. The shell of *F. flava* is much more rounded anterioventrally than that of *T. truncata*. Also, the shell of *F. flava* is heavier and its pseudocardinal teeth less compressed than those of *T. truncata*.

General Distribution

In the Great Lakes drainage *Truncilla truncata* has been reported only from lakes Erie, St. Clair and Michigan (Goodrich and van der Schalie, 1932; La Rocque and Oughton, 1937). In the Mississippi Basin it is known from southern Minnesota (Dawley, 1947) south to Louisiana (Vidrine, 1993) and from headwaters of the Ohio River drainage in western Pennsylvania (Ortmann, 1919) west to the Missouri River drainage of South Dakota (Backlund, 2000) and Kansas (Murray and Leonard, 1962). *Truncilla truncata* is widespread in the Cumberland River drainage downstream of Cumberland Falls (Cicerello et al., 1991; Parmalee and Bogan, 1998). It is widespread in the Tennessee River drainage, where it occurs from headwaters in southwestern Virginia downstream to the mouth of the Tennessee River (Ahlstedt, 1992a; Parmalee and Bogan, 1998). *Truncilla truncata* occurs in Gulf Coast drainages west of the Mississippi River from the Sabine River drainage in Louisiana and Texas to the San Jacinto River drainage in Texas (Vidrine, 1993; Howells et al., 1996).

Alabama and Mobile Basin Distribution

Truncilla truncata historically occurred in the Tennessee River across northern Alabama and in some tributaries.

Truncilla truncata is known to remain extant in the Tennessee River in tailwaters of Wilson Dam, Paint Rock River and a short reach of Bear Creek in Colbert County.

Ecology and Biology

Truncilla truncata occurs primarily in flowing water of creeks and small to medium rivers. However, it can also be found in large rivers, including tailwaters of dams, at depths exceeding 6 m. *Truncilla truncata* may sometimes be found in overbank habitats of reservoirs in areas with firm substrates. It may occur in a variety of substrates, including sand, gravel and occasionally firm mud.

Truncilla truncata has been reported to release sperm aggregated as hollow spermatozeugmata. Each spermatozeugmata is composed of 8,000 to 9,000 spermatozoa and has an inside diameter of 76 μm (Waller and Lasee, 1997). Heads of spermatozoa are oriented inward, with tails arranged radially. Each spermatozoan is made up of a head, midpiece and flagellum. They measure 3.3 μm, excluding the flagellum (Waller and Lasee, 1997).

Truncilla truncata is a long term brooder, gravid from late summer or autumn to the following summer (Lefevre and Curtis, 1912). Wilson (1916) reported *Sander canadensis* (Sauger) (Percidae) and *Aplodinotus grunniens* (Freshwater Drum) (Sciaenidae) to serve as glochidial hosts of *T. truncata* but did not indicate how they were determined.

Current Conservation Status and Protection

Truncilla truncata was considered currently stable throughout its range by Williams et al. (1993). However, in Alabama it was deemed threatened by Stansbery (1976) and a species of special concern by Lydeard et al. (1999). It was designated a species of highest conservation concern in the state by Garner et al. (2004).

Remarks

Ortmann (1925) reported *Truncilla truncata* as common in the Tennessee River at Muscle Shoals prior to impoundment of the river. However, its only appearance in the archaeological record is a single specimen from Muscle Shoals (Hughes and Parmalee, 1999).

Truncilla truncata remains rare in Wilson Dam tailwaters but appears to be increasing and is encountered routinely during sampling. This species was not reported from that reach during the second half of the 1900s (Garner and McGregor, 2001).

Synonymy

Truncilla truncata Rafinesque, 1819, *nomen nudum*. Rafinesque, 1819:426

Truncilla truncata Rafinesque, 1820. Rafinesque, 1820:301

Type locality: Falls of the Ohio River. Lectotype, ANSP 20217, length 42 mm, designated by Johnson and Baker (1973).

Truncilla truncata var. *fusca* Rafinesque, 1820. Rafinesque, 1820:301

Truncilla truncata var. *vermiculata* Rafinesque, 1820. Rafinesque, 1820:301

Unio elegans Lea, 1831. Lea, 1831:83, pl. 9, fig. 13; Lea 1834b:93, pl. 9, fig. 13

Unio elegans var. *elagantopsis* De Gregorio, 1914. De Gregorio, 1914:41, pl. 4, fig. 6

Unio elegans var. *magnelegans* De Gregorio, 1914. De Gregorio, 1914:41, pl. 5, figs. 1a–c

Truncilla truncata var. *lacustris* F.C. Baker, 1928. F.C. Baker 1928a:227, pl. 78, figs. 1, 2

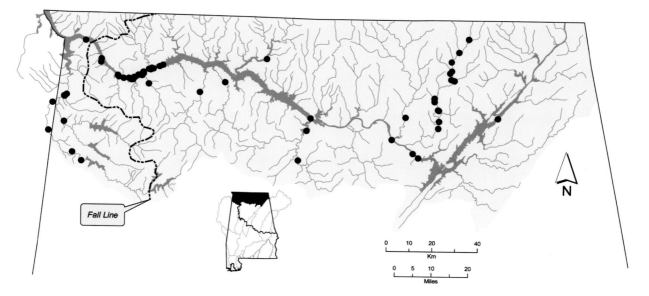

Distribution of *Truncilla truncata* in the Tennessee River drainage of Alabama.

Genus *Uniomerus*

Uniomerus Conrad, 1853, is found in the lower Mississippi Basin, Atlantic Coast drainages from Virginia to Florida, and Gulf Coast drainages from peninsular Florida to Texas. Turgeon et al. (1998) listed three *Uniomerus* species, one of which occurs in Alabama. An additional species, *Uniomerus columbensis* (Lea, 1857), is herein elevated from synonymy based on shell morphology and zoogeographic patterns. *Uniomerus carolinianus* (Bosc, 1801) has been reported from the Apalachicola Basin (e.g., Brim Box and Williams, 2000), but this species appears to be confined to Atlantic Coast drainages. *Uniomerus columbensis* is herein applied to *Uniomerus* of the Apalachicola Basin. *Uniomerus declivis* (Say, 1831) has been previously reported from Alabama (e.g., McGregor and Haag, 2004) but is not herein recognized as occurring in the state. In the Mobile Basin, variation in *Unio tetralasmus* Say, 1831, shell morphology appears to include individuals that have been identified as *U. declivis*. However, the two cannot be reliably distinguished, and further study is needed to determine if *U. declivis* is part of the Alabama fauna.

The taxonomic relationships of species assigned to *Uniomerus* are in a state of confusion. Simpson (1900b) incorrectly designated *Unio tetralasmus* Say, 1831, as the type species of *Uniomerus*, but it was not one of the taxa included in the original description (Conrad, 1853). Clench and Turner (1956) designated *Unio excultus* Conrad, 1838, as the type species of *Uniomerus* and noted it was a synonym of *Unio tetralasmus* Say, 1831. Johnson (1970) erroneously accepted Simpson's incorrect type designation and listed the correct type designation of Clench and Turner (1956) as invalid. Davis (1983) used morphological and electrophoretic techniques to examine relationships of *Uniomerus* but had difficulty separating *Uniomerus* from *Elliptio*.

Type Species
Unio excultus Conrad, 1838 (= *Unio tetralasmus* Say, 1831)

Diagnosis
Shell elongate rhomboid; dorsal and ventral margins parallel; umbo low, umbo sculpture distinct curved concentric ridges; periostracum cloth-like, yellow to brown or black, generally without rays; hinge plate well-developed but thin; pseudocardinal teeth compressed; nacre white to gray.

Incurrent aperture with arborescent papillae; mantle bridge separating excurrent and supra-anal apertures long; inner lamellae of inner gills connected to visceral mass only anteriorly; outer gills marsupial; glochidia held across entire gill; glochidium without styliform hooks (Conrad, 1853; Ortmann, 1912a; Simpson, 1914; Haas, 1969b; Britton and Fuller, 1980).

Synonymy
None recognized.

Uniomerus columbensis (Lea, 1857)
Apalachicola Pondhorn

Uniomerus columbensis – Length 41 mm, UF 366678. Mercers Mill Pond, unnamed tributary to Mill Creek, County Road 12, about 7 air miles south of Oakfield, Worth County, Georgia, 10 June 1992. © Richard T. Bryant.

Shell Description

Length to 125 mm; moderately thin; somewhat inflated; outline subrhomboidal; posterior margin obliquely truncate to rounded; anterior margin rounded; dorsal margin convex; ventral margin straight to slightly convex or concave; posterior ridge low to high, rounded; posterior slope flat to slightly concave, moderately steep dorsally, posterior terminus oblique; umbo broad, flat, may be slightly inflated, elevated slightly above hinge line, umbo sculpture irregular concentric ridges; periostracum cloth-like, dull, brown to black, young individuals may be weakly rayed.

Pseudocardinal teeth moderately thick, compressed, crests oriented almost parallel to shell margin, 2 in left valve, 1 in right valve; lateral teeth short, moderately thin, straight to slightly curved, 2 in left valve, 1 in right valve; interdentum long, very narrow; umbo cavity shallow, open; nacre bluish white to pale purple.

Soft Anatomy Description

Mantle creamy white to tan or light brown; visceral mass creamy white to tan, may be slightly darker anterioventrally; foot creamy white to tan. Lea (1863d) reported the mantle to be thickened marginally. The mantle margin just ventral to the incurrent aperture was observed to be slightly crenulate in one individual, but no distinct fold or flap was evident (J.T. Garner, personal observation).

Gills tan; dorsal margin straight to slightly sinuous, ventral margin elongate convex; gill length 55–65% of shell length; gill height 30–40% of gill length; outer gill height 65–85% of inner gill height; inner lamellae of inner gills connected to visceral mass only anteriorly.

No gravid females were available for marsupium description. Lea (1863d) reported outer gills to be marsupial, with glochidia held across the entire gill.

Labial palps creamy white to pale tan; straight dorsally, dorsal margin may be completely connected, straight to slightly convex ventrally, bluntly pointed distally, may be somewhat upswept; palp length 15–25% of gill length; palp height 50–70% of palp length; distal 20–55% of palps bifurcate.

Supra-anal aperture usually longer than incurrent and excurrent apertures; incurrent aperture usually slightly longer than excurrent aperture.

Incurrent aperture length 10–15% of shell length; tan within, may have scattered grayish brown areas, brown to grayish brown basal to papillae, some individuals with irregular creamy white band basal to papillae; papillae some combination of simple, bifid, trifid and arborescent; papillae light tan to brown or grayish brown, often creamy white distally.

Excurrent aperture length 5–15% of shell length; tan within, marginal color band grayish brown to dark brown; margin smooth to crenulate or with small simple papillae; papillae creamy white. Lea (1858e) reported the excurrent aperture to have small brown papillae.

Supra-anal aperture length 15–20% of shell length; creamy white to pale tan within, may have thin dark brown line marginally; margin smooth; mantle bridge separating supra-anal and excurrent apertures imperforate, length 15–35% of supra-anal length.

Specimens examined: Muckalee Creek, Lee County, Georgia (n = 1); Wamble Creek, Early County, Georgia (n = 3).

Glochidium Description

Glochidium unknown.

Similar Species

Uniomerus columbensis resembles several species of *Elliptio*, including *E. fumata* and *E. pullata*, but can be distinguished by its cloth-like periostracum. Some species of *Elliptio* have rays, but adult *U. columbensis* consistently lack rays. The umbo sculpture of *U. columbensis* is oblique to the hinge line, but it is not in species of *Elliptio*. Typically the incurrent aperture papillae of *U. columbensis* are arborescent, but may be bifid or trifid in some individuals. The papillae of most Apalachicola Basin *Elliptio* spp. are generally simple, with a few bifid papillae in some individuals. The exception to this is *Elliptio crassidens*, which may have some arborescent papillae.

General Distribution

The distribution of *Uniomerus columbensis* is unclear. It is herein considered an Apalachicola endemic, but further study may reveal its presence in adjacent river drainages along the eastern Gulf Coast.

Alabama and Mobile Basin Distribution

Uniomerus columbensis historically occurred in the Chattahoochee River and some tributaries as well as headwater streams of the Chipola River.

Uniomerus columbensis is extant in isolated tributaries of the Chattahoochee River and headwaters of the Chipola River.

Ecology and Biology

Uniomerus columbensis occurs in creeks, rivers and floodplain lakes in sand and sandy clay. It can also be found in holes and crevices of limestone substrates.

Uniomerus columbensis is presumably a long-term brooder, gravid from late summer or autumn to the following summer. Brim Box and Williams (2000) reported a gravid female during May. Glochidial hosts of *U. columbensis* are unknown.

Current Conservation Status and Protection

Uniomerus columbensis has previously been considered synonymous with *Uniomerus carolinianus*, a widespread and locally common species of the Atlantic Coast that was reported currently stable throughout its range by Williams et al. (1993). In Alabama *U. columbensis* (as *U. carolinianus*) was considered currently stable by Lydeard et al. (1999) and a species of moderate conservation concern by Garner et al. (2004).

Remarks

Apalachicola Basin *Uniomerus* have variously been recognized as *U. carolinianus* (e.g., Heard, 1979b; Brim Box and Williams, 2000) and/or *U. declivis* (e.g., Heard, 1979b; Davis, 1983). However, *U. carolinianus* may be confined to Atlantic Coast drainages, and *U. declivis* appears to be confined to lower reaches of the Mississippi Basin (A.E. Bogan, unpublished data). Because of the taxonomic uncertainty, a species of *Uniomerus* described from the Apalachicola Basin was herein recognized. *Unio columbensis* is the earliest available name. It differs conchologically from *Uniomerus tetralasmus* of Gulf Coast drainages west of the Apalachicola Basin. *Uniomerus columbensis* is more broadly rounded posteriorly than *U. tetralasmus*, which is often bluntly pointed to narrowly rounded.

Synonymy

Unio columbensis Lea, 1857. Lea, 1857a:31; Lea, 1858e:75, pl. 14, fig. 55; Lea, 1858h:75, pl. 14, fig. 55
Type locality: Creeks, near Columbus, [Muscogee County,] Georgia. Lectotype, USNM 85360, length 88 mm, designated by Johnson (1974).

Unio plantii Lea, 1857. Lea, 1857g:171; Lea, 1859f:192, pl. 21, fig. 76; Lea, 1859g:10, pl. 21, fig. 76

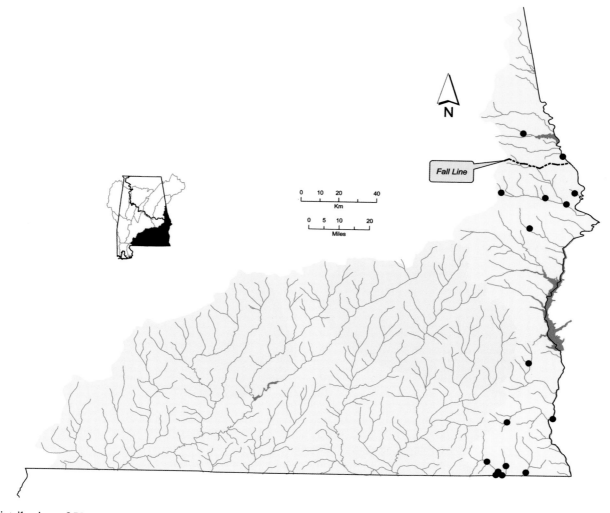

Distribution of *Uniomerus columbensis* in the eastern Gulf Coast drainages of Alabama.

Uniomerus tetralasmus (Say, 1830)
Pondhorn

Uniomerus tetralasmus – Upper figure: length 102 mm, AUM 1489. Calebee Creek, Macon County, Alabama. Lower figure: length 39 mm, UF 374093. Yellow River, below a spring tributary, north of U.S. Highway 84, Santa Rosa County, Florida, 19 August 1996. © Richard T. Bryant.

Shell Description

Length to 135 mm; moderately thin; somewhat inflated; outline elliptical to subrhomboidal; posterior margin obliquely truncate to narrowly rounded or bluntly pointed; anterior margin rounded; dorsal margin convex; ventral margin straight to slightly convex or slightly concave; posterior ridge low, rounded; posterior slope flat to slightly concave, moderately steep dorsally, less steep and broader ventrally, posterior terminus angular; umbo broad, flat, may be slightly inflated, elevated slightly above hinge line, umbo sculpture irregular concentric ridges; periostracum dull, brown to black, young individuals may be weakly rayed.

Pseudocardinal teeth moderately thick, compressed, crests oriented along margin, 2 in left valve, 1 in right valve; lateral teeth short, moderately thin, straight to slightly curved, 2 in left valve, 1 in right valve; interdentum long, narrow to very narrow; umbo cavity shallow, open; nacre bluish white.

Soft Anatomy Description

Mantle creamy white to tan or light brown, may have golden cast, light brown or rusty brown outside of apertures; visceral mass creamy white to pearly white, often with tan areas; foot creamy white to tan, may be darker distally.

Gills tan to gold; dorsal margin straight to slightly sinuous, ventral margin gently convex to almost straight; gill length 45–55% of shell length; gill height 35–50% of gill length; outer gill height 65–80% of inner gill height; inner lamellae of inner gills connected to visceral mass only anteriorly. Utterback (1915) reported gills to be brown.

No gravid females were available for marsupium description. Ortmann (1912a) and Utterback (1915) reported outer gills to be marsupial. Utterback (1915) reported gravid gills to be "rather padiform, distended at center, but not near the ventral edges".

Labial palps creamy white to tan; straight dorsally, convex ventrally, bluntly pointed distally; palp length 30–40% of gill length; palp height 40–65% of palp length; distal 25–35% of palps bifurcate.

Incurrent aperture longer than excurrent aperture; supra-anal aperture usually longer than incurrent aperture, occasionally of equal length.

Incurrent aperture length 10–15% of shell length; creamy white to tan within, may have rusty cast, with some combination of tan, rusty brown, dark brown and gray basal to papillae, often with bands perpendicular to margin; papillae arborescent, with branches lying mostly on 1 plane, papillae usually simple and bifid near aperture ends; papillae tan, rusty brown or grayish brown.

Excurrent aperture length approximately 10% of shell length; creamy white to light tan within, marginal color band usually uniform dark brown, may have irregular tan patches; margin with very small simple papillae; papillae tan to dark brown or rusty brown. Ortmann (1912a) and Utterback (1915) reported the excurrent aperture margin to have fine crenulations.

Supra-anal aperture length 15–20% of shell length; creamy white to light tan within, occasionally with sparse gray marginally; margin smooth; mantle bridge separating supra-anal and excurrent apertures rarely perforate, length 20–45% of supra-anal length. Ortmann (1912a) reported the mantle bridge to be "rather long", shorter than the supra-anal aperture but longer than the excurrent aperture.

Specimens examined: Brushy Creek, Conecuh County (n = 2); Opintlocco Creek, Russell County (n = 2); Pea River (n = 5).

Glochidium Description

Length 150–175 μm; height 200–225 μm; without styliform hooks; outline subelliptical; dorsal margin straight, ventral margin rounded, anterior and posterior margins convex.

Similar Species

Uniomerus tetralasmus may resemble several species of *Elliptio*, including *E. fumata*, *E. mcmichaeli* and *E. pullata*. However, *U. tetralasmus* has a well-defined angle at the junction of the dorsal and posterior margins and often has a doubled posterior ridge, which may be absent in *Elliptio* species. Shell nacre of *U. tetralasmus* is white, whereas nacre of many *Elliptio* species is usually purple or salmon. In live individuals the flattened arborescent papillae distinguish *U. tetralasmus* from all species of *Elliptio*.

General Distribution

Uniomerus tetralasmus occurs in the Mississippi Basin from the Ohio River drainage south to the Gulf of Mexico. It is found in Gulf Coast drainages from the Choctawhatchee River drainage in Alabama and Florida west to the Nueces River drainage in Texas (Howells et al., 1996). One specimen of *U. tetralasmus* was recently recovered from a prehistoric shell midden along the lower Tennessee River near Shiloh, Hardin County, Tennessee, but there are no historical records of this species from the Tennessee River drainage (Parmalee, 2005).

Alabama and Mobile Basin Distribution

Uniomerus tetralasmus is widespread in the Coastal Plain portion of the Mobile Basin and in Gulf Coast drainages from the Choctawhatchee River west to the Escambia River. There are several records from well above the Fall Line in the Coosa River drainage.

Uniomerus tetralasmus remains widespread in much of its historical range. It may be locally common.

Ecology and Biology

Uniomerus tetralasmus inhabits headwater streams, ponds and floodplain lakes of rivers in mud or sand substrates. It can tolerate high turbidity, periods of low dissolved oxygen and even prolonged stranding. *Uniomerus tetralasmus* often occurs in temporary water bodies, where it is the only freshwater mussel, and may be locally abundant. Frierson (1903a) noted that *U. tetralasmus* was able to live in localities "where, from three to six months at a time, there is absolutely no water; in fact living shells have been thrown out by the plowshare, and hundreds have been observed to be killed by fire sweeping over the dried-up ponds".

The brooding period of *Uniomerus tetralasmus* is unclear. Utterback (1915) reported individuals brooding mature glochidia in May and August in Missouri. However, Stern and Felder (1978) reported gravid individuals in February and March in Louisiana. Conglutinates were described by Utterback (1915) as white, "sole-shaped" and undivided, with "regular thin transparent areas arranged cross-wise made by the thickening of the septa at regular intervals". The only reported glochidial host of *U. tetralasmus* is *Notemigonus crysoleucas* (Golden Shiner) (Cyprinidae), based on observations of natural infestations (Stern and Felder, 1978).

Current Conservation Status and Protection

Uniomerus tetralasmus was considered currently stable throughout its range by Williams et al. (1993) and in Alabama by Lydeard et al. (1999). Garner et al. (2004) designated it a species of low conservation concern in the state.

Remarks

Systematic relationships of various forms within the genus *Uniomerus* are poorly understood. Conchological characters appear to be insufficient to delineate species, so it is unclear how many species exist. *Uniomerus declivis* appears to represent a valid species, at least in the lower Mississippi Basin (A.E. Bogan, unpublished data). Some individuals from

Alabama closely resemble *U. declivis* and it has been reported from the state (e.g., McGregor and Haag, 2004). However, in large series there are individuals that are intermediate between *Uniomerus tetralasmus* and *U. declivis*. Detailed phylogenetic analyses are needed to delineate distributional boundaries of species within the genus.

Synonymy

Unio tetralasmus Say, 1830. Say, 1830a:[no pagination], pl. 23
Type locality: Bayou St. John, New Orleans[, Louisiana]. Type specimen not found.

Unio camptodon Say, 1832. Say, 1832:[no pagination], pl. 42
Unio geometricus Lea, 1834. Lea, 1834a:38, pl. 4, fig. 10; Lea, 1834b:150, pl. 4, fig. 10
Unio excultus Conrad, 1838. Conrad, 1838:99, pl. 54, fig. 1
Unio sayanus Conrad, 1838. Conrad, 1838:102, pl. 56, fig. 2
Type locality: Small stream in Greene County, Alabama. Type specimen not found, reported lost by Johnson and Baker (1973). Measurement of figured shell not given in original description (Conrad, 1838).
Comment: Conrad appears to have assumed that the description of this species by Ward was going to be published prior to the appearance of his 1838 monograph. However, Conrad's description and figure appeared first, resulting in him being the author of this taxon.

Unio sayi Ward *in* Tappan, 1839. Tappan, 1839:268, pl. 3, fig. 1
Unio parallelus Conrad, 1841. Conrad, 1841:20
Unio symmetricus Lea, 1845. Lea, 1845:164; Lea, 1848a:73, pl. 4, fig. 11; Lea, 1848b:73, pl. 4, fig. 11
Unio rivularis Conrad, 1853, *nomen nudum*. Conrad, 1853:257
Unio rivularis Conrad, 1854. Conrad, 1854:296
Type locality: A small creek in Greene County, Alabama.
Comment: Conrad (1836) identified this species as *Unio declivis* Say, 1832. Subsequently he described it as *Unio rivularis*, designating the figured shell below (Conrad, 1836, pl. 23, fig. 1) as the type. Lectotype, ANSP 42852, designated by Johnson and Baker (1973).

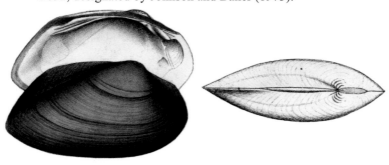

Unio porrectus Conrad, 1854. Conrad, 1854:296, pl. 26, fig. 7
Unio subcroceus Conrad, 1854. Conrad, 1854:297, pl. 27, fig. 1
Unio manubius Gould, 1856. Gould, 1856:229
Unio jamesianus Lea, 1857. Lea, 1857c:84; Lea, 1858e:52, pl. 6, fig. 35; Lea, 1858h:52, pl. 6, fig. 35
Unio electrinus Reeve, 1865. Reeve, 1865:[no pagination], pl. 25, fig. 121

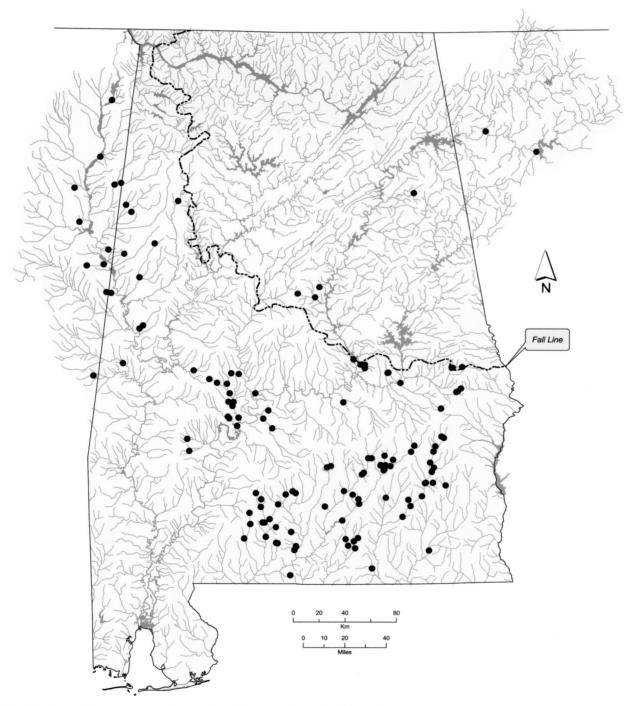

Distribution of *Uniomerus tetralasmus* in Alabama and the Mobile Basin.

Genus *Utterbackia*

Utterbackia Baker, 1927, occurs in the Great Lakes and Mississippi basins and Atlantic Coast drainages from Delaware to peninsular Florida. In Gulf Coast drainages it occurs from Florida to Texas and Mexico (Johnson, 1999). Turgeon et al. (1998) listed three species in this genus, two of which occur in Alabama. One species, *Utterbackia imbecillis* (Say, 1829) has been introduced outside of its native range (e.g., Fuller, 1980a).

Utterbackia Baker, January 1927, with the type species *Anodonta imbecillis* Say, 1829, has priority over *Utterbackiana* Frierson, April 1927, with the type species *Anodonta suborbiculata* Say, 1831, which makes this genus name a junior synonym of *Anodonta* Lamarck, 1799.

Utterbackia and all species currently recognized as belonging to it were previously considered to belong to *Anodonta*. Using morphological and allozymic data, Hoeh (1990) produced a phylogeny of North American *Anodonta* and included the type species *Anodonta cygnea* (Linneaus, 1758). The genus *Anodonta* was divided into three clades. Based on these phylogenetic analyses, Hoeh (1990) elevated *Pyganodon* Crosse and Fischer *in* Fischer and Crosse, 1894, and *Utterbackia* from subgenus to genus status.

Type Species
Anodonta imbecillis Say, 1829

Diagnosis
Shell inflated; thin; disk without sculpture; umbo not elevated above hinge line, umbo sculpture faint double-looped ridges; periostracum smooth, often shiny, green to brown often with numerous green rays; hinge teeth absent; nacre bluish white.

Excurrent aperture smooth; mantle bridge separating excurrent and supra-anal apertures long; inner lamellae of inner gills connected to visceral mass only anteriorly; outer gills marsupial; glochidia held across entire gill; marsupium well-padded when gravid, not extended beyond original gill ventral edge; secondary water tubes present when gravid; rusty brown when gravid; glochidium with styliform hooks (Baker, 1927, 1928a; Haas, 1969b; Britton and Fuller, 1980).

Synonymy
Utterbachia Baker, 1927; spelling corrected to *Utterbackia* by Baker (1928a)

Utterbackia imbecillis (Say, 1829)
Paper Pondshell

Utterbackia imbecillis – Length 63 mm, UMMZ 163354. Patsaliga River [Creek], Horton's Lake, 10 miles north of Searight, Crenshaw County, Alabama, July 1915. © Richard T. Bryant.

Shell Description

Length to 120 mm; thin; somewhat inflated; outline elliptical; posterior margin bluntly pointed to narrowly rounded; anterior margin rounded; dorsal margin straight, slightly concave in subadults, may be expanded into weak dorsal wing that often disappears with age; ventral margin convex; posterior ridge low, rounded; posterior slope moderately steep, flat; umbo broad, moderately inflated in adults, typically not elevated above hinge line, umbo of very large individuals may extend slightly above hinge line, umbo sculpture irregular ridges, slightly sinuous or double-looped ventrally; periostracum smooth, shiny, yellowish green to greenish brown, may be weakly rayed, especially in small individuals.

Pseudocardinal and lateral teeth absent; umbo cavity shallow, open; nacre white to bluish white.

Soft Anatomy Description

Mantle creamy white to light tan, outside of apertures tan with dark brown mottling; visceral mass creamy white; foot creamy white to tan.

Gills creamy white to tan; dorsal margin straight to slightly sinuous or slightly concave, ventral margin convex to elongate convex; gill length 45–50% of shell length; gill height 45–50% of gill length; outer gill height 80–90% of inner gill height; inner lamellae of inner gills connected to visceral mass only anteriorly.

Outer gills marsupial; glochidia held across most of gill; marsupium padded when gravid; light brown when brooding mature glochidia. Schwartz and Dimock (2001) reported gills to have secondary water tubes while gravid.

Labial palps light tan; straight dorsally, convex ventrally, bluntly pointed distally; palp length 25–35% of gill length; palp height 55–80% of palp length; distal 35–50% of palps bifurcate.

Incurrent aperture longer than excurrent and supra-anal apertures; supra-anal aperture longer or equal to excurrent aperture.

Incurrent aperture length 10–15% of shell length; creamy white within, tan to rusty brown or pale orange basal to papillae, may have brown lines extending from edges of papillae onto aperture wall, some lines converge proximally; papillae in 2–3 rows, long, simple, slender or thick; papillae creamy white to tan, rusty orange or gray, often with dark brown edges basally.

Excurrent aperture length 5–10% of shell length; creamy white within, marginal color band often dull orange with irregular lines or markings perpendicular to margin, some individuals with irregular lines but no distinct background coloration; margin smooth but undulating.

Supra-anal aperture length 5–10% of shell length; creamy white to pale gray within, margin may have irregular sparse brown or black; margin smooth; mantle bridge separating supra-anal and excurrent apertures imperforate, length 125–150% of supra-anal length.

Specimens examined: Tennessee River (n = 4).

Glochidium Description

Length 250–313 µm; height 289–310 µm; with styliform hooks; outline subtriangular, length may be greater than height; dorsal margin straight, ventral margin bluntly pointed, anterior and posterior margins convex (Ortmann, 1912a; Surber, 1912; Hoggarth, 1999; Schwartz and Dimock, 2001).

Similar Species

Utterbackia imbecillis closely resembles *Utterbackia peggyae* but has a more elongate shell. The shell of *U. peggyae* is higher posteriorly than that of *U. imbecillis*. *Utterbackia peggyae* usually has fine green rays, which are usually weak or absent on *U. imbecillis*. The mantle margin of *U. peggyae* is pigmented with small dark spots that give it a peppered appearance, whereas that of *U. imbecillis* is not.

Utterbackia imbecillis can be easily distinguished from *Pyganodon cataracta* and *Pyganodon grandis* because the umbo of those species is inflated and elevated above the hinge line.

Utterbackia imbecillis also resembles *Anodonta suborbiculata* and *Anodonta* sp. (Cypress Floater), but those species are much more round or oval in outline.

General Distribution

Utterbackia imbecillis is known from the Great Lakes Basin in lakes St. Clair, Erie and Ontario (Clarke, 1981a; Strayer and Jirka, 1997). In the Mississippi Basin it occurs from headwaters in Minnesota (Dawley, 1947; Graf, 1997) south to Louisiana (Vidrine, 1993) and from headwaters of the Ohio River drainage in western New York and Pennsylvania west to South Dakota (Backlund, 2000) and Nebraska (Hoke, 2000). *Utterbackia imbecillis* is present in the Cumberland River drainage downstream of Cumberland Falls (Cicerello et al., 1991). It is known from throughout the Tennessee River drainage but appears to have been uncommon historically. In that drainage Ortmann (1925) reported it only from the Paint Rock River. *Utterbackia imbecillis* occurs in Gulf Coast drainages from the Florida peninsula west to the Rio Grande and south into Nuevo Leon, Mexico (Contreras Arquieta, 1995; Howells et al., 1996). In Atlantic Coast drainages *U. imbecillis* occurs from the Delaware River drainage in Pennsylvania south to Florida. Its presence east of the Suwannee River and along the Atlantic Coast south of the Altamaha River drainage appears to be due to recent introductions and/or natural range expansion.

Alabama and Mobile Basin Distribution

Utterbackia imbecillis occurs in all of the major drainages of Alabama and the Mobile Basin. It does not occur in the Blackwater or Perdido River drainages and its presence in the Escatawpa River drainage is apparently due to a recent unintentional introduction.

Utterbackia imbecillis remains in most of its historical range and appears to be colonizing new areas through unintentional introductions from fish stockings.

Ecology and Biology

Utterbackia imbecillis usually inhabits slackwater areas of creeks and rivers, natural and artificial ponds and reservoirs, usually in mud or sand. Juveniles can be found under rocks or well-buried in gravel substrates in swift water, but adults are rarely found there. *Utterbackia imbecillis* was one of the first mussel species reported to colonize overbank habitats when the

Tennessee River was impounded by Kentucky Dam (Bates, 1962).

The reproductive biology of *Utterbackia imbecillis* is peculiar. It is one of few unionids in North America consistently reported to be a simultaneous hermaphrodite (Sterki, 1898b; van der Schalie, 1966, 1970; Heard, 1975a; Kotrla, 1988). However, at least one Florida panhandle population of *U. imbecillis* contains female individuals in addition to hermaphrodites (Heard, 1975a; Kat, 1983c; Kotrla, 1988). Higher ratios of male to female tissues have been reported in gonads of individuals from dense populations in slackwater areas compared to those from sparse populations in creeks and rivers (Kat, 1983c). *Utterbackia imbecillis* sperm are discharged as spermatozeugmata, which were described by Utterback (1931) as "hollow globular masses of sperm revolving clock-wise in the water by means of flagella thrust through a matrix from hundreds of individual sperm-cells".

Utterbackia imbecillis is a long-term brooder and may be found gravid at any time of the year (Utterback, 1915, 1916a; Ortmann, 1919; Baker, 1928a; Heard, 1975a; Howells, 2000). It may produce multiple broods per year (Allen, 1924) and possibly act as a short-term brooder during warm months and a long-term brooder during winter (Gordon and Layzer, 1989). The peculiar reproductive strategy of *U. imbecillis* has been referred to as "ultra-tachytictic" (Heard and Guckert, 1970). *Utterbackia imbecillis* may begin to reproduce as early as its second year (Allen, 1924). *Utterbackia imbecillis* glochidia are discharged in loose masses (Ortmann, 1910c).

Utterbackia imbecillis is one of few unionids reported to undergo transformation from glochidium to juvenile stage within the marsupium of the parent, bypassing the parasitic stage (Howard, 1914a, 1915). Though the report of Howard (1914a) was convincing, a study by Tucker (1928) did not duplicate the results, and there have been no further reports of this phenomenon in *U. imbecillis*. It has been shown to utilize a wide range of native fishes as well as a number of nonindigenous aquarium fishes and even amphibians as glochidial hosts. Native fishes that have been demonstrated to serve as glochidial hosts of *U. imbecillis* in laboratory trials include *Ambloplites rupestris* (Rock Bass), *Lepomis cyanellus* (Green Sunfish), *Lepomis gibbosus* (Pumpkinseed), *Lepomis macrochirus* (Bluegill), *Lepomis megalotis* (Longear Sunfish), *Micropterus salmoides* (Largemouth Bass) and *Pomoxis nigromaculatus* (Black Crappie) (Centrarchidae); *Cyprinella spiloptera* (Spotfin Shiner) and *Notemigonus crysoleucas* (Golden Shiner) (Cyprinidae); *Fundulus diaphanus* (Banded Killifish) (Fundulidae); *Ictalurus punctatus* (Channel Catfish) (Ictaluridae); and *Etheostoma lepidum* (Greenthroat

Darter) and *Perca flavescens* (Yellow Perch) (Percidae) (Tucker, 1928; Parker et al., 1980; Trdan and Hoeh, 1982; Hove et al., 1995; Howells, 1997; Keller and Ruessler, 1997). Additional species reported to serve as glochidial hosts of *U. imbecillis* based on observations of natural infestations include *Lepomis gulosus* (Warmouth) and *Lepomis marginatus* (Dollar Sunfish) (Centrarchidae); and *Gambusia affinis* (Western Mosquitofish) (Poeciliidae) (Stern and Felder, 1978). *Semotilus atromaculatus* (Creek Chub) (Cyprinidae) also possibly serves as a glochidial host of *U. imbecillis* (Clarke and Berg, 1959). Watters and O'Dee (1998) reported transformation of *U. imbecillis* glochidia on 26 species of nonindigenous aquarium fishes and 4 species of amphibian. Ortmann (1925) commented that "this species possesses exceptional means of dispersal".

Current Conservation Status and Protection

Utterbackia imbecillis was considered currently stable throughout its range by Williams et al. (1993) and in Alabama by Lydeard et al. (1999). Garner et al. (2004) designated it a species of lowest conservation concern in the state.

Remarks

The range of *Utterbackia imbecillis* appears to have expanded over the past century, presumably by introduction of sport and bait fishes infested with glochidia and natural expansion into habitats created by impoundment of rivers. For example, it was not reported historically from Florida east of the Suwannee River (Clench and Turner, 1956; Heard, 1979b) but now occurs in most of peninsular Florida. However, records from the early 1900s from all parts of its range in Alabama and the Mobile Basin suggest that it is native there. The abundance of *U. imbecillis* also appears to have increased with impoundment of most large rivers.

Synonymy

Anodonta imbecillis Say, 1829. Say, 1829:355
Type locality: Wabash River. Neotype, SMF 4301, length 58 mm, designated by Haas (1930); type locality information not given.
Anodonta incerta Lea, 1834. Lea, 1834a:46, pl. 6, fig. 16; Lea, 1834b:158, pl. 6, fig. 16
Anodon horda Gould, 1855. Gould, 1855:229
Anodonta phalena De Gregorio, 1914. De Gregorio, 1914:64, pl. 11, figs. 3a–c
Utterbackia imbecillis fusca F.C. Baker, 1927. F.C. Baker, 1927:222

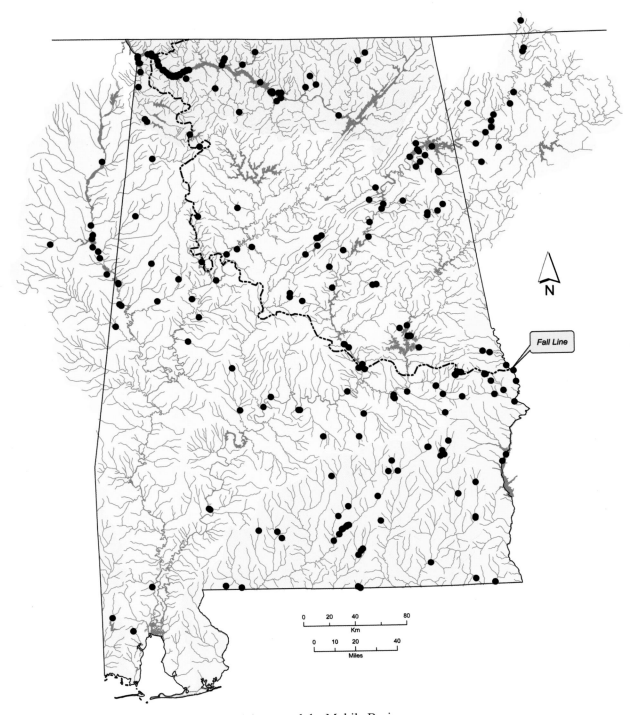

Distribution of *Utterbackia imbecillis* in Alabama and the Mobile Basin.

Utterbackia peggyae (Johnson, 1965)
Florida Floater

Utterbackia peggyae – Length 57 mm, UF 1950. Chipola River, Dead Lake at Chipola Park, 20 miles south of Blountstown, Calhoun County, Florida, 3 September 1954. © Richard T. Bryant.

Shell Description

Length to 86 mm; thin; moderately inflated; outline subelliptical, higher posteriorly; posterior margin rounded to narrowly truncate; anterior margin rounded; dorsal margin straight, with low dorsal wing; ventral margin convex; posterior ridge rounded, may be doubled; posterior slope flat to slightly concave; umbo broad, flat, not elevated above hinge line, umbo sculpture weakly double-looped ridges; periostracum smooth, shiny, green to greenish brown, typically with numerous fine green rays, darker on posterior slope.

Pseudocardinal and lateral teeth absent; umbo cavity shallow, open; nacre bluish white.

Soft Anatomy Description

Mantle creamy white to tan or light brown, outside of apertures usually rusty brown; visceral mass creamy white or pearly white to tan; foot creamy white to tan, may be darker distally.

Gills tan, may have rusty cast in some areas; dorsal margin straight to slightly sinuous, ventral margin elongate convex; gill length 55–70% of shell length; gill height 30–45% of gill length; outer gill height 80–100% of inner gill height; inner lamellae of inner gills connected to visceral mass only anteriorly.

Outer gills marsupial; glochidia held across entire gill; marsupium padded when gravid; tan to light brown. Brim Box and Williams (2000) reported gravid gills to be reddish.

Labial palps creamy white to tan or light brown; straight to slightly convex or concave dorsally, convex ventrally, bluntly pointed distally; palp length 20–35% of gill length; palp height 30–60% of palp length; distal 30–40% of palps bifurcate, occasionally less.

Incurrent aperture longer than excurrent aperture; incurrent aperture usually longer than supra-anal

aperture, occasionally equal in length; excurrent aperture longer, shorter or equal to supra-anal aperture.

Incurrent aperture length 10–20% of shell length; creamy white to tan within, may have rusty brown basal to papillae; papillae in 2–3 rows, inner row often sparse, simple, may be short or long; papillae usually some combination of creamy white, tan and rusty brown, may change color distally.

Excurrent aperture length 5–15% of shell length; creamy white to tan within, usually with marginal color band in some combination of tan, rusty brown and dark brown, often solid rusty brown proximally, developing an open reticulated pattern distally; margin smooth, may be undulating.

Supra-anal aperture length 5–20% of shell length; creamy white to tan within, may have very thin rusty brown band marginally; margin smooth; mantle bridge separating supra-anal and excurrent apertures imperforate, length 180–330% of supra-anal length, rarely absent.

Specimens examined: Apalachicola River, Florida (n = 1); Attapulga Creek, Gadsden County, Florida (n = 3); Escambia River, Florida (n = 1); Flint River, Georgia (n = 3) (specimens previously preserved).

Glochidium Description

Length 287–312 μm; height 300–325 μm; with styliform hooks; outline subtriangular; dorsal margin straight to slightly convex, ventral margin bluntly pointed, anterior and posterior margins convex, anterior margin slightly more convex than posterior margin.

Similar Species

Utterbackia peggyae closely resembles *Utterbackia imbecillis* but is less elongate and higher posteriorly. *Utterbackia peggyae* also has fine green rays that are usually very weak or absent on *U. imbecillis*. The mantle margin of *U. peggyae* is pigmented with small dark spots that give it a peppered appearance, whereas that of *U. imbecillis* is not.

General Distribution

Utterbackia peggyae occurs in large Gulf Coast drainages from the Ochlockonee River drainage in Florida and Georgia west to the Escambia River drainage in Florida.

Alabama and Mobile Basin Distribution

Utterbackia peggyae is found only in the Choctawhatchee River drainage and headwaters of the Chipola River. It is known from Florida reaches of the Yellow and Escambia River drainages, but there are no records from Alabama reaches of those systems.

The current distribution of *Utterbackia peggyae* in Alabama is unclear. It is presumably extant, but uncommon, in isolated areas.

Ecology and Biology

Utterbackia peggyae inhabits sluggish waters of creeks, sloughs of rivers, ponds and some man-made reservoirs in areas with little or no current and substrates composed of sand and mud.

Utterbackia peggyae is a long-term brooder, gravid from late summer or autumn to the following summer. It is dioecious, unlike the closely related *Utterbackia imbecillis*, which has been reported to be a simultaneous hermaphrodite (Heard, 1975a; Hoeh et al.,

1995). Heard (1979b) suggested that one population of *U. peggyae* in Florida produced two consecutive broods in a year. Glochidial hosts of *U. peggyae* are unknown.

Current Conservation Status and Protection

Utterbackia peggyae was considered currently stable throughout its range by Williams et al. (1993) and in Alabama by Lydeard et al. (1999). Garner et al. (2004) designated it a species of moderate conservation concern in the state.

Remarks

Utterbackia peggyae was formerly thought to occur from the Suwannee River system south to the Hillsborough River system (Johnson, 1965, 1972b). These records were subsequently recognized as a new species, *Utterbackia peninsularis*, which is restricted to peninsular Florida (Bogan and Hoeh, 1995). *Utterbackia peggyae* and *U. peninsularis* are geographically isolated and were found to differ electrophoretically and in the configuration of the right lateral sorting area in the stomach floor. Kat (1983b) contrasted stomach anatomy of *U. peggyae* with that of *Utterbackia imbecillis*. Bogan and Hoeh (1995) pointed out that Kat's illustration of the stomach anatomy of *U. peggyae* was actually an illustration of *U. peninsularis*. Bogan and Hoeh (1995) figured the stomach anatomy of *U. peggyae*.

Utterbackia peggyae was named for R.I. Johnson's wife Peggy.

Synonymy

Anodonta peggyae Johnson, 1965. Johnson, 1965:1, pl. 2, figs. 1–3
Type locality: Southeast shore of Lake Talquin, formed by a dam on the Ochlockonee River, Leon County public
    fishing ground, Leon County, Florida. Holotype, MCZ 251040, length 71mm (left); paratypes, MCZ 251041,
    lengths 54 mm (center) and 46 mm (right).

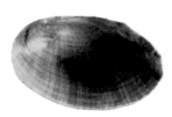

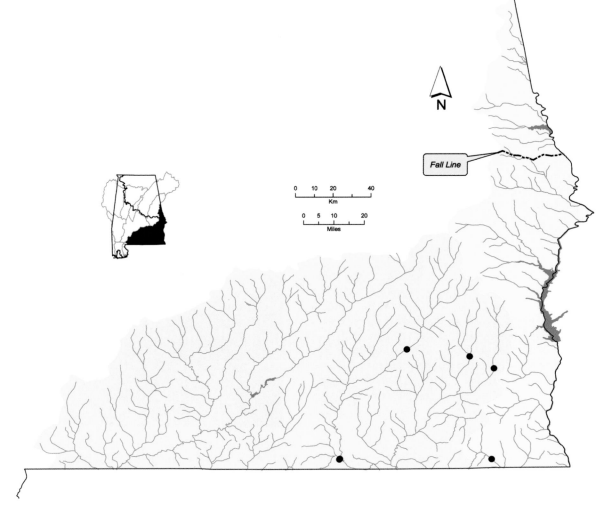

Distribution of *Utterbackia peggyae* in the eastern Gulf Coast drainages of Alabama.

Genus *Villosa*

Villosa Frierson, 1927, occurs in the Great Lakes and Mississippi basins and Atlantic Coast drainages from northern Virginia to Florida. It is found in Gulf Coast drainages from peninsular Florida to Texas. One purported *Villosa* species, *V. gundlachi* (Dunker, 1858), occurs in Cuba (Johnson, 1981). Turgeon et al. (1998) recognized 17 species of *Villosa*. One species, *Villosa vanuxemensis* (Lea, 1838), included two subspecies, *V. v. vanuxemensis* (Lea, 1838) and *V. v. umbrans* (Lea, 1857). *Villosa v. umbrans* is herein elevated to species status based on differences in periostracum color (Ortmann, 1924a), preliminary genetic analysis (W.R. Haag, personal communication) and their geographical isolation. This brings the number of *Villosa* species in Alabama to 11.

Villosa is the *Micromya* Agassiz, 1852, *non* Rondani, 1840, found in historical literature. Buhay et al. (2003) found that *Villosa*, as used today, is not a monophyletic group, based on a molecular phylogenetic analysis. *Villosa* includes as many as six distinct clades.

Species of *Villosa* have modified mantle folds located just ventral to their incurrent apertures. These folds, which are pigmented and have papillate margins, are believed to function as host-attracting lures (Figure 13.43).

Type Species
Unio villosus B.H. Wright, 1898

Diagnosis
Shell small; inflated; usually moderately thin; elliptical to oval; umbo located anteriorly; posterior ridge indistinct or evenly rounded; females swollen posteriorly, posterior margin often becoming truncate.

Mantle margin with well-developed papillate fold just ventral to incurrent aperture, fold rudimentary in males; excurrent aperture crenulate; mantle bridge separating excurrent and supra-anal apertures variable; inner lamellae of inner gills completely connected to visceral mass or connected only anteriorly; outer gills marsupial; glochidia held in posterior portion of gill; glochidium without styliform hooks (Frierson, 1927; Haas, 1969b; Britton and Fuller, 1980; Brim Box and Williams, 2000).

Figure 13.43. *Villosa vibex* female with papillate mantle fold displayed. Photograph by W.N. Roston.

Synonymy
Micromya Agassiz, 1852, *non* Rondani, 1840 (Insecta)

Villosa choctawensis Athearn, 1964
Choctaw Bean

Villosa choctawensis – Upper figure: female, length 30 mm, UF 57076. Patsaliga Creek, 7.6 miles north-northwest of Dozier, Crenshaw County, Alabama, 24 April 1970. Lower figure: male, length 25 mm, UMMZ 197257. East Fork Choctawhatchee River, 8 miles west of Abbeville, Henry County, Alabama. © Richard T. Bryant.

Shell Description

Length to 49 mm; moderately thin; somewhat inflated, females slightly more inflated posteriorly than males; outline oval; posterior margin narrowly to broadly rounded or obliquely truncate, females more broadly rounded or truncate than males; anterior margin rounded, may be slightly truncate; dorsal margin convex; ventral margin straight to slightly convex; posterior ridge low, rounded; posterior slope moderately steep, flat; umbo broad, flat, elevated slightly above hinge line, umbo sculpture unknown; periostracum shiny, tawny to greenish brown, typically with thin green rays, often obscure in darker individuals.

Pseudocardinal teeth moderately thick, triangular, 2 divergent teeth in left valve, 1 tooth in right valve, often with accessory denticle anteriorly; lateral teeth, straight to slightly curved, 2 in left valve, 1 in right valve; interdentum moderately long, narrow; umbo cavity wide, shallow; nacre bluish white.

Soft Anatomy Description

Mantle creamy white to tan, may have slight golden cast, outside of apertures pale orange, tan or pale rusty brown, usually with dark brown or gray mottling or blotches, mottling sometimes oriented perpendicular to margin; visceral mass creamy white to tan; foot tan to pinkish orange, sometimes darker ventrally.

Females with well-developed but shallow papillate mantle fold just ventral to incurrent aperture; fold length 10–20% of shell length; margin brown to rusty brown with heavy black mottling; papillae small, simple, may be numerous. Males with rudimentary mantle fold; often with little marginal coloration, usually with crenulate margin, occasionally with small, widely spaced, simple papillae.

Gills tan to light brown, may have golden cast; dorsal margin straight to slightly sinuous, ventral margin convex, female outer gill may be somewhat bilobed; gill length 45–55% of shell length; gill height 55–75% of gill length; outer gill height 60–75% of inner gill height; inner lamellae of inner gills connected to visceral mass only anteriorly.

Outer gills marsupial; glochidia held in posterior 30–50% of gill; marsupium outline oval, padded when gravid; creamy white to light tan.

Labial palps tan; straight to slightly concave dorsally, convex ventrally, bluntly pointed distally; palp length 30–40% of gill length; palp height 50–65% of palp length; distal 20–45% of palps bifurcate.

Incurrent aperture longer than excurrent aperture; supra-anal aperture and incurrent apertures usually of similar length, but supra-anal aperture occasionally longer or shorter; supra-anal aperture usually longer than excurrent aperture

Incurrent aperture length 15–20% of shell length; creamy white to tan within, usually with some combination of orange, dark tan, rusty brown, dark brown gray and black basal to papillae, may be mottled or in form of crude lines perpendicular to margin; papillae in 1–3 rows, long, slender, simple, inner row larger; papillae some combination of creamy white, tan, pale orange, rusty orange, brown and gray, often with dark brown or black edges basally.

Excurrent aperture length 10–15% of shell length; creamy white to tan within, marginal color band rusty brown or gray, usually with irregular dark brown or black lines or oblong blotches oriented perpendicular to margin; margin with small simple papillae; papillae tan to pale orange or rusty brown.

Supra-anal aperture length 10–20% of shell length; creamy white to tan within, may have slight golden cast, may have sparse gray marginally; margin smooth; mantle bridge separating supra-anal and excurrent apertures imperforate, length 30–75% of supra-anal length.

Specimens examined: Conecuh River (n = 1); Pea River (n = 4); West Fork Choctawhatchee River (n = 4).

Glochidium Description

Length 175–200 μm; height 225–237 μm; without styliform hooks; outline subelliptical; dorsal margin straight, ventral margin rounded, anterior and posterior margins convex.

Similar Species

Villosa choctawensis may resemble some individuals of *Villosa lienosa*. However *V. choctawensis* is typically less elongate, with thin green rays on the periostracum, which are usually absent or weakly developed in *V. lienosa*. Shell nacre is always white in *V. choctawensis*, but is usually some shade of purple or salmon in *V. lienosa*.

Villosa choctawensis may resemble small *Pleurobema strodeanum*, especially when the periostracum is covered by mineral deposits. However, the posterior ridge of *P. strodeanum* is much better developed and terminates in a blunt point posterioventrally, giving it an overall more angular outline than that of *V. choctawensis*. The shell of *V. choctawensis* is thinner than that of *P. strodeanum*, and its pseudocardinal teeth are more delicate. The periostracum of *V. choctawensis* is typically adorned with green rays, whereas *P. strodeanum* is seldom rayed.

General Distribution

Villosa choctawensis is restricted to Gulf Coast drainages from the Choctawhatchee River west to the Escambia River in southeastern Alabama and western Florida.

Alabama and Mobile Basin Distribution

Villosa choctawensis is known from the Choctawhatchee, Yellow and Escambia River drainages.

Villosa choctawensis is extant in isolated populations in the Choctawhatchee, Yellow and Escambia River drainages.

Ecology and Biology

Villosa choctawensis inhabits large creeks and small rivers. It occurs in silty sand to sandy clay substrates in moderate current.

Villosa choctawensis is presumably a long-term brooder, gravid from late summer or autumn to the following summer. Its glochidial hosts are unknown.

Current Conservation Status and Protection

Villosa choctawensis was considered threatened throughout its range by Williams et al. (1993) and imperiled in Alabama by Lydeard et al. (1999). Garner et al. (2004) designated it a species of high conservation concern in the state. In 2003 *V. choctawensis* was elevated to a candidate for protection under the federal Endangered Species Act.

Remarks

There is no published phylogenetic information on *Villosa choctawensis*. However, conchologically it more closely resembles *Villosa* from Atlantic Coast drainages than species in Gulf Coast drainages.

Villosa choctawensis was named for the Choctaw, a tribe of Native Americans from the southeast (Athearn, 1964).

Synonymy

Villosa choctawensis Athearn, 1964. Athearn, 1964:137, pl. 9, figs. c–h

Type locality: Choctawhatchee River, 2 miles southwest of Caryville, about 1 mile downstream of U.S. Highway 90, Holmes County, Florida, 28 November 1958. Holotype, CMNML 20096, length 37 mm.

Comment: Figures c–f in the original description are paratypes from the type locality, and g–h are paratypes from an additional locality; the holotype was not figured.

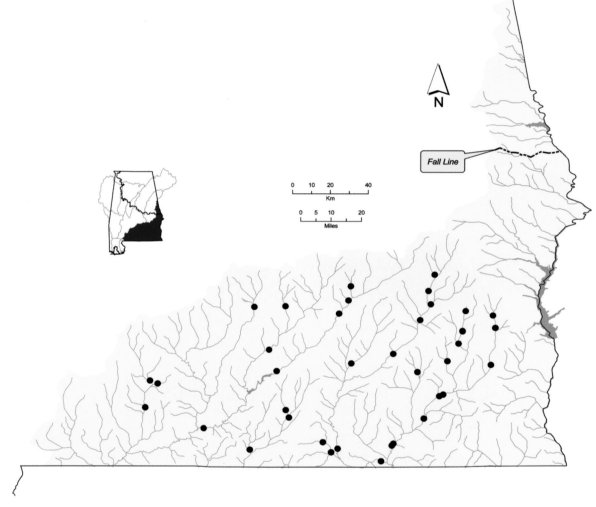

Distribution of *Villosa choctawensis* in the eastern Gulf Coast drainages of Alabama.

Villosa fabalis (Lea, 1831)
Rayed Bean

Villosa fabalis – Upper figure: female, length 27 mm, NCSM 45035. Nolichucky River, Hale Bridge, Bewleys Chapel Road, 3 air miles SE of Warrensburg, Greene County, Tennessee, 12 September 1964. Lower figure: male, length 32 mm, USNM 504859. Duck River, Columbia, Tennessee, 1891. © Richard T. Bryant.

Shell Description

Length to 40 mm, males often larger than females; moderately thick; males somewhat compressed, females moderately inflated; male outline elliptical, female outline oval; posterior margin narrowly rounded to bluntly pointed; anterior margin rounded; dorsal margin convex; ventral margin straight to slightly convex; posterior ridge low, rounded; posterior slope moderately steep; umbo broad, somewhat compressed, elevated slightly above hinge line, umbo sculpture double-looped ridges, interrupted in middle, nodulous posteriorly; periostracum olive to brown, with numerous thin, wavy, green rays.

Pseudocardinal teeth low, triangular, 2 divergent teeth in left valve, 1 tooth in right valve, usually with blade-like accessory denticle anteriorly; lateral teeth short, thick, straight, 2 in left valve, 1 in right valve; interdentum short, narrow; umbo cavity shallow, open; nacre white to bluish white.

Soft Anatomy Description

No material was available for soft anatomy description, but Lea (1863d) provided a brief description. The mantle is thin, dark brown marginally. The mantle fold just ventral to the incurrent aperture is simply described as a "fringed", spotted margin. Outer gills are marsupial, with glochidia held in the posterior portion. Inner lamellae of inner gills are connected to the visceral mass only anteriorly. Labial palps are "somewhat transverse, oval". The incurrent aperture has small brown papillae, and the excurrent aperture has numerous small papillae. The supra-anal aperture is large and dark brown, with spots marginally. Ortmann (1912a) added the following details based on material from the Ohio River drainage of western Pennsylvania. The mantle is white but has a "brownish black" edge, darkest posteriorly. Ortmann's description of the female mantle fold is more detailed. The fold is papillate, with papillae subconical and not closely spaced. There is "a streak of black pigment" on the adjacent mantle. Males have "very small" papillae in this region. Inner lamellae of inner gills are connected to the visceral mass for 25–

50% of their length. The excurrent aperture is crenulate, instead of papillate. Excurrent and supra-anal apertures are separated by a mantle bridge "of moderate length".

Glochidium Description

Length 170 µm; height 200 µm; without styliform hooks; outline subspatulate (Ortmann, 1919).

Similar Species

Villosa fabalis resembles *Villosa trabalis* but is less elongate and has a more rounded posterior margin. The umbo of *V. fabalis* is broader than that of *V. trabalis*. *Villosa fabalis* is also less inflated than *V. trabalis*.

Villosa fabalis resembles *Villosa iris* but has a thicker shell. Pseudocardinal teeth of *V. fabalis* are larger and more triangular than those of *V. iris*. The periostracum of *V. iris* from Alabama populations is generally more yellow, with less wavy rays than those of *V. fabalis*.

Villosa fabalis may resemble small *Villosa vanuxemensis*, but *V. fabalis* is more elongate, with a more pointed posterior margin and thicker shell. The periostracum of *V. fabalis* typically has more distinctive rays than that of *V. vanuxemensis*, which may lack rays altogether. Shell nacre of *V. fabalis* is white, whereas that of *V. vanuxemensis* is purple or salmon.

Villosa fabalis may superficially resemble *Toxolasma lividum*, but the periostracum of *V. fabalis* is typically rayed, whereas that of *T. lividum* is usually without rays. Male *V. fabalis* are usually less inflated than *T. lividum* of comparable size. Shell nacre of *V. fabalis* is white, whereas that of *T. lividum* is purple. Live *V. fabalis* can be distinguished from *T. lividum* by the presence of papillate mantle folds just ventral to the incurrent aperture. The folds are replaced by caruncles in *T. lividum*. These structures are rudimentary in males.

General Distribution

In the Great Lakes Basin *Villosa fabalis* is known only from Lake St. Clair and nearby Lake Erie (Clarke, 1981a). It is widespread in the Ohio River drainage, from headwaters in western New York and Pennsylvania downstream to near the mouth of the Ohio River (Ortmann, 1919; Cummings and Mayer, 1992; Strayer and Jirka, 1997). There are no records of *V. fabalis* from the Cumberland River drainage. It is widespread in the upper Tennessee River drainage of southwestern Virginia and eastern Tennessee as well as upper reaches of the Elk River (Parmalee and Bogan, 1998). The only historical record of *V. fabalis* from middle and lower reaches of the Tennessee River drainage is from the Duck River (Ortmann, 1925; Parmalee and Bogan, 1998).

Alabama and Mobile Basin Distribution

There are no historical records of *Villosa fabalis* from Alabama. However, its distribution includes upper reaches of the Tennessee River drainage as well as the upper Elk River in Tennessee, and it presumably occurred in intervening reaches of northern Alabama. There is one prehistoric record of *V. fabalis* from the Tennessee River near Bridgeport (Warren, 1975).

Ecology and Biology

Villosa fabalis occurs primarily in flowing water of small to large streams but may also be found in small or medium rivers and occasionally in natural lakes, including Lake Erie. In lakes it is usually found in areas that are subject to frequent wave action. It usually occurs in sand and gravel substrates, often in and around roots of aquatic plants. With regard to the presence of *V. fabalis* in beds of *Justicia americana* (Waterwillow), Ortmann (1919) stated that "it distinctly prefers these plants, in riffles, and is deeply buried in the sand and gravel bound together by their roots and rhizomes. By pulling up the plant it sometimes was brought to light in goodly numbers."

Villosa fabalis is a long-term brooder, gravid from late summer or autumn to the following summer. Gravid females have been reported in July and August (Ortmann, 1909b) and in May (Ortmann, 1919). Female *V. fabalis* display papillate mantle folds at certain times, presumably to attract glochidial hosts. The display involves the folds of each opposing valve undergoing rhythmic undulations simultaneously, showing white marsupial gills underneath (J.W. Jones, personal communication).

The only potential glochidial host reported for *Villosa fabalis* is *Etheostoma tippecanoe* (Tippecanoe Darter) (Percidae), based on observations of natural infestation in French Creek, Pennsylvania (White et al., 1996). This was determined by comparison of genetic material of glochidia removed from host fish with that of adult mussels.

Current Conservation Status and Protection

Villosa fabalis was considered a species of special concern throughout its range by Williams et al. (1993). Stansbery (1976) listed it as endangered in Alabama, and Lydeard et al. (1999) later listed it as extirpated from the state. *Villosa fabalis* was not included in Garner et al. (2004).

Remarks

The only *Villosa fabalis* in the archaeological record of Alabama is a single specimen from a prehistoric shell midden along the Tennessee River near Bridgeport (Warren, 1975).

Synonymy
Unio fabalis Lea, 1831. Lea, 1831:86, pl. 10, fig. 16; Lea, 1834b:96, pl. 10, fig. 16
Type locality: Ohio River. Lectotype, USNM 85270a, length 24 mm, designated by Johnson (1974).

Unio capillus Say, 1831. Say, 1831c:528
Unio lapillus Say, 1832. Say, 1832c:[no pagination], pl. 41
Unio donacopsis De Gregorio, 1914. De Gregorio, 1914:60–61, pl. 10, figs. 5a–d

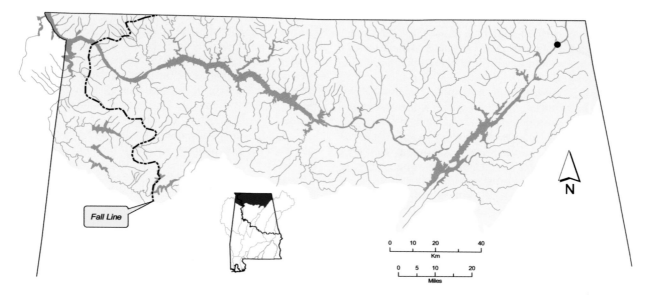

Distribution of *Villosa fabalis* in the Tennessee River drainage of Alabama.

Villosa iris (Lea, 1829)
Rainbow

Villosa iris – Upper figure: female, length 51 mm, UMMZ 85366. Lower figure: male, length 45 mm, UMMZ 85366. Spring River [Creek], Tuscumbia, Colbert County, Alabama. © Richard T. Bryant.

Shell Description

Length to 65 mm; thin; somewhat compressed, females more inflated posteriorly than males; outline subelliptical to oval; posterior margin narrowly to broadly rounded, more broadly rounded in females; anterior margin rounded; dorsal margin convex; ventral margin straight to convex; posterior ridge low, rounded; posterior slope low, flat to slightly concave; umbo broad, flat, elevated slightly above hinge line, umbo sculpture coarse ridges, becoming weakly double-looped ventrally; periostracum tawny to greenish brown, typically with well-developed, variable green rays.

Pseudocardinal teeth small, compressed, approximately parallel to hinge line, 2 in left valve, 1 in right valve, may have accessory denticle anteriorly; lateral teeth thin, moderately short, straight to slightly curved, 2 in left valve, 1 in right valve; interdentum long, narrow to very narrow; umbo cavity shallow, open; nacre usually bluish white, rarely salmon.

Soft Anatomy Description

Mantle tan to light brown, outside of apertures usually dull orange, mottled with gray, brown or black;

visceral mass creamy white, may be pearly white adjacent to foot; foot pale orange.

Females with well-developed, papillate mantle fold just ventral to incurrent aperture; fold length 10–30% of shell length; margin tan to dull orange or reddish brown, with heavy black mottling; papillae thick, blunt, simple, occasionally triangular and flattened, may be widely spaced, with crenulations between papillae. Males with rudimentary mantle fold; margin crenulate or irregularly papillate. Ortmann (1912a) reported females with smaller papillae between primary papillae, instead of crenulations. Lea (1863d) reported females to have the mantle adjacent to the fold to be "covered with whitish clouded spots".

Gills tan to light brown; dorsal margin slightly sinuous, ventral margin convex to elongate convex; gill length 50–60% of shell length; gill height 40–60% of gill length; outer gill height 70–85% of inner gill height; inner lamellae of inner gills completely connected to visceral mass or connected only anteriorly. Lea (1863d) reported "a thin whitish line along the lower edge" of inner gills.

Outer gills marsupial; glochidia held in posterior 35–50% of gill except extreme posterior end;

marsupium distended when gravid; creamy white, distal margin black.

Labial palps creamy white to tan; straight to slightly concave dorsally, curved ventrally, bluntly pointed distally; palp length 10–20% of gill length; palp height 50–100% of palp length; distal 50–75% of palps bifurcate, occasionally less.

Incurrent aperture longer or equal to excurrent aperture; incurrent aperture longer than supra-anal aperture; excurrent aperture usually longer than supra-anal aperture, occasionally shorter.

Incurrent aperture length 15–25% of shell length; creamy white to tan within, may have grayish tan to reddish brown or dark brown basal to papillae, often with sparse or irregular black mottling; papillae in 1–2 rows, long, slender, simple; papillae some combination of white, tan, pale orange, rusty brown, dark brown and black, often lighter or darker distally, may have dark brown or black edges basally.

Excurrent aperture length 10–20% of shell length; creamy white to gold or grayish tan within, marginal color band dull orange to rusty brown with dark brown or black mottling, occasionally solid brown; margin with small simple papillae, may be little more than crenulations; papillae dull orange or tan to brown or black.

Supra-anal aperture length 10–20% of shell length; creamy white to tan within, often with narrow brown, gray or black marginal band; margin smooth; mantle bridge separating supra-anal and excurrent apertures rarely perforate, length 10–45% of supra-anal length. Lea (1863d) reported the supra-anal aperture to be "spotted within".

Specimens examined: Estill Fork, Jackson County (n = 8); Larkin Fork, Jackson County (n = 1).

Glochidium Description

Length 196–263 μm; height 259–333 μm; without styliform hooks; outline subspatulate; dorsal margin straight, ventral margin rounded, anterior and posterior margins convex, slightly asymmetrical (Ortmann, 1912a, 1921; Surber, 1912; Utterback, 1916b; Zale and Neves, 1982b; Hoggarth, 1999).

Similar Species

Villosa iris resembles *Villosa taeniata* but generally has a thinner, more elongate shell with narrower green rays. Rays of *V. iris* are often not interrupted, as are those of *V. taeniata*. Pseudocardinal teeth of *V. iris* are more compressed and delicate than those of *V. taeniata*.

Villosa iris resembles *Villosa fabalis* but has a thinner shell, and the periostracum is more yellow and rays are not as wavy, at least in Alabama populations. Pseudocardinal teeth of *V. iris* are smaller and more compressed than those of *V. fabalis*.

Villosa iris also resembles *Villosa trabalis* but has a thinner, more compressed shell with a lower umbo. The periostracum of *V. iris* from Alabama populations is generally more yellow than that of *V. trabilis* and its rays tend to be less wavy. Pseudocardinal teeth of *V. iris* are smaller and more compressed than those of *V. trabalis*.

Villosa iris may resemble small *Actinonaias pectorosa*. However, *V. iris* is more elongate, has a thinner shell, shallower umbo cavity and more compressed pseudocardinal teeth.

General Distribution

Villosa iris is known from the Great Lakes drainage in parts of lakes Ontario, Erie, St. Clair, Huron and Michigan as well as their tributaries (Goodrich and van der Schalie, 1932; La Rocque and Oughton, 1937; Clarke, 1981a; Strayer et al., 1992). It is absent from upper reaches of the Mississippi Basin but occurs in the Ohio, Cumberland and Tennessee River drainages as well as several tributaries of the Mississippi River in Missouri and Arkansas (Cummings and Mayer, 1992; Oesch, 1995). In the Ohio River drainage *V. iris* occurs from headwaters in western Pennsylvania downstream to near the mouth of the Ohio River (Ortmann, 1919; Cicerello et al., 1991). In the Cumberland River drainage it occurs only downstream of Cumberland Falls (Cicerello et al., 1991; Parmalee and Bogan, 1998). *Villosa iris* is widespread in the Tennessee River drainage, occurring from headwaters in southwestern Virginia, western North Carolina and eastern Tennessee downstream to Muscle Shoals, with disjunct populations in the Buffalo and Duck River drainages, Tennessee (Ahlstedt, 1992a, 1992b; Parmalee and Bogan, 1998).

Alabama and Mobile Basin Distribution

Villosa iris was historically widespread in the Tennessee River drainage, occurring in the Tennessee River proper and many tributaries.

Villosa iris is extant in the Paint Rock River system as well as several other Tennessee River tributaries.

Ecology and Biology

Villosa iris occurs in flowing water in a variety of stream sizes, ranging from small creeks to medium rivers as well as northern lakes. There are a few records from the Tennessee River, primarily at Muscle Shoals, indicating that it can live in large rivers under some conditions. *Villosa iris* is more commonly found in cobble and boulder substrates than sand and gravel, often occurring under large rocks (Layzer and Madison, 1995). It can often be found among emergent vegetation along stream margins. Ortmann (1919) described *V. iris* in Lake Erie as occurring in sandy substrates, "often among scanty growths of rushes". Under laboratory conditions, Gatenby et al. (1996) reported juvenile *V. iris* to survive and grow better in cultures to which silt had been added compared to those without silt.

Villosa iris is a long-term brooder. It is dioecious, but van der Schalie (1969) reported a single hermaphroditic individual. Both sexes reach sexual maturity at age three (Zale and Neves, 1982b). In Big Moccasin Creek, Virginia, gametogenesis occurs between April and mid-August, at which time spawning takes place (Zale and Neves, 1982b). In most populations glochidia mature by early October and females remain gravid through the following July (Ortmann, 1919, 1921; Zale and Neves, 1982b). In Big Moccasin Creek glochidia of *Villosa* species occurred in stream drift from October to May, and those in spring were believed to be *V. iris* based on host-infestation patterns (Zale and Neves, 1982b). Female *V. iris* have a modified mantle fold just ventral to the incurrent aperture. This fold is pigmented and has a papillate margin (Figure 13.44) and presumably functions as a host-attracting lure. Mantle folds have been found to differ in morphology and pigment patterns among some populations. It is unclear whether or not these represent cryptic species. Zale and Neves (1982a) reported female *V. iris* to lie exposed on the substrate with valves agape and foot extended during the glochidial release period, but there was no mention of a mantle display.

Figure 13.44. *Villosa iris* female with papillate mantle fold displayed. Photograph by W.N. Roston.

Fishes reported to serve as glochidial hosts of *Villosa iris* in laboratory trials include *Ambloplites rupestris* (Rock Bass), *Lepomis cyanellus* (Green Sunfish), *Micropterus dolomieu* (Smallmouth Bass), *Micropterus punctulatus* (Spotted Bass) and *Micropterus salmoides* (Largemouth Bass) (Centrarchidae); *Cottus bairdii* (Mottled Sculpin) (Cottidae); *Erimystax dissimilis* (Streamline Chub) and *Luxilus chrysocephalus* (Striped Shiner) (Cyprinidae); *Etheostoma blennioides* (Greenside Darter), *Etheostoma caeruleum* (Rainbow Darter), *Etheostoma camurum* (Bluebreast Darter) and *Perca flavescens* (Yellow Perch) (Percidae); and *Gambusia affinis* (Western Mosquitofish) (Poeciliidae) (Zale and Neves, 1982a; Neves et al., 1985; O'Beirn et al., 1998; O'Dee and Watters, 2000; Watters et al., 2005). Two allopatric species, *Micropterus notius* (Suwannee Bass) (Centrarchidae) (Neves et al., 1985) and *Betta splendins* (Siamese Fighting Fish) (Belontiidae) (Watters and O'Dee, 1998) have also been found to serve as glochidial hosts under laboratory conditions. Natural infestations of *A. rupestris*, *Lepomis auritus* (Redbreast Sunfish) (Centrarchidae) and *M. dolomieu* with *V. iris* glochidia were reported to begin in April, peak in June and continue through August in Big Moccasin Creek, Virginia (Zale and Neves, 1982a). However, *V. iris* glochidia did not transform on *L. auritus* in three laboratory trials (Zale and Neves, 1982a).

In a study of mussel diet, juvenile *Villosa iris* survived best on a mixture of three species of algae along with fine sediment (Gatenby et al., 1997).

Trematode-infested *Villosa iris* were frequently encountered in Big Moccasin Creek, Virginia, by Zale and Neves (1982b). Older individuals were more frequently and more heavily parasitized than younger individuals.

Current Conservation Status and Protection

Villosa iris was reported to be currently stable throughout its range by Williams et al. (1993) and in Alabama by Lydeard et al. (1999). Garner et al. (2004) designated it a species of moderate conservation concern in the state.

Remarks

Villosa iris has long been believed to represent a species complex (Parmalee and Bogan, 1998). It is wide-ranging and shows a high degree of variability in shell morphology among different drainages and regions within its range. Also, mantle margins exhibiting different pigment and papilla patterns have been observed in the species both within and among populations. Resolution of the taxonomy of this group will likely require detailed comparative studies of shell morphology, soft anatomy, life history traits and genetic analyses among populations throughout its range.

There is a paucity of *Villosa iris* in the archaeological record from Alabama. It is represented by only two specimens collected from middens at Muscle Shoals (Hughes and Parmalee, 1999).

Synonymy

Unio iris Lea, 1829. Lea, 1829:439, pl. 11, fig. 18; Lea, 1834b:53, pl. 11, fig. 18

Type locality: Ohio. Syntype not found, but length of figured shell in original description reported as 41 mm.

Comment: The name *Unio nebulosus* Conrad, 1834, has been variously used with the *Villosa iris* complex. Herein all of the described taxa from the Ohio, Tennessee and Cumberland River systems were included as synonyms of *Villosa iris* and *Villosa nebulosa* and synonyms were restricted to headwaters of the Mobile Basin. *Villosa iris* is often recognized as a species complex. However, this complex will not be resolved on conchological grounds (e.g., Gordon, 1995).

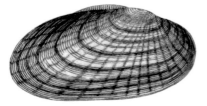  

Margarita (*Unio*) *notatus* Lea, 1836, *nomen nudum*. Lea, 1836:26
Margarita (*Unio*) *obscurus* Lea, 1836, *nomen nudum*. Lea, 1836:26
Margarita (*Unio*) *zeiglerianus* Lea, 1836, *nomen nudum*. Lea, 1836:26
Margarita (*Unio*) *cumberlandianus* Lea, 1836, *nomen nudum*. Lea, 1836:27
Margarita (*Unio*) *mühlfeldianus* Lea, 1836, *nomen nudum*. Lea, 1836:27
Margarita (*Unio*) *creperus* Lea, 1836, *nomen nudum*. Lea, 1836:28
Margarita (*Unio*) *glaber* Lea, 1836, *nomen nudum*. Lea, 1836:28
Margarita (*Unio*) *simus* Lea, 1836, *nomen nudum*. Lea, 1836:29
Unio obscurus Lea, 1838. Lea, 1838a:7, pl. 3, fig. 7; Lea, 1838c:7, pl. 3, fig. 7
Unio cumberlandicus Lea, 1838. Lea, 1838a:25–26, pl. 7, fig. 19; Lea, 1838c:25, pl. 7, fig. 19
Unio simus Lea, 1838. Lea, 1838a:26, pl. 8, fig. 20; Lea, 1838c:26, pl. 8, fig. 20
Unio notatus Lea, 1838. Lea, 1838a:28, pl. 8, fig. 22; Lea, 1838c:28, pl. 8, fig. 22
Unio zeiglerianus Lea, 1838. Lea, 1838a:32, pl. 10, fig. 27; Lea, 1838c:32, pl. 10, fig. 27
Unio creperus Lea, 1838. Lea, 1838a:33, pl. 10, fig. 28; Lea, 1838c:33, pl. 10, fig. 28
Unio glaber Lea, 1838. Lea, 1838a:34, pl. 10, fig. 29; Lea, 1838c:34, pl. 10, fig. 29
Unio mühlfeldianus Lea, 1838. Lea, 1838a:41, pl. 12, fig. 36; Lea, 1838c:41, pl. 12, fig. 36
Unio novi-eboraci Lea, 1838. Lea, 1838a:104, pl. 24, fig. 114; Lea, 1838c:104, pl. 24, fig. 114
Unio amoenus Lea, 1840. Lea, 1840:286; Lea, 1842b:200, pl. 10, fig. 12; Lea, 1842c:38, pl. 10, fig. 12
Unio tener Lea, 1840. Lea, 1840:286; Lea, 1842b:198, pl. 10, fig. 10; Lea, 1842c:36, pl. 10, fig. 10
Unio dactylus Lea, 1840. Lea, 1840:287; Lea, 1842b:196, pl. 9, fig. 7; Lea, 1842c:34, pl. 9, fig. 7
Unio fatuus Lea, 1840. Lea, 1840:287; Lea, 1842b:201, pl. 11, fig. 14; Lea, 1842c:39, pl. 11, fig. 14
Unio regularis Lea, 1841. Lea, 1841b:82; Lea, 1842b:243, pl. 25, fig. 59; Lea, 1842c:81, pl. 25, fig. 59
Unio puniceus Haldeman, 1842. Haldeman, 1842:201
Unio discrepans Lea, 1860. Lea, 1860a:92; Lea, 1860d:340, pl. 55, fig. 165; Lea, 1860g:22, pl. 55, fig. 165

Type locality: North Alabama, Professor Tuomey. Lectotype, USNM 85267, length 46 mm, designated by Johnson (1974).

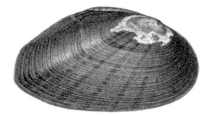  

Unio planicostatus Lea, 1860. Lea, 1860a:92; Lea, 1860d:354, pl. 59, fig. 179; Lea, 1860g:36, pl. 59, fig. 179
Type locality: Tuscumbia, [Colbert County,] Alabama, L.B. Thornton, Esq. Lectotype, USNM 84975, length 56
   mm, designated by Johnson (1974), was not found.

Unio scitulus Lea, 1860. Lea, 1860a:93; Lea, 1860d:342, pl. 55, fig. 167; Lea, 1860g:24, pl. 55, fig. 167
Type locality: Tuscumbia, [Colbert County,] Alabama, L.B. Thornton, Esq. Lectotype, USNM 85084, length 52 mm
   (male), designated by Johnson (1974).

Unio perpictus Lea, 1860. Lea, 1860c:306; Lea, 1860d:350, pl. 58, fig. 175; Lea, 1860g:32, pl. 58, fig. 175
Unio opalina Anthony, 1866. Anthony, 1866:146, pl. 7, fig. 2
Unio dispansus Lea, 1871. Lea, 1871:191; Lea, 1874c:19, pl. 6, fig. 16; Lea, 1874e:23, pl. 6, fig. 16
Comment: Simpson (1914) put *dispansus* in synonymy of *vanuxemensis*, but Ortmann (1918) placed it here.

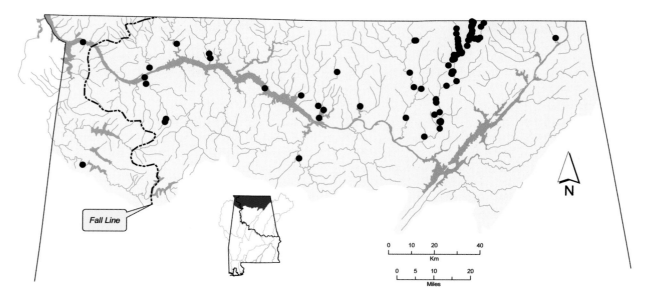

Distribution of *Villosa iris* in the Tennessee River drainage of Alabama.

Villosa lienosa (Conrad, 1834)
Little Spectaclecase

Villosa lienosa – Upper figure: female, length 47 mm, UF 68396. Dry Creek, Brantley, Crenshaw County, Alabama, July 1915. Lower figure: male, length 44 mm, UF 68364. Hunter's Creek, 8 miles west of Evergreen, Conecuh County, Alabama, May 1910. © Richard T. Bryant.

Shell Description

Length to 75 mm; moderately thin; moderately inflated, females more inflated posteriorly than males; outline subelliptical to oval; male posterior margin bluntly pointed to narrowly rounded, female posterior margin broadly rounded to obliquely truncate; anterior margin rounded; dorsal margin convex; ventral margin straight to convex, may be concave in mature females; posterior ridge low, rounded; posterior slope moderately steep; umbo broad, flat, elevated above hinge line, umbo sculpture weakly double-looped bars; periostracum tawny to olive brown, becoming dark brown to black with age, subadults often with variable green rays, rays often disappear with age.

Pseudocardinal teeth moderately small, 2 approximately parallel teeth in left valve, usually somewhat compressed, anterior tooth crest often parallel to margin, 1 compressed to knob-like tooth in right valve, often with accessory denticle anteriorly; lateral teeth moderately short, straight to slightly curved, 2 in left valve, 1 in right valve; interdentum long, narrow to very narrow; umbo cavity shallow, open; nacre usually some shade of purple, occasionally salmon, rarely white.

Soft Anatomy Description

Mantle creamy white to tan or light brown, may have golden cast, outside of apertures usually pale orange, tan or brown, mottled with gray, dark brown or black, mottling may be oriented perpendicular to margin; visceral mass pearly white to creamy white or tan, may have golden cast; foot creamy white to tan, dull orange or gold, often darker distally.

Females with well-developed papillate mantle fold just ventral to incurrent aperture; fold length 20–35% of shell length; margin gray, brown or black, often in thin band, may be absent, marginal coloration usually ends abruptly at anterior end of fold; papillae often widely spaced but occasionally numerous, usually short, simple, blunt, increasing in size with anterior progression. Males with rudimentary mantle fold.

Gills creamy white to gold, tan or light brown; dorsal margin straight to slightly sinuous, ventral margin convex to elongate convex, female outer gills may be folded anterior of marsupium; gill length 50–65% of shell length, occasionally less; gill height 40–65% of gill length, occasionally greater; outer gill height 60–95% of inner gill height; inner lamellae of inner gills connected to visceral mass only anteriorly.

Outer gills marsupial; glochidia held in posterior 35–65% of gill; marsupium outline may be reniform, greatly distended when gravid; creamy white, sometimes with opaque white band on distal margin.

Labial palps creamy white to tan, gold or light brown; straight to slightly concave or convex dorsally, slightly convex ventrally, bluntly pointed to rounded distally; palp length 25–45% of gill length; palp height 50–80% of palp length; distal 20–40% of palps bifurcate, occasionally greater.

Incurrent aperture usually considerably longer than excurrent and supra-anal apertures; excurrent aperture longer, shorter or equal to supra-anal aperture.

Incurrent aperture length 15–25% of shell length; creamy white to tan within, usually with some combination of tan, orange, brown, gray and black basal to papillae, often mottled; papillae in 1–2 rows, long, slender, simple, rarely with few bifid; papillae usually some combination of creamy white, tan, dull orange, brown, gray and black, often changing color distally, often with dark brown or black edges basally.

Excurrent aperture length 10–20% of shell length, occasionally less; creamy white to tan within, marginal color band some combination of rusty orange, gray, brown and black, often in form of irregular lines perpendicular to margin, some lines converging proximally, may also have oblong spots perpendicular to margin, irregular spots, open reticulated pattern or irregular mottling; margin with small simple papillae, often well-developed, occasionally little more than crenulations; papillae tan, dull orange, gray or brown.

Supra-anal aperture length 10–20% of shell length; creamy white within, sometimes with very narrow gray or brown marginal band; margin smooth; mantle bridge separating supra-anal and excurrent apertures usually imperforate, length highly variable, 15–85% of supra-anal length.

Specimens examined: Big Creek, Houston County (n = 2); Bogue Chitto, Dallas County (n = 2); Burnt Corn Creek, Conecuh County (n = 1); Chewacla Creek, Macon County (n = 1); Conecuh River (n = 3); Eightmile Creek, Walton County, Florida (n = 3); Pea River (n = 1); Red Creek, Washington County (n = 2); Sepulga River (n = 3); West Fork Choctawhatchee River (n = 3).

Glochidium Description

Length 200–220 μm; height 270–288 μm; without styliform hooks; outline subelliptical; dorsal margin straight, ventral margin rounded, anterior and posterior margins convex (Ortmann, 1912a, 1916b; Utterback, 1916b).

Similar Species

Villosa lienosa resembles *Villosa vibex*. However, *V. vibex* has a thinner shell, more broadly rounded posterior margin and more prominent dark green rays. *Villosa vibex* pseudocardinal teeth are thinner and more blade-like than those of *V. lienosa*. *Villosa lienosa* nacre is typically some shade of purple or salmon, only rarely white, whereas that of *V. vibex* is typically bluish white. Live individuals from drainages east of the Mobile Basin can be distinguished by pigmentation along the ventral mantle margin. In those *V. lienosa* the dark pigmentation ends abruptly just anterior of the mantle fold, whereas in *V. vibex* the pigmentation extends along the entire length of the ventral mantle margin but tapers anteriorly.

Villosa lienosa closely resembles *Villosa umbrans* but usually has a thicker shell and darker periostracum. *Villosa umbrans* often has more prominent green rays posteriorly. The nacre of *V. umbrans* is often coppery purple. *Villosa lienosa* nacre is usually various shades of purple but is not coppery. These two species occur sympatrically only in the Coosa River drainage.

Villosa lienosa resembles *Villosa villosa* but is less elongate and has thicker pseudocardinal teeth. *Villosa villosa* nacre is always bluish white, whereas that of *V. lienosa* is typically purple or salmon, only rarely white. Live individuals can be distinguished by pigmentation along the ventral mantle margin. In *V. lienosa* the dark pigmentation ends abruptly just anterior of the mantle fold, whereas in *V. villosa* the pigmentation extends along the entire length of the mantle but tapers anteriorly.

Villosa lienosa may resemble *Villosa choctawensis*, but is more elongate. *Villosa choctawensis* typically has a yellowish to greenish yellow periostracum with green rays, whereas that of *V. lienosa* is brown and rays usually do not persist into adulthood. *Villosa choctawensis* nacre is typically white, whereas that of *V. lienosa* is usually purple or salmon, only rarely white.

Villosa lienosa may resemble *Villosa nebulosa* but typically has a dark olive to black periostracum with obscure rays, whereas the periostracum of *V. nebulosa* is yellowish, usually with interrupted green rays. *Villosa lienosa* usually has nacre that is some shade of purple or pink, only rarely white. Nacre of *V. nebulosa* is always white but may have a salmon tint in the umbo cavity.

Male *Villosa lienosa* may superficially resemble *Hamiota australis* but are less elongate and have a less shiny periostracum. *Hamiota australis* nacre is bluish white, whereas that of *V. lienosa* is typically purple or salmon, only rarely white. *Villosa lienosa* has papillate mantle folds just ventral to the incurrent aperture, but *H. australis* has unmodified mantle margins.

Small *Villosa lienosa* may resemble *Toxolasma corvunculus* or *Toxolasma paulum*, but small *V. lienosa* are more compressed than similar-sized individuals of those species. Also, *V. lienosa* nacre is often darker purple than that of *T. corvunculus* and *T. paulum*. The periostracum of *T. corvunculus* tends to be more greenish than that of *V. lienosa*. Live *V. lienosa* can be distinguished by the presence of papillate mantle folds just ventrally to the incurrent aperture. The folds are replaced by caruncles in *T. corvunculus* and *T. paulum*. These structures are rudimentary in males.

General Distribution

Villosa lienosa occurs in the Mississippi Basin from southern Missouri south to the Gulf Coast and in the Ohio and Cumberland River drainages (Cicerello et al., 1991; Cummings and Mayer, 1992). In Gulf Coast drainages it occurs from the Suwannee River drainage, Florida, west to the San Jacinto River drainage, Texas (Williams and Butler, 1994; Howells et al., 1996).

Alabama and Mobile Basin Distribution

Villosa lienosa is widespread in the Mobile Basin above and below the Fall Line. It occurs in eastern Gulf Coast streams of Alabama, including the Chattahoochee, Chipola, Choctawhatchee, Yellow and Conecuh River drainages and Red Creek of the Pascagoula River drainage.

Villosa lienosa is extant in widespread populations, where it may be locally common.

Ecology and Biology

Villosa lienosa usually inhabits streams ranging in size from small creeks to large rivers. It is generally found in stable sand, sandy mud and gravel substrates in slow to moderate current but may also occur in rocky substrates in moderate to swift current. *Villosa lienosa* is most common in creeks and smaller rivers and is often the only mussel found in small headwater streams. *Villosa lienosa* has also colonized overbank habitat in at least one reservoir (Lake Martin, Tallapoosa River).

Villosa lienosa is a long-term brooder, gravid from late summer or autumn to the following summer (Ortmann, 1912a, 1924a). Conglutinates of *V. lienosa* are large and club-shaped (Utterback, 1916b). Fishes reported to serve as glochidial hosts for *V. lienosa* in laboratory trials include *Lepomis macrochirus* (Bluegill) and *Micropterus salmoides* (Largemouth Bass) (Centrarchidae); and *Ameiurus nebulosus* (Brown Bullhead) and *Ictalurus punctatus* (Channel Catfish) (Ictaluridae) (Keller and Ruessler, 1997).

Current Conservation Status and Protection

Villosa lienosa was considered currently stable throughout its range by Williams et al. (1993) and in Alabama by Lydeard et al. (1999). Garner et al. (2004) designated it a species of lowest conservation concern in the state.

Remarks

This species is becoming rare in the Mississippi Basin. A thorough taxonomic study is needed to determine if those populations are taxonomically distinct from those along the Gulf Coast.

Synonymy

Unio lienosus Conrad, 1834. Conrad, 1834a:339, pl. 1, fig. 4

Type locality: Small streams in south Alabama. Length of figured shell in original description reported as about 64 mm.

Comment: There is some confusion surrounding type material of *Villosa lienosa*. A specimen labeled "Type" and with a notation in the shell "Type *U. lienosus* Con. Big Pr. [Prairie] Creek, Marengo Co., Alabama" is in the Florida Museum of Natural History, UF 174330 (male), length 59 mm. This specimen was formerly in the collection of the Alabama Museum of Natural History. Clench and Turner (1956) referred to a collection, ANSP 9747, as the holotype. Johnson and Baker (1973) reported ANSP 9747 as syntypes but did not designate a lectotype.

Unio saxeus Conrad, 1838. Conrad, 1838:Part 11, back cover; Conrad, 1840:109, pl. 60, fig. 1
Type locality: Alabama River, Claiborne, [Monroe County, Alabama]. Lectotype not found, also not found by
  Johnson and Baker (1973). Length of figured shell in original description reported as about 51mm. Johnson
  and Baker (1973) erroneously reported ANSP 9747 as a syntype of *Unio saxeus*, but this number was also
  reported for syntypes of *Unio lienosus*.

Unio caliginosus Lea, 1845. Lea, 1845:165; Lea, 1848a:79, pl. 7, fig. 21; Lea, 1848b:79, pl. 7, fig. 21
Unio gouldii Lea, 1845. Lea, 1845:165; Lea, 1848a:76, pl. 6, fig. 16; Lea, 1848b:50, pl. 6, fig. 16
Type locality: [Black Warrior River,] Tuscaloosa, [Tuscaloosa County,] Alabama, R.E. Griffith, M.D. Holotype,
  USNM 85160, length 50 mm (male).

Unio nigerrimus Lea, 1852, *nomen nudum*. Lea, 1852a:251
Unio proximus Lea, 1852, *nomen nudum*. Lea, 1852a:252
Unio nigerrimus Lea, 1852. Lea, 1852b:268, pl. 18, fig, 23; Lea, 1852c:24, pl. 18, fig. 23
Unio proximus Lea, 1852. Lea, 1852b:271, pl. 20, fig. 27; Lea, 1852c:27, pl. 20, fig. 27
Unio fuligo Reeve, 1856. Reeve, 1856:[no pagination], pl. 30, fig. 159
Unio subellipsis Lea, 1856. Lea, 1856d:262; Lea, 1858e:62, pl. 10, fig. 44; Lea, 1858h:62, pl.10, fig. 44
Unio concestator Lea, 1857. Lea, 1857a:31; Lea, 1858e:66, pl. 12, fig. 48; Lea, 1858h:66, pl. 12, fig. 48
Type locality: Creeks near Columbus, Georgia, Bishop Elliott. Lectotype, USNM 85102, length 60 mm, designated
  by Johnson (1974).

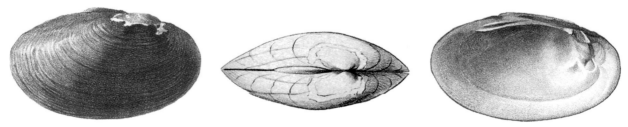

Unio apicinus Lea, 1857. Lea, 1857a:32; Lea, 1858e:76, pl. 14, fig. 56; Lea, 1858h:76, pl. 14, fig. 56
Type locality: Othcalooga Creek, Gordon County, Georgia, Bishop Elliott. Type specimen not found, also not found
    by Johnson (1974). Length of figured shell in original description reported as about 30 mm.

Unio fallax Lea, 1857. Lea, 1857a:32; Lea, 1858e:79, pl. 15, fig. 59; Lea, 1858h:79, pl. 15, fig. 59
Unio intercedens Lea, 1857. Lea, 1857a:32; Lea, 1858e:77, pl. 15, fig. 57; Lea, 1858h:77, pl. 15, fig. 57
Type locality: Streams near Columbus, Georgia, Bishop Elliott. Lectotype, USNM 85122, length 48 mm (male),
    designated by Johnson (1974).

Unio obfuscus Lea, 1857. Lea, 1857g:172; Lea, 1859f:197, pl. 22, fig. 80; Lea, 1859g:15, pl. 22, fig. 80
Unio prattii Lea, 1858. Lea, 1858c:166; Lea, 1858d:47, pl. 5, fig. 8; Lea, 1858h:47, pl. 5, fig. 8
Unio dispar Lea, 1860. Lea, 1860b:305; Lea, 1860d:327, pl. 51, fig. 153; Lea, 1860g:9, pl. 51, fig. 153
Type locality: Columbus, [Muscogee County,] Georgia, Bishop Elliott and G. Hallenbeck. Lectotype, USNM
    85101, length 55 mm, designated by Johnson (1974).

Unio linguaeformis Lea, 1860. Lea, 1860b:305; Lea, 1860d:345, pl. 56, fig. 170; Lea, 1860g:27, pl. 56, fig. 170
Type locality: Columbus, [Muscogee County,] Georgia. Lectotype, USNM 85259, length 50 mm, designated by
    Johnson (1974).

Unio contiguus Lea, 1861. Lea, 1861d:392; Lea, 1862c:199, pl. 28, fig. 268; Lea, 1863a:21, pl. 28, fig. 268
Unio bicaelatus Reeve, 1865. Reeve, 1865:[no pagination], pl. 26, fig. 130
Unio fontanus Conrad, 1866. Conrad, 1866b:279, pl. 15, fig. 13
Unio unicostatus B.H. Wright, 1899. B.H. Wright, 1899:69

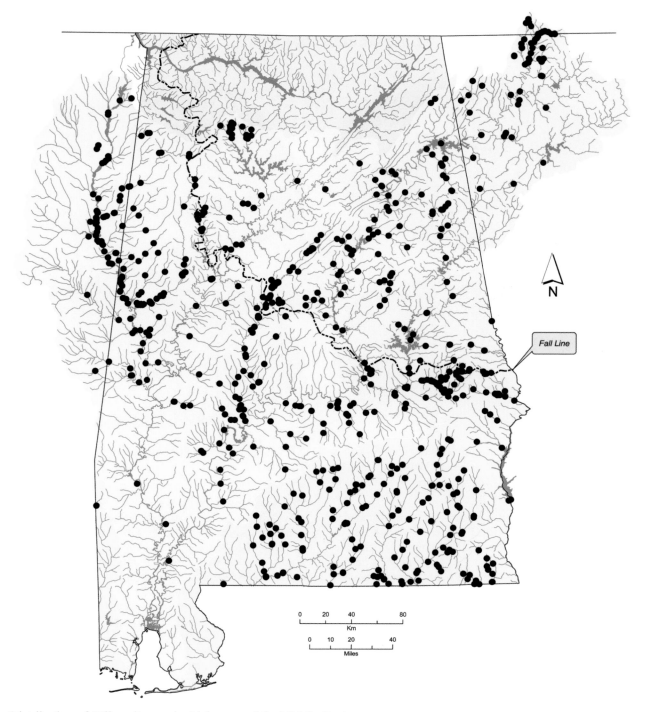

Distribution of *Villosa lienosa* in Alabama and the Mobile Basin.

Villosa nebulosa (Conrad, 1834)
Alabama Rainbow

Villosa nebulosa – Upper figure: female, length 39 mm, CMNH 61.744. Lower figure: male, length 46 mm, CMNH 61.744. Cahaba River, Bibb County, Alabama. © Richard T. Bryant.

Shell Description

Length to 71 mm; moderately thin; somewhat compressed; outline elliptical; posterior margin bluntly pointed to narrowly rounded or obliquely truncate, females higher than males posteriorly; anterior margin rounded; dorsal margin convex; ventral margin usually straight, may be slightly convex; posterior ridge low, rounded; posterior slope flat; mature females with slight marsupial swelling posteriorly; umbo low, not inflated, elevated slightly above hinge line, umbo sculpture coarse ridges, may be slightly knobby; periostracum tawny to greenish yellow, typically with numerous, usually interrupted, green rays, especially on posterior half.

Pseudocardinal teeth small, compressed, 2 slightly divergent teeth in left valve, 1 tooth in right valve, may have blade-like accessory denticle along margin anteriorly; lateral teeth moderately long, straight to slightly curved, 2 in left valve, 1 in right valve; interdentum short, narrow; umbo cavity shallow, open; nacre white to bluish white, may have pale salmon tint in umbo.

Soft Anatomy Description

Mantle creamy white to grayish tan, outside of apertures usually rusty brown with dark brown or black mottling oriented perpendicular to margin, may be mostly black with sparse tan mottling; visceral mass light tan, may have dark brown or black area posteriorly; foot pale orange.

Females with papillate mantle fold just ventral to incurrent aperture; fold length approximately 35% of shell length; adjacent mantle rusty brown or black; mantle outside of fold black to dark grayish brown; papillae sparse, widely separated, thick, simple. Males with rudimentary mantle fold; margin smooth, crenulate or with sparse papillae. Haag and Warren (2000) reported the mantle display of females to be "pale to pure white" in live individuals.

Gills tan; dorsal margin slightly sinuous, ventral margin convex to elongate convex; gill length 50–60% of shell length; gill height 50–65% of gill length; outer gill height 75–85% of inner gill height; inner lamellae of inner gills connected to visceral mass only anteriorly.

Outer gills marsupial; glochidia held in posterior 35–50% of gill; marsupium somewhat oval in outline, distended when gravid; creamy white, distal margin dark gray. Lea (1859f) reported the distal margin to be "purplish".

Labial palps creamy white to tan; straight dorsally, convex ventrally, bluntly pointed distally; palp length 20–25% of gill length; palp height 60–80% of palp length; distal 40–50% of palps bifurcate.

Incurrent aperture longer than excurrent and supra-anal apertures; supra-anal aperture longer or equal to excurrent aperture.

Incurrent aperture length 15–20% of shell length; creamy white within, may have slight grayish cast, with some combination of rusty brown, grayish brown and dark brown basal to papillae, usually mottled; papillae in 2 rows, long, slender, simple; papillae some combination of creamy white, rusty brown, grayish brown and gray, often with white tips.

Excurrent aperture length approximately 15% of shell length; creamy white to gray within, may have slight rusty cast, marginal color band rusty brown mottled with dark brown, mottling may be oriented perpendicular to margin; margin usually crenulate, may have very small, widely spaced papillae; papillae rusty brown to black.

Supra-anal aperture length 15–20% of shell length; creamy white within, may have faint grayish cast, some individuals with thin rusty brown and dark brown or black marginal band; margin smooth; mantle bridge separating supra-anal and excurrent apertures imperforate, length 20–30% of supra-anal length.

Specimens examined: Flannagin Creek, Lawrence County (n = 4).

Glochidium Description

Length 225–237 μm; height 287–300 μm; without styliform hooks; outline subelliptical; dorsal margin straight, ventral margin rounded, anterior and posterior margins convex.

Similar Species

Villosa nebulosa resembles *Villosa vibex*, but the periostracum of *V. nebulosa* usually has thinner rays that are interrupted. The periostracum of *V. vibex* is often more greenish or brownish than the yellowish periostracum of *V. nebulosa*. Pseudocardinal teeth of *V. nebulosa* are thicker than those of *V. vibex*, which are thin and often blade-like. The posterior margin of *V. nebulosa* is narrower than that of *V. vibex*.

Villosa nebulosa superficially resembles *Villosa umbrans* but usually has interrupted rays, compared to the uninterrupted rays of *V. umbrans*. Rays of *V. umbrans* are more often confined to the posterior half of the shell than those of *V. nebulosa*. The nacre of *V. nebulosa* is usually white, whereas the nacre of *V. umbrans* is typically coppery purple, only occasionally white.

Villosa nebulosa may superficially resemble *Villosa lienosa* but has a lighter periostracum, usually with interrupted green rays. The rays of *V. lienosa* are often obscure. *Villosa nebulosa* always has white nacre, sometimes with a salmon tint in the umbo, whereas *V. lienosa* is usually some shade of pink or purple only rarely white.

General Distribution

Villosa nebulosa is endemic to the Mobile Basin in Alabama, Georgia and Tennessee above the Fall Line but is absent from the Tallapoosa River system.

Alabama and Mobile Basin Distribution

Villosa nebulosa was historically widespread in the Black Warrior, Cahaba and Coosa River drainages above the Fall Line.

Villosa nebulosa is extant in a few isolated tributary populations, including headwaters of Sipsey Fork in Bankhead National Forest and scattered tributaries of the Coosa River.

Ecology and Biology

Villosa nebulosa occurred historically in habitats ranging from small creeks to large rivers. However, it was extirpated from large rivers with their impoundment. *Villosa nebulosa* occurs in flowing water in substrates of various combinations of sand and gravel, and may be found in fine sediments among cobble and boulders.

Villosa nebulosa is a long-term brooder, gravid from late summer or autumn to the following summer. Females display a modified mantle margin in the form of a papillate fold that is used to attract glochidial hosts. The long, simple, tentacle-like papillae, which number approximately 15, are pulsated in rapid bursts that last 2 to 4 seconds. The display occurs primarily at night (Haag and Warren, 1997).

Fishes reported to serve as glochidial hosts in laboratory trials include *Lepomis megalotis* (Longear Sunfish), *Micropterus coosae* (Redeye Bass), *Micropterus punctulatus* (Spotted Bass) and *Micropterus salmoides* (Largemouth Bass) (Centrarchidae) (Haag and Warren, 1997).

Current Conservation Status and Protection

Villosa nebulosa was reported as threatened throughout its range by Williams et al. (1993), but Lydeard et al. (1999) considered it currently stable in Alabama. It was designated a species of moderate conservation concern in the state by Garner et al. (2004).

Remarks

Villosa nebulosa closely resembles some forms of *Villosa iris* that occur in the Mississippi and Great

Lakes basins. The taxonomic relationships among various forms of *V. iris* and *V. nebulosa* remain unclear.

The name *Unio nebulosus* Conrad, 1834, has been included with the *Villosa iris* complex in some literature. All of the described taxa from the Ohio, Tennessee and Cumberland River systems were herein included as synonyms of *Villosa iris* and *Villosa nebulosa* was considered restricted to the headwaters of the Mobile Basin. The relationship between *V. nebulosa* and the *V. iris* complex will not be resolved on conchological grounds (e.g., Gordon, 1995).

Synonymy

Unio nebulosus Conrad, 1834. Conrad, 1834b:28, pl. 3, fig. 7
Type locality: Mountainous region of Alabama, in the Black Warrior River. Type specimen not found, also not found by Johnson and Baker (1973). Measurement of figured shell not given in original description (Conrad, 1834b).

Unio radians Lea, 1857. Lea, 1857a:32; Lea, 1859f:201, pl. 23, fig. 84; Lea, 1859g:19, pl. 23, fig. 84
Type locality: Othcalooga Creek, Gordon County, Georgia, Bishop Elliott. Lectotype, USNM 85156, length 45 mm, designated by Johnson (1974).

Unio jonesii Lea, 1859. Lea, 1859d:171; Lea, 1860d:339, pl. 54, fig. 164; Lea, 1860g:21, pl. 54, fig. 164
Type locality: Euharlee Creek, [Bartow and Polk counties,] Georgia, J. Postell. Lectotype, USNM 85257, length 48 mm, designated by Johnson (1974).

Unio plancus Lea, 1860. Lea, 1860c:307; Lea, 1862d:81, pl. 10, fig. 229; Lea, 1862e:85, pl. 10, fig. 229
Type locality: Coosa River, Watumpka [Wetumpka], [Elmore County,] Alabama, E.R. Showalter, M.D. Lectotype, USNM 85345, length 41 mm, designated by Johnson (1974).

Unio sparus Lea, 1868. Lea, 1868a:143; Lea, 1868c:306, pl. 47, fig. 119; Lea, 1869a:66, pl. 47, fig. 119
Type locality: Swamp Creek, north Georgia, Major T.C. Downie. Lectotype, USNM 85082, length 46 mm,
   designated by Johnson (1974).

Unio difficilis Lea, 1868. Lea, 1868a:144; Lea, 1868c:311, pl. 49, fig. 124; Lea, 1869a:71, pl. 49, fig. 124
Type locality: Swamp Creek, Georgia, Major T.C. Downie; Holston River, Washington County, [Virginia,]
   Professor Cope. Lectotype, USNM 85158, length 38 mm, designated by Johnson (1974), is from Swamp
   Creek, Georgia.

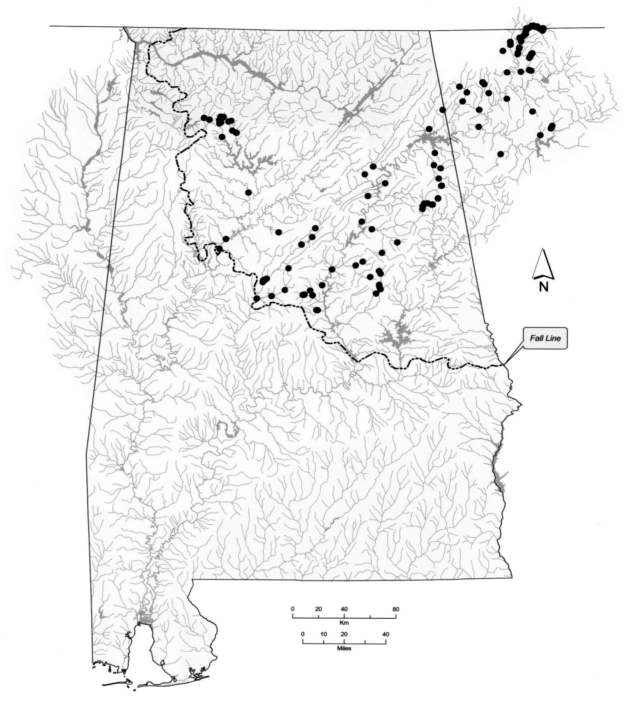

Distribution of *Villosa nebulosa* in Alabama and the Mobile Basin.

Villosa taeniata (Conrad, 1834)
Painted Creekshell

Villosa taeniata – Upper figure: female, length 54 mm, USNM 451955. Lower figure: male, length 66 mm, USNM 451955. South Harpeth River, Tennessee. © Richard T. Bryant.

Shell Description

Length to 80 mm; moderately thin; somewhat compressed, females more inflated posteriorly than males; outline subelliptical to oval; posterior margin bluntly pointed to broadly rounded, more broadly rounded in females than males; anterior margin rounded; dorsal margin straight to convex; ventral margin convex; posterior ridge low, rounded; posterior slope low, flat to slightly concave; umbo low, broad, flat, elevated slightly above hinge line, umbo sculpture double-looped ridges; periostracum tawny to yellowish green or brown, typically with numerous interrupted green rays.

Pseudocardinal teeth small, triangular, 2 divergent teeth in left valve, 1 tooth in right valve, may have very weak accessory denticle anteriorly; lateral teeth short, moderately thick, straight to slightly curved, 2 in left valve, 1 in right valve; interdentum moderately long, narrow; umbo cavity shallow, open; nacre bluish white, often with pale salmon tint in umbo cavity.

Soft Anatomy Description

Mantle creamy white to tan, outside of apertures pale orange or rusty tan with gray or dark brown mottling, mottling may form irregular bands perpendicular to margin; visceral mass pearly white to creamy white or light tan; foot pale orange to pinkish orange.

Females with mantle fold just ventral to incurrent aperture; fold length 25–40% of shell length; marginal coloration some combination of rusty tan, rusty orange, gray, brown and black, often with heavy mottling; usually brown or black under fold; margin smooth, crenulate or papillate; papillae thick, blunt, increasing in size anteriorly, may be irregularly spaced. Males with rudimentary mantle fold.

Gills tan; dorsal margin slightly sinuous, ventral margin elongate convex, female outer gills may be slightly bilobed; gill length 50–65% of shell length; gill height usually 45–55% of gill length, occasionally greater; outer gill height 50–90% of inner gill height; inner lamellae of inner gills usually completely connected to visceral mass, occasionally connected only anteriorly.

Outer gills marsupial; glochidia held across entire gill except anterior and posterior ends; marsupium distended when gravid; creamy white, with dark brown or black distal margin.

Labial palps creamy white to tan; straight to slightly concave or convex dorsally, convex ventrally, bluntly pointed to narrowly rounded distally; palp length 15–25% of gill length; palp height 55–85% of palp length; distal 45–60% of palps bifurcate, occasionally greater.

Incurrent aperture longer than excurrent and supra-anal apertures; excurrent and supra-anal aperture length variable, either may be longer than the other.

Incurrent aperture length 15–20% of shell length; creamy white to pale tan within, occasionally pale orange, usually with some combination of pale orange, rusty tan, golden brown, dark brown and black basal to papillae, usually with lighter background and irregular dark mottling; papillae usually in 2 rows, long, slender, simple; papillae some combination of creamy white, tan, gray, golden brown and rusty brown, often changing color distally, sometimes with irregular mottling.

Excurrent aperture length 10–15% of shell length; creamy white to pale tan within, marginal color band some combination of creamy white, pale orange, rusty tan, golden brown, dark brown and black, often with lighter background and darker mottling oriented perpendicular to margin; margin crenulate or with short simple papillae; papillae tan to golden brown.

Supra-anal aperture length 10–15% of shell length; creamy white to pinkish orange within, sometimes with thin pale orange, golden brown or brown marginal band; margin smooth; mantle bridge separating supra-anal and excurrent apertures may be perforate, length 5–45% of supra-anal length.

Specimens examined: Duck River, Tennessee (n = 3); Estill Fork, Jackson County (n = 4).

Glochidium Description

Length 220–240 μm; height 270–300 μm; without styliform hooks; outline subelliptical; dorsal margin straight, ventral margin broadly rounded, anterior and posterior margins slightly convex (Ortmann, 1912a; Surber, 1915).

Similar Species

Villosa taeniata resembles *Villosa iris* but generally has a thicker, less elongate shell with broader green rays. Rays of *V. iris* are sometimes not interrupted, but those of *V. taeniata* generally are. Pseudocardinal teeth of *V. taeniata* are thicker and often more triangular than those of *V. iris*, which may be very thin and blade-like.

Villosa taeniata may resemble small *Actinonaias pectorosa* but has a shallower umbo cavity. Rays on the periostracum of *V. taeniata* are usually interrupted, whereas those on *Actinonaias ligamentina* are not.

General Distribution

Villosa taeniata is endemic to the Cumberland and Tennessee River drainages in Alabama, Kentucky and Tennessee.

Alabama and Mobile Basin Distribution

Most records of *Villosa taeniata* are from Tennessee River tributaries, but a few are from the Tennessee River proper at Muscle Shoals.

Villosa taeniata is extant in a few Tennessee River tributaries in Alabama, including the Paint Rock River system and Cypress Creek, Lauderdale County.

Ecology and Biology

Villosa taeniata is primarily a species of creeks and small rivers. However, it was occasionally collected from large rivers (e.g., Tennessee River at Muscle Shoals) historically. It is usually found in substrates composed of gravel with interstitial sand in areas with moderate to swift current.

Villosa taeniata is a long-term brooder, gravid from late summer or autumn to the following summer. *Ambloplites rupestris* (Rock Bass) (Centrarchidae) has been reported to serve as a glochidial host in laboratory trials (Gordon et al., 1994). Kirk and Layzer (1997) reported transformation of small numbers of juveniles on *Cottus carolinae* (Banded Sculpin) (Cottidae) after injecting the fish with the immunosuppressant cortisol. *Villosa taeniata* has been reported to live as long as 44 years (Houslet and Layzer, 1997).

Current Conservation Status and Protection

Villosa taeniata was considered currently stable throughout its range by Williams et al. (1993). In Alabama Stansbery (1976) reported it as endangered, but Lydeard et al. (1999) considered it currently stable. Garner et al. (2004) designated *V. taeniata* a species of moderate conservation concern in the state.

Remarks

The type locality of *Villosa taeniata* was given as Flint River, Morgan County, Alabama (Conrad, 1834b). On modern maps it is called Flint Creek. This is sometimes confused with Flint River, which is a northern tributary of the Tennessee River in Madison County, Alabama.

Villosa taeniata is uncommon in prehistoric shell middens along the Tennessee River in northern Alabama (Morrison, 1942; Warren, 1975; Hughes and Parmalee, 1999).

Synonymy

Unio taeniatus Conrad, 1834. Conrad, 1834b:26, pl. 4, fig. 2

Type locality: Flint River, Morgan County, Alabama. Lectotype, ANSP 56441, length 52 mm, designated by Johnson and Baker (1973).

Unio pictus Lea, 1834. Lea, 1834a:73, pl. 11, fig. 32; Lea, 1834b:185, pl. 11, fig. 32

Margarita (*Unio*) *interruptus* Lea, 1836, *nomen nudum*. Lea, 1836:24

Margarita (*Unio*) *pulcher* Lea, 1836, *nomen nudum*. Lea, 1836:25

Unio interruptus Lea, 1838, *non* Rafinesque, 1820, *non* Say, 1831. Lea, 1838a:15, pl. 6, fig. 15; Lea, 1838c:15, pl. 6, fig. 15

Unio latiradiatus Conrad, 1838. Conrad, 1838:96–97, pl. 53 [replacement name for *Unio interruptus* Lea, 1838]

Unio pulcher Lea, 1834. Lea, 1838a:6, pl. 3, fig. 6; Lea, 1838c:6, pl. 3, fig. 6

Unio menkianus Lea, 1838. Lea, 1838a:76, pl. 19, fig. 59; Lea, 1838c:76, pl. 19, fig. 59

Unio tennesseensis Lea, 1840. Lea, 1840:288; Lea, 1842b:199, pl. 10, fig. 11; Lea, 1842c:37, pl. 10, fig. 11

Unio camelopardilis Lea, 1860. Lea, 1860a:92; Lea, 1860d:355, pl. 59, fig. 180; Lea, 1860g:37, pl. 59, fig. 180

Type locality: North Alabama, Professor Tuomey. Lectotype, USNM 85662 (male), length 50 mm, designated by Johnson (1974).

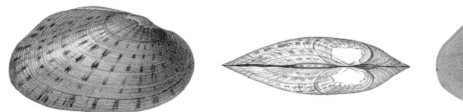

Unio fucatus Lea, 1860. Lea, 1860a:92; Lea, 1860d:353, pl. 59, fig. 178; Lea, 1860g:35, pl. 59, fig. 178

Type locality: North Alabama, Professor Tuomey; Tuscumbia, [Colbert County, Alabama,] L.B. Thornton, Esq. Type specimen not found, also not found by Johnson (1974).

Unio lindsleyi Lea, 1860. Lea, 1860c:306; Lea, 1860d:351, pl. 58, fig. 176; Lea, 1860g:33, pl. 58, fig. 176

Unio punctatus Lea, 1865. Lea, 1865:89; Lea, 1868b:261, pl. 32, fig. 76; Lea, 1869a:21, pl. 32, fig. 76

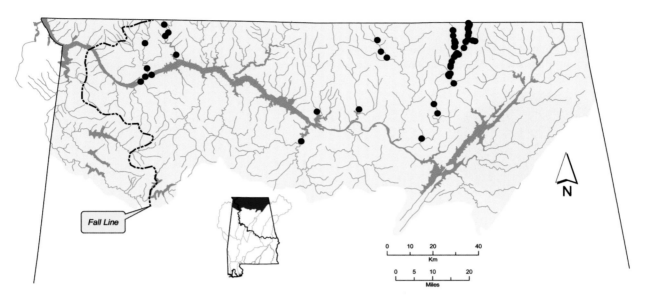

Distribution of *Villosa taeniata* in the Tennessee River drainage of Alabama.

Villosa trabalis (Conrad, 1834)
Cumberland Bean

Villosa trabalis – Upper figure: female, length 47 mm, USNM 84702. Tennessee. Lower figure: male, length 42 mm, NCSM 45080. Hiwasaee River at Apalachia Powerhouse, Polk County, Tennessee, 15 October 1994. © Richard T. Bryant.

Shell Description

Length to 55 mm; moderately thick; moderately inflated; outline oval to elliptical, sexual dimorphism subtle, males slightly narrower centrally than females; posterior margin narrowly rounded; anterior margin rounded; dorsal margin convex; ventral margin straight to convex; posterior ridge rounded; posterior slope moderately steep; umbo broad, elevated above hinge line, umbo sculpture unknown; periostracum greenish brown to dark brown, typically with thin, somewhat wavy green rays over most of shell.

Pseudocardinal teeth low, triangular, 2 divergent teeth in left valve, 1 tooth in right valve, usually with blade-like accessory denticle anteriorly and posteriorly; lateral teeth short to moderately long, straight, 2 in left valve, 1 in right valve; interdentum short to moderately long, narrow to moderately wide; umbo cavity shallow, open; nacre white.

Soft Anatomy Description

Mantle creamy white, outside of apertures grayish brown and rusty brown; visceral mass pale tan; foot pale tan.

Females with well-developed papillate mantle fold just ventral to incurrent aperture; fold length approximately 30% of shell length; adjacent mantle with very narrow, rusty brown and grayish brown marginal band; mantle outside of fold rusty brown and grayish brown, darker than band inside of margin; papillae simple, long, slender, widely spaced; papillae brown basally, gray distally. Ortmann (1912a) reported "a few smaller" papillae between the primary papillae.

Gills tan; dorsal margin straight, ventral margin elongate convex; gill length approximately 60% of shell length; gill height approximately 50% of gill length; outer gill height approximately 85% of inner gill height; inner lamellae of inner gills completely connected to visceral mass.

Outer gills marsupial; glochidia held across most of gill; marsupium greatly distended when gravid; creamy white with gray distal margin. Ortmann (1912a) reported glochidia to be held in posterior portion of outer gills only and the distal margin of marsupia to be "intensely black".

Labial palps tan; slightly convex dorsally, convex ventrally, bluntly pointed distally; palp length approximately 25% of gill length; palp height approximately 50% of palp length; distal 30% of palps bifurcate.

Incurrent aperture longer than excurrent and supra-anal apertures; supra-anal aperture slightly longer than excurrent aperture.

Incurrent aperture length approximately 20% of shell length; creamy white within, with irregular grayish brown basal to papillae, may form crude lines in somewhat reticulated pattern; papillae in 2 rows, long, slender, simple; papillae rusty tan and creamy white, with dark brown edges basally.

Excurrent aperture length approximately 10% of shell length; creamy white within, marginal color band grayish brown; margin with simple papillae, short but well-developed, in 2 rows on central part of aperture; papillae rusty brown and dark brown basally, white distally.

Supra-anal aperture length approximately 15% of shell length; creamy white within, with thin dark brown marginal band; margin minutely crenulate, becoming smooth dorsally; mantle bridge separating supra-anal and excurrent apertures imperforate, length approximately 10% of supra-anal length.

Specimen examined: Big South Fork Cumberland River [Tennessee?] (n = 1) (specimen previously preserved).

Glochidium Description

Length 193–220 μm; height 255–280 μm; without styliform hooks; outline subelliptical; dorsal margin short, straight, ventral margin rounded, anterior and posterior margins convex (Ortmann, 1912a; Surber, 1912; Hoggarth, 1993, 1999).

Similar Species

Villosa trabalis resembles *Villosa fabalis* but is more elongate and pointed posteriorly with a narrower umbo. Also, *V. trabalis* is more inflated than *V. fabalis*.

Villosa trabalis also resembles *Villosa iris* but has a thicker, more inflated shell with a higher umbo. Also, the periostracum of *V. trabalis* tends to be darker brown and the rays are usually wavier than those of *V. iris* from Alabama populations. Pseudocardinal teeth of *V. trabalis* are larger and more triangular than those of *V. iris*.

Villosa trabalis superficially resembles small *Villosa vanuxemensis* but is more elongate and more inflated with a narrower umbo. The periostracum of *V. trabalis* typically has more distinctive, wavier rays than that of *V. vanuxemensis*, the adults of which are often

without rays. Shell nacre of *V. trabalis* is white, whereas that of *V. vanuxemensis* is typically purple or salmon.

Villosa trabalis may superficially resemble *Toxolasma lividum* but is more elongate with a more pointed posterior margin. The periostracum of *V. trabalis* is typically rayed, whereas that of *T. lividum* is usually without rays. Shell nacre of *V. trabalis* is white, but that of *T. lividum* is purple. Live *V. trabalis* may be distinguished by the presence of papillate mantle folds just ventral to the incurrent aperture. The folds are replaced by caruncles in *T. lividum*. These structures are rudimentary in males.

General Distribution

Villosa trabalis is endemic to the Cumberland and Tennessee River drainages. In the Cumberland River drainage it is known only from downstream of Cumberland Falls (Parmalee and Bogan, 1998). In the Tennessee River drainage *V. trabalis* historically occurred from headwaters in southwestern Virginia and eastern Tennessee downstream to Muscle Shoals, Alabama (Parmalee and Bogan, 1998).

Alabama and Mobile Basin Distribution

Villosa trabalis historically occurred in the Tennessee River across northern Alabama as well as some large tributaries.

Villosa trabalis may be extirpated from Alabama. The most recent report of this species was from Hurricane Creek, tributary to the Paint Rock River in Jackson County, in 1980.

Ecology and Biology

Villosa trabalis occurs primarily in creeks and small rivers, usually in flowing water. However, records from preimpoundment Tennessee River proper indicate that it can live in large rivers under some conditions. Its preferred substrates appear to be gravel or mixtures of sand and gravel.

Villosa trabalis is a long-term brooder, gravid from late summer or autumn to the following summer. Gravid female *V. trabalis* have been observed lying exposed or partially exposed on the substrate from December to February, a behavior that is presumably associated with glochidial discharge or host attraction (Layzer and Madison, 1995).

Fishes reported to serve as glochidial hosts of *V. trabalis* in laboratory trials include *Cottus baileyi* (Black Sculpin) (Cottidae); and *Etheostoma blennioides* (Greenside Darter), *Etheostoma caeruleum* (Rainbow Darter), *Etheostoma flabellare* (Fantail Darter), *Etheostoma kennicotti* (Stripetail Darter), *Etheostoma nigrum* (Johnny Darter), *Etheostoma obeyense* (Barcheek Darter), *Etheostoma olivaceum* (Sooty Darter), *Etheostoma sagitta* (Arrow Darter),

Etheostoma simoterum (Snubnose Darter) and *Etheostoma virgatum* (Striped Darter) (Percidae) (Layzer *in* Parmalee and Bogan, 1998; Jones et al., 2003). However, *E. caeruleum* appears to be a secondary host (J.B. Layzer, personal communication).

Current Conservation Status and Protection

Villosa trabalis was considered endangered throughout its range by Stansbery (1970a) and Williams et al. (1993) and in Alabama by Lydeard et al. (1999). Garner et al. (2004) included *V. trabalis* on a list of species extirpated from Alabama. It was listed as endangered under the federal Endangered Species Act in 1976.

In 2001 *Villosa trabalis* was included on a list of species approved for a Nonessential Experimental Population in Wilson Dam tailwaters on the Tennessee River. However, no reintroductions had taken place as of 2007.

Remarks

There has been some confusion over the type locality of *Villosa trabalis*, which was not included with the original description (Conrad, 1834b). Locality data associated with the figured type gives "Flint River, Alabama". The map used by Conrad (Tanner, 1830) includes two Tennessee River tributaries that are labeled "Flint River", one on the south side of the Tennessee River in Morgan County and one on the north side in Madison County. Conrad's travels in northern Alabama are believed to have included only the river in Morgan County (Wheeler, 1935). On modern maps, the Flint River in Morgan County is labeled Flint Creek.

Villosa trabalis is represented in the archaeological record of Alabama by a single specimen collected from a prehistoric midden at Muscle Shoals (Hughes and Parmalee, 1999).

Synonymy

Unio trabalis Conrad, 1834. Conrad, 1834b:27, pl. 3, fig. 5

Type locality: Type locality not stated in original description, but locality data for figured lectotype is Flint River, Alabama. Holotype, ANSP 56481, length 42 mm.

Unio troostensis Lea, 1834. Lea, 1834a:71, pl. 10 fig. 30; Lea, 1834b:183, pl. 10, fig. 30
Unio troostii Lea, 1866. Lea, 1866:47 [unjustified emendation of *Unio troostensis* Lea, 1834]

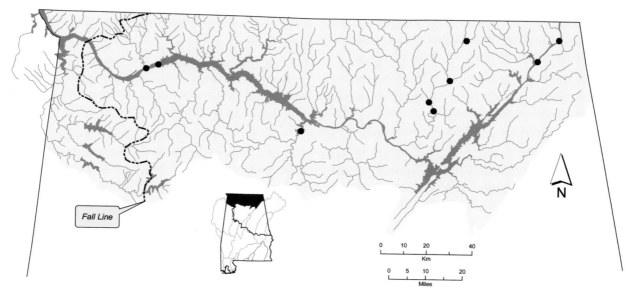

Distribution of *Villosa trabalis* in the Tennessee River drainage of Alabama.

Villosa umbrans (Lea, 1857)
Coosa Creekshell

Villosa umbrans – Upper figure: female, length 34 mm, CMNH 61.8139. Choccolocco Creek, Jackson Shoals, Talladega County, Alabama. Lower figure: male, length 49 mm, CMNH 61.8130. Coahulla Creek, Herndon's Mill, Whitfield County, Georgia. © Richard T. Bryant.

Shell Description

Length to 70 mm; moderately thin; moderately inflated, females more inflated posteriorly than males; outline elliptical to irregularly oval; male posterior margin narrowly rounded, female posterior margin bluntly pointed, posterior terminus positioned near or above midline, may have slight concavity posterioventrally; anterior margin rounded; dorsal margin convex; ventral margin straight to convex; posterior ridge low, rounded; posterior slope moderately steep; umbo broad, elevated slightly above hinge line, umbo sculpture slightly double-looped, coarse ridges; periostracum yellowish green to tawny or brown, usually with thin but prominent green rays posteriorly.

Pseudocardinal teeth compressed, 2 teeth in left valve, 1 tooth in right valve, often with small blade-like accessory denticle along margin anteriorly; lateral teeth thin, short, straight to slightly curved, 2 in left valve, 1 in right valve; interdentum long, very narrow; umbo cavity shallow, open; nacre typically coppery purple, often lighter marginally, occasionally white.

Soft Anatomy Description

Mantle tan, outside of apertures rusty orange, with sparse grayish brown mottling oriented perpendicular to margin; visceral mass creamy white to tan; foot dull orange. Lea (1863d) reported the "color of the mass [to be] light salmon".

Females with well-developed papillate mantle fold just ventral to incurrent aperture; fold length approximately 20% of shell length; adjacent mantle with dark grayish brown color band; papillae short, simple, blunt, closely spaced; papillae creamy white, may be grayish brown basally. Mantle fold of males with slightly reduced papillae.

Gills tan, may have slight grayish cast; dorsal margin slightly sinuous, ventral margin convex; gill length 50–65% of shell length; gill height 45–60% of gill length; outer gill height 80–100% of inner gill height; inner lamellae of inner gills completely connected to visceral mass.

Outer gills marsupial; glochidia held in posterior 50% of gill; marsupium greatly distended when gravid; creamy white, distal margin brown. Lea (1863d) reported the marsupia to occupy "nearly the whole length" of the gill.

Labial palps creamy white to tan; straight dorsally, convex ventrally, narrowly rounded to bluntly pointed distally; palp length 15–25% of gill length; palp height 70–80% of palp length; distal 50–60% of palps bifurcate.

Incurrent aperture longer than excurrent aperture; incurrent aperture longer or equal to supra-anal aperture; supra-anal aperture longer or shorter than excurrent aperture.

Incurrent aperture length approximately 20% of shell length; creamy white within, rusty orange with sparse dark brown mottling basal to papillae; papillae in 1–2 rows, long, slender, mostly simple, few bifid; papillae rusty orange, often creamy white distally. Lea (1863d) reported incurrent aperture papillae to be dark brown.

Excurrent aperture length approximately 15% of shell length; creamy white to golden tan within, marginal color band rusty orange with crude dark brown lines perpendicular to margin, some lines converge proximally; margin with short simple papillae, may be little more than crenulations; papillae creamy white to rusty orange.

Supra-anal aperture length 10–20% of shell length; creamy white to pale tan within, with sparse irregular grayish brown marginally; margin smooth; mantle bridge separating supra-anal and excurrent apertures imperforate, length 35–100% of supra-anal length.

Specimens examined: Holly Creek, Murray County, Georgia (n = 3).

Glochidium Description

Length 225–250 µm; height 300–325 µm; without styliform hooks; outline subelliptical; dorsal margin straight, ventral margin rounded, anterior and posterior margins convex.

Similar Species

Villosa umbrans closely resembles *Villosa lienosa* but usually has a thinner shell and lighter periostracum with more prominent green rays posteriorly. The nacre of *V. umbrans* is usually coppery purple. *Villosa lienosa* nacre is usually various shades of purple but is not coppery.

Villosa umbrans superficially resembles *Villosa nebulosa* but has rays that are uninterrupted and mostly confined to the posterior part of the shell. *Villosa nebulosa* usually has interrupted rays that are more numerous and wider, covering a greater portion of the shell. The nacre of *V. umbrans* is usually coppery purple, but the nacre of *V. nebulosa* is typically white.

Villosa umbrans superficially resembles *Villosa vibex* but has a narrower posterior margin and less prominent rays. *Villosa umbrans* has thicker pseudocardinal teeth and coppery purple nacre. The nacre of *V. vibex* is always white or bluish white.

Villosa umbrans superficially resembles *Anodontoides radiatus*. However, *A. radiatus* usually has rays across the shell disk, whereas *V. umbrans* has rays more prominent on the posterior half. Pseudocardinal and lateral teeth of *A. radiatus* are rudimentary to absent, but are well-developed in *V. umbrans*.

Small *Villosa umbrans* may also resemble *Toxolasma corvunculus*, but *V. umbrans* is longer and less inflated than similar-sized *T. corvunculus*. The periostracum of *T. corvunculus* is often greener than that of *V. umbrans*. Live *V. umbrans* may be distinguished by the presence of papillate mantle folds just ventral to the incurrent aperture. The folds are replaced by caruncles in *T. corvunculus*. These structures are rudimentary in males.

General Distribution

Villosa umbrans is endemic to the Coosa River drainage in Alabama, Georgia and Tennessee.

Alabama and Mobile Basin Distribution

Villosa umbrans is known only from the Coosa River drainage above the Fall Line.

Villosa umbrans is extant in small isolated populations in Coosa River tributaries.

Ecology and Biology

Villosa umbrans is primarily a species of small creeks to medium rivers. However, there are a few records from the Coosa River proper prior to its impoundment. It is found in mixtures of sand, gravel and cobble substrates in moderate current.

Villosa umbrans is a long-term brooder, gravid from September through May or June of the following summer (Gangloff, 2003). *Villosa umbrans* utilizes species of *Lepomis* (Centrarchidae) and *Cottus* (Cottidae) as glochidial hosts. Host use varies among individuals and populations, with some using either *Lepomis* or *Cottus* and some using both (W.R. Haag and P.D. Johnson, personal communication).

Current Conservation Status and Protection

Villosa umbrans was considered a species of special concern throughout its range by Williams et al. (1993) and in Alabama by Lydeard et al. (1999). Garner et al. (2004) designated *V. umbrans* a species of high conservation concern in the state.

Remarks

Villosa umbrans has been considered a subspecies of *Villosa vanuxemensis* by many authors. Preliminary genetic analyses suggest that *V. umbrans* of the Mobile Basin and *V. vanuxemensis* of the Cumberland and Tennessee River drainages are distinct species (W.R. Haag, personal communication). Ortmann (1924a) discussed periostracum and nacre color differences between shells of *V. umbrans* and *V. vanuxemensis*. Shells of *V. umbrans* tend to be lighter in color than those of *V. vanuxemensis*, but there is some overlap. Ortmann (1924a) stated that "light shells from the Tennessee [*V. vanuxemensis*] cannot be distinguished from the normal Coosa-form [*V. umbrans*]; and the dark shells of the latter cannot be told apart from the normal Tennessee-form".

Synonymy

Unio umbrosus Lea, 1857, *non* Lea, 1856. Lea, 1857a:32 [*Unio umbrosus* changed to *Unio umbrans* Lea, 1857]
Unio umbrans Lea, 1857. Lea, 1857f:104; Lea, 1858e:72, pl. 13, fig. 53; Lea, 1858h:72, pl. 13, fig. 53
Type locality: Othcalooga Creek, Gordon County, Georgia, Bishop Elliott. Lectotype, USNM 85121, length 35 mm (female), designated by Johnson (1974).

 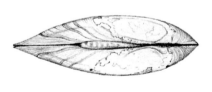 

Unio tenebricus Lea, 1857. Lea, 1857g:171; Lea, 1858e:83, pl. 17, fig. 63; Lea, 1858h:83, pl. 17, fig. 63
Type locality: Etowah River, Georgia, Bishop Elliott and Reverend G. White. Lectotype, USNM 85143, length 47 mm (male), designated by Johnson (1974).

Unio fabaceus Lea, 1861. Lea, 1861a:38; Lea, 1862b:90, pl. 13, fig. 238; Lea, 1862d:94, pl. 13, fig. 238
Type locality: Oostanaula River, Georgia, Bishop Elliott. Holotype, USNM 85165, length 32 mm.

Unio porphyreus Lea, 1861. Lea, 1861c:60; Lea, 1862b:80, pl. 10, fig. 228; Lea, 1862d:84, pl. 10, fig. 228
Type locality: Coosa River, Alabama, E.R. Showalter, M.D. Holotype, USNM 84851, length 46 mm.

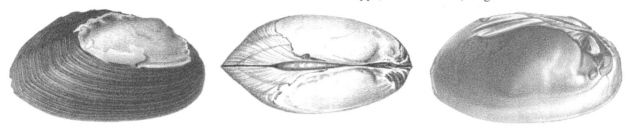

Unio proprius Lea, 1865. Lea, 1865:89; Lea, 1868b:256, pl. 31, fig. 70; Lea, 1869a:16, pl. 31, fig. 70
Type locality: Lafayette, [Walker County,] Georgia, Reverend G. White. Lectotype, USNM 85141, length 39 mm,
    designated by Johnson (1974).

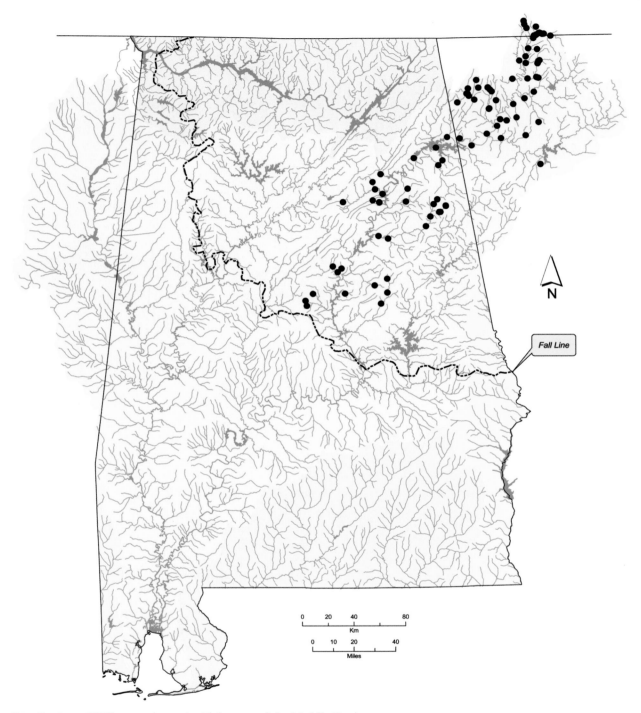

Distribution of *Villosa umbrans* in Alabama and the Mobile Basin.

Villosa vanuxemensis (Lea, 1838)
Mountain Creekshell

Villosa vanuxemensis – Upper figure: female, length 36 mm, CMNH 61.7040. Lower figure: male, length 53 mm, CMNH 61.7040. Paint Rock River, Paint Rock, Jackson County, Alabama. © Richard T. Bryant.

Shell Description

Length to 70 mm; moderately thin; moderately inflated, females more inflated posteriorly than males; outline oval to elliptical; male posterior margin narrowly rounded, female posterior margin broadly rounded to obliquely truncate, posterior terminus at or above midline of shell, may have slight concavity posterioventrally; anterior margin rounded; dorsal margin convex; ventral margin straight to convex; posterior ridge low, rounded; posterior slope moderately steep; umbo broad, elevated slightly above hinge line, umbo sculpture a few somewhat double-looped ridges; periostracum tawny to dark brown, sometimes with thin green rays posteriorly.

Pseudocardinal teeth small, triangular, 2 slightly compressed teeth in left valve, 1 tooth in right valve, less compressed than those of left valve, often with small blade-like accessory denticle along margin anteriorly; lateral teeth thin, short, straight to slightly curved, 2 in left valve, 1 in right valve; interdentum short to moderately long, very narrow; umbo cavity shallow, open; nacre typically some shade of purple, may be salmon or white.

Soft Anatomy Description

Mantle tan to light brown or rusty brown, outside of apertures and mantle fold dull orange to tan or light brown, with heavy black mottling, mottling may be oriented perpendicular to margin; visceral mass pearly white to creamy white or tan; foot creamy white to tan or pale orange.

Females with well-developed papillate mantle fold just ventral to incurrent aperture; fold length 20–30% of shell length; mantle adjacent to fold usually with black marginal color band, occasionally brown or rusty tan; papillae simple, thick, blunt, widely spaced; papillae dull orange, tan, brown or black, color may vary within an individual. Males with rudimentary mantle fold. Ortmann (1912a) described papillae of females as irregular, subconical and closely spaced. Ortmann (1915) reported males to have crenulations in place of the papillate fold of females.

Gills usually tan, may be creamy white, reddish brown or grayish brown; dorsal margin straight to slightly sinuous, ventral margin convex to elongate convex; gill length 45–60% of shell length; gill height 40–60% of gill length, occasionally greater; outer gill height 75–80% of inner gill height, occasionally greater; inner lamellae of inner gills usually completely connected to visceral mass.

Outer gills marsupial; glochidia held in posterior 35–50% of gill, but not extending completely to posterior end; marsupium outline oval, greatly distended when gravid; creamy white to tan, distal margin rusty brown to dark brown. Ortmann (1912a) reported the marsupium to sometimes occupy slightly more than half of the outer gill.

Labial palps creamy white to tan; straight to slightly convex dorsally, convex ventrally, narrowly rounded to bluntly pointed distally; palp length 15–25% of gill length; palp height 60–80% of palp length; distal 40–60% of palps bifurcate, occasionally less.

Incurrent aperture longer than excurrent and supra-anal apertures; excurrent aperture longer, shorter or equal to supra-anal aperture.

Incurrent aperture length 15–20% of shell length; creamy white within, typically with some combination of tan, rusty tan, rusty brown, dark brown and black basal to papillae, usually as mottling, occasionally as crude lines perpendicular to margin; papillae in 1–3 rows, long, slender, mostly simple, with occasional bifid; papillae creamy white to tan, grayish tan or dull orange, often with brown edges basally.

Excurrent aperture length 10–15% of shell length, occasionally less; creamy white within, marginal color band creamy white to tan or dull orange with brown or black mottling, lines in open reticulated pattern or as bands perpendicular to margin; margin with small simple papillae; papillae creamy white to tan, light brown or dull orange.

Supra-anal aperture length 10–15% of shell length, occasionally less; creamy white within, often with thin tan, rusty tan or dark brown marginal color band; margin smooth; mantle bridge separating supra-anal and excurrent apertures often with 1 or more small perforations, length 50–125% of supra-anal length.

Specimens examined: Estill Fork, Jackson County (n = 3); Fowler Creek, Madison County (n = 1); Limestone Creek, Limestone County (n = 3); Paint Rock River (n = 1).

Glochidium Description

Length 207–240 μm; height 277–325 μm; without styliform hooks; outline subelliptical; dorsal margin straight, ventral margin rounded, anterior and posterior margins convex (Ortmann, 1912a, 1915, 1921; Zale and Neves, 1982b).

Similar Species

Small *Villosa vanuxemensis* may resemble *Villosa trabalis*. However, *V. trabalis* is more elongate and inflated, with a narrower umbo. The periostracum of *V. trabalis* typically has more distinctive rays than that of *V. vanuxemensis*, which may be without rays. Shell nacre of *V. vanuxemensis* is typically purple or salmon, whereas that of *V. trabalis* is white.

Small *Villosa vanuxemensis* may resemble *Villosa fabalis*, but the periostracum of *V. fabalis* typically has more distinctive rays than that of *V. vanuxemensis*, which may be without rays. Shell nacre of *V. fabalis* is white, whereas that of *V. vanuxemensis* is typically purple or salmon.

Small *Villosa vanuxemensis* may resemble *Toxolasma lividum*. However, *V. vanuxemensis* has a proportionally thinner shell and is less inflated overall, with a less inflated umbo. Live *V. vanuxemensis* can be distinguished from *T. lividum* by the presence of papillate mantle folds just ventral to the incurrent aperture, which are replaced by caruncles in *T. lividum*. These features are rudimentary in males.

Small *Villosa vanuxemensis* may resemble *Toxolasma cylindrellus*. However, *T. cylindrellus* typically has a more yellowish periostracum, without rays. *Villosa vanuxemensis* may or may not have rays. Live *V. vanuxemensis* can be distinguished from *T. cylindrellus* by the presence of papillate mantle folds just ventral to the incurrent aperture, which are replaced by caruncles in *T. cylindrellus*. These features are rudimentary in males.

General Distribution

Villosa vanuxemensis is endemic to the Cumberland and Tennessee River drainages. In the Cumberland River drainage it is known from the Stones River downstream to the mouth of the Cumberland River (Cicerello et al., 1991; Parmalee and Bogan, 1998). *Villosa vanuxemensis* is more widespread in the Tennessee River drainage, occurring from headwaters in southwestern Virginia, western North Carolina and eastern Tennessee downstream to Muscle Shoals, with disjunct populations in the Buffalo and Duck River drainages, Tennessee (Ahlstedt, 1992a, 1992b; Parmalee and Bogan, 1998).

Alabama and Mobile Basin Distribution

Villosa vanuxemensis was historically widespread in the Tennessee River and tributaries across northern Alabama.

Villosa vanuxemensis is known to be extant in the Paint Rock River system and several smaller Tennessee River tributaries across northern Alabama. It can occasionally be collected from the Tennessee River in tailwaters of Wilson Dam.

Ecology and Biology

Villosa vanuxemensis is primarily a species of medium creeks to medium rivers. However, it can occasionally be found in very small creeks as well as large rivers. *Villosa vanuxemensis* is often found in riffles but can also be found in runs and pools. Though very rare, it can still be found in tailwaters of some Tennessee River dams in water exceeding 6 m in depth. It usually occurs in substrates composed of sand and gravel and can sometimes be found under large rocks.

Villosa vanuxemensis is a long-term brooder. Both sexes reach sexual maturity at age three (Zale and Neves, 1982b). Gametogenesis occurs during the four months just prior to spawning in late July (Zale and Neves, 1982b). Glochidia are mature by September and brooded until the following May (Ortmann, 1921; Zale and Neves, 1982b). In Big Moccasin Creek, Virginia, glochidia of *Villosa* species occurred in stream drift from October to May, and those in autumn and winter were believed to be *V. vanuxemensis* based on host-infestation patterns (Zale and Neves, 1982b). Female *V. vanuxemensis* have a well-developed mantle fold just ventral to the incurrent aperture. The papillate, pigmented fold is believed to function as a host-attracting lure.

Fishes reported to serve as glochidial hosts of *Villosa vanuxemensis* in laboratory trials include *Cottus baileyi* (Black Sculpin), *Cottus bairdii* (Mottled Sculpin), *Cottus carolinae* (Banded Sculpin) and *Cottus cognatus* (Slimy Sculpin) (Cottidae) (Zale and Neves, 1982a; Neves et al., 1985). Natural infestations of *C. carolinae* were observed in Big Moccasin Creek, Virginia, from autumn through late spring (Zale and Neves, 1982a).

Trematode-infested *Villosa vanuxemensis* were frequently encountered in Big Moccasin Creek, Virginia, by Zale and Neves (1982b). Older individuals were more frequently and more heavily parasitized than younger individuals.

Current Conservation Status and Protection

Villosa vanuxemensis was considered a species of special concern throughout its range by Williams et al. (1993) and in Alabama by Lydeard et al. (1999). Garner et al. (2004) designated it a species of low conservation concern in the state.

Remarks

Villosa vanuxemensis has previously been considered to consist of two subspecies, nominal *V. vanuxemensis* and *Villosa vanuxemensis umbrans*. The latter is endemic to the Mobile Basin. Preliminary genetic analyses suggest that the two are distinct species (W.R. Haag, personal communication). Ortmann (1924a) discussed periostracum and nacre color differences between shells of *V. vanuxemensis* and *Villosa umbrans*. Shells of *V. vanuxemensis* tend to be darker in color than those of *V. umbrans*, but there is some overlap. Ortmann (1924a) stated that "light shells from the Tennessee [*V. vanuxemensis*] cannot be distinguished from the normal Coosa-form [*V. umbrans*]; and the dark shells of the latter cannot be told apart from the normal Tennessee-form".

Only two specimens of *Villosa vanuxemensis* have been reported from archaeological sites in Alabama, one from Muscle Shoals and one from a site near Bridgeport (Warren, 1975; Hughes and Parmalee, 1999).

Lea (1838) named *Villosa vanuxemensis* for his friend, Professor L. Vanuxem.

Synonymy

Margarita (*Unio*) *vanuxemensis* Lea, 1836, *nomen nudum*. Lea, 1836:26

Unio vanuxemensis Lea, 1838. Lea, 1838a:36, pl. 11, fig. 31; Lea, 1838c:36, pl. 11, fig. 31

Type locality: Cumberland River, Tennessee. Lectotype, USNM 85146, length 48 mm, designated by Johnson (1974).

Unio nitens Lea, 1840. Lea, 1840:288; Lea, 1842b:205, pl. 12, fig. 19; Lea, 1842c:43, pl. 12, fig. 19

Unio pybasii Lea, 1858. Lea, 1858a:40; Lea, 1862b:67, pl. 6, fig. 216; Lea, 1862c:71, pl. 6, fig. 216
Type locality: Tennessee River at Florence, [Lauderdale County,] Alabama, B. Pybas. Lectotype, USNM 85138,
        length 57 mm (female), designated by Johnson (1974).

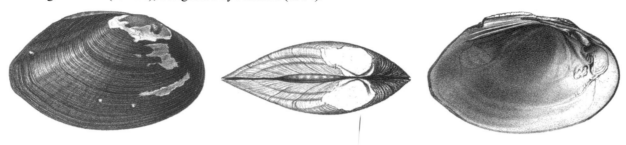

Unio vanuxemii Lea, 1858. Lea, 1858e:83 [unjustified emendation of *Unio vanuxemensis* Lea, 1838]
Unio copei Lea, 1868. Lea, 1868a:144; Lea, 1868c:307, pl. 47, fig. 120; Lea, 1869a:67, pl. 47, fig. 120

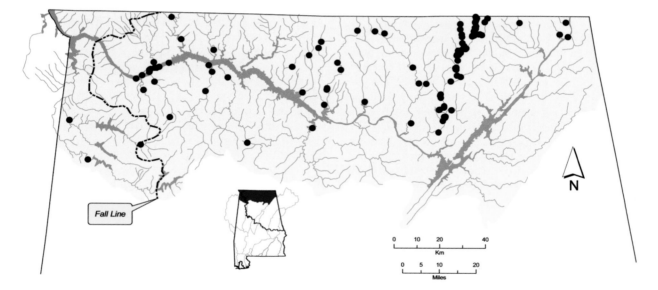

Distribution of *Villosa vanuxemensis* in the Tennessee River drainage of Alabama.

Villosa vibex (Conrad, 1834)
Southern Rainbow

Villosa vibex – Upper figure: female, length 75 mm, UF 358676. Uphapee Creek, State Highway 81, 2.5 miles north of Tuskegee, Macon County, Alabama, 24 June 1973. Lower figure: male, length 68 mm, UF 373811. Conecuh River, County Road 6, about 0.7 miles east-southeast of Glenwood, Crenshaw County, Alabama, 26 July 1995. © Richard T. Bryant.

Shell Description

Length to 100 mm; thin; slightly inflated, females more inflated posteriorly than males; outline elongate oval; posterior margin rounded, females more broadly rounded posteroventrally than males; anterior margin rounded; dorsal margin straight; ventral margin straight to slightly convex, may be slightly concave just anterior of posteroventral swelling in females; posterior ridge low, rounded; posterior slope flat to slightly concave; umbo broad, flat, elevated slightly above hinge line, umbo sculpture a few double-looped ridges; periostracum shiny, tawny to olive, typically with numerous dark green rays of variable width and intensity, rays usually wider and darker posteriorly.

Pseudocardinal teeth small, compressed, blade-like, crests approximately parallel to margin, 2 nearly parallel teeth in left valve, anterior tooth larger than posterior tooth, 1 tooth in right valve, often with accessory denticle anteriorly; lateral teeth short, thin, straight to slightly curved, 2 in left valve, 1 in right valve; interdentum long, narrow to very narrow; umbo cavity shallow, open; nacre bluish white.

Soft Anatomy Description

Mantle opaque, creamy white to tan, outside of apertures creamy white to dull orange, with grayish blotches, mantle outside of ventral mantle margin pigmented; visceral mass creamy white to tan; foot creamy white to pale orange or pinkish orange.

Females with papillate mantle fold just ventral to incurrent aperture; fold length 30–40% of shell length; adjacent mantle with 1–3 distinct color bands of various shades of brown or black, bands replaced by irregular mottling in some individuals, subadults often without marginal coloration; papillae simple, usually thick, blunt, widely spaced, often increasing in size with anterior progression. Males with rudimentary mantle fold; sparse reddish brown, mottled with dark brown or black; papillae usually sparse, often small, occasional individuals with papillae replaced by crenulate margin. Haag and Warren (2000) reported the female mantle display to be "inky black to rusty orange with numerous fine black spots" in live individuals.

Gills tan to light brown; dorsal margin straight to slightly sinuous, ventral margin convex to elongate convex; gill length 45–60% of shell length; gill height

45–65% of gill length, occasionally greater; outer gill height 70–90% of inner gill height; inner lamellae of inner gills connected to visceral mass only anteriorly. Lea (1863d) reported the inner lamellae of inner gills to be completely connected to the visceral mass.

Outer gills marsupial; glochidia held in posterior 35–50% of gill; marsupium distended when gravid; creamy white to tan, distal margin black.

Labial palps creamy white to tan, often with irregular opaque patch dorsally; straight to slightly concave dorsally, convex ventrally, usually bluntly pointed distally, occasionally rounded; palp length 15–40% of gill length; palp height 25–50% of palp length; distal 45–70% of palps bifurcate.

Incurrent aperture usually longer than excurrent and supra-anal apertures, occasionally shorter or equal in length; excurrent aperture length shorter, longer or equal to supra-anal aperture length.

Incurrent aperture length 10–20% of shell length; creamy white to light tan or gold within, usually with some shade of rusty orange or brown basal to papillae; papillae in 2 rows, long, slender, simple, rarely with a few bifid; papillae creamy white to rusty orange or brown, often differently colored distally.

Excurrent aperture length 10–20% of shell length; creamy white to golden tan within, marginal color band dull orange to dark brown or black, usually mottled; margin with small simple papillae, occasionally replaced by deep crenulations; papillae rusty orange to dark brown or black. Ortmann (1924a) described the excurrent aperture margin as crenulate.

Supra-anal aperture length 10–20% of shell length; creamy white within, with thin brown or gray marginal band, often irregular and sparse; margin smooth; mantle bridge separating supra-anal and excurrent apertures rarely perforate, length 20–65% of supra-anal length. Lea (1963a) reported the inner edge of the supra-anal aperture to be "maculate" (i.e., spotted).

Specimens examined: Burnt Corn Creek, Conecuh County (n = 1); Conecuh River (n = 1); Eightmile Creek, Walton County, Florida (n = 2); Flat Creek, Geneva County (n = 1); Little Cedar Creek, Conecuh County (n = 3); Murder Creek, Conecuh County (n = 2); Pauls Creek, Barbour County (n = 3); Red Creek, Washington County (n = 1); Sepulga River (n = 3).

Glochidium Description

Length 210–239 μm; height 270–304 μm; without styliform hooks; outline subspatulate; dorsal margin straight, ventral margin rounded, anterior and posterior margins convex, slightly asymmetrical (Ortmann, 1924a; Hoggarth, 1999).

Similar Species

Villosa vibex is similar to *Villosa lienosa* and *Villosa villosa*, but usually has a more broadly rounded posterior margin. *Villosa vibex* has more prominent green rays, a thinner shell and thinner, more blade-like pseudocardinal teeth than *V. lienosa* and *V. villosa*. *Villosa vibex* nacre is always bluish white, whereas that of *V. lienosa* is typically a shade of purple or salmon, only rarely white. Live *V. vibex* can be distinguished from live *V. lienosa* by coloration of the ventral mantle margin in some populations. In *V. vibex* dark pigmentation extends the ventral length of the mantle, whereas in *V. lienosa* from drainages east of the Mobile Basin it ends abruptly just anterior of the papillate mantle folds.

General Distribution

In eastern Gulf Coast drainages *Villosa vibex* occurs from the Hillsborough River drainage of peninsular Florida west to the Amite River system of the Lake Pontchartrain drainage, Louisiana (Vidrine, 1993). It occurs in Mississippi River tributaries as far north as western Tennessee (Parmalee and Bogan, 1998). As currently recognized, the range of *V. vibex* includes drainages of the southern Atlantic Coast as well as many Gulf Coast drainages. On the Atlantic Coast it occurs from the Cape Fear River drainage, North Carolina, south to the St. Marys River drainage, Georgia and Florida (Johnson, 1970; Butler, 1990).

Alabama and Mobile Basin Distribution

Villosa vibex occurs throughout the Mobile Basin above and below the Fall Line and along the Gulf Coast in the Chattahoochee, Chipola, Choctawhatchee, Yellow and Conecuh River drainages and Red Creek of the Pascagoula River drainage.

Villosa vibex is extant in much of its historical range, with the exception of large rivers that have been impounded or channelized.

Ecology and Biology

Villosa vibex is found in small creeks to large rivers usually in slight to moderate current. It typically occurs in substrates of various combinations of sand, clay and gravel. *Villosa vibex* appears to have some tolerance of silty conditions but is typically not found in pure mud substrates.

Villosa vibex is a long-term brooder, gravid from late summer or autumn to the following summer. In the Escambia River drainage, gravid *V. vibex* have been observed as early as August brooding embryos. Glochidia are usually mature by October (Ortmann, 1924a). *Villosa vibex* females display a papillate mantle fold as a host-attracting behavior (Figure 13.45). When displaying, females extend approximately 15 long, simple, tentacle-like papillae that may be more than 1 cm long. The papillae pulsate in bursts lasting 2 to 4 seconds to give an overall effect of a clump of small worms. The display may be performed with the female upright in the substrate or lying prone on the substrate. *Villosa vibex* displays its mantle lure primarily during the day (Haag and Warren, 1999, 2000).

Figure 13.45. *Villosa vibex* female with papillate mantle fold displayed. Flint River drainage, Georgia, May 1995. Photograph by N.M. Burkhead.

Fishes reported to serve as glochidial hosts of *Villosa vibex* in laboratory trials include *Lepomis*

cyanellus (Green Sunfish), *Micropterus coosae* (Redeye Bass), *Micropterus punctulatus* (Spotted Bass) and *Micropterus salmoides* (Largemouth Bass) (Centrarchidae) (Haag and Warren, 1999). However, *L. cyanellus* and *M. salmoides* appear to be secondary hosts.

Current Conservation Status and Protection

Villosa vibex was considered currently stable throughout its range by Williams et al. (1993) and in Alabama by Lydeard et al. (1999). Garner et al. (2004) designated *V. vibex* a species of lowest conservation concern in the state.

Remarks

Studies are needed to evaluate the taxonomic relationship between *Villosa vibex* from Atlantic and Gulf Coast drainages. The type locality of *V. vibex* is in the Black Warrior River drainage, Blount County, Alabama (Conrad, 1834b). If populations in Atlantic Coast drainages prove to be distinct from those in Gulf Coast drainages, the name *V. vibex* will be retained by the Gulf Coast species. The earliest available name for the Atlantic Coast form is *Villosa modioliformis* (Lea, 1834).

Synonymy

Unio vibex Conrad, 1834. Conrad, 1834b:31, pl. 4, fig. 3
Type locality: Black Warrior River, south of Blount's [Blount] Springs, [Blount County,] Alabama. Lectotype, ANSP 56488, length 59 mm, designated by Johnson and Baker (1973).

Unio exiguus Lea, 1840. Lea, 1840:287; Lea, 1842b:191, pl. 7, fig. 1; Lea, 1842c:29, pl. 7, fig. 1
Type locality: Chattahoochee River, Columbus, [Muscogee County,] Georgia, Dr. Boykin. Lectotype, USNM 84974, length 45 mm, designated by Johnson (1974).

Margaron (*Unio*) *prevostianus* Lea, 1852, *nomen nudum*. Lea 1852a:29
Margaron (*Unio*) *nigrinus* Lea, 1852, *nomen nudum*. Lea, 1852a:39
Unio nigrinus Lea, 1852. Lea, 1852a:252; Lea, 1852b:284, pl. 24, fig. 44; Lea, 1852c:40, pl. 24, fig. 44

Unio prevostianus Lea, 1852. Lea, 1852b:269, pl. 19, fig. 24; Lea, 1852c:25, pl. 19, fig. 24
Type locality: Eutowah [Etowah] River, Georgia, C.M. Wheatley. Holotype not found, also not found by Johnson
  (1974). Length of figured shell in original description reported as about 58 mm.

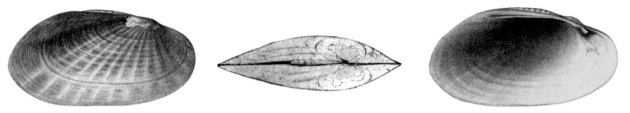

Unio rutilans Lea, 1856. Lea, 1856d:262; Lea, 1858e:59, pl. 9, fig. 41; Lea, 1858h:59, pl. 9, fig. 41
Type locality: Othcalooga Creek, Gordon County, Georgia, Bishop Elliott. Lectotype, USNM 85093, length 66 mm,
  designated by Johnson (1974).

Unio sudus Lea, 1857. Lea, 1857g:170; Lea, 1859f:194, pl. 21, fig. 77; Lea, 1859g:12, pl. 21, fig. 77

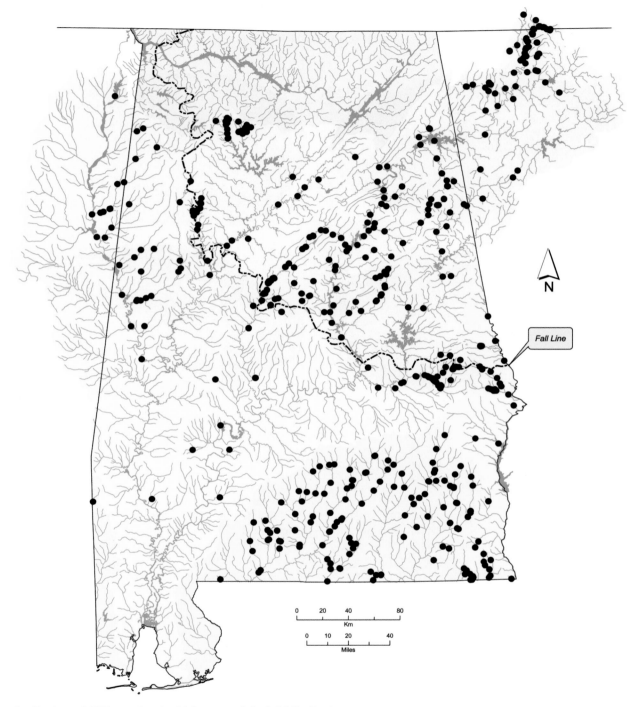

Distribution of *Villosa vibex* in Alabama and the Mobile Basin.

Villosa villosa (Wright, 1898)
Downy Rainbow

Villosa villosa – Upper figure: female, length 52 mm, UF 267468. Suwannee River, off State Route 26 in campground at Fanning Springs, Gilchrist County, Florida. Lower figure: male, length 55 mm, UF 375669. Mercers Mill Pond, off unnamed tributary to Mill Creek, County Route 12, about 7 air miles south of Oakfield, Worth County, Georgia, 10 June 1992. © Richard T. Bryant.

Shell Description

Length to 90 mm; moderately thin; inflated; outline elliptical; male posterior margin bluntly pointed, female posterior margin broadly rounded to obliquely truncate; anterior margin rounded; dorsal margin slightly convex; ventral margin straight to convex; posterior ridge low, rounded; posterior slope moderately steep; umbo broad, moderately inflated, elevated slightly above hinge line, umbo sculpture fine corrugations, may be broken; periostracum usually cloth-like, dark green to brown or black, may be covered with wide dark green rays.

Pseudocardinal teeth small, compressed, 2 in left valve, anterior tooth typically larger than posterior tooth, 1 tooth in right valve, usually with very thin blade-like accessory denticle along margin; lateral teeth moderately long, thin, straight to slightly curved, 2 in left valve, 1 in right valve; interdentum moderately long, very narrow; umbo cavity shallow, open; nacre bluish white.

Soft Anatomy Description

Mantle tan, sometimes with golden cast, outside of apertures dull orange to rusty tan or rusty brown with dark brown mottling oriented perpendicular to margin; visceral mass creamy white to tan, may have golden cast; foot pale tan to dull orange.

Females with papillate mantle fold just ventral to incurrent aperture; fold length 30–55% of shell length; adjacent mantle margin with irregular tan, rusty brown or black color band; mantle under and outside of fold with very dense dark brown mottling, may be almost solid brown; papillae simple, thick, blunt, may be larger anteriorly; papillae may be darkly pigmented. Males with rudimentary mantle flap; marginal coloration reduced or absent; papillae small, simple.

Gills tan, may have slight golden cast; dorsal margin straight to slightly sinuous, ventral margin elongate convex, female outer gills may be bilobed; gill length 50–70% of shell length; gill height 45–55% of gill length; outer gill height 60–90% of inner gill height; inner lamellae of inner gills completely connected to visceral mass or connected only anteriorly.

Outer gills marsupial; glochidia held in posterior 50–65% of gill; marsupium outline may be oval, greatly distended when gravid; tan with rusty brown distal margin.

Labial palps tan; straight to slightly concave dorsally, convex ventrally, bluntly pointed distally; palp length 25–35% of gill length; palp height 50–70% of palp length; distal 25–40% of palps bifurcate.

Incurrent, excurrent and supra-anal apertures of variable relative lengths, any of the three may be longer than the others.

Incurrent aperture length 10–20% of shell length; creamy white to tan within, may have slight golden cast, rusty tan to rusty brown or dark brown basal to papillae, may be mottled; papillae usually in 2 rows, long, slender, simple; papillae tan to rusty brown or dark grayish brown.

Excurrent aperture length 10–20% of shell length; light gray to tan or grayish brown within, marginal color band creamy white to tan or rusty brown; margin with short simple papillae or crenulations; papillae dark brown to rusty brown.

Supra-anal aperture length 10–20% of shell length; creamy white to tan within, may have thin grayish brown marginal band; margin smooth; mantle bridge separating supra-anal and excurrent apertures imperforate, length variable, 15–120% of supra-anal length, occasionally absent.

Specimens examined: Eightmile Creek, Walton County, Florida (n = 2); Four Mile Creek, Decatur County, Georgia (n = 3) (some specimens previously preserved).

Glochidium Description

Length 240–250 μm; height 296–308 μm; without styliform hooks; outline subelliptical; dorsal margin straight, ventral margin rounded, anterior and posterior margins convex (Hoggarth, 1999).

Similar Species

Villosa villosa resembles *Villosa lienosa* but has a thinner, more elongate shell. Pseudocardinal teeth of *V. villosa* are often thinner than those of *V. lienosa*. Shell nacre of *V. villosa* is always bluish white, whereas that of *V. lienosa* is typically some shade of purple or salmon, only rarely white. Live individuals can be distinguished by pigmentation along the ventral mantle margin, with dark pigmentation extending the length of the mantle or tapering anteriorly in *V. villosa*, but ending abruptly just anterior of the papillate mantle folds in *V. lienosa*.

Villosa villosa resembles *Villosa vibex*, but *V. villosa* is more elongate and more inflated with a slightly thicker shell. Green rays on the periostracum of *V. vibex* are typically prominent, but those of *V. villosa*

are absent or much weaker, somewhat obscured by the overall darker periostracum.

Villosa villosa also resembles *Hamiota australis*. However, *H. australis* has a shiny, instead of cloth-like, periostracum and almost always has more prominent green rays of variable widths on the posterior slope. Pseudocardinal teeth of *H. australis* are thicker than those of *V. villosa*.

Villosa villosa may resemble *Ptychobranchus jonesi*, but *P. jonesi* always has a doubled posterior ridge and is more inflated posteriorly. Pseudocardinal teeth of *P. jonesi* are thicker than those of *V. villosa*. The two can be easily distinguished if soft parts of gravid females can be examined. Gills of *P. jonesi* females are folded when gravid, whereas those of *V. villosa* are not. Also, *P. jonesi* lacks papillate mantle folds just ventral to the incurrent aperture.

Male *Villosa villosa* may resemble *Pleurobema pyriforme*. However, the bluntly pointed posterior terminus of *V. villosa* is positioned near the midline of the shell, whereas that of *P. pyriforme* is located more posterioventrally. The periostracum of *V. villosa* typically has green rays, but *P. pyriforme* is usually without rays.

General Distribution

In Atlantic Coast drainages *Villosa villosa* is known from the St. Marys River drainage, Georgia and Florida, south in the St. Johns River drainage in peninsular Florida. It occurs in most of the larger Gulf Coast drainages from the Myakka River, southwestern Florida, to the Escambia River drainage in western Florida and southeastern Alabama. However, there are no records of *V. villosa* from the Yellow River drainage.

Alabama and Mobile Basin Distribution

Villosa villosa is known from Little Uchee Creek, the Chattahoochee River and Chipola River headwaters near the Alabama and Florida state line. It is also known from a tributary of the Conecuh River in the Escambia River drainage. There are records from the Choctawhatchee River drainage in Florida but none from Alabama reaches of that drainage.

The current distribution of *Villosa villosa* in Alabama is unclear. It is presumably extant, but uncommon, in isolated tributary populations.

Ecology and Biology

Villosa villosa may be found in a wide range of habitats, from small, spring-fed creeks to large rivers as well as backwater sloughs and some reservoirs. It occurs in clear, tannic-stained and turbid water. *Villosa villosa* is typically found in substrates of sand and mud in slow current, but also occurs in sand and fine gravel in moderate current.

Villosa villosa is dioecious (Kotrla, 1988). It is a long-term brooder, gravid from late summer or autumn to the following summer. Females use a papillate mantle fold to lure glochidial hosts. Papillae may approach 1 cm in length during display. Fishes found to serve as glochidial hosts of *V. villosa* in laboratory trials include *Lepomis macrochirus* (Bluegill) and *Micropterus salmoides* (Largemouth Bass) (Centrarchidae) (Keller and Ruessler, 1997).

Current Conservation Status and Protection

Villosa villosa was considered a species of special concern throughout its range by Williams et al. (1993). It was not included in a status review of Alabama mussels by Lydeard et al. (1999) but was designated a species of highest conservation concern in the state by Garner et al. (2004).

Remarks

Villosa villosa was first reported from the Escambia River by Simpson (1900b), but it has apparently never been common there or in the Choctawhatchee River drainage, based on a paucity of museum records. It is much more common in the Apalachicola Basin and drainages to the east. *Villosa villosa* appears to become less common with an inland progression. *Villosa villosa* was not reported from the Chipola River by van der Schalie (1940), but some of his museum records labeled *Carunculina vesicularis* (Lea, 1872) appear to be misidentifications of *V. villosa*.

Individuals from populations in drainages west of the Apalachicola Basin have a much shinier periostracum than those of eastern populations, in which the periostracum is cloth-like. Given the zoogeographical boundary between the Apalachicola Basin and the Choctawhatchee River drainage, it is not implausible that populations in the Conecuh and Choctawhatchee River drainages represent an undescribed species.

Synonymy

Unio villosus B.H. Wright, 1898. B.H. Wright, 1898d:32; Simpson, 1900a:77, pl. 1, fig. 1
Type locality: Suwanee River, [Luraville,] Suwannee County, Florida. Lectotype, USNM 150503, length 48 mm, designated by Johnson (1967b).

Lampsilis wrightiana Frierson, 1927. Frierson, 1927:81; Frierson, 1928:139, pl. 2, fig. 3
Type locality: Volusia County, Florida.

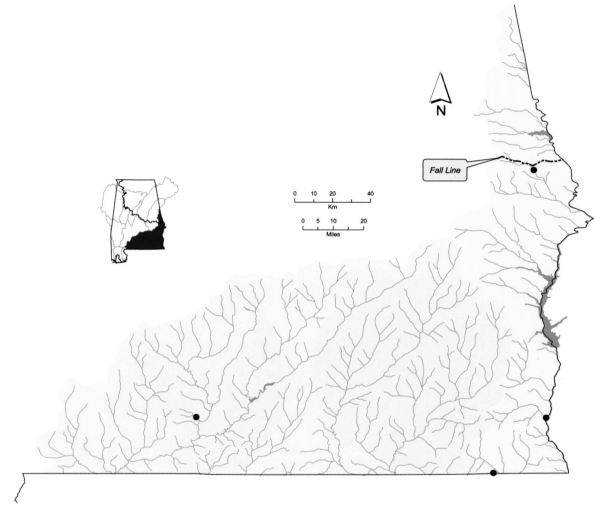

Distribution of *Villosa villosa* in the eastern Gulf Coast drainages of Alabama.

Chapter 14
Species of Hypothetical Occurrence

There are several mussel species that have never been reported from the state of Alabama, or have unsubstantiated reports but occur in rivers that flow into or out of the state. Three of these—*Amblema neislerii* (Lea, 1858), *Anodonta heardi* Gordon and Hoeh, 1995, and *Quadrula refulgens* (Lea, 1868)—may have inhabited Alabama streams historically, but their occurrence has not been documented. Two species, *Fusconaia flava* (Rafinesque, 1820) and *Quadrula nodulata* (Rafinesque, 1820), appear to be expanding their ranges up the Tennessee River and may eventually be found in Alabama. Abbreviated accounts are provided for these five species.

Amblema neislerii (Lea, 1858)
Fat Threeridge

Amblema neislerii – Length 70 mm, UF 369. Chipola River, Dead Lake at Chipola Park, 20 miles south of Blountstown, Calhoun County, Florida, 13 September 1954. © Richard T. Bryant.

Shell Description

Length to 90 mm; moderately thick; moderately to highly inflated; outline oval to quadrate; posterior margin obliquely truncate to broadly rounded; anterior margin truncate to convex; dorsal margin straight to convex; ventral margin straight to convex; posterior ridge poorly developed, rounded, obscured by plications; posterior slope broad, slightly concave; disk adorned with slightly oblique, parallel plications, extending from anterior portion of shell to posterior margin; umbo low, broad, elevated slightly above hinge line, umbo sculpture unknown; periostracum dark brown to black, without rays.

Pseudocardinal teeth thick, rough, triangular, 2 divergent teeth in left valve, 1 tooth in right valve, may have accessory denticle anteriorly and/or posteriorly; lateral teeth moderately long, somewhat thin, straight to slightly curved, 2 in left valve, 1 in right valve; interdentum short to moderately long, narrow to moderately wide; umbo cavity somewhat shallow, open; nacre white to bluish white.

General Distribution

Amblema neislerii is endemic to the Apalachicola Basin in Florida and Georgia, where it is known from the Apalachicola, Chipola and Flint rivers.

Reason for Inclusion

No records of *Amblema neislerii* exist for the Chattahoochee River. However, that river was one of the first in which habitat was destroyed by human endeavors (Brim Box and Williams, 2000). *Amblema neislerii* probably occurred in the Chattahoochee River upstream to the Fall Line but was extirpated before it could be collected by early naturalists.

Amblema neislerii has been reported from the Apalachicola River very near the mouth of the Chattahoochee River, Jackson County, Florida. This is a distance of approximately 48 km downstream of the Alabama state line.

Remarks

In the original description of *Pleurobema strodeanum*, Wright (1898c) erroneously reported *Amblema neislerii* from the Escambia River in western Florida.

Amblema neislerii was listed as endangered under the federal Endangered Species Act in 1998.

Anodonta heardi Gordon and Hoeh, 1995
Apalachicola Floater

Anodonta heardi – Length 113 mm, UF 358656. Harrison Creek, north side of bend, 29°52.80' W, 85°02.54' N, Franklin County, Florida, 7 September 1991. © Richard T. Bryant.

Shell Description

Length to 140 mm; thin; moderately inflated; outline elliptical; posterior margin narrowly rounded; anterior margin rounded; dorsal margin straight to slightly convex; ventral margin convex; posterior ridge rounded, indistinct; posterior slope flat to slightly convex; umbo broad, inflated, elevated slightly above hinge line, umbo sculpture low, irregular ridges; periostracum smooth, shiny, greenish yellow to tawny, darkening with age, usually without rays.

Pseudocardinal and lateral teeth absent; umbo cavity wide, shallow; nacre white to bluish white.

General Distribution

Anodonta heardi is known only from Coastal Plain reaches of the Apalachicola Basin in Florida and Georgia (Brim Box and Williams, 2000).

Reasons for Inclusion

Anodonta heardi is known from the Chattahoochee River near its junction with the Flint River, approximately 32 km downstream of the Alabama and Florida state line. Brim Box and Williams (2000) reported a single record of *A. heardi* from the Chattahoochee River drainage of Alabama. However, upon reevaluation of the material it was found to be a conchologically atypical *Utterbackia imbecillis*.

Remarks

Prior to its description in 1995, *Anodonta heardi* was included with *Anodonta couperiana* Lea, 1840, a southern Atlantic Coast and peninsular Florida species (Heard, 1979b), or with *Pyganodon gibbosa* (Say, 1824), which is endemic to the Altamaha River drainage of the southern Atlantic Coast (Clench and Turner, 1956).

Anodonta heardi was named for Dr. William H. Heard, Department of Biological Sciences, Florida State University, in recognition of his significant contributions to malacology. The volume of *Walkerana* in which *A. heardi* was described is dated 1993–1994. However, the volume was not copyrighted, printed and distributed until 1995. Therefore, the recognized date of authorship is 1995 (ICZN, 1999: Articles 21 and 22).

Fusconaia flava (Rafinesque, 1820)
Wabash Pigtoe

Fusconaia flava – Length 66 mm, NCSM 30111. Tennessee River, river mile 106.7, [0.8 km east of Rockport,] Humphries County, Tennessee, 1 December 1997. © Richard T. Bryant.

Shell Description

Length to 101 mm; thick; inflated; outline subquadrate to subtriangular; posterior margin bluntly pointed to obliquely truncate; anterior margin almost straight to broadly rounded; dorsal margin slightly convex; ventral margin convex; posterior ridge well defined, sharp dorsally; posterior slope steep, flat to slightly concave; umbo high, inflated, elevated well above hinge line, oriented anteriorly, umbo sculpture concentric ridges, angled on posterior ridge; periostracum light brown to black, typically without rays.

Pseudocardinal teeth thick, erect, triangular, often deeply serrate, 2 divergent teeth in left valve, 1 tooth in right valve; lateral teeth thick, slightly curved, 2 in left valve, 1 in right valve, right valve may have rudimentary lateral tooth ventrally; interdentum very short, moderately wide; umbo cavity deep; nacre white, sometimes with pink or salmon cast.

General Distribution

Fusconaia flava occurs in portions of the Hudson Bay and Great Lakes drainages (Clarke, 1981a). It also occurs through much of the Mississippi Basin, including the Ohio, Cumberland and lower Tennessee River drainages (Parmalee and Bogan, 1998).

Reason for Inclusion

Fusconaia flava occurs in lower reaches of the Tennessee River proper in western Tennessee. It has been reported from tailwaters of Pickwick Dam, Hardin County, Tennessee, approximately 13 km from the Alabama state line (Parmalee and Bogan, 1998). It is likely that *F. flava* will eventually appear in Alabama reaches of the Tennessee River. Other Interior Basin species (e.g., *Arcidens confragosus*, *Fusconaia ebena* and *Quadrula quadrula*) have been documented as expanding their ranges up the Tennessee River following its impoundment (Garner and McGregor, 2001).

Remarks

Morrison (1942) reported collecting *Fusconaia undata trigona* (= *Fusconaia flava*) from along the Tennessee River in drift at "Little Slough" on the northern side of Sevenmile Island, Lauderdale County, Alabama, prior to completion of Pickwick Dam. Specimens were also recovered from "button-shell discard piles" in Hardin County, Tennessee, near the Alabama and Tennessee state line, but no museum specimens to verify these reports were found.

Fusconaia flava is a reservoir-tolerant species. If it occurred in Alabama reaches of what is now Pickwick Reservoir, it would likely not have been extirpated with impoundment of the river. However, none have been encountered during several hundred hours of dive time in the vicinity of Sevenmile Island between 1995 and 2007 (J.T. Garner, personal observation).

Fusconaia flava was inadvertently included on Alabama's original list of commercially harvestable mussel species in 1993 (McGregor et al., 1996). It was subsequently removed from the list.

A total of 229 individuals of *Fusconaia flava*, collected from lower reaches of the Tennessee River, were introduced into Shoal Creek, Lauderdale County, Alabama, and Lawrence County, Tennessee, from 1994 to 1996 (Morgan, 1996; Morgan et al., 1997). The species apparently did not become established.

Quadrula nodulata (Rafinesque, 1820)
Wartyback

Quadrula nodulata – Length 59 mm, UF 370410. Tennessee River, river mile 93.0, [Benton County, Tennessee,] 1972. © Richard T. Bryant.

Shell Description

Length to 70 mm; moderately thick; moderately inflated; outline quadrate; posterior margin truncate, often slightly concave; anterior margin rounded; dorsal margin straight to convex; ventral margin convex; posterior ridge low but prominent, usually with a row of widely spaced pustules; posterior slope moderately steep, flattened into a short dorsal wing, usually with small pustules or plications distally; disk with sparse, widely spaced, elongate pustules, often oriented in one or more radial rows; umbo somewhat narrow, inflated, elevated above hinge line, umbo sculpture short nodulous ridges; periostracum shiny to satiny, tawny to greenish brown, subadults may have weak green rays.

Pseudocardinal teeth heavy, rough, triangular, 2 divergent teeth in left valve, 1 tooth in right valve, may have accessory denticle anteriorly and/or posteriorly; lateral teeth moderately long, somewhat thick, slightly curved, 2 in left valve, 1 in right valve, right valve may have rudimentary second lateral tooth ventrally; interdentum short, narrow to moderately wide; umbo cavity deep; nacre white.

General Distribution

Quadrula nodulata is found through much of the Mississippi Basin, including the Ohio, Cumberland and lower Tennessee River drainages (Parmalee and Bogan, 1998).

Reason for Inclusion

Quadrula nodulata occurs in lower reaches of the Tennessee River proper in western Tennessee (Parmalee and Bogan, 1998). The reported upstream limit of *Q. nodulata* in the Tennessee River is near the mouth of Beech River, Decatur County, Tennessee, approximately 125 km downstream of the Alabama state line (Parmalee and Bogan, 1998). It would not be surprising for this species to eventually appear in the Alabama reaches of the Tennessee River. Other Interior Basin species (e.g., *Arcidens confragosus*, *Fusconaia ebena* and *Quadrula quadrula*) have been documented as expanding their ranges up the Tennessee River following its impoundment (Garner and McGregor, 2001).

Remarks

Quadrula nodulata was inadvertently included on Alabama's original list of commercially harvestable mussel species in 1993 (McGregor et al., 1996). It was subsequently removed from the list. Stansbery (1976) erroneously reported *Q. nodulata* from the Tennessee

River drainage in Alabama and Ahlstedt (1998) erroneously reported it from the Paint Rock River.

A total of 22 individuals of *Quadrula nodulata* were introduced into Shoal Creek, Lawrence County, Tennessee, in 1995 and 1996 (Morgan, 1996; Morgan et al., 1997). Shoal Creek flows through Lauderdale County, Alabama, before emptying into the Tennessee River. The species apparently did not become established.

Quadrula refulgens (Lea, 1868)
Purple Pimpleback

Quadrula refulgens – Length 43 mm, UF 294893. Escatawpa River at Alabama and Mississippi state line, downstream of U.S. Highway 98, Mobile County, Alabama, and Jackson County, Mississippi, 29 August 2000. © Richard T. Bryant.

Shell Description

Length to 62 mm; moderately thick; moderately inflated; outline round to oval; posterior margin broadly rounded, may be somewhat truncate; anterior margin rounded; dorsal margin convex; ventral margin convex; posterior ridge rounded; posterior slope moderately steep, corrugations may extend from posterior ridge toward shell margin; disk with variable pustules, primarily on posterior half of shell; umbo high, inflated, elevated above hinge line, umbo sculpture coarse, nodulous ridges; periostracum reddish brown to black, without rays.

Pseudocardinal teeth thick, erect, triangular, 2 divergent teeth in left valve, 1 tooth in right valve, may have accessory denticle anteriorly and/or posteriorly; lateral teeth moderately thick, slightly curved, 2 in left valve, 1 in right valve; interdentum short, moderately wide; umbo cavity deep; nacre white to pink or purple.

General Distribution

Quadrula refulgens is endemic to the Gulf Coast of Mississippi and Louisiana, from the Pascagoula River drainage west to the Lake Pontchartrain drainage (Vidrine, 1993; Jones et al., 2005).

Reason for Inclusion

Quadrula refulgens is known at least as far east as the Pascagoula River drainage. The easternmost tributaries of this drainage lie in Choctaw, Washington and Mobile counties, Alabama, so this species may have historically occurred in the state. The most likely place for its occurrence is in Red Creek, Washington County. This stream has been poorly studied, but several species of freshwater mussels are known to occur there. The Escatawpa River is larger than Red Creek but is not known to harbor any bivalves (including *Corbicula fluminea*) or operculate snails, with the exception of its extreme lower reach near the confluence with the Pascagoula River in Mississippi.

Remarks

Shells of two *Quadrula refulgens* have been collected from Alabama on the banks of the Escatawpa River, a tributary of the Pascagoula River, at the U.S. Highway 98 crossing on the Alabama and Mississippi state line. This locality is a popular canoe launch. Since no bivalves or any other mollusks are known to occur in the entire reach of the Escatawpa River in Alabama, it is likely that the shells were discarded by a canoeist.

Chapter 15
Other Bivalves in Inland Waters

In addition to the Margaritiferidae and Unionidae, four other families of bivalve mollusks are known to occur in inland waters of Alabama. The family Corbiculidae is represented in the state by two taxa, *Polymesoda caroliniana*, a native species that occurs only in brackish and tidally influenced fresh waters, and *Corbicula fluminea*, a nonindigenous species that occurs in most inland waters of Alabama and the Mobile Basin (Figure 15.1). The family Dreissenidae is also represented in Alabama by two taxa, *Mytilopsis leucophaeata*, a native species that occurs primarily in brackish and tidally influenced fresh waters, and *Dreissena polymorpha*, a nonindigenous species that occurs throughout the Tennessee River proper. The family Mactridae is represented by one species, *Rangia cuneata*, that occurs only in brackish and tidally influenced fresh waters. The family Sphaeriidae is a group of small, fragile bivalves that occur statewide and are often abundant but poorly known. Four genera, *Eupera*, *Musculium*, *Pisidium* and *Sphaerium*, are known to occur in Alabama. However, the number of species in the state is unknown. None of these taxa closely resemble unionoids and can be distinguished from them based on shell morphology. Abbreviated accounts are provided for the representative species of these four families.

Figure 15.1. *Corbicula fluminea* (with siphons extended). Little Cahaba River, Bibb County, Alabama, October 2005. Photograph by A.D. Huryn.

FAMILY CORBICULIDAE – MARSH CLAMS

There are four species of the family Corbiculidae found in the U.S. Two are native brackish water species, *Polymesoda caroliniana* (Bosc, 1801) and *Polymesoda maritima* (d'Orbigny, 1842), and two are nonindigenous *Corbicula* species. *Corbicula fluminea* (Müller, 1774) is widespread in the U.S. The second species, identified by Hillis and Patton (1982), was reported from southwestern Texas, southwestern Arizona and southeastern California but was not given a specific epithet. McMahon and Bogan (2001) provided a distribution map for the two *Corbicula* species.

Corbicula fluminea (Müller, 1774)
Asian Clam
INTRODUCED

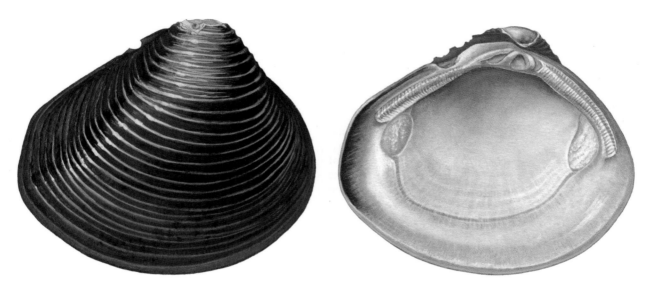

Corbicula fluminea – Length 30 mm, specimen not retained. Choctawhatchee River, 1.5 air miles south of Interstate Highway 10 bridge, Holmes and Washington counties, Florida, 3 June 1998. Illustration by S. Trammell.

Description

Length to 50 mm; shell moderately thick; outline round to ovate in young individuals, becoming more triangular with age; umbo moderately inflated, elevated above hinge line, centrally located; external surface sculptured with numerous evenly spaced, thin, concentric ridges; periostracum yellowish to tawny or dark olive, becoming dark brown to black with age. Internally, hinge plate with 3 cardinal teeth in each valve, lateral teeth straight to slightly curved, serrated, 2 on each side of cardinal teeth in right valve, 1 on each side of cardinal teeth in left valve; pallial line without pallial sinus; interior white to purple. Soft anatomy differs from that of unionoids by the presence of true siphons (McMahon and Bogan, 2001) (Figure 15.1).

Range and Habitat

Corbicula fluminea has been widely introduced in the U.S., beginning in the 1920s (Counts, 1986), but was not reported from Alabama until the 1960s (Sinclair and Ingram, 1961; Hubricht, 1963). It quickly spread to most of the state and now occurs in most fresh waters of Alabama and the Mobile Basin except for the Blackwater, Perdido and Escatawpa rivers and some direct tributaries of Mobile Bay.

Corbicula fluminea occurs in a variety of natural and man-made habitats including creeks, rivers, canals, ponds and reservoirs. It is found in mud, sand and gravel substrates as well as under cobble and boulders and in cracks and depressions in bedrock. *Corbicula fluminea* tolerates low salinity waters up to 10 ppt for short periods. In the Mobile Delta, Swingle and Bland (1974) reported the most downstream occurrence of live *Corbicula* near the junction of the Tensaw and

Blakeley rivers, approximately 10 km upstream of the Interstate Highway 10 bridge, in areas where salinity ranged from 0 to 11.4 ppt and averaged about 1.4 ppt.

Remarks

The serrated lateral teeth of *Corbicula fluminea* distinguish it from the Sphaeriidae. There has been much debate regarding the number of *Corbicula* species introduced into the U.S. and their relationships to Asian populations (Morton, 1979; Hillis and Patton, 1982; Siripattrawan et al., 2000; McMahon and Bogan, 2001; Lee et al., 2005). The only *Corbicula* species recognized as occurring in Alabama and the Mobile Basin is *C. fluminea*.

Polymesoda caroliniana (Bosc, 1801)
Carolina Marsh Clam

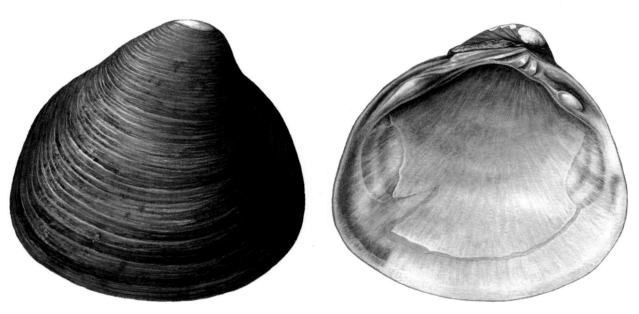

Polymesoda caroliniana – Length 43 mm, UF 181786. Mouth of Suwannee River, Dixie County, Florida, 1987. Illustration by S. Trammell.

Description

Length to 90 mm; shell moderately thick; outline ovate to subtriangular; posterior ridge absent; periostracum thin, brown, with scaly concentric ridges; external hinge ligament long. Internally, hinge plate with 3 cardinal teeth and 1 lateral tooth in each valve; pallial line with deep, narrow pallial sinus; interior white to purple. Soft anatomy differs from that of unionoids by the presence of true siphons (Abbott, 1974; Rehder, 1981; Porter and Houser, 1998).

Range and Habitat

Polymesoda caroliniana occurs along the Atlantic Coast from Virginia to northern Florida and in the Gulf of Mexico from northwestern Florida to Texas (Abbott,1974; Rehder, 1981; Porter and Houser, 1998).

In Alabama *Polymesoda caroliniana* occurs along the Gulf Coast and in the Mobile Delta. It is found on sandy mud substrates of estuaries and in low-salinity mouths of rivers and creeks, including tidally influenced fresh waters of the Mobile Delta (Swingle and Bland, 1974). The most inland record of *P. caroliniana* is from the junction of the Tensaw and Blakeley rivers in the Mobile Delta, approximately 10 km upstream of the Interstate Highway 10 bridge (Swingle and Bland, 1974).

Remarks

Polymesoda caroliniana inhabits intertidal areas with salinity ranging from 1 ppt to 15 ppt (Swingle and Bland, 1974). *Polymesoda caroliniana* may be distinguished from *Rangia cuneata* by its thinner shell, the presence of an external hinge ligament, rounder outline and absence of a posterior ridge.

FAMILY DREISSENIDAE – FALSEMUSSELS AND ZEBRA MUSSELS

In the U.S. the family Dreissenidae is represented by four species. Two are native, *Mytilopsis leucophaeata* (Conrad, 1831) and *Mytilopsis sallei* (Recluz, 1849), and two are nonindigenous, *Dreissena bugensis* Andrusov, 1897, and *Dreissena polymorpha* (Pallas, 1771). Nomenclature of *Dreissena* was examined by Rosenberg and Ludyansky (1994). Two species, *M. leucophaeata* and *D. polymorpha*, occur in Alabama.

Dreissena polymorpha (Pallas, 1771)
Zebra Mussel
INTRODUCED

Dreissena polymorpha – Length 23 mm, specimen not retained. Kentucky Lake in lock chamber at Kentucky Dam, Livingston County, Kentucky, 5 September 1992. Illustration by S. Trammell.

Description

Length to 40 mm; shell thin, inflated; outline subtriangular to lachrymous; posterior margin rounded to obliquely truncate; dorsal margin broadly rounded; anterior margin bluntly pointed; ventral margin nearly straight, with a prominent ridge extending the shell length; ventral surface flat; umbo forming blunt point at anterior end; hinge ligament short, internal, located anteriodorsally; periostracum light tan to brown, often marked with dark vertical zigzag lines; byssal threads present on ventral margin. Internally, hinge plate, cardinal and lateral teeth absent; shell with small septum across extreme anterior end, without apophysis; pallial sinus absent; interior bluish white. Soft anatomy differs from that of unionoids by the presence of true siphons (Parmalee and Bogan, 1998; McMahon and Bogan, 2001).

Range and Habitat

Dreissena polymorpha was introduced into the U.S. from Europe in the mid-1980s and has spread through most of the Great Lakes and St. Lawrence Basin, as well as the Mississippi Basin.

In Alabama it is found only in the Tennessee River. It has not been reported from the Mobile Basin, though the Tennessee Tombigbee Waterway has opened an avenue for its invasion. *Dreissena polymorpha* occurs primarily attached to hard surfaces by means of byssal threads.

Remarks

Dreissena polymorpha can be distinguished from *Mytilopsis leucophaeata* by the sharply angled ridge extending the shell length, flattened ventral surface and absence of an apophysis on the septum.

Mytilopsis leucophaeata (Conrad, 1831)
Dark Falsemussel

Mytilopsis leucophaeata – Length 22 mm, UF 388524. St. Johns River, Florida, June 2002. Illustration by S. Trammell.

Description

Length to 25 mm; shell thin; outline lachrymous; posterior margin rounded; anterior margin bluntly pointed, usually slightly ventrally hooked; ventral margin straight to slightly concave or convex; dorsal margin broadly rounded; ventral surface not flat; umbo forming blunt point at anterior end; hinge ligament short, internal, located anteriodorsally; periostracum thin, tawny to brown, may have weak vertical zigzag lines, often disappear ventrally, eroded individuals may appear bluish; byssal threads present on ventral margin. Internally, hinge plate, cardinal and lateral teeth absent; shell with small septum across extreme anterior end of shell, with apophysis on the side facing the internal ligament; interior bluish white. Soft anatomy differs from that of unionoids by the presence of true siphons (Abbott, 1974; Rehder, 1981; McMahon and Bogan, 2001).

Range and Habitat

Mytilopsis leucophaeata is found in brackish and tidally influenced fresh waters from New York south to Florida and west along the Gulf Coast to Mexico (Abbott, 1974; Rehder, 1981).

In Alabama it is locally abundant in upper Mobile Bay and parts of the Mobile Delta. It is occasionally found far inland in the Tennessee River and Mobile Basin, presumably dispersed by barges. There is no evidence that *Mytilopsis leucophaeata* reproduces in fresh water in Alabama. *Mytilopsis leucophaeata* occurs primarily attached to hard surfaces by means of byssal threads.

Remarks

Mytilopsis leucophaeata can be distinguished from *Dreissena polymorpha* by the absence of a sharply angled ridge extending the shell length, a ventral surface that is not distinctly flattened and the presence of an apophysis on the septum.

FAMILY MACTRIDAE – SURF CLAMS

In the U.S. the family Mactridae is represented by 9 genera and 19 species. All are native and occur in marine habitats, with the exception of one species, *Rangia cuneata*. This species occurs in estuaries and lower reaches of some rivers.

Rangia cuneata (Sowerby, 1831)
Atlantic Rangia

Rangia cuneata – Length 50 mm, UF 360321. Choctawhatchee River at junction with Sister River, Walton County, Florida, 22 March 2000. Illustration by S. Trammell.

Description

Length to 85 mm; shell thick; outline triangular to ovate; posterior ridge often well-developed; posterior slope steep; umbo inflated, elevated well above hinge line; periostracum tawny to grayish brown. Internally, hinge plate moderately wide, with well-developed anterior and posterior lateral teeth; hinge ligament fully internal, positioned in chondrophore; pallial line with small but moderately deep, distinct pallial sinus; interior white to bluish white. Soft anatomy differs from that of unionoids by the presence of true siphons (Abbott, 1974; Rehder, 1981; Porter and Houser, 1998).

Range and Habitat

Rangia cuneata is known coastally from New York south to Florida and west to Texas (Abbott, 1974; Rehder, 1981; Carlton, 1992; Porter and Houser, 1998). On the Atlantic Coast living individuals have been found north of Georgia only since the late 1950s (Wells, 1961; Godwin, 1968). *Rangia cuneata* is also known from Pleistocene deposits in that area (Hopkins and Andrews, 1970).

In Alabama *Rangia cuneata* is found in brackish water with 12 ppt to 20 ppt salinity from Mobile Bay upstream to near the confluence of the Tombigbee and Alabama rivers (Swingle and Bland, 1974). This species usually occurs in mud or muddy sand substrates, where it may be found in dense beds (Swingle and Bland, 1974). In upper reaches of the Mobile Delta *R. cuneata* often occurs sympatrically with several unionid species, such as *Glebula rotundata*, *Plectomerus dombeyanus*, *Pyganodon grandis* and *Quadrula apiculata*, as well as the native dreissenid *Mytilopsis leucophaeata* (Swingle and Bland, 1974).

Remarks

Rangia cuneata has an internal hinge ligament located in the chondrophore, which is absent in *Polymesoda caroliniana*. *Rangia cuneata* is typically more triangular in outline, with a more prominent posterior ridge and thicker shell than *P. caroliniana*.

FAMILY SPHAERIIDAE – PEA CLAMS OR FINGERNAIL CLAMS

In the U.S. the family Sphaeriidae includes 4 genera (*Eupera*, *Musculium*, *Pisidium* and *Sphaerium*) and 39 species, 4 of which are nonindigenous (Turgeon et al., 1998). A typical family representative, *Musculium partumeium*, is illustrated below.

Musculium partumeium – Length 11 mm, UF 187891. First lateral canal west of Lake Osborn, Lantana, Palm Beach County, Florida, 1936. Illustration by S. Trammell.

Family Description

Length to 25 mm; shell very thin, fragile; outline round to oval or subquadrate; periostracum thin, yellow to brown. Internally, hinge plate with cardinal teeth and anterior and posterior lateral teeth, lateral teeth smooth; pallial line without pallial sinus. Soft anatomy differs from that of unionoids by the presence of true siphons and only the inner gills are used to brood developing offspring (Heard, 1965; Burch, 1975b; McMahon and Bogan, 2001).

Range and Habitat

Sphaeriids are found in lakes, ponds, creeks and rivers, including ephemeral ponds and intermittent streams, through most of the U.S. One species, *Eupera cubensis* (Prime, 1865), uses byssal threads to attach to hard surfaces.

Sphaeriids are widespread in Alabama but diversity and distribution are poorly known (Burch, 1975b; McMahon and Bogan, 2001). All four genera occur in the state.

Remarks

Large populations of sphaeriids may be found in wetlands and shallow lakes, where they constitute important forage organisms for some species of waterfowl and fish.

Chapter 16
Spurious Record

Unio decumbens Lea, 1861

Unio decumbens – Length 84 mm, USNM 86150. Alabama. Photograph by J.D. Williams.

Lea (1861a) described a new species, *Unio decumbens*, based on a single specimen, allegedly from somewhere in Alabama. Simpson (1900b, 1914) considered it a species of *Margaritana* (= *Margaritifera*), suggesting a possible relationship with *Cumberlandia monodonta,* and remarked that it is "quite different from the other species, yet having the characters of a *Margaritana*." Simpson also noted that "it is a little strange that only the type is known of this curious compressed, rather thin, trapezoidal species." Frierson (1922, 1927) considered it to be a deformed specimen of *Elliptio complanata*. Without discussion Johnson (1970) placed *decumbens* in the synonymy of *E. complanata*. Smith (2001) reexamined the type and concluded that it is possibly a unionid.

Based on examination of the type, the shell appears most similar to *Trapezoideus exolescens* (Gould, 1843), an Asian unionid from Thailand, Myanmar, Laos, Cambodia and Vietnam. It does not resemble any known Alabama species and appears to be a specimen with incorrect locality data.

Synonymy
Unio decumbens Lea, 1861. Lea, 1861a:40; Lea, 1862b:87, pl. 12, fig. 236; Lea, 1862e:91, pl. 12, fig. 236
Type locality: Alabama, E.R. Showalter, M.D. Holotype, USNM 86150, length 84 mm.

Appendix
North American Unioniform Type Catalogs

The taxonomy and nomenclature of freshwater mussels is difficult to address without consulting original descriptions and examining type material, some of which is in foreign museums. This list of type catalogs is provided to assist in locating original descriptions and type material. These catalogs may contain assumptions and designations of types and restrictions of type localities which are incorrect or unnecessary. Some catalogs may not be complete, so a researcher should not be completely dependent on them. The International Code of Zoological Nomenclature (ICZN, 1999) should be consulted to clarify questions of validity of type material.

Boyko, C.B., and W.E. Sage, III. 1996. Catalog of recent type specimens in the Department of Invertebates, American Museum of Natural History. II Mollusca Part I (Classes Aplacohpora, Polyplacophora, Gastropoda [Subclass Opisthobranchia], Bivalvia and Scaphopoda). American Museum Novitates, Number 3170. 50 pages, 12 figures.

Brooks, S.T., and B.W. Brooks. 1931. List of types of Pelecypoda in the Carnegie Museum on January 1, 1931. Annals of Carnegie Museum 20(2):171–177.

Clarke, A.H. 1981. The tribe Alasmidontini (Unionidae: Anodontinae), Part I. *Pegias, Alasmidonta,* and *Arcidens.* Smithsonian Contributions to Zoology Number 326. 101 pages.

Clarke, A.H. 1985. The tribe Alasmidontini (Unionidae: Anodontinae), Part II. *Lasmigona* and *Simpsonaias.* Smithsonian Contributions to Zoology Number 399. 75 pages.

Franzen, D.S. 1956. Types of mollusks described by F.C. Baker, Part I, University of Illinois. The Nautilus 70(1):21–27.

Franzen, D.S. 1957. Types of mollusks described by F.C. Baker, Part II, University of Wisconsin. The Nautilus 71(1):30–35.

Franzen, D.S. 1958. Types of mollusks described by F.C. Baker, Part III, Chicago Academy of Sciences. The Nautilus 72(1):30–34.

Haas, F. 1930. Über nord- und mittelamerikanische Najaden. Senckenbergiana Biologica 12(6):317–330.

Johnson, R.I. 1952. A study of Lamarck's types of Unionidae and Mutelidae. The Nautilus 66(2):63–67.

Johnson, R.I. 1953. A study of Lamarck's types of Unionidae and Mutelidae (continued). The Nautilus 66(3):90–95.

Johnson, R.I. 1956. Types of naiades (Mollusca: Unionidae) in the Museum of Comparative Zoology. Bulletin of the Museum of Comparative Zoology 115(4):102–142, plates 1–2.

Johnson, R.I. 1959. The Charles M. Wheatley Collections. The Nautilus 73(2):72–74.

Johnson, R.I. 1967. Illustrations of all the mollusks describe by Berlin Hart and Samuel Hart Wright. Museum of Comparative Zoology, Harvard University, Occasional Papers on Mollusks 3(35):1–35, plates 1–13.

Johnson, R.I. 1969. Illustrations of Lamarck's types of North American Unionidae mostly in the Paris Museum. The Nautilus 83(2):52–61.

Johnson, R.I. 1969. The Unionacea of William Irvin Utterback. The Nautilus 82(4):132–135.

Johnson, R.I. 1970. The systematics and zoogeography of the Unionidae (Mollusca: Bivalvia) of the southern Atlantic Slope Region. Bulletin of the Museum of Comparative Zoology 140(6):263–449.

Johnson, R.I. 1971. The types and figured specimens of Unionacea (Mollusca: Bivalvia) in the British Museum (Natural History). Bulletin of the British Museum (Natural History) 20(3):75–108, plates 1–10.

Johnson, R.I. 1972. Illustrations of all of the mollusks described by Lorraine Screven Frierson. Museum of Comparative Zoology, Harvard University, Occasional Papers on Mollusks 3(41):137–173, plates 22–32.

Johnson, R.I. 1973. The types of Unionidae (Mollusca:Bivalvia) described by C.S. Rafinesque in the Museum National d'histoire Naturelle, Paris. Journal de Conchyliologie 110(2):35–36, plate 1.

Johnson, R.I. 1974. Lea's unionid types or recent and fossil taxa of Unionacea and Mutelacea introduced by Isaac Lea, including the location of all the extant types. Museum of Comparative Zoology, Harvard University, Special Occasional Papers Number 2. 159 pages.

Johnson, R.I. 1974. Marshall's unionid types or types of recent and fossil Unionacea and Mutelacea introduced by William B. Marshall, including a bibliography of all his writings on mollusks. Museum of Comparative Zoology, Harvard University, Special Occasional Papers Number 3. 14 pages.

Johnson, R.I. 1975. R. Ellsworth Call with a bibliography of his works on mollusks and a catalog of his taxa. Museum of Comparative Zoology, Harvard University, Occasional Papers on Mollusks 4(54)133–144.

Johnson, R.I. 1975. Simpson's unionid types and miscellaneous unionid types in the National Museum of Natural History. Museum of Comparative Zoology, Harvard University, Special Occasional Papers Number 4. 56 pages.

Johnson, R.I. 1975. William A. Marsh, his introduced taxa of Unionidae or freshwater mussels. Museum of Comparative Zoology, Harvard University, Occasional Papers on Mollusks 4(54):145–147.

Johnson, R.I. 1977. Monograph of the genus *Medionidus* (Bivalvia: Unionidae) mostly from the Apalachicolan region, southeastern United States. Museum of Comparative Zoology, Harvard University, Occasional Papers on Mollusks 4(56):161–187.

Johnson, R.I. 1977. Norman Macdowell Grier, a bibliography of his work on mollusks, with a catalogue of his unionid taxa. Museum of Comparative Zoology, Harvard University, Occasional Papers on Mollusks 4(57):226–227.

Johnson, R.I. 1977. Arnold Edward Ortmann, a bibliography of his work on mollusk, with a catalogue of his recent molluscan taxa. Museum of Comparative Zoology, Harvard University, Occasional Papers on Mollusks 4(58):229–241, plate 27.

Johnson, R.I. 1978. Systematics and zoogeography of *Plagiola* (=*Dysnomia* =*Epioblasma*), an almost extinct genus of freshwater mussels (Bivalvia: Unionidae) from middle North America. Bulletin of the Museum of Comparative Zoology 148(6):239–320.

Johnson, R.I. 1979. The types of Unionacea (Mollusca: Unionidae) in the Museum of Zoology, the University of Michigan. Malacological Review 12(1–2):29–36.

Johnson, R.I. 1980. The types of Unionacea (Mollusca: Bivalvia) in the Academy of Natural Sciences of Philadelphia: Additions and corrections. Proceedings of the Academy of Natural Sciences of Philadelphia 132:277–278.

Johnson, R.I. 1981. Recent and fossil Unionacea and Mutelacea (Freshwater Bivalves) of the Caribbean Islands. Museum of Comparative Zoology, Harvard University, Occasional Papers on Mollusks 4(60):269–288.

Johnson, R.I. 1998. Addenda to be included with Lea's Figured Unionid Types (1974). Department of Mollusks, Museum of Comparative Zoology, Harvard University, Cambridge, Massachusetts. 2 pages.

Johnson, R.I., and H.B. Baker. 1973. The types of Unionacea (Mollusca: Bivalvia) in the Academy of Natural Sciences of Philadelphia. Proceedings of the Academy of Natural Sciences of Philadelphia 125(9):145–186, plates 1–10.

Kabat, A.R., and K.J. Boss. 1992. An indexed catalogue of publications on Molluscan type specimens. Museum of Comparative Zoology, Harvard University, Occasional Papers on Mollusks 5(69):157–336.

Knudsen, J., K.R. Jensen, C. Nielsen and R.I. Johnson. 2003. Lorentz Spengler's description of freshwater mussel (Mollusca: Unionacea): translation and notes. Steenstrupia 27(2):263–279.

Mikkelsen, P.M., and A. Bradford. 1997. Annotated catalog of the type specimens in the malacological collections of the Delaware Museum of Natural History. Part II. Additions and corrections to Part I (Prosobranchs, Heterostropha, and Opisthobranchia), plus Bivalvia, Scaphopoda, and Polyplacophora. Nemouria, Occasional Papers of the Delaware Museum of Natural History Number 41. 76 pages.

Parodiz, J.J. 1967. Types of North American Unionidae in the collection of Carnegie Museum. Sterkiana 28:21–30.

Parodiz, J.J., and J.J. Tripp. 1988. Types of Mollusca in the collection of the Carnegie Museum of Natural History. Part 1. Bivalvia and Gastropoda (Prosobranchia and Opisthobranchia). Annals of the Carnegie Museum 57(5):111–154.

Rehder, H.A. 1967. Valid zoological names of the Portland Catalogue. Proceedings of the United States National Museum 121(3579):1–51.

Richards, M.C., and W.E. Old, Jr. 1969. A catalog of the molluscan type specimens in the Department of Living Invertebrates, the American Museum of Natural History. Copies on file at AMNH Division of Invertebrate Zoology. 147 pages.

Scudder, N.P. 1885. The published writings of Isaac Lea, LL.D. Biographies of American Naturalists II. Bulletin of the United States National Museum Number 23. 278 pages.

Trew, A. 1987. The Melvill-Tomlin Collection, Part 44, Unionacea. Handlists of the molluscan collections in the Department of Zoology, National Museum of Wales, Cardiff. 130 pages.

Vanatta, E.G. 1915. Rafinesque's types of *Unio*. Proceedings of the Academy of Natural Sciences of Philadelphia 67(1915):549–559.

Walker, B. 1916. The Rafinesque-Poulson Unios. The Nautilus 30(4):43–47.

Wheeler, M.J. 1963. Type of *Unio luteolus* Lamarck, 1819. The Nautilus 77(2):58–61.

Zilch, A. 1967. Die typen und typoide des Natur-Museums Senckenberg, 39(1). Archiv für Molluskenkunde 97(1/6):45–154.

Zilch, A. 1983. Die typen und typoide des Natur-Museums Senckenberg, 72(1). Archiv für Molluskenkunde 114(1/3):77–92.

Glossary

absolute tautonymy – identical spelling of generic and specific names of one of the species originally included in a genus description (e.g., *Villosa villosa*).

accessory denticle – a small, compressed or triangular prominence on the hinge plate anterior or posterior to pseudocardinal teeth.

acinus (acini, pl.) – a small compartment of the gonad in which gametes are produced; may also be referred to as an alveolus (alveoli, pl.) or follicle.

acute – tapering to a sharp point.

adductor muscle – a large muscle attached internally to valves, located dorsally, one each near the anterior and posterior ends of the hinge plate, which serve to draw the two valves together.

adductor muscle scar – an impression in the nacreous shell layer that is the attachment site for an adductor muscle.

adventive – not native, referring to an organism transported to, and becoming established in, a new area by human or natural means.

alate – having a wing or wing-like part; in unionids a thin, flat, dorsal extension of the shell.

allometry – relative growth of a structure in relation to size of the remainder of the organism, or the measurement or study of such growth.

allopatric – distributional relationship in which the ranges of two species do not overlap (see **sympatric**).

allotype – a specimen in the type series of the opposite sex from the holotype.

allozyme data – results from protein electrophoresis, separated first by electric charge and then by molecular weight, useful in analysis of phylogenetic relationships.

Ambleminae, amblemine – a subfamily of the Unionidae recognized by some authors (or used as a separate family by others, Amblemidae); typically characterized by well-developed hinge dentition, no mantle modification anterior to the incurrent aperture, marsupia formed by all four or only the outer two gills, glochidia without styliform hooks. Type genus *Amblema*.

amoebocyte – a motile cell, capable of changing shape by cytoplasmic flow, often functions in excretion and assimilation.

angular, angulate – having a corner where two edges meet; not rounded.

annulus (annuli, pl.) – a ring; a line in the shell or periostracum, sometimes referred to as a growth line or growth rest, often visible as a dark line in the periostracum, often presumed to be annual in occurrence.

Anodontinae, anodontine – a subfamily of the Unionidae recognized by some authors; typically characterized by reduced or absent hinge dentition, no mantle modification anterior to the incurrent aperture, marsupia restricted to outer gills and extended below ventral gill margin when gravid, glochidia with styliform hooks. Type genus *Anodonta*.

anteriad, anterior – of or relating to the front or forward portion, typically the end nearer the umbo in unioniforms.

anteriodorsal – of or relating to the upper portion of the anterior end.

anterior shell margin – herein refers to the portion of shell lying forward of an imaginary vertical line extending along the posterior edge of the anterior adductor muscle scar, and perpendicular to an imaginary horizontal line extending along the dorsal margins of the anterior and posterior adductor muscle scars.

anterioventral – of or relating to the lower portion of the anterior end.

anthropogenic – of, relating to or involving the impact of humans on nature.

aperture – an opening between the mantle of the two valves of a unionoid, through which water is drawn or expelled from the mantle cavity; sometimes erroneously referred to as a siphon.

apophysis – a small, triangular, downward protruding shell structure on the anterior septum of *Mytilopsis* valves, which serves as a muscle attachment.

aragonite – one of two crystalline forms of calcium carbonate.

arborescent – divided distally into multiple branches; tree-like in form.

arcuate – curved or bent; in the shape of a bow.

artificial infestation – subjection of potential hosts to glochidia for determination of glochidial hosts or propagation of juveniles; may be used interchangeably with artificial infection.

basin – the entire area of land drained by the largest rivers and lakes and their tributaries (e.g., Mobile, Mississippi and Great Lakes basins) (see **drainage**).

bedrock – solid rock.

biangular, biangulate – having two angles.

bifid – divided distally into two approximately equal parts.

bifurcation – division into two parts.

binomen – a polynomial consisting of a genus and species.

bioturbation – disturbance of sediments by biological activity.

Bivalvia – formerly Pelecypoda, a class of the phylum Mollusca, in which the primary diagnostic characteristic is two opposing valves.

brackish water – water with salinity content intermediate between that of fresh and sea water.

bradytictic – having a brooding period that encompasses much of the year, typically from late summer or autumn to the following summer; long-term brooder.

brood – to maintain eggs, embryos and/or glochidia in the marsupia.

brooding period – the interval during which eggs, embryos and glochidia are held in the marsupia; also called gravid period.

byssal thread, byssus – a tough proteinaceous thread used to anchor a bivalve to a hard substrate, secreted by the byssal gland in the base of the foot; often occur in bundles (e.g., *Dreissena polymorpha*), but are single in unionoid bivalves; not homologous to larval threads of glochidia.

calcareous, calcified – consisting of or containing calcium carbonate.

canalization – the process of excavating a canal or converting a stream into a canal for the purpose of navigation.

cardinal teeth – hinge teeth positioned between two sets of lateral teeth, which serve to stabilize opposing valves; found in some marine and freshwater families, including Corbiculidae and Sphaeriidae.

caruncle – a fleshy protuberance found on the posterioventral mantle margins of female *Toxolasma*, apparently used to attract glochidial hosts.

channelization – the process of straightening and removing obstructions in the course of a stream for more efficient drainage of a watershed.

chevron – a marking shaped like a wide-angled "V", usually inverted.

chitin, chitinous – a hard, amorphous polysaccharide that comprises the covering of some invertebrates and is the material forming the ligament of bivalve mollusks.

chondrophore – a deep pit or depression on the hinge plate ventral to the umbo.

cilia – small, hair-like processes extending from cells, such as those on gills, labial palps, foot and mantle surfaces.

ciliary action – the process of transporting water or small particles along a ciliated surface by repeated rhythmic movements of cilia.

clade – a natural, monophyletic group of two or more taxa.

clinal variation – gradual changes in morphological characteristics over geographic space; in unionoid bivalves typically exhibited by change in shell characteristics from headwaters to lower reaches of a river.

cloth-like – having a surface texture of fabric; satin-like.

commensal – an organism that benefits from another organism without injury to the host.

commercial species – mussel species which are harvested for economic benefit; typically for use in production of cultured pearl nuclei or jewelry, historically those used in the pearl button industry.

compressed – laterally flattened.

concave – curved or rounded inward.

concentric – curving about a common center.

conchological – of or relating to shell characteristics.

conchology – the science dealing with mollusk shells.

congeneric – belonging to the same genus.

conglutinate – a mass of glochidia bound together in a gelatinous or mucous mass, which may mimic food items of glochidial hosts.

conspecific – belonging to the same species.

convex – curved or rounded outward.

corrugation – one of a complex of small, roughly parallel ridges and grooves.

cotype – a term formerly used for either syntype or paratype, not recognized by ICZN (see **syntype**).

crenulate – having a margin with fine notches or scallops.

cryptic species – two or more species that appear so similar as to be confused as a single species.

crystalline style – the gelatinous rod of glyco-protein that extends from the style sac into the stomach, which aids in digestion.

cultured pearl – a pearl produced by deliberate insertion of foreign material into a bivalve.

Cumberlandian fauna – the assemblage of freshwater mussel species which appears to have its origin in the region encompassing the upper Cumberland and Tennessee River drainages, an area often referred to as the Cumberlandian region.

currently stable – a term used to describe the conservation status of a species which is not declining through much of its range and is not in imminent danger of extinction.

cyst – a capsule around a glochidium attached to a host, formed from epithelial tissues or a combination of epithelial and connective tissues of the host.

cytolysis – the dissolution or degeneration of cells.

decorticate – to remove the outer covering, such as loss of the periostracum by erosion.

demibranch – a gill; one of the paired, membranous organs situated between the mantle and visceral mass of a bivalve.

denticle, denticulation – a small, tooth-like projection.

depressed – dorsoventrally flattened.

detritivore – an organism whose diet consists primarily of decomposing organic material.

diaphragm – in unionids the separation between the incurrent and excurrent apertures formed by a connection between the posteriodorsal surface of the gills and the mantle.

diaphragmatic septa – in margaritiferids the posterior portions of the mantle which expand to meet the posterior end of the gill plate to form the diaphragm.

digestive gland – one of three glands external to the stomach, connected to the stomach by ducts which end in blind sacs (i.e., digestive diverticula); the location of extracellular and intercellular breakdown and absorption of food.

dioecious – having separate sexes; not hermaphroditic.

discharge – to expel; liberation of glochidia and/or conglutinates from marsupia in unionoids.

disjunct – separated or isolated.

disk – the central portion of the exterior of a shell valve.

distal – away from the point of attachment or origin, as opposed to proximal.

distended – enlarged from internal pressure.

dorsal, dorsum – of or relating to the back or top; represented by the hinge line in unioniforms.

dorsal muscles – small muscles located in the umbo region, often in the umbo cavity, which serve to attach the mussel mantle to the shell.

dorsal shell margin – herein refers to the shell periphery above an imaginary horizontal line extending along the dorsal margins of the anterior and posterior adductor muscle scars, located between imaginary vertical lines extending along the proximal margins of the anterior and posterior adductor muscle scars, excluding the umbo.

dorsal wing – a thin, flat, dorsal projection of a bivalve shell (see **alate**).

double-looped – in the form of dual arcs, connected centrally.

doubled posterior ridge – the condition of having two ridges between the umbo and posterioventral margin, which diverge distally.

drainage – the entire area of land drained by a river and its tributaries; herein primarily used to refer to large tributaries within basins (e.g., Tennessee River drainage within the Mississippi Basin and Tombigbee River drainage within the Mobile Basin) (see **basin**).

ecophenotypic variation – nonhereditary divergence in shell morphology resulting from varying environmental conditions.

ectobranchous – the condition in which only outer gills are used as marsupia.

edentulous – without hinge teeth.

electrophoresis – a process of separating proteins suspended in a gelatinous medium using an electric current with differential protein movement based on their size and charge, useful in analysis of phylogenetic relationships.

elliptical – elongate oval.

elongate – long in proportion to height.

emarginate – having the margin notched.

embayment – the permanently flooded mouth of a tributary.

embryo – an organism in the early stages of development; in Unioniformes the period from fertilization to the point where the glochidium is fully developed.

encyst – to become enclosed in a cyst; the process of a glochidium attaching to and becoming surrounded by tissues of a gill filament or fin of a host.

endangered – in danger of extinction through all or a significant portion of its range; an official designation under the federal Endangered Species Act of 1973.

endemic – having a native range restricted to a particular area, region or drainage.

endobranchous – the condition in which only inner gills are used as marsupia; a condition that does not occur in Unionidae or Margaritiferidae.

epithelium, epithelial – thin layer of cells covering a tissue or lining a cavity.

Etheriidae, etheriid – a monotypic family of Unioniformes, represented by the genus *Etheria*; individuals with one valve cemented to a solid substrate, both adductor muscles are present and the larval stage is unknown (formerly contained two other genera, subsequently removed, see Bogan and Hoeh, 2000); restricted primarily to sub-Saharan Africa, the Nile River above the cataracts and northwest Madagascar.

excurrent aperture – the posterior external opening between the mantle of the two valves of a unionoid, positioned between the incurrent and supra-anal apertures, through which water is expelled from the mantle cavity; sometimes referred to as anal or exhalent aperture; erroneously referred to as a siphon.

excurrent aperture length – the distance from the ventral terminus of the excurrent aperture to its dorsal terminus, measured in a straight line.

excyst – the process of a juvenile unionid escaping from its cyst on a glochidial host.

extant – not extinct or regionally extirpated.

exterior of mantle – the external portion of the mantle lying next to the interior surface of the shell.

extinct – taxa which have been eliminated from their entire ranges, with no living representatives.

extirpated – eliminated from a particular area, but populations are extant elsewhere.

extrapallial swelling – the posterioventral swelling of females of the genus *Epioblasma*, which is filled with a spongy pad, possibly used to attract glochidial hosts; erroneously referred to as marsupial swelling.

fabelliform – bean-shaped.

facultative parasite – an organism with the capacity to undergo direct development or pass through a parasitic larval stage on a host.

falcate, falciform – curved or sickle-shaped.

Fall Line – the boundary between the Coastal Plain and upland physiographic provinces along the Atlantic and Gulf of Mexico coasts of eastern North America.

family – a group of related taxa ranking above genus and below order.

female hermaphrodite – see **hermaphrodite**.

flange – the curved, inward projecting, ventral margin of some glochidia (e.g., *Potamilus*).

foot – the large, muscular organ ventrally located in bivalves, used for locomotion and anchorage in the substrate.

fluvial, fluviatile – riverine; of or pertaining to a river or stream; lotic.

gamete – a mature germ cell (ovum or spermatozoan).

gametogenesis – production of gametes.

ganglion (ganglia, pl.) – a mass of nerve tissue.

gape – the space between two valves where they do not meet when closed.

gastric shield – the cuticular plate on the posterior portion of the stomach wall, against which the crystalline style rotates during digestion.

genetic analysis – comparison of hereditary material among individuals and populations in an effort to determine phylogenetic relationships.

genus – a group of taxa with common characteristics, ranking above species and below family.

gill – an organ used to obtain oxygen from water; in bivalves, they occur as paired, membranous organs situated between the visceral mass and mantle on each side of the body.

gill height – the distance between dorsal and ventral margins of a gill, measured approximately centrally, perpendicular to the dorsal margin.

gill length – the distance between anterior and posterior margins of a gill, measured along the dorsal margin.

gill septa – the vertical dividing membranes within the gill, separating water tubes, situated between the inner and outer lamellae of a gill.

glochidial hook – an acute projection located on the ventral terminus or lateral edges of valves of some glochidia to aid host attachment.

glochidial host – an organism on which transformation of a glochidium to the juvenile stage is possible, typically a fish.

glochidial infestation – the presence of glochidia on, or encysted in, tissues of a host; sometimes referred to as glochidial infection.

glochidium (glochidia, pl.) – the bivalved larval stage of Unionidae, Margaritiferidae or Hyriidae (see Wächtler et al., 2001).

glochidium height – the distance between dorsal and ventral margins of a glochidium shell, measured approximately centrally, perpendicular to the dorsal margin.

glochidium length – the distance between anterior and posterior margins of a glochidium shell, measured parallel to the dorsal margin.

glycogen – a polysaccharide that is the primary carbohydrate storage in animals.

gonad – the organ in which gametes are formed.

gonochorism – the condition of having separate sexes; dioecious.

gravid – containing unborn young; with regard to unioniforms, brooding embryos or glochidia in marsupia (charged gills).

gravid period – interval during which a unioniform broods eggs, embryos and glochidia; also called brooding period.

growth rest – see **annulus**.

habitat fragmentation – division of suitable habitat into smaller portions by destruction or impairment of habitat in intervening areas.

haustorium (haustoria, pl.) – the parasitic larval stage of the Iridinidae, consisting of an elongate body with a very long larval thread (Fryer, 1959, 1961); subsequently used as the second stage of development of the lasidium in the Iridinidae (Wächtler et al., 2001).

headcut – stream bed and bank destabilization that proceeds upstream from a point of disturbance such as sediment removal or an increase in stream gradient due to channelization or canalization.

headwaters – streams comprising the source of a river.

hemolymph – the fluid in tissues, circulatory vessels and hemocoels of mollusks; blood.

hermaphrodite – an individual capable of producing both eggs and sperm. **Female hermaphrodite** – a hermaphroditic individual that primarily produces eggs. **Male hermaphrodite** – a hermaphroditic individual that primarily produces sperm.

Sequential hermaphrodite – an individual that changes sex (male to female or female to male) over a period of time. **Simultaneous hermaphrodite** – an individual that produces both eggs and sperm at the same time.

high conservation concern – as defined in Mirarchi et al. (2004) for taxa in Alabama that are imperiled because of three of four of the following: rarity; very limited, disjunct, or peripheral distribution; decreasing population trend/population viability problems; specialized habitat needs/habitat vulnerability due to natural/human-caused factors.

highest conservation concern – as defined in Mirarchi et al. (2004) for taxa in Alabama that are critically imperiled and at risk of extinction/extirpation because of extreme rarity, restricted distribution, decreasing population trend/population viability problems and specialized habitat needs/habitat vulnerability due to natural/human-caused factors.

hinge ligament – the elongate, elastic, proteinaceous structure that unites the dorsal margins of opposing valves, which acts to open valves when adductor muscles are relaxed.

hinge line – a hypothetical line passing through the hinge area from the anterior end of the pseudocardinal teeth to the posterior end of the lateral teeth, or the homologous area in shells with teeth reduced or absent.

hinge plate – the dorsal portion of the interior of a unionoid bivalve shell, including pseudocardinal and lateral teeth and interdentum if present.

historical – relating to the period of recorded history, following arrival of Europeans to North America.

Holarctic – a biogeographical province encompassing the northern hemisphere.

holotype – the single specimen designated by the author as the type of a species or subspecies in the original description.

homologous – similarity attributed to common origin.

homonym – species, genus or family names having the same spelling, differing only in suffix and denoting different taxa. For species, two or more available names having the same spelling and established for different nominal taxa, either originally (primary homonyms) or subsequently (secondary homonyms) combined with the same generic name.

hypolimnion – the cold layer below the thermocline in a thermally stratified body of water.

hypostracum – a layer of shell deposited only beneath muscle attachments.

hypoxia – deficiency in the amount of oxygen reaching tissues of an organism.

Hyriidae, hyriid – a family of Unioniformes with 17 genera; the larval form is a glochidium but may not be homologous to the unionoid glochidium (Wächtler et al., 2001); restricted to South America, Australia, New Guinea and New Zealand.

imperforate – without holes (perforations).

impoundment – a body of water created by damming a stream; reservoir.

incertae sedis – uncertain seat; denoting uncertain taxonomic position or placement of a group or category.

incurrent aperture – the posterior external opening between the mantle of the two valves of a unionoid, positioned ventral to the excurrent aperture, through which water is drawn into the mantle cavity; sometimes referred to as an inhalant aperture; erroneously referred to as a siphon.

incurrent aperture length – the distance from the ventral terminus of the incurrent aperture to its dorsal terminus, measured in a straight line.

inflated – swollen; expanded.

inflation – a measure of shell width in relation to height.

interdental projection – a small protuberance on the interdentum.

interdentum – the area between the posterior end of the pseudocardinal teeth and the anterior end of the lateral teeth.

interlamellar gill connection – see **gill septa**.

interrupted – not continuous; broken.

interstitial – of or relating to small spaces among substrate particles.

invagination – an outer surface folded inward to become the inner surface.

Iridinidae, iridinid – a family of Unioniformes with six genera; the larval stage is the haustorium (Wächtler et al., 2001); restricted to the Nile Basin and sub-Saharan Africa. Formerly recognized as Mutelidae.

junior synonym – see **synonym**.

juvenile – a sexually immature individual.

knob – a rounded protuberance; with regard to shell sculpture, usually refers to a large protuberance, as opposed to smaller tubercles or pustules.

labial palp distal bifurcation – the unfused portion of the dorsal margin of a pair of palps.

labial palp height – the distance from the dorsal margin to the ventral margin of the palp, measured at the dorsal end of the palp connection to the visceral mass.

labial palp length – the distance from the distal terminus of the palp to the midpoint of its basal connection, measured in a straight line.

labial palps – small, paired organs located near the anterioventral end of the gills which serve to separate food from nonfood particles and transfer them to the oral groove.

lachrymose – in the shape of a teardrop; broadly rounded on one end, tapered on the other.

lamella (lamellae, pl.) – a membranous structure, such as one of the two gill layers.

lamellar, lamellate – thin; membranous.

Lampsilinae, lampsiline – a subfamily of the Unionidae recognized by some authors (or used as a tribe by others, Lampsilini); typically characterized by well-developed hinge dentition, a mantle modification anterior to the incurrent aperture, marsupia restricted to the outer gills and extended below ventral gill margin when gravid, glochidia without styliform hooks. Type genus *Lampsilis*.

lanceolate – narrow; tapering to a point at one or both ends.

lapsus calami – an unintentional error; a slip of the pen or printer error.

larval thread – an elongate strand produced between the valves of some glochidia; not homologous to byssal threads.

lasidium (lasidia, pl.) – the larval stage of Mycetopodidae and Iridinidae with an elongate body and a very long larval thread. The larva of Iridinidae, sometimes referred to as a haustorium, has a different developmental history (see Wächtler et al., 2001).

lateral – of or relating to the side, as opposed to medial.

lateral teeth – in unioniform bivalves the linear, raised, interlocking processes positioned along the hinge line of a valve, located posterior to the pseudocardinal teeth and interdentum, which function to stabilize opposing valves.

lectotype – a specimen designated from a syntypic series, upon which a species is based, subsequent to the original description.

lentic – relating to still waters (e.g., lakes and reservoirs).

ligament – the chitinous structure on the dorsal edge of bivalves, in unionids posterior to the umbo, which functions to open the valves when the adductor muscles are relaxed.

ligamental notch – an indentation in the dorsal margin of the shell at the posterior end of the ligament.

ligulate – dorsoventrally elongate, truncate on one end (dorsally in glochidia) and round on the other end (ventrally in glochidia); shaped like a tongue or strap.

long-term brooder – a species of unionoid bivalve which broods glochidia for much of the year, typically from late summer or autumn to the following summer; bradytictic.

lotic – relating to flowing waters (e.g., rivers and creeks).

low conservation concern – as defined in Mirarchi et al. (2004) for taxa in Alabama that are secure, yet conservation concerns exist because of one of the following: relative abundance; limited, disjunct, or peripheral distribution; decreasing population trend/population viability problems; specialized habitat needs/increasing habitat vulnerability due to natural/human-caused factors.

lowest conservation concern – as defined in Mirarchi et al. (2004) for taxa in Alabama that are demonstrably secure, with size of populations stable/increasing, geographical distribution stable/expanding, population trend/population viability stable/increasing and relatively limited habitat vulnerability due to natural/human-caused factors.

lumen (lumina, pl.) – a hollow cavity or passageway.

lunule – the falcate or semicircular impression just anterior to the umbo, one-half being on each valve.

maculate, maculation – spotted.

malacology – branch of zoology which deals with the study of mollusks.

male hermaphrodite – see **hermaphrodite**.

mantle – the membranous organ attached to each valve that envelops the soft tissues of a bivalve, responsible for shell and periostracum development.

mantle attachment scars – a series of small pits near the center of the inside of the valve, characteristic of the Margaritiferidae, that are sites of mantle attachment to the shell; sometimes erroneously referred to as mantle muscle scars.

mantle bridge – the connection between mantle tissues of right and left sides of unionoid bivalves, located posteriodorsally and separating excurrent and supra-anal apertures, variable in length, absent in some species.

mantle flap – the specialized mantle modification of *Lampsilis* and some *Hamiota*, located just ventral to the incurrent aperture, in the form of a variable expansion of tissue, typically mimics prey items of potential glochidial hosts, thought to serve as lures.

mantle fold – the specialized modification of the mantle margin of some species, principally some long-term brooders (e.g., *Medionidus*, *Ligumia* and *Villosa*), generally not as deep or well-developed as the mantle flap of *Lampsilis* and some *Hamiota*, thought to serve as lures for glochidial hosts.

mantle margin – generally the outer edge of the mantle which actively secretes the shell and periostracum.

mantle muscle scars – see **mantle attachment scars**.

Margaritiferidae, margaritiferid – a family of Unioniformes with three genera; the larval stage is a glochidium (Wächtler et al., 2001); Holarctic in distribution with a single representative in Morocco, northwest Africa.

marginal glochidial host – a species on which glochidia are capable of transforming to the juvenile stage, but the success rate is much lower relative to species identified as primary hosts; sometimes referred to as secondary glochidial host.

marsupial swelling – the posterior inflated area of the shells of long-term brooding females; the term is often incorrectly applied to the posterioventral shell expansion in species of *Epioblasma*, which accommodates mantle modifications instead of marsupia.

marsupium – the portion of bivalve gill in which embryos and glochidia are held.

medial – of or relating to the middle, as opposed to lateral.

mesocosm – experimental enclosure or habitat designed to approximate natural conditions in which environmental factors can be regulated.

microlure – a very small mantle projection located just ventral to the incurrent aperture, serving as a lure in some species of *Epioblasma*, displayed against the mantle pad located within the extrapallial swelling.

micropoint – a small projection on the ventral margin of some glochidia.

microstylet – a very small styliform projection on the ventral margin or the hook of some glochidia.

microvillus (microvilli, pl.) – a finger-like extension on the surface of a cell, which often functions to increase surface area and enhance absorptive capacity.

midden – a pile of debris left by predators or humans often comprised of mollusk shells; refuse heap.

mitochondrion (mitochondria, pl.) – a membrane-bound organelle in which adenosine triphosphate (ATP) is produced by aerobic respiration of glucose, providing energy for cellular activities.

mitotype – a distinct DNA sequence that differs from other sequences by one or more nucleotide positions.

moderate conservation concern – as defined in Mirarchi et al. (2004) for taxa in Alabama with conservation problems because of insufficient data or because of two of four of the following: small populations; limited, disjunct, or peripheral distribution; decreasing population trend/popu-lation viability problems; specialized habitat needs/habitat vulnerability due to natural/human-caused factors.

monoecious – having male and female sexes in the same individual; hermaphroditic.

monophyletic – a group consisting of a common ancestor and all descendent taxa.

monotypic, monotypy – represented by a single taxon.

mottling – colored spots or blotches.

mucilagenous – being moist and viscid.

multinucleated inclusion – a cluster of DNA surrounded by a thin layer of cytoplasm; also known as a sperm morula.

muscle scar – the site of attachment for one of various muscles on the inside of a valve.

mushroom body – cells comprising the mantle of encysted glochidia, enlarged and projected into the mantle cavity, involved in digestion of larval adductor muscle and tissues of the host trapped between the glochidial valves.

Mycetopodidae, mycetopodid – a family of Unioniformes with 11 genera; the larval stage is a lasidium (Wächtler et al., 2001); restricted to South America and Central America as far north as central west Mexico.

nacre, nacreous layer – the inner layer of unioniform bivalve shells, composed of calcium carbonate (aragonite) deposited in an organic matrix.

naiad, naiade – a former rank in the classification of mollusks, approximately equal to unionoids.

natural infestation – the presence of glochidia on hosts observed in nature, which is often used to suggest which species serve as hosts. Natural infestation of a particular species does not necessarily indicate that successful transformation will occur.

neotype – the type of a species or subspecies designated subsequent to its original description to replace the original type if it is lost, destroyed or deemed inadequate.

nephridium (nephridia, pl.) – one of the tubules of the excretory organ (kidney).

nodule, nodulous – a small, irregular rounded mass on the shell surface.

nomen conservatum – a name that is retained by authorization of the ICZN, though the name strictly contravenes one or more provisions of the code.

nomen dubium (*nomina dubia*, pl.) – a name of unknown or doubtful application.

nomen novum (*nomina nova*, pl.) – a new replacement name.

nomen nudum (*nomina nuda*, pl.) – a nude name, one that was published without description or figure and is unavailable as published.

nomen oblitum (nomina oblita, pl.) – a forgotten name, a senior synonym or homonym that does not take precedence over a younger synonym or homonym in prevailing use.

nomen protectum (nomina protecta, pl.) – a protected name which has been given precedence over its unused senior synonym or senior homonym that has been relegated to status of *nomen oblitum*.

nominal – a taxon which is denoted by an available name, and in the family, genus and species group is based on a name-bearing type.

nominate – formerly used for nominotypical taxa.

nominotypical – the nominal taxon at a subordinate rank with the family, genus or species group that contains the name-bearing type of a subdivided taxon of that group (e.g., *Epioblasma torulosa torulosa*).

nonindigenous – a species not native to an area; introduced.

obligate parasite – an organism that requires a parasitic stage.

oblique – slanting; neither perpendicular nor parallel.

obscure – faint; lacking prominence; indistinct.

Ohioan fauna – the assemblage of freshwater mussels which has its presumed origin in the region encompassing the Ohio River drainage, exclusive of the Tennessee and Cumberland River systems.

oocyte – a developing egg; ovum.

oogenesis – the formation and maturation of female gametes (ova).

oogonium (oogonia, pl.) – a primordial female germ cell that gives rise to an oocyte.

oral groove – the ciliated channel extending from the anterior end of the labial palps to the mouth.

osmoregulation – control of osmotic pressure within cells of an organism.

ostium (ostia, pl.) – a small opening.

overbank – the area of permanently inundated floodplain created by impoundment of a stream.

oval, ovate – oblong and curvilinear; broadly elliptical.

ovoviviparous – producing eggs which are retained within the body of the female until a larval or juvenile stage is attained.

ovum (ova, pl.) – a female gamete; egg.

padded – slightly thickened or inflated.

pallial line – a thin, indented line consisting of a row of small pallial muscle scars, approximately concentric to the ventral margin of a valve, between the anterior and posterior adductor muscle scars.

pallial muscles – a narrow row of small muscles approximately concentric to the ventral margin of the mantle, which serve to attach the mantle to its respective valve and retract the mantle from the edge of the shell.

pallial sinus – an indentation in the posterior portion of the pallial line marking the space into which the siphons are drawn upon closure of the valves; not present in unionids or margaritiferids.

papilla (papillae, pl.) – a small projection, typically finger-like in form; in unioniforms, present along margins of the incurrent aperture and sometimes the excurrent aperture, also found on modified mantle margins anterior to incurrent aperture in some species.

papillate – having papillae.

paralectotype – each specimen of a former syntype series remaining after the designation of a lectotype.

paraphyletic – a group consisting of a common ancestor and some but not all of its descendant taxa.

parasite – an organism that lives on or in another species, from which it obtains nutriment, usually to the detriment of the host.

paratype – a specimen of the original type series, other than the holotype, designated in the original description of the taxon.

patronym – a name of a taxon derived from the name of a person, used in honor of that person.

pearl – a mineralized deposit within soft tissues of mollusks in the form of a concretion of concentric layers of nacre; usually an abnormal growth around a nucleus of extraneous material.

pearl nucleus – extraneous material around which concentric layers of nacre are laid to form a pearl.

pedal feeding – ingestion of food particles accumulated during sweeping motions of the ciliated foot of juvenile mussels.

pedal protractor muscle – small muscle used to extend the foot, located just posterioventral to the anterior adductor muscle.

pedal retractor muscle – one of two small muscles located near the anterior and posterior adductor muscles, which functions to withdraw the foot into the shell.

Pelecypoda – the former name used for the class Bivalvia.

perforation – a small hole or slit.

pericardium – the sac surrounding the heart and lower portion of the intestine.

periostracum – the thin, uncalcified, proteinaceous outer layer of bivalve shells produced by the mantle margin, which serves to protect underlying layers from erosion and dissolution.

phylogeny – the evolutionary history of a group of organisms.

plication – one of a series of parallel ridges; herein used to refer to relatively large ridges, as opposed to smaller corrugations.

polyphyletic – a group in which the most recent common ancestor is assigned to some other group and not the group itself.

posteriodorsal – of or relating to the upper part of the posterior end.

posterior – at or near the back or rear portion; in bivalves tends to be distant from the umbo.

posterior ridge – the radially oriented crest on the exterior of a bivalve shell, extending from the umbo to the posterioventral margin, variously developed among species ranging from angular to rounded.

posterior shell margin – herein refers to the portion of shell lying behind an imaginary vertical line extending along the anterior edge of the posterior adductor muscle scar, and perpendicular to an imaginary horizontal line extending along the dorsal margins of the anterior and posterior adductor muscle scars.

posterior slope – area of the external surface of a valve between the posterior ridge and the posteriodorsal shell margin.

posterioventral – of or relating to the lower part of the posterior end.

prehistoric – relating to times antedating recorded history, prior to arrival of Europeans to North America.

preoccupied – a term used in zoological nomenclature for the scientific name of a taxon that has been used earlier as the name of a different taxon.

primary glochidial host – a species on which glochidia are capable of transforming to the juvenile stage with optimal success rate.

primary synonymy – a list of all scientific names applied to a given species subsequent to its description, but not all of the subsequent generic combinations.

prismatic layer – a layer in freshwater mussel shells comprised of minute calcium carbonate prisms, oriented perpendicular to the shell surface.

produced – disproportionately elongate.

protandry – assumption of a functional male state by an individual before reversal to a functional female state.

proteinaceous – comprised primarily or entirely of proteins.

protractor – a muscle that extends a body part.

proximal – toward the point of attachment or origin, as opposed to distal.

pseudocardinal teeth – in unioniform bivalves the raised, interlocking processes positioned near the anterior end of the hinge line of a valve, typically triangular or blade-like, located anterior to the lateral teeth and interdentum, which serve to stabilize opposing valves.

pseudofeces – in suspension or deposit feeding bivalves, nonfood particles separated from food, bound in mucus and rejected ventrally from the gills.

pseudotaxodont – dentition in the form of numerous irregular short teeth transverse to the hinge; found in the Iridinidae.

pustule – sculpture in the form of a small prominence on the surface of a shell, may be round or oblong.

pyriform – broad on one end and narrow on the other; pear-shaped.

quadrate – having four sides, approximately square or rectangular.

radial – proceeding out from a central point; from the umbo in bivalves.

range – the geographic area in which an organism occurs.

ray – a radial line of pigment on the periostracum, originating on the umbo, usually some shade of green, may be interrupted or continuous.

relative abundance – the number of individuals of a particular taxon present in a given area compared to numbers of other species.

reniform – kidney-shaped.

reservoir – an artificial lake created by impoundment of a stream.

restriction of type locality – any modification of the type locality subsequent to its original publication, usually a clarification or selection of a single location as the type locality when multiple localities are involved.

reticulated – an interconnected pattern of lines or ridges.

retractor – a muscle that withdraws a body part.

rhomboid, rhomboidal – an outline in which opposite sides and angles are equal, but adjacent sides and angles are unequal.

riparian – pertaining to or situated on the bank of a stream or other water body.

riverine – found in or characteristic of rivers.

rotund – round or nearly round in outline.

rounded – having an evenly curved contour.

rudimentary – underdeveloped; poorly developed.

satiny – having the soft, lustrous texture of satin; used to describe the texture of periostracum on some species of unionoid bivalves.

sculpture – impressed or raised structures on the surface of a shell.

secondary glochidial host – a species on which glochidia are capable of transforming to the juvenile stage, but the success rate is much lower relative to species identified as primary hosts; sometimes referred to as marginal host.

secondary water tubes – small tubes formed by accessory partitions in the gravid marsupial portion of anodontine gills, present just before and persisting until just after the gravid period.

sediment – geological material deposited by water, wind or glaciers.

senior synonym – see **synonym**.

septum – a dividing wall or membrane.

sequential hermaphrodite – see **hermaphrodite**.

serrated – notched or toothed on the edge.

sexually dimorphic – males and females of the same species differing in shell morphology (e.g., *Epioblasma* and *Lampsilis*).

shear stress – force exerted by flowing water on a stream bed or bank.

shell height – the distance between dorsal and ventral margins, measured near the midpoint of the hinge ligament, perpendicular to the shell length.

shell length – the distance between the anterior and posterior margins, measured parallel to the hinge ligament.

shell morphology – form and structure of a shell.

shell width – the maximum distance between the outer surfaces of the paired valves.

short-term brooder – a species of unionoid bivalve which broods glochidia from late winter or spring to summer; tachytictic.

simple papillae – papillae that are not divided or branched distally.

simultaneous hermaphrodite – see **hermaphrodite**.

sinuous, sinuate – having a serpentine or wavy form; broadly undulating.

siphon – a tubular organ formed by the fusion of mantle margins that aids in the intake or discharge of fluid; apertures of unionids and margaritiferids are often erroneously referred to as siphons.

slack water – water with little or no current.

spatulate – oblong with an attenuated base opposite a rounded end.

spawning – discharge of gametes in aquatic organisms.

special concern – conservation term used to describe species which are currently not imperiled, but are thought to be declining in parts of their ranges, requiring diligent monitoring of their status.

species – a group of interbreeding natural populations that are reproductively isolated from all other such groups.

sperm morula (sperm morulae, pl.) – a cluster of DNA surrounded by a thin layer of cytoplasm; also known as a multinucleated inclusion.

spermatid – a cell derived from division of a spermatocyte, which differentiates into a spermatozoan.

spermatocyte – a cell derived from a spermatogonium that ultimately produces four spermatids.

spermatogenesis – formation and maturation of male gametes (spermatozoa).

spermatogonium (spermatogonia, pl.) – a primordial male germ cell that gives rise to a primary spermatocyte.

spermatozeugmata – unencapsulated sperm aggregates held together by an extracellular matrix, as opposed to spermatophores, which are encapsulated.

spermatozoan (spermatozoa, pl.) – a male gamete, usually motile.

statocyst – a sensory organ of equilibrium.

striate, striae – fine, raised lines.

styliform – having parallel sides and a pointed apex.

sub – a prefix meaning almost or nearly.

subadult – a sexually immature individual that approaches the size and shape of an adult.

subspecies – a morphologically distinguishable and geographically isolated group of organisms capable of successfully interbreeding with other subspecies of the same species.

substrate – the surface on which an organism lives; material at the bottom of a body of water.

sulcus – a groove, furrow or channel; typically radially oriented in bivalves.

superconglutinate – the packet of glochidia containing the entire annual reproductive output of an individual, tethered to the excurrent aperture of the producing female by a hollow mucus tube, which mimics food of glochidial hosts; known to occur only in *Hamiota*.

supernumerary hook – one of the small, straight, sharply pointed projections sparsely distributed on the ventral margin of *Epioblasma* glochidia.

supra-anal aperture – the posteriodorsal opening just dorsal to the excurrent aperture.

supra-anal aperture length – the distance from the ventral terminus of the supra-anal aperture to its dorsal terminus, measured in a straight line.

suprabranchial chamber – the narrow, elongate cavity dorsal to the connection of the gills; sometimes referred to as the epibranchial cavity.

sympatric – distributional relationship in which two or more species occur in the same area (see **allopatric**).

synonym – one of two or more names for the same organism. **Junior synonym** – a name for an organism published after the recognized name. **Senior synonym** – a name for an organism published before the currently recognized name.

synonymy – a list of synonyms for a given taxon.

syntype – one of a series of specimens of the same taxon, with equal rank, upon which a species was described with no designation of a holotype by the author in the original publication.

system – a stream or river and all of its tributaries; herein refers primarily to tributaries of drainages (e.g., Paint Rock River system within the Tennessee River drainage and Sipsey River system within the Tombigbee River drainage).

tachytictic – having a brooding period that lasts only a few months, typically from late winter or spring to summer; short-term brooder.

tautonymy – see **absolute tautonymy**.

tawny – yellowish brown.

taxon – the name applied to any group of organisms in a formal system of nomenclature.

taxonomy – study of the classification of organisms based on natural relationships.

tetragenous – the condition in which all four gills are used as marsupia.

thermocline – the middle layer in a thermally stratified water body, characterized by an abrupt decline in temperature.

threatened – likely to become endangered in all or a significant portion of its range during the foreseeable future; an official designation under the federal Endangered Species Act of 1973.

topotype – a specimen collected from the type locality subsequent to the taxon's original description.

transformation – the process of changing structure; herein refers to the transition of a glochidium to a free-living juvenile.

trapezoid, trapezoidal – an outline with four distinct sides, with only two sides parallel.

tributary – a stream feeding a larger stream or lake.

trifid – divided distally into three approximately equal parts.

trinomen – a polynomial consisting of genus, species and subspecies.

tripartite – divided into three parts by membranes, as in tripartite water tubes of anodontine gills.

truncate – ending abruptly, more or less squarely.

tubercle – sculpture in the form of a small prominence on the surface of a shell, may be round or oblong.

type – a specimen upon which a taxon is described.

type locality – the geographical location of the collection of the name-bearing type of a species or subspecies.

type species – the species designated to characterize a genus or subgenus.

typhlosole – a longitudinal fold on the floor of the stomach in bivalves that assists in sorting food particles; also a longitudinal fold in the ventral wall of the intestine which functions to increase surface area.

umbo – the raised, rounded, often inflated portion of a bivalve shell, located on the dorsal margin near the anterior end of the hinge ligament; the oldest part of a bivalve shell.

umbo cavity – the depression located inside each valve, beneath the umbo.

umbo sculpture – the diminutive, natural, raised markings found on umbos of shells, generally in the form of small ridges or corrugations.

undulate – having a wavy appearance or outline.

Unionacea, unionacean – an inappropriate term formerly used for Unionoidea.

Unionidae, unionid – a family of Unioniformes with about 120 genera; the larval stage is a glochidium (Wächtler et al., 2001); found in North America, Europe, Africa, northwest Madagascar and Asia.

Unioniformes, unioniform – the order comprised of 6 freshwater families: Etheriidae, Hyriidae, Iridinidae, Margaritiferidae, Mycetopodidae and Unionidae; all with an obligate parasitic larval stage; the order contains about 180 genera and approximately 800 species and occur on 6 continents; formerly recognized as Unionoida.

Unioninae, unionine – a subfamily of the Unionidae recognized by some authors; typically characterized by well-developed hinge dentition, no mantle modification anterior to the incurrent aperture, marsupia formed by all four or just the outer gills, marsupia are not distended when gravid and have a sharp ventral edge, glochidia with styliform hooks. Type genus *Unio*.

Unionoida – see **Unioniformes**.

Unionoidea, unionoid – the superfamily containing Margaritiferidae and Unionidae.

valve – one of the opposing shells of a bivalved mollusk.

veliger – a pelagic larval mollusk with a ciliated swimming membrane.

ventral – relating to or situated near the bottom or underside.

ventral shell margin – herein refers to the lower shell periphery between imaginary vertical lines extending along the proximal margins of the anterior and posterior adductor muscle scars.

viscera, visceral mass – the collective assemblage of internal organs, including the digestive system, kidney and reproductive tissues, which are held within a thin sheath of muscle.

vitelline membrane – the thin casing that encloses an egg (ovum).

water tube – one of a number of canals in the intralamellar space of a mussel gill, extending from the dorsal to the ventral margin.

wrinkle – a small furrow or ridge on a shell surface; herein refers to sculpture that is irregularly arranged.

Literature Cited

Abbott, R.T. 1974. American Seashells. The Marine Mollusca of the Atlantic and Pacific Coasts of North America. Second edition. Van Nostrand Reinhold Company, New York. 663 pages.

Ackerman, E.A. 1950. Ten Rivers in America's Future. The report of the Presidents Water Resources Policy Commission, Volume 2. U.S. Government Printing Office, Washington, District of Columbia. 801 pages.

Adams, W.F., J.M. Alderman, R.G. Biggins, A.G. Gerberich, E.P. Keferl, H.J. Porter and A.S. Van Devender. 1990. A report on the conservation status of North Carolina's freshwater and terrestrial molluscan fauna. North Carolina Wildlife Resources Commission, Raleigh, North Carolina. 246 pages, Appendix A, 37 pages.

Agassiz, L. 1842–1846. Nomenclator zoologicus, continenes nomina systematica generum animalium, tam viventium quam fossilium. Jent and Gassman, Solothurn, Switzerland. Two sections, 98 pages and 393 pages.

Agassiz, L. 1852. Ueber die Gattungen unter den nordamerikanischen Najaden. Archiv für Naturgeschichte 18(1):41–52.

Ahlstedt, S.A. 1983. The molluscan fauna of the Elk River in Tennessee and Alabama. American Malacological Bulletin 1:43–50.

Ahlstedt, S.A. 1992a. Twentieth century changes in the freshwater mussel fauna of the Clinch River (Tennessee and Virginia). Walkerana (for 1991) 5(13):73–122.

Ahlstedt, S.A. 1992b. Cumberlandian Mollusk Conservation Program. Activity 1: mussel surveys in six Tennessee Valley streams. Walkerana (for 1991) 5(13):123–160.

Ahlstedt, S.A. 1998. Status survey for federally listed endangered freshwater mussel species in the Paint Rock River system, northeastern Alabama, U.S.A. Walkerana (for 1995–1996) 8(19):63–80.

Ahlstedt, S.A., S. Bakaletz, M.T. Fagg, D. Hubbs, M.W. Treece and R.S. Butler. 2005. Current status of freshwater mussels (Bivalvia: Unionidae) in the Big South Fork National River and Recreation Area of the Cumberland River and recreation area of the Cumberland River, Tennessee and Kentucky. Evidence of faunal recovery. Walkerana (for 2003–2004) 14(31):33–77.

Ahlstedt, S.A., and J.J. Jenkinson. 1993. Distribution and abundance of *Potamilus capax* and other freshwater mussels in the St. Francis River system, Arkansas and Missouri, U.S.A. Walkerana (for 1991) 5(14):225–261.

Ahlstedt, S.A., and T.A. McDonough. 1993. Quantitative evaluation of commercial mussel populations in the Tennessee River portion of Wheeler Reservoir, Alabama. Pages 38–49. In: K.S. Cummings, A.C. Buchanan and L.M. Koch (editors). Conservation and Management of Freshwater Mussels. Proceedings of a UMRCC Symposium, 12–14 October 1992, St. Louis, Missouri. Upper Mississippi River Conservation Committee, Rock Island, Illinois.

Ahlstedt, S.A., and C. Saylor. 1998. Status survey of the Little-wing Pearly Mussel, *Pegias fabula* (Lea, 1838). Walkerana (for 1995–1996) 8(19):81–105.

Allen, E. 1924. The existence of a short reproductive cycle in *Anodonta imbecillis*. Biological Bulletin 46:88–94.

Allen, W.R. 1914. The food and feeding habits of freshwater mussels. Biological Bulletin 27(2):127–147.

Allen, W.R. 1921. Studies of the biology of freshwater mussels. Experimental studies of the food relations of certain Unionidae. Biological Bulletin 40(4):21–241.

Allen, W.R. 1922. Studies of the biology of freshwater mussels. III. Distribution and movements of Winona Lake mussels. Proceedings of the Indiana Academy of Science 1922:227–238.

Alyakrinskaya, I.O. 2001. The dimensions, characteristics and functions of the crystalline style of molluscs. Biology Bulletin 28(5):523–535. [Translated from Izvestiya Akademii Nauk, Seriya Biologicheskaya 5:613–627.]

Amyot, J.-P., and J.A. Downing. 1997. Seasonal variation in vertical and horizontal movement of the freshwater bivalve *Elliptio complanata* (Mollusca: Unionidae). Freshwater Biology 37(2):345–354.

Amyot, J.-P., and J.A. Downing. 1998. Locomotion in *Elliptio complanata* (Mollusca: Unionidae): a reproductive function? Freshwater Biology 39(2):351–358.

Andrusov, N.I. 1897. Fossil and living Dreissenidae of Eurasia. Trudy Sanktpeterburgskogo Obshchestva Estestvoispytateley. Otdel Geologii i Mineralogii [Proceedings of the Saint Petersburg Society of Naturalists, Department of Geology and Mineralogy] 25:285–286.

Anthony, J.G. 1841. On the byssus of *Unio*, with notes by J.E. Gray, Esq. Annals of Natural History 6:77.

Anthony, J.G. 1865. Descriptions of new species of North American Unionidae. American Journal of Conchology 1(2):155–164, 5 plates.

Anthony, J.G. 1866. Descriptions of new American fresh-water shells. American Journal of Conchology 2(2):144–147, plates 6–7.

Anthony, J.L., and J.A. Downing. 2001. Exploitation trajectory of a declining fauna: a century of freshwater mussel fisheries in North America. Canadian Journal of Fisheries and Aquatic Sciences 58(10):2071–2090.

Apgar, A.C. 1887. The musk rat and the *Unio*. The Zoologist (Series 3) 11:425–426.

Appleton, C.C. 1996. Freshwater Molluscs of Southern Africa. University of Natal Press, Pietermaritzburg, South Africa. 64 pages.

Arey, L.B. 1921. An experimental study on glochidia and the factors underlying encystment. Journal of Experimental Zoology 33(2):463–499, plates 1–3.

Arey, L.B. 1924. Glochidial cuticulae, teeth and the mechanics of attachment. Journal of Morphology and Physiology 39(2):332, 1 plate.

Arey, L.B. 1932a. A microscopical study of glochidial immunity. Journal of Morphology 53(2):367–379.

Arey, L.B. 1932b. The formation and structure of the glochidial cyst. Biological Bulletin 62(2):212–221.

Arey, L.B. 1932c. The nutrition of glochidia during metamorphosis. Journal of Morphology 53(1):201–221.

Arnold, W.H. 1965. A glossary of a thousand-and-one terms used in conchology. The Veliger. Volume 7 Supplement. 50 pages.

Athearn, H.D. 1964. Three new unionids from Alabama and Florida and a note on *Lampsilis jonesi*. The Nautilus 77(4):134–139.

Athearn, H.D. 1970. Discussion of Dr. Heard's paper. Symposium on Endangered Mollusks. Malacologia 10(1):28–31.

Bába, K. 2000. An area-analytical zoogeographical classification of Palearctic Unionaceae species. Bollettino Malacologico, Roma 36(5–8):133–140.

Backlund, D.C. 2000. Summary of current known distribution and status of freshwater mussels (Unionoida) in South Dakota. Central Plains Archeology 8(1):69–77.

Bailey, R.M., and F.B. Cross. 1954. River sturgeons of the American genus *Scaphirhynchus*: characters, distribution, and synonymy. Papers of the Michigan Academy of Science, Arts, and Letters 39:169–208.

Baird, M.S. 2000. Life history of the Spectaclecase, *Cumberlandia monodonta*, (Say 1829) (Bivalvia, Unionoidea, Margaritiferidae). Masters thesis, Southwest Missouri State University, Springfield. 108 pages.

Baker, F.C. 1898a. The molluscan fauna of western New York. Transactions of the Academy of Science of St. Louis 8(5):71–94, plate 10.

Baker, F.C. 1898b. The Mollusca of the Chicago area, Part I: The Pelecypoda. Bulletin of the Chicago Academy of Science 3(1):1–130, 27 plates.

Baker, F.C. 1922. New species and varieties of Mollusca from Lake Winnebago, Wisconsin, with new records from this state. The Nautilus 35(4):130–133; 36(1):19–21.

Baker, F.C. 1926. The naiad fauna of the Rock River System: a study of the law of stream distribution. Transactions of the Illinois State Academy of Science 19:103–112.

Baker, F.C. 1927. On the division of the Sphaeriidae into two subfamilies: and the description of a new genus of Unionidae, with descriptions of new varieties. American Midland Naturalist 10(7):220–223.

Baker, F.C. 1928a. The fresh water Mollusca of Wisconsin. Part II. Pelecypoda. Bulletin of the University of Wisconsin, Serial Number 1527. Bulletin of the Wisconsin Geological and Natural History Survey 70(2):1–495.

Baker, F.C. 1928b. The Mollusca of Chautauqua Lake, New York, with descriptions of a new variety of *Ptychobranchus* and of *Helisoma*. The Nautilus 42(2):48–60.

Baker, H.B. 1964a. Some of Rafinesque's unionid names. The Nautilus 77(4):140–142.

Baker, H.B. 1964b. *Dromus* not a homonym. The Nautilus 77(4):142.

Baker, H.B. 1964c. *Carunculina* (Lampsilinae). The Nautilus 78(1):33.

Baker, H.B. 1964d. *Elliptio* feminine. The Nautilus 78(1):33.

Baker, H.B., Y. Kondo and J.B. Burch. 1998. H. Burrington Baker's anatomical nomenclator. Walkerana (for 1995–1996) 8(19):1–30.

Baker, S.M., and D.J. Hornbach. 1997. Acute physiological effects of zebra mussel *(Dreissena polymorpha)* infestation on two unionid mussels, *Actinonaias ligamentina* and *Amblema plicata*. Canadian Journal of Fisheries and Aquatic Science 54(3):512–519.

Baker, S.M., and J.S. Levinton. 2003. Selective feeding by three native North American freshwater mussels implies food competition with zebra mussels. Hydrobiologia 505(1–3):97–105.

Baldwin, C.S. 1973. Changes in the freshwater mussel fauna in the Cahaba River over the past forty years. Masters thesis, Tuskegee Institute, Tuskegee, Alabama. 45 pages.

Ball, G.H. 1922. Variation in fresh-water mussels. Ecology 3(2):93–121, figures 1–6, tables i–iv.

Barfield, M.L., and G.T. Watters. 1998. Non-parasitic life cycle in the Green Floater, *Lasmigona subviridis* (Conrad, 1835). Triannual Unionid Report 16:22.

Barnes, D.W. 1823. On the genera *Unio* and *Alasmodonta*; with introductory remarks. American Journal of Science and Arts 6(1):107–127; 6(2):258–280, 13 plates.

Barnhart, C., F. Riusech and M. Baird. 1998. Hosts of salamander mussel (*Simpsonaias ambigua*) and snuffbox (*Epioblasma triquetra*) from the Meramec River system, Missouri. Triannual Unionid Report 16:34.

Barnhart, C., F.A. Riusech and A.D. Roberts. 1997. Fish hosts of the federally endangered pink mucket, *Lampsilis abrupta*. Triannual Unionid Report 13:35.

Barnhart, C., and A. Roberts. 1996. When clams go fishing. Missouri Conservationist 57(2):22–25.

Barnhart, C., and A. Roberts. 1997. Reproduction and fish hosts of unionids from the Ozark uplift. Pages 16–20. In: K.S. Cummings, A.C. Buchanan, C.A. Mayer and T.J. Naimo (editors). Conservation and Management of Freshwater Mussels II: Initiatives for the Future. Proceedings of a UMRCC Symposium, 16–18 October 1995, St. Louis, Missouri. Upper Mississippi River Conservation Committee, Rock Island, Illinois.

Bates, J.M. 1962. The impact of impoundment on the mussel fauna of Kentucky Reservoir, Tennessee River. The American Midland Naturalist 68(1):232–236.

Bauer, G. 1992. Variation in the life span and size of the freshwater pearl mussel. The Journal of Animal Ecology 61(2):425–436.

Bauer, G. 1994. The adaptive value of offspring size among freshwater mussels (Bivalvia; Unionoidea). The Journal of Animal Ecology 63(4):933–944.

Bauer, G., S. Hochwald and W. Silkenat. 1991. Spatial distribution of freshwater mussels: the role of host fish and metabolic rate. Freshwater Biology 26(3):377–386.

Beams, H.W., and S. Sekhon. 1966. Electron microscope studies on the oocyte of the fresh-water mussel (*Anodonta*), with special reference to the stalk and mechanism of yolk deposition. Journal of Morphology 119(4):477–501.

Beck, K., and R.J. Neves. 2003. An evaluation of selective feeding by three age-groups of the Rainbow Mussel *Villosa iris*. North American Journal of Aquaculture 65(3):203–209.

Beck, W.M., Jr. 1965. The streams of Florida. Bulletin of the Florida State Museum, Biological Sciences 10(3):91–126.

Beninger, P.G., J.E. Ward, B.A. MacDonald and R.J. Thompson. 1992. Gill function and particle transport in *Placopecten magellanicus* (Mollusca: Bivalvia) as revealed using video endoscopy. Marine Biology (Berlin) 114(2):281–288.

Benz, G.W., and D.E. Collins (editors). 1997. Aquatic Fauna in Peril. The Southeastern Perspective. Southeast Aquatic Research Institute. Special Publication 1. Lenz Design and Communications, Decatur, Georgia. 553 pages.

Benz, G.W., and S. Curren. 1997. Results of an ongoing survey of metazoan symbionts of freshwater mussels (Unionidae) from Kentucky Lake, Tennessee. Pages 39–66. In: A.F. Scott, S.W. Hamilton, E.W. Chester and D.S. White (editors). Proceedings of the Seventh Symposium on the Natural History of the Lower Tennessee and Cumberland River Valleys. The Center for Field Biology, Austin Peay State University, Clarksville, Tennessee.

Bequaert, J.C., and W.B. Miller. 1973. The Mollusks of the Arid Southwest, with an Arizona Check List. The University of Arizona Press, Tucson. 271 pages.

Bisbee, G.D. 1984. Ingestion of phytoplankton by two species of freshwater mussels, the Black Sandshell, *Ligumia recta*, and the Three Ridger, *Amblema plicata*, from the Wisconsin River in Oneida County, Wisconsin. Bios 55(4):219–225.

Bishai, H.M., A.M. Ibrahim and M.T. Khalil. 1999. Freshwater Molluscs of Egypt. Egyptian Environmental Affairs Agency, Department of Nature Protection. Publication of National Biodiversity Unit, Number 10. 145 pages.

Blalock-Herod, H.N., J.J. Herod and J.D. Williams. 2002. Evaluation of conservation status, distribution, and reproductive characteristics of an endemic Gulf Coast freshwater mussel, *Lampsilis australis* (Bivalvia: Unionidae). Biodiversity and Conservation 11(10):1877–1887.

Blalock-Herod, H.N., J.J. Herod, J.D. Williams, B.N. Wilson and S.W. McGregor. 2005. A historical and current perspective of the freshwater mussel fauna (Bivalvia: Unionidae) of the Choctawhatchee River drainage in Alabama and Florida. Alabama Museum of Natural History Bulletin 24:1–26.

Blankenship, S. 1971. Notes on *Alasmidonta fabula* (Lea) in Kentucky (Unionidae). The Nautilus 85(2):60–61.

Blystad, C.N. 1923. Significance of larval mantle of freshwater mussels during parasitism, with notes on a new mantle condition exhibited by *Lampsilis luteola*. Bulletin of the U.S. Bureau of Fisheries 39(1923–1924):203–219. [Issued separately as U.S. Bureau of Fisheries Document 950.]

Bogan, A.E. 1980. A comparison of Late Prehistoric Dallas and Overhill Cherokee subsistence in the Little Tennessee River Valley. Doctoral dissertation, University of Tennessee, Knoxville. 209 pages.

Bogan, A.E. 1990. Stability of Recent unionid (Mollusca: Bivalvia) communities over the past 6000 years. Pages 112–136. In: W. Miller (editor). Paleocommunity temporal dynamics: the long-term development of multispecies assemblages. The Paleontological Society. Special Publications, Number 5. Knoxville, Tennessee.

Bogan, A.E. 1992. Anal structures as a new source of anatomical characters in freshwater bivalves (Mollusca: Bivalvia: Unionoida). Unitas Malacologica, Abstracts of the Eleventh International Malacological Congress, Siena, Italy. Pages 14–16.

Bogan, A.E. 1993. Freshwater bivalve extinctions: search for a cause. American Zoologist 33(6):599–609.

Bogan, A.E. 1997. A resolution of the nomenclatural confusion surrounding *Plagiola* Rafinesque, *Epioblasma* Rafinesque, and *Dysnomia* Agassiz (Mollusca: Bivalvia: Unionidae). Malacological Review 30(1):77–86.

Bogan, A.E. 1998. Freshwater molluscan conservation in North America: problems and practices. Pages 223–230. In: I.J. Killeen, M.B. Seddon and A.M. Holmes (editors). Molluscan conservation: a strategy for the 21st Century. Journal of Conchology, Special Publication, Number 2.

Bogan, A.E. 2004. Freshwater bivalves: diversity and distribution of the Unionoida. Journal of the Egyptian German Society of Zoology 44(d):111–120.

Bogan, A.E. 2006. Conservation and extinction of the freshwater molluscan fauna of North America. Pages 373–384. In: C.F. Sturm, T.A. Pearce and A. Valdés (editors). The Mollusks: A Guide to Their Study, Collection, and Preservation. American Malacological Society.

Bogan, A.E. In press. Global diversity of freshwater mussels (Mollusca, Bivalvia) in freshwater. Hydrobiologia.

Bogan, A.E., S.A. Ahlstedt and P.W. Parmalee. 2002. Exotic freshwater bivalves found in the Nolichucky River, East Tennessee. Ellipsaria 4(3):9.

Bogan, A.E., and C.M. Bogan. 2002. The development and evolution of Isaac Lea's publications on the Unionoida. Pages 363–375. In: M. Falkner, K. Groh and M.C.D. Speight (editors). Collectanea Malacologica: Festschrift für Gerhard Falkner. ConchBooks and Verlag der Friedrich-Held-Gesellschaft, Hackenheim and Munich, Germany.

Bogan, A.E., and W.R. Hoeh. 1995. *Utterbackia peninsularis*, a newly recognized freshwater mussel (Bivalvia: Unionidae: Anodontinae) from Peninsular Florida, USA. Walkerana (for 1993–1994) 7(17–18):275–287.

Bogan, A.E., and W.R. Hoeh. 2000. On becoming cemented: evolutionary relationships among the genera in the freshwater bivalve family Etheriidae (Bivalvia: Unionoida). Pages 159–168. In: E.M. Harper, J.D. Taylor and J.A. Crane (editors). The Evolutionary Biology of the Bivalvia. Geological Society, London, Special Publication 177.

Bogan, A.E., and P.W. Parmalee. 1983. Tennessee's Rare Wildlife. Volume II: The Mollusks. Tennessee Wildlife Resources Agency and Tennessee Department of Conservation, Nashville, Tennessee. 123 pages.

Bogan, A.E., J.M. Pierson and P. Hartfield. 1995. Decline in the freshwater gastropod fauna in the Mobile Bay Basin. Pages 249–252. In: E.T. LaRoe, G.S. Farris, C.E. Puckett, P.D. Doran and M.J. Mac (editors). Our living resources: a report to the nation on the distribution, abundance, and health of U.S. plants, animals and ecosystems. U.S. Department of the Interior, National Biological Service, Washington, District of Columbia. 530 pp.

Bogan, A.E., and R.R. Polhemus. 1987. Faunal Analysis. Pages 971–1112. In: R.R. Polhemus (editor). The Toqua Site, a Late Mississippian Dallas Phase town. Volume II. University of Tennessee, Department of Anthropology, Report of Investigations, Number 41; Tennessee Valley Authority Publications in Anthropology, Number 44.

Bogan, A.E., and R.W. Portell. 1995. Freshwater bivalves (Bivalvia: Unionidae) from the Early Pleistocene Leisey Shell Pits, Hillsborough County, Florida. Leisey Shell Pits Volume. Bulletin of the Florida State Museum, Biological Sciences 37(Part I) (6):171–182.

Bogan, A.E., and J.D. Williams. 1986. *Megalonaias* Utterback, 1915 (Mollusca, Bivalvia): proposed conservation by the suppression of *Magnonaias* Utterback, 1915. Z.N.(S.) 2512. Bulletin of Zoological Nomenclature 43(3):273–276.

Bogan, A.E., J.D. Williams and S.L.H. Fuller. 1990. Comments on the proposed conservation of *Proptera* Rafinesque, 1819 (Mollusca, Bivalvia). Bulletin of Zoological Nomenclature 47(3):206–207.

Bogan, A.E., J.D. Williams and J.T. Garner. 2006. Case 3353. *Obovaria* Rafinesque, 1819 (Mollusca, Bivalvia): proposed conservation of usage by designation of *Unio retusa* Lamarck, 1819 as the type species of *Obovaria*. Conservation. Bulletin of Zoological Nomenclature 63(4):226–230.

Bogatov, V.V. 2001. New data on Unioniformes of Sakhalin Island. Byulleten Dal'nevostochnogo Malakologicheskogo Obshchestva 5:71–77 [in Russian].

Bogatov, V.V., E.M. Sayenko and Ya. I. Starobogatov. 2002. On taxonomic position of the genus *Kunashiria* (Bivalvia, Unioniformes). Zoologicheskii Zhurnal 81(5):521–528 [in Russian].

Bogatov, V.V., and M.N. Zatravkin. 1988. New species of the order Unioniformes (Mollusca Bivalvia) from the south part of Soviet Far East. Trudy Zoologicheskogo Instituta 187:155–168 [in Russian].

Bonaventura, C., and J. Bonaventura. 1983. Respiratory pigments: structure and function. Pages 1–50. In: P.W. Hochachka (editor). The Mollusca, Volume 2, Environmental Chemistry and Physiology. Academic Press, New York.

Bosc, L.A.G. 1801–1804. Histoire naturelle des coquilles, contenant leur description, les moeurs des animaux qui les habitent et leurs usages. Avec figures dessinées d'après nature. Volumes 1–5. Deterville, Paris, France.

Boschung, H.T., Jr., and R.L. Mayden. 2004. Fishes of Alabama. Smithsonian Books, Washington, District of Columbia. 736 pages.

Böttger, O. 1878. Systematisches Verzeichniss der lebenden Arten der Landschnecken-Gattung *Clausilia* Drap. mit ausführlicher Angabe der geographischen Verbreitung der einzelnen Species. Berichte des Offenbacher Vereins für Naturkunde, Offenbach 17/18:18–100.

Bourguignat, J.R. 1880–1881. Matériaux pour server à l'historoire des Mollusques Acephales du système européen. Fascicle 1:1–96, 1 plate (May 1880), plates 12–15; Fascicle 2:97–387, and title page (May 1881). Poissy (S. Lejay).

Bovbjerg, R.V. 1957. Feeding related to mussel activity. Proceedings of the Iowa Academy of Science 64:650–653.

Bowen, Z.H. 1993. Evaluation of the commercial freshwater mussel fishery on Wheeler Reservoir, Alabama. Masters thesis, Auburn University, Alabama. 50 pages.

Bowen, Z.H., S.P. Malvestuto, W.D. Davies and J.H. Crance. 1994. Evaluation of the mussel fishery in Wheeler Reservoir, Tennessee River. Journal of Freshwater Ecology 9(4):313–319.

Boyko, C.B., and W.E. Sage, III. 1996. Catalog of recent type specimens in the Department of Invertebrates, American Museum of Natural History. II. Mollusca Part I (Classes Aplacophora, Polyplacophora, Gastropoda [Subclass Opisthobranchia], Bivalvia and Scaphopoda). American Museum Novitates, Number 3170. 50 pages, 12 figures.

Brady, T., M. Hove, C. Nelson, R. Gordon, D. Hornbach and A. Kapuscinski. 2004. Suitable host fish determined for hickorynut and pink heelsplitter. Ellipsaria 6(1):14–15.

Brand, A.R. 1972. The mechanism of blood circulation in *Anodonta anatina* (L.) (Bivalvia, Unionidae). Journal of Experimental Biology 56(2):361–379.

Brim Box, J., and J. Mossa. 1999. Sediment, land use, and freshwater mussels: prospects and problems. Journal of the North American Benthological Society 18(1):99–117.

Brim Box, J., and J.D. Williams. 2000. Unionid mollusks of the Apalachicola Basin in Alabama, Florida, and Georgia. Alabama Museum of Natural History Bulletin 21:1–143.

Britton, J.C., and S.L.H. Fuller. 1980. The freshwater bivalve Mollusca (Unionidae, Sphaeriidae, Corbiculidae) of the Savannah River Plant, South Carolina. Savannah River Plant, U.S. Department of Energy, SRO-NERP-3. 37 pages.

Brown, C.J.D., C. Clark and B. Gleissner. 1938. The size of certain naiades from western Lake Erie in relation to shoal exposure. The American Midland Naturalist 19(3):682–701.

Brück, A. 1914. Die muskulatur von *Anodonta cellensis* Schröt. Ein beitrag zur anatomi und histology der muskelfasern. Zeitschrift für wissenschaftlishe Zoologie 110:33–619.

Bruenderman, S.A., and R.J. Neves. 1993. Life history of the endangered Fine-rayed Pigtoe, *Fusconaia cuneolus* (Bivalvia: Unionidae) in the Clinch River, Virginia. American Malacological Bulletin 10(1):83–91.

Bryan, P., and L.F. Miller. 1953. Changes in the commercial fishery on the Alabama portion of the Tennessee River. The Progressive Fish Culturist 15(2):75–77.

Buchanan, A.C. 1980. Mussels (Naiades) of the Meramec River Basin, Missouri. Missouri Department of Conservation, Aquatic Series 17:1–68.

Buchner, O. 1910. Ueber individuel Formverschiedenheiten bei Anodonten. Jahreshelfte des Vereins für Väterlandischer Naturkunde in Wütemburg. Fünfundsechziegster Jahrgang 56:60–223.

Buhay, J.E., and W.R. Haag. 2003. Molecular systematics of the freshwater mussel genus *Villosa* (Bivalvia: Unionidae). Third Biennial Symposium, Freshwater Mollusk Conservation Society, 16–19 March 2003, Meeting Program and Abstracts, pages 15–16.

Buhay, J.E., W.R. Haag, C. Lydeard and M.L. Warren, Jr. 2003. Something's fishy with *Villosa vanuxemensis, V. lienosa*, and *V. ortmanni* (Bivalvia: Unionidae): fish host usage and phylogeographic analysis of morphologically similar species. Third Biennial Symposium, Freshwater Mollusk Conservation Society, 16–19 March 2003, Meeting Program and Abstracts, page 18.

Burch, J.B. 1973. Freshwater Unionacean Clams (Mollusca: Pelecypoda) of North America. Biota of Freshwater Ecosystems. Identification Manual 11, U.S. Environmental Protection Agency, Washington, District of Columbia. 176 pages.

Burch, J.B. 1975a. Freshwater Unionacean Clams (Mollusca: Pelecypoda) of North America. Revised edition. Malacological Publications, Hamburg, Michigan. 204 pages.

Burch, J.B. 1975b. Freshwater Sphaeriacean Clams (Mollusca: Pelecypoda) of North America. Malacological Publications, Hamburg, Michigan. 96 pages.

Burch, J.B. 1993. Glossary for North American freshwater malacology. I. Gastropoda. Walkerana (for 1991) 5(14):263–288.

Burkhead, N.M., S.J. Walsh, B.J. Freeman and J.D. Williams. 1997. Status and restoration of the Etowah River, an imperiled southern Appalachian ecosystem. Pages 375–444. In: G.W. Benz and D.E. Collins (editors). Aquatic Fauna in Peril: The Southeastern Perspective. Southeast Aquatic Research Institute, Special Publication 1. Lenz Design and Communications, Decatur, Georgia.

Butler, R.S. 1990. Distributional records for freshwater mussels (Bivalvia: Unionidae) in Florida and south Alabama, with zoogeographic and taxonomic notes. Walkerana (for 1989) 3(10):239–261.

Byrne. M. 2000. Calcium concretions in the interstitial tissues of the Australian freshwater mussel *Hyridella depressa* (Hyriidae). Pages 329–337. In: E.M. Harper, J.D. Taylor and J.A. Crame (editors). The Evolutionary Biology of the Bivalvia. Geological Society of London, Special Publication 177.

Cahn, A.R. 1936. Three reports dealing with the clam-shell industry of the Tennessee River Valley. Unpublished report, Tennessee Valley Authority, Norris, Tennessee.

Call, R.E. 1880. Polymorphous anodontae. American Naturalist 14(7):529–530.

Call, R.E. 1896. A revision and synonymy of the *parvus* group of Unionidae. Proceedings of the Indiana Academy of Science for 1895:109–119, 6 plates.

Call, R.E. 1900. A descriptive illustrated catalogue of the Mollusca of Indiana. Indiana Department of Geology and Natural Resources, 24th Annual Report, 1899:335–535, 1013–1017 [index unpaginated], plates 1–78.

Campbell, D.C., J.M. Serb, J.E. Buhay, K.J. Roe, R.L. Minton and C. Lydeard. 2005. Phylogeny of North American amblemines (Bivalvia, Unionoida): prodigious polyphyly proves pervasive across genera. Invertebrate Biology 124(2):131–164.

Carlton, J.T. 1992. Introduced marine and estuarine mollusks of North America. An end-of-the-20th-Century perspective. Journal of Shellfish Research 11(2):489–505.

Casey, J.L. 1986. The prehistoric exploitation of Unionacean bivalve mollusks in the Lower Tennessee-Cumberland-Ohio River valleys in western Kentucky. Masters thesis, Department of Archaeology, Simon Fraser University, British Columbia, Canada. 176 pages.

Chamberlain, T.K. 1931. Annual growth of fresh-water mussels. Bulletin of the U.S. Bureau of Fisheries 46(1930):713–739. [Issued separately as U.S. Bureau of Fisheries Document 1130.]

Chamberlain, T.K. 1934. The glochidial conglutinates of the Arkansas fanshell, *Cyprogenia aberti* (Conrad). Biological Bulletin 66(1):55–61.

Checa, A. 2000. A new model for periostracum and shell formation in the Unionidae (Bivalvia, Mollusca). Tissue and Cell 32(5):405–416.

Chittick, B., M. Stoskopf, M. Law, R. Overstreet and J. Levine. 2001. Evaluation of potential health risks to eastern elliptio (*Elliptio complanata*) (Mollusca: Bivalvia: Unionioda: Unionidae) and implications for sympatric endangered freshwater mussel species. Journal of Aquatic Ecosystem Stress and Recovery 9(1):35–42.

Christian, A.D., B.N. Smith, D.J. Berg, J.C. Smoot and R.M. Findlay. 2004. Trophic position and potential food sources of 2 species of unionid bivalves (Mollusca: Unionidae) in 2 small Ohio streams. Journal of the North American Benthological Society 23(1):101–113.

Churchill, E.P., Jr., and S.I. Lewis. 1924. Food and feeding in fresh-water mussels. Bulletin of the U.S. Bureau of Fisheries 39(1923–1924):439–471. [Issued separately as U.S. Bureau of Fisheries Document 963.]

Cicerello, R.R., and E.L. Laudermilk. 2001. Distribution and status of freshwater mussels (Bivalvia: Unionoidea) in the Cumberland River basin upstream from Cumberland Falls, Kentucky. Transactions of the Kentucky Academy of Science 62(1):26–34.

Cicerello, R.R., and G.A. Schuster. 2003. A guide to the freshwater mussels of Kentucky. Kentucky State Nature Preserves Commission, Scientific and Technical Series, Number 7. 62 pages.

Cicerello, R.R., M.L. Warren, Jr., and G.A. Schuster. 1991. A distributional checklist of the freshwater unionids (Bivalvia: Unionoidea) of Kentucky. American Malacological Bulletin 8(2):113–129.

Claassen, C. 1994. Washboards, pigtoes, and muckets: historic musseling in the Mississippi watershed. Historical Archaeology 28(2):1–145.

Clarke, A.H. 1973. The freshwater molluscs of the Canadian Interior Basin. Malacologia 13(1–2):1–509.

Clarke, A.H. 1981a. The freshwater molluscs of Canada. National Museum of Natural Sciences, National Museum of Canada, Ottawa, Canada. 446 pages.

Clarke, A.H. 1981b. The tribe Alasmidontini (Unionidae: Anodontinae). Part I: *Pegias, Alasmidonta,* and *Arcidens.* Smithsonian Contributions to Zoology, Number 326. 101 pages.

Clarke, A.H. 1985. The tribe Alasmidontini (Unionidae: Anodontinae). Part II: *Lasmigona* and *Simpsonaias.* Smithsonian Contributions to Zoology, Number 399. 75 pages.

Clarke, A.H. 1986. Competitive exclusion of *Canthyria* (Unionidae) by *Corbicula fluminea* (Müller). Malacology Data Net, Ecosearch Series, Number 1:3–10.

Clarke, A.H. 1988. Aspects of Corbiculid-Unionid sympatry in the United States. Malacology Data Net, Ecosearch Series 2(3–4):57–99.

Clarke, A.H. 1990. Comments on the proposed conservation of *Proptera* Rafinesque, 1819 (Mollusca, Bivalvia). (Case 2558). Bulletin of Zoological Nomenclature 47(3):205–206.

Clarke, A.H., and C.O. Berg. 1959. The freshwater mussels of central New York with an illustrated key to the species of northeastern North America. Cornell University Agricultural Experiment Station, New York State College of Agriculture, Ithaca, New York, Memoir Number 367. 79 pages.

Clarke, A.H., Jr., and W.J. Clench. 1965. *Amblema* Rafinesque, 1820 (Lamellibranchiata): proposed addition to the official list and proposed suppression of *Amblema* Rafinesque, 1819. Bulletin of Zoological Nomenclature 22(3):196–197.

Clavero, M., and E. Garcia-Berthou. 2005. Invasive species are a leading cause of animal extinctions. TRENDS in Ecology and Evolution 20(3):110.

Clench, W.J. 1955. A freshwater mollusk survey of North Florida rivers. The Nautilus 68(3):95–98.

Clench, W.J. 1974. Mollusca from Russell Cave. Pages 86–90. In: J.W. Griffin (editor). Investigations in Russell Cave, Russell Cave National Monument, Alabama. National Park Service, Publications in Archaeology, Number 13.

Clench, W.J., and R.D. Turner. 1956. Freshwater mollusks of Alabama, Georgia, and Florida from the Escambia to the Suwannee River. Bulletin of the Florida State Museum, Biological Sciences 1(3):97–239, plates 1–9.

Cliff, M., M. Hove and M. Haas. 2001. Creeper glochidia appear to be host generalists. Ellipsaria 3(1):19–20.

Coe, W.R., and H.J. Turner. 1938. Development of the gonads and gametes in the Soft-shell Clam (*Mya arenaria*). Journal of Morphology 62(1):91–111.

Coker, R.E. 1914a. The protection of fresh-water mussels. Appendix VIII to the Report of the U.S. Commissioner of Fisheries for 1912:1–23, 2 plates. [Issued separately U.S. Bureau of Fisheries Document 793.]

Coker, R.E. 1914b. Water-power development in relation to fishes and mussels of the Mississippi. Appendix VIII to the Report of the U.S. Commissioner of Fisheries for 1913:1–28, 6 plates. [Issued separately as U.S. Bureau of Fisheries Document 805.]

Coker, R.E. 1916. The utilization and preservation of fresh-water mussels. Transactions of the American Fisheries Society 46:39–49.

Coker, R.E. 1919. Fresh-water mussels and mussel industries of the United States. Bulletin of the U.S. Bureau of Fisheries 36(1917–1918):13–89, 46 plates. [Issued separately as U.S. Bureau of Fisheries Document 865.]

Coker, R.E. 1921. The fisheries biological station at Fairport, Iowa. Appendix I to the Report of the U.S. Commissioner of Fisheries for 1920:3–12, 3 plates. [Issued separately as U.S. Bureau of Fisheries Document 895.]

Coker, R.E., A.F. Shira, H.W. Clark and A.D. Howard. 1921. Natural history and propagation of fresh-water mussels. Bulletin of the U.S. Bureau of Fisheries 37(1919–1920):77–181, 17 plates. [Issued separately as U.S. Bureau of Fisheries Document 893.]

Coker, R.E., and J.B. Southall. 1915. Mussel resources in tributaries of the upper Missouri River. Appendix IV to the Report of the U.S. Commissioner of Fisheries for 1914:1–17, 1 plate. [Issued separately as U.S. Bureau of Fisheries Document 812.]

Coker, R.E., and T. Surber. 1911. A note on the metamorphosis of the mussel *Lampsilis lævissimus.* Biological Bulletin 20(3):179–182, 1 plate.

Collins, K.A. 1971. Gametogenesis in the fresh-water mussel, *Elliptio dilatatus,* of the Rockcastle River, Rockcastle County, Kentucky. Masters thesis, Eastern Kentucky University, Richmond. 43 pages.

Conner, C.H. 1905. Glochidia of *Unio* on fishes. The Nautilus 18(12):142–143.

Conner, C.H. 1909. Supplementary notes on the breeding seasons of the Unionidae. The Nautilus 22(10):111–112.

Conrad, T.A. 1831. Description of fifteen new species of recent and three of fossil shells, chiefly from the coast of the United States. Journal of the Academy of Natural Sciences of Philadelphia 6:256–268, plate 11.

Conrad, T.A. 1834a. Descriptions of some new species of fresh water shells from Alabama, Tennessee, etc. American Journal of Science and Arts 25(2):338–343, 1 plate. [Reprinted in Sterkiana 9(1963).]

Conrad, T.A. 1834b. New Freshwater Shells of the United States, with lithographic illustrations; and a monograph of the genus *Anculotus* of Say; also a synopsis of the American naiades. J. Dobson, Philadelphia. May 3, 1834. 76 pages, 8 plates.

Conrad, T.A. 1835a. Additions to, and corrections of, the catalogue of species of American naiades, with descriptions of new species and varieties of fresh water shells. Pages 1–8, plate 9. Appendix to Conrad, 1834b. New Freshwater Shells of the United States, with lithographic illustrations; and a monograph of the genus *Anculotus* of Say; also a synopsis of the American naiades. J. Dobson, Philadelphia. May 3, 1834. 76 pages, 8 plates.

Conrad, T.A. 1835b–1840. Monography of the family Unionidae, or naiades of Lamarck, (fresh water bivalve shells) of North America, illustrated by figures drawn on stone from nature. J. Dobson, Philadelphia. Part 1(1835):1–12 [pages 13–16 not published], plates 1–5; Part 2(1836):17–24, plates 6–10; Part 3(1836):25–32, plates 11–15; Part 4(1836):33–40, plates 16–20; Part 5(1836):41–48, plates 21–25; Part 6(1836):49–56, plates 26–30; Part 7(1836):57–64, plates 32–36; Part 8(1837):65–72, plates 36–40; Part 9(1837):73–80, plates 41–45; Part 10(1838):81–94, plates 46–51; Part 11(1838):95–102, plates 52–57; Part 12(1840):103–110, plates 58–60; Part 13[1840, part 13 not dated]:111–118, plates 61–65.

Conrad, T.A. 1841. [Descriptions of three new species of *Unio* from the rivers of the United States.] Proceedings of the Academy of Natural Sciences of Philadelphia 1(2):19–20.

Conrad, T.A. 1849. Descriptions of new fresh water and marine shells. Proceedings of the Academy of Natural Sciences of Philadelphia 4(7):152–155. [Subsequently published in 1849 in The Annals and Magazine of Natural History 4(22):300–303.]

Conrad, T.A. 1850. Descriptions of new fresh water and marine shells. Journal of the Academy of Natural Sciences of Philadelphia 1 (New Series):275–280, plates 37–39.

Conrad, T.A. 1853. A synopsis of the family of Naïades of North America, with notes, and a table of some of the genera and sub-genera of the family, according to their geographical distribution, and descriptions of genera and sub-genera. Proceedings of the Academy of Natural Sciences of Philadelphia 6(7):243–269.

Conrad, T.A. 1854. Descriptions of new species of *Unio*. Journal of the Academy of Natural Sciences of Philadelphia (New Series) 2(4):295–298, plates 26–27.

Conrad, T.A. 1855. Descriptions of three new species of *Unio*. Proceedings of the Academy of Natural Sciences of Philadelphia 7(7):256.

Conrad, T.A. 1856. On a new species of *Unio*. American Journal of Science and Arts 21 (Second Series) (62):172.

Conrad, T.A. 1866a. Description of a new species of *Unio*. American Journal of Conchology 2(2):107, plate 10.

Conrad, T.A. 1866b. Descriptions of American fresh-water shells. American Journal of Conchology 2(3):278–279, plate 15.

Contreras Arquieta, A. 1995. Chapter 16. Capitulo 10. Mollusks. Pages 141–149. In: S. Contreras Balderas, S.F. Gonzalez Saldivar, D. Lazcano Villarreal and A. Contreras Arquieta (editors). Listado Preliminar de la Fauna Silvestre del Estado de Nuevo León, Mexico. Consejo Consultivo Estatal para la Preservatión y Fomento de la Flora y Fauna Silvestre de Nuevo León, Mexico.

Convey, L.E., J.M. Hanson and W.C. MacKay. 1989. Size-selective predation on unionid clams by muskrats. Journal of Wildlife Management 53(3):654–657.

Cooper, W. 1834. List of shells collected by Mr. Schoolcraft in western and northwestern territory. Appendix, pages 153–156. In: H.B. Schoolcraft. Narrative of Expedition Through Upper Missouri to Itasca Lake, etc., Under the Direction of Henry B. Schoolcraft. Harper and Brothers, New York.

Corey, C.A., R. Dowling and D.L. Strayer. 2006. Display behavior of *Ligumia* (Bivalvia: Unionidae). Northeastern Naturalist 13(2):319–332.

Counts, C.L., III. 1986. The zoogeography and history of the invasion of the United States by *Corbicula fluminea* (Bivalvia: Corbiculidae). American Malacological Bulletin, Special Edition, Number 2:7–39.

Cragin, F.W. 1887. A new species of *Unio* from Indian Territory. Bulletin of the Washburn College Laboratory of Natural History 1(8):6.

Crownhart, A., B. Sietman, M. Hove and N. Rudh. 2006. *Quadrula metanevra* glochidia metamorphose on select minnow species. *Ellipsaria* 8(3):6–7.

Culp, J.J., A.C. Shepard and M.A. McGregor. 2006. New Host Fish Identifications for the Pyramid Pigtoe, *Pleurobema rubrum*. *Ellipsaria* 8(3):5–6.

Cummings, K.S., and C.A. Mayer. 1992. Field Guide to Freshwater Mussels of the Midwest. Illinois Natural History Survey, Manual 5. 194 pages.

Cummings, K.S., and C.A. Mayer. 1993. Distribution and host species of the federally endangered freshwater mussel, *Potamilus capax* (Green, 1832), in the lower Wabash River, Illinois and Indiana. Illinois Natural History Survey, Center for Biodiversity, Technical Report 1993(1):1–29.

Curole, J.P., and T.D. Kocher. 2002. Ancient sex-specific extension of the cytochrome *c* oxidase II gene in bivalves and the fidelity of doubly-uniparental inheritance. Molecular Biology and Evolution 19(8):1323–1328.

Curren, C.B., Jr., E. Reitz and J. Walden. 1977. Faunal remains, bone and shell artifacts. Pages 173–192. In: E.M. Futato. The Bellefonte Site 1JA300. Office of Archaeological Research, University of Alabama, Research Series, Number 2.

Currier, A.O. 1868. List of the shell-bearing Mollusca of Michigan, especially of Kent and adjoining counties. Kent Scientific Institute, Miscellaneous Publications, Number 1. 12 pages.

Cvancara, A.M. 1963. Clines in three species of *Lampsilis* (Pelecypoda: Unionidae). Malacologia 1(2):215–225.

Cvancara, A.M. 1970. Mussels (Unionidae) of the Red River valley in North Dakota and Minnesota, U.S.A. Malacologia 10(1):57–92.

Cvancara, A.M. 1983. Aquatic mollusks of North Dakota. North Dakota Geological Survey, Report of Investigation, Number 78. Kaye's Incorporated, Fargo, North Dakota. 141 pages.

Daget, J. 1998. Catalogue raisonné des Mollusques bivalves d'eau douce africains. Backhuys Publishers, Leiden; Ostrom, Paris. 329 pages.

Dall, W.H. 1889. On the hinge of pelecypods and its development, with an attempt toward a better subdivision of the group. American Journal of Science and Arts (Series 3) 38(228):445–462.

Danglade, E. 1914. The mussel resources of the Illinois River. Appendix VI to the Report of the U.S. Commissioner of Fisheries for 1913:1–48, 6 plates. [Issued separately as U.S. Bureau of Fisheries Document 804.]

Daniels, L.E. 1902. A new species of *Lampsilis*. The Nautilus 16(2):13–14, plate 2.

Daniels, L.E. 1909. Records of Minnesota mollusks. The Nautilus 22(11):119–121.

Davis, G.M. 1983. Relative roles of molecular genetics, anatomy, morphometrics and ecology in assessing relationships among North American Unionidae (Bivalvia). Pages 193–222. In: G.S. Oxford and D. Rollinson (editors). Protein Polymorphism: Adaptive and Taxonomic Significance. Systematics Association, Special Volume 24, Academic Press, London.

Davis, G.M., and S.L.H. Fuller. 1981. Genetic relationships among Recent Unionacea (Bivalvia) of North America. Malacologia 20(2):217–253.

Dawley, C. 1947. Distribution of aquatic mollusks in Minnesota. The American Midland Naturalist 38(3):671–697.

De Gregorio, A. 1914. Su taluni molluschi di acqua dolce di America [American fresh water shells of America]. Il Naturalista Siciliano 22(2–3):31–72, plates 3–12.

De Kay, J.E. 1843. Zoology of New York, or the New York Fauna; comprising detailed descriptions of all the animals hitherto observed within the state of New York; with brief notices of those occasionally found near its borders: and accompanied by appropriate illustrations. Part 5. Mollusca. Carroll and Cook, Printers to the Assembly, Albany, New York. 271 pages, 40 plates.

DeCamp, W.H. 1881. List of the shell-bearing Mollusca of Michigan. Kent Scientific Institute, Miscellaneous Publication, Number 5. 15 pages, plate 1.

Delessert, B. 1841. Recueil des Coquilles decrites par Lamarck, dans son Histoire naturelle des Animaux sans Vertébres et non encore figurées. 40 folios, colored plates.

Deshayes, D.P. 1830. Mollusques. Volume 2. Pages 471–553. In: Encyclopédie Méthodique, Histoire des Vers par Bruguière et Lamarck, complétée par Deshayes. Panckoucke, Paris.

Dickens, R.S., Jr. 1971. Archaeology in the Jones Bluff Reservoir of Central Alabama. Journal of Alabama Archaeology 17(1):1–107.

Dietz, T.H. 1985. Ionic regulation in freshwater mussels: a brief review. American Malacological Bulletin 3(2):233–242.

Dietz, T.H., A.S. Udoetok, J.S. Cherry, H. Silverman and R.A. Bryne. 2000. Kidney function and sulfate uptake and loss in the freshwater bivalve *Toxolasma texasensis*. Biological Bulletin 199(1):14–20.

Diggins, T.P., and K.M. Stewart. 2000. Evidence of large change in unionid mussel abundance from selective muskrat predation, as inferred by shell remains left on shore. International Review of Hydrobiology 85(4):505–520.

D'Orbigny, A. 1840–1853. Mollusques. In: R. de la Sagra. Histoiria Físicia. Política y Natural de la Isla de Cuba. Volume 1 (1840–1842). Volume 2 (1842–1853). Atlas [1842?]. Paris.

Downing, J.A., J.-P. Amyot, M. Pérusse and Y. Rochon. 1989. Visceral sex, hermaphroditism and protandry in a population of the freshwater bivalve *Elliptio complanata*. Journal of the North American Benthological Society 8(1):92–99.

Downing, J.A., Y. Rochon and M. Pérusse. 1993. Spatial aggregation, body size, and reproductive success in the freshwater mussel *Elliptio complanata*. Journal of the North American Benthological Society 12(2):148–156.

Downing, W.L., J. Shostell and J.A. Downing. 1992. Non-annual external annuli in the freshwater mussels *Anodonta grandis grandis* and *Lampsilis radiata siliquoidea*. Freshwater Biology 28(3):309–317.

Dudgeon, D., and B. Morton. 1983. The population dynamics and sexual strategy of *Anodonta woodiana* (Bivalvia: Unionacea) in Plover Cove Reservoir, Hong Kong. Journal of Zoology (London) 201(1983):161–183.

Dundee, D.S. 1953. Formed elements of the blood of certain fresh-water mussels. Transactions of the American Microscopical Society 72(3):254–264.

Dunker, W. 1858. Einige neue species der Naiaden. Malakozoologische Blätter 5:225–229.

Dyke, S. 2000. Freshwater mussel management in North Dakota. Central Plains Archeology 8(1):99–102.

Eager, R.M.C. 1978. Shape and function of the shell: a comparison of some living and fossil bivalve molluscs. Biological Reviews of the Cambridge Philosophical Society 53(2):169–210.

Edgar, A.L. 1965. Observations on the sperm of the pelecypod *Anodontoides ferussacianus* (Lea). Transactions of the American Microscopical Society 84(2):228–230.

Edwards, D.D., and R.V. Dimock, Jr. 1995. Life history characteristics of larval *Unionicola* (Acari: Unionicolidae) parasitic on *Chironomus tentans* (Diptera: Chironomidae). Journal of Natural History 29(5):1197–1208.

Ellis, M.M. 1929. The artificial propagation of freshwater mussels. Transactions of the American Fisheries Society 59:217–223.

Ellis, M.M. 1931a. A survey of conditions affecting fisheries in the upper Mississippi River. U.S. Bureau of Fisheries, Fishery Circular, Number 5. 18 pages.

Ellis, M.M. 1931b. Some factors affecting the replacement of the commercial fresh-water mussels. U.S. Bureau of Fisheries, Fishery Circular, Number 7. 10 pages.

Ellis, M.M. 1936. Erosion silt as a factor in aquatic environments. Ecology 17(1):29–42.

Ellis, M.M. 1937. Detection and measurement of stream pollution. Bulletin of the U.S. Bureau of Fisheries 48(1940):365–437, figures 1–22. [Issued separately as U.S. Bureau of Fisheries Bulletin, Number 22.]

Ellis, M.M. 1941. Fresh-water impoundments. Transactions of the American Fisheries Society 71:80–93.

Ellis, M.M., and M.D. Ellis. 1926. Growth and transformation of parasitic glochidia in physiological nutrient solutions. Science 64(1667):579–580.

Ellis, M.M., and M. Keim. 1918. Notes on the glochidia of *Strophitus edentulus pavonis* (Lea) from Colorado. The Nautilus 32(1):17–18.

Ellis, M.M., A.D. Merrick and M.D. Ellis. 1931. The blood of North American fresh-water mussels under normal and adverse conditions. Bulletin of the U.S. Bureau of Fisheries 46(1930):509–542. [Issued separately as U.S. Bureau of Fisheries Document 1097.]

Etnier, D.A., and W.C. Starnes. 1993. The Fishes of Tennessee. University of Tennessee Press, Knoxville. 681 pages.

Evermann, B.W., and H.W. Clark. 1918. The Unionidae of Lake Maxinkukee. Proceedings of the Indiana Academy of Science 1917:251–285.

Evermann, B.W., and H.W. Clark. 1920. Lake Maxinkuckee: a Physical and Biological Survey. Indiana Department of Conservation, Publication Number 7, Volume 2. W.B. Burford Printer, Indianapolis, Indiana. 512 pages.

Farris, J.L., S.E. Belanger, D.S. Cherry and J. Cairns, Jr. 1989. Cellulytic activity as a novel approach to assess long-term zinc stress to *Corbicula*. Water Research 23(10):1275–1283.

Farris, J.L, J.L. Grudzien, S.E. Belanger, D.S. Cherry and J. Cairns, Jr. 1994. Molluscan cellulolytic activity responses to zinc exposure in laboratory and field stream comparisons. Hydrobiologia 287(2):161–178.

Ferriss, J.H. 1900. A new *Lampsilis* from Arkansas. The Nautilus 14(4):38–39.

Fischer, P. 1880–1887. Manuel de Conchyliologie et de Paléontologie conchyliologique ou Histoire Naturelle des Mollusques vivants et fossiles. Librairie F. Savy, Paris. 369 pages.

Fischer, P.H., and H. Crosse. 1894. Études sur les mollusques terrestres et fluviatiles du Mexique et du Guatemala. 2(15):489–576, colored plates 59–62. In: P.H. Fischer and H. Crosse (editors). Mission Scientifique au Mexique et dans l'Amérique Centrale. Reserches Zoologiques pour server à l'histoire de la faune de l'Amerique Centrale et du Mexique. Imprimerie Nationale, Paris.

Fisher, G.R., and R.V. Dimock, Jr. 2002a. Morphological and molecular changes during metamorphosis in *Utterbackia imbecillis* (Bivalvia: Unionidae). Journal of Molluscan Studies 68(2):159–164.

Fisher, G.R., and R.V. Dimock, Jr. 2002b. Ultrastructure of the mushroom body: digestion during metamorphosis of *Utterbackia imbecillis* (Bivalvia: Unionidae). Invertebrate Biology 121(2):126–135.

Foscue, V.O. 1989. Place Names in Alabama. The University of Alabama Press, Tuscaloosa. 175 pages.

Frierson, L.S. 1900. A new Alabama *Unio*. The Nautilus 13(10):109–110, plate 2.

Frierson, L.S. 1902. Collecting Unionidae in Texas and Louisiana. The Nautilus 16(4):37–40.

Frierson, L.S. 1903a. The specific value of *Unio declivus*, Say. The Nautilus 17(5):49–51, plate 3.

Frierson, L.S. 1903b. Observations on the byssus of Unionidae. The Nautilus 17(7):76–77.

Frierson, L.S. 1904. Observations on the genus *Quadrula*. The Nautilus 17(10):111–112.

Frierson, L.S. 1905. New Unionidae from Alabama. The Nautilus 19(2):13–14, plate 1, figures 1–4.

Frierson, L.S. 1908. Description of a new *Pleurobema*. The Nautilus 22(3):27–28, plate 3, figures 3–4.

Frierson, L.S. 1910. Description of a new species of *Anodonta*. The Nautilus 23(9):113–114, plate 10.

Frierson, L.S. 1911. Remarks on *Unio varicosus, cicatricosus* and *Unio compertus*, new species. The Nautilus 25(5):51–54, plates 2–3.

Frierson, L.S. 1912. *Unio (Obovaria) jacksonianus*, new species. The Nautilus 26(2):23–24, plate 3.

Frierson, L.S. 1914. Remarks on classification of the Unionidae. The Nautilus 28(1):6–8.

Frierson, L.S. 1916. Observations on the *Unio cor*, of Conrad. The Nautilus 29(9):102–104.

Frierson, L.S. 1922. Observations on the genus *Margaritana* with a new subgenus. The Nautilus 36(2):42–44.

Frierson, L.S. 1924. Interesting facts in the history of *Unio orbiculatus* Hildreth and *U. abruptus* Say. The Nautilus 37(4):135–137.

Frierson, L.S. 1927. A Classification and Annotated Check List of the North American Naiades. Baylor University Press, Waco, Texas. 111 pages.

Frierson, L.S. 1928. Illustrations of Unionidae. The Nautilus 41(4):138–139, plates 1–3.

Fryer, G. 1959. Development in a mutelid lamellibranch. Nature 183(4671):1342–1343.

Fryer, G. 1961. The developmental history of *Mutela bourguignati* (Ancey) Bourguignat (Mollusca: Bivalvia). Philosophical Transactions of the Royal Society of London, Series B, Biological Sciences 244(711):259–298.

Fuller, S.L.H. 1971. A brief guide to the fresh-water mussels (Mollusca: Bivalvia: Unionacea) of the Savannah River system. ASB [Association of Southeastern Biologists] Bulletin 18(4):137–146, 1 plate.

Fuller, S.L.H. 1974. Clams and mussels (Mollusca: Bivalvia). Pages 215–273. In: C.W. Hart, Jr., and S.L.H. Fuller (editors). Pollution Ecology of Freshwater Invertebrates. Academic Press, New York.

Fuller, S.L.H. 1978. Fresh-water mussels (Mollusca: Bivalvia: Unionidae) of the Upper Mississippi River: Observations at selected sites within the 9-foot channel navigation project on behalf of the United States Army Corps of Engineers. The Academy of Natural Sciences of Philadelphia, Division of Limnology and Ecology, Report Number 78-33. 401 pages.

Fuller, S.L.H. 1980a. *Anodonta imbecillis* Say (Bivalvia: Unionidae) in the Delaware River Basin. The Nautilus 94(1):4.

Fuller, S.L.H. 1980b. Freshwater mussels (Mollusca: Bivalvia: Unionidae) of the Upper Mississippi River: Observations at selected sites within the 9-foot navigation channel project for the St. Paul District, United States Army Corps of Engineers, 1977–1979. Academy of Natural Sciences of Philadelphia, Report 175. 1179 pages, Appendix, 441 pages.

Fuller, S.L.H. 1980c. Historical and current distributions of fresh-water mussels (Mollusca: Bivalvia: Unionidae) in the Upper Mississippi River. Pages 72–119. In: J.L. Rassmussen (editor). Proceedings of the UMRCC Symposium on Upper Mississippi River Bivalve Mollusks. Upper Mississippi River Conservation Committee, Rock Island, Illinois.

Fuller, S.L.H., and D.J. Bereza. 1973. Recent additions to the naiad fauna of the eastern Gulf drainage (Bivalvia: Unionoida: Unionidae). ASB [Association of Southeastern Biologists] Bulletin 20(2):53.

Fuller, S.L.H., and D.J. Bereza. 1974. The value of anatomical characters in naiad taxonomy (Bivalvia: Unionacea). Bulletin of the American Malacological Union, Inc. 1974:21–22.

Fuller, S.L.H., and M.J. Imlay. 1976. Spatial competition between *Corbicula manilensis* (Philippi), the Chinese clam (Corbiculidae), and fresh-water mussels (Unionidae) in the Waccamaw River Basin of the Carolinas (Mollusca: Bivalvia). ASB [Association of Southeastern Biologists] Bulletin 23(2):60.

Gagnon, P.M., S.W. Golladay, W.K. Michener and M.C. Freeman. 2004. Drought responses of freshwater mussels (Unionidae) in Coastal Plain tributaries of the Flint River Basin, Georgia. Journal of Freshwater Ecology 19(4):667–679.

Gangloff, M.M. 2003. The status, physical habitat associations, and parasites of freshwater mussels in the upper Alabama River drainage, Alabama. Doctoral dissertation, Auburn University, Alabama. 237 pages.

Gangloff, M.M., and D.L. Gustafson. 2000. The freshwater mussels (Bivalvia: Unionoida) of Montana. Central Plains Archeology 8(1):121–130.

Gangloff, M.A., J.D. Williams and J.W. Feminella. 2006. A New Species of Freshwater Mussel (Bivalvia: Unionidae), *Pleurobema athearni*, from the Coosa River Drainage of Alabama, USA. Zootaxa 1118:43-56.

Gardiner, D.B., H. Silverman and T.H. Dietz. 1991. Musculature associated with the water canals in freshwater mussels and response to monoamines in vitro. Biological Bulletin 180(3):453–465.

Garner, J.T. 1993. Reproductive cycle of *Quadrula metanevra* (Unionidae) in Kentucky Reservoir, Tennessee River, Hardin Co., Tennessee. Masters thesis, University of Alabama in Huntsville. 56 pages.

Garner, J.T., H. Blalock-Herod, A.E. Bogan, R.S. Butler, W.R. Haag, P.D. Hartfield, J.J. Herod, P.D. Johnson, S.W. McGregor and J.D. Williams. 2004. Freshwater mussels and snails. Pages 13–58. In: R.A. Mirarchi (editor). Alabama Wildlife. Volume 1. A Checklist of Vertebrates and Selected Invertebrates: Aquatic Mollusks, Fishes, Amphibians, Reptiles, Birds, and Mammals. The University of Alabama Press, Tuscaloosa.

Garner, J.T., T.M. Haggerty and R.F. Modlin. 1999. Reproductive cycle of *Quadrula metanevra* (Bivalvia: Unionidae) in the Pickwick Dam Tailwater of the Tennessee River. The American Midland Naturalist 141(2):277–283.

Garner, J.T., and S.W. McGregor. 2001. Current status of freshwater mussels (Unionidae, Margaritiferidae) in the Muscle Shoals area of the Tennessee River in Alabama (Muscle Shoals revisited again). American Malacological Bulletin 16(1–2):155–170.

Gatenby, C.M., R.J. Neves and B.C. Parker. 1996. Influence of sediment and algal food on cultured juvenile freshwater mussels. Journal of the North American Benthological Society 15(4):597–609.

Gatenby, C.M., B.C. Parker and R.J. Neves. 1997. Growth and survival of juvenile mussels, *Villosa iris* (Lea, 1829) (Bivalvia: Unionidae), reared on algal diets and sediment. American Malacological Bulletin 14(1):57–66.

Geiser, S.W. 1911. The correct name for *Tritogonia tuberculata*. The Academician 1(1):15.

Gillis, P.L., and G.L. Mackie. 1994. Impact of the zebra mussel, *Dreissena polymorpha*, on populations of Unionidae (Bivalvia) in Lake St. Clair. Canadian Journal of Zoology 72(7):1260–1271.

Girod, P. 1889. Manipulations de Zoologie. Guide pour les travaux pratiques de dissection. Animaux Invertébrés. Librairie J.B. Baillière et Fils. Paris. 140 pages, 25 plates.

Gledhill, T., and M.F. Vidrine. 2002. Two new sympatric water-mites (Acari: Hydrachnidia: Unionicolidae) from the mutelid bivalve, *Aspatharia sinuata* (von Martens) in Nigeria with some data on unionicoline-bivalve relationships. Journal of Natural History 36(11):1351–1381.

Glenn, L.C. 1911. Denudation and erosion in the Southern Appalachian Region and the Monongahela Basin. U.S. Geological Survey, Professional Paper 72. 137 pages, 21 plates.

Gmelin, J.F. 1791. Caroli Linnaeus, Systema naturæ per regna tria naturæ: secundum classes, ordines, genera, species, cum characteribus, differentiis, synonymis, locis. Lipsiae, Impensis Georg. Emanuel. Beer. 13th edition. Volume 1(Part 6) (Vermes):3021–3910.

Godwin, W.F. 1968. The distribution and density of the brackish water clam, *Rangia cuneata*, in the Altamaha River, Georgia. Georgia Department of Natural Resources, Contribution Series, Number 5. 10 pages.

Golladay, S.W., P. Gagnon, M. Kearns, J.M Battle and D.W. Hicks. 2004. Response of freshwater mussel assemblages (Bivalvia: Unionidae) to a record drought in the Gulf Coastal Plain of southwestern Georgia. Journal of the North American Benthological Society 23(3):494–506.

Gooch, C.H., W.J. Pardue and D.C. Wade. 1979. Recent mollusk investigations on the Tennessee River: 1978. Draft Report, Tennessee Valley Authority, Division of Environmental Planning, Water Quality and Ecology Branch, Muscle Shoals, Alabama, and Chattanooga, Tennessee. 126 pages.

Goodrich, C. 1931. *Pleurobema aldrichianum,* a new naiad. Occasional Papers of the Museum of Zoology, University of Michigan, Number 229. 4 pages, 1 plate.

Goodrich, C., and H. van der Schalie. 1932. I. On an increase in the naiad fauna of Saginaw Bay, Michigan. II. The naiad species of the Great Lakes. Occasional Papers of the Museum of Zoology, University of Michigan, Number 238. 14 pages.

Gordon, M.E. 1990. Case 2558. *Proptera* Rafinesque, 1819 (Mollusca, Bivalvia): proposed conservation. Bulletin of Zoological Nomenclature 47(1):19–21.

Gordon, M.E. 1993. Freshwater mussel investigations in the Little Tennessee River. Part 2: Glochidial hosts of *Lasmigona holstonia* (Bivalvia: Unionidae: Anodontinae). Unpublished report submitted to U.S. Forest Service, Asheville, North Carolina. 7 pages.

Gordon, M.E. 1995. *Venustaconcha sima* (Lea), an overlooked freshwater mussel (Bivalvia: Unionoidea) from the Cumberland River basin of central Tennessee. The Nautilus 108(3):55–60.

Gordon, M.E., and W.R. Hoeh. 1995. *Anodonta heardi*, a new species of freshwater mussel (Bivalvia: Unionidae) from the Apalachicola River system of the southeastern United States. Walkerana (for 1993–1994) 7(17–18):265–273.

Gordon, M.E., and J.B. Layzer. 1989. Mussels (Bivalvia: Unionoidea) of the Cumberland River. Review of life histories and ecological relationships. U.S. Fish and Wildlife Service, Biological Report 89(15):1–99.

Gordon, M.E., J.B. Layzer and L.M. Madison. 1994. Glochidial host of *Villosa taeniata* (Mollusca: Unionidae). Malacological Review 27(1–2):113–114.

Gordon, M.E., and D.G. Smith. 1990. Autumnal reproduction in *Cumberlandia monodonta* (Unionoidea: Margaritiferidae). Transactions of the American Microscopical Society 109(4):407–411.

Goudreau, S.E., R.J. Neves and R.J. Sheehan. 1993. Effects of wastewater treatment plant effluents on freshwater mollusks in the upper Clinch River, Virginia, U.S.A. Hydrobiologia 252(3):211–230.

Gould, A.A. 1843. [Descriptions of shells from Tavoy, British Burmah.] Proceedings of the Boston Society of Natural History 1(17):139–141.

Gould, A.A. 1850. [The following shells from the United States exploring expedition were described.] Proceedings of the Boston Society of Natural History 3(19):292–296.

Gould, A.A. 1855. New species of land and fresh-water shells from western (N.) America. (Cont.). Proceedings of the Boston Society of Natural History 5(15):228–229.

Gould, A.A. 1856. [Descriptions of shells.] Proceedings of the Boston Society of Natural History 6(1):11–16.

Gove, P.B. (editor in chief). 1961. Webster's Seventh New Collegiate Dictionary. G.C. Merriam Company, Springfield, Massachusetts. 1220 pages.

Graf, D.L. 1997. Distribution of unionoid (Bivalvia) faunas in Minnesota, U.S.A. The Nautilus 110(2):45–54.

Graham, A. 1949. The molluscan stomach. Transactions of the Royal Society of Edinburgh 61(3):737–778, figures 1–23.

Grande, C., R. Araujo and M.A. Ramos. 2001. The gonads of *Margaritifera auricularia* (Spengler, 1793) and *M. margaritifera* (Linnaeus, 1758) (Bivalvia: Unionoidea). Journal of Molluscan Studies 67(1):27–35.

Gray, J.E. 1840. [Mollusca.] Pages 86–89, 106–151. [In:] Synopsis of the contents of the British Museum. Forty-second edition. G. Woodfall and Son, London.

Gray, J.E. 1843. Catalogue of the species of Mollusca and their shells, which have hitherto been recorded as found at New Zealand, with the description of some lately discovered species. Appendix 4, pages 228–265. In: E. Dieffenbach. Travels in New Zealand; with contributions to the geography, geology, botany, and natural history of that country. Volume 2. Murray, London.

Gray, J.E. 1847. A list of the genera of recent Mollusca, their synonyms and types. Proceedings of the Zoological Society of London 1847(15):129–219.

Green, J. 1827. Some remarks on the Unios of the United States, with a description of a new species. Contributions of the Maclurian Lyceum to the Arts and Sciences 1(2):41–47, plate 3.

Green, R.H. 1972. Distribution and morphological variation of *Lampsilis radiata* (Pelecypoda, Unionidae) in some central Canadian lakes: a multivariate statistical approach. Journal of the Fisheries Research Board, Canada 29(11):1565–1570.

Grier, N.M. 1918. New varieties of naiades from Lake Erie. The Nautilus 32(1):9–12.

Grier, N.M. 1920. Morphological features of certain mussel-shells found in Lake Erie, compared with those of the corresponding species found in the drainage of the upper Ohio. Annals of the Carnegie Museum 13(1–2):145–182, plates 2–3.

Grier, N.M., and J.F. Mueller. 1926. Further studies in correlation of shape and station in fresh water mussels. Bulletin of the Wagner Free Institute of Science of Philadelphia 1(2–3):11–28.

Guérin-Méneville, F.E. 1828–1844. Iconographie du regne animal du G. Cuvier: ou Representation d'apres nature de l'une des especies les plus remarquables, et souvent non enore figurees, de chaque genre d'animaux. Avec un texte descriftif mis au courant de la science. Ouvarage pouvant sevir d'atlas a tous les traites de zoologie. J. B. Baiaere, Paris. Volume 2. Planches des animaux invertébrés. [No pagination, 218 plates.]

Gurevitch, J., and D.K. Padilla. 2004. Are invasive species a major cause of extinctions? TRENDS in Ecology and Evolution 19(9):471–474.

Guthiel, F. 1912. Über Darmkanal und die Mitteldarmdrüse von *Anodonta cellensis* Schröt. Zeitschrift für Wissenschaftliche Zoologie 99:444–538.

Haag, W.R., D.J. Berg, D.W. Garton and J.L. Farris. 1993. Reduced survival and fitness in native bivalves in response to fouling by the introduced zebra mussel (*Dreissena polymorpha*) in western Lake Erie. Canadian Journal of Fisheries and Aquatic Science 50(1):13–19.

Haag, W.R., R.S. Butler and P.D. Hartfield. 1995. An extraordinary reproductive strategy in freshwater bivalves: prey mimicry to facilitate larval dispersal. Freshwater Biology 34(3):471–476.

Haag, W.R., and J.L. Staton. 2003. Variation in fecundity and other reproductive traits in freshwater mussels. Freshwater Biology 48(12):2118–2130.

Haag, W.R., and M.L. Warren, Jr. 1997. Host fishes and reproductive biology of 6 freshwater mussel species from the Mobile Basin, U.S.A. Journal of the North American Benthological Society 16(3):576–585.

Haag, W.R., and M.L. Warren, Jr. 1998. Role of ecological factors and reproductive strategies in structuring freshwater mussel communities. Canadian Journal of Fisheries and Aquatic Sciences 55(2):297–306.

Haag, W.R., and M.L. Warren, Jr. 1999. Mantle displays of freshwater mussels elicit attacks from fish. Freshwater Biology 42(1):35–40.

Haag, W.R., and M.L. Warren, Jr. 2000. Effects of light and presence of fish on lure display and larval release behaviours in two species of freshwater mussels. Animal Behaviour 60(6):879–886.

Haag, W.R., and W.L. Warren, Jr. 2001. Host fishes and reproductive biology of freshwater mussels in the Buttahatchee River, Mississippi. Final report submitted to Mississippi Wildlife Heritage Fund. 41 pages.

Haag, W.R., and M.L. Warren, Jr. 2003. Host fishes and infection strategies of freshwater mussels in large Mobile Basin streams, U.S.A. Journal of the North American Benthological Society 22(1):78–91.

Haag, W.R., M.L. Warren, Jr., and M. Shillingsford. 1999. Host fishes and host-attracting behavior of *Lampsilis altilis* and *Villosa vibex* (Bivalvia: Unionidae). The American Midland Naturalist 141(1):149–157.

Haag, W.R., M.L. Warren, Jr., K. Wright and L. Shaffer. 2002. Occurrence of the rayed creekshell, *Anodontoides radiatus*, in the Mississippi River Basin: implications for conservation and biogeography. Southeastern Naturalist 1(2):169–178.

Haas, F. 1910. *Pseudunio*, neues genus für *Unio sinuatus* Lam. Nachrichtsblatt der Deutschen Malakozoologischen Gesellschaft 42(4):181–183.

Haas, F. 1910–1920. Die Unioniden. Systematisches Conchylien-Cabinet von Martini und Chemnitz. Küster edition. 9(Part 2, Section 2):1–344, plates 1–73, supplemental plate 12a [plates 74–75 not published but exist; work completed in Haas, 1924].

Haas, F. 1924. Beiträge zu einer Mongraphie der asiatischen Unioniden. Abhandlungen der Senckenbergishen Naturfoschenden Gesellschaft 38(2):129–203, plates 15–16.

Haas, F. 1929–1941. Dr. H.G. Bronns Klassen und Ordnungen des Tierreichs. Band 3 Mollusken und Tunikaten. Abt. 3 Bivalvia. Lief. 1, 3–7, 984 pages. Leipzig: Akademische Verlagsgesellschaft.

Haas, F. 1930. Über nord-und mittelamerikanische Najaden. Senckenbergiana Biologica 12(6):317–330.

Haas, F. 1940. A tentative classification of the Palearctic unionids. Field Museum of Natural History, Zoological Series 24(11):115–141.

Haas, F. 1969a. Superfamilia Unionacea. Das Tierreich (Berlin) 88. 663 pages.

Haas, F. 1969b. Superfamily Unionacea. Pages N411–N470. In: R.C. Moore (editor). Treatise on Invertebrate Paleontology. Geological Society of America and the University of Kansas, Part N, Volume 1 [of 3]. Mollusca 6. Bivalvia.

Haas, F., and E. Schwarz. 1913. Die Unioniden des Gebietes zwischen Mainz und deutschen Donau in tiergeographischer Hinsicht. Abhandlungen der Mathematisch-Physikalischen Classe der Königlich Bayerischen Akademie der Wissenschaften 26(7):1–34, plates 1–3.

Haggerty, T.M., and J.T. Garner. 2000. Seasonal timing of gametogenesis, spawning, brooding and glochidia discharge in *Potamilus alatus* (Bivalvia: Unionidae) in the Wheeler Reservoir, Tennessee River, Alabama, U.S.A. Invertebrate Reproduction and Development 38(1):35–41.

Haggerty, T.M., J.T. Garner, G.H. Patterson and L.C. Jones, Jr. 1995. A quantitative assessment of the reproductive biology of *Cyclonaias tuberculata* (Bivalvia: Unionidae). Canadian Journal of Zoology 73(1):83–85.

Haggerty, T.M., J.T. Garner and R. Rogers. 2005. Reproductive phenology in *Megalonaias nervosa* (Bivalvia: Unionidae) in Wheeler Reservoir, Tennessee River, Alabama, U.S.A. Hydrobiologia 539(2005):131–136.

Hakenkamp, C.C., and M.A. Palmer. 1999. Introduced bivalves in freshwater ecosystems: the impact of *Corbicula* on organic matter dynamics in a sandy stream. Oecologia 119(3):445–451.

Haldeman, S.S. 1842. Description of five new species of North American fresh-water shells. Journal of the Academy of Natural Sciences of Philadelphia 8(2):200–202.

Haldeman, S.S. 1846. Description of *Unio abacoides*, a new species. Proceedings of the Academy of Natural Sciences of Philadelphia 3(3):75. [Also published verbatim in 1846 in American Journal of Science and Arts (Series 2), 52(5):274 and Annals and Magazine of Natural History 18(121):430.]

Hall, B.M., and M.R. Hall. 1916. Second Report on the Water Powers of Alabama. Geological Survey of Alabama Bulletin 17. 448 pages.

Hall, R.O., Jr., J.L. Tank and M.F. Dybdahl. 2003. Exotic snails dominate nitrogen and carbon cycling in a highly productive stream. Frontiers in Ecology and the Environment 1(8):407–411.

Hanley, R.W. 1982. An analysis of the mollusk remains from 22CI814. Pages 153–155. In: C. Solis and R. Walling. Archaeological investigations at the Yarborough Site (22CI814), Clay County, Mississippi. Office of Archaeological Research, University of Alabama, Tuscaloosa, Report of Investigations, Number 30. Submitted to the U.S. Army Corps of Engineers, Mobile District.

Hanley, R.W. 1983. Mollusk Remains. Pages 78–84. In: E.M. Futato and C. Solis. Archaeology at Site 1JA78, the B.B. Comer Bridge Site, Jackson County, Alabama. Journal of Alabama Archaeology 29(1):1–123.

Hanley, R.W. 1984a. Invertebrate animal remains [1HA19]. Pages 103–105. In: C. Curren. The Protohistoric Period in central Alabama. Alabama Tombigbee Regional Commission.

Hanley, R.W. 1984b. Invertebrate remains [1TU4]. Pages 162, 164–165. In: C. Curren. The Protohistoric Period in Central Alabama. Alabama Tombigbee Regional Commission.

Hannibal, H. 1912. A synopsis of the Recent and Tertiary freshwater Mollusca of the Californian Province, based upon an ontogenetic classification. Proceedings of the Malacological Society of London 10(2):112–166; 10(3):167–211.

Hardison, B.S., and J.B. Layzer. 2001. Relations between complex hydraulics and the localized distribution of mussels in three regulated rivers. Regulated Rivers: Research and Management 17:77–84.

Harms, W. 1907. I. Wissenschaftliche Mitteilungen. 1. Über die postembryonale Entwicklung von *Anodonta piscinalis*. Zoologischer Anzeiger 31(25):801–814.

Harris, J.L., and M.E. Gordon. 1990. Arkansas Mussels. Arkansas Game and Fish Commission, Little Rock, Arkansas. 32 pages.

Harris, S.C. 1990. Preliminary considerations on rare and endangered invertebrates in Alabama. Journal of the Alabama Academy of Science 61(2):64–92.

Hartfield, P. 1990. Mussels. Pages 76–79. In: J.W. O'Hear. Archaeological investigations at the Sanders Site (22CI917), an Alexander midden on the Tombigbee River, Clay County, Mississippi. Mississippi State University, Cobb Institute of Archaeology, Report of Investigations, Number 6.

Hartfield, P. 1993. Headcuts and their effect on freshwater mussels. Pages 131–141. In: K.S. Cummings, A.C. Buchanan and L.M. Koch (editors). Conservation and Management of Freshwater Mussels. Proceedings of a UMRCC Symposium, 12–14 October 1992, St. Louis, Missouri. Upper Mississippi River Conservation Committee, Rock Island, Illinois.

Hartfield, P., and R.S. Butler. 1997. Observations on the release of superconglutinates by *Lampsilis perovalis*. Pages 11–14. In: K.S. Cummings, A.C. Buchanan, C.A. Mayer and T.J. Naimo (editors). Conservation and Management of Freshwater Mussels II: Initiatives for the Future. Proceedings of a UMRCC Symposium, 16–18 October 1995, St. Louis, Missouri. Upper Mississippi River Conservation Committee, Rock Island, Illinois.

Hartfield, P.D., and E. Hartfield. 1996. Observations on the conglutinates of *Ptychobranchus greeni* (Conrad, 1834) (Mollusca: Bivalvia: Unionoidea). The American Midland Naturalist 135(2):370–375.

Hartfield, P.D., and R.L. Jones. 1989. Population status of endangered mussels in the Tombigbee River at Gainesville Bendway, Alabama. Mississippi Department of Wildlife, Fish and Parks, Museum Technical Report, Number 6. 30 pages.

Hartfield, P.D., and R. Jones. 1990. Population status of endangered mussels in the Buttahatchee River, Mississippi and Alabama. Mississippi Museum of Natural Science, Museum Technical Report, Number 9. 35 pages.

Havlik, M.E., and D.H. Stansbery. 1977. The naiad mollusks of the Mississippi River in the vicinity of Prairie du Chien, Wisconsin. Bulletin of the American Malacological Union, Inc. 1977:9–12.

Hazay, J. 1881. Die Mollusken-fauna von Budapest. mit besonderer Rücksichtnahme auf die embryonalen und biologischen Verhältnisse ihrer Vorkommnisse. Malakozoologische Blätter, Neue Folge 3:1–183, plates 1–9; 4:43–224, plates 1–7. [Die Najaden 4:132–208.]

Hazleton, B.J., and G.R. Isenberg. 1977. Formed blood elements of the mussel *Elliptio complanatum*. Proceedings of the Pennsylvania Academy of Science 51(1):54–56.

Healy, J.M. 1989. Spermiogenesis and spermatozoa in the relict bivalve genus *Neotrigonia*: relevance to trigonioid relationships, particularly Unionoidea. Marine Biology (Berlin) 103(1):75–85.

Healy, J.M. 1996. Spermatozoan ultrastructure in the trigonioid bivalve *Neotrigonia margaritacea* Lamarck (Mollusca): comparison with other bivalves, especially Trigonioida and Unionoida. Helgoländer Meeresuntersuchungen 50:259–264.

Heard, W.H. 1964. *Corbicula fluminea* in Florida. The Nautilus 77(3):105–107.

Heard, W.H. 1965. Comparative life histories of North American pill clams (Sphaeriidae: *Pisidium*). Malacologia 2(3):381–411.

Heard, W.H. 1970a. Eastern freshwater mollusks (II), the South Atlantic and Gulf drainages. Malacologia 10(1):23–31.

Heard, W.H. 1970b. Hermaphroditism in *Margaritifera falcata* (Gould) (Pelecypoda: Margaritiferidae). The Nautilus 83(3):113–114.

Heard, W.H. 1974. Anatomical systematics of freshwater mussels. Malacological Review 7(1):41–42.

Heard, W.H. 1975a. Sexuality and other aspects of reproduction in *Anodonta* (Pelecypoda: Unionidae). Malacologia 15(1):83–103.

Heard, W.H. 1975b. Determination of the endangered status of freshwater clams of the Gulf and southeastern states. Department of Biological Sciences, Florida State University, Tallahassee. Final Report prepared for the Office of Endangered Species, Bureau of Sport Fisheries and Wildlife, U.S. Department of Interior, Contract Number 14-16-000-8905. 31 pages.

Heard, W.H. 1979a. Hermaphroditism in *Elliptio* (Pelecypoda: Unionidae). Malacological Review 12:21–28.

Heard, W.H. 1979b. Identification manual of the freshwater clams of Florida. Florida Department of Environmental Regulation, Technical Series 4(2). 83 pages.

Heard, W.H. 1998a. Brooding patterns in freshwater mussels. Malacological Review, Supplement 7, Bivalvia 1:105–121.

Heard, W.H. 1998b. A history of the study of generation of unionoidean bivalves: Aristotle to Houghton (1862). Malacological Review, Supplement 7, Bivalvia 1:123–136.

Heard, W.H. 2000. Glochidial larvae of freshwater mussels and their fish hosts: early discoveries and interpretations of the association. Malacological Review, Supplement 8, Freshwater Mollusca 1:83–88.

Heard, W.H., and G. Dinesen. 1999. A history of the controversy about *Glochidium parasiticum* Rathke, 1797 (Palaeoheterodonta: Unionoida: Unionoidea). Malacological Review, Supplement 8, Freshwater Mollusca 1:89–106.

Heard, W.H., and R.H. Guckert. 1970. A re-evaluation of the recent Unionacea (Pelecypoda) of North America. Malacologia 10(2):333–355.

Hebert, P.D.N., C.C. Wilson, M.M. Murdoch and R. Lazar. 1991. Demography and ecological impacts of the invading mollusc *Dreissena polymorpha*. Canadian Journal of Zoology 69(2):405–409.

Heinricher, J.R., and J.B. Layzer. 1999. Reproduction by individuals of a nonreproducing population of *Megalonaias nervosa* (Mollusca: Unionidae) following translocation. The American Midland Naturalist 141(1):140–148.

Henderson, J. 1929. Non-marine Mollusca of Oregon and Washington. The University of Colorado Studies 17(2):47–190.

Hendrix, S.S., M.F. Vidrine and R.H. Hartenstine. 1985. A list of records of freshwater aspidogastrids (Tremotoda) and their hosts in North America. Proceedings of the Helminthological Society, Washington 52(2):289–296.

Herbers, K. 1914. Entwicklungsgeschichte von *Anodonta cellensis* Schröt. Zeitschrift für Wissenshaftliche Zoologie 108:1–174.

Herrmannsen, A.N. 1846–1852. Indicis generum Malacozoorum primordia. Nomina subgenerum, familiarum, tribuum, ordinum, classium; adjectis auctoribus, temporibus, locis systematicis atque literariis, etymis, synonymis. Praetermittuntur Cirripedia, Tunicata et Rhizopoda. Volume 1 (1846):1–232; (1847):233–637. Volume 2 (1847):1–352; (1848):353–492; (1849):493–717. Supplement (1852):1–140.

Hickman, M.E. 1937. A contribution to the knowledge of the molluscan fauna of east Tennessee. Masters thesis, University of Tennessee, Knoxville. 165 pages, 104 plates.

Higgins, F. 1858. A catalogue of the shell-bearing species of Mollusca, inhabiting the vicinity of Columbus, Ohio, with some remarks thereon. Pages 548–555. In: Twelfth annual report of the Ohio State Board of Agriculture with an abstract of the Proceedings of the County Agricultural Societies, to the General Assembly of Ohio: for the Year 1857.

Hildreth, S.P. 1828. Observations on, and descriptions of the shells found in the waters of the Muskingum River, Little Muskingum, and Duck Creek, in the vicinity of Marietta, Ohio. American Journal of Science and Arts 14(2):276–291, 2 plates. [Reprinted in Sterkiana 8(1962).]

Hill, D.M. 1986. Cumberland Mollusk Conservation Program, Activity 3: Identification of fish hosts. Tennessee Valley Authority, Office of Natural Resources and Economic Development, Norris, Tennessee. 57 pages.

Hillegass, K.R., and M.C. Hove. 1997. Suitable fish hosts for glochidia of three freshwater mussels: Strange Floater, Ellipse, and Snuffbox. Triannual Unionid Report 13:25.

Hillis, D.M., and J.C. Patton. 1982. Morphological and electrophoretic evidence for two species of *Corbicula* (Bivalvia: Corbiculidae) in North America. The American Midland Naturalist 108(1):74–80.

Hinch, S.G., R.C. Bailey and R.H. Green. 1986. Growth of *Lampsilis radiata* (Bivalvia: Unionidae) in sand and mud: a reciprocal transplant experiment. Canadian Journal of Fisheries and Aquatic Sciences 43(3):548–552.

Hinkley, A.A. 1904. List of Alabama shells collected in October and November, 1903. The Nautilus 18(4):37–45; 18(5):54–57.

Hinkley, A.A. 1906. Some shells from Mississippi and Alabama. The Nautilus 20(3):34–36; 20(4):40–44; 20(5):52–55.

Hoeh, W.R. 1990. Phylogenetic relationships among eastern North American *Anodonta* (Bivalvia: Unionidae). Malacological Review 23(1–2):63–82.

Hoeh, W.R., M.B. Black, R. Gustafson, A.E. Bogan, R.A. Lutz and R.C. Vrijenhoek. 1998. Testing alternative hypotheses of *Neotrigonia* (Bivalvia: Trigonioida) phylogenetic relationships using cytochrome *c* oxidase Subunit 1 DNA sequences. Malacologia 40(1–2):267–278.

Hoeh, W.R., A.E. Bogan, K.S. Cummings and S.I. Guttman. 2002. Evolutionary relationships among the higher taxa of freshwater mussels (Bivalvia: Unionoida): inferences on phylogeny and character evolution from analyses of DNA sequence data. Malacological Review 31–32(2):123–141.

Hoeh, W.R., A.E. Bogan and W.H. Heard. 2001. A phylogenetic perspective on the evolution of morphological and reproductive characteristics in the Unionoida. Pages 257–280. In: G. Bauer and K. Wächtler (editors). Ecology and Evolution of the Freshwater Mussels Unionoida. Ecological Studies, Volume 145. Springer-Verlag, Berlin, Germany.

Hoeh, W.R., K.S. Frazer, E. Naranjo-Garcia and R.J. Trdan. 1995. A phylogenetic perspective on the evolution of simultaneous hermaphroditism in a freshwater mussel clade (Bivalvia: Unionidae: *Utterbackia*). Malacological Review 28(1–2):25–42.

Hoeh, W.R., D.T. Stewart and S.I. Guttman. 2002. High fidelity of mitochondrial genome transmission under the doubly uniparental mode of inheritance in freshwater mussels (Bivalvia: Unionoidea). Evolution 56(11):2252–2261.

Hoeh, W.R., D.T. Stewart, B.W. Sutherland and E. Zouros. 1996. Multiple origins of gender-associated mitochondrial DNA lineages in bivalves (Mollusca: Bivalvia). Evolution 50(6):2276–2286.

Hoggarth, M.A. 1980. A study of the distinguishing characteristics of *Obovaria jacksoniana* Frierson (1912) [sic] and *Obovaria unicolor* (Lea, 1845). Masters thesis, University of North Alabama, Florence. 74 pages.

Hoggarth, M.A. 1993. Glochidial functional morphology and rarity in the Unionidae. Pages 76–80. In: K.S. Cummings, A.C. Buchanan and L.M. Koch (editors). Conservation and Management of Freshwater Mussels. Proceedings of a UMRCC Symposium, 12–14 October 1992, St. Louis, Missouri. Upper Mississippi River Conservation Committee, Rock Island, Illinois.

Hoggarth, M.A. 1998. The Unionidae (Mollusca: Bivalvia) of the Walhonding River, Coshocton County, Ohio, including the federally endangered catspaw (*Epioblasma obliquata obliquata*), fanshell (*Cyprogenia stegaria*), and clubshell (*Pleurobema clava*) mussels. Walkerana (for 1995–1996) 8(20):149–176.

Hoggarth, M.A. 1999. Description of some of the glochidia of the Unionidae (Mollusca: Bivalvia). Malacologia 41(1):1–118.

Hoggarth, M.A., and A.S. Gaunt. 1988. Mechanics of glochidial attachment (Mollusca: Bivalvia: Unionidae). Journal of Morphology 198(1):71–81.

Hoke, E. 2000. A critical review of the unionoid mollusks reported for Nebraska by Samuel Aughey (1877). Central Plains Archeology 8(1):12–23.

Holland-Bartels, L.E. 1990. Physical factors and their influence on the mussel fauna of a main channel border habitat of the upper Mississippi River. Journal of the North American Benthological Society 9(4):327–335.

Holland-Bartels, L.E., and T.W. Kammer. 1989. Seasonal reproductive development of *Lampsilis cardium*, *Amblema plicata plicata* and *Potamilus alatus* (Pelecypoda: Unionidae) in the upper Mississippi River. Journal of Freshwater Ecology 5(1):87–92.

Hopkins, S.H., and J.D. Andrews. 1970. *Rangia cuneata* on the East Coast: thousand mile range extension, or resurgence? Science 167(3919):868–869.

Horn, K. 1983. Orientation of the freshwater mussel *Lampsilis radiata luteola* (Lam. 1819) in an eastern Kentucky stream. Masters thesis, Marshall University, Huntington, West Virginia. 54 pages.

Horn, K.J., and H.J. Porter. 1981. Correlations of shell shape of *Elliptio waccamawensis*, *Leptodea ochracea* and *Lampsilis* sp. (Bivalvia: Unionidae) with environmental factors in Lake Waccamaw, Columbus County, North Carolina. The Bulletin of the American Malacological Union, Inc. 1981:1–3.

Hornbach, D.J., J.G. March, T. Deneka, N.H. Troelstrup, Jr., and J.A. Perry. 1996. Factors influencing the distribution and abundance of the endangered Winged Mapleleaf Mussel *Quadrula fragosa* in the St. Croix River, Minnesota and Wisconsin. The American Midland Naturalist 136(2):278–286.

Horne, F.R., and S. McIntosh. 1979. Factors influencing distribution of mussels in the Blanco River of central Texas. The Nautilus 93(4):119–133.

Houghton, W. 1862. On the parasitic nature of the fry of *Anodonta cygnea*. Quarterly Journal of Microscopical Science (New Series) 2:162–168, plate 7.

Houslet, B.S., and J.B. Layzer. 1997. Difference in growth between two populations of *Villosa taeniata* in Horse Lick Creek, Kentucky. Pages 37–44. In: K.S. Cummings, A.C. Buchanan, C.A. Mayer and T.J. Naimo (editors). Conservation and Management of Freshwater Mussels II: Initiatives for the Future. Proceedings of a UMRCC Symposium, 16–18 October 1995, St. Louis, Missouri. Upper Mississippi River Conservation Committee, Rock Island, Illinois.

Hove, M.C. 1995. Suitable fish hosts of the lilliput, *Toxolasma parvus*. Triannual Unionid Report 8:9.

Hove, M.C. 1997. Ictalurids serve as suitable hosts for the purple wartyback. Triannual Unionid Report 11:4.

Hove, M.C., R.A. Engelking, M.E. Peteler and E.M. Peterson. 1995. *Anodontoides ferussacianus* and *Anodonta imbecillis* host suitability tests. Triannual Unionid Report 6:22.

Hove, M.C., R.A. Engelking, M.E. Peteler, E.M. Peterson, A.R. Kapuscinski, L.A. Sovell and E.R. Evers. 1997. Suitable fish hosts for glochidia of four freshwater mussels. Pages 21–25. In: K.S. Cummings, A.C. Buchanan, C.A. Mayer and T.J. Naimo (editors). Conservation and Management of Freshwater Mussels II: Initiatives for the Future. Proceedings of a UMRCC Symposium, 16–18 October 1995, St. Louis, Missouri. Upper Mississippi River Conservation Committee, Rock Island, Illinois.

Hove, M.C., R.A. Engelking, M.E. Peteler and L. Sovell. 1994. Life history research on *Ligumia recta* and *Lasmigona costata*. Triannual Unionid Report 4:[23].

Hove, M.C., J.E. Kurth and A.R. Kapuscinski. 1998. Brown bullhead suitable host for *Tritogonia verrucosa*; *Cumberlandia monodonta* host(s) remain elusive. Triannual Unionid Report 15:13.

Hove, M.C., and R.J. Neves. 1994. Life history of the endangered James spinymussel *Pleurobema collina* (Conrad, 1837) (Mollusca: Unionidae). American Malacological Bulletin 11(1):29–40.

Howard, A.D. 1913. The catfish as a host for fresh-water mussels. Transactions of the American Fisheries Society 42:65–70.

Howard, A.D. 1914a. A new record in rearing fresh-water pearl mussels. Transactions of the American Fisheries Society 44:45–47.

Howard, A.D. 1914b. A second case of metamorphosis without parasitism in the Unionidae. Science 40(1027):353–355.

Howard, A.D. 1914c. Some cases of narrowly restricted parasitism among commercial species of fresh water mussels. Transactions of the American Fisheries Society 44:41–44.

Howard, A.D. 1914d. Experiments in propagation of fresh-water mussels of the *Quadrula* group. Appendix IV to the Report of the U.S. Commissioner of Fisheries for 1913:1–52, 6 plates. [Issued separately as U.S. Bureau of Fisheries Document 801.]

Howard, A.D. 1915. Some exceptional cases of breeding among the Unionidae. The Nautilus 29(1):4–11.

Howard, A.D. 1922. Experiments in the culture of freshwater mussels. Bulletin of the U.S. Bureau of Fisheries 38(1924):63–89. [Issued separately as U.S. Bureau of Fisheries Document 916.]

Howard, A.D. 1951. A river mussel parasitic on a salamander. The Chicago Academy of Sciences, Natural History Miscellanea, Number 77. 6 pages.

Howard, A.D., and B.J. Anson. 1922. Phases in the parasitism of the Unionidae. Journal of Parasitology 9(2):68–82, 2 plates.

Howard, C.S. 1999. Efficacy of shell growth, behavior, and tissue condition to indicate effects of suspended sediment in the freshwater mussel, *Amblema plicata*. Masters thesis, Auburn University, Alabama. 144 pages.

Howells, R.G. 1995. Rio Grande Bleufer. Info-Mussel Newsletter 3(1):1.

Howells, R.G. 1996. Pistolgrip and Gulf Mapleleaf hosts. Info-Mussel Newsletter 4(3):3.

Howells, R.G. 1997. New fish hosts for nine freshwater mussels (Bivalvia: Unionidae). The Texas Journal of Science 49(3):255–258.

Howells, R.G. 2000. Reproductive seasonality of freshwater mussels (Unionidae) in Texas. Pages 35–48. In: R.A. Tankersley, D.I. Warmolts, G.T. Watters and B.J. Armitage (editors). Freshwater Mollusk Symposia Proceedings, Part I. Proceedings of the Conservation, Captive Care and Propagation of Freshwater Mussels Symposium, 1998. Ohio Biological Survey, Columbus, Ohio.

Howells, R.G., R.W. Neck and H.D. Murray. 1996. Freshwater Mussels of Texas. Texas Parks and Wildlife Department, Inland Fisheries Division, Austin, Texas. 218 pages.

Hubricht, L. 1963. *Corbicula fluminea* in the Mobile River. The Nautilus 77(1):31.

Hudson, R.G., and B.G. Isom. 1984. Rearing juveniles of the freshwater mussels (Unionidae) in a laboratory setting. The Nautilus 98(4):129–137.

Hueber, L.V. 1871. Zur Naturgeschiste der Unionin. Jahrbuch des naturhistorischen Landes-Museum von Kärnthen 10:150–157.

Huebner, J. 1810. Epistle. 1 page. ["This was a printed letter that apparently does not exist any more. Gerhard Falkner has looked for it in all sorts of libraries in Germany and it just seems gone." P. Bouchet, personal communication, 23 November 2005.]

Huff, S.W., D. Campbell, D.L. Gustafson, C. Lydeard, C.R. Altaba and G. Giribet. 2004. Investigations into the phylogenetic relationships of freshwater pearl mussels (Bivalvia: Margaritiferidae) based on molecular data: implications for their taxonomy and biogeography. Journal of Molluscan Studies 70:379–388.

Hughes, M.H., and P.W. Parmalee. 1999. Prehistoric and modern freshwater mussel (Mollusca: Bivalvia: Unionoidea) faunas of the Tennessee River: Alabama, Kentucky, and Tennessee. Regulated Rivers: Research and Management 15:25–42.

Hunter, R.D., and J.F. Bailey. 1992. *Dreissena polymorpha* (zebra mussel): colonization of soft substrata and some effects on unionid bivalves. The Nautilus 106:60–67.

Hurd, J.C. 1974. Systematics and zoogeography of the unionacean mollusks of the Coosa River drainage of Alabama, Georgia and Tennessee. Doctoral dissertation, Auburn University, Alabama. 240 pages.

Imlay, M.J. 1968. Environmental factors in activity rhythms of the freshwater clam *Elliptio complanatus catawbensis* (Lea). The American Midland Naturalist 80(2):508–528.

Imlay, M.J. 1972. Greater adaptability of freshwater mussels to natural rather than to artificial displacement. The Nautilus 86(2–4):76–79.

Imlay, M.J., and M.L. Paige. 1972. Laboratory growth of freshwater sponges, unionid mussels, and sphaeriid clams. Progressive Fish Culturist 34(4):210–216.

International Commission on Zoological Nomenclature (ICZN). 1926. Opinion 94. *Anodonta*. Smithsonian Miscellaneous Collections 73(4):12–13.

International Commission on Zoological Nomenclature (ICZN). 1957. Opinion 495. Designation under the plenary powers of a type species in harmony with accustomed usage for the nominal genus "*Unio*" Philipsson, 1788 (Class Pelecypoda) and validation under the same powers of the family-group name "Margaritiferidae" Haas, 1940. Opinions and Declarations Rendered by the International Commission on Zoological Nomenclature 17(17):287–322.

International Commission on Zoological Nomenclature (ICZN). 1959. Opinion 561. Protection under the Plenary Powers of the generic name *Anodonta* Lamarck, 1799 (Class Pelecypoda), a name placed on the Official list of Generic Names in Zoology in 1926 by the Ruling given in Opinion 94. Opinions and Declarations Rendered by the International Commission on Zoological Nomenclature 20(28):303–310.

International Commission on Zoological Nomenclature (ICZN). 1968. Opinion 840. *Amblema* Rafinesque, 1820 (Lamellibranchiata), validated under the plenary powers. Bulletin of Zoological Nomenclature 24:339–340.

International Commission on Zoological Nomenclature (ICZN). 1988. Opinion 1487. *Megalonaias* Utterback, 1915 (Mollusca, Bivalvia): conserved. Bulletin of Zoological Nomenclature 45(2):159.

International Commission on Zoological Nomenclature (ICZN). 1992. Opinion 1665. *Potamilus* Rafinesque, 1818 (Mollusca, Bivalvia): not suppressed. Bulletin of Zoological Nomenclature 49(1):81–82.

International Commission on Zoological Nomenclature (ICZN). 1999. International Code of Zoological Nomenclature. Fourth edition. International Trust for Zoological Nomenclature, London. 306 pages.

Irwin, E.R., G. Kowalski and D. Buckmeier. 1998. Distribution of endemic and threatened aquatic fauna in the upper Tallapoosa River. Unpublished final report by Alabama Cooperative Fish and Wildlife Research Unit, Auburn University, Alabama, submitted to U.S. Fish and Wildlife Service, Daphne, Alabama. 38 pages.

Isely, F.B. 1911. Preliminary note on the ecology of the early juvenile life of the Unionidae. Biological Bulletin 20(2):77–80.

Isely, F.B. 1914. Experimental study of the growth and migration of freshwater mussels. Appendix III to the Report of the U.S. Commissioner of Fisheries for 1913:1–24, 3 plates. [Issued separately as U.S. Bureau of Fisheries Document 792.]

Isom, B.G. 1968. The naiad fauna of Indian Creek, Madison County, Alabama. The American Midland Naturalist 79(2):514–516.

Isom, B.G. 1969. The mussel resource of the Tennessee River. Malacologia 7(2–3):397–425.

Isom, B.G. 1972. Mussels in the unique Nickajack Dam construction site, Tennessee River, 1965. Malacological Review 5(1):4–6.

Isom, B.G., C.H. Gooch and S.D. Dennis. 1979. Rediscovery of a presumed extinct river mussel, *Dysnomia sulcata* (Unionidae). The Nautilus 93(2–3):84.

Isom, B.G., and R.G. Hudson. 1982. In vitro culture of parasitic freshwater mussel glochidia. The Nautilus 96(4):147–151.

Isom, B.G., and P. Yokley, Jr. 1968. Mussels of Bear Creek watershed, Alabama and Mississippi, with a discussion of the area geology. The American Midland Naturalist 79(1):189–196.

Isom, B.G., and P. Yokley, Jr. 1973. The mussels of the Flint and Paint Rock River systems of the southwest slope of the Cumberland Plateau in north Alabama – 1965 and 1967. The American Midland Naturalist 89(2):442–446.

Isom, B.G., P. Yokley, Jr., and C.H. Gooch. 1973. Mussels of the Elk River basin in Alabama and Tennessee 1965–1967. The American Midland Naturalist 89(2):437–442.

Israel, W.V. 1910. Die Najadeen des Weidagebietes. Nachrichtsblatt der Deutschen Malakozoologischen Gesellschaft, Beilage Number 4:49–56.

Israel, W.V. 1911. Najadologische Miscellen. Nachrichtsblatt der Deutschen Malakozoologischen Gesellschaft 43(1):10–17.

Israel, W.V. 1913. Biologie der europäischen Süsswassermuscheln. Thüringer Lehrerverein für Naturkunde. Stuttgart. 93 pages, 18 plates.

Jackson, H.H., III. 1995. Rivers of History: Life on the Coosa, Tallapoosa, Cahaba, and Alabama. The University of Alabama Press, Tuscaloosa. 300 pages.

Jackson, H.H., III. 1997. Putting "Loafing Streams" to Work. The University of Alabama Press, Tuscaloosa. 230 pages.

Jansen, W.A. 1991. Seasonal prevalence, intensity of infestation, and distribution of glochidia of *Anodonta grandis simpsoniana* Lea on yellow perch, *Perca flavescens*. Canadian Journal of Zoology 69:964–972.

Jansen, W.A., and J.M. Hanson. 1991. Estimates of the number of glochidia produced by clams (*Anodonta grandis simpsoniana* Lea), attaching to yellow perch (*Perca flavescens*), and surviving to various ages in Narrow Lake, Alberta. Canadian Journal of Zoology 69:973–977.

Jass, J., and J. Glenn. 1999. Measuring the degree of variation in Wisconsin *Pyganodon grandis* (Say, 1821) (Mollusca: Bivalvia: Unionidae). Transactions of the Wisconsin Academy of Sciences, Arts and Letters 87:105–110.

Jay, J.C. 1839. A catalogue of the shells arranged according to the Lamarckian system; together with descriptions of new or rare species, contained in the collection of John C. Jay, M.D. Third edition. Wiley and Putnam, New York. 125 pages.

Jegla, T.C., and M.J. Greenberg. 1968. Structure of the bivalve rectum II. Notes on cell types and enervation. The Veliger 10:314–319, plates 45–48.

Jenkinson, J.J. 1973. Distribution and zoogeography of the Unionidae (Mollusca: Bivalvia) in four creek systems in east-central Alabama. Masters thesis, Auburn University, Alabama. 96 pages.

Jenkinson, J.J. 1976. Chromosome numbers of some North American naiads. Bulletin of the American Malacological Union, Inc. 1976:16–17.

Jenkinson, J.J. 1980. The Tennessee Valley Authority Cumberlandian Mollusk Conservation Program. Bulletin of the American Malacological Union, Inc. 1980:62–63.

Jenkinson, J.J. 1982. Cumberlandian Mollusk Conservation Program. Pages 95–103. In: A.C. Miller (compiler). Report of Freshwater Mollusks Workshop, 19–20 May 1981, U.S. Army Corps of Engineers, Waterways Experiment Station, Vicksburg, Mississippi.

Jirka, K.J., and R.J. Neves. 1992. Reproductive biology of four species of freshwater mussels (Mollusca: Unionidae) in the New River, Virginia and West Virginia. Journal of Freshwater Ecology 7(1):35–44.

Johnson, F.F. 1934. Aquatic shell industries. U.S. Bureau of Fisheries, Fishery Circular 15:1–17.

Johnson, J.A. 1997. The mussel, snail, and crayfish species of the Tallapoosa River drainage, with an assessment of their distribution in relation to chemical and physical habitat characteristics. Masters thesis, Auburn University, Alabama. 232 pages.

Johnson, J.A., and D.R. DeVries. 2002. The freshwater mussel and snail species of the Tallapoosa River Drainage, Alabama, U.S.A. Walkerana (for 1997–1998) 9(22):121–138.

Johnson, P.D., and R.R. Evans. 2000. A contemporary and historical database of freshwater mollusks in the Conasauga River Basin. A report to the U.S. Geological Survey Species-at-Risk Program Contract Number 98HQAG-2154. 293 pages.

Johnson, R.I. 1952. A study of Lamarck's types of Unionidae and Mutelidae. The Nautilus 66(2):63–67.

Johnson, R.I. 1953. A study of Lamarck's types of Unionidae and Mutelidae (cont). The Nautilus 66(3):90–95.

Johnson, R.I. 1956. Types of naiades (Mollusca: Unionidae) in the Museum of Comparative Zoology. Bulletin of the Museum of Comparative Zoology 115(4):102–142, plates 1–2.

Johnson, R.I. 1965. A hitherto overlooked *Anodonta* (Mollusca: Unionidae) from the Gulf drainage of Florida. Breviora, Number 213:1–7.

Johnson, R.I. 1967a. Additions to the unionid fauna of the Gulf drainage of Alabama, Georgia and Florida (Mollusca: Bivalvia). Brevoria, Number 270. 21 pages.

Johnson, R.I. 1967b. Illustrations of all the mollusks described by Berlin Hart and Samuel Hart Wright. Museum of Comparative Zoology, Harvard University, Occasional Papers on Mollusks 3(35):1–35, plates 1–13.

Johnson, R.I. 1968. *Elliptio nigella*, overlooked unionid from Apalachicola River system. The Nautilus 82(1):22–24.

Johnson, R.I. 1969a. Further additions to the unionid fauna of the Gulf drainage of Alabama, Georgia and Florida. The Nautilus 83(1):34–35.

Johnson, R.I. 1969b. Illustrations of Lamarck's types of North American Unionidae mostly in the Paris Museum. The Nautilus 83(2):52–61, figures 1–14.

Johnson, R.I. 1970. The systematics and zoogeography of the Unionidae (Mollusca: Bivalvia) of the southern Atlantic Slope Region. Bulletin of the Museum of Comparative Zoology 140(6):263–449.

Johnson, R.I. 1971. The types and figured specimens of Unionacea (Mollusca: Bivalvia) in the British Museum (Natural History). Bulletin of the British Museum (Natural History) 20(3):75–108, plates 1–10.

Johnson, R.I. 1972a. Illustrations of all of the mollusks described by Lorraine Screven Frierson. Museum of Comparative Zoology, Harvard University, Occasional Papers on Mollusks 3(41):137–173, plates 22–32.

Johnson, R.I. 1972b. The Unionidae (Mollusca: Bivalvia) of Peninsular Florida. Bulletin of the Florida State Museum, Biological Sciences 16(4):181–249.

Johnson, R.I. 1973. The types of Unionidae (Mollusca: Bivalvia) described by C.S. Rafinesque in the Museum national d'Histoire naturelle, Paris. Journal de Conchyliologie 110(2):35–37, plate 1.

Johnson, R.I. 1974. Lea's unionid types or Recent and fossil taxa of Unionacea and Mutelacea introduced by Isaac Lea, including the location of all the extant types. Museum of Comparative Zoology, Harvard University, Special Occasional Publication, Number 2. 159 pages.

Johnson, R.I. 1975. Simpson's unionid types and miscellaneous unionid types in the National Museum of Natural History. Museum of Comparative Zoology, Harvard University, Special Occasional Publication, Number 4. 56 pages, 3 plates.

Johnson, R.I. 1977. Monograph of the genus *Medionidus* (Bivalvia: Unionidae) mostly from the Apalachicolan region, southeastern United States. Museum of Comparative Zoology, Harvard University, Occasional Papers on Mollusks 4(56):161–187.

Johnson, R.I. 1978. Systematics and zoogeography of *Plagiola* (=*Dysnomia* =*Epioblasma*), an almost extinct genus of freshwater mussels (Bivalvia: Unionidae) from middle North America. Bulletin of the Museum of Comparative Zoology 148(6):239–320.

Johnson, R.I. 1979. The types of Unionacea (Mollusca: Unionidae) in the Museum of Zoology, the University of Michigan. Malacological Review 12(1–2):29–36.

Johnson, R.I. 1980. The types of Unionacea (Mollusca: Bivalvia) in the Academy of Natural Sciences of Philadelphia: Additions and corrections. Proceedings of the Academy of Natural Sciences of Philadelphia 132:277–278.

Johnson, R.I. 1981. Recent and fossil Unionacea and Mutelacea (freshwater bivalves) of the Caribbean Islands. Museum of Comparative Zoology, Harvard University, Occasional Papers on Mollusks 4(60):269–288.

Johnson, R.I. 1983. *Margaritifera marrianae*, a new species of Unionacea (Bivalvia: Margaritiferidae) from the Mobile-Alabama-Coosa and Escambia River systems, Alabama. Museum of Comparative Zoology, Harvard University, Occasional Papers on Mollusks 4(62):299–304, plate 41.

Johnson, R.I. 1998. Addenda to be included with Lea's Figured Unionid Types (1974). Museum of Comparative Zoology, Harvard University. 2 pages.

Johnson, R.I. 1999. Unionidae of the Rio Grande (Rio Bravo del Norte) system of Texas and New Mexico. Museum of Comparative Zoology, Harvard University, Occasional Papers on Mollusks 6(77):1–65.

Johnson, R.I. 2006. Conchology at the Lyceum of Natural History of New York: 1817–1876. Sporadic Papers on Mollusks 1:3–53.

Johnson, R.I., and H.B. Baker. 1973. The types of Unionacea (Mollusca: Bivalvia) in the Academy of Natural Sciences of Philadelphia. Proceedings of the Academy of Natural Sciences of Philadelphia 125(9):145–186, plates 1–10.

Johnson, R.L., F.Q. Liang, C.D. Milam and J.L. Farris. 1998. Genetic diversity and cellulolytic activity among several species of unionid bivalves in Arkansas. Journal of Shellfish Research 17(5):1375–1382.

Jokela, J., and P. Mutilainen. 1995. Effect of size-dependant muskrat (*Ondatra zibethica*) predation on the spatial distribution of a freshwater clam, *Anodonta piscinalis* Nilsson (Unionidae, Bivalvia). Canadian Journal of Zoology 73(6):1085–1094.

Jones, J.W. 2004. A holistic approach to taxonomic evaluation of two closely related endangered freshwater mussel species, the Oyster Mussel (*Epioblasma capsaeformis*) and Tan Riffleshell (*Epioblasma florentina walkeri*) (Bivalvia: Unionidae). Masters thesis, Virginia Polytechnic Institute and State University, Blacksburg. 178 pages.

Jones, J.W., R. Mair and R.J. Neves. 2003. Annual progress report for 2002: life history and artificial culture of endangered mussels. Report submitted to Tennessee Wildlife Resources Agency, Nashville, Tennessee. 80 pages.

Jones, J.W., and R.J. Neves. 2000. Annual progress report for 1999: life history and artificial culture of endangered mussels. Report submitted to Tennessee Wildlife Resources Agency, Nashville, Tennessee. 57 pages.

Jones, J.W., and R.J. Neves. 2001. Annual progress report for 2000: life history and artificial culture of endangered mussels. Report submitted to Tennessee Wildlife Resources Agency, Nashville, Tennessee. 66 pages.

Jones, J.W., and R.J. Neves. 2002a. Annual progress report for 2001: life history and artificial culture of endangered mussels. Report submitted to Tennessee Wildlife Resources Agency, Nashville, Tennessee. 90 pages.

Jones, J.W., and R.J. Neves. 2002b. Life history and propagation of the endangered fanshell pearlymussel, *Cyprogenia stegaria* Rafinesque (Bivalvia: Unionidae). Journal of the North American Benthological Society 21(1):76–88.

Jones, J.W., R.J. Neves, S.A. Ahlstedt and R.A. Mair. 2004. Life history and propagation of the endangered dromedary pearlymussel (*Dromus dromas*) (Bivalvia: Unionidae). Journal of the North American Benthological Society 23(3):515–525.

Jones, R.L. 1991. Population status of endangered mussels in the Buttahatchee River, Alabama and Mississippi. Segment 2. 1990. Mississippi Museum of Natural Science, Museum Technical Report, Number 14. 36 pages.

Jones, R.L., C.L. Knight and T.C. Majure. 1996. Endangered mussels of the Tombigbee River tributaries: the Noxubee River. Mississippi Museum of Natural Science, Museum Technical Report, Number 39. 21 pages.

Jones, R.L., and T.C. Majure. 1999. Endangered mussels of the Tombigbee River tributaries: Bull Mountain Creek. Mississippi Museum of Natural Science, Jackson, Mississippi, Museum Technical Report, Number 65. 10 pages.

Jones, R.L., W.T. Slack and P.D. Hartfield. 2005. The freshwater mussels (Mollusca: Bivalvia: Unionidae) of Mississippi. Southeastern Naturalist 4(1):77–92.

Jones, R.O. 1950. Propagation of fresh-water mussels. The Progressive Fish-Culturist 12(1):13–25.

Jones, W.B. 1938. Conservation of our natural and wildlife resources. Geological Survey of Alabama, Circular 11. 14 pages.

Jousseaume, F.P. 1875.Coquilles de la famille des marginelles. Monographie. Revue et Magasin de Zoologie (Series 3) 3:164–271, 429–435, plates 7–8.

Kandl, K.L., H.-P. Liu, R.S. Butler, W.R. Hoeh and M. Mulvey. 2001. A genetic approach to resolving taxonomic ambiguity among *Pleurobema* (Bivalvia: Unionidae) of the eastern Gulf Coast. Malacologia 43(1–2):87–101.

Kat, P.W. 1982. Effects of population density and substratum type on growth and migration of *Elliptio complanata* (Bivalvia: Unionidae). Malacological Review 15(1–2):119–127.

Kat, P.W. 1983a. Morphologic divergence, genetics, and speciation among *Lampsilis* (Bivalvia: Unionidae). Journal of Molluscan Studies 49(2):133–145.

Kat, P.W. 1983b. Genetic and morphological divergence among nominal species of North American *Anodonta* (Bivalvia: Unionidae). Malacologia 23(2):362–374.

Kat, P.W. 1983c. Sexual selection and simultaneous hermaphroditism among the Unionidae (Bivalvia: Mollusca). Journal of Zoology (London) 201(3):395–416.

Kat, P.W. 1984. Parasitism and the Unionacea (Bivalvia). Biological Review 59(2):189–207.

Kays, W.T., H. Silverman and T.H. Dietz. 1990. Water channels and water canals in the gill of the freshwater mussel, *Ligumia subrostrata*: ultrastructure and histochemistry. The Journal of Experimental Zoology 254(2):256–269.

Keeler, J.E. (editor). 1972. Rare and Endangered Vertebrates of Alabama. Alabama Department of Conservation and Natural Resources, Montgomery, Alabama. 92 pages.

Keller, A.E., and D.S. Ruessler. 1997. Determination or verification of host fish for nine species of unionid mussels. The American Midland Naturalist 138(2):402–407.

Keller, A.E., D.S. Ruessler and C.M. Chaffee. 1998. Testing the toxicity of sediments contaminated with diesel fuel using glochidia and juvenile mussels (Bivalvia, Unionidae). Aquatic Ecosystem Health and Management 1(1998):37–47.

Keller, A.E., and S.G. Zam. 1990. Simplification of in vitro culture techniques for freshwater mussels. Environmental Toxicology and Chemistry 9(10):1291–1296.

Kesler, D.H., and R.C. Bailey. 1993. Density and ecomorphology of a freshwater mussel (*Elliptio complanata*, Bivalvia: Unionidae) in a Rhode Island lake. Journal of the North American Benthological Society 12(3):259–264.

Kesler, D.H., and J.A. Downing. 1997. Internal shell annuli yield inaccurate growth estimates in the freshwater mussels *Elliptio complanata* and *Lampsilis radiata*. Freshwater Biology 37(2):325–332.

Khym, J.R., and J.B. Layzer. 2000. Host fish suitability for glochidia of *Ligumia recta*. The American Midland Naturalist 143(1):178–184.

Kilias, R. 1956. Über die kieme der teichmuschel (*Anodonta* Lam.) (Ein beitrag zur anatomie). Mitteilungen aus dem Zoologischen Museum in Berlin 32(1):151–174.

Kirk, S.G., and J.B. Layzer. 1997. Induced metamorphosis of freshwater mussel glochidia on nonhost fish. The Nautilus 110(3):102–106.

Kirtland, J.P. 1840. Fragments of natural history. Number I. Habits of naiades. American Journal of Science and Arts 39(1):164–168.

Kitchel, H.E. 1985. Life history of the endangered Shiny Pigtoe mussel, *Fusconaia edgariana*, in the North Fork Holston River, Virginia. Masters thesis, Virginia Polytechnic Institute and State University, Blacksburg. 124 pages.

Kokai, F.L. 1974. Variation in the incurrent and excurrent apertures of *Quadrula quadrula* Rafinesque, 1820 and *Quadrula pustulosa* (Lea, 1831). Bulletin of the American Malacological Union, Inc. 1974:32–34.

Kokai, F.L. 1976. Variations in aperture characteristics of eighteen species of Unionidae from Lake Erie. Bulletin of the American Malacological Union, Inc. 1976:12–15.

Kotrla, B. 1988. Gametogenesis and gamete morphology of *Anodonta imbecilis* [sic], *Elliptio icterina*, and *Villosa villosa* (Bivalvia: Unionidae). Doctoral dissertation, Florida State University, Tallahassee. 165 pages.

Kovalak, W.P., G.D. Longton and R.D. Smithee. 1993. Infestation of power plant water systems by the Zebra Mussel (*Dreissena polymorpha* Pallas). Pages 359–380. In: T.F. Nalepa and D.W. Schloesser (editors). Zebra Mussels: Biology, Impacts, and Control. Lewis Publishers, Boca Raton, Florida.

Kraemer, L.R. 1967. The distribution of the posterior nerves in *Lampsilis ventricosa* (Barnes). The American Malacological Union, Inc. Annual Reports for 1967:42.

Kraemer, L.R. 1970. The mantle flap in three species of *Lampsilis* (Pelecypoda: Unionidae). Malacologia 10(1):225–282.

Kraemer, L.R. 1978. Discovery of two distinct kinds of statocysts in freshwater bivalved mollusks: some behavioral implications. The Bulletin of the American Malacological Union, Inc. 1978:24–28.

Kraemer, L.R. 1979. *Corbicula* (Bivalvia: Sphaeriacea) *vs.* indigenous mussels (Bivalvia: Unionacea) in U.S. rivers: a hard case for interspecific competition. American Zoologist 19(4):1085–1096.

Kraemer, L.R. 1984. Aspects of the functional morphology of some fresh-water bivalve nervous systems: effects on reproductive processes and adaptation of sensory mechanisms in the Sphaeriacea and Unionacea. Malacologia 25(1):221–239.

Krakow, K.K. 1975. Georgia Place-names. First edition. Winship Press, Macon, Georgia. 172 pages.

Krebs, R.A. 2004. Combining paternally and maternally inherited mitochondrial DNA for analysis of population structure in mussels. Molecular Ecology 13(6):1701–1705.

Krug, C. 1922. Morphologie und Histologie des Herzens und Pericards von *Anodonta cellensis*. Zeitschrift für wissenschaftliche Zoologie 119:155–246.

Kunz, G.F. 1898a. A brief history of the gathering of fresh-water pearls in the United States. U.S. Fish Commission Bulletin 17(1897):321–330.

Kunz, G.F. 1898b. The fresh-water pearls and pearl fisheries of the United States. U.S. Fish Commission Bulletin 17(1897):373–426, 22 plates.

Küster, H.C. 1838–1842, 1853, 1873–1876. Die Gattung *Anodonta* nebst den übregen Najaden mit unvollkomeuem Schloss. In: Systematisches Conchylien-Cabinet von Martini und Chemnitz. Second edition. 9(1):1–288, plates 1–87. Nürnberg. Verlag von Bauer and Raspe.

Küster, H.C. 1839, 1848–1862. Die flussperlmuscheln (*Unio* et *Hyria*) in Abbildungen nach der Natur. In: Systematisches Conchylien-Cabinet von Martini und Chemnitz. Second edition. 9(2, Section 1):1–318, plates 1–100. Nürnberg. Verlag von Bauer and Raspe.

Lamarck, J.B.P.A. 1799. Prodrome d'une nouvelle classification des coquilles, comprenent une rédaction appropriée des caractères génériques, et l'établissement d'un grand nombre de genres nouveaux. Memoirs de la Societie d'Histoire Naturelle, Paris 1:63–91.

Lamarck, J.B.P.A. 1815–1822. Histoire naturelle des Animaux sans Vertébres. 8 volumes. [Les nayades, 1819. 5:67–100.]

Landman, N.H., P.M. Mikkelsen, R. Bieler and B. Bronson. 2001. Pearls: a Natural History. Harry N. Abrams, Incorporated, New York, in association with The American Museum of Natural History, New York, and the Field Museum, Chicago, Illinois. 232 pages.

La Rocque, A., and J. Oughton. 1937. A preliminary account of the Unionidae of Ontario. Canadian Journal of Research 15(8):147–155.

Lasee, B.A. 1991. Histological and ultrastructural studies of larval and juvenile *Lampsilis* (Bivalvia) from the upper Mississippi River. Doctoral dissertation, Iowa State University, Ames. 146 pages.

Lavrentyev, P.J., W.S. Gardner, J.F. Cavaletto and J.R. Beaver. 1995. Effects of the zebra mussel (*Dreissena polymorpha* Pallas) on protozoa and phytoplankton from Saginaw Bay, Lake Huron. Journal of Great Lakes Research 21(4):545–557.

Layzer, J.B., and R.M. Anderson. 1992. Impacts of the coal industry on rare and endangered aquatic organisms of the upper Cumberland River Basin. Final report to Kentucky Department of Fish and Wildlife Resources, Frankfurt, Kentucky, and Tennessee Wild Resources Agency, Nashville, Tennessee. 118 pages.

Layzer, J.B., and L.M. Madison. 1995. Microhabitat use by freshwater mussels and recommendations for determining their instream flow needs. Regulated Rivers: Research and Management 10(2–4):329–345.

Lea, I. 1828. Description of six new species of the genus *Unio*, embracing the anatomy of the oviduct of one of them, together with some anatomical observations on the genus. Transactions of the American Philosophical Society 3 (New Series) (3):259–273, plates 3–6, colored.

Lea, I. 1829. Description of a new genus of the family of naïades, including eight species, four of which are new; also the description of eleven new species of the genus *Unio* from the rivers of the United States: with observations on some of the characters of the naïades. Transactions of the American Philosophical Society 3 (New Series) (4):403–457, plates 7–14.

Lea, I. 1831. Observations on the naïades, and descriptions of new species of that and other families. Transactions of the American Philosophical Society 4 (New Series) (1):63–121, plates 3–18.

Lea, I. 1834a. Observations on the naïades; and descriptions of new species of that, and other families. Transactions of the American Philosophical Society 5 (New Series) (1):23–119, plates 1–19.

Lea, I. 1834b. Observations on the genus *Unio*, together with descriptions of new genera and species in the families Naïades, Conchae, Colimacea, Lymnaeana, Melaniana and Peristomiana: consisting of four memoirs read before the American Philosophical Society from 1827 to 1834, and originally published in their Transactions. James Kay, Jun. and Company, Philadelphia. Volume 1. 4 leaves, 233 pages, plates 3–14, 3–18, 1–19.

Lea, I. 1836. A synopsis of the family of Naïades. Cary, Lea and Blanchard, Philadelphia; John Miller, London. 59 pages, 1 plate, colored.

Lea, I. 1838a. Description of new freshwater and land shells. Transactions of the American Philosophical Society 6 (New Series) (1):1–154, plates 1–24, colored.

Lea, I. 1838b. A synopsis of the family of Naïades. Second edition, enlarged and improved. Philadelphia. 44 pages. [This publication represents pages 113–152 of Lea (1838a) repaginated and bound separately.]

Lea, I. 1838c. Observations on the genus *Unio*, together with descriptions of new genera and species in the families Naïades, Conchae, Colimacea, Lymnaeana, Melaniana and Peristomiana. Printed for the author, Philadelphia. Volume 2. 152 pages, plates 1–24, colored.

Lea, I. 1840. Descriptions of new fresh water and land shells. Proceedings of the American Philosophical Society 1(13):284–289.

Lea, I. 1841a. Continuation of paper [On fresh water and land shells]. Proceedings of the American Philosophical Society 2(17):30–35.

Lea, I. 1841b. Continuation of Mr. Lea's Paper [On fresh water and land shells]. Proceedings of the American Philosophical Society 2(19):81–83.

Lea, I. 1842a. Continuation of paper [On new fresh water and land shells]. Proceedings of the American Philosophical Society 2(23):224–225.

Lea, I. 1842b. Description of new fresh water and land shells. Transactions of the American Philosophical Society 8 (New Series) (Part 2):163–250, plates 5–27.

Lea, I. 1842c. Observations on the genus *Unio*, together with descriptions of new species in the families Naïades, Conchae, Colimacea, Lymnaeana, Melaniana and Peristomiana. Printed for the author, Philadelphia. Volume 3. 88 pages, plates 5–27.

Lea, I. 1843. Description of twelve new species of Uniones. 1 page. [Read by Isaac Lea before the American Philosophical Society on 18 August 1843, and privately published on 19 August 1843, Philadelphia.]

Lea, I. 1845. Descriptions of new fresh water and land shells. Proceedings of the American Philosophical Society 4(33):162–168.

Lea, I. 1846. Description of new fresh water and land shells. Transactions of the American Philosophical Society 9 (New Series) (Part 2): 275–282, plates 39–42.

Lea, I. 1848a. Description of new fresh water and land shells. Transactions of the American Philosophical Society 10 (New Series) (Part 1):67–101, plates 1–9.

Lea, I. 1848b. Observations on the genus *Unio*, together with descriptions of new species in the families Naïades, Conchae, Colimacea, Lymnaeana, Melaniana and Peristomiana. Printed for the author, Philadelphia. Volume 4. 101 pages, plates 39–42, 1–9.

Lea, I. 1852a. A synopsis of the family of naïades. Third edition, greatly enlarged and improved. Blanchard and Lea, Philadelphia. 88 pages.

Lea, I. 1852b. Descriptions of new species of the family Unionidae. Transactions of the American Philosophical Society 10 (New Series) (Part 2):253–294, plates 12–29.

Lea, I. 1852c. Observations on the genus *Unio*, together with descriptions of new species in the families Unionidae, Colimacea, and Melaniana. Printed for the author, Philadelphia. Volume 5. 62 pages, plates 12–30.

Lea, I. 1854. Rectification of Mr. T.A. Conrad's "Synopsis of the Family of Naiades of North America," published in the "Proceedings of the Academy of Natural Sciences of Philadelphia, February, 1853." Proceedings of the Academy of Natural Sciences of Philadelphia 7:236–249. [Issued separately as a pamphlet.]

Lea, I. 1856a. Description of twenty-five new species of exotic uniones. Proceedings of the Academy of Natural Sciences of Philadelphia 8(2):92–95.

Lea, I. 1856b. Description of four new species of exotic uniones. Proceedings of the Academy of Natural Sciences of Philadelphia 8(3):103.

Lea, I. 1856c. Description of the byssus in the genus *Unio*. Proceedings of the Academy of Natural Sciences of Philadelphia 8(5):213–214.

Lea, I. 1856d. Description of eleven new species of uniones, from Georgia. Proceedings of the Academy of Natural Sciences of Philadelphia 8(6):262–263.

Lea, I. 1857a. Description of thirteen new species of uniones, from Georgia. Proceedings of the Academy of Natural Sciences of Philadelphia 9(2):31–32.

Lea, I. 1857b. Description of six new species of uniones from Alabama. Proceedings of the Academy of Natural Sciences of Philadelphia 9(1857):83.

Lea, I. 1857c. Description of eight new species of naïades from various parts of the United States. Proceedings of the Academy of Natural Sciences of Philadelphia 9(1857):84.

Lea, I. 1857d. Description of twelve new species of naïades from North Carolina. Proceedings of the Academy of Natural Sciences of Philadelphia 9(1857):85–86.

Lea, I. 1857e. Description of six new species of fresh water and land shells of Texas and Tamaulipas, from the collection of the Smithsonian Institution. Proceedings of the Academy of Natural Sciences of Philadelphia 9(1857):101–102.

Lea, I. 1857f. [Proposed change of the names of *Unio umbrosus* and *Unio wheatleyi* to *Unio umbrans* and *Unio catawbensis* respectively.] Proceeding of the Academy of Natural Sciences of Philadelphia 9(1857):104.

Lea, I. 1857g. Descriptions of twenty-seven new species of uniones from Georgia. Proceedings of the Academy of Natural Sciences of Philadelphia 9(1857):169–172.

Lea, I. 1858a. Descriptions of new species of *Unio*, from Tennessee, Alabama and North Carolina. Proceedings of the Academy of Natural Sciences of Philadelphia 10(1858):40–41.

Lea, I. 1858b. Descriptions of seven new species of Margaritanae, and four new species of Anodontae. Proceedings of the Academy of Natural Sciences of Philadelphia 10(1858):138–139.

Lea, I. 1858c. Descriptions of twelve new species of uniones and other fresh-water shells of the United States. Proceedings of the Academy of Natural Sciences of Philadelphia 10(1858):165–166.

Lea, I. 1858d. Descriptions of the embryonic forms of thirty-eight species of Unionidae. Journal of the Academy of Natural Sciences of Philadelphia 4 (New Series) (1):43–50, plate 5.

Lea, I. 1858e. New Unionidae of the United States. Journal of the Academy of Natural Sciences of Philadelphia 4 (New Series) (1):51–95, plates 6–20.

Lea, I. 1858f. Descriptions of exotic genera and species of the family Unionidae. Journal of the Academy of Natural Sciences of Philadelphia 3 (New Series) (4):289–321, plates 21–33.

Lea, I. 1858g. Observations on the genus *Unio*, together with descriptions of new species in the family Unionidae. Printed for the author, Philadelphia. Volume 6(1):9–41, plates 21–33.

Lea, I. 1858h. Observations on the genus *Unio*, together with descriptions of new species, their soft parts, and embryonic forms, in the family Unionidae. Printed for the author, Philadelphia. Volume 6(2):43–95 [index unpaginated], plates 5–20.

Lea, I. 1858i. Observations on the genus *Unio*, together with descriptions of new species, their soft parts, and embryonic forms, in the family Unionidae. Printed for the author, Philadelphia. Volume 6. 97 pages, plates 5–33.

Lea, I. 1859a. Descriptions of eight new species of Unionidae, from Georgia, Mississippi and Texas. Proceedings of the Academy of Natural Sciences of Philadelphia 11(1859):112–113.

Lea, I. 1859b. Descriptions of Two New Species of Uniones, from Georgia. Proceedings of the Academy of Natural Sciences of Philadelphia 11(1859):154.

Lea, I. 1859c. Descriptions of seven new species of Uniones from South Carolina, Florida, Alabama and Texas. Proceedings of the Academy of Natural Sciences of Philadelphia 11(1859):154–155.

Lea, I. 1859d. Descriptions of twelve new species of uniones, from Georgia. Proceedings of the Academy of Natural Sciences of Philadelphia 11(1859):170–172.

Lea, I. 1859e. [Change of name of *Margaritana etowahensis* to *M. georgiana*.] Proceedings of the Academy of Natural Sciences of Philadelphia 11(1859):280.

Lea, I. 1859f. New Unionidae of the United States. Journal of the Academy of Natural Sciences of Philadelphia 4 (New Series) (2):191–233, plates 21–32.

Lea, I. 1859g. Observations on the genus *Unio*, together with descriptions of new species, their soft parts, and embryonic forms, in the family Unionidae. Printed for the author, Philadelphia. Volume 7(1):9–51, plates 21–32.

Lea, I. 1860a. Descriptions of five new species of uniones from north Alabama. Proceedings of the Academy of Natural Sciences of Philadelphia 12(1860):92–93.

Lea, I. 1860b. Descriptions of two new species of uniones from Georgia. Proceedings of the Academy of Natural Sciences of Philadelphia 12(1860):305.

Lea, I. 1860c. Descriptions of seven new species of Unionidae from the United States. Proceedings of the Academy of Natural Sciences of Philadelphia 12(1860):306–307.

Lea, I. 1860d. New Unionidae of the United States and northern Mexico. Journal of the Academy of Natural Sciences of Philadelphia 4 (New Series) (4):327–374, plates 51–66.

Lea, I. 1860e. Observations on the genus *Unio*, together with descriptions of new species, their soft parts, and embryonic forms, in the family Unionidae. Printed for the author, Philadelphia. Volume 7(2):53–91 [index unpaginated], plates 33–45.

Lea, I. 1860f. Observations on the genus *Unio*, together with descriptions of new species, their soft parts, and embryonic forms, in the family Unionidae. Volume 7. Frontispiece, 93 pages, plates 21–45.

Lea, I. 1860g. Observations on the genus *Unio*, together with descriptions of new species, their soft parts, and embryonic forms, in the family Unionidae. Printed for the author, Philadelphia. Volume 8(1):9–56, plates 51–66.

Lea, I. 1861a. Descriptions of twenty-five new species of Unionidae from Georgia, Alabama, Mississippi, Tennessee and Florida. Proceedings of the Academy of Natural Sciences of Philadelphia 13(1861):38–41.

Lea, I. 1861b. Descriptions of two new species of Anodontae, from Arctic America. Proceedings of the Academy of Natural Sciences of Philadelphia 13(1861):56.

Lea, I. 1861c. Descriptions of twelve new species of Uniones, from Alabama. Proceedings of the Academy of Natural Sciences of Philadelphia 13(1861):59–60.

Lea, I. 1861d. Descriptions of eleven new species of the genus *Unio* from the United States. Proceedings of the Academy of Natural Sciences of Philadelphia 13(1861):391–393.

Lea, I. 1862a. Descriptions of ten new species of Unionidae of the United States. Proceedings of the Academy of Natural Sciences of Philadelphia 14(1862):168–169.

Lea, I. 1862b. New Unionidae of the United States. Journal of the Academy of Natural Sciences of Philadelphia 5 (New Series) (1):53–109, plates 1–18.

Lea, I. 1862c. New Unionidae of the United States and Arctic America. Journal of the Academy of Natural Sciences of Philadelphia 5 (New Series) (2):187–216, plates 24–33.

Lea, I. 1862d. Observations on the genus *Unio*, together with descriptions of new species, their soft parts, and embryonic forms, in the family Unionidae. Printed for the author, Philadelphia. Volume 8(2):57–113 [index unpaginated], plates 1–18.

Lea, I. 1862e. Observations on the genus *Unio*, together with descriptions of new species, their soft parts, and embryonic forms, in the family Unionidae. Printed for the author, Philadelphia. Volume 8. 115 pages, plates 51–66, 1–18.

Lea, I. 1863a. Observations on the genus *Unio*, together with descriptions of new species, their soft parts, and embryonic forms, in the family Unionidae and descriptions of new genera and species of the Melanidae. Printed for the author, Philadelphia. Volume 9. 180 pages, plates 24–39.

Lea, I. 1863b. Descriptions of eleven new species of exotic Unionidae. Proceedings of the Academy of Natural Sciences of Philadelphia 15(1863):189–190.

Lea, I. 1863c. Descriptions of twenty-four new species of Unionidae of the United States. Proceedings of the Academy of Natural Sciences of Philadelphia 15(1863):191–194.

Lea, I. 1863d. Descriptions of the soft parts of one hundred and forty-three species and some embryonic forms of Unionidae of the United States. Journal of the Academy of Natural Sciences of Philadelphia 5 (New Series) (4):401–456.

Lea, I. 1863e. Observations on the genus *Unio*, together with descriptions of new species, their soft parts, and embryonic forms, in the family Unionidae. Printed for the author, Philadelphia. Volume 10. 94 pages, plates 41–50.

Lea, I. 1865. Descriptions of eight new species of *Unio* of the United States. Proceedings of the Academy of Natural Sciences of Philadelphia 17(1865):88–89.

Lea, I. 1866. New Unionidae, Melanidae, etc., chiefly of the United States. Journal of the Academy of Natural Sciences of Philadelphia 6 (New Series) (1):5–65, plates 1–21.

Lea, I. 1867a. Descriptions of five new species of Unionidae and one *Paludina* of the United States. Proceedings of the Academy of Natural Sciences of Philadelphia 19(1867):81.

Lea, I. 1867b. Observations on the genus *Unio*, together with descriptions of new species in the family Unionidae, and descriptions of new species of the Melanidae, Limneidae, Paludinae, and Helicidae. Printed for the author, Philadelphia. Volume 11. 146 pages, plates 1–24.

Lea, I. 1867c. Index to Volume I to XI of Observations on the genus *Unio*, together with description of new species of the family Unionidae. And descriptions of new species of the Melanidae, Paludinae, Helicidae, *etc.* Volume I. Printed for the author by T.K. Collins, Philadelphia. 63 pages.

Lea, I. 1868a. Description of sixteen new species of the genus *Unio* of the United States. Proceedings of the Academy of Natural Sciences of Philadelphia 20(1868):143–145.

Lea, I. 1868b. New Unionidae, Melanidae, *etc.*, chiefly of the United States. Journal of the Academy of Natural Sciences of Philadelphia 6 (New Series) (3):249–302, plates 29–45.

Lea, I. 1868c. New Unionidae, Melanidae, *etc.*, chiefly of the United States. Journal of the Academy of Natural Sciences of Philadelphia 6 (New Series) (4):303–343, plates 46–54.

Lea, I. 1869a. Observations on the genus *Unio*, together with descriptions of new species in the family Unionidae, and descriptions of new species of the Melanidae, and Paludinae. Printed for the author, Philadelphia. Volume 12. 105 pages, plates 29–54.

Lea, I. 1869b. Index to Volume XII and supplementary index to Volumes I to XI of Observations on the genus *Unio*, together with description of new species of the family Unionidae, and description of new species of the Melanidae, Paludinae, Helicidae, *etc.* Volume II. Printed for the author by T.K. Collins, Philadelphia. 23 pages.

Lea, I. 1870. A synopsis of the family Unionidae. Fourth edition, very greatly enlarged and improved. Henry C. Lea, Philadelphia. 184 pages.

Lea, I. 1871. Descriptions of twenty new species of uniones of the United States. Proceedings of the Academy of Natural Sciences of Philadelphia 23(3):189–193.

Lea, I. 1872a. Rectification of T. A. Conrad's Synopsis of the family of Naïades of North America, published in the Proceedings of the Academy of Natural Sciences of Philadelphia, February, 1853. New edition. Collins, Printer, Philadelphia. 45 pages.

Lea, I. 1872b. Descriptions of twenty-nine species of Unionidae from the United States. Proceedings of the Academy of Natural Sciences of Philadelphia 24(2):155–161.

Lea, I. 1874a. Description of seven new species of Unionidae of the United States. Proceedings of the Academy of Natural Sciences of Philadelphia 25(3):422–423.

Lea, I. 1874b. Description of three new species of Uniones of the United States. Proceedings of the Academy of Natural Sciences of Philadelphia 25(3):424.

Lea, I. 1874c. Descriptions of fifty-two species of Unionidae. Journal of the Academy of Natural Sciences of Philadelphia 8 (New Series) (1):5–54, plates 1–18.

Lea, I. 1874d. Supplement to Isaac Lea's paper on Unionidae. Journal of the Academy of Natural Sciences of Philadelphia 8 (New Series) (1):55–69, plates 19–22.

Lea, I. 1874e. Observations on the genus *Unio*, together with descriptions of new species in the family Unionidae, and descriptions of embryonic forms and soft parts, also, new species of Strepomatidae, Limnaeidae. Printed for the author, Philadelphia. Volume 13. 75 pages, plates 1–22.

Lea, I. 1874f. Index to Volume I to XIII Observations on the genus *Unio*, together with descriptions of new species of the family Unionidae, and descriptions of new species of the Melanidae, Paludinae, Helicidae, *etc*. Volume III. Printed for the author by Collins Printer, Philadelphia. 29 pages.

Lea, I. 1876. A catalogue of the published works of Isaac Lea, LL.D. From 1817 to 1876. Collins, Printer, Philadelphia. 22 pages.

Leach, W.E. 1814, 1815, 1817. The Zoological Miscellany: being descriptions of new, or interesting animals; illustrated with coloured figures. E. Nodder and Son, London. Volume 1 (1814), 144 pages. Volume 2 (1815), 154 pages. Volume 3 (1817), 151 pages.

Leach, W.E. 1847. The classification of the British Mollusca. Annals and Magazine of Natural History (Series 1) 20:267–273.

Lee, H.G. 2006. Musings on a local specimen of *Toxolasma paulum* (I. Lea, 1840), the Iridescent Lilliput. The Shell-O-Gram 47(5):3-6.

Lee, T., S. Siripattrawan, C.F. Ituarte and D.Ó Foighil. 2005. Invasion of the clonal clams: *Corbicula* lineages in the New World. American Malacological Bulletin 20(1–2):113–122.

Lefevre, G., and W.C. Curtis. 1910a. The marsupium of the Unionidæ. Biological Bulletin 19(1):31–34.

Lefevre, G., and W.C. Curtis. 1910b. Reproduction and parasitism in the Unionidae. Journal of Experimental Zoology 99(1):79–115, 4 plates.

Lefevre, G., and W.C. Curtis. 1910c. Experiments in the artificial propagation of fresh-water mussels. Bulletin of the U.S. Bureau of Fisheries 28(1908):615–626. [Issued separately as U.S. Bureau of Fisheries Document 671.]

Lefevre, G., and W.C. Curtis. 1911. Metamorphosis without parasitism in the Unionidæ. Science 33(857):863–865.

Lefevre, G., and W.C. Curtis. 1912. Studies on the reproduction and artificial propagation of fresh-water mussels. Bulletin of the U.S. Bureau of Fisheries 30(1910):105–201, 12 plates. [Issued separately as U.S. Bureau of Fisheries Document 756. Reprinted in Sterkiana 47, 48 (1972); 49, 51 (1973); 57 (1975) and 61, 63, 64 (1976).]

Leff, L.G., J.L. Burch and J.V. McArthur. 1990. Spatial distribution, seston removal, and potential competitive interactions of the bivalves *Corbicula fluminea* and *Elliptio complanata*, in a coastal plain stream. Freshwater Biology 24(2):409–416.

Lellis, W.A., and T.L. King. 1998. Release of metamorphosed juveniles by the green floater, *Lasmigona subviridis*. Triannual Unionid Report 16:23.

Lewis, J. 1876. Fauna of Alabama: fresh water and land shells. Alabama Geological Survey Report of Progress for 1876:61–97.

Lewis, J. 1877. Unionidae of Ohio and Alabama. Proceedings of the Academy of Natural Sciences of Philadelphia 29(1):26–36.

Lewis, J.B. 1985. Breeding cycle of the freshwater mussel *Anodonta grandis* Say. Canadian Journal of Zoology 63(10):2477–2478.

Lightfoot, J. 1786. A catalogue of the Portland Museum, lately the property of the Dutchess Dowager of Portland, deceased: which will be sold by auction by Mr. Skinner and Co. on Monday the 24th of April, 1786, and the thirty-seven following days at her late dwelling-house, in Privy-garden, Whitehall: by order of the acting executrix. Skinner, London. Pages [3]–194.

Lillie, F.R. 1895. The embryology of the Unionidae. Journal of Morphology 10(1):1–100, 6 plates.

Lillie, F.R. 1901. The organization of the egg of *Unio*, based on a study of its maturation, fertilization, and cleavage. Journal of Morphology 17(2):227–292, plates 26–27.

Lincoln, R.J., G.A. Boxshall and P.F. Clark. 1982. A Dictionary of Ecology, Evolution and Systematics. Cambridge University Press. 298 pages.

Linnaeus, C. 1758. Systema Naturae. Edition X. (Systema naturae per regna tria naturae, secundum classes, ordines, genera, species cum characteribus, differentiis, synonymis, locis. Tomus I. Edtio decima, reformata.) Holmiae. Volume 1:1–824.

Liu, H.-P, J.B. Mitton and S.K. Wu. 1996. Paternal mitochondrial DNA differentiation far exceeds maternal mitochondrial DNA and allozyme differentiation in the freshwater mussel, *Anodonta grandis grandis*. Evolution 50(2):952–957.

Lomte, V.S. 1973. Digestive enzymes in the freshwater mussel, *Parreysia corrugata*. Broteria 42(3–4):97–110.

Lomte, V.S., and M.L. Jadhav. 1980. A study on the crystalline style of the freshwater mussel, *Parreysia corrugata*. Hydrobiologia 69(1–2):175–178.

Loosanoff, V.L. 1937. Seasonal gonadal changes of adult clams, *Venus mercenaria*. Biological Bulletin 72(3):406–416.

Luo, M. 1993. Host fishes of four species of freshwater mussels and development of an immune response. Masters thesis, Tennessee Technological University, Cookeville. 32 pages.

Lydeard, C., R.H. Cowie, W.F. Ponder, A.E. Bogan, P. Bouchet, S.A. Clarke, K.S. Cummings, T.J. Frest, O. Gargominy, D.G. Herbert, R. Hershler, K.E. Perez, B. Roth, M. Seddon, E.E. Strong and F.G. Thompson. 2004. The global decline of nonmarine mollusks. BioScience 54(4):321–330.

Lydeard, C., J.T. Garner, P. Hartfield and J.D. Williams. 1999. Freshwater mussels in the Gulf region: Alabama. Gulf of Mexico Science 1999(2):125–134.

Lydeard, C., and R.L. Mayden. 1995. A diverse and endangered aquatic ecosystem of the southeast United States. Conservation Biology 9(4):800–805.

Lydeard, C., R.L. Minton and J.D. Williams. 2000. Prodigious polyphyly in imperiled freshwater pearly-mussels (Bivalvia: Unionidae): a phylogenetic test of species and generic designations. Pages 145–158. In: E.M. Harper, J.D. Taylor and J.A. Crane (editors). The Evolutionary Biology of the Bivalvia. Geological Society Special Publication, Number 177.

Lydeard, C., M. Mulvey and G.M. Davis. 1996. Molecular systematics and evolution of reproductive traits of North American freshwater unionacean mussels (Mollusca: Bivalvia) as inferred from 16SrRNA gene sequences. Philosophical Transactions of the Royal Society of London 351(1347):1593–1603.

Lyell, C., Sir. 1849. A second visit to the United States of North America. Harper and Brothers, New York; J. Murray, London. Volume 1, 368 pages. Volume 2, 385 pages.

Lynn, J.W. 1994. The ultrastructure of sperm and motile spermatozeugmata released from the freshwater mussel *Anodonta grandis* (Mollusca, Bivalvia, Unionidae). Canadian Journal of Zoology 72(8):1452–1461.

Mackie, G.L. 1991. Biology of the exotic zebra mussel, *Dreissena polymorpha*, in relation to native bivalves and its potential impact in Lake St. Clair. Hydrobiologia 219:251–268.

Mackie, G.L., and J.M. Topping. 1988. Historical changes in the unionid fauna of the Sydenham River Watershed and downstream changes in shell morphometrics of three common species. Canadian Field-Naturalist 102(4):617–626.

Maio, J.D., and L.D. Corkum. 1997. Patterns of orientation in unionids as a function of rivers with differing hydrological variability. Journal of Molluscan Studies 63(4):531–539.

Mandahl-Barth, G. 1988. Studies on African freshwater bivalves. Danish Bilharziasis Laboratory, Charlottenlund, Denmark. 161 pages.

Marsh, W.A. 1880. [Description of *Unio upsoni* Marsh. Read before the Mercer County, Illinois, Scientific and Historical Society, March 2, 1880.] Copies privately printed and distributed, Aledo, Illinois.

Marsh, W.A. 1891. Description of two new species of *Unio* from Arkansas. The Nautilus 5(1):1–2.

Marsh, W.A. 1901. Description of a new *Unio* from Missouri. The Nautilus 15(7):74–75.

Marsh, W.A. 1902a. Description of a new *Unio* from Tennessee. The Nautilus 15(10):115–116.

Marsh, W.A. 1902b. Description of a new *Unio* from Tennessee. The Nautilus 16(1):7–8, plate 1.

Marshall, W.B. 1916. A new genus and species of naiad from the James River at Huron, South Dakota. The Nautilus 29(12):133–135, plate 4.

Martel, A.L., D.A. Pathy, J.B. Madill, C.B. Renaud, S.L. Dean and S.J. Kerr. 2001. Decline and regional extirpation of freshwater mussels (Unionidae) in a small river system invaded by *Dreissena polymorpha*: the Rideau River, 1993–2000. Canadian Journal of Zoology 79(12):2181–2191.

Mathiak, H.A. 1979. A River Survey of the Unionid Mussels of Wisconsin 1973–1977. Sand Shell Press, Horicon, Wisconsin. 75 pages.

Matteson, M.R. 1948. Life history of *Elliptio complanata* (Dillwyn, 1817). The American Midland Naturalist 40(3):690–723.

Matteson, M.R. 1955. Studies on the natural history of the Unionidae. The American Midland Naturalist 53(1):126–145.

McCall, P.I., M.J.S. Tevesz and S.F. Scheigen. 1979. Sediment mixing by *Lampsilis radiata siliquoidea* (Mollusca) from western Lake Erie. Journal of Great Lakes Research 5(2):105–111.

McCullagh, W.H., J.D. Williams, S.W. McGregor, J.M. Pierson and C. Lydeard. 2002. The unionid (Bivalvia) fauna of the Sipsey River in northwestern Alabama, an aquatic hotspot. American Malacological Bulletin 17(1–2):1–15.

McGregor, M.A., J.T. Garner and Z. Bowen. 1996. Commercial mussels in Alabama. ADCNR Educational Booklet. Auburn University Printing Services, Alabama. 32 pages.

McGregor, S.W. 1992. A mussel survey of the streams draining Bankhead National Forest and the Okmulgee Division of Talladega National Forest, Alabama. 1992. Biological Resources Division, Geological Survey of Alabama. Challenge Cost Share Agreement Number CCS SO 92-1. 30 pages, Appendix, 37 pages.

McGregor, S.W., and J.T. Garner. 2004. Changes in the freshwater mussel (Bivalvia: Unionidae) fauna of the Bear Creek system of northwest Alabama and northeast Mississippi. American Malacological Bulletin 18(1–2):61–70.

McGregor, S.W., and W.R. Haag. 2004. Freshwater mussels (Bivalvia: Unionidae) and habitat conditions in the upper Tombigbee River system, Alabama and Mississippi, 1993–2001. Geological Survey of Alabama, Bulletin 176. 75 pages.

McGregor, S.W., P.E. O'Neil and J.M. Pierson. 2002. Status of the freshwater mussel (Bivalvia: Unionidae) fauna in the Cahaba River system, Alabama. Walkerana (for 2000) 11(26):215–237.

McGregor, S.W., and J.M. Pierson. 1999. Recent freshwater mussel (Bivalvia: Unionacea) records from the North River system, Fayette and Tuscaloosa counties, Alabama. Journal of the Alabama Academy of Science 70(4):153–162.

McGregor, S.W., and D.N. Shelton. 1995. A qualitative assessment of the unionid fauna of the headwaters of the Paint Rock and Flint rivers of north Alabama and adjacent areas of Tennessee, 1995. Geological Survey of Alabama, Open-File Report to Alabama Department of Conservation and Natural Resources. 23 pages.

McGregor, S.W., T.E. Shepard, T.D. Richardson and J.F. Fitzpatrick, Jr. 1999. A survey of the primary tributaries of the Alabama and lower Tombigbee rivers for freshwater mussels, snails and crayfish. Geological Survey of Alabama Circular 196:1–29.

McMahon, R.F. 1982. The occurrence and spread of the introduced Asiatic freshwater clam, *Corbicula fluminea* (Müller), in North America: 1924–1982. The Nautilus 96(4):134–141.

McMahon, R.F. 1999. Invasive characteristics of the freshwater bivalve *Corbicula fluminea*. Pages 315–343. In: R. Claudi and J. Leach (editors). Nonindigenous Freshwater Organisms: Vectors, Biology and Impacts. CRC Press, Boca Raton, Florida.

McMahon, R.F., and A.E. Bogan. 2001. Mollusca: Bivalvia. Pages 331–429. In: J.H. Thorpe and A.P. Covich (editors). Ecology and Classification of North American Freshwater Invertebrates. Second edition. Academic Press, New York.

McRae, S.E., J.D. Allen and J.B. Burch. 2004. Reach- and catchment-scale determinants of the distribution of freshwater mussels (Bivalvia: Unionidae) in south-eastern Michigan, U.S.A. Freshwater Biology 49(2):127–142.

Menke, C.T. 1828. Synopsis methodica molluscorum generum omnium et specierum earum quae in museo menkeano adservantur; cum synonymia critica et novarum specierum diagosibus. 91 pages.

Merrick, A.D. 1930. Some quantitative determinations of glochidia. The Nautilus 43(3):89–91.

Miles, M. 1861. A catalogue of the mammals, birds, reptiles and mollusks, of Michigan. Pages 219–241. In: A. Winchell (editor). First biennial report of the progress of the Geological Survey of Michigan, embracing observations on the geology, zoölogy and botany of the Lower Peninsula. Homer and Kerr, Lansing, Michigan.

Miller, A.C. 1994. A survey of the Black Warrior and Tombigbee rivers, Alabama, for the threatened Inflated Heelsplitter Mussel *Potamilus inflatus*, May 1993. U.S. Army Corps of Engineers, Waterways Experiment Station, Vicksburg, Mississippi, Technical Report EL-94-13. 78 pages.

Miller, A.C. 2000. An investigation of the mussel resource at selected locations along the Alabama River, Alabama, 1999. U.S. Army Corps of Engineers, Engineer Research and Development Center, Environmental Laboratory, Vicksburg, Mississippi, ERDC/EL TR-00-16. 39 pages.

Miller, A.C. 2001. An analysis of freshwater mussels (Unionidae) along Luxapallila Creek, Mississippi, 1999 Studies. U.S. Army Corps of Engineers, Engineer Research and Development Center, Environmental Laboratory, Vicksburg, Mississippi, ERDC/EL TR-01-26. 30 pages.

Miller, A.C., B.S. Payne, T.J. Naimo and W.D. Russell-Hunter. 1987. Gravel bar mussel communities: a community model. U.S. Army Corps of Engineers, Waterways Experiment Station, Vicksburg, Mississippi, Technical Paper EL-87-13. 71 pages.

Mirarchi, R.E. (editor). 2004. Alabama Wildlife, Volume 1. A Checklist of Vertebrates and Selected Invertebrates: Aquatic Mollusks, Fishes, Amphibians, Reptiles, Birds and Mammals. The University of Alabama Press, Tuscaloosa. 209 pages.

Mirarchi, R.E., M.A. Bailey, J.T. Garner, T.M. Haggerty, T.L. Best, M.F. Mettee and P.E. O'Neil (editors). 2004. Alabama Wildlife, Volume 4. Conservation and Management Recommendations for Imperiled Wildlife. The University of Alabama Press, Tuscaloosa. 221 pages.

Mirarchi, R.E., M.A. Bailey, T.M. Haggerty and T.L. Best (editors). 2004. Alabama Wildlife, Volume 3. Imperiled Amphibians, Reptiles, Birds and Mammals. The University of Alabama Press, Tuscaloosa. 225 pages.

Mirarchi, R.E., J.T. Garner, M.F. Mettee and P.E. O'Neil (editors). 2004. Alabama Wildlife, Volume 2. Imperiled Aquatic Mollusks and Fishes. The University of Alabama Press, Tuscaloosa. 255 pages.

Mitchell, R.D. 1955. Anatomy, life history, and evolution of the mites parasitizing fresh-water mussels. University of Michigan, Museum of Zoology, Miscellaneous Publications, Number 89. 28 pages, 6 plates.

Modell, H. 1942. Das natüruliche system der Najaden. Archiv für Molluskenkunde 74(5–6):161–191.

Modell, H. 1949. Das natüruliche system der Najaden. 2. Archiv für Molluskenkunde 78(1–3):29–46.

Modell, H. 1964. Das natüruliche system der Najaden. 3. Archiv für Molluskenkunde 93(3–4):71–126.

Moneymaker, B.C. 1941. Subriver solution cavities in the Tennessee Valley. Journal of Geology 49(1):74–86.

Monroe, W.H. 1941. Notes on deposits of Selma and Ripley Age in Alabama. Geological Survey of Alabama Bulletin 48:1–150, plates 1–2.

Mörch, O.A.L. 1852–1853. Catalogus conchyliorum quae reliquit D. Alphonso D'Aguirra & Gadea, Comes de Yoldi, regis daniae cubiculariorum princeps, ordinis dannebrogici in prima classe & ordinis caroli tertii eques. Hafniae, Typis Ludovici Kleini, Copenhagen. Part 1 (1852), 170 pages; Part 2 (1853), 74 pages.

Morgan, A. 1996. Feasibility of reintroducing freshwater mussels into Shoal Creek, Alabama and Tennessee. Masters thesis, Tennessee Technological University, Cookeville. 58 pages.

Morgan, A., N.J. Welker and J.B. Layzer. 1997. Feasibility of reintroducing threatened and endangered mussels into Shoal Creek in Alabama and Tennessee. Pages 196–204. In: K.S. Cummings, A.C. Buchanan, C.A. Mayer and T.J. Naimo (editors). Conservation and Management of Freshwater Mussels II: Initiatives for the Future. Proceedings of a UMRCC Symposium, 16–18 October 1995, St. Louis, Missouri. Upper Mississippi River Conservation Committee, Rock Island, Illinois.

Morris, T.J., and L.D. Corkum. 1999. Unionid growth patterns in rivers of differing riparian vegetation. Freshwater Biology 42(1):59–68.

Morrison, J.P.E. 1942. Preliminary report on mollusks found in the shell mounds of the Pickwick Landing Basin in the Tennessee River Valley. Pages 337–392. In: W.S. Webb and D.L. DeJarnette (editors). An archaeological survey of Pickwick Basin in the adjacent portions of the states of Alabama, Mississippi and Tennessee. Bureau of American Ethnology, Bulletin 129.

Morrison, J.P.E. 1969. The earliest names for North American naiads. The Annual Report for 1969 of the American Malacological Union, Inc. 36:22–24.

Morrison, J.P.E. 1973. The families of the pearly freshwater mussels. The American Malacological Union, Inc. Bulletin for 1972 38:45–46.

Morton, B. 1979. *Corbicula* in Asia. Pages 16–38. In: J.C. Britton (editor). Proceedings, First international *Corbicula* Symposium. Texas Christian University Research Foundation, Fort Worth.

Müller, O.F. 1773–1774. Vermium terrestrium et fluviatilium, seu animalium Infusoriorum, Helminthicorum et Testaceorum, non marinorum, succincta historia, &c. Havniæ; Lipsiæ, 2 volumes.

Mulvey, M., C. Lydeard, D.L. Pyer, K. Hicks, J. Brim-Box, J.D. Williams and R.S. Butler. 1997. Conservation genetics of North American freshwater mussels *Amblema* and *Megalonaias*. Conservation Biology 11(4):868–878.

Murphy, G. 1942. Relationship of the fresh-water mussel to trout in the Truckee River. California Fish and Game 28(2):89–102.

Murray, H.D., and A.B. Leonard. 1962. Handbook of unionid mussels in Kansas. Museum of Natural History, University of Kansas, Miscellaneous Publication, Number 28. 184 pages.

Mutvei, H., and T. Westermark. 2001. How environmental information can be obtained from naiad shells. Pages 367–379. In: G. Bauer and K. Wächtler (editors). Ecology and Evolution of Freshwater Mussels Unionoida. Ecological Studies, Volume 145. Springer-Verlag, Berlin, Germany.

Myers, P.R., and D.S. Franzen. 1970. Histological studies of the nephridium and pericardial lining of *Quadrula nodulata*. The Nautilus 83(4):139–144.

Nalepa, T.F. 1994. Decline of native unionid bivalves in Lake St. Clair after infestation by the zebra mussel, *Dreissena polymorpha*. Canadian Journal of Fisheries and Aquatic Science 51(10):2227–2233.

Nalepa, T.F., and D.W. Schloesser (editors). 1993. Zebra Mussels: Biology, Impacts, and Control. Lewis Publishers, Ann Arbor, Michigan. 810 pages.

Narain, A.S. 1972. Formed elements of the red blood of the freshwater mussel. Journal of Morphology 137(1):63–69.

Narain, A.S., and K. Singh. 1990. Unionid rectal structure in relation to function and with particular reference to the piercing of heart by gut. Proceedings of the National Academy of Science, India 60 (Biological Sciences) 3:245–252.

National Native Mussel Conservation Committee. 1998. National strategy for the conservation of native freshwater mussels. Journal of Shellfish Research 17(5):1419–1428.

Neel, J.K. 1941. A taxonomic study of *Quadrula quadrula* (Rafinesque). Occasional Papers of the Museum of Zoology, University of Michigan, Number 448. 8 pages, 1 plate.

Nelson, T.C. 1918. On the origin, nature, and function of the crystalline style of lamellibranchs. Journal of Morphology 31(1):53–111.

Neves, R.J. 1991. Mollusks. Pages 251–320. In: K. Terwilliger (editor). Virginia's Endangered Species. Proceedings of a Symposium, Virginia Department of Game and Inland Fisheries. McDonald and Woodward, Blacksburg, Virginia.

Neves, R.J., A.E. Bogan, J.D. Williams, S.A. Ahlstedt and P.W. Hartfield. 1997. Status of aquatic mollusks in the southeastern United States: a downward spiral of diversity. Pages 43–85. In: G.W. Benz and D.E. Collins (editors). Aquatic Fauna in Peril: The Southeastern Perspective. Southeast Aquatic Research Institute, Special Publication 1. Lenz Design and Communications, Decatur, Georgia.

Neves, R.J., and S.N. Moyer. 1988. Evaluation of techniques for age determination of freshwater mussels (Unionidae). American Malacological Bulletin 6(2):179–188.

Neves, R.J., and M.C. Odum. 1989. Muskrat predation on endangered freshwater mussels in Virginia. Journal of Wildlife Management 53(4):934–941.

Neves, R.J., L.R. Weaver and A.V. Zale. 1985. An evaluation of host fish suitability for glochidia of *Villosa vanuxemi* and *V. nebulosa* (Pelecypoda: Unionidae). The American Midland Naturalist 113(1):13–19.

Neves, R.J., and J.C. Widlak. 1987. Habitat ecology of juvenile freshwater mussels (Bivalvia: Unionidae) in a headwater stream in Virginia. American Malacological Bulletin 5(1):1–7.

Nezlin, L.P., R.A. Cunjak, A.A. Zotin and V.V. Ziuganov. 1994. Glochidium morphology of the freshwater pearl mussel (*Margaritifera margaritifera*) and glochidiosis of Atlantic salmon (*Salmo salar*): a study by scanning electron microscopy. Canadian Journal of Zoology 72(1):15–21.

Nichols, S.J., and D. Garling. 2000. Food-web dynamics and trophic-level interactions in a multispecies community of freshwater unionids. Canadian Journal of Zoology 78(5):871–882.

Nico, L.G., J.D. Williams and H.L. Jelks. 2005. Black Carp: Biological Synopsis and Risk Assessment of an Introduced Fish. American Fisheries Society, Special Publication 32. 337 pages.

Nilsson, S. 1822. Historia molluscorum Sveciae terrestrium et fluviatilium breviter delineate. Lundae, Sumptibus J.H. Schubothii. 124 pages.

Nitzsch, C.L. 1820. *Anodonta*. Pages 189–190. In: J.S. Ersch and J.G. Gruber (editors). Allgemeine Encyclopädie der Wissenschaften und Künste 4:1–475. Leipzig.

Novak, S.F. 2004. Current and historic freshwater mollusk distributions of the upper Coosa River basin. Masters thesis, University of Tennessee at Chattanooga. 318 pages.

O'Beirn, F.X., R.J. Neves and M.B. Steg. 1998. Survival and growth of juvenile freshwater mussels (Unionidae) in a recirculating aquaculture system. American Malacological Bulletin 14(2):165–171.

O'Brien, C.A., and J. Brim-Box. 1999. Reproductive biology and juvenile recruitment of the Shinyrayed Pocketbook, *Lampsilis subangulata* (Bivalvia: Unionidae) in the Gulf Coastal Plain. The American Midland Naturalist 142(1):129–140.

O'Brien, C.A., and J.D. Williams. 2002. Reproductive biology of four freshwater mussels (Bivalvia: Unionidae) endemic to eastern Gulf Coastal Plain drainages of Alabama, Florida, and Georgia. American Malacological Bulletin 17(1–2):147–158.

O'Brien, C.A., J.D. Williams and M.A. Hoggarth. 2003. Morphological variation in glochidia shells of six species of *Elliptio* from Gulf of Mexico and Atlantic Coast drainages in the southeastern United States. Proceedings of the Biological Society of Washington 116(3):719–731.

O'Connell, M.T., and R.J. Neves. 1999. Evidence of immunological responses by a host fish (*Ambloplites rupestris*) and two non-host fishes (*Cyprinus carpio* and *Carassius auratus*) to glochidia of a freshwater mussel (*Villosa iris*). Journal of Freshwater Ecology 14(1):71–78.

O'Dee, S.H., and G.T. Watters. 2000. New or confirmed host identifications for ten freshwater mussels. Pages 77–82. In: R.A. Tankersley, D.I. Warmolts, G.T. Watters and B.J. Armitage (editors). Freshwater Mollusk Symposia Proceedings, Part I. Proceedings of the Conservation, Captive Care and Propagation of Freshwater Mussels Symposium, 1998. Ohio Biological Survey, Columbus, Ohio.

Oesch, R.D. 1995. Missouri Naiades. A Guide to the Mussels of Missouri. Missouri Department of Conservation, Jefferson City, Missouri. 271 pages.

Opler, P.A. 1977. The parade of passing species: a survey of extinctions in the U.S.A. The Science Teacher 44:30–34.

Ortmann, A.E. 1909a. A preliminary list of the Unionidae of western Pennsylvania, with new localities for species from eastern Pennsylvania. Annals of the Carnegie Museum 5(2–3):178–210.

Ortmann, A.E. 1909b. The breeding season of Unionidae in Pennsylvania. The Nautilus 22(9):91–95; 22(10):99–103.

Ortmann, A.E. 1909c. The destruction of the fresh-water fauna in western Pennsylvania. Proceedings of the American Philosophical Society 48(191):90–110.

Ortmann, A.E. 1910a. The discharge of the glochidia in the Unionidae. The Nautilus 24(8):94–95.

Ortmann, A.E. 1910b. A new system of the Unionidae. The Nautilus 23(9):114–120.

Ortmann, A.E. 1910c. The marsupium of the Anodontinae. Biological Bulletin 19(3):217.

Ortmann, A.E. 1911a. A monograph of the najades of Pennsylvania. Parts I and II. Memoirs of the Carnegie Museum 4(6):279–347, 4 plates.

Ortmann, A.E. 1911b. The anatomical structure of certain exotic naiades compared with that of the North American forms. The Nautilus 24(9):103–108, 2 plates; 24(10):114–120; 24(11):127–131.

Ortmann, A.E. 1912a. Notes upon the families and genera of the najades. Annals of the Carnegie Museum 8(2):222–365, plates 18–20.

Ortmann, A.E. 1912b. *Cumberlandia*, a new genus of naiades. The Nautilus 26(2):13–14.

Ortmann, A.E. 1913a. The Alleghenian Divide, and its influence upon the freshwater fauna. Proceedings of the American Philosophical Society 52(210):287–390, plates 12–14.

Ortmann, A.E. 1913b. Studies in najades. The Nautilus 27(8):88–91.

Ortmann, A.E. 1914. Studies in najades (continued). The Nautilus 28(2):20–22; 28(3):28–34; 28(4):41–47; 28(5[sic]):65–69.

Ortmann, A.E. 1915. Studies in najades (continued). The Nautilus 28(9):106–108; 28(11):129–131; 28(12):141–143; 29(6):63–67.

Ortmann, A.E. 1916a. The anatomy of *Lemiox rimosus* (Rafinesque). The Nautilus 30(4):39–41.

Ortmann, A.E. 1916b. Studies in najades (concluded). The Nautilus 30(5):54–57.

Ortmann, A.E. 1917. A new type of the naiad-genus *Fusconaia*. Group of *F. barnesiana* Lea. The Nautilus 31(2):58–64.

Ortmann, A.E. 1918. The nayades (freshwater mussels) of the upper Tennessee drainage. With notes on synonymy and distribution. Proceedings of the American Philosophical Society 57(6):521–626.

Ortmann, A.E. 1919. A monograph of the naiades of Pennsylvania. Part III: systematic account of the genera and species. Memoirs of the Carnegie Museum 8(1):1–384, 21 plates.

Ortmann, A.E. 1920. Correlation of shape and station in freshwater mussels (Naiades). Proceedings of the American Philosophical Society 59(4):269–312.

Ortmann, A.E. 1921. The anatomy of certain mussels from the upper Tennessee. The Nautilus 34(3):81–91.

Ortmann, A.E. 1923a. The anatomy and taxonomy of certain Unioninae and Anodontinae from the Gulf drainage. The Nautilus 36(3):73–84; 36(4):129–132.

Ortmann, A.E. 1923b. Notes on the anatomy and taxonomy of certain Lampsilinae from the Gulf Drainage. The Nautilus 37(2):56–60.

Ortmann, A.E. 1924a. Notes on the anatomy and taxonomy of certain Lampsilinae from the Gulf drainage. The Nautilus 37(3):99–105; 37(4):137–144.

Ortmann, A.E. 1924b. Mussel Shoals. Science 60(1564):565–566.

Ortmann, A.E. 1925. The naiad fauna of the Tennessee River system below Walden Gorge. The American Midland Naturalist 9(8):321–372.

Ortmann, A.E., and B. Walker. 1912. A new North American naiad. The Nautilus 25(9):97–100, plate 8.

Ortmann, A.E., and B. Walker. 1922a. On the nomenclature of certain North American naiades. Occasional Papers of the Museum of Zoology, University of Michigan, Number 112. 75 pages.

Ortmann, A.E., and B. Walker. 1922b. A new genus and species of American naiades. The Nautilus 36(1):1–6, plate 1, figures 1–4.

Pallary, P. 1933. Résultats généraux d'une prospection Malacologique effectuée en Syrie de 1929 à 1932. Bulletin du Muséum National d'Histoire Naturelle Paris (Series 2) 5(2):148–154.

Pallas, P.S. 1771, 1773, 1776. Reise durch verschiedene Provinzen des russischen Reichs. St. Petersburg, Gedruckt bey der kayserlichen Academie der Wissenschaften.

Palmer, S. 1985. Some extinct mollusks of the U.S.A. ATALA, the Journal of Invertebrate Conservation 13(1):1–7.

Parker, R.S., C.T. Hackney and M.F. Vidrine. 1984. Ecology and reproductive strategy of a south Louisiana freshwater mussel, *Glebula rotundata* (Lamarck) (Unionidae: Lampsilini). Freshwater Invertebrate Biology 3(2):53–58.

Parker, R.S., M.F. Vidrine and C.T. Hackney. 1980. A new centrarchid host for the paper pond shell, *Anodonta imbecillis* Say (Bivalvia: Unionidae). ASB [Association of Southeastern Biologists] Bulletin 27(2):54–55.

Parmalee, P.W. 1967. The fresh-water mussels of Illinois. Illinois State Museum, Popular Science Series, Volume 8. 108 pages.

Parmalee, P.W. 1994. Freshwater mussels from Dust and Smith Bottom Caves, Alabama. Journal of Alabama Archaeology 40(1–2):135–162.

Parmalee, P.W. 2005. A Prehistoric Record of the Pond Mussel (*Uniomerus tetralasmus*) from the Tennessee River, Hardin County, Tennessee. *Ellipsaria* 7(1):7–8.

Parmalee, P.W., and A.E. Bogan. 1986. Molluscan remains from aboriginal middens at the Clinch River Breeder Reactor Plant Site, Roane County, Tennessee. American Malacological Bulletin 4(1):25–37.

Parmalee, P.W., and A.E. Bogan. 1998. The Freshwater Mussels of Tennessee. The University of Tennessee Press, Knoxville. 328 pages.

Parmalee, P.W., and W.E. Klippel. 1974. Freshwater mussels as a prehistoric food resource. American Antiquity 39(3):421–434.

Parmalee, P.W., W.E. Klippel and A.E. Bogan. 1980. Notes on the prehistoric and present status of the naiad fauna of the middle Cumberland River, Smith County, Tennessee. The Nautilus 94(3):93–105.

Parmalee, P.W., W.E. Klippel and A.E. Bogan. 1982. Aboriginal and modern freshwater mussel assemblages (Pelecypoda: Unionidae) from the Chickamauga Reservoir, Tennessee. Brimleyana 8:75–90.

Parodiz, J.J., and A.A. Bonetto. 1963. Taxonomy and zoogeographic relationships of the South American naiades (Pelecypoda: Unionacea and Mutelacea). Malacologia 1(2):179–213.

Passamonti, M., and V. Scali. 2001. Gender-associated mitochondrial DNA heteroplasmy in the venerid clam *Tapes philippinarum* (Mollusca Bivalvia). Current Genetics 39(2):117–124.

Patch, D.C. 1976. An analysis of the archaeological shell of freshwater mollusks from the Carlston Annis Shellmound west central Kentucky. Bachelors honors thesis, Washington University, St. Louis, Missouri. 76 pages.

Patch, D.C. 2005. The freshwater molluscan fauna: identification and interpretation for archaeological research. Pages 257–278. In: W.H. Marquardt and P.J. Watson (editors). Archaeology of the Middle Green River Region, Kentucky.

Patterson, C.C. 1984. A technique for determining apparent selective filtration in the fresh-water bivalve *Elliptio complanata* (Lightfoot). The Veliger 27(2):238–241.

Patterson, C.C. 1986. Particle-size selectivity in the freshwater bivalve *Elliptio complanata* (Lightfoot). The Veliger 29(2):235–237.

Payne, B.S., and A.C. Miller. 1987. Effects of current velocity on the freshwater bivalve *Fusconaia ebena*. American Malacological Bulletin 5(2):177–179.

Peacock, E. 1998. Freshwater mussels as indicators of prehistoric human environmental impact in the southeastern United States. Doctoral dissertation, University of Sheffield, United Kingdom. 435 pages.

Peacock, E. 2000. Assessing bias in archaeological shell assemblages. Journal of Field Archaeology 27(2):183–196.

Peck, R.H. 1877. The minute structure of the gills of lamellibranch Mollusca. Quarterly Journal of Microscopical Science 17(65):43–66, plates 4–7.

Pekkarinen, M., and I. Valovirta. 1996. Anatomy of the glochidia of the freshwater pearl mussel, *Margaritifera margaritifera* (L.). Archiv für Hydrobiologia 137(3):411–423.

Pelseneer, P. 1935. Essai d'éthologie zoologique d'aprés l'étude des mollusques. Bruxelles, Palais des Académies. 662 pages.

Penn, G.H. 1939. A study of the life cycle of the freshwater mussel, *Anodonta grandis*, in New Orleans. The Nautilus 52(3):99–101.

Pepi, V.E., and M.C. Hove. 1997. Suitable fish hosts and mantle display behavior of *Tritogonia verrucosa*. Triannual Unionid Report 11:5.

Percy, G.W. 1976. Salvage investigations at the Scholz Steam Plant Site (8JA104), a middle Weeden Island habitation site in Jackson County, Florida. Bureau of Historic Sites and Properties, Division of Archives, History, and Records Management, Department of State, Tallahassee, Florida, Miscellaneous Projects Report Series, Number 35. 150 pages.

Perles, S.J., A.D. Christian and D.J. Berg. 2003. Vertical migration, orientation, aggregation, and fecundity of the freshwater mussel *Lampsilis siliquoidea*. Ohio Journal of Science 103(4):73–78.

Petit, H., W.L. Davis, R.G. Jones and J.K. Hagler. 1980. Morphological studies on the calcification process in the fresh-water mussel *Amblema*. Tissue and Cell 12(1):13–28.

Pharris, G.L., C.C. Chandler and J.B. Sickel. 1982. Range extension for *Plectomerus dombeyanus* (Bivalvia: Unionidae) into Kentucky. (Abstract). Transactions of the Kentucky Academy of Science 43(1–2):95–96.

Picha, P.R., and F.E. Swenson. 2000. Freshwater shell tool/ornament production and resource use in the middle Missouri subarea of North Dakota. Central Plains Archeology 8(1):102–120.

Pierson, J.M. 1991a. A survey of the Sipsey River, Alabama, for *Quadrula stapes* (Lea, 1831) and *Pleurobema taitianum* (Lea, 1834). Unpublished report, Calera, Alabama. 61 pages.

Pierson, J.M. 1991b. Status survey of the Southern Clubshell, *Pleurobema decisum* (Lea, 1831). Unpublished report, Calera, Alabama. 51 pages.

Pilarczyk, M.M., P.M. Stewart, D.N. Shelton, H.N. Blalock-Herod and J.D. Williams. 2006. Current and recent historical freshwater mussel assemblages in the Gulf Coastal Plains. Southeastern Naturalist 5(2): 205-226.

Pilarczyk, M.M., P.M Stewart, D.N. Shelton, W.H. Heath and J.M. Miller. 2005. Contemporary survey and historical freshwater mussel assemblages in southeast Alabama and northwest Florida and life history and host fish identification of two candidate unionids (*Quincuncina burkei* and *Pleurobema strodeanum*). A report to the U.S. Fish and Wildlife Service, Panama City, Florida, Contract Number 401214G049. 93 pages, appendices 11 pages.

Pilsbry, H.A. 1892. New and unfigured Unionidae. Proceedings of the Academy of Natural Sciences of Philadelphia 44:131–132, plates 7–8.

Pilsbry, H.A., and J. Bequaert. 1927. The aquatic mollusks of the Belgian Congo, with a geographical and ecological account of Congo malacology. Bulletin of the American Museum of Natural History 53(2):69–602.

Pinder, M.J., E.S. Wilhelm and J.W. Jones. 2003. Status survey of the freshwater mussels (Bivalvia: Unionidae) in the New River drainage, Virginia. Walkerana 12(29–30):189–223.

Poole, K.E., and J.A. Downing. 2004. Relationship of declining mussel biodiversity to stream-reach and watershed characteristics in an agricultural landscape. Journal of the North American Benthological Society 23(1):114–125.

Porter, H.J., and L. Houser. 1998. Seashells of North Carolina. North Carolina Sea Grant Program, North Carolina State University, Raleigh. UNC-SG-97-03. 132 pages.

Pratt, W.H. 1876. Description of a *Unio* shell found on the south bank of the Mississippi River, opposite the Rock Island Arsenal, in 1870. Proceedings of the Davenport Academy of Natural Sciences 1:167–168, 1 plate.

Prentice, J. 1994. Student wins science fair with mussel project. Pages 5–6. In: R.G. Howells. Info-Mussel Newsletter 2(3):5–6.

Rafinesque, C.S. 1818. Discoveries in natural history, made during a journey through the western region of the United States, by Constantine Samuel Rafinesque, Esq. Addressed to Samuel L. Mitchill, President and the other members of the Lyceum of Natural History, in a letter dated at Louisville, Falls of Ohio, 20th July 1818. The American Monthly Magazine and Critical Review 3(1):354–356.

Rafinesque, C.S. 1819. Prodrome de 70 nouveaux Genres d'Animaux découverts dans l'intérieur des États-Unis d'Amérique, durant l'année 1818. Journal de Physique, de Chimie, d'Histoire Naturelle et des Arts 88:417–429.

Rafinesque, C.S. 1820. Monographie des coquilles bivalves fluviatiles de la Rivière Ohio, contenant douze genres et soixante-huit espèces. Annales générales des sciences Physiques, a Bruxelles 5(5):287–322, plates 80–82.

Rafinesque, C.S. 1831. Continuation of a monograph of the bivalve shells of the River Ohio, and other rivers of the western states. By Prof. C.S. Rafinesque. (Published at Brussels, September 1820.) Containing 46 species, from Number 76 to Number 121. Including an appendix on some bivalve shells of the rivers of Hindustan, with a supplement on the fossil bivalve shells of the Western states, and the Tulosites, a new genus of fossils. Philadelphia. 8 pages.

Rafinesque, C.S. 1832. *Odatelia* N.G. of N. American bivalve fluviatile shell. Atlantic Journal and Friend of Knowledge 4:154.

Raikow, D.F., and S.K. Hamilton. 2001. Bivalve diets in a midwestern U.S. stream: a stable isotope enrichment study. Limnology and Oceanography 46(3):514–522.

Read, W.A. 1937. Indian Place Names in Alabama. Louisiana State University Studies, Number 29. Baton Rouge. 84 pages. [Revised edition by J.B. McMillan. 1984. The University of Alabama Press, Tuscaloosa. 107 pages.]

Reardon, L. 1929. A contribution to our knowledge of the anatomy of the fresh-water mussel of the District of Columbia. Proceedings of the U.S. National Museum 75(11):1–12, plates 1–5.

Récluz, C.A. 1849. Descriptions de quelque novelles espèces de coquilles. Revue et Magasin de Zoologie 2:64–71.

Reeve, L. 1856, 1864–1868. Monograph of the genus *Unio*. Volume 16. 96 colored plates. In: Conchologia Iconica: or Illustrations of the shells of molluscous animals. L. Reeve and Company, London.

Rehder, H.A. 1981. The Audubon Society Field Guide to North American Seashells. Alfred A. Knopf, New York. 894 pages.

Reitz, E.J. 1987. Vertebrate and invertebrate fauna. Pages 193–214. In: E.M. Futato. Archaeological investigations at Shell Bluff and White Springs, two Late Woodland sites in the Tombigbee River Multi Resource District. University of Alabama, Office of Archaeological Research, Report of Investigations 50.

Remington, P.S., Jr., and W.J. Clench. 1925. Vagabonding for shells. The Nautilus 38(4):127–143.

Remington, W.C., and T.J. Kallsen. 1999. Historical Atlas of Alabama, Volume 1, Historical Locations by County. Second edition. Department of Geography, College of Arts and Sciences, University of Alabama, Tuscaloosa. 383 pages.

Reuling, F.H. 1919. Acquired immunity to an animal parasite. Journal of Infectious Diseases 24:337–347.

Retzius, A.J. 1788. Dissertatio historico-naturalis sistens nova testaceorum genera. Quam venia ampliss. Facult. Philosophicae preaeside D.M. Andr. J. Retzio ... Ad publicum examen defert Laurentius Munter Philipsson scanus. Ad Diem X. Decembris MDCCLXXXVIII. L.H.S. Lundae, Typis Berlingianis. 23 pages.

Rhoads, S.N. 1899. On a recent collection of Pennsylvanian mollusks from the Ohio River system below Pittsburg. The Nautilus 12(12):133–138.

Ricciardi, A., R.J. Neves and J.B. Rasmussen. 1998. Impending extinctions of North American freshwater mussels (Unionidae) following the zebra mussel (*Dreissena polymorpha*) invasion. Journal of Animal Ecology 67(4):613–619.

Ricciardi, A., F.G. Whoriskey and J.B. Rasmussen. 1996. Impact of the *Dreissena* invasion on native unionid bivalves in the upper St. Lawrence River. Canadian Journal of Fisheries and Aquatic Science 53(6):1434–1444.

Richard, P.E., T.H. Dietz and H. Silverman. 1991. Structure of the gill during reproduction in the unionids *Anodonta grandis*, *Ligumia subrostrata* and *Carunculina parva texasensis*. Canadian Journal of Zoology 69(7):1744–1754.

Roback, S.S., D.J. Bereza and M.F. Vidrine. 1980. Description of an *Ablabesmyia* (Diptera: Chironomidae: Tanypodinae) symbiont of unionid fresh-water mussels (Mollusca: Bivalvia: Unionacea), with notes on its biology and zoogeography. Transactions of the American Entomology Society 105(4):577–620.

Robison, N.D. 1983. Archeological records of naiad mussels along the Tennessee-Tombigbee Waterway. Pages 115–129. In: A.C. Miller (compiler). Report of Freshwater Mollusks Workshop, 26–27 October 1982, U.S. Army Corps of Engineers, Waterways Experimental Station, Vicksburg, Mississippi.

Rocha, E., and C. Azevedo. 1990. Ultrastructural study of the spermatogenesis of *Anodonta cygnea* L. (Bivalvia, Unionidae). Invertebrate Reproduction and Development 18(3):169–176.

Roe, K.J., and P.D. Hartfield. 2005. *Hamiota*, a new genus of freshwater mussel (Bivalvia: Unionidae) from the Gulf of Mexico drainages of the southeastern United States. The Nautilus 119(1):1–10.

Roe, K.J., P.D. Hartfield and C. Lydeard. 2001. Phylogenetic analysis of the threatened and endangered superconglutinate-producing mussels of the genus *Lampsilis* (Bivalvia: Unionidae). Molecular Ecology 10(9):2225–2234.

Roe, K.J., and W.R. Hoeh. 2003. Systematics of freshwater mussels (Bivalvia: Unionoida). Pages 91–122. In: C. Lydeard and D.R. Lindberg (editors). Molecular Systematics and Phylogeography of Mollusks. Smithsonian Books, Washington, District of Columbia.

Roe, K.J., and C. Lydeard. 1998. Molecular systematics of the freshwater mussel genus *Potamilus* (Bivalvia: Unionidae). Malacologia 39(1–2):195–205.

Roe, K.J., A.M. Simons and P. Hartfield. 1997. Identification of a fish host of the inflated heelsplitter *Potamilus inflatus* (Bivalvia: Unionidae) with a description of its glochidium. The American Midland Naturalist 138(1):48–54.

Rogers, C.L., and D.V. Dimock, Jr. 2003. Acquired resistance of bluegill sunfish *Lepomis macrochirus* to glochidia larvae of the freshwater mussel *Utterbackia imbecillis* (Bivalvia: Unionidae) after multiple infections. The Journal of Parasitology 89(1):51–56.

Rogers, S.O., B.T. Watson and R.J. Neves. 2001. Life history and population biology of the endangered tan riffleshell (*Epioblasma florentina walkeri*) (Bivalvia: Unionidae). Journal of the North American Benthological Society 20(4):582–594.

Ropes, J.W., and A.P. Stickney. 1965. Reproductive cycle of *Mya arenaria* in New England. Biological Bulletin 128(2):315–327.

Rosenberg, G., and M.L. Ludyanskiy. 1994. A nomenclatural review of *Dreissena* (Bivalvia: Dreissenidae), with identification of the quagga mussel as *Dreissena bugensis*. Canadian Journal of Fisheries and Aquatic Sciences 51(7):1474–1484.

Rothra, E.O. 1995. Florida's pioneer naturalist. The life of Charles Torrey Simpson. The University Press of Florida, Gainesville. 232 pages.

Rummel, R. 1980. Appendix D. Mussel shells from Kellogg Site features. Pages 342–345. In: J.R. Atkinson, J.C. Phillips and R. Walling (editors). The Kellogg Village Site investigations, Clay County, Mississippi. Department of Anthropology, Mississippi State University, Starkville, Report to the U.S. Army Corps of Engineers, Mobile District.

Saarinen, M., and J. Taskinen. 2003. Burrowing and crawling behavior of three species of Unionidae in Finland. Journal of Molluscan Studies 69(1):81–86.

Salmon, A., and R.H. Green. 1983. Environmental determinants of unionid clam distribution in the Middle Thames River, Ontario. Canadian Journal of Zoology 61(4):832–838.

Savazzi, E., and P. Yao. 1992. Some morphological adaptations in freshwater bivalves. Lethaia 25(2):195–209.

Say, T. 1817. Article Conchology. [No pagination, 14 pages, plates 1–4]. In: W. Nicholson (editor). American Edition of the British Encyclopedia or Dictionary of Arts and Sciences, Comprising an Accurate and Popular View of the Present Improved State of Human Knowledge. Volume 2. First edition. Samuel A. Mitchel and Horace Ames, Philadelphia.

Say, T. 1818. Description of a new genus of fresh water bivalve shells. Journal of the Academy of Natural Sciences of Philadelphia 1(11):459–460.

Say. T. 1824. Appendix. Part 1. Natural History. [Section] 1. Zoology. Pages 254–267, plates 14–15. In: W.H. Keating. Narrative of an expedition to the source of St. Peter's River, Lake Winnepeek, Lake of the Woods, etc.: Performed in the year 1823, by order of the Hon. J.C. Calhoun, Secretary of War, under the command of Stephen H. Long, Major U.S.T.E. Compiled from the notes of Major Long, Messrs. Say, Keating and Colhoun [sic]. In 2 volumes. H.C. Carey and I. Lea, Philadelphia.

Say, T. 1825. Descriptions of some new species of fresh water and land shells of the United States. Journal of the Academy of Natural Sciences of Philadelphia 5(3–4):119–131.

Say, T. 1829. Descriptions of some new terrestrial and fluviatile shells of North America. The Disseminator of Useful Knowledge; containing hints to the youth of the United States, from the School of Industry, New Harmony, Indiana 2(19):291–293, 23 September 1829; 2(20):308–310, 7 October 1829; 2(21):323–325, 21 October 1829; 2(22):339–341, 4 November 1829; 2(23):355–356, 18 November 1829.

Say, T. 1830a–1834. American Conchology, or descriptions of the shells of North America. Illustrated by colored figures from original drawings executed from nature. School Press, New Harmony, Indiana. Part 1 (1830); Part 2 (April 1831); Part 3 (September 1831); Part 4 (March 1832); Part 5 (August 1832); Part 6 (April 1834); Part 7 (1834?, published after Say's death, edited by T.A. Conrad).

Say, T. 1830b–1831b. New terrestrial and fluviatile shells of North America (continued). The Disseminator (Second Series). New Harmony, Indiana. 1(27) (28 December 1830); 1(29) (15 January 1831); 1(31) (29 January 1831) [no pagination].

Say, T. 1831c. Descriptions of several new species of shells and of a new species of *Lumbricus*. Transylvania Journal of Medicine and the Associated Sciences 4(4):525–528.

Scammon, R.E. 1906. The Unionidae of Kansas. Part I. An illustrated catalogue of the Kansas Unionidae. The Kansas University Science Bulletin 3(9):279–373; 3(10), plates 52–85.

Scarlato, O.A., and Ya.I. Starobogatov. 1979. Osnovnye Cherty Evolyutsii I sistema Klassa Bivalvia. Pages 5–38. In: O.A. Scarlatao (editor). Morfologiya, sistematika i Filogeniya Molliuskov [Morphology, systematics and phylogeny of the molluscs]. Trudy Zoologischeskogo Instituta, Akademiia Nausk SSSR 80:1–126 [in Russian]. [Translated by K.J. Boss and M.K. Jacobson (1985). Department of Mollusks, Harvard University, Special Occasional Publication, Number 5. 77 pages.]

Schierholz, C. 1889. Uber entwicklung der Unioniden. Denkschriften der Kaiserlichen Akademie der Wissenschaften, Wien [Vienna]. Mathematisch-Naturwissenschaftliche Classe. 55:183–214, 4 plates.

Schlüter, F. 1838. Kurzgefasstes systematisches Verzeichniss meiner Conchyliensammlung nebst Andeutung aller bis jetzt von mir bei Halle gefundenen: Land- und Flussconchylien zur Erleichterung des Tausches für Freunde der Conchyliologie. Halle Druck der Gebauerschen Buchdr. 40 pages.

Schnell, F.T., V.J. Knight, Jr., and G.S. Schnell. 1981. Cemochechobee. Archaeology of a Mississippian ceremonial center on the Chattahoochee River. University Presses of Florida, Gainesville. 290 pages.

Schumacher, C.F. 1816. [*Margaritifera*.] Oversigt over Kongelige Danske Videnskabernes Selskabs Forhandlinger og dets Medlemmers Arbeider i de sidste to Aar 1816:7.

Schwartz, M.L., and R.V. Dimock, Jr. 2001. Ultrastructural evidence for nutritional exchange between brooding unionid mussels and their glochidia larvae. Invertebrate Biology 120(3):227–236.

Schwebach, M., D. Schriever, N. Dillon, M. Hove, M. McGill, C. Nelson, F. Thomas and A. Kapuscinski. 2002. Channel catfish is a suitable host species for mapleleaf glochidia. Ellipsaria 4(3):12–13.

Schwanecke, H. 1913. Das Blutgefäßsystem von *Anodonta cellensis* Schröt. Zeitschrift für Wissenschaftliche Zoologie 107(1):42–70.

Scruggs, G.D., Jr. 1960. Status of freshwater mussel stocks in the Tennessee River. U.S. Fish and Wildlife Service, Special Scientific Report 370. 41 pages.

Scudder, N.P. 1885. The published writings of Isaac Lea, LL.D. Biographies of American Naturalists II. Bulletin of the U.S. National Museum, Number 23. Government Printing Office, Washington, District of Columbia. 278 pages [a biographical sketch of Isaac Lea is included on pages vii–ix].

Scudder, S.H. 1882. Nomenclator Zoologicus. an alphabetical list of all generic names that have been employed by naturalists for recent and fossil animals from the earliest times to the close of the year 1879. Part I, Supplemental List. Bulletin of the U.S. National Museum 19. 376 pages.

Sell, H. 1908. Biologische Beobachtungen an Najades. Archiv für Hydrobiologie und Planktonkunde 1908:179–188.

Serb, J.M., J.E. Buhay and C. Lydeard. 2003. Molecular systematics of the North American freshwater bivalve genus *Quadrula* (Unionidae: Ambleminae) based on mitochondrial ND1 sequences. Molecular Phylogenetics and Evolution 28(1):1–11.

Servain, G. 1890. Des Acéphales lamellibranches fluviatiles du système européen. Bulletins de la Société malacologique de France 7:281–323, plates 5–7.

Shelton, D.N. 1997. Observations on the life history of the Alabama pearl shell, *Margaritifera marrianae* R.I. Johnson, 1983. Pages 26–29. In: K.S. Cummings, A.C. Buchanan, C.A. Mayer and T.J. Naimo (editors). Conservation and Management of Freshwater Mussels II: Initiatives for the Future. Proceedings of a UMRCC Symposium, 16–18 October 1995, St. Louis, Missouri. Upper Mississippi River Conservation Committee, Rock Island, Illinois.

Shively, S.H., and M.F. Vidrine. 1984. Fresh-water mollusks in the alimentary tract of a Mississippi Map Turtle. Proceedings of the Louisiana Academy of Sciences 47:27–29.

Sickle, J.B. 1980. Correlation of unionid mussels with bottom sediment composition in the Altamaha River, Georgia. Bulletin of the American Malacological Union, Inc. 1980:10–13.

Silverman, H., E.C. Achberger, J.W. Lunn and T.H. Dietz. 1995. Filtration and utilization of laboratory-cultured bacteria by *Dreissena polymorpha*, *Corbicula fluminea* and *Carunculina texasensis*. Biological Bulletin 189(3):308–319.

Silverman, H., W.T. Kays and T.H. Dietz. 1987. Maternal calcium contribution to glochidial shells in freshwater mussels (Eulamellibranchia: Unionidae). Journal of Experimental Zoology 242(2):137–146.

Silverman, H., S.J. Nichols, J.S. Cherry, E. Archberger, J.W. Lynn and T.H. Dietz. 1997. Clearance of laboratory-cultured bacteria by freshwater bivalves: differences between lentic and lotic unionids. Canadian Journal of Zoology 75(11):1857–1866.

Silverman, H., P.E. Richard, R.H. Goddard and T.H. Dietz. 1989. Intracellular formation of calcium concretions by phagocytic cells in freshwater mussels. Canadian Journal of Zoology 67(1):198–207.

Silverman, H., W.L. Steffens and T.H. Dietz. 1983. Calcium concretions in the gills of a freshwater mussel serve as a calcium reservoir during periods of hypoxia. Journal of Experimental Zoology 227(2):177–189.

Silverman, H., W.L. Steffens and T.H. Dietz. 1985. Calcium from extracellular concretions in the gills of freshwater unionid mussels is mobilized during reproduction. Journal of Experimental Zoology 236(2):137–147.

Simpson, C.T. 1893. On the relationship and distribution of the North American Unionidae, with notes on the west coast species. American Naturalist 27(316):353–358.

Simpson, C.T. 1896. The classification and geographical distribution of the pearly fresh-water mussels. Proceedings of the U.S. National Museum 18(1068):295–343, 1 map.

Simpson, C.T. 1899. The pearly fresh-water mussels of the United States; their habits, enemies, and diseases, with suggestions for their protection. Bulletin of the U.S. Fish Commission 18(1898):279–288.

Simpson, C.T. 1900a. New and unfigured Unionidae. Proceedings of the Academy of Natural Sciences of Philadelphia 52(1):74–86, plates 1–5.

Simpson, C.T. 1900b. Synopsis of the naiades, or pearly fresh-water mussels. Proceedings of the U.S. National Museum 22(1205):501–1044.

Simpson, C.T. 1914. A Descriptive Catalogue of the Naiades, or Pearly Fresh-water Mussels. Parts I–III. Bryant Walker, Detroit, Michigan. 1540 pages.

Simpson, G.B. 1884. Anatomy and physiology of *Anodonta fluviatilis*. New York State Museum of Natural History, Annual Report 35:169–191.

Simpson, J.C. 1956. A Provisional Gazetteer of Florida Place-names of Indian Derivation, Either Obsolete or Retained Together with Others of Recent Application. Florida Geological Survey, Special Publications, Number 1. 158 pages.

Sinclair, R.M., and W.M. Ingram. 1961. A new record for the Asiatic clam in the United States, the Tennessee River. The Nautilus 74(3):114–118.

Siripattrawan, S., J.-K. Park and D. Ó Foighil. 2000. Two lineages of the introduced Asian freshwater clam *Corbicula* occur in North America. Journal of Molluscan Studies 66(3):423–429.

Skelton, P.W., and M.J. Benton. 1993. Mollusca: Rostroconchia, Scaphopoda and Bivalvia. Pages 237–263. In: M.J. Benton (editor). The Fossil Record 2. Chapman and Hall, London.

Skibinski, D.O.F., C. Gallagher and C.M. Beynon. 1994a. Mitochondrial DNA inheritance. Nature 368(6474):817–818.

Skibinski, D.O.F., C. Gallagher and C.M. Beynon. 1994b. Sex limited mitochondrial DNA transmission in the marine mussel *Mytilus edulis*. Genetics 138(3):801–809.

Smith, D.G. 1980. Anatomical studies on *Margaritifera margaritifera* and *Cumberlandia monodonta* (Mollusca: Pelecypoda: Margaritiferidae). Zoological Journal of the Linnaean Society 69(3):257–270.

Smith, D.G. 1983. On the so-called mantle muscle scars on the shells of the Margaritiferidae (Mollusca, Pelecypoda), with observations on mantle-shell attachment in the Unionoida and Trigonoida. Zoologica Scripta 12(1):67–71.

Smith, D.G. 1986. The stomach anatomy of some eastern North American Margaritiferidae (Unionoida: Unionacea). American Malacological Bulletin 4(1):13–19.

Smith, D.G. 1991. Comment on proposed conservation of *Proptera* Rafinesque, 1819 (Mollusca: Bivalvia). (Case 2558). Bulletin of Zoological Nomenclature 48(2):142–143.

Smith, D.G. 2000. Investigations on the byssal gland in juvenile unionids. Pages 103–107. In: R.A. Tankersley, D.I. Warmolts, G.T. Watters and B.J. Armitage (editors). Freshwater Mollusk Symposia Proceedings, Part 1. Proceedings of the Conservation, Captive Care and Propagation of Freshwater Mussels Symposium. Ohio Biological Survey, Columbus, Ohio.

Smith, D.G. 2001. Systematics and distribution of the recent Margaritiferidae. Pages 33–49. In: G. Bauer and K. Wächtler (editors). Ecology and Evolution of Freshwater Mussels Unionoida. Ecological Studies, Volume 145. Springer-Verlag, Berlin, Germany.

Smith, D.G., and W.P. Wall. 1984. The Margaritiferidae reinstated: a reply to Davis and Fuller (1981), genetic relationships among Recent Unionacea (Bivalvia) of North America. Museum of Comparative Zoology, Harvard University, Occasional Papers on Mollusks 4:321–330.

Smith, H.M. 1899. The mussel fishery and pearl-button industry of the Mississippi River. U.S. Fish Commission Bulletin 18(1898):289–314, plates 65–85.

Soucek, D.J., D.S. Cherry and C.E. Zipper. 2003. Impacts of mine drainage and other nonpoint source pollutants on aquatic biota in the upper Powell River System, Virginia. Human and Ecological Risk Assessment 9(4):1059–1073.

Sowerby, G.B. (I). 1820–1834. The Genera of Recent and Fossil Shells, for the use of Students in Conchology and Geology Commenced by J. Sowerby, and Continued by G.B. Sowerby. Sowerby, London. [Without title page or index, plates not numbered, genera and plates arranged alphabetically and bound in two volumes.]

Sowerby, G.B. (II). 1839. A Conchological Manual. G.B. Sowerby, London. 130 pages, 24 plates, 2 folded tables.

Sowerby, G.B. (II). 1867–1870. Monograph of the genus *Anodon*. In: L. Reeve and G.B. Sowerby (editors). Conchologia Iconica 17:1–57, 37 plates.

Spengler, L. 1793. Beskrivelse over et nyt Slægt af de toskallede Konkylier, forhen af mig kaldet *Chaena*, saa og over det Linnéiske Slægt *Mya*, hvilket nøiere bestemmes, og inddeles i tvende Slægter. Skrivter af Naturhistorie-Selskabet. København 3:16–69, plate 2.

Splittstößer, P. 1913. Zur Morphologie des Nervensystems von *Anodonta cellensis* Schröt. Zeitschrift für wissenschaftliche Zoologie 104(3):388–470.

Stansbery, D.H. 1964. The Mussel (Muscle) Shoals of the Tennessee River revisited. (Abstract). American Malacological Union, Inc. Annual Reports for 1964 31:25–28.

Stansbery, D.H. 1970a. 2. Eastern freshwater mollusks. I. The Mississippi and St. Lawrence River systems. American Malacological Union Symposium on Rare and Endangered Mollusks. Malacologia 10(1):9–22.

Stansbery, D.H. 1970b. A study of the growth rate and longevity of the naiad *Amblema plicata* (Say, 1817) in Lake Erie (Bivalvia: Unionidae). American Malacological Union, Inc. Annual Reports for 1970:78–79.

Stansbery, D.H. 1971. Rare and endangered freshwater mollusks in eastern United States. Pages 5–18f, 50 figures. In: S.E. Jorgensen and R.E. Sharp (editors). Proceedings of a Symposium on Rare and Endangered Mollusks (naiads) of the U.S. Region 3, Bureau Sport Fisheries and Wildlife, U.S. Fish Wildlife Service, Twin Cities, Minnesota.

Stansbery, D.H. 1976. Naiad mollusks. Pages 42–52. In: H.T. Boschung (editor). Endangered and Threatened Plants and Animals of Alabama. Alabama Museum of Natural History Bulletin 2.

Stansbery, D.H., and C.B. Stein. 1976. Changes in the distribution of *Io fluvialis* (Say, 1825) in the upper Tennessee River system (Mollusca: Gastropoda: Pleuroceridae). Bulletin of the American Malacological Union, Inc. 1976:28–33.

Starnes, L.B., and W.C. Starnes. 1980. Discovery of a new population of *Pegias fabula* (Lea) (Unionidae). The Nautilus 94(1):5–6.

Starobogatov, Ya.I. 1970. Fauna mollyuskov i zoogeographicheskoe raionirovanie kontinental'nykh vodoemov zemnogo shara [Mollusk fauna and zoogeographical partitioning of continental water reservoirs of the world]. Akademiya Nauk SSSR. Zoologischeskii Instituti Nauka, Leningrad. 372 pages, 39 figures, 12 tables [in Russian].

Starobogatov, Ya.I. 1991. Problems in the nomenclature of higher taxonomic categories. Bulletin of Zoological Nomenclature 48(1):6–18.

Starobogatov, Ya.I. 1992. Morphological basis for phylogeny and classification of Bivalvia. Ruthenica 2(1):1–25.

Starobogatov, Ya.I. 1995. The pearly freshwater mussels (Mollusca, Unionoida, Margaritiferidae) of Russia. Pages 109–112. In: Proceedings of the Ninth International Colloquium of the European Invertebrate Survey, Helsinki, Finland, 3–4 September 1993. WWF [World Wildlife Fund] Finland, Report Number 7.

Starobogatov, Ya.I, L.A. Prozorova, V.V. Bogatov and E.M. Sayenko. 2004. Mollusks. Pages 9–491. In: S.J. Tsalolikhin (editor). Key to Freshwater Invertebrates of Russia and Adjacent Lands. Volume 6. Molluscs, Polychaetes, Nemerteans. Nauka, St. Petersburg, Russia [in Russian].

Steg, M.B. 1998. Identification of host fishes and experimental culture of juveniles for selected species of freshwater mussels in Virginia. Masters thesis, Virginia Polytechnic Institute and State University, Blacksburg. 101 pages.

Steg, M.B., and R.J. Neves. 1997. Fish host identification for Virginia listed unionids in the upper Tennessee River drainage. Triannual Unionid Report 13:34.

Stein, C.B. 1968. Studies in the life history of the naiad, *Amblema plicata* (Say, 1817). (Abstract). American Malacological Union, Inc. Annual Reports for 1968 35:46–47.

Stein, C.B. 1969. Gonad development in the three-ridge naiad, *Amblema plicata* (Say, 1817). (Abstract). American Malacological Union, Inc. Annual Reports for 1969 36:30.

Sterki, V. 1891a. A byssus in *Unio*. The Nautilus 5(7):73–74.

Sterki, V. 1891b. On the byssus of Unionidae. II. The Nautilus 5(8):90–91.

Sterki, V. 1895. Some notes on the genital organs of Unionidae, with the reference to systematics. The Nautilus 9(8):91–94.

Sterki, V. 1898a. Some observations on the genital organs of Unionidae with reference to classification. The Nautilus 12(2):18–21; 12(3):28–32.

Sterki, V. 1898b. *Anodonta imbecillis*, hermaphroditic. The Nautilus 12(8):87–88.

Sterki, V. 1911. Notes on the anatomy and physiology of the Unionidae. The Ohio Naturalist 11(6):331–334.

Stern, E. 1983. Depth distribution and density of freshwater mussels (Unionidae) collected by SCUBA from the lower Wisconsin and St. Croix Rivers. The Nautilus 97(1):36–42.

Stern, E.M., and D.L. Felder. 1978. Identification of host fishes for four species of freshwater mussels (Bivalvia: Unionidae). The American Midland Naturalist 100(1):233–236.

Stimpson, W. 1851. Shells of New England. A Revision of the Synonymy of the Testaceous Mollusks of New England. Phillips, Sampson and Company, Boston, Massachusetts. 66 pages, 11 plates.

Stoliczka, F. 1870–1871. The Pelecypoda, with a review of all known genera of this class; fossil and recent. In: Palaeontologia Indica, being figures and descriptions of the organic remains procured during the progress of the geological survey of India. Cretaceous fauna of southern India. Memoirs of the Geological Survey of India 3. 409 pages, plates 1–50.

Strayer, D.L. 1981. Notes on the microhabitats of unionid mussels in some Michigan streams. The American Midland Naturalist 106(2):411–415.

Strayer, D.L. 1983. The effects of surface geology and stream size on freshwater mussel (Bivalvia: Unionidae) distribution in southeastern Michigan, U.S.A. Freshwater Biology 13(3):253–264.

Strayer, D.L. 1993. Macrohabitats of freshwater mussels (Bivalvia: Unionacea) in the streams of the northern Atlantic Slope. Journal of the North American Benthological Society 12(3):236–246.

Strayer, D.L. 1999a. Effects of alien species on freshwater mollusks in North America. Journal of the North American Benthological Society 18(1):74–98.

Strayer, D.L. 1999b. Use of flow refuges by unionid mussels in rivers. Journal of the North American Benthological Society 18(4):468–476.

Strayer, D.L., J.A. Downing, W.R. Haag, T.L. King, J.B. Layzer, T.J. Newton and S.J. Nichols. 2004. Changing perspectives on pearly mussels, North America's most imperiled animals. BioScience 54(5):429–439.

Strayer, D.L., D.C. Hunter, L.C. Smith and C.K. Borg. 1994. Distribution, abundance, and roles of freshwater clams (Bivalvia: Unionidae) in the freshwater tidal Hudson River. Freshwater Biology 31(2):239–248.

Strayer, D.L., and K.J. Jirka. 1997. The pearly mussels of New York state. New York State Museum Memoir 26:1–113, 27 plates.

Strayer, D.L., K.J. Jirka and K.J. Schneider. 1992. Recent collections of freshwater mussels (Bivalvia: Unionidae) from western New York. Walkerana (for 1991) 5(13):63–72.

Strayer, D.L., and J. Ralley. 1993. Microhabitat use by an assemblage of stream-dwelling unionaceans (Bivalvia), including two rare species of *Alasmidonta*. Journal of the North American Benthological Society 12(3):247–258.

Strayer, D.L., and L.C. Smith. 1996. Relationships between zebra mussels (*Dreissena polymorpha*) and unionid clams during the early stages of the zebra mussel invasion in the Hudson River. Freshwater Biology 36(3):771–779.

Strecker, J.K., Jr. 1931. The distribution of the naiades or pearly fresh-water mussels of Texas. Baylor University Museum, Special Bulletin, Number 2. 71 pages.

Stringfellow, R.C. 1997. A descriptive survey of freshwater Unionidae bivalves in five creeks located in west central Georgia. Masters thesis, Columbus State University, Georgia. 77 pages.

Subba Rao, N.V. 1989. Handbook Freshwater Molluscs of India. Zoological Survey of India, Calcutta, India. 289 pages.

Surber, T. 1912. Identification of the glochidia of freshwater mussels. Report and Special Papers of the U.S. Bureau of Fisheries 1912:1–10, plates 1–3. [Issued separately as U.S. Bureau of Fisheries Document 771.]

Surber, T. 1913. Notes on the natural hosts of fresh-water mussels. Bulletin of the U.S. Bureau of Fisheries 32(1912):101–116, plates 29–31. [Issued separately as U.S. Bureau of Fisheries Document 778.]

Surber, T. 1915. Identification of the glochidia of fresh-water mussels. Appendix V to the Report of the U.S. Commissioner of Fisheries for 1914:3–9, 1 plate. [Issued separately as U.S. Bureau of Fisheries Document 813.]

Swainson, W. 1823. The specific characters of several undescribed shells. The Philosophical Magazine and Journal 62:401–403.

Swainson, W. 1824. Description of two new remarkable freshwater shells, *Melania setosa* and *Unio gigas*. Quarterly Journal of Science 17:13–17.

Swainson, W. 1820–1823, 1829–1833. Zoological illustrations, or Original figures and descriptions of new, rare, or interesting animals, selected chiefly from the classes of ornithology, entomology, and conchology, and arranged on the principles of Cuvier and other modern zoologists. London, Baldwin, Cradock, & Joy [etc.]. Six volumes, first series 1820-1823, second series 1829-1833.

Swainson, W. 1840. A Treatise on Malacology or the Natural Classification of Shells and Shell-fish. Printed for Longman, Orme, Brown, Green and Longmans [etc.], London. 419 pages.

Swainson, W. 1841. Exotic conchology; or figures and descriptions of rare, beautiful, or undescribed shells, drawn on stone from the most select specimens; the descriptions systematically arranged on the principles of the natural system. Second Edition. Edited by Sylvanus Hanley. Henry G. Bohn, York Street, Covent Garden. London. 339 pages, 48 plates.

Swingle, H.A., and D.G. Bland. 1974. Distribution of the estuarine clam *Rangia cuneata* Gray in coastal waters of Alabama. Alabama Marine Resources Bulletin 10:9–16.

Šyvokienė, J. 1988. The amount and biomass of water and digestive tract bacteria of Unionacea. Acta Hydrobiologica Lituanica 7:63–68 [in Russian].

Šyvokienė, J., D. Sinevičienė and B. Šalčiūtė. 1987. Microflora of the mussel *Anodonta piscinalis* digestive system depending on feeding. Lietuvos TSR Moksly Akamijos darbai. C. serija 2(98):48–54 [in Russian].

Šyvokienė, J., D. Sinevičienė and B. Šalčiūtė. 1988. Peculiarities of group composition of symbiotic microflora of Unionacea digestive tract depending on their nutrition. Acta Hydrobiologica Lituanica 7:69–76 [in Russian].

Tankersley, R.A. 1996. Multipurpose gills: effect of larval brooding on the feeding physiology of freshwater unionid mussels. Invertebrate Biology 115(3):243–255.

Tankersley, R.A., and R.V. Dimock, Jr. 1992. Quantitative analysis of the structure and function of the marsupial gills of the freshwater mussel *Anodonta cataracta*. Biological Bulletin 182(1):145–154.

Tankersley, R.A., and R.V. Dimock, Jr. 1993a. The effect of larval brooding on the respiratory physiology of the freshwater unionid mussel *Pyganodon cataracta*. The American Midland Naturalist 130(1):146–163.

Tankersley, R.A., and R.V. Dimock, Jr. 1993b. Endoscopic visualization of the functional morphology of the ctenidia of the unionid mussel *Pyganodon cataracta*. Canadian Journal of Zoology 71(4):811–819.

Tanner, D.P. 1970. Five additions to southwestern Pennsylvania naiad fauna. Sterkiana 37:27–29. [Reprinted in Pittsburgh Shell Club Bulletin Number 8:16–17.]

Tanner, H.S. 1830. The traveller's pocket map of Alabama, with its roads and distances from place to place, along the stage and steamboat routes. H.S. Tanner, Philadelphia.

Tappan, B. 1839. Description of some new shells. American Journal of Science and Arts 35(2):268–270, plate 3.

Taylor, J.D., W.J. Kennedy and A. Hall. 1969. The shell structure and mineralogy of the Bivalvia: introduction, Nuculacea - Trigonacea. Bulletin of the British Museum (Natural History), Zoology, Supplement 3:3–125, plates 1–29.

Tennessee Valley Authority. 1949. Geology and foundation treatment, Tennessee Valley Authority projects, Knoxville, Tennessee. Tennessee Valley Authority, Technical Report 22. 548 pages.

Tennessee Valley Authority. 1964. Progress Report for August 1964. Unpublished report, Tennessee Valley Authority, Division of Forestry Development, Knoxville, Tennessee. 8 pages.

Tennessee Valley Authority. 1966. The mussel resource of the Tennessee River. Tennessee Valley Authority, Fish and Wildlife Branch, Norris, Tennessee. 32 pages.

Tennessee Valley Authority. 1986. Cumberlandian Mollusk Conservation Program. Activity 3: identification of fish hosts. Tennessee Valley Authority, Office of Natural Resources and Economic Development, Knoxville, Tennessee. 57 pages.

Tepe, W.C. 1943. Hermaphroditism in *Carunculina parva*, a fresh water mussel. The American Midland Naturalist 29(3):621–623.

The Nature Conservancy. 1997. Freshwater Initiative. The Nature Conservancy, Arlington, Virginia. 33 pages.

Thiele, J. 1929–1935. Handbuch der systematischen Weichtierkunde. In four parts. 1154 pages. Gustav Fischer, Jena. 1(1):1–376 (1929); 1(2):377–778 (1931); 2(3):779–1022 (1934); 2(4):1023–1154 (1935). [Reprinted (1963), A. Asher and Company, Amsterdam, The Netherlands. Translated by R. Bieler and P.M. Mikkelsen. Handbook of Systematic Malacology. Smithsonian Institution Libraries and the National Science Foundation, Washington, District of Columbia, Parts 1 and 2 (1992); Parts 3 and 4 (1998).]

Threlfall, W. 1986. Seasonal occurrence of *Anodonta cataracta* Say, 1817, glochidia on three-spined sticklebacks, *Gasterosteus aculeatus* Linnaeus. The Veliger 29(2):231–234.

Thurston, W.N. 1973. The Apalachicola-Chattahoochee-Flint River water route system in the nineteenth century. The Georgia Historical Quarterly 57(2):200–212.

Trautman, M.B. 1981. The fishes of Ohio with illustrated keys. Second edition. Ohio State University Press, Columbus. 782 pages.

Trdan, R.J., and W.R. Hoeh. 1982. Eurytopic host use by two congeneric species of freshwater mussels (Pelecypoda: Unionidae: *Anodonta*). The American Midland Naturalist 108(2):381–388.

Trimble, J.J., III, and D. Gaudin. 1975. Fine structure of the sperm of the freshwater clam *Ligumia subrostrata* (Say, 1831). The Veliger 18(1):34–36, figures 4–10.

Troschel, F.H. 1847. Ueber die Brauchbarkeit der Mundlappen und Kiemen zur Familieuntescheidung und über die Familie der Najaden. Archiv für Naturgeschichte 13(1): 257–274, plate 6.

Trueman, E.R. 1968. The locomotion of the freshwater clam *Margaritifera margaritifera* (Unionacea: Margaritiferidae). Malacologia 6(3):401–410.

Tucker, M.E. 1927. Morphology of the glochidium and juvenile of the mussel *Anodonta imbecillis*. Transactions of the American Microscopical Society 46(4):286–293.

Tucker, M.E. 1928. Studies on the life cycle of two species of fresh-water mussels belonging to the genus *Anodonta*. Biological Bulletin 54(2):117–127.

Turgeon, D.D., A.E. Bogan, E.V. Coan, W.K. Emerson, W.G. Lyons, W.L. Pratt, C.F.E. Roper, A. Scheltema, F.G. Thompson and J.D. Williams. 1988. Common and Scientific Names of Aquatic Invertebrates from the United States and Canada: Mollusks. American Fisheries Society, Special Publication 16. 277 pages, 12 plates.

Turgeon, D.D., J.F. Quinn, A.E. Bogan, E.V. Coan, F.G. Hochberg, W.G. Lyons, P. Mikkelsen, R.J. Neves, C.F.E. Roper, G. Rosenberg, B. Roth, A. Scheltema, F.G. Thompson, M. Vecchione and J.D. Williams. 1998. Common and Scientific Names of Aquatic Invertebrates from the United States and Canada: Mollusks. Second edition. American Fisheries Society, Special Publication 26. 526 pages.

Tyrrell, M., and D.J. Hornbach. 1998. Selective predation by muskrats on freshwater mussels in two Minnesota rivers. Journal of the North American Benthological Society 17(3):301–310.

U.S. Army Corps of Engineers. 1985a. Alabama-Mississippi stream mileage tables with drainage areas. U.S. Army Corps of Engineers, Mobile District, Mobile, Alabama. 276 pages.

U.S. Army Corps of Engineers. 1985b. Florida-Georgia stream mileage tables with drainage areas. U.S. Army Corps of Engineers, Mobile District, Mobile, Alabama. 233 pages.

U.S. Environmental Protection Agency. 1990. The quality of our nation's water: a summary of the 1988 National Water Quality Inventory. EPA Report 440/4-90-005. U.S. Environmental Protection Agency, Washington, District of Columbia. 23 pages.

U.S. Fish and Wildlife Service. 1976. Endangered and threatened wildlife and plants; endangered status for 159 taxa of animals. Federal Register 41(115):24062–24064.

U.S. Fish and Wildlife Service. 1984. Endangered and threatened wildlife and plants; review of invertebrate wildlife for listing as endangered or threatened species. Federal Register 49(100):21664–21675.

U.S. Fish and Wildlife Service. 1989a. Endangered and threatened wildlife and plants; animal notice of review. Federal Register 54(4):554–579.

U.S. Fish and Wildlife Service. 1989b. Recovery plan for Little-wing Pearl Mussel (*Pegias fabula*). U.S. Fish and Wildlife Service, Atlanta, Georgia. 26 pages.

U.S. Fish and Wildlife Service. 1991. Endangered and threatened wildlife and plants; animal candidate review for listing as endangered or threatened species. Federal Register 56(225):58804–58836.

U.S. Fish and Wildlife Service. 1994. Endangered and threatened wildlife and plants; animal candidate review for listing as endangered or threatened species. Federal Register 59(219):58982–59028.

U.S. Fish and Wildlife Service. 2000. Mobile River Basin aquatic ecosystem recovery plan. U.S. Fish and Wildlife Service, Atlanta, Georgia. 128 pages.

U.S. Fish and Wildlife Service. 2003. Recovery plan for endangered Fat Threeridge (*Amblema neislerii*), Shinyrayed Pocketbook (*Lampsilis subangulata*), Gulf Moccasinshell (*Medionidus penicillatus*), Ochlockonee Moccasinshell (*Medionidus simpsonianus*), Oval Pigtoe (*Pleurobema pyriforme*); and Threatened Chipola Slabshell (*Elliptio chipolaensis*), and Purple Bankclimber (*Elliptoideus sloatianus*). U.S. Fish and Wildlife Service, Atlanta, Georgia. 142 pages.

U.S. Fish and Wildlife Service. 2004. Recovery plan for Cumberland Elktoe (*Alasmidonta atropurpurea*), Oyster Mussel (*Epioblasma capsaeformis*), Cumberland Combshell (*Epioblasma brevidens*), Purple Bean (*Villosa perpurpurea*), and Rough Rabbitsfoot (*Quadrula cylindrica strigillata*). U.S. Fish and Wildlife Service, Atlanta, Georgia. 167 pages.

Uthaiwan, K., N. Noparatnaraporn and J. Machado. 2001. Culture of glochidia of the freshwater pearl mussel *Hyriopsis myersiana* Lea, 1856 in artificial media. Aquaculture 195(1–2):61–69.

Uthaiwan, K., P. Pakkong, N. Noparatnaraporn, L. Vilarinho and J. Machado. 2002. Study of a suitable fish plasma for in vitro culture of glochidia *Hyriopsis myersiana*. Aquaculture 209(1):197–208.

Utterback, W.I. 1915. The naiades of Missouri. The American Midland Naturalist 4(3):41–53, 4(4):97–152, 4(5):181–204, 4(6):244–273.

Utterback, W.I. 1916a. Breeding record of Missouri mussels. The Nautilus 30(2):13–21.

Utterback, W.I. 1916b. The naiades of Missouri (continued). The American Midland Naturalist 4(7):311–327, 4(8):339–354, 4(9):387–400, 4(10):432–464, plates 1–28.

Utterback. W.I. 1917. Naiad geography of Missouri. The American Midland Naturalist 5(1):26–30.

Utterback, W.I. 1931. Sex behavior among naiades. West Virginia Academy of Science 5:43–45.

Valenciennes, A. 1827. Coquilles fluviatiles bivalves du Nouveau-Continent, recueillies pendant le voyage de MM. De Humboldt et Bonpland. In: A. von Humboldt and A.J.A. Bonpland. Recueil d'observations de zoologie et d' anatomie compare, faites dans l'ocean Atlantique, dans l'intérieur du nouveau continent et dans la mer du sud pendant les années 1799, 1800, 1801, 1802 et 1803; par Al. de Humbodt et A. Bonpland. J. Smith and Gide, Paris. 2(13):225–237, colored plates 48, 50, 53, 54.

Valentine, B.D., and D.H. Stansbery. 1971. An introduction to the naiades of the Lake Texoma region, Oklahoma, with notes on the Red River fauna (Mollusca: Unionidae). Sterkiana 42:1–40.

van Damme, D. 1984. The Freshwater Mollusca of Northern Africa. Distribution, Biogeography and Palaeoecology. Developments in Hydrobiology, Volume 25. Dr. W. Junk Publishers, The Netherlands. 164 pages.

van der Schalie, H. 1933. Notes on the brackish water bivalve, *Polymesoda caroliniana* (Bosc). Occasional Papers of the Museum of Zoology, University of Michigan, Number 258. 8 pages, 1 plate.

van der Schalie, H. 1934. *Lampsilis jonesi*, a new naiad from southeastern Alabama. The Nautilus 47(4):125–127, plate 15.

van der Schalie, H. 1938a. The naiades (freshwater mussels) of the Cahaba River in northern Alabama. Occasional Papers of the Museum of Zoology, University of Michigan, Number 392. 29 pages.

van der Schalie, H. 1938b. Contributing factors in the depletion of naiades in the eastern United States. Basteria 3(4):51–57.

van der Schalie, H. 1938c. The naiad fauna of the Huron River, in southeastern Michigan. University of Michigan, Museum of Zoology, Miscellaneous Publications, Number 40. 83 pages.

van der Schalie, H. 1938d. The taxonomy of naiades inhabiting a lake environment. Journal of Conchology 21(8):246–253.

van der Schalie, H. 1939a. *Medionidus mcglameriae*, a new naiad from the Tombigbee River, with notes on other naiades of that drainage. Occasional Papers of the Museum of Zoology, University of Michigan, Number 407. 6 pages.

van der Schalie, H. 1939b. Additional notes on the naiades (fresh-water mussels) of the lower Tennessee River. The American Midland Naturalist 22(2):452–457.

van der Schalie, H. 1940. The naiad fauna of the Chipola River, in northwestern Florida. Lloydia 3(3):191–208.

van der Schalie, H. 1966. Hermaphroditism among North American freshwater mussels. Malacologia 5(1):77–78.

van der Schalie, H. 1969. Two unusual unionid hermaphrodites. Science 163(3873):1333–1334.

van der Schalie, H. 1970. Hermaphroditism among North American freshwater mussels. Malacologia 10(1):93–112.

van der Schalie, H. 1981a. Past, present and future status of the Mollusca of the upper Tombigbee River. Sterkiana 71:8–11.

van der Schalie, H. 1981b. Perspective on North American malacology, I: Mollusks in the Alabama River drainage; past and present. Sterkiana 71:24–40.

van der Schalie, H., and F. Locke. 1941. Hermaphroditism in *Anodonta grandis*, a fresh-water mussel. Occasional Papers of the Museum of Zoology, University of Michigan, Number 432. 7 pages, 3 plates.

van der Schalie, H., and P.W. Parmalee. 1960. Animal remains from the Etowah Site, Mound C, Bartow County, Georgia. Florida Anthropologist 13(2–3)37–54.

van Snik Gray, E., W.A. Lellis, J.C. Cole and C.S. Johnson. 1999. Hosts of *Pyganodon cataracta* (eastern floater) and *Strophitus undulatus* (squawfoot) from the upper Susquehanna River basin, Pennsylvania. Triannual Unionid Report 18:6.

van Snik Gray, E., W.A. Lellis, J.C. Cole and C.S. Johnson. 2002. Host identification for *Strophitus undulatus* (Bivalvia: Unionidae), the Creeper, in the upper Susquehanna River basin, Pennsylvania. The American Midland Naturalist 147(1):153–161.

Vanatta, E.G. 1915. Rafinesque's types of *Unio*. Proceedings of the Academy of Natural Sciences of Philadelphia 67(1915):549–559.

Vannote, R.L., and G.W. Minshall. 1982. Fluvial processes and local lithology controlling abundance, structure, and composition of mussel beds. Proceedings of the National Academy of Sciences, United States of America 79(13):4103–4107.

Vaughn, C.C. 1997. Regional patterns of mussel species distributions in North American rivers. Ecography 20(2):107–115.

Vaughn, C.C., K.B. Gido and D.E. Spooner. 2004. Ecosystem processes performed by unionid mussels in stream mesocosms: species roles and effects of abundance. Hydrobiologia 527(1):35–47.

Vaughn, C.C., and C.C. Hakenkamp. 2001. The functional role of burrowing bivalves in freshwater ecosystems. Freshwater Biology 46(11):1431–1446.

Vaughn, C.C., and C.M. Taylor. 1999. Impoundment and the decline of freshwater mussels: a case study of an extinction gradient. Conservation Biology 13(4):912–920.

Vaughn, C.C., and C.M. Taylor. 2000. Macroecology of a host-parasite relationship. Ecography 23(1):11–20.

Vicentini, H. 2005. Unusual spurting behavior of the freshwater mussel *Unio crassus*. Journal of Molluscan Studies 71(4):409–410.

Vidrine, M.F. 1993. The Historical Distributions of Freshwater Mussels in Louisiana. Gail Q. Vidrine Collectibles, Eunice, Louisiana. 225 pages, 20 color plates, 7 tables, 136 maps.

Vidrine, M.F. 1996a. North American *Najadicola* and *Unionicola*: photomicrographs. Gail Q. Vidrine Collectibles, Eunice, Louisiana. 205 pages.

Vidrine, M.F. 1996b. North American *Najadicola* and *Unionicola*: collections and communities. Gail Q. Vidrine Collectibles, Eunice, Louisiana. 259 pages.

Vidrine, M.F. 1996c. North American *Najadicola* and *Unionicola*: systematics and coevolution. Gail Q. Vidrine Collectibles, Eunice, Louisiana. 145 pages.

Vidrine, M.F. 1996d. North American *Najadicola* and *Unionicola*: diagnoses and distributions. Gail Q. Vidrine Collectibles, Eunice, Louisiana. 365 pages.

Vidrine, M.F. 1996e. North American *Najadicola* and *Unionicola*: I. Diagnoses of genera and subgenera. II. Key. III. List of reported hosts. Gail Q. Vidrine Collectibles, Eunice, Louisiana. 180 pages.

Vogt, R.C. 1981. Food partitioning in three sympatric species of map turtle, genus *Graptemys* (Testudinata, Emydidae). American Midland Naturalist 105(1):102–111.

von Ihering, H. 1891. *Anodonta* und *Glabaris*. Zoologischer Anzeiger 14(380):474–484.

von Ihering, H. 1901. The Unionidae of North America. The Nautilus 15(4):37–39; 15(5):50–53.

von Martens, E. 1890–1901. Biologia Centrali-Americana. Land and Freshwater Mollusca. 675 pages.

Wächtler, K., M.C. Dreher-Mansur and T. Richter. 2001. Larval types and early postlarval biology in naiads (Unionoida). Pages 93–125. In: G. Bauer and K. Wächtler (editors). Ecology and Evolution of Freshwater Mussels Unionoida. Ecological Studies, Volume 145. Springer-Verlag, Berlin, Germany.

Wade, D.C., R.G. Hudson and A.D. McKinney. 1993. Comparative response of *Ceriodaphnia dubia* and juvenile *Anodonta imbecillis* to selected complex industrial whole effluents. Pages 109–112. In: K.S. Cummings, A.C. Buchanan and L.M. Koch (editors). Conservation and Management of Freshwater Mussels. Proceedings of a UMRCC Symposium, 12–14 October 1992, St. Louis, Missouri. Upper Mississippi River Conservation Committee, Rock Island, Illinois.

Walker, B. 1905a. List of shells from northwestern Florida. The Nautilus 18(12):133–136, plate 9, figures 6 and 7.

Walker, B. 1905b. A new species of *Medionidus*. The Nautilus 18(12):136–137, plate 9, figures 4–5.

Walker, B. 1910a. Description of a new species of *Truncilla*. The Nautilus 24(4):42–44, plate 3.

Walker, B. 1910b. Notes on *Truncilla*, with a key to the species. The Nautilus 24(7):75–81.

Walker, B. 1915. *Pleurobema missouriensis* Marsh. The Nautilus 28(12):140–141, plate 5.

Walker, B. 1918. Notes on North American naiades. I. Occasional Papers of the Museum of Zoology, University of Michigan, Number 49. 6 pages.

Wallengren, H. 1905. Zur biologie der Muscheln. I. Die Wasserströmungen. Acta Universitatis Lundensis (Series 2) 1(2):1–64; 1(3):1–59.

Waller, D.L., and B.A. Lasee. 1997. External morphology of spermatozoa and spermatozeugmata of the freshwater mussel *Truncilla truncata* (Mollusca: Bivalvia: Unionidae). The American Midland Naturalist 138(1):220–223.

Waller, D.L., and L.G. Mitchell. 1989. Gill tissue reactions in walleye *Stizostedion vitreum vitreum* and common carp *Cyprinus carpio* to glochidia of the freshwater mussel *Lampsilis radiata siliquoidea*. Diseases of Aquatic Organisms 6(2):81–87.

Ward, J.E., B.A. MacDonald, R.J. Thompson and P.G. Benninger. 1993. Mechanisms of suspension feeding in bivalves: resolution of current controversies by means of endoscopy. Limnology and Oceanography 38(2):265–272.

Warren, M.L., and W.R. Haag. 2005. Spatio-temporal patterns of the decline of freshwater mussels in the Little South Fork Cumberland River, U.S.A. Biodiversity and Conservation 14(6):1383–1400.

Warren, R.E. 1975. Prehistoric unionacean (Freshwater Mussel) utilization at the Widows Creek Site (1JA305), Northeast Alabama. Masters thesis, University of Nebraska, Department of Anthropology, Lincoln. 245 pages.

Warren, R.E. 1991. Freshwater mussels as paleoenvironmental indicators: a quantitative approach to assemblage analysis. Pages 23–66. In: J.R. Purdue, W.E. Klippel and B.W. Styles (editors). Beamers, Bobwhites, and Blue-points: Tributes to the Career of Paul W. Parmalee. Illinois State Museum Scientific Papers, Volume 23 and The University of Tennessee, Department of Anthropology Report of Investigations, Number 52.

Watson, B.T., and R.J. Neves. 1998. Fish host identification for two federally endangered unionids in the upper Tennessee River drainage. Triannual Unionid Report 14:7.

Watters, G.T. 1992. Unionids, fishes and the species-area curve. Journal of Biogeography 19(5):481–490.

Watters, G.T. 1993a. A guide to the freshwater mussels of Ohio. Revised edition. Division of Wildlife, Ohio Department of Natural Resources. 106 pages.

Watters, G.T. 1993b. Mussel diversity as a function of drainage area and fish diversity: management implications. Pages 113–116. In: K.S. Cummings, A.C. Buchanan and L.M. Koch (editors). Conservation and Management of Freshwater Mussels. Proceedings of a UMRCC Symposium, 12–14 October 1992, St. Louis, Missouri. Upper Mississippi River Conservation Committee, Rock Island, Illinois.

Watters, G.T. 1994a. An annotated bibliography of the reproduction and propagation of the Unionoidea (primarily of North America). Ohio Biological Survey, Miscellaneous Contributions, Number 1. 158 pages.

Watters, G.T. 1994b. Form and function of unionoidean shell sculpture and shape (Bivalvia). American Malacological Bulletin 11(1):1–20.

Watters, G.T. 1995. Sampling freshwater mussel populations: the bias of muskrat middens. Walkerana (for 1993–1994) 7(17–18):63–69.

Watters, G.T. 1996a. Hosts for the northern riffle shell (*Epioblasma torulosa rangiana*). Triannual Unionid Report 10:14.

Watters, G.T. 1996b. Small dams as barriers to freshwater mussels (Bivalvia, Unionoida) and their hosts. Biological Conservation 75(1):79–85.

Watters, G.T. 1999. Morphology of the conglutinate of the kidneyshell freshwater mussel, *Ptychobranchus fasciolaris*. Invertebrate Biology 118(3):289–295.

Watters, G.T. 2001. The evolution of the Unionacea in North America, and its implications for the Worldwide fauna. Pages 281–307. In: G. Bauer and K. Wächtler (editors). Ecology and Evolution of the Freshwater Mussels Unionoida. Ecological Studies, Volume 145. Springer-Verlag, Berlin, Germany.

Watters, G.T. 2002. The kinetic conglutinates of the Creeper freshwater mussel, *Strophitus undulatus* (Say, 1817). Journal of Molluscan Studies 68(2):155–158.

Watters, G.T., and K. Kuehnl. 2004. Ohio pigtoe host suitability trials. Ellipsaria 6(2):10.

Watters, G.T., T. Menker, S. Thomas and K. Kuehnl. 2005. Host identifications or confirmations. Ellipsaria 7(2):11–12.

Watters, G.T., and S.H. O'Dee. 1996. Shedding of untransformed glochidia by fishes parasitized by *Lampsilis fasciola* Rafinesque, 1820 (Mollusca: Bivalvia: Unionidae): evidence of acquired immunity in the field? Journal of Freshwater Ecology 11(4):383–389.

Watters, G.T., and S.H. O'Dee. 1997a. Surrogate hosts: transformation on exotic and non-piscine hosts. Triannual Unionid Report 11:35.

Watters, G.T., and S.H. O'Dee. 1997b. Potential hosts for *Villosa iris* (Lea, 1829). Triannual Unionid Report 12:7.

Watters, G.T., and S.H. O'Dee. 1998. Metamorphosis of freshwater mussel glochidia (Bivalvia: Unionidae) on amphibians and exotic fishes. American Midland Naturalist 139(1):49–57.

Watters, G.T., S.H. O'Dee and S. Chordas, III. 1998. Infective and non-infective glochidia in *Lasmigona costata*? Triannual Unionid Report 15:29.

Watters, G.T., S.H. O'Dee and S. Chordas, III. 2001. Patterns of vertical migration in freshwater mussels (Bivalvia: Unionoida). Journal of Freshwater Ecology 16(4):541–549.

Watters, G.T., S.H. O'Dee, S. Chordas, III, and D. Glover. 1999. Seven potential hosts for *Ligumia recta* (Lamarck, 1819). Triannual Unionid Report 18:5.

Way, C.M., D.J. Hornbach, T. Deneka and R.A. Whitehead. 1989. A description of the ultrastructure of the gills of freshwater bivalves, including a new structure, the frontal cirrus. Canadian Journal of Zoology 67(2):357–362.

Way, C.M., A.C. Miller and B.S. Payne. 1989. The influence of physical factors on the distribution and abundance of freshwater mussels (Bivalvia: Unionidae) in the lower Tennessee River. The Nautilus 103(3):96–98.

Weaver, L.R., G.B. Pardue and R.J. Neves. 1991. Reproductive biology and fish hosts of the Tennessee Clubshell *Pleurobema oviforme* (Mollusca: Unionidae) in Virginia. The American Midland Naturalist 126(1):82–89.

Webb, W.F. 1942. United States Mollusca. Ideal Printing Company, St. Petersburg, Florida. 224 pages.

Weiss, J.L., and J.B. Layzer. 1993. Seasonal and spatial variation in glochidial infections of fish in the Barren River, Kentucky. Pages 72–75. In: K.S. Cummings, A.C. Buchanan and L.M. Koch (editors). Conservation and Management of Freshwater Mussels. Proceedings of a UMRCC Symposium, 12–14 October 1992, St. Louis, Missouri. Upper Mississippi River Conservation Committee, Rock Island, Illinois.

Weiss, J.L., and J.B. Layzer. 1995. Infestations of glochidia on fishes in the Barren River, Kentucky. American Malacological Bulletin 11(2):153–159.

Wells, H.W. 1961. The fauna of oyster beds, with special reference to the salinity factor. Ecological Monographs 31(3):239–266.

Wesler, K.W. 2001. Excavations at Wickliffe Mounds. University of Alabama Press, Tuscaloosa. 178 pages.

Westerlund, C.A. 1890. Fauna der in der Paläaretischen Region lebenden Binnencouchylien. Volume 7. Malcozoa Acephala. H. Ohlsson's Buchdr. Lund. 319 pages, plates 1–15.

Wheeler, H.E. 1935. Timothy Abbott Conrad, with particular reference to his work in Alabama one hundred years ago. Bulletins of American Paleontology 23(77):1–157.

White, L.R., B.A. McPheron and J.R. Stauffer, Jr. 1996. Molecular genetic identification tools for the unionids of French Creek, Pennsylvania. Malacologia 38(1–2):181–202.

Wiles, M. 1975. The glochidia of certain Unionidae (Mollusca) in Nova Scotia and their fish hosts. Canadian Journal of Zoology 53(1):33–41.

Wiley, E.O. 1981. Phylogenetics: the Theory and Practice of Phylogenetic Systematics. John Wiley and Sons, New York. 439 pages.

Willem, V., and A. Minne. 1899. Recherches expérimentales sur la circulation sanguine chez Anodonte. Mémoiores couoronnés et Mémoires de savants, étrangers, Bruxelles 57:28, plates 1–2.

Williams, J.C., and G.A. Schuster. 1989. Freshwater mussel investigations of the Ohio River, mile 317.0 to mile 981.0. Kentucky Department of Fish and Wildlife Resources, Division of Fisheries, Frankfort, Kentucky. 57 pages.

Williams, J.D. 1965. Studies on the fishes of the Tallapoosa River system in Alabama and Georgia. Masters thesis, University of Alabama, Tuscaloosa. 135 pages.

Williams, J.D. 1976. A review of the Endangered Species Act of 1973. ASB [Association of Southeastern Biologists] Bulletin 23(3):138–141.

Williams, J.D. 1982. Distribution and habitat observations of selected Mobile Basin unionid mollusks. Pages 61–85. In: A.C. Miller (compiler). Report of Freshwater Mollusks Workshop, U.S. Army Corps of Engineers, Waterways Experiment Station, Vicksburg, Mississippi.

Williams, J.D., and R.S. Butler. 1994. Freshwater bivalves. Pages 53–128. In: M. Deyrup and R. Franz (editors). Rare and Endangered Biota of Florida. Volume IV. University Press of Florida, Gainesville.

Williams, J.D., and A. Fradkin. 1999. *Fusconaia apalachicola*, a new species of freshwater mussel (Bivalvia: Unionidae) from pre-Columbian archeological sites in the Apalachicola Basin of Alabama, Florida, and Georgia. Tulane Studies in Zoology 31(1):51–62.

Williams, J.D., S.L.H. Fuller and R. Grace. 1992. Effects of impoundments on freshwater mussels (Mollusca: Bivalvia: Unionidae) in the main channel of the Black Warrior and Tombigbee rivers in western Alabama. Alabama Museum of Natural History Bulletin 13:1–10.

Williams, J.D., M.L. Warren, Jr., K.S. Cummings, J.L. Harris and R.J. Neves. 1993. Conservation status of the freshwater mussels of the United States and Canada. Fisheries 18(9):6–22.

Wills, D. 1995. Alabama Mollusks. Freshwater Mussels of the Tennessee River near Decatur, Alabama. Privately published by D. Wills, Hartselle, Alabama. 99 pages.

Wilson, C.B. 1916. Copepod parasites of fresh-water fishes and their economic relations to mussel glochidia. Bulletin of the U.S. Bureau of Fisheries 34(1914):331–374, plates 60–74. [Issued separately as U.S. Bureau of Fisheries Document 824.]

Wilson, C.B., and H.W. Clark. 1912. The mussel fauna of the Maumee River. Report and Special Papers of the Commissioner of Fisheries 1911 (Document Number 757):1–72, plates 1–2. [Issued separately as U.S. Bureau of Fisheries Document 757.]

Wilson, C.B., and H.W. Clark. 1914. The mussels of the Cumberland River and its tributaries. Report and Special Papers of the Commissioner of Fisheries 1912 (Document Number 757):1–63, figures 1–2. [Issued separately as U.S. Bureau of Fisheries Document 781.]

Wilson, C.B., and E. Danglade. 1914. The mussel fauna of central and northern Minnesota. Appendix V to the Report of the U.S. Commissioner of Fisheries for 1913:1–26, 1 map. [Issued separately as U.S. Bureau of Fisheries Document 803.]

Winslow, M.L. 1918. *Pleurobema clava* (Lam.) and *Planorbis dilatatus buchanensis* (Lea) in Michigan. Occasional Papers of the Museum of Zoology, University of Michigan, Number 51. 6 pages, 1 plate.

Womochel, D.R. 1982. Mollusks from 1AU139 and the Ivy Creek Locality, Autauga County, Alabama. Pages 354–365. In: J.W. Cottier (editor). The Archaeology of Ivy Creek. Auburn University Archaeological Monograph 3.

Wood, E.M. 1974. Development and morphology of the glochidium larva of *Anodonta cygnea* (Mollusca: Bivalvia). Journal of Zoology (London) 173(1):1–13.

Wood, K.G., E. Reitz and M. Russo. 1987. Zooarchaeology. Pages 117–144. In: T.H. Gresham, K.G. Wood, R.J. Ledbetter, D.O. Bradley and P.S. Gardner. Archaeological testing at 1MN30, Eureka Landing, Monroe County, Alabama. Report prepared by Southeastern Archeological Services, Incorporated, for the U.S. Army Corps of Engineers, Mobile District.

Wood, W. 1828. Index Testaceologicus: or a Catalogue of Shells: with the Latin and English names, references to figures and places where found, &c. Supplement, &c. (References from Lamarck's "Animaux sans Vertebres," adapted to the figures in the ... "Index", &c.) London. 59 pages, 8 plates.

Wood, W. 1856. Index testaceologicus, an illustrated catalogue of British and foreign shells, containing about 2,800 figures accurately coloured after nature, by W. Wood, a new and entirely revised edition, with ancient and modern appellations, synonyms, localities, etc. Edited by S. Hanley. Willis and Sotheran, London. 234 pages.

Woodrick, A. 1981. An analysis of the faunal remains from the Gainesville Lake area. Pages 91–168. In: G.M. Caddell, A. Woodrick and M.C. Hill. Biocultural studies in the Gainesville Lake area. University of Alabama, Office of Archaeological Research, Report of Investigations, Number 14.

Woodrick, A. 1983. Molluscan remains and shell artifacts. Pages 391–429. In: C. Peebles (editor). Studies of material remains from the Lubbub Creek archaeological locality. Volume II of prehistoric agricultural communities in west central Alabama. Report to Heritage, Conservation, and Recreation Service, Interagency Archaeological Services, Atlanta, Georgia.

Woody, C.A., and L. Holland-Bartels. 1993. Reproductive characteristics of a population of the washboard mussel *Megalonaias nervosa* (Rafinesque 1820) in the Upper Mississippi River. Journal of Freshwater Ecology 8(1):57–66.

Wright, B.H. 1888. Check List of North American Unionidae and Other Fresh Water Bivalves. Dore and Cook, Book and Job Printers, Portland, Oregon. [No pagination, 8 pages.]

Wright, B.H. 1896. New American Unionidae. The Nautilus 9(12):133–135, plate 3.

Wright, B.H. 1897. A new plicate *Unio*. The Nautilus 11(8):91–92.

Wright, B.H. 1898a. A new undulate *Unio* from Alabama. The Nautilus 11(9):101–102.

Wright, B.H. 1898b. New varieties of Unionidae. The Nautilus 11(11):123–124.

Wright, B.H. 1898c. New Unionidae. The Nautilus 12(1):5–6.

Wright, B.H. 1898d. A new *Unio*. The Nautilus 12(3):32–33.

Wright, B.H. 1899. New southern Unios. The Nautilus 13(1):6–8; 13(2):22–23; 13(3):31; 13(4):42–43; 13(6):69; 13(7):75–76; 13(8):89–90.

Wright, B.H. 1900. New Southern Unios. The Nautilus 13(12):138–139.

Wright, B.H., and B. Walker. 1902. Check List of North American Naiades. Privately published, Detroit, Michigan. 19 pages.

Wright, S.H. 1891. Unionidae of Ga., Ala., S.C., and La., in south Florida. The Nautilus 4(11):125.

Wright, S.H. 1897. Contributions to a knowledge of United States Unionidae. The Nautilus 10(12):136–139; 11(1):4–5.

Wurtz, C.B., and S.S. Roback. 1955. The invertebrate fauna of some Gulf Coast rivers. Proceedings of the Academy Natural Sciences of Philadelphia 107:167–206.

Yeager, B.L., and R.J. Neves. 1986. Reproductive cycle and fish hosts of the rabbit's foot mussel, *Quadrula cylindrica strigillata* (Mollusca: Bivalvia: Unionidae) in the upper Tennessee River drainage. The American Midland Naturalist 116(2):329–340.

Yeager, B.L., and C.F. Saylor. 1995. Fish hosts for four species of freshwater mussels (Pelecypoda: Unionidae) in the upper Tennessee River drainage. The American Midland Naturalist 133(1):1–6.

Yeager, M.M., D.S. Cherry and R.J. Neves. 1994. Feeding and burrowing behaviors of juvenile rainbow mussels, *Villosa iris* (Bivalvia: Unionidae). Journal of the North American Benthological Society 13(2):217–222.

Yokley, P., Jr. 1968. A study of the anatomy of the naiad *Pleurobema cordatum* (Rafinesque, 1820) (Mollusca: Bivalvia: Unionoida). Doctoral dissertation, Ohio State University, Columbus. 139 pages.

Yokley, P., Jr. 1972. Life history of *Pleurobema cordatum* (Rafinesque, 1820) (Bivalvia: Unionacea). Malacologia 11(2):351–364.

Young, D. 1911. The implantation of the glochidium on the fish. University of Missouri Bulletin, Science Series 2(1):1–16, 3 plates.

Young, M.R., and J. Williams. 1984. The reproductive biology of the freshwater pearl mussel *Margaritifera margaritifera* (Linn.) in Scotland. I. Field studies. Archiv für Hydrobiologie 99(4):405–422.

Zale, A.V., and R.J. Neves. 1982a. Fish hosts of four species of lampsiline mussels (Mollusca: Unionidae) in Big Moccasin Creek, Virginia. Canadian Journal of Zoology 60(11):2535–2542.

Zale, A.V., and R.J. Neves. 1982b. Reproductive biology of four freshwater mussel species (Mollusca: Unionidae) in Virginia. Freshwater Invertebrate Biology 1(1):17–28.

Zale, A.V., and R.J. Neves. 1982c. Identification of a fish host for *Alasmidonta minor* (Mollusca: Unionidae). The American Midland Naturalist 107(2):386–388.

Zatravkin, M.N., and V.V. Bogatov. 1987. Large bivalve molluscs in fresh and marine waters of the Far East of the U.S.S.R.: keys to identification. Akademiya Nauk SSSR, Vladivostok. 153 pages [in Russian].

Zatravkin, M.N., and A.L. Lobanov. 1987. Morphometrical limits for the species of the genus *Unio* (Bivalvia, Unioniformes) in the fauna of the U.S.S.R. Byulleten Moskovskogo Obshchestva Ispytatelei Prirody Otdel Biologicheskii 92(6):42–51 [in Russian].

Zatravkin, M.N., and A.L. Lobanov. 1989. The experience of the creation of data base on taxonomy and distribution of the contemporary and Paleogene-Quaternary mollusks belonging to order Unioniformes in fauna of the U.S.S.R. Byulleten Moskovskogo Obshchestva Ispytatelei Prirody Otdel Biologicheskii 94(4):59–63 [in Russian].

Zatravkin, M.N., and Ya.I. Starobogatov. 1984. New species of the superfamily Unionoidea (Bivalvia, Unioniformes) from the Soviet Far East. Zoologicheskii Zhurnal 63(12):1785–1791 [in Russian].

Zimmerman, L.L., and R.J. Neves. 2002. Effects of temperature on duration of viability for glochidia of freshwater mussels (Bivalvia: Unionidae). American Malacological Bulletin 17(1–2):31–35.

Zimmerman, L.L., R.J. Neves and D.G. Smith. 2003. Control of predacious flatworms *Macrostomum* sp. in culturing juvenile freshwater mussels. North American Journal of Aquaculture 65(1):28–32.

Zouros, E., A.O. Ball, C. Saavedra and K.R. Freeman. 1994a. Mitochondrial DNA inheritance. Nature 368:817–818.

Zouros, E., A.O. Ball, C. Saavedra and K.R. Freeman. 1994b. A new type of mitochondrial DNA inheritance in the blue mussel *Mytilus*. Proceedings of the National Academy of Sciences, United States of America 91(16):7463–7467.

Ziuganov, V., E. San Miguel, R.J. Neves, A. Longa, C. Fernández, R. Amaro, V. Beletsky, E. Popkovitch, S. Kalluzhin and T. Johnson. 2000. Life span variation of the freshwater pearl mussel: a model species for testing longevity mechanisms in animals. Ambio 29(2):102–105.

Zuiganov, V., A. Zotin, L. Nezlin and V. Tretiakov. 1994. The Freshwater Pearl Mussels and Their Relationships with Salmonid Fish. VNIRO Publishing House, Moscow. 104 pages.

Zurawski, A. 1978. Summary appraisals of the Nation's ground-water resources; Tennessee region. U.S. Geological Survey Professional Paper 813-L. 34 pages.

Index